Glycopeptides and Related Compounds

Glycopeptides and Related Compounds

Synthesis, Analysis, and Applications

edited by

David G. Large

Liverpool John Moores University
Liverpool, England

Christopher D. Warren

Shriver Center for Mental Retardation
Waltham, Massachusetts

MARCEL DEKKER, INC.   NEW YORK · BASEL · HONG KONG

Library of Congress Cataloging-in-Publication Data

Glycopeptides and related compounds: synthesis, analysis, and
  applications / edited by David G. Large, Christopher D. Warren.
      p. cm.
   Includes index.
   ISBN 0-8247-9531-8 (hardcover : alk. paper)
   1. Glycopeptides. 2. Glycoproteins. I. Large, David G.
II. Warren, Christopher D.
QP552.G59G597 1997
547'.754—dc21

                                                              97-10206
                                                                  CIP

The publisher offers discounts on this book when ordered in bulk quantities.
For more information, write to Special Sales/Professional Marketing at the
address below.

This book is printed on acid-free paper.

MARCEL DEKKER, INC.
270 Madison Avenue, New York, New York 10016
http://www.dekker.com

Current printing (last digit):
10 9 8 7 6 5 4 3 2 1

PRINTED IN THE UNITED STATES OF AMERICA

Preface

Much of the successful research on glycoproteins and glycopeptides has involved a fruitful partnership among physicians, biochemists, and chemists. The aim of this book is to review the many contributions to this joint effort made by chemists, and to discuss state-of-the-art methods for synthesis, analysis, and conformational investigation, with appropriate pointers to future directions for research and development. The intention is to present a balanced picture of the chemistry of this exciting field, with emphasis on smaller molecules of defined structure. We feel that this will complement the growing literature devoted to studies of the isolation and biological roles of glycoproteins.

Glycopeptides and their larger cousins, glycoproteins, lie chemically at the interface of two important areas of natural molecules: carbohydrates and peptides/proteins. During the last four decades, peptide chemistry has received much study, as have the intricate interactions that drive a protein toward its active conformation. One testimony to this effort is the fact that the synthesis of peptides, and even of small proteins, is now a relatively routine procedure. By contrast, carbohydrate chemistry as a whole, and oligo/polysaccharide chemistry in particular, has received much less attention, especially by a majority of mainstream organic chemists. For many years researchers and graduate students have often regarded the study of complex saccharides as simultaneously complex yet intellectually less challenging than other groups of biomolecules. Whereas nucleic acids, polypeptides, prostaglandins, and steroids all had obviously interesting and important biological properties, the humble carbohydrate seemed to serve only as a structural material or energy source, and hence seemed so much less worthy of study. Even the protein chemist regarded the "contaminating" carbohydrate as a hindrance to the real work of protein separation and isolation.

But slowly and painstakingly during the past 20 or 30 years there has been a quiet revolution in attitudes toward carbohydrates. Careful isolation of glycoproteins from mainly mammalian systems, their structural determination, and a growing understanding of their biological roles have created an intense interest in these "hybrid" molecules. This interest in protein glycosylation stems from current knowledge of the ability of a glycan to significantly modify the folding, stability, or biological activity of a peptide, and to express informational markers, such as sialyl Lewis[x], vital for cell–cell adhesion in inflammatory processes.

The biological importance of the oligosaccharides attached to glycoproteins arises primarily from their ability to convey recognitional information. This power originates from the enormous variety of structures that can result from permutations of monosaccharide composition, anomeric configuration, sequence and branching patterns, and modifications such as phosphorylation and sulfation. This far exceeds the informational capability of RNA, with its genetic code of just 64 "letters." In addition, the glycans are usually heterogeneous, with several different variants being found at any one glycosylation site. As a result, glycoproteins are usually large and very complex molecules. This great complexity presents a severe challenge to researchers who want to work with glycoproteins. This book is intended to assist glycobiologists, graduate students, pharmaceutical scientists, and other practitioners in organic chemistry and biochemistry in their response to that challenge.

The chapters cover glycopeptides, oligosaccharides, glycosylphosphatidylinositol anchors, dolichol intermediates, and substrates and inhibitors of glycosyltransferases and glycosidases. These topics were chosen because the compounds are all tools for the study of the structure, biosynthesis, biological function, and metabolism of glycoproteins. In Chapter 1, Christopher Warren and Hari Garg discuss the history of glycopeptide synthesis and briefly review the problems it still presents to chemists today. After a discussion of the therapeutic applications of glycopeptides, the future of glycopeptide research is considered from the point of view of both synthesis and analysis. This chapter is intended to introduce the reader to glycopeptides and related compounds, and many of the topics covered are dealt with in much more detail in later chapters.

Chapters 2–7, 9, and 10 all deal with different aspects of synthesis. In Chapter 2, Horst Kunz and Michael Schultz discuss recent advances in glycopeptide synthesis. After reviewing protecting group strategies for both peptide and carbohydrate components, synthesis of biologically important oligosaccharides, and the methods for the formation of sugar–peptide linkages, the chapter concludes with a section ("Selective Deblocking and Chain Extension of Glycosylated Peptides") illustrated by several examples of the assembly of small glycopeptides into clustered units, to provide important O- and N-linked glycopeptide antigens. The methods described employ both solution and solid-phase chemistry, but in this context the solid support functions primarily as a vehicle for peptide synthesis, glycosylated molecules being incorporated as modified amino acids or short peptides. This is in contrast to the solid-phase technology described in Chapter 5 (see below), where the support is the anchor for elongating the saccharide chain. However, for both of these the common quest is for an anchor resistant to the conditions for repetitive coupling reactions and deprotection steps, and for a final cleavage reaction that must leave sensitive glycosidic linkages unscathed.

Enzymatic synthesis, with its advantages of speed, convenience, and built-in regio- and stereospecificity, has established itself as a worthy partner of chemical synthesis. In Chapter 3, Yoshitaka Ichikawa, who with Chi-Huey Wong pioneered several outstanding studies of the enzymatic synthesis of oligosaccharides and glycopeptides, reviews this topic. The glycosyltransferases are each considered with regard to their occurrence, purification, or cloning, their properties (especially donor and acceptor substrate specificity), and applications. In each case a typical enzymatic glycosylation reaction is described. Also covered are the application of glycosidases to oligosaccharide synthesis, and the synthesis of glycopeptides using both glycosyltransferases and glycosidases. Directly following, in Chapter 4, is an account of the chemical synthesis of the same classes of compounds, by Sabine Flitsch and Gregory Watt. These authors make a complex topic readable and interesting by addressing the important issues and illustrating them with clear, concise formula schemes. Included are: regioselectivity and protecting groups, the chemistry of glycosidic linkage formation, stereoselectivity, strategies for block synthesis, and the introduction of phosphate and sulfate groups at selected positions on the carbohydrate moiety. As expected, the emphasis is on the more recent, innovative, synthetic approaches to glycoprotein glycans, with special attention given to the "difficult" linkages such as β-D-mannosyl and 2-acetamido-2-deoxy-α-D-glucopyranosyl. (Readers interested in detailed, step-by-step descriptions of other state-of-the-art oligosaccharide syntheses will find several outstanding examples in Chapter 7.)

In Chapter 5, a recent development in oligosaccharide synthesis, the application of sold-phase technologies, is comprehensively reviewed by Samuel Danishefsky and Jacques Roberge. General aspects of the topic—nature of the support (solid or soluble), linkage to the support, nature of the activation reactions, coupling reactions, deprotections, capping procedures, cleavage from the support—are all discussed in terms of (a) enzymatic synthesis and (b) chemical synthesis, for both oligosaccharides and glycopeptides. The account of chemical glycosylation emphasizes the application of the glycal assembly method, which has been successfully pioneered by Professor Danishefsky's team. Completing the set of chapters on different aspects of glycopeptide synthesis is Chapter 6, by David Large and Ian Bradshaw, who review the strategies and methodologies successfully used in the synthesis of the peptide component of glycopeptides.

In Chapter 7, Roy Gigg and Jill Gigg review the synthesis of glycosylphosphatidylinositol (GPI) anchors, starting from a subject of their own special expertise, the preparation of specifically protected chiral derivatives of *myo*-inositol. Subsequent sections describe the addition of glucosamine, stepwise assembly of the oligosaccharide, and addition of the diacylated glycerol phosphate and ethanolamine phosphate moieties, showing examples of the synthetic strategies employed by the different investigators. The descriptions of syntheses of complex structures containing up to seven sugar residues, plus inositol and different types of phosphodiesters, using the very latest protection, glycosidation, and phosphorylation strategies, dramatically illustrate just how far glycoconjugate synthesis has come in the last decade or so.

GPI anchor–containing protozoan parasites such as trypanosomes and *Leishmania* cause worldwide debilitating disease. These organisms are unique in possessing an unusual plasma membrane glycoconjugate, lipophosphoglycan (LPG), anchored to the membrane by a GPI, and critical for survival in the host. Malcolm McConville is an expert on the biosynthesis and structure of LPG and GPI, and in Chapter 8 he and Julie Ralton review the methodology for analysis of these related molecules. After an introductory section on the structure and identification of GPI anchors, their analysis is treated in logical stages: (a) as the protein or peptide-linked molecule, (b) as the GPI glycolipid released from peptide, (c) as the purified glycan moiety and its side chains, and (d) as the released PI glycolipid.

The biosynthesis of *N*-linked glycans differs fundamentally from that of other glycoconjugates in that a large oligosaccharide with a very specific structure must be preassembled on a dolichol diphosphate carrier before being transferred to protein.

In Chapter 9, Vladimir Shibaev and Leonid Danilov address the special challenges posed by the chemical synthesis of polyisoprenyl sugar intermediates. After an account of the structure and synthesis of polyprenols, the authors review synthetic approaches to polyprenyl phosphates, glycosyl phosphates, and, most importantly, glycosyl polyprenyl phosphates and diphosphates, an area in which their own laboratory has made a major contribution. The chapter concludes by discussing the important role of synthetic dolichol intermediates in studies of glycoprotein biosynthesis.

Inhibitors of glycosidases and glycosyltransferases are versatile tools for glycobiologists, especially in studies of oligosaccharide function; their inhibitory activities and synthesis are reviewed by J. Michael Williams in Chapter 10. Among the important classes of glycosidase inhibitors described are the lactones, piperidines (e.g., deoxynojirimycin), pyrrolidines, indolizidines (e.g., swainsonine and castanospermine), and pyrrolizidines. This review is followed by a discussion of the design of donor and acceptor analogs as glycosyltransferase inhibitors.

In general for natural products, isolation and analysis precedes synthesis, and this has been true for glycoconjugates. Instrumental and other analytical techniques have continued to develop in recent years at a rapid rate, partly in response to the need of chemists and biochemists to work at the nanomole scale or smaller. Chapters 11–14 all deal with modern methods of glycoconjugate analysis and show the great power of these methods for these complex biomolecules. Carbohydrate linkage and sequence analysis by permethylation, gas–liquid chromatography, and mass spectrometry have been standard methodology for many years. In Chapter 11, Steven Levery focuses on recent refinements of practical importance for glycobiologists and places these developments in a historical context to show how the technique has evolved. Particular attention is paid to the challenging problem of intramolecular esterification in glycoconjugates containing sialic acids. Recent developments in mass spectrometry are reviewed in Chapter 12, by David Harvey. After a survey of methods for ionization, mass separation, and fragmentation, with some discussion of instrument design, Dr. Harvey focuses in detail on two areas of his special expertise, the application of electrospray and matrix-assisted laser desorption ionization (MALDI) to intact glycoproteins, and the determination of composition and structure of oligosaccharides released by chemical or enzymatic procedures from specific glycosylation sites of proteins, by application of GC/MS, fast atom beam, electrospray, and MALDI.

In Chapter 13, in the first of two linked reviews of NMR methods, Elizabeth Hounsell and David Bailey describe the application of modern NMR methods to several classes of glycoconjugates and illustrate their utility by reference to studies of glycosaminoglycans, O-glycans, human milk oligosaccharides, glycosphingolipids, lipopolysaccharides, and intact glycoproteins. Chapter 14, also on NMR, is by Trevor Rutherford. In the first part the focus is on the value of NMR for conformational studies. After an introduction to two-dimensional NMR methods, the author describes correlation spectroscopy (COSY and its variants), heteronuclear shift correlation spectroscopy, and multidimensional NMR. Then there follows a detailed treatment of conformationally sensitive NMR parameters, including heteronuclear coupling constants, nuclear Overhauser effects (NOE), and NOESY and its variants. The second part of the chapter, molecular modeling, is concerned with force fields, potential-energy maps, and molecular dynamics simulations, after which the two main topics of the chapter come together in a discussion of the relationship between observed NOEs and those calculated from restrained models.

A volume on glycopeptides in the '90s would be incomplete without a chapter on glycobiology and medicine. Changes in the surface glycans of cells that accompany malignant transformation have been known for many years, but their significance was often unclear. In Chapter 15, authors Rao Koganty, Mark Reddish, and B. Michael Longenecker show how alterations in the structures of mucins elaborated by epithelial cells expose novel epitopes that are the basis for vaccines against cancer. The authors show how correctly designed synthetic glycopeptides that express the cancer-associated antigens can be used to direct the action of the immune system against the parent glycoprotein.

In the areas of glycoconjugate research considered in this book, great strides have been made within the recent past, and whether it is in synthesis, analysis, or biological activity, the very complexity of these molecules continues to open doors to important new areas of research. Furthermore, we believe that this challenging subject possesses a high degree of topical interest and commercial potential and that for these crucially important biomolecules, the chemical future is likely to be even more exciting than the past.

David G. Large
Christopher D. Warren

Contents

Contributors

David Bailey, M.Sc., Ph.D.   Department of Biochemistry and Molecular Biology, University College London, London, England

Ian J. Bradshaw, Ph.D.   School of Pharmacy and Chemistry, Liverpool John Moores University, Liverpool, England

Leonid L. Danilov, Ph.D.   N. D. Zelinsky Institute of Organic Chemistry, Russian Academy of Sciences, Moscow, Russia

Samuel J. Danishefsky, Ph.D.   Kettering Chair and Director, Laboratory for Bioorganic Chemistry, Memorial Sloan-Kettering Institute for Cancer Research, and Professor, Department of Chemistry, Columbia University, New York, New York

Sabine L. Flitsch, D.Phil.   Chemistry Department, The University of Edinburgh, Edinburgh, Scotland

Hari G. Garg, Ph.D., D.Sc.   Pulmonary and Critical Care Unit, Massachusetts General Hospital, and Department of Medicine, Harvard Medical School, Boston, Massachusetts

Jill Gigg, Ph.D.   Division of Lipid and General Chemistry, National Institute for Medical Research, London, England

Roy Gigg, Ph.D., D.Sc., C.Chem., F.R.S.C.   Division of Lipid and General Chemistry, National Institute for Medical Research, London, England

David J. Harvey, Ph.D., D.Sc., C.Chem., F.R.S.C.   Department of Biochemistry, Glycobiology Institute, University of Oxford, Oxford, England

Elizabeth F. Hounsell, Ph.D., F.R.S.C.   Head, MRC Glycoprotein Structure/Function Group, Department of Biochemistry and Molecular Biology, University College London, London, England

Yoshitaka Ichikawa, Ph.D.   Assistant Professor, Department of Pharmacology and Molecular Sciences, The Johns Hopkins University School of Medicine, Baltimore, Maryland

R. Rao Koganty, Ph.D.   Director of Chemistry Research and Development, Biomara Inc., Edmonton, Alberta, Canada

Horst Kunz, Ph.D. Professor, Institut für Organische Chemie, Johannes Guten-berg-Universität, Mainz, Germany

David G. Large, Ph.D. School of Pharmacy and Chemistry, Liverpool John Moores University, Liverpool, England

Steven B. Levery, Ph.D. Co-Technical Director, NIH Resource for Biomedical Complex Carbohydrates, Complex Carbohydrate Research Center, University of Georgia, Athens, Georgia

B. Michael Longenecker, Ph.D. Senior Vice President, Research and Development, Biomira Inc., Edmonton, Alberta, Canada

Malcolm J. McConville, Ph.D. Department of Biochemistry and Molecular Biology, University of Melbourne, Parkville, Victoria, Australia

Julie E. Ralton, Ph.D. Department of Biochemistry and Molecular Biology, University of Melbourne, Parkville, Victoria, Australia

Mark A. Reddish, M.Sc. Director of Immunology, Research and Development, Biomira Inc., Edmonton, Alberta, Canada

Jacques Y. Roberge, Ph.D. Research Investigator, Combinatorial Drug Discovery, Bristol-Myers Squibb, Princeton, New Jersey

Trevor J. Rutherford, Ph.D. Center for Biomolecular Sciences, School of Chemistry, University of St. Andrews, Fife, Scotland

Michael Schultz, Ph.D. Institut für Organische Chemie, Johannes Gutenberg-Universität, Mainz, Germany

Vladimir N. Shibaev, Ph.D., D.Sc. N. D. Zelinsky Institute of Organic Chemistry, Russian Academy of Sciences, Moscow, Russia

Christopher D. Warren, Ph.D. Department of Biomedical Sciences, Shriver Center for Mental Retardation, Waltham, Massachusetts

Gregory Michael Watt, Ph.D. Chemistry Department, The University of Edinburgh, Edinburgh, Scotland

J. Michael Williams, Ph.D., C.Chem., F.R.S.C. Chemistry Department, University of Wales Swansea, Swansea, Wales

1

Glycopeptides and Glycoproteins: Their Past, Present, and Future

Christopher D. Warren
Shriver Center for Mental Retardation, Waltham, Massachusetts

Hari G. Garg
Massachusetts General Hospital and Harvard Medical School, Boston, Massachusetts

It is now widely accepted that oligosaccharides covalently bound to proteins can have important effects on their folding, stability, immunogenicity, and biological activity [1]. In addition, the structural relationship of the carbohydrate and polypeptide moieties in a glycoprotein is an example of the broader phenomenon of carbohydrate−protein interactions now known to be vital for such diverse biological processes as the role of the selectins in inflammation [2] and that of host lectins in the infectivity of pathogenic microorganisms [3]. It is therefore very important to understand the mechanisms by which sugars and peptides interact. But these are very difficult to study in a glycoprotein because of the daunting molecular size and complexity. As a conjugate of two different classes of macromolecule, intact glycoproteins are refractory to direct study, even by most of the latest and most powerful spectroscopic technologies. Also, methods for the study of oligosaccharides are less well developed than those for polypeptides because of their structural complexity and the lack, until recently, of an appreciation for their biological importance.

Glycopeptides, which retain the critical carbohydrate−peptide linkage region of a glycoprotein but lack its size and complexity, are much more amenable to study, e.g., by modern nuclear magnetic resonance (NMR) techniques and molecular dynamics simulations [4,5]. Also, though smaller and simpler, they may retain important properties, such as antigenicity, and for this reason may be the basis of vaccines [6,7]. Therefore glycopeptides are important models for glycoproteins. Importantly, the methods developed for glycopeptide synthesis can also be applied to artificial protein glycosylation for purposes such as altering stability to proteolysis or routing a recombinant protein drug to its target site by making use of receptors like the macrophage or placenta mannose receptor [8,9]. Therefore the study of glycopeptides is important for basic research and for the development of diagnostic and therapeutic agents.

I. COMPOUNDS RELATED TO GLYCOPEPTIDES AND GLYCOPROTEINS

Many important physical and chemical properties of glycopeptides can be investigated by studies of their attached oligosaccharides, or glycans. An example would be the recently characterized molecular motions derived from torsional oscillations about glycosidic linkages [5,10,11]. Also, the capability for biological recognition is often resident in the glycan, so that relatively small oligosaccharide analogs can function as mimics of natural glycoprotein ligands such as those of the selectins [12,13,10,14,15], or act as decoys for pathogenic organisms that are dependent on a cell surface lectin for their initial attachment to host tissues [6,16]. Furthermore, oligosaccharides with the structures of glycoprotein glycans are vital intermediates for glycopeptide synthesis, especially by the more recently developed direct, or "convergent," approach (see detailed discussion later). Therefore such oligosaccharides are important for a variety of reasons and are the most obviously interesting category of related compounds. In this context the interest in peptides is focused mainly on their value as intermediates for glycopeptide synthesis.

A. Biosynthetic Intermediates and Protein Folding

For researchers concerned with the biosynthesis of glycoproteins and its regulation, biosynthetic intermediates attract a high level of interest. This is especially the case for the asparagine-linked or *N*-linked class of glycoproteins, because their glycans are preassembled on a lipid carrier prior to their transfer to protein [17]. The multiplicity of steps leading to the synthesis of Glc$_3$Man$_9$GlcNAc$_2$-*PP*-dolichol, each dependent on a specific glycosyltransferase, that constitute the so-called dolichol pathway offer enzymologists and molecular biologists many opportunities for investigation. The synthetic chemist has the important role of providing pure and structurally defined intermediates to serve as exogenous acceptors for these transferases [18]. The complexity of the dolichol pathway does not end with transfer of the preassembled oligosaccharide intermediate to protein. Recent studies on protein folding [19] have revealed an important new role for the transferred oligosaccharide. One of the glucose residues, whether one of the original three or a new one added by a special glycosyltransferase in the endoplasmic reticulum (ER), serves to facilitate binding of the newly formed glycoprotein to the molecular chaperone calnexin. The resulting complex is held in the ER until folding of the polypeptide has been satisfactorily completed, after which the glycoprotein is released from the complex and proceeds to the Golgi apparatus for further processing. In cases where a mutant protein fails to fold correctly, such as the ΔF 508 form of the cystic fibrosis transmembrane conductance regulator [20], the malformed glycoprotein never escapes from the chaperone complex and is retained in the ER for eventual disposal. This role in regulating chaperone-mediated folding is a newly recognized general function for *N*-linked glycosylation, unconnected with the specific biological functions of fully processed glycans [1]. After the satisfactory folding of normal proteins, there follows an extensive series of processing reactions, in which many of the original monosaccharide residues are removed and then replaced by new ones to form the mature glycans. These hydrolytic and transfer reactions are catalyzed in turn by a series of processing glycosidases and a further set of glycosyltransferases.

B. Glycosidase Inhibitors

The glycohydrolases, whether they are involved in glycan biosynthesis or degradation, are susceptible to inhibition by sugar analogs in which the ring oxygen atom has been replaced by nitrogen [21,22]. These inhibitors, which differ widely in their degree of anomeric or monosaccharide specificity, are either naturally occurring compounds, often alkaloids such as swainsonine or castanospermine, or mainly synthetic, like deoxynojirimycin and its variants and derivatives. The α-mannosidases, which perform important functions in both biosynthesis and degradation of the *N*-linked glycans, can be mechanistically divided into two classes based on their inhibition by *pyranose* or *furanose* analogs [23]. Because these inhibitors can be employed to alter the processing of glycoprotein glycans and hence their structures, they are valuable for studies of biological activity, and for the development of antiviral agents, because pathogenic viruses such as HIV are dependent on the correct synthesis of the glycan chains of the envelope glycoprotein for viral maturation and replication [24,25]. Therefore the glycosidase inhibitors are also, in an important functional sense, compounds related to glycoproteins.

C. Glycosylphosphatidylinositol Anchors

Glycosylphosphatidylinositol membrane anchors consist of an oligosaccharide linked through an ethanolamine phosphate diester bridge to the *C*-terminus of a protein at the "nonreducing" end, and through myoinositol to a moiety closely resembling phosphatidylinositol at the "reducing" end [26]. They can therefore be considered as a specialized class of glycoprotein, although the anchor itself is actually a glycophospholipid. Methods for their isolation, analysis, and identification are heavily dependent on hydrolytic chemical and enzymatic procedures specific for phosphate esters in general, the phosphate diester group or fatty acyl groups of phosphatidylinositol, or glycosides of glucosamine [27,28]. GPI anchors have attracted the attention of synthetic chemists, presumably because of their novelty as a previously unknown class of glycoconjugate, but the synthesized compounds will also have value as reference compounds for degradative procedures.

II. SYNTHESIS OF GLYCOPEPTIDES AND RELATED COMPOUNDS

Historically, structure determination and chemical synthesis have often been undertaken in parallel because the synthesis of a pure, structurally defined compound provides an ideal standard for comparison with the natural material. However, in the early days of glycoprotein research (roughly from the 1930s onward) the targets for synthesis were often poorly defined because of uncertainties about glycoprotein structures. In several instances compounds were synthesized with either the wrong linkages or the wrong component sugars. Thus Link and co-workers [29,30] and previously Bergmann and Zervas [31] prepared glucosamine derivatives acylated with glycine, methionine, glutamic acid, or lysine through the N-acetyl group. Later, Micheel and Kochling [32] and Jones et al. [33] synthesized N-acetylglucosamine β-linked to serine, presumably because the involvement of N-acetyl galactosamine α-linked to serine or threonine had yet to be revealed [34] and because galactosamine was not available. With the benefit of hindsight it is not too surprising that the synthesized compounds were often unstable and did not resemble natural materials. It is a little ironic that some 30 years after the Jones work GlcNAc-Ser (Thr) would be discovered as a component of nuclear and cytoplasmic proteins [35,36].

Widespread application of synthesis for structure confirmation did not occur until improved methods of isolation, separation, and purification had been devised, particularly the introduction of chromatography: first paper, then ion exchange, gel filtration, adsorption, and thin layer. Together with increasingly sophisticated analytical methodologies such as Smith degradation, gas-liquid chromatography, and later early mass spectrometry (and hence gc/ms and methylation analysis), and in the 1960s early applications of NMR, a large number of structures were determined (some were later shown to be incorrect), and many targets became available to the synthetic chemist.

Though today synthesis is viewed mainly as a means of obtaining pure material for further experimentation, either chemical, spectroscopic, enzymatic, or drug related, it is apparent that in the past much synthetic work, especially at earlier times when extensive justification for basic research was not needed, was often viewed as an end in itself. Compounds were synthesized because, like Mount Everest, "they were there" and provided a challenging target. However, it is also true that the work was often undertaken to advance synthetic chemistry as a branch of science and to introduce new and improved methods and reagents.

For glycopeptides and oligosaccharides, the major explosion of chemical, and more recently enzymatic, synthesis occurred during the 1970s, '80s and '90s, in parallel with the introduction of modern NMR and mass spectrometry and the advent of HPLC. However, for the dolichol intermediates [37,17] the stimulus came mainly from advances in biochemical techniques, notably those employing radiolabeling, and advances in the techniques for the study of membrane-bound glycosyltransferases in cell-free systems. The contribution of chemical synthesis was especially pronounced for the early steps of the dolichol pathway, where it was able to provide useful amounts of pure compounds for which important structural details, such as anomeric configuration, were defined [38]. Then the synthetic compounds could be carefully exposed to the kind of diagnostic treatments that were being employed to characterize the radiolabeled biosynthetic intermediates [39]. For example, α and β anomers of dolichyl mannosyl phosphate were subjected to hydrolysis under mild acidic and basic conditions [40]. The products of these treatments were compared with those from the labeled intermediate, and from the results it was possible to conclude without any doubt that the biosynthetic Man-P-Dol was β linked [41].

The major traditional roles of chemical synthesis in the study of glycopeptides, glycoproteins, and related compounds can be summarized as follows:

1. The intellectual satisfaction of making something previously available only from nature
2. The development of new and improved methods of synthesis
3. To establish or confirm the structure of a natural compound
4. To provide pure reference compounds for chromatography and spectroscopy
5. To provide exogenous substrates for glycosyl transferases and glycosidases in biosynthetic studies

In the recent past, a new role for synthesis, which is rapidly assuming great importance, has emerged and is likely to dominate future efforts; this is synthesis not to mimic nature but to improve on it. In this new role the compounds made will be smaller and simpler than the natural materials, while equaling them in their important biological activities or even excelling in this capacity. The example that immediately comes to mind is that of a synthetic ligand that binds to a receptor with superior avidity or specificity than the natural one. The obvious importance of this is in the development of carbohydrate-based drugs like the previously mentioned and very topical small molecular mimics of selectin ligands, for treatment of inflammatory conditions [12,13,15].

III. THE CHALLENGE OF GLYCOPEPTIDE SYNTHESIS

Loosely speaking, any molecule containing a monosaccharide covalently bound to an amino acid could in a minimal sense be called a glycopeptide. In reality, such a compound would attract minimal interest, except, perhaps, as a model for the chemistry of linkage formation. To be useful for most purposes a glycopeptide should contain an oligosaccharide and an oligopeptide, both of sufficient length to facilitate the kinds of binding (hydrogen bonding, hydrophobic bonding, van der Vaal's contacts, and possibly electrostatic bonding) that routinely occur between carbohydrate and peptide in a glycoprotein. Therefore the challenge of glycopeptide synthesis is twofold: (1) to promote efficient linkage formation with the correct regio- and stereo-specificity, and (2) to devise a successful strategy for oligomerization of the saccharide and peptide moieties, either before or after linkage formation. With these facts in mind, it is possible to elaborate the basic principles of glycopeptide synthesis, as is done in the succeeding subsections.

A. Linkage Formation

Although glycopeptides may be N-linked or O-linked, general features of linkage formation are similar. Thus every case involves the condensation of a reducing terminal monosaccharide residue with the side chain of an amino acid, with the elimination of one mole of water. The sugar component is always a glycoside, either a 2-acetamido-2-deoxy-β-D-glucopyranosylamine for the N-linked, or an oxazoline, orthoacetate, or glycosyl halide for the O-linked. For N-linked glycopeptides, the amino acid side chain is the β-carboxyl group of aspartic acid, and the coupling is most often induced by a carboxyl activating reagent such as a carbodiimide or one of the more recently introduced reagents like 2-(1H-benzotriazol-1-yl)-1,1,3,3-tetra-methyluronium hexafluorophosphate. Alternatively, the acid can be activated by prior conversion into the pentafluorophenyl ester. Obviously all other COOH and NH_2 groups, and other reactive groups except for hydroxyl, must be protected during the coupling reaction. Therefore protective group strategy is an important part of experimental design, especially since any subsequent deprotection steps must not affect the newly formed N-(aspart-4-oyl)-glucopyranosyl linkage or any of the interresidue glycosidic linkages in an oligosaccharide. The coupling reaction results in the net conversion of aspartic acid to asparagine and therefore does not mimic natural synthesis, in which an oligosaccharide is transferred to a preexisting aparagine residue.

For O-linked glycopeptides the amino acid side chain is the OH group of serine, threonine, hydroxyproline, or hydroxylysine, and linkage formation is an example of cis glycoside synthesis, because the anomeric configuration is α. Therefore the reaction conditions are those designed to favor formation of α glycosides of GalNAc, GlcNAc, Gal, or Fuc. For the N-acetyl hexosamines this presents a serious problem, since only β-glycosides are readily formed. The usual solution to this has been to use a 2-azido derivative as the glycosyl donor, which is then converted into GalNAc or GlcNAc after linkage formation. Obviously all other OH groups, and any other reactive groups, must be protected during coupling, and therefore protective group considerations are very important, as in N-glycopeptide synthesis. Unlike the latter, however, O-linkage formation is analogous to the natural process.

B. Oligomerization Strategy

A glycopeptide can theoretically be constructed in at least five ways:

   a. By synthesizing a monosaccharide-amino acid core (e.g., GlcNAc-Asn or GalNAc-Ser) and then extending both sugar and amino acid parts
   b. By synthesizing a monosaccharide-peptide and then extending the sugar into an oligosaccharide

c.   By synthesizing an oligosaccharide-amino acid and then extending the amino acid into an oligopeptide by adding more residues to the *N*- or *C*-terminus

d.   By synthesizing an oligopeptide and then incorporating a presynthesized core (e.g., GlcNAc-Asn or GalNAc-Ser) into the chain as a modified amino acid, then completing the elongation of the peptide

e.   By coupling a preformed oligosaccharide to a preformed oligopeptide (the direct, or "convergent," approach)

The last method (e), in which linkage formation takes place after the oligo-merization of both components, is conceptually the simplest way of forming a gly-copeptide, and is attractive because it minimizes protection and deprotection prob-lems, including that of cleavage of the peptide from a resin when solid phase synthesis is employed. Another advantage is that a valuable oligosaccharide is sub-jected to a minimal number of synthetic reactions (i.e., one only), during which loss of yield can occur. Also, for *N*-glycopeptide synthesis, the carbohydrate need not, and indeed must not, be protected by *O*-acetyl groups, which means that any sugar, including oligosaccharides, can be very conveniently converted into the glycosyl-amine by reaction with excess aqueous ammonium hydrogen carbonate and then immediately coupled with a peptide [42–45]. The only requirement is that the reduc-ing terminal monosaccharide residue be an *N*-acetyl hexosamine, which is, of course, the case for the *N*-glycopeptides. The disadvantages of direct synthesis (e) are that steric hindrance between the large oligosaccharide and peptide components may make the coupling difficult, and, for *N*-glycopeptides, that an unwanted side reaction, namely, an intramolecular cyclization resulting from attack of the neighboring residue NH on the activated aspartic acid β carboxyl group to form an aspartimide, occurs to a greater or lesser extent, depending on the identity of the X residue in Asp-X-Ser (Thr). The first problem has been overcome by the application of recently intro-duced highly potent coupling reagents, and the second problem by careful optimi-zation of the coupling conditions [42,43].

C.   Experimental Considerations

Whereas linkage formation always employs chemical synthesis, elongation of the sugar and peptide moieties can be by chemical or enzymatic procedures, and the former can employ liquid- or solid-phase technique, although the latter is only now being successfully developed for oligosaccharide synthesis [46–48].

In practice, glycopeptides have in the past usually been made by methods (c) and (d), and the sugar has almost always been a mono- or disaccharide. Recently, much more interesting oligosaccharides, with structures similar to those in glycopro-teins, have become available [49] and have been incorporated into glycopeptides by method (e) [43–45]. Another very recent development is the introduction of enzy-matic methodology for the extension of both oligosaccharide and oligopeptide chains, for *N*-glycopeptide synthesis by method (a) [47,48].

D.   Position of Linkage in *N*-Glycopeptides

The regiospecificity of the linkage of GlcNAc to asparagine is a problem that has occupied researchers over a span of more than three decades. In the early 1960s

Neuberger and associates [50] assigned the linkage to the β carboxyl of Asn, but the method used to arrive at this has been criticized as not truly conclusive. The approach was classical; enzymatic and chemical degradation of a glycopeptide derived from a readily available glycoprotein, ovalbumin, down to the core containing the linkage, i.e. GlcNAc-Asn, then comparison with chemically synthesized reference compounds in which the linkage was through either the α- or β-carboxyl group, respectively. Unfortunately, the strong acid hydrolysis employed in the degradation step caused low yields of GlcNAc-Asn, so that although the natural core corresponded to 2-acetamido-N-(L-aspart-4-oyl)-2-deoxy-β-D-glucopyranosylamine, this was not a good formal proof that the natural linkage was through the β position. Only recently has a more definitive proof been established, as part of a study of the enzyme that transfers preassembled oligosaccharide $Glc_3Man_9GlcNAc_2$ to nascent polypeptide during protein glycosylation by the dolichol pathway. Coward and coworkers [51] employed oligosaccharyl transferase from yeast to glycosylate in vitro the synthetic glycotripeptide N-benzoyl-Asn-Leu-Thr-NH_2. The product was degraded by endo-N-acetyl-β-glucosaminidase H to yield $Glc_3Man_9GlcNAc$ and N-benzoyl-Asn(GlcNAc)-Leu-Thr-NH_2, which was exhaustively purified and compared by HPLC with synthetic GlcNAc-tripeptides in which the linkage was via the α- or β-carboxyl groups of asparagine, respectively. The ^{3}H-labeled biosynthetic product was clearly identical to the β compound, and none of the alternative compound was observed. Thus the regiospecificity of the linkage was finally established.

IV. BIOMEDICAL APPLICATIONS OF GLYCOPEPTIDES AND GLYCOPROTEINS

In contrast to much of the past work on glycopeptides, which emphasized structure and synthesis, a major present effort is directed toward applications of glycopeptides and glycoproteins in drug development. The beginning of this trend was coincident with the recent explosion of interest in glycobiology, when it was recognized that glycoconjugates have exciting biological activities and are not merely structural components of tissues, dietary materials, or "contaminants" of proteins. The speed of change in attitudes toward carbohydrates will accelerate in the near future as the range of their known biological activities widens. Significantly, these developments are a direct outgrowth of the results of basic research on the structure, biosynthesis, antigenicity, or lectin-binding properties of glycopeptides and glycoproteins.

A. Anticancer Therapy

One of the most rapidly developing areas, with tremendous potential, is that of cancer therapy. One approach, immunotherapy of tumors, is presently undergoing clinical trials and showing signs of success [6,7]. This makes use of the fact that the glycan chains of the mucins that line the surfaces of the cells of the breast, lung, colon, pancreas, and ovary are truncated and aberrant, so abnormal epitopes are exposed. The immune response, by cytotoxic T cells, to these tumor-specific antigens is usually too weak to be clinically useful. However, it can be greatly magnified by synthesizing a glycopeptide expressing the antigen, binding to keyhole limpet hemocyanin, and conjugating the product to an adjuvant, thus formulating a tumor-specific vaccine that can be used to immunize patients against their own tumors [7].

A different approach to cancer therapy employs the α-mannosidase inhibitor swainsonine [23]. This compound blocks the normal processing of N-linked glycans because it is a potent inhibitor of Golgi α-mannosidase II, which normally hydrolyzes 3- and 6-linked mannose residues from the α-$(1\rightarrow6)$-linked mannose in the N-glycan core. This results in hybrid chains being formed instead of complex glycans. When cells undergo malignant transformation, and especially when the resulting tumor cells attain metastatic potential, an enhanced activity of GlcNAc transferase V is observed [52]. This enzyme transfers GlcNAc to the C-6 position of the α-$(1\rightarrow6)$-linked core mannose to form 2,6-branched mannose, but it cannot act if this branch is unprocessed. Also, the 2,6-branched mannose residue is the one that most often carries polylactosaminyl chains, the increased incidence of which also correlates with advanced carcinoma and metastasis. Interference with the action of GlcNAc transferase V and the formation of polylactosaminyl chains may explain the antitumor and antimetastatic results of therapy with swainsonine, which have been demonstrated in several animal trials [53].

B. Anti-infective Therapy

Many infectious microbial agents rely on lectin-carbohydrate interactions for their initial attachment to host tissues, and the abrogation of this process is the basis of several drugs under development. The principle is to flush out the organism with a decoy oligosaccharide that resembles a glycan exposed on the surfaces of host epithelial cells. The diseases targeted in work in progress include *Helicobacter pylori*–induced gastric and duodenal ulcers, pneumonia, urinary tract infections, ear infections, and bacterial diarrhea [16,6]. In the case of a particularly devastating form of the latter, hemolytic uremic syndrome caused by *E. coli* 0157-H7, a different kind of therapy also under development involves oligosaccharides that bind to the bacterial toxins and remove them from circulation [54]. A different but related approach is to modify the host glycan and interfere with the binding of the organism's lectin. This is being done with sialidase, against the influenza virus [55]. A very different approach employs the lectin-carbohydrate interaction in the opposite way, i.e., to target an agent to its intended site of action. A conjugate of a glycoprotein containing galactose-terminated glycans with a polycation that can bind antisense oligonucleotides is used to ferry these to the surface of parenchymal cells in the liver, via binding to the asialoglycoprotein receptor. This mode of attack is currently being used against hepatitis B, but in due course will be used for other liver diseases and perhaps eventually cancer.

In an effort to develop an anti-HIV agent that differs in its mechanism of action from most of the others being tested, the synthesis of the envelope glycoprotein gp 120, is being targeted. Derivatives of the potent glucosidase inhibitor deoxynojirimycin (DNJ), such as the N-butyl compound, which can penetrate infected cells, interfere with processing of the gp 120 glycans by inhibiting the removal of glucose residues from the newly formed oligosaccharides, and in some way disrupt viral replication or maturation; the exact mechanism is not known [56,25], but the possibilities include a disruption of chaperone-mediated protein folding [19]. The problems with such a systemic approach to disease therapy are obvious—not only are other glucosidases like those involved in digestion inhibited, but the processing of host glycoproteins is affected also. Therefore a major part of the effort involved in developing such a drug is that directed toward the mitigation of side effects. Successes against digestive complications have been reported that employ a modified N-butyl DNJ as a prodrug.

Recently it has been shown that N-butyl DNJ can also inhibit glycolipid formation. When human cells were cultured in the presence of N-butyl DNJ, the synthesis of glucosylceramide, and hence that of more complex glycolipids derived from it, was markedly curtailed [57a]. An analog of N-butyl DNJ, N-butyldeoxygalactonojirimycin (N-butyl DGNJ), was an equally good inhibitor of glycolipid synthesis and had the advantage of not affecting N-glycan processing glucosidase I, or lysosomal β-glucocerebrosidase [57b]. These results opened up the possibility of a new approach to the therapy of Gaucher disease, in which an inherited deficiency of glucocerebrosidase leads to massive storage of glucosylceramide in patient's cells, especially in liver and spleen, where macrophages become engorged with lipid [57c]. Therefore Gaucher patients could be treated with N-butyl DGNJ to impair glycolipid synthesis without the risk of side effects due to the inhibitor's affecting glycoprotein processing or the activities of intestinal glucosidases.

C. Enzyme Replacement Therapy

This can also employ a receptor-ligand interaction to direct an agent, in this case an enzyme, to a location where it can perform its function. The glycan chains of β-glucocerebrosidase were modified by digestion with exoglycosidases to yield an enzyme in which the glycans consisted of the trimannosyl core, which targets it for efficient uptake by the macrophage mannose receptor [8,57c]. There the enzyme is able to perform the breakdown of glycolipids stored in Gaucher disease and alleviate manifestations of the disease. This is the traditional approach to storage disease therapy, in contrast to the novel approach of inhibiting biosynthesis (see above).

D. Anti-inflammatory Therapy

This is currently the most discussed application of glycoprotein-related compounds in therapeutics and the one with the most obvious potential for commercial reward. The selectins are adhesion proteins that mediate recruitment of neutrophils and other white blood cells to sites of inflammation and tissue injury [2,3]. For example, E-selectin is expressed on the surface of endothelial cells and binds to glycans on the surface of leukocytes that terminate with the tetrasaccharide sialyl Lewis x, Sialylα(2→3)-Galβ(1→4)[Fucα(1→3)]-GlcNAc [15]. The resulting tendency of the leukocytes to stick to the endothelium initiates a rolling motion that slows the cells down and enables a second level of much tighter binding involving integrins. Many investigators and companies are competing in a race to develop molecular mimics of sialyl Lewis x that will bind to the receptor on the blood vessel wall and hinder the ability of the blood cells to bind [13,12]. In one especially interesting approach, a pharmacophore was constructed from knowledge of the structural elements necessary for selectin binding, as derived from conformational (nuclear Overhauser effect, or NOE) and structure-function studies [14]. This pharmacophore was employed to search a database for potential inhibitors of selectin binding. This revealed the inhibitory activity of the triterpene fucoside glycyrrhizin, a component of licorice long used as an antiinflammatory in Chinese medicine. It also made possible the synthesis of related fucose derivatives in which inhibitory activity and stability were optimized in a simpler molecule. These derivatives had superior antiinflammatory activity in a mouse model.

The successful design of potent inhibitors of selectin binding will be further stimulated by the recent determination by X-ray crystallography of the E selectin carbohydrate recognition domain [58] because structures of oligosaccharides and other ligands as determined by NMR can be tailored for optimal binding. However, sophisticated as these studies were, they did not take into account the importance of ligand multivalency. When monovalent and multivalent forms of sialy Lewis x were conjugated to bovine serum albumin and tested for binding affinity to immobilized E selectin, the multivalent form was roughly 1,000 times superior [59]. Conversely, selectin density on the cell membrane has also been shown to be an important parameter [60]. This was demonstrated in a 1992 binding-affinity study in which structurally defined oligosaccharides were conjugated to phosphatidylethanolamine dipalmitate (neoglycolipids) and E selectin was expressed from the cloned cDNA transfected into Chinese hamster ovary cells. In a very recent review, it has been hypothesized that neither multivalency in ligand nor selectin is alone sufficient for high-affinity binding in vivo and that ''clustured oligosaccharide patches'' are critical [61].

Despite the uncertainties surrounding the nature of true ligand–selectin interactions, reduction of tissue injury has already been demonstrated in several animal trials of selectin ligand mimics. The diseases targeted by this effort include acute conditions like ARDS, septic shock and reperfusion injury (as after heart attack, stroke, or organ transplant), and chronic conditions such as psoriasis and rheumatoid arthritis. In many cases a major thrust of the research is to search for smaller, relatively simple molecules (''glycomimetics'') that can substitute for the natural molecule, which is difficult to synthesize. One example of this is to replace the sialic acid residue by sulfate [62], but other materials are being developed in which most of the sugar residues are replaced by space filling structures. However, the recent reports, mentioned above, highlighting the importance of ligand multivalency and clustering [59,61] might seem to cast something of a cloud over the prospects for these efforts.

Closely following the identification of carbohydrate ligands for the selectins have been the first reports of patients with an inherited defect in selectin ligand synthesis. Children with leukocyte adhesion deficiency, Type 2, [63] had severe mental retardation, abnormal physical characteristics, and frequent infections resulting from a marked defect in neutrophil motility. They were unable to synthesize sialyl Lewis x, and their Bombay blood phenotype showed that the defect was not limited to a $(1{\rightarrow}3)$-fucosyltransferase, suggesting a more general problem with the fucosylation of oligosaccharide precursors.

A new group of inherited metabolic diseases has recently been identified and named carbohydrate-deficient glycoprotein (CDG) syndromes [64]. The symptoms are similar to those of glycoprotein storage disorders, involving mental retardation, severe neuromotor disability, and often the failure of major organ systems [65]. Diagnosis depends on the identification of serum transferrin lacking one or both of its sialylated biantennary glycans [66]. This implies a general defect in *N*-glycosylation of secreted proteins. The most prevalent form of CDG, Type 1, has been identified in more than 120 patients, in 14 nations. Previously, biochemical studies suggested that Type 1 CDG probably involves one of the steps in the synthesis or transfer to protein of the dolichol intermediate, oligosaccharide-*PP*-dolichol. Recently, the primary defect has been identified as a deficiency in the activity of phosphomannomutase [67], which converts mannose 6-phosphate into mannose 1-phosphate, a key step in the synthesis of GDP-mannose. From our knowledge of glycoconjugate biosynthesis, this would promote a cascade of events as follows: The defective supply of GDP-mannose would lead to a deficiency of dolichol-*P*-mannose, leading in turn to a deficient supply of oligosaccharide-*PP*-dolichol, so that some *N*-glycosylation sites in proteins would fail to be glycosylated. This could lead to chaperone-mediated protein folding problems [19] and to secondary effects due to defective glycosylation of a wide range of enzymes, hormones, neurotransmitters, etc. and their receptors.

If the deficiency of dolichol-*P*-mannose is severe enough, it could also affect the synthesis of glycosylphosphatidylinositol protein anchors [26–28]. In addition, because GDP-fucose is normally synthesized from GDP-mannose, fucosylation would also be affected, leading to altered glycan chains in both glycoproteins and glycolipids and to impaired synthesis of vital ligands such as sialyl Lewis x [2,3,12–15]. The end result of all these abnormalities would be multiple defects in maintenance of cellular function at different stages of development in a fetus affected by CDG, Type 1, and multisystemic failures of major organ systems in affected children. Therefore the identification of the primary defect in this inherited metabolic disease can, in theory, fully explain the complexities of the observed phenotype.

Interestingly, aberrations of glycosylation have sometimes been valuable for detecting abnormalities of protein function where the actual lesion does not primarily involve glycosylation per se. An example of this is the cystic fibrosis transmembrane regulator (CFTR). In the disease resulting from the most common mutation in cystic fibrosis patients, termed ΔF 508 [20], in which the protein has a deletion of a single phenylalanine residue, the mutant protein is trapped in the endoplasmic reticulum (ER) by a quality control process that monitors protein synthesis and assembly and retains for disposal those that are faulty [68]. The first valuable clue that this was happening came from the observation that the glycans of CFTR were not processed by any of the enzymes present in the Golgi apparatus; i.e., they remained as immature, high-mannose types [20]. Thus it was realized that the lack of CFTR function in these patients was due to defective transport of the protein to the cell surface, not to an aberration of protein structure that disrupted biological activity. Interestingly, the ΔF 508 mutant protein retains some activity as a chloride channel, so that some aspects of cystic fibrosis pathogenicity may actually result from aberrant channel activity in the ER.

V. THE FUTURE OF GLYCOPEPTIDE RESEARCH

Many of the future trends are already evident, and changes are occurring rapidly, especially in the area of drug research. The overwhelming need will be for procedures, both synthetic and analytical, that are simple, convenient, rapid, and inexpensive. Similarly, too many analytical instruments, though they may offer high-resolution chromatographic separations or highly sophisticated spectroscopic information, are very expensive, and so technically demanding that they need full-time skilled technicians to operate them and are often "down" because of their complexity of operation.

A. Future Trends in Synthesis

The example of selectin ligands discussed earlier provides a good case for illustration. The best-known ligand, sialyl Lewis x, is a tetrasaccharide, a relatively simple molecule compared with many other biologically important carbohydrates. Yet the chemical synthesis is a complex and expensive undertaking, requiring an extended period of intense effort by highly skilled individuals. Two types of solutions to this problem have evolved: (a) to make the synthesis simpler by replacing some of the complicated saccharides by simpler molecules that mimic them, and (b) to make the process simpler and faster by replacing chemical steps with enzymatic ones [10]. For the latter, the enzymes are most often glycosyltransferases, but sometimes it is possible to make use of glycosidases, in which case it is necessary to shift the equilibrium toward product formation and away from hydrolysis [69].

In some ways, enzymatic synthesis is the opposite of chemical synthesis. The latter employs glycoside-forming reactions that are nonspecific in terms of position of the linkage formed, this regio-specificity being installed via protection of the other positions. In enzymatic synthesis, no protection is necessary, because regio-specificity is determined by the choice of enzyme, as is stereochemistry. Thus the complications of protection and deprotection, and problems of separating stereoisomers, are all avoided. These are the major factors that make enzymatic synthesis potentially much faster and less labor intensive. Another difference is that chemical synthesis employs volatile, toxic solvents, hazardous reagents, and sometimes harsh conditions (temperature, pH), whereas enzymatic reactions are usually performed in aqueous media at or near room temperature and near neutral pH. They are therefore much more convenient and less likely to harm either the compound under study or the person doing the work. Furthermore, efficient chemical synthesis often requires rigorous conditions that necessitate complicated apparatus, which makes scaling up difficult and expensive. Scaling up enzymatic synthesis presents different problems, notably the questionable availability of the enzymes (see below), the high expense of the sugar nucleotide donors, and inhibition of the transfer reactions by the nucleoside phosphates formed as a second product. The enzymes are becoming available via molecular cloning and overexpression, and the other problems have been solved through in situ cofactor regeneration [10,69].

Purely in terms of versatility, chemical synthesis cannot be excelled. In theory almost any molecule can be constructed, though this may involve many steps and a low overall yield, especially in cases with unfavorable stereochemistry, such as the formation of β-mannosides or α-glycosides of the N-acetyl hexosamines. In contrast, some reactions cannot be performed enzymatically because a glycosyltransferase with the required specificity has not yet been purified or cloned or simply does not exist. This might be due to a peculiarity of biosynthesis. A good example of this is the pentasaccharide core of the N-glycopeptides, which is formed via the dolichol pathway of protein glycosylation. Because an oligosaccharide is initially assembled on a dolichol intermediate, the glycosyltransferases involved will not transfer

GlcNAc and mannose residues to nonlipid acceptors. Therefore, until or unless novel transferases, with unnatural substrate specificities, are developed by molecular biology techniques, the N-glycopeptides and their oligosaccharides will remain accessible only by chemical synthesis. Unfortunately, the syntheses of such molecules, though a monumental achievement from an academic standpoint, are all too often so complicated, labor intensive, and intellectually demanding that they are very difficult to reproduce, yield small amounts of the target compound, and hence are of little practical use.

One solution to the conundrum of having chemical methods that can produce exactly the right compound in low yields after a formidable investment of time and effort, and enzymatic methods that may not be able to approach the target so accurately but are much faster, simpler, and more convenient, is to design molecular mimics that are more readily accessible. For example, instead of a mannose residue, for which none of the available mannosyltransferases can provide the required linkage, it may be possible to utilize an analog with a galactose residue instead [48]. Or, instead of a minor structural variation, more radical molecular substitution may be employed, in which a relatively small molecule displays an active center or determinant that is naturally part of a larger one, or where the critical reactive groups are carried by a polymer skeleton that reproduces spatial characteristics of a natural ligand without having to imitate its complexity. Other factors that are making enzymatic synthesis more generally applicable are the increasing number of transferases being made available in purified form via molecular cloning, and innovations involving unnatural applications of enzymes. An exciting example of the latter is the recent use of the subtilisins in the enzymatic extension of the peptide parts of glycopeptides [47,48].

Another factor that is tilting the balance in favor of enzymatic synthesis is that of broad substrate tolerance. Although reactions catalyzed by glycosyltransferases are kinetically most favored when the donor and acceptor substrates have exactly the same structure as those in the natural biosynthesis, the reactions will often proceed, though less efficiently, when the structures are not exactly "correct." For example, β-(1→4)-galactosyltransferase can transfer glucose, 4-deoxyglucose, arabinose, and N-acetylgalactosamine and N-acetylglucosamine from their UDP derivatives and will tolerate GlcNAc and glucose derivatives as acceptors [70]. Acceptor specificity can also be loosened by lowering pH of the incubations and by the presence of α-lactalbumin [71]. Although the reactions may be slow, and yields low, the method is so much easier and faster than the chemical synthesis of the same compounds that the enzymatic method is preferable.

Therefore substrate tolerance significantly broadens the scope of the enzymatic approach. However, it should be noted that in most of the recently reported examples of successful oligosaccharide synthesis using glycosyltransferases, a combination of chemical and enzymatic methodologies is employed. Often, chemical synthesis has been used to fabricate a core or primer that is then extended by enzymatic transfer [72–77]. In a particularly interesting "hybrid" approach, the loose acceptor specificity of Lewis a-(1→3/4)-fucosyltransferase for human milk was taken advantage of. A radically modified GDP-fucose analog was chemically synthesized in which the bulky human blood group B trisaccharide antigen was linked to C-6 of the fucose residue through an 8-carbon "spacer." The partially purified transferase was capable of transferring the preassembled oligosaccharide antigen to acceptor conjugates containing β-Gal(1→3)GlcNAc [78]. Another way in which hybrid chemical/enzymatic methods can be used, this time in glycopeptide synthesis, is the use of hydrolytic enzymes to effect the removal of protecting groups during chemical synthesis [79].

Another recent development with far-reaching implications for future glyco-peptide research is the introduction of a practical solid-phase synthesis of oligosac-charides [46]. This work differs fundamentally from previous efforts to utilize solid-phase methodology in that the glycosyl donor is attached to the solid support and the incoming sugar residues are glycosyl acceptors, instead of vice versa. The se-quential glycosidations are performed by the epoxidation of glycals in which the OH groups are protected. The critical features of the method are the highly stereoselective epoxidation and susequent nucleophilic attack on the epoxide by the glycosyl ac-ceptor. A very important point is that the problem of contamination of the product with deletion products, a bugbear that haunts solid-phase technology, is effectively solved by their elimination during the synthesis because any unreacted epoxides are hydrolyzed during the workup. Although the method has so far been demonstrated only for the most favorable case of oligosaccharide synthesis, that in which all the sugar residues are neutral hexoses and all the linkages are β-(1→6)- and trans, ex-tension to hexosamines and other linkage types is in progress. It is possible that in the distant future such technology may eventually lead to the automated synthesis of oligosaccharides and glycopeptides, especially when it is combined with enzy-matic synthesis.

In a development that combines elements of solid-phase synthesis, enzymatic synthesis, and affinity chromatography [16], different saccharides are immobilized by conjugation to affinity supports and treated with a tissue homogenate that contains a glycosyltransferase with acceptor specificity for the immobilized sugar. The idea is that the enzyme binds to the column and is then released when the correct sugar nucleotide is provided, so that a highly enriched preparation of the glycosyltrans-ferase elutes from the column. When all the required transferases have been partially purified in this way they are employed sequentially for oligosaccharide synthesis. The most frequent criticisms of this approach are that the enzymes will not bind tightly enough and that more than one will bind to each acceptor, resulting in a mixture of activities. However, it is likely that some useful enzymes will be obtained in this way, whereas others will be made available by conventional purification tech-niques, and increasingly, others will be obtained by molecular cloning and expression in eukaryotic cells. Probably the latter is the most attractive in the long term because it will provide enzymes uncontaminated by other activities.

Just as oligosaccharides can be synthesized by reversing the action of glyco-sidases, peptides can be efficiently synthesized with proteolytic enzymes, especially when engineered to enhance favorable characteristics, such as enhancing the bond-forming activity over the hydrolytic, as was achieved in a variant of thiolsubtilisin [80]. Limitations of this approach have led to an exploration of monoclonal anti-bodies for peptide synthesis [81]. Perhaps it is not too wildly imaginative to suggest that antibodies could eventually be developed that could catalyze the coupling of unprotected sugar residues to form oligosaccharides. A technique that has been called "native chemical ligation," recently developed for the synthesis of proteins [82,83], has the potential for the synthesis of "large" glycopeptides (glycoproteins?), i.e., those containing at least 200 amino acid residues. In this technique, a thiol ester group added to the carboxyl end of a peptide reacts with the sulfide of a cysteine residue at the amino terminus of a second peptide. The thioester intermediate, initially formed rapidly, rearranges to yield a peptide bond at the ligation site. An enzymatic version of this employs another variant of subtilisin, called subtiligase [84].

B. Future Trends in Analysis

Well-established technologies such as capillary electrophoresis and high-performance (or pressure) liquid chromatography (HPLC), especially the anion exchange method with pulsed amperometric detection [85–89], will achieve even greater resolution and will become easier to use. In contrast to synthesis, where the prospect of automation is still a remote future possibility, the use of instruments for analytical operations previously performed manually is already becoming routine in many glycobiology laboratories. Thus many investigators are successfully employing a machine in which hydrazinolysis is used to achieve the quantitative release of glycans from glycoproteins, with selectivity for O- and N-linked glycans [90]. Another instrument performs the next step—glycan "mapping," by carefully calibrated gel permeation chromatography, with sizing in terms of glucose units. Sequencing is achieved by radiolabeling, digestions with mixtures of exoglycosidases, and high-resolution gel-permeation chromatography (GPC). The resulting profile is analyzed by a computer, which, armed with knowledge of the activity and specificity of each enzyme, generates a structure from measurements of peak elution positions and intensities. Interestingly, this is an example of a trend toward the modern application of old technology, a new level of sophistication being introduced at the level of profile analysis by clever application of software. The overwhelming emphasis is on simplicity and convenience of operation and on the ability to process many samples with minimal effort. Many of the users of such instrumentation are interested in rapid profiling, e.g., of glycosylated recombinant proteins produced under different culture conditions. This activity has been greatly stimulated by the magnitude of the business of producing recombinant protein therapeutics (some $350 million predicted in 1994 alone) and the government-regulated requirement for "correct" glycosylation [91]. Glycan analysis will also be important for the characterization of proteins secreted into the milk of transgenic animals, because the glycans of such proteins will be expected to show species-, tissue-, or cell-specific structural differences from those formed in the usual human organ or tissue.

The need for rapid profiling of multiple samples has also stimulated some new technologies, such as fluorophore-assisted carbohydrate electrophoresis (FACE), in which glycans are released enzymatically, labeled with a fluorescent tag by reductive amination, and separated by polyacrylamide gel electrophoresis [92,93]. The gel profiles are analyzed by an imaging system, and sequencing is achieved by digestion with exoglycosidases followed by reanalysis. As in mapping by GPC, the strength of the method is derived from an ingenious application of a software program for analyzing band (peak) positions, intensities, and shifts. An additional advantage of FACE is that it is low tech and therefore affordable. The same cannot be said for the extremely high-tech methodologies, such as matrix-assisted laser desorption mass spectrometry and electrospray [94], which are currently achieving ever-higher resolution and sensitivity and in some cases the unique ability to handle unseparated mixtures of glycoproteins. Perhaps even more demanding in terms of instrumentation, computer power, and budget are the investigations of oligosaccharide conformation and molecular dynamics in aqueous solution, via multidimensional NMR spectroscopy and NOE measurements, and theoretical calculations correlating potential energy minima with variations in glycosidic torsion angles [4,5,11].

C. Perspectives and Conclusions

The ability to investigate structures in three dimensions is accelerating a trend toward viewing molecules as surfaces on which active sites are displayed, rather than as formulas. There is also a realization that the complementarity of any two interacting surfaces must be defined in both structural and thermodynamic parameters. The new spectroscopic techniques, availability of 3-D search programs [95], and other novel phenomena, such as virtual reality, will provide much more powerful ways of looking at receptor–ligand interactions, such as those between lectins and oligosaccharides, and a refinement of attitudes toward the importance of molecular motion. An example of the latter of prime interest to glycobiologists concerns the extent of torsional oscillation about glycosidic linkages in oligosaccharides. The importance of these phenomena to both academic researchers and pharmaceutical scientists will lead to a blurring of the distinction between academic research and drug development, because both will need the technology and it will be too expensive for either one to pursue alone. If this leads to more collaboration and financial interdependence between industry and universities, there would be advantages for all. Companies could utilize the vast pool of skilled and cheap labor provided by graduate students and postdoctoral fellows, while academic investigators would be less dependent on government handouts (grants) for their survival. Also of immense benefit to everyone will be the new generation of drugs with unprecedented potency and specificity developed as a result of the new understanding of molecular interactions.

In the past several years, as the full experimental potential of molecular biology has emerged, there has been a headlong rush to apply the methods to every imaginable problem and situation. In glycobiology the efforts have mostly concerned the molecular cloning and expression of glycosidases and glycosyltransferases. As discussed earlier, the successful efforts along those lines are already very valuable for the rapid development of enzymatic oligosaccharide and glycopeptide synthesis. A side effect, however, has been that the traditional disciplines based on organic chemistry and enzymology have become eclipsed. This has had a negative impact on the success rate of grant applications to fund worthwhile research that employs those techniques, because they are viewed as unfashionable. It is to be hoped that once the novelty of molecular biology as something overwhelmingly new and exciting has worn off with time, it will take its place beside the older disciplines as one of several, equally valuable experimental approaches. This would allow a renaissance of interest in chemistry and biochemistry, which will be needed to characterize the reactions catalyzed by the cloned enzymes and to identify the products of their reactions.

REFERENCES

1.   A. Varki, Biological roles of oligosaccharides: All of the theories are correct, *Glycobiology 3*: 97 (1993).

2.   L. A. Lasky, Selectins: Interpreters of cell-specific carbohydrate information during inflammation, *Science 258*: 964 (1992).

3.   H. Lis and N. Sharon, Protein glycosylation: Structural and functional aspects, *Eur. J. Biochem. 218*: 1 (1993).

4.   S. W. Homans, Homonuclear three-dimensional NMR methods for the complete assignment of proton NMR spectra of oligosaccharides—application to Galβ1-4(Fucα1-3)GlcNAcβ1-3Galβ1-4Glc, *Glycobiology 2*: 153 (1992).

5.   S. W. Homans, Conformation and dynamics of oligosaccharides in solution, *Glycobiology 3*: 551 (1993).

6.   J. F. Wong, Carbohydrates may form the next wave of interest in biotech, *Gen. Eng. News,* July 1993, p. 22.

7.   J. Taylor-Papadimitriou and A. E. Epenetos, Exploiting altered glycosylation patterns in cancer, *Tibtech 12*: 227 (1994).

8.   M. E. Taylor, K. Bezouska, and K. Drickamer, Contribution to ligand binding by multiple carbohydrate-recognition domains in the macrophage mannose receptor, *J. Biol. Chem. 267*: 1719 (1992).

9.   V. Kery, J. J. F. Krepinsky, C. D. Warren, P. Capek, and P. D. Stahl, Ligand recognition by purified human mannose receptor, *Arch. Biochem. Biophys. 298*: 49 (1992).

10. Y. Ichikawa, Y-C. Lin, D. P. Dumas, G-J. Shen, E. Garcia-Junceda, M. A. Williams, R. Bayer, C. Ketchum, L. E. Walker, J. C. Paulson, and C-H. Wong, Chemical-enzymatic synthesis and conformational analysis of sialyl Lewis^x and derivatives, *J. Am. Chem Soc. 114*: 9283 (1992).

11. T. J. Rutherford and S. W. Homans, Restrained vs free dynamics simulations of oligosaccharides: Application of solution dynamics of biantennary and bisected biantennary N-linked glycans, *Biochemistry 33*: 9606 (1994).

12. R. Sherman-Gold, Companies pursue therapies based on complex cell adhesion molecules, *Gen. Eng. News*, July 1993, p. 6.

13. S. Borman, Race is on to develop sugar-based anti-inflammatory, antitumor drugs, *Chem. Eng. News*, December 1992, p. 25.

14. B. N. Narasinga Rao, M. B. Anderson, J. H. Musser, J. H. Gilbert, M. E. Schaefer, C. F. Foxall, and B. K. Brandley, Sialyl Lewis X mimics derived from a pharmacophore search are selectin inhibitors with anti-inflammatory activity, *J. Biol. Chem. 269*: 19663 (1994).

15. Y. Ichikawa, R. L. Halcombe, and C-H. Wong, Sticky solutions, *Chemistry in Britain*, February 1994, p. 117.

16. S. Borman, Glycotechnology drugs begin to emerge from the lab, *Chem. Eng. News*, June 1993, p. 27.

17. R. Kornfeld and S. Kornfeld, Assembly of asparagine-linked oligosaccharides, *Ann. Rev. Biochem. 54*: 631 (1985).

18. C. D. Warren, Y. Goussault, B. Bugge, S. Nakabayashi, and R. W. Jeanloz, The application of glycosyl polyprenyl phosphate intermediates to the study of glycoconjugate biosynthesis, *Chemica Scripta 27*: 121 (1987).

19. A. Helenius, How N-linked oligosaccharides affect glycoprotein folding in the endoplasmic reticulum, *Mol. Biol. Cell 5*: 253 (1994).

20. S. H. Cheng, R. J. Gregory, J. Marshall, S. Paul, D. W. Souza, G. A. White, C. R. O'Riordan, and A. E. Smith, Defective intracellular transport and processing of CFTR is the molecular basis of most cystic fibrosis, *Cell 63*: 827 (1990).

21. A. D. Elbein, Glycosidase inhibitors: inhibitors of N-linked oligosaccharide processing, *FASEB J. 5*: 3055 (1991).

22. B. Winchester and G. W. J. Fleet, Amino-sugar glycosidase inhibitors: Versatile tools for glycobiologists, *Glycobiology 2*: 199 (1992).

23. P. F. Daniel, B. Winchester, and C. D. Warren, Mammalian α-mannosidases—Multiple forms but a common purpose?, *Glycobiology 4*: 551 (1994).

24. T. Feizi and M. Larkin, AIDS and glycosylation, *Glycobiology 1*: 17 (1990).
25. G. B. Karlsson, T. D. Butters, R. A. Dwek, and F. M. Platt, Effects of the imino sugar *N*-butyldeoxynojirimycin on the *N*-glycosylation of recombinant gp 120, *J. Biol. Chem. 268*: 570 (1993).
26. M. G. Low, The glycosyl-phosphatidylinositol anchor of membrane proteins, *Biochim. Biophys. Acta 988*: 427 (1989).
27. M. A. J. Ferguson, S. W. Homans, R. A. Dwek, and T. W. Rademacher, Glycosyl-phosphatidylinositol moiety that anchors *Trypanosoma brucei* variant surface glyco-protein to the membrane, *Science 239*: 753 (1988).
28. J. R. Thomas, R. A. Dwek, and T. W. Rademacher, Structure, biosynthesis, and function of glycosylphosphatidylinositols, *Biochemistry 29*: 5413 (1990).
29. D. G. Doherty, E. A. Popenoe, and K. P. Link, Amino acid derivatives of D-glucosa-mine, *J. Am. Chem. Soc. 75*: 3466 (1953).
30. E. A. Popenoe, D. G. Doherty, and K. P. Link, Interaction of 1,3,4,6-tetraacetyl-β-D-glucosamine with acyl amino acid azides, *J. Am. Chem. Soc. 75*: 3469 (1953).
31. M. Bergmann and L. Zervas, Uber die synthese von glucopeptiden des *d*-glucosamins (*N*-glycyl-*d*-glucosamin und *N*-*d*-alanyl-*d*-glucosamin), *Chem. Ber. 65*: 1201 (1932).
32. F. Micheel and H. Kochling, Darstellung von Glycosiden des D-Glucosamins mit ali-phatischen und aromatischen Alcoholen und mit serin nach der Oxazolin-Methode, *Chem. Ber. 91*: 673 (1958).
33. J. K. N. Jones, M. B. Perry, B. Shelton, and D. J. Walton, The carbohydrate-protein linkage in glycoproteins. Part 1. The syntheses of some model substituted amides and an L-seryl-D-glucosaminide, *Can. J. Chem. 39*: 1005 (1960).
34. R. D. Marshall and A. Neuberger, Aspects of the structure and metabolism of glyco-proteins, *Adv. Carbohydr. Chem. Biochem. 25*: 407 (1970).
35. G. W. Hart, G. D. Holt, and R. S. Haltiwanger, Nuclear and cytoplasmic glycosylation: Novel saccharide linkages in unexpected places, *Trends Biochem. Sci. 13*: 380 (1988).
36. S. P. Jackson and R. Tjian, *O*-glycosylation of eukaryotic transcription factors: Impli-cations for mechanisms of transcriptional regulation, *Cell 55*: 125 (1988).
37. S. C. Hubbard and R. J. Ivatt, Synthesis and processing of asparagine-linked oligosac-charides, *Annu. Rev. Biochem. 50*: 555 (1981).
38. C. D. Warren and R. W. Jeanloz, Chemical synthesis of dolichyl glycosyl phosphates and pyrophosphates or dolichol intermediates, *Methods Enzymol. 50 (Part C)*: 122 (1978).
39. C. D. Warren and R. W. Jeanloz, Synthesis of P^1-dolichyl P^2α-D-mannopyranosyl py-rophosphate. The acid and alkaline hydrolysis of polyisoprenyl α-D-mannopyranosyl mono- and pyrophosphate diesters, *Biochemistry 14*: 412 (1975).
40. C. D. Warren, I. Y. Liu, A. Herscovics, and R. W. Jeanloz, The synthesis and chemical properties of polyisoprenyl β-D-mannopyranosyl phosphates, *J. Biol. Chem. 250*: 8069 (1975).
41. A. Herscovics, C. D. Warren, and R. W. Jeanloz, The anomeric configuration of the dolichyl D-mannosyl phosphate formed in calf pancreas microsomes, *J. Biol. Chem. 250*: 8079 (1975).
42. S. T. Anisfeld and P. T. Lansbury, A convergent approach to the chemical synthesis of aparagine-linked glycopeptides, *J. Org. Chem. 55*: 5560 (1990).
43. S. T. Cohen-Anisfeld and P. T. Lansbury, A practical, convergent method for glyco-peptide synthesis, *J. Am. Chem. Soc. 115*: 10531 (1993).
44. C. D. Warren and H. G. Garg, Glycopeptide synthesis by the direct glycosylation of a peptide, *Glycobiology 3*: 540 (1993).
45. S. Y. C. Wong, G. R. Guile, T. W. Rademacher, and R. A. Dwek, Synthetic glycosyl-ation of peptides using unprotected saccharide β-glycosylamines, *Glycoconjugate J. 10*: 227 (1993).

46.   S. J. Danishefsky, K. F. McClure, J. T. Randolph, and R. B. Ruggeri, A strategy for
       solid-phase synthesis of oligosaccharides, *Science 260*: 1307 (1993).
47.   C-H. Wong, K. K-C. Liu, T. Kajimoto, L. Chen, Z. Zhong, D. P. Dumas, J. L-C. Liu,
       Y. Ichikawa, and G. J. Shen, Enzymes for carbohydrate and peptide synthesis, *Ind. J.
       Chem. 32B*: 135 (1993).
48.   C-H. Wong, M. Schuster, P. Wang, and P. Sears, Enzymatic synthesis of *N*- and *O*-
       linked glycopeptides, *J. Am. Chem. Soc. 115*: 5894 (1993).
49.   C. D. Warren, P. F. Daniel, B. Bugge, J. E. Evans, L. F. James, and R. W. Jeanloz,
       The structures of oligosaccharides excreted by sheep with swainsonine toxicosis, *J.
       Biol. Chem. 263*: 15041 (1988).
50.   P. G. Johansen, R. D. Marshall, and A. Neuberger, The preparation and some of the
       properties of a glycopeptide from hen's egg albumin, *Biochem. J. 78*: 518 (1961).
51.   R. S. Clark, S. Banerjee, and J. K. Coward, Yeast oligosaccharyltransferase: Glyco-
       sylation of peptide substrates and chemical characterization of the glycopeptide prod-
       uct, *J. Org. Chem. 55*: 6275 (1990).
52.   J. W. Dennis, S. Laferte, C. Waghorne, M. L. Breitman, and R. S. Kerbel, β1-6 branch-
       ing in Asn-linked oligosaccharides is directly associated with metastasis, *Science 236*:
       582 (1987).
53.   J. W. Dennis, Oligosaccharides in carcinogenesis and metastasis, *Glyco News*, Oxford
       Glycosystems, Inc., 1992, p. 1.
54.   M. J. Pramik, New biotech treatment binds toxins from *E. coli*, *Gen Eng News*, June
       1993, p. 1.
55.   T. Corfield, Tailor-made sialidase inhibitors home in on influenza virus, *Glycobiology
       3*: 413 (1993).
56.   G. W. J. Fleet, A. Karpas, R. A. Dwek, L. E. Fellows, A. S. Tyms, S. Petursson, S.
       K. Namgoong, N. G. Ramsden, P. W. Smith, J. C. Son, F. Wilson, D. R. Witty, G. S.
       Jacob, and T. W. Rademacher, Inhibition of HIV replication by amino sugar derivatives,
       FEBS Lett. 237: 128 (1988).
57a.   F. M. Platt, G. R. Neises, R. A. Dwek, and T. D. Butters, *N*-Butyldeoxynojirimycin is
       a novel inhibitor of glycolipid biosynthesis, *J. Biol. Chem. 269*: 8362 (1994).
57b.   F. M. Platt, G. R. Neises, G. B. Karlsson, R. A. Dwek, and T. D. Butters, *N*-Butyl-
       deoxygalactonojirimycin inhibits glycolipid biosynthesis but does not affect *N*-linked
       oligosaccharide processing, *J. Biol. Chem. 269*: 27108 (1994).
57c.   E. Beutler, Gaucher disease as a paradigm of current issues regarding single gene
       mutations of humans, *Proc. Nat. Acad. Sci. USA 90*: 5384 (1993).
58.   K. Drickamer, Three-dimensional view of a selectin cell adhesion molecule, *Glyco-
       biology 4*: 245 (1994).
59.   J. K. Welply, S. Zaheer Abbas, P. Scudder, J. L. Keene, K. Broschat, S. Casnocha, C.
       Gorka, C. Steininger, S. C. Howard, J. J. Schmuke, M. Graneto, J. M. Rotsaert, I. D.
       Manger, and G. S. Jacob, Multivalent sialyl-Le-X; Potent inhibitors of E-selectin-
       mediated cell adhesion; reagent for staining activated endothelial cells, *Glycobiology
       4*: 259 (1994).
60.   M. Larkin, T. J. Ahern, M. S. Stoll, M. Shaffer, D. Sako, J. O'Brien, C-T. Yuen,
       A. M. Lawson, R. A. Childs, K. M. Barone, P. R. Langer-Safer, A. Hasegawa, M.
       Kiso, G. R. Larsen, and T. Feizi, Spectrum of sialylated and nonsialylated fuco-oligo-
       saccharides bound by endothelial-leukocyte adhesion molecule E-selectin, *J. Biol.
       Chem. 267*: 13661 (1992).
61.   A. Varki, Selectin ligands, *Proc. Nat. Acad. Sci. USA 91*: 7390 (1994).
62.   B. K. Brandley, M. Kiso, S. Abbas, P. Nikrad, O. Srivastava, C. Foxall, Y. Oda, and
       A. Hasegawa, Structure–function studies on selectin carbohydrate ligands. Modifica-
       tions to fucose, sialic acid and sulphate as a sialic acid replacement, *Glycobiology 3*:
       633 (1993).

63. A. Etzioni, M. Frydman, S. Pollack, I. Avidor, M. L. Philips, J. C. Paulson, and R. Gershoni-Baruch, Brief report: Recurrent severe infections caused by a novel leukocyte adhesion deficiency, *New Eng. J. Med. 327*: 1789 (1992).
64. J. Jaeken, H. Carchon, and H. Stibler, The carbohydrate-deficient glycoprotein syndromes: Pre-Golgi and Golgi disorders, *Glycobiology 3*: 423 (1993).
65. B. A. Hagberg, G. Blennow, B. Kristiansson, and H. Stibler, Carbohydrate-deficient glycoprotein syndromes: Peculiar group of new disorders, *Pediatr. Neurol. 9*: 255 (1993).
66. K. Yamashita, H. Ideo, T. Ohkura, K. Fukushima, I. Yuasa, K. Ohno, and K. Takashita, Sugar chains of serum transferrin from patients with carbohydrate-deficient glycoprotein syndrome, *J. Biol. Chem. 268*: 5783 (1993).
67. E. Van Schaftingen and J. Jaeken, Phosphomannomutase deficiency is a cause of carbohydrate-deficient glycoprotein syndrome type 1, *FEBS Lett. 377*: 318 (1995).
68. R. D. Klausner and R. Sitia, Protein degradation in the endoplasmic reticulum, *Cell 62*: 611 (1990).
69. Y. Ichikawa, G. C. Look, and C-H. Wong, Enzyme-catalyzed oligosaccharide synthesis, *Anal. Biochem. 202*: 215 (1992).
70. M. M. Palcic, O. P. Srivastava, and O. Hindsgaul, Transfer of D-galactosyl groups to 6-O-substituted 2-acetamido-2-deoxy-D-glucose residues by use of bovine D-galactosyl transferase, *Carbohydr. Res. 159*: 315 (1987).
71. E. Yoon and R. A. Laine, Synthesis of four novel trisaccharides by induction of loose acceptor specificity in Galβ1-4 transferase (EC 2.4.1.22):Galp(β1-4)Glcp(X)Glc where X = β1-3: β1-4: β1-6: α1-4, *Glycobiology 2*: 161 (1992).
72. A. Lubineau, C. Auge, and P. Francois, The use of porcine liver (2-3)-α-sialyltransferase in the large-scale synthesis of α-Neup5Ac-(2-3)-β-D-Galp-(1-3)-D-GlcpNAc, the epitope of the tumor-associated carbohydrate antigen CA 50, *Carbohydr. Res. 228*: 137 (1992).
73. M. M. Palcic, A. P. Venot, R. M. Ratcliffe, and O. Hindsgaul, Enzymic synthesis of oligosaccharides terminating in the tumor-associated sialyl-Lewisª determinant, *Carbohydr. Res. 190*: 1 (1989).
74. K. J. Kaur, G. Alton, and O. Hindsgaul, Use of N-acetylglucosaminyltransferases I and II in the preparative synthesis of oligosaccharides, *Carbohydr. Res. 210*: 145 (1991).
75. K. J. Kaur and O. Hindsgaul, Combined chemical-enzymic synthesis of a dideoxypentasaccharide for use in a study of the specificity of N-acetylglucosaminyl transferase-III, *Carbohydr. Res. 226*: 219 (1992).
76. R. Oehrlein, O. Hindsgaul, and M. M. Palcic, Use of the ''core-2''-N-acetylglucosaminyltransferase in the chemical-enzymatic synthesis of sialyl-Leˣ-containing hexasaccharide found on O-linked glycoproteins, *Carbohydr. Res. 244*: 149 (1993).
77. M. A. Kashem, C. Jiang, A. P. Venot, and G. R. Alton, Combined chemical-enzymic synthesis of an internally monofucosylated hexasaccharide corresponding to the CD-65/VIM-2 epitope: Use of a terminal α2-6-linked N-acetylneuraminic acid as a temporary blocking group, *Carbohydr. Res. 230*: C7 (1992).
78. G. Srivastava, K. J. Kaur, O. Hindsgaul, and M. M. Palcic, Enzymatic transfer of a preassembled trisaccharide antigen to cell surfaces using a fucosyltransferase, *J. Biol. Chem. 267*: 22356 (1992).
79. H. Waldmann, A. Heuser, P. Braun, and H. Kunz, New enzymatic protecting group techniques for peptide and carbohydrate chemistry, *Ind. J. Chem. 31B*: 799 (1992).
80. L. Abrahmsen, J. Tom, J. Burnier, K. A. Butcher, A. Kossiakoff, and J. A. Wells, Engineering subtilisin and its substrates for efficient ligation of peptide bonds in aqueous solution, *Biochemistry 30*: 4151 (1991).
81. R. Hirschmann, A. B. Smith III, C. M. Taylor, P. A. Benkovic, S. D. Taylor, K. M. Yager, P. A. Sprengeler, and S. J. Benkovic, Peptide synthesis by an antibody containing a binding site for variable amino acids, *Science 265*: 234 (1994).

82. C-F. Liu and J. P. Tam, Chemical ligation approach to form a peptide bond between unprotected peptide segments. Concept and model study, *J. Am. Chem. Soc 116*: 1135 (1994).
83. P. E. Dawson, T. W. Muir, I. Clark-Lewis, and S. B. H. Kent, Synthesis of proteins by native chemical ligation, *Science 266*: 776 (1994).
84. D. Y. Jackson, J. Burnier, C. Quan, M. Stanley, J. Tom, and J. A. Wells, A designed peptide ligase for total synthesis of ribonuclease A with unnatural catalytic residues, *Science 266*: 243 (1994).
85. M. R. Hardy and R. R. Townsend, Separation of positional isomers of oligosaccharides and glycopeptides by high-performance anion-exchange chromatography with pulsed amperometric detection, *Proc. Nat. Acad. Sci USA 85*: 3289 (1988).
86. L-M. Chen, M-G. Yet, and M-C. Shao, New methods for rapid separation and detection of oligosaccharides from glycoproteins, *FASEB J. 2*: 2819 (1988).
87. R. R. Townsend, M. R. Hardy, D. A. Cumming, J. P. Carver, and B. Bendiak, Separation of branched sialylated oligosaccharides using high-pH anion-exchange chromatography with pulsed amperometric detection, *Anal. Biochem. 182*: 1 (1989).
88. G. P. Reddy and C. A. Bush, High-performance anion-exchange chromatography of neutral milk oligosaccharides and oligosaccharide alditols derived from mucin glycoproteins, *Anal. Biochem. 198*: 278 (1991).
89. E. Watson, A. Bhide, W. C. Kenney, and F-K. Lin, High-performance anion-exchange chromatography of asparagine-linked oligosaccharides, *Anal. Biochem. 205*: 90 (1992).
90. T. Merry, Carrying out routine glycosylation analysis of monoclonal antibodies, *Gen. Eng. News*, July 1994, p. 16.
91. K. Seamon, Evaluation of recombinant glycoproteins, *Glyco News*, Oxford Glycosystems, Inc., 1991, p. 5.
92. C. M. Starr, R. I. Masada, and C. Hague, Characterization of the glycosylation of recombinant gp120 using FACE: Fluorophore-assisted carbohydrate-electrophoresis, *Glycobiology 3*: 511 (1993).
93. R. I. Masada and C. M. Starr, Sequencing the *N*-linked oligosaccharides on rgp120 using FACE, *J. NIH Res. 5*: 78 (1993).
94. V. N. Reinhold, B. B. Reinhold, and S. Chan, Carbohydrate sequence analysis, (ES-MS/CID/MS): High resolution separation of biological macromolecules, *Methods Enzymol. 271*: 377 (1996).
95. S. Borman, New 3-D search and de novo design techniques aid drug development, *Chem. Eng. News*, August 1992, p. 18.

2

Recent Advances in the Synthesis of Glycopeptides

Horst Kunz and Michael Schultz
Institut für Organische Chemie, Johannes Gutenberg-Universität, Mainz, Germany

I. INTRODUCTION

For a long time carbohydrates and polysaccharides on the one hand and peptides and proteins on the other had been considered separate classes of natural products. Emil Fischer and his co-workers had developed the fundamental chemical methodology for the synthesis of both saccharides [1] and peptides [2]. However, the analytical techniques available at this time were insufficient for a detailed analysis of complex biomacromolecules, and the general importance of glycoconjugates, in particular of glycoproteins, remained undiscovered. In the first decades of the century, synthetic conjugates between saccharide and peptide components of quite artificial structure were prepared, for example, in the nonselective diazo coupling of p-diazoniumphenyl glycosides with amino acid side chains of proteins [3].

Exciting results of cell biological, biochemical, and immunological research obtained during the past three decades have shown that glycosylation is a very common posttranslational modification of proteins in eukaryotic cells. Actually, most natural proteins of multicellular organisms are glycoproteins; that is, they contain covalently linked saccharide portions [4]. Besides their influence on physicochemical properties and on the preferred conformation, carbohydrate parts of glycoproteins can also play crucial functions in biological recognition processes, e.g., in the selective uptake of serum components into cells [5], in infectious processes [6], in cell adhesion [7], or in the regulation of cell growth [8].

Corresponding to the increasing interest in the investigation of biological recognition phenomena, the synthesis of glycopeptides representing partial structures of glycoproteins is receiving more and more attention.

A. Linkage Regions Between Glycans and the Peptide Backbone

Despite the large number of different glycoproteins, the linkage between the peptide and the carbohydrate portions shows only limited variation. Most frequently found are N-glycoproteins, which contain a N-glycosidic bond between N-acetylglucosamine and the amide group of an asparagine (**1**) [9]. The sites of N-glycosylation in N-glycoproteins are characterized by sequons Asn-X-Ser/Thr. The hydroxyl group of serine or threonine, respectively, apparently plays a crucial role in the biosynthetic transfer of the glycan chain from the glycolipid precursor to the amide group of asparagine [10,11].

Among the O-glycoproteins with important biological functions in mammalians, the mucin-type glycoproteins are characterized by α-O-glycosidic linkages between N-acetylgalactosamine and threonine or serine (**2**) [12,13]. The typical linkage structure of proteoglycans consists of the β-xylosyl-serine unit **3** [14].

Early efforts in synthetic glycoconjugate chemistry were aimed at the synthesis of such glycosylated amino acid derivatives, which served as model compounds in proving the proposed structure of the linkage regions of glycoproteins. For example, conjugates of type **1** were synthesized by condensation of N-benzyloxycarbonyl aspartic acid α-benzyl ester **6** with O-protected 1-amino-N-acetylglucosamine derivatives (**5**) [15,16]. The β-xylosyl serine structure **3** was first obtained according to the Koenigs-Knorr methodology [17].

In contrast to the structures **1**, **2**, and **3**, O-(N-acetylglucosamine)serine and threonine conjugates **4** [18,19] and even glycopeptides of this type [20,21] had been

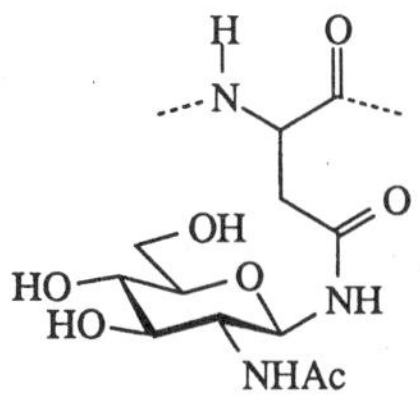

1

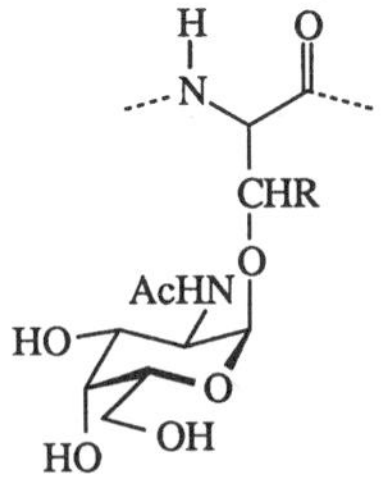

2 R= H, CH_3

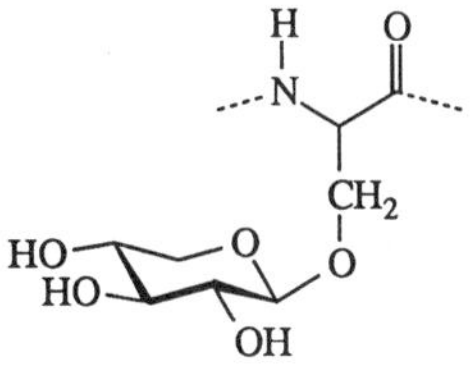

3

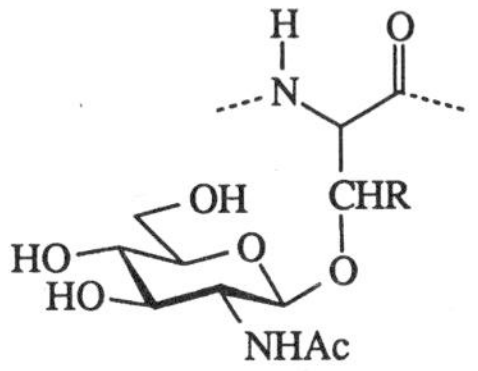

4 R= H, CH_3

Scheme 1 Typical linkage regions between glycans and the peptide backbone.

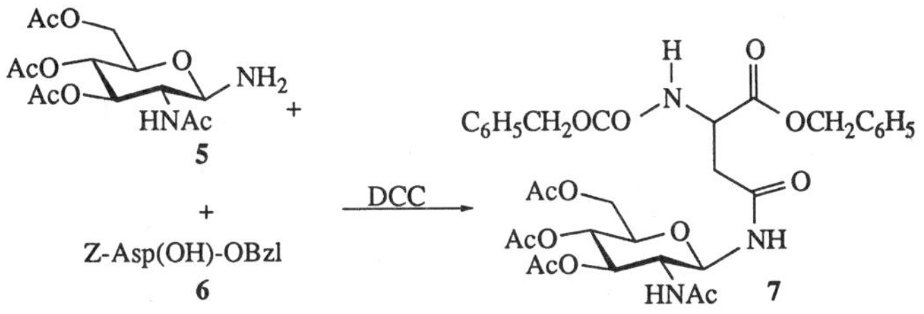

Scheme 2 Synthesis of the characteristic *N*-glycoprotein linkage region between aspartic acid and glucosamine.

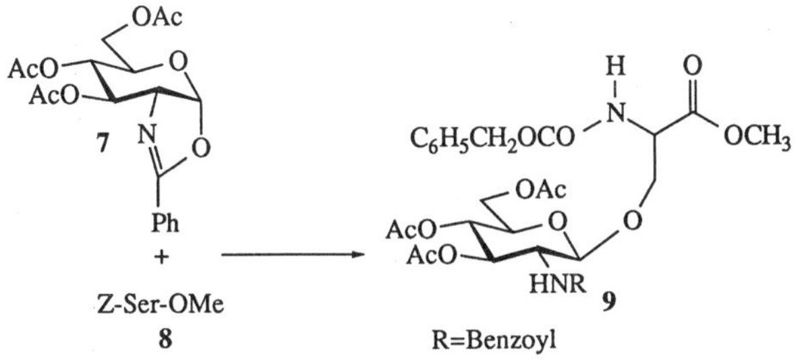

Scheme 3 Synthesis of a widely occurring *O*-glycoprotein linkage region between serine and glucosamine

synthesized before the occurence of structure **4** in ubiquitous mammalian glycoproteins was discovered [22].

The synthesis of the fully protected glucosamine serine ester **9** from the corresponding glucosamine-derived oxazolidine **7** by Micheel et al. marks the beginning of the chemical synthesis of biologically interesting glycan-peptide conjugates. However, at this time and until the early '80s problems concerning the selective deprotection of glycosyl amino acid derivatives, the chain extension to give glycopeptides of defined sequence, the complete deprotection of these complex compounds, and their coupling to proteins had not been solved [23]. Since the investigation of biological recognition phenomena and the elucidation of structure-function relationships of enzymes and lectins as well as the immunodifferentiation by antibodies requires glycoproteins of exactly specified structure, the development of a selective and efficient synthetic methodology was required. In addition to the progress in oligosaccharide synthesis [24,25] chemoselective protecting group strategies had to be elaborated that allow the selective deblocking of one functional group of these polyfunctional conjugates without affecting the other protecting groups and sensitive structures, e.g., glycosidic bonds [26–28].

In the following chapter a brief summary of protecting groups relevant to glycopeptide chemistry is given. Subsequently, the formation of *N*- and *O*-glycosidic linkages and the conversion of these key building blocks into complex glycopeptides will be illustrated on selected examples.

II. PROTECTING GROUPS IN GLYCOPEPTIDE CHEMISTRY

Depending on the chosen synthetic strategy, the number of protecting groups required for the different functionalities (Scheme 4) varies. The most common strategy of glycopeptide synthesis involves the preparation of a more or less complex saccharide, which is subsequently coupled to the corresponding amino acid derivative (Asp, Ser, Thr). For several reasons, it is advantageous to have only one type of carbohydrate protecting group (usually acyl type) prior to the attachment to the amino acid.

For this approach, at least four different types of protecting groups are necessary in order to accomplish the protection and deprotection of the amino, the carboxylic, and the side-chain functions of the peptide portion and of the hydroxyl groups within the carbohydrate. Moreover, in order to facilitate the extension of the peptide chain, the amino and carboxylic protections must be removable independently.

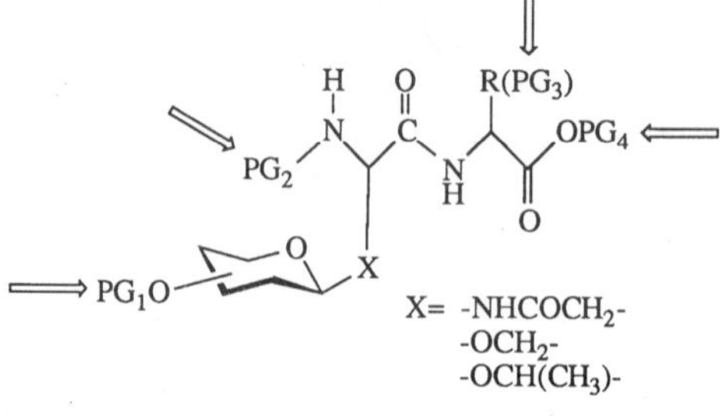

Scheme 4   Functional groups in glycopeptides that require suitable protection.

As a rule, the reaction conditions applied during the protecting group manipulations and the coupling steps must take into consideration the lability of the glycosidic linkages, which are generally sensitive to acids (acetal cleavage, anomerization). In addition, O-glycosyl serine and threonine derivatives are also sensitive to bases, which can cause β-elimination of the carbohydrate portion. Of course, treatment with bases may also result in epimerization of amino acid units. Therefore, an efficient protecting group strategy should combine complete orthogonality of the protecting groups with mild reaction conditions.

Examination of the available protecting groups shows eight major types of cleavage reactions applicable to glycopeptide synthesis. They are compiled in Table 1 together with the corresponding cleavage reagents. The discussed cleavage mechanisms and possible combinations of these protecting groups are briefly outlined below, except for the enzymatic methods, which will be introduced in Section V.A.

A. Benzyl-Type Protection

The hydrogenolytic cleavage of the N-benzyloxycarbonyl group (Scheme 5), benzyl esters, and benzyl ethers [29] is quite suitable for selective deprotections in glycopeptide synthesis. Actually, it was successfully applied in early preparations of completely deblocked glycosyl amino acids [17].

Table 1 Types of Protecting Groups Reliable in Glycopeptide Chemistry

Cleavage principle	Reagents	Protecting group	Ref.
Hydrogenolysis	H_2/palladium	benzyloxycarbonyl (Z) benzyl ether (Bzl) benzyl ester (OBzl) methoxybenzyl ether	29, 30
Acidolysis	Trifluoroacetic acid Formic acid, HCl/ether	tert-butyloxycarbonyl (Boc) tert-butyl ether (tBu) tert-butylester (OtBu) methoxybenzyl ether (Mpm)	31
Base-promoted cleavage	Morpholine, piperidine NaOMe/MeOH, NH_2NH_2/MeOH	9-fluorenylmethoxycarbonyl (Fmoc) O-acetyl (OAc)	32
Reductive cleavage	Zn/acetic acid	2,2,2-trichloroethoxycarbonyl (Tcoc)	33
Oxidative cleavage	Ceric ammonium nitrate 5,6-Dichloro-2,3-dicyano- 1,4-benzo-quinone (DDQ)	methoxybenzyl ether (Mpm)	34, 35
Metal-complex catalyzed cleavage	[(Ph$_3$P)$_4$Pd^0]/nucleophile	allyl ester (OAll) allyloxycarbonyl (Aloc)	36, 37
Photolysis	hν	o-nitrobenzyl (Nb)	38
Enzymatic cleavage	Lipases, esterases, proteases	alkyl and alkoxyalkylesters	39, 40

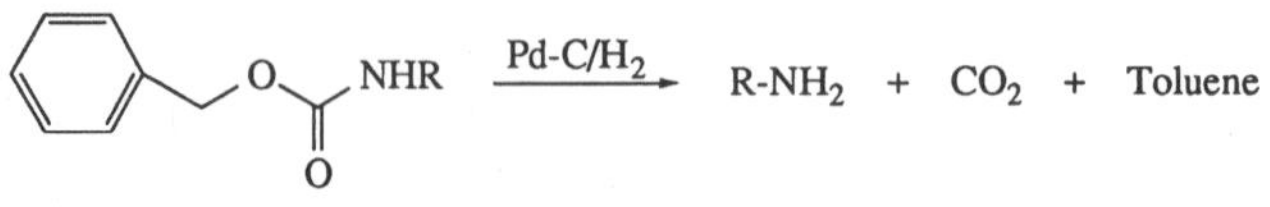

Scheme 5   Hydrogenolysis of the benzyloxycarbonyl group.

Combinations as benzyl/methoxybenzyl, benzyl/trityl, or benzyloxycarbonyl/allyl do not exhibit orthogonal stability since both benzyl-type groups and allyl moieties are attacked under hydrogenolytic conditions. Fluorenylmethoxycarbonyl (Fmoc, see next subsection) in combination with benzyl esters has successfully been used in peptide [41] and glycopeptide [42] synthesis, although the Fmoc group was reported to be not completely stable towards hydrogenolysis, in particular under hydrogen transfer conditions [41]. Hence, these reactions have to be performed carefully.

The deprotection of complex perbenzylated molecules is sometimes afflicted with problems due to incomplete conversions even at high temperatures and hydrogen pressures. Strongly acidic conditions, e.g., HBr in acetic acid, that have been used for the detachment of the benzyloxycarbonyl group must be avoided in glycopeptide synthesis, since glycosidic bonds are cleaved or anomerized under these circumstances.

B. Fluorenylmethyloxycarbonyl (Fmoc) Group

The fluorenylmethoxycarbonyl (Fmoc) group [32] can be rapidly cleaved under moderately basic conditions. This method was shown to be very valuable for glycopeptide synthesis.

Application of the weak base morpholine (pk_a 8.33) results in the selective and complete removal of this protecting group. The intrinsic basicity of morpholine is sufficient to initiate the E1cB cleavage of the Fmoc group but is obviously too low to cause β-elimination of the carbohydrate from O-glycosyl serine and threonine derivatives [42]. More recent results [43,44] suggest that stronger bases than piper-

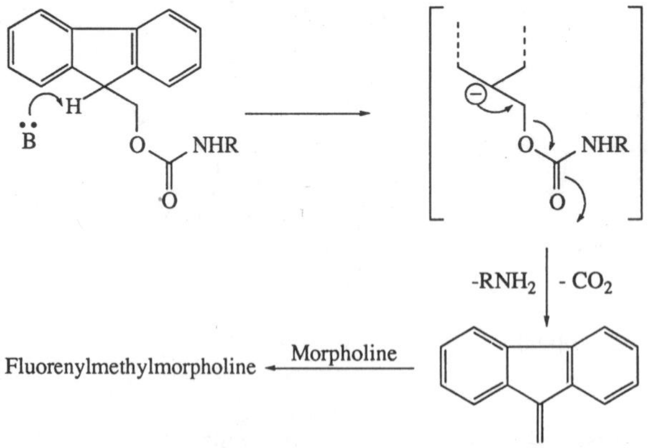

Scheme 6   Mechanism of the base-promoted removal of the Fmoc group.

idine (pk$_a$ 11.03) [45] and very dilute sodium methoxide in methanol can be applied, provided that the amino acid is bound as an amide (not as an ester) and that water is excluded.

A successful use of the Fmoc group demands that a potent nucleophile that efficiently traps the formed fulvene be present during the cleavage process. Both morpholine and piperidine simultaneously serve as cleaving bases and trapping nucleophiles.

The Fmoc group can be applied in glycopeptide synthesis in combination with the benzyl ester [42], the *tert*-butyl ester [45], and the allyl ester [46,47]. Due to the mild and selective removal under homogenous reaction conditions, which do not affect acid-sensitive and allylic protecting groups, the Fmoc group is successfully used as the temporary amino protection in the solid phase synthesis of glycopeptides [48–50].

C.   *tert*-Butyl-Type Protection

The slightly acidic conditions required for the removal of *tert*-butyl groups usually allow the use of the *tert*-butyloxycarbonyl (Boc) group as well as that of *tert*-butyl esters and ethers in the synthesis of glycopeptides [51]. However, the presence of acetalic glycosidic bonds within glycopeptides, not only in the form of the glycosidic linkage to the amino acid but also as the intersaccharidic bonds, demands that special preconditions be fulfilled if the application of the acidolytic cleavage of the *tert*-butyl-type groups is applied. For example, the removal of the *tert*-butyl ester from a fucosyl-chitobiose-asparagine conjugate that contained an *O*-benzyl-protected α-fucosyl moiety resulted in complete cleavage of the fucoside bond [52,53] even if formic acid was applied. But it was demonstrated in this study that the exchange of the *O*-benzyl for *O*-acyl protection in the fucoside results in a marked stabilization of the glycosidic bonds towards acids [52,53]. This stabilization is illustrated by the successful removal of the *N*-Boc or *tert*-butyl ester protection from Lewis^x-type glycopeptides containing exclusively *O*- and *N*-acetyl groups within the carbohydrate portion. Even by application of HCl in diethyl ether, the glycoside bonds remained unaffected [54].

A particular problem of the acidolytic cleavage of groups producing intermediate stabilized cations consists in alkylations as side reactions. Usually, trapping nucleophiles, e.g., thiols, have to be added [55].

The introduction of the Boc group via the pyrocarbonate [56,57] and the formation of *tert*-butyl esters [58] is well established. The selective formation of *tert*-butyl esters of serine and threonine in the presence of the hydroxyl group, however, requires special conditions, e.g. the application of *tert*-butanol, DCC and CuCl under specified conditions [59–61].

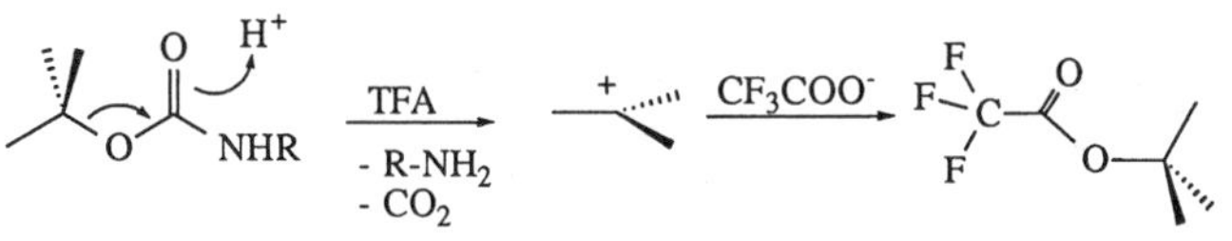

Scheme 7   Acidolysis of the Boc group using TFA.

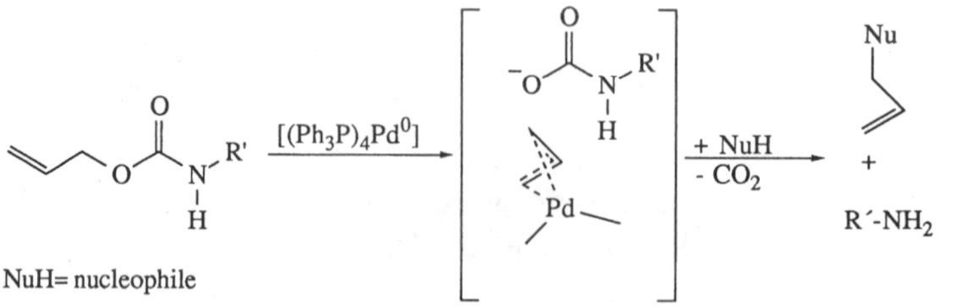

Scheme 8   Pd(0)-catalyzed removal of the Aloc group.

D.   Allyl Esters and the Allyloxycarbonyl (Aloc) Group

Palladium(0)-catalyzed allyl transfer to nucleophiles is very popular in synthetic organic chemistry as a C—C-bond forming reaction. Its application in peptide and glycopeptide synthesis [37,62] opened up the way to amino [37] and carboxylic [62] protecting groups, e.g., allylurethanes (Aloc) and allylesters, which are stable to both acids and bases and can be removed under practically neutral conditions.

However, a prerequisite for a successful application of the Aloc group and the allyl ester in peptide chemistry is that the allylic moiety be irreversibly trapped from the π-allyl palladium complex. Suitable reagents for this aim are morpholine [62], dimedone [37], N,N'-dimethyl barbituric acid [63] or other weakly basic nucleophiles [64,65].

The use of carboxylates as trapping nucleophiles successfully applied in β-lactam chemistry [66] results in the formation of allylic esters which themselves are reversibly cleavable by the same mechanism and finally cause undesired allylations of the peptide and glycopeptide functional groups.

Allyl-type esters have also been used as a novel anchoring principle in solid-phase glycopeptide synthesis and therein facilitate an exceptionally mild detachment of the target molecule from the polymeric support [48,67,68].

It should be mentioned that the transition-metal-catalyzed cleavage of allyl ethers follows a different reaction mechanism via an intermediate enol ether that is solvolyzed under slightly acidic conditions. All types of allylic protecting groups are affected under hydrogenolytic conditions, which limits their combination with benzyl-type protecting groups.

E.   Trichloroethoxycarbonyl (Tcoc) Group

The application of reducing agents such as zinc in acetic acid [33,69] or cobalt(I) phthalocyanine [70] induces the reductive β-elimination of dichloroethylene from trichloroethyl esters [69] and trichloroethoxycarbonyl (Tcoc) [33] derivatives and,

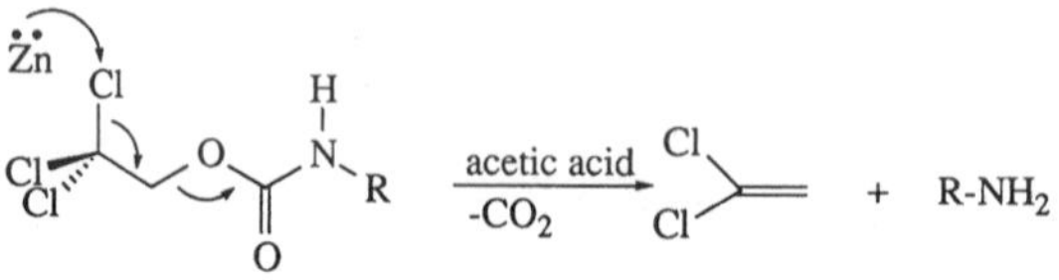

Scheme 9   Reductive elimination of the Tcoc group using zinc in acetic acid.

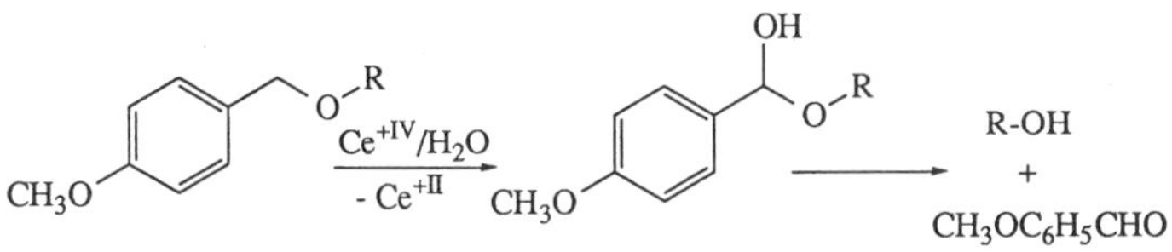

Scheme 10 Oxidative cleavage of Mpm ethers.

thus, results in the deblocking of the carboxylic or amino groups, respectively. This reaction is mild and selective. The Tcoc group is stable towards hydrogenolysis and can also be used as an amino-protecting group for amino sugars. However, a limitation, arises from its limited stability towards basic media [71]. For instance, in the presence of the Tcoc group, the O-deacetylation of saccharides has to be carried out with hydrazine in methanol [42] instead of sodium methoxide/methanol [50].

F. 4-Methoxybenzyl (Mpm) Ethers

Like benzyl ethers, methoxybenzyl (Mpm) ethers are valuable protecting groups in carbohydrate chemistry. They can be oxidized with ceric ammonium nitrate ($[Ce(NH_4)_2(NO_2)_6]$ [34] or 5,6-dichloro-2,3-dicyano-1,4-benzoquinone (DDQ) [35]. The intermediate hemi-acetal rapidly hydrolyzes and sets the hydroxyl group free (Scheme 10).

Beyond the reaction just outlined above, Mpm ethers can be cleaved by acid-olysis and by hydrogenation. Protecting-group combinations such as Mpm/tBu and Mpm/Bzl only allow a one-directional selective deblocking, that is, the oxidative removal of the Mpm group.

G. Acetal-Type Protection

Acetal-type protecting groups are routinely used in the synthesis of the saccharide portions of glycopeptides. The described selectivity between methoxybenzyl and benzyl ethers can be transferred to the corresponding diol-protecting benzylidene (Bzd) and p-methoxybenzylidene (Pmb) groups. However, the acid lability of the acetals is significantly enhanced in comparison to the corresponding ethers. Under reducing conditions and depending on the reagents, the regioselective opening of the cyclic acetal structure can be accomplished. The use of sodium cyanoborohydride and tri-fluoroacetic acid [72] leads to the corresponding 6-O-protected products, while treat-

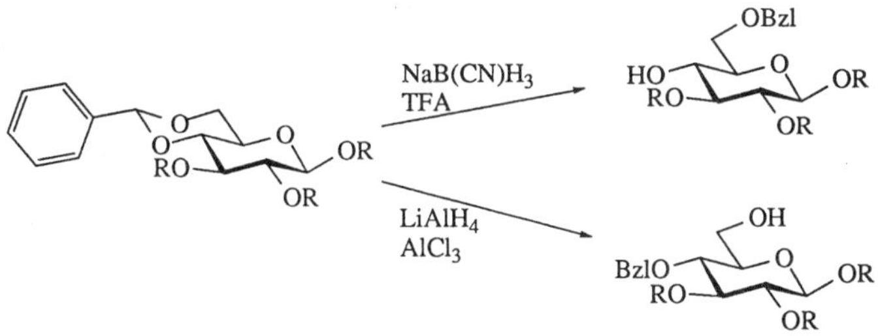

Scheme 11 Possibilities for a regioselective opening of 4,6-benzylidene acetals.

ment with lithium aluminum hydride and aluminum trichloride [73] gives the 4-O-benzyl ether.

H. O-Acyl-Type Protection

The use of acetyl-, benzoyl-, or pivaloyl-protected carbohydrates is inevitable for the construction of 1,2-*trans*-glycosides, in particular, β-linked saccharides, since these neighboring groups cause the formation of an intermediate 1,2-acyloxonium cation that is subsequently attacked by nucleophiles from the trans-(β-)side (**I.**a) [24]. The formation of the undesired orthoester from intermediate **I**(b.) (Scheme 12) can often be observed in basic reaction media. In the absence of a 2-O-acyl protecting group, the *cis*-(α-)product is generally favored, due to the anomeric effect, and is usually obtained as the major product.

   The acid-stability of glycosidic linkages can be significantly enhanced if the carbohydrate hydroxyl groups are peracylated [52]. During deprotections that require acidic conditions (e.g., cleavage of *tert*-butyl esters with trifluoro acetic acid), ester groups exhibit a stronger inductive (-I) effect, and ester carbonyl oxygens are partly protonated and, thus, exert a Coulomb repulsion towards proton attack at the acetal linkages (for an example, see Section II.C. *tert*-Butyl-Type Protection).

   The introduction of acyl groups using common acylating reagents as anhydrides or acylchlorides in combination with pyridine proceeds under mild conditions in high yield. In contrast, special attention has to be paid to the base-promoted cleavage of these esters in the presence of O-glycosidic linkages to serine or threonine. In these cases, β-elimination of the carbohydrate portion (Scheme 12, **II**) can occur. Such undesired side reactions are avoided, for example, if catalytic amounts of sodium methoxide in methanol [50] or, even milder, hydrazine in methanol [42] are applied.

   The combined application of the briefly described protecting groups ensures that selective N- or, alternatively, C-terminal deprotection of glycosyl amino acids and glycopeptides and subsequent chain-extending glycopeptide synthesis can nowadays be carried out with high efficiency. It should be mentioned that side-chain functional groups of amino acids must also be blocked reversibly. With this aim, further types of protecting groups are needed during the synthesis of these poly-functional molecules. First of all, however, the stereo- and regioselective construction

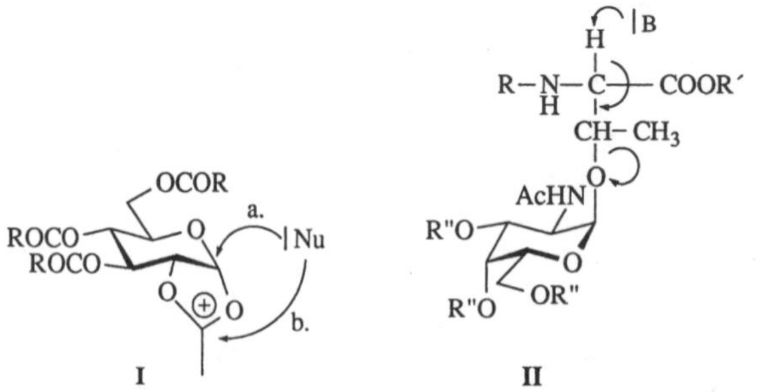

Scheme 12   Orthoester formation during O-glycosylation and β-elimination of the glycosyl portion under basic conditions.

of oligosaccharides remains a special challenge and prerequisite of an efficient synthesis of biologically interesting glycopeptides.

III. OLIGOSACCHARIDE PORTIONS OF BIOLOGICAL RELEVANCE

Some significant problems occurring during the synthesis of important glycoprotein-related saccharide structures will be outlined in this section. In addition, strategies that facilitate an efficient coupling of the synthetic oligosaccharides to the peptide backbone will be dealt with. With regard to the large number of required reaction steps and the complexity of the target compounds caused by their polyfunctionality, the synthesis of oligosaccharides often is by far the most laborious part of glycopeptide synthesis [24–26].

A. Lewis-Type Oligosaccharides

Glycoconjugates containing Lewis-type structures, especially Lewis[a] and Lewis[x], are of particular interest, since their crucial functions in biological recognition processes have been elucidated only recently [74–77]. In connection with at least two pathogenic phenomena of pharmacological interest—namely, tumor genesis [78] and inflammatory diseases [79], Lewis-type antigens (Scheme 13) and structures derived thereof play most important roles. It has been discovered [80] that neutrophils while patrolling through the blood circulation carry the sialyl Lewis X (SLe[x]) antigen on their surface. If inflamed tissue is encountered, this ligand is recognized by cytokine-activated vascular endothelial cells. The so-called E-selectin (ELAM-1) has been identified as the receptor on these cells specifically interacting with the SLe[x] molecules on lymphocytes. After ligand-receptor binding has occurred, a cascade of adhesion processes is initiated that finally enables the neutrophils to pass between the endothelial cells into the inflamed surrounding tissue [76]. If the process of lymphocyte-endothelium recognition can be efficiently inhibited, a promising idea for the treatment of diseases such as rheumatoid arthritis will become realizable.

Sialyl Le[a], a regio-isomer of SLe[x], as well as the asialo compounds also show affinity to E-selectin. The binding of the asialo compounds is significantly weaker [81]. However, sulfated Lewis-type antigens have recently been shown to display increased affinity [82]. A sulfated Le[a] antigen was recently reported to be the most potent ligand known for E-selectin [83].

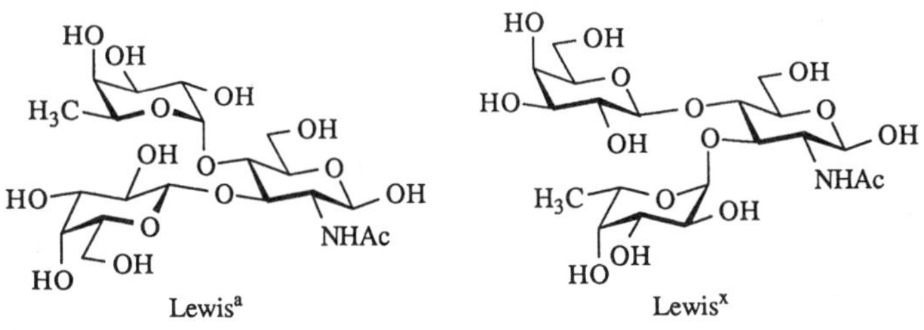

Scheme 13   The regioisomeric Lewis[a] and Lewis[x] structure.

With respect to the important recognition effects, the synthesis of Lewis-type antigens and their linkage to lipids and peptides is a challenging task [84]. For the synthesis of Lewis[a] and Lewis[x] glycopeptides, different strategies compared to those applied in "pure" carbohydrate chemistry are required. Particular problems arising in syntheses of glycopeptides that contain Lewis and sialyl-Lewis antigen saccharides are:

1.   The stereoselective formation of α-fucosides is only possible using ether-type-protected fucosyl donors. The resulting ether-type-protected fucoside bonds, however, are sensitive to acids, even to formic acid, usually applied in glycopeptide synthesis [52,53].
2.   An anomeric protecting group is needed, which facilitates a mild and stereoselective sequence of deblocking, activation, and coupling to amino acid and peptide components [52–54,85].
3.   Sialic acid contains a further carboxylic function demanding a chemoselective differentiation from the functional groups of the peptide [86].

Successful solutions of these difficulties will be illustrated in the following sections.

Syntheses of Lewis-Type Side Chains Using Glycosyl Azides

Lewis[a] Saccharide Azide.   Glycosylamines are commonly used for the construction of the N-glycosidic bond to asparagine, the typical linking element of N-glycoproteins [27]. Moreover, they are generally applicable for the stereoselective coupling of the synthesized saccharide units to various types of carriers, e.g., polymers, peptides or dendrimers. Glycosyl azides constitute anomerically protected forms of saccharides that, at the same time, are percursors of glycosylamines, later on required for the construction of N-glycosidic bonds.

According to this theory, a suitably protected N-acetylglucosamine was considered the key building block for the synthesis of the Lewis[a] saccharide side-chain [85]. To facilitate the amide bond formation between saccharide and aspartic acid at a later stage, the anomeric position is masked as glycosyl azide. This group can efficiently be converted into the amine by a hydrogenation catalyzed by (neutral) Raney-nickel. Of course, hydrogenations of benzyl ethers must be avoided during the construction of such oligosaccharide azides.

The formation of 2-acetamido-3,4,6-tri-O-acetyl-2-deoxy-β-D-glucopyranosyl azide **2** from the corresponding glycosyl chloride **1** was first achieved using sodium azide in formamide [87]. But this procedure proved to be not generalizable. Already for the chitobiose series, hazardous silver azide had to be applied [88]. A useful alternative was found in the reaction of glycosyl halides like **1**, with sodium azide under phase transfer conditions [89]. After removal of the O-acetyl groups, glycosyl azide **2** was reacted with methoxybenzaldehyde to give compound **3** [90] prepared for the subsequent selective attachment of the galactose at the 3-position. However, glycosylation of the 3-OH of **3** with tetraacetylgalactosyl bromide and silver triflate/collidine failed and instead gave the corresponding orthoester in high yield [85].

Collidine, added in order to trap the strong acid generated during the glycosylation and suspected to affect the acid-sensitive methoxybenzylidene group, favored the charge-controlled attack of the 3-OH of the glycosyl acceptor at C-2′ of the dioxolenium (acyloxonium) ion. If the Helfrich promotor ($Hg(CN)_2$) was used,

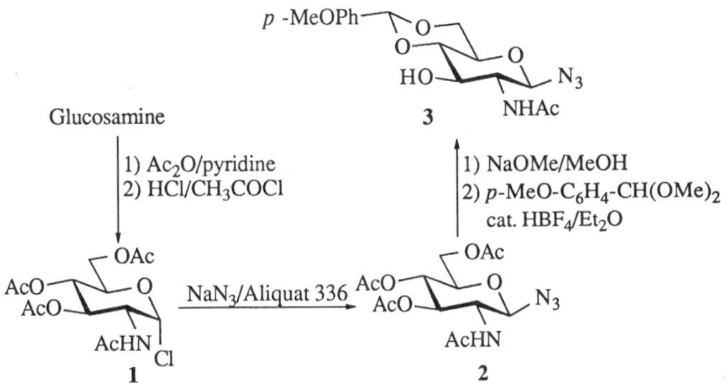

p -MeOPh

O

O

HO N$_3$

NHAc

3

Glucosamine

1) Ac$_2$O/pyridine
2) HCl/CH$_3$COCl

1) NaOMe/MeOH
2) p-MeO-C$_6$H$_4$-CH(OMe)$_2$
   cat. HBF$_4$/Et$_2$O

OAc

AcO
AcO
O

AcHN
Cl

1

NaN$_3$/Aliquat 336

OAc

AcO
AcO
O
N$_3$

AcHN

2

Scheme 14 Synthesis of the key glucosamine building block for the construction of Lewis-type molecules.

this reaction was suppressed and the desired disaccharide was obtained in a yield of 81% [85].

Deprotection of the 4-OH to give **4** was now achieved by regioselective opening of the 4,6-acetal with sodium cyanoborohydride and trifluoroacetic acid [72]. The fucosyl donor **5** required for the construction of the Lewis^a saccharide was synthesized from the fucosyl thioglycoside [91] after exchange of the acetyl for the Mpm ether protecting groups and treatment with bromine and was then activated according to the in situ anomerization procedure [92]: Tetrabutylammonium bromide caused an equilibrium between the α- and β-fucosyl bromide from which the much more reactive β-anomer rapidly and selectively formed the α-fucoside **6** [85]. As for the transformation of the thiofucoside, the sterically hindered base 2,6-di-*tert*-butyl pyridine [93] was added in order to prevent the cleavage of the acid-labile methoxybenzyl ethers. In comparison to the commonly applied benzyl ether, the Mpm ether opens up the way for an exchange of the ether type for an acyl-type protection of the fucoside portion in the presence of the anomeric azido group [52,53,85]. This was realized by oxidative cleavage of the Mpm ethers and subsequent peracetylation, which finally gave the Le^a azide **7** [85].

Lewis^x Saccharide Azide.   According to the strategy described above, access to the series of Le^x antigen structures is achieved starting from the methoxybenzylidene-protected azide **3** [54]. The straightforward route involving fucosylation of the 3-OH of **3** and regioselective reductive acetal opening gave glycosyl acceptor **8**. However, the galactosylation of the 4-OH failed under all conditions examined (Scheme 16) [94].

The 4-hydroxy group of the fucosyl-glucosamine derivative is quite unreactive. Its glycosylation demanded harsh conditions that caused the formation of unacceptable amounts of by-products. A conformational analysis of the acceptor disaccharide **9** suggests that electronic rather than steric reasons are responsible for the decreased reactivity of this 4-OH [94]. Alternatively, the lactosamine derivative **9** was synthesized first. With this aim, the 3-OH of **3** was allylated with allyl bromide in the presence of barium hydroxide [95]. Other reagents, for instance, sodium hydride and allyl bromide, predominantly led to decomposition of the starting material. After the regioselective opening of the acetal, glycosylation with peracetylated galactosyl bromide was carried out in the presence of silver triflate and di-*tert*-butyl pyridine.

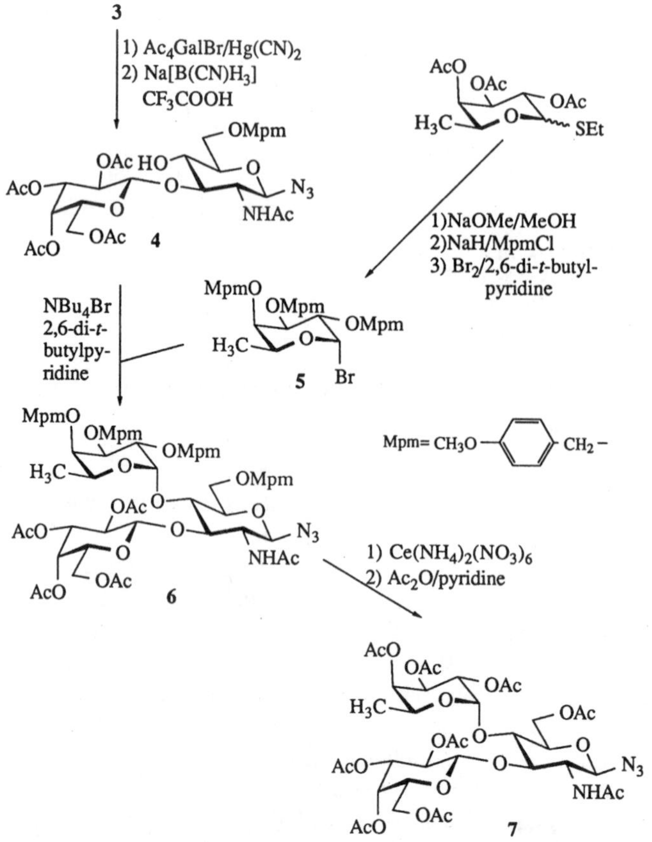

3

Scheme 15   Synthesis of peracetylated Lewisᵃ azide.

Attempts to use mercury cyanide as the promotor, efficient in the synthesis of the Lewisᵃ analog [85] (Scheme 15), surprisingly resulted in the formation of the orthoester in a yield of 82% [54].

To facilitate the introduction of the fucoside in the next step, the allyl ether had to be removed. Side reactions at the azido group were avoided when palladium dichloride in acetic acid/sodium acetate [96] was used in order to generate the propenyl ether, which instantaneously hydrolyzed under these conditions. However,

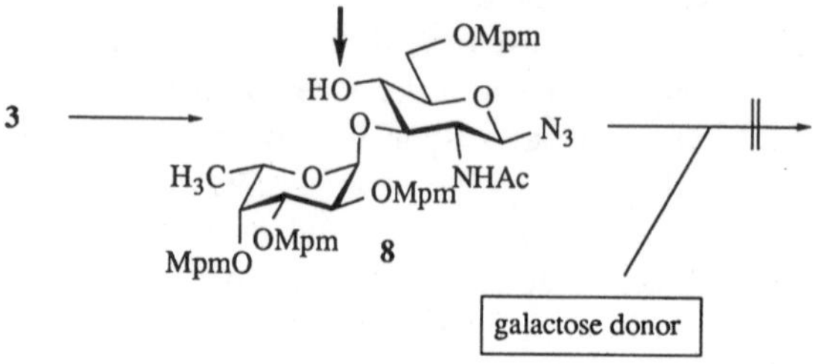

Scheme 16   4-OH of the Lewisˣ azide precursor **8** is unreactive.

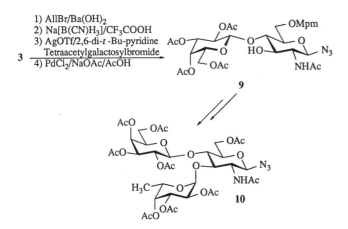

1) AllBr/Ba(OH)$_2$
2) Na[B(CN)H$_3$]/CF$_3$COOH
3) AgOTf/2,6-di-t-Bu-pyridine
   Tetraacetylgalactosylbromide
4) PdCl$_2$/NaOAc/AcOH

3 →

Scheme 17 Synthesis of the peracetylated Lewis^x azide.

cleavage of the methoxybenzyl group occurred and disaccharide **9** was obtained in a yield of only 58%. Exchange of the acid-labile 6-Mpm group for benzoyl, prior to the allyl ether cleavage, is possible but requires two more steps and does not improve the overall yield. Fucosylation of the acceptor **9** and exchange of the ether type for acetyl protection proceeded by analogy to the reactions described for the Lewis^a saccharide and gave the Lewis^x trisaccharide azide **10** in high yield [54].

The syntheses of both Lewis antigen regioisomers emphasize the potential of the anomeric azide methodology. The incompatibility of the azido function with the hydrogenolysis of benzyl ether groups can be efficiently circumvented by the use of methoxybenzyl ethers, which are selectively removable by oxidation. This strategy allows a convenient standardization of the protecting group pattern before the attachment of the saccharide to the aspartic acid moiety is performed. Peracetylated compounds such as **7** or **10** are quite stable toward treatment with acids [54,85]. O- and N-acyl protections markedly stabilize the intersaccharidic linkages by inductive (-I) and Coulomb repulsion effects, particularly resulting from protonation of the carbonyl oxygens. This property is most important for the deblocking steps carried out during the synthesis of glycosylated peptides.

Sialyl Lewis^x Azides. Sialyl Lewis^x saccharides, being attractive ligands for investigations of cell adhesion phenomena, can also be synthesized by applying the anomeric azide, which later on can serve for the coupling to peptides [86]. In this sense, the selectively protected 3-O-(α-fucosyl)-glucosamine azide **11** obtained by analogy to the synthesis of **8** was glycosylated using the O-(2,3,4-tri-O-acetyl-6-benzyl-α,β-D-galactopyranosyl)-trichloroacetimidate according to the method of R. R. Schmidt [97] to give the Lewis^x azide [86]. The low reactivity of the 4-OH of **11** demands a particularly efficient glycosylation, which is most advantageously accomplished with trichloroacetimidates.

The protecting-group pattern of the intermediate product allowed the deacetylation of the secondary hydroxyl functions of the galactose unit to give **12**. Sialylation of this acceptor using sialic acid methyl ester methylthioglycoside **13** [98,99] that was activated with methylsulfenyl triflate [100] resulted in a stereoselective and completely regioselective formation of the desired sialyl Lewis^x tetrasaccharide. The

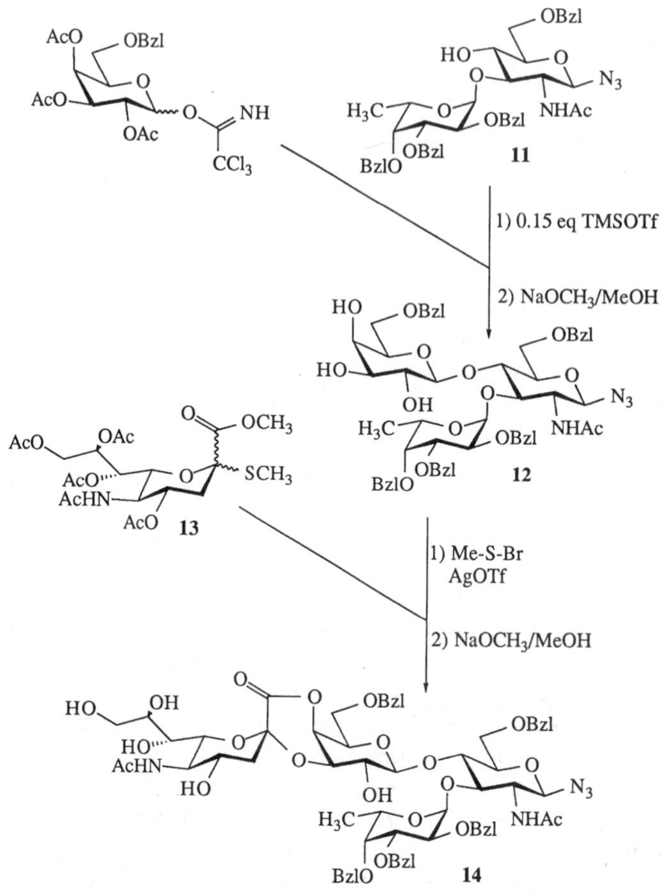

Scheme 18   Construction of the protected sialyl Lewis^x azide **14**.

stability of the anomeric azide allowed the subsequent removal of the O-acetyl and the methyl ester protection. Upon treatment with acidic ion exchange resin, the sialic acid unit was converted into the 4'-lactone [101] in **14**, which served as an internal protection during glycopeptide formation. The azido funcion of **14** was reduced to give the sialyl Lewis^x amine useful for the coupling to amino acid or peptide components.

Activation of Glycosyl Azides and Their Use in O-Glycosylations

The protection of the anomeric center by the azido function is not only favorable for the construction of N-glycosides, but can also be utilized in O-glycoside and saccharide syntheses [102,103]. The potential of this concept was demonstrated in the construction of neolacto-tetraose. The N-phthaloyl protected glucosamine azide **15** [53] was deacetylated, subsequently converted into the 4,6-benzylidene derivative, and then allylated at O-3 to give **16**. The latter was subjected to regioselective reductive opening of the acetal to furnish acceptor **17**. During all of these manipulations, the anomeric azido group remained unaffected [103]. Galactosylation of **17**

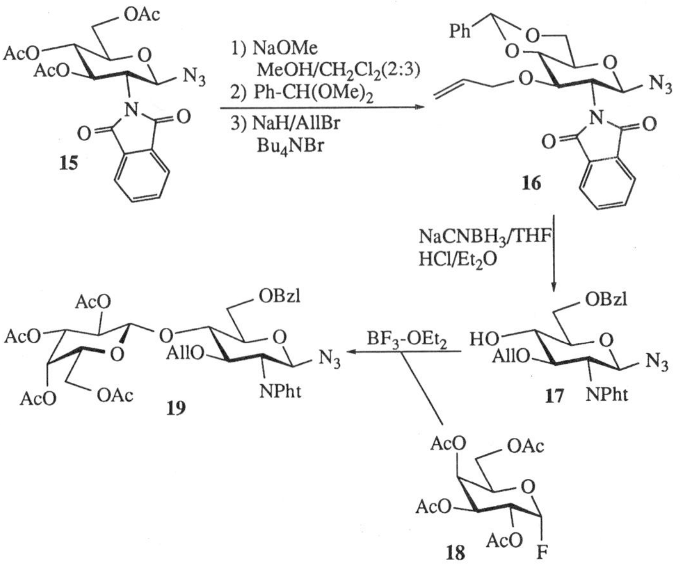

Scheme 19   Synthesis of a suitably protected lactosaminyl azide.

was achieved using the O-acetylated galactosyl fluoride **18** [104] in the presence of boron trifluoride [105,106] and gave the lactosamine azide **19** in a yield of 76%.

The azido function utilized up to this point as the anomeric protection was now subjected to a 1,3-dipolar cycloaddition with di-*tert*-butyl-acetylene-dicarboxylate to furnish the glycosyl triazole **20**. Treatment of **20** with hydrogen fluoride pyridine gave the disaccharide fluoride **21**. Activation of the glycosyl fluoride using BF_3 afforded an efficient coupling to the selectively deblocked benzyl lactoside **22** [107] and resulted in formation of the neolactotetraose **23** in high yield [103]. It should be noticed that **23** is differentiated in the protection of O-3 of the glucosamine unit and, thus, accessible to a later introduction of an α-fucoside unit at this position. Of course, the use of glycosyl azides for O-glycoside synthesis, illustrated in the synthesis of **23**, can also be adapted for the synthesis of O-glycopeptides [102].

From the various other successful attempts in the synthesis of important Lewis-type antigens, two at least should be outlined in further detail. And although the oligosaccharides produced from Schemes 21 and 22 have not yet been converted into glycopeptides, the approaches used do hold great potential for such a possibility.

Synthesis of Sulfated Lewis[x] [84]

The application of glycosyl fluorides in the syntheses of Lewis antigens was carried out with particular success in the field of sulfated derivatives. These compounds exhibit significantly higher affinity toward the relevant receptors (E-selectin) than sialyl Lewis[x] itself. Interestingly, a mixture of sulfated Le[a]/Le[x] has been isolated from an ovarian cystadenoma glycoprotein [82]. A linear synthesis of the sulfated Le[x] started with the formation of disaccharide **24**. After exchange of the N-phthaloyl for an N-acetyl group, the silyl ether in the 4′-position was removed. The selectively deblocked product was reacted with a suitably protected galactosyl fluoride with

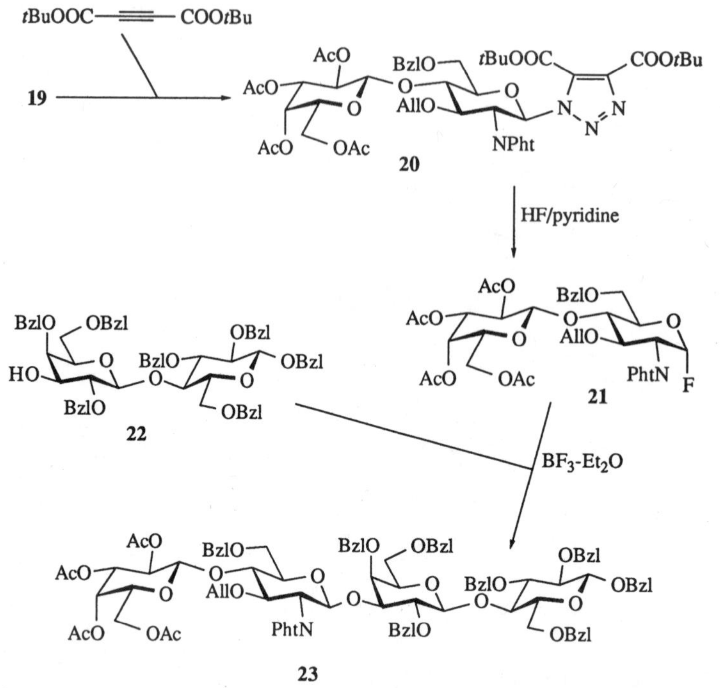

Scheme 20   Construction of a neolactotetraose via activation of the azido group.

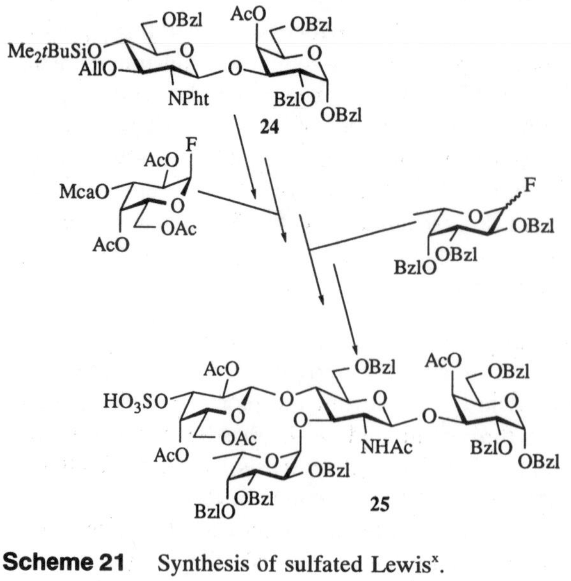

Scheme 21   Synthesis of sulfated Lewis^x.

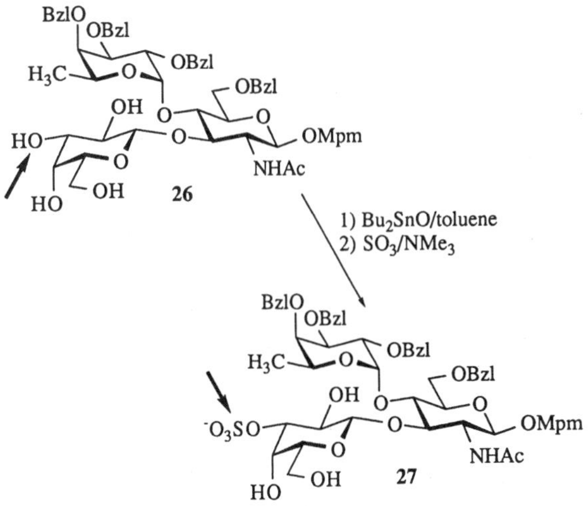

Scheme 22   Regioselective sulfation of a partially protected Lewis^a derivative.

promotion by SnCl$_2$/AgClO$_4$ [108]. Fucosylation was similarly carried out subsequent to the cleavage of the allyl ether in the 3-position of the GlcNAc portion. In the resulting tetrasaccharide, the chloroacetyl group (Mca) was selectively removed from the 3-hydroxy group using thiourea/lutidine and converted into the sulfate **25** by treatment with SO$_3$/NMe$_3$.

A simplification of the regioselective sulfation was published for the Le^a analog **27** [109]. In this case, the terminal galactose in **26** was completely deprotected and transformed into the stannylene derivative, which was regioselectively attacked by SO$_3$/NMe$_3$. Introduction of the sulfate after protection of the primary 6-hydroxy group significantly increased the yield.

Synthesis of Lewis^y Using Glycals

An elegant synthesis of the Lewis^y antigen saccharides employing glycals as glycosyl donors was published recently [110]. The crucial step of the synthetic strategy was the transformation of the tetrasaccharide glycal **28** into the 2-iodo-1-sulfonamide **29**. This product reacted with another glycal as the acceptor, presumably via an intermediate aziridine intermediate, to form the 2-sulfonamido glycoside **30**. The latter compound was converted into the natural N-acetyl derivative.

Danishefsky et al. have also reported the use of glycals for a promising polymer-supported oligosaccharide synthesis [111].

B.   Core-Region Glycans of Natural Glycoproteins

Almost all glycans of naturally occurring N-glycoproteins contain a central ("core") pentasaccharide **31**. Many are further branched to form antennas [112,113]. Unusual antennas often are of biological relevance. An additional (1-6)-fucosylation (**31**) of the asparagine linked N-acetylglucosamine, for instance, is often found on virus envelope glycoproteins. The occurrence of this structure in membrane glycoproteins of cancer cells has been correlated with metastasis [114]. For the investigation of

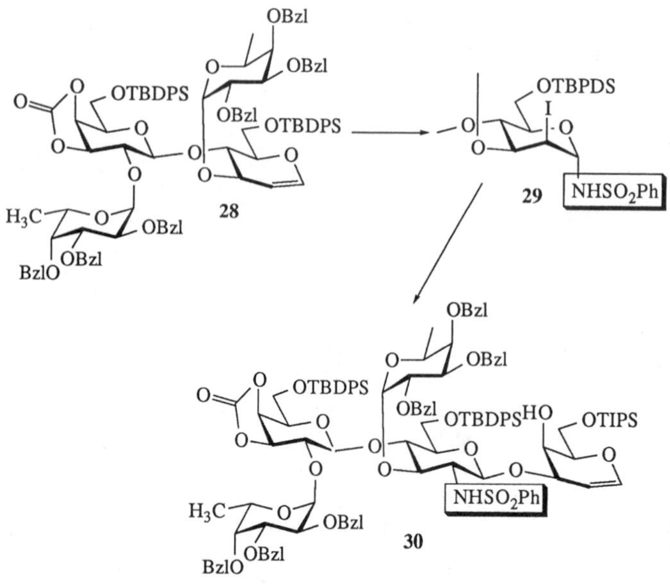

Scheme 23   Synthesis of Lewis^y via the activation of glycals.

the effects of these membrane components, model compounds with partial structures of such *N*-glycoproteins are required.

Synthesis of a Fucosylated Core-N-Glycan

The synthesis of glycans from this series concerned the fucosylated core-*N*-glycan **35**, which constitutes an instructive example for the access to this type of saccharide [52,53]. As outlined above, the azido group was employed as the anomeric protection, thus allowing the subsequent *N*-glycosidic attachment of the saccharide to the peptide backbone.

In order to ensure a subsequent introduction of the fucose at the correct position, the chitobiosyl azide has to be equipped with a suitable protecting group pattern. Starting from the 4,6-methoxybenzylidene-protected glucosaminyl azide **8**, the 3-OH group was acetylated and the benzylidene acetal opened regioselectively. The 4-OH set free was then glycosylated with the *N*-phthaloyl-protected glycosyl bromide **32** [115] to give chitobiosyl derivative **33**. After oxidative removal of the methoxybenzyl group [34], the fucose was coupled stereoselectively to the 6-position using the Mpm-protected fucosyl chloride [52,53] under in situ anomerization conditions [92]. The trisaccharide was subsequently subjected to exchange of the protecting groups.

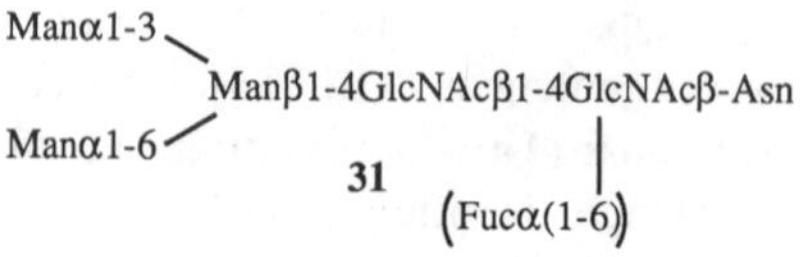

Scheme 24   Core pentasaccharide of *N*-glycoproteins.

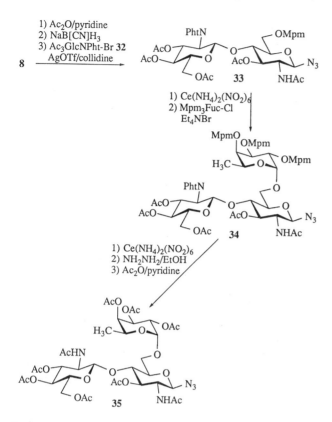

Scheme 25   Synthesis of a 1-6 fucosylated chitobiosyl azide.

The Mpm groups were removed selectively by oxidation with ceric ammonium nitrate. The azido group remained unaffected. Then the phthaloyl group was cleaved using hydrazine in ethanol. Acetylation gave peracetylated derivative **35**. It should be noted that an analogous tri-O-benzyl-fucosyl chitobiose derivative had been synthesized in a previous approach [53] and was linked to N-protected asparagine *tert*-butyl ester. However, under the acidic conditions required for the cleavage of the *tert*-butyl ester during subsequent glycopeptide synthesis, the fucosidic linkage was completely destroyed. This side reaction can be avoided if the hydroxyl groups of the fucose residue have been acetylated prior to the acidolysis of the *tert*-butyl ester, which is a prerequisite for extension of the peptide chain [52,53] (see also Section II.C.).

N-*Glycans Containing the β-Mannoside Linkage*

A particular problem during the construction of the core pentasaccharide **31** consists in the stereoselective formation of the β-mannoside linkage to chitobiose. In contrast to glucosylation or galactosylation, the utilization of neighboring groups such as acetyl in the 2-position of a mannosyl donor leads to the formation of α-glycosides. Moreover the formation of the α-mannosides is also favored by the anomeric effect.

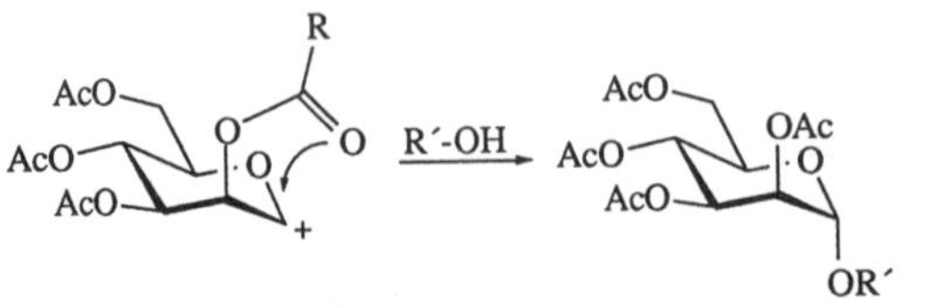

Scheme 26   Neighboring groups lead to α-mannosides.

A stereoselective formation of β-mannosides was achieved with tetra-*O*-benzyl-mannosyl bromide in the presence of the heterogeneous promotor silver silicate [116]. However, depending on the protecting group pattern of the mannose donor and the reactivity of the acceptor, the undesired α-mannosides are also obtained in significant amounts.

A multistep methodology was developed by Barresi and Hindsgaul [117], who took advantage of a preformed donor-acceptor conjugate **36** that, upon activation of the anomeric position, gave the desired β-mannosyl glycosides.

In a similar strategy, Stork and Kim [118] have used the mixed dialkylalkoxysilane to furnish β-mannoside **38** after activation of an intermediate anomeric sulfoxide **37**. Recently, this approach has been varied by the application of a mixed *p*-methoxybenzylidene acetal in the 2-position, which also served as the source for the glycosyl acceptor while the anomeric position was activated [119].

The required β-mannoside configuration can alternatively be established by transformation of the stereoselectively accessible gluco derivatives. For instance, in

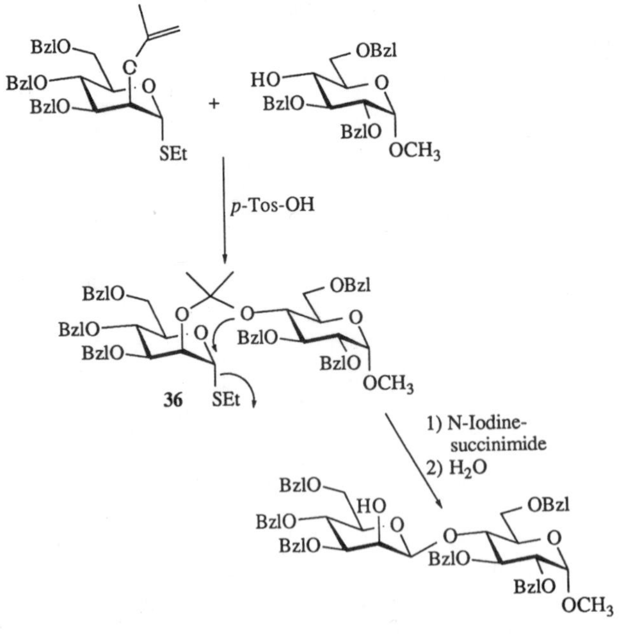

Scheme 27   Synthesis of a β-mannoside via intramolecular aglycon delivery.

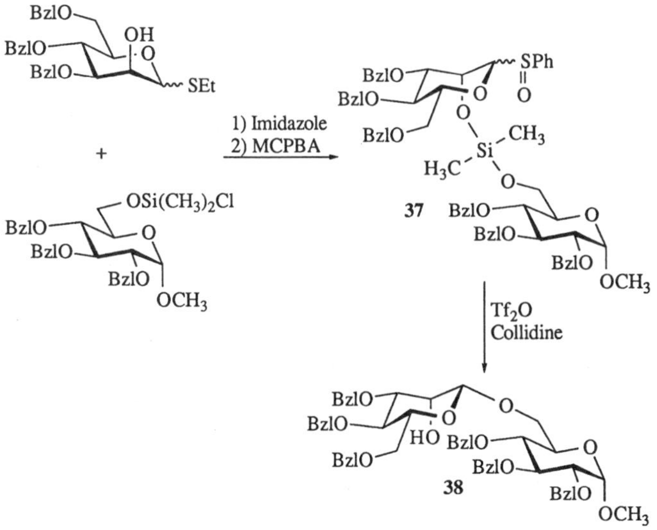

Scheme 28   A bisglycosylalkoxysiloxane as β-mannoside precursor.

an oxidation/reduction sequence of β-glucosides selectively deblocked at the 2-position, β-mannosides were obtained in high yield [120].

Another stereoselective access to β-mannosides can be realized by S_N2-type reactions at the 2-position of β-glucosides. Generally, it is thought that the S_N2 substitution in the 2-position of glucosides is unfavorable because of the repulsion of the approaching nucleophile by the lone pairs at the ring oxygen. But it was shown that the inversion of configuration at the 2-position can efficiently be achieved via an intramolecular S_N2 reaction [121,122]. To this end, a leaving group was introduced at the 2-hydroxy group of disaccharide **39**, which contained a phenylcarbamoyl group in the 3-position. Stirring of this compound in N,N'-dimethylformamide/pyridine at 40–50°C resulted in the intramolecular S_N2 reaction at C-2, to give the 2,3-cis-carbonate with manno configuration. This strategy allows the differentiation between the 3- and the 6-positions of the mannoside unit and opens up the way to the synthesis of 3,6-mannosylated core-region saccharides [123].

It is interesting to note that the thioglycoside **39** remained stable during these manipulations. Its subsequent activation allowed the extension to the core trisaccharide, which was finally converted into the isothiocyanide **40**. In this case, all attempts to introduce an anomeric azido group failed, in contrast to descriptions for analogous reactions in the literature [124]. Therefore, a special reaction of the intermediate oxazolidine (not shown) with potassium thiocyanide under phase transfer conditions

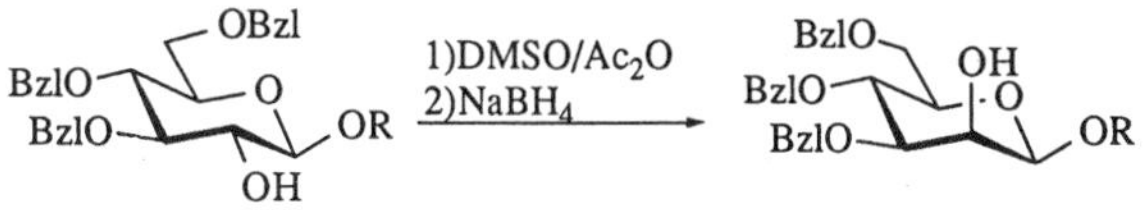

Scheme 29   β-Mannoside formation via an oxidation/reduction sequence.

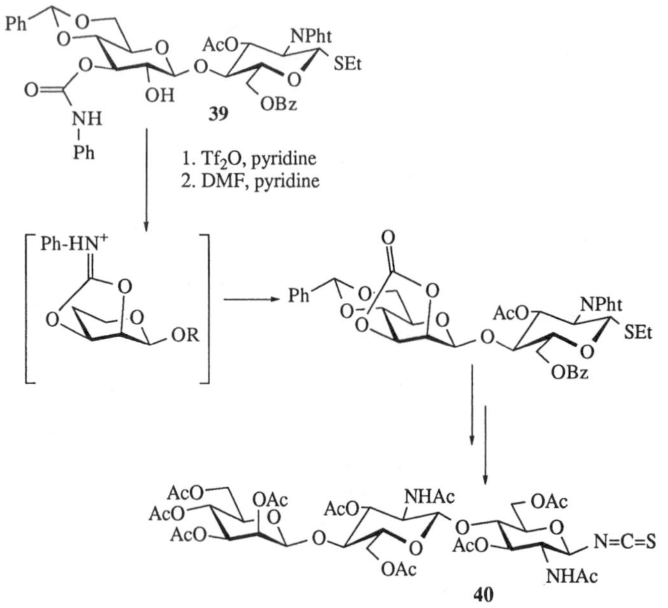

Scheme 30   β-Mannoside formation via intramolecular S_N2 reaction: Synthesis of the core trisaccharide.

had to be applied in order to furnish **40**, which is a precursor for *N*-glycopeptide formations [123].

The potential of this methodology was recently demonstrated in the synthesis of the branched core-region heptasaccharide **43** [125]. In order to circumvent the problems of introducing the azide function at a late stage of the synthesis, the chitobiosyl azide **41** was used as acceptor for the attachment of the 3-*O*-phenylcarbamoyl-protected glucosyl residue. The above-described intramolecular epimerization sequence led to the corresponding manno-2,3-carbonate. After alkaline hydrolysis of this intermediate, the 2″-/3″-*cis*-diol was regioselectively coupled at 3″-OH with a glucosaminyl mannose donor to give the pentasaccharide **42** in high yield. After subsequent cleavage of the benzylidene group of **42**, a further regioselective glycosylation at 6″-OH gave, under carefully optimized reaction conditions, heptasaccharide **43**.

These examples illustrate the potential of the intramolecular S_N2-type transformation of β-glucosides to β-mannosides for the construction of the complex *N*-glycan structures of *N*-glycoproteins.

C.   *O*-Glycopeptide Side Chains and Their Coupling to Serine or Threonine

The biosynthesis of *O*-glycoproteins proceeds in a way fundamentally different from that of *N*-glycoproteins [126]. Glycosyl nucleotides, e.g., UDP-glucose, UDP-galactose, etc., are activated by specific transferases [127], resulting in the glycosylation of the hydroxy group of serine or threonine [128]. Further monosaccharide residues are subsequently coupled in the same manner.

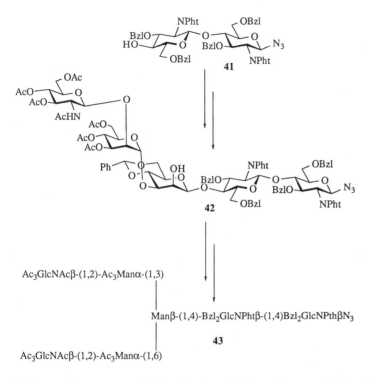

Scheme 31 Total synthesis of the branched core heptasaccharide **43** using the intramolecular S_N2 reaction.

Not much is yet known about the controlling mechanisms of this process or whether there are peptide signal sequences for *O*-glycosylation analogous to the -Asn-Xaa-Ser/Thr- sequon in the biosynthesis of *N*-glycoproteins. There is some evidence, however, that at least in mucins (see next subsection) *O*-glycosylation starting with the attachment of *N*-acetylgalactosamine proceeds preferentially in proline-rich regions of the peptide backbone [129,130]. Important linkage regions of *O*-glycoproteins are depicted in scheme 1 (**2–4**).

Carbohydrate Side Chains of Mucin-Type Glycoproteins

Mucins are highly glycosylated (~50% of their molecular weight) and large macromolecules (molecular weight $\sim 10^7$) [131]. They are produced by a wide variety of epithelial and endothelial cells in order to protect these sensitive tissues. During the development of tumors, however, marked alterations of the mucin carbohydrate side chains have been observed. In particular, mucins with Thomsen−Friedenreich antigen side chains have been described as tumor-associated antigens [132,133]. Owing to these reports, increased efforts were aimed at the synthesis of GalNAcα-Ser/Thr **2** (T_n antigen) and its derivatives with extended saccharide chains, e.g., T antigen (Galβ(1,3)-GalNAcα-Ser/Thr) or sialyl-T_n (NeuAcα(2,6)-GalNAcα-Ser/Thr). Synthetic approaches concerning these glycan side chains will be described next.

Synthesis of the T_n Antigen. Investigations using monoclonal antibodies have proven the T_n antigen (GalNAcα-Ser/Thr) to be expressed in over 70% of carcinomas in lung, colon, and stomach [133,134]. Therefore, T_n antigen derivatives are receiving increasing interest as target molecules for diagnosis and immunotherapy of cancers [135].

A major synthetic problem during the preparation of **2** is the stereoselective construction the α-glycosidic linkage between *N*-acetylgalactosamine and serine or threonine. An efficient solution to this problem was provided by Paulsen et al. [136,137], who used 2-azido-2-deoxy-galactosyl halides conveniently prepared via the azidonitration procedure introduced by Lemieux [138]: Peracetylated galactose can be converted directly into the glycosyl bromide. Reductive elimination resulted in the formation of tri-*O*-acetyl-galactal **44**, which was treated with ceric ammonium nitrate in acetonitrile to give the 2-azido-1-nitro compound **45** as a mixture of anomers. Upon treatment of **45** with LiBr, the corresponding α-bromide **46** is accessible [138]. By preparing a thioglycoside **47** from the α-azido nitrate **45** via the corresponding xanthate and subsequent elimination of carbonoxysulfide [139], the more stable glycosyl donor **47** was obtained. Its use in glycosylation reactions avoids the need for silver or mercury salts.

For glycosylations of serine and threonine derivatives, e.g., the *N*-benzyloxy-carbonyl serine heptylester **48** (see Section V.) with **47** as the glycosyl donor, dimethyl(methylthio)sulfonium triflate (DMTST) [140] was the reagent of choice since it proved to be more thiophilic than methyl triflate [141]. Previous to these syntheses the α-(2-azido)galactosyl bromide was used for analogous glycosylations and activated with silver carbonate/silver perchlorate in a mixture of toluene/dichloromethane [136,142,143]. This procedure provided an efficient and in most cases stereoselective formation of the desired T_n antigen conjugates.

Conversion of the 2-azido group into the natural 2-acetamido function was achieved by reduction with sodium boranate/nickel chloride [144,145], trioctylphosphine/acetic acid [46,146], hydrogen sulfide in pyridine [147], or, already including the acetylating step, with thioacetic acid [148] (Scheme 34).

Suitable protecting-group combinations in the amino acid portion as Fmoc/OBzl [42,143], Fmoc/OAll [46,47], Fmoc/O*t*Bu [137], Aloc/O*t*Bu, or Z/OHep [40] mentioned above allowed selective amino or carboxy deprotection and the applica-

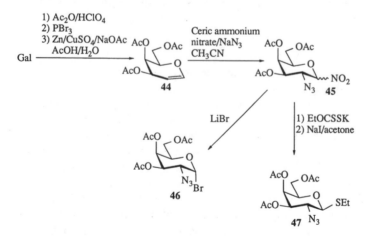

Scheme 32   Preparation of 2-azido-2-deoxy-galactopyranosyl donors.

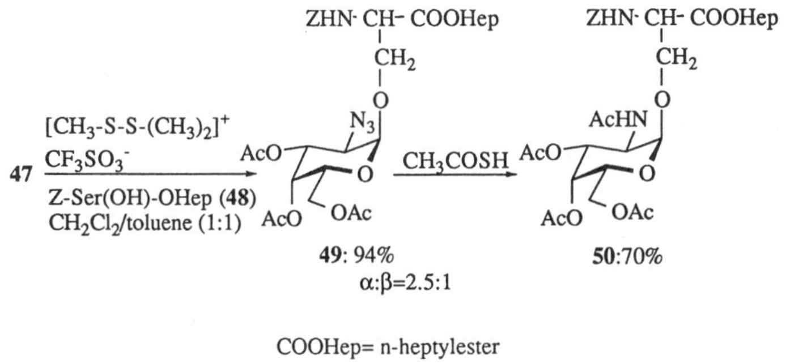

COOHep= n-heptylester

Scheme 33 Synthesis of a protected Tn antigen derivative.

tion of the obtained carboxy components in glycopeptide synthesis either in solution or on solid phase. The glycosylation of Fmoc serine and threonine pentafluorophenyl esters [149] affords building blocks that can be employed directly in peptide synthesis [49]. Examples will be shown in Section V.

Synthesis of the T Antigen. A successful strategy reported for the construction of the Galβ(1,3)GalNAcα-Ser/Thr structure involved the formation of the Galβ(1,3)GalN₃ disaccharide [150]. After its transformation into the corresponding glycosyl bromide, the disaccharide was coupled with a serine or threonine acceptor

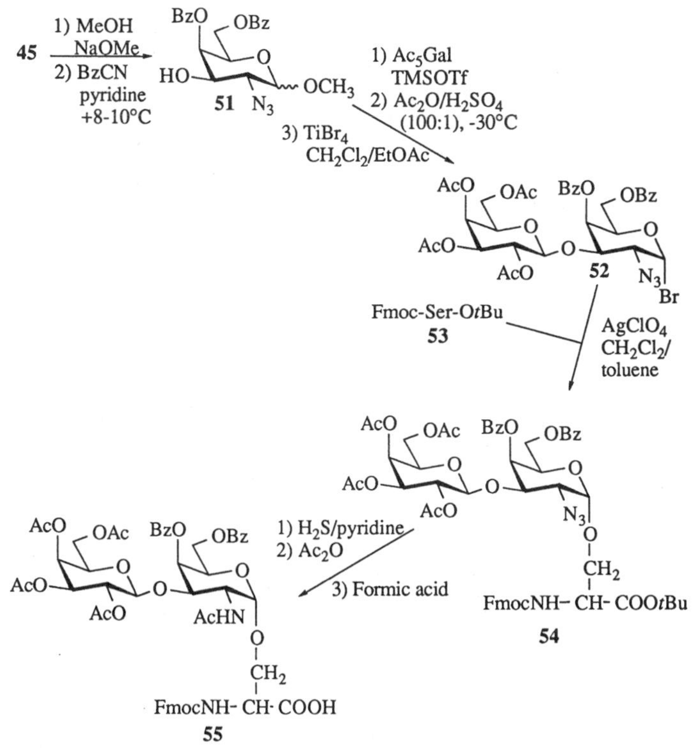

Scheme 34 Synthesis of a protected T antigen derivative.

in order to form the desired T antigen conjugates. For this purpose the azidonitrate **45** was again used as a suitable precursor. It was converted into the methyl glycoside, which was regioselectively benzoylated in the 4- and 6-positions to give **51** [150]. Galactose is introduced at the 3-position of **51** via a trimethylsilyl triflate-mediated activation [151] of penta-*O*-acetyl-galactose. After acetolysis and reaction with titanium tetrabromide in dichloromethane/ethyl acetate (10:1), the desired disaccharide bromide **52** was obtained. The selectively deblocked disaccharide-amino acid conjugate **55** [152] results from a silver perchlorate promoted glycosylation of Fmoc-Ser-O*t*Bu **53** (**54**, 54%, α:β = 3:1), subsequent reduction of the azido group with hydrogen sulfide in pyridine, acetylation, and cleavage of the *tert*-butyl ester using formic acid.

Synthesis of Sialyl T$_n$ Antigen. The sialyl T$_n$ antigen (Neu5Acα(2,6)Gal-NAcα-Ser/Thr) has also been described as a tumor-associated antigen [153,154]. Glycopeptides bearing clustered oligosaccharide structures of the sialyl T$_n$ type especially exhibit increased immunoreactivity [153], and their application as vaccines directed toward carcinomas is therefore under intensive investigation.

Syntheses of sialyl T$_n$-amino acid conjugates can either involve preformed Neu5Acα(2,6)GalNAc components [155] or start from a suitably protected GalNAcα-Ser/Thr moiety that is subsequently glycosylated with a sialic acid donor [44]. The latter strategy advantageously uses the valuable neuraminic acid at a late stage of the synthesis.

In any case, the sialylation reaction has to be controlled carefully in order to furnish a maximum excess of the desired α-sialoside. One possibility to ensure the stereoselective formation of the α-sialoside **58** (scheme 35) involves the effect of a temporary neighboring group, e.g., phenylthio, in the 2-position of the NeuAc donor **56** [156], which forces the attack of the glycosyl acceptor **57** from the opposite (α) side. Neither a regio- nor a stereoisomeric product was obtained by the reaction of this designed sialyl donor.

Another strategy [44] employs the more readily accessible NeuAc xanthate **59** [157] and the galactosamine-threonine acceptor **60** containing three unprotected hy-

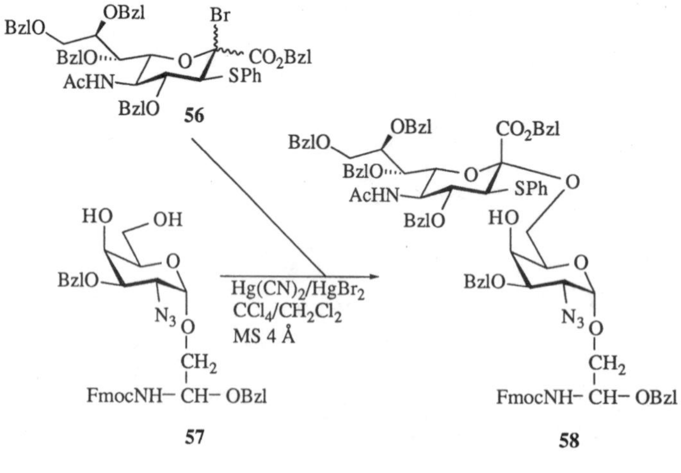

Scheme 35   Synthesis of sialyl Tn antigen using a neighboring group at the NeuAc donor.

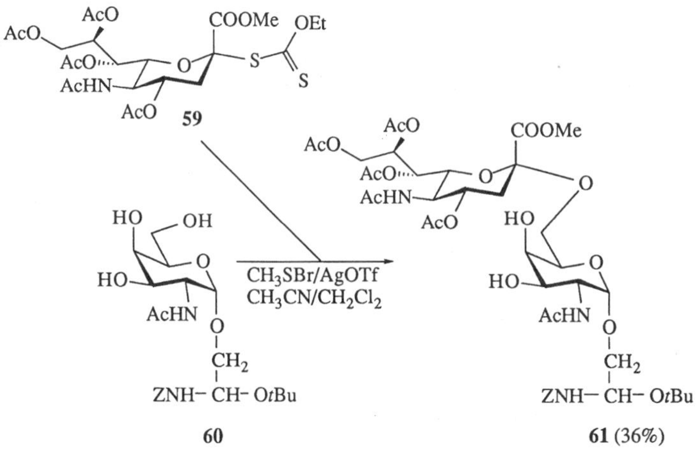

Scheme 36 Synthesis of sialyl Tn antigen using a trihydroxygalactosyl threonine acceptor and a NeuAc xanthate donor.

droxyl groups in the 3-, 4-, and 6-positions. The stereoselectivity of the glycosylation reaction was $\alpha:\beta$ 4:1; but the reactive donor led to high yields of the isolated sialyl T_n antigen derivative (71%). After separation by RP-HPLC chromatography, the pure α-sialyl T_n antigen conjugate **61** was obtained in a yield of 36%.

IV. PREPARATION OF GLYCOSYLATED ASPARAGINE DERIVATIVES

In general, the synthesis of glycosylated asparagine derivatives can conveniently be accomplished via an amide formation between a suitably protected aspartic acid and the corresponding glycosylamine. While different aspartic acid derivatives are commercially available or easily accessible in a few steps [158,159], the preparation of precursors of glycosylamines usually is much more laborious, as was illustrated in Sections III.A. and B.

A. Synthesis of Glycosylamines

Glycosylamines are accessible from glycosyl azides by catalytic reduction. The azido function itself can be part of the protecting group strategy, as was outlined earlier [52,53]. It was also reported [124] that the azide can be introduced via an intermediate oxazoline **62** that is reacted with trimethyl silyl azide in the presence of tin tetrachloride (Scheme 37). The reduction of the azido function usually proceeds in high yield without anomerization or acyl migration if a suitable solvent (isopropanol, dioxane) and proper catalysts, such as Raney-Nickel or Lindlar catalyst, are employed. Special attention has to be paid to benzyl-type protecting groups present in the molecules. But the selective reduction of such azides was also successfully realized [52,53].

    Alternatively, glycosyl amines are obtained after thorough treatment of reducing saccharides with saturated ammonium bicarbonate [160]. By this method, oli-

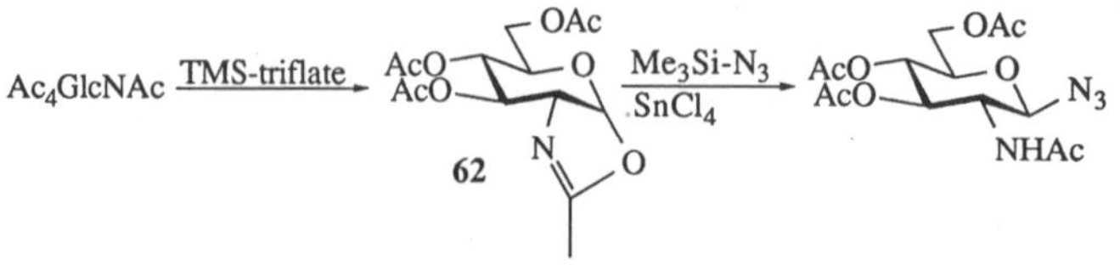

Scheme 37   Conversion of glucosaminyl oxazolidinone into *N*-acetylglucosaminyl azide using TMS-azide and tin tetrachloride.

gosaccharides isolated from natural sources such as **63** could be converted into gly-cosylamines as **64**. However, protected saccharides, e.g., *O*-acetylated saccharides, are not as useful in this process [161]. In any case, this reaction has to be controlled indirectly by an HPLC analysis, which determines a product formed in a subsequent standard coupling reaction of the amine. To this end, a sample of the reaction mixture containing the desired glycosylamine was reacted with Boc-Asp-OBzl. Moreover, excess of ammonia has to be removed carefully from the product to ensure satisfying yields during the subsequent amide formation [161].

Very recently the aforementioned methodology has been improved by the util-ization of equimolar amounts of ammonium bicarbonate in the presence of ammonia [162]. These optimized reaction conditions led to an efficient formation of the desired product. Again, this process can be realized only with unprotected *N*-acetylglucos-amine derivatives.

B.   Synthesis of the *N*-Glycosidic Linkage

Different methodologies for the acylation of glycosylamines with suitably protected amino acid derivatives have already been extensively reported [26,27,161]. There-fore, this section will refer only to the most efficient and recent developments.

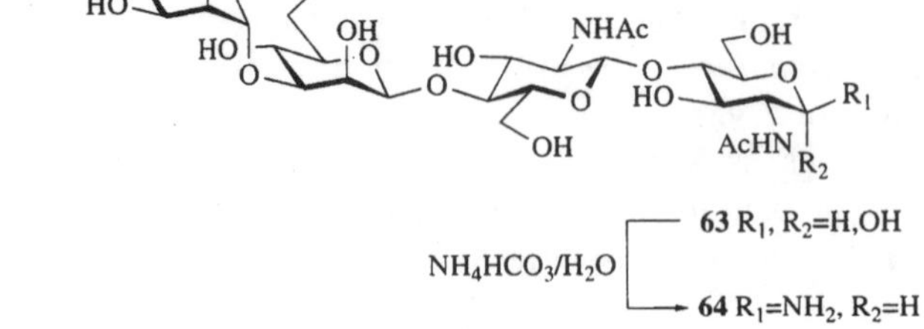

Scheme 38   Preparation of glycosylamines using aqueous ammonium carbonate.

Usually, the asparagine-glucosamine linkage, being the most important type occurring in natural N-glycoproteins, is formed with the aid of acylating reagents as dicyclohexyl carbodiimide in the presence of additives like 1-hydroxy-benzotriazole [163] or 2-ethoxy-1-ethoxycarbonyl-1,2-dihydroquinoline (EEDQ) [164]. Subsequently, the amino acid is selectively deprotected and peptide chain extension is carried out either in solution or on a solid support.

A study performed by Cohen and Lansbury [161] revealed promising results in coupling aspartic acid containing peptides with unprotected glycosylamines (see Section IV.A.) The outcome of the coupling reaction according to Scheme 39 depending on the reaction conditions and reaction partners is given in Table 2. Entries 1–3 and 7–10 in the table show that N-glycosidic linkages can be formed in high yield if the corresponding peracetylated glycosylamines are reacted with suitably protected aspartic acid derivatives and condensation reagents as EEDQ (ethyl-2-ethoxy-1,2-dihydroquinoline-1-carboxylate) [164], IIDQ (2-isobutoxy-1-isobutoxy-carbonyl-1,2-dihydroquinoline) [169], or the water-soluble carbodiimide EDC (1-ethyl-3-(3'-dimethylaminopropyl)-carbodiimide) [170,171]. The coupling reactions proceed efficiently even with more complex carbohydrate amines. Entry 12 shows that, starting from a protected sialyl Lewis^x amine (for its azido precursor **14**, see Scheme 17), the synthesis of the complex sialyl Lewis^x-RGDA N-glycopeptide was accomplished in high yield provided optimized conditions were applied [86].

For those glycans isolated from natural sources, the conversion into their glycosylamines using the improved ammonia/ammonium hydrogen carbonate system is obviously promising. For pharmaceutical purposes, however, a larger amount of material is usually required and, in most cases, will be accessible only by synthetic approaches. For this aim, the azido group has proved to be a valuable protection of the anomeric center that can be transformed into the required glycosylamine. The use of such a glycosylamine precursor thus seems to be straighforward for the formation of glycan-asparagine conjugates required for a subsequent solid-phase glycopeptide synthesis. In order to establish a sufficient acid stability of the glycan part during the glycopeptide synthesis, the protecting-group pattern in the carbohydrate moiety should permit an exchange of ether type for acyl-type protection in the presence of the azido group. The use of the methoxybenzyl protecting groups (see Section III.A.1.) instead of benzyl groups in the carbohydrate portion facilitates this protecting group exchange without affecting the anomeric azido function.

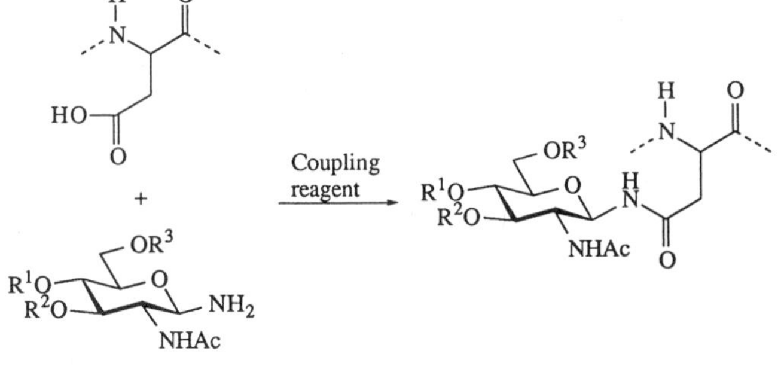

Scheme 39 Amide formation between aspartic acid and glycosylamines.

Table 2 The Formation of the *N*-Glycosidic Linkage Under Different Conditions

Glycosylamine	Amino acid/peptide	Reagents	Yield (%)	Ref.
1 R^1, R^2, R^3 = Acetyl	Boc-Asp-OAll	EEDQ	74	165
2 R^1, R^2, R^3 = Acetyl	Aloc-Asp-O*t*Bu	IIDQ	100	166
3 R^1, R^2, R^3 = Acetyl	Fmoc-Asp-O*t*Bu	IIDQ	68	94
4 R^1, R^2, R^3 = H	Ac-Asp-Leu-Phe-NH$_2$	HBTU/HOBT	95[a]	167
5 R^1 = βGlcNAc R^2, R^3 = H (2 eq)	Ac-Tyr-Asp-Leu-Thr-Ser-NH$_2$	HBTU (3 eq) DIEA (2 eq)	88	161
6 R^1 = βGlcNAc R^2, R^3 = Ac	Boc-Asp-OAll	EEDQ	91	62
7 R^1 = βAc$_4$Gal R^2 = Ac, R^3 = Bz	Fmoc-Asp-O*t*Bu	IIDQ	92	94
8 R^1 = βAc$_3$GlcNAc R^2 = Ac R^3 = αAc$_3$Fuc	Aloc-Asp-O*t*Bu	EEDQ	60	166
9 R^1 = αAc$_3$Fuc R^2 = βAc$_4$Gal R^3 = Ac	Tcoc-Asp-OAll	EDC/HOBT	80	168
10 R^1 = βAc$_4$Gal R^2 = αAc$_3$Fuc R^3 = Ac	Boc-Asp-OAll	IIDQ	84	54
11 **64**	Ac-Tyr-Asp-Leu-Thr-Ser-NH$_2$	HBTU/HOBt DIEA (5:5:2 eq)	55	161
12 R^1 = αNeu5Ac(2,6)-βBzlGal[b] R^2 = αBzl$_3$Fuc R^3 = Bzl	Z-Arg(Z)$_2$-Gly-Asp(OBzl)- Ala-OH[c] (1.25 eq)	TBTU/HOBT DIPEA (3:1:0.25 eq)	65	86

[a]Yields estimated by HPLC analysis.
[b]The carboxy group of Neu5Ac is protected as internal lactone with the 4-OH of galactose.
[c]This linkage is formed by condensation with the carboxy group of alanine.

V. SELECTIVE DEBLOCKING AND CHAIN EXTENSION OF GLYCOSYLATED PEPTIDES

A. Glycopeptide Synthesis in Solution

Glycosylated amino acid derivatives are sterically demanding amino or carboxylic components in peptide condensations. The construction of oligopeptides containing one or more glycan side chains is, therefore, favorably carried out in solution. This holds true, in particular, for the synthesis of *N*-/*O*-glycopeptides with a relatively high number of glycan side chains compared to the overall length of the peptide. Such clustered glycopeptide structures are receiving increasing interest because they should display stronger immunoreactivity compared to monovalent antigens. In addition, multivalent glycoconjugates have also been demonstrated to exhibit strongly enhanced binding to receptors [172].

A successful synthesis of such a synthetic antigen, the divalent Lewis^x containing hexapeptide **72** [54], started with *N*-Boc-protected α-allyl-aspartate **65** [165],

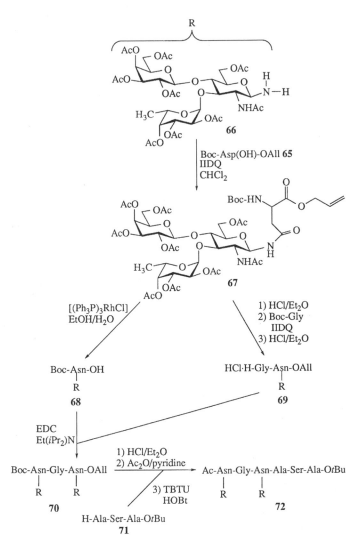

Scheme 40   Synthesis of a bivalent *N*-glycohexapeptide carrying two Lewis[x] residues.

which was coupled with the peracetylated trisaccharide amine **66** in a reaction promoted by IIDQ (Table 2, entry 10). The Boc group of the formed *N*-glycosyl asparagine conjugate **67** was selectively removed using HCl/diethylether, and the deprotected amino function subsequently condensed with Boc-glycine. Again the Boc group was removed from the glycosylated dipeptide to give **69**. On the other hand, Rh(I)-catalyzed isomerization and cleavage of the allyl ester [165] of **67** provided compound **68** selectively deblocked at the carboxylic function. Coupling of **69** and **68** employing water-soluble carbodiimide (EDC) [170,171] gave a tripeptide **70** that carries two Lewis[x] saccharides [54]. Replacement of the *N*-terminal Boc by the biocompatible *N*-acetyl group, cleavage of the allyl ester, and condensation of the resulting carboxyl function with tripeptide ester **71** led to the glycohexapeptide **72** carrying two Lewis^x chains, simulating a biantennary presentation of the carbohydrate antigen [54].

For complete deblocking, the *tert*-butyl ester was cleaved with formic acid and the *O*-acetyl groups were removed using sodium methoxide in methanol. To facilitate immunological investigations, conjugation of the deprotected *bis*-Lewis^x hexapeptide **73** to bovine serum albumin (BSA) [173] or keyhole limpet hemocyanin (KLH) [174,175] was carried out with water-soluble carbodiimide/*N*-hydroxysuccinimide and furnished neoglycoproteins **74** and **75**. The particularly mild and selective application of enzymes promises further improvements for the synthesis of complex glycopeptides in solution, e.g., of sensitive *O*-glycopeptides of the mucin type [131]. Thus, the twofold glycosylated decapeptide **83** from the tandem repeat of MUC1 [176], a highly glycosylated protein found on the surface of breast and pancreatic tumor cells [177], was prepared by repeated application of a lipase-catalyzed cleavage of heptyl esters [40,178,179].

Starting from the 2-azido-2-deoxygalactopyranosyl serine conjugate **76**, the heptyl ester was removed enzymatically. The resulting Fmoc-protected glycosyl serine derivative **77** was coupled to the *O*-glycosylated threonyl-alanine dipeptide heptyl ester **78** using EDC/HOBt. Since the various lipases tested did not accept 2-acetamido galactopyranosyl amino acid heptyl esters (e.g., **50**) as substrates, the azido functions could only be reduced and acetylated employing thioacetic acid after the enzymatic reaction had been completed. Probably due to an increased distance between the 2-acetamido group and the heptyl ester, the 2-acetamido groups of *N*-acetylgalactosamine in C-terminally extended glycopeptides did not inhibit the cleavage of the heptyl esters. With carboxylic compound **79**, further chain extension employing tetrapeptide fragment **80** led to the glycohexapeptide **81**, from which the Fmoc group was removed with morpholine. Condensation with tripeptide fragment **82** furnished the desired glycopeptide sequence of MUC1. Lipase-catalyzed cleavage of the heptyl ester and removal of the *O*-acetyl groups with sodium methoxide in methanol gave the target molecule **83** [40]. It should be noted that during the lipase-

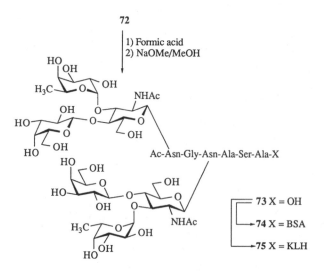

Scheme 41 Preparation of immunorelevant conjugates from **72**.

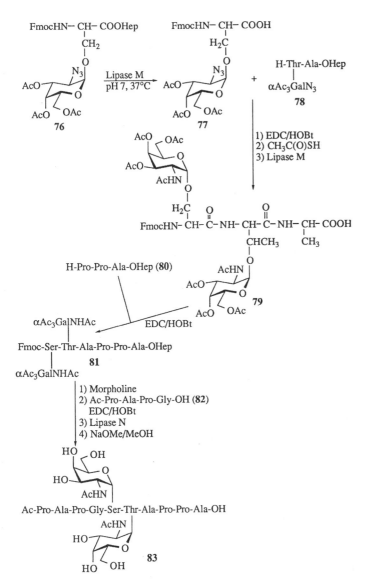

Scheme 42 Synthesis of a decapeptide from the repeating unit of MUC1 carrying a Tn antigen side chain.

catalyzed hydrolyses, no undesired cleavage of *O*-acetyl groups or effect on other functionalities was observed.

Meanwhile, the limitations in the substrate quality of heptyl esters, in some cases probably arising from their insufficient wetability in phosphate buffer/acetone systems, have been overcome by the introduction of methoxyethoxyethyl (MEE) esters [39]. These polar esters even facilitated the enzyme-catalyzed hydrolysis of such distinctly hydrophobic and inert substrates as **84**.

As the concluding example in this section, the preparation of a bivalent T antigen derivative, representing a partial structure from glycophorin A, a glycoprotein

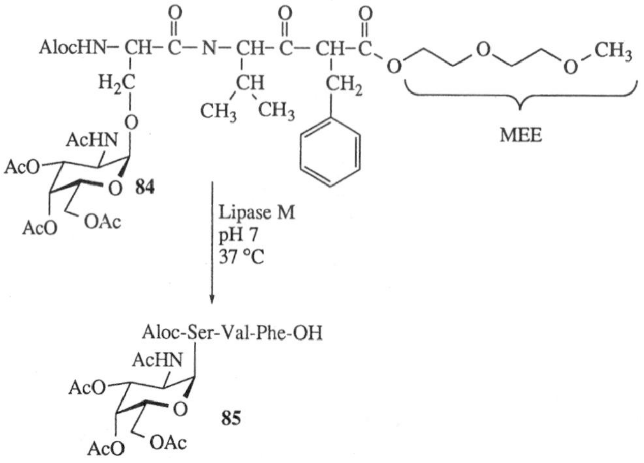

Scheme 43 Lipase-catalyzed cleavage of MEE esters of O-glycopeptides.

from the erythrocyte surface, and its conversion into an immunorelevant carrier-hapten conjugate will be described [143,180]. Starting from the key disaccharide bromide **52** and the Fmoc-protected serine and threonine benzyl esters **86** and **87**, the glycosylated amino acids **88** and **89** were obtained by the procedure outlined earlier (Scheme 34) using silver perchlorate/silver carbonate. After reduction of the azido function employing nickel dichloride and sodium borohydride [145], subsequent acetylation, and the mild and selective removal of the Fmoc group of **88** using morpholine [42], the glycosylated dipeptide benzyl ester was furnished by condensation of the N-terminally deblocked component with the 2-(4-pyridyl)-ethoxycarbonyl (4-Pyoc)-protected serine derivative **90** [181]. The protecting-group pattern in the coupling product allowed the selective hydrogenolysis of the C-terminal benzyl ester and thus the further elongation of **92** with the glycosylated threonine benzyl ester **91.**

After conversion into the C-H acidic pyridinium form using methyliodide/dichloromethane, the 4-Pyoc group of **92** was removed by treatment with morpholine/dichloromethane. The N-terminus was then acetylated, the benzyl ester cleaved by hydrogenolysis, and the O-acyl protections of the carbohydrate moiety were removed using hydrazine in methanol [42]. For the preparation of the desired synthetic T antigen, bovine serum albumin was used as the carrier protein [173]. Attachment of the glycosylated tripeptide **93** was achieved using water soluble carbodiimide and HOBt [143,180]. Immunological studies using the resulting neoglycoprotein **94** afforded monoclonal antibodies specific for the T antigen structure.

B. Solid-Phase Glycopeptide Synthesis

The application of the solid-phase peptide methodology developed by Merrifield [182,183] in the field of glycopeptide synthesis promises simplified access to large partial sequences of glycoproteins. The advantages of the solid-phase method, e.g., rapid syntheses, possible automation, easy removal of surplus reagents, and the circumvention of intermittent purification procedures, have stimulated intensive re-

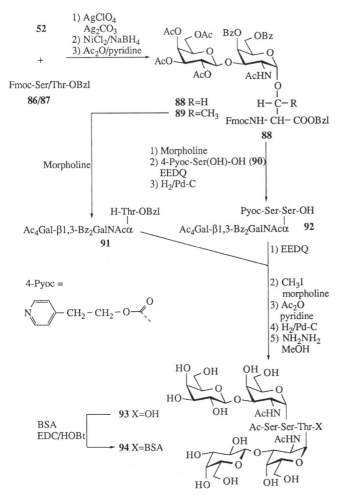

Scheme 44   Synthesis of an immunorelevant twofold glycosylated glycotripeptide carrying two T antigen side chains.

search to find reaction conditions and polymer-anchoring systems adapted to the enhanced sensitivity of glycosylated components during the intermediate and final coupling, cleavage, and deprotection steps.

The choice of the anchoring structure, in particular, is of great importance for a successful solid-phase glycopeptide synthesis. The anchor, connecting the growing peptide chain to the polymer backbone, has to be inert toward the conditions applied for coupling and removal of the temporary *N*-terminal protections, e.g., Boc or Fmoc groups. Moreover, the final cleavage of the glycopeptide–anchor linkage has to proceed without affecting the sensitive glycosidic structures.

Often, the anchoring principle used for the synthesis of glycopeptides is based on acid-labile linkers [49,50,184]. Typically, these linkers are methoxy-substituted benzyloxy or benzylamino structures, e.g., **95** ("SASRIN") [185] and **96** ("RINK") [186], which allow the isolation of glycopeptides with free C-terminus or of glycopeptide amides after a cleavage with trifluoroacetic acid (TFA). Linkers such as **96**, with a higher degree of methoxy substitution, additionally permit the retention of acid-labile side-chain protections, e.g., *tert*-butyl or trityl ethers, since diluted TFA is sufficient in these cases.

The RINK-linker **96** has been used for the solid-phase synthesis of glycopeptide amides using the ''multiple-column'' technique [187]. With the aid of this method, a variety of GalNAc-, GalNAcα-(1,3)-GalNAc-, and GalNAcα-(1,6)-GalNAc-carrying glycopeptides with systematically varied peptide sequence have been obtained on a small scale. Starting from the Fmoc-Gly-RINK-norleucine-poly-(dimethylacrylamide) conjugate **98**, assembly of the peptide chain was achieved with Fmoc-protected amino acid pentafluorophenyl esters. 3-Hydroxy-4-oxo-3,4-dihydro-1,2,3-benzotriazine (Dhbt-OH) [49,188] served as auxiliary nucleophile during the coupling conditions and facilitated the estimation of turnover since the initially yellow color of the Dhbt-OH ammonium salt disappears in the course of the reaction. After the last coupling step, the *N*-terminus of **99** was acetylated, the azido functions in the carbohydrate portion were reduced and acetylated, and the glycopeptide amide was cleaved from the resin with TFA. Finally the *O*-acetyl groups were removed using very dilute sodium methoxide in methanol.

A total circumvention of acidic or basic conditions during the cleavage of the glycopeptide-linker structure was achieved by the introduction of the allylic anchoring principle **97** [48,189]. With 4-hydroxycrotonic acid in **101** serving as the key element of the ''HYCRAM™'' linker **103**, a mild and selective cleavage of the assembled glycopeptide from the resin using the Pd(0)-catalyzed allyl transfer reaction becomes possible. The HYCRAM™ resin was further improved by insertion of β-alanine as standard amino acid between hydroxycrotonic acid and the polymer backbone (''β-HYCRAM™'') [67,190]. This spacer allows convenient quantitative control of the coupling reactions by amino acid analysis.

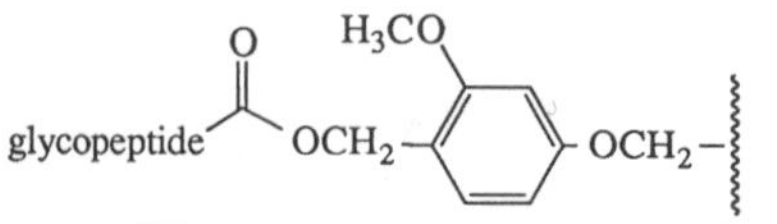

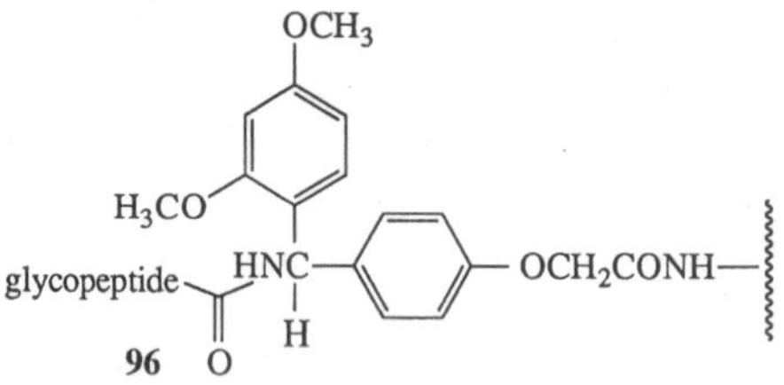

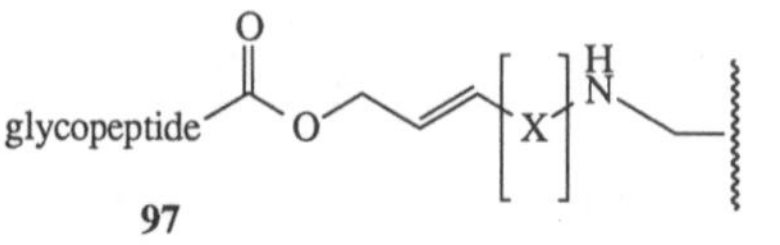

Scheme 45   Different types of anchoring systems suitable for solid-phase glycopeptide synthesis.

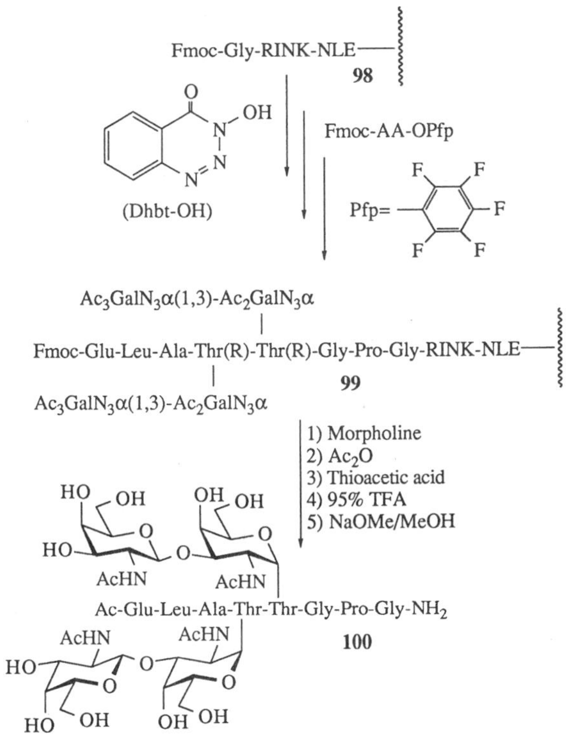

Scheme 46 Synthesis of O-glycopeptide amides carrying GalNAc-GalNAc side chains.

The applicability of this anchoring methodology was demonstrated in the solid-phase synthesis of partial structures of the HIV surface glycoprotein gp 120 on a gram-scale [67]. The glycosylated Peptide T [191], carrying lactosamine (Galβ-(1,4)GlcNAc) N-glycosidically linked to asparagine, was synthesized starting from the preformed conjugate **101** already bearing the C-terminal amino-acid [67]. Further Fmoc-protected amino acids, including Fmoc-Asn(Ac$_4$Galβ1,3Ac$_2$GlcNAc)-OH, were coupled according to a standard protocol (coupling: diisopropylcarbodiimide(DIC)/HOBt; Fmoc-cleavage: morpholine/dichloromethane 1:1). After the final coupling step using Boc-Ala-OH, release of lactosamine-Peptide T from the polymer conjugate **104** was achieved by the Pd^0-catalyzed allyl transfer to N-methylaniline in DMSO. Finally, tert-butyl group protections were removed using TFA. Subsequent deprotection of the disaccharide portion using diethylmethylamine yielded lactosamine-Peptide T **105** [67].

However, as further investigations showed, this anchoring system could still be improved. The α,β-unsaturated carbonyl part in **103** seemed to be responsible for undesired side reactions if the Fmoc group was applied as temporary protection—in particular, on the level of the resin-linked dipeptide [192]. As a consequence, out of the allylic anchors [189] the HYCRON-linker **106** was developed, in which 4-hydroxycrotonic acid was replaced by 1,4-dihydroxybut-2-ene combined with a flexible and polar spacer [68]. In fact, this linker diminished the tendency for aminolysis and allowed a distinct improvement in the yields of the solid-phase synthesis involving the Pd^0-catalyzed cleavage.

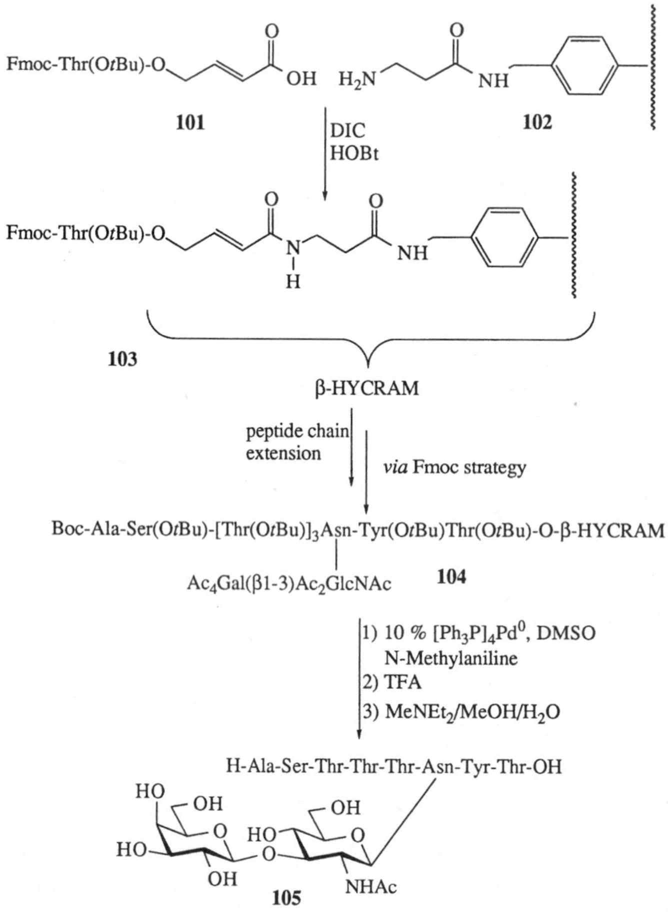

Scheme 47 Solid-phase synthesis of glycosylated peptide T using the β-HYCRAM methodology.

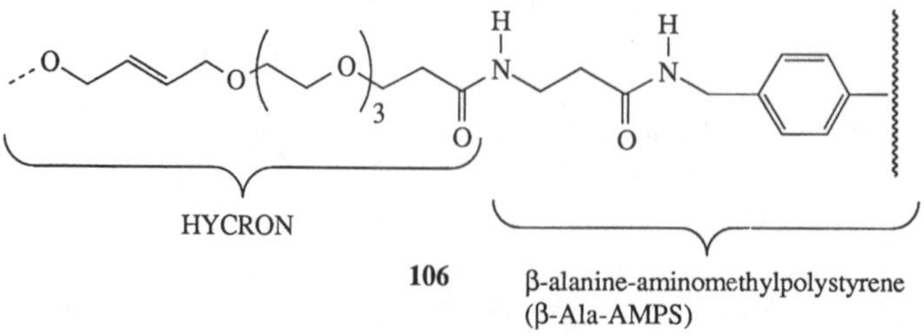

Scheme 48 The HYCRON linker attached to the polymer backbone.

The benefits of the HYCRON linker were demonstrated in the synthesis of the glycosylated partial structure **110** from MUC1 (see Section V.A.). Starting from the polymer-linked Fmoc-glycine, the *N*-terminus was deblocked using morpholine. The resin-linked amino compound was then condensed with Boc-proline. Since a protonated *N*-terminus resulted from the acidolytic removal of the *tert*-butyloxycarbonyl group, the formation of a diketopiperazine [193] prior to the coupling of the next amino acid was avoided. For further coupling steps, Fmoc-protected amino acids were activated with DIC/HOBt, except for the glycosylated building block and the immediately following amino acid, which were coupled using *O*-(1-benzotriazol-1-yl)-*N,N,N',N'*-tetramethyluronium tetrafluoroborate (TBTU) [194] and *N*-methylmorpholine. After acetylation of the terminal alanine, the glycononapeptide was detached from the resin by Pd⁰-catalyzed allyl transfer to give the partially deblocked glycononapeptide (purity > 95%) in an overall yield (18 steps) of 95% relative to the loaded resin **107**. Deprotection of the amino acid side chains delivered the target compound in an isolated overall yield of 85%, referring to the initial loading of the starting amino acid (determined by amino acid analysis) [68].

The presented results prove the allylic anchoring methodology to be an efficient tool in solid-phase glycopeptide synthesis: Both Fmoc and Boc strategies are applicable and sensitive structural elements like *tert*-butyl side-chain protections remain unaffected during the removal of the glycopeptide from the resin. Thus, larger and selectively protected fragments can be obtained and subjected to further coupling reactions.

The stability of the allylic anchor toward acids not only permits the selective removal of the *N*-terminal Boc group from resin-linked peptides, e.g., **108**, but also allows the selective cleavage of acid-sensitive side chain protections, e.g., the *tert*-butyl ether protection from resin-bound tripeptide **111** to form the polymer-bound

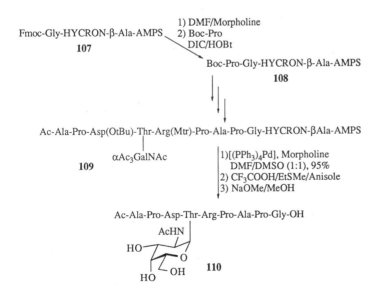

Scheme 49   Solid-phase synthesis of a partial structure of MUC1 using the HYCRON linker.

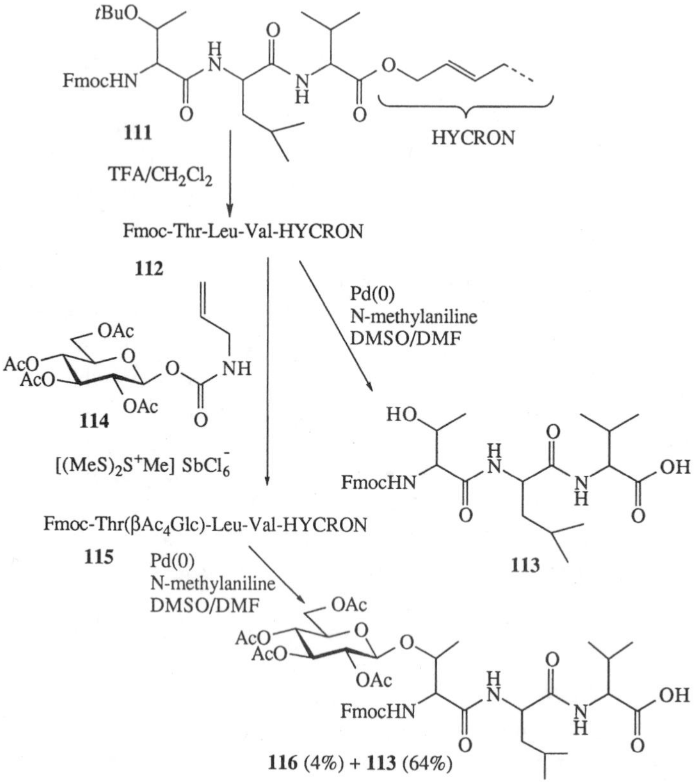

Scheme 50   Side-chain deprotection and *O*-glycosylation of a threonine peptide linked to HYCRON resin.

tripeptide **112**. Palladium(0)-catalyzed cleavage of the allylic anchor afforded the side-chain-deblocked tripeptides **113** in high overall yield related to the resin-loaded amount of valine [195]. According to this result, the removal of the side chain *tert*-butyl ether from **111** had occurred almost quantitatively and with complete selectivity. Attempts to subject the resin-bond tripeptide **112** to an *O*-glycosylation by treatment with the *O*-glycosyl allylcarbamate **114** activated with methyl bisme-thylthiosulfonium salt [196] were of very limited success. After Pd⁰-catalyzed release, the *O*-glycosylated Fmoc tripeptide **116** was isolated in very low yield together with a major amount of the unglycosylated peptide **113** [195,197]. This observation is in contradiction to the successful *O*-glycosylation of immobilized serine-containing peptides achieved with a glucosamine-derived oxazoline [198]. The low yield in the formation of **115** may be traced back to general problems concerning glycosylations of acceptors linked to solid phases, e.g., steric hindrance and incomplete removal of water. But also the glycosylation method and the neighboring bulky Fmoc group in **112** could be responsible for the incomplete reaction.

Promising results have been reported for the formation of *N*-glycopeptides formed in a condensation of resin-linked glutamic acid- and aspartic-acid-containing peptides with glycosylamines [199]. The method is based on the convergent synthesis of *N*-glycopeptides from preformed aspartic acid-containing peptides and glycosyl-amines [161,167], already outlined in Section IV (see Schemes 38 and 39 and Table 2).

The resin-linked [leu]enkephaline *N*-terminally substituted with glutamic acid **117** was synthesized using SASRIN resin [185] and temporary Fmoc protection [199]. The allyl ester in the side chain of glutamic can selectively be removed by palladium(0)-catalyzed allyl transfer [200]. The free γ-carboxy group of **118** was

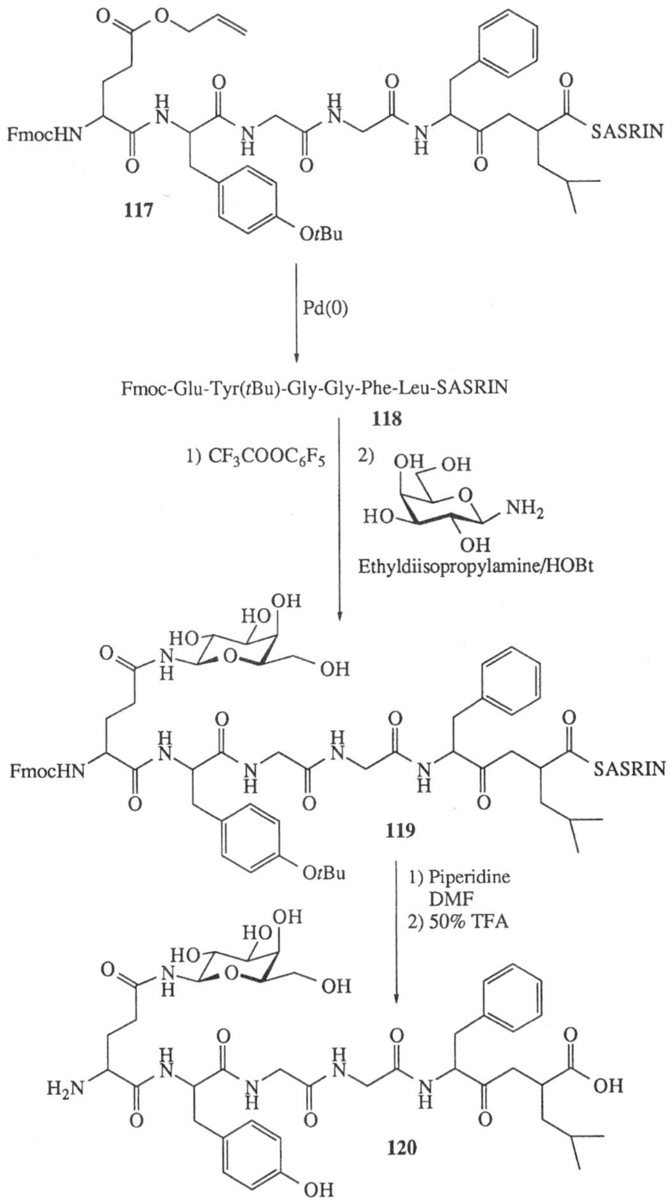

Scheme 51 Convergent solid-phase synthesis of *N*-glycopeptides linked to SASRIN resin.

converted into the β-galactosylamine in the presence of HOBt to give the *N*-galac-
tosyl glutamine glycopeptide **119** [189]. Special attention has to be paid in this
condensation reaction to the case of the corresponding aspartic acid-containing peptides, which are prone to easy aspartimide formation and subsequent isomerisation
(see Scheme 39). In particular, excess of base should be omitted [167]. It is also
noteworthy, that *O*-acyl-protected glycosylamines do not react under these conditions
[167], whereas *O*-benzyl-protected saccharides, e.g., the Sialyl Lewis^x lactone amine,
can be subjected to this type of condensations [86]. After base-catalyzed removal of
the Fmoc group from **119** and acidolytic cleavage of the *tert*-butyl ether and the
SASRIN anchor, the free *N*-glycopeptide **120** was obtained in an overall yield of
58% after purification by preparative HPLC. More than 20 *N*-glycosylated glutamine-
and asparagine-containing peptides have been synthesized by this convergent methodology [199]. The final application of acidolysis did not affect the glycosidic bonds
within the *N*-glycopeptides except for the fucose-containing glycopeptide, which underwent cleavage of the fucoside bond to a greater extent. Nevertheless, the combination of the solid-phase peptide synthesis with selectively removable side-chain
protection and subsequent condensation with glycosylamines opens up ways to the
construction of glycopeptide libraries.

VI. OUTLOOK

The selective methods developed during the past 15 years allow the construction of
glycopeptides of biological relevancy in preparative amounts. Model glycopeptides
thus available can be used to demonstrate and clarify the role of carbohydrates in
biological processes. For example, it has been shown that *N*-glycosylation can break
helices and thus changes secondary structures of peptides [201]. Recently, it was
reported that glycosylation of immunogenic peptides can affect the binding to MHC
II molecules [202] and also influence the T-cell stimulatory activity of epitopes on
virus glycoproteins [203].

Glycosylation also affects the activity of transporter proteins [204,205], protects
peptides and proteins from proteolysis, and has various other interesting effects on
the biological activity of glycoprotein factors [206,207]. In this chapter, major selective protecting-group methodologies for glycopeptide synthesis have been compiled. Their use was demonstrated on examples, in particular on the synthesis of T
antigen *O*-glycopeptides, and Lewis antigen glycopeptides. Of course, efficient and
stereoselective oligosaccharide syntheses are inevitable preconditions for the construction of biologically relevant glycopeptides. Since such biologically important
glycans often contain sensitive bonds, e.g., fucoside units, the demands for a suitable
protecting-group methodology become even more challenging.

The solid-phase synthesis provides versatile access to glycopeptides of various
structures. Some examples are shown in Section V.B. Promising are convergent synthesis, either on solid phase [199] or in solution [86,161,167]. For the synthesis of
preparative amounts, fragment condensations are particularly efficient [54,85]. A valuable tool for the construction of glycopeptides are enzymatic reactions. Glycosyltransferase-catalyzed saccharide chain extensions have successfully been achieved
with *N*-glycopeptides [208] and *O*-glycopeptides [209], and recently were successfully applied to solid-phase synthesis of glycopeptides [210,211]. Supplemented by

enzymatic condensation reactions of glycopeptide fragments [212], enzymatic methods constitute promising alternatives to the chemical methods used for the construction of glycopeptides.

ABBREVIATIONS

Aloc	Allyloxycarbonyl
AMPS	Aminomethylpolystyrene
Boc	*tert*-Butyloxycarbonyl
Bzd	Benzylidene
Bz	Benzoyl
Bzl	Benzylether
CAN	Ceric ammonium nitrate
DCC	Dicyclohexylcarbodiimide
DDQ	5,6-Dichloro-2,3-dicyano-1,4-benzoquinone
Dhbt	3-Hydroxy-4-oxo-3,4-dihydro-1,2,3-benzotriazine
DIC	Diisopropylcarbodiimide
DIEA	Diisopropylethylamine
EDC	1-Ethyl-3-(3′-dimethyl-aminopropyl)carbodiimide hydrochloride
EEDQ	1-Ethoxycarbonyl-2-ethoxy-1,2-dihydroquinoline
Fmoc	9-Fluorenylmethoxycarbonyl
HBTU	2-(1*H*-benzotriazol-1-yl)-1,1,3,3-tetramethyluronium hexafluorphosphate
HOBt	1-Hydroxybenzotriazole
HYCRAM	Hydroxycrotonylaminomethylpolystyrene
HYCRON	(*E*)-17-Hydroxy-4,7,10,13-tetraoxy-15-heptadecenoyl
IIDQ	1-Isobutoxycarbonyl-2-isobutoxy-1,2-dihydroquinoline
Mpm	Methoxybenzylether
OAll	Allylester
OBzl	Benzylester
OHep	*n*-Heptylester
OMEE	Methoxyethoxyethylester
O*t*Bu	*tert*-Butylester
OTf	Trifluoromethanesulfonic acid ester
Pmb	*para*-Methoxybenzylidene
Pyoc	2-(4-Pyridyl)ethoxycarbonyl
SLe$^{x/a}$	Sialyl Lewis$^{x/a}$
TBDPS	*tert*-Butyldiphenylsilylether
TBTU	*O*-(1-Benzotriazole-1-yl)-*N*,*N*,*N*′,*N*′-tetramethyluronium tetrafluoroborate
*t*Bu	*tert*-Butylether
Tcoc	Trichloroethoxycarbonyl
TFA	Trifluoroacetic acid
TMS	Trimethylsilyl
Z	Benzyloxycarbonyl

REFERENCES

1. E. Fischer, Untersuchungen über Kohlenhydrate und Fermente I+II (1884–1919), *E. Fischer: Gesammelte Werke* (M. Bergmann, ed.), Springer, Berlin, 1922.
2. E. Fischer, Untersuchungen über Aminosäuren, Polypeptide und Proteine I+II (1899–1907), *E. Fischer, Gesammelte Werke* (M. Bergmann, ed.), Springer, Berlin, 1923.
3. W. F. Goebel and O. T. Avery, Chemo-immunological studies on conjugated carbohydrate—proteins, I. The synthesis of *p*-aminophenol β-glucosides, *p*-aminophenol β-galactosides and their coupling with serum globulin, II. Immunological specificity of synthetic sugar-protein antigens, III. Active and passive anaphylaxis with synthetic sugar-proteins, *J. Exp. Med. 50*: 521–567 (1929).
4. H. Lis and N. Sharon, Protein glycosylation. Structural and functional aspects, *Eur. J. Biochem. 218*: 1 (1993).
5. G. Ashwell and A. G. Morell, Membrane glycoproteins and recognition phenomena, *Adv. Enzymol. 41*: 99 (1977).
6. I. Ofek and N. Sharon, Lectinophagocytosis: A molecular mechanism of recognition between cell surface sugars and lectins in the phagocytosis of bacteria, *Infect. Immun. 56*: 539 (1988).
7. R. P. McEver, K. L. Moore, and R. D. Cummings, Leukocyte trafficking mediated by selectin-carbohydrate interactions, *J. Biol. Chem. 270*: 11,025 (1995).
8. R. J. Wieser and F. Oesch, Contact inhibition of growth by complex carbohydrates, *Trends Glycosci. Glycobiol. 4*: 160–167 (1992).
9. P. G. Johansen, R. D. Marshall, and A. Neuberger, Preparation and some of the properties from hen-egg albumin, *Biochem. J. 78*: 518 (1961).
10. E. Bause and G. Legler, The role of the hydroxy amino acid in the triplet sequence Asn-Xaa-Thr(Ser) for the *N*-glycosylation step during glycoprotein biosynthesis, *Biochem. J. 195*: 639 (1981).
11. L. Kasturi, J. R. Eshleman, W. H. Wunner, and S. H. Shakin-Eshleman, The hydroxy amino acid in an Asn-X-Ser/Thr sequon can influence N-linked core glycosylation efficiency and the level of expression of cell surface glycoprotein, *J. Biol. Chem. 270*: 14,756 (1995).
12. V. P. Bhavanandan, E. Buddecke, R. Carubelli, and A. Gottschalk, Complete enzymic degradation of glycopeptides containing *O*-seryl- and *O*-threonyl-linked carbohydrates, *Biochem. Biophys. Res. Commun. 16*: 333 (1964).
13. E. Buddecke, H. Schauer, E. Werries, and A. Gottschalk, Characterization of *O*-seryl-*N*-acetylgalactosaminide hydrolysases as a-*N*-acetylgalactosaminidase, *Biochem. Biophys. Res. Commun. 34*: 517 (1969).
14. U. Lindahl and L. Rodén, The role of galactose and xylose in the linkage of heparin to protein, *J. Biol. Chem. 240*: 2821 (1965).
15. G. S. Marks and A. Neuberger, Synthetic studies relating to the carbohydrate—protein linkage in egg albumin, *Biochem. J. 85*: 15p–16p (1962).
16. C. H. Bolton and R. W. Jeanloz, Amino sugars. XXXVI. Synthesis of a glucosamine asparagine compound.—Benzyl N^2-benzyloxycarbonyl-N^4-(2-acetamido-3,4,6-tri-*O*-acetyl-2-deoxy-β-*O*-glucopyranosyl)-L-asparaginate, *J. Org. Chem. 28*: 3228 (1963).
17. B. Lindberg and B. G. Silvander, Synthesis of *O*-β-D-xylopyranosyl-L-serine, *Acta Chem. Scand. 19*: 530 (1965).
18. F. Micheel and H. Köchling, Über die Reaktionen des Glucosamins, *Chem. Ber. 91*: 673–676 (1958).
19. J. R. Vercellotti and A. E. Luetzow, β-Elimination of α glycoside monosaccharide from 3-*O*-(2-amino-2-deoxy-D-glucopyranosyl)serine—evidence for an intermediate in glycoprotein hydrolysis, *J. Org. Chem. 31*: 825–830 (1966).

20. H. G. Garg and R. W. Jeanloz, The synthesis of N-(benzyloxycarbonyl)-3-O-[3,4,6-tri-O-acetyl-2-deoxy-2-(2,4-dinitroanilino)-β-D-glucopyranosyl]-L-serine methyl ester, and its condensation with activated esters of amino acids, *Carbohydr. Res. 49*: 482–488 (1976).

21. H. Kauth and H. Kunz, Synthesis of protected serine glycopeptides via N-terminal elongation of the peptide chain, *Liebigs Ann. Chem. 1983*: 360–366 (1983).

22. C. R. Torres and G. W. Hart, Topography and polypeptide distribution of terminal N-acetylglucosamine residues on the surfaces of intact lymphocytes. Evidence for O-linked GlcNAc, *J. Biol. Chem. 259*: 3308 (1984).

23. H. G. Garg and R. W. Jeanloz, Synthetic N- and O-glycosyl derivatives of L-asparagine, L-serine and L-threonine, *Adv. Carbohydr. Chem. Biochem. 43*: 135–201 (1985).

24. H. Paulsen, Progress in the selective chemical synthesis of complex oligosaccharides, *Angew. Chem. Int. Ed. Engl. 21*: 155 (1982).

25. R. R. Schmidt, New methods of glycoside and oligosaccharide synthesis—Are there alternatives to the Koenigs–Knorr method? *Angew. Chem. Int. Ed. Engl. 25*: 212 (1986).

26. H. Kunz, Synthesis of glycopeptides. Partial structures of biological recognition components, *Angew. Chem. Int. Ed. Engl. 26*: 294–308 (1987).

27. H. G. Garg, K. v. d. Bruch, and H. Kunz, Developments in the synthesis of glycopeptides containing glycosyl L-asparagine, L-serine, and L-threonine, *Adv. Carbohydr. Chem. Biochem. 50*: 277–310 (1994).

28. H. Kunz, Glycopeptides of biological interest: A challenge for chemical synthesis, *Pure Appl. Chem. 65*: 1223–1232 (1993).

29. M. Bergmann and L. Zervas, Über ein allgemeines Verfahren der Peptid-Synthese, *Ber. Dtsch. Chem. Ges. 65*: 1192 (1932).

30. W. H. Hartung and R. Simonoff, Hydrogenolysis of benzyl groups attached to O, N or S, *Org. React. (N. Y.) 7*: 263 (1953).

31. L. A. Carpino, Oxidative reactions of hydrazines. II. Isophthalimides. New protective groups on nitrogen, *J. Am. Chem. Soc. 79*: 98 (1957).

32. L. A. Carpino, The 9-fluorenylmethyloxycarbonyl family of base-sensitive amino-protecting groups, *Acc. Chem. Res. 20*: 401 (1987).

33. T. B. Windholz and D. B. R. Johnston, Trichloroethoxycarbonyl. A generally applicable protecting group, *Tetrahedron Lett. 27*: 2555 (1967).

34. R. Johansson and B. Samuelsson, Regioselective reductive ring-opening of 4-methoxybenzylidene acetals of hexopyranosides. Access to a novel protecting group strategy, *J. Chem. Soc., Perkin Trans. 1 1984*: 2371 (1984).

35. K. Horita, T. Yoshioka, T. Tanaka, Y. Oikawa, and O. Yonemitsu, On the selectivity of deprotection of benzyl, MPM (4-methoxybenzyl) and DMPM (3,4-dimethoxy benzyl) protecting groups for hydroxy functions, *Tetrahedron 42*: 3021 (1986).

36. H. Kunz and H. Waldmann, The allyl group as mildly and selectively removable carboxy-protecting group for the synthesis of labile O-glycopeptides, *Angew. Chem. Int. Ed. Engl. 23*: 71–72 (1984).

37. H. Kunz and C. Unverzagt, Allyloxycarbonyl (Aloc) moiety—conversion of an unsuitable into a useful amino protecting group for peptide synthesis, *Angew. Chem. Int. Ed. Engl. 23*: 436–437 (1984).

38. U. Zehavi, B. Amit, and A. Patchornik, Light-sensitive glycosides. I. 6-Nitroveratryl-β-D-glucopyranoside and 2-nitrobenzyl-β-D-glucopyranoside, *J. Org. Chem. 37*: 2281 (1972).

39. H. Kunz, D. Kowalczyk, P. Braun, and G. Braum, Enzymatic hydrolysis of hydrophilic diethylenglycol and polyethylenglycol esters of peptides and glycopeptides by lipases, *Angew. Chem. Int. Ed. Engl. 33*: 336 (1994).

40. P. Braun, H. Waldmann, and H. Kunz, Chemoenzymatic synthesis of O-glycopeptides carrying the tumor-associated Tn antigen structure, *Bioorg. Med. Chem. 1*: 197–207 (1993).

41. M. Bodanszky and A. Bodanszky, *The practice of peptide synthesis*, Vol. XVII, Springer-Verlag, Berlin, 1984, p. 284.

42. P. Schultheiss-Reimann and H. Kunz, O-Glycopeptide synthesis by 9-fluorenylme-thoxycarbonyl(Fmoc)-protected blocks, *Angew. Chem. Int. Ed. Engl. 22*: 62 (1983).

43. M. Meldal, T. Bielfeldt, S. Peters, K. J. Jensen, H. Paulsen, and K. Bock, Susceptibility of glycans to β-elimination in Fmoc-based O-glycopeptide synthesis, *Int. J. Peptide Protein Res. 43*: 529 (1994).

44. B. Liebe and H. Kunz, Synthesis of sialyl-Tn antigen. Regioselective sialylation of a galactosamine threonine conjugate unblocked in the carbohydrate portion, *Tetrahedron Lett. 35*: 8777–8778 (1994).

45. B. Ferrari and A. A. Pavia, Antigénes de groupe sanguin. Synthese de glycopeptides Tn representant la partie N-terminae de la glycophorine, humaine A^N et A^M, *Tetrahedron 41*: 1939 (1985).

46. M. Ciommer and H. Kunz, Synthesis of glycopeptides with partial structure of human glycophorin using the fluorenylmethoxycarbonyl/allyl protecting group combination, *Synlett 1991*: 593 (1991).

47. S. Rio, J.-M. Beau, and J.-C. Jaquinet, Synthesis of glycopeptides from the carbohy-drate linkage region of proteoglycans, *Carbohydr. Res. 219*: 71 (1991).

48. H. Kunz and B. Dombo, Solid-phase synthesis of peptides and glycopeptides on polymer supports carrying allylic anchoring groups, *Angew. Chem. Int. Ed. Engl. 27*: 711–713 (1988).

49. S. Peters, T. Bielfeldt, M. Meldal, K. Bock, and H. Paulsen, Multiple-column solid-phase glycopeptide synthesis, *J. Chem. Soc., Perkin Trans. 1 1992*: 1163 (1992).

50. H. Paulsen, G. Merz, and U. Weichert, Solid-phase synthesis of O-glycopeptide se-quences, *Angew. Chem. Int. Ed. Engl. 27*: 1365 (1988).

51. V. V. Bencomo and P. Sinay, Synthesis of glycopeptides having clusters of O-glycosylic disaccharide chains [β-D-Gal-(1-3)-α-D-GalNAc] located at vicinal amino acid residues of the peptide chain, *Carbohydr. Res. 116*: C9 (1983).

52. H. Kunz and C. Unverzagt, Protective group dependent stability of intersaccharidic bonds. Synthesis of fucosylchitobiose-glycopeptides, *Angew. Chem. Int. Ed. Engl. 27*: 1697–1699 (1988).

53. C. Unverzagt and H. Kunz, Synthesis of glycopeptides and neoglycoproteins containing the fucosylated linkage region of N-glycoproteins, *Bioorg. Med. Chem. 2*: 1189–1201 (1994).

54. K. v. d. Bruch and H. Kunz, Synthesis of N-glycopeptide clusters with Lewis[x] antigen side-chains and their coupling to carrier proteins, *Angew. Chem. Int. Ed. Engl. 33*: 101–103 (1994).

55. B. F. Lundt and N. L. Johansen, Removal of t-butyl and t-butyloxycarbonyl protecting groups with trifluoroacetic acid. Mechanisms, byproduct formation and evaluation of scavengers, *Int. J. Peptide Protein Res. 12*: 258 (1978).

56. D. S. Tarbell, Y. Yamamoto, and B. M. Pope, New method to prepare N-*tert*-butoxy-carbonyl derivatives and the corresponding sulfur analogs from di-*tert*-butyl-di-carbon-ate or di-*tert*-butyl-di-thiolodicarbonates and amino acids, *Proc. Natl. Acad. Sci. USA 69*: 730 (1972).

57. L. Moroder, A. Hallett, E. Wünsch, O. Keller, and G. Wersin, Di-*tert*-butyl-dicarbonate, a useful reagent for the introduction of the *tert*-butyloxycarbonyl protecting group, *Hoppe-Seyler's Z. Physiol. Chem. 357*: 1651 (1976).

58. R. W. Roeske, Amino acid *tert*-butyl esters, *Chem. Ind. (London) 1959*: 1251 (1959).

59. E. Vowinkel, Darstellung von Carbonsäureestern mittels *O*-Alkyl-*N*,*N'*-dicyclohexyli-soharnstoffen, *Chem. Ber. 100*: 16 (1967).

60. G. Braum, Dissertation, Universität Mainz, (1991).

61. M. Schultz and H. Kunz, Synthetic *O*-glycopeptides as model substrates for glycosyl-transferases, *Tetrahedron Asymmetry 4*: 1205 (1993).

62. H. Kunz and H. Waldmann, The allyl group as selectively cleavable carboxy-protecting group for the synthesis of sensitive *O*-glycopeptides, *Angew. Chem. Int. Ed. Engl. 23*: 71 (1984).

63. H. Kunz and J. März, The *p*-nitro cinnamyloxycarbonyl(Noc) group—An acid stable amino protecting group for peptide and glycopeptide synthesis—cleavable under neutral conditions, *Angew. Chem. Int. Ed. Engl. 27*: 1375–1377 (1988).

64. F. Guibe, O. Dangles, G. Balavoine, and A. Loffet, Use of an allylic anchor group and of its palladium catalyzed hydrostannolytic cleavage in the solid phase synthesis of protected peptide fragments, *Tetrahedron Lett. 30*: 2641 (1989).

65. J. P. Genet, E. Blart, M. Savignac, S. Lemeune, and J.-M. Lemeune, Palladium catalyzed reaction in aqueous media. An efficient removal of allyloxycarbonyl protecting group from oxygen and nitrogen, *Tetrahedron Lett. 34*: 4189 (1993).

66. P. D. G. Jeffrey and S. W. McCombie, Homogeneous, palladium(0)-catalyzed exchange deprotection of allylic ester, carbonates and carbamates, *J. Org. Chem. 47*: 587 (1982).

67. W. Kosch, J. März, and H. Kunz, Synthesis of glycopeptide derivatives of Peptide T on a solid phase using an allylic linkage, *Reactive Polymers 22*: 181–194 (1994).

68. O. Seitz and H. Kunz, A novel allylic anchor for solid-phase synthesis—Synthesis of protected and unprotected mucin-type *O*-glycopeptides, *Angew. Chem. Int. Ed. Engl. 34*: 803–805 (1995).

69. R. B. Woodward, K. Heusler, J. Gosteli, P. Naegeli, P. Oppolzer, R. Ramage, S. Ranganathan, and H. Vorbrüggen, Total synthesis of cephalosporin C, *J. Am. Chem. Soc. 88*: 852 (1966).

70. H. Eckert and I. Ugi, New protective group technique—Cleavage of β-halogenated alkyl esters with super nucleophilic cobalt(II)phthalocyanine, *Liebigs Ann. Chem. 1979*: 278 (1979).

71. V. Poszgay, Synthesis of oligosaccharides related to plant, vertebrate, and bacterial cell-wall glycans. *Carbohydrates—Synthetic Methods and Application in Medicinal Chemistry* (H. Ogawa, A. Hasegawa, and T. Suami, eds.), Verlag Chemie, Weinheim, 1992, pp. 188–227.

72. P. J. Garegg, H. Hultberg, and S. Wallin, A novel ring-opening of carbohydrate benzylidene acetals, *Carbohydr. Res. 108*: 97–101 (1982).

73. A. Liptak, J. Imre, J. Harangi, P. Nánási, and A. Neszmélyi, Chemo-, stereo-, and regioselective hydrogenolysis of carbohydrate benzylidene acetals. Synthesis of benzyl ethers of benzyl α-D-, methyl β-D-mannopyranoside and α-D-rhamnopyranosides by ring cleavage of benzylidene derivatives with the lithium aluminum hydride–aluminum trichloride reagent, *Tetrahedron 38*: 3721 (1982).

74. T. A. Springer, Adhesion receptors of the immune system, *Nature 346*: 425 (1990).

75. L. A. Lasky, An endothelial ligand for L-selectin is a novel mucin-like molecule, *Science 258*: 964 (1992).

76. L. A. Lasky and S. D. Rosen, Further characterization of the interaction between L-selectin and its endothelial ligands, *Glycobiol. 2*: 373 (1992).

77. A. Varki, Selectins and other mammalian sialic acid-binding lectins, *Curr. Op. Cell Biol. 4*: 257 (1992).

78. S. H. Itzkowitz, M. Yuan, Y. Fukushi, A. Palekar, P. C. Phelps, A. M. Shamsuddin, B. F. Trump, S.-i. Hakomori, and Y. S. Kim, Lewis X and sialylated Lewis X related antigen expression in human malignant and nonmalignant colonic tissues, *Cancer Res. 46*: 2627–2632 (1986).

79. T. A. Springer, Traffic signals for lymphocyte recirculation and leukocyte emigration: The multistep paradigm, *Cell 76*: 301 (1994).

80. M. L. Phillips, E. Nudelman, F. C. A. Gaeta, M. Perez, A. K. Singhal, S.-i. Hakomori, and J. C. Paulson, ELAM 1 mediates cell adhesion by recognition of a carbohydrate ligand. Sialyl Lewis X, *Science 250*: 1130–1132 (1990).

81. M. Larkin, T. J. Ahern, M. S. Stoll, M. Shaffer, D. Sako, J. O'Brien, C.-T. Yuen, A. M. Lawson, R. A. Childs, K. M. Barone, P. R. Langer-Safer, A. Hasegawa, M. Kiso, G. R. Larsen, and T. Feizi, Spectrum of sialylated and nonsialylated fuco-oligosaccharides bound by the endothelial-leukocyte adhesion molecule E-selectin. Dependence of the carbohydrate binding activity on E-selectin density, *J. Biol. Chem. 267*: 13661 (1992).

82. C.-T. Yuen, A. M. Lawson, W. Chai, M. Larkin, M. S. Stoll, A. C. Stuart, F. X. Sullivan, T. J. Ahern, and T. Feizi, Sulfated blood group Lewis A, *Biochemistry 31*: 9126 (1992).

83. A. Lubineau, J. L. Gallic, and R. Lemoine, First synthesis of the 3'-sulfated Lewis^a pentasaccharide, the most potent human E-selectin ligand so far, *Bioorg. Med. Chem. 2*: 1143 (1994).

84. K. C. Nicolaou, N. J. Bockovich, and D. R. Carcanague, Total synthesis of sulfated Le^x and Le^a-type oligosaccharide lectin ligands, *J. Am. Chem. Soc. 115*: 8843 (1993).

85. H. Kunz and J. März, Synthesis of selectively deprotectable asparagine glycoconjugates with a Lewis A antigen side chain, *Synlett 1992*: 589 (1992).

86. U. Sprengard, G. Kretzschmar, E. Bartnik, C. Hüls, and H. Kunz, Synthesis of a RGD-sialyl-Lewis X-glycoconjugate: A new and highly potent ligand for P-selectin, *Angew. Chem. Int. Ed. Engl. 34*: 990 (1995).

87. A. Yamamoto, C. Miyashita, and H. Tsukamoto, Amino sugars (II). Preparation of 2-acetamido-1-N[L-α(and β) aspartyl]-2-deoxy-β-D-glucosylamino and 2-acetamido-2-deoxy-N-[L-γ-glutamyl]-β-D-glucosylamine, *Chem. Pharm. Bull. 13*: 1041 (1965).

88. M. Spinola and R. W. Jeanloz, The synthesis of a di-N-acetylchitobiose asparagine derivative, 2-acetamido-4-O-(2-acetamido-2-deoxy-β-D-glucopyranose)-1-N-(4-L-aspartyl)-2-deoxy-β-D-glucopyranosylamine, *J. Biol. Chem. 245*: 4158 (1970).

89. H. Kunz and H. Waldmann, Construction of disaccharide N-glycopeptides—Synthesis of the linkage region of the transmembrane neuraminidase of an influenza virus, *Angew. Chem. Int. Ed. Engl. 24*: 883 (1985).

90. H. Kunz and C. Unverzagt, Synthesis of fucosyl glycosides and disaccharides using 4-methoxybenzyl (Mpm) protected fucosyl donors, *J. prakt. Chem. 334*: 579 (1992).

91. H. Lönn, Synthesis of a tetra- and a nonasaccharide which contain a-fucopyranosyl groups and are part of the complex type of the carbohydrate moiety of glycoproteins, *Carbohydr. Res. 139*: 105 (1985).

92. R. U. Lemieux and J. I. Haymi, The mechanism of the anomerization of the tetra-O-acetyl-D-glucopyranosyl chloride, *Can. J. Chem. 43*: 2162 (1965).

93. M. Nilsson and T. Norberg, Synthesis of dimeric Lewis X hexasaccharide derivative corresponding to a tumor-associated glycolipid, *Carbohydr. Res. 183*: 71 (1988).

94. K. v. d. Bruch, Dissertation, Universität Mainz, (1993).

95. J. C. Jacquinet and P. Sinay, Synthesis of blood group substances. 6. Synthesis of O-α-L-fucopyranosyl-(1-2)-O-β-galactopyranosyl-1(1-4)-O-[α-L-fucopyranosyl-(1-3)]-2-acetamido-2-deoxyglucopyranose—The postulated Le^d antigenic determinant, *J. Org. Chem. 42*: 720 (1977).

96. R. Boss and R. Scheffold, Cleavage of allyl ethers using Pd/C, *Angew. Chem. Int. Ed. Engl. 15*: 558 (1976).

97. R. R. Schmidt and J. Michel, Simple synthesis of α- and β-O-glycosyl imidates; preparation of glycosides and disaccharides, *Angew. Chem. Int. Ed. Engl. 19*: 731 (1980).

98. T. Murase, H. Ishida, M. Kiso, and A. Hasegawa, A facile regio- and stereoselective synthesis of α-glycosides of N-acetylneuraminic acid, *Carbohydr. Res. 184*: C1-C4 (1988).

99. A. Hasegawa, H. Ohki, T. Nagahama, H. Ishida, and M. Kiso, Synthetic studies on sialo glycoconjugates. Part 19. A facile large-scale preparation of the methyl-2-thioglycoside of *N*-acetylneuraminic acid , and its usefulness for the a-stereoselective synthesis of sialoglycosides, *Carbohydr. Res. 212*: 277 (1991).

100. F. Dasgupta and P. J. Garegg, Alkyl sulfenyl triflate as activator in the thioglycoside-mediated formation of 3-glycosidic linkages during oligosaccharide synthesis, *Carbohydr. Res. 177*: C13–C17 (1988).

101. K. C. Nicolaou, C. W. Hummel, and Y. Iwabuchi, Total synthesis of dimeric Le^x, *J. Am. Chem. Soc. 114*: 3126 (1992).

102. W. Bröder and H. Kunz, Glycoside and saccharide synthesis using *N*-glycosyl triazoles as hydrolytically stable glycosyl donors, *Synlett 1990*: 251 (1990).

103. W. Bröder and H. Kunz, A new method of anomeric protection and activation based on the conversion of glycosyl azides to glycosyl fluorides, *Carbohydr. Res. 249*: 221 (1993).

104. L. D. Hall, J. F. Manville, and N. S. Bhacca, Specifically fluorinated carbohydrates. I. Nuclear magnetic resonance studies of hexopyranosyl fluoride derivatives, *Can. J. Chem. 47*: 1 (1969).

105. K. C. Nicolaou, R. E. Dolle, D. P. Papahatjis, and J. L. Randall, Practical synthesis of oligosaccharides. Partial synthesis of avermectin B1a, *J. Am. Chem. Soc. 106*: 4189 (1984).

106. H. Kunz and W. Sager, Stereoselective glycosylations of alcohols and silyl ethers using glycosyl fluorides and boron trifluoride etherate, *Helv. Chem. Acta 68*: 285 (1985).

107. K. H. Jung, M. Hoch, and R. R. Schmidt, Glycosyl imidates 42. Selectively protected lactose and 2-azido lactose building blocks for glycoside synthesis, *Liebigs Ann. Chem. 1989*: 1099 (1989).

108. T. Mukaiyama, Y. Muria, and T. Shoda, An efficient method for glycosylation of hydroxy compounds using glucopyranosyl fluorides, *Chem. Lett. 1981*: 431 (1981).

109. A. Lubineau, J. L. Gallic, and R. Lemoine, First synthesis of the 3′-sulfated Lewis^a trisaccharide, putative ligand for the leukocyte homing receptor, *J. Chem. Soc., Chem. Commun. 1993*: 1419 (1993).

110. V. Behar and S. Danishefsky, Highly convergent synthesis of the blood group determinant Lewis Y in conjugate forming shape, *Angew. Chem. Int. Ed. Engl. 33*: 1468 (1994).

111. S. Danishefsky, K. F. McClure, J. T. Randolph, and R. B. Ruggeri, A strategy for the solid-phase synthesis of oligosaccharides, *Science 260*: 1307 (1993).

112. J. Montreuil, Primary structure of glycoprotein glycans, *Advances in Carbohydrate Chemistry and Biochemstry* (R. S. Tipson and D. Horton, eds.), Vol. 37, Academic Press, New York, 1980, pp. 157–223.

113. A. Varki, Biological roles of oligosaccharides: All of the theories are correct, *Glycobiol. 3*: 97 (1993).

114. J. W. Dennis and S. Laferte, Oncodevelopmental expression of -GlcNAcβ1-6Manα1-6Manβ1—Branched asparagine-linked olisosaccharides in murine tissues and breast carcinomas, *Cancer Res. 49*: 945 (1989).

115. R. U. Lemieux, T. Takeda, and B. Y. Chung, Synthesis of 2-Amino-2-deoxy-β-D-glucopyranosides, *Am. Chem. Symp. Ser. 39*: 90 (1976).

116. H. Paulsen and O. Lockhoff, Neue, effektive b-Glycosidsynthese für Mannose-Glycoside. Synthese von Mannose-haltigen Oligosacchariden, *Chem. Ber. 114*: 3102 (1981).

117. F. Barresi and O. Hindsgaul, Synthesis of β-mannopyranosides by intramolecular aglycon delivery, *J. Am. Chem. Soc. 113*: 9376 (1991).

118. G. Stork and G. Kim, Stereocontrolled synthesis of disaccharides via the temporary silicon connection, *J. Am. Chem. Soc. 114*: 1087 (1992).

119. Y. Ito and T. Ogawa, A new approach for the stereoselective synthesis of β-mannosides, *Angew. Chem. Int. Ed. Engl. 33*: 1765 (1994).

120. C. Augé, C. C. Warren, R. W. Jeanloz, M. Kiso, and L. Anderson, The synthesis of *O*-β-D-Mannopyranosyl-(1→4)-*O*-(2-acetamido-2-deoxy-β-D-glucopyranosyl)-(1→4)-2-acetamido-2-deoxy-D-glucopyranose, *Carbohydr. Res. 82*: 71 (1980).

121. W. Günther and H. Kunz, Synthesis of a β-mannosyl chitobiosyl asparagine conjugate—A central element of the core region of *N*-glycoproteins, *Angew. Chem. Int. Ed. Engl. 29*: 1068 (1990).

122. H. Kunz and W. Günther, β-Mannoside synthesis via inversion of configuration on β-glucosides by intramolecular nucleophilic substitution, *Angew. Chem. Int. Ed. Engl. 27*: 1188 (1988).

123. W. Günther and H. Kunz, Synthesis of β-D-mannosides from β-D-glucosides via an intramolecular S_N2 reaction at C-2, *Carbohydr. Res. 228*: 217 (1992).

124. S. Nakabayashi, C. D. Warren, and R. W. Jeanloz, Amino sugars 135. The preparation of a partially protected heptasaccharide-asparagine intermediate for glycopeptide synthesis, *Carbohydr. Res. 174*: 219 (1988).

125. C. Unverzagt, Synthesis of a branched heptasaccharide by regioselective glycosylations, *Angew. Chem. Int. Ed. Engl. 33*: 1102 (1994).

126. H. Schachter and D. Williams, Biosynthesis of mucus glycoproteins, *Mucus in Health and Disease* (E. N. Chantler, J. B. Elder and M. Elstein, eds.), Vol. 2, Plenum Press, New York, 1982, pp. 3–28.

127. T. A. Beyer, J. E. Sadler, J. J. Rearick, J. C. Paulson, and R. Hill, Glycosyltransferases and their use in assessing oligosaccharide structure and structure–function relationships, *Adv. Enzymol. 52*: 23–175 (1981).

128. L. F. Leloir, Two decades of research on the biosynthesis of saccharides, *Science 172*: 1299 (1971).

129. J. E. Hansen, O. Lund, J. Engelbrecht, H. Bohr, J. O. Nielsen, J.-E. S. Hansen, and S. Brunak, Prediction of *O*-glycosylation of mammalian proteins: Specificity patterns of UDP-GalNAc:polypeptide *N*-acetylgalactosaminyl transferase, *Biochem. J. 308*: 801–813 (1995).

130. R. A. Dwek, Glycobiology: Towards understanding the function of sugars, *Biochem. Soc. Trans. 23*: 1 (1995).

131. N. Jentoft, Why are proteins *O*-glycosylated? *TIBS 15*: 291 (1990).

132. G. Uhlenbruck, G. I. Pardoe, and G. W. G. Bird, On the specificity of lectins with a broad agglutination spectrum, *Z. Immun. Forsch. 138*: 423 (1969).

133. G. F. Springer, T and Tn, general carcinoma autoantigens, *Science 224*: 1198 (1984).

134. S. H. Itzkowitz, M. Yuan, C. K. Montgomery, T. Kjeldsen, H. K. Takahashi, W. L. Bigbee, and Y. S. Kim, Expression of Tn, sialosyl Tn, and T antigen in human colon cancer, *Cancer Res. 49*: 197 (1989).

135. T. Toyokuni, S.-i. Hakomori, and A. K. Singhal, Synthetic carbohydrate vaccines: Synthesis and immunogenicity of Tn antigen conjugates, *Bioorg. Med. Chem. 2*: 1119 (1994).

136. H. Paulsen and J.-P. Hölck, Building blocks of oligosaccharides. Part XV. Synthesis of *O*-β-D-galactopyranosyl-(1-3)-*O*-(α-D-2-acetamido-2-deoxy-α-D-galactopyranosyl)-(1-3)-L-serine and -L-threonine glycopeptides, *Carbohydr. Res. 109*: 89 (1982).

137. H. Paulsen, Syntheses, conformations and X-ray structure analyses of the saccharide chains from the core region of glycoproteins, *Angew. Chem. Int. Ed. Engl. 29*: 823 (1990).

138. R. U. Lemieux and R. M. Ratcliffe, The azidonitration of tri-*O*-acetylgalactal, *Can. J. Chem. 57*: 1244 (1979).

139. H. Paulsen, W. Rauwald, and U. Weichert, Building units of oligosaccharides. LXXXVI. Glycosidation of oligosaccharide thioglycosides to *O*-glycoproteins segments, *Liebigs Ann. Chem. 1988*: 75.

140. M. Ravenscroft, R. M. G. Roberts, and J. G. Tillett, The reaction of some cyclic and open-chain disulfides with methyl trifluoromethanesulfonic acid, *J. Chem. Soc., Perkin Trans. 1982*: 1568 (1982).

141. P. Fügedi, P. J. Garegg, H. Lönn, and T. Norberg, Thioglycosides as glycosylating agents in oligosaccharide synthesis, *Glycoconjugate J. 4*: 97 (1987).

142. B. Ferrari and A. A. Pavia, The synthesis of derivatives of 3-*O*-(2-acetamido-2-deoxy-α-D-galactopyranosyl)-L-serine and -L-threonine, *Carbohydr. Res. 79*: C1 (1980).

143. H. Kunz, S. Birnbach, and P. Wernig, Synthesis of glycopeptides with Tn and T antigen structure and their linkage to bovine serum albumin, *Carbohydr. Res. 202*: 207 (1990).

144. H. C. Brown and C.A. Brown, The reaction of sodium borohydride with nickel acetate in aqueous solution—A convenient synthesis of an active nickel hydrogenation catalyst of low isomerizing tendency, *J. Am. Chem. Soc. 85*: 1003 (1963).

145. K. Heyns, D. Feldmann, D. Hadamczyk, J. Schwertner, and J. Thiem, Ein Verfahren zur Synthese von α-Glycosiden der 3-Amino-2,3,6-tridesoxyhexopyranosen aus Glycalen, *Chem. Ber. 114*: 232 (1981).

146. S. Friedrich-Bochnitschek, H. Waldmann, and H. Kunz, Allyl esters as carboxy protecting groups in the synthesis of *O*-glycopeptides, *J. Org. Chem. 54*: 751 (1989).

147. T. Adachi, Y. Yamada, and J. Inoue, An alternative method for the selective reduction of unsaturated nucleoside azides to the amines, *Synthesis 1977*: 45 (1977).

148. T. Rosen, I. M. Lico, and D. T. W. Chu, A convenient and highly chemoselective method for the reductive acetylation of azides, *J. Org. Chem. 53*: 1580 (1988).

149. E. Atherton, R. R. Cameron, and R. C. Sheppard, Peptide Synthesis. Part 10. Use of pentafluorophenyl esters of fluorenylmethoxycarbonylamino acids in solid phase peptide synthesis, *Tetrahedron 44*: 843 (1988).

150. H. Paulsen and M. Paal, Synthesis of oligosaccharides. LVI. Block synthesis of *O*-glycopeptides and other T-antigen structures, *Carbohydr. Res. 135*: 71 (1984).

151. T. Ogawa, K. Beppu, and S. Nakabayashi, Trimethylsilyl trifluoromethane sulfonate as an effective catalyst for glycoside synthesis, *Carbohydr. Res. 93*: C6 (1981).

152. H. Paulsen and K. Adermann, Synthese von *O*-Glycopeptid-Sequenzen des N-Terminus von Interleukin-2, *Liebigs Ann. Chem. 1989*: 751

153. A. Kurosaka, H. Kitagawa, S. Fukui, Y. Numata, H. Nakada, I. Funakoshi, T. Kawasaki, T. Ogawa, H. Iijima, and I. Yamashina, A monoclonal antibody that recognizes a cluster of a disaccharide, NeuAcα2-6GalNAc, in mucin-type glycoproteins, *J. Biol. Chem. 263*: 8724 (1988).

154. T. Kjeldsen, H. Clausen, S. Hirohashi, T. Ogawa, H. Iijima, and S.-i. Hakomori, Preparation and characterization of monoclonal antibodies directed to the tumor-associated *O*-linked sialosyl-2-6-α-*N*-acetylgalactosaminyl (sialosyl-Tn) antigen, *Cancer Res. 48*: 2214 (1988).

155. H. Iijima and T. Ogawa, Synthetic studies on cell-surface glycans. Part 52. Total synthesis of 3-*O*-[2-acetamido-6-*O*-(*N*-acetyl-α-D-neuraminyl)-2-deoxy-α-D-galactosyl]-L-serine and a stereoisomer, *Carbohydr. Res. 172*: 183 (1988).

156. Y. Nakahara, H. Iijima, S. Sibayama, and T. Ogawa, A highly stereoselective synthesis of di- and trimeric sialosyl-Tn epitope—A partial structure of glycophorin A1, *Tetrahedron Lett. 31*: 6897 (1990).

157. A. Marra and P. Sinay, A novel stereoselective synthesis of *N*-acetyl-α-neuraminosyl-galactose disaccharide derivatives using anomeric *S*-glycosyl xanthates, *Carbohydr. Res. 187*: 35 (1989).

158. W. A. R. v. Heswijk, M. J. D. Eenink, and J. Feijen, An improved method for the preparation of γ-esters of glutamic acid and β-esters of aspartic acid esters, *Synthesis 1982*: 744 (1982).

159. C.-D. Chang, M. Waki, M. Ahmad, J. A. Meienhofer, E. O. Lundell, and J. D. Haug, Preparation and properties of N2-9-fluorenylmethyloxycarbonyl amino acids bearing *tert*-butyl side chain protection, *Int. J. Pep. Prot. Res. 15*: 59 (1980).

160. L. M. Likhosherstov, O. S. Novikova, V. A. Derevitskaja, and N. K. Kochetkov, A new simple synthesis of amino-sugar β-D-glycosylamines, *Carbohydr. Res. 146*: C1 (1986).

161. S. T. Cohen-Anisfeld and P. T. Lansbury, A practical, convergent method for glyco-peptide synthesis, *J. Am. Chem. Soc. 115*: 10531 (1993).

162. A. Lubineau, J. Augé, and B. Drouillat, Improved synthesis of glycosylamines and a straightforward preparation of N-acylglycosylamines as carbohydrate-based detergents, *Carbohydr. Res. 266*: 211 (1995).

163. W. König and R. Geiger, New method for the synthesis of peptides: Activation of the carboxyl group with dicyclohexylcarbodiimide by using 1-hydroxybenzotriazoles as additives, *Chem. Ber. 103*: 788 (1970).

164. B. Belleau and G. Malek, New convenient reagent for peptide syntheses, *J. Am. Chem. Soc. 90*: 1651 (1968).

165. H. Waldmann and H. Kunz, Allylester als selektiv abspaltbare Carboxyschutzgruppe in der Peptid- und N-Glycopeptidsynthese, *Liebigs Ann. Chem. 1983*: 1712.

166. C. Unverzagt, Dissertation, 1988, Universität Mainz.

167. S. T. Cohen-Anisfeld and P. T. Lansbury, A convergent approach to the chemical syn-thesis of asparagine-linked glycopeptides, *J. Org. Chem. 55*: 5560 (1990).

168. J. März and H. Kunz, Synthesis of selectively deprotectable asparagine glycoconjugates with a Lewis[a] antigen side chain, *Synlett 1992*: 589 (1992).

169. Y. Kiso and H. Yajima, 2-Isobutoxy-N(isobutoxycarbonyl)-1,2-dihydroquinoline as a coupling reagent in peptide synthesis, *J. Chem. Soc., Chem. Commun. 1972*: 942 (1972).

170. J. C. Sheehan and S. L. Ledis, Total synthesis of monocyclic peptide lactone antibiotic, etamycin, *J. Am. Chem. Soc. 95*: 875 (1973).

171. M. K. Dhaon, R. K. Olsen, and R. Ramasamy, Esterification of N-protected α-amino acids with alcohol/carbodiimide/4-(dimethylamino)pyridine. Racemization of aspartic and glutamic acid derivatives, *J. Org. Chem. 47*: 1962 (1982).

172. R. T. Lee and Y. C. Lee, Preparation of cluster glycosides of N-acetylgalactosamine that have subnanomolar binding constants towards the mammalian hepatic Gal/GalNAc specific receptor, *Glycoconjugate J. 4*: 317 (1987).

173. C. P. Stowell and Y. C. Lee, The binding of glucosyl-neoglycoproteins to the hepatic asialoglycoprotein receptor, *Adv. Carbohydr. Protein Res. 37*: 225 (1980).

174. E. T. Adman, Copper protein structures, *Advances in Protein Chemistry* (C. B. Anfin-sen, J. T. Edsall, F. M. Richards, and D. S. Eisenberg, eds.), Vol. 42, Academic Press, San Diego, 1991, pp. 145–192.

175. C. P. Stowell and Y. C. Lee, Neoglycoproteins, and the preparation and application of synthetic glycoproteins, *Adv. Carbohydr. Chem. Biochem.* (R. S. Tipson and D. Horton, eds.), Academic Press, New York, 1980, pp. 225–281.

176. S. Gendler, J. Taylor-Papadimitriou, T. Duhig, J. Rothbard, and J. Burchell, A highly immunogenic region of human polymorphic epithelial mucin expressed by carcinomas is made up of tandem repeats, *J. Biol. Chem. 263*: 12820 (1988).

177. J. Hilkens, M. J. L. Ligtenberg, H. L. Vos, and S. L. Litvinov, Cell membrane associated mucins and their adhesion-modulating property, *TIBS 17*: 359 (1992).

178. P. Braun, H. Waldmann, W. Vogt, and H. Kunz, Selective enzymatic removal of pro-tecting functions: Heptyl esters as carboxy protecting groups in peptide synthesis, *Syn-lett 1990*: 105 (1990).

179. P. Braun, H. Waldmann, W. Vogt, and H. Kunz, Selektive enzymatische Schutzgrup-penabspaltungen: Der n-Heptylester als Carboxylschutzgruppe in der Peptidsynthese, *Liebigs Ann. Chem. 1991*: 165 (1991).

180. H. Kunz and S. Birnbach, Synthesis of O-glycopeptides of the tumor associated Tn and T antigen type and their linkage to bovine serum albumin, *Angew. Chem. Int. Ed. Engl. 25*: 360 (1986).

181. S. Birnbach and H. Kunz, Peptide synthesis using the hydrophilic-, acid and base stable 2-(4-pyridyl)ethoxycarbonyl protecting group, *Tetrahedron Lett. 25*: 3567 (1984).

182. R. B. Merrifield, Solid-phase peptide synthesis. I. The synthesis of a tetrapeptide, *J. Am. Chem. Soc. 85*: 2149 (1963).

183. G. Barany, N. Kneib-Cordonier, and D. G. Mullen, Solid-phase peptide synthesis: A silver anniversary report, *Int. J. Peptide Protein Res. 30*: 705 (1987).

184. B. Lüning, T. Norberg, and J. Tejbrant, Synthesis of mono- and disaccharide amino acid derivatives for use in solid phase peptide synthesis, *Glycoconjugate J. 6*: 5 (1989).

185. M. Mergler, R. Tanner, J. Gosteli, and P. Grogg, Peptide synthesis by a combination of solid-phase and solution methods. I. A new very acid-labile anchor group for the solid-phase synthesis of fully protected fragments, *Tetrahedron Lett. 29*: 4009 (1988).

186. H. Rink, Solid-phase synthesis of protected peptide fragments using a trialkoxy-diphenyl-methyl ester resin, *Tetrahedron Lett. 28*: 3787 (1987).

187. S. Rio-Anneheim, H. Paulsen, M. Meldal, and K. Bock, Synthesis of the building blocks Nα-Fmoc-O-[α-D-Ac$_3$GalN$_3$p-(1-3)α-D-Ac$_2$GalN$_3$p]-Thr-OPfp and Nα-Fmoc-O-[α-D-Ac$_3$GalN$_3$p-(1-6)α-D-Ac$_2$GalN$_3$p]-Thr-OPfp and their application in the solid-phase glycopeptide synthesis of core 5 and core 7 mucin O-glycopeptides, *J. Chem. Soc., Perkin Trans. 1 1995*: 1071 (1995).

188. E. Atherton, J. L. Holder, M. Meldal, R. C. Sheppard, and R. C. Valerio, Peptide synthesis. Part 12. 3,4-Dihydro-4-oxo-1,2,3-benzotriazin-3-yl esters of fluorenylmethoxycarbonyl amino acids as self-indicating reagents for solid-phase peptide synthesis, *J. Chem. Soc., Perkin Trans. 1 1988*: 2887 (1988).

189. H. Kunz and B. Dombo, Polymer-linked allylic protecting and anchoring systems in solid-phase peptide and glycopeptide synthesis, *DEP 3720269.3 (1987), Orpegen GmbH, Heidelberg, US Pat. 4929671 (1990)*: (1990).

190. H. Kunz, W. Kosch, and J. März, New allyl esters and their application for the construction of solid-phase systems for solid-phase reactions, *P 3938850.6*: (1989).

191. A. Motta, D. Picone, D. A. Temussi, M. Marastoni, and R. Tomatis, Conformational analysis of peptide T and of its C-pentapeptide fragment, *Biopolymers 28*: 479 (1989).

192. W. Kosch, Dissertation, Universität Mainz, (1992).

193. B. F. Gisin and R. B. Merrifield, Carboxyl-catalyzed intramolecular aminolysis. Side reactions in solid-phase peptide synthesis, *J. Am. Chem. Soc. 94*: 3102 (1972).

194. R. Knorr, A. Trzeciak, W. Bannworth, and D. Gillessen, New coupling reagents in peptide chemistry, *Tetrahedron Lett. 30*: 1927 (1989).

195. H. Kunz and O. Seitz, unpublished Dissertation, 1995, Universität Mainz.

196. H. Kunz and J. Zimmer, Glycoside synthesis via electrophile-induced activation of N-allyl carbamates, *Tetrahedron Lett. 34*: 2907 (1993).

197. H. Kunz, J. Zimmer, and O. Seitz, unpublished results; J. Zimmer, Dissertation 1994, Universität Mainz.

198. M. Hollósi, E. Kollát, I. Laczkó, K. F. Medziradszky, J. Thurin, and L. Otvos Jr., Solid-phase synthesis of glycopeptides: Glycosylation of resin-bound serine peptides by 3,4,6-tri-O-acetyl-D-glucose-oxazoline, *Tetrahedron Lett. 32*: 1531 (1991).

199. D. Vetter, D. Tumelty, S. K. Singh, and M. A. Gallop, A versatile solid-phase synthesis of N-linked glycopeptides, *Angew. Chem. Int. Ed. Engl. 34*: 60 (1995).

200. H. Kunz, H. Waldmann, and C. Unverzagt, Allyl ester as temporary protecting group for the β-carboxy function of aspartic acid, *Int. J. Peptide Protein Res. 26*: 493 (1985).

201. L. Otvos Jr., J. Thurin, E. Kollat, H. H. Mautsch, and M. Hollósi, Glycosylation of synthetic peptides breaks helices, *Int. J. Peptide Protein Res. 38*: 476 (1991).

202. S. Mourissen, M. Meldal, I. Christiansen-Brams, E. Elsner, and O. Werdelin, Attachment of oligosaccharides to peptide antigen profoundly affects binding to major histocompatibility complex class II molecules and peptide immunogenicity, *Eur. J. Immunol. 24*: 1066 (1994).

203. L. Otvos jr, G. R. Krivulka, L. Urge, G. I. Szendrei, L. Nagy, Z. Q. Xiang, and H. C. J. Ertl, Comparison of the effect of amino acid substitutions and β-*N*-vs. α-*O*-glycosylation on the T-cell stimulatory activity and conformation of an epitope on the rabies virus glycoprotein, *Biochim Biophys. Acta 1267*: 55 (1995).

204. K. Fiedler and K. Simons, The role of N-glycans in the secretory pathway, *Cell 81*: 309 (1995).

205. W. A. Lubas, M. Smith, C. M. Starr, and J. A. Hanover, Analysis of nuclear pore protein p62 glycosylation, *Biochemistry 34*: 1686 (1995).

206. H. Lis and N. Sharon, Protein glycosylation—Structural and functional aspects, *Eur. J. Biochem. 218*: 1 (1993).

207. P. M. Rudd, R. J. Woods, M. R. Wormald, G. Opdenakker, A. K. Downing, I. D. Campbell, and R. A. Dwek, The effects of variable glycosylation on the functional activities of ribonuclease, plasminogen and tissue plasminogen activator, *Biochim. Biophys. Acta 1248*: 1 (1995).

208. C. Unverzagt, H. Kunz, and J. C. Paulson, High-efficiency synthesis of sialooligosaccharides and sialoglycopeptides, *J. Am. Chem. Soc. 112*: 9308 (1990).

209. M. Schultz and H. Kunz, Synthetic *O*-glycopeptides as model substrates for glycosyltransferases, *Tetrahedron Asymm. 4*: 1205 (1993).

210. M. Schuster, P. Wang, J. C. Paulson, and C.-H. Wong, Solid-phase chemical-enzymatic synthesis of glycopeptides and oligosaccharides, *J. Am. Chem. Soc. 116*: 1135 (1994).

211. C.-H. Wong, R. L. Halcomb, Y. Ishikawa, and T. Kajimoto, Enzymes in organic synthesis: Application to the problem of carbohydrate recognition, *Angew. Chem. Int. Ed. Engl. 34*: 521 (1995).

212. C.-H. Wong, M. Schuster, P. Wang, and P. Sears, Enzymatic synthesis of *N*- and *O*-linked glycopeptides, *J. Am. Chem. Soc. 115*: 5893 (1993).

3

Enzymatic Synthesis of Oligosaccharides and Glycopeptides

Yoshitaka Ichikawa
The Johns Hopkins University School of Medicine, Baltimore, Maryland

I. INTRODUCTION

Enzymes have come into frequent use in synthetic organic chemistry because they are recognized as "easy-to-use reagents" that show high diastereo- and regioselectivity [1–6]. In the field of glycobiology (carbohydrate-related biology), these biological catalysts have been analyzed with respect to their specificity and kinetics. In 1980, Nunez and Barker reported the enzymatic synthesis of N-acetyllactosamine (Galβ(1→4)GlcNAc) using β(1,4)galactosyltransferase on a preparative scale [7]. This landmark paper stimulated synthetic chemists' interest in examining and developing enzymes (glycosyltransferases and glycosidases) as tools for the synthesis of complex oligosaccharides [8–14].

Three typical oligosaccharide structures are shown in Figure 1: a common N-linked triantennary complex-type oligosaccharide, a sialyl Lewis x-bearing O-linked glycan with a core 3 structure, and a tumor antigen glycolipid G_{D3} structure [15–18]. Since all these carbohydrates are synthesized in vivo by a series of glycosyltransferases [19–22], we can assume that these highly complicated structures could be synthesized in the laboratory in 10, 9, and 4 steps, respectively, if we were able to use all the necessary enzymes.

Enzymatic oligosaccharide synthesis has a great advantage over chemical synthesis: The reactions proceed in a highly stereo- and regioselective manner without the need to protect the sugars' functional groups [9,12,13]. Four types of enzymes are known to catalyze the formation of glycosidic bonds: phosphorylases, glycosidases, transglycosidases, and glycosyltransferases (Figure 2).

Phosphorylases utilize a sugar 1-phosphate as a donor substrate and transfer it to an acceptor molecule. One such enzyme is sucrose phosphorylase (EC 2.4.1.7), which is derived from the bacterium *Pseudomonas saccharophila* and catalyzes the following reaction (Scheme 1) [23]:

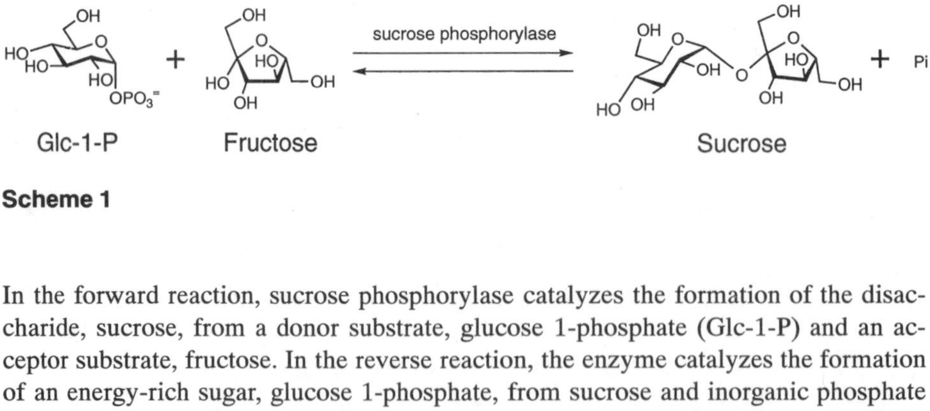

Glc-1-P   Fructose   Sucrose

Scheme 1

In the forward reaction, sucrose phosphorylase catalyzes the formation of the disaccharide, sucrose, from a donor substrate, glucose 1-phosphate (Glc-1-P) and an acceptor substrate, fructose. In the reverse reaction, the enzyme catalyzes the formation of an energy-rich sugar, glucose 1-phosphate, from sucrose and inorganic phosphate (Pi). The enzyme has been studied with respect to the acceptor substrate specificity and has been found to transfer a glucosyl residue to 2-keto-sugars [23].

Glycosidases normally catalyze the hydrolysis of the glycosidic bond; however, under appropriate conditions the enzymes can also catalyze the formation of a glycosidic bond (see Section IV on glycosidase-catalyzed oligosaccharide synthesis). Since glycosidases are easily obtained enzymes, their application to the synthesis of oligosaccharides has been well studied [10,11,13,14]. The reaction proceeds stereo-selectively but not efficiently with respect to regioselectivity.

Transglycosidases are a unique class of enzymes that transfer a particular car-bohydrate residue from an oligosaccharide to an acceptor oligosaccharide. One such enzyme is the recently isolated trans-sialidase from *Trypanosoma cruzi* (see Section IV on glycosidase-catalyzed oligosaccharide synthesis).

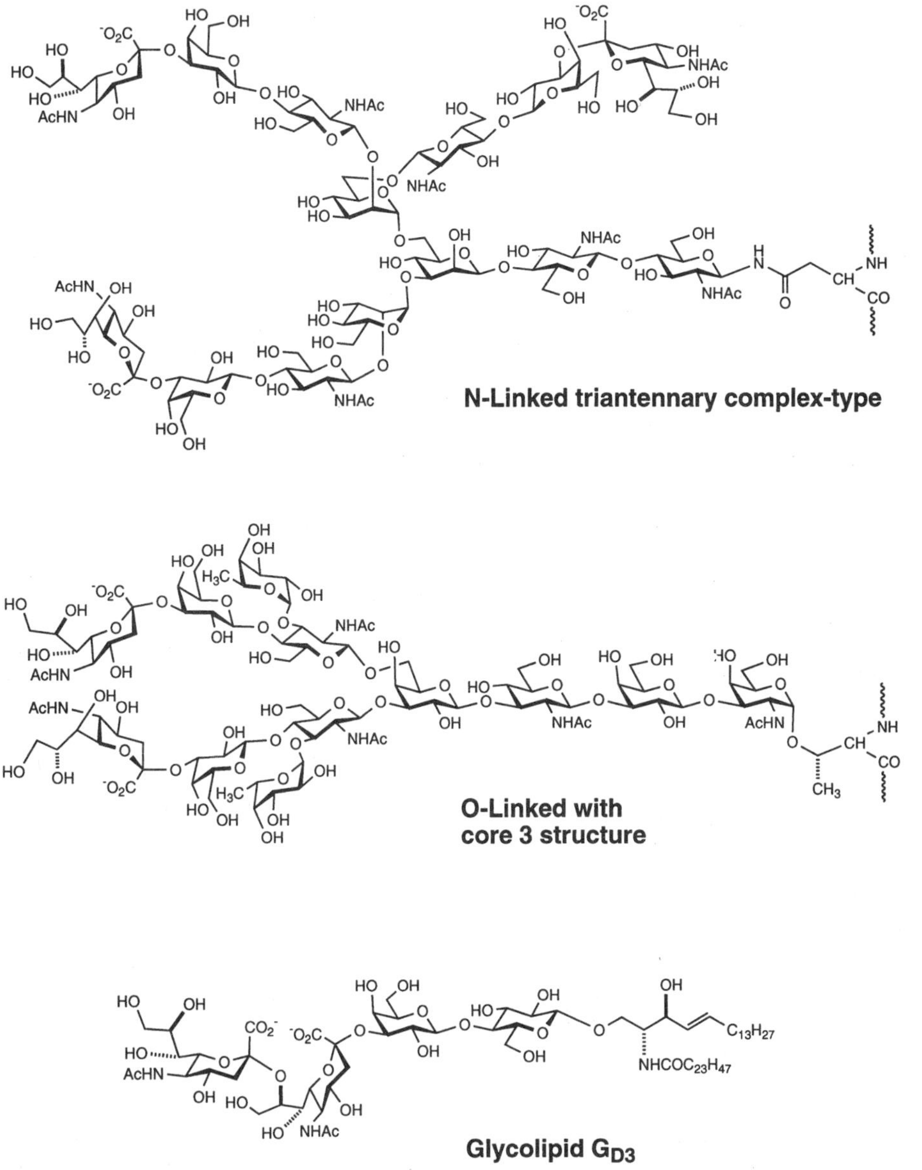

N-Linked triantennary complex-type

O-Linked with core 3 structure

Glycolipid G$_{D3}$

Figure 1

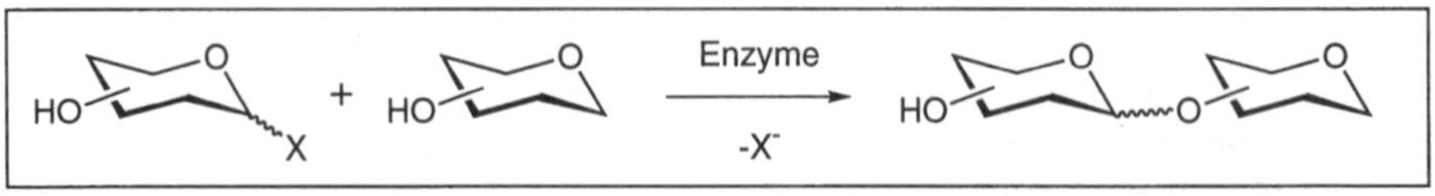

Enzyme	X (Leaving group)
Phosphorylase	-O-PO$_3^=$
Glycosidase	-OR, -F, or -OH
Transglycosidase	-O-Sugar or -OR
Glycosyltransferase	-O-UDP, -O-GDP, or -O-CMP

Figure 2

Glycosyltransferases require sugar nucleotides as donor substrates and are responsible for constructing most of the carbohydrate structures [8,9]. Because their reactions are stereo- and regioselective, glycosyltransferases are the most attractive, among all these glycosidic bond-forming enzymes, for use in synthesizing complex oligosaccharides [12,13].

In this chapter, recent advances in the enzymatic synthesis of oligosaccharides and glycopeptides will be discussed, with emphasis on the following subjects: (II) the properties and specificity of the most commonly used glycosyltransferases, (III) an enzymatic method for large-scale synthesis of oligosaccharides, (IV) glycosidase-catalyzed oligosaccharide synthesis, and (V) the enzymatic preparation of glycopeptides.

II. GLYCOSYLTRANSFERASES

In this section, we will discuss several glycosyltransferase reactions according to the type of sugar transferred, and examine their substrate specificity and the preparation of their donor substrates, sugar nucleotides.

A. Fucosyltransferases

Fucosylation is often the final modification of oligosaccharide structures in their biosynthetic pathway, and therefore levels of fucosyltransferase activity have been thought to play a regulatory role in intercellular recognition [24]. In fact, many biologically important oligosaccharides, such as blood-group substances and tumor antigens, are fucosylated, and high levels of fucosyltransferase activity are observed in the serum of cancer patients [25–27].

Fucosyltransferase catalyzes the transfer of L-fucose from guanidine 5'-diphospho-β-L-fucose (GDP-fucose) to an acceptor molecule (Scheme 2). The reaction is often carried out in Na cacodylate (dimethylarsinic acid sodium salt), MOPS (3-[N-morpholino]propanesulfonic acid), or HEPES (N-[2-hydroxyethyl]piperazine-N'-[2-ethanesulfonic acid]) buffer at 200 mM (pH 6 to 7.5) and usually requires Mn^{2+} for activity. One unit of the enzyme will transfer one mmol of L-fucose from GDP-fucose per hour at pH 6.2, in the presence of 5 mM ATP and 20 mM MnCl$_2$.

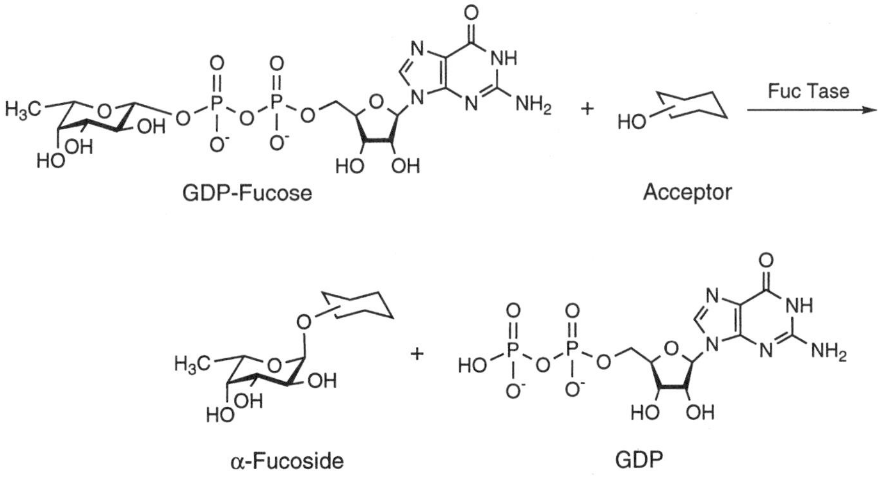

Scheme 2

Although GDP-fucose is commercially available, it is very expensive (currently $82 per mg). Chemical synthesis of GDP-fucose has been reported by several laboratories [28–34]: L-fucose β-1-phosphate is prepared and subsequently coupled to an activated guanosine 5'-monophosphate derivative such as GMP-morpholidate to give GDP-fucose (Scheme 3); however, the coupling reaction usually gives low yields (20–40%) because of the poor solubility of fucose 1-phosphate, the instability of fucose 1-phosphate and GDP-fucose, and the requirement for strict conditions (absolutely anhydrous) for the coupling reactions.

Two enzymatic syntheses of GDP-fucose have been reported (Methods A and B). Method A (Scheme 4) is a normal biosynthetic pathway for GDP-fucose, in which GDP-mannose is transformed into GDP-fucose [35]. (In Scheme 4, the carbon atoms that are modified by enzyme are highlighted with asterisks.) D-Mannose α-1-

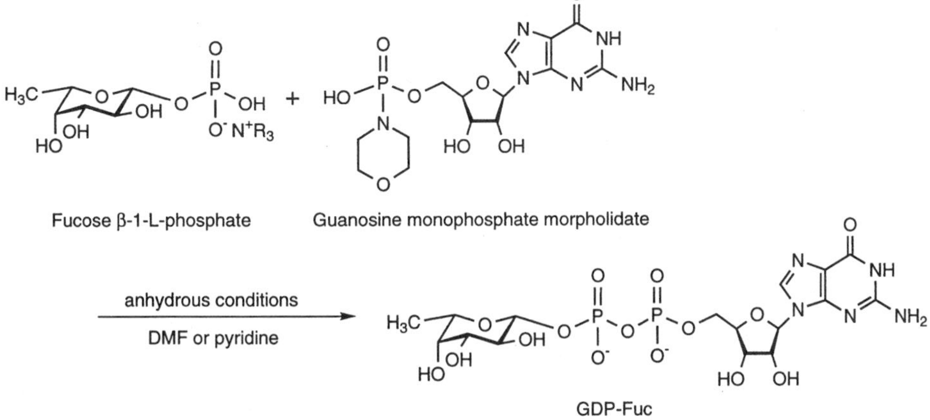

Scheme 3

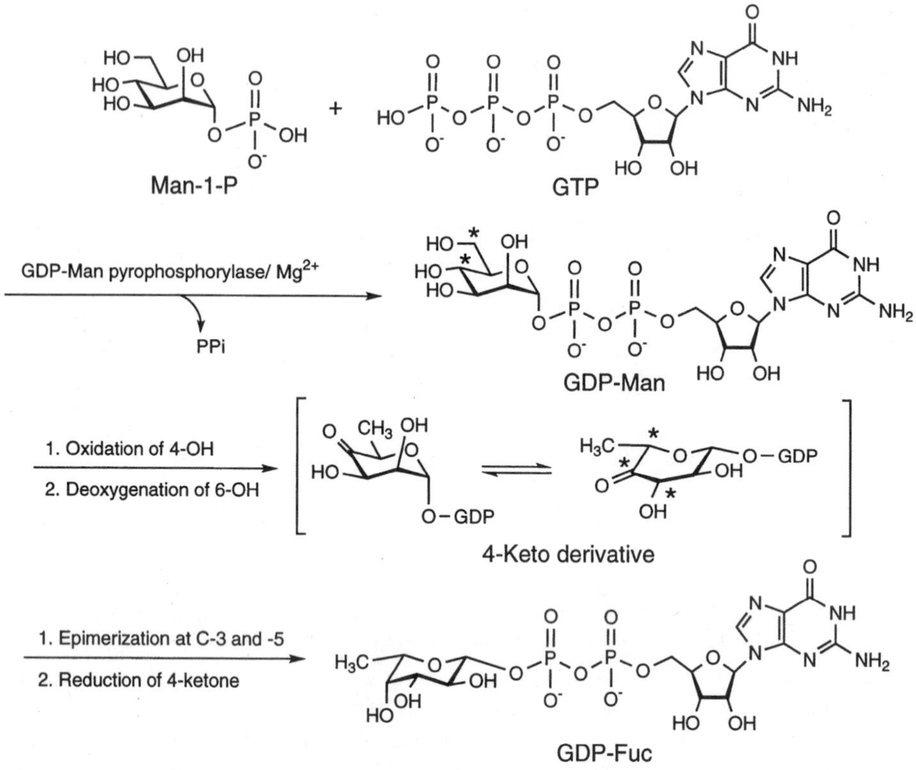

Scheme 4

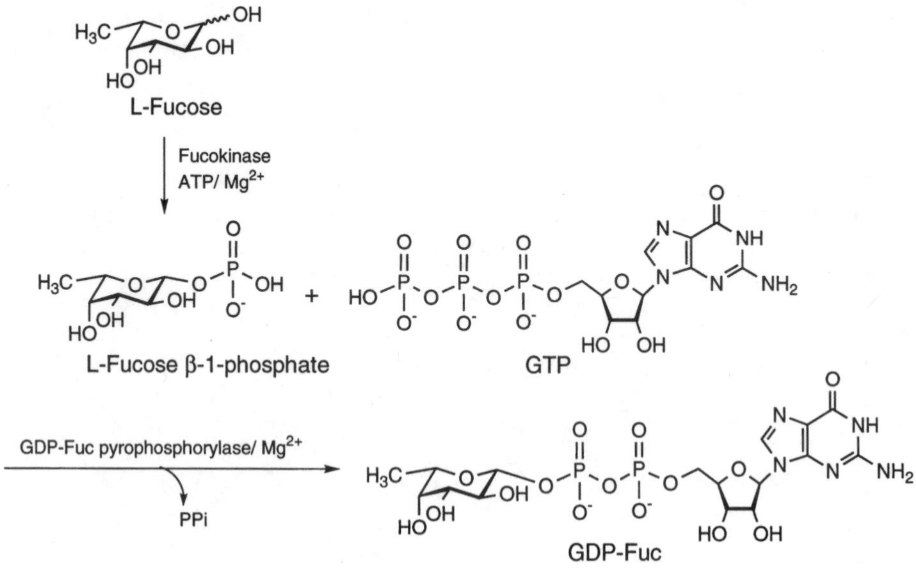

Scheme 5

phosphate reacts with guanosine 5'-triphosphate (GTP) in the presence of Mg^{2+} and GTP-mannose pyrophosphorylase (EC 2.7.7.22), which is easily obtained from yeast. GDP-mannose is subjected to oxidation of the 4-OH of its mannose residue by unknown NADP-dependent enzyme(s), and subsequently to deoxygenation of the 6-OH group to give a 4-keto intermediate. The 4-keto intermediate is further modified by epimerization of the stereochemistry at C-3 and C-5 (both of which are next to the 4-keto position), and subsequent hydride reduction of the 4-keto group. Overall, GDP-α-D-mannose is converted to GDP-β-L-fucose. The enzyme(s) involved in such a conversion has (have) been partially purified from the bacterium *Klebsiella pneumonia* [35], but the enzymatic reaction mechanism has not been yet studied.

The other enzymatic method, Method B, is a salvage pathway for GDP-fucose [36–42] (Scheme 5). When a supply of GDP-mannose is too low, L-fucose is used directly for the formation of GDP-fucose. Fucose is phosphorylated with ATP and fucokinase (EC 2.7.1.52) to give L-fucose β-1-phosphate, which reacts with GTP in the presence of GDP-fucose pyrophosphorylase (EC 2.7.7.30) to give GDP-fucose. These enzymes have been partially purified from porcine thyroid [39,43–45].

α(1,2)Fucosyltransferase (EC 2.4.1.69): Blood group H α(1,2)fucosyltransferase

Blood groups are defined by antigenic differences in particular glycoproteins or glycolipids presented in the membrane of human erythrocytes. These antigenic differences can be identified by the use of specific antibodies that recognize subtle differences in their oligosaccharide structure (the process known as blood typing).

In humans, the blood group determinants A, B, and H are synthesized from the inactive precursor, Galβ(1→3)GlcNAcβ1→R, where R represents an underlying oligosaccharide-containing protein or lipid structure (Scheme 6). Attachment of a fucose residue to the terminal galactose residue via an α(1→2)-linkage is catalyzed

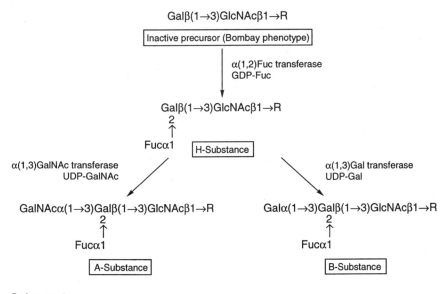

Scheme 6

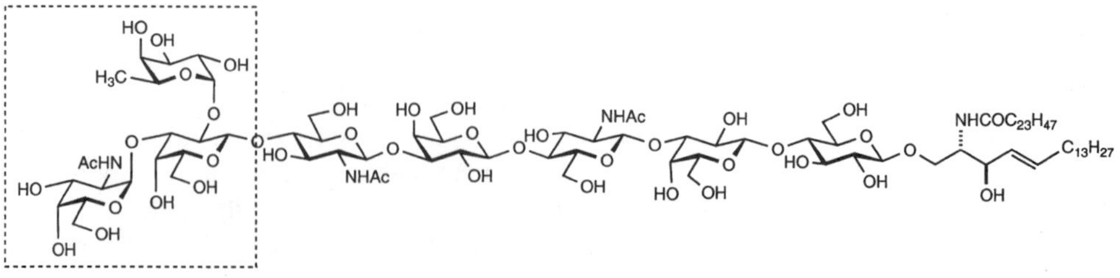

Blood group A antigen

Figure 3

by the enzyme α(1,2)fucosyltransferase (EC 2.4.1.69). In rare instances, individuals may be lacking this enzyme, and hence no A, B, or H antigenicity is expressed on their red blood cells or in their secreted glycoproteins. This condition was first identified in India, and therefore is called the "Bombay" phenotype. The H-positive determinants may be converted into either A- or B-positive determinants by subsequent glycosylation with N-acetylgalactosamine (GalNAc) or galactose (Gal) residues, respectively. The A, B, and H determinants are expressed on the termini of O-linked oligosaccharides in the secreted glycoproteins; however, the majority of the antigenic sequences are present on glycolipids in the erythrocyte membrane (Figure 3).

The enzyme GDP-L-fucose: β-D-galactoside 2-α-L-fucosyltransferase is responsible for the synthesis of the Fucα(1→2)Galβ-linkage (Figure 4). This enzyme has been purified from porcine submaxillary glands [46–48], and the gene encoding the enzyme has been cloned [49,50]. The enzymatic reaction is usually carried out in Na cacodylate buffer (25 mM, pH 6.0) and requires Mn^{2+} for enzyme activity.

The Barker group has employed a purified α(1,2)fucosyltransferase to synthesize blood group H trisaccharide from its Type 2 disaccharide (Galβ(1→4)GlcNAc) precursor and has used the enzymatically prepared trisaccharide to carry out an NMR-based conformational study of the blood group substance in solution [51].

Hindsgaul and co-workers have studied the acceptor substrate specificity of a recombinant α(1,2)fucosyltransferase (H transferase) using structural analogs of octyl β-D-galactopyranoside, which is the minimum acceptor substrate of the enzyme [52] (Figure 5). The examined compounds included those in which the hydroxyl groups at C-3, -4, and -6 were replaced, independently, with deoxy, fluoro, O-methyl, amino, and acetamido groups. Modification of the C-6 OH group with a noncharged group

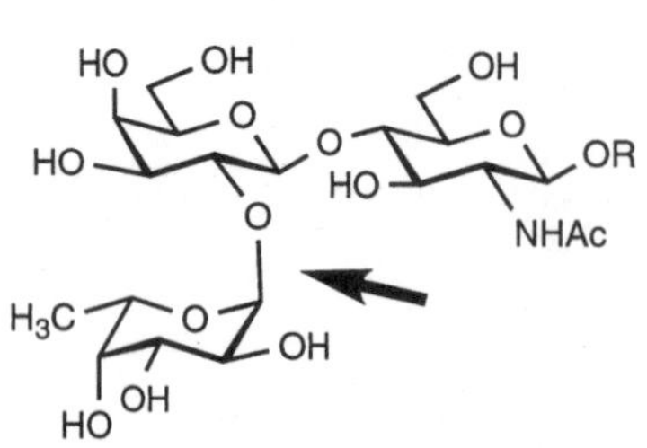

Figure 4

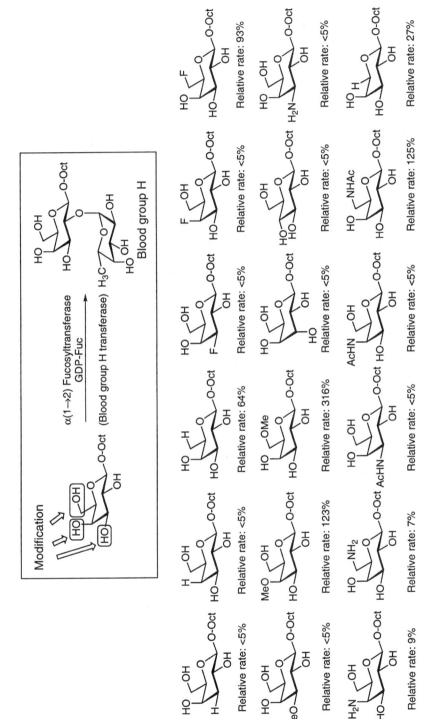

Figure 5

such as deoxy, fluoro, or acetamido was only tolerably successful, with relative enzymatic reaction rates in the range of 27% to 316%; other modifications at the C-3 and -4 positions were not tolerated [52].

α(1,3/4)Fucosyltransferase (EC 2.4.1.65): Blood group Lewis α(1,3/4)fucosyltransferase

A biologically important class of carbohydrate sequences, the Lewis blood group antigens are named for the patient, Ms. Lewis, in which they were first identified [53]. The Lewis blood group oligosaccharide sequences, Lewis x (Galβ(1→4)-[Fucα(1→3)]GlcNAc) and Lewis a (Galβ(1→3)[Fucα(1→4)]GlcNAc), and their sialylated forms, sialyl Lewis x (Siaα(2→3)Galβ(1→4)[Fucα(1→3)]GlcNAc) and sialyl Lewis a (Siaα(2→3)Galβ(1→3)[Fucα(1→4)]GlcNAc), are often present in antigenic oligosaccharide structures, and they have been found to play a key role in intercellular recognition [24]. The Lewis x structure is expressed on glycolipids only during embryonic development (hence its other name, SSEA-1 [stage-specific embryonic antigen-1]), and its expression has been thought to be developmentally regulated [17]. In addition, the Lewis x structures have been reported to bind to other Lewis x structures on adjacent cells in a Ca^{2+}-dependent manner [54]. Sialyl Lewis x (Figure 6) has been found to serve as a ligand for the adhesion molecule, ELAM-1 (endothelial leukocyte adhesion molecule-1), which was later named E-selectin [55–72]; this interaction was the first example of a carbohydrate acting as an initiator of intercellular recognition.

Sialyl Lewis a (Figure 6) has also been identified as a ligand of selectins [63,68]. Furthermore, the sulfated trisaccharides, 3'-O-sulfated Lewis x (Fucα(1→3)-[3-O-(SO$_3^-$)-Galβ(1→4)]GlcNAc) and 3'-O-sulfated Lewis a (Fucα(1→4)[3-O-(SO$_3^-$)-Galβ(1→3)]GlcNAc) (Figure 6), in which sulfate replaces the sialic acid residue, have been recognized as ligands of E-selectins [73,74]. Very recently, a nonsialylated and nonsulfated trisaccharide has been reported as a more potent ligand

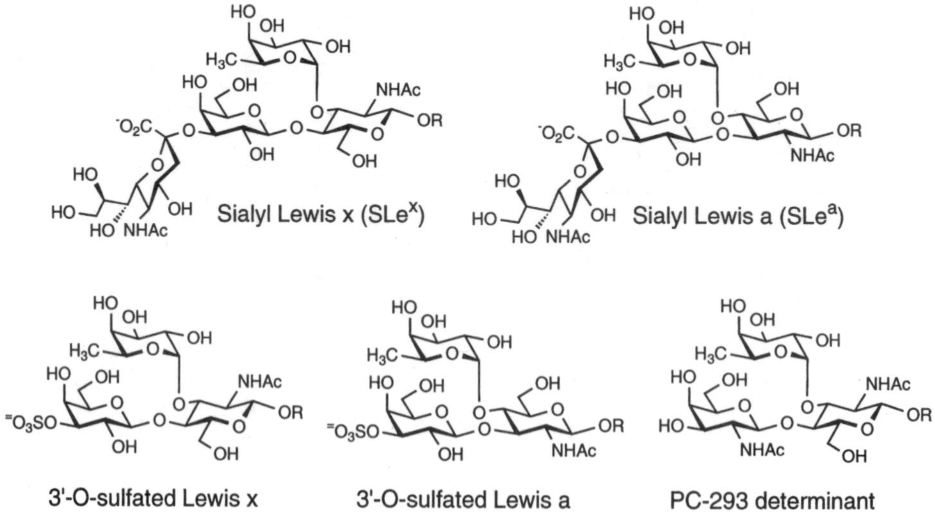

Sialyl Lewis x (SLe^x)

Sialyl Lewis a (SLe^a)

3'-O-sulfated Lewis x

3'-O-sulfated Lewis a

PC-293 determinant

Figure 6

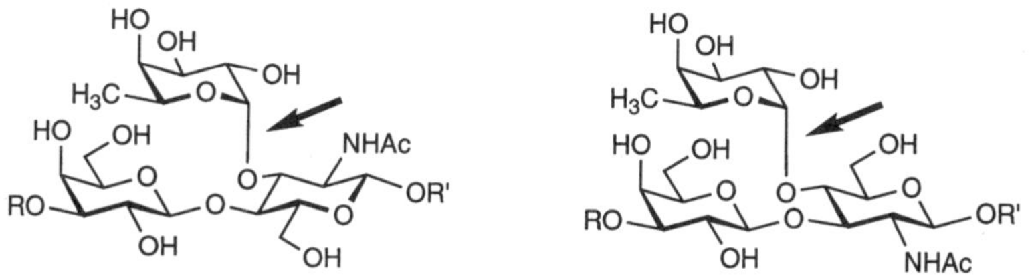

R= H, sialic acid, or other sugar   R= H, sialic acid, or other sugar

Figure 7

of E-selectin [75,76]. This ligand is the PC-293 determinant, a GalNAcβ(1→4)-
[Fucα(1→3)]GlcNAc trisaccharide sequence (Figure 6) that is expressed on the terminus of the N-linked oligosaccharide of the human anticoagulant factor glycoprotein, Protein C. The affinity of this trisaccharide for E-selectin is reported to be about
ten times higher than that of sialyl Lewis x [76]. Structure–activity relationships
have been extensively studied by use of oligosaccharide analogs synthesized chemically and enzymatically on the basis of the structures of sialyl Lewis x and sialyl
Lewis a [71,72,74].

The enzyme GDP-L-fucose: β-D-galactosyl-N-acetyl-D-glucosaminide 3/4-α-L-
fucosyltransferase is responsible for the biosynthesis of Lewis x (Galβ(1→4)-
[Fucα(1→3)]GlcNAc), Lewis a (Galβ(1→3)[Fucα(1→4)]GlcNAc), sialyl Lewis x
(Siaα(2→3)Galβ(1→4)[Fucα(1→3)]GlcNAc), and sialyl Lewis a (Siaα(2→3)
Galβ(1→3)[Fucα(1→4)]GlcNAc) structure (Figure 7).

Biosynthetically, sialyl Lewis x is thought to be synthesized from the terminal
N-acetylglucosamine residue by sequential glycosyltransferase reactions [55]
(Scheme 7): First, β(1,4)galactosyltransferase (β(1,4)GalTase) introduces a galactose
residue onto the 4-OH of N-acetylglucosamine via a β(1→4)-linkage. Second,
α(2,3)sialyltransferase (α(2,3)SiaTase) adds sialic acid onto the 3-OH of the galactose via an α(2→3)-linkage. Finally, α(1,3)fucosyltransferase (α(1,3)FucTase) transfers a fucose residue onto the 3-OH of the N-acetylglucosamine via an α(1→3)-
linkage to furnish sialyl Lewis x. Based on this proposed biosynthetic pathway, sialyl
Lewis x has been synthesized on a preparative scale [45,77].

α(1,3/4)Fucosyltransferase has been purified from various sources, including
human milk, mouse kidney, and human leukocytes [78–82]. Enzyme activity also
has been demonstrated in colon carcinoma cells and embryonic brain tissues. Genes
encoding the Lewis α1,3- and α1,3/4-fucosyltransferases have been cloned [83–90].
Purified recombinant α1,3/4-fucosyltransferase has been studied extensively with respect to its acceptor substrate specificity [91–109] (see below).

Differential expression of the α(1,3)fucosyltransferase during human embryonic development and in adult tissues (stomach, intestine, lung, thymus, heart, liver,
and brain) has been studied by use of synthetic oligosaccharide acceptors, and four
main α(1,3)fucosyltransferase patterns have been defined [93] (Table 1): (1) In myeloid cells, granulocytes, monocytes, and lymphoblasts, the fucosyltransferase transfers fucose to the 3-OH group of the GlcNAc residue of H blood-group Type 2
oligosaccharide (Fucα(1→2)Galβ(1→4)GlcNAcβ→R), with Mn^{2+} as an activator.
The enzyme from brain tissue shows the same specificity and uses Co^{2+} as an acti-

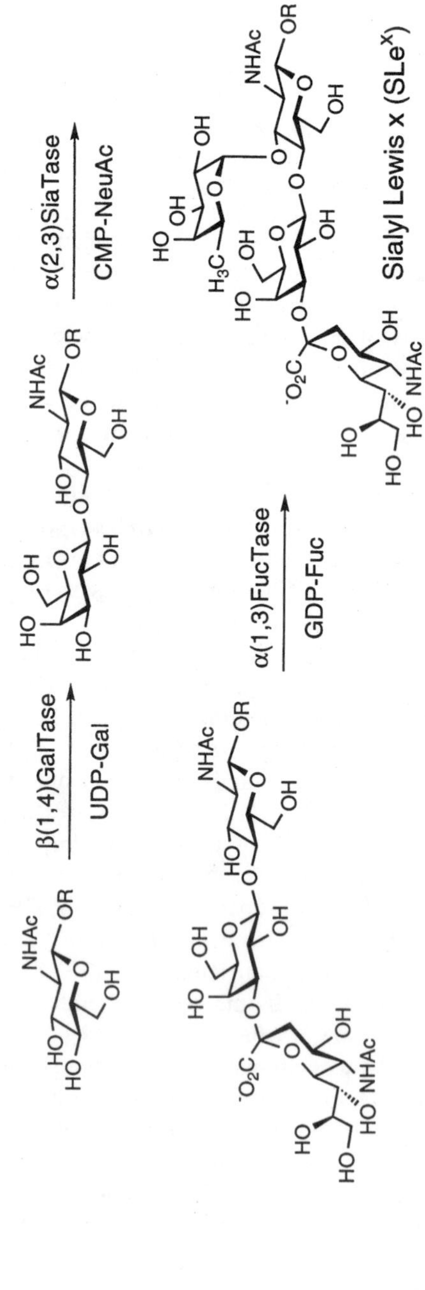

Scheme 7

Structure	Expressed in					
	Myeloid granulocytes / Monocytes / Lymphoblasts / Brain	Plasma / Liver	Intestine / Gall bladder / Kidney / Milk	Stomach mucosa	Early-stage development	Tumor epithelial cell
1 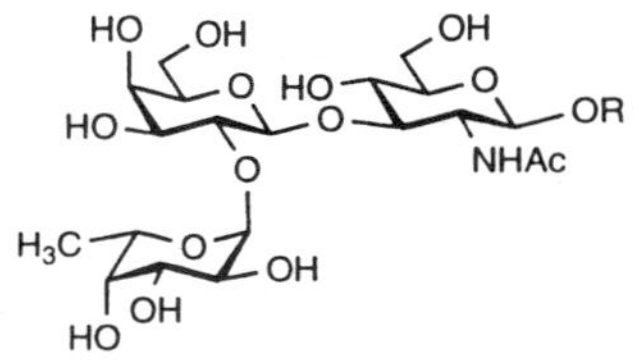			O	O		
2	O	O	O	O	O	O
3			O	O		
4		O	O			

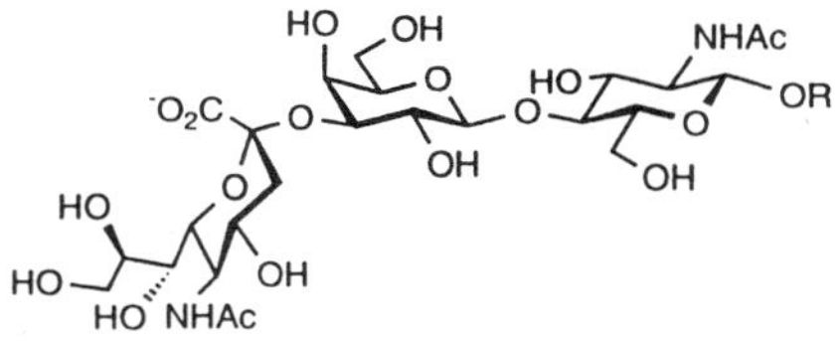

Note: R represents an underlying oligosaccharide [93].

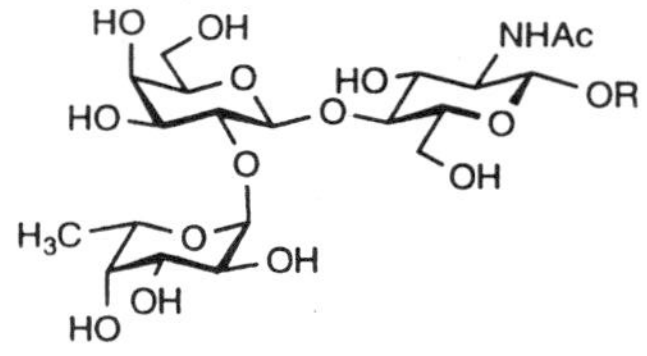

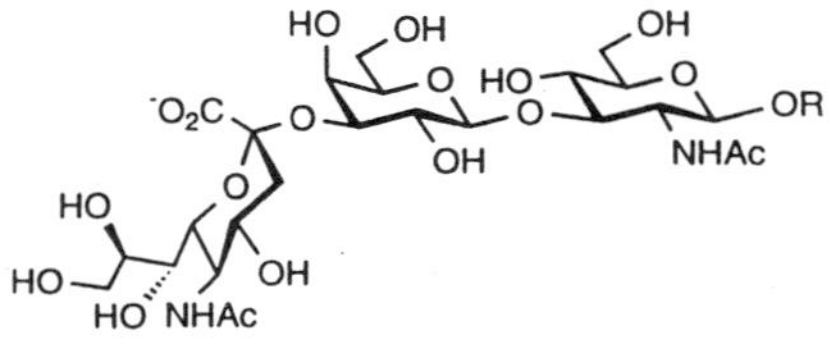

vator. (2) In plasma and liver tissue, the enzyme uses H blood-group Type 2 and sialyl N-acetyllactosamine (sialyl Type 2 oligosaccharide) (Siaα(2→3)Galβ-(1→4)GlcNAcβ→R) as acceptor oligosaccharides. (3) The fucosyltransferase from intestine, gall bladder, kidney, and milk shows the same specificity as that from plasma, and also transfers fucose to the 4-OH of the GlcNAc residue of the H blood-group Type 1 (Fucα(1→2)Galβ(1→3)GlcNAcβ→R) and sialyl Type 1 (Siaα(2→3)-Galβ(1→3)GlcNAcβ→R) oligosaccharides. (4) The enzyme from stomach mucosa uses Type 1 and Type 2 but not sialylated Type 2 as acceptors. The myeloidlike pattern of α(1,3)fucosyltransferase occurs at early stages of development. Tumor epithelial cell lines express the myeloidlike pattern of α(1,3)fucosyltransferase found in normal embryonic tissues as well as α(1,3/4)fucosyltransferase.

An α(1,3)fucosyltransferase that catalyzes the difucosylation of asparagine-bound GlcNAc [110,111] has been found in honey bee (*Apis mellifica*) venom glands [111]. This enzyme was defined as GDP-fucose: β-N-acetylglucosamine (Fuc to [Fucα1→6GlcNAc]-Asn-peptide) α1,3-fucosyltransferase.

A plant α(1,4)fucosyltransferase has been purified from green mung beans and used for the synthesis of various fucosylated oligosaccharides [106]. The enzyme is specific for Type 1 disaccharide (Galβ1→3GlcNAc) and forms the α(1→4)-fucose linkage on the GlcNAc residue; it does not utilize Type 2 disaccharide (Galβ1→4GlcNAc) as an acceptor. The Hindsgaul group has developed two very sensitive assay procedures for the fucosyltransferase activity: One employs an enzyme-linked immunosorbent (ELISA)-type assay [108], and the other uses a syn-thetic chromophore-containing oligosaccharide acceptor and identifies the fucosy-lated product by capillary zone electrophoresis and laser-induced fluorescence detec-tion [97] (Scheme 8). This second assay method has been reported to detect 100 molecules of the fucosylated product. This α(1→4)-fucosyltransferase reaction is carried out in Na cacodylate buffer (250 mM, pH 7.0) and requires Mn^{2+} for the enzyme activity.

Hindsgaul and his co-workers have demonstrated that the Lewis α(1,3/4)fu-cosyltransferase from human milk can catalyze transfer of preassembled trisaccha-ride, linked via a spacer to the 6-position of L-galactose in a GDP-L-galactose de-rivative (or an L-fucose derivative, because L-fucose is 6-deoxy-L-galactose), to an oligosaccharide acceptor [105] (Scheme 9). This reaction transformed blood group O Le^{a-b-} red blood cells, in which Type 1 (Galβ1→3GlcNAc) or Type 2 (Galβ1→4GlcNAc) disaccharide sequences (both acceptor substrates for the Lewis α(1,3/4)fucosyltransferase) are exposed on the cell surface, into blood group B red blood cells by introducing a blood group B trisaccharide-carrying fucose residue onto the N-acetylglucosaminyl moiety of the terminal cell surface oligosaccharide sequences.

The Hindsgaul group also used chemically modified fucose-containing GDP derivatives to study the donor substrate specificity of the Lewis α(1,4)fucosyltrans-ferase from human milk [109] (Table 2). The relative rates of GDP-arabinose (GDP-6-normethyl-L-fucose) and GDP-3-deoxy-L-fucose were reported to be 5.9% and 2.3%, respectively, of that for GDP-Fuc [109].

Because of the importance of the Lewis α(1,3/4)fucosyltransferase, the acceptor substrate specificity of purified and recombinant enzymes has been extensively stud-ied. The acceptor substrate specificity of purified Lewis α(1,3/4)fucosyltransferase from human milk has been studied with a variety of oligosaccharides [100] (Figure

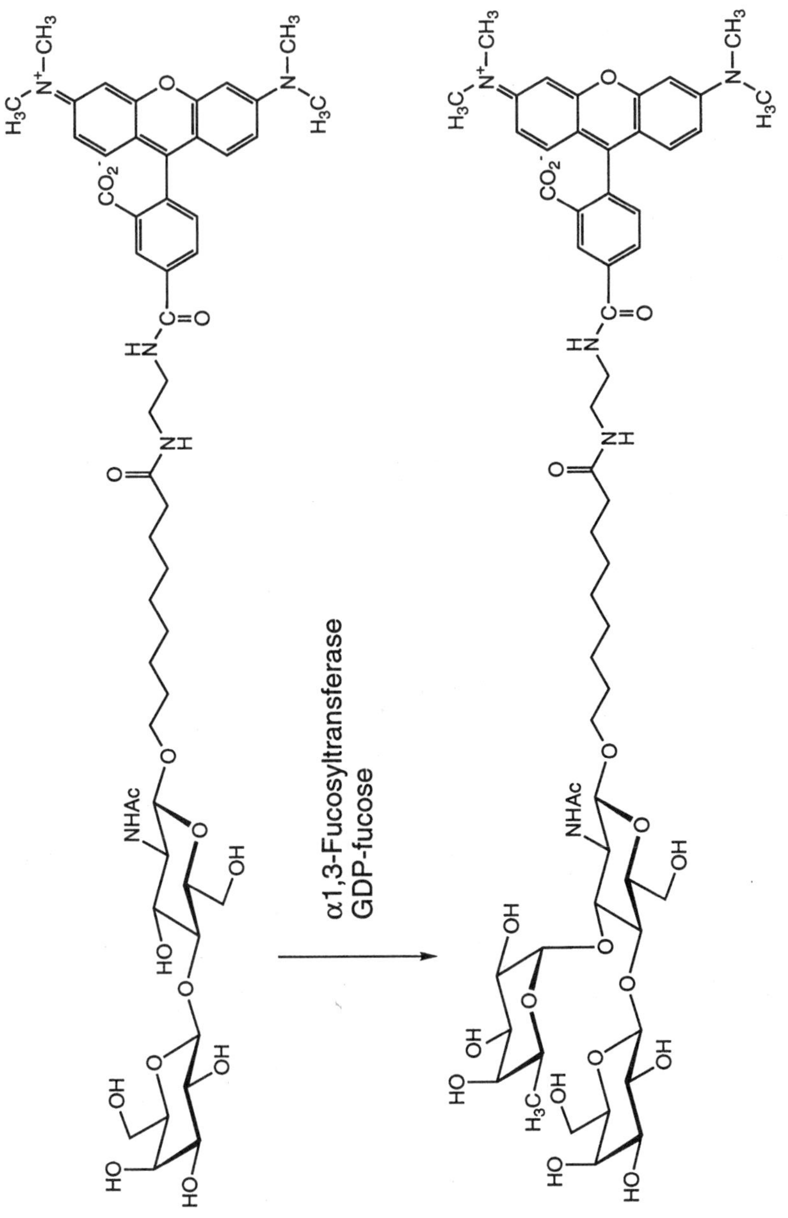

α1,3-Fucosyltransferase
GDP-fucose

Scheme 8

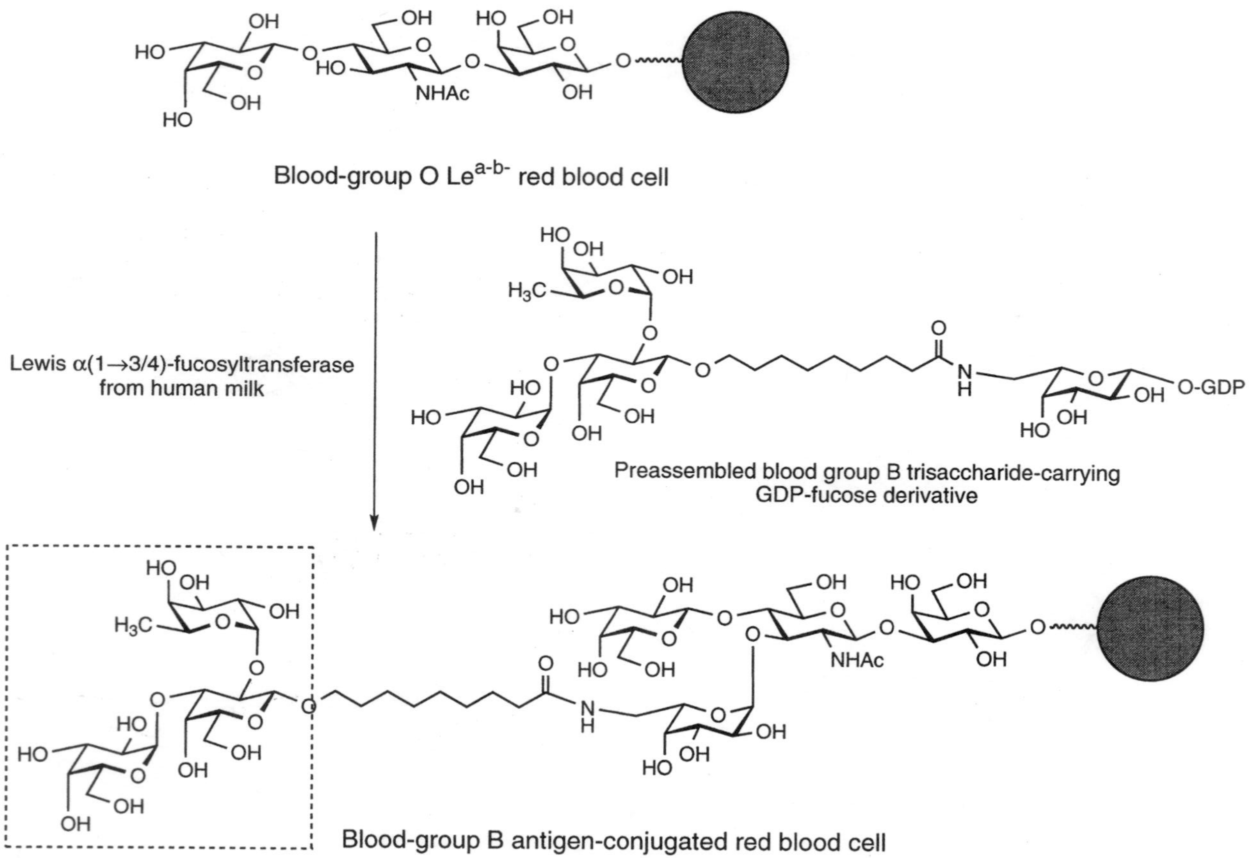

Blood-group O Le^{a-b-} red blood cell

Preassembled blood group B trisaccharide-carrying GDP-fucose derivative

Lewis α(1→3/4)-fucosyltransferase from human milk

Blood-group B antigen-conjugated red blood cell

Scheme 9

Table 2 Donor Substrate Specificity of Lewis
$\alpha(1,4)$Fucosyltransferase from Human Milk [109]

GDP-modified L-fucose	Relative rate (%) for Lewis $\alpha(1{\rightarrow}4)$fucosyltransferase from human milk
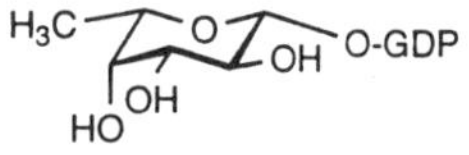 GDP-L-Fucose	100
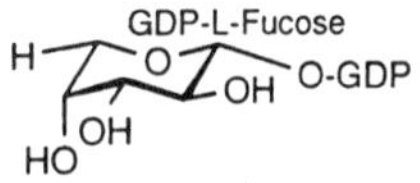 GDP-D-Arabinose	5.9
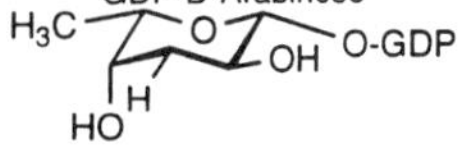 GDP-3-deoxy-L-Fucose	2.3

8). Both Type 1 (Galβ1$\rightarrow$3GlcNAc) and Type 2 (Galβ1$\rightarrow$4GlcNAc) disaccharides were good substrates, which showed V/K_m values of 41.7 and 53.0, respectively. When a hydrophobic aglycone (—O—(CH$_2$)$_8$CO$_2$Me) was introduced into both types of disaccharides, they became extremely good substrates, with a V/K_m of 655 for Type 1 and 450 for Type 2 disaccharide glycosides. The H blood group trisaccharide was also good substrate ($V/K_m = 188$), as were sialyl Lewis x precursor trisaccharide (NeuAcα2$\rightarrow$3Galβ1$\rightarrow$4GlcNAc) ($V/K_m = 293$) and the sialyl Lewis a precursor-containing pentasaccharide (NeuAcα2$\rightarrow$3Galβ1$\rightarrow$3GlcNAcβ1$\rightarrow$3Gal-β1$\rightarrow$4Glc) ($V/K_m = 176$).

A recombinant human Lewis $\alpha(1,3)$fucosyltransferase has also been studied with respect to its acceptor substrate specificity [92] (Figure 9). This enzyme efficiently utilized Type 2 disaccharide ($V/K_m = 2.86$) but not Type 1 disaccharide ($V/K_m = 0.22$). The sialyl Lewis x precursor trisaccharide was a very good substrate ($V/K_m = 6.20$). Surprisingly, sialyl lactal (NeuAcα2$\rightarrow$3Galβ1$\rightarrow$4Glucal) was a good acceptor ($V/K_m = 5.16$), and sialyl Lewis x glycal was synthesized using this recombinant human Lewis $\alpha(1,3)$fucosyltransferase [45,92].

$\alpha(1,6)$Fucosyltransferase (EC 2.4.1.68): Glycoprotein 6-α-L-fucosyltransferase

An L-fucosyl residue, linked via an $\alpha(1{\rightarrow}6)$-linkage, is also found on N-acetyl-glucosamines of N-linked oligosaccharides [112,114]. The enzyme extracted from porcine liver has been named GDP-L-fucose: glycoprotein (L-fucose asparagine-linked N-acetylglucosamine of N-acetyl-β-D-glucosaminyl-1$\rightarrow$2-α-D-mannosyl-1$\rightarrow$4-N-acetylglucosaminyl) asparagine 6-α-L-fucosyltransferase because its substrate requires a terminal $\beta(1{\rightarrow}2)$-linked GlcNAc residue on the $\alpha(1{\rightarrow}3)$-linked mannose branch of the core. The enzyme activity has also been demonstrated in an extract from human skin fibroblasts [112]. The substrate specificity of the partially purified $\alpha(1,6)$fucosyltransferase from rat liver Golgi membranes has been studied

Galβ1,3GlcNAc
K_m = 2.4 mM, V= 100

Galβ1,4GlcNAc
K_m = 2.0 mM, V= 106

Galβ1,3GlcNAc-OR
K_m = 0.2 mM, V= 131

Galβ1,4GlcNAc-OR
K_m = 0.3 mM, V= 135

Fucα1,2Galβ1,4GlcNAc
K_m = 0.5 mM, V= 94

NeuAcα2,3Galβ1,4GlcNAc
K_m = 0.3 mM, V= 88

NeuAcα2,6Galβ1,4GlcNAc
K_m = *, V= 0

GalNAcα1,3Galβ1,4GlcNAc
K_m = 1.0 mM, V= 105

Galβ1,3GlcNAcβ1,3Galβ1,4Glc
K_m = 2.4 mM, V= 123

NeuAcα2,3Galβ1,3GlcNAcβ1,3Galβ1,4Glc
K_m = 0.5 mM, V= 88

Galβ1,4GlcNAcβ1,3Galβ1,4Glc
K_m = 1.5 mM, V= 141

Fucα1,2Galβ1,3GlcNAcβ1,3Galβ1,4Glc
K_m = 0.5 mM, V= 126

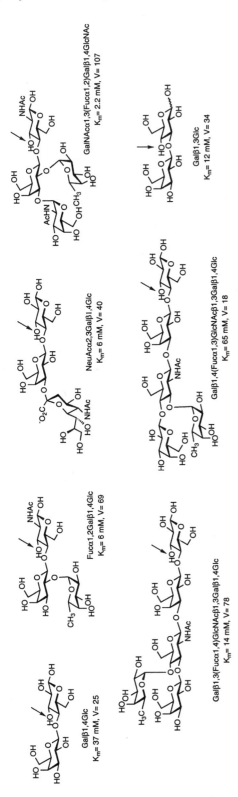

Figure 8

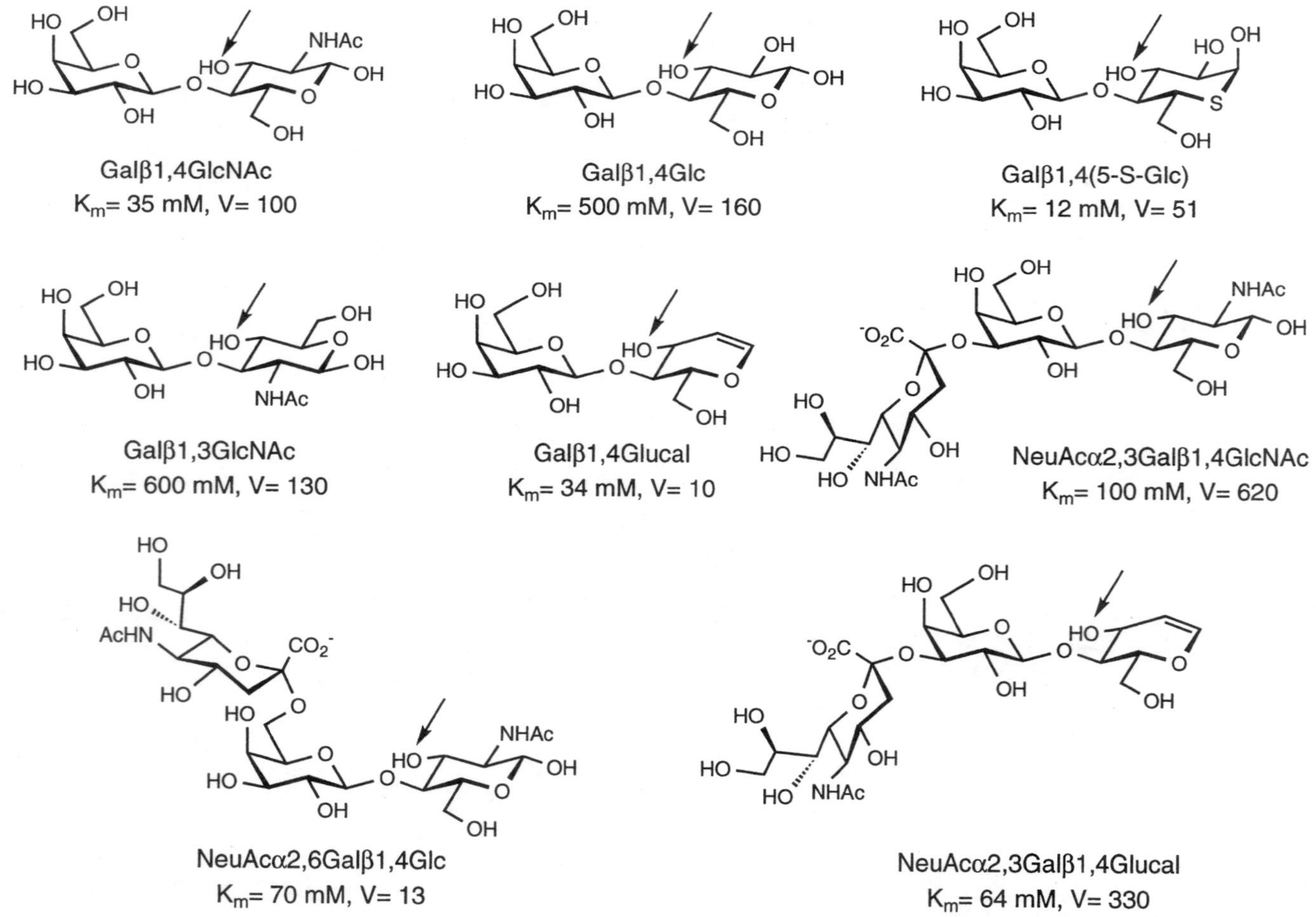

Galβ1,4(5-S-Glc)
K_m= 12 mM, V= 51

Galβ1,4Glc
K_m= 500 mM, V= 160

Galβ1,4GlcNAc
K_m= 35 mM, V= 100

NeuAcα2,3Galβ1,4GlcNAc
K_m= 100 mM, V= 620

Galβ1,4Glucal
K_m= 34 mM, V= 10

Galβ1,3GlcNAc
K_m= 600 mM, V= 130

NeuAcα2,3Galβ1,4Glucal
K_m= 64 mM, V= 330

NeuAcα2,6Galβ1,4Glc
K_m= 70 mM, V= 13

Figure 9

with respect to the 3-dimensional conformations of the substrate oligosaccharides [113].

A typical enzymatic fucosylation reaction: *Synthesis of* Neu5Acα(2→3)Galβ(1→4)[Fucα(1→3)]GlcNAcβ(1→6)[Galβ(1→3)]-GalNAcα1-O-(CH₂)₈CO₂Me [114]

A solution of Neu5Acα(2→3)Galβ(1→4)GlcNAcβ(1→6)[Galβ(1→3)]GalNAcα1-O-(CH$_2$)$_8$CO$_2$Me (3.3 mg, 2.8 μmol), GDP-fucose (3.9 μmol), ATP (5 mM; to prevent degradation of GDP-fucose by contaminating phosphatase because the α(1,3)fucosyltransferase was partially purified from mouse kidney), BSA (2%), MnCl$_2$ (20 mM), and α(1,3)fucosyltransferase (EC 2.4.1.65; 25 mU, partially purified from mouse kidney) in HEPES buffer (1 mL, 250 mM, pH 7.0) was stirred for 6 h at room temperature. TLC showed that the starting material (R_f = 0.02 in CHCl$_3$/MeOH/H$_2$O 13:7:1) was converted into one product (R_f = 0.30 in CHCl$_3$/MeOH/H$_2$O 5:4:1). The reaction mixture was diluted to 9 mL with water, and applied onto a C$_{18}$ Sep-Pak cartridge. The cartridge was washed with water (40 mL), and the product was eluted with MeOH (50 mL). The MeOH fraction was concentrated, and the residue was lyophilized from water. The resulting powder was chromatographed on silica gel with CHCl$_3$/MeOH/H$_2$O (12:7:1). Fractions containing the product were pooled and concentrated. The residue was dissolved in water, passed through a 0.22-μm filter, and lyophilized to give 3.3 mg of the α(1→3)fucosylated product (89% yield).

B. Sialyltransferases

Sialic acid (Sia) is the common terminal carbohydrate residue of *N*- and *O*-linked oligosaccharides and of glycolipid structures [15]. Eight different sialic acid (Sia) linkages have been identified: Siaα(2→6)Gal, Siaα(2→3)Gal, Siaα(2→6)GalNAc, Siaα(2→4)Gal, Siaα(2→6)GlcNAc, Siaα(2→4)GlcNAc, Siaα(2→6)Man [115], and Siaα(2→8)Sia. The functions of sialic acid-carrying oligosaccharides (sialosides) are diverse [24,116,117]: they serve as tumor-specific antigens [17,118] and are directly involved, as ligands, in cellular recognition of cell adhesion molecules [68], viruses [119,120], mycoplasma [121,122], animal and plant lectins [123,124], and bacterial toxins [24]. Terminal sialic acids also function in protecting the underlying carbohydrate residues from carbohydrate-specific recognition that results in clearance of the oligosaccharides by galactose-binding protein [125,126] or endoglycosidase digestion [116].

The structural variation in sialic acid has been well studied [127]: variations in the *O*-acetylation of the hydroxyl groups (at the C-7, -8, and/or -9 positions) of the side chain and in the *N*-glycolyl derivatives have been characterized. An α(2→8)-linked sialic acid polymer [128–131] is expressed as a component of colominic acid in the outer membrane of microorganisms and also in early stages of embryonic development, during which it has been suggested to play a role in cell adhesion.

Since the chemical sialylation reaction is one of the most difficult reactions in synthetic carbohydrate chemistry, and often results in poor stereoselectivity or low chemical yields [132–134], sialylation with sialyltransferases is a valuable alternative method. Sialyltransferase (SiaTase) transfers sialic acid (Neu5Ac) from cytidine 5′-

monophospho sialic acid (CMP-Neu5Ac) onto the specific hydroxyl group of the acceptor sugar (Scheme 10).

Two chemical syntheses of CMP-sialic acid have been reported (Scheme 11). The Schmidt group has employed sialyl phosphite as a sialyl donor and performed a coupling reaction with a partially protected cytidine monophosphate derivative [135]. The coupling reaction was driven by the acidity of the CMP derivative itself, and the coupled product was deprotected to give CMP-NeuAc. Hata and co-workers have coupled a cytidine 5'-phosphoramidite derivative with a β-2-OH sialic acid derivative protected with allyloxy (AOC) group [136]. After the coupling, the phosphite group was oxidized to phosphate, and deprotection gave CMP-NeuAc. Very recently, CMP-Neu5Acα(2→8)Neu5Ac has been synthesized [137]; however, the enzymatic transfer of such a novel CMP-sialic acid has not been yet reported.

Enzymatic synthesis of CMP-Neu5Ac has been reported by several groups [138–144]. These syntheses involved the reaction of sialic acid with cytidine 5'-triphosphate (CTP), catalyzed by CTP: N-acetylneuraminate cytidylyltransferase (CMP-sialic acid synthetase) (EC 2.7.7.43) (Scheme 12). This enzyme has been purified from natural sources [145–147]. Gene encoding the enzyme has been cloned in E. coli, and the protein has been overexpressed in large quantity [148–153]. The enzyme reaction mechanism has been studied by NMR with ^{13}C-enriched sialic acid [154].

The activity of the CMP-sialic acid synthetase from calf brain has been found to be dependent on metal concentration and pH [147]. The best enzyme activity was obtained at pH 10 in the presence of 25 mM Mg^{2+} and at pH 8.0 in the presence of 6 mM Mn^{2+}. However, the maximum stimulated activity in the presence of Mg^{2+} at pH 9.5 was 1.5 times higher than that in the presence of Mn^{2+} at pH 7.0. Higa and Paulson have prepared CMP-NeuAc, -9-O-Ac-NeuAc, and -NeuGc and examined these derivatives as donor substrates for six different sialyltransferases [147]: In addition to Galβ(1→3/4)GlcNAc-specific α(2,3)sialyltransferase from rat liver, they also studied two Galβ(1→4)GlcNAc-specific α(2,6)sialyltransferases from rat liver

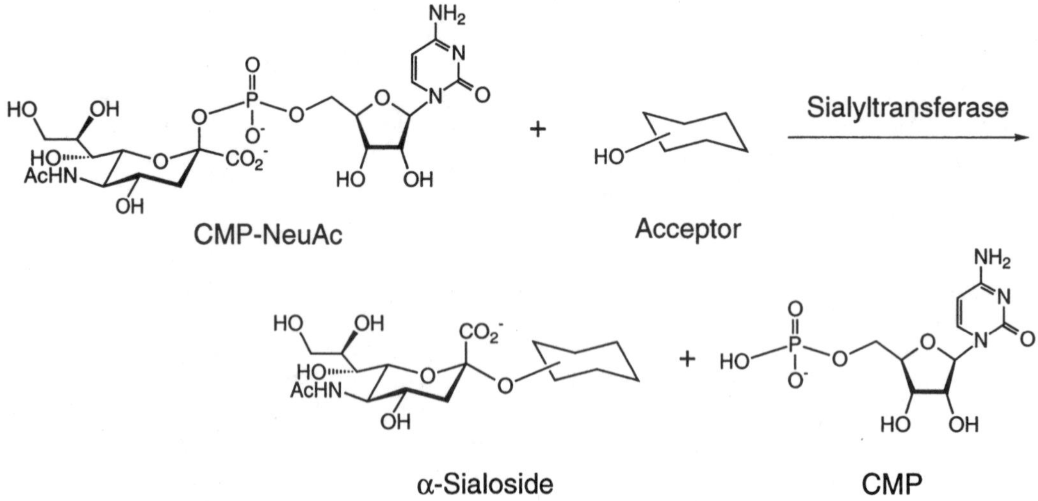

CMP-NeuAc   Acceptor

α-Sialoside   CMP

Scheme 10

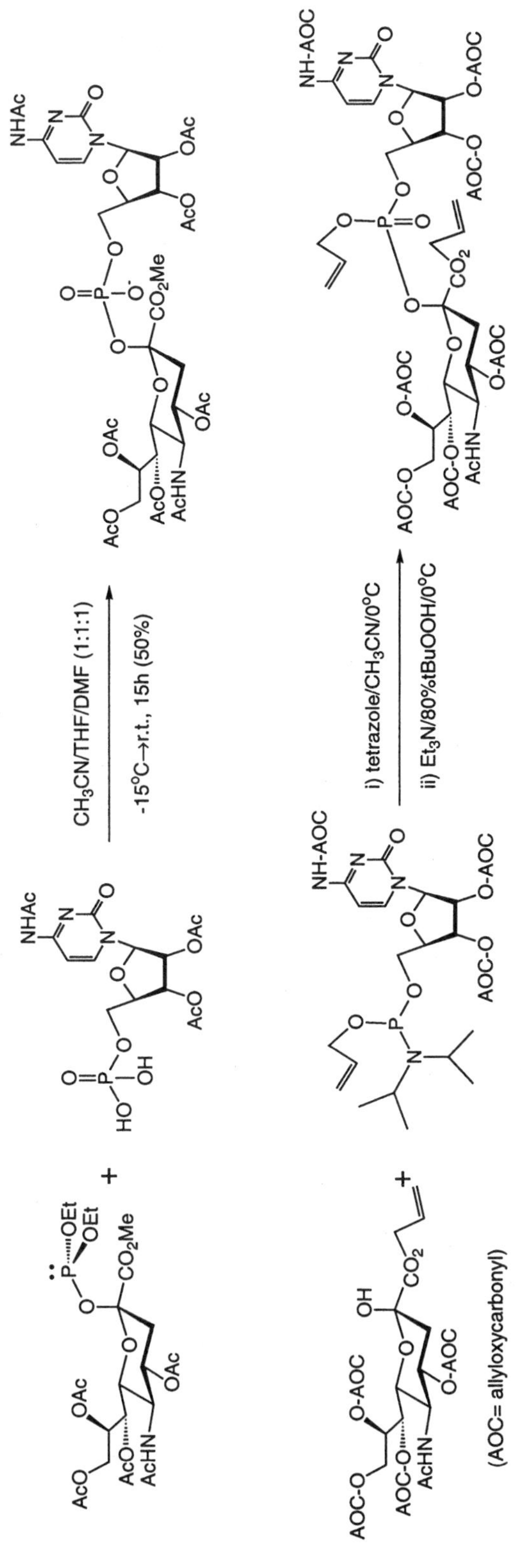
CH3CN/THF/DMF (1:1:1)
-15°C→r.t., 15h (50%)
i) tetrazole/CH3CN/0°C
ii) Et3N/80%tBuOOH/0°C
(AOC= allyloxycarbonyl)

Scheme 11

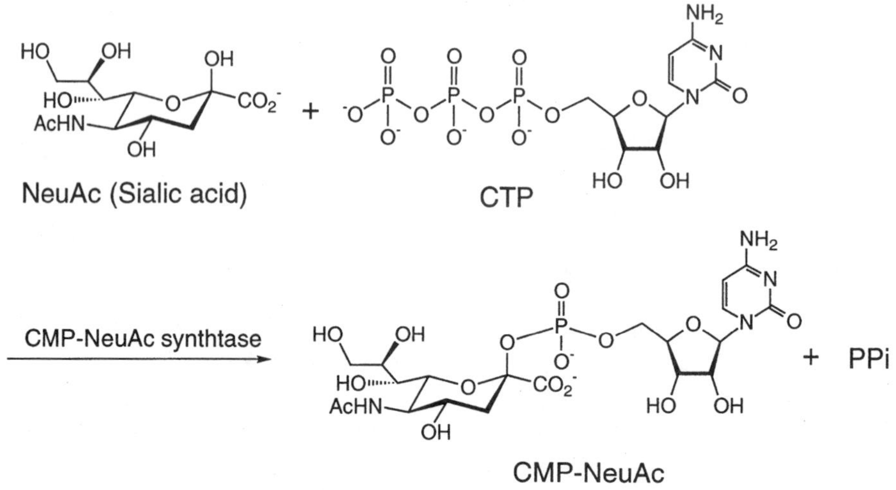

NeuAc (Sialic acid)   CTP

CMP-NeuAc synthtase
⟶

                                                                    + PPi

CMP-NeuAc

Scheme 12

and bovine colostrum (α1-acid asialoglycoprotein with N-linked oligosaccharides containing the Galβ(1→4)GlcNAc terminal sequence was used as an acceptor). This group also characterized two GalNAc-specific α(2,3)sialyltransferases prepared from porcine and bovine submaxillary glands (ovine asialosubmaxillary mucin with a simple O-linked oligosaccharide, GalNAcαThr/Ser, as an acceptor) and Galβ(1→3)-GalNAc-specific α(2,3)sialyltransferase from porcine submaxillary gland (with antifreeze glycoprotein with O-linked oligosaccharides bearing Galβ(1→3)GalNAcαThr/Ser as an acceptor). In general, the incorporation rates for CMP-NeuGc were equal to or higher than those observed for CMP-NeuAc. When CMP-9-O-Ac-NeuAc was tested as a substrate, for the sialyltransferases that utilize N-linked oligosaccharides, the incorporation rates were 50% to 75% of those obtained for CMP-NeuAc; for the sialyltransferases that use O-linked glycans the incorporation rates were 10% to 25% of those for CMP-NeuAc.

Biosynthesis of CMP-N-glycolyl sialic acid normally involves hydroxylation of the N-acetamido moiety of CMP-N-acetyl sialic acid in a reaction catalyzed by N-hydroxylase (monooxygenase) [155–157]. However, CMP-sialic acid synthetase has also been found to act on the N-glycolyl sialic acid to produce CMP-N-glycolyl sialic acid [147]. Several CMP-9-modified sialic acid derivatives have been prepared using purified enzymes from rat liver, then utilized as donor substrates to introduce the modified sialic acid residues onto the termini of the oligosaccharides (see below).

Very recently, CMP-KDN (deaminoneuraminic acid) synthetase from rainbow trout testis has been identified and characterized; this enzyme is involved in the biosynthesis of a novel KDN-containing glycolipid [158] (Figure 10).

α(2,6)Sialyltransferase (EC 2.4.99.1): β-Galactoside (galβ(1→4)GlcNAc)-specific

In 1973 the Roseman group reported the isolation of an enzyme from rat mammary glands that could transfer the sialic acid of CMP-NeuAc onto β-galactosides such

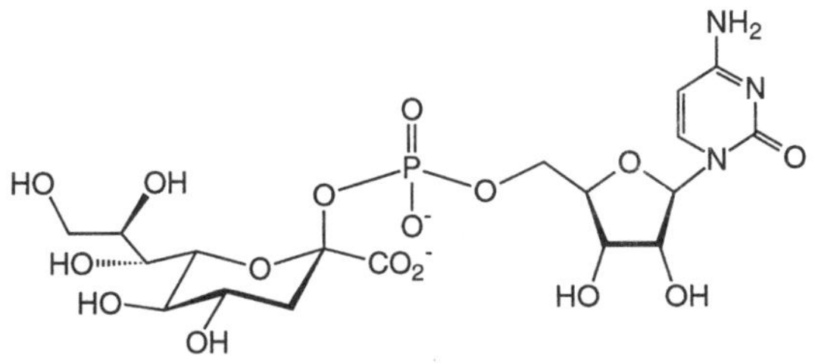

Figure 10

as lactose to form Neu5Acα(2→6)Galβ(1→ structures [159–161]. Since then, sia-lyltransferase has been actively studied because of the biological importance of sia-lylated oligosaccharides.

Neu5Acα(2→6)Galβ(1→4)GlcNAcβ(1→ is a common terminal structure in the N-linked glycans of glycoproteins, and the expression of this sequence has been suggested to play a regulatory role in intercellular recognition [162,163]. The enzyme CMP-N-acetylneuraminate: β-D-galactosyl-1,4-N-acetyl-β-D-glucosamine α-2,6-N-acetylneuraminyltransferase is responsible for the synthesis of the Siaα(2→6)Galβ(1→4)GlcNAc structure (Figure 11).

This enzyme has been purified from natural sources, such as bovine colostrum, rat liver, and human liver, and its substrate specificity has been extensively studied [164–169]. The enzyme from rat liver is commercially available from Sigma and Boehringer Mannheim: The current price is $140 per 0.1 unit; one unit of enzyme will transfer 1 mmol NeuAc from CMP-NeuAc to asialoglycoprotein (the Galβ(1→4)GlcNAc structure is exposed at the oligosaccharide terminus) per min at pH 6.5 and 37°C.

The α(2,6)sialyltransferase gene from rat liver has been cloned [170]. The gene from chick embryo has also been cloned and compared with the predicted sequence of the rodent enzyme; the chick sequence was 58% identical to that of rat liver α(2,6)sialyltransferase [171].

In terms of the donor structure, some modification of the 9-hydroxyl group of Neu5Ac in CMP-NeuAc can be tolerated (e.g., substitution with O-acetyl, azido, deoxy or fluorescent groups [172–175] (Figure 12). These 9'-modified CMP-NeuAc derivatives are relatively good substrates for α(2,6)sialyltransferase: Their relative rates are comparable to that of CMP-NeuAc (see below).

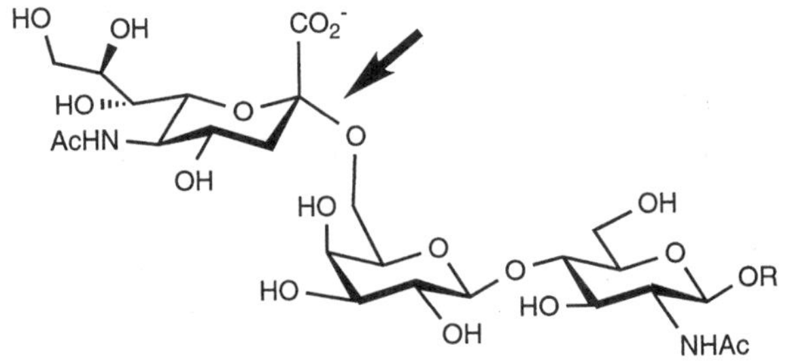

Figure 11

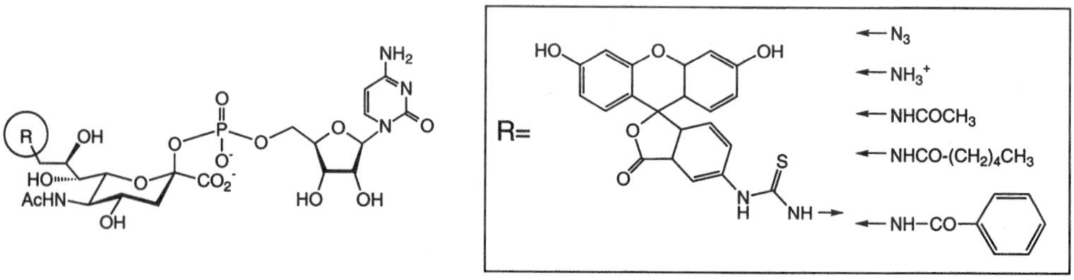

Figure 12

The Brossmer group has employed sialyltransferase to introduce 9-modified sialic acid derivatives into oligosaccharide acceptors [167,172–175]. They used two types of α(2,6)sialyltransferases specific for Galβ(1→4)GlcNAc, one from human liver and the other from rat liver; and α₁-acid asialo GP (glycoprotein), which has an abundantly exposed Galβ(1→4)GlcNAc sequence, was used as a substrate (Table 3). For both enzymes, CMP-fluoresceinyl-NeuAc was a seven- to eightfold better substrate than was CMP-unmodified NeuAc. When a positive charge was generated on the C-9 position (CMP-9-amino-NeuAc), this substrate was 10 to 18 times less active than CMP-NeuAc. In general, hydrophobicity on the C-9 position was preferred by the α(2,6)sialyltransferase [167].

Purified α(2,6)sialyltransferases from rat liver and bovine colostrum, which are specific for Type 2 (Galβ(1→4)GlcNAc) but not Type 1 disaccharides (Galβ(1→3)-GlcNAc), have been studied with respect to their donor and acceptor specificities [165] (Table 4). The enzyme has been used for the synthesis of sialosides [177–179].

The enzyme from rat liver accepted CMP-NeuAc derivatives such as CMP-9-O-Ac-NeuAc and CMP-NeuGc as good substrates, with relative rates of 210% and 70%, respectively [147]. It has recently been demonstrated that α(2→6)sialyltransferase transfers a sialic acid residue onto the Man residues of Manβ(1→4)GlcNAc and Manβ(1→4)GlcNAcβ(1→4)GlcNAc via α(2→6)-linkages in yields of 4% and 27%, respectively; however, no kinetic data were provided [176].

α(2,6)Sialyltransferase from bovine colostrum showed rather strict acceptor substrate specificity: It preferentially sialylates the Galβ(1→4)GlcNAc-terminating structure. It also accepted CMP-NeuAc derivatives such as CMP-9-O-Ac-NeuAc and CMP-NeuGc as good substrates with relative rates of 150% and 50%, respectively [147]. Recently, GalNAcβ(1→4)GlcNAc was found to be sialylated by this enzyme, although its relative acceptor activity (V_{max}/K_m) was low (0.45) when compared to the value of 1.16 obtained for Galβ(1→4)GlcNAc [168].

α(2,6)Sialyltransferase (EC 2.4.99.3): α-N-Acetylgalactosaminide-specific

The carbohydrate structure Siaα(2→6)[Galβ(1→3)]GalNAcα(1→Ser/Thr is typical of O-linked glycans in glycoproteins. The α(2→6)sialyltransferase that forms the Siaα(2→6)GalNAcα1→ linkage has been purified and studied with respect to its substrate specificity (Table 5). The enzyme from porcine submaxillary gland very effectively accepts CMP-9-O-Ac-NeuAc as a donor (relative rate: 90%), but not the N-glycolyl derivative (relative rate: 10%) [147]. The enzyme from bovine submaxillary gland showed similar specificity for the CMP-9-O-Ac and N-glycolyl deriva-

Table 3 Kinetic Study of CMP-9-Modified NeuAc Derivatives as Donor Substrates for Galβ(1→4)GlcNAc-Specific α(2,6)Sialyltransferases from Rat Liver and Human Liver [167]

		Substrates		Kinetics		
Enzyme	Source	Donor	Acceptor	K_m (μM)	Rel. rate (%)	V/K_m (1/μM)
α2,6ST Linkage NeuAcα(2,6)Galβ(1,4)GlcNAc	Rat liver	CMP-NeuAc	α₁-Acid asialo-GP	50	100	2
		CMP-9-fluoresceinyl-NeuAc	α₁-Acid asialo-GP	7	100	14.2
		CMP-9-amino-NeuAc	α₁-Acid asialo-GP	720	90	0.12
		CMP-9-acetamido-NeuAc	α₁-Acid asialo-GP	120	110	0.92
		CMP-9-benzamino-NeuAc	α₁-Acid asialo-GP	30	110	3.6
		CMP-9-hexanoylamino-NeuAc	α₁-Acid asialo-GP		112	
		CMP-9-azido-NeuAc	α₁-Acid asialo-GP		139	
α2,6ST Linkage NeuAcα(2,6)Galβ(1,4)GlcNAc	Human liver	CMP-NeuAc	α₁-Acid asialo-GP	20	100	5
		CMP-9-fluoresceinyl-NeuAc	α₁-Acid asialo-GP	2	80	40
		CMP-9-amino-NeuAc	α₁-Acid asialo-GP	150	70	0.47
		CMP-9-acetamido-NeuAc	α₁-Acid asialo-GP	84	110	1.31
		CMP-9-benzamino-NeuAc	α₁-Acid asialo-GP	10	130	13

Table 4 Substrate Specificity of Galβ(1→4)GlcNAc-Specific α(2,6)Sialyltransferases from Rat Liver and Bovine Colostrum [147,165,168,176]

Enzyme	Source	Substrates		Kinetics		
		Donor	Acceptor	K_m (mM)	V_{max}	Rel. rate (%)
α2,6ST	Rat liver	CMP-NeuAc	$α_1$-Acid asialo-GP	50		100
Linkage		CMP-9-O-Ac-NeuAc	$α_1$-Acid asialo-GP			210
NeuAcα(2,6)Galβ(1,4)GlcNAc		CMP-NeuGc	$α_1$-Acid asialo-GP			70
		CMP-NeuAc	Manβ(1,4)GlcNAc			
		CMP-NeuAc	Manβ(1,4)GlcNAcβ(1,4)GlcNAc			
α2,6ST	Bovine colostrum	CMP-NeuAc	$α_1$-Acid asialo-GP			100
Linkage		CMP-9-O-Ac-NeuAc	$α_1$-Acid asialo-GP			150
NeuAcα(2,6)Galβ(1,4)GlcNAc		CMP-NeuGc	$α_1$-Acid asialo-GP			50
		CMP-NeuAc	Galβ(1,4)GlcNAc	12	4.8	100
		CMP-NeuAc	Galβ(1,3)GlcNAc	1780	0.29	<1
		CMP-NeuAc	Galβ(1,4)Glc	390	4.90	<1
		CMP-NeuAc	Galβ-OMe	1040	1.30	<1
		CMP-NeuAc	Araβ-OMe	1160	0.10	<1
				K_m (mM)	V_{max}	V/K_m
		CMP-NeuAc	Galβ(1,4)GlcNAc	7	8.14	1.16
		CMP-NeuAc	GalNAcβ(1,4)GlcNAc	13	5.92	0.45

tives [147]. For the CMP-9-modified NeuAc derivatives, the enzyme from porcine submaxillary gland accepted CMP-9-azido NeuAc very well (relative rate: 150%) but did not tolerate other modifications of the C-9 position of the NeuAc in CMP-NeuAc (relative rates were in the range of 0% to 10%) [174].

More than two types of α(2,6)sialyltransferase genes that sialylate the 6-OH of α-linked N-acetylgalactosaminide have been cloned [180,181]. The enzyme from chicken embryos showed a 12% sequence homology with the α(2,6)sialyltransferase for β-galactoside and was demonstrated to sialylate GalNAc-Ser. The enzyme was specific for GalNAcα(1→Ser/Thr, and when the GalNAc residue was galactosylated via a β(1→3)-linkage, it became a better substrate (Table 6). The fact that GalNAcα(1→Ser/Thr was an acceptor substrate but Galβ(1→3)GalNAc (which lacks peptide portion) was not indicates that the enzyme required not only the terminal carbohydrate but also a peptide moiety [180].

A second sialyltransferase, from chicken testes, has been overexpressed in COS cells and shown to require the presence of a β(1,3)-linked galactose on the acceptor GalNAc residue [181].

α(2,3)Sialyltransferase (EC 2.4.99.4): Galβ(1→3)GalNAc-specific

The Neu5Acα(2→3)Galβ(1→3)GalNAcα(1→ linkage is a common terminal structure in O-linked oligosaccharides of glycoproteins and glycolipids. CMP-N-

Table 5 Substrate Specificity of GalNAc-Specific α(2,6)Sialyltransferases from Porcine Submaxillary Gland (SG) and Bovine Submaxillary Gland (SG) [147,174]

| Enzyme | Source | Substrates | | Kinetics | | |
		Donor	Acceptor	K_m (mM)	V_{max}	Rel. rate (%)
α2,6ST Linkage NeuAcα(2,6)GalNAcα-Thr/Ser	Porcine SG	CMP-NeuAc	Ovine asialo mucin	0.13	1.0	100
		CMP-9-O-Ac-NeuAc	Ovine asialo mucin	0.21	1.3	90
		CMP-NeuGc	Ovine asialo mucin	0.08	0.09	10
		CMP-NeuAc	Antifreeze GP			100
		CMP 9-amino-NeuAc	Antifreeze GP			0
		CMP-9-acetamido-NeuAc	Antifreeze GP			15
		CMP-9-benzamido-NeuAc	Antifreeze GP			9
		CMP-9-hexanolyamino-NeuAc	Antifreeze GP			7
		CMP-9-azido-NeuAc	Antifreeze GP			165
α2,6ST Linkage NeuAcα(2,6)GalNAcα-Thr/Ser	Bovine SG	CMP-NeuAc	Ovine asialo mucin	0.27	1.0	100
		CMP-9-O-Ac-NeuAc	Ovine asialo mucin	0.52	1.6	150
		CMP-NeuGc	Ovine asialo mucin	0.10	0.10	60

Table 6 Acceptor Substrate Specificity of a Recombinant GalNAc-Specific α(2,6)Sialyltransferase [180,181]

Enzyme	Source	Donor	Acceptor [abundant terminal structure]	Activity (pmol/h/10 μL media)
α2,6ST	Cloned from chick embryo	CMP-NeuAc	Fetuin [Galβ(1,3)NAcα-R]	142
Linkages		CMP-NeuAc	Asialo fetuin [Galβ(1,4)GlcNAc or Galβ(1,3)GalNAcα-R]	96
NeuAcα(2,6)GalNAcα-peptides		CMP-NeuAc	α₁-Acid GP [Siaα(2,3)Galβ(1,4)GlcNAcβ-R]	6
		CMP-NeuAc	α₁-Acid asialo GP [Galβ(1,4)GlcNAcβ-R]	4
		CMP-NeuAc	Bovine submaxillary mucin [Siaα(2,3)Galβ(1,3)GalNAcα-R]	15
		CMP-NeuAc	Bovine submaxillary asialo mucin [Galβ(1,3)GalNAcα-R]	186
		CMP-NeuAc	Ovomucoid [Siaα(2,3)Galβ(1,4)GlcNAcβ-R]	7
		CMP-NeuAc	Asialoovomucoid [Galβ(1,4)GlcNAcβ-R]	0
		CMP-NeuAc	Galβ(1,3)GlcNAcβ(1,3)Galβ(1,4)Glc	0
		CMP-NeuAc	Galβ(1,4)GlcNAc	0
		CMP-NeuAc	Galβ(1,3)GalNCa	0
		CMP-NeuAc	GalNAcβ(1,4)Gal	0
		CMP-NeuAc	Galβ(1,4)Glc	0
		CMP-NeuAc	GalNAcα-Ser(NAc)	4
		CMP-NeuAc	GalNAcα-benzyl	2

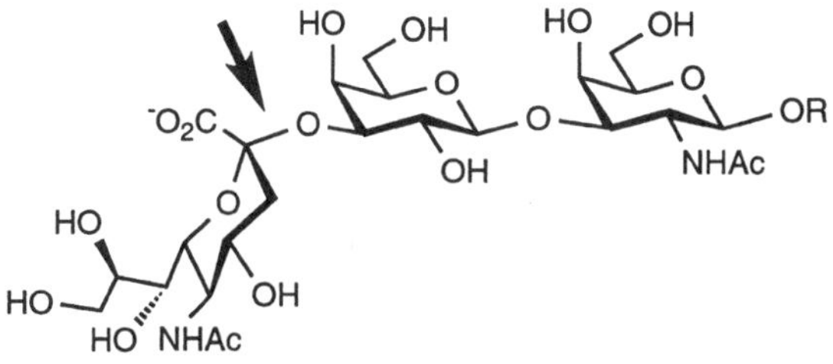

Figure 13

acetylneuraminate: β-D-galactosyl-1,3-N-acetyl-β-D-galactosamine α-2,3-N-acetyl-neuraminyltransferase is responsible for the synthesis of the Siaα(2→3)-Galβ(1→3)GalNAc structure (Figure 13). This enzyme has been purified from porcine liver and is commercially available ($157 to $205 per 0.02 unit); one unit of enzyme will transfer 1 μmol NeuAc from CMP-Neu5Ac to asialoglycoprotein (Galβ(1→3)GalNAc is exposed in the oligosaccharide terminus) per min at pH 6.5 and 37°C.

The acceptor substrate specificity of α(2,3)sialyltransferase (Galβ(1→3)-GalNAc-specific) has been well studied [147,182,183] (Table 7). The enzyme purified from porcine submaxillary glands accepted CMP-NeuAc derivatives such as 9-O-Ac (relative rate: 90%) and N-glycolyl (relative rate: 20%) [147]. The 9'-modified CMP-NeuAc derivatives were accepted equally well; However, if a positive charge was present on the C-9 position, the CMP-9-amino-NeuAc was not a substrate [174].

The enzyme showed remarkable specificity for the Galβ(1→3)GalNAc structure with a V_{max}/K_m value of 44,500 when compared to either Type 1 (Galβ-(1→3)GlcNAc) or Type 2 (Galβ(1→4)GlcNAc) disaccharides (V_{max}/K_m: 67.7 and 2.6, respectively) [183].

The enzyme purified from human placenta showed a similar specificity to that of the one from porcine submaxillary glands [184] (Table 8). It preferentially sialylated Galβ(1→3)GalNAc rather than other typical disaccharide sequences found in N- and O-linked oligosaccharides. The enzyme also sialylated Type 1 disaccharides, although the relative rate was low (6.2%).

The Paulson group has cloned two types of Galβ(1→3)GalNAc-specific α(2,3)sialyltransferase genes [185,186] (Table 9). The one designated ST3O, which was derived from the porcine enzyme, is specific for a Galβ(1→3)GalNAc structure and does not sialylate the Type 2 structure (Galβ(1→4)GlcNAc) [185]. The other, designated STZ, is expressed from COS-1 cells and is a Galβ(1→3)GalNAc-specific α(2,3)sialyltransferase; although, it sialylated the Type 2 structure with a relative rate of 39%, it did not sialylate the Type 1 structure [186]. Other Galβ(1→3)GalNAc-specific α(2,3)sialyltransferase genes have also been cloned from mouse brain [187,188].

α(2,3)Sialyltransferase (EC 2.4.99.6): Galβ(1→3/4)GlcNAc-specific

The NeuAcα(2→3)Galβ(1→3/4)GlcNAcβ→ linkage is a common terminal structure in the N- and O-linked oligosaccharides of glycoproteins and glycolipids, and the expression of this sequence is often found in biologically important oligosaccharide structures such as sialyl Lewis x and sialyl Lewis a structures. CMP-N-acetylneu-

Table 7 Substrate Specificity of Galβ(1→3)GalNAc-Specific α(2,3)Sialyltransferase from Porcine Submaxillary Gland (SG) [147,174,183]

Substrates				Kinetics		
Enzyme	Source	Donor	Acceptor	K_m (μM)	V_{max}	Rel. rate (%)
α2,3ST	Porcine SG	CMP-NeuAc	Antifreeze GP			100
Linkage		CMP-9-O-Ac-NeuAc	Antifreeze GP			90
NeuAcα(2,3)Galβ(1,3)GalNAc		CMP-NeuGc	Antifreeze GP			20
		CMP-9-amino-NeuAc	Antifreeze GP			0
		CMP-9-acetamido-NeuAc	Antifreeze GP			78
		CMP-9-benzamido-NeuAc	Antifreeze GP			80
		CMP-9-hexanoylamino-NeuAc	Antifreeze GP			100
		CMP-9-azido-NeuAc	Antifreeze GP			84
				K_m (μM)	V_{max}	V_{max}/K_m (1/mM)
		CMP-NeuAc	Galβ(1,3)GalNAc	0.2	8.9	44500
		CMP-NeuAc	Galβ(1,3)GlcNAc	65	4.4	67.7
		CMP-NeuAc	Galβ(1,4)GlcNAc	42	0.11	2.62
		CMP-NeuAc	Galβ(1,6)GlcNAc	29	0.24	8.3
		CMP-NeuAc	Galβ(1,4)Glc	180	0.62	3.4

Table 8 Acceptor Substrate Specificity of Galβ(1→3)GalNAc-Specific α(2,3)Sialyltransferase from Human Placenta [184]

Enzyme	Sources	Substrates		Kinetics		
		Donor	Acceptor [abundant terminal structure]	K_m (mM)	V_{max}	Rel. rate (%)
α2,3ST	Human placenta	CMP-NeuAc	Galβ(1,3)GalNAc	0.27	0.484	100
Linkages		CMP-NeuAc	Galβ(1,3)GalNAcαThr			60.6
NeuAcα(2,3)Galβ(1,3)GalNAc		CMP-NeuAc	Galβ(1,3)GlcNAcβ-PNP			28.6
		CMP-NeuAc	Galβ(1,3)GlcNAcβ(1,3)Galβ(1,4)Glc			10.4
		CMP-NeuAc	Galβ(1,3)GlcNAc	220	0.12	6.2
		CMP-NeuAc	Galβ(1,4)Glc			0.2
		CMP-NeuAc	Galβ(1,4)GlcNAc			<0.1
		CMP-NeuAc	Antifreeze GP [Galβ(1,3)GalNAcα-R]			100
		CMP-NeuAc	Porcine submaxillary asialo/afuco mucin [Galβ(1,3)GalNAcα-R]			30
		CMP-NeuAc	Ovine submaxillary asialo mucin [GalNAcα-R]			0.9
		CMP-NeuAc	Asialo fetuin [Galβ(1,4)GlcNAc or Galβ(1,3)GalNAcα-R]			13
		CMP-NeuAc	α1-Acid asialo GP [Galβ(1,4)GlcNAcβ-R]			1.5
		CMP-NeuAc	Asialo transferrin [Galβ(1,4)GlcNAcβ-R]			0.2

Table 9 Acceptor Substrate Specificity of a Recombinant Galβ(1→3)GalNAc-Specific α(2,3)Sialyltransferase [185,186]

Enzyme	Sources	Substrates Donor	Acceptor	Rel. rate (%) ST30	STZ
α2,3ST	Cloned from porcine	CMP-NeuAc	Galβ(1,3)GalNAc	100	100
Linkages		CMP-NeuAc	Galβ(1,3)GlcNAc	2	0
NeuAcα(2,3)Galβ(1,3)GalNAc		CMP-NeuAc	Galβ(1,4)GlcNAc	0	9
		CMP-NeuAc	Galβ(1,4)Glc	0	8
		CMP-NeuAc	Galβ(1,3)GlcNAcβ(1,3)Galβ(1,4)Glc	1	0
		CMP-NeuAc	Galβ(1,4)GlcNAcβ(1,3)Galβ(1,4)Glc	0	39

raminate: β-D-galactosyl-1,4-N-acetyl-β-D-glucosamine α-2,3-N-acetylneuraminyl-transferase is responsible for the synthesis of Siaα(2→3)Galβ(1→3/4)GlcNAc structure(s) (Figure 14).

The enzyme has been purified from rat liver [189,190], and its gene has also been cloned [191–194]; but the enzyme is not yet commercially available. This sialyltransferase preferentially sialylates Galβ(1→3)GlcNAc (the Type 1 structure) but also acts on Galβ(1→4)GlcNAc (the Type 2 structure) to form the Siaα(2→3)Gal linkage. The purified and recombinant enzymes have been used for the enzymatic synthesis of a variety of sialosides [12,13,45,77,95,107,114,142,195–198].

The enzyme purified from rat liver has been studied with respect to its donor specificity (Table 10). It accepted CMP-NeuAc derivatives such as 9-O-Ac and N-glycolyl (NeuGc) as donor substrates, and it utilized, as acceptor, asialo α₁-acid GP (glycoprotein), which has the Type 2 disaccharide as a major oligosaccharide component of its structure, with relative rates of 150% and 60%, respectively [147].

CMP-9-modified NeuAc derivatives were also relatively good substrates, for which the relative rates were in the range of 27%–147% (Table 10). Unlike other sialyltransferases, this enzyme seemed not to accept well a bulky substituent on the C-9 of the NeuAc of CMP-NeuAc [174].

The acceptor substrate specificity of this α(2,3)sialyltransferase from rat liver has also been studied by the use of naturally occurring glycoproteins whose oligosaccharide structures have been modified by glycosidases such as sialidase (to asialo) and galactosidase (to agalacto) [190] (Table 11). The enzyme preferentially sialylated

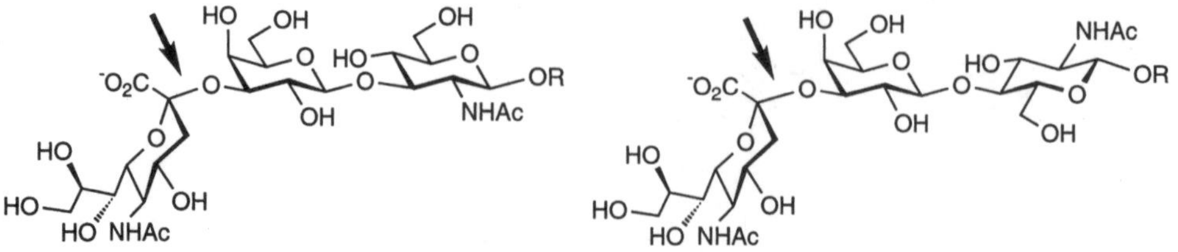

NeuAcα(2→3)Galβ(1→3)GlcNAcβ→OR   NeuAcα(2→3)Galβ(1→4)GlcNAcβ→OR

Figure 14

Table 10 Donor Substrate Specificity of Galβ(1→3/4)GalNAc-Specific α(2,3)Sialyltransferase from Rat Liver [147,174]

		Substrates		Kinetics	
Enzyme	Source	Donor	Acceptor	K_m (μM)	Rel. rate (%)
α2,3ST	Rat liver	CMP-NeuAc	α$_1$-Acid asialo GP	70	100
Linkage		CMP-9-O-NeuAc	α$_1$-Acid asialo GP		150
NeuAcα(2,3)Galβ(1,3/4)GlcNAc		CMP-NeuGc	α$_1$-Acid asialo GP		60
		CMP-NeuAc	Galβ(1,4)GlcNAc		N.A.
		CMP-NeuAc	Galβ(1,3)GlcNAc		N.A.
		CMP-9-amino-NeuAc	α$_1$-Acid asialo GP		10
		CMP-9-acetamido-NeuAc	α$_1$-Acid asialo GP	245	53
		CMP-9-benzamido-NeuAc	α$_1$-Acid asialo GP	220	62
		CMP-9-hexanoylamino-NeuAc	α$_1$-Acid asialo GP		27
		CMP-9-azido-NeuAc	α$_1$-Acid asialo GP		147

Table 11 Acceptor Substrate Specificity of Galβ(1→3/4)GlcNAc-Specific α(2,3)Sialyltransferase from Rat Liver [190]

Enzyme	Source	Substrates		Rel. rate (%)
		Donor	Acceptor [abundant terminal structure]	
α2,3ST	Rat liver	CMP-NeuAc	Asialo α₁-acid GP [Galβ(1,4)GlcNAcβ-R]	48
Linkage		CMP-NeuAc	Asialo transferrin [Galβ(1,4)GlcNAcβ-R]	11
NeuAcα(2,3)Galβ(1,3/4)GlcNAc		CMP-NeuAc	Asialo prothrombin [Galβ(1,4)GlcNAcβ-R and Galβ(1,4)GlcNAcβ-R]	97
		CMP-NeuAc	Asialoagalacto prothrombin [Galβ(1,3)GlcNAcβ-R]	100
		CMP-NeuAc	Antifreeze GP [Galβ(1,3)GalNAcα-R]	0
		CMP-NeuAc	Asialo ovine submaxillary mucin [GalNAcα-R]	0

Table 12   Acceptor Substrate Specificity of a Recombinant Galβ(1→3/4)GlcNAc-Specific α(2,3)Sialyltransferase [193]

		Substrates		Rel. rate
Enzyme	Source	Donor	Acceptor	(%)
α2,3ST	Cloned from	CMP-NeuAc	Galβ(1,3)GalNAc	3
	porcine			
Linkages		CMP-NeuAc	Galβ(1,3)GlcNAc	49
NeuAcα(2,3)Galβ(1,3/4)GlcNAc		CMP-NeuAc	Galβ(1,4)GlcNAc	19
		CMP-NeuAc	Galβ(1,4)Glc	7
		CMP-NeuAc	Galβ(1,3)GlcNAcβ(1,3)Galβ(1,4)Glc	100
		CMP-NeuAc	Galβ(1,4)GlcNAcβ(1,3)Galβ(1,4)Glc	4

the Type 1 structure rather than the Type 2 structure (relative rate: 48% for asialo α_1-acid glycoprotein).

Kitagawa and Paulson have cloned an α(2,3)sialyltransferase gene that is specific for Galβ(1→3/4)GlcNAc designated ST3N [193]. Like other known α(2,3)sialyltransferases, this enzyme preferentially sialylated the Type 1 (relative rate: 49%) rather than the Type 2 disaccharide (relative rate: 19%) (Table 12).

Ito and Paulson have reported that lactose, a poor substrate, became a better substrate for a recombinant Galβ(1→3/4)GlcNAc α(2,3)sialyltransferase (ST3N) after chemical modification [197]. With the introduction of a hydrophobic aglycone into lactose, the resulting trimethylsilylethyl lactoside became an effective substrate ($V_{max}/K_m = 0.15$ mM^{-1}). In addition, if the 2-OH of the glucose residue was acylated with a pivaloyl group, the 2-O-modified lactose became a better acceptor substrate ($V_{max}/K_m = 1.6$ mM^{-1}) for the enzyme (Table 13).

Table 13   Chemical Modification of Lactose Changed the Kinetics of a Recombinant Galβ(1→3/4)GlcNAc-Specific α(2,3)Sialyltransferase [197]

	K_m (mM)	V_{max}	V_{max}/K_m
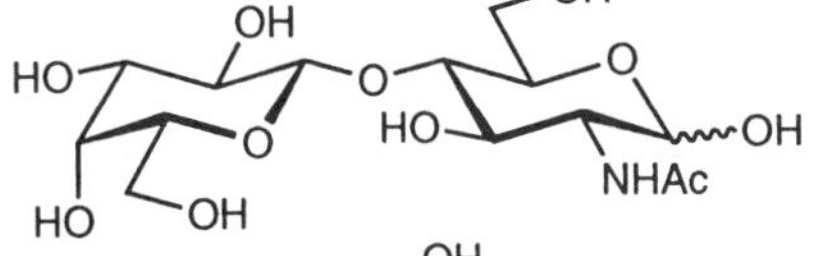	1.6	1.0	0.63
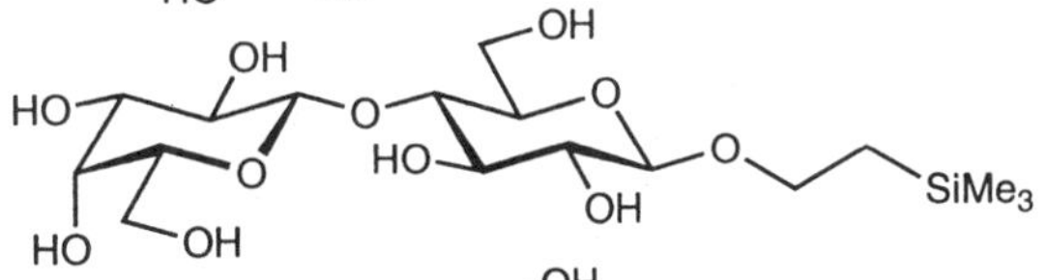	9.8	1.4	0.15
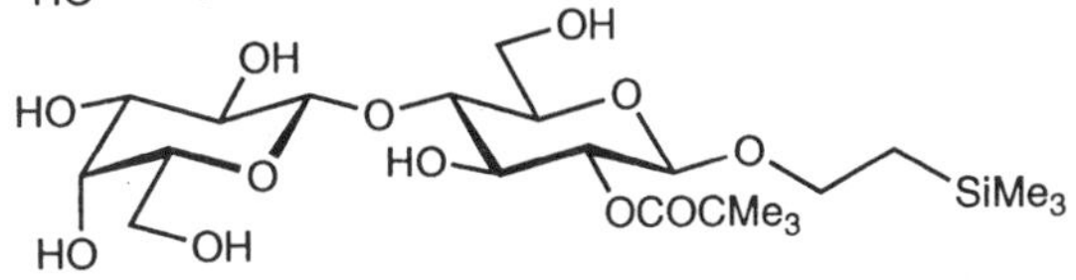	0.8	1.3	1.60

Table 14 Acceptor Substrate Specificity of Lac-Cer-Specific α(2,3)Sialyltransferase (G_{M3} Synthase) from Rat Brain [203]

Substrate		Rel. activity (%)
Galβ(1→4)Glc	Lac	0
Glcβ(1→1')Cer	GlcCer	0
Galβ(1→1')Cer	GalCer	22.6
Galβ(1→4)Glcβ(1→1')Cer	LacCer	100
GalNAcβ(1→4)Galβ(1→4)Glcβ(1→1')Cer	G_{A2}	14.0
Galα(1→4)Galβ(1→4)Glcβ(1→1')Cer	G_{b3}Cer	0
Galβ(1→3)GalNAcβ(1→4)[NeuAcα(2→3)]Galβ(1→4)Glcβ(1→1')Cer	G_{M1}	0

α(2,3)Sialyltransferase (EC 2.4.99.9): G_{M3}-synthase

G_{M3}, NeuAcα(2→3)Galβ(1→4)Glcβ(1→Cer, is the first and simplest acidic glyco-sphingolipid (ganglioside) and is one of the major glycosphingolipids in most extra-neuronal tissues. The expression of the G_{M3} molecule has been suggested to be related to cell development and differentiation [24].

The enzyme that catalyzes the transfer of NeuAc from CMP-NeuAc onto lac-tosylceramide is CMP-N-acetylneuraminate: lactosylceramide α(2,3)sialyltransferase (EC 2.4.99.9) [199–203]. This enzyme has been purified from several natural sources, including rat brain. Its activity has been found to be dependent on the structure of the ceramide portion of the substrate [202].

The acceptor substrate specificity has been studied by the use of several gly-colipid structures [203] (Table 14). Although lactose (Lac) shares the same carbo-hydrate structure as that of LacCer, Lac was not a substrate for this sialyltransferase. This observation suggests that the enzyme recognizes not only the terminal carbo-hydrate but also the aglycone moiety. GalCer was a weak substrate (relative rate: 23%) for the enzyme. When the galactose residue of LacCer was glycosylated with GalNAc via a β(1→4)-linkage, the resulting G_{A2} was also a weak substrate (relative rate: 14%).

α(2,8)Sialyltransferase (EC 2.4.99.8)

The glycolipid G_{D3}, Siaα(2→8)Siaα(2→3)Galβ(1→4)Glcβ(1→Cer, is a minor ganglioside in adult normal tissues; however, it is highly expressed, as one of the major gangliosides, on the cell surface of human tumors of neuroectodermal origin, such as melanoma, gliomas, and neuroblastomas. G_{D3} is also a major ganglioside in the central nervous system of the embryo but shows a progressive decline in its expression during development [24]. CMP-N-acetylneuraminate: α-N-acetylneur-aminyl-2,3-β-D-galactoside α-2,8-N-acetylneuraminyltransferase is responsible for the synthesis of the Siaα(2→8)Siaα(2→3)Galβ(1→ structure (Figure 15). The sia-lyltransferase gene from a human melanoma cell line has been cloned [204–206]; this enzyme is specific for the G_{M3} structure to produce G_{D3} and does not form polysialic acid structures [206].

The enzyme activity responsible for the synthesis of polysialic acid has been demonstrated in fetal rat brain and newborn rat brain by the use of bacteriophage

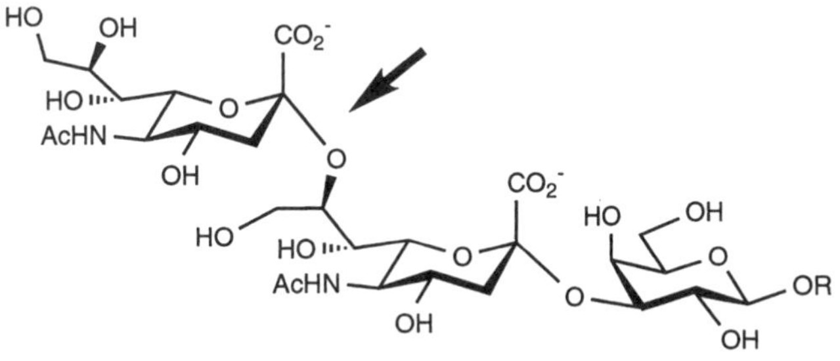

Figure 15

endoneuraminidase, but the enzyme has not yet been fully characterized [128,129]. The enzyme responsible for the biosynthesis of the $\alpha(2\rightarrow8)$-sialic acid polymer has been speculated to be associated with cell adhesion molecules (NCAM) [128–130].

Typical enzymatic sialylation reaction: Synthesis of
Neu5Acα(2→6)Galβ(1→4)GlcNAc [196]

A solution of an acceptor (LacNAc, Galβ(1→4)GlcNAc; 8.3 mg, 23 μmol), CMP-Neu5Ac sodium salt (20 mg, 30 μmol), NaN$_3$ (0.065 mg, 10 μmol), and bovine serum albumin (BSA, 1.4 mg) in Na cacodylate buffer (50 mM, pH 7.7; 2.3 mL) was adjusted to pH 6.5. To the solution were added $\alpha(2,6)$sialyltransferase (100 mU) and calf intestinal alkaline phosphatase (EC 3.1.3.1; 6 U), and the reaction mixture was stirred for 2 days at 37°C. TLC showed that the starting materials, LacNAc (R_f = 0.63 in 1 M NH$_4$OAc/2-propanol 1:2.4, v/v) and CMP-Neu5Ac (R_f = 0.19) were converted into the sialoside (R_f = 0.30). The mixture was concentrated and chromatographed on a Sephadex G-25 column (2.3 × 32 cm), with 0.1 M NH$_4$HCO$_3$, to give Neu5Acα(2→6)Galβ(1→4)GlcNAc in almost quantitative yield.

C. Galactosyltransferase

Galactose is a common constituent sugar in *N*- and *O*-linked oligosaccharides and glycolipids, and is usually capped with sialic acid. Once the sialic acid is removed, the resulting galactose-exposed oligosaccharide-containing glycoproteins are immediately removed from the circulation because these glycans with exposed galactose residues are recognized as aberrant oligosaccharide structures by the body. This so-called clearance is mediated by galactose-binding proteins expressed on hepatocytes [125,126].

Galactosyltransferase expressed on the cell surface is thought to participate in cell–cell adhesion by binding to its specific acceptor carbohydrate on an adjacent cell surface or in the extracellular matrix [207–215]. In addition, a galactosyltransferase present on the head of the sperm cells has been postulated to mediate binding to the egg in fertilization, by binding to the oligosaccharide residues in the egg coat glycoprotein [208–210,213,214].

Galactosyltransferase transfers a galactose residue from uridine 5′-diphospho galactose (UDP-Gal) to a specific hydroxyl group on the acceptor, usually glucose or *N*-acetylglucosamine (Scheme 13).

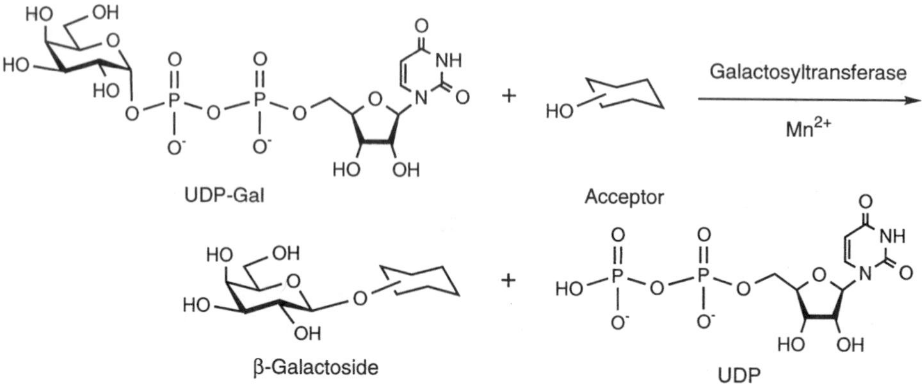

Scheme 13

The donor substrate UDP-galactose is prepared by the following two enzymatic methods (A and B) [28,216–221]: Method A utilizes the interconversion of UDP-glucose and UDP-galactose (Scheme 14). Glucose 1-phosphate reacts with UTP (uridine 5'-triphosphate) in the presence of UDP-glucose pyrophosphorylase (EC 2.7.7.9) to give UDP-glucose. The UDP-glucose is then converted to UDP-galactose by an equilibrium reaction catalyzed by UDP-galactose 4-epimerase (EC 5.1.3.2), although this equilibration favors UDP-glucose [216,221]. Glucose 1-phosphate is prepared from glucose: Glucose is phosphorylated by ATP and hexokinase (EC 2.7.1.1) to give glucose 6-phosphate, which is converted to glucose 1-phosphate by an equilibrium reaction catalyzed by phosphoglucomutase (EC 5.4.2.2) in the presence of a small amount of glucose 1,6-diphosphate.

Method B utilizes a galactose 1-phosphate uridyltransferase reaction (Scheme 15). Galactose 1-phosphate undergoes an exchange reaction, catalyzed by galactose 1-phosphate uridyltransferase (EC 2.7.7.12), between galactose 1-phosphate and UDP-glucose, to give UDP-galactose and glucose 1-phosphate [217–221]. Galactose 1-phosphate is prepared from galactose and ATP in a reaction catalyzed by galactokinase (EC 2.7.1.6). All these enzymes are commercially available from Sigma.

β(1,4)Galactosyltransferase: (EC 2.4.1.22, EC 2.4.1.38, EC 2.4.1.90)

N-Acetyllactosamine (Galβ(1→4)GlcNAc) and lactose (Galβ(1→4)Glc) sequences are abundant in N- and O-linked oligosaccharides and glycolipids [15]. UDP-Galactose: D-glucose 4-β-D-galactosyltransferase is responsible for the synthesis of the β(1→4)-galactosyl linkage (Figure 16). Three EC numbers have been given to this enzyme [222]. The enzyme EC 2.4.1.22 is defined as UDP-galactose: D-glucose 4-β-D-galactosyltransferase, which is a complex of two proteins: A and B [223–231]. Protein B is an α-lactalbumin [228–231]; in the absence of protein B the enzyme catalyzes the transfer of galactose from UDP-galactose to N-acetylglucosamine, which is identical to the activity of enzyme EC 2.4.1.90. Enzyme EC 2.4.1.38 is defined as UDP-galactose: N-acetyl-D-glucosamine-glycopeptide 4-β-D-galactosyltransferase, which catalyzes the formation of the LacNAc sequence in glycopeptide oligosaccharides and is a component (protein A) of the enzyme EC 2.4.1.22 [225]. The enzyme EC 2.4.1.90 is defined as UDP-galactose: N-acetyl-D-

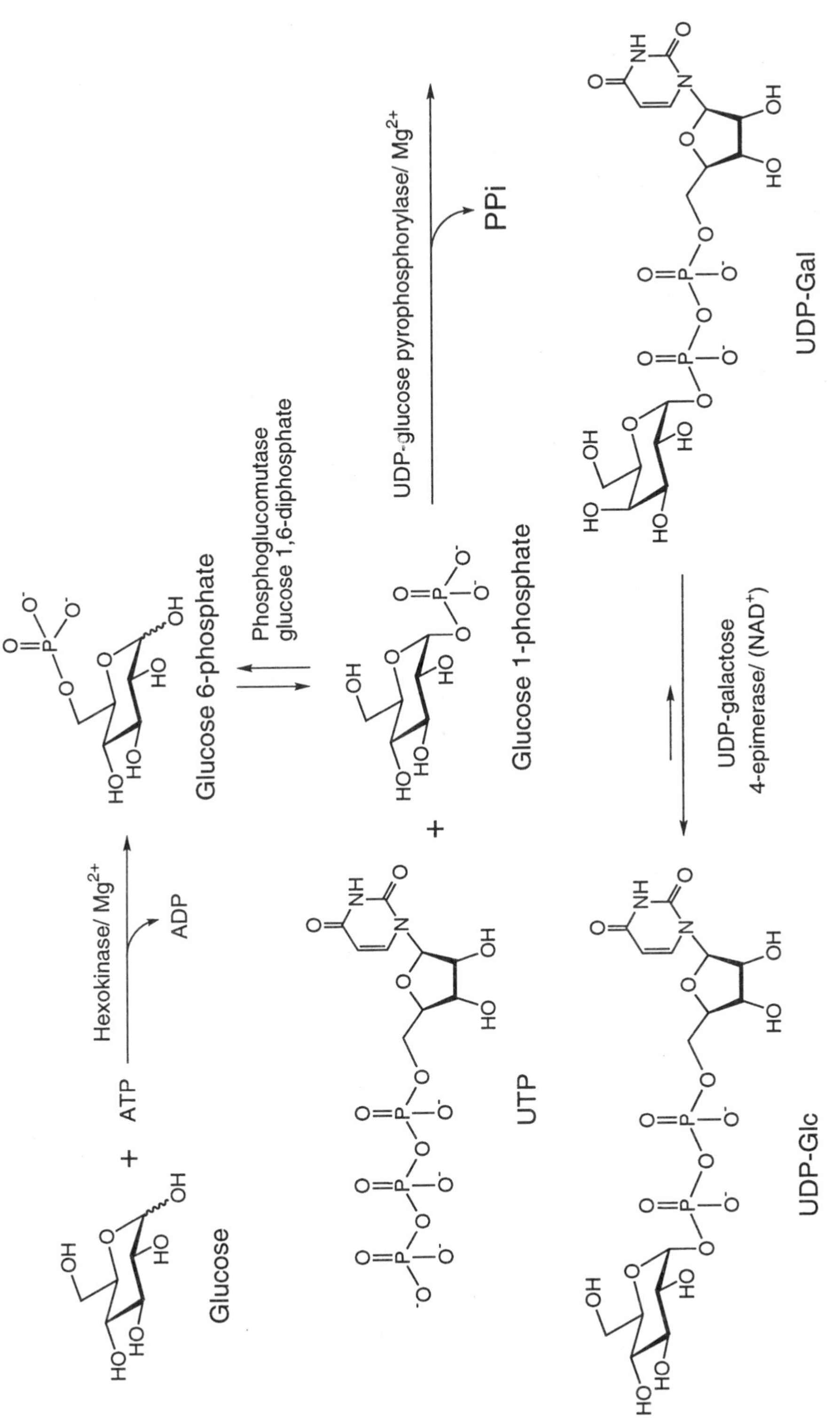

Scheme 14

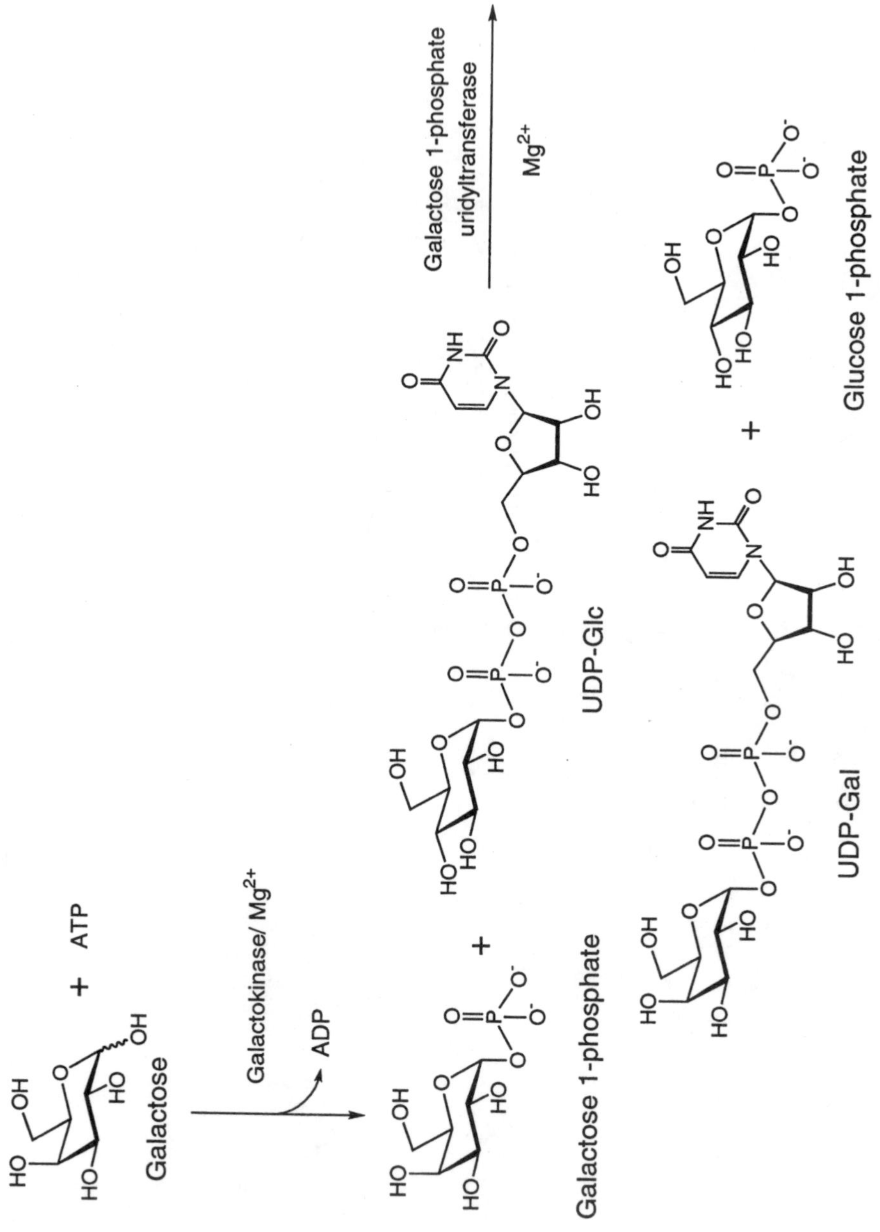

Galactose + ATP

Galactose

Galactokinase/ Mg²⁺ → ADP

Galactose 1-phosphate + UDP-Glc

Galactose 1-phosphate uridyltransferase, Mg²⁺

UDP-Gal + Glucose 1-phosphate

Scheme 15

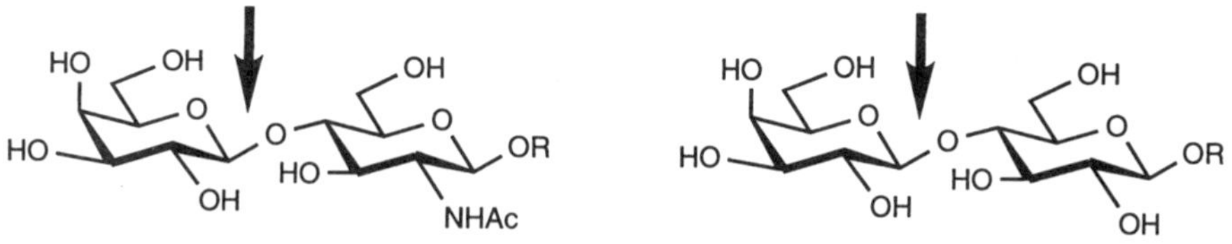

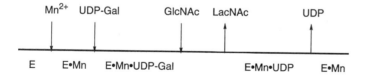

Figure 16

glucosamine 4-β-D-galactosyltransferase, and the reaction catalyzed by this enzyme is the same as that catalyzed by the component of EC 2.4.1.22 (protein A) and EC 2.4.1.38. As mentioned above, the acceptor specificity of this enzyme is changed by α-lactalbumin: In the presence of α-lactalbumin the enzyme accepts D-glucose and forms lactose; on the other hand, in the absence of adequate amount of α-lactalbumin the enzyme accepts N-acetyl-D-glucosamine and forms LacNAc [231].

This enzyme has been purified from bovine milk [223], and its gene has been cloned [232–243]. The enzyme is commercially available from Sigma (EC 2.4.1.22 from bovine milk, currently \$102.50 for 5 units; EC 2.4.1.22 from human milk, \$61 for 0.2 units) and Boehringer Mannheim (EC 2.4.1.38 from human milk, \$70 for 0.2 units). One unit of the enzyme will transfer 1.0 μmol of galactose from UDP-galactose to D-glucose per minute at pH 8.4 and 30°C, in the presence of 0.2 mg of α-lactalbumin per mL reaction mixture. The enzyme reaction is usually carried out in Na cacodylate buffer (50 mM, pH 7.4) and requires Mn^{2+}.

The enzyme kinetics have been thoroughly studied, and the enzyme reaction proceeds by an ordered mechanism [244,245] (Figure 17). A divalent cation (Mn^{2+}) binds to the enzyme's active site (E) first, followed by UDP-Gal (a donor substrate) to form a complex of E•Mn•UDP-Gal. GlcNAc (an acceptor substrate) then comes in, and galactosyl transfer reaction occurs. LacNAc (the product) comes off and leaves an E•Mn•UDP complex. After UDP is released, the enzyme's active site still holds Mn^{2+}, and another round of galactosyl transfer takes place.

Donor Substrate Specificity. Donor substrate specificity has been well studied by Berliner and Robinson and by the Hindsgaul group (Figure 18). Hindsgaul and co-workers have synthesized UDP-galactose analogs in which the galactose residue was chemically modified by reactions such as deoxygenation, and they have evaluated these analogs as donor substrates for β(1,4)galactosyltransferase (EC 2.4.1.22) from bovine milk. When the 2-OH of the galactose of UDP-galactose was converted to 2-NHAc (UDP-GalNAc), the relative rate of the enzyme reaction was 4% of that for UDP-galactose [247]. The enzyme apparently does not accept UDP-glucosamine derivatives such as UDP-GlcNAc and UDP-GlcNH₂, since their relative rates were 0% and 0.09%, respectively [247]. When the 4-OH of the galactose was

```
Mn²⁺   UDP-Gal          GlcNAc   LacNAc          UDP
 |        |                |        ↑              ↑
 |        |                |        |              |
_|_____|_____|_____|_____|_____

E      E•Mn     E•Mn•UDP-Gal            E•Mn•UDP       E•Mn
```

Figure 17

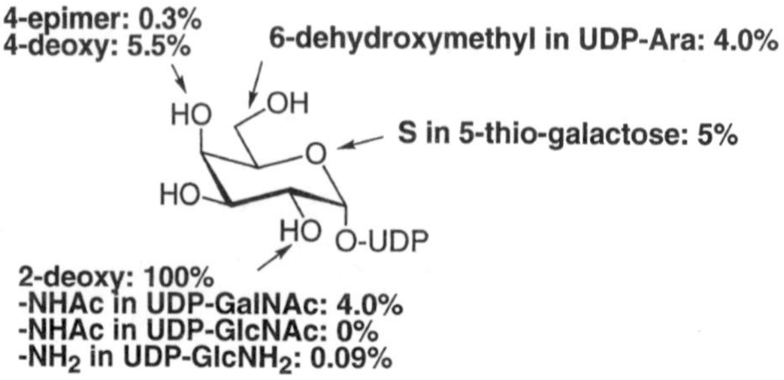

4-epimer: 0.3%
4-deoxy: 5.5%   6-dehydroxymethyl in UDP-Ara: 4.0%

S in 5-thio-galactose: 5%

O-UDP

2-deoxy: 100%
-NHAc in UDP-GalNAc: 4.0%
-NHAc in UDP-GlcNAc: 0%
-NH₂ in UDP-GlcNH₂: 0.09%

Figure 18

epimerized or removed, i.e., as in UDP-glucose or UDP-4-deoxy-glucose, the relative rates were 0.3% and 5.5%, respectively [246].

Replacing the ring oxygen of the galactose with a sulfur atom gave UDP-5-thio-galactose, which had a relative rate of 5% [248] (Figure 18). Several modified galactose-containing N-acetyllactosamine derivatives were synthesized by the enzymatic reaction [249].

Acceptor Substrate Specificity.   The acceptor substrate specificity of this enzyme has been extensively studied by several groups [250–255] (Figure 19). Modification of the C-1 position of D-glucosamine or D-glucose did not affect the relative rate significantly (relative rates were in the range of 20%–100%). However, the presence of a lipophilic aglycone dramatically enhanced the relative rate to 263% for R = O(CH₂)₈CO₂Me [252]. The enzyme tolerated the modification of the acyl group of the NHAc of N-acetyl-D-glucosamine (~80%) or deoxygenation of the 2-OH of D-glucose (60%) [251]; however, the enzyme did not accept the C-2 epimer or D-mannose as an acceptor substrate (relative rate: 1%) [251]. D-Glucal was a very poor acceptor substrate (0.4%) [253].

Modification of the 3-OH group was almost ineffective: 3-Deoxy-GlcNAc was a very poor acceptor (relative rate: 1%), and masking the 3-OH of the GlcNAc abolished its activity [253]. However, after modification of the 3-OH with lactic acid,

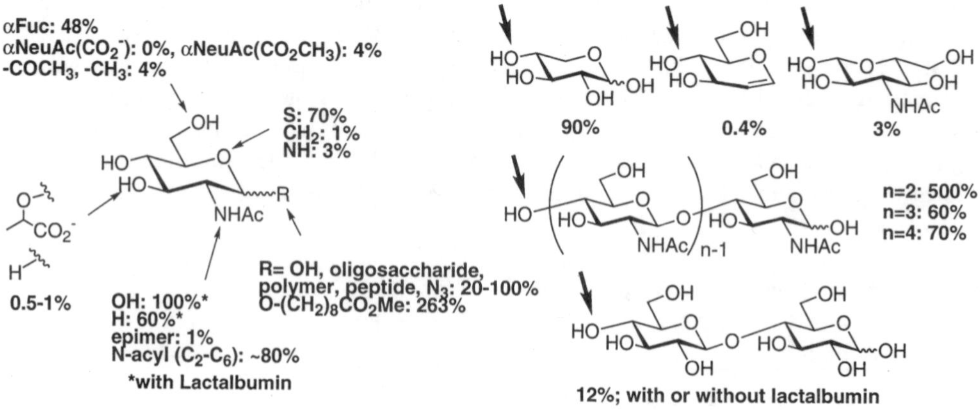

Figure 19

the resulting N-acetylmuramic acid was a moderate acceptor (relative rate: 44%). These finding suggests that the 3-OH group may be a key functional group for enzyme recognition (via hydrogen bonding?).

The 6-OH groups could be modified to some extent: Fucα(1→6)GlcNAcOR was a moderate acceptor (relative rate: 48%), whereas NeuAcα(1→6)GlcNAcOR was not an acceptor. When the carboxylate of the NeuAc was blocked with a methyl group (CO_2Me), the resulting NeuAc(CO_2Me)α(1→6)GlcNAcOR was a poor acceptor (relative rate: 4%) [252]. It is interesting that when the 6-hydroxymethyl group was removed entirely, the resulting sugar (D-xylose) was a good acceptor substrate (relative rate: 90%) [250].

The ring oxygen could be replaced with a sulfur atom (70%) [253]; however, replacement with a methylene (—CH_2—) or imine (NH) decreased the relative rate dramatically (1%–3%) [253]. Oligomers of N-acetyl-D-glucosamine were good acceptor substrates: The relative rate for di-N-acetylchitobiose (GlcNAcβ(1→4)-GlcNAc) (n = 2) was 500% [250]. A glucose oligomer, cellobiose (Glcβ(1→4)Glc), was a moderate acceptor (relative rate: 12%), and the reaction rate was found not to be affected by α-lactalbumin [250].

It is worth mentioning that an N-acetyl-D-glucosamine analog, 3-acetamido-3-deoxy-D-glucopyranose, was an acceptor substrate for β(1,4)galactosyltransferase, although the relative rate was low (3%); however, the enzyme transferred galactose to the C-1 OH group of 3-acetamido-3-deoxy-D-glucopyranose to form D-galactopyranosyl-β(1→1)-3-acetamido-3-deoxy-β-D-glucopyranoside   [254,255] (Figure 19). The enzyme has been used for enzymatic synthesis of β(1→4)galactose-terminated oligosaccharides [12,13,177,196,256–261].

Another β(1,4)galactosyltransferase is the enzyme responsible for the biosynthesis of lactosyl ceramide (Lac-Cer). This enzyme transfers a galactose residue to the glucose residue of glucosyl ceramide (Glc-Cer). The enzyme has been purified and its gene has been cloned [262,263].

α(1,3)Galactosyltransferase (EC 2.4.1.37): Blood group B transferase

An α(1→3)-galactosyl linkage is found in the blood group B substance. UDP-galactose: α-L-fucosyl-(1,2)-β-D-galactose 3-α-D-galactosyltransferase is responsible for the synthesis of the Galα(1→3)[Fucα(1→2)]Gal linkage. In contrast to the group B structure, the blood group A substance has an α(1→3)-N-acetylgalactosamine residue. The genes for both glycosyltransferases that introduce α(1→3)GalNAc and α(1→3)Gal onto the H substance (Fucα(1→2)Gal), referred to as A and B transferase, respectively, have been cloned by Hakomori and his co-workers [264]. This group has found that the amino acid sequences deduced from the two are identical except for four amino acid residues. They also performed an interesting gene reconstruction experiment in which they showed that these four amino acid substituents are crucial for nucleotide sugar-binding specificity [264b]. In addition, they reconstructed a new enzyme that catalyzed the transfer of both GalNAc and Gal residues from the UDP-GalNAc and UDP-Gal, respectively, onto the H substance [264b].

The Hindsgaul group has studied the acceptor substrate specificity using synthetic structural analogs of the disaccharide, Fucα(1→2)Galβ-O-Octyl, which is a minimum acceptor sequence for the glycosyltransferase: α(1,3)galactosyltransferase (EC 2.4.1.37) (B-transferase) for blood groups [265,266] (Figure 20). The com-

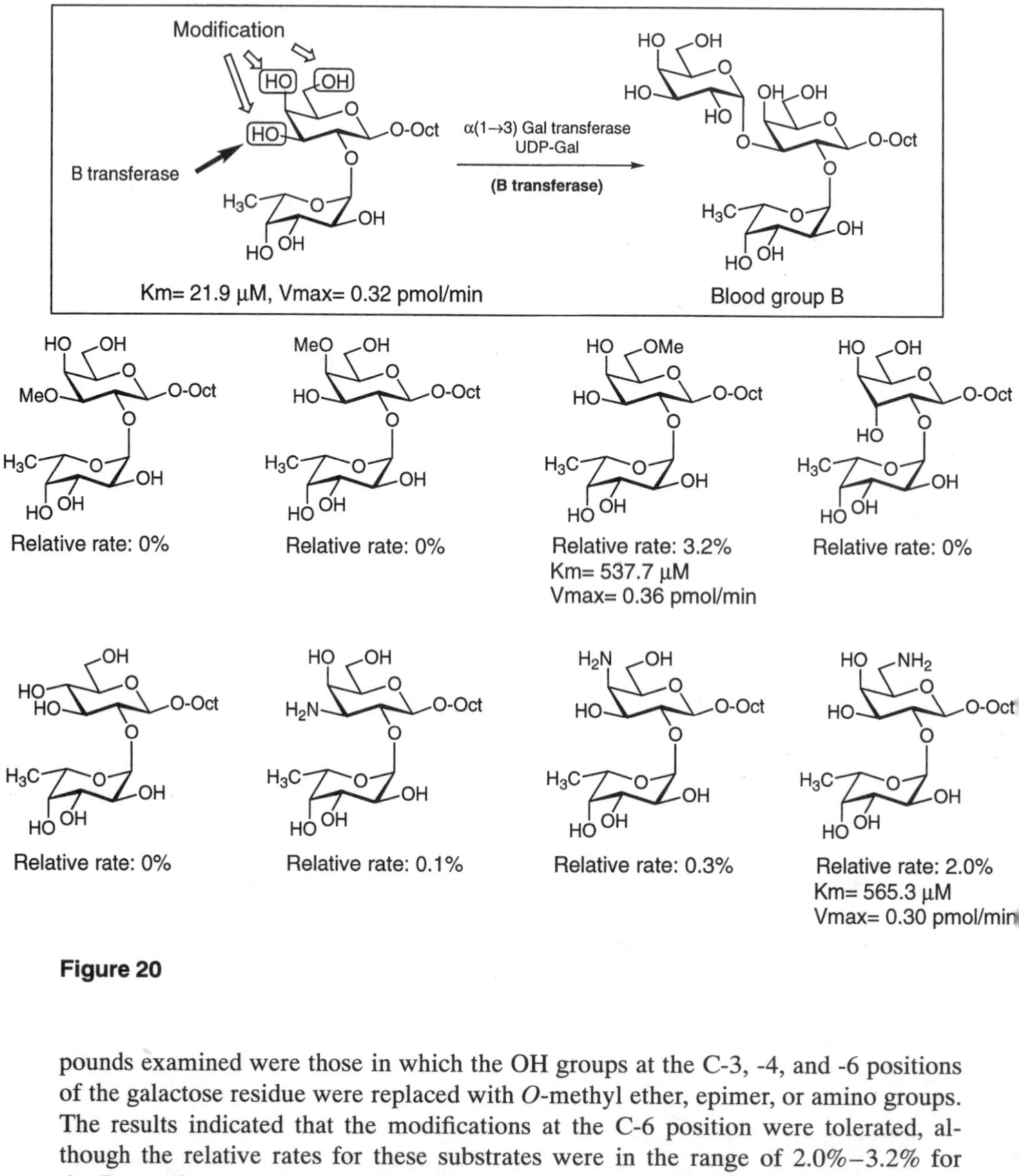

Figure 20

pounds examined were those in which the OH groups at the C-3, -4, and -6 positions of the galactose residue were replaced with *O*-methyl ether, epimer, or amino groups. The results indicated that the modifications at the C-6 position were tolerated, although the relative rates for these substrates were in the range of 2.0%–3.2% for the B-transferase. Other modifications at C-3 and -4 totally abolished the activity.

α(1,3)Galactosyltransferase (EC 2.4.1.151)

UDP-galactose: β-D-galactosyl-1,4-*N*-acetyl-D-glucosaminide 3-α-D-galactosyltransferase is responsible for the synthesis of the Galα(1→3)Galβ(1→3/4)GlcNAc linkage (Figure 21). This α(1→3)-galactosyl linkage is found at the terminus of *N*-acetyllactosamine-type glycans, and as such is a noncharged alternative to oligosaccharide chain termination by sialic acid. Since α(1,3)galactosyltransferase is widely expressed in a variety of nonprimate mammals and in New World monkeys but not in Old World monkeys or humans, the production of this enzyme has

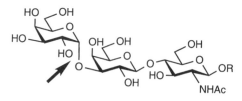

Figure 21

been suggested to be developmentally regulated [267,268]. In addition, the Galα(1→3)Galβ→ determinant is highly immunogenic in man, and many individuals show significant levels of circulating IgG specific for this determinant [267,268].

The enzyme has been purified from calf thymus, and its gene has been cloned [269]. The enzyme reaction is carried out in Na cacodylate buffer (100 mM, pH 6.5) and requires Mn^{2+} for enzyme activity. Joziasse and co-workers have examined the acceptor specificity of two types of the enzymes: One was purified from calf thymus, and the other was a recombinant bovine α(1,3)galactosyltransferase (Table 15) [270]. Both enzymes showed similar specificity: In addition to the Type 2 disaccharide sequence, the enzymes also accepted lactose, with a relative rate of 20%. Blood group H substance was a very poor acceptor substrate for either of the enzymes [270].

The Joziasse group has synthesized an "α3-galactosyl-Le^{x}" tetrasaccharide using a recombinant α(1,3)galactosyltransferase and α(1,3)fucosyltransferase partially purified from human milk [271] (Scheme 16). N-Acetyllactosamine was galactosylated by α(1,3)galactosyltransferase and UDP-Gal to form a Galα(1→3)-Galβ(1→4)GlcNAc, which was then fucosylated by α(1,3)fucosyltransferase and GDP-fucose to furnish the Galα(1→3)Galβ(1→4)[Fucα(1→3)]GlcNAc structure, which they proposed to be a potential E-selectin ligand; however, no biological data have since been published [271].

Table 15 Acceptor Substrate Specificity of a Purified (from Calf Thymus) and a Recombinant α(1,3)Galactosyltransferase [270]

Acceptor substrate [abundant terminal structure]	Rel. rate (%)	
	α(1,3)GalT calf thymus	α(1,3)GalT recombinant
Asialo α₁-Acid GP [Galβ(1,4)GlcNAcβ-R]	77	50
Asialo fetuin [Galβ(1,4)GlcNAcβ-R or Galβ(1,4)GalNAcα-R]	71	40
Galβ(1,4)GlcNAcβ-R	100	100
Galβ(1,4)Glc	20	24
Fucα(1,2)Galβ(1,4)GlcNAc	3	1

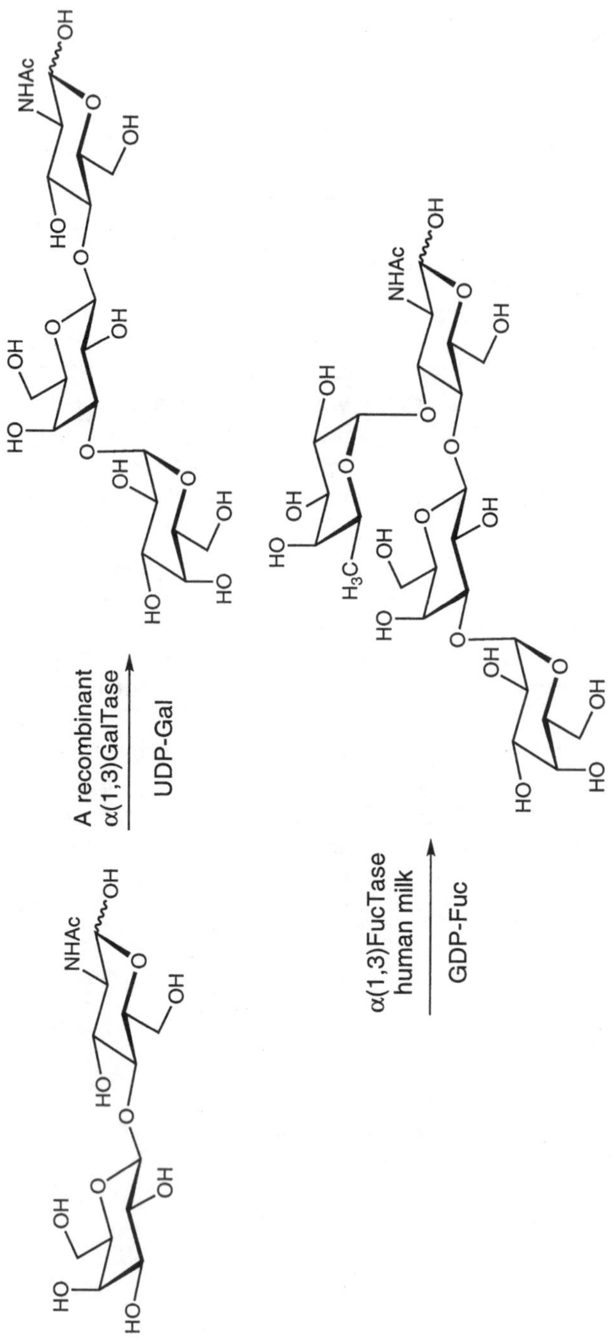

A recombinant
α(1,3)GalTase
UDP-Gal

α(1,3)FucTase
human milk
GDP-Fuc

Scheme 16

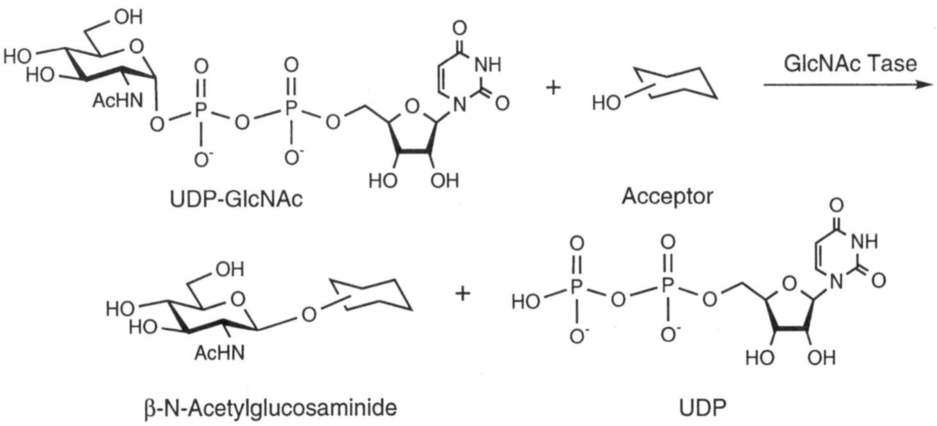

Scheme 17

A typical enzymatic galactosylation reaction: Synthesis of Galβ(1→4)-GlcNAcβ(1→6)[Galβ(1→3)]GalNAcα1-O-(CH₂)₈CO₂Me [114]

A solution of GlcNAcβ(1→6)[Galβ(1→3)]GalNAcα1-O-(CH$_2$)$_8$CO$_2$Me (19 mg, 25 μmol), UDP-galactose (30 μmol), MnCl$_2$ (14 mM), bovine serum albumin (BSA, 1%), calf intestine alkaline phosphatase (12 U), and β(1,4)-galactosyltransferase (EC 2.4.1.22, 2 U) in Na cacodylate buffer (0.62 mL; 50 mM, pH 7.4) was stirred for 1 h at room temperature. TLC showed that the starting material (R_f = 0.28 in CHCl$_3$/MeOH/H$_2$O 13:7:1) was converted into a product (R_f = 0.19). The reaction mixture was diluted to 9 mL with water, and applied onto a C$_{18}$ Sep-Pak cartridge. The cartridge was washed with water (40 mL), and the product was eluted with MeOH (50 mL). The MeOH fraction was concentrated, and the residue was lyophilized from water. The resulting powder was chromatographed on silica gel, with CHCl$_3$/MeOH/H$_2$O (12:7:1). Fractions containing the product were pooled and concentrated. The residue was dissolved in water, passed through a 0.22-μm filter, and lyophilized to give the β(1→4)-galactosylated product (22.7 mg, 100%).

D. *N*-Acetylglucosaminyltransferase

The enzyme responsible for the synthesis of the *N*-acetyl-D-glucosaminyl linkage is *N*-acetylglucosaminyltransferase. This enzyme utilizes uridine 5′-diphospho-*N*-acetyl-D-glucosamine (UDP-GlcNAc) as a donor substrate (Scheme 17).

All *N*-linked oligosaccharides share a common core structure: Manα(1→6)-[Manα(1→3)]Manβ(1→4)GlcNAcβ(1→4)GlcNAcβ1-Asn. Formation of antennary or branched structures begins by the addition of *N*-acetylglucosamine residues to the mannose residues of this core structure. Such branching is initiated by at least five different Golgi *N*-acetylglucosaminyltransferases [272] (Figure 22):

GnT I	(forms GlcNAcβ(1→2)Manα(1→3)-linkages)
GnT II	(forms GlcNAcβ(1→2)Manα(1→6)-linkages)
GnT IV	(forms GlcNAcβ(1→4)Manα(1→3)-linkages)

GnT V   (forms GlcNAcβ(1→6)Manα(1→6)-linkages)
GnT VI   (forms GlcNAcβ(1→4)Manα(1→6)-linkages)

In addition, bisected N-linked complex-type glycans have an N-acetylglucosaminyl residue attached to the β(1→4)-linked mannose residue via a β(1→4)-linkage:

GnT III   (forms GlcNAcβ(1→4)Manβ(1→4)-linkages)

The sites of potential glycosylation with GnT I–VI on the mannose residues of the N-linked trimannosyl core structure are shown in Figure 22, and the oligosaccharide structures formed are shown in Figure 23.

In O-linked glycans, at least five classes of core structures are known, and attachment of an N-acetylglucosaminyl residue also plays a key role in diversifying the O-linked oligosaccharide structures [15] (Figure 24). Taken together, N-acetyl-glucosaminyltransferases can regulate the branching of oligosaccharide structures in both N- and O-linked glycans, and thereby exert a controlling influence on carbohydrate-mediated intercellular recognition processes [272].

UDP-N-Acetylglucosamine has been chemically prepared by coupling N-acetyl-D-α-glucosamine 1-phosphate to an activated UMP-derivative such as UMP-morpholidate [28] (Scheme 18).

The donor substrate, UDP-GlcNAc, can also be enzymatically prepared from UTP and N-acetyl-α-D-glucosamine 1-phosphate (GlcNAc-1-P) in a reaction catalyzed by UDP-N-acetylglucosamine pyrophosphorylase (EC 2.7.7.23) (Scheme 19). This enzyme has been partially purified from yeast [28]; whole yeast cells can also be used for the preparation of UDP-GlcNAc [273].

An alternative chemo-enzymatic method of preparation has been reported by the Whitesides group [274]. They employed chemical acetylation of enzymatically prepared UDP-glucosamine (Scheme 20). In this approach, hexokinase (EC 2.7.1.1) phosphorylates D-glucosamine in the presence of ATP to give glucosamine 6-phosphate. Phosphoglucomutase (EC 5.4.2.2) then catalyzes the isomerization of the glucosamine 6-phosphate to the corresponding 1-phosphate. The resulting GlcNH$_2$-1-P reacts with UTP in the presence of UDP-glucose pyrophosphorylase (EC 2.7.7.9) to give UDP-glucosamine. The last step is N-acetylation of the UDP-glucosamine with N-acetyloxysuccinimide to give UDP-N-acetylglucosamine.

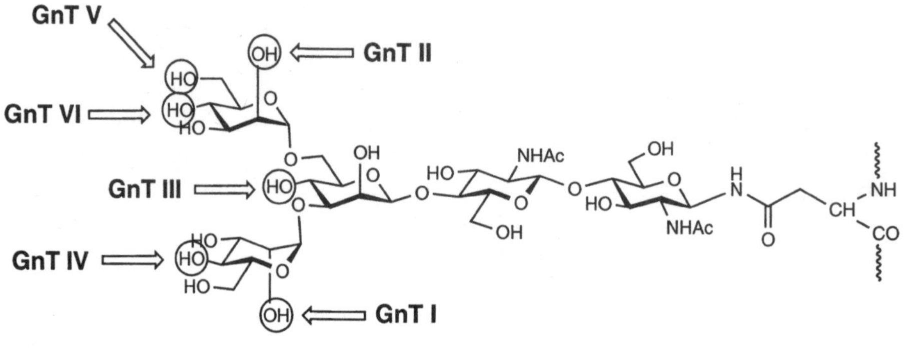

Figure 22

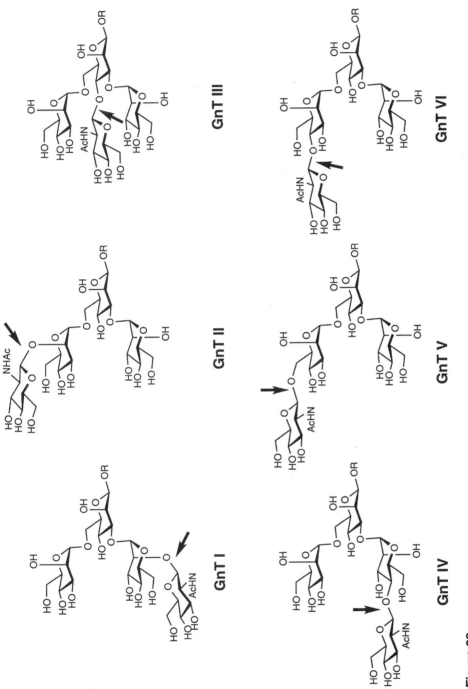

Figure 23

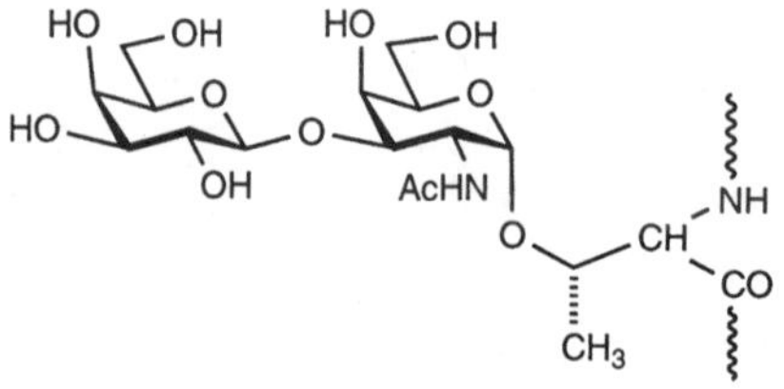

Core class 1 structure

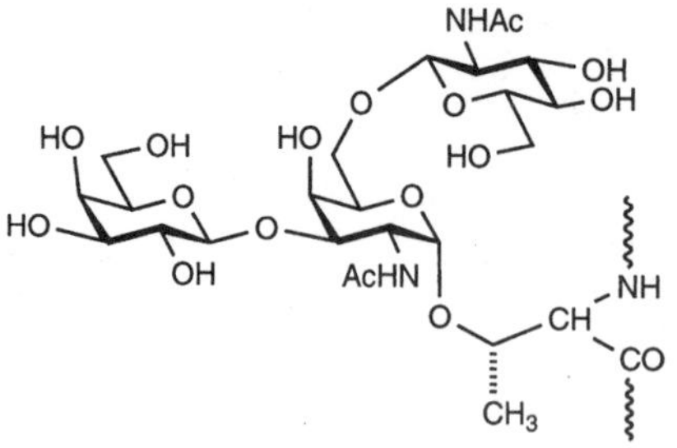

Core class 2 structure

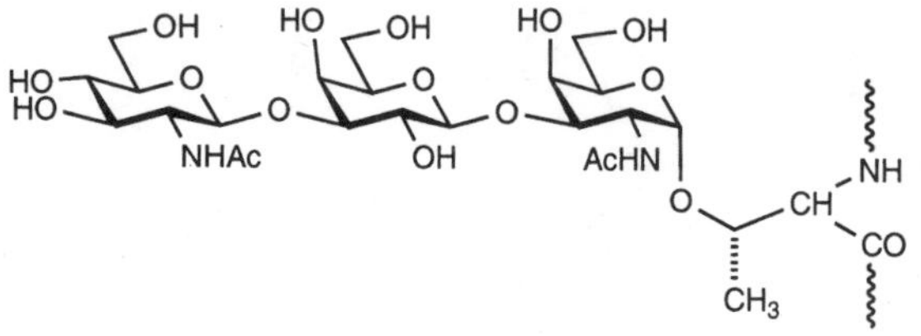

Core class 3 structure

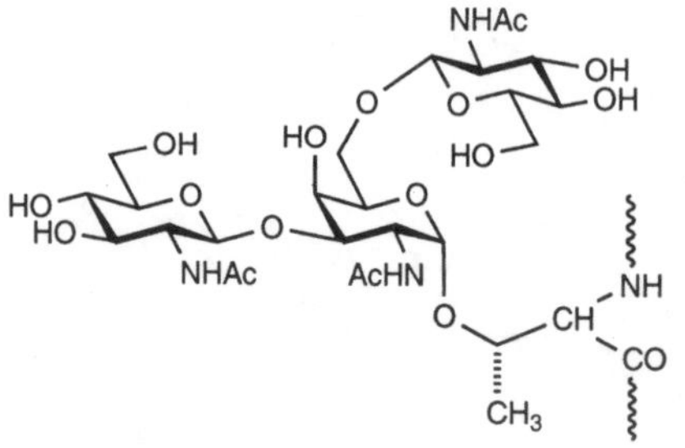

Core class 4 structure

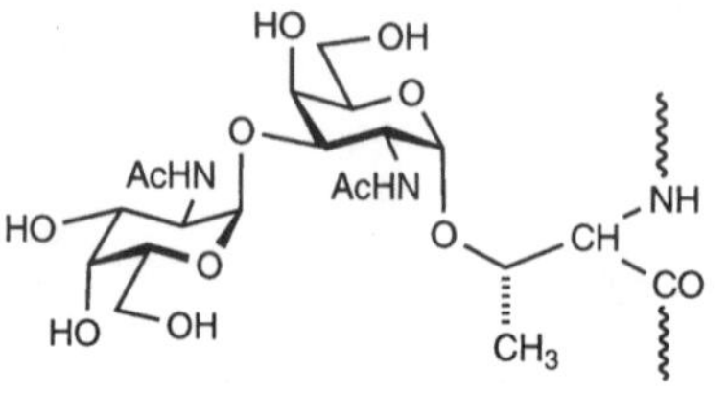

Core class 5 structure

Figure 24

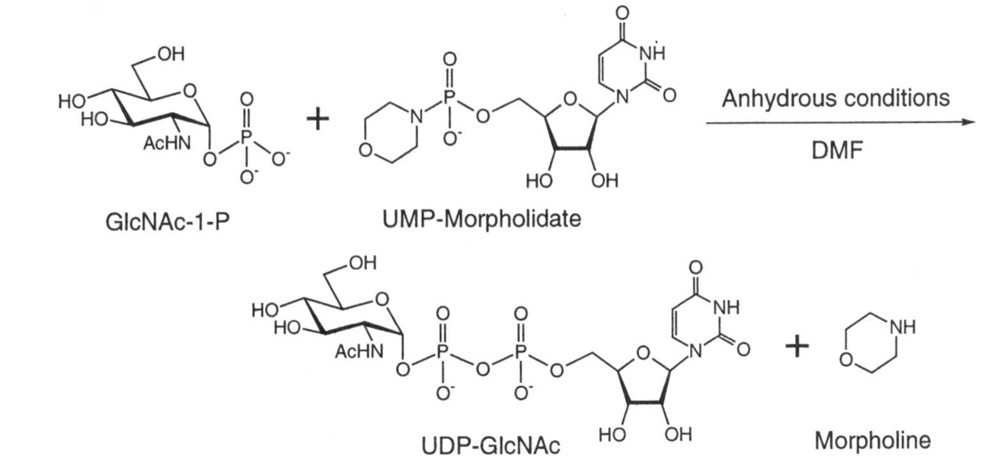

GlcNAc-1-P   UMP-Morpholidate   Anhydrous conditions

DMF

UDP-GlcNAc   Morpholine

Scheme 18

β(1,2)N-Acetylglucosaminyltransferase (EC 2.4.1.101): GnT I

UDP-*N*-Acetylglucosamine: α-D-mannosyl-(1→3)-β-D-mannoside β(1→2)*N*-acetyl-D-glucosaminyltransferase is responsible for the synthesis of the GlcNAcβ(1→2)-Manα(1→3)Manβ(1→4) linkage. The enzyme has been purified from rat liver, and its gene has been cloned [275–279]. The substrate specificity of the enzyme has been also studied [280,281].

Hindsgaul and co-workers have studied the donor substrate specificity of the enzyme, using *N*-acetylglucosaminyltransferase (GnT I) purified from human milk and synthetic UDP-*N*-acetylglucosamine analogs in which the OH groups at the C-3, -4, and -6 positions of the GlcNAc residue of UDP-GlcNAc had been independently deoxygenated [281] (Figure 25). The kinetic parameters of these deoxygenated UDP-GlcNAc derivatives with GnT I were not provided; however, both the 3″- and

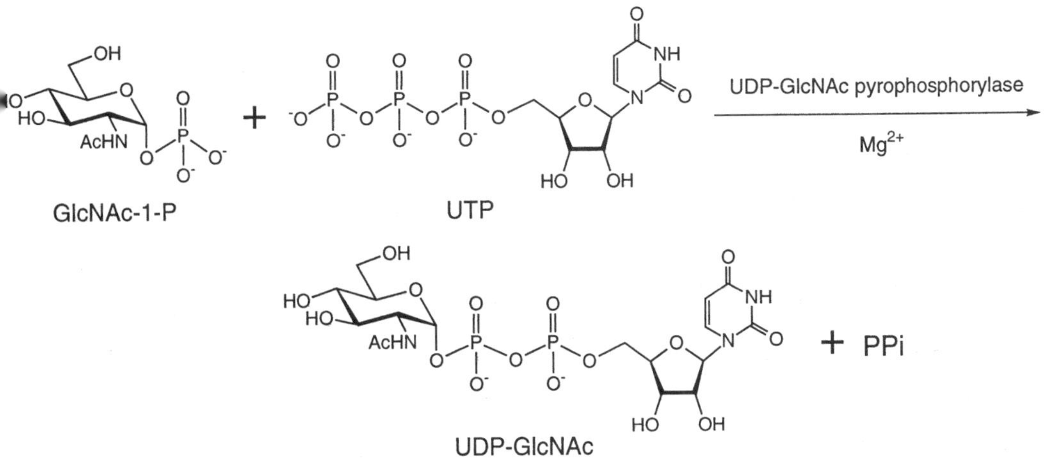

GlcNAc-1-P   UTP   UDP-GlcNAc pyrophosphorylase

Mg^{2+}

UDP-GlcNAc   + PPi

Scheme 19

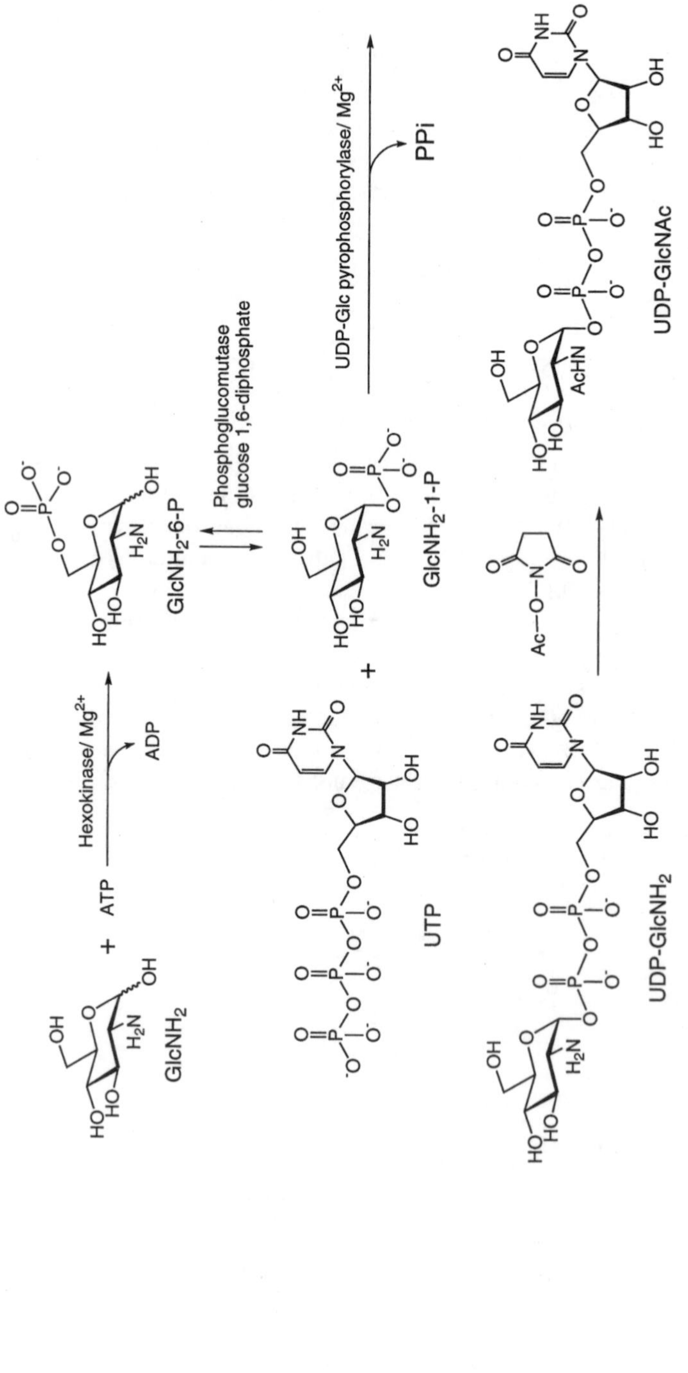
GlcNH$_2$
+ ATP
Hexokinase/ Mg^{2+}
ADP
GlcNH$_2$-6-P
Phosphoglucomutase
glucose 1,6-diphosphate
UTP
+
GlcNH$_2$-1-P
UDP-Glc pyrophosphorylase/ Mg^{2+}
PPi
UDP-GlcNH$_2$
UDP-GlcNAc

Scheme 20

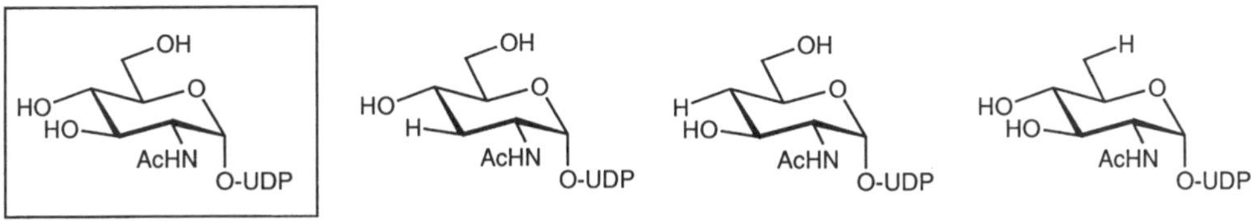

Figure 25

6″-deoxy derivatives of UDP-GlcNAc were reasonably reactive as donor substrates for the enzyme, but the 4″-deoxy derivative was much less reactive.

β(1,2)N-Acetylglucosaminyltransferase (EC 2.4.1.143): GnT II

UDP-*N*-Acetylglucosamine: α-D-mannosyl-(1→6)-β-D-mannoside β(1→2)*N*-acetyl-D-glucosaminyltransferase is responsible for the synthesis of the GlcNAcβ(1→2)-Manα(1→6)Manβ(1→4) linkage. The enzyme has been purified from rat liver [282,283].

The Hindsgaul group has studied the acceptor substrate specificity of partially purified GnT II from rat liver, using deoxygenated oligosaccharide acceptor analogs prepared enzymatically from chemically synthesized UDP-deoxygenated-GlcNAc and an acceptor, Manα(1→3)[Manα(1→6)]Manβ-*O*-(CH₂)₈CO₂Me [283] (Figure 26). The results indicated that GnT II recognizes a fairly large part of the acceptor structure and that deoxygenation of the 4- and 6-OH groups of the GlcNAc residue of the acceptor causes a modest decrease in acceptor activity, whereas removal of the 3-OH totally abolishes its activity.

β(1,4)N-Acetylglucosaminyltransferase (EC 2.4.1.144): GnT III

UDP-*N*-Acetylglucosamine: β-(1→4)-D-mannoside β(1→4)*N*-acetyl-D-glucosaminyltransferase is responsible for the synthesis of the GlcNAcβ(1→4)Manβ(1→4) linkage. This enzyme has been purified from rat liver, and its gene has been cloned [284,285]; however, its substrate specificity has not yet been reported [286].

β(1,6)N-Acetylglucosaminyltransferase (EC 2.4.1.155): GnT V

UDP-*N*-Acetylglucosamine: α-D-mannosyl-(1→6)-β-D-mannoside β(1→6)*N*-acetyl-D-glucosaminyltransferase is responsible for the synthesis of the GlcNAcβ(1→6)-Manα(1→6)Manβ(1→4) linkage. Current interest in this enzyme is based on reports that the β(1→6)*N*-acetylglucosaminyl-branching in *N*-linked oligosaccharide structures increase when baby hamster kidney (BHK) cells are transformed by polyoma or Rous sarcoma virus and that the GnT V activity may correlate directly with the metastatic potential of some transformed cell lines [277,287]. The enzyme has been purified from natural sources [288,289], and its gene has been cloned [290]. A sensitive ELISA assay method for GnT V activity has been developed by the Hindsgaul group [291].

This group has also evaluated the critical role of the 4-OH group of the GlcNAc residue β(1→2)-linked to the Man residue [292] (Figure 27). They prepared several structural analogs in which the 4-OH group of the GlcNAc residue of the acceptor

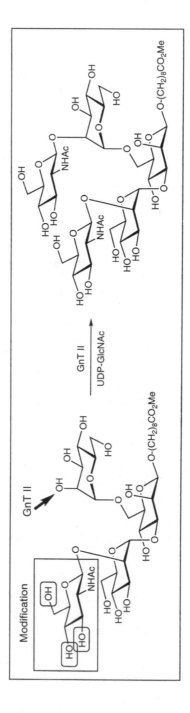

Figure 26

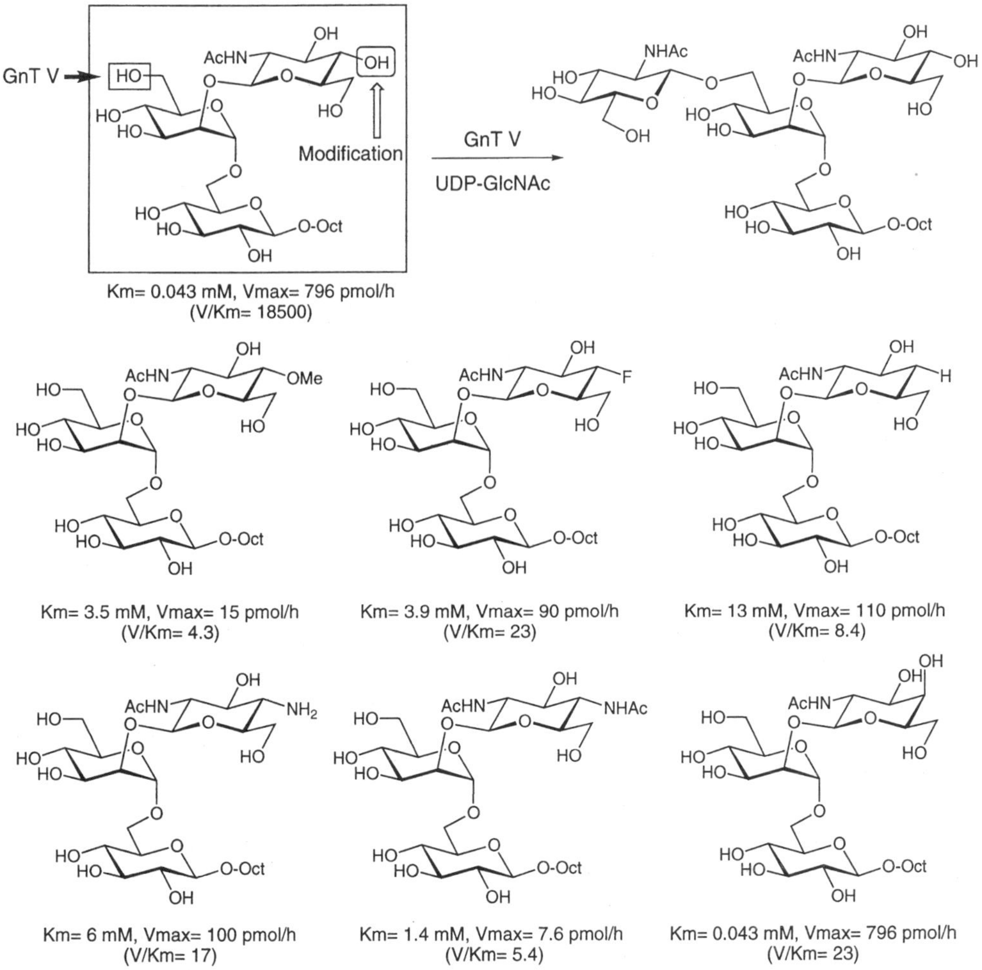

Figure 27

was replaced independently with an *O*-methyl, fluoro, deoxy, amino, or acetamido group or with their respective epimers. Modification of the 4-OH group of the GlcNAc residue abolished its acceptor activity, with V_{max}/K_m values in the range of 4.3–23 (compared to a V_{max}/K_m of 18,500 for the unmodified substrate). The activity of GnT V was found to be highly dependent on the acceptor structure, in which the core $\beta(1\rightarrow4)$-linked mannose was replaced by an octyl β-glucopyranoside residue.

Hindsgaul and his co-workers have also addressed an interesting question [293,294] (Figures 28 and 29): Which rotamer, with respect to the Man$\alpha(1\rightarrow6)$-Man$\beta(1\rightarrow$ bond, of the acceptor substrate does GnT V preferentially glycosylate? They prepared acceptor analogs whose free rotation around the Man$\alpha(1\rightarrow6)$-Man$\beta(1\rightarrow$ bond was somewhat restricted by the additional 6-methyl group (Figure 28). The results indicated that the enzyme (GnT V) recognizes one rotamer and preferentially transfers GlcNAc onto a **gt** conformer [293].

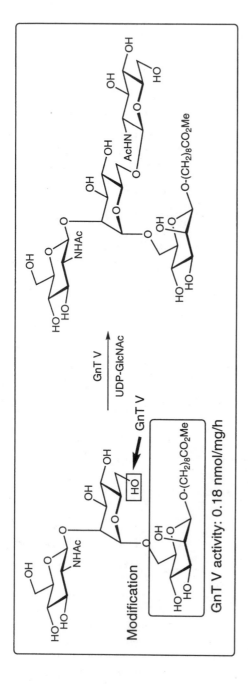

GnT V activity: 0.18 nmol/mg/h

GlcNAcβ(1→2)Manα(1→O
GnT V activity: 0.22 nmol/mg/h

GlcNAcβ(1→2)Manα(1→O
GnT V activity: 0.14 nmol/mg/h

GlcNAcβ(1→2)Manα(1→O
GnT V activity: 0.21 nmol/mg/h

GlcNAcβ(1→2)Manα(1→O
GnT V activity: 0.52 nmol/mg/h

GlcNAcβ(1→2)Manα(1→O
GnT V activity: 0.41 nmol/mg/h

GlcNAcβ(1→2)Manα(1→O
GnT V activity: 0.05 nmol/mg/h

Figure 28

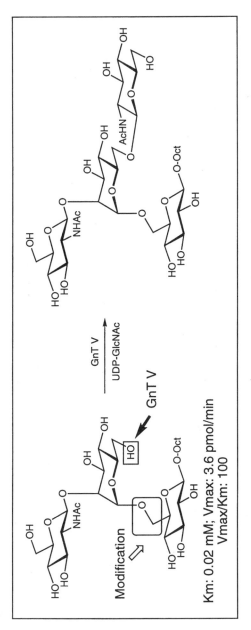

Figure 29

The Hindsgaul group also prepared two conformationally restricted trisaccharide analogs of GlcNAcβ(1→2)Manα(1→6)Glcβ→O-Oct as acceptor substrates for GnT V; in these structures, the rotamer is fixed, with a pyran ring fused to the 4-OH and C-6 carbon of the glucose residue [294] (Figure 29). Their experiments produced a striking result: The enzyme (GnT V) preferentially accepted the **gt**-rotamer analog of the trisaccharide substrate, GlcNAcβ(1→2)Manα(1→6)Glcβ→O-Oct, with a V_{max}/K_m value of 227 as compared to 0.05 for the **gg**-rotamer. They also found that the enzyme did not require any of the hydroxyl groups of the Glc residue in a separate experiment [295].

β(1,6)N-Acetylglucosaminyltransferase: O-Linked oligosaccharide "core-2"-GlcNAc transferase (EC 2.4.1.102)

A branching enzyme of mucin-type (O-linked) oligosaccharide biosynthesis is UDP-N-acetyl-D-glucosamine: O-glycosyl-glycoprotein (N-acetyl-D-glucosamine to N-acetyl-D-galactosamine of β-D-galactosyl-1,3-N-acetyl-D-galactosaminyl-R) β-1,6-N-acetyl-D-glucosaminyltransferase [20]. The enzyme has been highly purified from bovine tracheal and swine tracheal epithelium [296–300], and its gene has also been cloned [301–303]. The enzyme has been utilized for the preparative synthesis of mucin-type trisaccharide, GlcNAcβ(1→6)[Galβ(1→3)]GalNAcα→OR, with regeneration of UDP-GlcNAc [273].

The Hindsgaul group has demonstrated an enzymatic synthesis of a sialyl Lewis x-containing mucin-type oligosaccharide [114]. They purified the enzyme from a commercially available (Sigma) mouse kidney acetone powder (1 g) by a single step of affinity chromatography on UDP-hexanolamine Sepharose, and obtained 383 mU of core-2 GlcNAc transferase [114] (Scheme 21). GlcNAc was introduced, together with this partially purified enzyme and UDP-GlcNAc, into the core 2 disaccharide, Galβ(1→3)GalNAcα(1→OR, to give the trisaccharide GlcNAcβ(1→6)[Galβ(1→3)]-GalNAcα(1→OR. This trisaccharide was then sequentially glycosylated by galactosylation (with galactosyltransferase and UDP-Gal), sialylation (with sialyltransferase and CMP-NeuAc), and fucosylation (with fucosyltransferase and GDP-Fuc) to give a sialyl Lewis x-carrying mucin-type oligosaccharide.

β(1,6)N-Acetylglucosaminyltransferase: Involved in the biosynthesis of blood group I and i antigens

The carbohydrate structures known as blood group antigens I and i are often expressed on the erythrocyte cell surface [304,305] (Figure 30). The i-active form is a linear structure that is extended from the terminal galactose residue of the polylactosaminoglycan in a reaction catalyzed by β(1,3)N-acetylglucosaminyltransferase. The I-active form is a branched structure that is formed from the internal N-acetylglucosamine residue of the polylactosaminoglycan in a reaction catalyzed by β(1,6)N-acetylglucosaminyltransferase [306].

During the transition from fetal to adult erythrocytes, the linear i-active polylactosaminoglycans are converted to the branched I-active glycans: This conversion is mediated by the blood group I-specific β(1,6)N-acetylglucosaminyltransferase. β(1,6)N-Acetylglucosaminyltransferase has been purified from several sources but is not yet well characterized [306].

Two biosynthetic pathways for the expression of the I-active carbohydrate structures have been proposed in response to the discovery of this β(1,6)N-acetylglucosaminyltransferase [306] (Scheme 22). One pathway starts with a β(1,6)

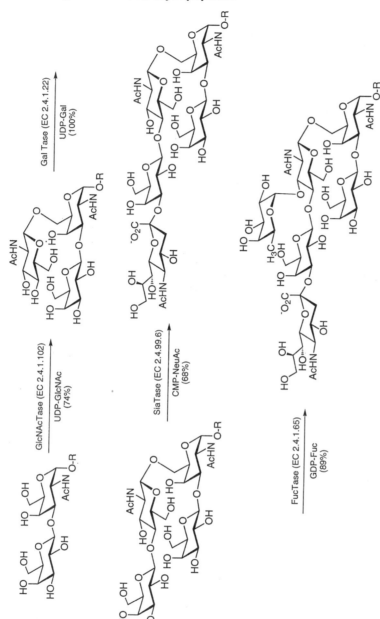

Scheme 21

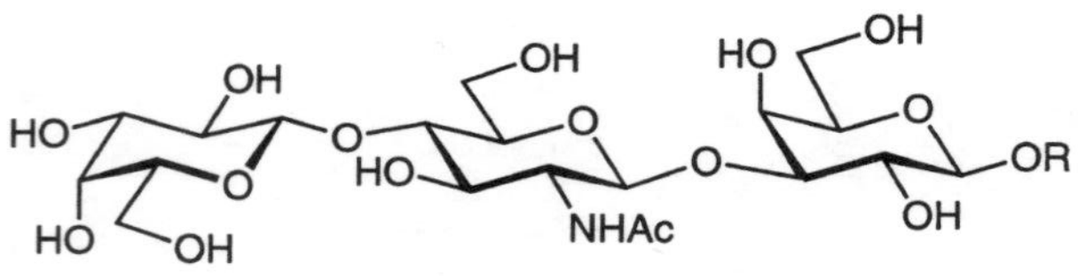

Blood group I antigen

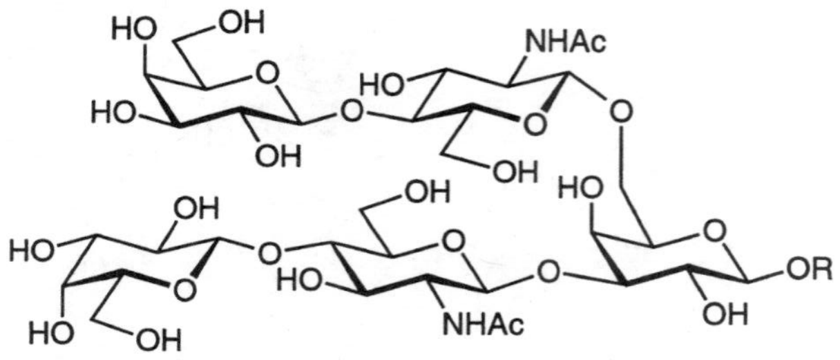

Blood group i antigen

Figure 30

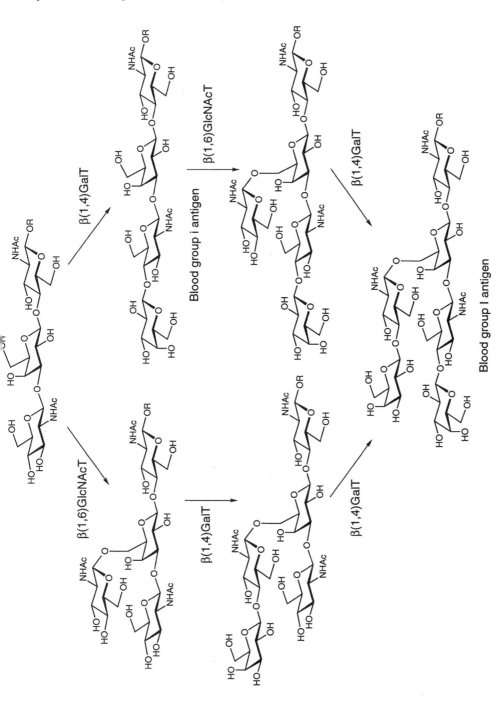

Scheme 22

N-acetylglucosaminyltransferase-mediated reaction of a GlcNAcβ(1→3)Gal exposed on a polylactosaminoglycan chain; this reaction generated a GlcNAcβ(1→6)-[GlcNAcβ(1→3)]Gal structure. Two consecutive β(1,4)galactosyltransferase reactions furnish the I antigen structure. In the other proposed pathway, blood group i antigen is first formed by a β(1,4)galactosyltransferase reaction. The i antigen is further glycosylated by the β(1,6)N-acetylglucosaminyltransferase [306] to give a branched structure, and subsequently galactosylated with β(1,4)galactosyltransferase to give the I antigen structure. A β(1,6)N-acetylglucosaminyltransferase that utilizes the i antigen structure has been detected in rat intestine [306].

Another interesting N-acetylglucosaminyltransferase reaction is the formation of GlcNAcα-O-Ser linkages. This sequence is often found in nuclear and endo-plasmic proteins. The gene encoding this enzyme has been cloned [307].

A typical enzymatic N-acetylglucosaminylation reaction with GnT I and II: Synthesis of GlcNAcβ(1→2)Manα(1→6)[GlcNAcβ(1→2)Manα(1→3)]Manβ1-O-(CH₂)₈CO₂Me [282]

A solution of Manα(1→6)[Manα(1→3)]Manβ1-O-(CH$_2$)$_8$CO$_2$Me (80 mg, 0.12 mmol), UDP-N-acetylglucosamine (710 mg, 1.0 mmol), and partially purified GnT I and II from rat liver acetone powder in Na cacodylate buffer (35 mL; 50 mM, pH 6.5) was gently stirred for 3 days. More UDP-GlcNAc (320 mg, 0.46 mmol) was added, and the mixture was stirred for another 3 days. TLC showed that the starting material (R_f = 0.4 in CH$_2$Cl$_2$/MeOH/H$_2$O 60:35:6) was converted to the product (R_f = 0.14). The mixture was applied onto a C$_{18}$ column (3.5 × 25 cm) and eluted with MeOH/H$_2$O (35:65, v/v) at 1 mL/min. The appropriate fractions were pooled, concentrated, and lyophilized from water to give the product (102 mg, 80%).

E.   Mannosyltransferase

Mannose is a major component of glycoprotein N-linked oligosaccharides and the microbial cell wall [15]. All the N-linked oligosaccharides are biologically synthesized from a high mannose-type structure, with three α-linked glucose residues [Glcα(1→2)Glcα(1→3)Glcα(1→3)] as a precursor [19,21] (Figure 31). Terminal 6-O-phosphorylated mannose residues of high mannose-type N-linked oligosaccharides are specific ligands for the mannose 6-phosphate receptor, which plays a key role in the intracellular trafficking of newly synthesized lysosomal enzymes from the trans Golgi to lysosome [308] (Figure 31).

Mannosyltransferase transfers D-mannose from guanidine 5'-diphosphomannose (GDP-Man) to an appropriate acceptor (Scheme 23).

The donor substrate, GDP-mannose, is commercially available from Sigma (currently $107 for 50 mg). GDP-Mannose is enzymatically synthesized from mannose 1-phosphate and guanosine 5'-triphosphate (GTP) in the presence of GDP-mannose pyrophosphorylase (EC 2.7.7.22), which has been purified from yeast [28,309] (Scheme 24).

α(1,2)Mannosyltransferase (EC 2.4.1.131)

GDP-mannose: α-D-mannoside α(1,2)-mannosyltransferase is responsible for the synthesis of Manα(1→2)Manα(1→ linkage (Figure 32); and its gene has been cloned

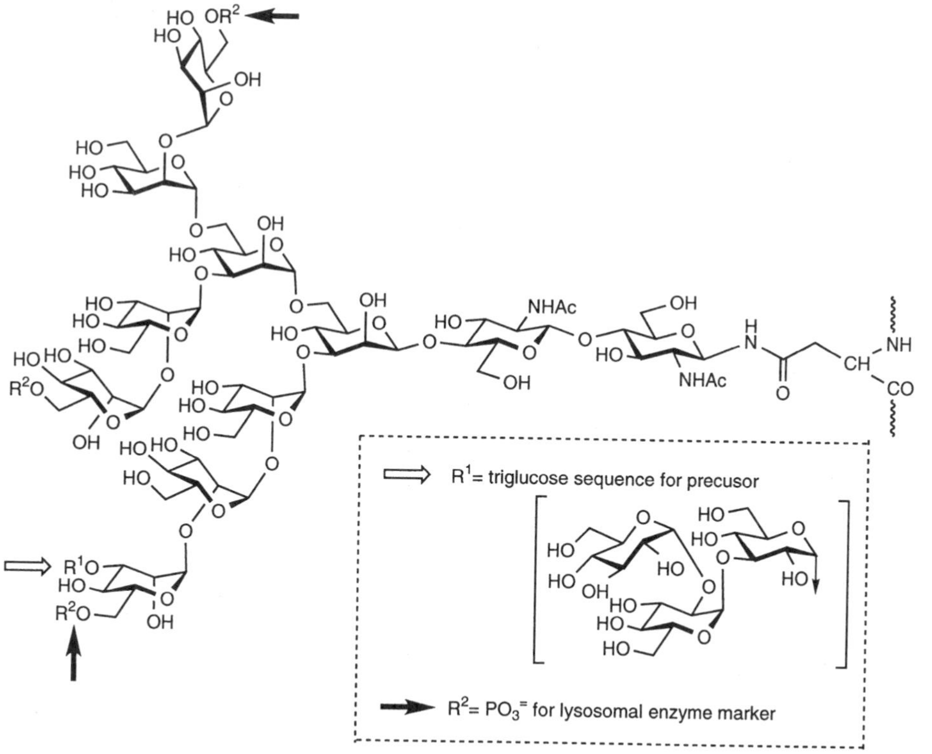

Figure 31

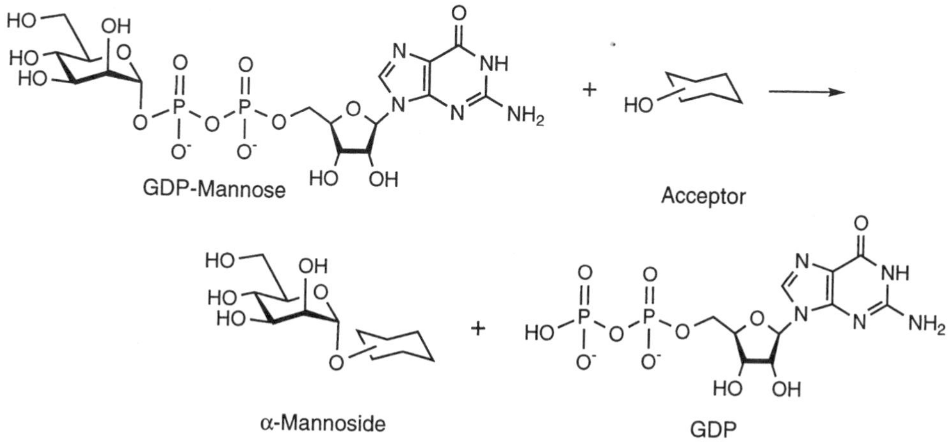

Scheme 23

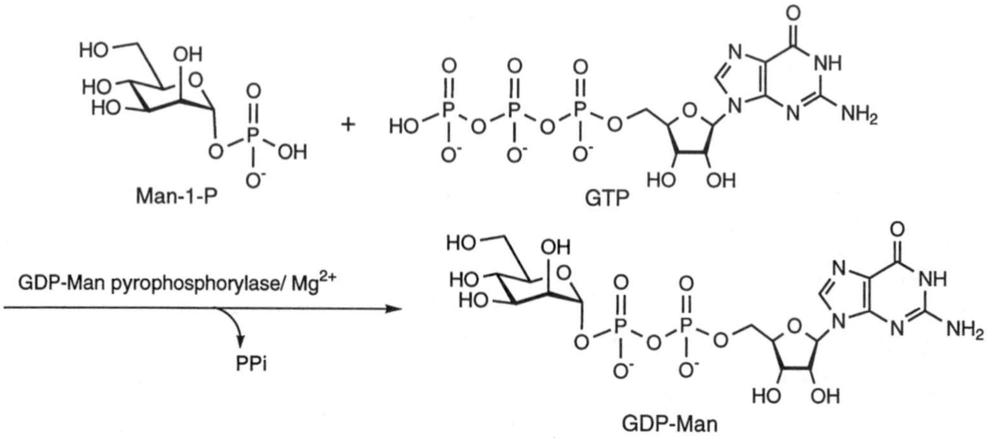

Scheme 24

[310–312]. The enzyme reaction is carried out in Tris buffer (100 mM, pH 7.5), and requires Mn^{2+} for the enzyme activity.

Studies of acceptor substrate specificity have been carried out using a recombinant $\alpha(1,2)$mannosyltransferase [311] (Table 16). In terms of relative velocity (relative rates), α-methyl mannoside was the best acceptor substrate among the compounds tested. However, in terms of the efficiency of the acceptor substrate (V/K_m), peptide-containing mannosides were normally better acceptor substrates than mannose derivatives terminating with a simple alkyl glycoside [311].

F. N-Acetylgalactosaminyltransferase

N-Acetylgalactosamine is a unique carbohydrate residue because it is often found in glycolipid structures as a terminal carbohydrate. It is also a common linkage carbohydrate in O-linked glycans linked to the serine or threonine residues of peptides [15,18].

UDP-N-Acetylgalactosamine has been prepared by various enzymatic and chemo-enzymatic methods [28,221]. Galactosamine is phosphorylated by an enzyme obtained from *Klyveromyces fragilis* in the presence of ATP to give galactosamine

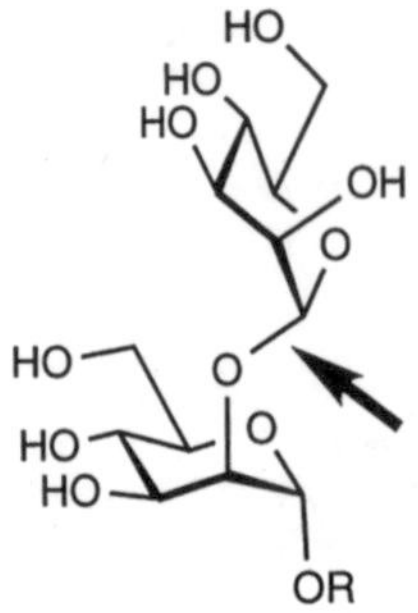

Figure 32

α-1-phosphate (GalN-1-P). GalN-1-P undergoes an exchange reaction with UDP-Glc, catalyzed by galactose 1-phosphate uridyltransferase (EC 2.7.7.12), to give UDP-galactosamine (UDP-GalN). The amino group of UDP-GalN is then acetylated with N-acetoxysuccinimide to give UDP-N-acetylgalactosamine (UDP-GalNAc) (Scheme 25).

Table 16   Acceptor Substrate Specificity of a Recombinant α(1,2)Mannosyltransferase [311]

Acceptor (left)	K_m (mM)	Rel. rate (%)	V/K_m	Acceptor (right)	K_m (mM)	Rel. rate (%)	V/K_m
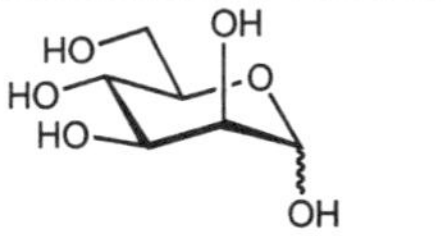 mannose (free anomeric OH)	193	46	0.24	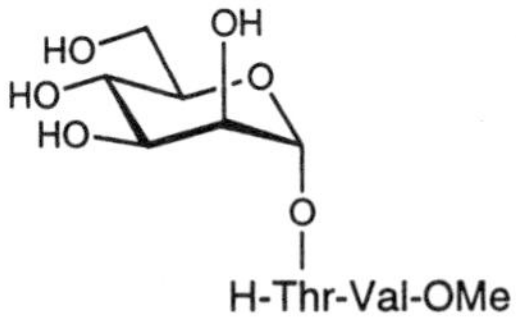 mannose-O-linked, H-Thr-Val-OMe		68	
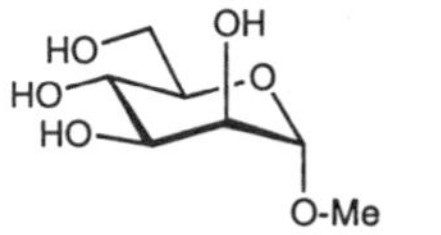 mannose O-Me	57	100	1.75	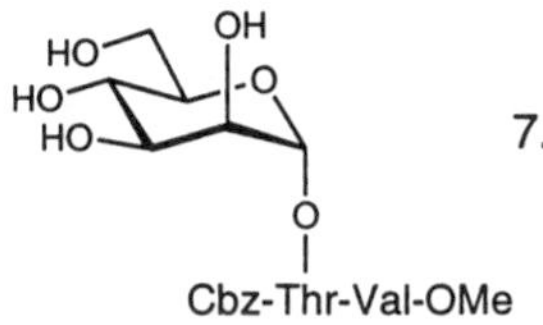 mannose-O-linked, Cbz-Thr-Val-OMe	7.8	71	9.1
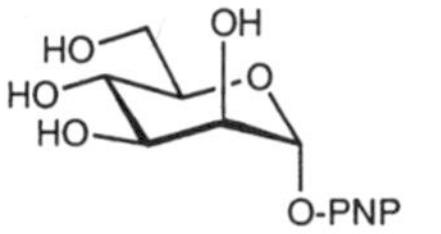 mannose O-PNP	0			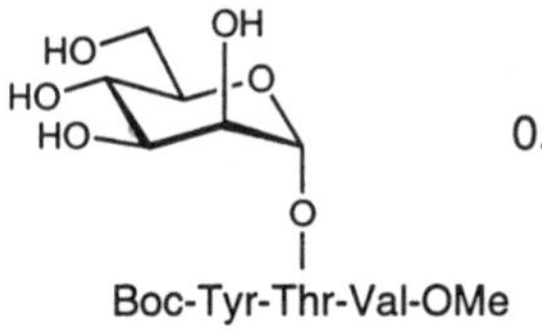 mannose-O-linked, Boc-Tyr-Thr-Val-OMe	0.7	35	50
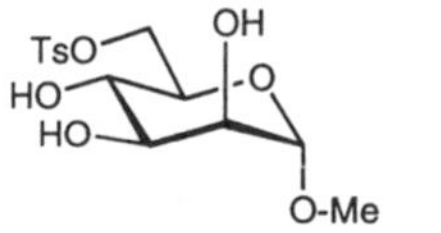 6-O-Ts mannose O-Me	0			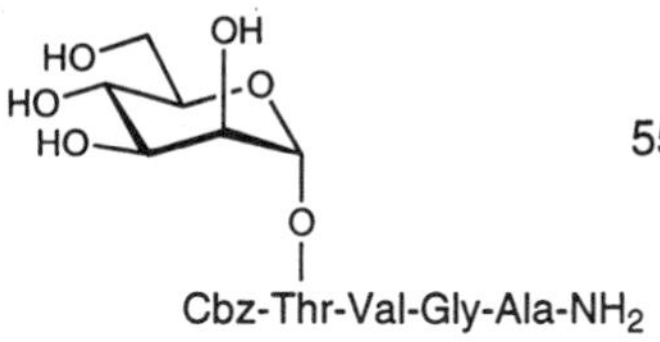 mannose-O-linked, Cbz-Thr-Val-Gly-Ala-NH₂		55	
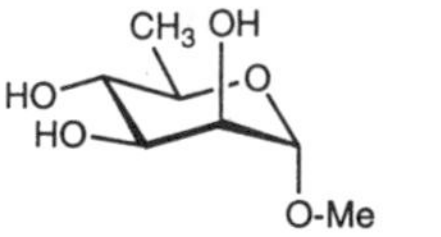 6-deoxy (CH₃) mannose O-Me		6		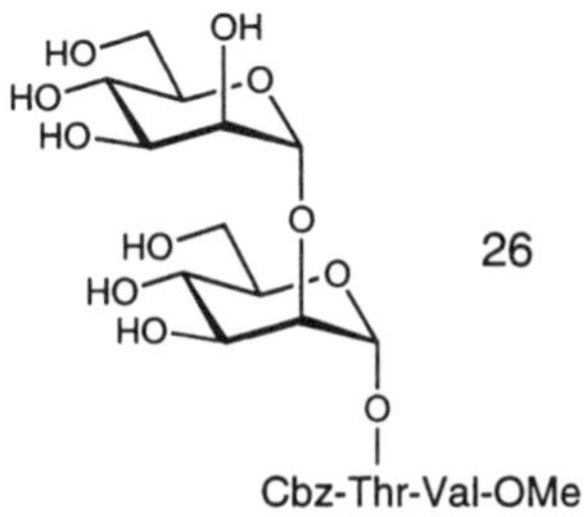 disaccharide-O-linked, Cbz-Thr-Val-OMe	26	17	0.65
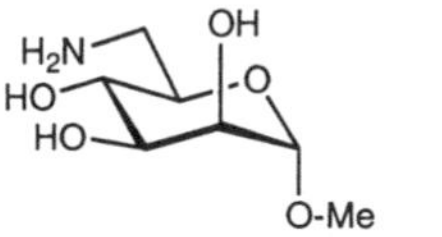 6-amino (H₂N) mannose O-Me		7					
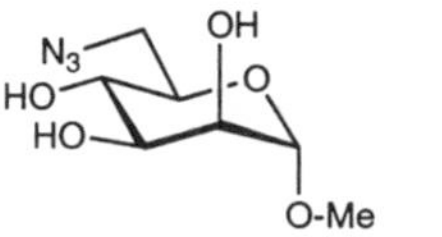 6-azido (N₃) mannose O-Me		2					
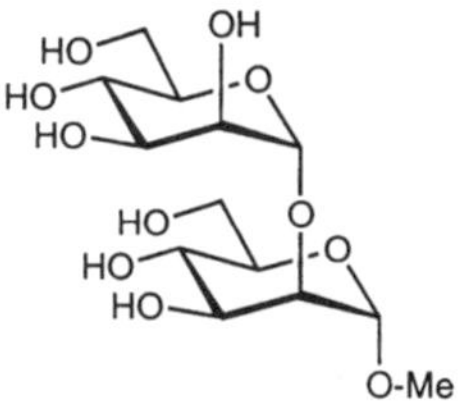 disaccharide O-Me	28	62	2.21				

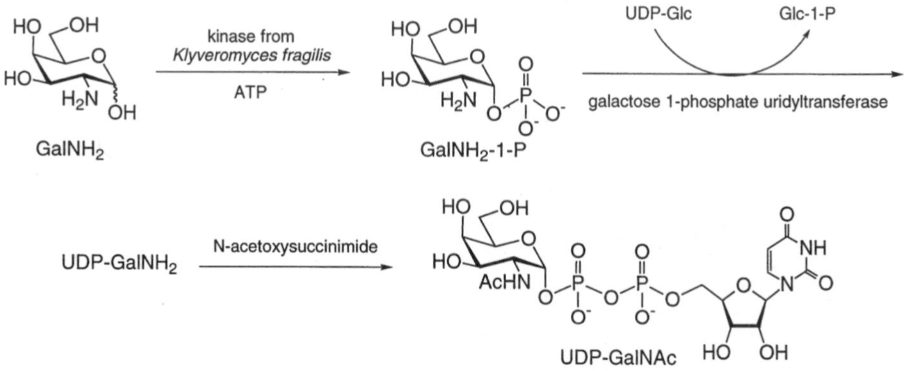

Scheme 25

α(1,3)N-Acetylgalactosaminyltransferase (EC 2.4.1.40): Blood group A transferase

Blood group A substance is synthesized in vivo by the addition of N-acetylgalactos-amine onto the galactose residue of the H substance (Fucα(1→2)Galβ(1→) via an α(1→3)-linkage) (Scheme 26).

The blood group A transferase has been purified from human tissues [313–315], and its gene has been cloned [264]. Hindsgaul and co-workers have studied the acceptor substrate specificity of the enzyme using synthetic structural analogs of the disaccharide Fucα(1→2)Galβ-O-Octyl, which is a minimum acceptor sequence for the glycosyltransferase: α(1,3)N-acetylgalactosaminyltransferase (EC 2.4.1.165) (A-transferase) [265,266] (Figure 33). They examined a series of compounds in which the OH groups at the C-3, -4, and -6 positions of the galactose residue were replaced with O-methyl ether, epimer, or amino groups. The results indicated that modifications at the C-6 position were tolerated, although the relative rates for these substrates were in the range of 4.7%–13.4% of that for the control disaccharide Fucα(1→2)Galβ-O-Octyl. Other modifications at C-3 and -4 totally abolished the activity (relative rates < 0.8%).

β(1,4)N-Acetylgalactosaminyltransferase (EC 2.4.1.165)

A ganglioside, G_{M2} (GalNAcβ(1→4)[NeuAcα(2→3)]Galβ(1→4)Glcβ(1→Cer), is observed to accumulate in gastric carcinoma [316,317]. In contrast, the glycolipids NGM (normal gastric mucosa)-1, -2, and -3 are detected in normal gastric mucosa, but they are completely lost in the carcinoma tissue. NGM-1 shares a terminal epi-tope with the G_{M2} ganglioside and with the Cad blood group antigen. The difference between G_{M2} and NGM-1 is that NGM-1 has a Type 2 disaccharide sequence in its structural backbone, whereas G_{M2} does not (Figure 34). The decrease in NGM content and the increase in G_{M2} levels, and therefore in the levels of β(1,4)N-acetylgalac-tosaminyltransferase, have been considered to be cancer-associated changes in gastric mucosa [314,315].

G_{M2} is biosynthesized from the sialylated precursor G_{M3} by the addition of N-acetylgalactosamine residue via a β(1→4)-linkage to the galactose residue of G_{M3} (Scheme 27). The enzyme responsible for this transformation is β(1,4)N-acetylga-

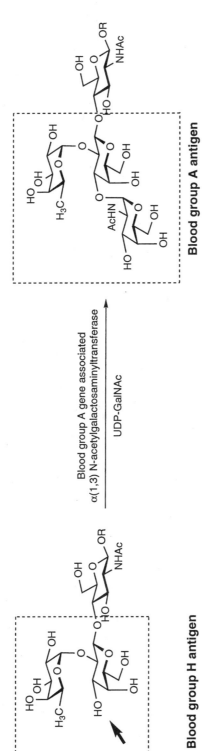

Scheme 26

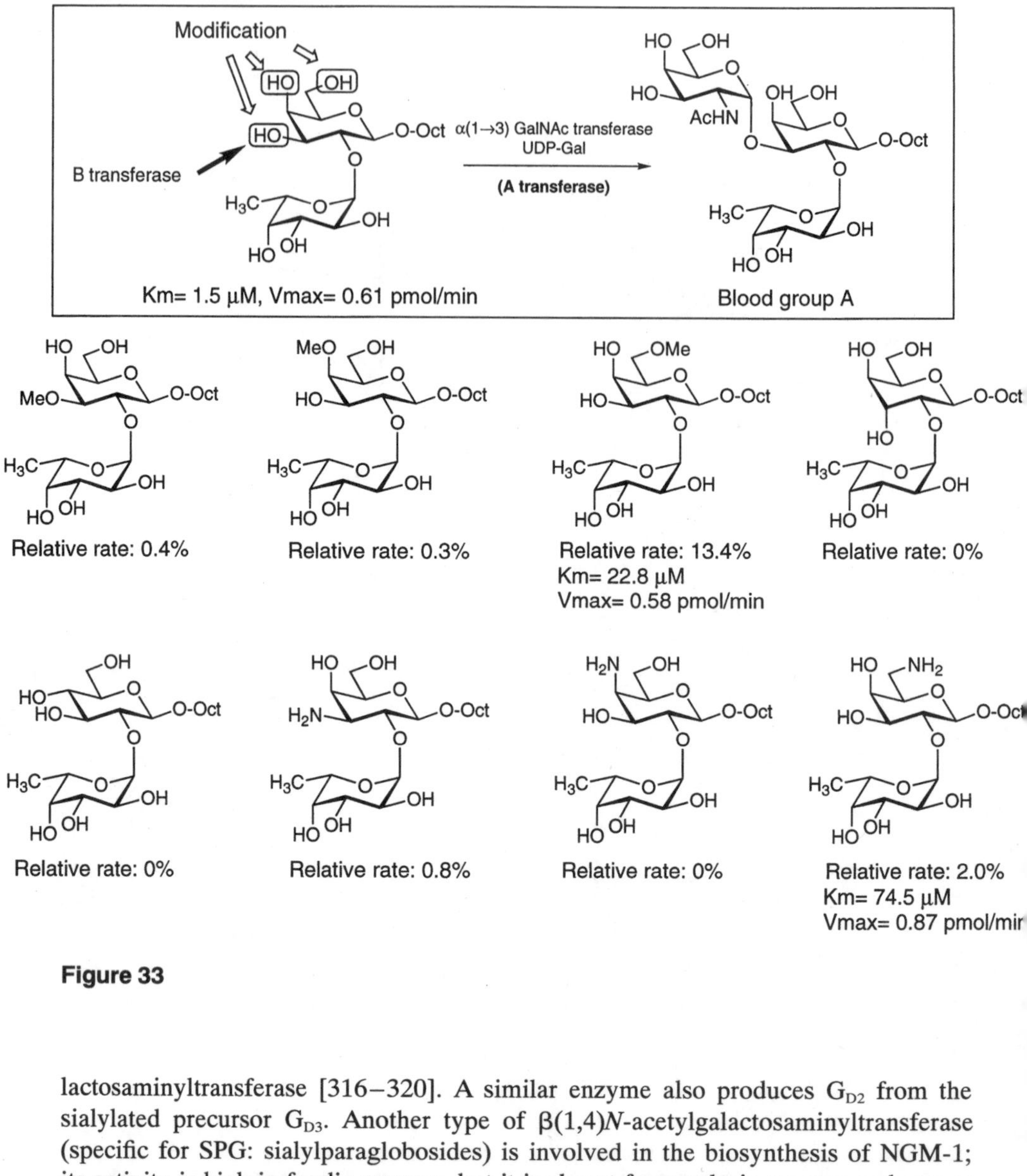

Figure 33

lactosaminyltransferase [316–320]. A similar enzyme also produces G_{D2} from the sialylated precursor G_{D3}. Another type of $\beta(1,4)N$-acetylgalactosaminyltransferase (specific for SPG: sialylparaglobosides) is involved in the biosynthesis of NGM-1; its activity is high in fundic mucosa, but it is absent from pyloric mucosa and cancer cells [317,318]. In contrast, an increase in the activity of $\beta(1,4)N$-acetylgalactosaminyltransferase for G_{M3} has been observed in cancer tissues and cancer cell lines [317,318].

The enzyme has been purified from mouse liver and studied with respect to its kinetics and substrate specificity [317] (Table 17). It exhibited a pH optimum of 7.5–7.9 and required 2.5–10 mM Mn^{2+} for the maximal activity. The K_m for UDP-GalNAc was 7 µM. Among the glycolipids tested as acceptor substrates, G_{M3} (NeuGc) and G_{D3} (NeuGc) were better acceptor substrates, with the V_{max}/K_m values of 0.024 and 0.027, respectively, than G_{M3} (NeuAc, $V_{max}/K_m = 0.006$) and G_{D3} (NeuAc, $V_{max}/K_m = 0.012$). The enzyme did not accept SPG (sialylparaglobosides)

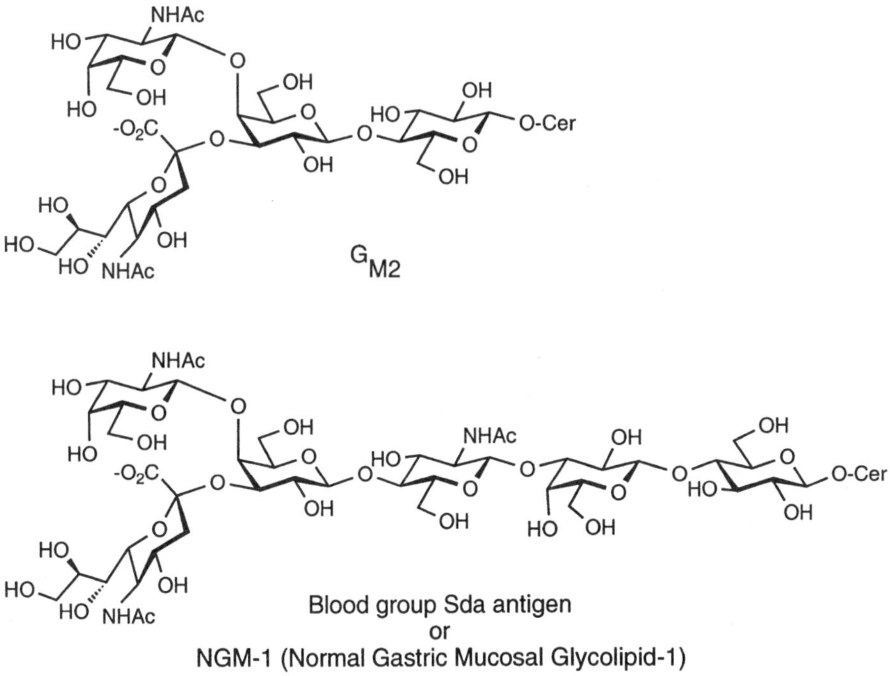

G_{M2}

Blood group Sda antigen
or
NGM-1 (Normal Gastric Mucosal Glycolipid-1)

Figure 34

as an acceptor substrate (relative rate < 2%). The $\beta(1,4)N$-acetylgalactosaminyltrans-
ferase gene has been cloned from human melanoma cell lines [318].

α-N-Acetylgalactosaminyltransferase (EC 2.4.1.41): Polypeptide: N-acetylgalactosaminyltransferase

The O-linked oligosaccharide core GalNAc-α-Ser/Thr linkage is formed by α-N-acetylgalactosaminyltransferase and UDP-GalNAc. The acceptor substrate specificity of this enzyme has been extensively studied because the enzyme catalyzes the first step of O-linked oligosaccharide biosynthesis [20]. It has been purified from several sources, including porcine, bovine, and ovine submaxillary glands, and the various forms have been found to show different acceptor substrate specificities [321–325].

The Hill group has studied the substrate specificity of αN-acetylgalactosami-nyltransferase from porcine submaxillary gland. By examining a series of synthetic peptides that are partial peptide sequence analogs of erythropoietin peptide, they have found that the enzyme activity (formation of GalNAc-α-O-Ser) was markedly influenced by the peptide sequence (PPDAASAAPLR) adjacent to the threonine or serine residues [324] (Table 18). Thus, PPDASSSAPLR and PPDVVSVVPLR were about 5- (V_{max}/K_m 0.08) and 30-fold (V_{max}/K_m 0.017) less active than the erythropoietin peptide, PPDAASAAPLR (V_{max}/K_m 0.51), respectively. Replacement of the Ala residue adjacent to Ser with a Gly residue to give PPDGGSGGPLR abolished the acceptor activity. A shorter peptide, DAASAAPL, was less active by 5-fold (V_{max}/K_m 0.11), and the pentapeptide AASAA was inactive.

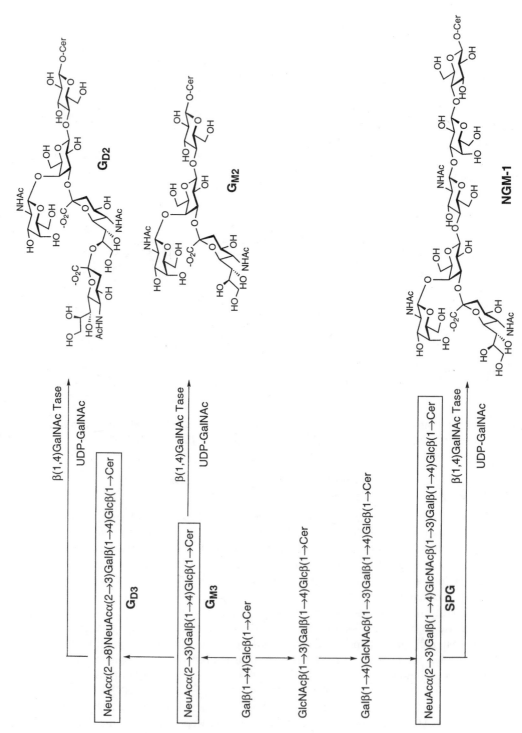

G_{D2}

G_{M2}

NGM-1

β(1,4)GalNAc Tase

UDP-GalNAc

NeuAcα(2→8)NeuAcα(2→3)Galβ(1→4)Glcβ(1→Cer

G_{D3}

β(1,4)GalNAc Tase

UDP-GalNAc

NeuAcα(2→3)Galβ(1→4)Glcβ(1→Cer

G_{M3}

Galβ(1→4)Glcβ(1→Cer

GlcNAcβ(1→3)Galβ(1→4)Glcβ(1→Cer

Galβ(1→4)GlcNAcβ(1→3)Galβ(1→4)Glcβ(1→Cer

NeuAcα(2→3)Galβ(1→4)GlcNAcβ(1→3)Galβ(1→4)Glcβ(1→Cer

SPG

β(1,4)GalNAc Tase

UDP-GalNAc

Scheme 27

Table 17 Acceptor Substrate Specificity of β(1,4)N-Acetylgalactosaminyltransferase from Mouse Liver [317]

Acceptor	Relative activity (%)	K_m (μM)	V_{max} (units/mg)
NeuGcα(2→3)Galβ(1→4)Glcβ(1→Cer G_{M3} (NeuGc)	100	160	3.8
NeuAcα(2→3)Galβ(1→4)Glcβ(1→Cer G_{M3} (NeuAc)	73	2100	12.9
Neucα(2→8)NeuGcα(2→3)Galβ(1→4)Glcβ(1→Cer G_{D3} (NeuGc)	66	27	0.7
NeuAcα(2→8)NeuAcα(2→3)Galβ(1→4)Glcβ(1→Cer G_{D3} (NeuAc)	76	350	4.3
Galβ(1→4)Glcβ(1→Cer Lac-Cer	<2		
NeuAcα(2→3)Galβ(1→4)GlcNAcβ(1→3)Galβ(1→4)Glcβ(1→Cer Sialylparagloboside (SPG) (NeuAc)	<2		

G. Glucuronic Acid Transferase

D-Glucuronic acid is a major component of extracellular matrix polysaccharides [such as heparin, heparan, chondroitin 4- and 6-sulfates, and hyaluronic acid (glycosaminoglycans)], microbial capsular polysaccharides, and a number of plant polysaccharides. It is also occurs in many low molecular weight metabolic products and serves as a detoxifying agent in mammals. Some poisonous compounds in metabolic products are excreted in the urine and bile as alkyl or aryl D-glucuronic acids.

β-D-Glucuronosyltransferase (EC 2.4.1.17)

A primary route for the detoxification of steroids and other endogenous lipophiles proceeds via enzyme-catalyzed glucuronidation. Toxic lipophilic compounds (acceptor substrates of the enzyme) are thereby converted to water-soluble glucuronides, which have enhanced rates of excretion from cells. UDP-Glucuronate: β-D-glucuronosyltransferase is responsible for the synthesis of alkyl and aryl β-D-glucuronides (Scheme 28). This enzyme (EC 2.4.1.17) accepts a wide range of substrates, includ-

Table 18 Acceptor Substrate Specificity of Peptide: α-N-Acetylgalactosaminyltransferase from Porcine Submaxillary Gland [324]

Acceptor substrate amino acid sequence	K_m (mM)	V_{max} (μmol/min/mg)	V_{max}/K_m
Pro-Pro-Asp-Ala-Ala-Ser*-Ala-Ala-Pro-Leu-Arg	4.5	2.3	0.51
Asp-Ala-Ala-Ser*-Ala-Ala-Pro-Leu	14.9	1.6	0.11
Pro-Pro-Asp-Ala-Ser-Ser*-Ser-Ala-Pro-Leu-Arg	18.0	1.6	0.08
Pro-Pro-Asp-Val-Val-Ser*-Val-Val-Pro-Leu-Arg	12.9	0.22	0.017
Pro-Pro-Asp-Ala-Ala-Ser*-Ala-Ala-Pro-Leu-Arg		Not a substrate	
Ala-Ala-Ser*-Ala-Ala		Not a substrate	

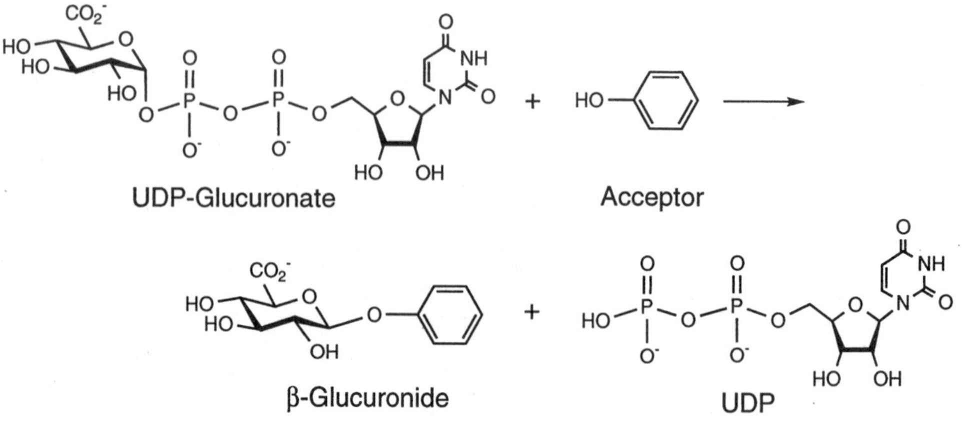

UDP-Glucuronate   Acceptor

β-Glucuronide   +   UDP

Scheme 28

ing phenols, alkyl alcohols, amines, and fatty acids, and utilizes uridine 5′-diphosphoglucuronic acid (UDP-GlcU) as a donor substrate.

The donor substrate UDP-GlcU is enzymatically prepared from UDP-glucose by UDP-glucose dehydrogenase (EC 1.1.1.22) in the presence of 2 equivalents of NAD [28,309] (Scheme 29), and is commercially available (currently $120 per 100 mg). The dehydrogenase is also available from Sigma.

The enzyme UDP-glucuronate: β-D-glucuronosyltransferase from rabbit liver and bovine liver is available from Sigma (currently $43 for 5 units). The gene encoding the human liver enzyme has been cloned, and the enzyme has been studied with respect to its substrate specificity [326] (Figure 35). The order of relative activity was as follows: 4-hydroxy-estrone > estriol > 2-hydroxy-estriol > 4-hydroxy-estradiol > 6α-hydroxy-estriol > 5α-androstane-3α,11β,17β-triol = 5β-androstane-3α,11β,17β-triol.

UDP-Glc   UDP-Glc dehydrogenase

2NAD

UDP-GlcUA

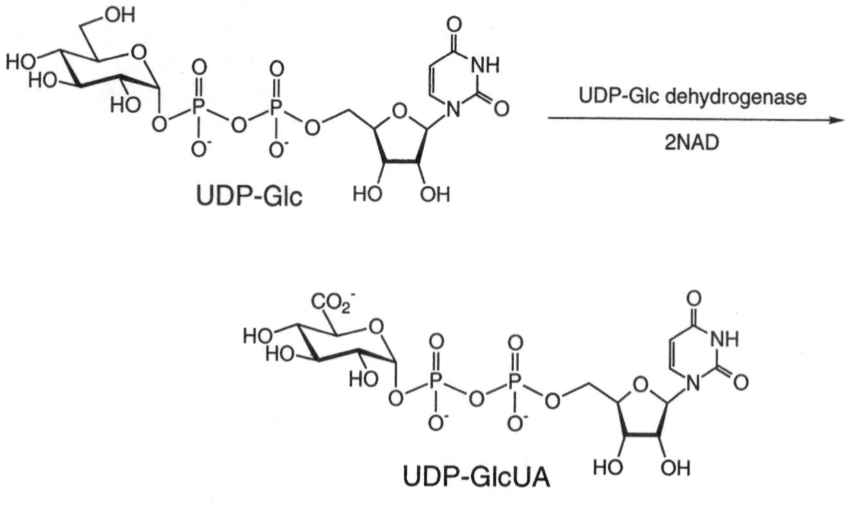

Scheme 29

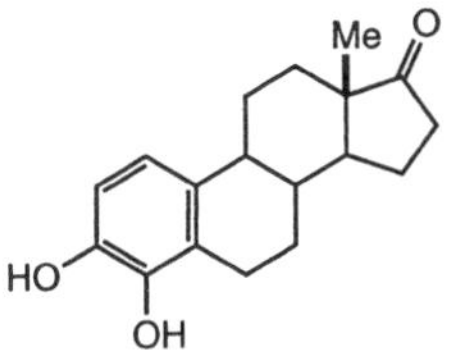

4-Hydroxy-estrone	Estriol	2-Hydroxy-estriol
4-Hydroxy-estradiol	6α-Hydroxy-estriol	5α-Androstane-3α,11β,17β-triol

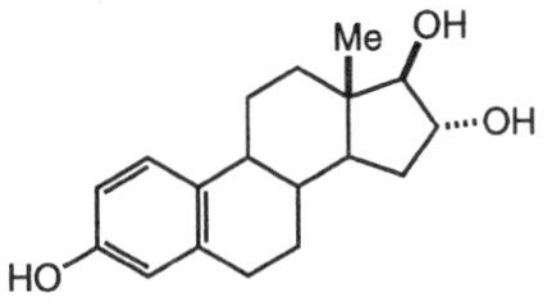

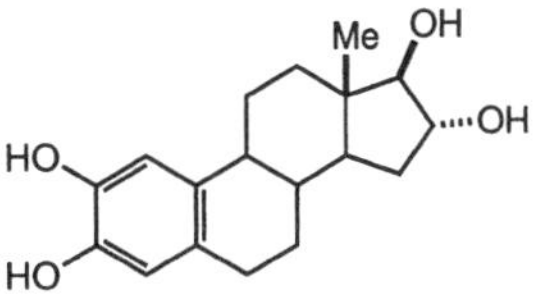

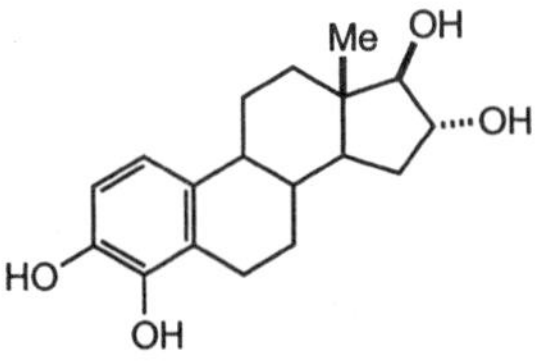

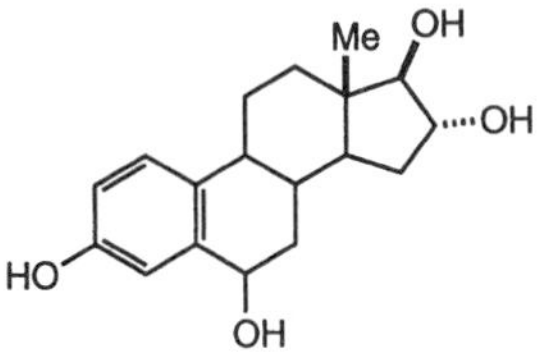

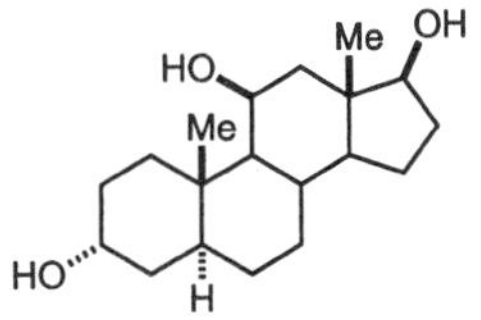

Figure 35

III. METHODS FOR LARGE-SCALE ENZYMATIC SYNTHESIS OF OLIGOSACCHARIDES

As discussed earlier, enzymatic methods for the synthesis of oligosaccharides have been well studied and are very attractive; however, they have a major drawback. The procedures based on the stoichiometric reaction of an acceptor and a donor substrate (sugar nucleotide) require a separate preparation of the sugar nucleotide. Sugar nucleotides are fairly unstable (because of the glycosylphosphate bond) and very expensive. In addition, the stoichiometric enzyme reactions often suffer from product inhibition caused by the released nucleoside di- or monophosphate.

The inhibition problem can be reduced by employing a phosphatase such as calf intestinal phosphatase to decompose the released nucleoside di- or monophosphate to yield a noninhibiting nucleoside [196]. However, stoichiometric amounts of the sugar nucleotides are needed. A practical solution to these problems is to use catalytic amounts of cofactors such as nucleoside phosphates. Then sugar nucleosides are continuously regenerated, by a series of in situ enzymatic reactions, during the glycosyltransferase reactions. Under these circumstances, the nucleoside di- or monophosphate is released only in a catalytic amount and does not inhibit the glycosylation reaction. Furthermore, there is no need for the sugar nucleotide(s) to be purchased or prepared separately [13,327].

A. Galactosylation

A UDP-galactose regeneration reaction employing UDP-galactose 4-epimerase

The usefulness of in situ cofactor regeneration in a glycosyltransferase reaction was first demonstrated by Wong et al. for the galactosylation reaction employing β(1,4)galactosyltransferase to give *N*-acetyllactosamine [328] (Scheme 30). In this scheme, glucose 1-phosphate (Glc-1-P), prepared in situ from glucose 6-phosphate in a reaction catalyzed by phosphoglucomutase (EC 5.4.2.2), reacted with uridine 5′-triphosphate (UTP) in the presence of UDP-glucose pyrophosphorylase (EC 2.7.7.9) to give UDP-glucose (UDP-Glc). The by-product, pyrophosphate (PPi), was decomposed by inorganic pyrophosphatase (EC 3.6.1.1) to 2 equivalents of ortho-phosphate (Pi). UDP-Glc was converted to UDP-galactose (UDP-Gal) under equilib-rium conditions, with UDP-galactose 4-epimerase (EC 5.1.3.2) as a catalyst and NAD$^+$ as a cofactor. The equilibrium is in favor of the formation of UDP-Glc; however, the UDP-Gal produced was irreversibly utilized by β(1,4)galactosyltrans-ferase (GalTase; EC 2.4.1.22) to give β(1→4)-galactose-terminated carbohydrates, so that UDP-Gal was continuously formed from UDP-Glc. The released uridine 5′-diphosphate (UDP) was converted to UTP with phospho(enol)pyruvate (PEP) in a reaction catalyzed by pyruvate kinase (EC 2.7.1.40). UTP reacted again with Glc-1-P to form UDP-Glc. The UDP-galactose regeneration reaction worked well, starting with a catalytic amount of UDP and equivalent amounts of Glc-1-P and PEP, to produce multigram quantities of *N*-acetyllactosamine in 85% yield [328]. The overall reaction did require a phosphorylated sugar, glucose 6-phosphate, as a starting material, although it can be separately prepared from glucose by hexokinase (EC 2.7.1.1) and adenosine 5′-triphosphate (ATP).

A similar in situ UDP-galactose regeneration system has been reported by the Augé group [329]. They synthesized branched penta- and heptasaccharides using β1,4-galactosyltransferase. Thiem and co-workers have employed a similar in situ

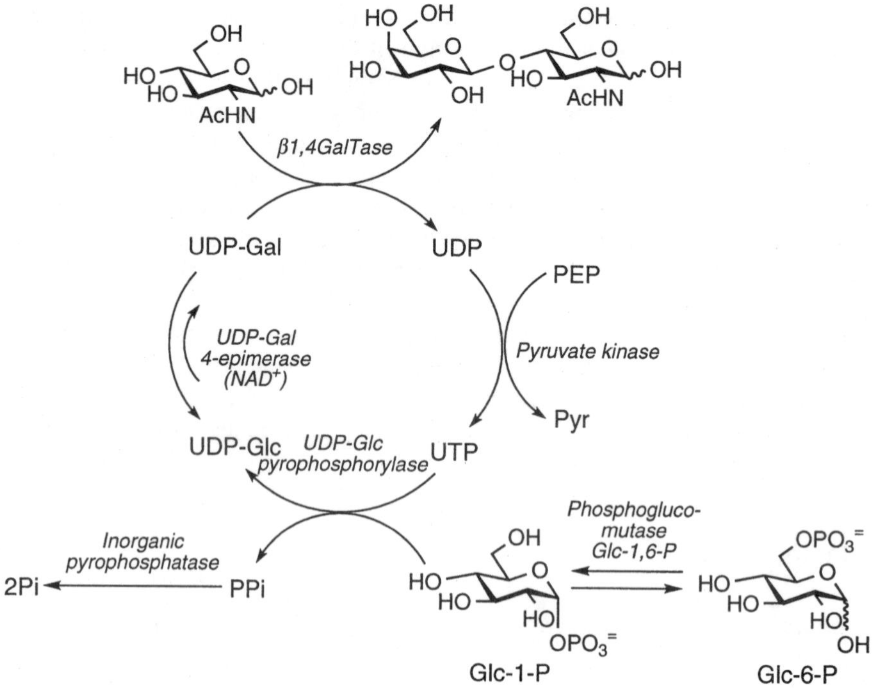

Scheme 30

cofactor regeneration system for β1,4-galactosyltransferase reaction in which they regenerated UDP-2-deoxy-galactose [258]. They started with 2-deoxy-glucose 1-phosphate and synthesized 2-deoxy-galactose-β(1→4)-glucose and 2-deoxy-galactose-β(1→4)-N-acetylglucosamine.

A typical galactosylation reaction involving an in situ cofactor regeneration system that employs UDP-glucose 4-epimerase: Synthesis of Galβ(1→4)GlcNAcβ1-O-allyl [45]

A solution of GlcNAcβ1-O-allyl (2.00 g, 7.65 mmol), glucose 1-phosphate (Glc-1-P) (2.74 g, 7.65 mmol), phospho(enol)pyruvate (PEP) K salt (1.60 g, 7.65 mmol; 95%), uridine 5'-diphosphate (UDP) (90 mg, 0.19 mmol), NAD$^+$ (193 mg, 0.25 mmol), dithiothreitol (DTT) (306 mg, 2.0 mmol), MgCl$_2$·6H$_2$O (162 mg, 0.80 mmol), MnCl$_2$·4H$_2$O (79 mg, 0.40 mmol), KCl (1.04 g, 15 mmol), and NaN$_3$ (20 mg, 0.31 mmol) in HEPES buffer (100 mM, pH 7.5; 200 mL) was carefully adjusted with 10 N and 1 N NaOH solution to pH 7.5. To the solution were added the enzymes: pyruvate kinase (EC 2.7.1.40; 100 units), inorganic pyrophosphatase (EC 3.6.1.1; 100 units), galactose 1-phosphate uridyltransferase (EC 2.7.7.12; 10 units), UDP-glucose pyrophosphorylase (EC 2.7.7.9; 20 units), and β(1,4)galactosyltransferase (EC 2.4.1.22; 5 units), and the reaction mixture was gently stirred under argon atmosphere for 3 days at 37°C. The mixture was concentrated and chromatographed on silica gel, with EtOAc/MeOH (2:1), to give a disaccharide, which was further purified with a Sephadex G-25 column, eluted with water, to give Galβ(1→4) GlcNAcβ1-O-allyl (1.70 g, 50%).

A UDP-galactose regeneration reaction employing galactokinase and galactose 1-phosphate uridyltransferase

Another UDP-galactose regeneration in situ system has been reported by Wong, Wang, and Ichikawa, in which an unphosphorylated monosaccharide, galactose, served as a starting material for UDP-galactose [330] (Scheme 31). Galactose was treated with galactokinase (EC 2.7.1.6) and ATP to give galactose 1-phosphate; the by-product, adenosine 5'-diphosphate (ADP), was converted back to ATP in a reaction catalyzed by pyruvate kinase (EC 2.7.1.40) and PEP. Galactose 1-phosphate uridyltransferase (EC 2.7.7.12) then catalyzed an exchange reaction between galactose 1-phosphate and a catalytic amount of UDP-Glc to give UDP-Gal and Glc-1-P. UDP-Gal was used by β(1,4)galactosyltransferase (EC 2.4.1.22) to form β(1→4)-galactose-terminated oligosaccharides. The released UDP was phosphorylated by pyruvate kinase (EC 2.7.1.40) and PEP to give UTP. UTP then reacted with the released Glc-1-P from the exchange reaction between the galactose 1-phosphate and UDP-Glc to give UDP-Glc; the by-product, pyrophosphate (PPi), was decomposed by inorganic pyrophosphatase (EC 3.6.1.1). The newly produced UDP-Glc again underwent an exchange reaction with galactose 1-phosphate to generate UDP-Gal and Glc-1-P. The overall reaction was quite effective, starting with catalytic amounts of UDP, Glc-1-P, and ATP, equimolar amounts of galactose, and two equivalent amounts of PEP as a phosphate source. Based on this regeneration system, several β(1→4)-galactose-terminated disaccharides were prepared [330].

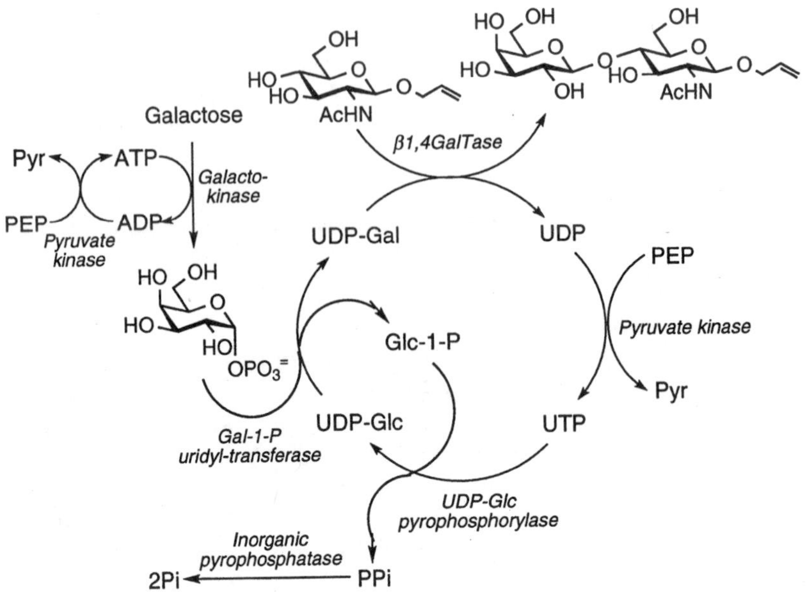

Scheme 31

A typical galactosylation reaction involving an in situ cofactor regeneration system that employs galactokinase: Synthesis of Galβ(1→4)GlcNAcβ1-O-allyl [45]

A solution of GlcNAcβ1-O-allyl (1.15 g, 4.4 mmol), galactose (800 mg, 4.4 mmol), phospho(enol)pyruvate (PEP) K salt (1.82 g, 8.8 mmol; 95%), uridine 5'-diphosphate (UDP) (90 mg, 0.19 mmol), adenosine 5'-triphosphate (ATP) (100 mg, 0.18 mmol), cysteine (116 mg, 0.96 mmol), dithiothreitol (DTT) (183 mg, 1.2 mmol), MgCl$_2$·6H$_2$O (244 mg, 1.2 mmol), MnCl$_2$·4H$_2$O (118 mg, 0.6 mmol), KCl (179 mg, 2.4 mmol), NaN$_3$ (12 mg, 0.18 mmol), and glucose 1-phosphate (Glc-1-P) (77 mg, 0.22 mmol) in HEPES buffer (100 mM, pH 7.5; 120 mL) was carefully adjusted with 10 N and 1 N NaOH solution to pH 7.5. To the solution were added the enzymes: galactokinase (EC 2.7.1.6; 10 units), pyruvate kinase (EC 2.7.1.40; 200 units), inorganic pyrophosphatase (EC 3.6.1.1; 10 units), galactose 1-phosphate uridyltransferase (EC 2.7.7.12; 10 units), UDP-glucose pyrophosphorylase (EC 2.7.7.9; 10 units), and β1,4-galactosyltransferase (EC 2.4.1.22; 10 units). The reaction mixture was gently stirred under argon atmosphere for 3 days at 37°C. The mixture was then concentrated and chromatographed on silica gel, with EtOAc/MeOH (2:1), to give a disaccharide, which was further purified on a Sephadex G-25 column, eluted with water, to give Galβ(1→4)GlcNAcβ1-O-allyl (1.06 g, 57%).

B. Sialylation

The chemical glycosylation is well developed, and chemical galactosylation has proved successful in preparing multigram quantities of β(1→4)-galactose-terminated oligosaccharides. However, stereoselective chemical synthesis of α-sialosides on a multigram scale has not yet been accomplished. Ichikawa et al. have reported an elegant enzymatic sialoside synthesis in which CMP-sialic acid is regenerated in situ in a multienzyme system [150] (Scheme 32).

In this scheme, cytidine 5′-monophosphate (CMP) was phosphorylated by myokinase (EC 2.7.4.3) or nucleoside monophosphate kinase (EC 2.7.4.4) in the presence of ATP to give cytidine 5′-diphosphate (CDP); the by-product, ADP, was converted back to ATP by pyruvate kinase and PEP. CDP was further phosphorylated to give cytidine 5′-triphosphate (CTP) in a reaction catalyzed by pyruvate kinase and PEP. CTP then reacted with sialic acid (Neu5Ac) in the presence of CMP-sialic acid synthetase (EC 2.7.7.43) to give CMP-sialic acid (CMP-NeuAc); the by-product, pyrophosphate, was decomposed by inorganic pyrophosphatase to yield orthophosphate. CMP-NeuAc was utilized by sialyltransferase (α(2,6)sialyltransferase; EC 2.4.99.1) to give the α-sialoside (Neu5Acα(2→6)Galβ(1→4)GlcNAc). The released CMP was again converted to CDP, CTP, and CMP-NeuAc. The overall reaction, starting with catalytic amounts of CMP and ATP and equimolar amounts of Neu5Ac and 2 equivalents of PEP, was very effective and produced a 98% yield of the product in 2 days.

Enzymatic sialylation with cofactor regeneration in situ has been further extended by introducing another in situ enzymatic reaction to generate Neu5Ac. Ichikawa et al. have introduced the in situ generation of sialic acid (Neu5Ac) from N-acetylmannosamine (ManNAc) and pyruvic acid generating as a by-product of the phosphorylation reaction of PEP. This Neu5Ac generation reaction is catalyzed by sialic acid aldolase (EC 4.1.3.3) [331] (Scheme 33). Although the reaction is reversible, the Neu5Ac produced was irreversibly incorporated into CMP-NeuAc,

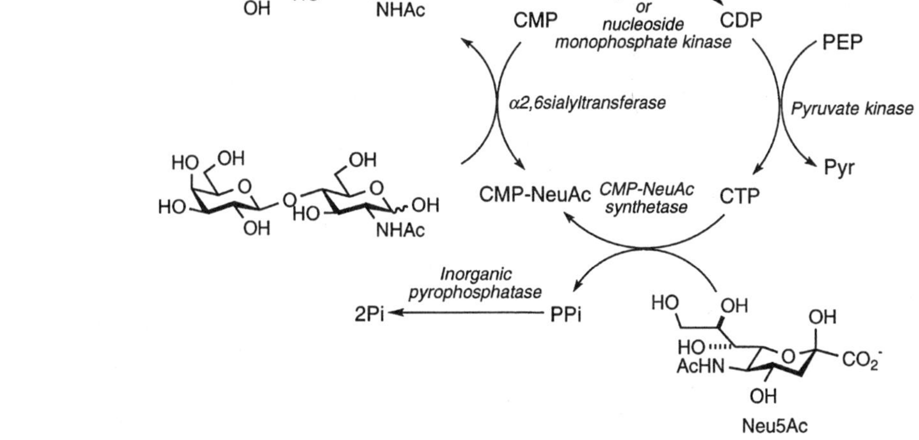

Scheme 32

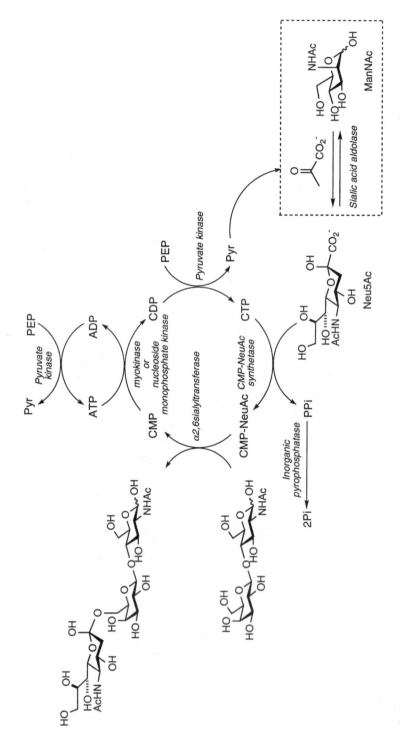

Scheme 33

which was utilized by the sialyltransferase. The reaction worked well, starting with an acceptor and ManNAc (no sialic acid), and produced the sialoside in 88% yield in 2 days.

This group also demonstrated that two glycosyltransferases with different substrate specificities can work together in a one-pot multienzyme system with cofactor regeneration in situ [331] (Scheme 34). A galactosyltransferase reaction generating *N*-acetyllactosamine (LacNAc), with cofactor regeneration in situ, was incorporated into the sialyltransferase reaction with in situ generation of NeuAc. This multienzyme system started with three monosaccharides (*N*-acetylglucosamine, glucose 1-phosphate, and *N*-acetylmannosamine) in the presence of two glycosyltransferases (β(1,4)galactosyltransferase: EC 2.4.1.22 and α(2,6)sialyltransferase: EC 2.4.99.1). β(1,4)Galactosyltransferase accepts only *N*-acetylglucosamine as an acceptor substrate, and neither *N*-acetyllactosamine nor *N*-acetylmannosamine nor glucose 1-phosphate existed in the reaction mixture. In addition, α(2,6)sialyltransferase accepts only *N*-acetyllactosamine as an acceptor substrate, and neither *N*-acetylmannosamine nor UDP-galactose generated in situ nor glucose 1-phosphate were present in the reaction mixture. The entire reaction, therefore, proceeded sequentially: Gal was added onto *N*-acetylglucosamine by β(1,4)galactosyltransferase to produce *N*-acetyllactosamine, which was then sialylated with α(2,6)sialyltransferase to give the sialoside.

A typical enzymatic sialylation reaction with CMP-NeuAc regeneration:
Neu5Acα(2→3)Galβ(1→4)GlcNAcβ1-O-allyl[45]

A solution of Galβ(1→4)GlcNAcβ1-*O*-allyl (210 mg, 0.50 mmol), sialic acid (Neu5Ac) (160 mg, 0.52 mmol), phospho(enol)pyruvate (PEP) Na$_3$ salt (120 mg, 0.51 mmol), MgCl$_2$·6H$_2$O (20 mg, 0.10 mmol), MnCl$_2$·4H$_2$O (4.9 mg, 0.025 mmol), KCl (7.5 mg, 0.10 mmol), cytidine 5′-monophosphate (CMP) (16 mg, 0.05 mmol), adenosine 5′-triphosphate (ATP) (2.7 mg, 0.005 mmol), and mercaptoethanol (0.34 mL) in HEPES buffer (200 mM, pH 7.5; 3.5 mL) was carefully adjusted to pH 7.5 with 1 N NaOH solution. To the solution were added the enzymes: nucleoside monophosphate kinase (EC 2.7.4.4) (5 units), pyruvate kinase (EC 2.7.1.40) (100 units), inorganic pyrophosphatase (EC 3.6.1.1) (10 units), CMP-sialic acid synthetase (EC 2.7.7.43) (0.4 units), and α2,3-sialyltransferase (0.1 unit), and the reaction mixture was gently stirred under argon atmosphere for 3 days at room temperature (25°C). The mixture was concentrated, and chromatographed on silica gel, with EtOAc/iPrOH/H$_2$O (2:2:1), to give a trisaccharide, which was further purified on a BioGel P-2 column, with water, to give Neu5Acα(2→3)Galβ(1→4)GlcNAcβ1-*O*-allyl (88 mg, 24%).

C. Fucosylation

A GDP-fucose regeneration system has not yet been optimized, but two approaches have been reported [45] (Schemes 35 and 36). One employed the enzymatic conversion of GDP-mannose to GDP-fucose, and the other generated GDP-fucose directly from fucose 1-phosphate.

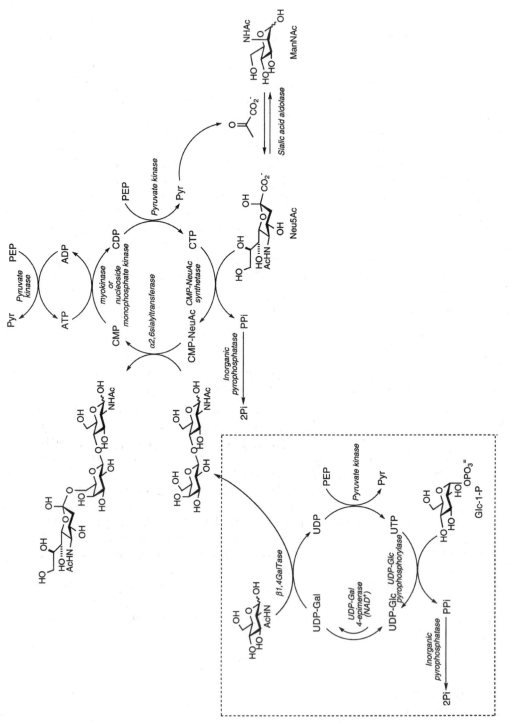

Scheme 34

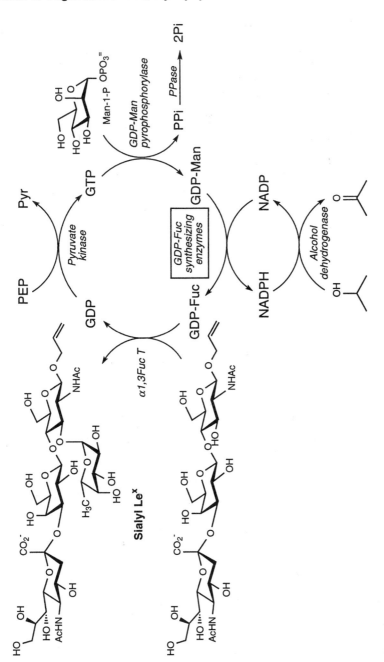

Sialyl Le^x

Scheme 35

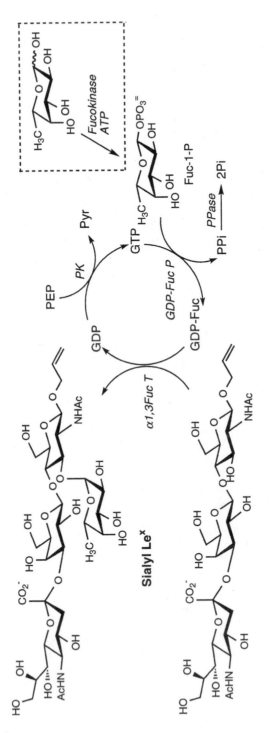

Sialyl Le^x

Scheme 36

GDP-Fucose from GDP-mannose

GDP-fucose was converted to GDP-mannose by "GDP-fucose synthesizing enzyme(s)," which is known to exist in the bacterium *Klebsiella pneumoniae*, but has been little studied. In the GDP-fucose regeneration system (Scheme 35), the reaction of mannose 1-phosphate (Man-1-P) with guanosine 5'-triphosphate (GTP) to give GDP-mannose (GDP-Man) was catalyzed by GDP-mannose pyrophosphorylase (EC 2.7.7.22) obtained from yeast. GDP-Man was converted to GDP-fucose (GDP-Fuc) by "GDP-fucose synthesizing enzyme(s)" obtained from *Klebsiella pneumoniae*, and the consumed NADP that acts as a cofactor in this conversion was regenerated from its product NADPH by *Thermoanaerobium brockii* alcohol dehydrogenase (EC 1.1.1.2). The GDP-Fuc produced was utilized by fucosyltransferase (α(1,3)fucosyltransferase) to give an $\alpha(1 \rightarrow 3)$-fucosylated product. The released GDP was phosphorylated to GTP by phospho(enol)pyruvate (PEP) and pyruvate kinase (EC 2.7.1.40). GTP again reacted with Man-1-P to give GDP-Man, which was then converted to GDP-Fuc.

GDP-Fucose from fucose 1-phosphate

An alternative route is to use GDP-fucose pyrophosphorylase (Scheme 36), which can be obtained from porcine liver [45]. Reaction of fucose 1-phosphate (Fuc-1-P) with GTP, catalyzed by GDP-fucose pyrophosphorylase (EC 2.7.7.30) from porcine thyroids, gave GDP-Fuc. Fuc-1-P could be obtained directly from fucose by treatment with ATP and fucokinase (EC 2.7.1.52) from porcine liver. The GDP-Fuc was utilized by fucosyltransferase (α1,3-fucosyltransferase) to give an α1,3-fucosylated product, and the released GDP was phosphorylated to GTP by phospho(enol)pyruvate (PEP) and pyruvate kinase (EC 2.7.1.40). The GTP again reacted with Fuc-1-P to give GDP-Fuc.

These in situ GDP-fucose regeneration systems have not yet succeeded in producing fucosylated oligosaccharides on a preparative scale (>1 g), most probably because key enzymes—GDP-mannose pyrophosphorylase, the "GDP-fucose synthesizing enzyme(s)" that convert GDP-mannose to GDP-Fuc, fucokinase, and GDP-fucose pyrophosphorylase—are very difficult to obtain in quantities and quality required for a large-scale synthesis of oligosaccharides.

D. Glucuronylation

Gygax and co-workers reported a glucuronosyltransferase reaction with in situ regeneration of UDP-glucuronic acid [332] (Scheme 37). In this UDP-GlcUA regeneration, UDP-Glc pyrophosphorylase (EC 2.7.7.9) catalyzed the reaction of glucose 1-phosphate with UTP to give UDP-Glc. UDP-Glc was oxidized by UDP-glucose dehydrogenase (EC 1.1.1.22) in the presence of 2 equimolar NAD to give UDP-glucuronic acid (UDP-GlcUA). UDP-GlcUA was utilized by β-glucuronosyltransferase (EC 2.4.1.17) to give the product. The released UDP was phosphorylated by pyruvate kinase and PEP to give UTP. UTP again reacted with Glc-1-P to give UDP-Glc. This system worked well and produced D-glucuronosylated steroids and phenols in 28%–65% yields.

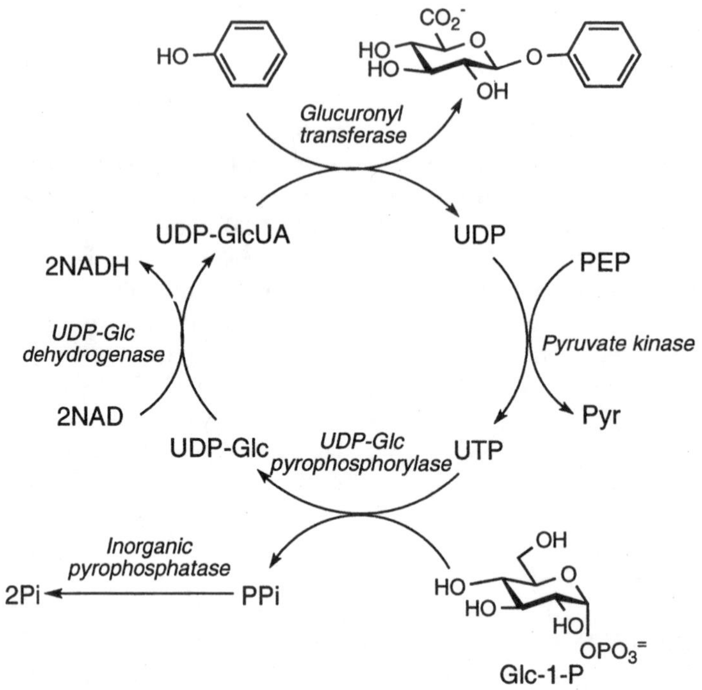

Scheme 37

E. *N*-Acetylglucosaminylation

An in situ regeneration system for UDP-*N*-acetylglucosamine (UDP-GlcNAc) has also been studied in which the regeneration is coupled with β(1,6)*N*-acetylglucosaminyltransferase reaction, which is specific for the *O*-linked core class 1 structure [273] (Scheme 38). In this system, *N*-acetylglucosamine 1-phosphate (GlcNAc-1-P) reacted with uridine 5′-triphosphate in the presence of UDP-*N*-acetylglucosamine pyrophosphorylase (EC 2.7.7.23) to give UDP-GlcNAc. UDP-GlcNAc was then consumed by the β(1,6)*N*-acetylglucosaminyltransferase (EC 2.4.1.148) from bovine liver in the reaction. The released UDP was converted to UTP by pyruvate kinase and PEP. The enzyme UDP-*N*-acetylglucosamine pyrophosphorylase (EC 2.7.7.23) has been partially purified from yeast; however, the cell-free extracts produced by this purification method contain phosphatases that degrade the phosphate-bearing compounds (UDP, UTP, UDP-GlcNAc, and GlcNAc-1-P) present in the regeneration system.

F. Mannosylation

A GDP-mannose regeneration system involving a recombinant α(1,2)mannosyltransferase has been studied [311] (Scheme 39). Mannose 1-phosphate reacted with GTP in the presence of GDP-mannose pyrophosphorylase (EC 2.7.7.22), which had been partially purified from yeast, to give GDP-mannose. GDP-Mannose was used by a recombinant α(1,2)mannosyltransferase to form a Manα(1→2)-linkage. The released

GDP was converted to GTP in the presence of PEP and pyruvate kinase. GTP was again transformed into GDP-mannose.

G. Glucosylation (Sucrose Synthetase)

A UDP-glucose regeneration system has been developed for the synthesis of sucrose and trehalose, which are α-D-glucopyranosyl-linked nonreducing disaccharides that are ubiquitous in plants and abundant in fungi (sucrose) and insects (trehalose). Haynie and Whitesides have explored the enzymatic synthesis of sucrose and trehalose with cofactor regeneration in situ [333] (Scheme 40).

UDP-Glucose was synthesized from glucose 1-phosphate (Glc-1-P) and UTP in the presence of UDP-glucose pyrophosphorylase (EC 2.7.7.9). The conversion of UDP-glucose to sucrose was catalyzed by sucrose synthetase, partially purified from wheat germ. The released UDP was transformed to UTP, in the presence of PEP and pyruvate kinase, and the UTP was then converted to UDP-glucose.

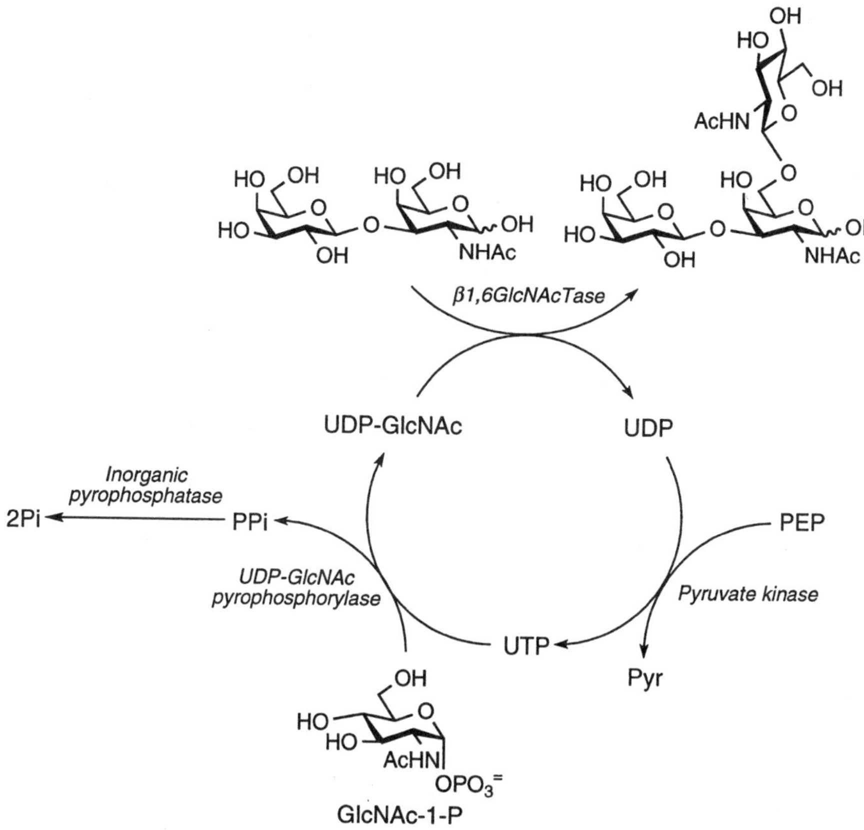

Scheme 38

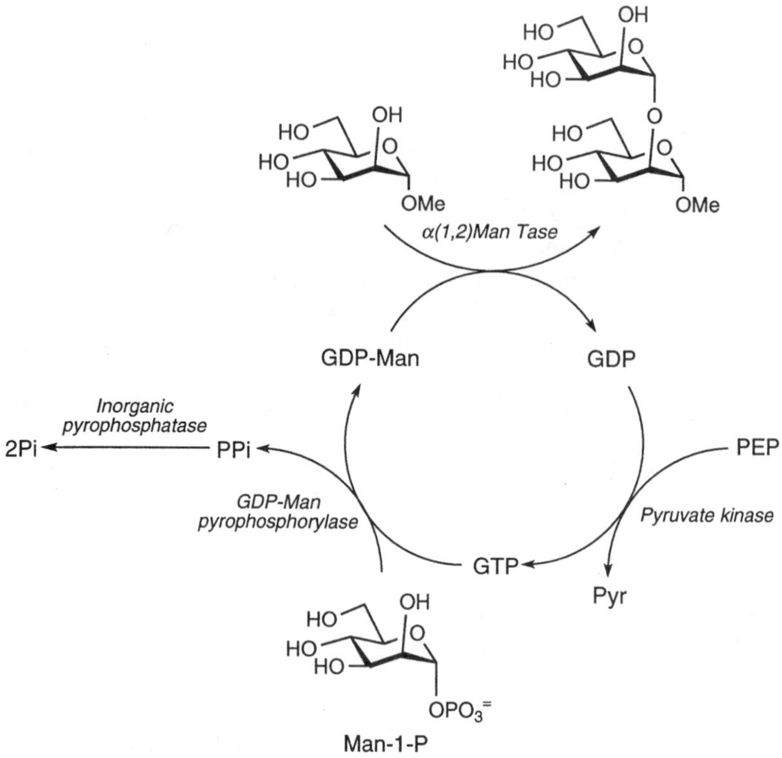

Scheme 39

IV. GLYCOSIDASE REACTIONS

Glycosidases are glycosylhydrolases, which means the enzymes hydrolytically cleave a glycosidic bond in nature. The glycosidases are divided into two groups: exoglycosidases, which cleave the glycosidic bond at the nonreducing end of an oligosaccharide, and endoglycosidases, which cleave an internal glycosidic bond [334,335]. A recent study of the glycosidase reaction showed that under appropriate conditions the enzyme can catalyze the formation of a glycosidic bond in a stereo- and regioselective manner. The advantage of utilizing glycosidases as catalysts for glycosyl-transfer reactions is that the reaction does not require expensive sugar nucleotides as donor substrates; in addition, the glycosidases are generally more readily available and less expensive than glycosyltransferases.

The mechanism of glycosidase-mediated glycosidic bond-cleaving reaction has generally been accepted as following the model proposed for lysozyme: acid-catalyzed hydrolysis of glycosidic bonds. In such a model (Scheme 41), there are two important carboxylic acid moieties in the enzyme active site: One is protonated (AH) and therefore functions as an acid (general acid), and one is ionized (B) and functions as a base (general base). The α- and β-glycosidases have the disposition of these acids reversed (Scheme 41). In general, the glycosidic oxygen atom is protonated by the general acid (AH); the glycosidic bond cleavage is thereby facilitated, leaving an unstable oxocarbenium ion structure that is stabilized by the general base

(B) through ionic interaction. Such an unstable intermediate is trapped by a nucleophile, in most cases water, to furnish glycosidic bond hydrolysis. If the nucleophile
is an alcohol or carbohydrate hydroxy group, the reaction is a "transglycosidation
reaction" or a glycosidic bond-forming reaction. For an α-glycosidase, a β-linked
enzyme-carbohydrate complex is formed and attacked by the nucleophile from the
opposite face (α side) to form an α-linked product. In the case of the β-glycosidase,
the α-linked enzyme-carbohydrate complex is formed and attacked by the nucleophile from the opposite face (β side) to form a β-linked product. By this means, the
stereoselectivity of glycosidase-mediated glycosidic bond-forming reactions is accomplished; however, the regioselectivity is poor. In addition, the enzyme reaction
is reversible, and therefore the yields of the reactions are, at best, moderate.

This Section discusses three examples of glycosidase-catalyzed reactions; the
reader is referred to other review articles for additional examples of glycosidase
reactions.

Glycals as acceptor substrates for β-galactosidase

Glycals have been employed as acceptor and donor substrates in glycosidase-
catalyzed glycosidic bond-forming reactions [336,337]. Bay and Cantacuzene have
reported that when a glucal functioned as both donor and acceptor for β-glucosidase
(from almond), a glucal-containing 2-deoxy-β-(1→3)-linked disaccharide formed as
a major product, rather than the corresponding β-(1→6)-linked disaccharide [337]

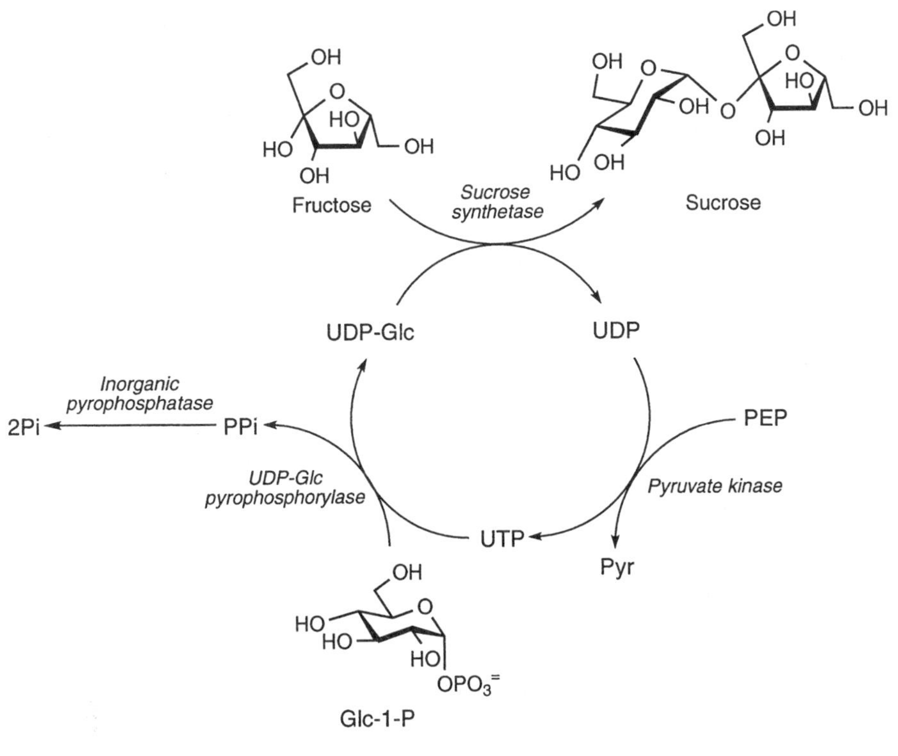

Scheme 40

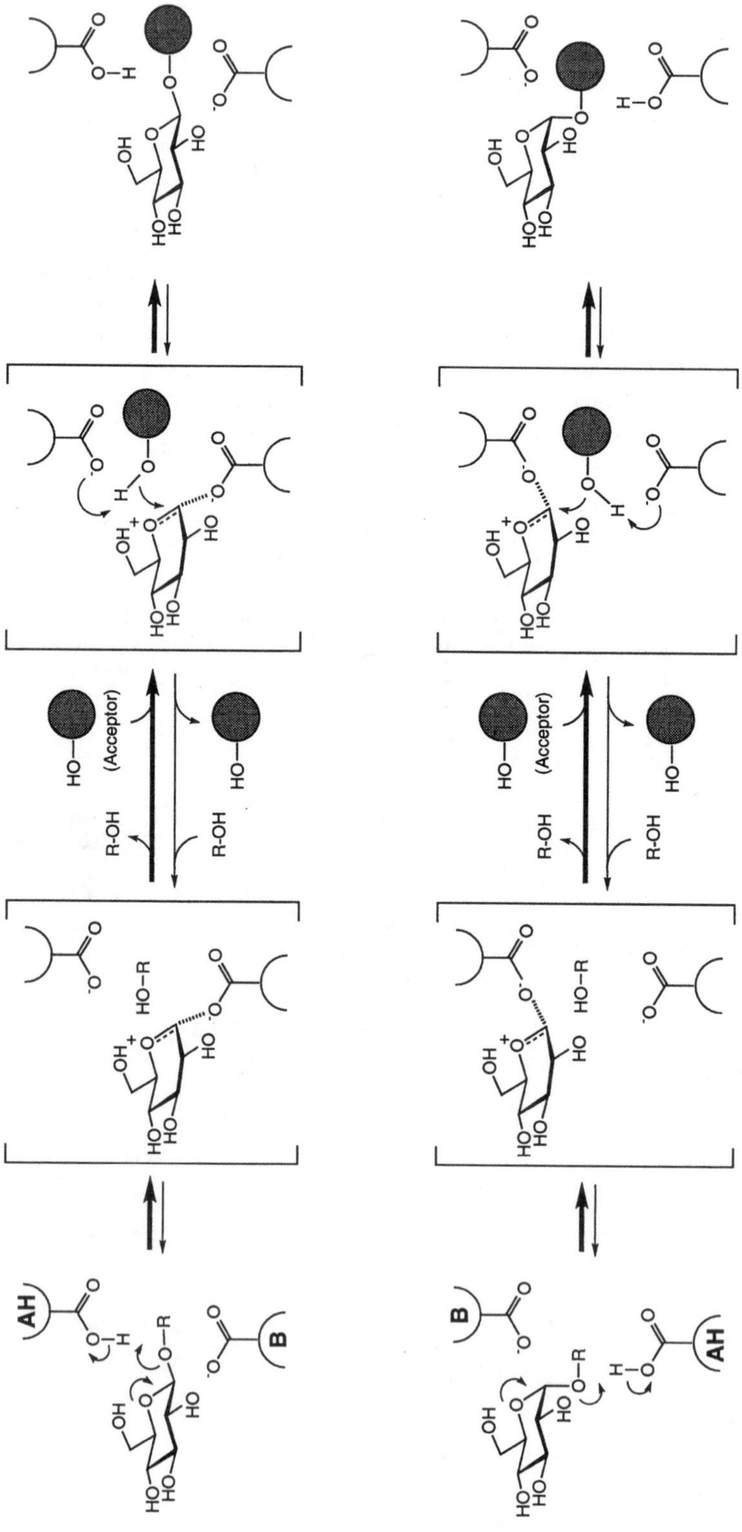

Scheme 41

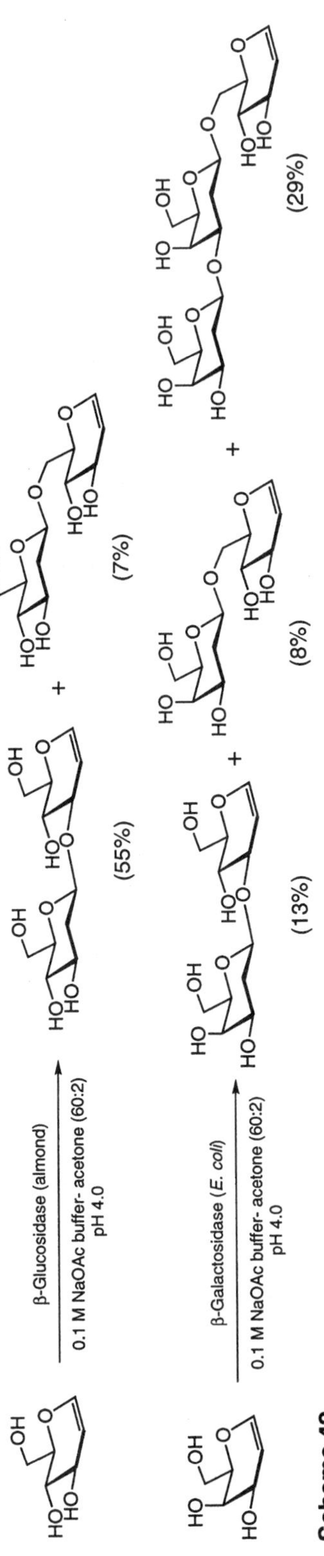

β-Glucosidase (almond)
0.1 M NaOAc buffer- acetone (60:2)
pH 4.0

(55%)

+

(7%)

(29%)

Scheme 42

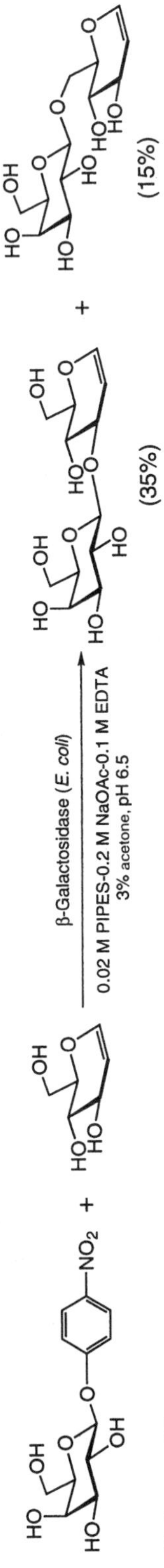

β-Galactosidase (E. coli)
0.1 M NaOAc buffer- acetone (60:2)
pH 4.0

(13%)

+

(8%)

+

β-Galactosidase (E. coli)
0.02 M PIPES-0.2 M NaOAc-0.1 M EDTA
3% acetone, pH 6.5

(35%)

Scheme 43

β-Galactosidase (E. coli)
0.02 M PIPES-0.2 M NaOAc-0.1 M EDTA
3% acetone, pH 6.5

(42%)

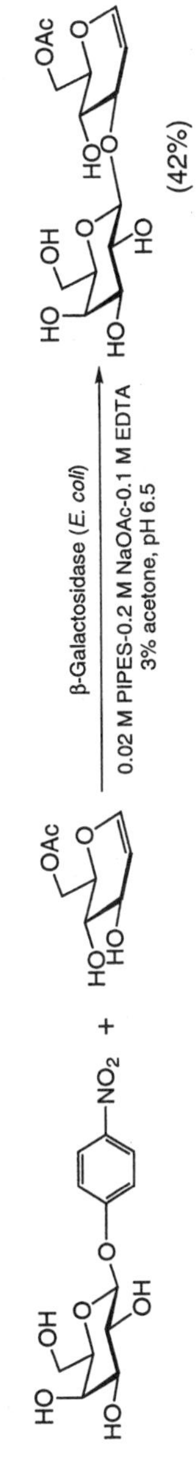

Scheme 44

(Scheme 42). Usually glycosidase-catalyzed glycosidic bond formation occurs on the most reactive hydroxyl group, the 6-OH (primary hydroxyl group), and therefore a (1→6)-linked glycosidic bond is preferentially formed. However, the glycal has an allylic hydroxy group and its reactivity (nucleophilicity) is greater than that of the ordinary secondary hydroxyl groups, and consequently the β-(1→3)-linked disaccharide is formed as a major additional product. In the case of galactal, the authors isolated a trisaccharide as a major product.

Look and Wong have examined the possibility that if a suitably protected carbohydrate such as glycal is recognized as an acceptor for glycosidase-mediated oligosaccharide formation, the regioselectivity can be improved [338]. They used a glycal as an acceptor and p-nitrophenyl β-D-galactoside as a donor substrate, with β-galactosidase from E. coli as a glycosidase. When D-glucal was employed as an acceptor, Galβ(1→3)glucal was produced as a major product in 35% yield, and the yield of the β(1→6)-linked disaccharide was 15% (Scheme 43). When 6-O-acetyl-glucal, prepared by means of an enzymatic regioselective acetylation catalyzed by subtilisin in the presence of vinyl acetate [339], was instead used as the acceptor, the reaction produced only one disaccharide, Galβ(1→3)-6-O-Ac-glucal in 42% yield (Scheme 44). Thus, the regioselectivity of glycosidase-mediated oligosaccharide synthesis was perfectly controlled [338].

trans-*Sialidase*

The *trans*-sialidase from *Trypanosoma cruzi* is a unique enzyme that catalyzes the equilibrium reaction between Neu5Acα(2→3)Galβ-O-R^1 and Galβ-O-R^2 to yield a new sialoside, Neu5Acα(2→3)Galβ-O-R^2 [340–343] (Scheme 45). The enzyme

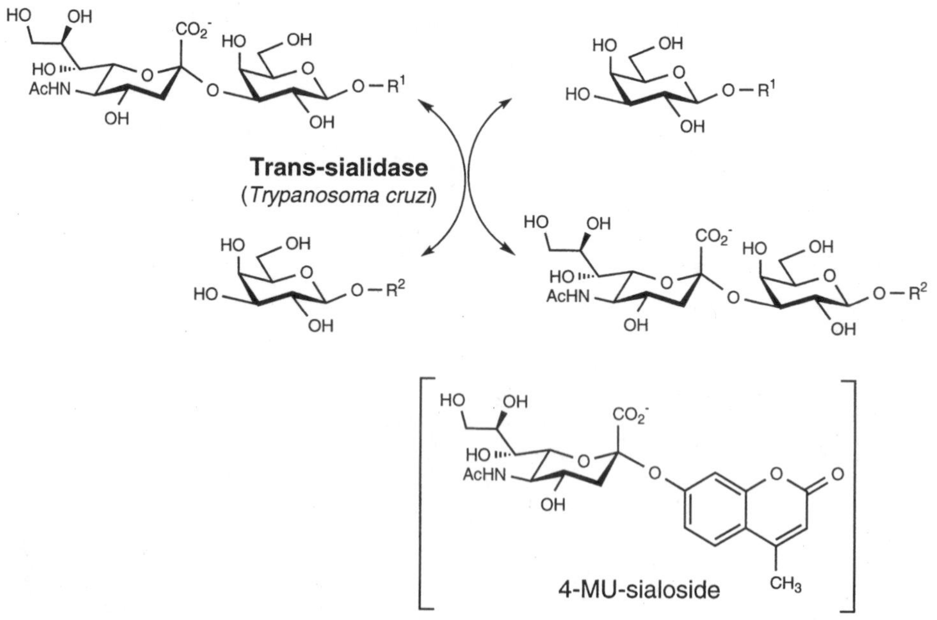

Scheme 45

shows a rather strict specificity for Neu5Acα(2→3)β-galactoside or 4-methylumbel-liferone-α-sialoside (4-MU-Neu5Ac).

Ito and Paulson have demonstrated the synthetic utility of the *trans*-sialidase reaction [344] (Scheme 46): The *trans*-sialidase reaction was coupled with a CMP-NeuAc regeneration system employing α(2,3)sialyltransferase to synthesize the dif-ficult-to-synthesize α(2→3)sialoside by α(2,3)sialyltransferase. An excellent acceptor substrate (in a catalytic amount) was used in the transient α(2,3)sialyltransferase reaction to form an α(2→3)-sialoside as a donor substrate for the *trans*-sialidase. The sialic acid moiety of the transient α(2→3)-sialoside was then transferred to the second acceptor, β-galactoside, which was not a good acceptor substrate for the α(2,3)sialyltransferase.

Lee and Lee showed another advantage of this *trans*-sialidase reaction [345] (Scheme 47): The enzyme utilizes the 4-MU-sialoside, in which the side chain of the sialic acid is chemically modified. Even though the conversion efficiency of the transfer reaction was not good, it was still satisfactory when compared to those reactions employing CMP-sialic acid synthetase and α(2,3)sialyltransferase. 4-MU-Sialoside derivatives were synthesized according to Scheme 47, which involved: (1) oxidative cleavage of the sialic acid side chain to generate an aldehyde group, and (2) reductive amination of the aldehyde and an amine component. Lee and Lee prepared a Neu5Acα(2→3)Galβ(1→4)Glc derivative in which the sialic acid side chain was attached to a dansyl group [345].

Cellulase

Cellulose is the most abundant organic substance, and some 10^{15} kg of cellulose are produced by photosynthesis each year. Synthesis of cellulose had been one of the most challenging problems in synthetic chemistry. The Kobayashi group has reported enzymatic synthesis of cellulose using a transglycosylation reaction [346,347] (Scheme 48). The reaction of β-D-cellobiosyl fluoride with cellulase in acetonitrile and acetate buffer gave cellulose with a degree of polymerization (DP) $\geq$ 22.

Glycosidase-glycosyltransferase combination

Glycosidase-mediated glycosidic bond formation is a reversible reaction. After a certain period of time, the formation of the glycosidic bond reaches a maximum, and then hydrolysis of the newly produced glycosidic bond becomes predominant. If the product can be removed, the equilibrium of the reaction can be shifted in favor of the glycosidic bond-forming reaction.

The Wong group has introduced a sialyltransferase reaction into a β-galactosidase-catalyzed reaction [348] (Scheme 49). β-Galactosidase from *Bacillus* is known to catalyze N-acetyllactosamine synthesis from N-acetylglucosamine and lactose (galactosyl donor) with good regioselectivity [349,350]. The LacNAc produced is a substrate for both enzymes, the α(2,6)sialyltransferase and the β-galactosidase. When the sialyltransferase sialylates the LacNAc, the resulting NeuAcα(2→6)LacNAc is no longer a substrate for the β-galactosidase. Since the reaction rate for the hydrolysis of the β-galactosidase was much faster than that of the sialyltransferase, the chemical yield of NeuAcα(2→6)LacNAc was not high; however, this is a promising new approach for the "next generation" of enzymatic oligosaccharide syntheses.

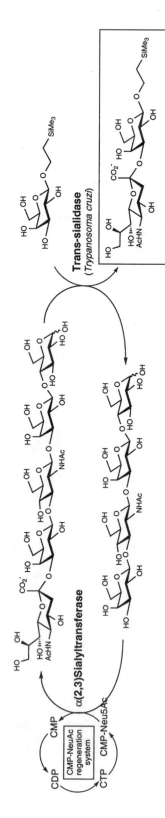

Scheme 46

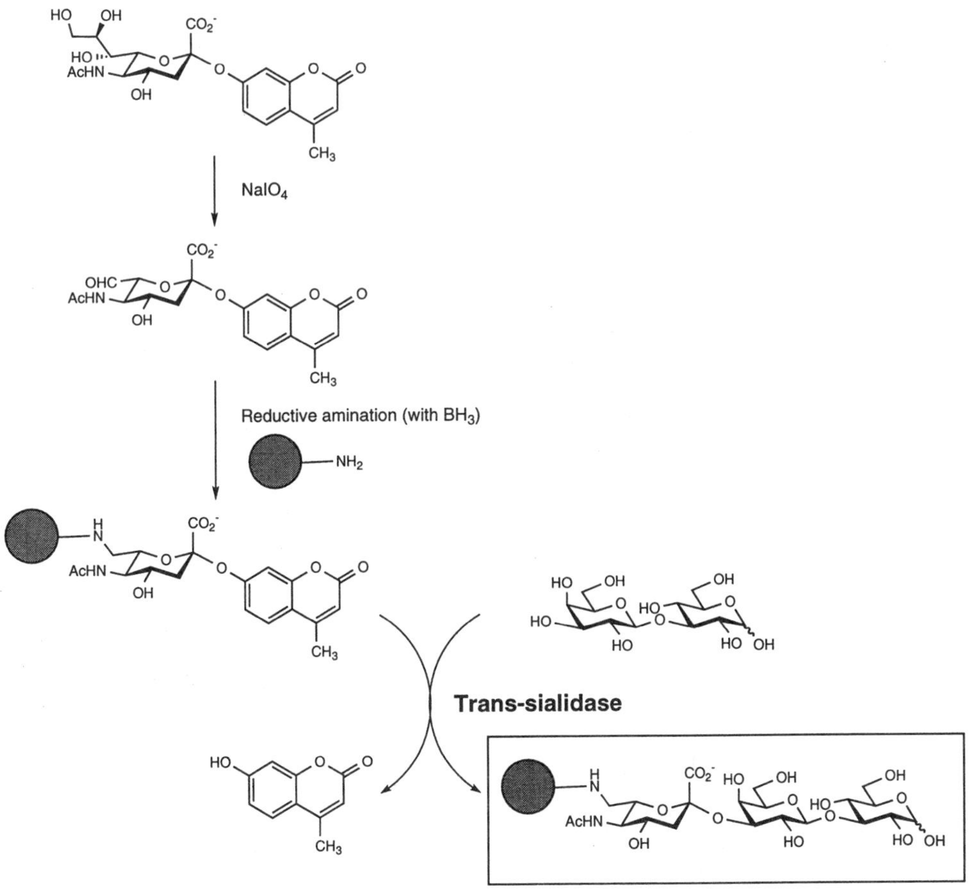

Scheme 47

V. ENZYMATIC SYNTHESIS OF GLYCOPEPTIDES

A. Application of a Glycosyltransferase Reaction to Glycopeptide Synthesis

Application of enzymes to the synthesis of glycopeptides has not yet been fully explored. Recently, Schaster et al. have reported one possible approach in which the solid-phase peptide backbone was synthesized chemically and the attached oligosaccharide chain was elongated enzymatically by a series of glycosyltransferase reactions [351] (Scheme 50). This group used a silica-based solid support, aminopropylsilica, which is compatible with both organic and aqueous solvents. Glycine hexamer linked to phenylalanine was chosen as an enzymatically cleavable spacer. This spacer also provided a proper distance from the solid support to the carbohydrate moiety, where the glycosyltransferase-catalyzed sugar chain elongation took place. The reducing terminal N-acetylglucosamine residue was introduced as GlcNAcβ-Asn onto the N-terminus of the Gly-Phe moiety of the peptide bond. Two consecutive

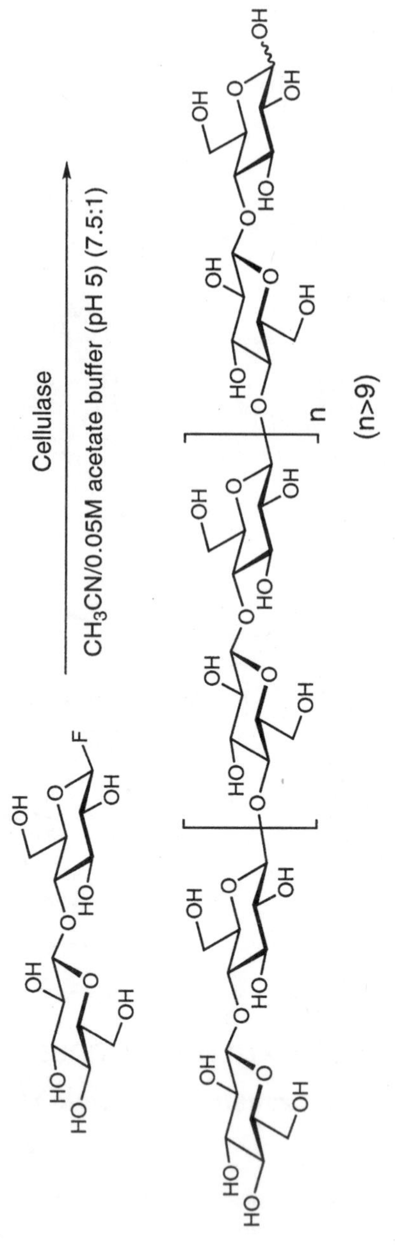

Scheme 48

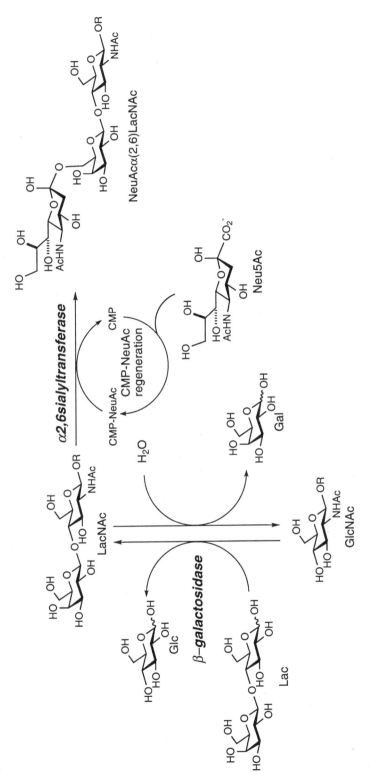

Scheme 49

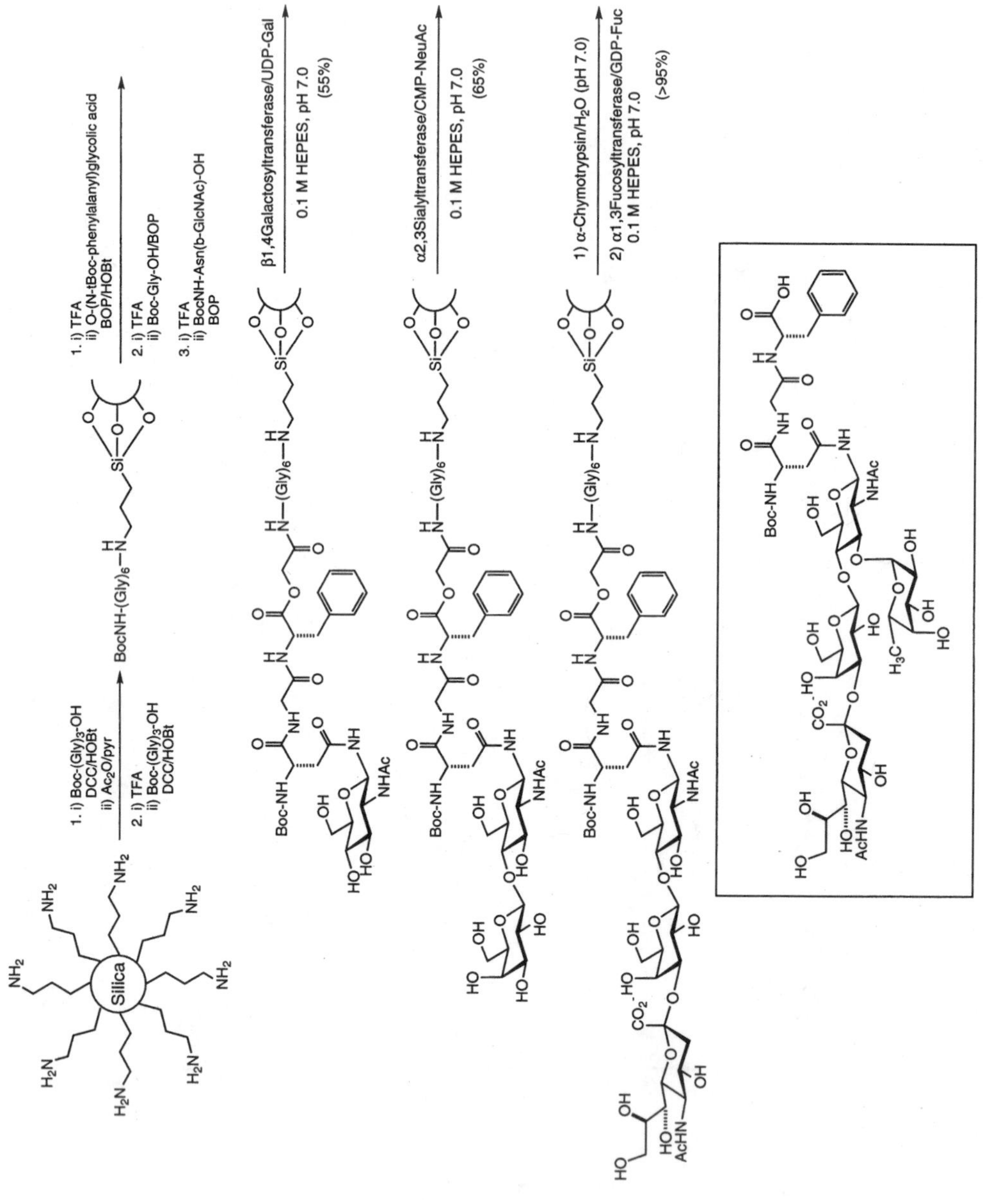
Silica
1. i) Boc-(Gly)3-OH DCC/HOBt ii) Ac2O/pyr
2. i) TFA ii) Boc-(Gly)3-OH DCC/HOBt
BocNH-(Gly)6-N
1. i) TFA ii) O-(N-tBoc-phenylalanyl)glycolic acid BOP/HOBt
2. i) TFA ii) Boc-Gly-OH/BOP
3. i) TFA ii) BocNH-Asn(b-GlcNAc)-OH BOP
β1,4Galactosyltransferase/UDP-Gal
0.1 M HEPES, pH 7.0
(55%)
α2,3Sialyltransferase/CMP-NeuAc
0.1 M HEPES, pH 7.0
(65%)
1) α-Chymotrypsin/H2O (pH 7.0)
2) α1,3Fucosyltransferase/GDP-Fuc
0.1 M HEPES, pH 7.0
(>95%)

Scheme 50

glycosyltransferase reactions, β(1,4)galactosyltransferase with UDP-galactose and α(2,3)sialyltransferase with CMP-NeuAc, gave the trisaccharide on the solid support; the peptide bond was then selectively cleaved by α-chymotrypsin to afford the glycopeptide. The released glycopeptide was purified by HPLC and subsequently fucosylated with GDP-Fuc to afford a sialyl Lex-containing glycopeptide. This study was preliminary in nature, and there is much to be improved, including the chemical yield, method of purification, and flexibility of the procedure.

Another approach has been reported by the Oxford group [352] (Scheme 51). They utilized the coupling reaction of iodo-(or bromo)acetamido glycosides of oligosaccharides with cysteine residue(s) of a peptide in a site-specific manner. The oligosaccharide was converted to the corresponding glycosylamine, which was then converted into the iodoacetamido derivative. The iodoacetamido group reacted with a sulfhydryl moiety of the cysteine residue(s) to form covalently linked glycopeptide derivatives. The linked oligosaccharide moiety was then further modified by introducing a sialic acid residue with sialyltransferase.

B. Glycopeptide Preparation by Glycosurgery with Glycosidases

Exoglycosidases have been used for the structural elucidation of oligosaccharide structures isolated from natural sources [353–356]. They cleave monosaccharides from their glycosidic linkages only when the monosaccharides are the terminal residues. Most exoglycosidases show good specificity for the sugar moiety and anomeric configuration of the glycosidic linkage. In addition, some of the exoglycosidases display linkage specificity. When a glycopeptide whose structure is unknown is treated with an exoglycosidase whose specificity is known, a change in chromatographic behavior means that the unidentified oligosaccharide structure has this specific monosaccharide at the terminal, in a specific glycosidic linkage. If no change is seen, the oligosaccharide structure has no such linkage. This sequence is repeated with another type of exoglycosidase, on the glycopeptide or oligosaccharide product of the previous exoglycosidase reaction. Several types of exoglycosidases are now commercially available for use in oligosaccharide sequence analysis.

Neuraminidase (EC 3.2.1.18)

This enzyme cleaves terminal sialic acid residues that are linked via an α2→3-, α2→6-, or α2→8-linkage to Gal, GlcNAc, GalNAc, NeuAc (N-acetyl-neuraminic acid), or NeuGly (N-glycolyl-neuraminic acid) residues of oligosaccharides, glycolipids, or glycoproteins (Table 19).

Neuraminidase from A. *ureafaciens* shows some linkage specificity. The order of susceptibility of the sialic acid linkages for the enzyme digestion is: α2→6 > α2→3 > α2→8. The initial rate of hydrolysis of the α2→6-linkage is twice that for the α2→3-linkage.

Neuraminidase from *Clostridia perfringens* has the broadest specificity of the available sialidases, but it still preferentially cleaves α2→3 linkages by a factor of 2.5–3.0. N-Glycolyl derivatives of sialic acid are effectively cleaved, although the enzyme shows higher K_m and lower V_{max} values for the glycolyl derivatives than for the N-acetyl derivatives. The enzyme can cleave 9-O-acetyl sialic acid but does not cleave the 4-O-acetyl derivative.

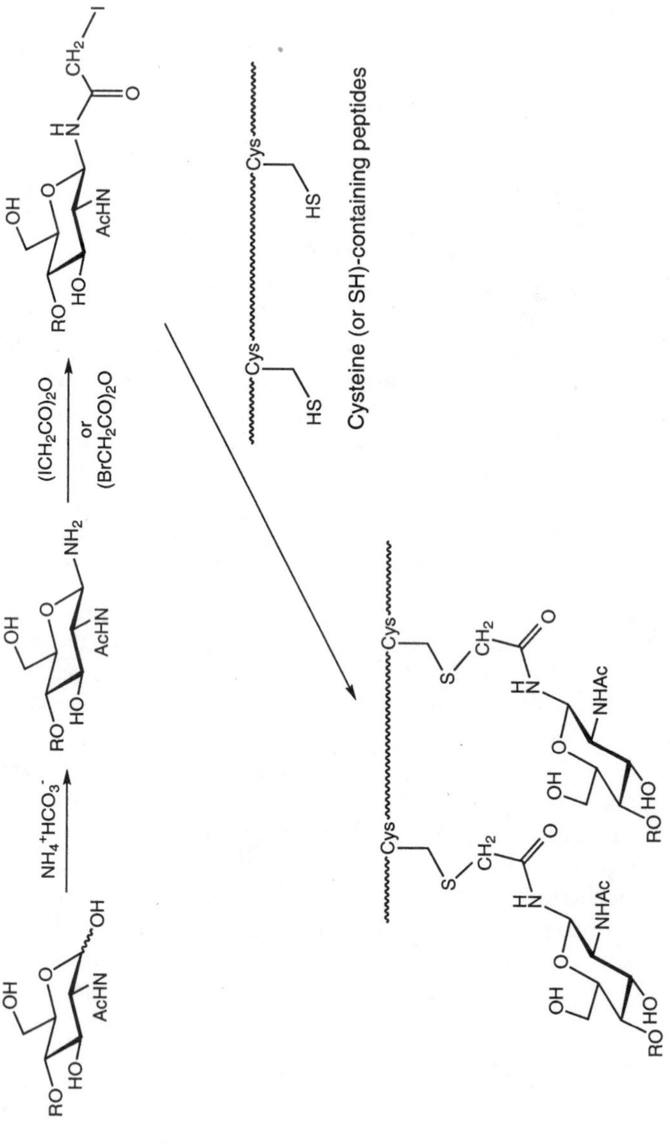

Scheme 51

Table 19 Linkage Specificity of Neuraminidases

Enzyme source	Linkage specificity
Neuraminidase from *Anthrobacter ureafaciens*	$\alpha2\rightarrow6 > \alpha2\rightarrow3 > \alpha2\rightarrow8$
Neuraminidase from *Clostridium perfringens*	$\alpha2\rightarrow3 > \alpha2\rightarrow8 \cong \alpha2\rightarrow6$
Neuraminidase from *Vibrio cholerae*	$\alpha2\rightarrow3 > \alpha2\rightarrow6 > \alpha2\rightarrow8$
Neuraminidase from *Salmonella typhimurium* (recombinant in *E. coli*)	$\alpha2\rightarrow3 > \alpha2\rightarrow6 >> \alpha2\rightarrow8$
Neuraminidase from Newcastle disease	$\alpha2\rightarrow3 > \alpha2\rightarrow8$; not $\alpha2\rightarrow6$

Neuraminidase from *Vibrio cholerae* shows a preference for $\alpha2\rightarrow3$ linkages and hydrolyzes the terminal *N*- or *O*-acyl-neuraminic acid. This enzyme cleaves virtually all the sialic acid linkages and is particularly useful for removal of sialic acid from cell-surface molecules such as receptors.

Neuraminidase from *Salmonella typhimurium* has a 260-fold kinetic preference for the $\alpha2\rightarrow3$ linked sialic acid linkage. This preference places this enzyme between the one from Newcastle disease virus, which is $\alpha2\rightarrow3$ specific, and those from *C. perfringens* and *V. cholerae*, which show a preference of approximately threefold for this linkage. *O*-Acylated sialic acid derivatives are poor substrates.

Neuraminidase from Newcastle disease virus is very specific and essentially hydrolyzes only the $\alpha2\rightarrow3$-linked sialic acid; it does cleave $\alpha2\rightarrow8$ sialic acid linkages but only very slowly. In contrast to the pattern observed for bacterial neuraminidases, this viral neurominidase does not hydrolyze $\alpha2\rightarrow6$-linked sialic acid linkages. *N*-Glycolyl, 4-*O*-acetyl-, and 9-*O*-acetyl-derivatives of sialic acid are poor substrates for the enzyme but are, nevertheless, hydrolyzed at significant rates.

Fucosidase (EC 3.2.1.51)

This enzyme cleaves α-linked terminal L-fucose residues. There are five types of α-L-fucosidases, with distinct linkage specificities (Table 20). The fucosidase from *Charonia lampas* shows linkage specificity and cleaves $\alpha1\rightarrow2$- or 6-linked fucose residues preferentially over $\alpha1\rightarrow3$- or 4-linked fucose residues. The fucosidase from almond emulsion-I specifically cleaves $\alpha1\rightarrow3$- or 4-linked fucose residues. In contrast, the fucosidase from almond emulsion-II cleaves only $\alpha1\rightarrow2$-linked fucose residues. The fucosidase from bovine epididymis cleaves $\alpha1\rightarrow6$-linkages preferentially over $\alpha1\rightarrow2$-, 3-, or 4-linked fucose residues. Chicken liver fucosidase cleaves preferentially $\alpha1\rightarrow2$-, 4-, or 6-linked fucose residues, but does not cleave $\alpha1\rightarrow3$-linked fucose residues.

α-*Galactosidase*

α-Galactosidase cleaves α-linked terminal galactose residues; two types of enzyme are known (Table 21). α-Galactosidase from recombinant *E. coli* cleaves all the α-

Table 20 Linkage Specificity of α-Fucosidases

Enzyme source	Linkage specificity
Fucosidase from *Charonia lampas*	α1→2 or 6 > α1→3 or 4
Fucosidase from almond emulsion-I	α1→3 or 4
Fucosidase from almond emulsion-II	α1→2
Fucosidase from bovine epididymis	α1→6 > α1→2, 3 or 4
Fucosidae from chicken liver	α1→2, 4 or 6

linked terminal galactose residues from oligosaccharides. α-Galactosidase from coffee beans cleaves only the α1→3- or 6-linked terminal galactose residues.

β-*Galactosidase*

This enzyme cleaves β-linked terminal galactose residues, and four types of enzyme are known (Table 22). β-Galactosidase from jack bean shows good linkage specificity, cleaving β1→4- or 6-linked galactose residues much better than β1→3-linked galactose residues. β-Galactosidase from *Streptococcus pneumoniae* also shows a good linkage specificity and cleaves β1→4-linked galactose residues; however, this specificity is evident only at low enzyme concentration (<100 mU/mL). At higher enzyme concentrations, hydrolysis of the β1→3-linkage also becomes evident. β-

Table 21 Linkage Specificity of α-Galactosidases

Enzyme source	Linkage specificity
α-Galactosidase from recombinant *E. coli*	α1→
α-Galactosidase from coffee bean	α1→3 or 6

Table 22 Linkage Specificity of β-Galactosidases

Enzyme source	Linkage specificity
β-Galactosidase from jack bean	β1→4 (or 6) >> β1→3
β-Galactosidase from *Streptococcus pneumoniae*	β1→4; not β1→3 or 6
β-Galactosidase from bovine tests	β1→3 ≥ β1→4 > β1→6
β-Galactosidase from chicken liver	β1→3 or 6
β-Galactosidase from *Bacillus circulans*	β1→4; not β1→3 or 6

Galactosidase from bovine testes cleaves β1→3-linked galactose more readily than β1→4- or 6-linked galactose residues. β-Galactosidase from chicken liver cleaves Galβ1→3GalNAc and Galβ1→4GlcNAc linkages specifically.

β-N-*Acetylhexosaminidase (EC 3.2.1.30)*

In addition to cleaving β-linked *N*-acetylglucosamine, this enzyme also cleaves *N*-acetylgalactosamine, but with less efficiency (Table 23). β-*N*-Acetylhexosaminidase from bovine kidney cleaves β-linked *N*-acetylglucosamine with a broad aglycone specificity. The enzyme also cleaves β-linked *N*-acetylgalactosamine at 20% of the reaction rate. β-*N*-Acetylhexosaminidase from jack bean cleaves β-linked *N*-acetylglucosamine and *N*-acetylgalactosamine with a broad aglycone specificity.

β-*N*-Acetylhexosaminidase from *Streptococcus pneumoniae* shows some linkage specificity. When incubated with *N*-linked oligosaccharides at low enzyme concentrations (<10 mU/mL), the enzyme can differentiate between GlcNAcβ1→2Man, GlcNAcβ1→4Man, and GlcNAcβ1→6Man linkages. Under such conditions, the enzyme cleaves essentially only β1→2-linked GlcNAc. The GlcNAcβ1→2Man linkage is not hydrolyzed if the linked mannose residue is further substituted at the C-6 position. If the β-linked mannose of the pentasaccharide core structure of an *N*-linked oligosaccharide is substituted with a bisecting GlcNAc residue, only the GlcNAcβ1→2Man linkage on the β1→3Man branch is cleaved. At high enzyme concentrations, β1→4- and 6-linked *N*-acetylglucosamine residues are also hydrolyzed.

The β-*N*-acetylglucosaminidase from *Diplococcus pneumoniae* displays a unique linkage specificity. It cleaves GlcNAcβ1→3Gal, GlcNAcβ1→6Gal, and most GlcNAcβ1→2Man; however, it does not cleave GlcNAc β1→4Man or GlcNAc β1→6Man.

The chicken liver α-*N*-acetylgalactosaminidase (EC 3.2.1.49) cleaves α-linked *N*-acetylgalactosamine residues of glycolipids and glycoproteins: GalNAcα1→3 and GalNAcα1→Ser/Thr.

Mannosidase *(EC 3.2.1.24)*

This enzyme cleaves α- or β-linked terminal mannose residues. α-Mannosidase from jack bean shows a rather broad aglycone specificity but displays good linkage specificity (Table 24). The enzyme hydrolyzes Manα1→2Man and Manα1→6Man at a rate of 100%; however, the rate is only 7% for the Manα1→3Man linkage. The enzyme does not cleave the single Manα1→6 residue that is linked to the core β-mannose, but it does cleave the single Manα1→3 linked to the core β-mannose.

α-Mannosidase from *Aspergillus phoenics* (saitoi) shows strict linkage specificity and hydrolyzes only the Manα1→2Man linkage. β-Mannosidase from *Achatina fulica* and *Helix pomatia* cleaves all the terminal β-mannosyl residues: Manβ1→4GlcNAc (or 2, 3, or 6).

Application of exoglycosidases to the preparation of glycopeptides

More than a thousand oligosaccharide structures on glycoproteins and glycolipids from natural sources have been purified and characterized, and recent advances in purification methods have made these glycopeptides obtainable on a semipreparative

Table 23 Linkage Specificity of *N*-Acetylhexosaminidases

Enzyme source	Linkage specificity
β-*N*-Acetylhexosaminidase from bovine kidney	GlcNAcβ1→ > GalNAcβ1→
β-*N*-Acetylhexosaminidase from jack bean	GlcNAc/GalNAcβ1→
β-*N*-Acetylhexosaminidase from *Streptococcus pneumoniae*	GlcNAc/GalNAcβ1→2 or 3
β-*N*-Acetylglucosaminidase from *Diplococcus pneumoniae*	GlcNAcβ1→3Gal, GlcNAcβ1→6Gal, and most GlcNAcβ1→2Man; not GlcNAcβ1→4Man or GlcNAcβ1→6Man
α-*N*-Acetylgalactosaminidase from chicken liver	GalNAcα1→3 or Serine/Threonine

scale. We can imagine that a glycopeptide with well-defined oligosaccharide structure can be constructed from these isolatable glycoproteins by employing a series of exoglycosidases (Scheme 52). In fact, this is what oligosaccharide sequence analysis involves.

The process could start with a typical triantennary *N*-linked complex-type structure, such as has been found in many glycoproteins. Protease digestion cleaves the peptide chain to relatively small fragments, some with oligosaccharide and some without. If the amino acid sequence of the polypeptide chain and the specificity of the protease are both known, the amino acid sequence of a small peptide that carries an oligosaccharide becomes predictable.

The intact oligosaccharide has terminal sialic acid α(2→3)-linked to galactose residues, as well as a terminal fucose residue linked to the inner core *N*-acetylglucosamine residue via an α(1→6)-linkage. The sialic acid moieties can be selectively removed using a neuraminidase (from virtually any source) to give a triantennary oligosaccharide-carrying glycopeptide with three exposed terminal galactose residues. The fucose residue can then be removed by using α-L-fucosidase from *Charonial lampas* or bovine epididymis.

β-Galactosidase can then be used to remove the galactose residues from the neuraminidase- and fucosidase-treated glycopeptide. All *N*-acetylglucosamine residues are removed by *N*-acetylhexosaminidase from bovine kidney or jack bean. If

Table 24 Linkage Specificity of α- and β-Mannosidases

Enzyme source	Linkage specificity
α-Mannosidase from jack bean	α1→2 or 6 >> α1→3
α-Mannosidase from *Aspergillus phoenics*	α1→2
β-Mannosidase from *Achatina fulica*	β1→4 (or 2, 3, or 6)
β-Mannosidase from *Helix pomatia*	β1→4 (or 2, 3, or 6)

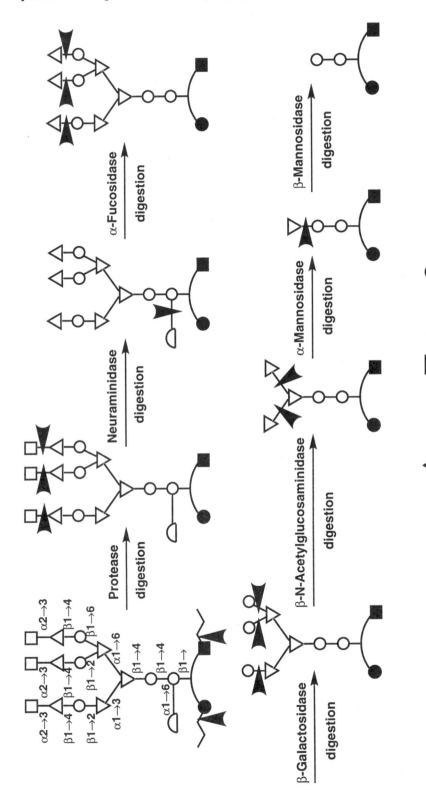

Scheme 52

the *N*-acetylhexosaminidase from *Streptococcus pneumoniae* is used, the enzyme digestion will leave the β(1→6)-linked *N*-acetylglucosamine intact on the mannose residue.

Mannose residues linked either α(1→6) or α(1→2) to the β-mannoside can be differentiated: α-Mannosidase from *Aspergillus phoenics* selectively removes the α(1→2)-linked mannose, leaving the α(1→6)-linked mannose residue on the oligosaccharide. Alternatively, α-mannosidase from jack bean removes both the α(1→6)- and α(1→2)-linked mannose residues and gives a glycopeptide with β-mannose exposed. β-Mannosidase from *Achatina fulica* can remove the β-mannosyl residue and leave a small glycopeptide, a short peptide chain with chitobiose (GlcNAcβ(1→4)GlcNAc) on its Asn residue.

VI. FUTURE OUTLOOK

A number of research papers have recently reported the identification of "novel carbohydrate structures" that serve as ligands in biologically important recognition systems. In addition, recent advances in chromatographic and analytical techniques have revealed complex carbohydrates to be realistic targets for synthetic chemists. If we are to obtain better understanding of such carbohydrate-mediated recognition processes at the molecular level, a good source of these complex carbohydrate molecules must be established.

Enzymes involved in the biosynthesis and metabolism of carbohydrates have been known for a long time and are now recognized as useful "reagents" for constructing complex carbohydrates. Glycosyltransferase reactions with cofactor regeneration in situ have paved the way to large-scale synthesis of complex carbohydrates. An increasing number of recombinant glycosyltransferases and other carbohydrate-related enzymes should provide enough of all the enzymes necessary for the synthesis. By combining enzymatic and chemical procedures, biologically active complex carbohydrates can be efficiently synthesized in large quantities.

REFERENCES

1. Enzymes in Organic Synthesis, *Symposium on Organic Synthesis Using Enzymes* (1984: Ciba Foundation), (R. Porter and C. Sarah, eds.), Pitman, London, 1985.
2. *Enzyme Catalysts in Organic Synthesis* (K. Drauz and H. Waldman, eds.), VCH Publisher, 1994.
3. D. G. Drueckhammer, W. J. Hennen, R. L. Pederson, C. F. Barbas, III, C. M. Gautheron, T. Krach, and C.-H. Wong, Enzyme catalysis in synthetic carbohydrate chemistry, *Synthesis* 499–525 (1991).
4. E. J. Toone, E. S. Simon, M. D. Bednarski, and G. M. Whitesides, Enzymes in carbohydrate chemistry, *Tetrahedron 45*: 5365–5422 (1989).
5. C.-H. Wong and G. M. Whitesides, Enzymes in Synthetic Organic Chemistry, *Tetrahedron Organic Chemistry Series*, Vol. 12, Pergammon, 1994.
6. C.-H. Wong, R. L. Halcomb, Y. Ichikawa, and T. Kajimoto, Enzymes in organic synthesis: Application to the problems of carbohydrate recognition, *Angew. Chem. Int. Ed. Engl. 34*: 412–432, 521–546 (1995).

7. H. A. Nunez and R. Barker, Enzymatic synthesis and carbon-13 nuclear magnetic resonance conformational studies of oligosaccharides containing β-D-galactopyranosyl and β-D-[β1-^{13}C]galactopyranosyl residues, *Biochemistry 19*: 485–495 (1980).

8. L. F. Leloir, Two decades of research on the biosynthesis of saccharides, *Science 172*: 1299–1303 (1971).

9. A. T. Bayer, J. E. Saddler, J. I. Rearick, J. C. Paulson, and H. L. Hill, Glycosyltransferases and their use in assembling oligosaccharide structure and structure-function relationships, *Adv. Enzymol. 52*: 23–175 (1981).

10. K. G. I. Nilsson, Enzymatic synthesis of oligosaccharides, *TIBTECH 6*: 256–264 (1988).

11. G. L. Cote and B. Y. Tao, Oligosaccharide synthesis by enzymatic transglycosylation, *Glycoconj. J. 7*: 145–162 (1990).

12. S. David, C. Augé, and C. Gautheron, Enzymatic methods in preparative carbohydrate chemistry, *Adv. Carbohydr. Chem. Biochem. 49*: 175–237 (1991).

13. Y. Ichikawa, G. C. Look, and C.-H. Wong, Enzyme-catalyzed oligosaccharide synthesis, *Anal. Biochem. 202*: 215–238 (1992).

14. R. A. Rastall and C. Bucke, Enzymatic synthesis of oligosaccharides, *Biotechnol. Genet. Eng. Rev. 10*: 253–281 (1992).

15. A. Kobata, The carbohydrates of glycoproteins, *Biology of Carbohydrates* (V. Ginsburg and P. W. Robins, eds.), Wiley, New York, 1984, pp. 87–161.

16. S. Hakomori, Tumor-associated carbohydrate antigens, *Ann. Rev. Immunol. 2*: 103–126 (1984).

17. T. Feizi, Demonstration by monoclonal antibodies that carbohydrate structures of glycoproteins and glycolipids are onco-developmental antigens, *Nature 314*: 53–57 (1985).

18. C. L. M. Stults, C. C. Sweeley, and B. A. Macher, Glycosphingolipids: Structure, biological source, and properties, *Methods in Enzymol. 179*: 167–214 (1989).

19. M. D. Snider, Biosynthesis of glycoproteins: Formation of *N*-linked oligosaccharides, *Biology of Carbohydrates* (V. Ginsburg and P. W. Robins, eds.), Wiley, New York, 1984, pp. 163–198.

20. J. E. Sadler, Biosynthesis of glycoproteins: Formation of *O*-linked oligosaccharides, *Biology of Carbohydrates* (V. Ginsburg and P. W. Robins, eds.), Wiley, New York, 1984, pp. 199–288.

21. R. Kornfeld and S. Kornfeld, Assembly of asparagine-linked oligosaccahrides, *Ann. Rev. Biochem. 54*: 631–664 (1985).

22. Y. T. Li and S. C. Li, Biosynthesis and catabolism of glycolipids, *Adv. Carbohydr. Chem. Biochem. 40*: 235–288 (1982).

23. W. Z. Hassid and M. Doudoroff, Enzymatic synthesis of sucrose and other disaccharides, *Adv. Carbohydr. Chem. Biochem. 5*: 29–48 (1950).

24. A. Varki, Biological roles of oligosaccharides: All of the theories are correct, *Glycobiology 3*: 97–130 (1993).

25. C. H. Bauer, W. G. Reutter, K. P. Erhart, E. Köttgen, and W. Gerok, Decrease of human serum fucosyltransferase as an indicator of successful tumor therapy, *Science 201*: 1232–1233 (1978).

26. S. Yazawa, R. Madiyalakan, H. Izawa, T. Asao, K. Furukawa, and K. L. Matta, Cancer-associated elevation of α(1→3)-L-fucosyltransferase activity in human serum, *Cancer 62*: 516–520 (1988).

27. S. Yazawa, R. Madiyalakan, R. K. Jain, N. Shimoda, and K. L. Matta, Use of benzyl 2-acetamido-2-deoxy-3-*O*-(2-*O*-methyl-β-D-galactosyl)-β-D-glucopyranoside [2'-*O*-methyllacto-*N*-biose 1βBn] as a specific acceptor for GDP-fucose: *N*-acetylglucosaminide α(1→4)-fucosyltransferase, *Anal. Biochem. 187*: 374–378 (1990).

28. J. E. Heidlas, K. W. Williams, and G. M. Whitesides, Nucleoside phosphate sugars: Synthesis on practical scales for use as reagents in the enzymatic preparation of oligosaccharides and glycoconjugates, *Acc. Chem. Res. 25*: 307–314 (1992).

29. H. A. Nunez, J. V. O'Connor, P. R. Rosevear, and R. Barker, The synthesis and characterization of α- and β-L-fucopyranosyl phosphates and GDP-fucose, *Can. J. Chem. 59*: 2086–2095 (1981).

30. U. B. Gokhale, O. Hindsgaul, and M. M. Palcic, Chemical synthesis of GDP-fucose analogs and their utilization by the Lewis α(1→4) fucosyltransferase, *Can. J. Chem. 68*: 1063–1071 (1990).

31. R. R. Schmidt, B. Wegmann, and K.-H. Jung, Stereospecific synthesis of α- and β-L-fucopyranosyl phosphates and of GDP-fucose via trichloroacetimidate, *Liebigs Ann. Chem.* 121–124 (1991).

32. G. H. Veeneman, H. J. G. Broxterman, G. A. van der Marel, and J. H van Boom, An approach towards the synthesis of 1,2-*trans*-glycosyl phosphates via iodonium ion assisted activation of thioglycosides, *Tetrahedron Lett. 32*: 6175–6178 (1991).

33. Y. Ichikawa, M. M. Sim, and C.-H. Wong, Efficient chemical synthesis of GDP-fucose, *J. Org. Chem. 57*: 2944–2946 (1992).

34. K. Adelhorst and G. M. Whitesides, Large-scale synthesis of β-L-fucopyranosyl phosphate and the preparation of GDP-β-L-fucose, *Carbohydr. Res. 242*: 69–76 (1993).

35. V. Ginsburg, Formation of guanosine diphosphate L-fucose from guanosine diphosphate D-mannose, *J. Biol. Chem. 235*: 2196–2201 (1960).

36. H. Ishihara, D. J. Massaro, and E. C. Heath, The metabolism of L-fucose. III. The enzymatic synthesis of β-L-fucose 1-phosphate, *J. Biol. Chem. 243*: 1103–1109 (1968).

37. H. Ishihara and E. C. Heath, The metabolism of L-fucose. IV. The biosynthesis of guanosine diphosphate L-fucose in porcine liver, *J. Biol. Chem. 243*: 1110–1115 (1968).

38. P. D. Yurchenco and P. H. Atkinson, Fucosyl-glycoprotein and precursor pools in HeLa cells, *Biochemistry 14*: 3107–3114 (1975).

39. R. Prohasko and H. Schenkel-Brunner, A simple and efficient method for the preparation of GDP-fucose, *Anal. Biochem. 69*: 536–544 (1975).

40. W. L. Richards and G. S. Serif, Canine thyroid fucokinase, *Biochim. Biophys. Acta 484*: 353–367 (1977).

41. R. D. Kilker, D. H. Shuey, and G. S. Serif, Isolation and properties of porcine thyroid fucokinase, *Biochim. Biophys. Acta 570*: 271–283 (1979).

42. W. Butler and G. S. Serif, Fucokinase, its anomeric specificity and mechanism of phosphate group transfer, *Biochim. Biophys. Acta 829*: 238–243 (1985).

43. P. Stangier and J. Thiem, Glycosylation by use of glycohydrolases and glycosyltransferases in preparative scale, *Enzymes in Carbohydrate Synthesis* (M. D. Bednarski and E. S. Simon, eds.), ACS Symposium Series, American Chemical Society, Washington, DC, 1991, Vol. 466, pp. 63–78.

44. R. Stiller and J. Thiem, Enzymatic synthesis of β-L-fucose-1-phosphate and GDP-fucose, *Liebigs Ann. Chem.* 467–471 (1992).

45. Y. Ichikawa, Y.-C. Lin, D. P. Dumas, G.-J. Shen, E. Garcia-Junceda, M. A. Williams, R. Bayer, C. Ketcham, L. E. Walker, J. C. Paulson, and C.-H. Wong, Chemical-enzymatic synthesis and conformational analysis of sialyl Lewis x and derivatives, *J. Am. Chem. Soc. 114*: 9283–9298 (1993).

46. A. Sarnesto, T. Kohlin, J. Thurin, and M. Blaszczyk-Thurin, Purification of H gene-encoded β-galactoside α1→2fucosyltransferase from human serum, *J. Biol. Chem. 265*: 15067–15075 (1990).

47. P. Kyprianou, A. Betteridge, A. S. Donald, and W. M. Watkins, Purification of the blood group H gene associated α-2-L-fucosyltransferase from human plasma, *Glycoconj. J. 7*: 573–588 (1990).

48. A. Sarnesto, T. Kohlin, O. Hindsgaul, J. Thurin, and M. Blaszczyk-Thurin, Purification of the secretor-type β-galactoside α1→2-fucosyltransferase from human serum, *J. Biol. Chem. 267*: 2737–2744 (1992).

49. L. K. Ernst, V. P. Rajan, R. D. Larsen, M. M. Ruff, and J. B. Lowe, Stable expression of blood group H determinants and GDP-L-fucose: β-D-galactoside 2-α-L-fucosyltransferase in mouse cells after transfection with human DNA, *J. Biol. Chem. 264*: 3436–3447 (1989).

50. R. D. Larsen, L. K. Ernst, R. P. Nair, and J. B. Lowe, Molecular cloning, sequence, and expression of a human GDP-L-fucose: β-D-galactoside 2-α-L-fucosyltransferase cDNA that can form the H blood group antigen, *Proc. Natl. Acad. Sci. USA 87*: 6674–6678 (1990).

51. P. R. Rosevear, H. A. Nunez, and R. Barker, Synthesis and solution conformation of the Type 2 blood group oligosaccharide α-L-Fuc(1→2)β-D-Gal(1→4)GlcNAc, *Biochemistry 21*: 1421–1431 (1982).

52. T. L. Lowary, S. J. Swiedler, and O. Hindsgaul, Recognition of synthetic analogues of the acceptor, β-D-Galp-OR, by the blood-group H gene-specified glycosyltransferase, *Carbohydr. Res. 256*: 257–273 (1994).

53. A. E. Mourant, A "new" human blood antigen of frequent occurrence, *Nature 158*: 237–238 (1946).

54. N. Kojima, B. A. Fenderson, M. R. Stroud, R. I. Goldberg, R. Habermann, T. Toyokuni, and S. Hakomori, Further studies on cell adhesion based on Le^x–Le^x interaction, with new approaches: Embryoglycan aggregation of F9 tetracarcinoma cells, and adhesion of various tumor cell based on Le^x expression, *Glycoconj. J. 11*: 238–248 (1994).

55. J. B. Lowe, L. M. Stoolman, R. P. Nair, R. D. Larsen, T. L. Berhend, and R. M. Marks, ELAM-1-dependent cell adhesion to vascular endothelium determined by a transfected human fucosyltransferase cDNA, *Cell 63*: 475–484 (1990).

56. M. L. Phillips, E. Nudelman, F. C. A. Gaeta, M. Perez, A. K. Shinghal, S. Hakomori, and J. C. Paulson, ELAM-1 mediates cell adhesion by recognition of a carbohydrate ligand, sialyl–Le^x, *Science 250*: 1130–1132 (1990).

57. G. Walz, A. Aruffo, W. Kolanus, M. Bevilacqua, and B. Seed, Recognition by ELAM-1 of the sialyl Lex determinant on myeloid and tumor cells, *Science 250*: 1132–1135 (1990).

58. M. J. Polly, M. E. Phillips, E. Wayner, E. Nudelman, A. K. Singhal, S. Hakomori, and J. C. Paulson, CD62 and endothelial cell-leukocyte adhesion molecule 1 (ELAM-1) recognize the same carbohydrate ligand, sialyl Lewis-x, *Proc. Natl. Acad. Sci. USA 88*: 6224–6228 (1991).

59. Y. Imai, M. S. Singer, C. Fennie, L. A. Lasky, and S. D. Rosen, Identification of a carbohydrate-based endothelial ligand for a lymphocyte homing receptor, *J. Cell Biol. 113*: 1213–1221 (1991).

60. E. C. Butcher, Leukocyte-endothelial cell recognition: Three (or more) steps to specificity and diversity, *Cell 67*: 1033–1036 (1991).

61. K. L. Moore, A. Varki, and R. P. McEver, GMP-140 binds to a glycoprotein receptor on human neutrophils: Evidence for a lectin-like interaction, *J. Cell Biol. 112*: 491–499 (1991).

62. M. Tiemeyer, S. J. Swiedler, M. Ishihara, M. Moreland, H. Schweingruber, P. Hirtzer, and B. K. Brandley, Carbohydrate ligands for endothelial-leukocyte adhesion molecule 1, *Proc. Nal. Acad. Sci. USA 88*: 1138–1142 (1991).

63. E. L. Berg, M. K. Robinson, O. Mansson, E. C. Butcher, and J. L. Magnani, A carbohydrate domain common to both sialyl Le^a and sialyl Le^x is recognized by the endothelial cell leukocyte adhesion molecule ELAM-1, *J. Biol. Chem. 266*: 14869–14872 (1991).

64. T. Feizi, Carbohydrate differentiation antigens: Probable ligands for cell adhesion molecules, *TIBS 16*: 84–86 (1991).

65. Y. Imai, L. A. Lasky, and S. D. Rosen, Further characterization of the interaction between L-selectin and its endothelial ligands, *Glycobiology 2*: 373–381 (1992).

66. D. V. Erde, B. A. Wolitzky, L. G. Presta, C. R. Norton, R. J. Ramos, D. K. Burns, J. M. Rumberger, B. N. N. Rao, C. Foxall, B. K. Brandley, and L. A. Lasky, Identification of an E-selectin region critical for carbohydrate recognition and cell adhesion, *J. Cell Biol. 119*: 215–227 (1992).

67. G. R. Larsen, D. Sako, T. J. Ahern, M. Shaffer, J. Erban, S. A. Sajer, R. M. Gibson, D. D. Wagner, B. C. Furie, and B. Furie, P-selectin and E-selectin, distinct but overlapping leukocyte ligand specificities, *J. Biol. Chem. 267*: 11104–11110 (1992).

68. A. Varki, Selectin ligands, *Proc. Natl. Acad. Sci. USA 91*: 7390–7397 (1994).

69. R. D. Cummings and D. F. Smith, The selectin family of carbohydrate-binding proteins: Structure and importance of cabohydrate ligands for cell adhesion, *BioEssays, 14*: 849–856 (1992).

70. D. V. Erve, S. R. Watson, L. G. Presta, B. A. Wolitzky, C. Foxall, B. K. Brandley, and L. A. Lasky, P- and E-selectin use common sites for carbohydrate ligand recognition and cell adhesion, *J. Cell Biol. 120*: 1227–1235 (1993).

71. D. Tyrrel, P. James, N. Rao, C. Foxall, S. Abbas, F. Dasgupta, M. Nashed, A. Hasegawa, M. Kiso, D. Asa, J. Kidd, and B. K. Brandley, Structural requirements for the carbohydrate ligand of E-selectin, *Proc. Natl. Acad. Sci. USA 88*: 10372–10376 (1991).

72. B. K. Brandley, M. Kiso, S. Abbas, P. Nikrad, O. Srivastava, C. Foxall, Y. Oda, and A. Hasegawa, Structure-function studies on selectin carbohydrate ligands. Modifications to fucose, sialic acid and sulfate group as a sialic acid replacement, *Glycobiology 3*: 633–639 (1993).

73. C.-T. Yuen, A. M. Lawson, W. Chai, M. Larkin, M. S. Stoll, A. C. Stuart, F. X. Sullivan, T. J. Ahern, and T. Feizi, Novel sulfated ligands for the cell adhesion molecule E-selectin revealed by the neoglycolipid technology among *O*-linked oligosaccharides on an ovarian cystadenoma glycoprotein, *Biochemistry 31*: 9126–9131 (1992).

74. C.-T. Yuen, K. Bezouska, J. O'Brien, M. Stoll, R. Lemoine, A. Lubineau, M. Kiso, A. Hasegawa, N. J. Bockovich, K. C. Nicolaou, and T. Feizi, Sulfated blood group Lewis[a], *J. Biol. Chem. 269*: 1595–1598 (1994).

75. S. B. Yan, Y. B. Chao, and H. van Halbeek, Novel Asn-linked oligosaccharides terminating in GalNAcβ(1→4)[Fucα(1→3)]GlcNAcβ(1→•) are present in recombinant human Protein C expressed in human kidney 293 cells, *Glycobiology 3*: 597–608 (1993).

76. B. W. Grinnel, R. B. Hermann, and S. B. Yan, Human Protein C inhibits selectin-mediated cell adhesion: Role of unique fucosylated oligosaccharide, *Glycobiology 4*: 221–225 (1994).

77. G. E. Ball, R. A. O'Neill, J. E. Schultz, J. B. Lowe, B. W. Weston, J. O. Nagy, E. G. Brown, C. J. Hobbs, and M. D. Bednarski, Synthesis and structural analysis using 2-D NMR of sialyl Lewis x (SLe^x) and Lewis x (Le^x) oligosaccharides: Ligands related to E-selectin [ELAM-1] binding, *J. Am. Chem. Soc. 114*: 5449–5450 (1992).

78. P. L. Sadowski and G. A. Strobel, Guanosine diphosphate-L-fucose glycopeptide fucosyltransferase activity in *Corynebacterium insidiosum, J. Bacteriol. 115*: 668–672 (1973).

79. P. H. Johnson, A. D. Yates, and W. M. Watkins, Human salivary fucosyltransferase: Evidence for two distinct α-3-L-fucosyltransferase activities, one of which is associated with the Lewis blood group Le gene, *Biochem. Biophys. Res. Commun. 100*: 1611–1618 (1981).

80. D. R. Howard, M. Fukuda, M. N. Fukuda, and P. Stanley, The GDP-fucose: *N*-acetyl-glucosaminide 3-α-L-fucosyltransferases of LEC11 and LEC12 Chinese hamster ovary

mutants exhibit novel specificity for glycolipid substrates, *J. Biol. Chem. 262*: 16830–16837 (1987).

81. M. Basu, J. W. Hawes, Z. Li, S. Ghosh, F. A. Khan, B. J. Zhang, and S. Basu, Biosynthesis in vivo of SA-Le x and SA-di-Le x by α1→3-fucosyltransferases from colon carcinoma cells and embryonic brain tissues, *Glycobiology 1*: 527–535 (1991).

82. A. Sarnesto, T. Kohlin, O. Hindsgaul, K. Vogele, M. Blaszczyk-Thurin, and J. Thurin, Purification of the β-*N*-acetylglucosaminide α1→3-fucosyltransferase from human serum, *J. Biol. Chem. 267*: 2745–2752 (1992).

83. S. E. Goelz, C. Hssion, D. Goff, B. Griffiths, R. Tizard, B. Newman, G. Chi-Rosso, and R. Lobb, ELFT: A gene that directs the expression of an ELAM-1 ligand, *Cell 63*: 1349–1356 (1990).

84. R. Kumar, B. Potvin, W. A. Muller, and P. Stanley, Cloning of a human α(1→3)-fucosyltransferase gene that encodes ELFT but does not confer ELAM-1 recognition on Chinese hamster ovary cell transfectants, *J. Biol. Chem. 266*: 21777–21783 (1991).

85. J. B. Lowe, J. F. Kulowska-Latallo, R. P. Nair, R. D. Larsen, R. M. Marks, B. A. Macher, R. J. Kelly, and L. K. Ernst, Molecular cloning of a human fucosyltransferase gene that determines expression of the sialyl Lewis x and VIM-2 epitopes but not ELAM-1 dependent cell adhesion, *J. Biol. Chem. 266*: 17467–17477 (1991).

86. B. W. Weston, R. P. Nair, R. D. Larsen, and J. B. Lowe, Isolation of a novel human α(1,3)fucosyltransferase gene and molecular comparison to the human Lewis blood group α(1,3/4)fucosyltransferase gene. Syntenic, homologous, nonallelic gene encoding enzymes with distinct acceptor substrate specificities, *J. Biol. Chem. 267*: 4152–4160 (1992).

87. B. W. Weston, P. L. Smith, R. J. Kelly, and J. B. Lowe, Molecular cloning of a fourth member of a human α(1,3)fucosyltransferase gene family. Multiple homologous sequences that determine expression of the Lewis x, sialyl Lewis x, and difucosyl sialyl Lewis x epitopes, *J. Biol. Chem. 267*: 24575–24584 (1992).

88. K. L. Koszdin and B. R. Bowen, The cloning and expression of a human α1,3-fucosyltransferase capable of forming the E-selectin ligand, *Biochem. Biophys. Res. Commun. 187*: 152–157 (1992).

89. S. Nakatsuka, K. M. Gersten, K. Zenita, R. Kannagi, and J. B. Lowe, Molecular cloning of a cDNA encoding a novel human leukocyte α-1,3-fucosyltransferase capable of synthesizing the sialyl Lewis x determinant, *J. Biol Chem. 269*: 16789–16794 (1994).

90. K. Sasaki, K. Kurata, K. Funayama, M. Nagata, E. Watanabe, S. Ohta, N. Hanai, and T. Nishi, Expression cloning of a novel α1,3-fucosyltransferase that is involved in biosynthesis of the sialyl Lewis x carbohydrate determinants in leukocytes, *J. Biol. Chem. 269*: 14730–14737 (1994).

91. R. Mollicone, A. Gibaud, A. Francois, M. Ratcliffe, and R. Oriol, Acceptor specificity and tissue distribution of three human α-3-fucosyltransferases, *Eur. J. Biochem. 191*: 169–176 (1990).

92. C.-H. Wong, D. P. Dumas, Y. Ichikawa, K. Koseki, S. J. Danishefsky, B. W. Weston, and J. B. Lowe, Specificity, inhibition, and synthetic utility of a recombinant human α-1,3-fucosyltransferase, *J. Am. Chem. Soc. 114*: 7321–7322 (1992).

93. R. Mollicone, J. J. Candelier, B. Mennesson, B. Mennesson, P. Couillin, A. P. Venot, and R. Oriol, Five specificity patterns of (1→3)-α-L-fucosyltransferase activity defined by use of synthetic oligosaccharide acceptors. Differential expression of the enzymes during human embryonic development and in adult tissues, *Carbohydr. Res. 228*: 265–276 (1992).

94. J. J. Candelier, R. Mollicone, B. Mennesson, A. M. Bergener, S. Henry, P. Coullin, and R. Oriol, α-3-Fucosyltransferases and their glycoconjugate antigen products in the developing human kidney, *Lab. Invest. 69*: 449–459 (1993).

95. M. A. Kashem, K. B. Wlasichuk, J. M. Gregson, and A. P. Venot, Chemoenzymic synthesis of sialylated and fucosylated oligosaccharides having an N-acetyllactosaminyl core, *Carbohydr. Res. 250*: 129–144 (1993).

96. M. Jezequel-Cuer, H. N'Guyen-Cong, D. Biou, and G. Durand, Oligosaccharide specificity of normal human hepatocyte α1–3fucosyltransferase, *Biochim. Biophys. Acta 1157*: 252–258 (1993).

97. J. Y. Zhao, N. J. Dovichi, O. Hindsgaul, S. Gosselin, and M. M. Palcic, Detection of 100 molecules of product formed in a fucosyltransferase reaction, *Glycobiology 4*: 239–242 (1994).

98. J.-P. Prieels, D. Monnom, M. Dolmans, T. A. Beyer, and R. L. Hill, Co-purification of the Lewis blood group N-acetylglucosaminide α1→4fucosyltransferase and an N-acetylglucosaminide α1→3fucosyltransferase from human milk, *J. Biol. Chem. 256*: 10456–10463 (1981).

99. P. H. Johnson and W. M. Watkins, Purification of the Lewis blood-group gene associated α-3/4-fucosyltransferase purified from human milk: An enzyme transferring fucose primarily to type 1 and lactose-based oligosaccharide chains, *Glycoconj. J. 9*: 241–249 (1992).

100. P. H. Johnson, A. S. Donald, J. Feeney, and W. M. Watkins, Reassessment of the acceptor specificity and general properties of the Lewis blood-group gene associated α-3/4-fucosyltransferase purified from human milk, *Glycoconj. J. 9*: 251–264 (1992).

101. E. H. Holmes, Human Lewis α1→3/4fucosyltransferase: Specificity of fucose transfer to GlcNAcβ1→3Galβ1→4Glcβ1→Cer, *Glycobiology 3*: 77–81 (1993).

102. P. V. Nikrad, M. A. Kashem, K. B. Wlasichuk, G. Alton, and A. P. Venot, Use of human milk fucosyltransferase in the chemoenzymic synthesis of analogues of the sialyl Lewis^x and sialyl Lewis^a tetrasaccharides modified at the C-2 position of the reducing unit, *Carbohydr. Res. 250*: 145–160 (1993).

103. T. de Vries and D. H. van den Eijnden, Biosynthesis of sialyl-oligomeric-Lewis x and VIM-2 epitopes: Site specificity of human milk fucosyltransferase, *Biochemistry 33*: 9937–9944 (1994).

104. D. P. Dumas, Y. Ichikawa, C.-H. Wong, J. B. Lowe, and R. P. Nair, Enzymatic synthesis of sialyl Lex and derivatives based on a recombinant fucosyltransferase, *Bioorg. Med. Chem. Lett. 1*: 425–428 (1991).

105. G. Srivastava, K. J. Kaur, O. Hindsgaul, and M. M. Palcic, Enzymatic transfer of a preassembled trisaccharide antigen to cell surface using a fucosyltransferase, *J. Biol. Chem. 267*: 22356–22361 (1992).

106. S. C. Crawley, O. Hindsgaul, R. M. Ratcliffe, L. R. Lamontagne, and M. M. Palcic, A plant fucosyltransferase with human Lewis blood-group specificity, *Carbohydr. Res. 193*: 249–256 (1989).

107. M. M. Palcic, A. P. Venot, R. M. Ratcliffe, and O. Hindsgaul, Enzymic synthesis of oligosaccharides terminating in the tumor-asociated sialyl-Lewis-a determinant, *Carbohydr. Res. 190*: 1–11 (1989).

108. M. M. Palcic, R. M. Ratcliffe, L. R. Lamontagne, A. H. Good, G. Alton, and O. Hindsgaul, An enzyme-linked immunosorbent assay for the measurement of Lewis blood-group α-(1→4)-fucosyltransferase activity, *Carbohydr. Res. 196*: 133–140 (1990).

109. U. B. Gokhale, O. Hindsgaul, and M. M. Palcic, Chemical synthesis of GDP-fucose analogs and their utilization by the Lewis α(1→4)fucosyltransferase, *Can. J. Chem. 68*: 1063–1071 (1990).

110. J. A. Voynow, T. F. Scanlin, and M. C. Glick, A quantitative method for GDP-L-Fuc: N-acetyl-β-D-glucosaminide α1→6fucosyltransferase activity with lectin affinity chromatography, *Anal. Biochem. 168*: 367–373 (1988).

111. E. Staudacher, F. Altmann, J. Glössl, L. März, H. Schachter, J. P. Kamerling, K. Hård, and J. F. G. Vliegenthart, GDP-fucose: β-*N*-acetylglucosamine (Fuc to (Fucα1→ 6GlcNAc)-Asn-peptide) α1,3-fucosyltransferase activity in honeybee (*Apis mellifica*) venom glands. The difucosylation of asparagine-bound *N*-acetylglucosamine, *Eur. J. Biochem. 199*: 745–751 (1991).

112. G. D. Longmore and H. Schachter, Product-identification and substrate-specificity studies of the GDP-L-fucose: 2-acetamido-2-deoxyβ-D-glucoside (Fuc→Asn-linked GlcNAc) 6-α-L-fucosyltransferase in a Golgi-rich fraction from porcine liver, *Carbohydr. Res. 100*: 365–392 (1982).

113. M. C. Shao, C. W. Sokolik, and F. Wold, Specificity studies of the GDP-L-fucose: 2-acetamido-2-deoxy-β-D-glucoside (Fuc→Asn-linked GlcNAc) 6-α-L-fucosyltransferase from rat-liver Golgi membranes, *Carbohydr. Res. 251*: 163–173 (1994).

114. R. Oehrlein, O. Hindsgaul, and M. M. Palcic, Use of the "core-2"-*N*-acetylglucos-aminyltransferase in the chemical-enzymatic synthesis of a sialyl-Le^x-containing hexa-saccharide found on *O*-linked glycoproteins, *Carbohydr. Res. 244*: 149–159 (1993).

115. J. van Pelt, L. Dorland, M. Duran, C. H. Hokke, J. P. Kamerling, and J. F. G. Vlie-genthart, Sialyl-α2,6-mannosyl-β1,4*N*-acetylglucosamine, a novel compound occurring in urine of patients with β-mannosidosis, *J. Biol. Chem. 265*: 19685–19689 (1990).

116. R. Schauer, Sialic acids, Chemistry, metabolism, and function, *Adv. Carbohydr. Chem. Biochem. 40*: 131–234 (1982).

117. Sialic acids 1988 (R. Schauer and T. Yamakawa, eds.), *Proceedings of Japanese-German Symposium on Sialic Acid*, Berlin, May 18–21, 1988.

118. W. B. De Lau, J. Kuipers, H. Voshol, H. Clevers, and B. J Base, HB4 antibody rec-ognizes a carbohydrate structure on lymphocyte surface proteins related to HB6, CDw75, and CD76 antigens, *J. Immunol. 150*: 4911–4919 (1993).

119. J. C. Paulson, Interactions of animal viruses with cell surface receptor, *The Receptor* (M. Conn, ed.), Vol. 2, Academic Press, New York, 1985, pp. 131–219.

120. J. Liekkonen, S. Haataja, K. Tikkanen, S. Kelm, and J. Finne, Identification of *N*-acetylneuraminyl α2→3 poly-*N*-acetyllactosamine glycans as the receptors of sialic acid-binding *Streptococcus suis* strains, *J. Biol. Chem. 267*: 21105–21111 (1992).

121. L. R. Glasgow and R. L. Hill, Interaction of Mycoplasma gallisepticum with sialyl glycoproteins, *Infec. Immun. 30*: 353–361 (1980).

122. L. M. Loomes, K. Vemura, R. A. Childs, J. C. Paulson, G. N. Rogers, P. R. Scudder, J. Michalski, E. F. Hounsell, D. Taylor-Robinson, and T. Feizi, Erythrocyte receptors for *Mycoplasma pneumoniae* are sialylated oligosaccharides of Ii antigen type, *Nature 307*: 560–563 (1984).

123. M. H. Ravindranath, H. H. Higa, E. L. Cooper, and J. C. Paulson, Purification and characterization of an *O*-acetylsialic acid-specific lectin from a marine crab *Cancer antennarius*, *J. Biol. Chem. 260*: 8850–8856 (1985).

124. J. Jancik and R. Schauer, Sialic acid—a determinant of the life-time of rabbit eryth-rocytes, *Hoppe Seylers Z. Physiol. Chem. 355*: 395–400 (1974).

125. G. Ashwell and A. G. Morell, The role of surface carbohydrates in the hepatic rec-ognition and transport of circulating glycoproteins, *Adv. Enzymol. Relat. Areas Mol. Biol. 41*: 99–128 (1974).

126. G. Ashwell and J. Harford, Carbohydrate-specific receptors of the liver, *Ann. Rev. Biochem. 51*: 531–554 (1982).

127. A. Varki, Diversity in the sialic acids, *Glycobiology 2*: 25–40 (1992).

128. R. D. McCoy, E. R. Vimr, and F. A. Troy, CMP-NeuAc: Poly-α-2,8-sialosyl sialyl-transferase and the biosynthesis of polysialosyl units in neural cell adhesion molecules, *J. Biol. Chem. 260*: 12695–12699 (1985).

129. A. Acheson, J. L. Sunshine, and U. Rutishauser, NCAM polysialic acid can regulate both cell–cell and cell–substrate interactions, *J. Cell Biol. 114*: 143–153 (1991).

130. J. Roth, U. Rutishauser, and F. Troy, *Polysialic Acids*, Birkhauser Verlag, Basel, 1992.
131. H. Baumann, J.-R. Brisson, F. Michon, R. Pon, and H. J. Jennings, Comparison of the conformation of the epitope of α(2→8)polysialic acid with its reduced and *N*-acyl derivatives, *Biochemistry 32*: 4007–4013 (1993).
132. Y. Ito and T. Ogawa, An efficient approach to stereoselective glycosylation of *N*-acetylneuraminic acid: Use of phenylselenyl group as a stereocontrolling auxiliary, *Tetrahedron Lett. 28*: 6221–6224 (1987).
133. K. Okamoto and T. Goto, Glycosidation of sialic acid, *Tetrahedron 45*: 5835–5857 (1989).
134. M. P. DeNinno, The synthesis and glycosidation of *N*-acetylneuraminic acid, *Synthesis* 583–593 (1991).
135. T. J. Martin and R. R. Schmidt, Convenient chemical synthesis of CMP-*N*-acetyl-neuraminic acid (CMP-Neu-5-Ac), *Tetrahedron Lett. 34*: 1765–1768 (1993).
136. S. Makino, Y. Ueno, M. Ishikawa, Y. Hayakawa, and T. Hata, Chemical synthesis of cytidine-5′-monophosphono-*N*-acetylneuraminic acid (CMP-Neu5Ac), *Tetrahedron Lett. 34*: 2775–2778 (1993).
137. Y. Kajihara, K. Koseki, T. Ebata, H. Kodama, H. Matsushita, and H. Hashimoto, Synthesis of a novel CMP-Neu5Ac analogue: CMP-[α-Neu5Ac-(2→8)-Neu5Ac], *Carbohydr. Res. 264*: C1–C5 (1994).
138. D. H. van den Eijnden and W. van Dijk, A convenient method for the preparation of cytidine 5′-monophospho-*N*-acetylneuraminic acid, *Hoppe-Seyler Z. Physiol. Chem. 353*: 1817–1820 (1972).
139. J. Haverkamp, J. M. Beau, and R. Schauer, Improved synthesis of CMP-sialates using enzymes from frog liver and equine submandibular gland, *Hoppe-Seyler Z. Physiol. Chem. 360*: 159–166 (1979).
140. J. Thiem and W. Treder, Synthesis of the trisaccharide Neu-5-Ac-α(2→6)Gal-β-(1→4)GlcNAc by the use of immobilized enzymes, *Angew. Chem. Int. Ed. Engl. 25*: 1096–1097 (1986).
141. C. Augé and C. Gautheron, An efficient synthesis of cytidine monophospho-sialic acids with four immobilized enzymes, *Tetrahedron Lett. 29*: 789–790 (1988).
142. E. S. Simon, M. D. Bednarski, and G. M. Whitesides, Synthesis of CMP-NeuAc from *N*-acetylglucosamine: Generation of CTP from CMP using adenylate kinase, *J. Am. Chem. Soc. 110*: 7159–7163 (1988).
143. J. Thiem and P. Stangier, Preparative-enzymatic formation of cytidine 5′-monophosphosialate by integrated cytidine 5′-triphosphate regeneration, *Liebigs Ann. Chem.* 1101–1105 (1990).
144. J. L.-C. Liu, G.-J. Shen, Y. Ichikawa, J. F. Rutan, G. Zapata, W. F. Vann, and C.-H. Wong, Overproduction of CMP-sialic acid synthetase for organic synthesis, *J. Am. Chem. Soc. 114*: 3901–3910 (1992).
145. E. L. Kean and S. Roseman, The sialic acids. X. Purification and properties of cytidine 5′-monophosphosialic acid synthetase, *J. Biol. Chem. 241*: 5643–5650 (1966).
146. E. L. Kean, Nuclear cytidine 5′-monophosphosialic acid synthetase, *J. Biol. Chem. 245*: 2301–2308 (1970).
147. H. H. Higa and J. C. Paulson, Sialylation of glycoprotein oligosaccharides with *N*-acetyl-, *N*-glycolyl-, and *N-O*-diacetylneuraminic acids, *J. Biol. Chem. 260*: 8838–8849 (1985).
148. W. F. Vann, R. P. Silver, C. Abeijon, K. Chang, W. Aronson, A. Sutton, C. W. Finn, W. Lindner, and M. Kotsatos, Purification, properties, and genetic location of *Escherichia coli* cytidine 5′-monophosphate *N*-acetylneuraminic acid synthetase, *J. Biol. Chem. 262*: 17556–17562 (1987).

149. G. Zapata, W. F. Vann, W. Aronson, M. S. Lewis, and M. Moos, Sequence of the cloned *Eschrichia coli* K1 CMP-*N*-acetylneuraminic acid synthetase gene, *J. Biol. Chem. 264*: 14769–14774 (1989).

150. Y. Ichikawa, G.-J. Shen, and C.-H. Wong, Enzyme-catalyzed synthesis of sialyl oligosaccharide with in situ regeneration of CMP-sialic acid, *J. Am. Chem. Soc. 113*: 4698–4700 (1991).

151. G.-J. Shen, J. L.-C. Liu, and C.-H. Wong, Cloning and overexpression of a tagged CMP-*N*-acetylneuraminic acid synthetase from *E. coli* using a lambda phage system and application of the enzyme to the synthesis of CMP-*N*-acetylneuraminic acid, *Biocatalysis 6*: 31–42 (1992).

152. S. L. Shames, E. S. Simon, C. W. Christopher, W. Schmid, and G. M. Whitesides, CMP-*N*-Acetylneuraminic acid synthetase of *Escherichia coli*: High-level expression, purification and use in the enzymatic synthesis of CMP-*N*-acetylneuraminic acid and CMP-*N*-acetylneuraminic acid derivatives, *Glycobiology 1*: 187–191 (1991).

153. S. Ganguli, G. Zapata, T. Wallis, C. Reid, G. Boulnois, W. F. Vann, and I. S. Roberts, Molecular cloning and analysis of genes for sialic acid synthesis in Neisseria meningitidis group B and purification of the meningococcal CMP-NeuAc synthetase enzyme, *J. Bacteriol. 176*: 4583–4589 (1994).

154. M. G. Ambrose, S. J. Freese, M. S. Reinhold, T. G. Warner, and W. F. Vann, ^{13}C NMR investigation of the anomeric specificity of CMP-*N*-acetylneuraminic acid synthetase from *Escherichia coli*, *Biochemistry 31*: 775–780 (1992).

155. J. Vamecq, N. Mestdagh, J.-P. Henichart, and J. Poupaert, Subcellular distribution of glycosyltransferases in rodent liver and their significance in special reference to the synthesis of *N*-glycolylneuraminic acid, *J. Biochem. 111*: 579–583 (1992).

156. W. Schlenzka, L. Shaw, P. Schneckenburger, and R. Schauer, Purification and characterization of CMP-*N*-acetylneuraminic acid hydroxylase from pig submandibular glands, *Glycobiology 4*: 675–683 (1994).

157. P. Schckenburger, L. Shaw, and R. Schauer, Purification and characterization of CMP-*N*-acetylneuraminic acid hydroxylase from mouse liver, *Glycoconj. J. 11*: 194–203 (1994).

158. T. Terada, S. Kitazume, K. Kitajima, S. Inoue, F. Ito, F. A. Troy, and Y. Inoue, Synthesis of CMP-deaminoneuraminic acid (CMP-KDN) using the CTP:CMP-3-deoxynonulosonate cytidylyltransferase from rainbow trout testis. Identification and characterization of a CMP-KDN synthetase, *J. Biol. Chem. 268*: 2640–2648 (1993).

159. D. M. Carlson, G. W. Jourdian, and S. Roseman, The sialic acids. XIV. Synthesis of sialyl-lactose by a sialyltransferase from rat mammary gland, *J. Biol. Chem. 248*: 5742–5750 (1973).

160. B. A. Bartholomew, G. W. Jourdian, and S. Roseman, The sialic acids. XV. Transfer of sialic acid to glycoproteins by a sialyltransferase from colostrum, *J. Biol. Chem. 248*: 5751–5762 (1973).

161. D. M. Carlson, E. J. McGuire, G. W. Jourdian, and S. Roseman, The sialic acids. XVI. Isolation of a mucin sialyltransferase from sheep submaxillary gland, *J. Biol. Chem. 248*: 5763–5773 (1973).

162. B. J. Bast, L. J. Zhou, G. J. Freeman, K. J. Colley, T. J. Erhst, J. M. Munro, and T. F. Tedder, The HB-6, CDw75, and CD76 differentiation antigens are unique cell-surface carbohydrate determinants generated by the β-galactoside α2,6-sialyltransferase, *J. Cell Biol. 116*: 423–435 (1992).

163. P. O. Skacel, A. J. Edwards, C. T. Harrison, and W. M. Watkins, Enzymatic control of the expression of the x determinant (CD15) in human myeloid cells during maturation: The regulatory role of 6-sialyltransferase, *Bood 78*: 1452–1460 (1991).

164. F. Dall'Olio, N. Malagolini, G. Di Stefano, M. Ciambella, and F. Serafini-Cessi, α2,6-Sialylation of *N*-acetyllactosaminic sequences in human colorectal cancer cell lines. Relationship with non-adherent growth, *Int. J. Cancer 47*: 291–297 (1991).

165. J. C. Paulson, J. I. Rearick, and R. L. Hill, Enzymatic properties of β-D-galactoside α2→6sialyltransferase from bovine colostrum, *J. Biol. Chem. 252*: 2363–2371 (1977).

166. J. Weinstein, U. de Souza-e-Silva, and J. C. Paulson, Purification of a Galβ1→4GlcNAc α2→6sialyltransferase and a Galβ1→3(4)GlcNAc α2→3sialyltransferase to homogeneity from rat liver, *J. Biol. Chem. 257*: 13835–13844 (1982).

167. U. Sticher, H. J. Gross, and R. Brossmer, Purification and characterization of α(2→6)-sialyltransferase from human liver, *Glycoconj. J. 8*: 45–54 (1991).

168. M. Nemansky and D. H. van den Eijnden, Bovine colostrum CMP-NeuAc: Galβ(1→4)GlcNAc-R α(2→6)-sialyltransferase is involved in the synthesis of the terminal NeuAcα(2→6)GalNAcβ(1→4)GlcNAc sequence occurring on *N*-linked glycans of bovine milk glycoproteins, *Biochem. J. 287*: 311–316 (1992).

169. M. Nemansky, H. T. Edzes, R. A. Wijnands, and D. H. van den Eijnden, The polypeptide part of human chorionic gonadotropin affects the kinetics of α6-sialylation of its *N*-linked glycans but does not alter the branch specificity of CMP-NeuAc: Galβ1-4GlcNAc-R α2-6-sialyltransferase, *Glycobiology 2*: 109–117 (1992).

170. J. Weinstein, E. U. Lee, K. McEntee, P. H. Lai, and J. C. Paulson, Primary structure of β-galactoside α2,6-sialyltransferase. Conversion of membrane-bound enzyme to soluble forms by cleavage of the NH₂-terminal signal anchor, *J. Biol. Chem. 262*: 17735–17743 (1987).

171. N. Kurosawa, M. Kawasaki, T. Hamamoto, T. Nakaoka, Y.-C. Lee, M. Arita, and S. Tsuji, Molecular cloning and expression of chick embryo Galβ1,4GlcNAc α2,6-sialyltransferase. Comparison with the mammalian enzyme, *Eur. J. Biochem. 219*: 375–381 (1994).

172. H. J. Gross, A. Bünsch, J. C. Paulson, and R. Brossmer, Activation and transfer of novel synthetic 9-substituted sialic acids, *Eur. J. Biochem. 168*: 595–602 (1987).

173. H. J. Gross and R. Brossmer, Enzymatic introduction of a fluorescent sialic acid into oligosaccharide chains of glycoproteins, *Eur. J. Biochem. 177*: 583–589 (1988).

174. H. J. Gross, U. Rose, J. M. Krause, J. C. Paulson, K. Schmist, R. E. Feeney, and R. Brossmer, Analogues to *N*- and *O*-linked glycoprotein glycans using four different mammalian sialyltransferases, *Biochemistry 28*: 7386–7392 (1989).

175. R. E. Kosa, R. Brossmer, and H. J. Gross, Modification of cell surfaces by enzymatic introduction of special sialic acid analogues, *Biochem. Biophys. Res. Commun. 190*: 914–920 (1993).

176. J. van Pelt, L. Dorland, M. Duran, C. H. Hokke, J. P. Kamerling, and J. F. G. Vliegenthart, Transfer of sialic acid in α2–6linkage to mannose in Manβ1–4GlcNAc and Manβ1–4GlcNAcβ1–4GlcNAc by the action of Galβ1–4GlcNAc α2–6-sialyltransferase, *FEBS Lett. 256*: 179–184 (1989).

177. S. Sabesan and J. C. Paulson, Combined chemical and enzymatic synthesis of sialyloligosaccharides and characterization by 500-MHz ¹H and ¹³C NMR spectroscopy, *J. Am. Chem. Soc. 108*: 2068–2080 (1986).

178. C. Augé, C. Gautheron, and H. Pora, Enzymatic synthesis of the sialylglycopeptide, α-D-Neu*p*Ac-(2→6)-β-D-Gal*p*-(1→4)-β-D-Glc*p*NAc-(1→4)-L-Asn, *Carbohydr. Res. 193*: 288–293 (1989).

179. C. Augé, R. Fernandez-Fernandez, and C. Gautheron, The use of immobilized glycosyltransferase in the synthesis of sialyloligosaccharides, *Carbohydr. Res. 200*: 257–268 (1990).

180. N. Kurosawa, T. Hashimoto, Y.-C. Lee, T. Nakaoka, N. Kojima, and S. Tsuji, Molecular cloning and expression of GalNAcα2,6-sialyltransferase, *J. Biol. Chem. 269*: 1402–1409 (1994).

181. N. Kurosawa, N. Kojima, M. Inoue, T. Hashimoto, and S. Tsuji, Cloning and expression of Galβ1,3GalNAc-specific GalNAcα2,6-sialyltransferase, *J. Biol. Chem. 269*: 19048–19053 (1994).

182. J. E. Sadler, J. I. Rearick, J. C. Paulson, and R. L. Hill, Purification to homogeneity of a β-galactoside α2→3-sialyltransferase and a partial purification of an α-*N*-acetylgalactosaminide α2→6-sialyltransferase from porcine submaxillay gland, *J. Biol. Chem. 254*: 4434–4443 (1979).

183. J. I. Rearick, J. E. Sadler, J. C. Paulson, and R. L. Hill, Enzymatic characterization of β-D-galactoside α2→3-sialyltransferase from porcine submaxillary gland, *J. Biol. Chem. 254*: 4444–4451 (1979).

184. D. H. Joziasse, M. L. E. Bergh, H. G. J. ter Hart, P. L. Koppen, G. J. M. Hooghwinkel, and D. H. Van den Eijnden, Purification and enzymatic characterization of CMP-sialic acid: β-galactosyl1→3-*N*-acetylgalactosaminide α2→3-sialyltransferase from human placenta, *J. Biol Chem. 260*: 4941–4951 (1985).

185. W. Gillespie, S. Kelm, and J. C. Paulson, Cloning and expression of the Galβ1,3GalNAc α2,3-sialyltransferase, *J. Biol. Chem. 267*: 21004–21010 (1992).

186. H. Kitagawa and J. C. Paulson, Cloning of a novel α2,3-sialyltransferase that sialylates glycoprotein and glycolipid carbohydrate groups, *J. Biol. Chem. 269*: 1394–1401 (1994).

187. Y.-C. Lee, N. Kurosawa, T. Hashimoto, and S. Tsuji, Molecular cloning and expression of Galβ1,3GalNAc α2,3-sialyltransferase from mouse brain, *Eur. J. Biochem. 216*: 377–385 (1993).

188. Y.-C. Lee, N. Kojima, E. Wada, N. Kurosawa, T. Nakaoka, T. Hashimoto, and S. Tsuji, Cloning and expression of cDNA for a new type of Galβ1,3GalNAc α2,3-sialyltransferase, *J. Biol. Chem. 269*: 10028–10033 (1994).

189. D. H. van den Eijnden and W. E. C. M. Schiphorst, Detection of β-galactosyl(1→4)*N*-acetylglucosaminide α(2→3)sialyltransferase activity in fetal calf liver and other tissues, *J. Biol. Chem. 256*: 3159–3162 (1981).

190. J. Weinstein, U. de Souza-e-Silva, and J. C. Paulson, Sialylation of glycoprotein oligosaccharides *N*-linked to asparagine. Enzymatic characterization of a Galβ1→3(4)GlcNAc α2→3sialyltransferase and a Galβ1→4GlcNAc α2→6sialyl-transferase from rat liver, *J. Biol. Chem. 257*: 13845–13858 (1982).

191. D. X. Wen, B. D. Livingston, K. F. Medzihradsky, S. Kelm, A. L. Burllingame, and J. C. Paulson, Primary structure of Galβ1,3(4)GlcNAc α2,3-sialyltransferase determined by mass spectroscopy sequence analysis and molecular cloning. Evidence for a protein motif in the sialyltransferase gene family, *J. Biol. Chem. 267*: 21011–21019.

192. B. D. Livingston and J. C. Paulson, Polymerase chain reaction cloning of a developmentally regulated member of the sialyltransferase gene family, *J. Biol. Chem. 268*: 11504–11507 (1993).

193. H. Kitagawa and J. C. Paulson, Cloning and expression of human Gaβ1,3(4)GlcNAc α2,3-sialyltransferase, *Biochem. Biophys. Res. Commun. 194*: 375–382 (1993).

194. K. Sasaki, E. Watanabe, K. Kawashima, S. Sekine, T. Dohi, M. Oshima, N. Hanai, T. Nishi, and M. Hasegawa, Expression cloning of a novel Galβ(1→3/4)GlcNAc α2,3-sialyltransferase using lectin resistance selection, *J. Biol. Chem. 268*: 22782–22787 (1993).

195. A. Lubineau, C. Augé, and P. François, The use of porcine liver (2→3)-α-sialyltransferase in the large-scale synthesis of α-Neup5Ac-(2→3)-β-D-Galp-(1→3)-D-GlcpNAc, the epitope of the tumor-associated carbohydrate antigen CA50, *Carbohydr. Res. 228*: 137–144 (1992).

196. C. Unverzagt, H. Kuntz, and J. C. Paulson, High-efficiency synthesis of sialyloligosaccharides and sialoglycopeptides, *J. Am. Chem. Soc. 112*: 9308–9306 (1990).

197. Y. Ito and J. C. Paulson, A novel strategy for synthesis of ganglioside G_{M3} using an enzymatically produced sialoside glycosyl donor, *J. Am. Chem. Soc. 115*: 1603–1605 (1993).

198. K. K.-C. Liu and S. Danishefsky, A striking example of the interfacing of glycal chemistry with enzymatically mediated sialylation: A concise synthesis of G_{M3}, *J. Am. Chem. Soc. 115*: 4933–4934 (1993).

199. G. K. Ostrander and E. H. Holmes, Characterization of a CMP-NeuAc: Lactosylceramide α2–3-sialyltransferase from rainbow trout hepatoma (RTH-149) cells, *Comp. Biochem. Physiol. [B] 98*: 87–95 (1991).

200. H. Iber, G. van Echten, and K. Sandhoff, Substrate specificity of α2–3-sialyltransferase in ganglioside biosynthesis of rat liver Golgi, *Eur. J. Biochem. 195*: 115–120 (1991).

201. L. J. Melkerson and C. C. Sweeley, Special consideration in the purification of the GM3 ganglioside-forming enzyme, CMP-sialic acid: Lactosamide α2–3-sialyltransferase (SAT-1): Effects of protease inhibitors on rat hepatic SAT-1 activity, *Biochem. Biophys. Res. Commun. 175*: 325–332 (1991).

202. L. J. Melkerson and C. C. Sweeley, Purification to apparent homogeneity by immunoaffinity chromatography and partial characterization of the G_{M3} ganglioside-forming enzyme, CMP-sialic acid: Lactosylceramide α2,3-sialyltransferase (SAT-1), from rat liver Golgi, *J. Biol. Chem. 266*: 4448–4457 (1991).

203. U. Preuss, X. Gu, T. Gu, and R. K. Yu, Purification and characterization of CMP-*N*-acetylneuraminic acid: Lactosylceramide α(2–3)sialyltransferase (G_{M3}-synthase) from rat brain, *J. Biol. Chem. 268*: 26273–26278 (1993).

204. X. B. Gu, T. J. Gu, and R. K. Yu, Purification to homogeneity of G_{D3} synthase and partial purification of G_{M3} synthase from rat brain, *Biochem. Biophys. Res. Commun. 166*: 387–393 (1990).

205. K. Nara, Y. Watanabe, K. Maruyama, K. Kasahara, Y. Nagai, and Y. Sanai, Expression cloning of a CMP-NeuAc: NeuAcα2–3Galβ1–4Glcβ1–1′Cer α2,8-sialyltransferase (G_{D3} synthetase) from human melanoma cells, *Proc. Natl. Acad. Sci. USA 91*: 7952–7956 (1994).

206. K. Sasaki, K. Kurata, N. Kojima, S. Ohta, N. Hanai, S. Tsuji, and T. Nishi, Expression cloning of a G_{M3}-specific α2,8-sialyltransferase (G_{D3} synthetase), *J. Biol. Chem. 269*: 15950–15956 (1994).

207. S. Roseman, The synthesis of carbohydrates by multiglycosyltransferase systems and their potential function in intercellular adhesion, *Chem. Phys. Lipids 5*: 270–297 (1970).

208. B. D. Shur and N. G. Hall, Sperm surface galactosyltransferase activities during in vitro capacitation, *J. Cell Biol. 95*: 567–573 (1982).

209. B. D. Shur and N. G. Hall, A role for mouse sperm surface galactosyltransferase in sperm binding to the egg zona pellucida, *J. Cell Biol. 95*: 574–579 (1982).

210. B. D. Shur, Expression and function of cell surface galactosyltransferase, *Biochem. Biophys. Acta 988*: 389–409 (1989).

211. C. P. Loeber and R. B. Runyan, A comparison of fibronectin, laminin, and galactosyltransferase adhesion mechanisms during embryonic cardiac mesenchymal cell migration in vitro, *Dev. Biol. 140*: 401–412 (1990).

212. A. Passaniti and G. W. Hart, Metastasis-associated murine melanoma cell surface galactosyltransferase: Characterization of enzyme activity and identification of the major surface substrates, *Cancer Res. 50*: 7261–7271 (1990).

213. D. A. Benau, E. J. McGuire, and B. T. Storey, Further characterization of the mouse sperm surface zona-binding site with galactosyltransferase activity, *Mol. Reprod. Dev. 25*: 393–399 (1990).

214. D. J. Miller, M. B. Macek, and B. D. Shur, Complementarity between sperm surface β1,4-galactosyltransferase and egg-coat ZP3 mediates sperm-egg binding, *Nature 357*: 589–593 (1992).

215. M. H. Barcellos-Hoff, Mammary epithelial reorganization on extracellular matrix is mediated by cell surface galactosyltransferase, *Exp. Cell Res. 201*: 225–234 (1992).

216. E. S. Maxwell, The enzymatic interconversion of uridine diphosphogalactose and uridine diphosphoglucose, *J. Biol. Chem. 229*: 139–151 (1957).

217. E. P. Anderson, E. S. Maxwell, and R. M. Burton, Enzymatic synthesis of C-14-labeled uridine diphosphoglucose, galactose 1-phosphate, and uridine diphosphogalactose, *J. Am. Chem. Soc. 81*: 6514–6517 (1959).

218. E. F. Neufeld, D. S. Feingold, and W. Z. Hassid, Phosphorylation of D-galactose and L-arabinose by extracts from *Phaseolus aureus* seedlings, *J. Biol. Chem. 235*: 906–909 (1960).

219. K. Kurahashi and A. Sugiura, Purification and properties of galactose 1-phosphate uridyl transferase from *Escherichia coli*, *J. Biol. Chem. 235*: 940–946 (1960).

220. J. Kim, F. Ruzicka, and P. A. Frey, Remodeling hexose-1-phosphate uridyltransferase: Mechanism-inspired mutation into a new enzyme, UDP-hexose synthetase, *Biochemistry 29*: 10590–10593 (1990).

221. J. E. Heidlas, W. J. Lees, and G. M. Whitesides, Practical enzyme-based syntheses of uridine 5′-diphosphogalactose and uridine 5′-diphospho-*N*-acetylgalactosamine on a gram scale, *J. Org. Chem. 57*: 152–157 (1992).

222. Enzyme nomenclature, *Recommendations of the Nomenclature Committee of the International Union of Biochemistry*, Academic Press, San Diego, 1992.

223. H. Babad and W. Z. Hassid, Soluble uridine diphosphate D-galactose: D-glucose β-4-D-galactosyltransferase from bovine milk, *J. Biol. Chem. 241*: 2672–2678 (1966).

224. U. Brodbeck and K. E. Ebner, Resolution of a soluble lactose synthetase into two protein components and solubilization of microsomal lactose synthetase, *J. Biol. Chem. 241*: 762–764 (1966).

225. I. P Trayer and R. L. Hill, The purification and properties of the A protein of lactose synthetase, *J. Biol. Chem. 246*: 6666–6675 (1971).

226. C. R. Geren, S. T. Magee, and K. E. Ebner, Hydrophobic chromatography of galactosyltransferase, *Arch. Biochem. Biophys. 172*: 149–155 (1976).

227. H. J. Hathaway, R. B. Runyan, S. Khounlo, and B. D. Shur, Purification and characterization of avian β1,4-galactosyltransferase: Comparison with the mammarian enzyme, *Glycobiology 1*: 211–221 (1991).

228. U. Brodbeck, W. L. Denton, N. Tanahashi, and K. E. Ebner, The isolation and identification of the B protein of lactose synthetase as α-lactalbumin, *J. Biol. Chem. 242*: 1391–1397 (1967).

229. A. Moore, L. Hall, and D. W. Hamilton, An 18-kDa androgen-regulated protein that modifies galactosyltransferase activity is synthesized by the rat caput epididymidis, but has no structural similarity to rat milk α-lactalbumin, *Biol. Reprod. 43*: 497–506 (1990).

230. J. A. Grobler, M. Wang, A. C. Pike, and K. Brew, Study by mutagenesis of the role of two aromatic clusters of α-lactalbumin in aspects of its action in the lactose system, *J. Biol. Chem. 269*: 5106–5114 (1994).

231. J. E. Bell, T. A. Beyer, and R. L. Hill, The kinetic mechanism of bovine milk galactosyltransferase: The role of α-lactalbumin, *J. Biol. Chem. 251*: 3003–3013 (1976).

232. H. Narimatsu, S. Sinha, K. Brew, H. Okayama, and R. K. Qasba, Cloning and sequencing of cDNA of bovine *N*-acetylglucosamine (β1–4)galactosyltransferase, *Proc. Natl. Acad. Sci. USA 83*: 4720–4724 (1986).

233. G. D'Agostaro, B. Bendiak, and M. Tropak, Cloning of cDNA encoding the membrane-bound form of bovine β1,4-galactosyltransferase, *Eur. J. Biochem. 183*: 211–217 (1989).

234. J. H. Shaper, G. F. Hollis, and N. L. Shaper, Evidence for two forms of murine β-1,4-galactosyltransferase based on cloning studies, *Biochimie 70*: 1683–1688 (1988).

235. K. Nakagawa, T. Ando, T. Kimura, and H. Narimatsu, Cloning and sequencing of a full-length cDNA of mouse *N*-acetylglucosamine (β1,4)-galactosyltransferase, *J. Biochem. (Tokyo) 104*: 165–168 (1988).

236. K. Nakazawa, K. Furukawa, A. Kobata, and H. Narimatsu, Characterization of a murine β1–4galactosyltransferase expressed in COS-1 cells, *Eur. J. Biochem. 196*: 363–368 (1991).

237. M. G. Humphrey-Beher, B. Bunnell, P. van Tuinen, D. H. Ledbetter, and V. J. Kidd, Molecular cloning and chromosomal localization of human 4-β-galactosyltransferase, *Proc. Natl. Acad. Sci. USA 83*: 8918–8922 (1986).

238. D. Aoki, H. E. Appert, D. Johnson, S. S. Wong, and M. N. Fukuda, Analysis of the substrate binding site of human galactosyltransferase by protein engineering, *EMBO J. 9*: 3171–3178 (1990).

239. S. K. Chatterjee, Molecular cloning of human β1,4-galactosyltransferase and expression of catalytic activity of the fusion protein in *Escherichia coli, Int. J. Biochem. 23*: 695–702 (1991).

240. T. Uejima, M. Uemura, S. Nozawa, and H. Narimatsu, Complementary DNA cloning for galactosyltransferase associated with tumor and determination of antigenic epitopes recognized by specific monoclonal antibodies, *Cancer Res. 52*: 6158–6163 (1992).

241. S. Basu, S. Ghosh, S. S. Basu, J. W. Kyle, Z. Li, and M. Basu, Regulation of expression of cell surface neolacto-glycolipids and cloning of embryonic chicken brain GalT-4 (UDP-Gal: GlcNAc-R β1-4galactosyltransferase), *Indian J. Biochem. Biophys. 30*: 315–323 (1993).

242. K. Nakazawa, K. Furukawa, H. Narimatsu, and A. Kobata, Kinetic study of human β-1,4-galactosyltransferase expressed in *E. coli, J. Biochem. (Tokyo) 113*: 747–753 (1993).

243. C. H. Krezdorn, G. Watzele, R. B. Kleene, S. X. Ivanov, and E. G. Berger, Purification and characterization of recombinant human β-1–4-galactosyltransferase expressed in Saccharomyces cerevisiae, *Eur. J. Biochem. 212*: 113–120 (1993).

244. J. F. Morrison and K. E. Ebner, Studies on galactosyltransferase: Kinetic investigations with *N*-acetylglucosamine as the galactosyl group acceptor, *J. Biol. Chem. 246*: 3977–3984 (1971).

245. J. F. Morrison and K. E. Ebner, Studies on galactosyltransferase: Kinetic investigations with glucose as the galactosyl group acceptor, *J. Biol. Chem 246*: 3985–3991 (1971).

246. L. J. Berliner and R. D. Robinson, Structure-function relationships in lactose synthetase. Structural requirement of the uridine 5′-diphospho galactose binding site, *Biochemistry 21*: 6340–6343 (1982).

247. M. M. Palcic and O. Hindsgaul, Flexibility in the donor substrate specificity of β1,4-galactosyltransferase: Application in the synthesis of complex carbohydrates, *Glycobiology 1*: 205–209 (1991).

248. H. Yuasa, O. Hindsgaul, and M. M. Palcic, Chemical-enzymatic synthesis of 5′-thio-*N*-acetyllactosamine: The first disaccharide with sulfur in the ring of the nonreducing sugar, *J. Am. Chem. Soc. 114*: 5891–5892 (1992).

249. G. Srivastava, O. Hindsgaul, and M. M. Palcic, Chemical synthesis and kinetic characterization of UDP-2-deoxy-D-lyxo-hexose (UDP-2-deoxy-D-galactose), a donor-substrate for β-(1→4)-D-galactosyltransferase, *Carbohydr. Res. 245*: 137–144 (1993).

250. F. L. Scanbacher and K. E. Ebner, Galactosyltransferase acceptor specificity of the lactose synthetase A protein, *J. Biol. Chem. 245*: 5057–5061 (1970).

251. L. J. Berliner, M. E. Davis, K. E. Ebner, T. A. Beyer, and J. E. Bell, The lactose synthetase acceptor site: A structural map derived from acceptor studies, *Mol. Cell. Biochem. 62*: 37–42 (1984).

252. M. M. Palcic, O. P. Srivastava, and O. Hindsgaul, Transfer of D-galactosyl group to 6-*O*-substituted 2-acetamido-2-deoxy-D-glucose residues by use of bovine D-galactosyltransferase, *Carbohydr. Res. 159*: 315–324 (1987).

253. C.-H. Wong, Y. Ichikawa, T. Krach, C. Gautheron-Le Narvor, D. P. Dumas, and G. C. Look, Probing the acceptor specificity of β-1,4-galactosyltransferase for the development of enzymatic synthesis of novel oligosaccharide, *J. Am. Chem. Soc. 113*: 8137–8145 (1991).

254. Y. Nishida, T. Wiemann, and J. Thiem, Transfer of galactose to the anomeric position of *N*-acetylgentosamine, *Tetrahedron Lett. 33*: 8043–8046 (1992).

255. Y. Nishida, T. Wiemann, V. Sinnwell, and J. Thiem, A new type of galactosyltransferase reaction: Transfer of galactose to the anomeric position of *N*-acetylkanosamine, *J. Am. Chem. Soc. 115*: 2536–2537 (1993).

256. C. Augé, S. David, C. Mathieu, and C. Gautheron, Synthesis with immobilized enzymes of two trisaccharides, one of them active as the determinant of A stage antigen, *Tetrahedron Lett. 25*: 1467–1470 (1984).

257. J. Thiem and T. Wiemann, Combined chemoenzymatic synthesis of *N*-glycoprotein building blocks, *Angew. Chem. Int. Ed. Engl. 29*: 80–82 (1990).

258. J. Thiem and T. Wiemann, Galactosyltransferase-catalyzed synthesis of 2′-deoxy-*N*-acetyllactosamine, *Angew. Chem. Int. Ed. Engl. 30*: 1163–1164 (1991).

259. R. Öhrlein, B. Ernst, and E. G. Berger, Galactosylation of non-natural glycosides with human β-D-galactosyltransferase on a preparative scale, *Carbohydr. Res. 236*: 335–338 (1992).

260. J. Thiem and T. Wiemann, Synthesis of galactose-terminated oligosaccharides by use of galactosyltransferase, *Synthesis* 141–145 (1992).

261. T. Wiemann, N. Taubken, U. Zehavi, and J. Thiem, Enzymic synthesis of *N*-acetyllactosamine on a soluble, light-sensitive polymer, *Carbohydr. Res. 257*: C1–C6 (1994).

262. S. Chatterjee, N. Ghosh, and S. Khurana, Purification of uridine diphospho-galactose:glucosyl ceramide, β-1–4-galactosyltransferase from human kidney, *J. Biol. Chem. 267*: 7148–7153 (1992).

263. S. Schulte and W. Stoffel, Ceramide UDP-galactosyltransferase from myelination of rat brain: Purification, cloning, and expression, *Proc. Natl. Acad. Sci. USA 90*: 10265–16269 (1993).

264. (a) F. Yamamoto, H. Clausen, T. White, J. Marken, and S. Hakomori, Molecular genetics basis of the histo-blood group ABO system, *Nature 345*: 229–233 (1990). (b) F. Yamamoto and S. Hakomori, Sugar-nucleotide donor specificity of histo-blood group A and B transferase is based on amino acid substituents, *J. Biol. Chem. 265*: 19257–19262 (1990).

265. T. L. Lowary and O. Hindsgaul, Recognition of synthetic deoxy and deoxyfluoro analogs of the acceptor α-L-Fuc*p*-(1→2)β-D-Gal*p*-O*R* by the blood-group A and B gene-specified glycosyltransferases, *Carbohydr. Res. 249*: 163–195 (1993).

266. T. L. Lowary and O. Hindsgaul, Recognition of synthetic *O*-methyl, epimeric, and amino analogues of the acceptor α-L-Fuc*p*-(1→2)β-D-Gal*p*-O*R* by the blood-group A and B gene-specified glycosyltransferases, *Carbohydr. Res. 251*: 33–67 (1994).

267. U. Galili, E. A. Rachmilewitz, A. Peleg, and I. Flechner, A unique natural human IgG with anti-α-galactosyl specificity, *J. Exp. Med. 160*: 1519–1531 (1984).

268. U. Galiki, M. R. Clark, S. B. Shohet, J. Buehler, and B. A. Macher, Evolutionary relationship between the natural anti-Gal antibody and the Galα1→3Gal epitope in primates, *Proc. Natl. Acad. Sci. USA 84*: 1369–1373 (1987).

269. (a) D. H. Joziasse, J. H. Shaper, D. H. van den Eijnden, A. J. van Tunen, and N. L. Shaper, Bovine α1→3-galactosyltransferase: Isolation and characterization of a cDNA clone, Identification of homologous sequences in human genomic DNA, *J. Biol. Chem.* *264*: 14290–14297 (1989). (b) R. D. Larsen, V. P. Rajan, M. M. Ruf, J. Kukowska-Latallo, R. D. Cummings, and J. B. Lowe, Isolation of a cDNA encoding a murine UDP-galactose: β-D-galactosyl-1,4-N-acetyl-D-glucosaminide α-1,3-galactosyltransferase: Expression cloning by gene transfer, *Proc. Natl. Acad. Sci. USA 86*: 8227–8231 (1989).

270. D. H. Joziasse, N. L. Shaper, L. S. Salyer, D. H. van den Eijnden, A. C. van der Spoel, and J. H. Shaper, α1→3-Galactosyltransferase: The use of recombinant enzyme for the synthesis of α-galactosylated glycoconjugates, *Eur. J. Biochem. 191*: 75–83 (1990).

271. D. H. Joziasse, W. E. Schiphorst, C. A. Koeleman, and D. H. van den Eijnden, Enzymatic synthesis of the α3-galactosyl-Le x tetrasaccharide: A potential ligand for selectin-type adhesion molecules, *Biochem. Biophys. Res. Commun. 194*: 358–367 (1993).

272. H. Schachter, Enzymes associated with glycosylation, *Curr. Opin. Struct. Biol. 1*: 755–765 (1991).

273. G. C. Look, Y. Ichikawa, G.-J. Shen, P.-W. Cheng, and C.-H. Wong, A combined chemical and enzyme strategy for the construction of carbohydrate-containing antigen core units, *J. Org. Chem. 58*: 4326–4330 (1993).

274. J. E. Heidlas, W. J. Lees, P. Pale, and G. M. Whitesides, Gram-scale synthesis of uridine 5'-diphospho-N-acetylglucosamine: Comparison of enzymatic and chemical routes, *J. Org. Chem. 57*: 146–151 (1992).

275. R. Kuar, J. Yang, R. D. Larsen, and P. Stanley, Cloning and expression of N-acetylglucosaminyltransferase I, the medial Golgi transferase that initiates complex N-linked carbohydrate formation, *Proc. Natl. Acad. Sci. USA 87*: 9948–9952 (1990).

276. M. Sarkar, E. Hull, Y. Nishikawa, R. J. Simpson, R. L. Moritz, R. Dunn, and H. Schachter, Molecular cloning and expression of cDNA encoding the enzyme that controls conversion of high-mannose to hybrid and complex N-glycans: UDP-N-acetylglucosamine: α-3-D-mannoside β-1,2-N-acetylglucosaminyltransferase I, *Proc. Natl. Acad. Sci. USA 88*: 234–238 (1991).

277. H. Schachter, E. Hull, M. Sarker, R. J. Simpson, R. L. Moritz, J. W. Hopperner, and R. Dunn, Molecular cloning of human and rabbit UDP-N-acetylglucosamine: α-3-D-mannoside β-1,2-N-acetylglucosaminyltransferase I, *Biochem. Soc. Trans. 19*: 645–648 (1991).

278. S. Pownall, C. A. Kozak, K. Schappert, M. Sarker, E. Hull, H. Schachter, and J. D. Marth, Molecular cloning and characterization of the mouse UDP-N-acetylglucosamine: α-3-D-mannoside β-1,2-N-acetylglucosaminyltransferase I gene, *Genomics 12*: 699–704 (1992).

279. R. Kumar, J. Yang, R. L. Eddy, M. G. Byers, T. B. Shows, and P. Stanley, Cloning and expression of the murine gene and chromosomal location of the human gene encoding N-acetylglucosaminyltransferase I, *Glycobiology 2*: 383–393 (1992).

280. G. Moller, F. Reck, H. Paulsen, K. J. Kaur, M. Sarkar, H. Schachter, and I. Brockhausen, Control of glycoprotein synthesis: Substrate specificity of rat liver UDP-GlcNAc: Manα3R β2-N-acetylglucosaminyltransferase I using synthetic substrate analogues, *Glycoconj. J. 9*: 180–190 (1992).

281. G. Srivastava, G. Alton, and O. Hindsgaul, Combined chemical-enzymic synthesis of deoxygenated oligosaccharide analogs: transfer of deoxygenated D-GlcpNAc residues from their UDP-GlcpNAc derivatives using N-acetylglucosaminyl-transferase I, *Carbohydr. Res. 207*: 259–276 (1990).

282. F. Reck, M. Springer, H. Paulsen, I. Brockhausen, M. Sarkar, and H. Schachter, Synthesis of tetrasaccharide analogues of the N-glycan substrate of β-(1→2)-N-acetylglu-

cosaminyltransferase II using trisaccharide precursors and recombinant β-(1→2)-*N*-acetylglucosaminyltransferase I, *Carbohydr. Res. 259*: 93–101 (1994).

283. K. J. Kaur, G. Alton, and O. Hindsgaul, Use of *N*-acetylglucosaminyltransferase I and II in the preparative synthesis of oligosaccharide, *Carbohydr. Res. 210*: 145–153 (1991).

284. A. Nishikawa, Y. Ihara, M. Hatakeyama, K. Kangawa, and N. Taniguchi, Purification, cDNA cloning, and expression of UDP-*N*-acetylglucosamine: β-D-mannoside β-1,4-*N*-acetylglucosaminyltransferase III from rat kidney, *J. Biol. Chem. 267*: 18199–18204 (1992).

285. Y. Ihara, A. Nishikawa, T. Tohma, S. Soejima, N. Nishikawa, and N. Taniguchi, cDNA cloning, expression, and chromosomal localization of human *N*-acetylglucosaminyl-transferase III (GnT-III), *J. Biochem. (Tokyo) 113*: 692–698 (1993).

286. K. J. Kaur and O. Hindsgaul, Combined chemical-enzymic synthesis of a dideoxypen-tasaccharide for use in a study of the specificity of *N*-acetylglucosaminyltransferase-III, *Carbohydr. Res. 226*: 219–231 (1992).

287. M. M. Palcic, J. Ripka, K. J. Kaur, M. Shoreibah, O. Hindsgaul, and M. Pierce, Reg-ulation of *N*-acetylglucosaminyltransferase V activity. Kinetic comparisons of parental, Rous sarcoma virus-transformed BHK, and L-phytohemagglutinin-resistant BHK cells using synthetic substrates and inhibitory substrate analog, *J. Biol. Chem. 265*: 6759–6769 (1990).

288. M. G. Shoreibah, O. Hindsgaul, and M. Pierce, Purification and characterization of rat kidney UDP-*N*-acetylglucosamine: α-6-D-Mannoside β-1,6-*N*-acetylglucosaminyltrans-ferase, *J. Biol. Chem. 267*: 2920–2927 (1992).

289. J. Gu, A. Nishikawa, N. Tsuruoka, M. Ohno, N. Yamaguchi, K. Kangawa, and N. Taniguchi, Purification and characterization of UDP-*N*-acetylglucosamine: α-6-D-man-noside β-1-6-*N*-acetylglucosaminyltransferase (*N*-acetylglucosaminyltransferase V) from a human lung cancer cell line, *J. Biochem. (Tokyo) 113*: 614–619 (1993).

290. H. Saito, A. Nishikawa, J. Gu, Y. Ihara, H. Soejima, Y. Wada, C. Sekiya, N. Niikawa, and N. Taniguchi, cDNA cloning and chromosomal mapping of human *N*-acetylglu-cosaminyltransferase V, *Biochem. Biophys. Res. Commun. 198*: 318–327 (1994).

291. S. C. Crawley, O. Hindsgaul, G. Alton, M. Pierce, and M. M. Palcic, An enzyme-linked immunosorbent assay for *N*-acetylglucosaminyltransferase-V, *Anal. Biochem. 185*: 112–117 (1990).

292. O. Kanie, S. C. Crawley, M. M. Palcic, and O. Hindsgaul, Acceptor-substrate recog-nition by *N*-acetylglucosaminyltransferase-V: Critical role of the 4″-hydroxyl group in β-D-GlcpNAc-(1→2)-α-D-Manp-(1→6)-β-D-Glcp-OR, *Carbohydr. Res. 243*: 139–164 (1993).

293. S. H. Khan, S. C. Crawley, O. Kanie, and O. Hindsgaul, A trisaccharide acceptor analog for *N*-acetylglucosaminyltransferase V which binds to the enzyme but sterically pre-cludes the transfer reaction, *J. Biol. Chem. 268*: 2468–2473 (1993).

294. I. Lindh and O. Hindsgaul, Synthesis and enzymatic evaluation of two conformationally restiricted trisaccharide analogues as substrates for *N*-acetylglucosaminyltransferase V, *J. Am. Chem. Soc. 113*: 216–223 (1991).

295. T. Linker, S. C. Crawley, and O. Hindsgaul, Recognition of the acceptor β-D-GlcpNAc-(1→2)-α-D-Manp-(1→6)-β-D-Glcp-OR by *N*-acetylglucosaminyltransferase V: None of the hydroxyl groups on the Glc-residue are important, *Carbohydr. Res. 245*: 323–331 (1993).

296. P.-W. Chen, W. E. Wingert, M. R. Little, and R. Wei, Mucin biosynthesis. Properties of a bovine tracheal mucin β-6-*N*-acetylglucosaminyltransferase, *Biochem. J. 227*: 405–412 (1985).

297. P. A. Ropp, M. R. Little, and P.-W. Cheng, Mucin biosynthesis: Purification and char-

acterization of a mucin β6*N*-acetylglucosaminyltransferase, *J. Biol. Chem. 266*: 23863–23871 (1991).

298. S. Yousefi, E. Higgins, Z. Daoling, A. Polles-Kruger, O. Hindsgaul, and J. W. Dennis, Increased UDP-GlcNAc: Galβ1–3GalNAc-O (GlcNAc to GalNAc) β-1,6-*N*-acetylglucosaminyltransferase activity in metastatic murine tumor cell lines. Control of polylactosamine synthesis, *J. Biol. Chem. 266*: 1772–1782 (1991).

299. S. Sangadala. S. Sivakami, and J. Mendicino, UDP-GlcNAc: Galβ3GalNAc-mucin: (GlcNAc→GalNAc) β6-*N*-acetylglucosaminyltransferase and UDP-GlcNAc: Galβ3-(GlcNAcβ6) GalNAc-mucin (GlcNAc→Gal)β6-*N*-acetylglucosaminyltransferase from swine trachea epithelium, *Mol. Cell. Biochem. 101*: 125–143 (1991).

300. P. A. Ropp, M. R. Little, and P. W. Cheng, Mucin biosynthesis: Purification and characterization of a mucin β6*N*-acetylglucosaminyltransferase, *J. Biol. Chem. 266*: 23863–23871 (1991).

301. M. F. Bierhuizen and M. Fukuda, Expression cloning of a cDNA encoding UDP-GlcNAc: Galβ1–3-GalNAc (GlcNAc to GalNAc) β1–6GlcNAc transferase by gene transfer into CHO cells expressing polyoma large tumor antigen, *Proc. Natl. Acad. Sci. USA 89*: 9326–9330 (1992).

302. M. F. Bierhuizen, M. G. Mattei, and M. Fukuda, Expression of the developmental I antigen by a cloned human cDNA encoding a member of a β-1,6-*N*-acetylglucosaminyltransferase gene family, *Genes Dev. 7*: 468–473 (1993).

303. M. F. Bierhuizen, K. Maemura, and M. Fukuda, Expression of a differentiation antigen and poly-*N*-acetyllactosaminyl *O*-glycans directed by a cloned core 2 β-1,6-*N*-acetylglucosaminyltransferase, *J. Biol. Chem. 269*: 4473–4479 (1994).

304. S. Hakomori, Blood group ABH and Ii antigens of human erythrocytes: Chemistry, polymorphism, and their developmental change, *Semin. Hematol. 18*: 39–62 (1981).

305. B. A. Fenderson, E. M. Eddy, and S. Hakomori, The blood group I antigen defined by monoclonal antibody C6 is a marker of early mesoderm during murine embryogenesis, *Differentiation 38*: 124–133 (1988).

306. J. Gu, A. Nishikawa, S. Fujii, S. Gasa, and N. Taniguchi, Biosynthesis of blood group I and i antigen in rat tissues. Identification of a novel β1–6-*N*-acetylglucosaminyltransferase, *J. Biol. Chem. 267*: 2994–2999 (1992).

307. R. S. Haltiwanger, M. A. Blomberg, and G. W. Hart, Glycosylation of nuclear and cytoplasmic proteins. Purification and characterization of a uridine diphospho-*N*-acetylglucosamine: Polypeptide β-*N*-acetylglucosaminyltransferase. *J. Biol. Chem. 267*: 9005–9013 (1992).

308. S. Kornfeld and I. Mellman, The biogenesis of lysosomes, *Ann. Cell Biol. 5*: 483–525 (1989).

309. E. S. Simon, S. Grabowski, and G. M. Whitesides, Convenient synthesis of cytidine 5′-triphosphate, guanosine 5′-triphosphate, and uridine 5′-triphosphate and their use in the preparation of UDP-glucose, UDP-glucuronic acid, and GDP-mannose, *J. Org. Chem. 55*: 1834–1841 (1990).

310. A. Hausler and P. W. Robbins, Glycosylation in *Saccharomyces cerevisiae*: Cloning and characterization of an α-1,2-mannosyltransferase structural gene, *Glycobiology 2*: 77–84 (1992).

311. P. Wang, G.-J. Shen, Y.-F. Wang, Y. Ichikawa, and C.-H. Wong, Enzymes in organic synthesis: Active domain overproduction, specificity study and synthetic use of an α1,2-mannosyltransferase with regeneration of GDP-Man, *J. Org. Chem. 58*: 3985–3990 (1993).

312. C. L. Yip, S. K. Welch, F. Klebl, T. Gilbert, P. Seidel, F. J. Grant, P. J. O'Hara, and V. L. MacKay, Cloning and analysis of the *Saccharomyces cerevisiae* MNN9 and

MNN1 genes required for complex glycosylation of secreted proteins, *Proc. Natl. Acad. Sci. USA 91*: 2723–2727 (1994).

313. A. Takeya, O. Hosomi, and M. Ishiura, Complete purification and characterization of α-3-*N*-acetylgalactosaminyltransferase encoded by the human blood group A gene, *J. Biochem. (Tokyo) 107*: 360–368 (1990).

314. H. Clausen, T. White, K. Takio, K. Titani, M. Stroud, E. Holmes, K. Karkov, L. Thim, and S. Hakomori, Isolation to homogeneity and partial characterization of a histo-blood group A defined Fucα1→2Gal α1→3-*N*-acetylgalactosaminyltransferase from human lung tissue, *J. Biol. Chem. 265*: 1139–1145 (1990).

315. N. Navaratnam, J. B. Findlay, J. N. Keen, and W. M. Watkins, Purification, properties and partial amino acid sequence of the blood-group-A-gene-associated α-3-*N*-acetyl-galactosaminyltransferase from human gut mucosal tissue, *Biochem. J. 271*: 93–98 (1990).

316. N. Malagolini, F. Dall'Olio, and F. Serafini-Cessi, UDP-GalNAc: NeuAcα2,3Galβ-R (GalNAc to Gal) β1,4-*N*-acetylgalactosaminyltransferase responsible for the Sda specificity in human colon carcinoma CaCo-2 cell line, *Biochem. Biopys. Res. Commun. 180*: 681–686 (1991).

317. T. Dohi, N. Hanai, K. Yamaguchi, and M. Ohshima, Localization of UDP-GalNAc: NeuAcα2,3Gal-R β1,4 (GalNAc to Gal) *N*-acetylgalactosaminyltransferase in human stomach. Enzymatic synthesis of a fundic gland-specific ganglioside and GM2, *J. Biol. Chem. 266*: 24038–24043 (1991).

318. T. Dohi, A. Nishikawa, I. Ishizaki, M. Totani, K. Yamaguchi, K. Nakagawa, O. Saitoh, S. Ohshiba, and M. Ohshima, Substrate specificity and distribution of UDP-GalNAc: Sialylparagloboside *N*-acetylgalactosaminyltransferase in the human stomach, *Biochem. J. 288*: 161–165 (1992).

319. Y. Nagata, S. Yamashiro, J. Yodoi, K. O. Lloyd, H. Shiku, and K. Furukawa, Expression cloning of β1,4-*N*-acetylgalactosaminyltransferase cDNA that determines the expression of G_{M2} and G_{D2} gangliosides, *J. Biol. Chem. 267*: 12082–12089 (1992).

320. Y. Hashimoto, M. Sekine, K. Iwasaki, and A. Suzuki, Purification and characterization of UDP-*N*-acetylgalactosamine G_{M3}/G_{D3} *N*-acetylgalactosaminyltransferase from mouse liver, *J. Biol Chem. 268*: 25857–25864 (1993).

321. Y. Wang, J. L. Abernethy, A. E. Eckhardt, and R. L. Hill, Purification and character-ization of a UDP-GalNAc: Polypeptide *N*-acetylgalactosaminyltransferase with speci-ficity for glycosylation of threonine residue, *J. Biol. Chem. 267*: 12709–12716 (1992).

322. A. P. Elhammer, R. A. Poorman, E. Brown, L. L. Maggiora, J. G. Hoogerheide, and F. J. Kezdy, The specificity of UDP-GalNAc: Polypeptide *N*-acetylgalactosaminyltrans-ferase as inferred from a database of in vivo substrates and from the in vitro glyco-sylation of proteins and peptides, *J. Biol. Chem. 268*: 10029–10038 (1993).

323. F. K. Hagen, B. Van Wuyckhuyse, and L. A. Tabak, Purification, cloning, and expres-sion of a bovine UDP-GalNAc: Polypeptide *N*-acetylgalactosaminyltransferase, *J. Biol. Chem. 268*: 18960–18965 (1993).

324. Y. Wang, N. Agrwal, A. E. Eckhardt, R. D. Stevens, and R. L. Hill, The acceptor specificity of porcine submaxillary UDP-GalNAc: Polypeptide *N*-acetylgalactosami-nyltransferase is dependent on the amino acid sequences adjacent to serine and thre-onine residue, *J. Biol. Chem. 268*: 22979–22983 (1993).

325. I. Nishimori, N. R. Johnson, S. D. Sanderson, F. Perini, K. Mountjoy, R. L. Cerny, M. L. Gross, and M. A. Hollingsworth, Influence of acceptor substrate primary amino acid sequence on the activity of human UDP-*N*-acetylgalactosamine: Polypeptide *N*-acetyl-galactosaminyltransferase. Studies with the MUC1 tandem repeat, *J. Biol. Chem. 269*: 16123–16130 (1994).

326. J. K. Ritter, Y. Y. Sheen, and I. S. Owens, Cloning and expression of human liver UDP-glucuronosyltransferase in COS-1 cells. 3,4-Catechol estrogens and estriol as primary substrates, *J. Biol. Chem. 265*: 7900–7906 (1990).

327. Y. Ichikawa, R. Wang, and C.-H. Wong, Regeneration of sugar nucleotide for enzymatic oligosaccharide synthesis, *Methods in Enzymol. 247*: 107–127 (1994).

328. C.-H. Wong, S. L. Haynie, and G. M. Whitesides, Enzyme-catalyzed synthesis of *N*-acetyllactosamine with in situ regeneration of uridine 5′-diphosphate glucose and uridine 5′-diphosphate galactose, *J. Org. Chem. 47*: 5416–5418 (1982).

329. C. Augé, C. Mathieu, and C. Mérienne, The use of an immobilized cyclic multi-enzyme system to synthesize branched penta- and hexa-saccharides associated with blood-group I epitope, *Carbohydr. Res. 151*: 147–156 (1986).

330. C.-H. Wong, R. Wang, and Y. Ichikawa, Regeneration of sugar nucleotide for enzymatic oligosaccharide synthesis: Use of Gal-1-phosphate uridyltransferase in the regeneration of UDP-galactose, *J. Org. Chem. 57*: 4343–4344 (1992).

331. Y. Ichikawa, J. L.-C. Liu, G.-J. Shen, and C.-H. Wong, A highly efficient multienzyme system for the one-step synthesis of a sialyl trisaccharide: In situ generation of sialic acid and *N*-acetyllactosamine coupled with regeneration of UDP-glucose, UDP-galactose, and CMP-sialic acid, *J. Am. Chem. Soc. 113*: 6300–6302 (1991).

332. D. Gygax, P. Spies, T. Winkler, and V. Pfaar, Enzymatic synthesis of β-D-glucuronide with in situ regeneration of uridine 5′-diphosphoglucuronic acid, *Tetrahedron 47*: 5119–5122 (1991).

333. S. L. Haynie and G. M. Whitesides, Enzyme-catalyzed organic synthesis of sucrose and trehalose with in situ regeneration of UDP-glucose, *Appl. Biochem. Biotech. 23*: 155–170 (1990).

334. M. J. Sinnott, Catalytic mechanisms of enzymatic glycosyltransfer, *Chem. Rev. 90*: 1171–1202 (1990).

335. A. J. Kirby, Mechanism and stereoelectronic effects in the lysozyme reaction, *CRC Crit. Rev. Biochem. 22*: 283–315 (1987).

336. J.-M. Petit, F. Paquet, and J.-M. Beau, Synthesis of β-2-deoxy-D-glycosides assisted by glycosidases, *Tetrahedron Lett. 32*: 6125–6128 (1991).

337. S. Bay and D. Cantacuzene, Glycosidase-catalyzed synthesis of 2-deoxy-β-glycosides, *Bioorg. Med. Chem. Lett. 2*: 423–426 (1992).

338. G. C. Look and C.-H. Wong, A facile enzymatic synthesis of Galβ1,3Glucal: A key intermediate for the synthesis of Le[a] and sialyl Le[a], *Tetrahedron Lett. 33*: 4253–4256 (1992).

339. E. W. Holla, Enzymatic synthesis of selectively protected glycals, *Angew. Chem. Int. Ed. Engl. 28*: 220–221 (1989).

340. S. Schenkman, M.-S. Jiang, G. W. Hart, and V. Nussenzweig, A novel cell surface *trans*-sialidase of *Trypanosoma cruzi* generates a stage-specific epitope required for invasion of mammalian cells, *Cell 65*: 1117–1125 (1991).

341. F. Vandekerckhove, S. Schenkman, L. P. de Carvalho, S. Tomlinson, M. Kiso, M. Yoshida, A. Hasegawa, and V. Nussenzweig, Substrate specificity of the *Trypanosoma cruzi trans*-sialidase, *Glycobiology 2*: 541–548 (1992).

342. S. Tomlinson, L. P. de Carvalho, F. Vandkerckhove, and V. Nussenzweig, Resialylation of silidase-treated sheep and human erythrocytes by *Trypanosoma cruzi trans*-sialidase: Restoration of complement resistance of desialylated sheep erythrocytes, *Glycobiology 2*: 549–551 (1992).

343. M. A. Ferrero-Garcia, S. E. Trombetta, D. O. Sanchez, A. Reglero, A. C. Frasch, and A. J. Parodi, The action of *Trypanozama cruzi trans*-sialidase on glycolipids and glycoproteins, *Eur. J. Biochem. 213*: 765–771 (1993).

344. Y. Ito and J. C. Paulson, Combined use of *trans*-sialidase and sialyltransferase for enzymatic synthesis of αNeuAc2→3βGal-*OR*, *J. Am. Chem. Soc. 115*: 7862–7863 (1993).

345. K. B. Lee and Y. C. Lee, Transfer of modified sialic acids by *Trypanosoma cruzi trans*-sialidase for attachment of functional groups to oligosaccharide, *Anal. Biochem. 216*: 358–364 (1994).

346. S. Kobayashi, K. Kashiwa, T. Kawasaki, and S. Shoda, Novel method for polysaccharide synthesis using an enzyme: The first in vitro synthesis of cellulose via a nonbiosynthetic path utilizing cellulase as catalyst, *J. Am. Chem. Soc. 113*: 3079–3084 (1991).

347. J. H. Lee, R. M. Brown, Jr., S. Kuga, S. Shoda, and S. Kobayashi, Assembly of synthetic cellulose I, *Proc. Natl. Acad. Sci. USA 91*: 7425–7429 (1994).

348. G. H. Herrmann, Y. Ichikawa, C. Wandrey, F. C. A. Gaeta, J. C. Paulson, and C.-H. Wong, A new multienzyme system for a one-pot synthesis of sialyloligosaccharide: Combined use of β-galactosidase and α(2,6)-sialyltransferase coupled with regeneration in situ of CMP-sialic acid, *Tetrahedron Lett. 34*: 3091–3094 (1993).

349. K. Sakai, R. Katsumi, H. Ohi, T. Usui, and Y. Ishido, Enzymatic synthesis of *N*-acetyllactosamine and *N*-acetylallolactosamine by the use of β-galactosidases, *J. Carbohydr. Chem. 11*: 553–565 (1992).

350. G. H. Herrmann, U. Kragl, and C. Wandrey, Continuous catalytic synthesis of *N*-acetyllactosamine, *Angew. Chem. Int. Ed. Engl. 32*: 1342–1343 (1993).

351. M. Schuster, P. Wang, J. C. Paulson, and C.-H. Wong, Solid-phase chemical-enzymatic synthesis of glycopeptides and oligosaccharides, *J. Am. Chem. Soc. 116*: 1135–1136 (1994).

352. S. Y. C. Wong, G. R. Guile, R. A. Dwek, and G. Arsequell, Synthetic glycosylation of proteins using *N*-(β-saccharide)iodoacetamides: Applications in site-specific glycosylation and solid-phase enzymic oligosaccharide synthesis, *Biochem. J. 300*: 843–850 (1994).

353. K. B. Lee, D. Loganathan, Z. M. Merchant, and R. J. Linhardt, Carbohydrate analysis of glycoproteins, *Appl. Biochem. Biotech. 23*: 53–80 (1990).

354. H. Watzlawick, M. T. Walsh, Y. Yoshioka, K. Schmid, and R. Brossmer, Structure of the *N*- and *O*-glycans of the A-chain of human plasma α₂HS-glycoprotein as deduced from the chemical compositions of the derivatives prepared by stepwise degradation with exoglycosidases, *Biochemistry 31*: 12198–12203 (1992).

355. Boehringer Mannheim Biochemicals, 1993 Catalog, *Exoglycosidases*, pp. 167–171.

356. Tools for Glycobiology, Oxford GlycoSystems Product Catalog, *ExoGlycosidases*, pp. 51–74.

4

Chemical Synthesis of
Glycoprotein Glycans

Sabine L. Flitsch and Gregory Michael Watt
The University of Edinburgh, Edinburgh, Scotland

I. INTRODUCTION

The oligosaccharide side chains of proteins have been challenging targets for chemical synthesis for some time. The availability of pure synthetic glycan oligosaccharides has been instrumental in understanding biological processes such as the biosynthetic pathway of glycans [1–3], the recognition and regulation of glycosyl transferases and glycosidases [4,5], and lectin specificity, in particular in cell–cell adhesion processes [6,7]. Synthetic studies have provided material for structural work, such as solution NMR [8] and X-ray measurements [9], and have provided vaccines and antigenic probes for generating antibodies for diagnostic use.

Despite their heterogeneity, different glycans can be classified according to common structural motifs. Firstly, N-glycans are linked to proteins via a glycosylamido linkage to asparagine residues, whereas O-glycans are linked to protein hydroxyl groups, mainly to serine or threonine residues. Both have distinctly different biosynthetic pathways, and each has particular structural features. Illustrations of typical N- and O-glycans are shown in Figure 1.

A striking structural feature of N-glycans is the common core pentasaccharide at the reducing end, consisting of three Man and two GlcNAc residues. Substructures of N-links can be classified according to further branching; for example, structure **1** in Figure 1 contains a tetraantennary structure with four branches off the nonreducing mannose residues of the pentasaccharide core. In mammals these branches can contain repeated N-acetyllactosamine units that are frequently capped by either α2–3- or α2–6-linked NeuNAc residues. An additional Fucα1–6 branch at the reducing GlcNAc residue of the N-glycan structure **1** has also been observed. O-Glycans are generally shorter oligomers than N-glycans and can be classified by a number of conserved cores; for example, the O-glycan structure **2** in Figure 1 contains a typical Core II trisaccharide at the reducing end [8].

The nonreducing sugars on oligosaccharides have been shown to be particularly important in the process of biological recognition, and cell–cell interactions can often be inhibited by small soluble oligosaccharide fragments [7], either single monosaccharide residues or oligomers. A recent important example is the tetrasaccharide Lewis^x structure shown in Figure 1, found both on glycolipids (gangliosides) and glycoproteins. This has been shown to be an important antigen with regard to cell adhesion [6] and can inhibit key cell adhesion processes.

The progress made in oligosaccharide synthesis over the past 15 years has been remarkable. Yet, although many biologically important glycan structures have now been prepared, there are still many challenges for the synthetic chemist. Oligosaccharide synthesis is difficult and time consuming, mainly due to the requirement of protecting-group strategies. Currently there is no universal method for the formation of glycosidic linkages. A comprehensive strategy must be carefully worked out prior to each novel synthesis, for oligosaccharide formation does not depend only on the size of the target structure, but also on its linkages and the composition of its building blocks. Thus, some decasaccharides may be obtained readily, whereas a trisaccharide with "difficult" linkages can present many problems. This chapter will try to present the state of the art of current synthetic methodology, highlighting some recent achievements and some of the problems that still exist. It is by no means comprehensive, and the reader is also referred to several excellent reviews that have been written on oligosaccharide synthesis [8,10–20]. More detailed reviews on some spe-

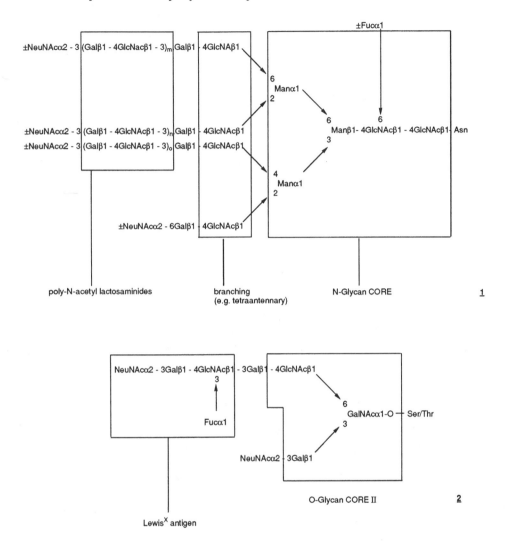

Figure 1 Typical *N*-linked glycan 1 and *O*-linked glycan 2 structures.

cific methods, including enzymatic synthesis and solid-phase synthesis, which are recent important developments in this area, are described in separate chapters in this book.

II. CHEMICAL SYNTHESIS OF THE OLIGOSACCHARIDES

The monosaccharide building blocks of glycans such as those shown in Figure 1 are generally readily available. *N*-Acetylneuraminic acid is still quite expensive, although a good chemo-enzymatic synthetic approach from mannose using commercially available aldolase has been developed [21]. Thus the major challenge in glycan synthesis is the formation of glycosidic linkages between a "glycosyl donor" and a "glycosyl acceptor" with high regio- and stereoselectivity. The reactivity of hydroxyl groups in partially protected glycosyl acceptors is influenced by steric factors and also by the size of the coupling components. The general order of reactivity in aldohexopyranosides containing equatorial hydroxyl groups is 6-OH >> 3-OH > 2-OH > 4-OH [20]. The stereoselectivity is harder to control in coupling reactions involving hydroxyl groups of low nucleophilicity, due to their low rate of reaction; this increases the possibility of side reactions, thus lowering the yield of the desired glycoside. The regioselectivity of glycosidation reactions is essentially ensured by the use of protecting groups, and their stereoselectivity is ensured by the choice of reagents and reaction conditions. The most commonly used protecting groups will be reviewed first.

A. Protecting Groups

In glycosidation reactions, regioselectivity is essentially obtained by selectively protecting the coupling components, although protecting groups may also influence the stereoselectivity. Before coupling, each individual building block needs to be unambiguously protected with a set of orthogonal groups such that each hydroxyl group is deprotected when required for glycosylation. Depending on the complexity of the target molecule, this might require up to four orthogonal hydroxyl-protecting groups for hexoses. The protection and deprotection steps are the main reasons why oligosaccharide synthesis is so time consuming, and, unfortunately, chemical methodologies are not selective enough to avoid them.

A selection of alcohol- and amino-protecting groups commonly employed in glycan synthesis is listed in Table 1. The most frequently used hydroxyl-protecting groups are acetate and benzoate esters and benzyl ethers, which are generally fully orthogonal, although primary benzyl ethers are sometimes cleaved by using typical ester hydrolysis conditions (e.g., sodium methoxide in methanol). The choice of protecting group depends on the overall synthetic strategy, where not only orthogonality has to be taken into account, but also the reactivity of synthetic intermediates. Thus the correct selection of protecting group may not only prevent undesirable side reactions, but can significantly influence the reactivity of coupling components during glycoside bond formation.

The C-2 substituent of the glycosyl donor is particularly important in influencing the stereoselectivity of the glycosidic linkage. A participating group such as an acetate, a benzoate, or an acetamide can direct the attack of the incoming nucleophile to produce, in general, the 1,2 trans configuration of the glycoside bond (Figure 2). An acetate-protecting group at C-2 can lead to undesired side products during glycosidation reactions, for example, ortho-esters (Figure 2). This may be reduced by using other acyl groups at C-2, such as benzoyl and pivaloyl groups.

In addition, protecting groups can influence the reactivity of both the glycosyl donor and the glycosyl acceptor, even if they are not directly involved in the coupling reaction. Acetates and benzoates are more electronegative than benzyl-protecting groups and can lower the nucleophilicity of free hydroxyl groups of partially protected sugars. For example, the C-4 hydroxyl group of tetrabenzyl glucoside 3 is much more nucleophilic than the C-4 hydroxyl group of the tetraacetate 4. The reactivity of the anomeric center of glycosyl donors can be markedly influenced by electronegative protecting groups through inductive effects. Several research groups

Table 1 Protecting Groups Used in Oligosaccharide Synthesis

hydroxyl protecting groups:

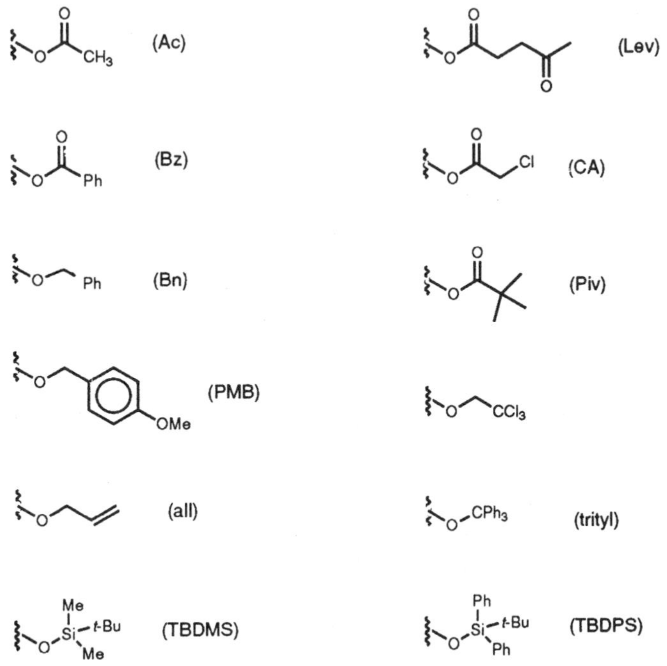

diol protecting groups:

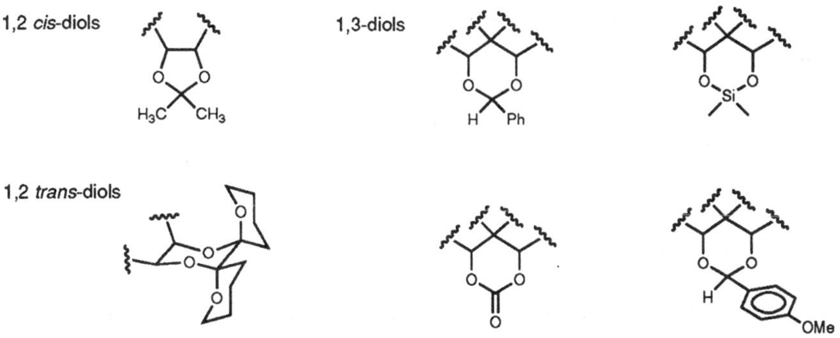

1,2 cis-diols

1,3-diols

1,2 trans-diols

amino-protecting groups

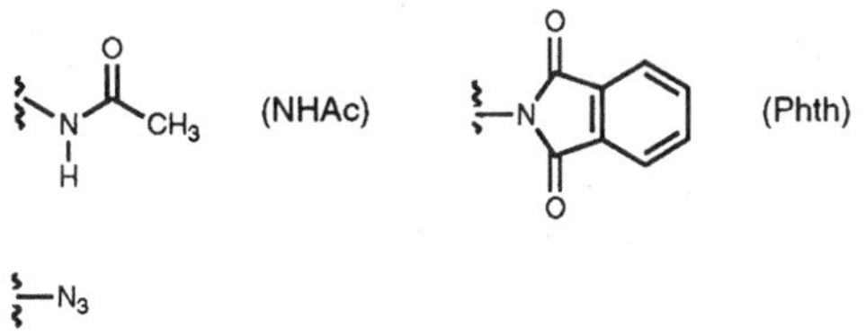

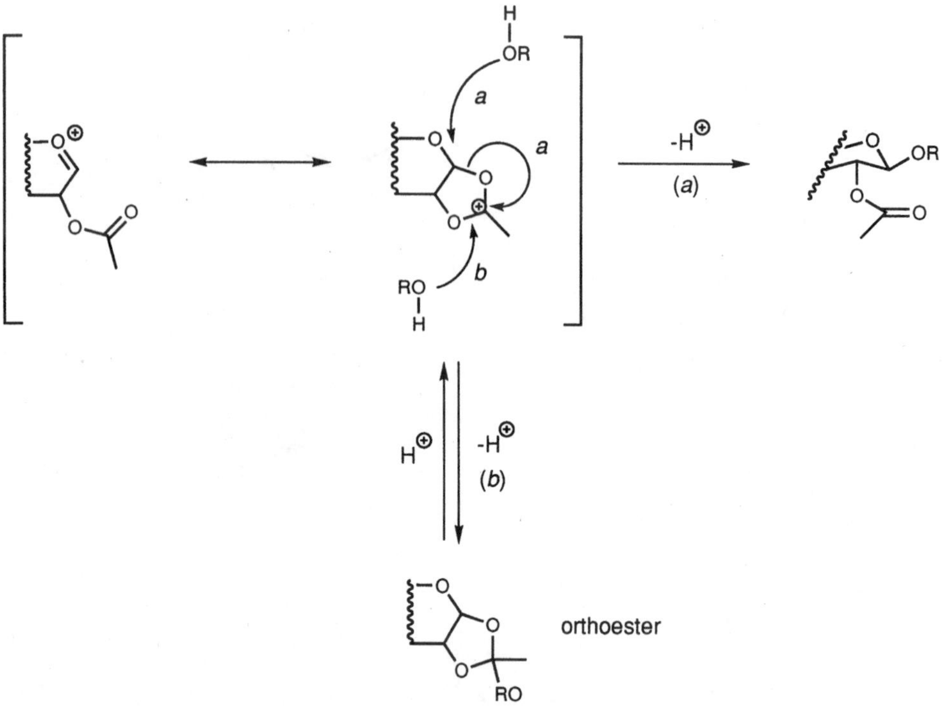

Figure 2 Proposed mechanism of glycosidation reactions involving neighboring-group participation. Path (a): formation of glycoside; path (b): formation of orthoester (reversible).

have taken advantage of this factor by using glycosylation strategies involving "active" and "dormant" glycosyl donors, which will be discussed in more detail later on.

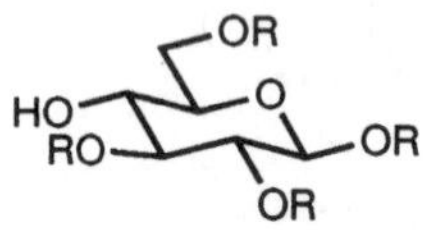

3 R= -Bn
4 R = -Ac

Selective Protection of Saccharides

The selective protection of sugars relies on subtle differences in reactivity between the different saccharide hydroxyl groups. This is demonstrated by the example shown, in Figure 3, where a glucoside is generated and then selectively protected. Protecting the anomeric hydroxyl is relatively straightforward. Direct C-1 protection of free sugars can be achieved by means of the Fischer reaction, which entails treatment with the corresponding alcohol of the intended glycoside (e.g., methyl or allyl) in the presence of acid, although generally both α- and β-anomers are formed. The Koenigs–Knorr method is a more stereoselective approach; for example, glucose pentaacetate **5** is converted into the glucosyl chloride **6**. Displacement of the chloride with an alcohol (e.g., R^1 = methyl or allyl) in the presence of a reaction promoter followed by deacetylation affords the C-1–protected β-glucoside **7**.

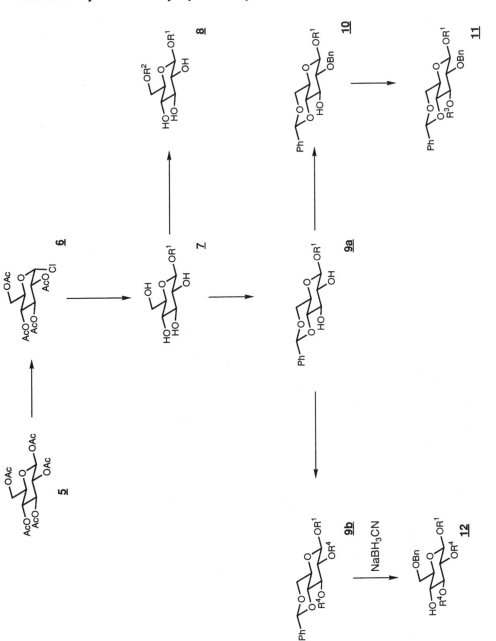

Figure 3

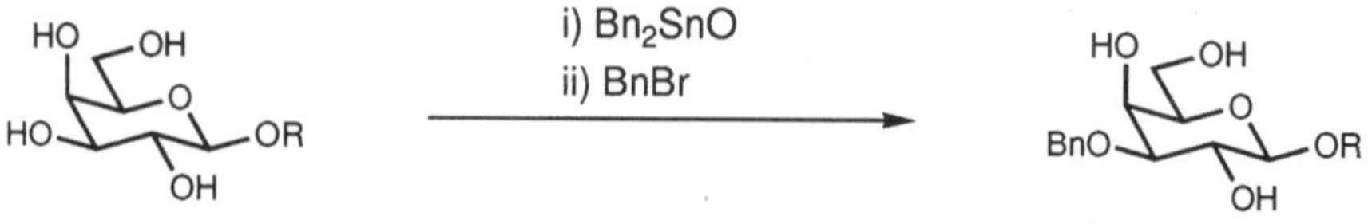

i) Bn₂SnO
ii) BnBr

13   **14**

Figure 4

    Straightforward protection of the primary hydroxyl at C-6 is possible because it reacts with bulky reagents faster than secondary sugar hydroxyls. Glucoside **7** is converted into the C-6–protected derivative **8** with the use of bulky protecting groups (e.g., R^2 = TBDMS or trityl).

    The selective protection of the secondary hydroxyl groups can present a more difficult problem. A breakthrough in this area has been the development of diol-protecting groups, such as acetals (Table 1). Benzylidene acetals are particularly useful for selectively protecting hexopyranose sugars; for example, the 4,6-protected glucopyranoside **9a** can be obtained using benzaldehyde and a Lewis acid promoter, or by an acyl-transfer reaction involving dimethylbenzyl acetal in the presence of an acid catalyst, whereby the methanol formed during the reaction can be removed to shift the equilibrium in favor of the products and thus drive the reaction to completion [22]. Compound **9a** can be selectively benzylated at the C-2 hydroxyl group under phase-transfer conditions (BnBr, NaOH, Bu₄NHSO₄, CH₂Cl₂) [23] to give the 2-O-benzyl ether **10**, which leaves the C-3 hydroxyl group free for orthogonal protection to give compound **11**. Alternatively, benzylidene acetal groups can be reductively cleaved with sodium cyanoborohydride, for example, in the transformation of the fully protected derivative **9b** to give **12**, whose C-4 hydroxyl group is left for further reaction [10].

    In the case of mannose and galactose, their unique cis-diol configuration can be effectively exploited by the formation of dialkyl-acetal derivatives. The use of

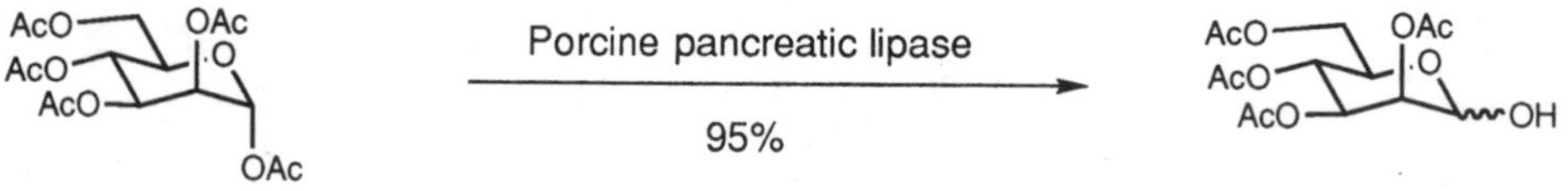

Porcine pancreatic lipase

95%

15   **16**

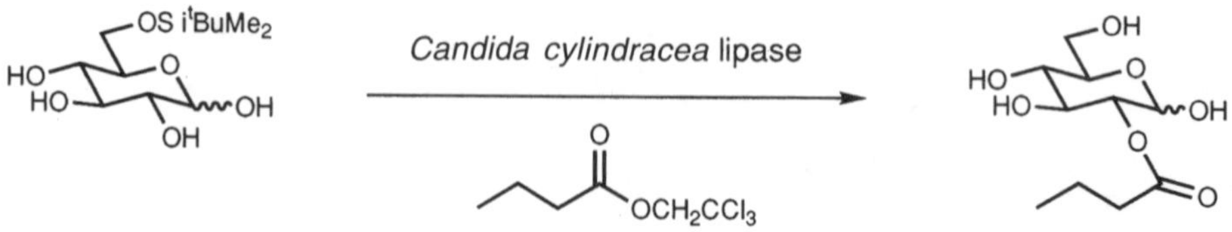

Candida cylindracea lipase

17   **18**

Figure 5

dispiroketal (dispoke) derivatives can enable the selective protection of vicinal trans-diols (Table 1). The regioselective protection of glucose derivatives, which contain both 2,3- and 3,4-trans-diols is possible by chiral recognition using enantiopure di-substituted-6,6'-bi-2H-pyrans [24].

The use of stannyl ethers and stannylene acetals have been very effective tools for the selective protection of sugars, for example, the selective C-3 protection of galactoside **13** via 3,4-stannylene acetal formation to give **14** (Figure 4) [25].

An interesting more recent approach is the use of lipases and esterases for selective carbohydrate protection [26]. These can be used either hydrolytically (e.g., **15** → **16**) or synthetically (e.g. **17** → **18**) (Figure 5).

B. Glycosidation Methods

The formation of glycosidic linkages is essentially the key step in oligosaccharide synthesis, and it should be emphasized that there are no universal methods for oli-gosaccharide synthesis, such as are available for peptides and oligonucleotides. The statement of Paulsen in his review of 1982 that ''each oligosaccharide remains an independent problem'' [20] still holds true; however, during the past 15 years a plethora of reactions have been developed that have allowed access to complex oligosaccharide structures.

By far the most common way of forming glycosides is by nucleophilic dis-placement of anomeric substituents on glycosyl donors by glycosyl acceptors. An-omeric hydroxyls are poor leaving groups and need to be activated, either by pro-tonation or by conversion into a good leaving group. This is particularly important since the hydroxyl group of the aglycon is generally a poor nucleophile, especially when it is a secondary ring hydroxyl group. Five glycosidation methods that are commonly used in oligosaccharide synthesis are listed in Table 2 and are discussed in the following sections.

Glycosyl Halides (Koenigs–Knorr)

The Koenigs–Knorr glycosylation method was the first approach to be used widely for the formation of glycosidic linkages [20]. Glycosyl chlorides and bromides are available from either glycosyl acetates (Figure 3) or hemiacetal sugars and are gen-erally stable enough to be purified and stored. Glycosyl halide formation is the last step carried out prior to the glycosylation reaction, for glycosyl halides are too labile to withstand the reaction conditions of, for example, protecting-group manipulations. Hence, they are ideally used for stepwise synthesis of oligosaccharides starting from the reducing end, but are relatively less useful for block syntheses. Although glycosyl halides can react with strong nucleophiles by direct displacement of the halide, re-action times and yields are generally greatly improved by halophilic activators such as silver or mercury salts (Table 2).

Trichloroacetimidates

The trichloroacetimidate reaction was developed in the early 1980s and is now con-sidered alongside the Koenigs–Knorr method as one of the most popular glycosi-dation methods. The use of trichloroacetimidates is well reviewed [13,18]. They are easily formed from hemiacetal sugars (e.g., **19**) by treatment with trichloroacetonitrile

Table 2 The Main Anomeric Activating Groups Used in Oligosaccharide Synthesis

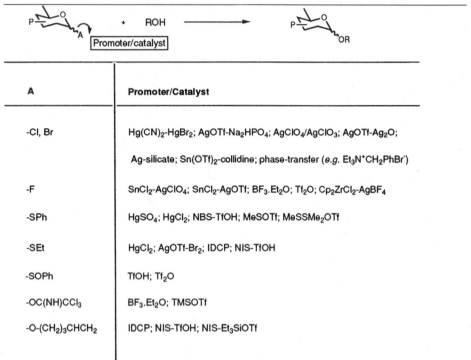

A	Promoter/Catalyst
-Cl, Br	Hg(CN)$_2$-HgBr$_2$; AgOTf-Na$_2$HPO$_4$; AgClO$_4$/AgClO$_3$; AgOTf-Ag$_2$O; Ag-silicate; Sn(OTf)$_2$-collidine; phase-transfer (*e.g.* Et$_3$N$^+$CH$_2$PhBr$^-$)
-F	SnCl$_2$-AgClO$_4$; SnCl$_2$-AgOTf; BF$_3$.Et$_2$O; Tf$_2$O; Cp$_2$ZrCl$_2$-AgBF$_4$
-SPh	HgSO$_4$; HgCl$_2$; NBS-TfOH; MeSOTf; MeSSMe$_2$OTf
-SEt	HgCl$_2$; AgOTf-Br$_2$; IDCP; NIS-TfOH
-SOPh	TfOH; Tf$_2$O
-OC(NH)CCl$_3$	BF$_3$.Et$_2$O; TMSOTf
-O-(CH$_2$)$_3$CHCH$_2$	IDCP; NIS-TfOH; NIS-Et$_3$SiOTf

in the presence of a base such as sodium hydride (NaH). These reaction conditions are generally compatible with most protecting groups and glycosidic linkages. The resulting glycosyl trichloroacetimidates are very reactive upon activation with Lewis acids such as boron trifluoride (see Table 2), generally giving excellent stereoselectivity. A further advantage of trichloroacetimidates is that they can be purified by column chromatography, and unlike glycosyl halides they do not require activation with toxic or expensive heavy-metal salts.

Preparation of both α- and β-glycosyl trichloroacetimidates is possible depending upon the reaction conditions [18], whereas β-glycosyl chlorides and bromides are often unobtainable. The in situ deprotonation of hemiacetals under basic conditions affords anomeric oxide ions that react with trichloroacetonitrile in a rapid and reversible process; due to the higher nucleophilicity of the β-anomeric oxide ion, the β-trichloroacetimidate is the kinetic product. This can then anomerize in a slow, base-catalyzed reaction to give the thermodynamically more stable α-trichloroacetimidate (Figure 6). The choice of conditions can sometimes allow the exclusive formation of one anomer; for example, the reaction of hemiacetal **19** with trichloroacetonitrile in dichloromethane using K$_2$CO$_3$ as the base gives the β-anomer (**20-β**) (kinetic product). The α-anomer (**20-α**) (thermodynamic product) can be selectively obtained by using NaH or 1,8-diazabicyclo[5,4,0]undec-7-ene (DBU) as the base. This is because, unlike K$_2$CO$_3$, NaH and DBU catalyze the anomerization of the β-anomer to generate the thermodynamically more stable α-anomer. Unfortu-

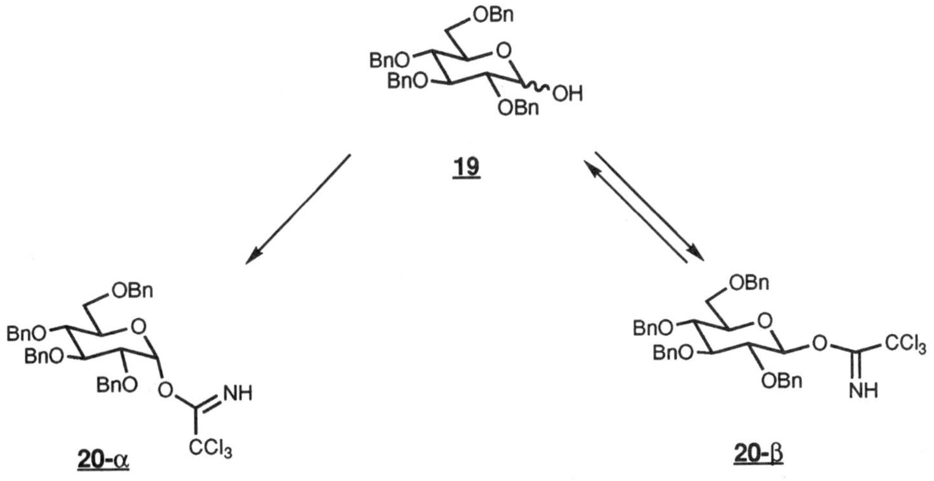

Figure 6

nately, the formation of β-trichloroacetimidates is sensitive to protecting-group substituents; for example, if the C-6 benzyl group of **19** is replaced with an acetyl group, the reaction using K_2CO_3 as the base is less selective (α/β = 1:3) [18].

Thioglycosides, Sulfoxides, and Selenides

Glycosylation via thioglycosides has gained greatly in popularity over the past years [10,11]. They can be obtained easily from glycosyl halides or peracetylated sugars on reaction with thiols. They are very stable until treated with specific thiophilic activators (Table 2) and can therefore be carried through several reactions before activation. This avoids the selective C-1 protection/deprotection/activation sequence required before coupling in the two previous methods and makes this method particularly useful for block synthesis. Activation of thioglycosides was first described by Ferrier [27] using mercury(II) sulfate and has been followed by a large number of alternative reagents (Table 2).

Glycosyl sulfoxides are more reactive than thioglycosides and are prepared by the oxidation of thioglycosides using *m*-chloroperbenzoic acid just before the glycosidation reaction. Activation of glycosyl sulfoxides is achieved under mild conditions using triflic anhydride in the presence of an acid scavenger such as 2,6-di-*t*-butyl-4-methylpyridine [28].

Glycosyl selenides can be selectively activated over thioglycosides using silver triflate in the presence of potassium carbonate. They have the added advantage that they are inert during the activation of glycosyl halides and trichloroacetimidates [29]. They can be prepared from peracetylated sugars or by azido-selenylation of glycals [30,31].

The reactivity of thioglycosides is sensitive to the type of protecting group used. Electron-withdrawing groups, such as acyl groups, are deactivating, and electron-donating groups (e.g., benzyl ethers) activate the thioglycoside. These effects can be exploited to the extent that a deactivated thioglycoside (such as **22** in Figure 7) can be used as a glycosyl acceptor when coupled with an activated thio-

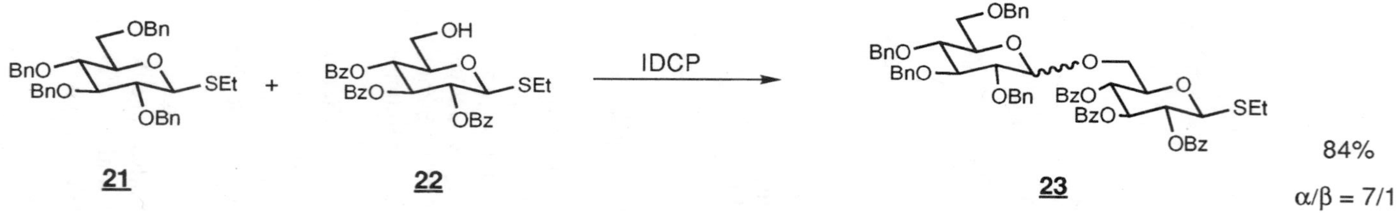

Figure 7

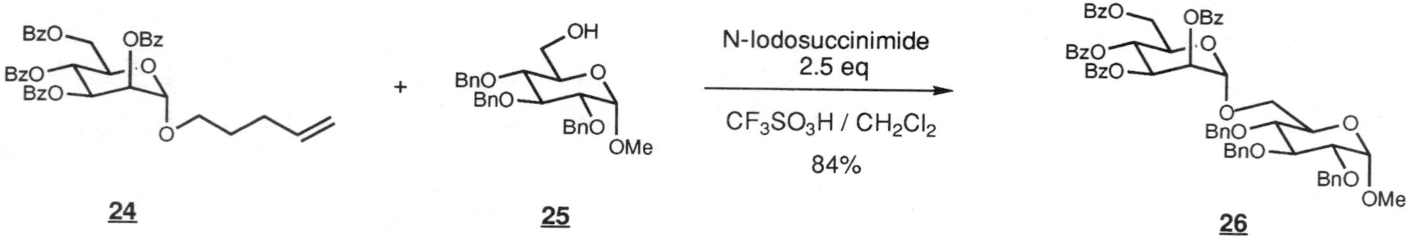

Figure 8

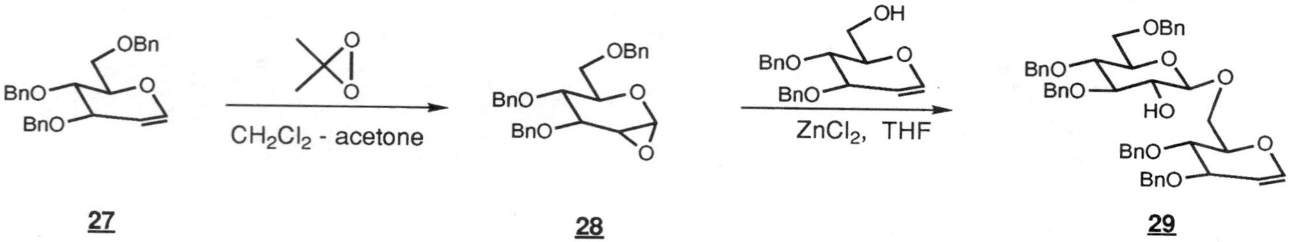

Figure 9

glycosyl donor (e.g., **21**) in the presence of iodonium dicollidine perchlorate (IDCP) to give as the predominant product one disaccharide (e.g., **23**) in good yield [32].

Pentenyl Glycosides

Pentenyl glycosides such as **24** (Figure 8) are activated in the presence of N-iodo-succinimide (NIS), whereby fragmentation of the resulting iodonium ion gives the oxocarbenium ion [33,34] that couples to the acceptor **25** to afford the disaccharide **26**. This reaction was discovered by Fraser-Reid and colleagues [35] and has been applied to the synthesis of N-glycans. Pentenyl glycosides have the advantage that they are stable and activated by mild reaction conditions that do not normally affect protected sugars. This strategy is therefore suitable for block synthesis and has been used by Fraser-Reid and colleagues for a number of oligosaccharide syntheses.

Glycals

Glycals such as **27** make very attractive building blocks for oligosaccharide synthesis, because there is no need for C-1 and C-2 protection before activation; thus the remaining hydroxyl groups can be selectively protected. Glycals are difficult to activate, which is the main reason their use is not widespread. However, recently Danishefsky and colleagues [36] have used 3,3-dimethyloxirane to generate α-epoxide intermediates (such as **28** in Figure 9), which can react with a nucleophile in the presence of $ZnCl_2$ to give saccharides (e.g., **29**) in good yields. More details on this method and its application to solid-phase synthesis are reported in Chapter 5 (pp. 256–271).

C. Diastereoselective Generation of Glycosidic Linkages

General Considerations

The stereoselective outcome of glycosylation reactions is sensitive to a range of different factors, such as the structure of the glycosyl donor and acceptor, the choice of protecting groups, the type of coupling method used, the choice of activator used, the solvent, and the reaction temperature. With this number of variables, the reaction conditions for each individual coupling reaction need to be optimized. However, with the help of systematic studies our understanding of coupling reactions has improved to the extent that it is sometimes possible to predict the effects of reaction conditions on the stereochemical outcome. The following is an attempt to present some of our present understanding.

The reaction scheme shown in Figure 10 is an illustration of a nucleophilic displacement at the anomeric center, where P denotes a fully protected sugar and **R** a nonparticipating substituent. From the initial hemiacetal **30**, two activated species **31**-α and **31**-β can be formed, where X is a good leaving group (Table 2). These can both form the oxocarbenium ion intermediate **32** via an S_N1-type displacement or lead to glycosides **33**-β and **33**-α by direct S_N2 displacement of X by the glycosyl acceptor R′OH. Once the oxocarbenium ion **32** is generated, product formation depends on whether the α- or the β-face is preferentially attacked by R′OH.

If R does not influence the reaction—i.e., if it is small and nonparticipating—formation of the α-anomer **33**-α is generally favored. This is because the axial (α)

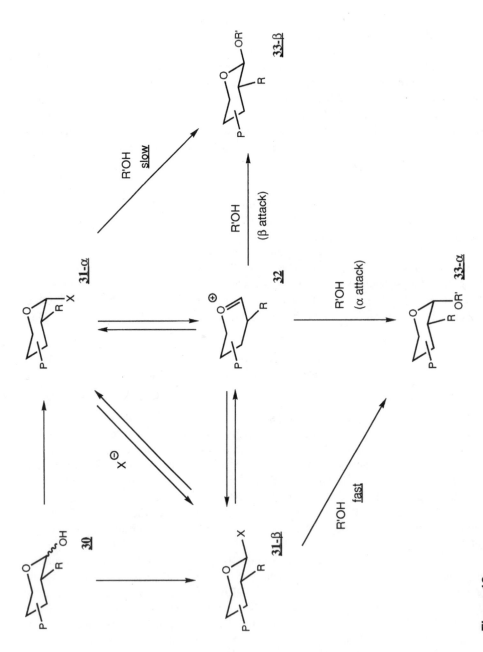

Figure 10

displacement of X on **31**-β by R′OH (**31**-β → **33**-α) and the axial (α) attack on **32** (**32** → **33**-α) are both faster than the corresponding equatorial (β) approaches. This has been explained using stereoelectronic arguments similar to those rationalizing the anomeric effect [37]. In Figure 10, **33**-β can only be formed in high yields from **31**-α by a direct S_N2-type displacement, avoiding the formation of the oxocarbenium ion **32**. High yields of the α-anomer **33**-α can be obtained by in situ anomerization of α-halides. For example, when **31**-α (X = Br) is coupled with R′OH in the presence of bromide ion, a rapid equilibrium between **31**-α and **31**-β is established; because the α-attack of R′OH on **31**-β is faster than the β-attack of R′OH on **31**-α, glycoside **33**-α is predominantly formed [20].

Neighboring-group participation. The protecting group at C-2 (R in Figure 10) can have a pronounced effect on the stereochemical outcome of glycosylation reactions. If it is a very bulky group (e.g., *O*-pivaloyl and *N*-phthalimido groups), the β-face of the oxocarbenium ion **32** may be preferentially attacked by R′OH, resulting in the predominant formation of **33**-β. Apart from providing steric bulk, the C-2−protecting group can also stabilize the oxocarbenium ion **32** by the formation of a dioxocarbenium ion (Figure 11). This neighboring-group stabilization competes with the reactions shown in Figure 10, but under the right conditions can dominate the stereochemical outcome of the glycosylation reaction. The most common participating protecting groups used in glycan synthesis are *O*-acetyl, *O*-benzoyl and *N*-phthalimido groups. Neighboring-group participation is a very effective way of generating glycosidic linkages with a 1,2 trans configuration, of either α- or β-glycosides, depending on the configuration of C-2 (Figure 11).

In discussing some general strategies for glycan synthesis it makes sense to classify glycosidic linkages of glycans into four different types: α-1,2 trans, β-1,2 trans, α-1,2 cis, and β-1,2 cis. In addition, sialic acids are structurally different from the other pyranoses and need to be considered separately, since they present particular synthetic problems.

α-1,2 Trans Linkages

The α-1,2 trans linkage is presented mainly by α-mannosides, which are common in *N*-linked glycans. From the mechanistic discussion in the previous section (Figures 10 and 11), this linkage would be expected to be relatively easy to obtain, which is essentially the case. Formation of the α-1,2 trans linkage is favored by neighboring-group effects and also by kinetic formation of the α-anomer. For example, mannosyl trichloroacetimidates afford predominantly α-products, even when starting from α-trichloroacetimidates [18]. α-Mannosides are frequently prepared using Koenigs−Knorr methodology because acetobromo mannose is readily available. A striking example of the selectivity of α-mannosylation is the one-pot synthesis of the core trisaccharide **36** (Figure 12), whereby the coupling of acetobromo mannose **34** and octyl β-D-mannoside **35** gave **36** in 17% yield, besides other regioisomers but no significant amount of stereoisomers [38]. The other regioisomers were removed by periodate oxidation, to which the desired compound **36** was not susceptible.

The trichloroacetimidate method is essentially a highly effective way of generating α-1,2 trans linkages with regard to block synthesis. For example, the two disaccharide building blocks **37** and **38** coupled in the presence of trimethylsilyl triflate to afford the tetrasaccharide **39** in good yield (Figure 13) [39]. *O*-Pentenyl

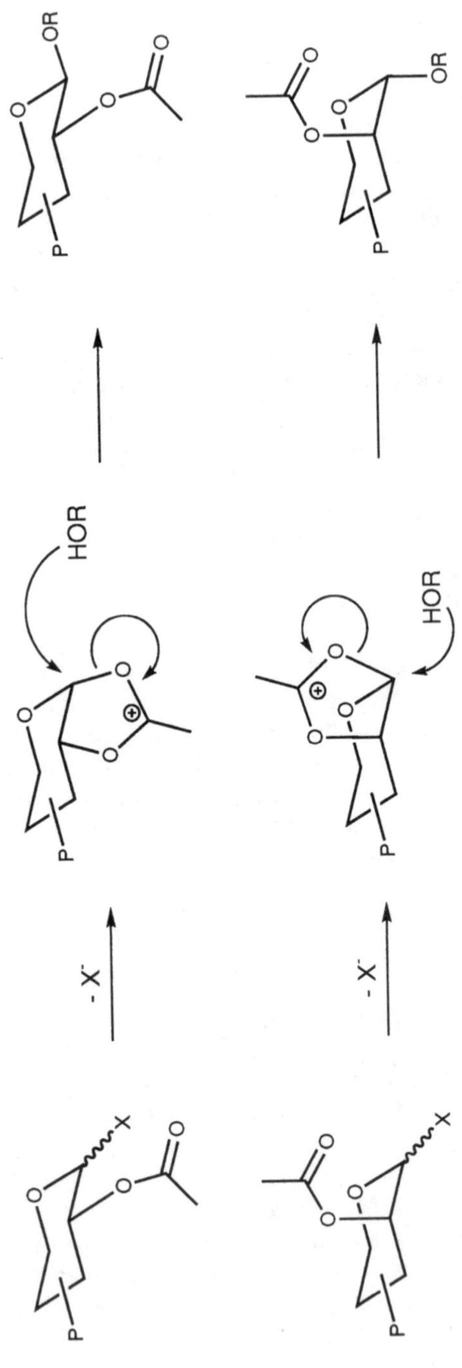

Figure 11

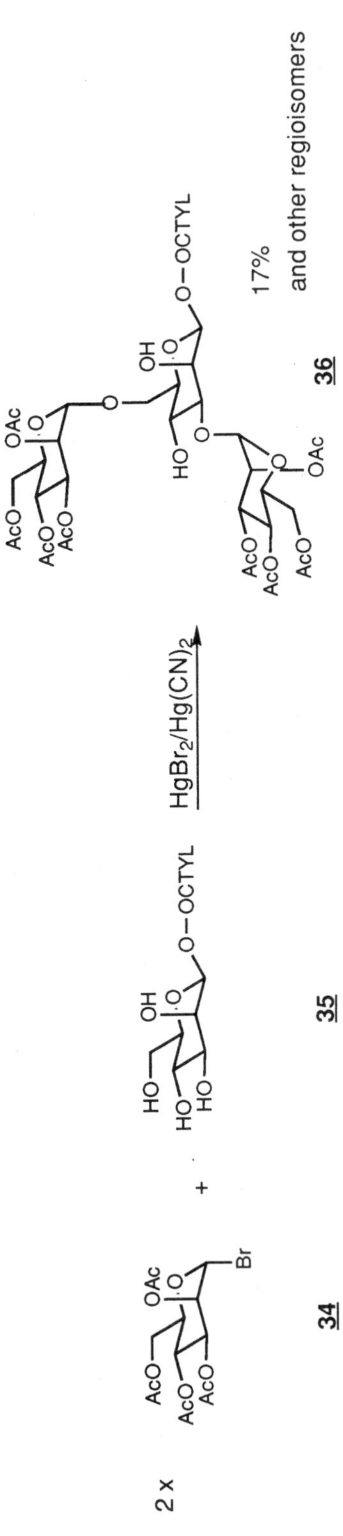

Figure 12

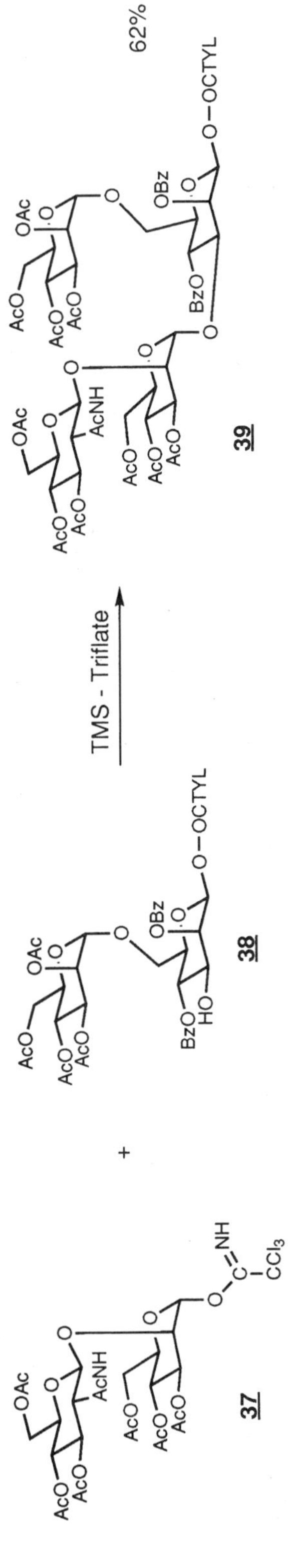

Figure 13

glycosides, such as **24** (Figure 8), are also effective at generating α-mannoside linkages. α-Mannosides can even be formed as the predominant anomer without a participating group at C-2, and some recent advances in this area have been reported [40,41].

α-1,2 Cis Linkages

A number of different α-1,2 cis linkages occur in glycans, the most important being α-glucosides, α-galactosides, 2-acetamido-2-deoxy-α-glucosides, 2-acetamido-2-deoxy-α-galactosides, and α-fucosides. The stereoselective synthesis of α-1,2 cis-glycosides is often troublesome; this is particularly true for 2-acetamido-2-deoxy-α-glycosides. From the earlier discussion on neighboring-group participation it follows that a nonparticipating group is required in the C-2 position of the glycosyl donor. Since the *N*-acetyl group provides good anchimeric assistance, it needs to be temporarily protected as a small group that is not participating. The best results have been achieved with 2-azido-2-deoxy-pyranosides, which allow good α-selectivity.

A very useful strategy for generating α-1,2 cis-glycosides is by in situ anomerization of glycosyl bromides. This method works well with fucosides, for example, the generation of the Fucα1-3GlcNAc disaccharide **42** from **40** and **41** in the presence of tetraethylammonium bromide (Figure 14) [42]. Thioglycosides have also been used successfully for the formation of α-1,2 cis-glycosides [43,44]. It has been shown [45] that reaction conditions such as temperature and solvent can greatly influence the stereoselectivity of glycosidation reactions and that probing of the reaction conditions can often lead to the desired anomeric configuration. The stereoselectivity of the trichloroacetimidate coupling method has been extensively investigated by Schmidt and colleagues [18]. β-Trichloroacetimidates usually result in the formation of the α-glycoside as the predominant anomer by an S_N2-type reaction, with inversion of configuration. For example, the α-anomer **45** is formed from the coupling of the β-trichloroacetimidate **43** and the acceptor **44** in ether in the presence of a strong Lewis acid catalyst (e.g., Me$_3$SiOTf) (Figure 15). The solvent appears to play an important role in determining the stereoselectivity of the reaction. Nonpolar solvents make the formation of the oxocarbenium ion unfavorable, thus encouraging the S_N2-type reaction pathway. Diethyl ether can also participate in the reaction by forming the more stable equatorial oxonium ions (e.g., **47**), which favor invertive attack and the formation of the thermodynamically more stable α-products (Figure 16).

2-Acetamido-2-deoxy-α-glucosides and 2-acetamido-2-deoxy-α-galactosides are synthesized using mostly 2-azido-2-deoxy-donors, which can be prepared by azido-nitration or azido-selenylation of glycals or by epoxide ring opening [17]. Both α- and β-glycosides can be obtained from 2-azido-2-deoxy glycosyl donors, depending on the choice of reaction conditions. Both the in situ anomerization of α-glycosyl bromides and the direct displacement of β-trichloroacetimidates in apolar solvents using a Lewis acid catalyst such as Me$_3$SiOTf [46] have been successful approaches for generating α-1,2 cis linkages.

β-1,2 Trans Linkages

The β-1,2 trans linkage is represented in glycans by β-galactosides, 2-acetamido-2-deoxy-β-glucosides, and 2-acetamido-2-deoxy-β-galactosides, and is best formed via neighboring-group participation. The most obvious choices are 2-NHAc and 2-OAc

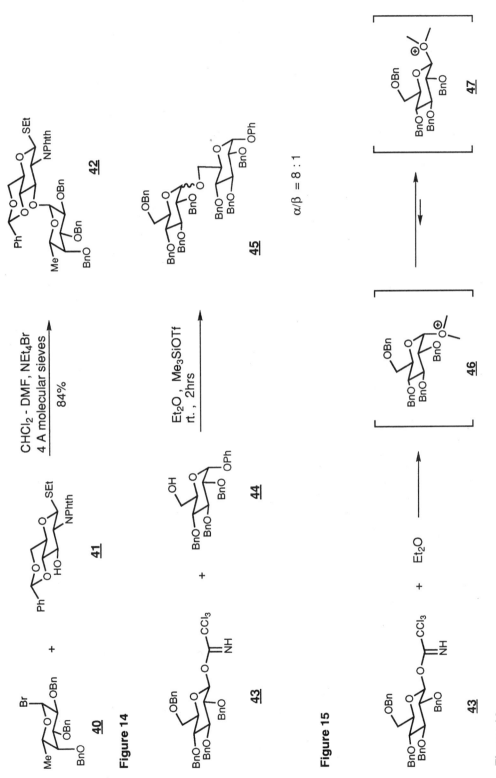

Figure 14

Figure 15

Figure 16

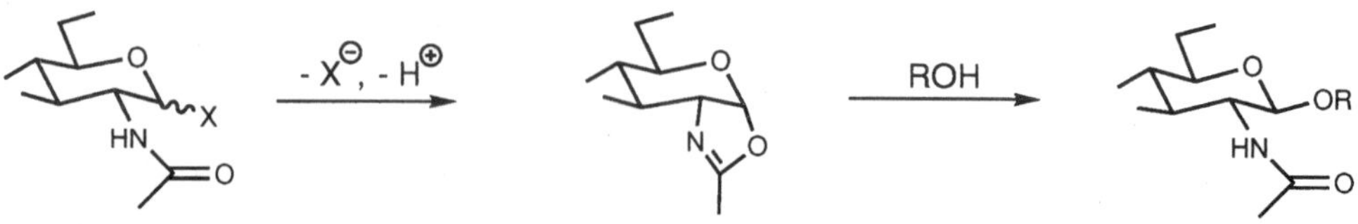

48   49   50

Figure 17

groups, but these can be problematic because the stabilized intermediates generally
react sluggishly. Orthoesters may be formed as possible side products (Figure 2),
although the trichloroacetimidate method has been reported to result in reduced orthoester formation, when compared with the Koenigs–Knorr method [18]. Oxazolines (e.g., **49**) can be formed readily from 2-acetamido-2-deoxy-β-glycosyl halides
(e.g., **48**; Figure 17) and have been useful β-directing intermediates, although they
generally react well only with reactive glycosyl acceptors.

Benzoates and pivaloates (Table 1) are usually better 2-O-participating groups
because they can also encourage β-glycoside formation through steric bulk, and
orthoesters are formed to a lesser extent. The phthalimido group is a very effective
2-N-protecting group, because it is relatively reactive and provides good stereoselectivity, both by anchimeric assistance and by providing steric bulk. Thus β-anomers
can generally be obtained in high yields even when using poor acceptors. A disadvantage of N-phthalimido protection is the requirement for several additional
protection and deprotection steps.

As an alternative, β-glycoside formation can also be achieved by S_N2-type
displacement of α-halides or trichloroacetimidates of glycosyl donors containing a
nonparticipating group at C-2 using nonpolar solvents and low temperatures to avoid
formation of the oxocarbenium ion. Acetonitrile appears to be a particularly good
solvent to favor β-glycoside formation. This is suggested [18] to be due to direct
solvent participation by the formation of a nitrilium-nitrile conjugate that can have
either an α- or a β-configuration, where the former is kinetically favored, thus resulting in preferential formation of β-glycosides.

β-1,2 Cis Synthesis of the β-Mannosidic Linkage

The only β-1,2 cis linkage found in glycans is the β-mannoside linkage of the N-
glycan pentasaccharide core (Figure 1). This linkage has presented a challenging
problem to chemists for a long time and is hence more extensively discussed in this
chapter. The methods discussed so far relating to stereochemistry, fail for β-mannoside synthesis and lead to α-mannosides as the predominant anomer, even under
conditions normally favorable for β-glycoside formation [18]. Thus, special glycosidation methods are required to solve this problem.

The first synthesis of β-mannosides involved an inversion of β-glucosides (e.g.,
52) at C-2 by means of an oxidation process (→**53**) and a reduction process (→ **54**)
(Figure 18) [47–50]. More recently, variations of this approach were reported, for
example, using 2-imidazylates [51]. A C-2 inversion via the neighboring participation
of an N-phenylurethane at C-3 (**55** → **56** → **57**; Figure 19) [52] has been applied
to the synthesis of the N-glycan pentasaccharide core [53]. Two different groups

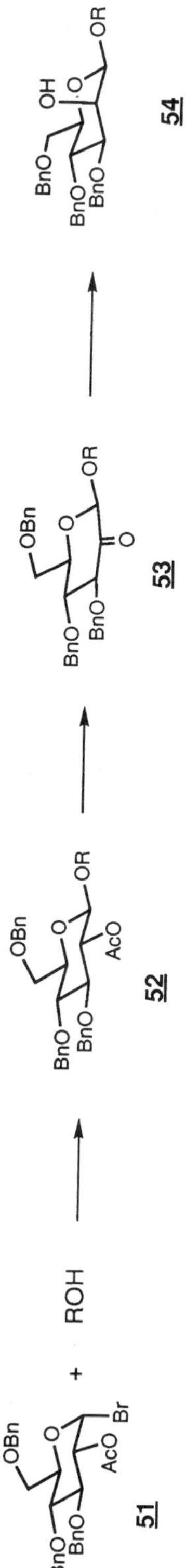

Figure 18

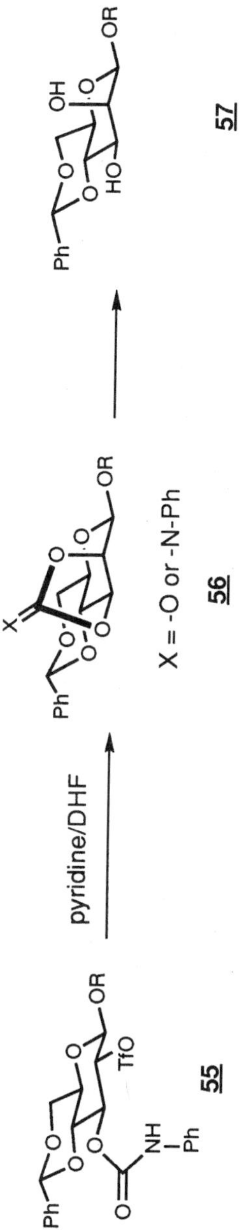

Figure 19

recently described an interesting radical-initiated inversion at the anomeric center (α → β) in monosaccharides [54,55], but this has not yet been applied to oligosaccharide synthesis.

The first direct method for β-mannoside formation developed by Paulsen and colleagues is carried out in the presence of a heterogeneous promoter (silver silicate) and requires both a reactive donor (e.g., **58**) and a reactive acceptor (e.g., **59**) [56,57] (Figure 20). The relatively high reactivity of the glycosyl acceptor **59** is attributed to the axial configuration of the 4-OH group, which is brought about by sugar ring inversion as a result of 1,6-anhydride formation. The β-anomer **60** was formed in a 7:1 ratio to the α-anomer, with a yield of 68%. This approach was applied in the successful synthesis of the N-glycan pentasaccharide core [57].

More recently several research groups have reported β-mannoside formation by the novel intramolecular aglycon delivery approach, where the acceptor is temporarily tethered to the C-2 position of a mannosyl donor and is then delivered to the anomeric center by means of a 1 → 2 cis transfer reaction. This was first carried out using acceptor **62**, which was attached to C-2 of the donor **61** via an acetal link to give **63** (Figure 21) [58–60]. Intramolecular aglycon delivery was then triggered by treatment with N-iodosuccinimide to give exclusively the β-anomer **64**. Unfortunately, this method has so far been unsuccessful in the synthesis of the N-glycan pentasaccharide core. A very similar procedure uses temporary silicon connection (**65** → **66**; Figure 22) but has so far only been successful for 1–6-linked oligosaccharides [61]. A recent improvement of the intramolecular aglycon delivery method using para-methoxy benzaldehydes [62] was applicable to the synthesis of the N-glycan pentasaccharide core [63].

Synthesis of Oligosaccharides Containing α-Neuraminic Acids

Neuraminic acid (or sialic acid) differs from the simple pyranoses discussed so far by having a carboxyl group attached to the anomeric center (C-2) and no substituent at C-3. This has important consequences for the synthesis of glycosides of neuraminic acids in that it presents two particular problems. Firstly, competing β-elimination is frequently observed, leading to a 2,3-unsaturated derivative (e.g., **70**). Secondly, stereoselectivities of coupling reactions are low [8] because there is no neighboring group to aid stereoselective glycoside formation. Most sialic acids in natural oligosaccharides have an α-configuration; thus the glycosidic linkages are equatorial. The first syntheses of sialosides resulted in epimeric mixtures that then had to be separated (Figure 23) [64]. This would dictate the overall synthetic strategy, in that sialic acid containing building blocks were synthesized first and the epimers separated and then subsequently coupled en bloc to other saccharides to avoid having to do the sialosyl coupling step at the end. Initial strategies would only work using unhindered nucleophiles, and hence only 1–6-linked sialosides could be formed.

The enormous interest in the biological activities of sialylated oligosaccharides such as the sialyl-Lewis antigens has led to increased activity in this area, and significant progress has been made over the past 10 years. The problem of neighboring-group participation was overcome by the use of temporary substituents at C-3 [65,66], such as OH-, Br-, PhS-, and PhSe-, which gave good stereoselectivity and prevented glycal formation. However, this approach lowered the total yield due to the extra steps required for introduction and removal of the 3-substituent.

Figure 20

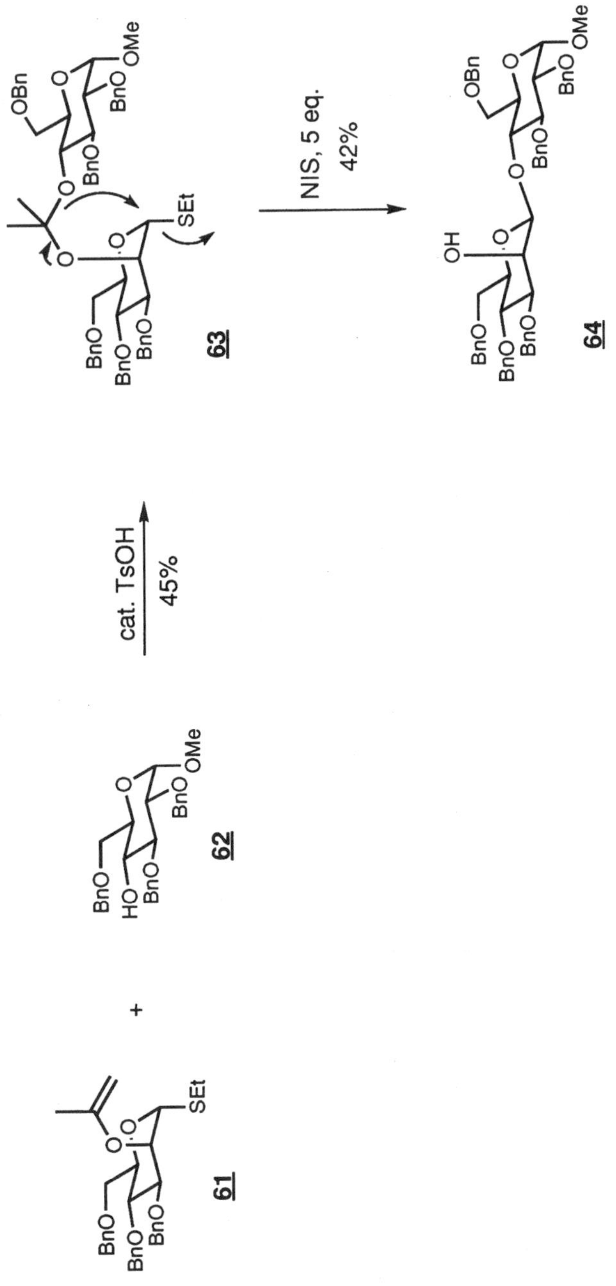

Figure 21

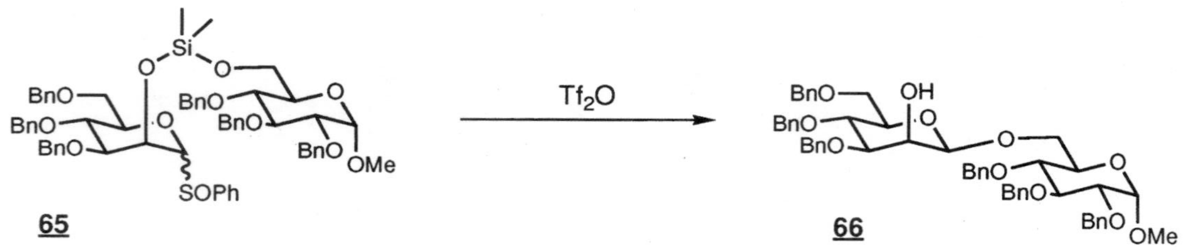

Figure 22

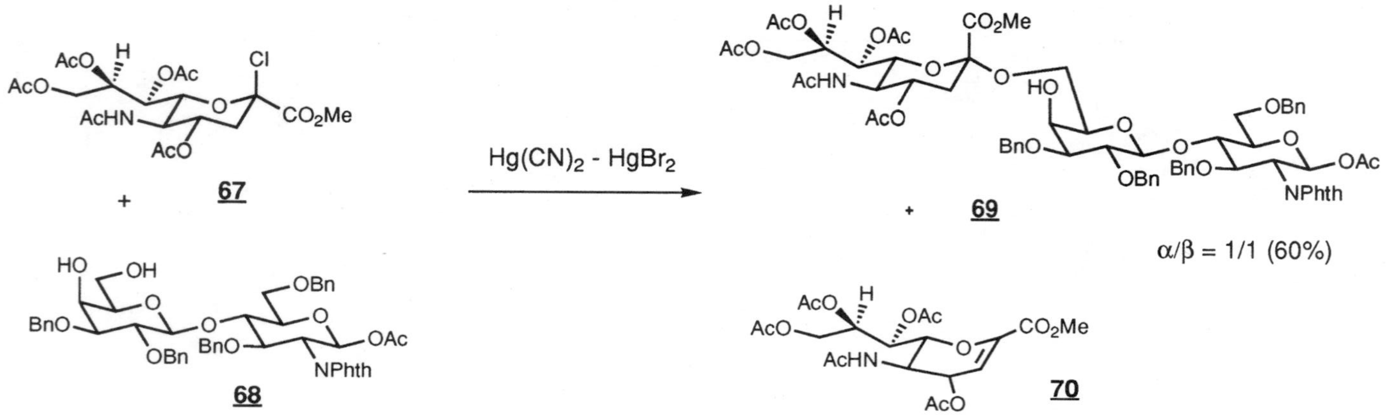

Figure 23

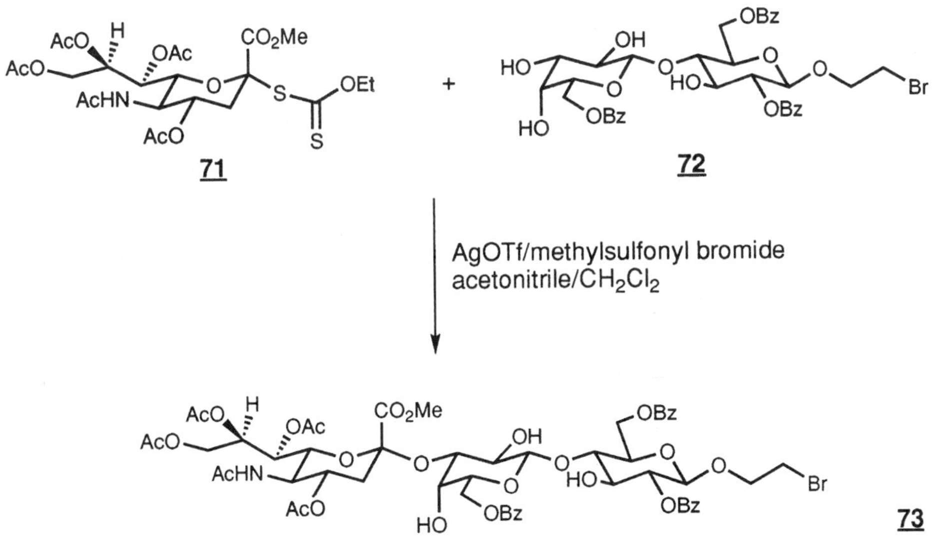

Figure 24

Using thioglycosides as activated donors resulted in better α/β selectivities, although the yields were still poor. However, yields increased dramatically by the use of the more reactive anomeric S-glycosyl xanthates [67]. The effect of temperature on sialic acid glycosylation using both thioglycosides and S-glycosyl xanthates has been studied systematically [68,69]. It was found that low temperatures favored α-glycoside formation in acetonitrile/dichloromethane and reduced glycal formation. Thus sialyl-α2-3-lactoside **73** was prepared from the xanthate **71** and lactoside **72** in 82% yield using a twofold excess of donor (Figure 24) [68]. S-Glycosyl xanthates have also been used by Kunz et al. in the successful synthesis of Sialyl-Tn antigen (NeuAcα2-6GalNAcα1-3-L-Ser/Thr) [70], where sialylation was carried out in 71% yield with an anomeric ratio of 4:1 in favor of the α-anomer. Another successful method of activating sialic acids is by way of their glycosyl phosphites (e.g., **74**), which result in good α-selectivity (Figure 25) [71–73].

D. Further Functionalization: Phosphates and Sulfates

Oligosaccharide chains of glycolipids and glycans can be derived further by the formation of either sulfate or phosphate esters. The recent discovery of the sulfated-Lewis antigens, mainly by mass-spectrometric analysis of isolated oligosaccharide fractions [74], has led to the publication of several syntheses [70,75–77]. Sulfate esters are generally formed at the very end of the oligosaccharide synthesis; i.e., the hydroxyl group that needs to be sulfated is temporarily protected with an orthogonal group that can be cleaved after all the coupling reactions have been accomplished. The reason for this is that no sufficiently stable protecting groups for sulfates are currently available.

The glycosyl phosphates of glycans containing mannose-6-phosphate at the nonreducing ends have been studied extensively and have been shown to be involved in lysosomal targeting. Since several phosphate-protecting groups are available, such

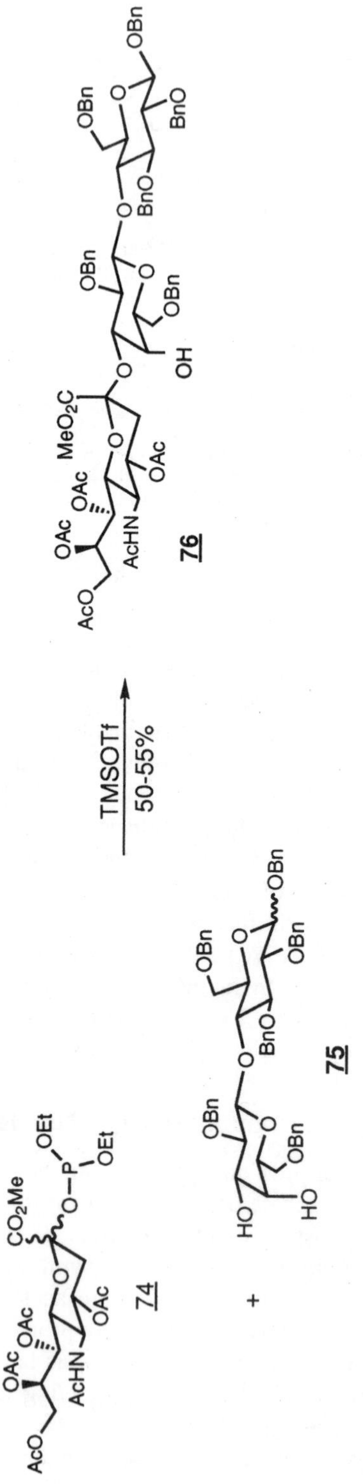
AcO
OAc
OAc
AcHN
OAc
CO2Me
O–P
OEt
OEt
74
+
HO
OBn
HO
OBn
BnO
OBn
OBn
OBn
75
TMSOTf
50-55%
AcO
OAc
OAc
AcHN
OAc
MeO2C
OH
OBn
OBn
BnO
BnO
OBn
OBn
76

Figure 25

as phenyl, benzyl, cyanoethyl, or trichloroethyl groups, phosphates can be introduced at either an early or a late stage in the oligosaccharide synthesis [78].

E. Total Synthesis of *N*-Glycans and *O*-Glycans by Block Synthesis

Several strategies are possible for glycan synthesis. A linear tetrasaccharide ABCD can be synthesized sequentially from either the reducing end (D → CD → BCD → ABCD) or the nonreducing end (A → AB → ABC → ABCD), by a mixed sequence (B → BC → ABC → ABCD) or by block synthesis (AB + CD, or A + BCD, or ABC + D). Which of these strategies is chosen depends on the target structure and the glycosylation methods employed.

Given the often poor-to-average yields and stereoselectivity of glycosidation steps, block synthesis is very attractive. However, a major problem with block synthesis is anomeric activation, which has to be carried out in the presence of labile glycosidic linkages. Block synthesis requires the use of temporary anomeric protecting groups that can be selectively cleaved in the presence of other protecting groups before the coupling reaction. This may require several steps on precious building-block saccharides. This problem can be overcome by "dormant" anomeric activation [10], whereby aglycons can be orthogonally activated during the synthesis. Thus, under certain coupling conditions one glycosyl donor may be activated in the presence of another more stable glycosyl donor that is used as the aglycon to form a glycosidic linkage, for example, the synthesis of **42** (Figure 14), where one disaccharide product was formed, despite the fact that both acceptor and donor carry an anomeric activating group. It would subsequently be possible directly to couple disaccharide **42** to another acceptor using a thiophilic promoter.

Because the block synthesis of oligosaccharides is such an attractive concept, several researchers have suggested strategies for "dormant" anomeric activation. In some cases the same coupling method may be used, which is illustrated by the elegant synthesis of the trisaccharide **80** (Figure 26) [79], which was assembled by an iterative protocol from the universal building block pentenyl mannoside **77**. This can be converted into the "dormant" dibromopentyl mannoside **78** by bromination of the pentenyl group. The dibromopentyl group is inert during pentenyl glycoside activation, but when the need arises it can be easily converted into the pentenyl group by reductive elimination. Thus, the "active" pentenyl glycoside **77** couples with the "dormant" dibromopentyl mannoside **78** to form the disaccharide **79**, which on selective deprotection followed by a further coupling with **77** forms the trisaccharide **80**. This is activated by conversion into the pentenyl glycoside, which on repeated (×2) coupling with the dormant trisaccharide **80** results in a nonamannan structure [79].

The synthesis of the bisected undecasaccharide structure **84** (Figure 27) was a milestone in the synthesis of large glycans and illustrates again the advantages of block synthesis [80]. The strategy involved the use of two building blocks, one a suitably protected form of the Manβ1-4GlcNPthβ1-4GlcNPth trisaccharide **82** and the other an activated tetrasaccharide precursor **81** of the side chains linked to the central mannose branch. The coupled product **83** was deprotected to give the bisected structure **84** in 15% yield.

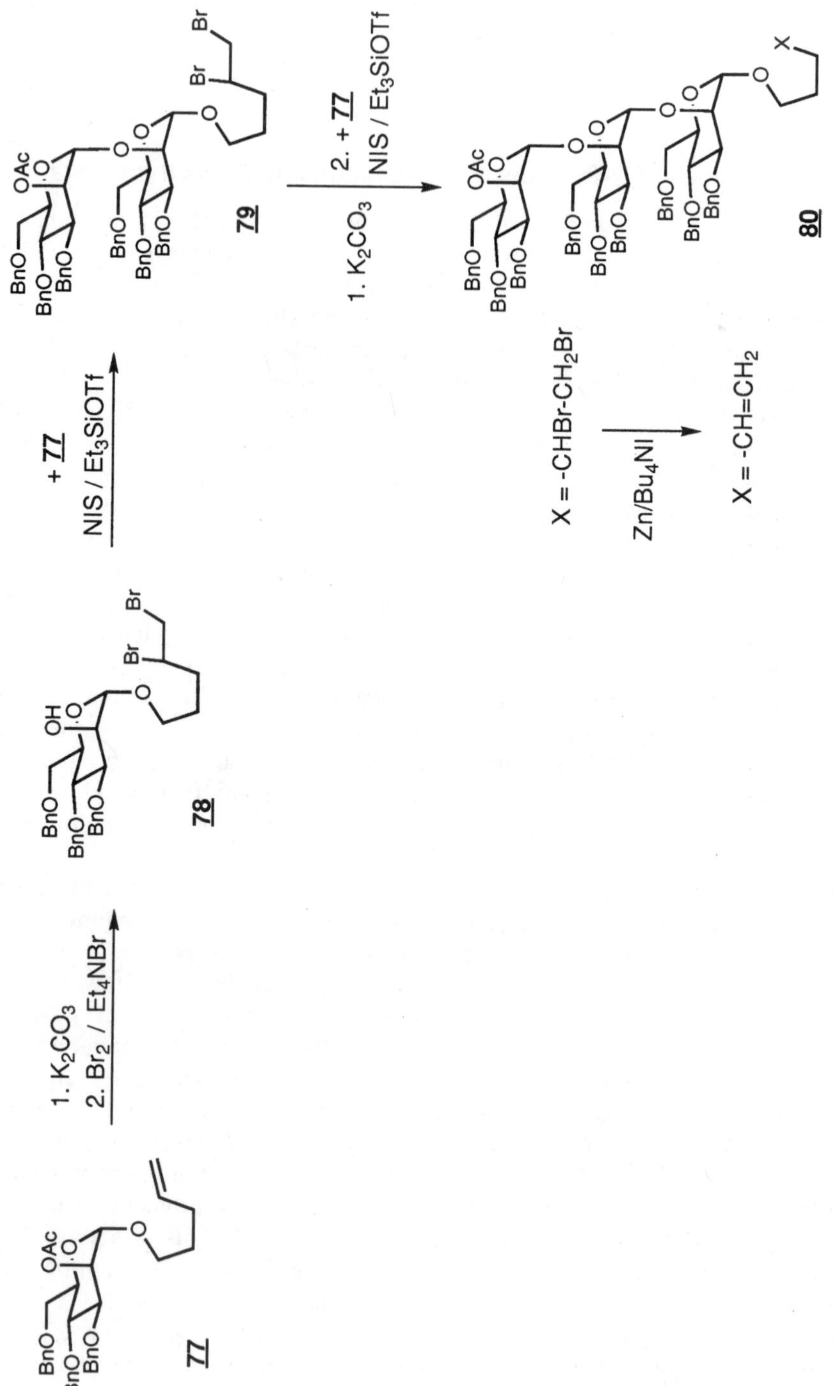

Figure 26

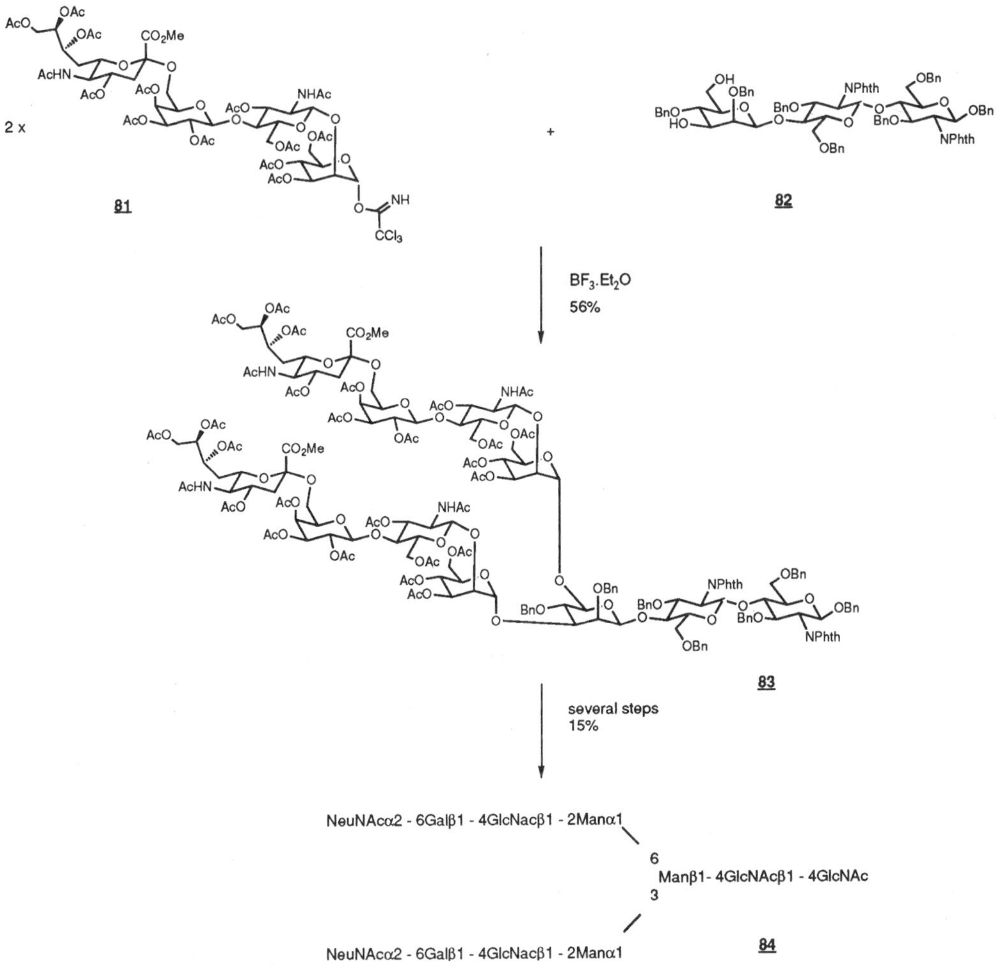

Figure 27

Building blocks **81** and **82** both contain a "difficult linkage": in **82** the β-mannoside, and in **81** the α-sialoside. The precursor of **82**, trisaccharide **87**, was formed by a Koenigs–Knorr coupling of the mannosyl bromide **85** to the aglycon **86** using silver silicate as the promoter to generate the β-mannoside linkage (Figure 28). The tetrasaccharide precursor **89** to building block **81** was constructed from the sialyl chloride **67** and the acceptor **88** as a mixture of epimers that were separated by chromatography (Figure 29). Thus glycosyl halides are used for the synthesis of the smaller building blocks, whereas the essentially more reactive trichloroacetimidate method is used for block-coupling.

Quite a number of *N*-linked core structures have now been synthesized, which cannot all be discussed here in detail. Table 3 shows some of the larger glycans that have been made so far [53,79,81–87], which demonstrate that total syntheses of glycans is feasible today.

87

86

85

silver silicate
40% β

Figure 28

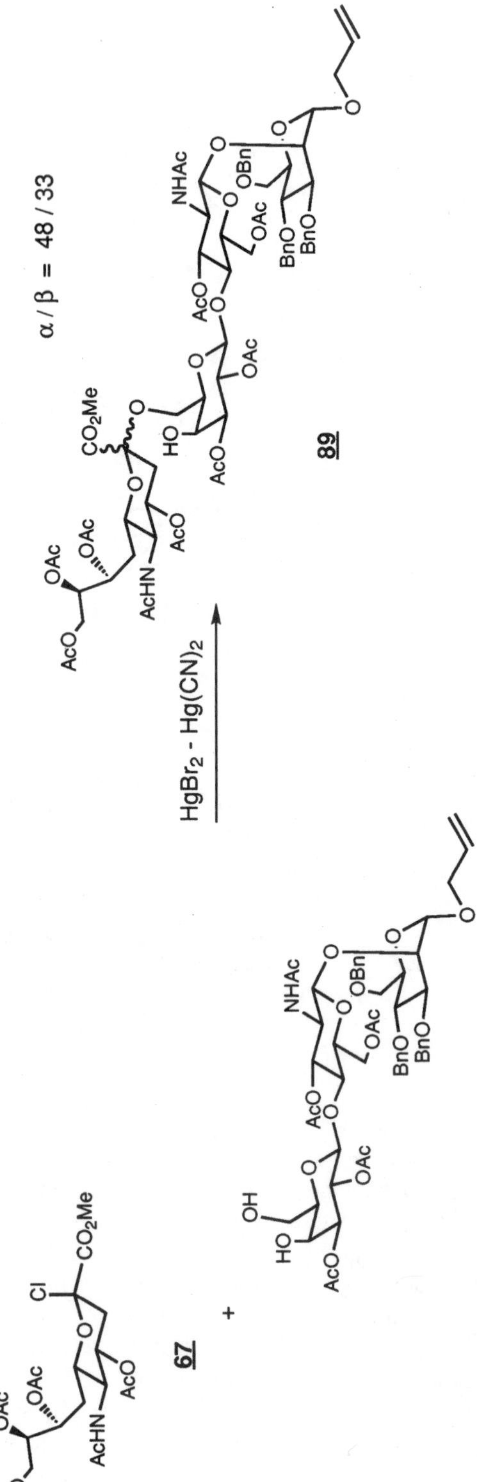

α / β = 48 / 33

89

HgBr₂ - Hg(CN)₂

67

88

Figure 29

Table 3   Some Large *N*-Linked Core Glycans that Have Been Synthesized

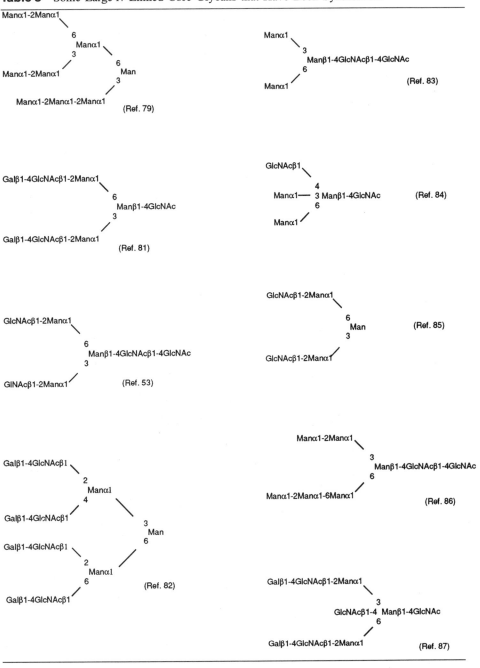

III. CONCLUSIONS/OUTLOOK

Current synthetic methodologies are able to give access to most of the glycan side chains of glycoproteins, even the larger and more complex N-linked glycans. However, most of the synthetic routes are still very long and low yielding, and we are far away from the custom synthesis that is possible for oligonucleotides and polypeptides. Glycan targets should still be considered a synthetic challenge; syntheses cannot be completed within days or weeks, but require months or in some cases years of effort. Thus, there is still much need for new methodologies in carbohydrate chemistry, in particular for glycosylation methods that may in the future be both regio- and stereoselective enough to avoid the use of protecting groups.

Oligosaccharide synthesis has come of age as the biological importance of oligosaccharides has become clear and the chemical tools of synthesis have become more sophisticated. In particular the last 10 years has seen the problems and challenges of selective glycosidic bond formation taken up by organic chemists and several good methods are now available. Block synthesis has become feasible by the use of orthogonal anomeric protecting groups or by using "dormant" anomeric activating groups. Many new developments are to be expected in the near future.

ABBREVIATIONS

Ac	acetyl
AgOTf	silver trifluoromethylsulfonate
All	allyl
Bn	benzyl
Bz	benzoyl
CA	chloroacetyl
DBU	1,8-diazabicyclo[5.4.0]undec-7-ene
Fuc	fucopyranose
Gal	galactopyranose
GlcNAc	N-acetylglucosamine
IDCP	iodonium dicollidine perchlorate
Lev	levulinoyl
Man	mannopyranose
NBS	N-bromosuccinimide
NeuNAc	N-acetylneuramic acid
NIS	N-iodosuccinimide
Phth	phthalimide
Piv	pivaloyl
PMB	p-methoxybenzyl
TBDMS	t-butyldimethylsilyl
TBDPS	t-butyldiphenylsilyl
Tf	trifluoromethanesulfonyl
TMSOTf	trimethylsilyl trifluoromethylsulfonate

REFERENCES

1. M. A. Kukuruzinska, M. L. E. Bergh, and B. J. Jackson, Protein glycosylation in yeast, *Ann. Rev. Biochem. 56*: 915–944 (1987).
2. R. Kornfeld and S. Kornfeld, Assembly of asparagine-linked oligosaccharides, *Ann. Rev. Biochem. 54*: 631–664 (1985).
3. C. D. Warren, M. L. Milat, C. Auge, and R. W. Jeanloz, The synthesis of a trisaccharide and a tetrasaccharide lipid intermediate P_1-dolichyl P_2-[*O*-β-D-mannopyranosyl-(1-4)-*O*-(2-acetamido-2-deoxy-β-D-glucopyranosyl)-(1-4)-2-acetamido-2-deoxy-α-D-glucopyranosyl] diphosphate and P_1-dolichyl P_2-[*O*-α-D-mannopyranosyl-(1-3)-*O*-β-D-mannopyranosyl-(1-4)-*O*-(2-acetamido-2-deoxy-β-D-glucopyranosyl-(1-4)-2-acetamido-2 -deoxy-α-D-glucopyranosyl] diphosphate, *Carbohydr. Res. 126*: 61–80 (1984).
4. F. Reck, E. Meinjohanns, M. Springer, R. Wilkins, J. A. L. M. Van Dorst, H. Paulsen, G. Moller, I. Brockhausen, and H. Schachter, Synthetic substrate analogs for UDP-GlcNAc-Man-α-1-6R β(1-2)-*N*-acetylglucosaminyltransferase II. Substrate specificity and inhibitors for the enzyme, *Glycoconj. J. 11*: 210–216 (1994).
5. H. Paulsen, F. Reck, and I. Brockhausen, Synthese von modifizierten oligosacchariden der *N*-glycoproteine als substrate für *N*-acetylglucosaminyltransferase I, *Carbohydr. Res. 236*: 39–71 (1992).
6. T. Feizi, Oligosaccharides that mediate mammalian cell–cell adhesion, *Current Opinion in Structural Biology 3*: 701–710 (1993).
7. K. Bezouska, C.-T. Yuen, J. O'Brien, R. A. Childs, W. Chai, A. Lawson, K. Drbal, A. Fiserova, M. Pospisil, and T. Feizi, Oligosaccharide ligands for NKR-P$_1$ protein activate NK cells and cytotoxicity, *Nature 372*: 150–157 (1994).
8. H. Paulsen, Syntheses, conformations and X-ray structure analyses of the saccharide chains from the core regions of glycoproteins, *Angew. Chem. Int. Ed. Engl. 29*: 823–839 (1990).
9. Y. Bourne, H. Van Tilbeurgh, and C. Cambillau, Protein–carbohydrate interactions, *Current Opinions in Structural Biology 3*: 681–686 (1993).
10. P. J. Garegg, Saccharides of biological importance: Challenges and opportunities for organic synthesis, *Acc. Chem. Res. 25*: 575–580 (1992).
11. K. Toshima and K. Tatsuta, Recent progress in *O*-glycosylation methods and its application to natural products synthesis, *Chem. Rev. 93*: 1503–1531 (1993).
12. O. Hindsgaul, Synthesis of carbohydrates for applications in glycobiology, *Seminars in Cell Biology 2*: 319–326 (1991).
13. R. R. Schmidt, Synthesis of glycosides, *Comprehensive Organic Synthesis 6*: 33–63 (1991).
14. M. Meldal, Recent developments in glycopeptide and oligosaccharide synthesis, *Current Opinion in Structural Biology 4*: 710–718 (1994).
15. O. Kanie and O. Hindsgaul, Synthesis of oligosaccharides, glycolipids and glycopeptides, *Current Opinion in Structural Biology 2*: 674–681 (1992).
16. R. L. Halcomb and C.-H. Wong, Synthesis of oligosaccharides, glycopeptides and glycolipids, *Current Opinion in Structural Biology 3*: 694–700 (1993).
17. J. Banoub, P. Boullanger, and D. Lafont, Synthesis of oligosaccharides of 2-amino-2-deoxy sugars, *Chem. Rev. 92*: 1167–1195 (1992).
18. R. R. Schmidt and W. Kinzy, Anomeric-oxygen activation for glycoside synthesis. The trichloroacetimidate method, *Adv. Carbohydr. Chem. Biochem. 50*: 21–123 (1994).
19. B. Fraser-Reid, U. E. Udodong, Z. Wu, H. Ottosson, J. R. Merritt, C. S. Rao, C. Roberts, and R. Madsen, *n*-Pentenyl glycosides in organic chemistry: A contemporary example of serendipity, *Synlett.* 927–942 (1992).
20. H. Paulsen, Advances in selective chemical syntheses of complex oligosaccharides, *Angew. Chem. Int. Ed. Engl. 21*: 155–173 (1982).

21. C. Wong, R. L. Halcomb, Y. Ichikawa, and T. Kajimoto, Enzymes in organic synthesis (Part 2): Application to the problems of carbohydrate recognition, *Angew. Chem. Int. Ed. Engl. 34*: 521–546 (1995).

22. T. W. Greene and P. G. M. Wuts, *Protective Groups in Organic Synthesis*, Wiley, New York, 1991.

23. P. J. Garegg, T. Iversen, and S. Oscarson, Monobenzylation of diols using phase-transfer catalysis, *Carbohydr. Res. 50*: C12–C14 (1976).

24. P. J. Edwards, D. A. Entwistle, C. Genicot, S. V. Ley, and G. Visentin, Dispiroketals in synthesis (Part 17): Regioselective protection of D-glucopyranoside, D-galactopyranoside and D-mannopyranoside substrates, *Tetrahedron: Asymmetry 5*: 2609–2632 (1994).

25. S. David and S. Hanessian, Regioselective manipulation of hydroxyl groups via organotin derivatives, *Tetrahedron 41*: 643–663 (1985).

26. D. G. Drueckhammer, W. J. Hennen, R. L. Pederson, C. F. Barbas, C. M. Gautheron, T. Krach, and C.-H. Wong, Enzyme catalysis in synthetic carbohydrate chemistry, *Synthesis* 499–525 (1991).

27. R. J. Ferrier, R. W. Hay, and N. Vethaviyasar, A potentially versatile synthesis of glycosides, *Carbohydr. Res. 27*: 55–61 (1973).

28. D. Kahne, S. Walker, Y. Cheng, and D. van Engen, Glycosylation of unreactive substrates, *J. Am. Chem. Soc. 111*: 6881–6882 (1989).

29. S. Mehta and B. M. Pinto, Novel glycosidation methodology. The use of phenyl selenoglycosides as glycosyl donors and acceptors in oligosaccharide synthesis, *J. Org. Chem. 58*: 3269–3276 (1993).

30. S. Czernecki, E. Ayadi, and D. Randriamandimby, New and efficient synthesis of protected 2-azido-2-deoxy-glucopyranoses from the corresponding glycal, *J. Chem. Soc., Chem. Commun.* 35–36 (1994).

31. F. Santoyo-Gonzalez, F. G. Calvo-Flores, P. Garcia-Mendoza, F. Hernandez-Mateo, J. Isac-Garcia, and R. Robles-Diaz, Synthesis of phenyl 2-azido-2-deoxy-1-selenoglycosides from glycals, *J. Org. Chem. 58*: 6122–6125 (1993).

32. G. H. Veeneman and J. H. van Boom, An efficient thioglycoside-mediated formation of α-glycosidic linkages promoted by iodonium dicollidine perchlorate, *Tetrahedron Lett. 31*: 275–278 (1990).

33. P. Konradsson, D. R. Mootoo, R. E. McDevitt, and B. Fraser-Reid, Iodonium ion generated in situ from *N*-iodosuccinimide and trifluoromethanesulphonic acid promotes direct linkage of "disarmed" pent-4-enyl glycosides, *J. Chem. Soc. Chem. Commun.* 270–272 (1990).

34. P. Konradsson, U. E. Udodong, and B. Fraser-Reid, Iodonium promoted reactions of disarmed thioglycosides, *Tetrahedron Lett. 31*: 4313–4316 (1990).

35. B. Fraser-Reid, J. R. Merritt, A. L. Handlon, and C. Andrews, The chemistry of *n*-pentenyl glycosides: Synthetic, theoretical and mechanistic ramifications, *Pure Appl. Chem. 65*: 779–786 (1993).

36. R. L. Halcomb and S. J. Danishefsky, On the direct epoxidation of glycals: Application of a reiterative strategy for the synthesis of β-linked oligosaccharides, *J. Am. Chem. Soc. 111*: 6661–6666 (1989).

37. R. U. Lemieux, K. B. Hendriks, R. V. Stick and K. James, Halide ion catalyzed glycosidation reactions. Synthesis of α-linked disaccharides, *J. Am. Chem. Soc. 97*: 4056–4062 (1975).

38. K. J. Kaur and O. Hindsgaul, A simple synthesis of octyl 3,6-di-*O*-(α-D-mannopyranosyl)-β-D-mannopyranoside and its use as an acceptor for the assay of *N*-acetylglucosaminyltransferase-I activity, *Glycoconj. J. 8*: 90–94 (1991).

39. H. Paulsen, E. Meinjohanns, F. Reck, and I. Brockhausen, Building units of oligosaccharides 107. Synthesis of modified oligosaccharides of *N*-glycoproteins for substrate

specificity studies of *N*-acetylglucosaminyl transferase—II, *Liebigs Ann. Chem.* 721–735 (1993).

40. K. Suzuki, H. Maeta, T. Suzuki, and T. Matsumoto, Cp2ZrC12-AgBF4 in benzene: A new reagent system for rapid and highly selective α-mannoside synthesis from tetra-*O*-benzyl-D-mannosyl fluoride, *Tetrahedron Lett. 30*: 6879–6882 (1989).

41. M. Nishizawa, Y. Kan, W. Shimomoto, and H. Yamada, α-Selective thermal glycosidation of rhamnosyl and mannosyl chlorides, *Tetrahedron Lett. 31*: 2431–2434 (1990).

42. H. Lonn, Synthesis of a tri- and a tetrasaccharide which contain α-L-fucopyranosyl groups and are part of the complex type of carbohydrate moiety of glycoproteins, *Carbohydr. Res. 139*: 105–113 (1985).

43. A. Kameyama, H. Ishida, M. Kiso, and A. Hasegawa, Total synthesis of sialyl lewis x, *Carbohydr. Res. 209*: C1–C4 (1991).

44. R. K. Jain and K. L. Matta, Methyl 3,4-*O*-isopropylidene-2-*O*-(4-methoxybenzyl)-1-thio-β-fucopyranoside: A novel, efficient glycosylating reagent for the synthesis of linear and other α-L-fucosyl oligosaccharides, *Tetrahedron Lett. 30*: 4325–4328 (1990).

45. H. Flowers, Synthesis of oligosaccharides of L-fucose containing α- and β-anomeric configurations in the same molecule, *Carbohydr. Res. 119*: 75–84 (1983).

46. G. J. P. H. Boons, M. Overhand, G. A. van der Marel, and J.-H. van Boom, Application of the dimethyl(phenyl)silyl group as a masked form of the hydroxy group in the synthesis of an L-*glycero*-α-D-*manno*-heptapyranoside-containing trisaccharide from the dephosphorylated inner core region of *Neisseria meningitidis, Angew. Chem. Int. Ed. Engl. 28*: 1504–1505 (1989).

47. M. A. E. Shaban and R. W. Jeanloz, Synthesis of 2-acetamido-2-deoxy-3-*O*-β-D-mannopyranosyl-D-glucose, *Carbohydr. Res. 52*: 103–114 (1976).

48. M. A. E. Shaban and R. W. Jeanloz, The synthesis of 2-acetamido-2-deoxy-4-*O*-β-D-mannopyranosyl-D-glucose, *Carbohydr. Res. 52*: 115–127 (1976).

49. G. Ekborg, B. Lindberg, and J. Lonngren, Synthesis of β-D-mannopyranosides, *Acta Chem. Scand. 26*: 3287–3292 (1972).

50. H. B. Boren, G. Ekborg, K. Eklind, P. J. Garegg, A. Pilotti, and C.-G. Swahn, Benzylated orthoesters in glycoside synthesis, *Acta Chem. Scand. 27*: 2639–2644 (1973).

51. S. David, A. Malleron, and C. Dini, Preparation of oligosaccharides with β-D-mannopyranosyl and 2-azido-2-deoxy-β-D-mannopyranosyl residues by inversion at C-2 after coupling, *Carbohydr. Res. 188*: 193–200 (1989).

52. H. Kunz and W. Gunther, β-Mannoside from β-glucosides by intramolecular nucleophilic substitution with inversion of configuration, *Angew. Chem. Int. Ed. Engl. 27*: 1086–1087 (1988).

53. C. Unverzagt, Synthesis of a biantennary heptasaccharide by regioselective glycosylations, *Angew. Chem. Int. Ed. Eng. 33*: 1102–1104 (1994).

54. J. Brunckova, D. Crich, and Q. W. Yao, Intramolecular hydrogen atom abstraction in carbohydrates and nucleosides: Inversion of an α- to β-mannopyranoside and generation of thymidine C-4′ radicals, *Tetrahedron Lett. 35*: 6619–6622 (1994).

55. N. Yamazaki, E. Eichenberger, and D. P. Curran, Synthesis of β-mannopyranosides from α-epimers by radical inversion, *Tetrahedron Lett. 35*: 6623–6626 (1994).

56. H. Paulsen, R. Lebuhn, and O. Lockhoff, Synthese des verzweigten Tetrasaccharid Bausteins der Schlusselsequenz von *N*-glycoproteinen, *Carbohydr. Res. 103*: C7–C11 (1982).

57. H. Paulsen and R. Lebuhn, Synthese von Tri- und Tetrasaccharid Sequenzen von *N*-glycoproteinen mit β-D-mannosidischer Verknupfung, *Liebigs Ann. Chem.* 1047–1072 (1983).

58. F. Barresi and O. Hindsgaul, Synthesis of β-mannopyranosides by intramolecular aglycon delivery, *J. Am. Chem. Soc. 113*: 9376–9377 (1991).

59. F. Barresi and O. Hindsgaul, Improved synthesis of β-mannopyranosides by intramolecular aglycon delivery, *Synlett.* 759–761 (1992).

60. F. Barresi and O. Hindsgaul, The synthesis of β-mannopyranosides by the intramolecular aglycon delivery: Scope and limitations of the existing methodology, *Can. J. Chem.* 72: 1447–1465 (1994).

61. G. Stork and G. Kim, Stereocontrolled synthesis of disaccharides via the temporary silicon connection, *J. Am. Chem. Soc. 114*: 1087–1088 (1992).

62. Y. Ito and T. Ogawa, A novel approach to the stereoselective synthesis of β-mannosides, *Angew. Chem. Int. Ed. Engl. 33*: 1765–1767 (1994).

63. A. Dan, Y. Ito, and T. Ogawa, Stereocontrolled synthesis of the pentasaccharide core structure of asparagine-linked glycoprotein oligosaccharide based on a highly convergent strategy, *Tetrahedron Lett. 36*: 7487–7490 (1995).

64. H. Paulsen and H. Tietz, Synthese eines *N*-acetylneuraminsaure-haltigen Syntheseblocks. Kupplung zum *N*-acetylneuraminsaure-tetrasaccharid mit trimethylsilyltriflat, *Carbohydr. Res. 144*: 205–229 (1985).

65. T. Ercegovic and G. Magnusson, A new *N*-acetylneuraminic acid donor for highly stereoselective α-sialylation, *J. Chem. Soc., Chem. Commun.* 831–832 (1994).

66. K. Okamoto and T. Goto, Glycosidation of sialic acid, *Tetrahedron 46*: 5835–5857 (1990).

67. A. Marra and P. Sinay, A novel stereoselective synthesis of *N*-acetyl-α-neuraminosyl-galactose disaccharide derivatives, using anomeric *S*-glycosyl xanthates, *Carbohydr. Res. 195*: 303–308 (1990).

68. H. Lonn and K. Stenvall, Exceptionally high yield in glycosylation with sialic acid. Synthesis of a GM3 glycoside, *Tetrahedron Lett. 33*: 115–116 (1992).

69. W. Birberg and H. Lonn, α-Selectivity and glycal formation are temperature dependent in glycosylation with sialic acid: Synthesis of a Neu5Ac-α(2-6) Gal thioglycoside building block, *Tetrahedron Lett. 32*: 7453–7456 (1991).

70. B. Liebe and H. Kunz, Synthesis of sialyl-Tn antigen. Regioselective sialylation of a galactosamine threonine conjugate unblocked in the carbohydrate portion, *Tetrahedron Lett. 35*: 8777–8778 (1994).

71. T. J. Martin and R. R. Schmidt, Efficient sialylation with phosphite as leaving group, *Tetrahedron Lett. 33*: 6123–6126 (1992).

72. T. J. Martin, R. Brescello, A. Toepfer, and R. R. Schmidt, Synthesis of phosphites and phosphates of neuraminic acid and their glycosyl donor properties. Convenient synthesis of GM(3), *Glycoconj. J. 10*: 16–25 (1993).

73. R. R. Schmidt, Chemical synthesis of sialylated glycoconjugates, *ACS Symposium Series 560*: 276–296 (1994).

74. C. T. Yuen, K. Bezouska, J. Obrien, M. Stoll, R. Lemoine, A. Lubineau, M. Kiso, A. Hasegawa, N. J. Bockovich, K. C. Nicolaou, and T. Feizi, Sulfated blood-group Lewis (A)—A superior oligosaccharide ligand for human E-selectin, *J. Biol. Chem. 269*: 1595–1598 (1994).

75. E. V. Chandrasekaran, R. K. Jain, and K. L. Matta, Ovarian cancer α1,3-L-fucosyltransferase, *J. Biol. Chem. 267*: 23806–23814 (1992).

76. K. C. Nicolaou, N. J. Bockovich, and D. R. Carcanague, Total synthesis of sulfated Le(x) and Le(a)-type oligosaccharide selectin ligands, *J. Am. Chem. Soc. 115*: 8843–8844 (1993).

77. R. K. Jain, R. Rampal, E. V. Chandrasekaran, and K. L. Matta, Total synthesis of 3'-*O*-sialyl, 6'-*O*-sulfo Lewis(x), NeuAc-α-2-]3(6-*O*-SO₃Na)Gal-β-1-]4(Fuc-α-1-]3)-GlcNAc-β-OMe—A major capping group of glycam-I, *J. Am. Chem. Soc. 116*: 12123–12124 (1994).

78. O. P. Srivastava and O. Hindsgaul, Synthesis of phosphorylated trimannosides corre-

sponding to end groups of the high-mannose chains of lysosomal enzymes, *Carbohydr. Res. 161*: 195–210 (1987).

79. J. R. Merritt, E. Naisang, and B. Fraser-Reid, *n*-Pentenyl mannoside precursors for synthesis of the nonamannan component of high-mannose glycoproteins, *J. Org. Chem. 59*: 4443–4449 (1994).

80. T. Ogawa, M. Sugimoto, T. Kitajima, K. K. Saozai, and T. Nukada, Total synthesis of an undecasaccharide: A typical carbohydrate sequence for the complex type of glycan chain of a glycoprotein, *Tetrahedron Lett. 27*: 5739–5742 (1986).

81. H. Paulsen, W. Rauwald, and R. Lebuhn, Synthese von unsymmetrischen Pentasac-charid-sequenzen der *N*-glycoproteine, *Carbohydr. Res. 138*: 29–40 (1985).

82. H. Lonn and J. Longren, Synthesis of a nona- and an undeca-saccharide that form part of the complex type of carbohydrate moiety of glycoproteins, *Carbohydr. Res. 120*: 17–24 (1983).

83. H. Paulsen and R. Lebuhn, Synthese der invarianten pentasaccharid-Core-Region der Kohlenhydrat-Ketten der *N*-glycoproteine, *Carbohydr. Res. 130*: 85–101 (1984).

84. H. Paulsen, M. Heume, Z. Gyorgydeak, and R. Lebuhn, Synthese einer verzweigten pentasaccharid-Sequenz der "bisected" Struktur von *N*-glycoproteinen, *Carbohydr. Res. 144*: 57–70 (1985).

85. J. Arnarp and J. Lonngren, Synthesis of hepta- and penta-saccharides, Part of the com-plex-type carbohydrate portion of glycoproteins, *J. Chem. Soc., Chem. Commun.* 1000 –1002 (1980).

86. T. Nukada, T. Kitajima, Y. Nakahara, and T. Ogawa, Synthesis of an octasaccharide fragment of high-mannose-type glycans of glycoproteins, *Carbohydr. Res. 228*: 157–170 (1992).

87. J. Arnarp, M. Haraldsson, and J. Lonngren, Synthesis of three oligosaccharides that form part of the complex type of carbohydrate moiety of glycoproteins containing in-tersecting *N*-acetylglucosamine, *J. Chem. Soc. Perkin Trans. I*: 535–539 (1985).

5

Syntheses of Oligosaccharides and Glycopeptides on Insoluble and Soluble Supports

Samuel J. Danishefsky
Memorial Sloan-Kettering Institute for Cancer Research and Columbia University, New York, New York

Jacques Y. Roberge
Bristol-Myers Squibb, Princeton, New Jersey

I. INTRODUCTION

Oligosaccharides and glycoconjugates [1,2] constitute important groups of biomolecules found in all living organisms. They are known to be involved in a myriad of biologically significant functions, which include serving as a source of energy and particularly as structural elements. Oligosaccharides and glycoconjugates are more complex than any other biopolymer. The structural and stereochemical diversity results in a massive information content. The information content is the key to the versatility of these biooligomers, which serve in hormones, transport molecules, blood clotting agents, antibodies, adhesion molecules, and immunological cell determinants [1,3–6].

Oligosaccharides and glycoconjugates are rarely isolated in homogeneous form from the living cell, since they exist as closely related mixtures [7]. The purification of these compounds to homogeneity is consequently tedious and often results in prohibitively small yields of end products that are difficult to characterize [8]. Though recent progress in the fields of separation technology and structure elucidation makes the isolation-purification process feasible for obtaining complex glycoconjugates [9], a need for the stepwise preparation of structurally defined entities still exists. In this way, the biological and medicinal potential of these molecules can be better exploited [10–17].

Synthesis of oligosaccharides and glycoconjugates presents a more difficult problem than syntheses of peptides or oligonucleotides. The structural complexity of oligosaccharides stems from a diversity of monomeric units with numerous connecting sites, resulting in a higher level of branching than in amino acids or nucleotides. Furthermore, the need to control the stereochemistry at the anomeric center in each glycosylation steps adds to the challenge of the synthesis of these molecules. Great progress has been made toward solving these synthetic problems by the development of a number of selective glycosylation reactions as well as through the refinement of protecting-group strategies [10–15,18–21].

The preparation of oligonucleotides [22] and peptides [23–28] has benefited significantly from the development of syntheses on solid supports. The application of the analogous solid-phase synthesis of oligosaccharides and glycoconjugates was reviewed by Fréchet in 1980 [29]. At that time, Fréchet concluded his review by saying:

> Although several approaches to the solid-phase synthesis of oligosaccharides have demonstrated its feasibility and have explored different strategies, numerous problems remain to be solved. These include mainly the problem of incomplete coupling and lack of complete stereoselectivity. Until these are solved, the solid-phase method will not be competitive with classical homogeneous-phase synthesis.

Since then, the progress of glycosylation chemistry, in terms of gross yields and stereoselectivity, has stimulated a renewed interest in the synthesis of oligosaccharides on solid supports. This review will focus on contributions in the area of solid-phase synthesis of oligosaccharides and glycopeptides since Fréchet's review. The discussion will also include the use of soluble supports, since this technique is conceptually related to the use of insoluble supports for synthesis. Our main focus will be on methods that create glycosidic linkages on the support, although some interesting syntheses of *O*- and *N*-linked glycopeptides using preformed glycosylated amino acids will also be discussed. The synthesis of oligonucleotides will be omitted from this chapter, since this area has been reviewed recently [22].

This work will summarize briefly what had been accomplished before Fréchet's review, and will then describe in greater detail the work that has been accomplished since then. The order of the discussion will follow as far as possible the chronological order of the various contributions. Chemoenzymatic glycosylations will be discussed first. Purely chemical glycosylation methods will follow, and we will conclude by reviewing the methods for the formation of glycopeptides and glycoconjugates on a polymeric support. Since these last methods do not usually involve fashioning glycosidic linkages on the support, this section will be brief and nonexhaustive.

II. EARLIEST DEVELOPMENTS

The initial reported work on the synthesis of oligosaccharides was accomplished by Fréchet and Schuerch, who glycosylated a polymer-bound glycosyl acceptor with anomeric bromides [30]. The substrates were attached to a polystyrene support at the reducing end of the growing saccharide via an ozone-sensitive allylic alcohol linker. Gagnaire and co-workers assembled disaccharides attached to a 0.2% cross-linked polystyrene with an ester linkage at the nonreducing end of 2-*N*-acetyl glucosamine [31,32]. These glycosylations were achieved with an oxazoline donor and *p*-toluenesulfonic acid (β1,6) [31] or with a 2-*N*-acetyl glucosyl chloride donor and mercuric cyanide (β1,3) [32]. The disaccharides were recovered from the support by solvolysis. The early contribution of Zehavi's group to the synthesis of oligosaccharides on a polymeric support was to introduce the use of a photo-removable linker [33]. Essentially the same chemistry and support as that used by Fréchet and Schuerch was used for the glycosylation, except that the α1,6-disaccharide was released from the support as the free reducing sugar upon UV irradiation.

In 1973, Guthrie and co-workers proposed the use of a sugar donor attached to a soluble polymeric support [34]. The copolymerization of styrene with a protected glycoside monomer, having a *p*-vinylbenzoyl [35] or *p*-vinylsulfonyl moiety at the C-6 hydroxyl, produced a polymer-bound donor after activation with hydrobromic acid. This material was glycosylated via the orthoester method, and the β1,6-disaccharide was recovered from the support by acetolysis. The use of an inorganic support has been explored with moderate success by the groups of Schuerch [36] and Holick [37].

Gagnaire and co-workers proposed the attachment of glycosides at the anomeric position to a 2% cross-linked polystyrene resin via a benzoyl propionyl ester linker. Using this support they accomplished the synthesis of a tetrasaccharide with β1,6-linkages [38] and two disaccharides (α1,3 and β1,3-linkages) [39]. Anderson and Chiu also contributed to the early development of the field. They described the formation of primarily α-linked disaccharides attached via thioglycosidic linkages to both soluble and insoluble polystyrene supports [40]. Such was, in brief, the state of the field of solid-phase synthesis of oligosaccharides at the time of Fréchet's review.

III. SYNTHESES OF OLIGOSACCHARIDES ON INSOLUBLE AND SOLUBLE SUPPORTS

A. General Aspects of the Solid-Support Synthesis of Oligosaccharides

The solid-phase synthetic approach toward the preparation of molecules made of repetitive units allows for the rapid assembly of compounds in a single flask. The relatively simple procedures used are suitable for the design of an automated synthesizer. One advantage of a synthesis where the growing oligomer is attached to insoluble beads is that the purification process between each discrete step is that of simple filtration and washing [27]. Excess reagents can be used, and the coupling steps can be repeated to drive the reaction to completion. The use of a single flask minimizes transfer and mechanical losses of materials [41]. Another advantage of this approach is that the solubility of support-bound oligomers is governed, in most part, by the physical properties of the support. The purpose of synthesizing oligosaccharides on a solid support goes beyond the goal of reaching the target molecule in free form. Hypothetically, the oligosaccharide assembled on a support can be used to prepare an affinity column for purification of oligosaccharide-binding proteins, for the preparation of combinatorial libraries for biological screening, or as a synthetic immunogenic material for vaccines.

Requirements for the synthesis of oligosaccharides on a solid support are essentially the same as those for the solid-phase synthesis of polypeptides or oligonucleotides. The first building block should be easily installed, the support must be compatible with the multiple deprotections, activations, couplings, and capping procedures, and the product must be retrievable in homogeneous form after cleavage from the support. The conditions necessary to form glycosidic linkages of oligosaccharides are usually more demanding (due to exposure to Lewis acid promoters) and more varied than those necessary to join amino acids or nucleotides. Thus, more stable supports and linker arms are necessary. The successful synthesis of an oligosaccharide on a polymeric support can be achieved only after careful examination of the type of polymer and linker to be used, careful selection of coupling methodology, and thoughtful planning of a protecting-group strategy. These aspects will be discussed briefly in each one of the examples presented. The physicochemical aspects of the use of solid supports for the synthesis of organic molecules has been previously discussed [27,29,41,42].

B.   Enzymatic Synthesis of Oligosaccharides

The use of glycosidases and glycosyl transferases for the synthesis of oligosaccharides has dramatically increased in the past 10 years. The major advantages in the use of enzymes for the formation of glycosidic linkages are their excellent substrate specificities, the high regio- and stereoselectivity of bond formation between unprotected sugars, and the mildness of the required reaction conditions. These advantages greatly simplify the assembly of complex oligosaccharides. Though great progress has been made in the preparation of enzymes by cloning and overexpression, the disadvantages of this approach are still the limited availability and high cost of the desired catalysts [14,43]. Furthermore, the intrinsic specificity of enzymes toward natural substrates can be a disadvantage when the goal is the preparation of non-natural analogs of oligosaccharides.

Attachment to a Solid Support via a Photolabile Linker

Zehavi and co-workers were the first to report the use of enzymes for the synthesis of oligosaccharides on a solid support. Their work was based on their previous attempts chemically to glycosylate a glucose derivative attached to a polymer via a photoremovable linker arm to produce the disaccharide isomaltose [33,44,45]. To achieve the first solid-phase chemoenzymatic synthesis of oligosaccharides they attached the initial glycosyl unit to a newly developed photolabile linker derived from methyl 4-hydroxymethyl-3-nitrobenzoate [46,47]. This preformed saccharide-linker unit **1** was coupled to the amino-functionalized water-compatible polyacrylamide-gel polymeric support **2** via the formation of an amide linkage to yield **3** (Figure 1). Though the formation of amide bonds is generally energetically favorable, a surprisingly small proportion of the available amino groups was functionalized. The polymer-bound glycoside **3** was glycosylated using D-galactosyl transferase in the presence of radio-labeled (^{14}C or ^{3}H) UDP-D-galactose to produce **4** after photolysis. The disaccharide **3** was also treated with lysozyme in the presence of chitin oligosaccharides **5** to obtain a mixture of tri- and tetrasaccharides **6** and **7**. In the first case, photolysis of the oligosaccharide was quantitative, because all radioactivity was released from the support upon irradiation. Although the photolabile linker performed well and the products were quantitatively removed from the polymer, Zehavi's yields of di-, tri-, and tetrasaccharides were disappointingly low. It was suggested that the yields were due to a low initial loading of the polymer and to inaccessibility of the

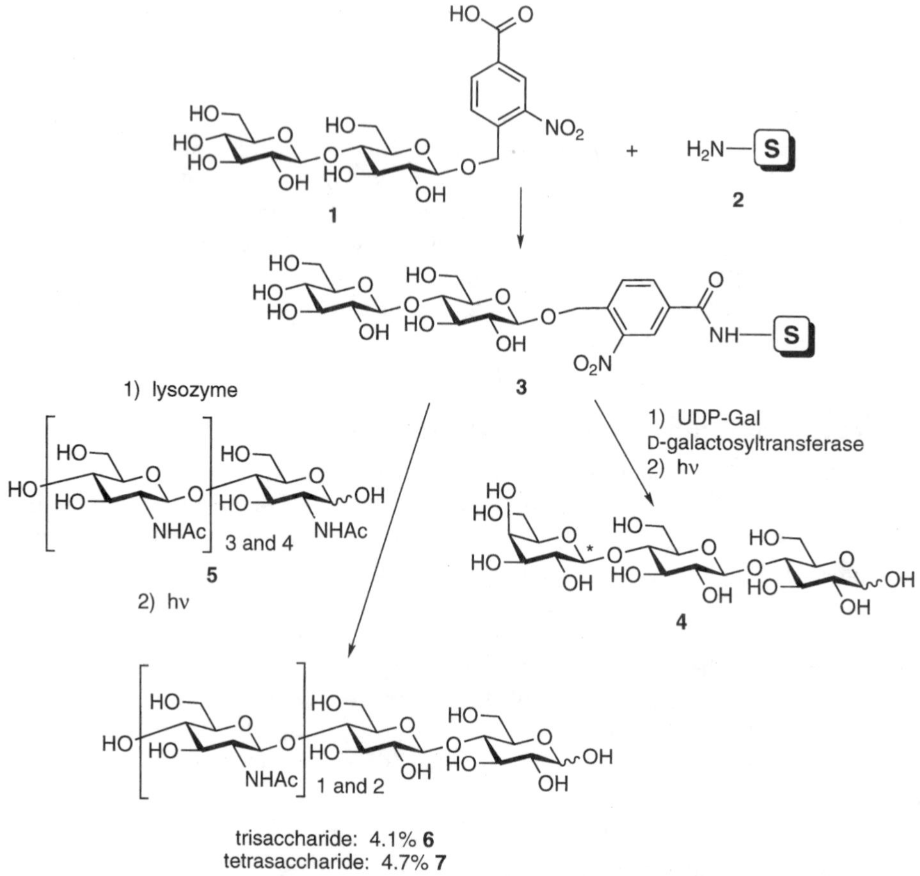

Figure 1 Synthesis of oligosaccharides via enzymatic glycosylation on a photolabile solid support. **S**, polyacrylamide-gel beads; *, radiolabel. (From Ref. 46.)

substrate to the enzyme because of steric interactions with the backbone of the polyacrylamide gel. Only the exposed surfaces of the polymer were directly accessible to the enzyme and thereby able to react. This same polymer and linker were later used to probe the acceptor specificity of the glycogen synthase enzymatic reaction [48].

Attachment to a Soluble Support via a Photolabile Linker

Zehavi and Herchman later proposed the use of a soluble polymeric support in an attempt to solve the problem of substrate accessibility [49]. A soluble polymeric support consists of a relatively small linear chain of monomers without cross-linking [50]. This type of support offers advantage in that it is fully solvated in solution and thus should not suffer from substrate inaccessibility to the extent of a cross-linked polymer. The physical properties of the attached oligosaccharide will be dominated by those of the polymer as long as the size of the growing oligomer remains much smaller than the support. The soluble polymer is usually precipitated after the re-

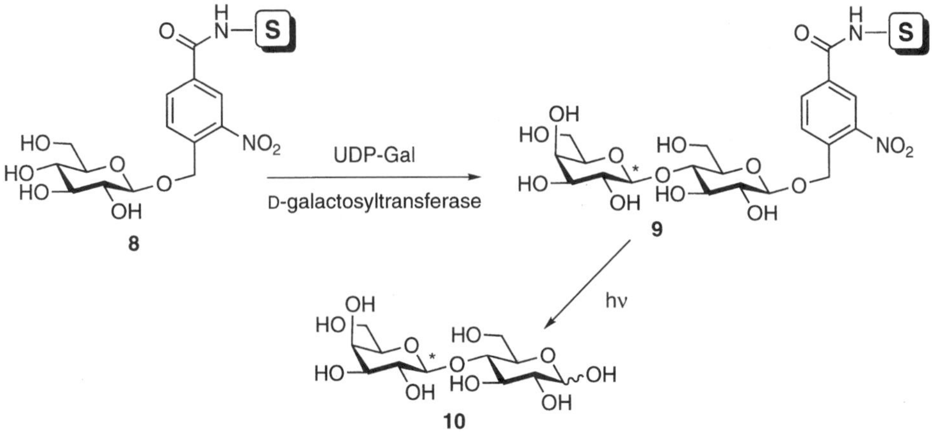

Figure 2 Synthesis of oligosaccharides via enzymatic glycosylation on a photolabile water-soluble polymeric support. **S**, poly(vinyl alcohol); *, radiolabel. (From Ref. 49.)

actions are completed and the impurities following washing are removed by simple filtration (thus, it is similar to a solid support). Alternatively, it can be purified by ultrafiltration (size exclusion chromatography or dialysis). Another feature of a soluble support is that progress can be monitored by common spectroscopic techniques such as NMR.

Zehavi and Herchman selected a substituted polyvinyl alcohol as the soluble support for the enzymatic synthesis of oligosaccharides. The use of this support, which is compatible with aqueous solvent, resulted in a more facile attachment of the first sugar and in greater accessibility of the substrate to the enzyme. The glycosylation of **8** was conducted as previously described using ^{14}C-labeled UDP-galactose in the presence of D-galactosyltransferase to give the disaccharide **9** in 34% yield as determined from the ^{14}C-radio-label (Figure 2). Upon UV irradiation of **9**, the measured release of radioactivity indicated that 88% of the polymer-bound lactose was detached from the support to give **10**. Although the use of a soluble support had indeed resulted in an increased yield of the desired product, some of the convenience and simplicity commonly associated with insoluble supports are compromised.

Glycopeptide Synthesis on a Solid Support Having a Chymotrypsin-Labile Linker

Wong and co-workers have accomplished an enzymatic synthesis of the sialyl Lewis X glycopeptide on a functionalized silica support [51]. The choice of a non-swelling support allowed for the use of organic solvents in the initial formation of the glycopeptide linkage. Aqueous buffers were successfully employed to mediate the enzymatic glycosylations [36,37]. A preformed N-acetyl glucosamine-asparagine glycopeptide was attached to the silica via the α-chymotrypsin-labile spacer-linker (GlyPhe(Gly)$_7$NH(CH$_2$)$_3$Silica) to give **11** with a loading of about 0.2 mmol/g of **12** (Figure 3). The use of a spacer unit was critical for rendering the substrate accessible to the enzymatic action. The glycopeptide **11** was galactosylated with UDP-

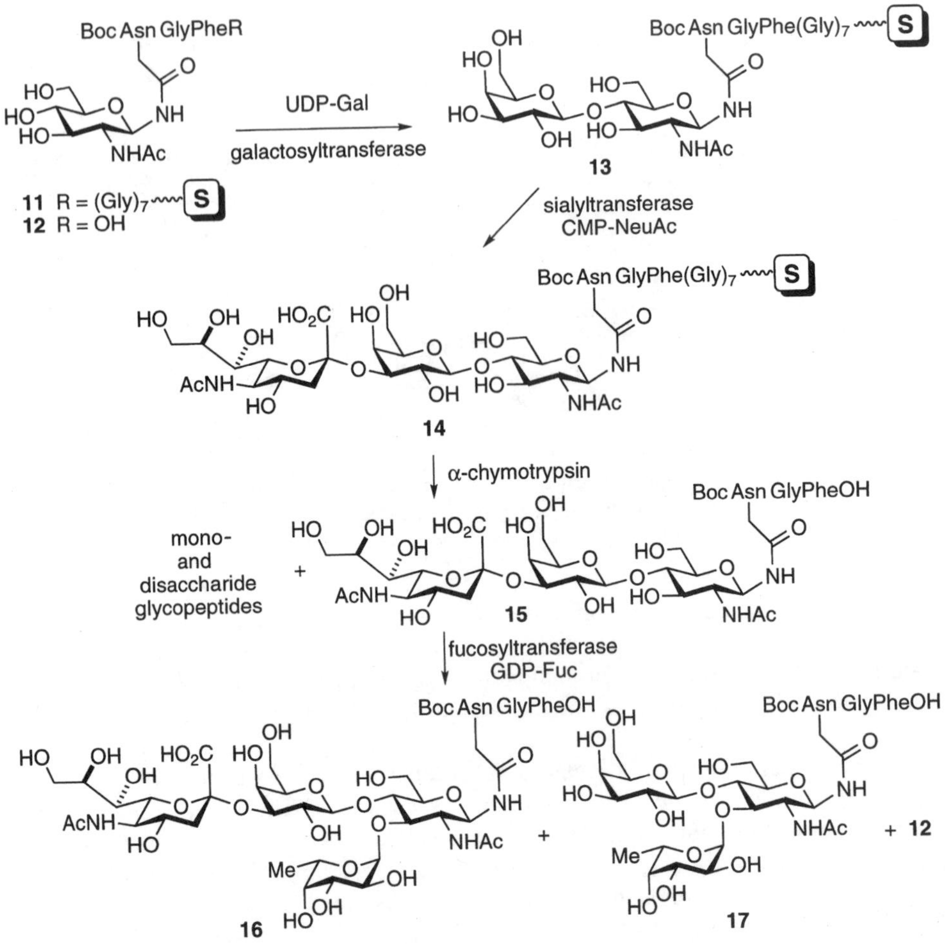

Figure 3 Synthesis of sialyl Lewis^x glycopeptide via enzymatic glycosylation on a solid support. **S**, aminopropyl silica. (From Ref. 51.)

galactose in the presence of β-galactosyltransferase to give **13**. The yield of disaccharide was evaluated by cleavage of the glycopeptide from an aliquot of the functionalized silica with α-chymotrypsin and found to be 55% of BocAsn(Galβ1,4-GlcNAcβ)GlyPheOH. Disaccharide **13** was then sialylated on the support in 65% yield with α-2,3-sialyltransferase and CMP-sialic acid to give **14**. Galactosylation and sialylation reactions were incomplete, and the products were released from the silica beads with α-chymotrypsin as a mixture of mono-, di-, and trisaccharide glycopeptides, which were fucosylated with α-1,3-fucosyltransferase and GDP-fucose. Purification of the reaction mixture by HPLC resulted in the isolation of the sialyl Lewis^x glycopeptide **16** (35% yield), the Lewis^x glycopeptide **17** (20%), and the glucosamine **12** (45%). Product **17** resulted from the fucosylation product of the unsialylated **13**, and the glucosamine **12** was the recovered unglycosylated starting material **11**. Even though the glycosylations did not all proceed to completion on the solid support, Wong's work provided critical demonstration of the feasibility of assembling complex oligosaccharides in acceptable yield using a solid-phase chemoenzymatic approach.

Glycopeptide Synthesis on a Solid Support Having an Acid-Labile Linker

More recently, Meldal and co-workers reported the use of a new polyethylene glycol/polyacrylamide copolymer for the enzymatic synthesis of glycopeptides on a solid support [52]. The copolymer had remarkably good swelling ability in organic as well as in aqueous solution and also possessed free amino groups for functionalization. The acid-labile 4-(α-amino-2′,4′-dimethoxybenzyl)phenoxyacetic acid (Rink's linker) was used to attach the glycopeptide to the support (Figure 4). The first N-acetyl glucosamine was installed on the support to give **18** with a loading of 0.16 mmol/g and was galactosylated, as in Wong's case, with galactosyltransferase in excellent yield. The final lactosamine glycopeptide **19** was recovered by TFA-mediated cleavage of the linker. The high isolated yield of products suggests that the swelling ability of the polymer enables the enzyme to mediate glycosylation of most reactive sites of the insoluble glycoside. However, the conditions used for glycopeptide retrieval may still be too acidic for some more sensitive glycosidic linkages. In principle, more readily cleaved linkers could perhaps be applicable in this approach (photo-, hydrogenolysis-, or enzyme-labile).

Attachment to a Soluble Support via a Linker That Is Cleaved by Hydrogenolysis

Nishimura and co-workers synthesized N-acetyl-lactosamine on a water-soluble polymeric support using galactosyltransferase [53]. The assembly of their supports used a radical-initiated copolymerization of acrylamide with glucosamine-acrylamide construct having two different spacer segments (see **20** and **21**, Figure 5). The benzylamido moiety was perceived to be labile to hydrogenolysis or DDQ oxidation. The copolymerization of each monomer was conducted in an aqueous medium in the

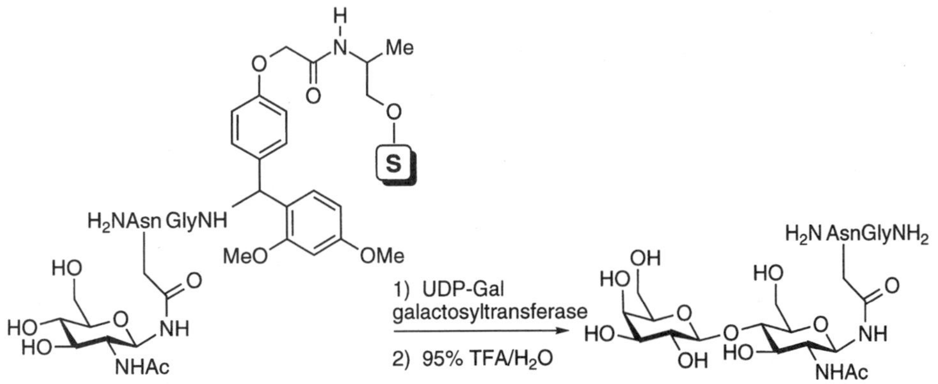

Figure 4 Synthesis of glycopeptides via enzymatic glycosylation on a solid support. S, polyethylene glycol/polyacrylamide copolymer. (From Ref. 52.)

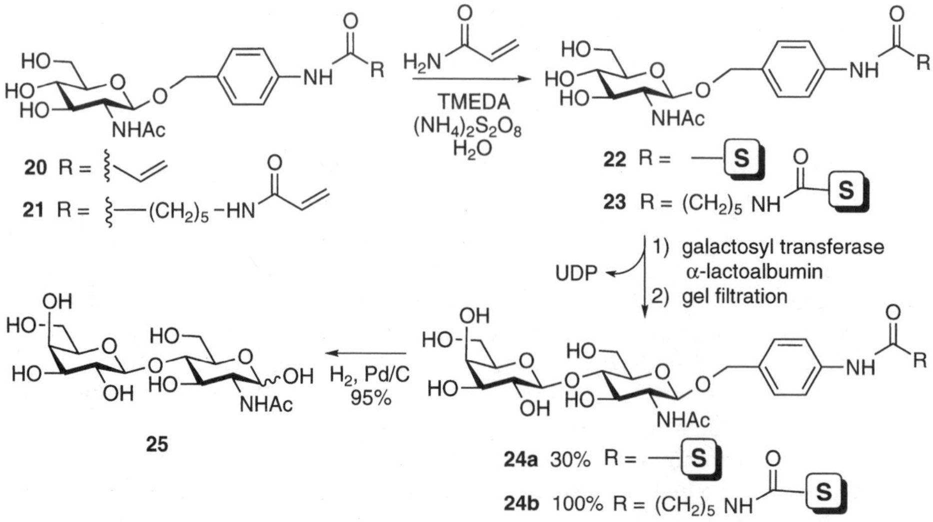

Figure 5 Synthesis of oligosaccharides via enzymatic glycosylation on a soluble polymeric support. **S**, polyacrylamide. (From Ref. 53.)

presence of acrylamide, tetramethylethylenediamine, and the radical promoter ammonium persulfate. The two resulting polyacrylamide polymers **22** and **23** were glycosylated separately with galactosyltransferase and α-lactoalbumin as the galactosyl donor to give polymer-bound lactosamine **24a** and **24b** after gel filtration and lyophilization. The yields for the formation of the disaccharides were established by NMR spectroscopy of the soluble polymer. A dramatic improvement in the glycosylation was observed in the case of **23**, in which the longer spacer arm was employed. Thus, it was found that the presence of a flexible spacer was crucial for performing enzymatic glycosylation of polymer-bound glucosamine. Lactosamine **25** was recovered in high yield from the catalytic hydrogenolysis of polymer **24b**.

Enzymatic Glycosylation of Water-Soluble Polymer-Bound N-Acetylglucosamine

A related approach has been used by Nishimura et al. to prepare polyacrylamide-bound lactosamine [54]. Their goal was not that of synthesizing free sugar but that of generating synthetic glycopolymer for use as biomimetic, diagnostic, and therapeutic ligands. As described in the previous section, polymer-bound glucosamine **26** was galactosylated in almost quantitative yield with UDP-galactose in the presence of bovine milk β-1,4-galactosyltransferase and α-lactoalbumin to give the disaccharide **27** (Figure 6). The reaction yield was measured by the weight increase of the soluble polymer and by NMR spectroscopy. This material was then incubated with *Trypanosoma cruzi trans*-sialidase and *p*-nitrophenyl sialic acid to give the heterogeneous glycopolymer **28** with partial sialylation of the available galactose. Selective trimming of the unsialylated substrates from the polymeric support using β-D-galactosidase and N-acetyl-D-glucosaminidase led to homogeneous **29**. This approach should be applicable to the preparation of oligosaccharides on polymeric support if a suitable, removable linker arm is introduced between the saccharide and polymer.

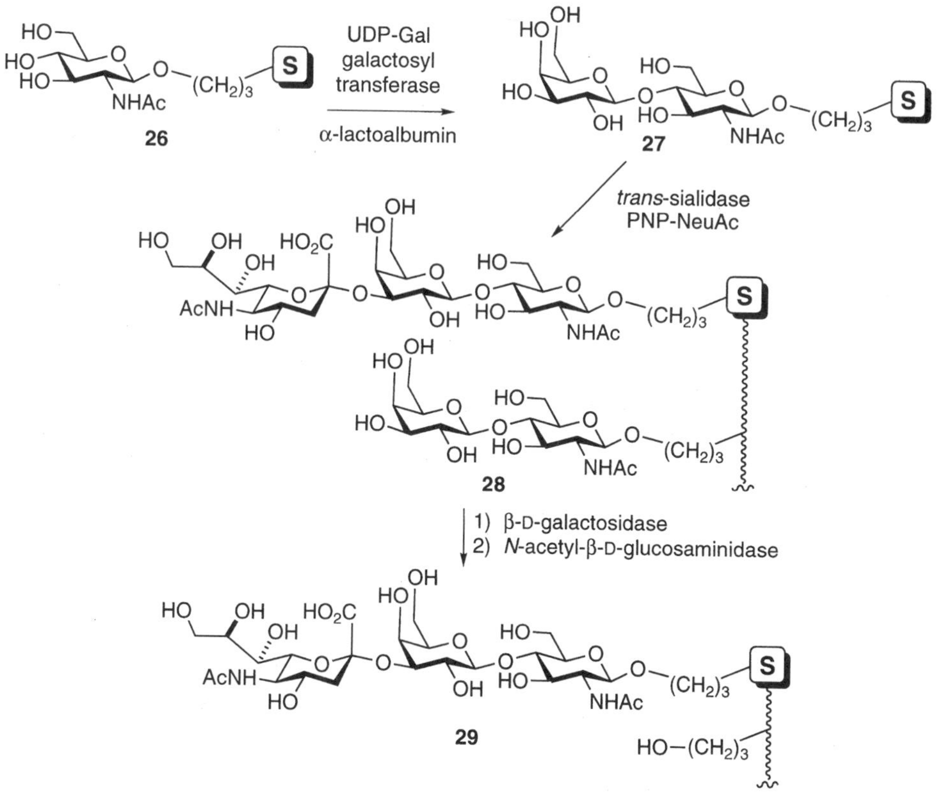

Figure 6 Synthesis of oligosaccharides via enzymatic glycosylation on a soluble polymeric support. **S**, polyacrylamide. (From Ref. 54.)

C. Chemical Glycosylation

Attachment to the Merrifield Resin via a Homoserine Linker

In a study directed toward the preparation of synthetic vaccines, van Boom and co-workers synthesized a naturally occurring β1,5-linked D-galactofuranosyl heptamer using solid-phase techniques [55]. The heptamer is an immunologically active component of the extracellular polysaccharide coat of *Aspergillus* and *Penicillium* species.

The first galactofuranosyl unit **30** was attached to Merrifield's resin via a homoserine linking arm suitable for the subsequent attachment of the polysaccharide to a macromolecular carrier (**31**, Figure 7). The presence of a participating group at the C-2 position of donor **30** resulted in the exclusive formation of the β-linkage using Helferich glycosylation conditions. The levulinoyl protecting group at the C-5 position of the polymer-bound saccharide was chosen to be selectively and efficiently removed with hydrazinium acetate in pyridine. The resulting deprotected material **32** was coupled with the anomeric chloride **30** (2 equivalents) to give **33** after acetylation. This capping step was necessary to prevent the formation of oligosaccharides with a shorter sequence (internal deletions) resulting from coupling

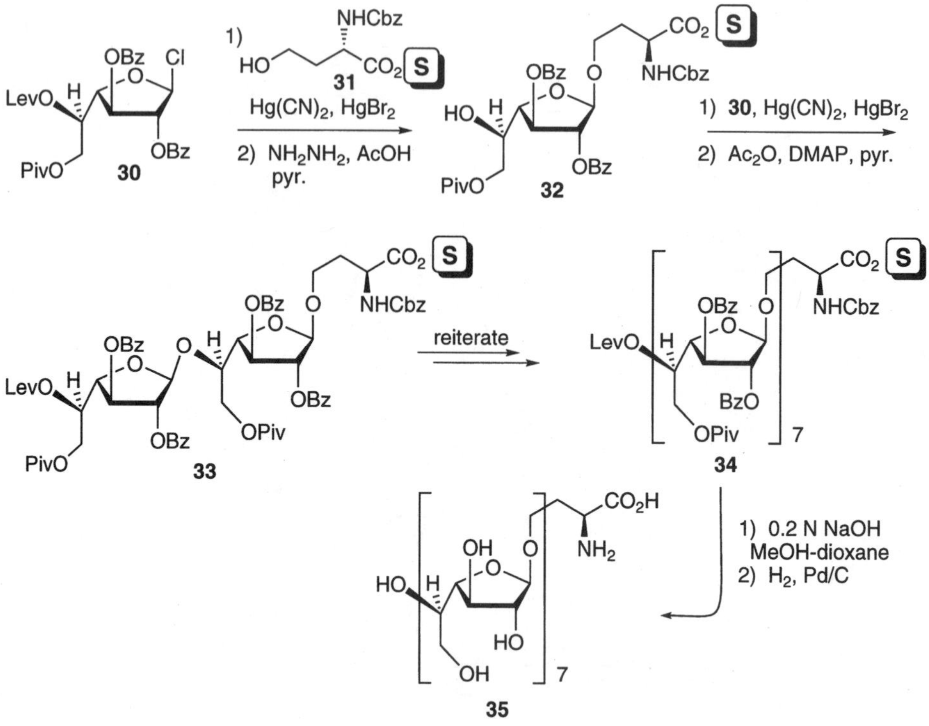

Figure 7 Solid-phase synthesis of a homoserine-linked heptasaccharide. S, Merrifield 1% cross-linked polystyrene. (From Ref. 55.)

the remaining uncoupled C-5 hydroxyls with the glycosylating reagent. The presence of such anomalous sequences would result in a very complex final mixture. Multiple iterations of the three-step sequence gave the polymer-bound heptasaccharide **34**, which was removed from the support by basic hydrolysis. Finally, the benzyloxy-carbonyl group was cleaved by hydrogenolysis. The yield for each coupling iteration was evaluated to be about 90%, and the overall isolated yield of the heptasaccharide **35** was 23%.

In a reaction of extracellular coat antibodies against *Aspergillus* and *Penicillium* species with the extracellular polysaccharide antigens, the heptamer-homoserine con-jugate was found to be the most effective inhibitor of a series of synthetic oligomers. van Boom's approach is well suited for the preparation of repeating polysaccharides on a solid support and provides the earliest reference to the use of a solid support for the synthesis of an oligosaccharide in a bioconjugatable form. The major incon-venience of this approach lies in the need selectively to manipulate protecting groups on the solid support, where monitoring of progress is difficult.

Application of the Glycal Assembly Method to the Synthesis of Oligosaccharides on a Solid Support

Glycal Assembly Methods. In a recent communication, Danishefsky et al. [56] presented an approach toward the solid-phase synthesis of oligosaccharides based on the glycal assembly method [57–64]. This method is described schematically in Figure 8 and consists of the activation of the double bond of glycal **36** with a chemical activation device **E**. The activated donor **37** can be reacted with an acceptor molecule and, by using a glycal **38** as the acceptor, yielding a disaccharide glycal **39**. The process can be reiterated, rapidly generating oligosaccharides (**40** + **41** → **42**). In principle, any hydroxyl of the glycal might be used to provide access to 1,6-, 1,4-, or 1,3-interglycosidic linkages. An attractive feature of the method is that the newly installed functionality **E** is a differentiated group that can be used to modify further the donor moiety.

The synthesis of polysaccharides from glycals has several advantages over the use of conventional pyranose coupling strategies. The hydroxyl groups of glycals are easier to protect differentially than those of equivalent saccharides. An assortment of activation protocols now exist to initiate the donating capacity of a glycal, lending access to diverse linkages upon reaction with a glycosyl acceptor.

Glycals are becoming widely used as efficient building blocks for the synthesis of a wide variety of complex carbohydrates and natural products. For example, ciclamycin 0 **43** (Figure 9) [60], allosamidin **44** [65], calicheamicin **45** [66,67], gangliosides GM_3 **46** [68], desgalactotigonin **47** (Figure 9) [57], the sialyl-Lewis x **48** [69,70], Lewis[b] **49** [71], and Lewis[y] **50** antigens [72] were some of the products recently synthesized in Danishefsky's laboratories using glycal-based strategies.

Glycals are readily epoxidized to give 1,2-anhydrosugars using 3,3-dimethyldioxirane [73]. The stereochemical outcome of the reaction has been studied and can be predicted for many glycals. The synthetic utility of 1,2-anhydrosugars is summarized in Figure 10 [59,74]. Treatment of 1,2-anhydrosugar **51** with an alcohol

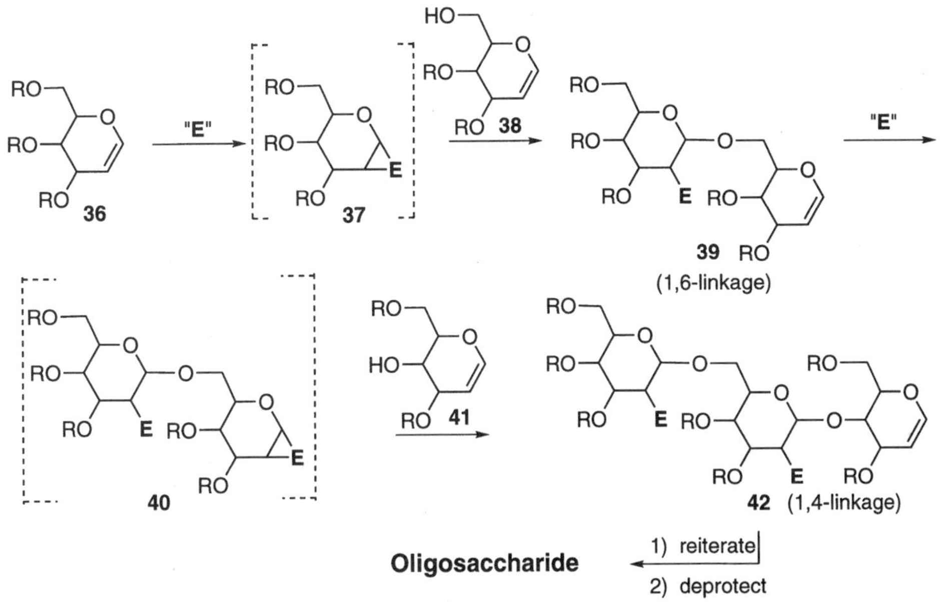

Figure 8 A general strategy for oligosaccharide synthesis using the glycal assembly method. (From Ref. 56.)

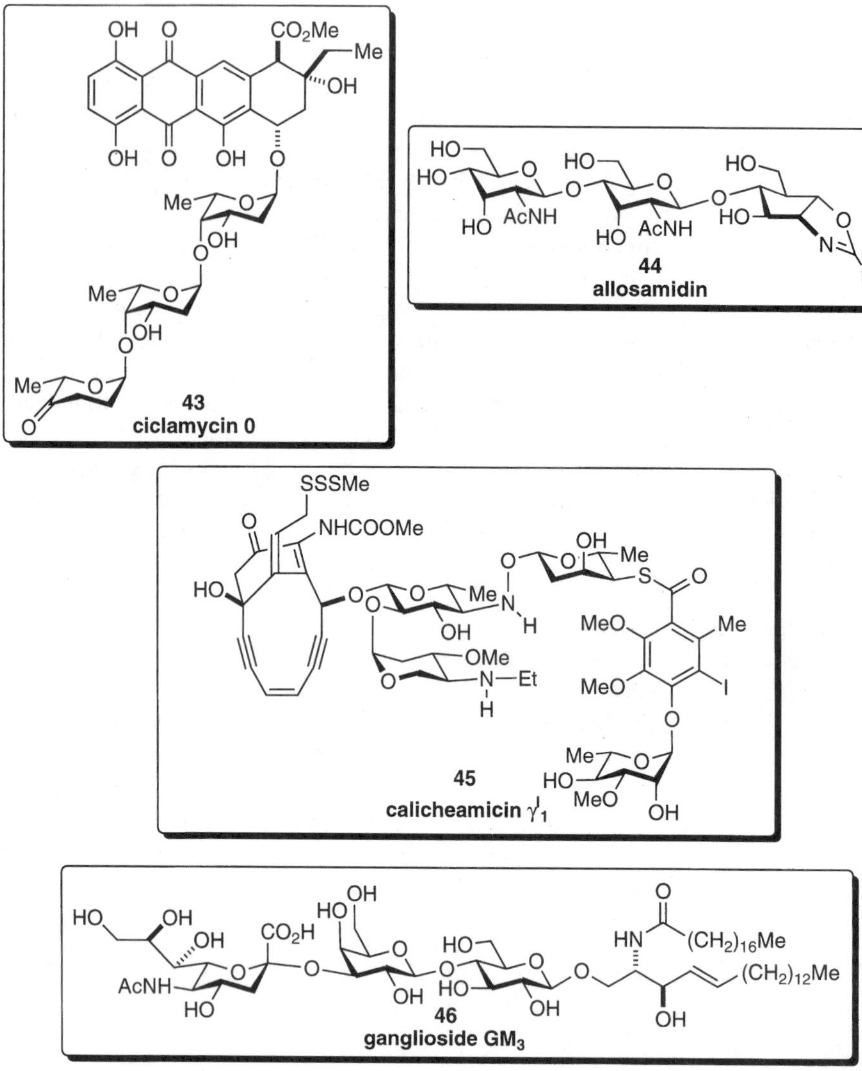

Figure 9 Molecules synthesized using the glycal assembly method.

glycosyl acceptor in the presence of a Lewis acid produces a new glycoside **52** [59]. Alternatively, 1,2-anhydrosugars have been used to prepare anomeric azides **53** and amines **54**. Several glycosides with known applicability as useful glycosyl donors have been prepared from 1,2-anhydrosugars. These include glycosyl fluorides **55**, 4-pentenyl glycosides **56**, phenylthio- **57** and selenoglycosides **58** [75], and 1,2-gly-cooxazoline **59** [76].

Glycals are known to be suitable for the preparation of a wide range of 2-deoxy sugars and sugar donors (Figure 11). Both azidonitration [77] and azido-phenylselenation [78] have been used to prepare 2-azido-2-deoxysugar donors (**60** and **61**, respectively). These products can serve as precursors for glycosides of *N*-

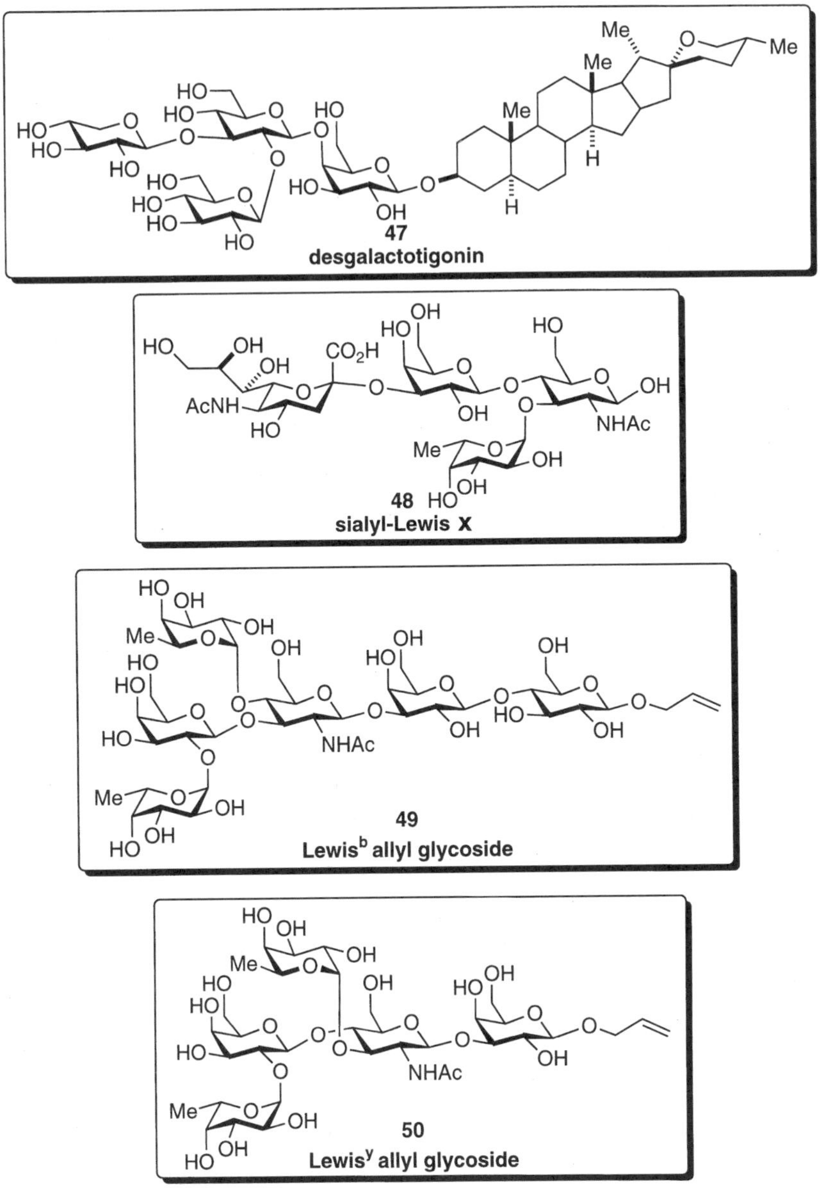

Figure 9 Continued

acetylglucosamine and *N*-acetylgalactosamine after simple modification of the azide and the anomeric functional groups. For example, azido nitrates **60** were readily transformed into azido trichloroacetamidates **62** [79] or azido xanthates **63**, which both act as potential donors to reach 2-azido-2-deoxy sugars [79]. Direct formation of 2-deoxyglycosides with or without Ferrier rearrangement, to provide **64** [80] or **65** [81], respectively, was accomplished by use of various proton or Lewis acid sources. 2-*N*-Acetyl-2-deoxy sugars were also accessible via the halogenosulfon-

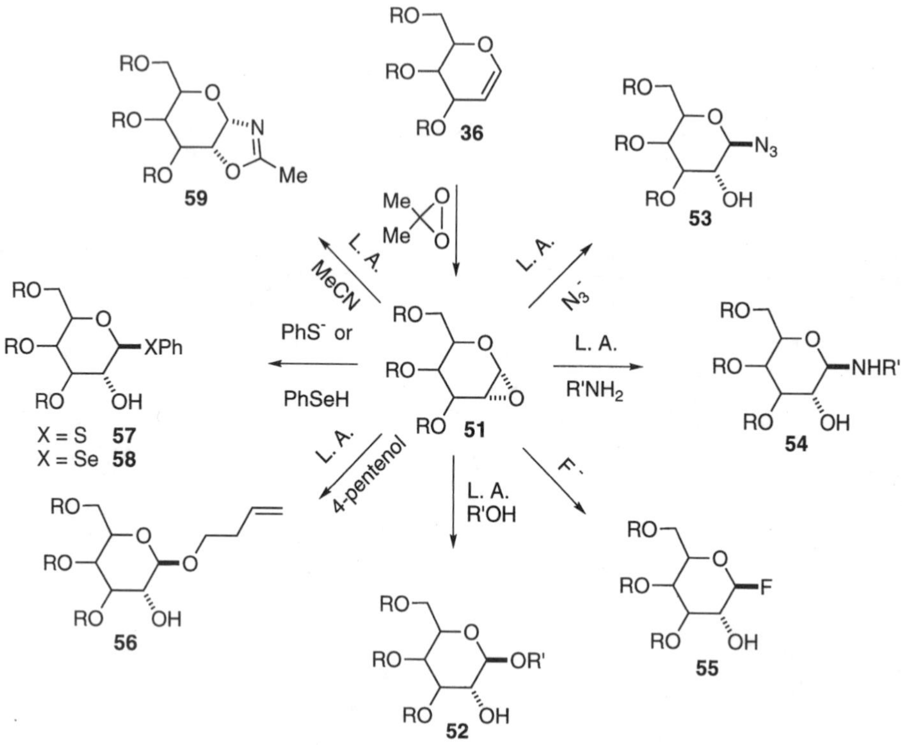

Figure 10 Reactions of glycal epoxides with nucleophiles.

amidation reaction of glycals [58]. Halogenosulfonamides **66** proved susceptible to attack by nucleophiles (N_3^- [82], or alkoxides [58]) to give 2-deoxy-2-sulfonamido saccharides **67**, which were ultimately converted to 2-N-acetyl-2-deoxy sugars **68** after transformation of the sulfonamide functional group into an acetamide. Halogenoglycosylation of glycals yielded 2-deoxy sugars **65** after reduction of the halide at the 2-position of glycoside **69** [10,62]. 2-Thiophenylglycosides **70** were obtained from the reaction of glycals with (phenylthio)sulfonium salts and stannyl alkoxides. The thiophenyl group also proved to be susceptible to reductive cleavage to yield 2-deoxyglycosides **65** [83]. Adducts **71**, obtained from the photoaddition of bis(trichloroethyl)azidocarboxylate to glycals, react with glycosyl acceptors in the presence of catalytic amounts of acid to yield glycosylated hydrazodicarboxylates **72**. These can serve as precursors to acetamides upon suitable reductive cleavage and acetylation [84]. Although not every possible type of glycosidic linkage is currently available from glycals, the versatility of these molecules, stemming from the large number of activation methods that can be employed to release their donating capacity, explains their increasing applicability to the assembly of carbohydrate constructs.

     The Glycal Assembly Method: Adaptability to Synthesis on the Solid Support. Solid-phase synthesis of oligosaccharides using a polymer-bound glycosyl donor has previously been intimidating because of the risk of incomplete activation and chain termination. In fact, only one report of the application of this approach

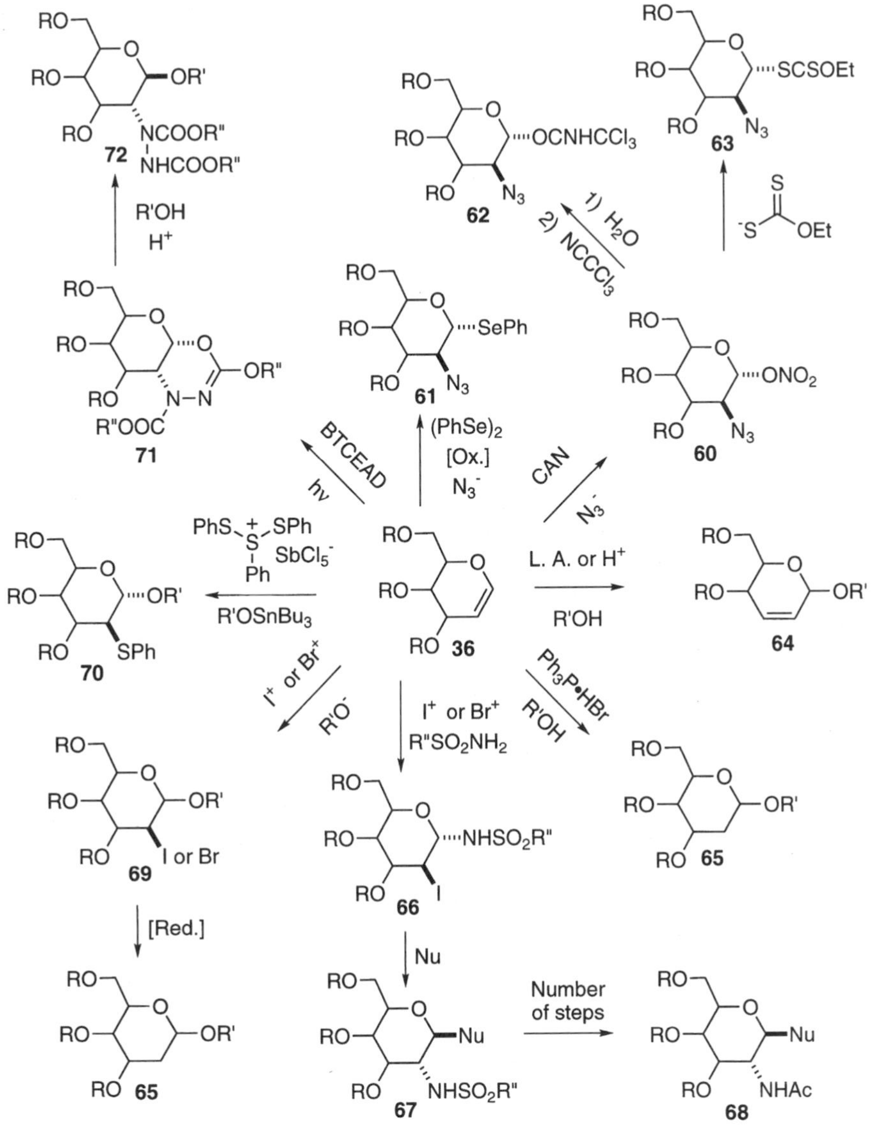

Figure 11 2-Deoxysugars synthesized from glycals.

has appeared to date [34]. However, the recent introduction of a number of stereo-selective, high-yielding activation methods for ''priming'' glycals to serve as gly-cosyl donors has made possible the use of such a strategy for the preparation of complex carbohydrates.

In their initial approach, Danishefsky et al. [56,85] attached glycal carbonate **75** (derived from D-galactal) to chlorotriarylsilyl-functionalized 1% cross-linked polystyrene **74**, prepared from polystyrene **73** according to a known procedure (Figure 12) [86,87]. The loading of the sugar donor **76** was calculated to be 0.5–0.6 mmol of glycal per gram of functionalized polymer after retrieval of the galactal

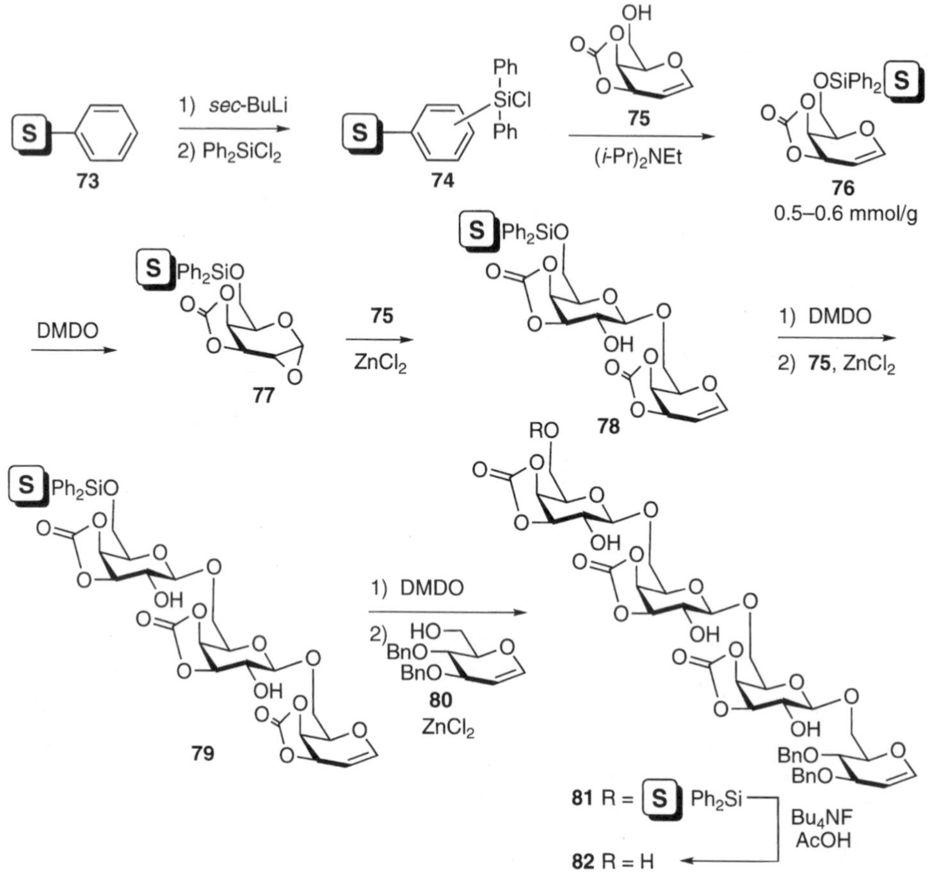

Figure 12 Synthesis of oligosaccharides on a solid support. **S**, 1% cross-linked polystyrene. (From Ref. 56.)

carbonate **75** from the resin using tetrabutylammonium fluoride buffered with acetic acid to prevent the removal of the carbonate.

The epoxidation of glycals with dimethyldioxirane followed by opening of the resulting 1,2-anhydrosugar with a glycosyl acceptor in the presence of $ZnCl_2$ is an effective method for glycosylation. Furthermore, since glycals can be used as acceptors with 1,2-anhydrosugars donors, this method can be used for an iterative approach to oligosaccharide assembly [59]. Application of this procedure to the polymer-bound donor **76** (Figure 12) produced the disaccharide **78** via opening of the oxirane **77**. Subsequent iterations of the oxidation-glycosylation sequence produced a tri- (**79** from reaction with the acceptor **75**) and a tetrasaccharide (**81** from reaction with the acceptor **80**). The yield of each iteration was found to be around 70%. Thus, the tetrasaccharide **82** was isolated in 32% overall yield from the polymer-bound donor **76** after fluoride-anion induced retrieval from the support.

Further experimentation demonstrated versatility in the use of 1,2-anhydrosugars for the solid-phase synthesis of oligosaccharides. Galactal (**76**) and glucal (**84**) donors attached to the solid support (Figure 13) were used to prepare oligosaccha-

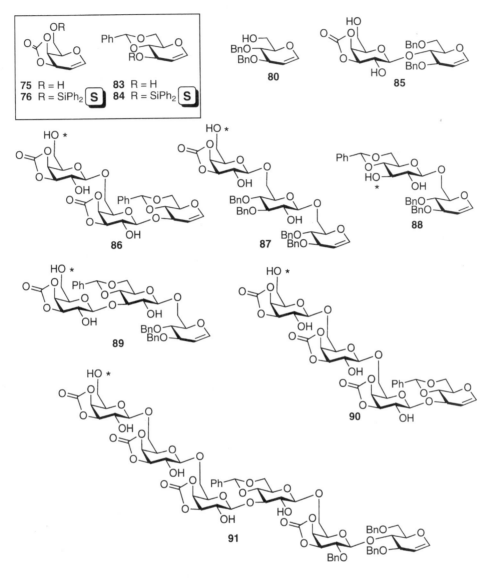

Figure 13 Oligosaccharides **86–91** assembled on solid support by the glycal assembly method from compounds **75**, **76**, **80**, and **83–85**. S, 1% cross-linked polystyrene; *, point of attachment to the polymer. (From Ref. 56.)

rides **86–91** by coupling with the sugar acceptors **75**, **80**, **83**, and **85**. The secondary allylic hydroxyl group at C-3 of glucal **83** was attached to the polymeric support to give the donor **84** and also acted as a glycosyl acceptor (see formation of **86, 89–91**). Lactal **85** was used as an acceptor and, since this compound was prepared on a solid support, this presented the possibility of making elaborate sugars by convergent solid-phase oligosaccharide synthesis analogous to convergent solid-phase [88] or solid-phase fragment condensation syntheses of peptides [89].

An interesting feature of the solid-phase glycal assembly method is the ease of purification of oligosaccharides upon release from the support. The desired products were isolated from more polar by-products by chromatography on silica gel. The apparent absence of products resulting from interior deletion of saccharide units is probably due to a "self-policing" feature of the method, in which uncoupled epoxide moieties are destroyed during the washing procedures following glycosylation. It is believed that a 1,2-diol is produced, which lacks glycosyl donating capacity under the reagent conditions employed. The resulting polar diols were easily separated from the final mixtures. This characteristically obviates the need for capping before subsequent glycosylations. Such capping is often necessary when using approaches in which the acceptor is attached to the polymer. Thus, internal deletions are virtually eliminated using the glycal epoxide assembly method. The unreacted glycosyl acceptors are easily recovered, in high yield, from the washings after each glycosylation.

Another advantageous feature of the method is that new, easily differentiable C-2 hydroxyls are introduced at each coupling iteration, yielding oligomers with a series of potentially unique functional groups. An exploitable feature of the methods is that it allows for branching directly on the solid support. The polymer-bound oligosaccharide becomes the acceptor in the branching process with minimal protecting group manipulations. The synthesis of branched oligosaccharides on a soluble polymeric support had been reported by Krepinsky et al. [90] (see Figure 22, Section III.C.).

Danishefsky et al. [91] prepared disaccharide 93 (Figure 14) by opening polymer-bound epoxide 77 with the diacetonide galactose acceptor 92. Disaccharide 93 was glycosylated with glycosyl fluoride 94 (derived from tribenzylglucal) in the presence of stannous triflate. The branched trisaccharide 96 was obtained in 53% overall yield from 95 after retrieval with fluoride anion. Similarly, fucosylated trisaccharide 99 was prepared from disaccharide 93 and fluorosugar 97 [92]. This method provides a straightforward solution to the problem of assembling oligosaccharide patterns that contain either α- or β-linked glycoside branches at C-2.

The Glycal Assembly Method: Solid-Phase Synthesis of Blood Group Determinants. Danishefsky et al. directed their efforts toward applying their solid-phase method to the synthesis of carbohydrate domains having blood group determining specificity. This class of molecules is found attached to both proteins and lipids and is widely distributed in erythrocytes, epithelial cells, and various secretions [93,94]. An early goal of studies of these branched sugars was to help establish their role in specifying blood types [95]. It was recently recognized that carbohydrate recognition is an important phase of cell adhesion and binding phenomena [96–100]. Novel carbohydrate patterns are often encountered as markers for the onset of various tumors [93,101]. An understanding of the origin of tumor-associated antigens may lead to new applications in cancer detection and treatment by immunotherapy [102].

In an attempt to assemble on solid phase a tetrasaccharide possessing H-type-2 specificity, Danishefsky et al. [91] treated the polymer-bound 1,2-anhydrosugar 77 in the usual manner using 3,6-dibenzylglucal 100 as a solution-based acceptor to produce the β1,4-disaccharide 101 (Figure 15). The newly introduced C-2 hydroxyl was fucosylated using a solution of 2,3,4-tribenzylfucosyl fluorides 97 [92] to give the branched trisaccharide 102, which was recovered from the support to give 103

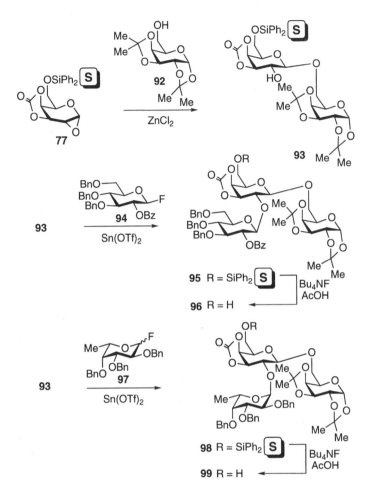

Figure 14 A strategy for the assembly of branched oligosaccharides on a solid support. S, 1% cross-linked polystyrene. (From Ref. 91.)

in 50% overall yield from **76**. The fucosylation was buffered with DTBP to preserve the acid-sensitive glycal moiety.

The polymer-bound trisaccharide glycal **102** was converted into the iodo benzenesulfonamide and iodo 2-(trimethylsilyl)ethanesulfonamide **104** and **105** in order to investigate the possibility of applying Danishefsky's sulfonamidoglycosylation procedure to the preparation of 2-acetamido glycosides (see Figure 11, **66** → **67** → **68**) [58]. However, this reaction sequence has not yet been successful on the solid support with an alkoxide acceptor. It is smoothly accomplished with an azide nucleophile (*vide infra*). The trisaccharide glycal **103** was released from the support and used to prepare the H-type-2 human blood group determinant in solution [91].

Upon further investigation it was discovered that the diphenylsilyl linker was labile to the basic and nucleophilic conditions of the sulfonamidoglycosylation reaction. This finding prompted the search for new, more stable, linkers. The linker **106** (Figure 16), involving a di-isopropylsilyl linkage, is more stable than the di-

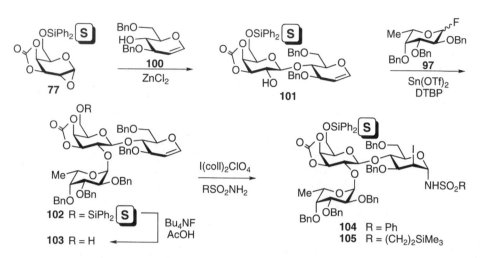

Figure 15 Solid-phase synthesis of a trisaccharide precursor to an H-type blood group determinant. **S**, 1% cross-linked polystyrene. (From Ref. 91.)

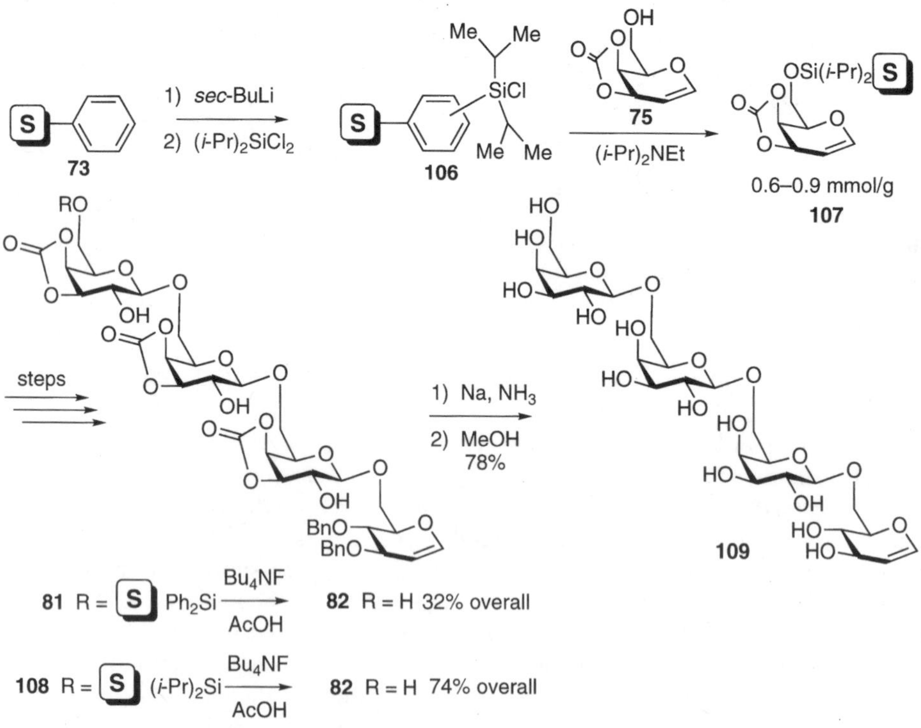

Figure 16 Synthesis of a new di-isopropylsilyl linker and preparation of a tetrasaccharide on the solid support. **S**, 1% cross-linked polystyrene. (From Ref. 103.)

phenyl counterpart. This linker was attached to the aromatic rings of the polystyrene beads **73** following the same procedure used previously [86], and the resulting polymer **106** provided the support for the preparation of polymer-bound galactal carbonate **107** with an improved loading of 0.6–0.9 mmol/g. The superiority of the diisopropylsilyl linker was demonstrated when the tetrasaccharide **82** was produced in 74% overall yield from the polymer-bound galactal carbonate **107** compared with an overall yield of 32% obtained for the diphenylsilyl-linked material **81**. The average per-coupling yield improved from ~70% to ~90% [103]. The tetrasaccharide **108** was released from the support to give **82**, which was recently fully deprotected to give **109** in 78% yield. Cleavage of the blocking groups was accomplished in a one pot, dissolving metal reduction followed by solvolysis of the carbonates [103].

Randolph and Danishefsky [71] recently used a combination of solid-phase and solution techniques to prepare neoglycoproteins with Lewis[b] specificity. The Le[b] determinant has been claimed to be a mediator for the attachment of *Helicobacter pylori* to human gastric epithelium [104], which is suspected to be a causative agent for chronic gastritis, gastric and duodenal ulcers, and gastric adenocarcinoma [105–108]. Furthermore, it was discovered that a soluble neoglycopeptide presenting the Le[b] antigen inhibited *H. pylori* binding to cells [104]. Since bacterial binding is a prerequisite for infection [109], soluble Le[b] analogs may prove to be effective therapeutic candidates against these ailments.

The core Le[b] tetrasaccharide was smoothly synthesized on the solid support using the glycal assembly method (Figure 17) [71]. The 1,2-anhydrosugar derived from galactal carbonate **107** was reacted with the monoprotected 6-TIPS glucal **110** to give the polymer-bound disaccharide-diol **111**. This reaction proceeded in a highly regioselective manner with selective glycosylation of the allylic position, C-3. This selectivity might have been difficult to achieve if the acceptor were attached to the polymer instead of the donor. Bisfucosylation of diol **111** with fluorides **97** followed by detachment of the glycal **112** from the support yielded the desired tetrasaccharide **113** in 40% overall yield from **107**. Saccharide **113** was used successfully to build Lewis[b] blood group determinant **49** via sulfonamide **114**. Hexasaccharide **49** was subsequently conjugated to human serum albumin (HSA) to give the macromolecule **115**, which is currently being evaluated as a determinant for binding with *Helicobacter pylori*.

The Glycal Assembly Method: Solid-Phase Synthesis of Glycopeptides. Roberge and Danishefsky [110] successfully synthesized *N*-linked glycopeptides using the glycal assembly method on the diphenylsilyl-functionalized solid support. The polymer-bound trisaccharide glycal **116** (Figure 18) was assembled as described in the previous section and used as the model substrate for their studies. The free C-2 hydroxyls were acetylated, and the glycal was converted to iodosulfonamide **117** by reaction with iodonium bis(*sym*-collidine) perchlorate and 2-trimethylsilylethylsulfonamide [70]. The iodosulfonamide **117** was treated with a solution of tetrabutylammonium azide [111] to give the anomeric azide **118** via nucleophilic opening of the aziridine intermediate [82]. The azide was reduced with propanedithiol [112] and the resultant amine **119** was coupled with the aspartic acid residue of protected tripeptide **120** by formation of the mixed anhydride with 2-isobutoxy-1-isobutoxycarbonyl-1,2-dihydroquinoline to give the polymer-bound glycoconjugate **121** [113]. Treatment with hydrogen fluoride·pyridine complex gave the glycopeptide **122** in 19% overall yield from galactal carbonate **76**. All attempts to convert or to replace

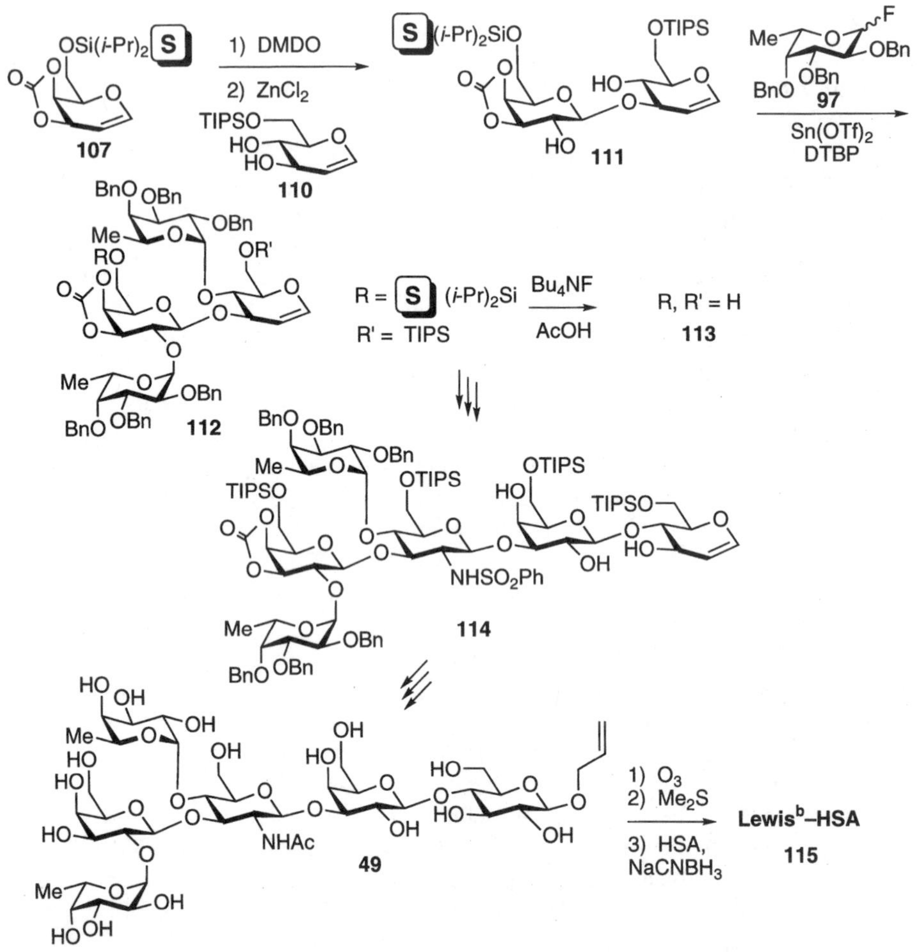

Figure 17 Synthesis of the Lewis[b]-HSA neoglycoprotein using a combined solid-support and solution approach. S, 1% cross-linked polystyrene. (From Ref. 71.)

the SES sulfonamide group by an *N*-acetyl function were unsuccessful. However, the preliminary work had demonstrated that it was possible to assemble glycopeptides, convergently, with the sugars directly attached to the polymeric support and a modified approach was developed.

The instability of the glycopeptide linkage complicated the transformation of the sulfonamide to the acetamide. It was proposed that the deprotection of the sulfonamide would be less difficult if the C-2 functionality were transformed into an acetamide prior to coupling with the peptide. A number of acetamides, carbamates, and sulfonamides were investigated, but only 9-anthracenyl sulfonamide [114] had all the desired properties. This approach has allowed for the successful synthesis of 2-*N*-acetyl anomeric azide intermediates.

For this work, the glycopeptide was prepared on the more stable, diisopropyl-silyl-functionalized polymeric support. The iodosulfonamide **124** (Figure 19), derived

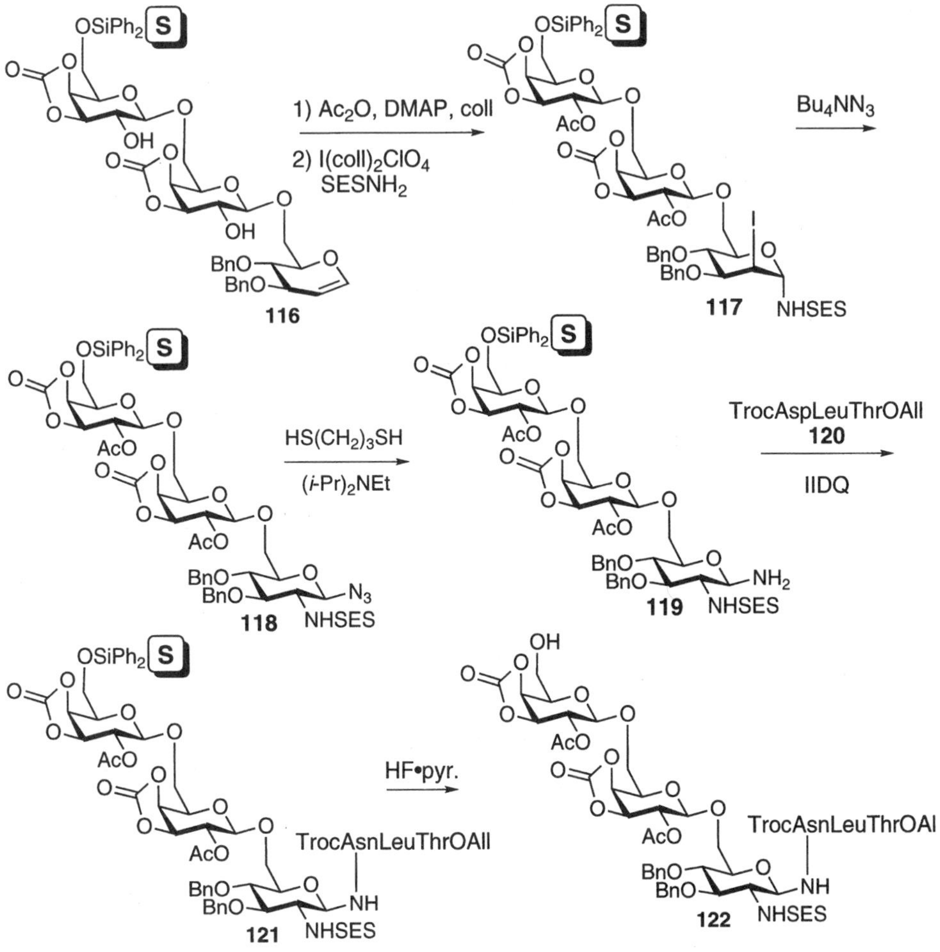

Figure 18 Solid-phase synthesis of N-linked 2-N-SES-glycopeptide by the glycal assembly method. S, 1% cross-linked polystyrene. (From Ref. 110.)

from glycal **123**, was treated with tetrabutylammonium azide, and the resulting sulfonamide was acylated with acetic anhydride in the presence of DMAP to give the anomeric azide **125**. Removal of the sulfonamide moiety was accomplished by treatment with thiophenol and Hünig's base. The resulting product, containing a 2-N-acetyl function and the anomeric azide was reduced to afford the anomeric amine. This compound was coupled with a protected tripeptide **120** as described previously to give the polymer-bound glycopeptide **126**. Reaction with HF·pyridine gave the glycopeptide **127** in about 20%–40% overall yield from the galactal carbonate **107**. This corresponded to an average yield of 90% for each of the 10 steps in the sequence. The glycopeptide **127** was fully deblocked. The sequence began with palladium-catalyzed removal of allyl ester in the presence of dimethylbarbituric acid to provide the free acid [115]. The latter was treated with zinc in acetic acid to yield glycopeptide **128** [116]. The benzyl-ether-protecting groups were removed by cata-

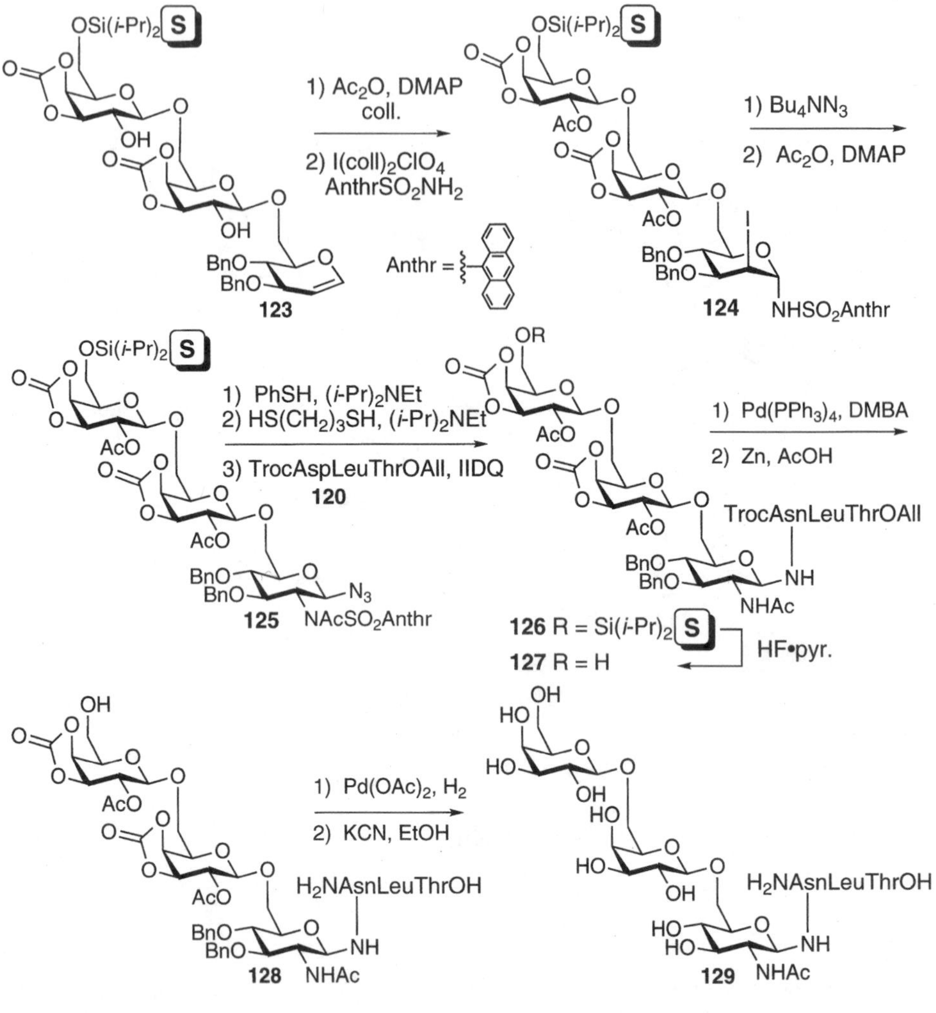

Figure 19 Solid-phase synthesis of *N*-linked glycopeptide by the glycal assembly method. **S**, 1% cross-linked polystyrene. (From Ref. 110.)

lytic hydrogenolysis with palladium black generated in situ [117], and the acetates and carbonates were cleaved by transesterification under mediation with potassium cyanide in ethanol [118]. In this way, **129** was obtained in 65%–75% yield from **127**.

The work described above has only begun to demonstrate the potential of the solid-phase version of the glycal assembly method for the synthesis of glycopeptides and oligosaccharides. The two major types of glycosidic linkages installed on the solid support were the 1,2-diequatorial (i.e., β-gluco- and β-galactoside) and the 1-axial, 2-equatorial (i.e., α-fuco-) linkage. The method accommodates primary and secondary hydroxyls, and all four hexose positions have been glycosylated (i.e., α-1,2; β-1,3; β-1,4; β-1,6) on the solid support. Although methods exist for the preparation of other types of linkage from glycals (see section III.C.), these methods have not yet been applied to the solid support. The convenience of the glycal strategy for manipulation of protecting groups was highlighted by the exploitation of the free hydroxyl at the C-2 position for synthesis of branched oligosaccharides. Also, as noted earlier, a major advantage of this solid-support approach is the ''self-policing'' feature that virtually eliminates deletion products, resulting in facile purification of the final material.

Attachment to the Merrifield Resin via a p-Hydroxythiophenol Linker

Recently Kahne and co-workers [119] reported a remarkable synthesis of oligosaccharides on Merrifield resin using an anomeric sulfoxide glycosylation method. The first unit was attached to the resin via an anomeric p-hydroxythiophenol linker that could be cleaved with aqueous mercuric trifluoroacetate [40]. The C-6 position of the polymer-bound thioglycoside **130** (Figure 20) was selectively protected with a triphenylmethyl group that was removed by treatment with trifluoroacetic acid. The free hydroxyl was glycosylated at low temperature with anomeric sulfoxide **131** in the presence of trifluoromethanesulfonic anhydride, buffered with 2,6-di-*tert*-butyl-4-methylpyridine, to produce disaccharide **132** and, after reiteration, trisaccharide **133** having β(1,6) linkages. Trisaccharide **134** was obtained in 52% overall yield from **130** after protecting group manipulations and cleavage from the support with mercuric trifluoroacetate.

Kahne and co-workers also explored the possibility of preparing partial structures of Lewis[x] and Lewis[a] blood group antigens on their solid support. In an approach towards the Le[x] blood group antigen, the polymer-bound p-hydroxythiophenyl glycoside of 2-azido-2-deoxy-4,6-benzylidene-D-glucose **135** was reacted with the

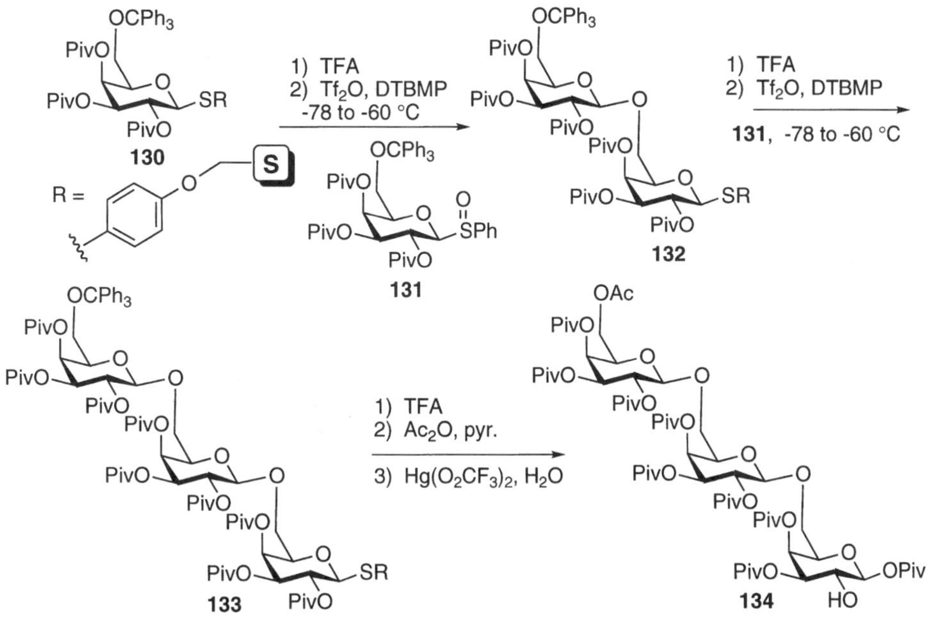

Figure 20 Trisaccharide synthesis on a solid support using anomeric sulfoxide donors. **S**, Merrifield's resin. (From Ref. 119.)

anomeric sulfoxide **136** at low temperature (Figure 21). The glycosylation was repeated to increase the yields of polymer-bound **137**. Disaccharide **138**, exclusively α-linked, was recovered from the support in the usual way in 67% yield. Repetitive glycosylation reactions gave the desired α-linked product in a yield superior to what is usually obtained in similar solution-based coupling reactions. The Le^a fragment was similarly assembled by coupling the identical acceptor **135** with the galactosyl donor **139**. The disaccharide **141** was isolated in 64% overall yield after treatment of **140** with the mercuric salt. The glycosidic linkage of the product was exclusively β due to the presence of the participating pivaloyl group at C-2.

Attaching the acceptor to the support has the obvious advantage of allowing for multiple coupling reactions to increase the yield of a difficult linkage. On the other hand, it may be difficult to recover the excess glycosyl donor because it may be destroyed in the process of activation. The low temperature required for coupling also increases the technical complexity of automating the procedure. When the acceptor is tied to the support, a method must be devised to remove selectively the protecting group of the polymer-bound glycoside to expose a particular hydroxyl group before the next coupling. If larger oligosaccharides are desired, it may also become necessary to include a capping step after each coupling in order to prevent interior deletions leading to the formation of a complex mixture of oligosaccharides.

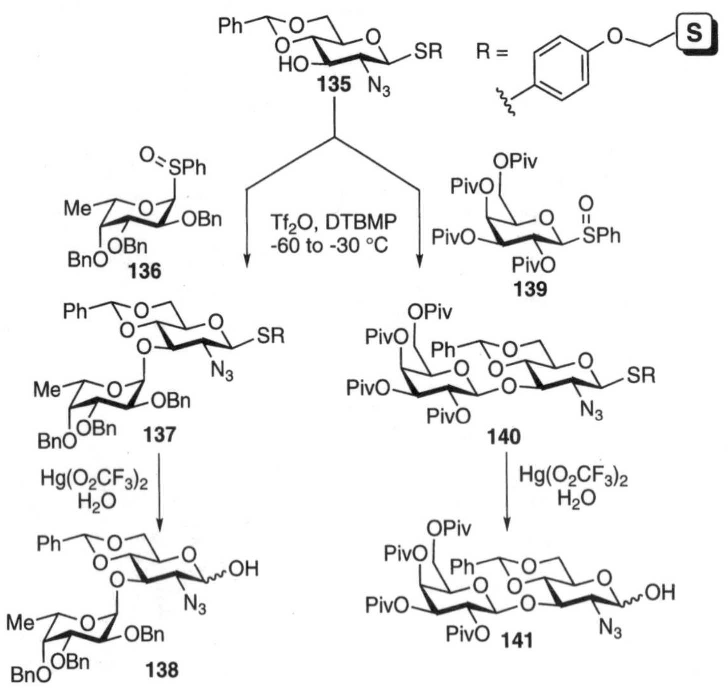

Figure 21 Disaccharide syntheses on a solid support using sulfoxide donors. **S**, Merrifield's resin. (From Ref. 119.)

Attachment to a Soluble Polymeric Support via a Succinate Diester

Krepinsky et al. [90] have used a soluble polyethylene glycol mono methyl ether polymer for the synthesis of di- and trisaccharides. Other researchers had previously exploited the more favorable kinetics of glycosidic couplings on a soluble polymeric support [34,40,49]. Krepinsky and co-workers attached their substrate to a polyethylene glycol (PEG) polymer via a succinate diester linkage [120]. Koenigs–Knorr glycosylation of polymer-bound diol **142** with anomeric bromide **143** resulted in formation of a mixture of disaccharide **144** and trisaccharide **145** in an undisclosed ratio (Figure 22). The presence of a single methyl group on the polymer served as an internal standard that allowed for monitoring of the reaction progress by NMR spectroscopy. After coupling, the polymer was precipitated with diethyl ether or *tert*-butyl methyl ether and recrystallized from ethanol. The products were released from the polymeric support by methanolysis or hydrazinolysis. The coupling reactions proceeded in approximately 85%–95% yield per coupling step.

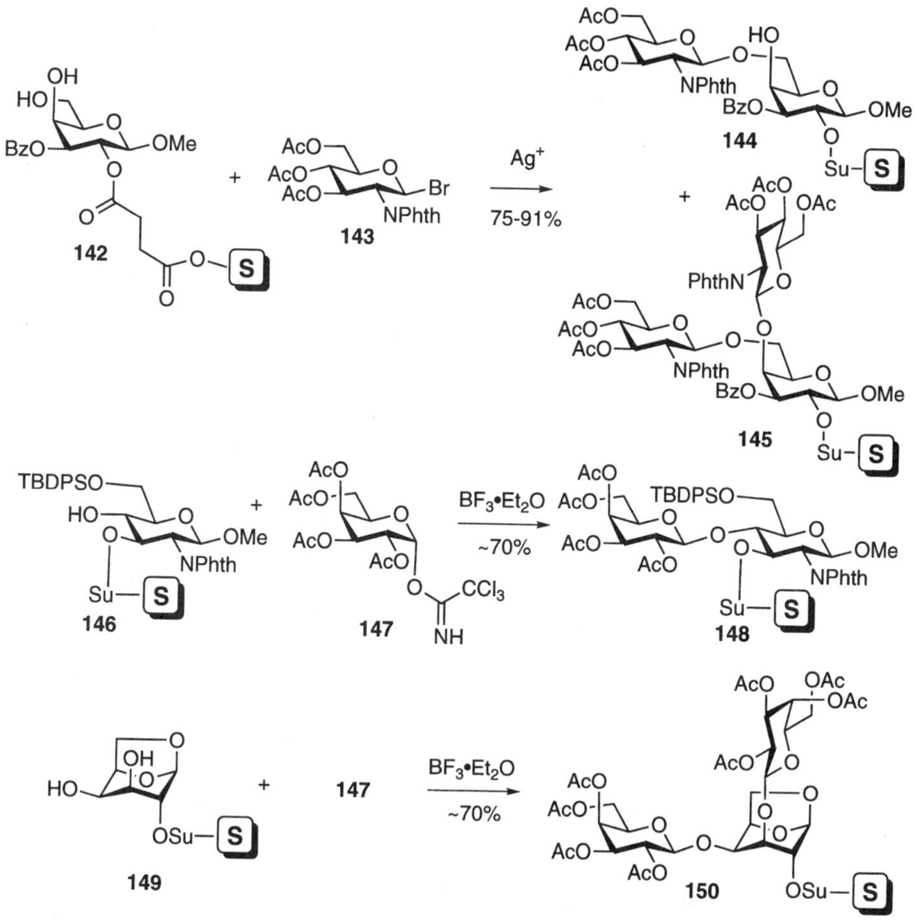

Figure 22 Synthesis of oligosaccharides via glycosylation on a soluble polymeric support. **S**, poly(ethylene glycol)monomethyl ether. (From Ref. 90.)

Krepinsky and co-workers also applied the Schmidt glycosylation protocol for the assembly of β-linked disaccharide **148** from polymer-bound glycoside **146** and trichloroacetamidate **147** bearing a participating acetate group at C-2. An identical donor was used to bis-galactosylate 1,6-anhydro acceptor **149**, thereby yielding trisaccharide **150**. Thus, Krepinsky et al. succeeded in a synthesis of branched oligosaccharide structures on a polymeric support.

van Boom et al. [121] also used the monomethyl ether polyethylene glycol/ succinoyl support to assemble a heptasaccharide, which has phytoalexin elicitor activity. The α-methylglycoside **151** (Figure 23), attached at the C-4 position to the polymer, was glycosylated with disaccharide **152** at the exposed C-6 hydroxyl using N-iodosuccinimide-triflic acid-promoted glycosylation of ethylthioglycosides [122].

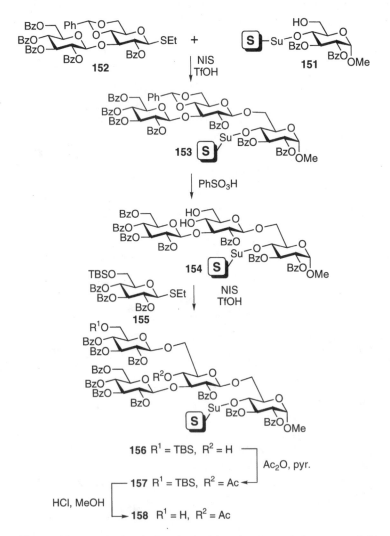

Figure 23 Synthesis of oligosaccharides via glycosylation on a soluble polymeric support. S, poly(ethylene glycol)monomethyl ether. (From Ref. 121.)

The resulting β-glycoside **153** was treated with benzenesulfonic acid to remove the benzylidene group, and the free primary hydroxyl of polymer-bound diol **154** was selectively coupled with monosaccharide **155** in the presence of the iodonium promoter to give **156**. The secondary hydroxyl of tetrasaccharide **156** was capped with an acetate group and the silyl ether of the resulting, fully protected product **157** was removed to produce oligosaccharide **158**, which was ready for further elongation by repetition of the ethylthioglycosylation procedure with donors **152** and **159** (Figure 24). The heptasaccharide **160** was deprotected and cleaved from the polymer support by the Zemplen solvolysis, and the resultant oligosaccharide **161** (R = H) was per-acetylated to simplify purification and characterization. The acetate functionalities of **162** (R = Ac) were easily removed by the Zemplen reaction, which produced the desired heptaglucoside **161** (R = H) in 18% overall yield.

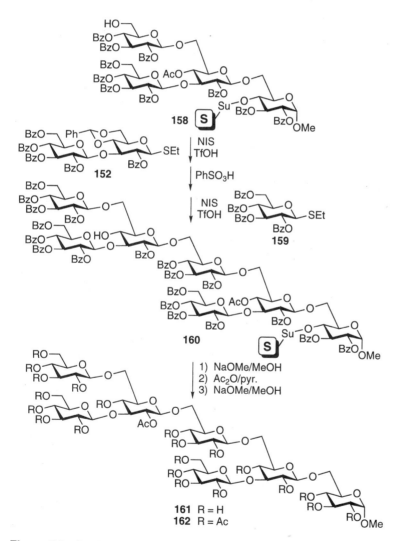

Figure 24 Synthesis of oligosaccharides via glycosylation on a soluble polymeric support. **S**, poly(ethylene glycol)monomethyl ether. (From Ref. 121.)

The use of a polymeric soluble support is advantageous for large-scale preparation of oligosaccharides because it simplifies the necessary purification processes while retaining reaction kinetics similar to those normally obtained in solution synthesis. Thus, precipitation and crystallization can be used as long as the oligosaccharide is small relative to the polymer. Conversely, the chemistry of the glycosidic coupling is limited because of the ability of the polyether backbone to complex metal ions. Other drawbacks of the soluble polymer are the material loss associated with the precipitation and crystallization steps. Furthermore, it would be difficult to adapt this procedure for an automated synthesizer.

IV. REITERATIVE FORMATION OF THE PEPTIDE BONDS OF GLYCOPEPTIDES ON A POLYMERIC SUPPORT

A. Solid-Phase Synthesis of Glycopeptides

The total synthesis of glycopeptides on a solid support is more complex than the synthesis of peptides or oligosaccharides, since this process must accommodate the protecting-group requirements of both peptides and carbohydrates [21,123,124]. Since the linkages of glycopeptides are sensitive to both acids and bases, all reaction conditions must be particularly mild [125]. With a few exceptions, no new glycosidic linkages have been formed on the support [51,52,126–128].

O-Linked Glycopeptide Synthesis on a Solid Support

The application of the solid-phase peptide synthesis (SPPS) to the synthesis of glycopeptides began in the early 80s when Lavielle and co-workers assembled a glycosylated analog of the tetradecapeptide neurohormone somatostatin **167** on a Merrifield resin (Figure 25) using standard Boc-based strategy [129]. The preformed glucopyranosylserine unit **163** was coupled to the polymer-supported cysteine **164** in the presence of dicyclohexylcarbodiimide (DCC) and hydroxybenzotriazole (HOBt). The other residues were sequentially installed, and the glycopeptide **166** was obtained from the resin **165** after treatment with hydrofluoric acid. The disulfide bridge was installed using the potassium ferricyanide oxidation method, and the somatostatin derivative **167** was obtained after deacetylation with saturated ammonia in methanol. Though these workers obtained the desired compound, the use of hydrofluoric acid for the cleavage of the glycopeptide from the support would probably be too harsh for a complex array of glycosidic linkages, which might suffer cleavage or anomerization.

    Kunz and co-workers introduced the use of a polymer-bound allyl alcohol linker for the solid-phase synthesis of peptides and glycopeptides [7,17,130]. This resin was based on the lability of the allyl ester to palladium(0)-catalyzed nucleophilic displacement, resulting in release of the substrate from the support. The reaction conditions employed for nucleophilic displacement were very mild, at near neutral pH, which is highly desirable for the synthesis of sensitive glycopeptides, and orthogonal with other protecting groups used in the major SPPS strategies (Cbz, Boc, Fmoc) [131]. For example, the glycosylated tripeptide **168** (Figure 26) was assembled on Kunz's HYCRAM resin (aminoacyl hydroxycrotonylaminomethyl

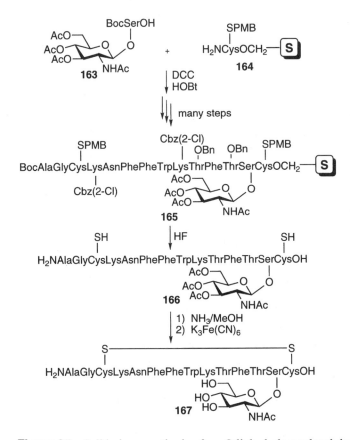

Figure 25 Solid-phase synthesis of an *O*-linked glycosylated derivative of somatostatin. **S**, polystyrene. (From Ref. 129.)

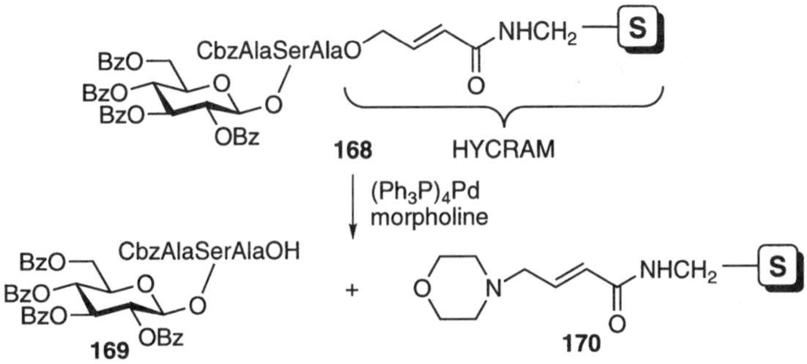

Figure 26 Solid-phase synthesis of an *O*-linked glycopeptide. **S**, polystyrene. (From Ref. 130.)

polystyrene). The glycopeptide **169** was smoothly released from the polymer, and the morpholine derivative **170** was formed upon exposure to Pd(0) and morpholine. Kunz also applied the use of the HYCRAM resin to the synthesis of N-linked glycopeptides (*vide infra*).

Most syntheses of O-linked glycopeptides used the Fmoc protecting group at the N-termini of glycosyl-serine/threonine moieties [16,132–151]. This protecting group was removed upon exposure to mild base—for example, morpholine—which does not promote the β-elimination of glycosides from serine and threonine residues.

In the context of O-linked glycopeptide syntheses, the Fmoc protecting group was most often used in conjunction with polymer linkers that were sensitive to mild acids. The Wang-type linker (**171**) (Figure 27) [126,132,133,136–138,141, 142,148,151] and the PAL linker (**172**) [139,140,143,144,146] were generally cleaved with 95% TFA. The RINK linker (**173**) [145,147,149,150,152,153] was cleaved with 30%–85% TFA, while super acid-sensitive resin SASRIN (**174**) [134,135] was labile with 1% TFA in dichloromethane. These acidic conditions were compatible with most mono- and disaccharide-containing glycopeptides on solid supports. In most cases the glycosidic linkages are usually further stabilized toward acids by peracetylation or perbenzoylation of the sugar hydroxyl groups [16,126,129,130,132–141,143–147,149–152,154]. This was especially true if the desired glycopeptide contained fucosyl moieties [155]. Nevertheless, Meldal and coworkers assembled a trisaccharide-glycopeptide on the solid support, with the sugar moieties free of protecting groups [148].

The use of the SASRIN linker had the advantage that following acidolysis from the support, the glycopeptide retained most protecting groups, thereby allowing for easy purification and further modification of the glycopeptide in solution [134,135].

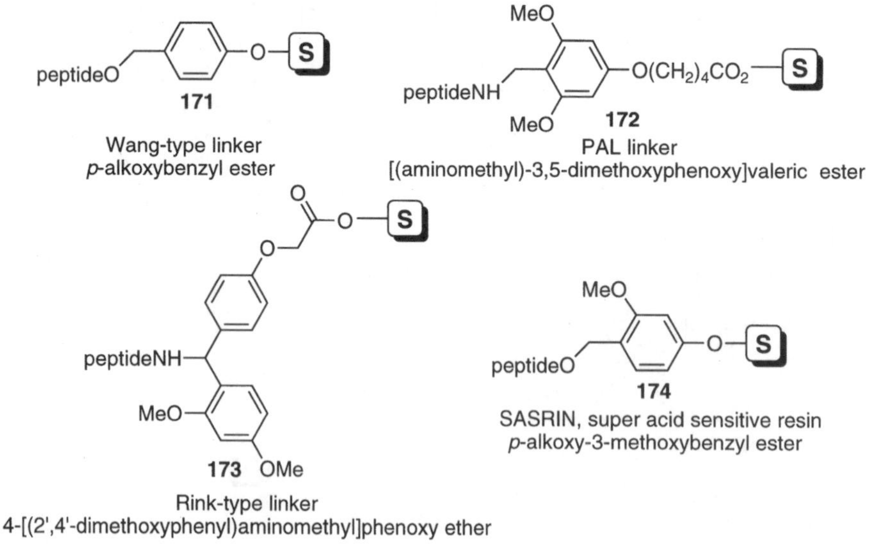

Figure 27 Acid-sensitive supports for solid-phase synthesis of glycopeptides. S, polystyrene, polydimethylacrylamide, or poly(ethyleneglycol)-polydimethylacrylamide copolymer.

Glycopeptides have been assembled manually as well as by using automated synthesizers on polystyrene [129,132–135,145,147,154,156], polydimethylacrylamide [126,136–138,141–144,146,148,151], poly(ethyleneglycol)-polydimethylacrylamide copolymer [149,150], or polystyrene-poly(ethyleneglycol) graft copolymer solid supports [16,152]. In a few syntheses, more than one glycosylated residue has been installed to mimic natural substrates [138,142–144,146,150].

Direct glycosylation of the serine moiety of a polymer-bound peptide was achieved by Hollósi and co-workers [126]. They added 17–25 equivalents of a per-acetylated 2-*N*-acetylglucose-oxazolinium trifluoromethanesulfonate complex to a polymer-bound peptide that possessed a serine moiety with a free hydroxyl. The sugar incorporation was relatively low (23% by HPLC) even though a large excess of donor was used.

A comparative study of the synthesis of *O*-linked glycopeptides using direct glycosylation versus the preformed building block approach was accomplished by Andrews and Seale [128]. They reported that the synthesis of mannosylated glycopeptides using preformed glycosylated serines or threonines was more efficient than the direct glycosylation of the free hydroxyl group of polymer-bound serine and threonine-containing peptides using the Koenig–Knorr glycosylation reaction.

In addition to Kunz et al., others have used an acid-stable linker. Trzupek et al. used the Kaiser–DeGrado oxime resin to give glycopeptides that were released from the solid support in fully protected form [154]. The acetate salt of perbenzoylated glucose **175** (Figure 28), attached to a serine residue protected with a benzyl group at the C-terminus, was used to displace the polymer-bound peptide **176** from their resin. This procedure generated C-terminal glycopeptide **177** in an original way.

In most cases the desired glycopeptides were fully deblocked after cleavage from the resin. Alternatively, protecting groups of the sugar moieties have been removed while the glycopeptides were still attached to the polymer [139,140,145,146]. Paulsen and co-workers successfully converted an azide group at C-2 to an acetamide directly on the solid support by treatment with thioacetic acid [146].

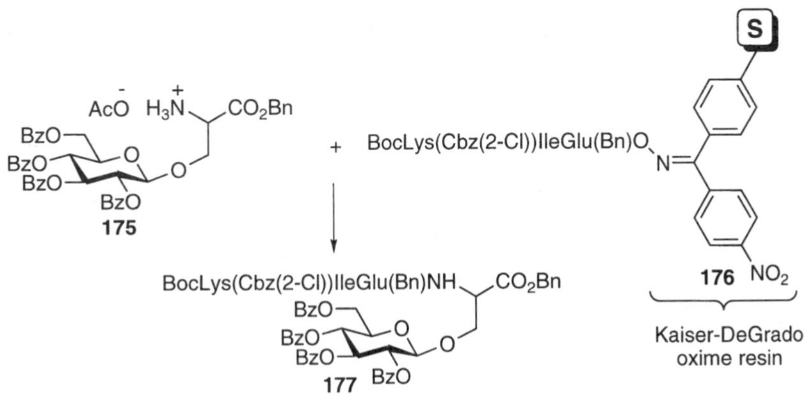

Figure 28 Synthesis of *O*-linked glycopeptide using the Kaiser–DeGrado oxime resin. **S**, polystyrene. (From Ref. 154.)

N-*Linked Glycopeptide Synthesis on a Solid Support*

Kunz and co-workers provided the major breakthrough on the solid-phase synthesis of *N*-linked glycopeptides. These researchers assembled glycopeptide **180** using pre-formed glycosylated asparagine **178** and Pd(0) removable allyl functionalized resin (Figure 29) [156]. Glycopeptide **179** was assembled on the HYCRAM resin using Boc-SPPS methodology. Glycopeptide **180** was liberated by palladium(0)-catalyzed nucleophilic displacement of the HYCRAM resin with morpholine. The usefulness of this method was demonstrated by the automated large-scale synthesis (0.5 g) of the fully deblocked *N*-linked disaccharide octapeptide **184** (Figure 30) [7,157]. The preformed lactosamine-asparagine unit **182** was coupled with a dipeptide attached to an improved version of the HYCRAM linker **181**. After extension of the peptide, the glycopeptide **183** was released upon reaction with Pd(0). The remaining protecting groups were removed to give glycoconjugate **184**, a partial structure of the HIV glycoprotein gp 120, which was involved in the initial steps of HIV virus binding to lymphocytes.

There are considerable similarities between the methods described for the solid-phase synthesis of *N*-linked glycopeptides. An automated synthesizer was used in most cases to assemble Fmoc-protected, preformed mono- and disaccharide-asparagine units on an acid-sensitive polyacrylamide support, and the final glyco-peptides were fully deprotected after removal from the resin [151,158–161]. The sugar-asparagine units were coupled to the preceding residue as the pentafluoro-phenyl (Pfp) ester or with DIC-HOBt [160]. Hepta- [160], octa- [161], undeca- [159], dodeca- [158], and hexadecaglycopeptides [151] have been synthesized for use in various biological and structural studies.

Albericio and co-workers [127] prepared *N*-linked glycopeptides by directly glycosylating the free carboxylic group of an aspartate or a glutamate residue of polymer-bound peptides with 1-amino-2-*N*-acetylglucosamine. In this manner, they obtained monoglycosylated tri- and tetrapeptides. The major side product observed in the case of the asparagine-containing glycopeptide was the formation of a small amount of aspartimide.

Recently, Otvos and co-workers assembled *N*-linked glycopeptides on a solid support using Fmoc-asparagine glycosylated with malto-oligosaccharides having two to seven sugar units [162]. The amino sugars were prepared from commercially

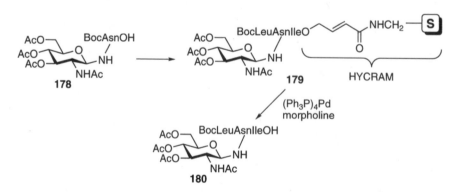

Figure 29   Solid-phase synthesis of *N*-linked glycopeptide. **S**, polystyrene. (From Ref. 156.)

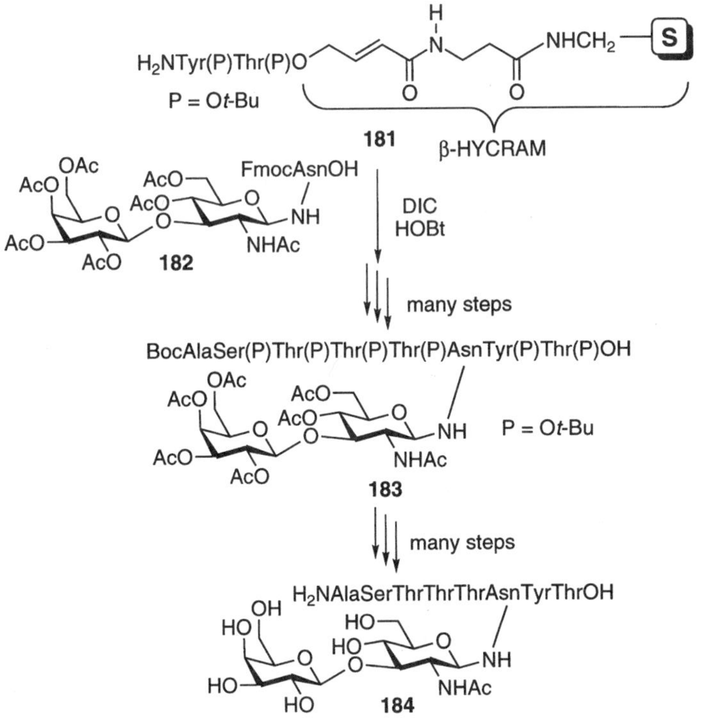

Figure 30 Solid-phase synthesis of *N*-linked glycopeptide T fragment. S, polystyrene. (From Ref. 157.)

available malto-oligosaccharides that were reacted, in the unprotected form, with ammonium bicarbonate. The resulting amino sugars were coupled with Fmoc-aspartate-O*t*Bu with the γ-carboxyl activated as the Pfp ester, and, after removal of the *tert*-butyl ester, the glycosylated asparagine residues were coupled at the terminal position of a polymer-bound peptide. The coupling yields between the aminosugars and the activated aspartates decreased as the number of sugar units increased. The same trend was also observed for the coupling of the glycosylated asparagine residues and the peptide. The stability of the glycopeptides was evaluated in basic and acidic conditions, and it was found that the glycopeptides were stable to 20% piperidine in DMF, slowly decomposed in 100% TFA, and were fairly labile with 5% water in TFA. The glycopeptide structures were studied by circular dichroism and by NMR spectroscopy. It was shown that the presence of sugars at the terminal position did not interfere with the α-helical structure of the peptide but had a size-dependent interference with disulfide bridge formation on proximal cysteine residues.

B. Dendrimeric Glycopeptide Synthesis

Roy and co-workers used the solid-phase approach to prepare inhibitors of influenza A virus haemagglutinins (HAs). HAs are the viral proteins that recognize the α-sialosides of the host cell surface and attach the virus to initiate the infection process [163]. An efficient inhibitor of the HAs must possess a large number of exposed α-sialosides and inhibit the virus sialidase enzymes that cleave the sialyl moieties. An inhibitor was prepared by attaching the α-thiosialoside **185** (Figure 31) to the dendrimeric polypeptide **186** via a thio-anomeric linkage to produce the hexadeca-α-thiosialoside **187** (the di-, tetra-, and octa-α-thiosialosides were also prepared). The dendritic skeleton should mimic the antenarry structure of the membrane oligosaccharides, and the α-thiosialoside should inhibit sialidase enzymes [164]. The structures presented to the virus were sialidase-resistant sugars in a "clustered" geometry that is known to enhance greatly the binding with oligosaccharide recognition molecules [164]. To synthesize the inhibitor, dendrimer **186** was assembled on the solid support by coupling N^α-N^ϵ-di-Fmoc-protected lysine benzotriazolyl esters with the Wang resin via a β-alanine spacer. The Fmoc groups were removed, exposing two free amine moieties that were coupled with two more di-Fmoc lysine moieties. The

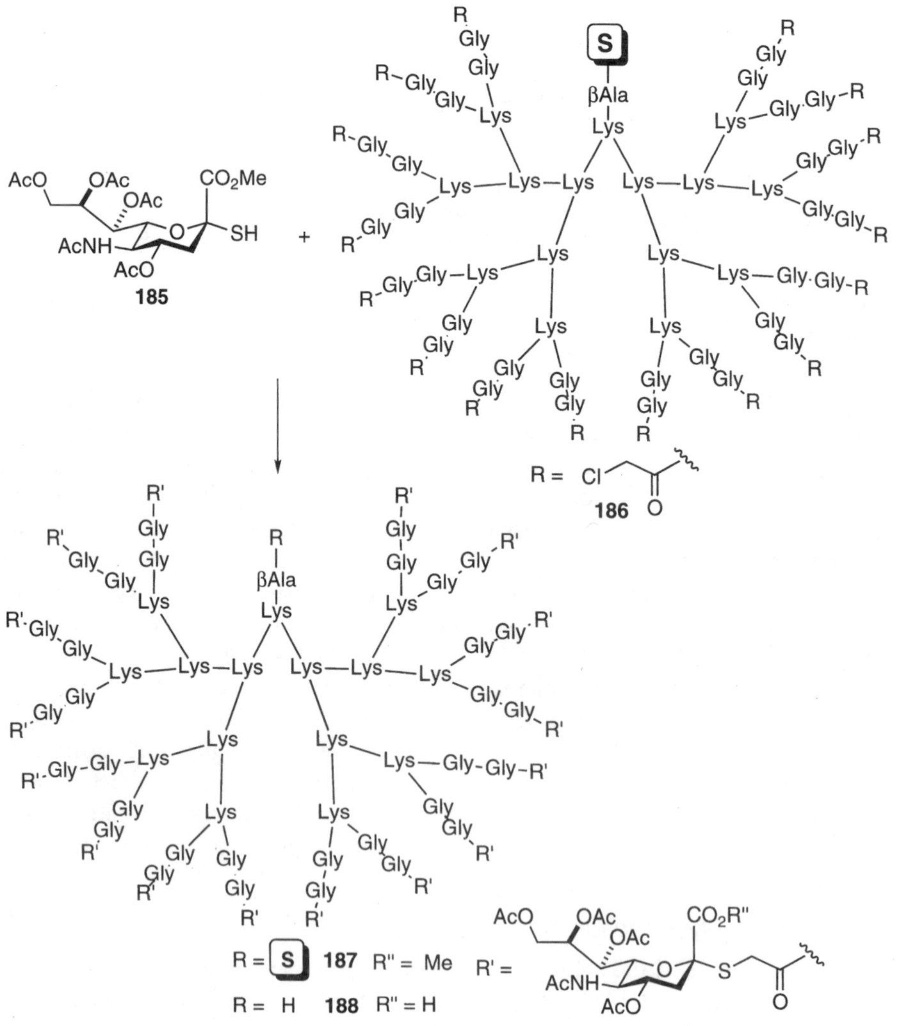

Figure 31 Synthesis of thiosialoside dendrimers. S, Wang resin. (From Ref. 163.)

deprotection and coupling steps were repeated twice, and the Fmoc groups were removed to give a dendrimer with 16 free amino groups. The amines were coupled with the benzotriazolyl ester of chloroacetylglycineglycine to yield the hexachloroacetyl derivative **186**. After coupling with the α-thiosialoside **185**, the dendrimer **188** was removed from the support, with 95% TFA in good yield, and was fully deblocked. The final product was shown to be an excellent inhibitor of the attachment of the influenza-A virus.

The capacity to synthesize glycopeptides on solid supports using preformed glycoside-amino acids units has improved over the past 13 years. There are now reliable methods for manual and automated syntheses of biologically relevant glycopeptides. However, considerable problems remain to be solved. Areas in which improvements are required include optimizing the method for glycoconjugates containing more than three sugars and increasing coupling yields for sterically bulky saccharide-amino acid moieties [148]. The years ahead may witness the interfacing of glycopeptide syntheses on solid supports with enzymatic techniques to modify the oligosaccharide moieties while attached to the polymer.

V. MISCELLANEOUS SOLID-PHASE OLIGOSACCHARIDE SYNTHESES

Solid-phase techniques for the synthesis of oligosaccharides were used to assemble bacterial cell wall fragments and capsular polysaccharides, with the goal of developing vaccines against infections. The solid-phase approach is well suited for the synthesis of such fragments, since they are assembled from repeating identical monomers.

An example can be found in the work of van Boom and co-workers, who prepared a fragment of the cell wall teichoic acid of *Bacillus licheniformis* ATCC 9945 [165]. The polymer-bound glycerol derivative **190**, attached to an amino-functionalized porous glass via a succinamide linker, was coupled with phosphite **189** (Figure 32). Disaccharide **191** was obtained after capping of uncoupled hydroxyls, oxidation with iodine, and removal of the dimethoxytrityl group. This sequence of reactions was repeated to generate pentasaccharide **192**. This compound was released from the support, with concomitant removal of the acetates and β-cyanoethyl groups. The desired cell wall fragment **193** was obtained upon hydrogenolysis of the benzyloxymethyl- and dimethoxytrityl-protecting groups in acceptable yields. A similar approach was applied to the syntheses of the capsular polysaccharides of *Haemophilus influenzae* type b [166,167] and *Haemophilus (Actinobacillus) Pleuropneumoniae* serotype 2 [168,169].

ACKNOWLEDGMENTS

We would like to thank Drs. Jana Pika, John T. Randolph, Mark T. Bilodeau, and Xenia Beebe for helpful comments on the manuscript. Financial support from the National Institutes of Health (#NIH-CA-28824) and from the Natural Sciences and Engineering Research Council of Canada (Postdoctoral Fellowship to J.Y.R.) is gratefully acknowledged.

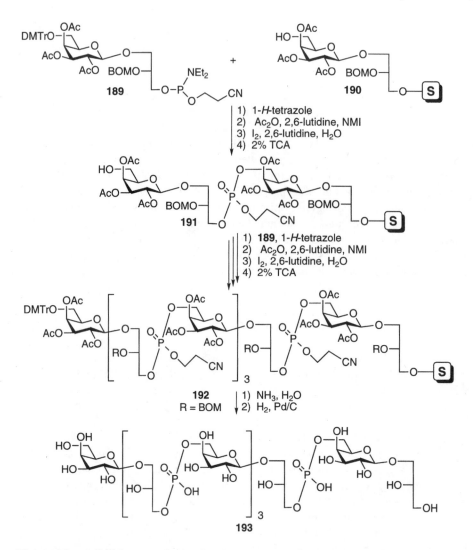

Figure 32 Solid-phase synthesis of a fragment of the cell wall teichoic acid of *Bacillus licheniformis* ATCC 9945. **S**, CO(CH$_2$)$_2$CONH-functionalized microporous glass. (From Ref. 165.)

ABBREVIATIONS

All	allyl
Anthr	9-anthracenyl
Bn	benzyl
Boc	*tert*-butyloxycarbonyl
BOM	benzyloxymethyl
BTCEAD	bis(trichloroethyl)azidocarboxylate
Bz	benzoyl

CAN	ceric ammonium nitrate
Cbz	benzyloxycarbonyl
Cbz(2-Cl)	2-chlorobenzyloxycarbonyl
CMP	cytidine-5′-monophosphate
coll	2,4,6-trimethylpyridine
DCC	dicyclohexylcarbodiimide
DDQ	2,3-dichloro-5,6-dicyano-1,4-benzoquinone
DIC	diisopropylcarbodiimide
DMAP	dimethylaminopyridine
DMBA	dimethylbarbituric acid
DMDO	dimethyldioxirane
DMTr	4,4′-dimethoxytrityl
DTBMP	2,6-di(*tert*-butyl)-4-methylpyridine
DTBP	2,6-di(*tert*-butyl)pyridine
E	electrophile
Fmoc	fluorenyl-9-methoxycarbonyl
Fuc	fucose
Gal	galactose
GDP	guanosine 5′-diphosphate
GlcNAc	2-*N*-acetyl-2-deoxy-glucosamine
HOBt	1-hydroxybenzotriazole
HSA	human serum albumin
HYCRAM	aminoacyl hydroxycrotonylaminomethyl polystyrene
IIDQ	2-isobutoxy-1-isobutoxycarbonyl-1,2-dihydroquinoline
L. A.	Lewis acid
Lev	levulinoyl ($CH_3COCH_2CH_2CO$)
NeuAc	sialic acid
NIS	*N*-iodosuccinimide
NMI	*N*-methylimidazole
NPhth	phthalimido
Nu	nucleophile
PAL	[(aminomethyl)-3,5-dimethoxyphenoxy]valeric ester
Pfp	pentafluorophenyl
Piv	pivaloyl
PMB	*p*-methoxybenzyl
PNP	*p*-nitrophenyl
pyr.	pyridine
SASRIN	superacid sensitive resin
SES	2-trimethylsilylethylsulfonyl
SPPS	solid-phase peptide synthesis
Su	succinate(-$COCH_2CH_2COO$-)
TBDPS	*tert*-butyldiphenylsilyl
TBS	*tert*-butyldimethylsilyl
TCA	trichloroacetic acid
Tf	trifluoromethanesulfonyl
TFA	trifluoroacetic acid
TIPS	triisopropylsilyl
TMEDA	tetramethylethylenediamine

Troc   2,2,2-trichloroethoxycarbonyl
UDP   uridine 5′-diphosphate
[Ox.]   oxidation
[Red.]   reduction

REFERENCES

1. H. J. Allen and E. C. Kisailus, eds., *Glycoconjugates: Composition, Structure, and Function*, Dekker, New York, 1992.
2. Y. C. Lee and R. T. Lee, eds., *Neoglycoconjugates: Preparation and Applications*, Academic Press, London, 1994.
3. A. Kobata, Glycobiology: An expanding research area in carbohydrate chemistry, *Acc. Chem. Res. 26*: 319 (1993).
4. G. Opdenakker, P. M. Rudd, C. P. Ponting, and R. A. Dwek, Concepts and principles of glycobiology, *FASEB J. 7*: 1330 (1993).
5. A. Varki, Selectins ligands, *Proc. Natl. Acad. Sci. USA 91*: 7390 (1994).
6. D. E. Levy, P. C. Tang, and J. H. Musser, Cell adhesion and carbohydrates, *Annual Reports in Medicinal Chemistry*, (W. K. Hagmann, ed.), Academic Press, San Diego, 1994, p. 215.
7. H. Kunz, Glycopeptides of biological interest: A challenge for chemical synthesis, *Pure & Appl. Chem. 65*: 1223 (1993).
8. R. A. Dwek, C. J. Edge, D. J. Harvey, M. R. Wormald, and R. B. Parekh, Analysis of glycoprotein-associated oligosaccharides, *Annu. Rev. Biochem. 62*: 65 (1993).
9. H. van Halbeek, NMR developments in structural studies of carbohydrates and their complexes, *Curr. Opin. Struct. Biol. 4*: 697 (1994).
10. K. Toshima and K. Tatsuta, Recent progress in *O*-glycosylation methods and its application to natural products synthesis, *Chem. Rev. 93*: 1503 (1993).
11. R. R. Schmidt, Synthesis of glycosides, *Comprehensive Organic Synthesis: Selectivity, Strategy and Efficiency in Modern Organic Chemistry*, (B. M. Trost and I. Fleming, eds.), Pergamon Press, Oxford, Eng., 1991, p. 33.
12. R. R. Schmidt, New methods for the synthesis of glycosides and oligosaccharides— Are there alternatives to the Koenigs–Knorr method? *Angew. Chem. Int. Ed. Engl. 25*: 212 (1986).
13. R. R. Schmidt, *Carbohydrates, synthetic methods and applications in medicinal chemistry*, VCH Publishers, Cambridge, Eng., 1992, Chap. 4.
14. E. J. Toone, E. S. Simon, M. D. Bednarski, and G. M. Whitesides, Tetrahedron Report Number 259: Enzyme-catalyzed synthesis of carbohydrates, *Tetrahedron 45*: 5365 (1989).
15. G. C. Look, C. H. Fotsch, and C.-H. Wong, Enzyme-catalyzed organic synthesis: Practical routes to aza sugars and their analogs for use as glycoprocessing inhibitors, *Acc. Chem. Res. 26*: 182 (1993).
16. A. H. Andreotti and D. Kahne, Effects of glycosylation on peptide backbone conformation, *J. Am. Chem. Soc. 115*: 3352 (1993).
17. H. Kunz, Synthesis of biologically interesting glycopeptides a problem of chemical selectivity, *Antibiotics and Antiviral Compounds: Chemical Synthesis and Modification*, (K. Krohn, H. Kirst, and H. Maas, eds.), VCH Publishers, Weinheim, Ger., 1993, p. 251.
18. H. Waldmann and D. Sebastian, Enzymatic protecting group techniques, *Chem. Rev. 94*: 911 (1994).
19. J. Banoub, P. Boullanger, and D. Lafont, Synthesis of oligosaccharides of 2-amino-2-deoxy sugars, *Chem. Rev. 92*: 1167 (1992).

20. R. L. Halcomb and C.-H. Wong, Synthesis of oligosaccharides, glycopeptides, and glycolipids, *Curr. Opin. Struct. Biol. 3*: 694 (1993).

21. M. Meldal, Glycopeptides synthesis, *Neoglycoconjugates: Preparation and Applications*, (Y. C. Lee and R. T. Lee, eds.), Academic Press, London, 1994, p. 145.

22. G. M. Blackburn and M. J. Gait, Synthesis of nucleosides, *Nucleic Acids in Chemistry and Biology*, (G. M. Blackburn and M. J. Gait, eds.), IRL Press at Oxford University Press, Oxford, Eng., 1990, p. 73.

23. J. Jones, *The chemical synthesis of peptides*, Clarendon Press, Oxford, Eng., 1991.

24. E. Atherton and R. C. Sheppard, *Solid phase peptide synthesis: A practical approach*, Practical Approach Series, (D. Rickwood and B. D. Hames, eds.), IRL Press at Oxford University Press, Oxford, Eng., 1989, p. 203.

25. G. Barany, N. Kneib-Cordonier, and D. G. Mullen, Solid-phase peptide synthesis: A silver anniversary report, *Int. J. Pept. Protein Res. 30*: 705 (1987).

26. M. Bodansky, Principles of peptide synthesis, *Reactivity and Structure Concepts in Organic Chemistry*, (K. Hafner, et al., eds.), Springer-Verlag, Vol. 16. New York, 1984, p. 307.

27. G. Barany and R. B. Merrifield, Special methods in peptide synthesis, Part A: Solid phase peptide synthesis, *The Peptides: Analysis, Synthesis, Biology*, Vol. 2, (E. Gross and J. Meienhofer, eds.), Academic Press, New York, 1980, p. 1.

28. C. Birr, *Aspects of the Merrifield Synthesis*, Vol. 8, Reactivity and Structure, Springer-Verlag, Berlin, 1978.

29. J. M. Fréchet, Polymer-supported synthesis of oligosaccharides, *Polymer-Supported Reactions in Organic Synthesis*, (P. Hodge and D. C. Sherrington, eds.), Wiley, New York, 1980, p. 407.

30. J. M. Fréchet and C. Schuerch, Solid-phase synthesis of oligosaccharides. I. Preparation of the solid support. Poly[*p*-(1-propen-3-ol-1-yl)styrene], *J. Am. Chem. Soc. 93*: 492 (1971).

31. G. Excoffier, D. Gagnaire, J. P. Utille, and M. Vignon, Solid-phase synthesis of oligosaccharides. II. Synthesis of 2-acetamido-6-*O*-(2-acetamido-2-deoxy-β-D-glucopyranosyl)-2-deoxy-D-glucose, *Tetrahedron Lett. 13*: 5065 (1972).

32. G. Excoffier, D. Gagnaire, J.-P. Utille, and M. Vignon, Synthèse d'oligosaccharides sur polymère support—IV, *Tetrahedron 31*: 549 (1975).

33. U. Zehavi and A. Patchornik, Oligosaccharide synthesis on a light-sensitive solid support. I. The polymer and synthesis of isomaltose (6-*O*-α-D-glucopyranosyl-D-glucose), *J. Am. Chem. Soc. 95*: 5673 (1973).

34. R. D. Guthrie, A. D. Jenkins, and G. A. F. Roberts, Synthesis of oligosaccharides on polymer supports. Part II. Synthesis of β-D-gentiobiose derivatives on soluble support copolymers of styrene and 6-*O*-(*p*-vinylbenzoyl) or 6-*O*-(*p*-vinylphenylsulphonyl) derivatives of D-glucopyranose, *J. Chem. Soc. Perkin Trans. 1*: 2414 (1973).

35. R. D. Guthrie, A. D. Jenkins, and J. Stehlícek, Synthesis of oligosaccharides on polymer supports. Part I. 6-*O*-(*p*-vinylbenzoyl) derivatives of glucopyranose and their copolymer with styrene, *J. Chem. Soc. C*: 2690 (1971).

36. R. Eby and C. Schuerch, Solid-phase synthesis of oligosaccharides. V. Preparation of an inorganic support, *Carbohydr. Res. 39*: 151 (1975).

37. S. A. Holick, *Systematic, Sequential Synthesis of Oligosaccharides. Investigation of Porous Glass as a Solid Support*, Ph. D. Thesis, 1974, University of Wisconsin, (Madison, Wis.).

38. G. Excoffier, D. Y. Gagnaire, and M. R. Vignon, Le groupe trichloroacétyl comme substituant temporaire; Synthèse du gentiotétraose, *Carbohydr. Res. 46*: 201 (1976).

39. G. Excoffier, D. Y. Gagnaire, and M. R. Vignon, Synthèse du nigérose et du laminarabiose sur polymère support, *Carbohydr. Res. 51*: 280 (1976).

40. S.-H. L. Chiu and L. Anderson, Oligosaccharide synthesis by the thioglycoside scheme on soluble and insoluble polystyrene supports, *Carbohydr. Res. 50*: 227 (1976).

41. K. Smith, ed., *Solid Supports and Catalysts in Organic Synthesis*, Ellis Horwood PTR Prentice Hall Organic Chemistry Series, (E. Horwood and J. Mellor, series eds.), Ellis Horwood, Chichester, Eng., 1992.

42. J. M. J. Fréchet, Tetrahedron Report Number 103: Synthesis and applications of organic polymers as supports and protecting groups, *Tetrahedron 37*: 663 (1981).

43. G. F. Herrmann, P. Wang, G.-J. Shen, and C.-H. Wong, Recombinant whole cells as catalyst for the enzymatic synthesis of oligosaccharides and glycopeptides, *Angew. Chem. Int. Ed. Engl. 33*: 1241 (1994).

44. For a review of the use of photoremovable protecting groups in synthesis, see: V. N. R. Pillai, Photoremovable protecting groups in organic synthesis, *Synthesis* : 1 (1980).

45. U. Zehavi, Applications of photosensitive protecting groups in carbohydrate chemistry, *Adv. Carbohydr. Chem. Biochem. 46*: 179 (1988).

46. U. Zehavi, S. Sadeh, and M. Herchman, Enzymic synthesis of oligosaccharides on a polymer support. Light-sensitive, substituted polyacrylamide beads, *Carbohydr. Res. 124*: 23 (1983).

47. U. Zehavi and M. Herchman, Probing acceptor specificity in the glycogen synthase reaction with polymer-bound oligosaccharides, *Carbohydr. Res. 151*: 371 (1986).

48. U. Zehavi and M. Herchman, Probing acceptor specificity in the glycogen synthase reaction with polymer-bond oligosaccharides, *Carbohydr. Res. 151*: 371 (1986).

49. U. Zehavi and M. Herchman, Enzymic synthesis of oligosaccharides on a polymer support, light-sensitive, water-soluble substituted poly(vinyl alcohol), *Carbohydr. Res. 128*: 160 (1984).

50. E. Bayer, Towards the chemical synthesis of proteins, *Angew. Chem. Int. Ed. Engl. 30*: 113 (1991).

51. M. Schuster, P. Wang, J. C. Paulson, and C.-H. Wong, Solid-phase chemical-enzymatic synthesis of glycopeptides and oligosaccharides, *J. Am. Chem. Soc. 116*: 1135 (1994).

52. M. Meldal, F.-I. Auzanneau, O. Hindsgaul, and M. M. Palcic, A PEGA resin for use in the solid-phase chemical-enzymatic synthesis of glycopeptides, *J. Chem. Soc. Chem. Commun.*: 1849 (1994).

53. S.-I. Nishimura, K. Matsuoka, and Y. C. Lee, Chemoenzymatic oligosaccharide synthesis on a soluble polymeric carrier, *Tetrahedron Lett. 35*: 5657 (1994).

54. S.-I. Nishimura, K. B. Lee, K. Matsuoka, and Y. C. Lee, Chemoenzymatic preparation of a glycoconjugate polymer having a sialyloligosaccharide: Neu5Acα(2→3)Galβ (1→4)GlcNAc, *Biochem. Biophys. Res. Commun. 199*: 249 (1994).

55. G. H. Veeneman, S. Notermans, R. M. J. Liskamp, G. A. van der Marel, and J. H. van Boom, Solid-phase synthesis of a naturally occurring β-(1→5)-linked D-galactofuranosyl heptamer containing the artificial linkage arm L-homoserine, *Tetrahedron Lett. 28*: 6695 (1987).

56. S. J. Danishefsky, K. F. McClure, J. T. Randolph, and R. B. Ruggeri, A strategy for the solid-phase synthesis of oligosaccharides, *Science 260*: 1307 (1993).

57. J. T. Randolph and S. J. Danishefsky, Application of the glycal assembly strategy to the synthesis of a branched oligosaccharide: The first synthesis of a complex saponin, *J. Am. Chem. Soc. 115*: 8473 (1993).

58. D. A. Griffith and S. J. Danishefsky, Sulfonamidoglycosylation of glycals. A route to oligosaccharides with 2-aminohexose subunits, *J. Am. Chem. Soc. 112*: 5811 (1990).

59. R. L. Halcomb and S. J. Danishefsky, On the direct epoxidation of glycals: Application of a reiterative strategy for the synthesis of β-linked oligosaccharides, *J. Am. Chem. Soc. 111*: 6661 (1989).

60. K. Suzuki, G. A. Sulikowski, R. W. Friesen, and S. J. Danishefsky, Application of substituent-controlled oxidative coupling of glycals in a synthesis and structural cor-

roboration of ciclamycin 0: New possibilities for the construction of hybrid anthracy-clines, *J. Am. Chem. Soc. 112*: 8895 (1990).

61. R. U. Lemieux and B. Fraser-Reid, The mechanisms of the halogenations and halo-genomethoxylations of D-glucal triacetate, D-galactal triacetate, and 3,4-dihydropyran, *Can. J. Chem. 43*: 1460 (1965).

62. R. U. Lemieux and S. Levine, Synthesis of alkyl 2-deoxy-α-D-glycopyranosides and their 2-deuterio derivatives, *Can. J. Chem. 42*: 1473 (1964).

63. J. Thiem and P. J. Ossowski, Studies of hexuronic acid ester glycals and the synthesis of 2-deoxy-β-glycoside precursors, *J. Carbohydr. Chem. 3*: 287 (1984).

64. J. Thiem, A. Prahst, and I. Lundt, Untersuchungen zur β-glycosylierung nach dem *N*-iodsuccinimid-verfahren: Synthese der terminalen disaccharideinheit von orthosomy-cinen, *Liebigs Ann. Chem.*: 1044 (1986).

65. D. A. Griffith and S. J. Danishefsky, Total synthesis of allosamidin: An application of the sulfonamidoglycosylation of glycals, *J. Am. Chem. Soc. 113*: 5863 (1991).

66. R. L. Halcomb, S. H. Boyer, and S. J. Danishefsky, Synthesis of the calicheamicin aryltetrasaccharide domain bearing a reducing terminus: Coupling of fully synthetic aglycone and carbohydrate domains by the Schmidt reaction, *Angew. Chem. Int. Ed. Engl. 31*: 338 (1992).

67. S. A. Hitchcock, S. H. Boyer, M. Y. Chu-Moyer, S. H. Olson, and S. J. Danishefsky, A convergent total synthesis of calicheamicin γ_1^I, *Angew. Chem. Int. Ed. Engl. 33*: 858 (1994).

68. K. K.-C. Liu and S. J. Danishefsky, A striking example of the interfacing of glycal chemistry with enzymatically mediated sialylation: A concise synthesis of GM_3, *J. Am. Chem. Soc. 115*: 4933 (1993).

69. S. J. Danishefsky, J. Gervay, J. M. Peterson, F. E. McDonald, K. Koseki, T. Oriyama, D. A. Griffith, C.-H. Wong, and D. P. Dumas, Remarkable regioselectivity in the chem-ical glycosylation of glycal acceptors: A concise solution to the synthesis of sialyl-Lewis X glycal, *J. Am. Chem. Soc. 114*: 8329 (1992).

70. S. J. Danishefsky, K. Koseki, D. A. Griffith, J. Gervay, J. M. Peterson, F. E. McDonald, and T. Oriyama, Azaglycosylation of complex stannyl alkoxides with glycal-derived iodo sulfonamides: A straightforward synthesis of sialyl-Lewis X antigen and other oligosaccharide domains, *J. Am. Chem. Soc. 114*: 8331 (1992).

71. J. T. Randolph and S. J. Danishefsky, An interactive strategy for the assembly of complex, branched oligosaccharide domains using solid-support methods: An appli-cation to a concise synthesis of the Lewis[b] domain in bioconjugatable form, *Angew. Chem. Int. Ed. Engl. 33*: 1470 (1994).

72. V. Behar and S. J. Danishefsky, A highly convergent synthesis of the Lewis[y] blood group determinant in a conjugatable form, *Angew. Chem. Int. Ed. Engl. 33*: 1468 (1994).

73. R. W. Murray and R. Jeyaraman, Dioxiranes: Synthesis and reactions of methyldioxi-ranes, *J. Org. Chem. 50*: 2847 (1985).

74. M. T. Bilodeau and S. J. Danishefsky, Coupling of glycals: A new strategy for the rapid assembly of oligosaccharides, in *Modern Methods in Carbohydrate Synthesis* (S. H. Khan and R. A. O'Neill, eds), Harwood Academic Publishers, 1996, p. 171–193.

75. D. M. Gordon and S. J. Danishefsky, Displacement reactions of a 1,2-anhydro-α-D-hexopyranose: Installation of useful functionality at the anomeric carbon, *Carbohydr. Res. 206*: 361 (1990).

76. D. M. Gordon and S. J. Danishefsky, Ritter-like reactions of 1,2-anhydropyranose de-rivatives, *J. Org. Chem. 56*: 3713 (1991).

77. R. U. Lemieux and R. M. Ratcliff, The azidonitration of tri-*O*-acetyl-D-galactal, *Can. J. Chem. 57*: 1979 (1979).

78. S. Czernecki and D. Randriamandimby, Azido-phenylselenylation of protected glycals, *Tetrahedron Lett. 34*: 7915 (1993).

79. A. Marra, F. Gauffeny, and P. Sinaÿ, A novel class of glycosyl donors: Anomeric *S*-xanthates of 2-azido-2-deoxy-D-galactopyranosyl derivatives, *Tetrahedron 47*: 5149 (1991).

80. R. J. Ferrier, Unsaturated sugars, *Adv. Carbohydr. Chem. Biochem. 24*: 199 (1969).

81. V. Bolit, C. Mioskowski, S.-G. Lee, and J. R. Falck, Direct preparation of 2-deoxy-D-glucopyranosides from glucals without Ferrier rearrangement, *J. Org. Chem. 55*: 5812 (1990).

82. F. E. McDonald and S. J. Danishefsky, A stereoselective route from glycals to asparagine-linked *N*-protected glycopeptides, *J. Org. Chem. 57*: 7001 (1992).

83. S. Ramesh, N. Kaila, G. Grewal, and R. W. Franck, Aureolic acid antibiotics: A simple method for 2-deoxy-β-glycosidation, *J. Org. Chem. 55*: 5 (1990).

84. Y. Leblanc and B. J. Fitzsimmons, [4+2] Cycloaddition reaction of bis(trichloroethyl)-azodicarboxylate and glycals: Preparation of a C1–C1 2-amino disaccharide, *Tetrahedron Lett. 30*: 2889 (1989).

85. S. Borman, New solid-phase synthesis of oligosaccharides developed, *C&EN*: 30 (1993).

86. T.-H. Chan and W.-Q. Huang, Polymer-anchored organosilyl protecting group in organic synthesis, *J. Chem. Soc. Chem. Commun.*: 909 (1985).

87. M. J. Farrall and J. M. J. Fréchet, Bromination and lithiation: Two important steps in the functionalization of polystyrene resins, *J. Org. Chem. 41*: 3877 (1976).

88. P. Lloyd-Williams, F. Albericio, and E. Giralt, Tetrahedron Report Number 347: Convergent solid-phase peptide synthesis, *Tetrahedron 49*: 11065 (1993).

89. H. Benz, The role of solid-phase fragment condensation (SPFC) in peptide synthesis, *Synthesis* : 337 (1994).

90. S. P. Douglas, D. M. Whitfield, and J. J. Krepinsky, Polymer-supported solution synthesis of oligosaccharides, *J. Am. Chem. Soc. 113*: 5095 (1991).

91. S. J. Danishefsky, J. T. Randolph, J. Y. Roberge, K. F. McClure, and R. B. Ruggeri, Recent applications of the glycal assembly method: Solid-phase synthesis of oligosaccharides and glycoconjugates, *The Schering Lecture Series, 26*: 7 (1995).

92. K. C. Nicolaou, T. J. Caulfield, H. Kataoka, and N. A. Stylianides, Total synthesis of the tumor-associated Le^x family of glycosphingolipids, *J. Am. Chem. Soc. 112*: 3693 (1990).

93. K. O. Lloyd, Blood group antigens as markers for normal differentiation and malignant change in human tissue, *Am. J. Clin. Path. 87*: 129 (1987).

94. J. B. Lowe, Red cell membrane antigens, *The Molecular Basis of Blood Diseases*, (G. Stamatoyannopoulos, et al., eds.), W. B. Saunders, Philadelphia, 1994, p. 293.

95. R. R. Race and R. Sanger, Blood Groups in Man, 6th ed., Blackwell, Oxford, England, 1975.

96. M. L. Phillips, E. Nudelman, F. C. A. Gaeta, M. Perez, A. K. Singhal, S.-i. Hakomori, and J. C. Paulson, ELAM-1 mediates cell adhesion by recognition of a carbohydrate ligand, sialyl-Le^x, *Science 250*: 1130 (1990).

97. Y. Hirabayashi, A. Hyogo, T. Nakao, K. Tsuchiya, Y. Suzuki, M. Matsumoto, K. Kon, and S. Ando, Isolation and characterization of extremely minor gangliosides, G_{M1b} and $G_{D1\alpha}$, in adult bovine brains as developmentally regulated antigens, *J. Biol. Chem. 265*: 8144 (1990).

98. M. Vandonselaar, L. T. Delbaere, U. Spohr, and R. U. Lemieux, Crystallization of the Lectin IV of *Griffonia simplicifolia* and its complexes with Lewis b and Y human blood group determinants, *J. Biol. Chem. 262*: 10848 (1987).

99. M. J. Polley, M. L. Phillips, E. Wayner, E. Nudelman, A. K. Singhal, S.-i. Hakomori, and J. C. Paulson, CD62 and endothelial cell-leukocyte adhesion molecule 1 (ELAM-

1) recognize the same carbohydrate ligand, sialyl-Lewis x, *Proc. Natl. Acad. Sci. USA* *88*: 6224 (1991).

100. O. Hindsgaul, T. Norberg, J. L. Pendue, and R. U. Lemieux, Synthesis of type 2 human blood-group antigenic determinants. The H, X, and Y haptens and variations of the H type 2 determinant as probes for the combining site of the lectin I of *Ulex europaeus*, *Carbohydr. Res. 109*: 109 (1982).

101. K. O. Lloyd, Humoral immune response to tumor associated carbohydrate antigens, *Seminars in Cancer Biol. 2*: 421 (1991).

102. T. Toyokuni, B. Dean, S. Cai, D. Boivin, S.-i. Hakomori, and A. K. Singhal, Synthetic vaccines: Synthesis of a dimeric Tn antigen-lipopeptide conjugate that elicits immune responses against Tn-expressing glycoproteins, *J. Am. Chem. Soc. 116*: 395 (1994).

103. J. T. Randolph, K. F. McClure, and S. J. Danishefsky, Major simplifications in oligo-saccharide syntheses arising from a solid-phase based method: An application to the synthesis of Lewis b antigen, *J. Am. Chem. Soc. 117*: 5712 (1995).

104. T. Borén, P. Falk, K. A. Roth, G. Larson, and S. Normark, Attachment of *Helicobacter pylori* to human gastric epithelium mediated by blood group antigens, *Science 262*: 1892 (1993).

105. J. I. Wyatt and M. F. Dixon, Chronic gastritis—A pathogenic approach, *J. Pathol. 154*: 113 (1988).

106. J. Alper, Ulcers as infectious disease, *Science 260*: 159 (1993).

107. D. Y. Graham, G. M. Lew, P. D. Klein, D. G. Evans, D. J. Evans, Z. A. Saeed, and H. M. Malaty, Effect of treatment of *Helicobacter pylori* infection on the long-term recurrence of gastric or duodenal ulcer. A randomized, controlled study, *Ann. Intern. Med. 116*: 705 (1992).

108. E. Hentschel, G. Brandstätter, B. Dragosics, A. M. Hirschl, H. Nemec, K. Schütze, M. Taufer, and H. Wurzer, Effect of ranitidine and amoxicillin plus metronidazole on the eradication of *Helicobacter pylori* and the recurrence of duodenal ulcer, *N. Engl. J. Med. 328*: 308 (1993).

109. N. Sharon, Bacterial lectins, *The Lectins: Properties, Functions and Applications in Biology and Medicine*, (I. E. Liener, N. Sharon, and I. J. Goldstein, eds.), Academic Press, Orlando, Fla., 1986, p. 494.

110. J. Y. Roberge and S. J. Danishefsky, A strategy for a convergent synthesis of *N*-linked glycopeptides on a solid support, *Science 269*: 202 (1995).

111. A. Brändström, B. Lam, and A. Palmertz, The use of tetrabutylammonium azide in the Curtius rearrangement, *Acta Chem. Scand. B 28*: 699 (1974).

112. S. T. Anisfeld and P. T. Lansbury Jr., A convergent approach to the chemical synthesis of asparagine-linked glycopeptides, *J. Org. Chem. 55*: 5560 (1990).

113. Y. Kiso and H. Yajima, 2-Isobutoxy-1-isobutoxycarbonyl-1,2-dihydroquinoline as a coupling reagent in peptide synthesis, *J. Chem. Soc. Chem. Commun.*: 942 (1972).

114. A. J. Robinson and P. B. Wyatt, Addition of a Reformatsky reagent to *N*-anthracene-9-sulfonyl and related imines: Synthesis of protected β-amino acids, *Tetrahedron 49*: 11329 (1993).

115. T. B. Windholz and D. B. R. Johnson, Trichloroethoxycarbonyl: A generally applicable protecting group, *Tetrahedron Lett. 8*: 2555 (1967).

116. H. Kunz and J. März, Synthesis of glycopeptides with Lewis[a] antigen side chain and HIV peptide T sequence using trichloroethoxycarbonyl/allyl ester protecting group combination, *Synlett.*: 591 (1992).

117. J. M. Schlatter, R. H. Mazur, and O. Goodmonson, Hydrogenation in solid phase peptide synthesis. I. Removal of product from the resin, *Tetrahedron Lett. 18*: 2851 (1977).

118. K. Mori and M. Sasaki, Synthesis of (±)-lineatin, the unique tricyclic pheromone of *Trypodendron lineatum* (olivier), *Tetrahedron Lett. 20*: 1329 (1979).

119. L. Yan, C. M. Taylor, R. Goodnow Jr., and D. Kahne, Glycosylation on the Merrifield resin using anomeric sulfoxides, *J. Am. Chem. Soc. 116*: 6953 (1994).

120. D. M. Whitfield, S. P. Douglas, and J. J. Krepinsky, Metathesis of oligosaccharides. Relative stabilities of activated and deactivated glycosides of polyethylene glycol, *Tetrahedron Lett. 33*: 6795 (1992).

121. R. Verduyn, P. A. M. van der Klein, M. Douwes, G. A. van der Marel, and J. H. van Boom, Polymer-supported solution synthesis of a heptaglucoside having phytoalexin elicitor activity, *Recl. Trav. Chim. Pays-Bas 112*: 464 (1993).

122. G. H. Veeneman, S. H. van Leeuwen, and J. H. van Boom, Iodonium promoted reactions at the anomeric centre. II. An efficient thioglycoside mediated approach toward the formation of 1,2-*trans* linked glycosides and glycosidic esters, *Tetrahedron Lett. 31*: 1331 (1990).

123. M. Meldal and K. Bock, Glycopinion mini-review: A general approach to the synthesis of *O*- and *N*-linked glycopeptides, *Glycoconjugate J. 11*: 59 (1994).

124. M. Meldal, Recent developments in glycopeptide and oligosaccharide synthesis, *Curr. Opin. Struct. Biol. 4*: 710 (1994).

125. K. Wakabayashi and W. Pigman, Synthesis of some glycodipeptides containing hydroxyamino acids, and their stabilities to acids and bases, *Carbohydr. Res. 35*: 3 (1974).

126. M. Hollósi, E. Kollát, I. Laczkó, K. F. Medzihradszky, J. Thurin, and L. Otvös Jr., Solid-phase synthesis of glycopeptides: Glycosylation of resin-bound serine-peptides by 3,4,6-tri-*O*-acetyl-D-glucose-oxazoline, *Tetrahedron Lett. 32*: 1531 (1991).

127. S. A. Kates, B. G. de la Torre, R. Erita, and F. Albericio, Solid-phase *N*-glycopeptide synthesis using allyl side-chain protected Fmoc-amino acids, *Tetrahedron Lett. 35*: 1033 (1994).

128. D. M. Andrews and P. W. Seale, Solid-phase synthesis of *O*-mannosylated peptides: Two strategies compared, *Int. J. Pept. Protein Res. 42*: 165 (1993).

129. S. Lavielle, N. C. Ling, R. Saltman, and R. C. Guillemin, Synthesis of a glycotripeptide and a glycosomatostatin containing the 3-*O*-(2-acetamido-2-deoxy-β-D-glucopyranosyl)-L-serine residue, *Carbohydr. Res. 89*: 229 (1981).

130. H. Kunz, B. Dombo, and W. Kosh, Solid phase synthesis of peptides and glycopeptides on resins with allylic anchoring groups, *Peptides 1988*, (G. Jung and E. Bayer, eds.), Walter de Gruyter, Berlin, 1989, p. 154.

131. G. Becker, H. Nguyen-Trong, C. Birr, B. Dombo, and H. Kunz, Evaluation of the new allylic anchor group HYCRAM in the Merrifield solid-phase peptide synthesis, *Peptides 1988*, (G. Jung and E. Bayer, eds.), Walter de Gruyter, Berlin, 1989, p. 157.

132. H. Paulsen, G. Merz, and U. Weichert, Solid-phase synthesis of *O*-glycopeptide sequences, *Angew. Chem. Int. Ed. Engl. 27*: 1365 (1988).

133. H. Paulsen, G. Merz, S. Peters, and U. Weichert, Festphasensynthese von *O*-gycopeptiden, *Liebigs Ann. Chem.*: 1165 (1990).

134. B. Lüning, T. Norberg, and J. Tejbrant, Solid-phase synthesis of mono- and disaccharide-containing glycopeptides, *J. Chem. Soc. Chem. Commun.* : 1267 (1989).

135. B. Lüning, T. Norberg, C. Rivera-Baeza, and J. Tejbrant, Solid-phase synthesis of the fibronectin glycopeptide V(Galβ3GalNAcα)THPGY, its β analogue, and the corresponding unglycosylated peptide, *Glycoconjugates J. 8*: 450 (1991).

136. A. M. Jansson, M. Meldal, and K. Bock, The active ester *N*-Fmoc-3-*O*-[Ac$_4$-α-D-Man$p$-(1→2)-Ac$_3$-α-D-Man$p$-1-]-threonine-*O*-Pfp as a building block in solid-phase synthesis of an *O*-linked dimannosyl glycopeptide, *Tetrahedron Lett. 31*: 6991 (1990).

137. A. M. Jansson, M. Meldal, and K. Bock, Solid-phase synthesis and characterization of *O*-dimannosylated heptadecapeptide analogues of human insulin-like growth factor 1 (IGF-1), *J. Chem. Soc. Perkin Trans. 1*: 1699 (1992).

138. F. Filira, L. Biondi, B. Scolaro, M. T. Foffani, S. Mammi, E. Peggion, and R. Rocchi, Solid phase synthesis and conformation of sequential glycosylated polytripeptide sequences related to antifreeze glycoproteins, *Int. J. Biol. Macromol. 12*: 41 (1990).

139. E. Bardají, J. L. Torres, P. Clapés, F. Albericio, G. Barany, and G. Valencia, Solid-phase synthesis of glycopeptide amides under mild conditions: Morphiceptin analogues, *Angew. Chem. Int. Ed. Engl. 29*: 291 (1990).

140. E. Bardají, J. L. Torres, P. Clapés, F. Albericio, G. Barany, R. E. Rodríguez, M. P. Sacristán, and G. Valencia, Synthesis and biological activity of *O*-glycosylated morphiceptin analogues, *J. Chem. Soc. Perkin Trans. 1*: 1755 (1991).

141. F. Filira, L. Biondi, F. Cavaggion, B. Scolaro, and R. Rocchi, Synthesis of *O*-glycosylated tuftsins by utilizing threonine derivatives containing an unprotected monosaccharide moiety, *Int. J. Pept. Protein Res. 36*: 86 (1990).

142. M. Meldal and K. Jansen, Pentafluorophenyl esters for the temporary protection of the α-carboxy group in solid phase glycopeptide synthesis, *J. Chem. Soc. Chem. Commun.*: 483 (1990).

143. S. Peters, T. Bielfeldt, M. Meldal, K. Bock, and H. Paulsen, Multiple column solid phase glycopeptide synthesis, *Tetrahedron Lett. 32*: 5067 (1991).

144. S. Peters, T. Bielfeldt, M. Meldal, K. Bock, and H. Paulsen, Multiple-column solid-phase glycopeptide synthesis, *J. Chem. Soc. Perkin Trans. 1*: 1163 (1992).

145. R. Polt, L. Szabó, J. Treiberg, Y. Li, and V. J. Hruby, General methods for α- or β-*O*-Ser/Thr glycosides and glycopeptides. Solid-phase synthesis of *O*-glycosyl cyclic enkephalin analogues, *J. Am. Chem. Soc. 114*: 10249 (1992).

146. T. Bielfeldt, S. Peters, M. Meldal, K. Bock, and H. Paulsen, A new strategy for solid-phase synthesis of *O*-glycopeptides, *Angew. Chem. Int. Ed. Engl. 31*: 857 (1992).

147. M. Elofsson, S. Roy, B. Walse, and J. Kihlberg, Solid-phase synthesis and conformational studies of glycosylated derivatives of helper-T-cell immunogenic peptides from hen-egg lysozyme, *Carbohydr. Res. 246*: 89 (1993).

148. K. B. Reimer, M. Meldal, S. Kusumoto, K. Fukase, and K. Bock, Small-scale solid-phase *O*-glycopeptide synthesis of linear and cyclized hexapeptides from blood-clotting factor IX containing *O*-(α-D-Xyl-1→3-α-D-Xyl-1→3-β-D-Glc)-L-Ser, *J. Chem. Soc. Perkin Trans. 1*: 925 (1993).

149. M. K. Christensen, M. Meldal, and K. Bock, Synthesis of mannose 6-phosphate-containing disaccharide threonine building blocks and their use in solid-phase glycopeptide synthesis, *J. Chem. Soc. Perkin Trans. 1*: 1453 (1993).

150. M. K. Christensen, M. Meldal, K. Bock, H. Cordes, S. Mouritsen, and H. Elsner, Synthesis of glycosylated peptide templates containing 6'-*O*-phosphorylated mannose disaccharides and their binding to the cation-independent mannose 6-phosphate receptor, *J. Chem. Soc. Perkin Trans. 1*: 1299 (1994).

151. M. Meldal, S. Mouritsen, and K. Bock, Synthesis and immunological properties of glycopeptide T-cell determinants, *Carbohydrate Antigens*, (P. J. Garegg and A. A. Lindberg, eds.), American Chemical Society, New York, 1993, p. 19.

152. J. Kihlberg and T. Vuljanic, Piperidine is preferable to morpholine for Fmoc cleavage in solid phase synthesis of *O*-linked glycopeptides, *Tetrahedron Lett. 34*: 6135 (1993).

153. G. Arsequell, L. Krippner, R. A. Dwek, and S. Y. C. Wong, Building blocks for solid-phase glycopeptide synthesis: 2-Acetamido-2-deoxy-β-D-glycosides of FmocSerOH and FmocThrOH, *J. Chem. Soc. Chem. Commun.*: 2383 (1994).

154. A. C. Bauman, J. S. Broderick, R. M. Dacus IV, D. A. Grover, and L. S. Trzupek, A convenient procedure for the preparation of fully-protected C-terminal glycopeptides, *Tetrahedron Lett. 34*: 7019 (1993).

155. H. Kunz and C. Unverzagt, Protecting-group-dependent stability of intersaccharide bonds—Synthesis of a fucosyl-chitobiose glycopeptide, *Angew. Chem. Int. Ed. Engl. 27*: 1697 (1988).

156. H. Kunz and B. Dombo, Solid phase synthesis of peptides and glycopeptides on polymeric supports with allylic anchor groups, *Angew. Chem. Int. Ed. Engl. 27*: 711 (1988).

157. W. Kosch, J. März, and H. Kunz, Synthesis of glycopeptide derivatives of Peptide T on a solid phase using an allylic linkage, *Reactive Polymers 22*: 181 (1994).

158. L. Otvos Jr., L. Urge, M. Hollowsi, K. Wroblewski, G. Graczyk, G. D. Fasman, and J. Thurin, Automated solid-phase synthesis of glycopeptides. Incorporation of unprotected mono- and disaccharide units of *N*-glycoprotein antennae into T cell epitopic peptides, *Tetrahedron Lett. 31*: 5889 (1990).

159. M. Meldal and K. Bock, Pentafluorophenyl esters for temporary carboxyl group protection in solid phase synthesis of *N*-linked glycopeptides, *Tetrahedron Lett. 31*: 6987 (1990).

160. R. J. Chadwick, J. S. Thompson, and G. Tomalin, Solid phase synthesis of monosaccharide-containing *N*-glycopeptides, *Biochem. Soc. Trans. 19*: 406S (1991).

161. I. Christiansen-Brams, M. Meldal, and K. Bock, Protected-mode synthesis of *N*-linked glycopeptides: Single-step preparation of building blocks as peracetyl glycosylated *N*^αFmoc asparagine OPfp esters, *J. Chem. Soc. Perkin Trans. 1*: 1461 (1993).

162. L. Urge, D. C. Jackson, L. Gorbics, K. Wroblewski, G. Graczyk, and L. Otvos Jr., Synthesis and conformational analysis of *N*-glycopeptides that contain extended sugar chains, *Tetrahedron 50*: 2373 (1994).

163. R. Roy, D. Zanini, S. J. Meunier, and A. Romanowska, Solid-phase synthesis of dendritic sialoside inhibitors of influenza A virus haemagglutinin, *J. Chem. Soc. Chem. Commun.* : 1869 (1993).

164. R. Roy, F. O. Andersson, G. Harms, S. Kelm, and R. Schauer, Synthesis of esterase-resistant 9-*O*-acetylated polysialoside as inhibitor of influenza C virus hemagglutinin, *Angew. Chem. Int. Ed. Engl. 31*: 1478 (1992).

165. P. Westerduin, G. H. Veeneman, Y. Pennings, G. A. van der Marel, and J. H. van Boom, Preparation of a fragment of the cell wall teichoic acid of *Bacillus licheniformis* ATCC 9945 via a solid phase approach, *Tetrahedron Lett. 28*: 1557 (1987).

166. C. J. J. Elie, H. J. Muntendam, H. van den Elst, G. A. van der Marel, P. Hoogerhout, and J. H. van Boom, Synthesis of fragments of the capsular polysaccharide of *Haemophilus influenzae* type b. Part III. A solid-phase synthesis of a spacer-containing ribosylribitol phosphate hexamer, *Recl. Trav. Chim. Pays-Bas 108*: 219 (1989).

167. S. Nilsson, M. Bengtsson, and T. Norberg, Solid-phase synthesis of a fragment of the capsular polysaccharide of *Haemophilus influenzae* type b using *H*-phosphonate intermediates, *J. Carbohydr. Chem. 11*: 265 (1992).

168. G. H. Veeneman, H. F. Brugghe, H. van den Elst, and J. H. van Boom, Solid-phase synthesis of a cell-wall component of *Haemophilus (Actinobacillus) pleuropneumoniae* serotype 2, *Carbohydr. Res. 195*: C1 (1990).

169. G. H. Veeneman, *Chemical Synthesis of Oligosaccharides in Solution and on a Solid Support*, Ph. D. Thesis, 1991, Leiden University (The Netherlands).

6

Chemical Synthesis of the Peptide Moiety of Glycopeptides

David G. Large and Ian J. Bradshaw
Liverpool John Moores University, Liverpool, England

I. INTRODUCTION

The methodology for the chemical synthesis of peptides and small proteins of defined amino acid sequence has been the subject of investigation for more than six decades. By contrast, the chemical synthesis of *glyco*peptides is relatively less well developed, and production of glycoproteins is still in its infancy. Nevertheless, significant advances in glycopeptide assembly have been made, and the pace of this development is accelerating, as problems of chemical selectivity are solved in this often-complex area of glycoconjugates. Sections II and III of Chapter 1 give an overview of this development, and in Chapter 2 a detailed and up-to-date examination is made of the total glycopeptide assembly process, in solution and by solid-phase methods.

The building of biologically relevant oligosaccharide moieties suitable for later glycopeptide synthesis is only now becoming possible, and the state of the art in this difficult synthetic area is reviewed in Chapters 4 and 5.

This review will highlight the methods used to produce the *peptide* portion of glycopeptides reported in the literature up to mid-1996. The methodologies described are those of synthesis in solution and by solid phase. Increasingly the latter methodology forms the majority of reported work. This coincides with the increasing interest in automated and semiautomated techniques to produce combinatorial libraries of glycopeptides for biological evaluation.

Naturally, the construction of the peptide moiety of a glycopeptide relies heavily upon the established chemistry and technology of pure peptide synthesis. But the chemical sensitivities of the carbohydrate must be taken into account, whether the sugar is present during the synthesis as only a "minimum" monosaccharide, such as in a simple glycosylated amino acid building block, or whether the carbohydrate moiety is to be added to the peptide later.

In order to place the construction of the peptide portion of the glycopeptide into the context of this review, the overall synthetic strategies represented in the literature for the production of the whole glycopeptide are presented next and in Figure 1 (see also Chapter 1 (Section III) and Ref. 82).

Strategy 1. The Building Block Approach. Here, a suitably protected derivative of the appropriate amino acid (aspartic acid for N-linked, serine or threonine for *O*-linked glycopeptides) is first linked to a mono- or disaccharide derivative. The resulting "building block" is then usually incorporated onto the N-terminus of a growing peptide chain, which may be attached to an insoluble resin (solid-phase synthesis, or SPS). Alternatively, the building block may undergo peptide chain extension in solution at its N- or C-terminus. The monosaccharide or disaccharide element of the original building block is not extended by this strategy. This approach has been the one most reported in the literature for the synthesis of *N*-glycopeptides in solution [1–12] or by SPS [13–20] and for the synthesis of *O*-glycopeptides by solution [21–30] or solid-phase methods [30–53].

Strategy 2. The glyco–amino acid building block, created as described in Strategy 1, then undergoes extension of the carbohydrate portion, followed by peptide chain synthesis. This has been the least described method to the present time [54–56].

Strategy 3. By this approach an oligosaccharide is first constructed, then coupled to a suitably protected derivative of an amino acid or short (di- or tri-) peptide, followed by peptide extension of this intermediate glycopeptide [57–62].

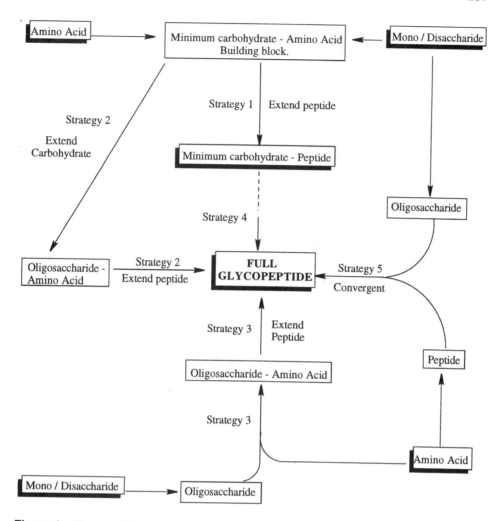

Figure 1   Five possible strategies for the synthesis of glycopeptides.

Strategy 4.   The "final" product of Strategy 1 undergoes carbohydrate lengthening, notably by enzymatic means [56,63–66].

Strategy 5.   Convergent Synthesis.   In this approach, the required carbohydrate chain and peptide are each built independently, and the glycopeptide link is created late in the synthesis [67–78,83,93].

The greatest effort in the development of synthetic methods for model glycopeptides has been directed toward synthesis of those having a relatively small carbohydrate component by Strategy 1. But in recent years, the need has arisen for glycopeptides having larger oligosaccharide moieties, for use in conformational, biological, or other studies. Perhaps this goal may be met by further development of Strategies 2–5, especially Strategies 4 and 5. For such complex macromolecules, it is of paramount importance that the often-different chemical methodologies and sensitivities of oligosaccharide synthesis and of peptide synthesis be carried out with the minimum of degradation of either portion of the glycopeptide (see Section IV).

II. SOLUTION SYNTHESIS OF THE PEPTIDE MOIETY OF GLYCOPEPTIDES

Historically, the solution synthesis of peptides preceded the solid-phase approach, and this is also true for the synthesis of the peptide portion of glycopeptides. Synthesis of peptides or glycopeptides in solution becomes increasingly difficult as the peptide chain length increases, because of the decreasing solubility in any suitable solvent of the (glyco)peptide with its many protecting groups. Pragmatically, the limit for solution synthesis is probably about 9–12 amino acid residues, although this can be sequence dependent and may also be affected if a substantial carbohydrate component is present.

Following the successful construction in the 1950s and 1960s of suitable building blocks for β-*N*-linked asparagine (**1**) [90,135] and β-*O*-linked-serine (**2**) [134] or threonine (**3**) [136] residues, successful peptide chain extension led to a range of monosaccharide di-, tri-, tetra-, and penta-peptides [1–5].

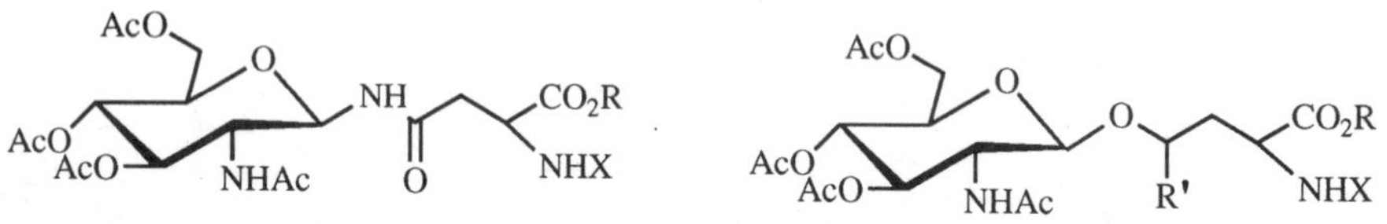

It was noted in the Introduction that the classical methodology of peptide synthesis is equally appropriate for synthesis of glycopeptides, provided that the chemistry of the carbohydrate component is taken into account when designing:

The overall protecting group strategy
The methods for coupling of protected glyco–amino acid to amino acid or
    peptide fragments
The final deprotection to give the finished glycopeptide

Thus the serine- or threonine-*O*-glycosyl link is sensitive to both acids and bases [84], and as a result the overall protecting-group strategy for *O*-glycopeptide synthesis is regarded as more demanding than that for *N*-glycopeptide assembly. This difficulty in the choice of suitable protection for synthesis of the peptide moiety is exacerbated if a more complex oligosaccharide structure is also present, since certain interglycosidic links are also particularly sensitive to acid (e.g., α-fuc links; sialic acid α 2,3- and 2,6-links), especially if the sugar hydroxyl groups are not masked [85]. This may partly explain why Strategies 2 and 3 have not been reported more frequently, since, for either methodology, the full oligosaccharide component must then survive the building of the peptide.

A. N- and C-Terminal Protection Employed in Solution Glycopeptide Synthesis

For the pioneering work of glycopeptide synthesis by the Jeanloz group in Boston during the 1970s [1–5], benzyloxycarbonyl- and p-nitrobenzyl ester were the usual protecting groups for N- and C-termini, respectively. Slightly later, in the early 1980s, the Kunz group in Mainz began a long exploration of synthetic methods in solution by examining alternative N-terminal protection with t-butoxycarbonyl-(Boc) (4) [7,9,10], removable by trifluoroacetic acid (TFA); and N-[2-(triphenylphos-phonio)ethoxycarbonyl]-(Peoc) (5) [86,6,23] removable with morpholine or diethyl-amine. Early in 1983, 9-fluorenylmethoxycarbonyl-(Fmoc) (6) amino acid N-protection was successfully applied to O-glycopeptide synthesis by the Kunz group [8] and later the same year also by Ferrari and Pavia in a study of human glycophorin A^M O-glycopentapeptides [21]. First proposed by Carpino and Han in 1970 [87], the Fmoc group has proved particularly valuable for O-glycopeptide assembly, by both solution [25–29,54,55,71] and solid-phase methods [Section III]. This is because of its facile removal under mild basic conditions, when piperidine or morpholine are used.

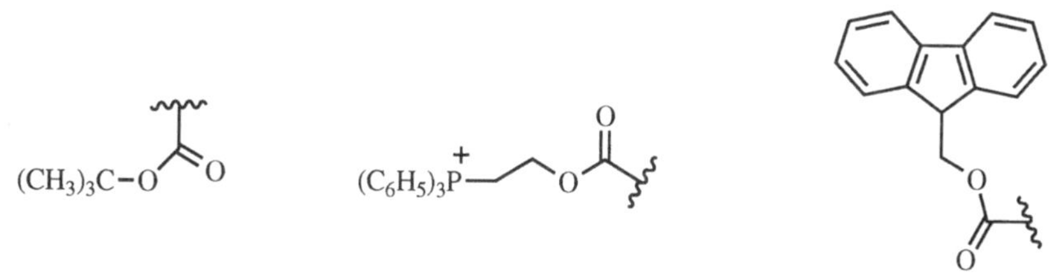

	4	5	6
	^{t}Boc	Peoc	Fmoc

^{t}Boc protection in solution synthesis of N-glycopeptides continues to be used [9,10,61,67], and the β-(1-4)-glycosidic link in chitobiosyl-glycopeptides remains unaffected by brief treatment with TFA [9], provided that the disaccharide hydroxyl groups are O-acylated. Similarly, ^{t}Boc-deprotection with HCl/diethyl ether was effected from a diglycosylated pentapeptide bearing fully O-acetylated Lewis^x trisaccharide groups [61]. Also used for amino acid N-blocking have been 2-(4-pyri-dyl)ethoxycarbonyl (Pyoc) 7 [100,57] (a useful solubilizing group) and allyloxycar-bonyl (Aloc) [27] 8, which can be removed selectively with the palladium complex (PPh$_3$)$_4$Pd when used in conjunction with ^{t}butyl ester protection. Recently, trichlo-roethoxycarbonyl (Teoc) 9 has also been used for glycopeptide synthesis [63].

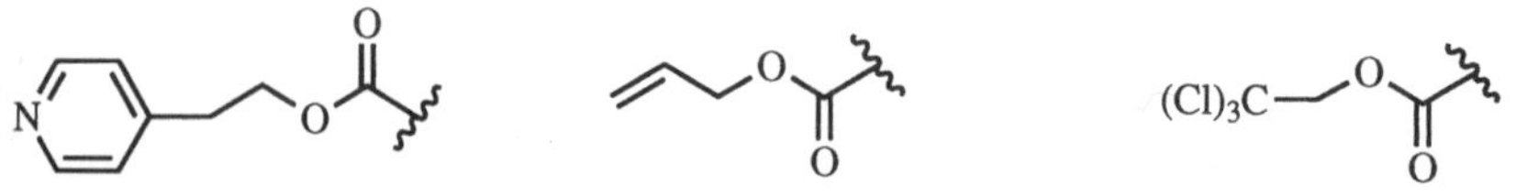

7   8   9

Pyoc   Aloc   Teoc

For peptide chain elongation, the C-terminal protection must be orthogonal with that used for the N-terminus and be semipermanent for the more usual C → N chain extension, or removable during the synthesis for the less common N → C approach. For semipermanent blocking, benzyl esters have been commonly used in conjunction with 'Boc [29,30,67a/b] or with Fmoc [8,25,57,66] protection. Other combinations used in glycopeptide chain extension have been Fmoc/'butyl ester [21,26,27] and benzyloxycarbonyl (Z)/'butyl ester [22,26,27,63]. Protection with the allyl ester group, which can be removed selectively by Pd(O)-catalyzed transfer to morpholine acting as nucleophile, was first reported by Kunz and Waldmann [24], and it has been much used by the Kunz group for temporary C-terminal masking in solution synthesis with 'Boc [7,9,10,12,61] or Fmoc [28] as N-terminal blocking groups. Other studies toward improved selectivity proposed by Kunz and co-workers include Peoc/benzyl ester [6]; Peoc/'butyl ester [6]; Aloc/'butyl ester [58], and Pyoc/benzyl ester [58].

In addition to the 'butyl and allyl esters just cited for temporary C-terminal protection, a novel alternative was successfully used for the temporary blocking of a glyco-asparagine intermediate during the synthesis of *N*-glycopeptides [88]: 1,3-dithian-2-ylmethyl (Dim) ester (**10**). The Dim group is cleaved under mild conditions at pH 8 after oxidation of the sulfur in the dithiane ring.

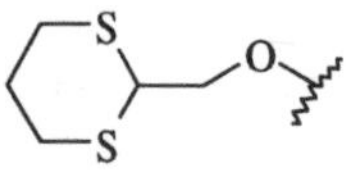

10

Dim-

B. Coupling Reagents for Peptide Bond Extension in Solution

Relatively few carboxyl-activation methods have been employed for extending the peptide from the initial glyco–amino acid building block, after removal of either the N- or C-terminal temporary masking group. In the early solution work of the Jeanloz group on *N*-glycopeptides [1–5], formation of the peptide bonds was mediated either by *N,N'*-dicyclohexylcarbodiimide **11** (DCCI) [79], *N*-ethoxycarbonyl-1,2-dihydroquinoline (**12**) (EEDQ) [80], or *N*-ethyl-5-phenylisoxazolium-3′-sulfonate (**13**) (WRK) [81] or by using preformed pentachlorophenyl active esters [92].

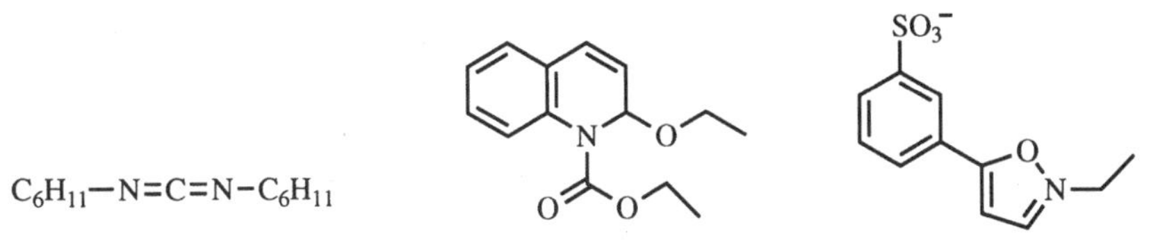

11	**12**	**13**
DCCI	EEDQ	WRK

In a comparison of DCCI and Woodward's Reagent K (WRK), Garg and Jeanloz [1] found that the latter was preferable, because of the easier procedure for product isolation when the troublesome dicyclohexylurea was absent. In this and in their later work it was found beneficial to couple short preformed peptides to the appropriately protected GlcNAc-β-*N*⁴-asparagine, rather than to extend the peptide in a stepwise manner. By this means, losses of the valuable carbohydrate-containing building block could be minimized.

EEDQ has been much employed in dichloromethane or dimethylformamide solution synthesis of the peptide component of both *N*- and *O*-glycopeptides. Quoted yields of isolated products have been in the 55%–85% range. The related 2-isobutoxy-1-isobutoxycarbonyl-1,2-dihydroquinoline (IIDQ) [108] has been used for the coupling of 1-glycosylamines to protected aspartyl residues in the process of creating *N*⁴-glycoasparagine building blocks [55,61] and also occasionally for peptide elongation [28].

EEDQ has been favored by the Kunz group for a number of years. For example, in the synthesis of glucosyl-β-*O*-serine glycotripeptides [8] and GlcNAc-β-*O*-serine glycopeptides [23], stepwise synthesis was used from the appropriate building block. For the building of chitobiosyl-β-*N*-asparagine peptides [9,10] and xylosyl-β-*O*-serine glycopeptides [24], extension of the peptide at the C-terminus of the relevant protected glycoamino acid was effected with dipeptide fragments and EEDQ as activating agent.

The Paulsen group has also employed EEDQ in the synthesis of interleukin-2-type *O*-glycopeptides [26,30] by stepwise peptide chain lengthening and for producing proline-containing (acetylated) GalNAc-α′-threonine glycopeptides in C-terminal chain extension. Other examples of Strategy 1 solution synthesis mediated by EEDQ include the heavily glycosylated *O*-glycopeptides related to the human T-antigen [22] and a glycohexapeptide sequence related to human oncofetal fibronectin [29]. In this last case a side reaction was observed in which the imidazole ring of a histidine residue was ethoxycarbonylated by the EEDQ.

Examples of the use of EEDQ in other strategies of glycopeptide synthesis include that by Ogawa and co-workers of a heavily sialylated glycopeptide [54,55]

(Strategy 2), *O*-glycotripeptides related to glycophorin A [57] (Strategy 3), and an *N*-glycopentapeptide related to vasoactive intestinal polypeptide [67] (Strategy 5).

The classical DCCI coupling reagent first proposed by Sheehan and Hess [79] has been little used in peptide chain extension during solution synthesis of glyco-peptides. This may reflect the later development of glycopeptide synthesis compared with peptide synthesis, which enabled the development of other methods and the recognition of certain problems with DCCI. The difficulty in removal of the by-product dicyclohexylurea is a serious obstacle to isolation of the product in a solution synthesis, and other carbodiimide condensation agents, in which the by-product is water soluble and separable from the glycopeptide product, have been preferred. Notable amongst these is 1-ethyl-3-(3′-dimethylaminopropyl)carbodiimide **14**, EDAC, or "water-soluble carbodiimide" (WSC) [58,61,67], always used with 1-hydroxybenzotriazole (HOBt) **15** for suppression of any possible racemization [119].

| **14** | **15** |
| WSC | HOBt |

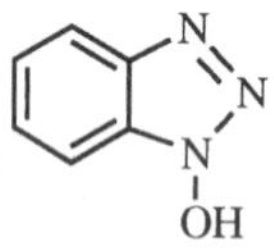

III. SOLID-PHASE SYNTHESIS (SPS) OF THE PEPTIDE MOIETY OF GLYCOPEPTIDES

The first synthesis of a glycopeptide on an insoluble resin was reported in 1981 [13], and 15 years later solid-phase methodology is the preferred route to both *N*- and *O*-glycopeptides. The inherent advantages of the technique are:

  Speed of peptide assembly
  Relative ease of separation of by-products from the required compound at each step
  Facile variation of the sequence of the peptide component

These are ideal features for the construction of a range of glycopeptide variants needed for biological studies or to produce larger protected glycopeptides, which are likely to be insoluble in a solution synthesis. Nevertheless, there are also certain disadvantages to solid-phase methodology, notably the need to work on a smaller scale, the difficulty of following the progress of the synthesis by many of the normal analytical procedures, and the necessity to ensure near-quantitative conversion in each reaction step. Each of these issues is discussed below.

Considerable development in solid-phase assembly of glycopeptides has taken place since the 1981 synthesis, which was carried out on the original Merrifield-type polystyrene resin [110]. These efforts have been directed mainly at the solution of some of the problems noted previously. Useful reviews of peptide or glycopeptide methods have been published [38,39,62,99,106]. In solvents such as dichloromethane or DMF, the original polystyrene resin, lightly cross-linked with divinyl benzene, provided a swollen gelatinous and insoluble matrix, with sufficient physical strength and accessibility for reagents. Attachment of the C-terminal amino acid by a benzyl ester linkage to the insoluble support then allows synthesis of the peptide (or glycopeptide) from the C- to the N-terminus using t-butoxycarbonyl ('Boc) as alpha-amino-group protection. The peptide bonds and any glycosidic linkages must both survive the following repetitive steps:

Deprotection of the 'Boc groups with TFA
Neutralization of the resulting N^α salt by weak base
Condensation with the next activated amino acid or N^α-protected peptide (fragment coupling)

In addition, the harsh acidic cleavage of the glycopeptide from the resin support, typically using liquid HF, HBr in acetic acid or trifluoromethanesulfonic acid, pose additional problems for glycopeptide synthesis compared with those for peptide assembly. These problems include the cleavage of sensitive glycosidic links or β-elimination of the whole saccharide entity. The use of different resin systems and ''linkers'' or ''anchors'' between resin and glycopeptide is described below, in relation to published glycopeptide syntheses.

A. Resin and Linker Systems Used for Solid-Phase Glycopeptide Assembly

In the 'Boc/polystyrene resin system, the crucial peptide-resin benzyl ester link must be stable to the TFA conditions used repetitively to deprotect the 'Boc groups. In practice some peptide chain cleavage does occur at each deprotection step, and this is difficult to control when both deprotection and cleavage from the resin employ the same chemical conditions. One solution to this problem was the development by Merrifield [115] of the 4-(hydroxymethyl)phenyl acetic acid (PAM) linker [A, Figure 2] connected to the modified polystyrene 16. Using this linker the growing peptide is again attached via a benzyl ester link and is cleaved after the synthesis by HF, but the peptide-PAM resin link is 100 times more stable during the repetitive TFA treatment than the peptide-resin link of the original Merrifield resin.

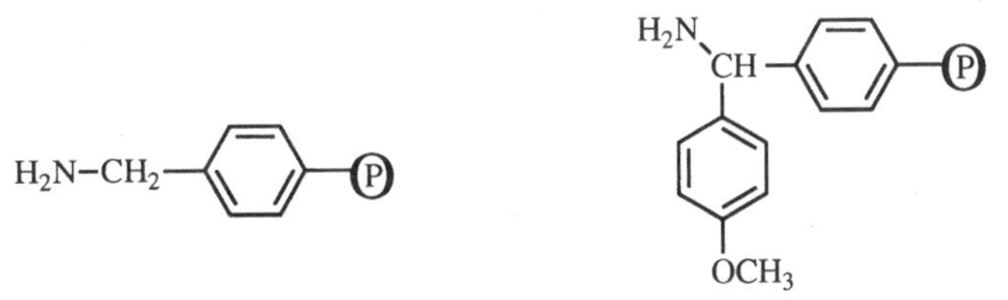

16   17

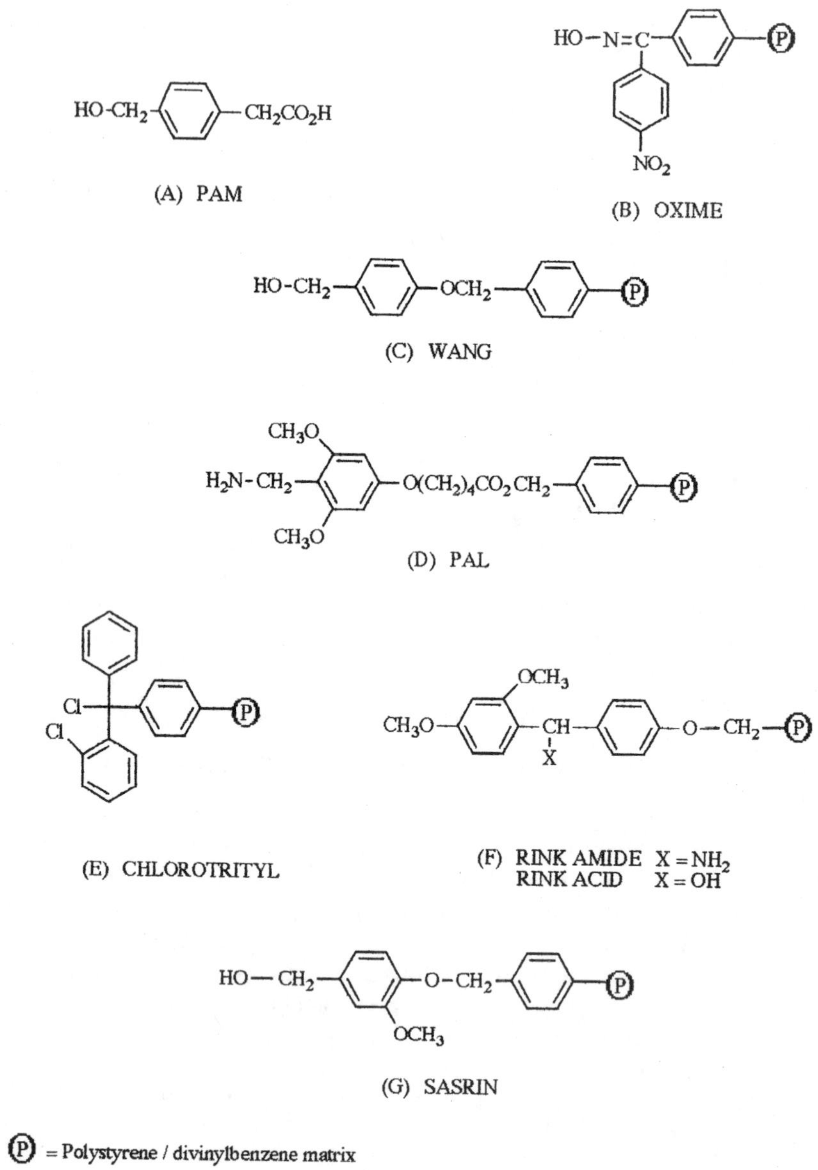

(A) PAM

(B) OXIME

(C) WANG

(D) PAL

(E) CHLOROTRITYL

(F) RINK AMIDE X = NH$_2$
    RINK ACID   X = OH

(G) SASRIN

(P) = Polystyrene / divinylbenzene matrix

Figure 2 Resin anchors used for glycopeptide synthesis. C, D, F and G may also be attached to polydimethylacrylamide or PEG-polydimethylacrylamide copolymer instead of polystyrene.

Unverzagt et al. used this resin system with ^{t}BOC amino protection and HF cleavage in an interesting example of N-glycopeptide synthesis [64], which overcame the dangers to the glycopeptide of such strong acid-cleavage conditions. Using a Strategy 4 approach, the acid-sensitive carbohydrate moiety was assembled onto the peptide *after* the HF cleavage, by enzymatic means in solution.

Cohen-Anisfeld and Lansbury, in their notable contribution to convergent glycopeptide synthesis (Strategy 5) [74], used a methylbenzyhydrylamine (MBHA)-

polystyrene resin **17**, which is a common resin support for pure peptide synthesis employing 'Boc chemistry. Although HF cleavage of the finished peptide is required with the MBHA resin, in this case also the carbohydrate was not exposed to these acid conditions, since glycosylation of the peptide occurred in solution, after cleavage from the resin.

Another polystyrene variant using 'Boc α-amino-protection that has been used for glycopeptide synthesis is the oxime resin, first proposed by Kaiser and DeGrado [126] (B, Figure 2). The advantage of this resin is that protected peptide fragments are obtained by cleavage from the resin with weak nucleophiles, such as acetate ion. This system was used by Anisfeld and Lansbury [72] and by Trzupek and co-workers [76]. In the latter work, the resin link was attached to the *penultimate* C-terminal amino acid and was cleaved by the amine salt of a glycosylated serine residue, which became C-terminal (**18**).

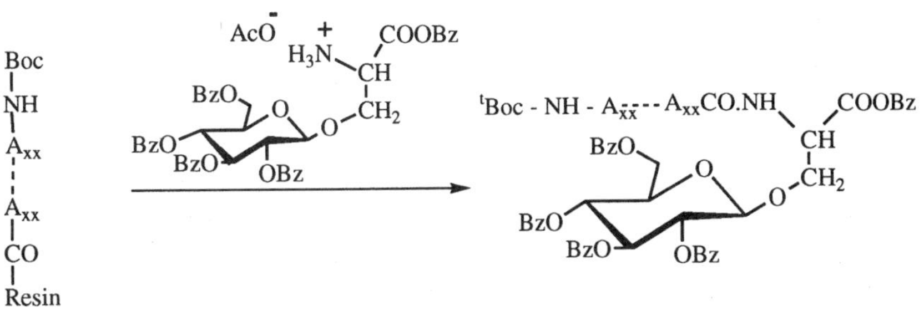

18

In addition to the above modified Merrifield resins, two other broad develop-ments in solid-phase peptide synthesis have been successfully applied to glycopeptide construction. First, the use of the base-labile Fmoc N^α-protection removes the need for repetitive acid treatment in the deprotection steps. Second, the use of Fmoc protection then makes it possible to increase the acid lability of the glycopeptide link to the resin, since the peptide-resin link is not now susceptible to cleavage during the repetitive amino-deprotection steps of the synthesis.

The development of the 4-hydroxymethylphenoxy-linker (C, Figure 2) by Wang [105] was one of the earliest modifications of the polystyrene resin to increase the acid lability of the peptide-resin bond. The Wang resin has been favored by the Paulsen group for synthesis of a number of GalNAc-α-O-serine/threonine glyco-hexapeptides [30–32] using Fmoc amino protection. The first amino acid was at-tached to the resin using its symmetrical anhydride, and subsequent amino acid or O-glycosyl serine/threonine building blocks were added in a stepwise manner (Strategy 1). Final cleavage from the Wang resin was achieved with TFA/water (95: 5) to produce the glycopeptide in the acid form. These examples, as well as those of the Dwek group [20] and of the Rocchi group [47], illustrate the solid-phase method in a batch synthesis, where the resin is filtered and washed free of reagents at each step. This can be automated if necessary. In the alternative ''continuous flow'' technology, the resin is usually contained within a permeable inorganic matrix (e.g.,

Kieselguhr), capable of withstanding continuous flow pressures. The reagents are circulated through the resin bed continuously, which allows efficient mixing and the possibility for monitoring the reactions if UV-absorbing reagents or by-products are involved (see Section III.C.). An example of this technique using Wang resin is the interesting comparison made by Andrews and Seale of the building block approach and the convergent strategy in their synthesis of a series of *O*-mannosylated glyco-peptides [75].

Other anchors utilized for glycopeptide assembly with Fmoc α-amino protec-tion include 5-(4-aminomethyl-3-5-dimethoxyphenoxy)-valeric acid, or PAL linker [104] (D, Figure 2); the 2-chlorotrityl chloride resin [114] (E, Figure 2); the 4-(2′,4′-dimethoxyphenylaminomethyl)-phenoxymethyl resin, or RINK linker [111] (F, Fig-ure 2); and the "super acid-sensitive" resin, or SASRIN [113] (G, Figure 2). A comparison of the typical acid conditions needed for cleavage from these anchors is shown in Table 1. These are guidelines only, since the amino acid sequence of the glycopeptide, as well as other factors, mean that optimization is usually needed.

With the PAL anchor, the final glycopeptide is obtained as the C-terminal am-ide. The RINK linker is capable of giving either the glycopeptide amide, from RINK amide handle, or the free acid from RINK acid handle. The chlorotrityl and SASRIN linkers deliver the C-terminal free acid following cleavage from the resin, while a fully protected glycopeptide is produced from the oxime resin. References to ex-amples of glycopeptide syntheses carried out with these modified polystyrene resins are shown in Table 1.

The Kunz group at Mainz has adopted another novel approach to the modifi-cation of polystyrene resins for glycopeptide synthesis, which makes it unnecessary to use any acid conditions to achieve final release from the solid phase. This has been achieved with a carefully designed allyl ester linkage between the growing glycopeptide and the resin anchor. It can be cleaved at the end of the synthesis by Pd(O)-catalyzed transfer to morpholine as nucleophile. Such a method increases the overall choice for orthogonality of protecting-group systems in glycopeptide synthe-sis. The two linker variations of this principle, HYCRAM™ and the more recent HYCRON, are described more fully in Chapter 2 (Section V.B.), with examples of their successful use.

In 1981 and in parallel with all of the preceding polystyrene modifications, Atherton, Sheppard, and co-workers proposed a quite different resin system [102,103]. They argued that a more polar solid-support system than polystyrene was needed for assembly of peptides, to allow better solvation for both peptide bond

Table 1 Literature Glycopeptide Syntheses Using Modified Polystyrene Resins, Showing Cleavage Conditions

Anchor	Typical acid cleavage conditions	Glycopeptide C-terminus	Refs.
PAL	70% TFA in CH_2Cl_2	Amide	35, 45
Chlorotrityl	AcOH/TFE/CH_2Cl_2; 0.5–2 hr	Acid	19, 93
RINK	10% AcOH in CH_2Cl_2; 4 hr	Acid/amide	37, 43, 95
SASRIN	1% TFA in CH_2Cl_2	Acid	34, 77, 93
Oxime	Acetate ion	Protected	72, 76

formation and deprotection steps during a stepwise synthesis from C- to N-terminus. The first successful copolymer was based upon *N,N*-dimethylacrylamide **19** and ac-ryloylsarcosine methyl ester **20**. Subsequent chemical modification then creates amino group termini to which the C-terminal amino acid is attached [38].

$$CH_2=CHCO.N\begin{smallmatrix}CH_3\\CH_3\end{smallmatrix} \qquad\qquad CH_2=CHCO.\underset{\underset{CH_3}{|}}{N}-CH_2-CO_2CH_3$$

$$\textbf{19} \qquad\qquad\qquad\qquad\qquad \textbf{20}$$

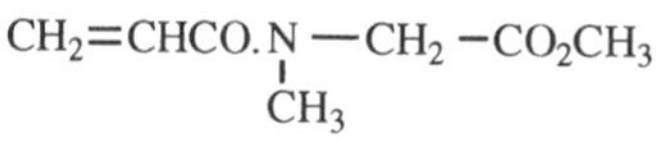

This and other related polyamide solid supports have been widely used in glycopeptide synthesis, with DMF as a solvating, dipolar, but aprotic solvent. The polyamide matrix may be polymerized within rigid macroporous Kieselguhr, and it is then mechanically stronger for the continuous-flow technique [124]. The orthogonal protection strategy consisting of N^α-Fmoc and mildly acid-cleavable side chain masking groups (e.g., ᵗButyl) has served to establish a valuable system for glycopeptide synthesis. Usually a linker molecule and a reference amino acid (norleucine) are interposed between the polyamide backbone and the growing peptide. For example, in a series of papers, the Bock and Meldal group in Copenhagen, jointly with the Paulsen group in Hamburg, used this approach to prepare multiglycosylated *O*-glycopeptides of the mucin type, using the PAL linker attached to a Sheppard-type polydimethylacrylamide resin [36,41,46,49]. In this work, a number of different gly-coamino acid building blocks were constructed, and these were introduced into a normal peptide assembly on the polyamide resin (Strategy 1). The relatively mild conditions of the Fmoc-polyamide system are very suitable for glycopeptide synthesis, as long as very acid-sensitive interglycosidic links are absent in the carbohydrate moiety. The Copenhagen and Hamburg groups have also explored the use of the acid-labile RINK linker for building mucin-type *O*-glycopeptides [51] and for a study of β-elimination in Fmoc *O*-glycopeptide synthesis [48]. Other groups have also used the RINK-polyamide resin combination [75,78].

The "standard" linker attachment to the polyamide resin, 4-hydroxymethyl-phenoxyacetic acid **21** [128,103,124], has been much used by the Bock and Meldal group [15,42,59,60] for both *O*- and *N*-glycopeptides and by the Rocchi group in Padua for glycosylated tuftsin-containing IgG fragments [17] and vespulakinin analogs [40].

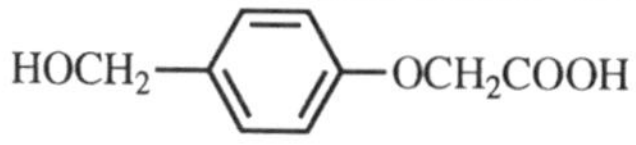

$$HOCH_2-\!\!\left\langle\!\!\bigcirc\!\!\right\rangle\!\!-OCH_2COOH$$

$$\textbf{21}$$

Other resin systems more polar than polystyrene have also been examined for glycopeptide synthesis. Kihlberg and Vuljanic [44] prepared an α-GalNAc-hepta-deca-glycopeptide using polyoxyethylene-polystyrene resin bearing a RINK amide linker. The Bock, Meldal, and Paulsen groups have successfully used for O-glyco-peptide synthesis a copolymer (PEGA) of poly(ethyleneglycol) and poly(N,N-di-methyacrylamide), also derivatized with the RINK linker [48,50,94]. This resin is "flowstable" for continuous-flow technology and allows faster peptide acylation than the normal polyamide resin [94].

B. Protecting Group Systems for Solid-Phase Glycopeptide Assembly

The choice of a temporary α-amino blocking group is most fundamental for solid-phase work. The carboxyl terminus is held by the resin, and its integrity during the synthesis is dependent upon the linker-peptide bond, as described in Section III.A. For polystyrene or linker-modified polystyrene synthesis, where HF cleavage is nec-essary, 'Boc has usually been used for α-amino protection. Orthogonal side-chain masking groups are chosen so that they remain in place during the repetitive 'Boc deprotections but are simultaneously removed when the glycopeptide is cleaved from the resin. Suitable groups for the side-chain protection are benzyloxycarbonyl (Z) and benzyl ethers and esters. The alternative Fmoc chemistry, described earlier, has become preferred in recent years, to avoid degradation of the carbohydrate moiety by its use of milder conditions and also in an attempt to reduce side reactions gen-erally. However, even with the use of Fmoc chemistry, TFA cleavage of the finished glycopeptide from the resin/linker with concurrent side-chain deprotection can give rise to stable cations that can modify sensitive amino acids such as tryptophan, methionine, tyrosine, and cysteine [132]. Cation scavengers are, therefore, routinely part of the resin cleavage "cocktail." From a practical point of view, no *single* protecting-group and resin-cleavage protocol has emerged. Orthogonal protection must be chosen with care, and deprotection of blocking groups must usually be carried out in stages and be optimized according to the peptide sequence.

C. Carboxyl Activation for Solid-Phase Glycopeptide Synthesis

For efficient and unambiguous formation of the peptide bond, activation of the N-α-protected amino acid or glyco-amino acid must be achieved under mild conditions with the avoidance of racemization. Furthermore, in solid-phase assembly, the acti-vated carboxyl group must then acylate the free amino group on the growing peptide with high coupling efficiency and without side reactions. Since purification does not occur at each step and normal analytical control of the synthesis is difficult, problems may be hidden until the synthesis is complete. Lastly, failure to acylate every avail-able growing chain for reasons of steric interaction or chain aggregation will lead to deletion sequences and difficulties in purification.

Carbodiimide Activation

Carbodiimides have been widely used for in situ activation in glycopeptide syntheses on solid phase, especially with the nonpolar polystyrene matrices and dichloro-methane as solvent and resin swelling agent. In such cases, dicyclohexylcarbodiimide

Table 2   Solid-Phase Glycopeptide Syntheses Using
Diisopropylcarbodiimide/HOBt

Resin matrix	Linker	Solvent	N^α	Ref
Polystyrene	PAM	NMP	'BOC	64
Polystyrene	RINK amide	DMF	Fmoc	43, 95
Polystyrene/PEG graft	RINK amide	DMF	Fmoc	44
Polystyrene/PEG graft	PAL	DMF	Fmoc	98
Polystyrene	RINK acid/amide	DMA	Fmoc	130
Aminomethyl-polystyrene	HYCRON	DCM	Fmoc	53

PEG = poly(oxyethylene).

has usually been employed [14,30,31,32,35,40,45,47]. 1-Hydroxybenzotriazole (HOBt) is invariably used as auxiliary reagent, and an activated hydroxybenzotriazolyl ester is formed in situ. As well as suppressing racemization at the adjacent alpha-carbon [119], HOBt also assists in preventing dehydration. The DCU byproduct can be removed more easily in a solid-phase synthesis, where the peptide is resin-bound, than in a solution synthesis. Nevertheless, N,N' diisopropylcarbodiimide, whose urea is more soluble, has also been used. This is illustrated in Table 2.

Carboxyl Activation by Symmetrical Anhydride Formation

Acyl amino acid symmetrical anhydrides are regarded as reactive carboxyl components for solid-phase peptide assembly [124]. They are relatively stable, but are usually preformed from two molar equivalents of the acyl amino acid and a carbodiimide in a nonpolar solvent, such as dichloromethane. The symmetrical anhydride, which is formed rapidly, can be filtered from the urea co-product, and the resin is not then exposed to the reactive carbodiimide, with the danger of side reactions. Separation of the urea from the resin at the next washing step is also avoided. The major disadvantage to the use of preformed symmetrical anhydrides is that only half of the acyl amino acid/glycosylated amino acid is incorporated into the growing peptide chain. This is usually unacceptably expensive, and few glycopeptide syntheses with this carboxyl activation method have been reported: [16], [17; polyamide/standard linker], [34; SASRIN resin]. All used Fmoc chemistry and employed symmetrical anhydrides only for some amino acids, together with other types of carboxyl activation such as activated esters. One report [76] uses 'Boc chemistry and the Kaiser–DeGrado oxime resin.

Activated Esters

Creation of the peptide bond by the aminolysis of methyl and ethyl esters was first proposed by Emil Fischer [E. Fischer and Y. Suzuki, *Chem. Ber. 38*, 4173 (1905)]. The subsequent development by Bodanszky in the 1950s [129] of nitrophenyl esters having a better leaving group for the nucleophilic reaction opened the way for effective peptide formation, using "active esters." Nitrophenyl and some other substituted-aromatic esters, e.g., pentachlorophenyl, have been used in solution synthesis of glycopeptides [2,21,25,91], but they react only slowly in solid-phase assembly [13,133], especially where sterically hindered amino acids are involved. However, pentafluorophenyl (Pfp) esters have been shown to be effective carboxyl-activating groups for both O- and N-glycopeptide synthesis. The acylation rate is enhanced by the addition of HOBt, and it then approaches that of symmetrical anhydrides [109,52]. First introduced into glycopeptide synthesis by Meldal and Jensen in 1990 [33], the Copenhagen group has since prepared a large number of glycosylated asparagine, serine, and threonine "building blocks" as their Pfp esters. The Pfp ester acts as a temporary masking group for the carboxyl function during the glycosylation and can then be purified prior to being introduced into a normal solid-phase peptide assembly. While the Pfp esters are "activated" carboxyl compounds from the point of view of peptide bond formation, they are stable enough for the glycosylation process to produce the building block and its purification, provided that they are kept scrupulously dry, to avoid hydrolysis [33,52].

Using the prepared and purified building blocks, Meldal, Bock, and co-workers have used the Pfp esters in an impressive series of glycopeptide syntheses using either polyamide or polyoxyethylene resins and either RINK, PAL, or "standard" polyamide anchors. Both N-glycopeptides [15,131] and O-glycopeptides have been prepared [33,36,41,42,45,48–51,59,60,94]. Other groups have also found the Pfp activation method valuable [17,18,52,78].

In most of the work just cited for the Copenhagen group, a further reagent is added at the acylation step: 3,4-dihydro-4-oxo-1,2,3-benzotriazine; Dhbt-OH; **22**). This increases the rate of coupling to the amino terminus on the resin, presumably through the formation of an even more activated Dhbt ester.

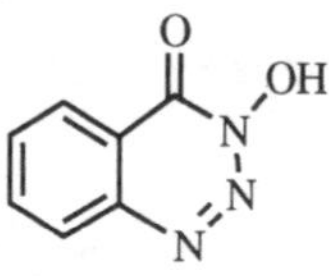

22

Dhbt - OH

First proposed for peptide synthesis [120,109], the Dhbt esters formed in situ from the Pfp esters have a valuable, if accidental, attribute for solid-phase synthesis. During the acylation with the Pfp ester in presence of Dhbt-OH, the resin (normally polyamide and off-white in color) is yellow, but the color disappears when the acylation is complete. The color is attributed to the formation of an ion pair between unacylated amino groups on the resin and ionized Dhbt-OH in the presence of di-isopropylethylamine (DIPEA) or to other bases usually present for the peptide acylation [109]. Thus, the Dhbt active esters act as a self-indicating method for the progress of the peptide bond formation. The Copenhagen group has used this finding to great effect in their glycopeptide syntheses, especially with continuous-flow conditions, because the resin color can be monitored at 440 nm [60].

Other Carboxyl Activation Methods Used for Glycopeptide Synthesis

Recently, a number of uronium [118] and phosphonium salts have been used to activate the carboxyl functions of protected amino acids or glycoamino acids for peptide bond formation when used in the presence of excess base, commonly DIPEA. For glycopeptide syntheses these have included Benzotriazol-l-yloxytris (dimethylamino)phosphonium hexafluorophosphate **23** (BOP or Castro's reagent [101]) in the presence of HOBt [29,37,66,72], *O*-(1*H*-benzotriazol-l-yl)-*N,N,N′,N′*-tetramethyluronium tetrafluoroborate **24** (TBTU), or the essentially similar hexafluorophosphate salt **25** (HBTU) [19,61,72,74,93,94]. These reagents produce active acylating species, in situ, presumably by generating the corresponding hydroxybenzotriazolyl esters. PyBOP **26** is now recommended as a replacement for BOP, since it does not give carcinogenic by-products [137].

23 BOP

24 $X^- = F_4B^-$ TBTU

25 $X^- = F_6P^-$ HBTU

26 PyBOP

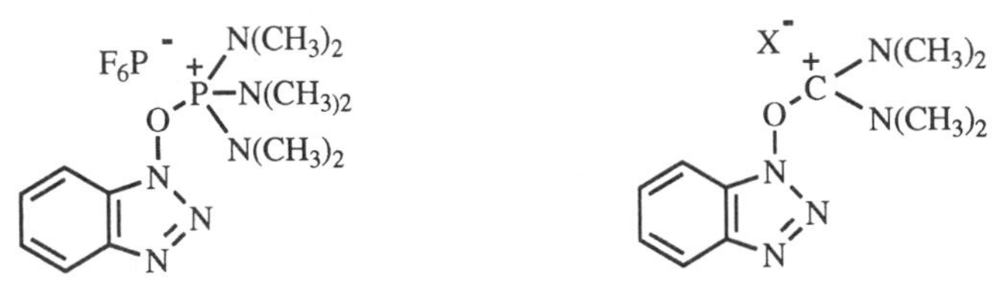

IV. SIDE REACTIONS AND OTHER PROBLEMS IN GLYCOPEPTIDE SYNTHESIS ARISING DURING PEPTIDE ELONGATION

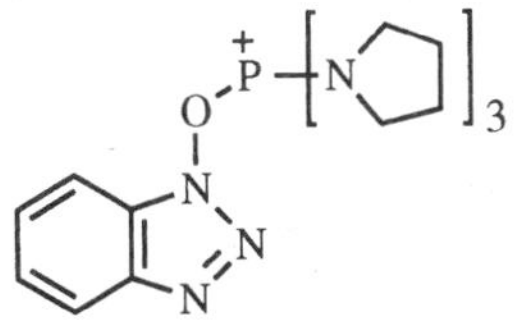

A. Sensitivity of Glycopeptides to Interglycosidic Bond Cleavage and β-Elimination

The interglycosidic links within an oligosaccharide are acetals in nature and therefore are sensitive to acid cleavage or to anomerization. In addition, glycosides of serine and threonine are reactive with bases above pH 11, with cleavage of the glycosidic link by β-elimination **27** [89,116].

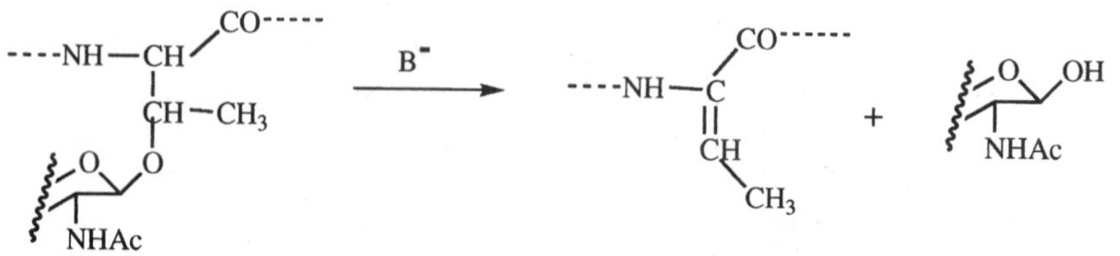

27

Such factors complicate the choice of protecting groups employed in glyco-peptide synthesis, either in solution or on solid phase. In the latter case, the method of cleavage of the finished glycopeptide from the resin is also a matter of concern. These issues have attracted much debate in the literature, and attempts have been made to define the limits for either acid or base conditions that will avoid damage to glycopeptides. In the solid-phase synthesis described by Lavielle and co-workers, the asparagine-GlcNAc link was subjected to 20-min treatments with 45% TFA for removal of 'Boc groups and to liquid HF at $-20°C$ and $0°C$ for 0.8 hr for resin cleavage, apparently without damage to the glycopeptide [13]. Although the GlcNAc-asparagine bond is regarded as perhaps the most stable, certainly in comparison with α-fucosyl and α-sialyl links, the use of 'Boc protection or 'butyl esters/ethers, both requiring TFA for their removal, seems to present some risk to sensitive glycosidic links. Thus Bencomo and Sinaÿ [22] found that the O-glycosyl links in a fully acetylated Gal(β1-3)-GlcNAc(α1-3)-threonyl derivative were stable to "brief" treatment with TFA. Urge, Otvos, and co-workers examined a number of glyco-sylated asparagine derivatives in which there were different sugar linkages [85]. These compounds were exposed to TFA to examine the stability of the N-glycosyl bond and the various O-glycosidic links under typical conditions for solid-phase synthesis. The lability of the various O-glycosidic links examined was found to be:

Gal(β1-3) < Gal(β1-4), Glc(β1-4) < GlcNAc(β1-4) < N-glycosyl bond

In an associated experiment using a Y-antigen oligosaccharide containing two α-Fuc links, TFA brought about 15% decomposition after 15 minutes and 80% after 1 hour. However, with TFA/DMF (1:1) no decomposition occurred after 1 hour.

The Kunz group found that the β1−4 link between GlcNAc units in a chito-biosyl peptide was resistant to TFA, but only if the sugar hydroxyls were acetylated [9,10]. This was also observed by other groups [125,20] and in other work by the Kunz group [58], where a sensitive Fuc(α1−6)-GlcNAc link was extensively dam-aged by TFA treatment during removal of a 'butyl ester group, when O-acetyl groups in the carbohydrate were omitted. Kunz ascribed the beneficial effect of the acetate ester groups to the protonation of their carbonyls with TFA, which effectively sup-presses the protonation of the glycosidic oxygen, a necessary first step in its rupture.

The O-acetylated glycosylamines have been reported to be less nucleophilic than their unprotected counterparts during the formation of GlcNAc- or GalNAc-amino acid building blocks [125,72], but pentafluorophenyl ester derivatives do not acylate unprotected sugar hydroxyl groups [16,18]. There is some evidence to suggest that nonacetylated glycoamino acid derivatives react more quickly with the growing peptide chain during automated solid-phase synthesis [16,18], but this is not universally accepted [131,36]. However, these problems may be avoided in a *solution* synthesis, as was demonstrated by the Kunz group [12]. In this work an acetylated trisaccharide, (Lewis[a])-octapeptide was synthesized; and in a final cleavage of [t]Boc and [t]butyl esters with TFA there was no damage to a sensitive α-fucosyl link.

As the development of solid-phase techniques has led to the use of resins and linkers requiring ever milder acid conditions for cleavage [37] (or no acid at all in the case of HYCRAM™ and HYCRON), more attention has been focused onto the use of the base-removable *N*-Fmoc protection. Removed by 10% piperidine (pK$_a$ 11.1) [44,95] and by morpholine (pKa 8.3) as well as by DBU, there remains concern at the possibility of β-elimination of the glycan during solid-phase glycopeptide synthesis by Strategy 1, 2, or 3. Though β-elimination can occur at pH 9 [116], the reaction depends upon the structure of the glycoconjugate [89]. Morpholine in DMF is safer [89] than piperidine, and DBU may be of assistance [48] if there is a danger of incomplete Fmoc removal [44]. The Kunz group advocate the safer Peoc group (5), since it is completely removable by morpholine (23) without β-elimination. An alternative strategy may be enzymatic removal of protecting groups, at least in the final stages of a solution-phase [97] or solid-phase synthesis [96]. By this means, exposure of a glycopeptide to either base or acid would be reduced.

B. Racemization and Epimerization During Glycopeptide Assembly

The use of a reliable carboxyl activation method and a urethane-protected amino acid (or glycoamino acid) is usually regarded as the safest approach for glycopeptide synthesis for the minimization of racemization. The danger of epimerization is greater if a *peptide* is activated, as in fragment condensation. In view of its potential importance for biological activity of synthetic products, surprisingly few reports [19] exist in the literature where specific assessment is made of the optical purity of synthetic glycopeptides.

The suppression of racemization in carbodiimide activation by HOBt and other substituted benzotriazoles [119–121] is well known, but there is still danger when fragment coupling is attempted [19]. The newer carboxyl activating reagent BOP [101] in conjunction with HOBt [117] has been shown to be relatively free from racemization when used to activate a protected amino acid, but Steinauer et al. [127] and the Rocchi group [29] have warned about its use for fragment coupling. Hindered bases, e.g., DIPEA, are safer than other amines normally used, but a combination of excess base and dichloromethane as solvent must be avoided [127]. Knorr and co-workers in a study of racemization found that TBTU/HOBt was much superior to BOP for carboxyl activation [118].

C. Aspartimide (Aminosuccinyl) Formation

The intramolecular cyclization of aspartyl peptides is a side reaction of significance in the synthesis of *N*-glycopeptides. It was shown by Bodanszky and Kwei to be peptide-sequence-dependent and to be acid- or base-catalyzed [123]. Sequences of Asp-Gly, Asp-Ser, and Asp-Thr are particularly susceptible, and Asp-β-benzyl esters are prone to nucleophilic attack by the amide nitrogen at the aspartyl C-terminus, as shown in **28** [112,67].

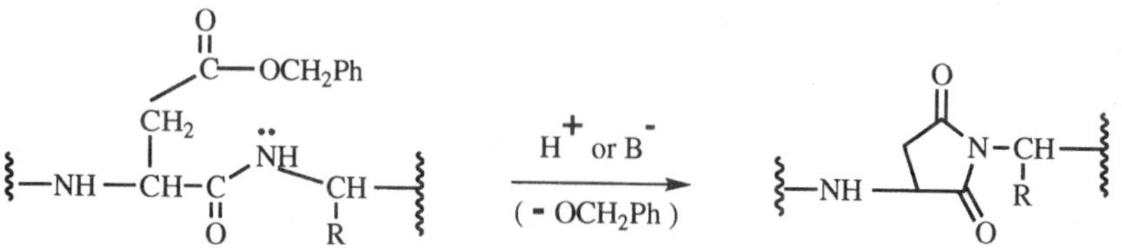

28

This damaging side reaction can lead to α/β-transpeptidation through the cyclic aspartimide, and some different approaches to its suppression have been proposed. Voelter and co-workers [19] prepared an N-terminally glycosylated 21-mer containing the aspartimide-susceptible sequence Asp-Gly- at positions 10 and 11. In the synthesis, residues 3–10 were assembled on a chlorotrityl resin and cleaved to produce a protected peptide with C-terminal Asp. This fragment was then linked to Gly(11) of the C-terminal sequence, which was attached to a chlorotrityl resin. Finally, the N-terminal glyco-dipeptide was added. No aspartimide formation was observed by this stratagem.

Aspartimide formation can also occur when a glycosylamine is reacted with the activated β-carboxyl group of an aspartyl peptide in a convergent solution synthesis [67]. This was studied by Anisfeld and Lansbury [72,74] in some detail, by varying the molar quantities of the amine, the base (DIPEA), the HBTU coupling reagent, and the added HOBt. It was found important to avoid excess base (aspartimide formation is base catalyzed). Also, increased amounts of HBTU and HOBt were helpful to accelerate the acylation reaction, which is in competition with the aspartimide formation. The β-carboxyl group of the peptide for glycosylation was protected as its bulky cyclohexyl ester, and this was effective in suppressing aspartimide formation during the solid-phase peptide synthesis itself.

In a recent elegant synthesis, Kunz and co-workers prepared the linear heptapeptide Ala-Asp-Ser-Asp-Ala-Asp-Gly, which contains the difficult -Asp-Gly- and -Asp-Ser- sequences. Each β-aspartyl carboxyl group was protected with the bulky ^{t}butyl ester group [93]. After assembly of the peptide on the chlorotrityl resin system, the released linear peptide was then cyclized in solution and the ^{t}butyl ester groups removed. Glycosylation of each aspartyl residue with the glycosylamine from a Lewis^x tetrasaccharide completed the convergent synthesis. No evidence of aspartimide formation was found throughout the synthesis. Presumably the ^{t}butyl groups prevented aspartimide ring formation during the peptide synthesis, and the cyclic heptapeptide was unable to cyclize further to aspartimide during the glycosylation. Another development in the prevention of aspartimide formation during the synthesis of N-glycopeptides is the proposal by Offer et al. to use the N-(2-acetoxy-4-methoxybenzyl) or AcHmb group to block temporarily the -aspartyl-CONH-X_{aa}-amide nitrogen during glycopeptide assembly [78]. By removal of the amide proton and concomitant decrease in the nucleophilicity of the nitrogen, aspartimide formation is again suppressed.

D. Miscellaneous Problems in Glycopeptide Synthesis

Intramolecular aminolysis of *N*-unprotected dipeptide esters, which results in the formation of cyclic diketopiperazines [112], has been reported in solution synthesis of glycopeptides at the dipeptide stage [6,9]. It is also a potential problem in solid-phase synthesis, resulting in premature cleavage of the peptide from the resin, as shown in **29**. For solid-phase synthesis, deprotection of the Fmoc group by base at the dipeptide stage liberates the free amino group, which could attack susceptible carbonyl groups at the peptide-resin junction. Formation of the stable six-membered diketopiperazine ring would result in the loss of peptide chains from the resin. Both glycine (as H_2N-Gly-X_{aa}-resin) and proline (as H_2N-Pro-X_{aa}-resin or H_2N-X_{aa}-Pro-resin) are vulnerable to this side reaction [122], and, therefore, immediate acylation by the third amino acid residue is desirable after Fmoc deprotection. Other methods of reducing the danger of this reaction include the use of the bulky chlorotrityl resin or, where proline is C-terminal or the penultimate amino acid, to couple a prolyl-dipeptide unit, as in the *O*-glycopeptide synthesis of Rocchi and co-workers [107].

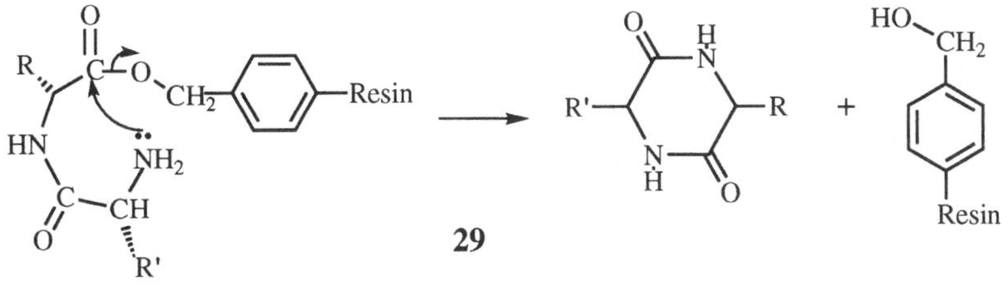

29

For successful solid-phase synthesis without deletion sequences, it is essential to obtain the highest possible chemical yield in each step of the glycopeptide elongation. For insoluble polymer supports, successful diffusion of reagents in solution is crucial, if quantitative yields are to be approached. To assist such diffusion in this essentially heterogeneous system, large excesses of reagents are commonly used, and repeated treatment may also be required if analytical checks show this to be necessary. Continuous-flow techniques can greatly reduce the need for such excesses of reagents [106], but this general feature of solid-phase methodology is an added difficulty for *glyco*peptide assembly when only small amounts of a valuable glycan may be available [42,72,74].

V. OUTLOOK

In the last few years great progress has been made in the chemical synthesis of glycopeptides, by exploiting both the technical advances in solid-phase assembly of the peptide portion and separation technology of the glycopeptide products by preparative HPLC and gel permeation chromatography. Many of the successful syntheses have been achieved by the use of glycosylated amino acid building blocks. This Strategy 1 approach is very versatile and valuable for the building of combinatorial libraries of glycopeptides using multicell synthesis machines. Multiglycosylated glycopeptides bearing different glycans at different positions are possible by this general method. It may be argued, at the present time, that it is *relatively* straightforward to build the peptide moiety. However, problems still remain to be solved, as is illustrated by the difficulties encountered by the Meldal and Bock group in the recent assembly of a glycosylated 25-mer [94]. These difficulties included glycopeptide aggregation on the resin and poor solubility after cleavage.

The chemical preparation of oligosaccharide precursors (with biological potential) for inclusion into glycopeptides remains costly in time and effort, and it is probable that quantities of suitable oligosaccharides for synthetic elaboration into glycopeptides will remain limited. The Strategy 3 approach to building the oligosaccharide on the solid phase, pioneered by the Danishefsky group [83], provides an avenue for progress in this respect. In addition, prebuilding the peptide with incorporation of a minimum carbohydrate ''handle,'' followed by enzymatic extension in solution of the carbohydrate moiety (Strategy 4), provides another route to glycopeptides that must surely be developed further [63–66]. By this approach and also in the fully convergent method [93,74] (Strategy 5), assembly of the peptide moiety must proceed in such a way that some protecting groups are retained following cleavage from the solid phase. It is therefore reasonable to expect that solid-phase assembly of such a peptide moiety will be further developed using mild methods for cleavage, exemplified currently by resins of the SASRIN, HYCRON, and chlorotrityl types and, perhaps, by more soluble and polar polymer supports [106,66]. Selective and mild methods are needed for the assembly of the sensitive glycopeptide conjugates, and further advancement in such synthetic methodology would be consistent with trends in other areas of natural product chemistry.

ABBREVIATIONS

Ac	CH_3CO-
AcHmb	*N*-(2-acetoxy-4-methoxybenzyl)
Aloc	allyloxycarbonyl
'Boc	*tert*-butoxycarbonyl
BOP	benzotriazole-1-yl-oxy-tris(dimethylamino)-phosphonium hexafluorophosphate
Bz	$PhCH_2-$
DBU	1,7-diaza-(5,4,0)-bicycloundec-1,8-ene
DCCI	*N,N'*-dicyclohexylcarbodiimide
DCU	dicyclohexylurea
DCM	dichloromethane
Dhbt-OH	3,4-dihydro-4-oxo-1,2,3-benzotriazine
Dim	1,3-dithian-2-ylmethyl
DIPEA	diisopropylethylamine
DMA	dimethylacetamide
DMF	dimethylformamide
EEDQ	*N*-ethoxycarbonyl-1,2-dihydroquinoline

Fmoc	9-fluorenylmethoxycarbonyl
GalNAc	*N*-acetyl-D-galactosamine
GlcNAc	*N*-acetyl-D-glucosamine
HBTU	2-(1*H*-benzotriazole-1-yl)-1,1,3,3-tetramethyluronium hexafluorophosphate
HOBt	1-hydroxybenzotriazole
HPLC	high-performance (-pressure) liquid chromatography
IIDQ	2-isobutoxy-1-isobutoxycarbonyl-1,2-dihydroquinoline
MBHA	methylbenzhydrylamine
NMP	*N*-methyl pyrrolidone
PAL	5-(4-aminomethyl-3,5-dimethoxy-3,5-dimethoxyphenoxy)-valeric acid
PAM	4-(hydroxymethyl)phenylacetamidomethyl
PEGA	copolymer of poly(ethylene glycol) and poly(*N*,*N*-dimethylacrylamide)
Peoc	*N*-[2-(triphenylphosphonio)ethoxycarbonyl]
Pfp	pentafluorophenyl
PyBOP	benzotriazole-1-yl-oxy-tris-pyrrolidino-phosphonium hexafluorophosphate
Pyoc	2-(4-pyridyl)ethoxycarbonyl)
RINK	4-(2′,4′-dimethoxyphenylaminomethyl)phenoxymethyl-
SASRIN	super acid-sensitive resin
TBTU	2-(1*H*-benzotriazole-1-yl)-1,1,3,3-tetramethyluronium tetrafluoroborate
Teoc	2,2,2-trichloroethoxycarbonyl
TFA	2,2,2-trifluoroethanoic acid
WRK	*N*-ethyl-5-phenylisoxazolium-3′-sulfonate
WSC	water-soluble carbodiimide
Z	benzyloxycarbonyl

ACKNOWLEDGMENT

We would like to express our appreciation to Dr. Janet Large for preparing the molecular structure drawings.

REFERENCES

1. H. G. Garg and R. W. Jeanloz, The synthesis of protected glycopeptides containing the amino acid sequences 34−37 and 34−38 of bovine ribonuclease B, *Carbohydr. Res. 32*, 37 (1974).
2. M. A. E. Shaban and R. W. Jeanloz, The synthesis of glycopeptide fragments of human plasma α_1-acid glycoproteins by sequential elongation at the terminal-amino group, *Carbohydr. Res. 43*, 281 (1975).
3. H. G. Garg and R. W. Jeanloz, Synthesis of substituted glycopeptides containing a 2-acetamido-2-deoxy-β-D-glucopyranosyl residue and the amino acid sequence 18−22 of bovine pancreatic deoxyribonuclease A, *J. Org. Chem. 41*, 2480 (1976).
4. H. G. Garg and R. W. Jeanloz, Synthesis of glycopeptide derivatives containing the 2-acetamido-*N*-(L-aspart-4-oyl)-2-deoxy-β-D-glucopyranosyl-amine linkage and having the amino acid sequences 32−34, 33−35, 33−37, and 33−38 of bovine ribonuclease, *Carbohydr. Res. 70*, 47 (1979).

5. H. G. Garg and R. W. Jeanloz, Synthesis of glycopeptides containing the amino acid sequence 17–23 of bovine pancreatic deoxyribonuclease, *Carbohydr. Res. 86*, 59 (1980).

6. H. Kunz and H. Kauth, Synthesis of protected asparagine glycopeptides via *N*-terminal elongation of the peptide chain—Partial sequences of bovine deoxyribonuclease A and of luteinizing hormone, *Liebigs Ann. Chem.*: 337 (1983). (In German).

7. H. Waldmann and H. Kunz, Allyl esters as selectively removable carboxy-protecting functions in peptide and *N*-glycopeptide syntheses, *Liebigs Ann. Chem.*: 1712 (1983). (In German).

8. H. Kunz and H. Waldmann, 1,3-dithian-2-yl methyl ester as two-step protecting group for the carboxy function in peptide synthesis, *Angew. Chem. Int. Ed. Engl. 22*, 62 (1983).

9. H. Kunz and H. Waldmann, Construction of disaccharide *N*-glycopeptides—Synthesis of the linkage region of the transmembrane-neuriminidase of an influenza virus, *Angew. Chem. Int. Ed. Engl. 24*, 883 (1985).

10. H. Kunz, H. Waldmann, and J. März, Synthesis of partial structures of *N*-glycopeptides representing the linkage regions of the transmembrane neuraminidase of an influenza virus and of factor B of the human complement system, *Liebigs Ann. Chem.*: 45 (1989). (In German).

11. R. S. Clark, S. Banerjee, and J. K. Coward, Yeast Oligosaccharyltransferase: Glycosylation of peptide substrates and chemical characterization of the glycopeptide product, *J. Org. Chem. 55*, 6275 (1990).

12. H. Kunz and J. März, Synthesis of glycopeptides with Lewis^a antigen side chain and HIV peptide T sequence using the trichloroethoxycarbonyl/allyl ester protecting group combination, *Synlett*, 591 (1992).

13. S. Lavielle, N. C. Ling, and R. C. Guillemin, Solid-phase synthesis of two glycopeptides containing the amino acid sequence 5 to 9 of somatostatin, *Carbohydr. Res. 89*, 221 (1981).

14. H. Kunz and B. Dombo, Solid phase synthesis of peptides and glycopeptides on polymeric supports with allylic anchor groups, *Angew. Chem. Int. Ed. Engl. 27*, 711 (1988).

15. M. Meldal and K. Bock, Pentafluorophenyl esters for temporary carboxyl group protection in solid-phase synthesis of *N*-linked glycopeptides, *Tetrahedron Lett. 31*, 6987 (1990).

16. L. Otvos, Jr., L. Urge, M. Hollosi, K. Wroblewski, G. Graczyk, G. D. Fasman, and J. Thurin, Automated solid-phase synthesis of glycopeptides. Incorporation of unprotected mono- and disaccharide units of *N*-glycoprotein antennae into T cell epitopic peptides, *Tetrahedron Lett. 41*, 5889 (1990).

17. L. Biondi, F. Filira, M. Gobbo, B. Scolaro, and R. Rocchi, Synthesis of glycosylated tuftsins and tuftsin-containing IgG fragment undecapeptide, *Int. J. Peptide Protein Res. 37*, 112 (1991).

18. L. Urge, D. C. Jackson, L. Gorbies, K. Wroblewski, G. Graczyk, and L. Otvos, Jr., Synthesis and conformational analysis of *N*-glycopeptides that contain extended sugar chains, *Tetrahedron 50*, 2373 (1994).

19. H. Zhang, Y. Wang, and W. Voelter, A new strategy for the synthesis of Asp-Gly units containing glycopeptides using Fmoc/Bzl protection, *Tetrahedron Lett. 36*, 8767 (1995).

20. G. Arsequell, J. S. Haurum, T. Elliott, R. A. Dwek, and A. C. Lellouch, Synthesis of major histocompatibility complex class I binding glycopeptides, *J. Chem. Soc. Perkin Trans. 1*, 1739 (1995).

21. B. Ferrari and A. A. Pavia, Blood group antigens: Synthesis of T_N glycopeptide-related to human glycophorin A^M, *Int. J. Peptide Protein Res. 22*, 549 (1983).

22. V. V. Bencomo and P. Sinay, Synthesis of glycopeptides having clusters of O-glyco-sylic disaccharide chains [β-D-Gal-(1 → 3)-α-D-GalNAc] located at vicinal amino acid residues of the peptide chain, *Carbohydr. Res. 116*, C9 (1983).

23. H. Kauth and H. Kunz, Synthesis of protected serine glycopeptides via N-terminal elongation of the peptide chain, *Liebigs Ann. Chem.*: 360 (1983). (In German).

24. H. Kunz and H. Waldmann, The allyl group as mildly and selectively removable carboxy-protecting group for the synthesis of labile O-glycopeptides, *Angew. Chem. Int. Ed. Engl. 23*, 71 (1984).

25. B. Ferrari and A. A. Pavia, Synthese de glycopeptide T_N representant la partie N-terminale de la glycophorine humaine A^N et A^{Mc}, *Tetrahedron 41*: 1939 (1985).

26. H. Paulsen and K. Adermann, Synthesis of varied O-glycopeptides of Interleukin-2, *Liebigs Ann. Chem.*: 771 (1989). (In German).

27. H. Paulsen, G. Merz, and I. Brockhausen, Synthesis of L-proline-containing O-gly-copeptides, *Liebigs Ann. Chem.*: 719 (1990). (In German).

28. M. Ciommer and H. Kunz, Synthesis of glycopeptides with partial structure of human glycophorin using the fluorenylmethoxycarbonyl/allyl ester protecting group combination, *Synlett*, 593 (1991).

29. M. Gobbo, L. Biondi, F. Filira, and R. Rocchi, Solution synthesis of the glyco-hex-apeptide sequence of the human oncofetal fibronectin defined by monoclonal antibody FDC-6, *Int. J. Peptide Protein Res. 38*, 417 (1991).

30. H. Paulsen, K. Adermann, G. Merz, M. Schultz, and U. Weichert, Synthesis of O-glycopeptides, *Starch/Stärke 40*, 465 (1988). (In German).

31. H. Paulsen, G. Merz, and U. Weichert, Solid-phase synthesis of O-glycopeptide se-quences, *Angew. Chem. Int. Ed. Engl. 27*, 1365 (1988).

32. H. Paulsen, G. Merz, S. Peters, and U. Weichert, Solid-phase synthesis of glycopep-tides, *Liebigs Ann. Chem.*: 1165 (1990). (In German).

33. M. Meldal and K. J. Jensen, Pentafluorophenyl esters for the temporary protection of the α-carboxy group in solid-phase glycopeptide synthesis, *J. Chem. Soc. Chem. Com-mun.*: 483 (1990).

34. B. Lüning, T. Norberg, C. Rivera-Baeza, and J. Tejbrant, Solid-phase synthesis of the fibronectin glycopeptide V(Galβ3GalNAcα)THPGY, its β analogue, and the corre-sponding unglycosylated peptide, *Glycoconjugate J. 8*, 450 (1991).

35. E. Bardaji, J. L. Torres, P. Clapés, F. Albericio, G. Barany, R. E. Rodriguez, M. P. Sacristan, and G. Valencia, Synthesis and biological activity of O-glycosylated mor-phiceptin analogues, *J. Chem. Soc. Perkin Trans. 1*, 1755 (1991).

36. S. Peters, T. Bielfeldt, M. Meldal, K. Bock, and H. Paulsen, Solid phase synthesis of mucin glycopeptides, *Tetrahedron Lett. 33*, 6445 (1992).

37. R. Polt, L. Szabó, J. Triberg, Y. Li, and V. J. Hruby, Synthesis of O-glycosyl cyclic enkephalins, *J. Am. Chem. Soc. 114*, 10,249 (1992).

38. E. Atherton and R. C. Sheppard, *Solid phase peptide synthesis—A practical approach*, IRL Press, Oxford/New York/Tokyo, 1989.

39. G. B. Fields and R. L. Noble, Solid phase peptide synthesis utilizing 9-fluorenyl-methoxycarbonyl amino acids, *Int. J. Peptide Protein Res. 35*, 161 (1990).

40. M. Gobbo, L. Biondi, F. Filira, B. Scolaro, R. Rocchi, and T. Piek, Synthesis and biological activity of the mono- and di-galactosyl-vespulakinin 1 analogues, *Int. J. Peptide Protein Res. 40*, 54 (1992).

41. T. Bielfeldt, S. Peters, M. Meldal, K. Bock, and H. Paulsen, A new strategy for solid-phase synthesis of O-glycopeptides, *Angew. Chem. Int. Ed. Engl. 31*, 857 (1992).

42. K. B. Reimer, M. Meldal, S. Kusumoto, K. Fukase, and K. Bock, Small-scale solid-phase O-glycopeptide synthesis of linear and cyclized hexapeptides from blood-clotting factor IX containing O-(α-D-Xyl-1 → 3-α-D-Xyl-1 → 3-β-D-Glc)-L-Ser, *J. Chem. Soc. Perkin Trans. 1*, 925 (1993).

43. M. Elofsson, S. Roy, B. Wasle, and J. Kihlberg, Solid-phase synthesis and confor-
     mational studies of glycosylated derivatives of helper-T-cell immunogenic peptides
     from hen-egg lysozyme, *Carbohydr. Res. 246*, 89 (1993).
44. J. Kihlberg and T. Vuljanic, Piperidine is preferable to morpholine for Fmoc cleavage
     in solid phase synthesis of *O*-linked glycopeptides, *Tetrahedron Lett. 34*, 6135 (1993).
45. H. Paulsen, T. Bielfeldt, S. Peters, M. Meldal, and K. Bock, A new strategy for the
     solid-phase synthesis of *O*-glycopeptides via 2-azido-glycopeptides, *Liebigs Ann.
     Chem.*: 369 (1994). (In German).
46. H. Paulsen, T. Bielfeldt, S. Peters, M. Meldal, and K. Bock, Application of the azido
     glycopeptide synthesis strategy for the multiple column solid phase synthesis of mucin
     O-glycopeptides, *Liebigs Ann. Chem.*: 381 (1994). (In German).
47. M. Zheng, M. Gobbo, L. Biondi, F. Filira, S. Hakomori, and R. Rocchi, Synthetic
     immunochemistry of glycohexapeptide analogues characteristic of oncofetal fibronec-
     tin, *Int. J. Peptide Protein Res. 43*, 230 (1994).
48. M. Meldal, T. Bielfeldt, S. Peters, K. J. Jensen, H. Paulsen, and K. Bock, Suscepti-
     bility of glycans to β-elimination in Fmoc-based *O*-glycopeptide synthesis, *Int. J.
     Peptide Protein Res. 43*, 529 (1994).
49. H. Paulsen, S. Peters, T. Bielfeldt, M. Meldal, and K. Bock, Synthesis of the glycosyl
     amino acids N^α-Fmoc-Ser[Ac$_4$-β-D-Gal *p*-(1 → 3)-Ac$_2$-α-D-GalN$_3$ *p*]-OPfp and N^α-
     Fmoc-Thr[Ac$_4$-β-D-Gal *p*-(1 → 3)-Ac$_2$-α-D-GalN$_3$ *p*]-OPfp and the application in the
     solid-phase peptide synthesis of multiply glycosylated mucin peptides with T^n and T
     antigenic structures, *Carbohydr. Res. 268*, 17 (1995).
50. E. Meinjohanns, A. Vargas-Berenguel, M. Meldal, H. Paulsen, and K. Bock, Com-
     parison of *N*-Dts and *N*-Aloc in the solid-phase syntheses of *O*-GlcNAc glycopeptide
     fragments of RNA-polymerase II and mammalian neurofilaments, *J. Chem. Soc. Per-
     kin Trans. 1*, 2165 (1995).
51. S. Rio-Anneheim, H. Paulsen, M. Meldal, and K. Bock, Synthesis of the building
     blocks N^α-Fmoc-*O*-[α-D-Ac$_3$GalN$_3$ *p*-(1 → 3)-α-D-Ac$_2$GalN$_3$ *p*]-Thr-OPfp and N^α-
     Fmoc-*O*-[α-D-Ac$_3$GalN$_3$*p*-(1 → 6)-α-D-Ac$_2$GalN$_3$*p*]-Thr-OPfP and their application in
     the solid phase glycopeptide synthesis of core 5 and 7 mucin *O*-glycopeptides, *J.
     Chem. Soc. Perkin Trans. 1*, 1071 (1995).
52. J. Rademann and R. R. Schmidt, Solid-phase synthesis of a glycosylated hexapeptide
     of human sialophorin, using the trichloroacetimidate method, *Carbohydr. Res. 269*,
     217 (1995).
53. O. Seitz and H. Kunz, A novel allylic anchor for solid-phase synthesis—Synthesis
     of protected and unprotected *O*-glycosylated mucin-type glycopeptides, *Angew. Chem.
     Int. Ed. Engl. 34*, 803 (1995).
54. H. Iijima, Y. Nakahara, and T. Ogawa, Synthesis of the N-terminal glycopentapeptides
     of human glycophorin A$_M$ and A$_N$ carrying trimeric sialosyl Tn epitope, *Tetrahedron
     Letters 33*, 7907 (1992).
55. Y. Nakamara, H. Iijima, and T. Ogawa, Stereocontrolled approaches to *O*-glycopeptide
     synthesis, *ACS Symp. Ser 560*, 249 (1994).
56. C-H Wong, M. Schuster, P. Wang, and P. Sears, Enzymatic synthesis of *N*- and *O*-
     linked glycopeptides, *J. Am. Chem. Soc. 115*, 5893 (1993).
57. H. Kunz and S. Birnbach, Synthesis of *O*-glycopeptides of the tumor-associated T$_N$-
     and T-antigen type and their binding to bovine serum albumin, *Angew. Chem. Int. Ed.
     Engl. 25*, 360 (1986).
58. H. Kunz and C. Unverzagt, Protecting-group-dependent stability of intersaccharide
     bonds of a fucosyl-chitobiose glycopeptide, *Angew. Chem. Int. Ed. Engl. 27*, 1697
     (1988).
59. A. M. Jansson, M. Meldal, and K. Bock, The active ester *N*-Fmoc-3-*O*-[Ac$_4$-α-D-
     Man*p*-(1 → 2)-Ac$_3$Man*p*-1-]-Threonine-*O*-Pfp as a building block in solid-phase

synthesis of an O-linked dimannosyl glycopeptide, *Tetrahedron Lett. 31*, 6991 (1990).

60. A. M. Jansson, M. Meldal, and K. Bock, Solid-phase synthesis and characterization of O-dimannosylated heptadecapeptide analogues of human insulin-like growth factor 1 (IGF-1), *J. Chem. Soc. Perkin Trans. 1*, 1699 (1992).

61. K. von dem Bruch and H. Kunz, Synthesis of N-glycopeptide clusters with Lewis[x] antigen side chains and their coupling to carrier proteins, *Angew. Chem. Int. Ed. Engl. 33*, 101 (1994).

62. J. Y. Roberge, X. Beebe, and S. J. Danishefsky, A strategy for a convergent synthesis of N-linked glycopeptides on a solid support, *Science 269*, 202 (1995).

63. M. Schultz and H. Kunz, Enzymatic Glycosylation of O-glycopeptides, *Tetrahedron Lett. 33*, 5319 (1992).

64. C. Unverzagt, S. Kelm, and J. C. Paulson, Chemical and enzymatic synthesis of multivalent sialoglycopeptides, *Carbohydr. Res. 251*, 285 (1994).

65. M. Meldal, F-I Auzanneau, O. Hindsgaul, and M. M. Palcic, A PEGA resin for use in the solid-phase chemical-enzymatic synthesis of glycopeptides, *J. Chem. Soc., Chem. Commun.*: 1849 (1994).

66. M. Schuster, P. Wang, J. C. Paulson and C-H. Wong, Solid-phase chemical-enzymatic synthesis of glycopeptides and oligosaccharides, *J. Am. Chem. Soc. 116*, 1135 (1994).

67a. R. P. Beckett, *Studies of Strategy in Glycopeptide Synthesis*, PhD Dissertation, Liverpool Polytechnic, (1985).

67b. R. P. Beckett and D. G. Large, A convergent route to solution synthesis of N-glycopeptides, VIII European Carbohydrate Symposium, Cracow, Poland (1993).

68. A. K. M. Anisuzzaman, L. Anderson, and (in part) J. L. Navia, Synthesis of a close analog of the repeating unit of the antifreeze glycoproteins of polar fish, *Carbohydr. Res. 174*, 265 (1988).

69. J. Lee and J. K. Coward, Enzyme-catalyzed glycosylation of peptides using a synthetic lipid disaccharide substrate, *J. Org. Chem. 57*, 4126 (1992).

70. F. E. McDonald and S. J. Danishefsky, A stereoselective route from glycals to asparagine-linked N-protected glycopeptides, *J. Org. Chem. 57*, 7001 (1992).

71. A. Rajca and M. Wiessler, Direct glycosylation of protected O-tritylserine esters and oligopeptides, *Carbohydr. Res. 274*, 123 (1995).

72. S. T. Anisfeld and P. T. Lansbury, Jr., A convergent approach to the chemical synthesis of asparagine-linked glycopeptides, *J. Org. Chem. 55*, 5560 (1990).

73. M. Hollosi, E. Kollat, I. Laczko, K. F. Medzihradszky, J. Thurin, and L. Otvos, Jr., Solid-phase synthesis of glycopeptides: Glycosylation of resin-bound serine-peptides by 3,4,6-tri-O-acetyl-D-glucose-oxazoline, *Tetrahedron Lett. 32*, 1531 (1991).

74. S. T. Cohen-Anisfeld and P. T. Lansbury, Jr., A practical, convergent method for glycopeptide synthesis, *J. Am. Chem. Soc. 115*, 10531 (1993).

75. D. M. Andrews and P. W. Seale, Solid-phase synthesis of O-mannosylated peptides: Two strategies compared, *Int. J. Peptide Protein Res. 42*, 165 (1993).

76. A. C. Bauman, J. S. Broderick, R. M. Dacus, IV, D. A. Grover, and L. S. Trzupek, A convenient procedure for the preparation of fully protected C-terminal glycopeptides, *Tetrahedron Lett. 34*, 7019 (1993).

77. D. Vetter, D. Tumelty, S. K. Singh, and M. A. Gallop, A versatile solid-phase synthesis of N-linked glycopeptides, *Angew. Chem. Int. Ed. Engl. 34*, 60 (1995).

78. J. Offer, M. Quibell, and T. Johnson, On-resin solid-phase synthesis of asparagine N-linked glycopeptides: Use of N-(2-acetoxy-4-methoxybenzyl) (AcHmb) aspartyl amide-bond protection to prevent unwanted aspartimide formation, *J. Chem. Soc. Perkin Trans. 1*, 175 (1996).

79. J. C. Sheehan and G. P. Hess, A new method of forming peptide bonds, *J. Am. Chem. Soc. 77*, 1067 (1955).

80. B. Belleau and G. Malek, A new reagent for peptide synthesis, *J. Am. Chem. Soc.* *90*, 1651 (1968).

81. R. B. Woodward, R. A. Olofson, and H. Mayer, A new synthesis of peptides, *J. Am. Chem. Soc. 83*, 1010 (1961).

82. M. Meldal and K. Bock, A general approach to the synthesis of *O*- and *N*-linked glycopeptides, *Glycoconjugate J. 11*, 59 (1994).

83. S. J. Danishefsky and J. Y. Roberge, Advances in the development of convergent schemes for the synthesis of biologically important glycoconjugates, *Pure Appl. Chem. 67*, 1647 (1995).

84. J. Montreuil, Primary structure of glycoprotein glycans: Basis for the molecular biology of glycoproteins, *Adv. Carbohydr. Chem. Biochem. 37*, 157 (1980).

85. L. Urge, L. Otvos, Jr., E. Lang, K. Wroblewski, I. Laczko, and M. Hollosi, Fmoc-protected, glycosylated asparagines potentially useful as reagents in the solid-phase synthesis of *N*-glycopeptides, *Carbohydr. Res. 235*, 83 (1993).

86. H. Kunz, The 2-(Triphenylphosphonio)ethoxycarbonyl group as an amino protective function in peptide chemistry, *Chem. Ber. 109*, 2670 (1976). (In German).

87. L. A. Carpino and G. Y. Han, The 9-Fluorenylmethoxycarbonyl function, a new base-sensitive amino-protecting group, *J. Am. Chem. Soc. 92*, 5718 (1970).

88. H. Waldmann and H. Kunz, 1,3-dithian-2-ylmethyl esters as two-step carboxy protecting groups in the synthesis of *N*-glycopeptides, *J. Org. Chem. 53*, 4172 (1988).

89. H. Kunz, Synthesis of glycopeptides, Partial structures of biological recognition components, *Angew. Chem. Int. Ed. Engl. 26*, 294 (1987).

90. G. S. Marks and A. Neuberger, Synthetic studies relating to the carbohydrate-protein linkage in egg albumin, *J. Chem. Soc.*, 4872 (1961).

91. H. G. Garg and R. W. Jeanloz, The synthesis, and study of the β-elimination reaction, of di- and tripeptides having a 3-*O*-(2-acetamido-3,4,6-tri-*O*-acetyl-2-deoxy-β-D-glucopyranosyl)-L-serine residue, *Carbohydr. Res. 76*, 85 (1979).

92. G. Kupryszewski and S. Pedagogiczna, Amino acid chlorophenyl esters. II. Synthesis of peptides by aminolysis of active *N*-protected amino acid 2,4,6-trichlorophenyl esters, *Roczniki Chem. 35*, 595 (1961). (*Chemical Abstracts 55*:27121).

93. U. Sprengard, M. Schudok, W. Schmidt, G. Kretzschmar, and H. Kunz, Multiple sialyl Lewis[x] *N*-glycopeptides: Effective ligands for E-selectin, *Angew. Chem. Int. Ed. Engl. 35*, 321 (1996).

94. A. M. Jansson, K. J. Jensen, M. Meldal, J. Lomako, W. M. Lomako, C. E. Olsen, and K. Bock, Solid-phase glycopeptide synthesis of tyrosine-glycosylated glycogenin fragments as substrates for glucosylation by glycogenin, *J. Chem. Soc. Perkin Trans. 1*, 1001 (1996).

95. P. Sjölin, M. Elofsson, and J. Kihlberg, Removal of acyl protective groups from glycopeptides: Base does not epimerize peptide stereocenters, and β-elimination is slow, *J. Org. Chem. 61*, 560 (1996).

96. J. Eberling, P. Braun, D. Kowalczyk, M. Schultz, and H. Kunz, Chemoselective removal of protecting groups from *O*-glycosyl amino acid and peptide (methoxyethoxy) ethyl esters using lipases and papain, *J. Org. Chem. 61*, 2638 (1996).

97. H. Kunz, D. Kowalczyk, P. Braun, and G. Braum, Enzymatic hydrolysis of hydrophilic diethyleneglycol and polyethyleneglycol esters of peptides and glycopeptides by lipases, *Angew. Chem. Int. Ed. Engl. 33*, 336 (1994).

98. K. J. Jensen, P. R. Hansen, D. Venugopal, and G. Barany, Synthesis of 2-acetamido-2-deoxy-β-D-glucopyranose *O*-glycopeptides from *N*-dithiasuccinoyl-protected derivatives, *J. Am. Chem. Soc. 118*, 3148 (1996).

99. M. Meldal, Glycopeptide synthesis, in *Neoglycoconjugates: Preparation and Application* (Y. C. Lee and R. T. Lee, eds.), p. 145, Academic Press, San Diego, 1994.

100. H. Kunz and S. Birnbach, Der 2-(4-pyridyl)ethoxycarbonyl-(4-Pyoc) rest eine hydrophile säure und basenstabile aminoschutzgruppe für die peptidsynthese, *Tetrahedron Lett. 25*, 3567 (1984).

101. B. Castro, J. R. Dormoy, G. Evin, and C. Selve, Reactifs de couplage peptidique IV (1)-l'hexafluorophosphate de benzotriazoyl *N*-oxytrisdimethylamino phosphonium (B.O.P.), *Tetrahedron Lett. 14*, 1219 (1975).

102. R. Arshady, E. Atherton, D. L. J. Clive, and R. C. Sheppard, Peptide synthesis. Part 1. Preparation and use of polar supports based on poly(dimethylacrylamide), *J. Chem. Soc. Perkin Trans. 1*, 529 (1981).

103. E. Atherton, C. J. Logan, and R. C. Sheppard, Peptide synthesis. Part 2. Procedures for solid-phase synthesis using N^α-fluorenylmethoxycarbonylamino-acids on polyamide supports. Synthesis of substance P and of acyl carrier protein 65–74 decapeptide, *J. Chem. Soc. Perkin Trans. 1*, 538 (1981).

104a. F. Albericio and G. Barany, An acid-labile anchoring linkage for solid-phase synthesis of *C*-terminal peptide amides under mild conditions, *Int. J. Peptide Protein Res. 30*, 206 (1987).

104b. F. Albericio, N. Kneib-Cordonier, S. Biancalana, L. Gera, R. I. Masada, D. Hudson, and G. Barany, Preparation and application of the 5-(4-(9-fluorenylmethoxylcarbonyl)aminomethyl-3,5-dimethoxyphenoxy)-valeric acid (PAL) handle for the solid-phase synthesis of C-terminal peptide amines under mild conditions, *J. Org. Chem. 55*, 3730 (1990).

105. S-S. Wang, *p*-Alkoxybenzyl alcohol resin and *p*-alkoxybenzyloxycarbonylhydrazide resin for solid phase synthesis of protected peptide fragments, *J. Am. Chem. Soc. 95*, 1328 (1973).

106. E. Bayer, Towards the synthesis of proteins, *Angew. Chem. Int. Ed. Engl. 30*, 113 (1991).

107. M. Gobbo, L. Biondi, F. Filira, R. Rocchi, and T. Piek, Synthesis and biological activity of some linear and cyclic kinin analogues, *Int. J. Peptide Protein Res. 44*, 1 (1994).

108. Y. Kiso and H. Yajima, 2-Isobutoxy-1-isobutoxycarbonyl-1,2-dihydroquinoline as a coupling reagent in peptide synthesis, *J. Chem. Soc. Chem. Comm.*, 942 (1972).

109. E. Atherton, J. L. Holder, M. Meldal, R. C. Sheppard, and R. M. Valerio, Peptide Synthesis. Part 12. 3,4-Dihydro-4-oxo-1,2,3-benzotriazin-3-yl esters of fluorenylmethoxycarbonylamino acids as self-indicating reagents for solid phase peptide synthesis, *J. Chem. Soc. Perkin Trans. 1*, 2887 (1988).

110. R. B. Merrifield, Solid phase peptide synthesis. I. The synthesis of a tetrapeptide, *J. Am. Chem. Soc. 85*, 2149 (1963).

111. H. Rink, Solid-phase synthesis of protected peptide fragments using a trialkoxy-diphenyl-methylester resin, *Tetrahedron Lett. 28*, 3787 (1987).

112. M. Bodanszky and J. Martinez, Side reactions in peptide synthesis, *Synthesis*: 333 (1981).

113. M. Mergler, R. Tanner, J. Gosteli, and P. Grogg, Peptide synthesis by a combination of solid phase and solution methods. I. A new very acid-labile anchor group for the solid-phase synthesis of fully protected fragments, *Tetrahedron Lett. 29*, 4005 (1988).

114a. K. Barlos, O. Chatzi, D. Gatos, and G. Stravropoulos, 2-chlorotrityl resin. Studies on anchoring of Fmoc-amino acids and peptide cleavage, *Int. J. Peptide Protein Res. 37*, 513 (1991).

114b. K. Barlos, D. Gatos, S. Kapolos, C. Poulos, W. Schäffer, and Y. Wenquing, Application of 2-chlorotrityl resin in solid phase synthesis of (Leu^{15})-gastrin I and unsulfated cholecystokinin, *Int. J. Peptide Protein Res. 38*, 555 (1991).

115. A. R. Mitchell, S. B. H. Kent, M. Engelhard, and R. B. Merrifield, A new synthetic

route to *tert*-butyloxycarbonylaminoacyl-4-(oxymethyl)phenylacetamidomethyl resin, an improved support for solid-phase peptide synthesis, *J. Org. Chem. 43*, 2845 (1978).

116. K. Wakabayashi and W. Pigman, Synthesis of some glycodipeptides containing hydroxyamino acids, and their stabilities to acids and bases, *Carbohydr. Res. 35*, 3 (1974).

117. A. Fournier, C. T. Wang, and A. M. Felix, Applications of BOP reagent in solid phase synthesis, *Int. J. Peptide Protein Res. 31*, 86 (1988).

118. R. Knorr, A. Trzeciak, W. Bannwarth, and D. Gillessen, New coupling reagents in peptide chemistry, *Tetrahedron Lett. 30*, 1927 (1989).

119. W. König and R. Geiger, Eine neue methode zur synthese von peptiden: Aktivierung der carboxylgruppe mit dicyclohexylcarbodiimide unter zusatz von l-hydroxy-benzotriazolen, *Chem. Ber. 103*, 788 (1970).

120. W. König and R. Geiger, Eine neue methode zur synthese von peptiden: Aktivierung der carboxylgruppe mit dicyclohexylcarbodiimid und 3-hydroxy-4-oxo-3,4-dihydro-1,2,3-benzotriazin, *Chem. Ber. 103*, 2034 (1970).

121. W. König and R. Geiger, Racemisierung bei peptidsynthesen, *Chem. Ber. 103*, 2024 (1970).

122. E. Pedroso, A. Grandas, X de las Heras, R. Eritja, and E. Giralt, Diketopiperazine formation in solid phase peptide synthesis using *p*-alkoxybenzyl ester resins and Fmoc-amino acids, *Tetrahedron Lett. 27*, 743 (1986).

123. M. Bodanszky and J. Z. Kwei, Side reactions in peptide synthesis. VII. Sequence dependence in the formation of aminosuccinyl derivatives from β-benzyl-aspartyl peptides, *Int. J. Peptide Protein Res. 12*, 69 (1978).

124. A. Dryland and R. C. Sheppard, Peptide Synthesis. Part 8. A system for solid-phase synthesis under low pressure continuous flow conditions, *J. Chem. Soc. Perkin Trans. 1*, 125 (1986).

125. S. Y. C. Wong, G. R. Guile, T. W. Rademacher, and R. A. Dwek, Synthetic glycosylation of peptides using unprotected saccharide β-glycosylamines, *Glycoconjugate J 10*, 227–234 (1993).

126. W. F. Degrado and E. T. Kaiser, Polymer-bound oxime esters as support for solid phase peptide synthesis, *J. Org. Chem. 45*, 1295 (1980).

127. R. Steinauer, F. M. F. Chen, and N. L. Benoiton, Studies on racemization associated with the use of benzotriazoyl-l-yl-tris (dimethylamino) phosphonium hexafluorophosphate (BOP), *Int. J. Peptide Protein Res. 34*, 295 (1989).

128. E. Atherton, M. J. Gait, R. C. Sheppard, and B. J. Williams, The polyamide method of solid phase peptide and oligonucleotide synthesis, *Bioorg. Chem. 8*, 351 (1979).

129a. M. Bodanszky, Synthesis of peptides by ammonolysis of nitrophenyl esters, *Nature 175*, 685 (1955).

129b. M. Bodanszky, Stepwise synthesis of peptides by the nitrophenyl-ester method, *Ann. N.Y. Acad. Sci. 88*, 655 (1960).

130. H. Rink and B. Ernst, Glycopeptide solid-phase synthesis with an acetic acid-labile trialkoxy-benzhydryl linker, in *Peptides* (E. Giralt and D. Andreu, (eds.), p. 418, ESCOM Science Publishers, Leiden, 1990.

131. I. Christiansen-Brams, M. Meldal, and K. Bock, Protected-mode synthesis of *N*-linked glycopeptides: Single-step preparation of building blocks as peracetyl glycosylated N^α-Fmoc asparagine OPfp esters, *J. Chem. Soc. Perkin Trans. 1*, 1461 (1993).

132. D. S. King, C. G. Fields, and G. B. Fields, A cleavage method which minimizes side reactions following Fmoc solid phase peptide synthesis, *Int. J. Peptide Protein Res. 36*, 255 (1990).

133. L. Otvos Jr., B. Dietzschold, and L. Kisfaludy, Solid-phase peptide synthesis using *tert*-butyloxycarbonoylamino acid pentafluorophenyl esters, *Int. J. Peptide Protein Res. 30*, 511 (1987).

134. F. Micheel and H. Köchling, Darstellung von glykosiden des D-glucosamins mit ali-phatischen und aromatischen alkoholen und mit serin nach der oxazolin, methode, *Chem. Ber. 91*, 673 (1958).

135. C. H. Bolton and R. W. Jeanloz, The synthesis of a glucosamine-asparagine compound benzyl N^2-carbobenzyloxy-*N*-(2-acetamido-3,4,6-tri-*O*-acetyl-2-deoxy-β-D-glucopyr-anosyl-L-asparaginate, *J. Org. Chem. 28*, 3228 (1963).

136. V. A. Derevitskaya, E. M. Klimov, and N. K. Kochetkov, Synthesis of *O*-glycosyl-threonines, *Carbohydr. Res. 7*, 7 (1968).

137. J. Coste, D. le Nguyen, and B. Castro, PyBOP: A new peptide coupling reagent devoid of toxic by-product, *Tetrahedron Lett. 31*, 205 (1990).

7

Synthesis of Glycosylphosphatidylinositol Anchors

Roy Gigg and Jill Gigg
National Institute for Medical Research, London, England

I. INTRODUCTION

A. Concerning Chiral *myo*-Inositol Derivatives

The chemical synthesis of the lipo-oligosaccharide that forms the "glycosylphos-phatidylinositol lipid anchor" involves several distinct phases, and in this article these will be considered in sequence. Firstly, and unusually for the synthetic car-bohydrate chemist (who is accustomed to utilizing the extensive chiral pool of mo-nosaccharides), a suitably substituted chiral derivative of *myo*-inositol is required.

Because *myo*-inositol (**2**, R = H) is a symmetrical molecule (it can be super-imposed upon its mirror image), positions 1 and 3 (and also 4 and 6) are equivalent in their chemical reactions (but not in enzymatic reactions), and this is often indicated by putting the alternative numbering in brackets, as in formula **2** (Scheme 1). De-rivatization at 1 (or 3) and 4 (or 6) will give an unsymmetrical molecule; thus **2** (R = alkyl) consists of equal portions of the two optical isomers (*enantiomers*) **1** and **3**. Because of a nomenclature rule ("lowest-number rule"), **2** is called 1-*O*-alkyl-*myo*-inositol rather than 3-*O*-alkyl-*myo*-inositol.

The enantiomer **1** is called 1-*O*-alkyl-D-*myo*-inositol [where the D indicates an optically active (*chiral*) molecule of defined configuration], and **3** is thus 3-*O*-alkyl-D-*myo*-inositol. But, on applying the "lowest-number rule," this is changed to 1-*O*-alkyl-L-*myo*-inositol, where the L indicates the opposite absolute configuration to D. **1** and **3** are nonsuperimposable mirror images and have identical physical properties

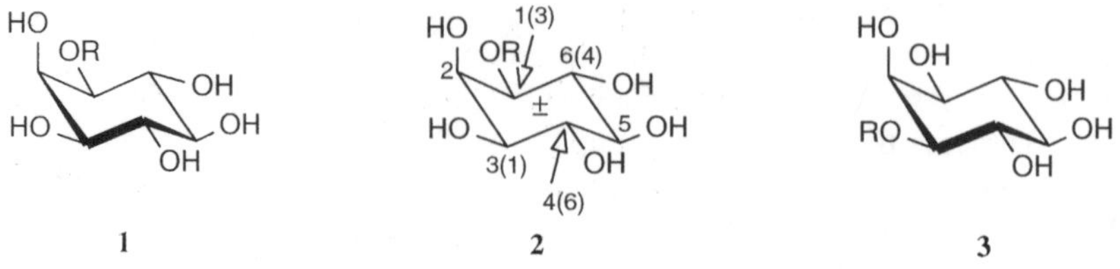

Scheme 1

(m.p., i.r., and n.m.r. spectra and solubility and chromatographic behavior) except for optical rotations, which are equal but opposite in sign. Therefore **2** (R = alkyl) is a *racemic* mixture of the enantiomers **1** and **3** and is referred to as (±)-1-*O*-alkyl-*myo*-inositol and will have some physical properties (particularly m.p. and solubility) different from **1** and **3** and will have zero optical activity (see Ref. 1 for an excellent discussion of the stereochemistry of *myo*-inositol derivatives and the problems of nomenclature).

In the formulae in this chapter, racemic *myo*-inositol derivatives are indicated with (±) in the ring, chiral inositol derivatives are drawn with thickened lines in the ring, and *meso*-compounds (i.e., symmetrical compounds like *myo*-inositol) are shown with neither of these modifications.

The separation of **1** and **3** from the mixture present in **2** is referred to as *resolution*. This can be achieved by chromatography on a suitable chiral column (i.e., the solid support in the column is optically active) or by chemical combination of the racemate **2** with an optically active compound that produces a mixture of two *diastereoisomers* (derived from **1** and **3**) differing in physical properties and that can sometimes be separated by crystallization or chromatography. Appropriate degradation of the separated diastereoisomers allows the enantiomers **1** and **3** to be recovered.

The preparation of pure enantiomers is often difficult but is essential for synthetic work with *myo*-inositol derivatives. Extensive recent synthetic studies [2] on the *myo*-inositol phosphates of the phosphatidylinositol cycle have produced numerous solutions to this problem: (a) resolutions of racemic *myo*-inositol derivatives with chiral reagents (particularly ω-camphanate esters), (b) the use of naturally occurring chiral inositol derivatives (other than *myo*-inositol) that can be converted into *myo*-inositol derivatives, and (c) the separation of the diastereoisomers prepared by condensing racemic derivatives of *myo*-inositol with derivatives of another chiral carbohydrate, e.g., 2-amino-2-deoxy-D-glucose. Unless otherwise stated, the optical rotations given in the text were recorded in chloroform at ambient temperatures.

B. Addition of the D-Glucosamine Residue

The preparation of α-linked glucosamine derivatives (such as occurs in the "lipid anchor") was for many years a difficult problem. But following the introduction by Paulsen of the 2-azido-2-deoxy-D-glucose derivatives, this problem has been completely solved. Nearly all of the researchers whose work is reviewed in this article have used this sugar derivative for the coupling of the glucosamine residue in α-linkage to the *myo*-inositol that represents the second stage of the synthetic sequence.

C. Preparation of Oligosaccharides

The third stage in the synthesis involves the addition of further sugars. Various relatively new methods, particularly those involving glycosyl-trichloracetimidates, glycosyl fluorides, and thioglycosides as glycosylating reagents, as well as the pentenyl glycoside approach of Fraser-Reid have been used. In order to couple the sugars in the correct positions, extensive use of protecting groups has been made. The benzyl ethers have been predominant as "persistent" protecting groups because in the final stage these can be removed by hydrogenolysis. Likewise the azido group is acting as a "persistent" form of protection for the amino group of glucosamine and is converted to the amino group in the final hydrogenolytic step. For "temporary" protecting groups, the allyl, *p*-methoxybenzyl and silyl ethers have been predominant.

D. Introduction of Phosphate Esters

The penultimate stage in the synthesis is the introduction of the phosphate esters: the diacylglycerol phosphate residue on the *myo*-inositol, and the ethanolamine phosphate residue (which is the site of attachment of the protein) on the terminal sugar. Again, recent progress in the field of phosphate ester synthesis (due primarily to the work of the nucleic acid chemists) has provided elegant methods for phosphorylation using either the phosphotriester approach or the *H*-phosphonate method. Syntheses of simple phosphatidylinositol derivatives by the phosphotriester method have been reviewed recently [3].

E. Deprotection of Synthetic Products

The final stage of the synthesis is the removal of all protecting groups. Acid hydrolysis removes acetal groups, and alkaline hydrolysis removes ester groups and the cyanoethyl protecting groups on phosphates. Because most of the synthetic routes use benzyl groups as "persistent" protecting groups, the last operation is usually hydrogenolysis to remove these.

II. CHIRAL *myo*-INOSITOL DERIVATIVES

A. Syntheses by the Durham Group

The Durham group [4] used the chiral penta-acetate **9** for their synthesis, which was prepared as shown in Scheme 2. The racemic 4-*O*-benzyl-1,2:5,6-di-*O*-cyclohexylidene-*myo*-inositol **4**, prepared as described by Garegg et al. [5], was resolved via the diastereoisomeric (−)-ω-camphanate esters, as described by the Merck group [6,7]. The diastereoisomer **5** (m.p. 153°; $[\alpha]_D$ − 37°) was obtained in 35% yield and was hydrolyzed with acid to give the diol **6** (m.p. 178°; $[\alpha]_D$ − 32°) which on saponification gave the triol **7** (m.p. 139°; $[\alpha]_D$ + 21°). Acidic hydrolysis of **7** and acetylation of the product gave **8**, which was hydrogenolyzed to give **9** (m.p. 161°; $[\alpha]_D$ + 15°).

In a new approach (Scheme 3) to a chiral *myo*-inositol derivative (designed to use both products from a tin-mediated alkylation) the Durham group [8] started from the diol **10**, which on tin-mediated *p*-methoxybenzylation gave a mixture of **11** (44%) and **13** (39%)—compare with Scheme 7 for a similar approach by the RIKEN group. Methoxymethylation of **11** gave **12**, which on acid hydrolysis (to give **15**) and subsequent benzylation gave **16**. Removal of the *p*-methoxybenzyl group [with cerium (IV) ammonium nitrate] gave **17**, which was converted into the diastereoisomeric mixture of (−)-ω-camphanates **18**. These were separated by chromatography to give **22** (36%, $[\alpha]_D$ − 16°), which was saponified to give **23** and this was allylated to

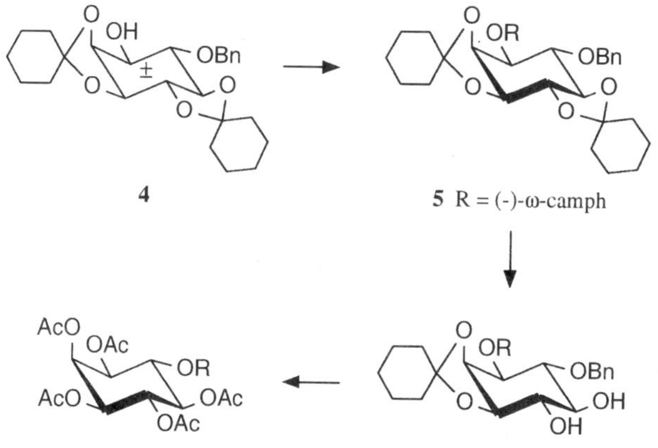

Scheme 2

give **24**. Removal of the methoxymethyl group by acid hydrolysis gave the required **27** ($[\alpha]_D - 9°$, cf $[\alpha]_D - 7°$ in Ref. 9).

The other product, **13**, from the tin-mediated p-methoxybenzylation was converted into the diastereoisomeric mixture of $(-)$-ω-camphanates **14**, which were separated to give **19** (36%, $[\alpha]_D - 37°$). Saponification of **19** (to give **20**) followed by allylation gave **21**, and this on acid hydrolysis gave **25**. Benzylation of the latter and subsequent removal of the p-methoxybenzyl group gave more of **27**.

B. Syntheses by the Mill Hill Group

The Mill Hill group [10] prepared and resolved the racemic 3,4,5-tri-O-benzyl-1,2-O-isopropylidene-myo-inositol **36**, as shown in Scheme 4. Because of the ease of the optical resolution and the substitution pattern, **36** is a very suitable intermediate for syntheses in the lipid anchor field. The di-O-isopropylidene derivative **28** of myo-inositol was preferentially benzylated at the 3 position to give **29** [11], which was converted into the p-methoxybenzyl ether **30** (m.p. 133°) or the allyl ether **31** (m.p. 120°). Partial hydrolysis of **30** and **31** gave the diols **32** (m.p. 149°) and **33** (m.p. 121°), respectively. Benzylation of **32** gave **34** (m.p. 77°) and the p-methoxybenzyl group was removed with dichlorodicyanoquinone to give **36** (m.p. 90°). Conversion of **36** into the $(-)$-ω-camphanate esters gave the readily separable diastereoisomers from which **37** (m.p. 179°; $[\alpha]_D - 47°$) was obtained. Saponification of **37** gave **38**, whose structure was established by conversion into the known chiral tetra-O-benzyl-myo-inositol **39**. Compound **33** was also converted via **35** into the diol **40** (m.p. 101°), which was also used to prepare chiral **38** via the triol **41** and the isopropylidene derivative **36**.

C. Syntheses by the Leiden Group

The Leiden group [9] converted the di-O-cyclohexylidene-myo-inositol **42** (Scheme 5) into the monobenzyl ether **43** (m.p. 138°), and this gave the p-methoxybenzyl

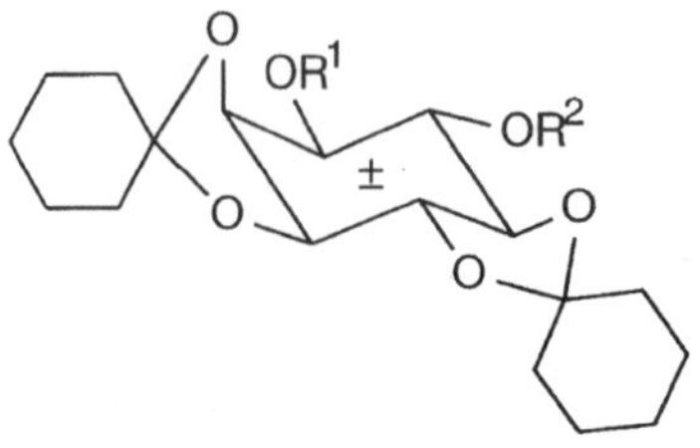

10 R¹ = R² = H

11 R¹ = pMB; R² = H

12 R¹ = pMB; R² = MOM

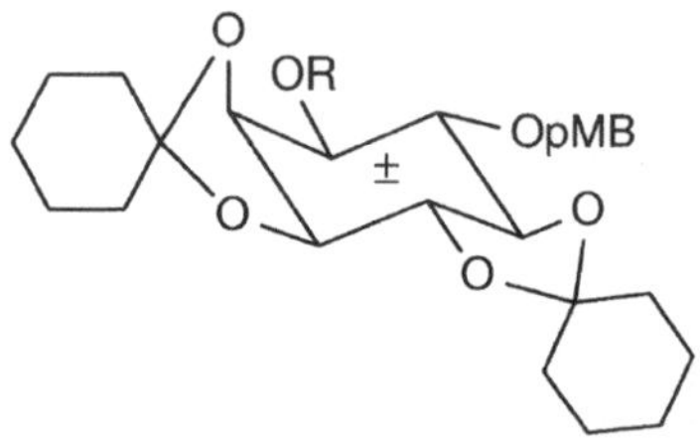

13 R = H

14 R = (-)-ω-camph

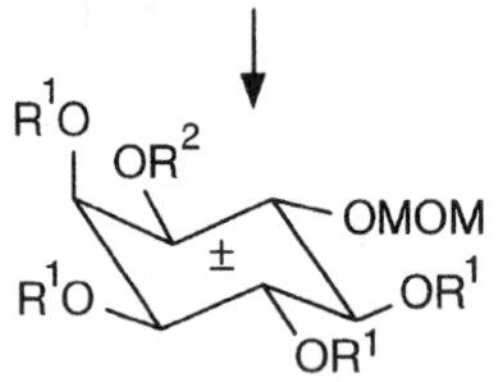

15 R¹ = H; R² = pMB

16 R¹ = Bn; R² = pMB

17 R¹ = Bn; R² = H

18 R¹ = Bn; R² = (-)-ω-camph

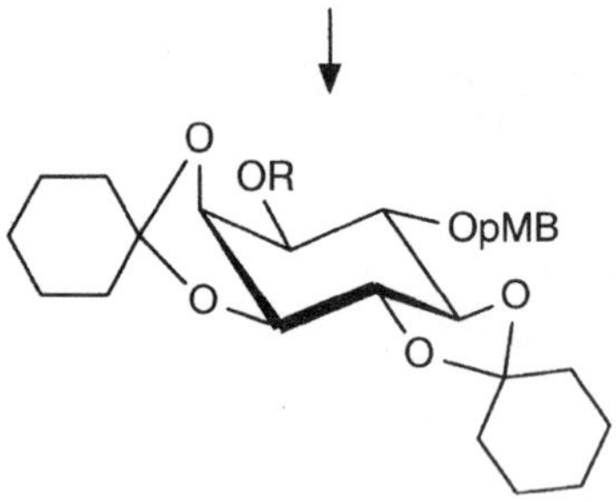

19 R = (-)-ω-camph

20 R = H

21 R = All

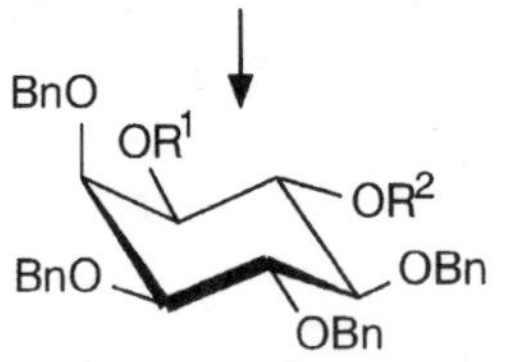

22 R¹ = (-)-ω-camph; R² = MOM

23 R¹ = H; R² = MOM

24 R¹ = All; R² = MOM

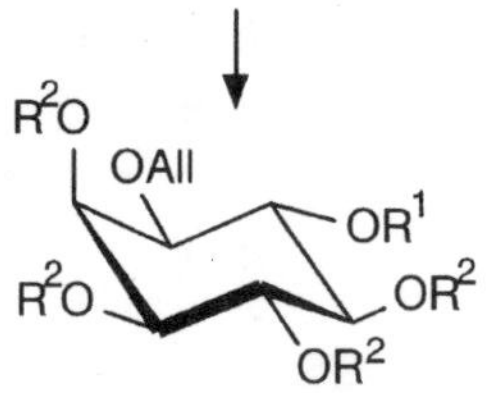

25 R¹ = pMB; R² = H

26 R¹ = pMB; R² = Bn

27 R¹ = H; R² = Bn

Scheme 3

ether **44** (m.p. 139°). Partial hydrolysis of **44** gave the diol **45** (m.p. 142°), and this was benzylated to give **46**. Removal of the *p*-methoxybenzyl group from **46** with dichlorodicyanoquinone gave the alcohol **47** (m.p. 103°; cf. **36**, Scheme 4). Compound **47** was resolved [12] via the diastereoisomeric (−)-menthoxyacetyl ester **48**. Chromatographic separation gave the less polar isomer **49** (m.p. 103°), which was saponified to give the alcohol **50** (cf. **38**, Scheme 4). The absolute configuration of **50** was confirmed by conversion into the tetra-*O*-benzyl ether **39** (Scheme 4).

The racemate **47** was glycosidated by the Leiden group, and the diastereoisomeric disaccharides were separated to achieve a resolution (see Section III.B).

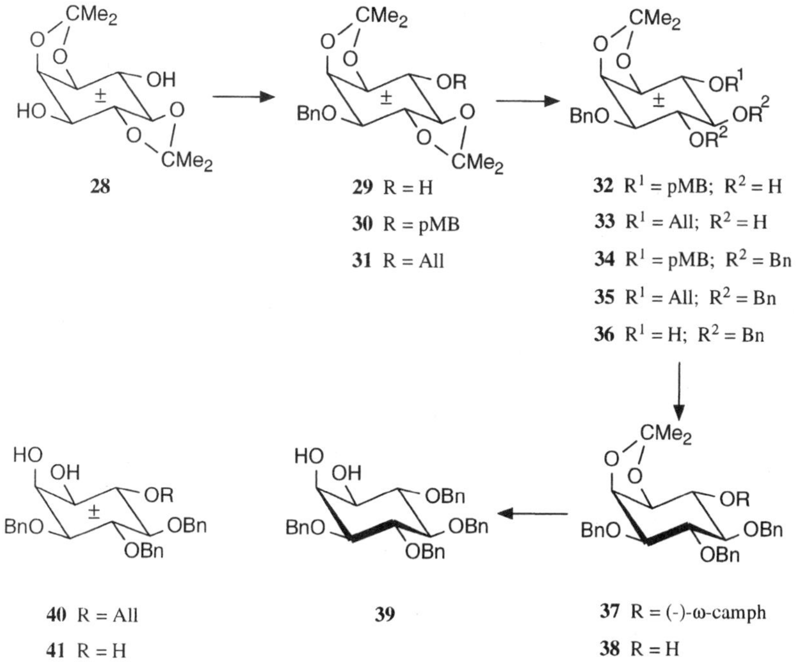

28

29 R = H
30 R = pMB
31 R = All

32 R^1 = pMB; R^2 = H
33 R^1 = All; R^2 = H
34 R^1 = pMB; R^2 = Bn
35 R^1 = All; R^2 = Bn
36 R^1 = H; R^2 = Bn

40 R = All
41 R = H

39

37 R = (-)-ω-camph
38 R = H

Scheme 4

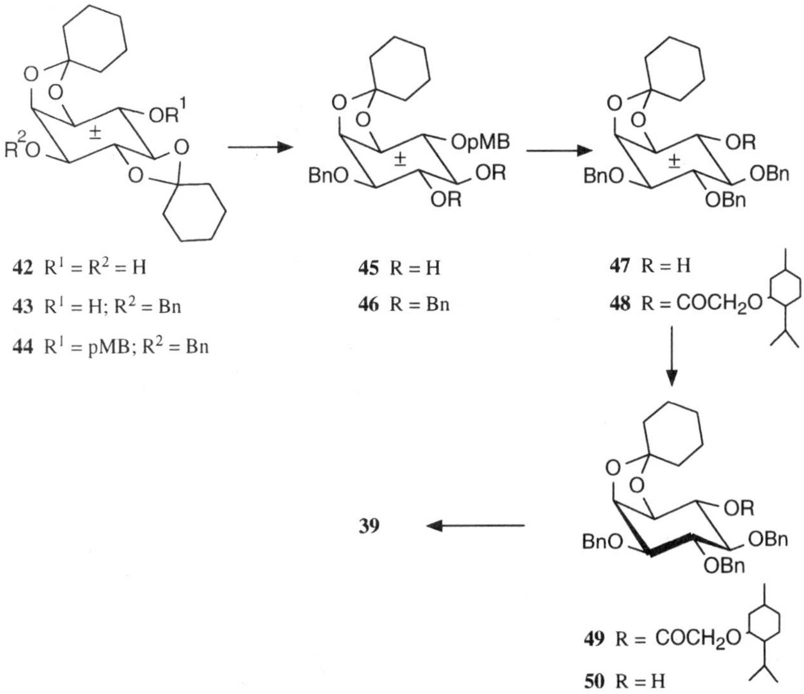

42 R^1 = R^2 = H
43 R^1 = H; R^2 = Bn
44 R^1 = pMB; R^2 = Bn

45 R = H
46 R = Bn

47 R = H
48 R = COCH$_2$O

39

49 R = COCH$_2$O
50 R = H

Scheme 5

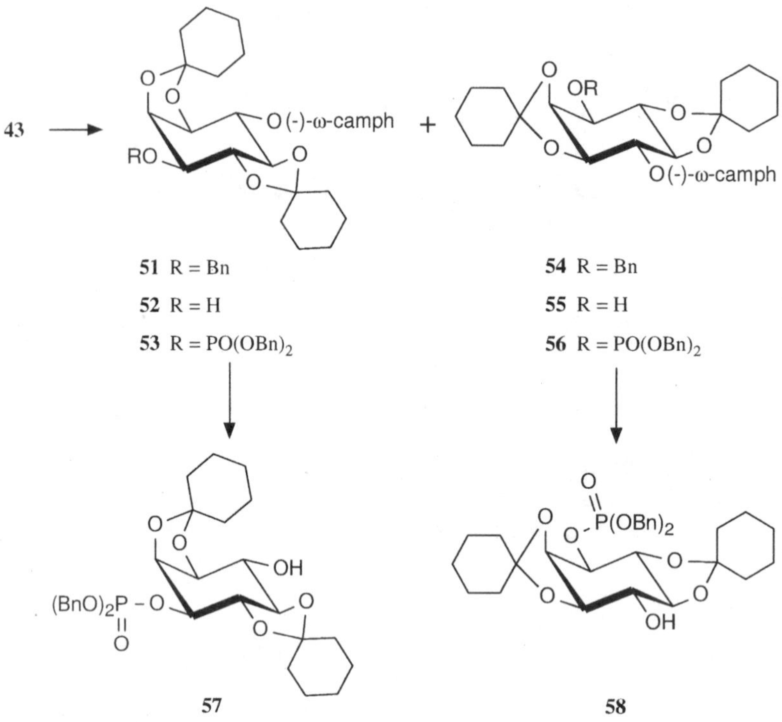

Scheme 6

D. Syntheses by the Charlottesville Group

The Charlottesville group [13] also started from **43** (Scheme 5), and this was resolved as described by Vacca et al. [7] via the (−)-ω-camphanates to give the diastereoisomers **51** and **54** (Scheme 6).

Shen [13] and Brimacombe [14] have drawn attention to the correction [7b] to the assignments previously made [7a] by the Merck group to **51** and **54**. This previous misassignment has led to an error in the synthetic work reported by Ogawa [15–17] (see Ref. 14 and Section V.A.).

The camphanate **51** (Scheme 6) was debenzylated by catalytic hydrogen transfer to give the alcohol **52** (m.p. 237°; $[\alpha]_D$ −26°), and this was phosphitylated with dibenzyloxy-*N,N*-diisopropylaminophosphine and subsequent oxidation of the phosphite to give **53** (m.p. 139°; $[\alpha]_D$ +4°). Similar treatment of **54** gave **55** (m.p. 225°; $[\alpha]_D$ +12°), which was phosphorylated to give **56** (m.p. 157°; $[\alpha]_D$ +5°). Saponification of **53** with methanolic ammonia gave **57** (m.p. 142°; $[\alpha]_D$ +10°). Similarly, **56** gave **58** (m.p. 141°; $[\alpha]_D$ −13°). The configurations of **57** and **58** were confirmed by their conversion into the known 1L-*myo*-inositol 1-phosphate and 1D-*myo*-inositol 1-phosphate, respectively.

E. Syntheses by the RIKEN Group

The RIKEN group [16] converted the diol **59** (=**10**) [5] (Scheme 7) by tin-mediated alkylation into the syrupy *p*-methoxybenzyl ether **60** in 60% yield, and this was

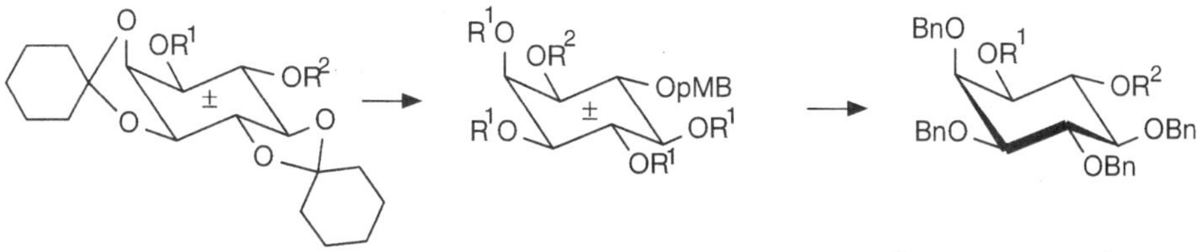

59 $R^1 = R^2 = H$

60 $R^1 = H; R^2 = pMB$

61 $R^1 = All; R^2 = pMB$

62 $R^1 = H; R^2 = All$

63 $R^1 = Bn; R^2 = All$

64 $R^1 = Bn; R^2 = H$

65 $R^1 = Bn; R^2 = (-)-\omega-camph$

66 $R^1 = (-)-\omega-camph; R^2 = pMB$

67 $R^1 = (-)-\omega-camph; R^2 = H$

68 $R^1 = (-)-\omega-camph; R^2 = CH(Me)OEt$

69 $R^1 = H; R^2 = CH(Me)OEt$

70 $R^1 = pMB; R^2 = CH(Me)OEt$

71 $R^1 = pMB; R^2 = H$

72 $R^1 = pMB; R^2 = Bn$

Scheme 7

converted into the allyl ether **61**. Acidic hydrolysis of this gave the tetraol **62**, which was benzylated to give **63**. Removal of the allyl group gave the alcohol **64**, which was converted into the diastereoisomeric mixture of $(-)-\omega$-camphanates **65**. The diastereoisomer **66** ($[\alpha]_D +10°$) was separated by chromatography and treated with cerium(IV) ammonium nitrate, which removed the p-methoxybenzyl group to give **67** ($[\alpha]_D +15°$), which was converted into the acetal **68** by reaction with ethyl vinyl ether. Saponification of the camphanoyl group gave the alcohol **69**, which was converted into the p-methoxybenzyl ether **70**. Acidic hydrolysis of **70** gave the alcohol **71** (m.p. 78°; $[\alpha]_D +9°$). In order to establish the absolute configuration, this was converted into the benzyl ether **72**, and the p-methoxybenzyl group was removed to give the chiral 2,3,4,5,6-penta-O-benzyl-myo-inositol ($[\alpha]_D +11°$). This was correlated with figures previously reported by Vacca et al. [7a] for this compound. However, these were later shown to be misassigned [7b,13,14,18]. It is concluded ([14]; see also Section V.A.) that the further synthesis of the lipid anchor reported by the RIKEN group [16,17] is based on the wrong isomer of myo-inositol.

F. Syntheses by the Tufts Group

The Tufts group [19] also used the chiral inositol derivative **5** (Scheme 2), which was saponified to give **73** (Scheme 8). This was converted into the t-butyldimethylsilyl ether **74**, which on hydrogenolysis gave **75**. In a further paper [20] these authors mention the use of compound **76** in synthesis, but the preparation has not yet been published (see their Ref. 29 in our Ref. 20). A further resolution of the racemic inositol derivative **59** (Scheme 7) was achieved [19] by separating the diastereoisomeric mixture obtained by condensation with a glucosamine derivative (see Section III.E.).

G. Syntheses by the Madrid Group

The Madrid group [21] described the conversion of myo-inositol into the hexaboronate **77** (Scheme 9), which was trans-metallated with di-n-butyltin-bis-acetylace-

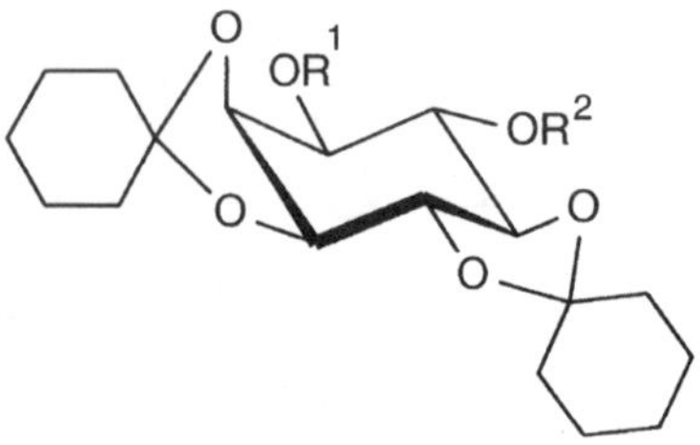

73 R^1 = H; R^2 = Bn

74 R^1 = TBDMS; R^2 = Bn

75 R^1 = TBDMS; R^2 = H

76 R^1 = (-)-ω-camph; R^2 = H

Scheme 8

tonate. The product was treated with (−)-menthylchloroformate and N-methylimi-dazole to give the chiral inositol derivative **78** (m.p. 173°; $[\alpha]_D$ −71°, MeOH) in 30% yield. This was converted into the mixture of di-O-isopropylidene derivatives **79** (40%; m.p. 186°; $[\alpha]_D$ −64°, MeOH) and **80** (51%, m.p. 64°; $[\alpha]_D$ −61°, MeOH), which were separated by chromatography. The absolute configuration of **79** was confirmed by conversion into 1D-1,4-di-O-methyl-*myo*-inositol (m.p. 222°; $[\alpha]_D$ −25°, H₂O), which is the known compound D-(−)-liriodendritol.

The Madrid group also used other methods of resolution [22] based on the condensation of glucosamine derivatives with racemic *myo*-inositol derivatives and

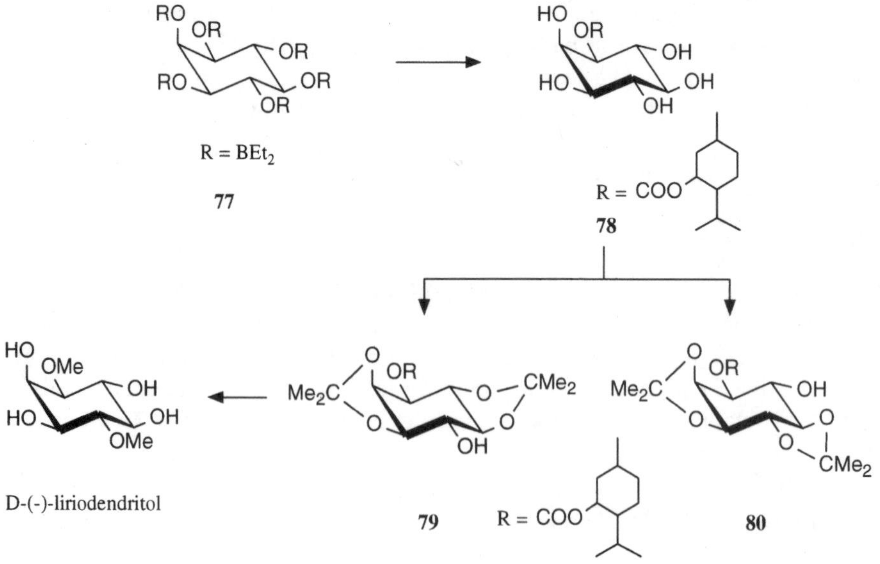

Scheme 9

separation of the diastereoisomers formed (see Section III.F.). But these methods were described [22] as "tedious, laborious and of limited practical use."

In a further approach [23–25] to chiral *myo*-inositol derivatives the Madrid group started from the unsaturated ketone **81** (m.p. 62°; $[\alpha]_D$ +64°, CH$_2$Cl$_2$) (Scheme 10), which had been prepared previously [26] by application of the Ferrier reaction to the 5,6-ene derived from methyl 2,3,4-tri-*O*-benzyl-α-D-glucopyranoside. Compound **81** was reduced in high yield with sodium borohydride in the presence of cesium(III) chloride to give **82** (m.p. 119°; $[\alpha]_D$ +117°), which was treated with 3-chloroperbenzoic acid to give the epoxide **83** (m.p. 150°; $[\alpha]_D$ +78°) in high yield. Acid-catalyzed opening of the epoxide ring in allyl alcohol gave the D-*chiro*-inositol derivative **86** ($[\alpha]_D$ +40°). Benzylation of **86** and subsequent deallylation, methylation, and debenzylation gave the known 1-*O*-methyl-D-*chiro*-inositol, thus establishing the stereochemistry of **86**.

Tin-mediated *p*-methoxybenzylation of **86** gave **85** ($[\alpha]_D$ +2°). Oxidation of **85** with pyridinium chlorochromate gave the ketone **84**, which was reduced with (R)-Alpine hydride to give the required *myo*-inositol derivative **87** (58%; m.p. 91°; $[\alpha]_D$ −8°) together with some of **85** (21%). Methylation of **87** and deprotection (removing allyl, *p*-methoxybenzyl, and benzyl groups) gave the known 1D-1-*O*-methyl-*myo*-inositol [(−)-bornesitol], thus establishing the structure of **87**.

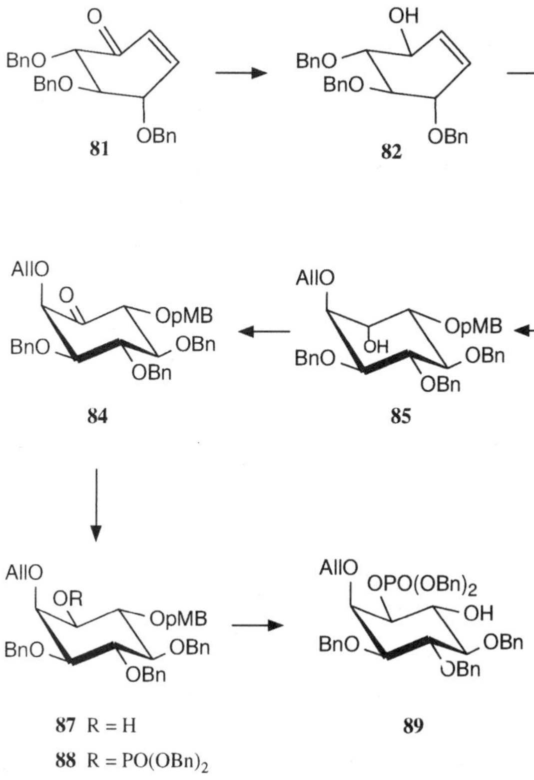

Scheme 10

Phosphorylation of **87** using phosphoramidite methodology gave the dibenzyl-phosphate **88** (m.p. 87°; $[\alpha]_D$ +6°), and this was de-*O*-*p*-methoxybenzylated with trifluoroacetic acid in dichloromethane to give the alcohol **89** (m.p. 104°; $[\alpha]_D$ +78°), ready for glycosidation, as described in Section III.F.

As a further approach to the preparation of a chiral *myo*-inositol derivative from the chiral pool of carbohydrates, the Madrid group [27] started from the di-epoxide **90** (Scheme 11), which is readily prepared from D-mannitol [28]. This was opened with benzyl alcohol to give **91** [29], which was converted into the bis-*t*-butyldiphenylsilyl ether **92**, and this was hydrogenolyzed to give the diol **93** ($[\alpha]_D$ +22°). Oxidation of the latter with oxalyl chloride-methyl sulfoxide-triethylamine gave the dialdehyde **94**. The crude reaction mixture was added to a solution of samarium(II) iodide [30] in *t*-butanol at −50° to give the *myo*-inositol derivative **95**

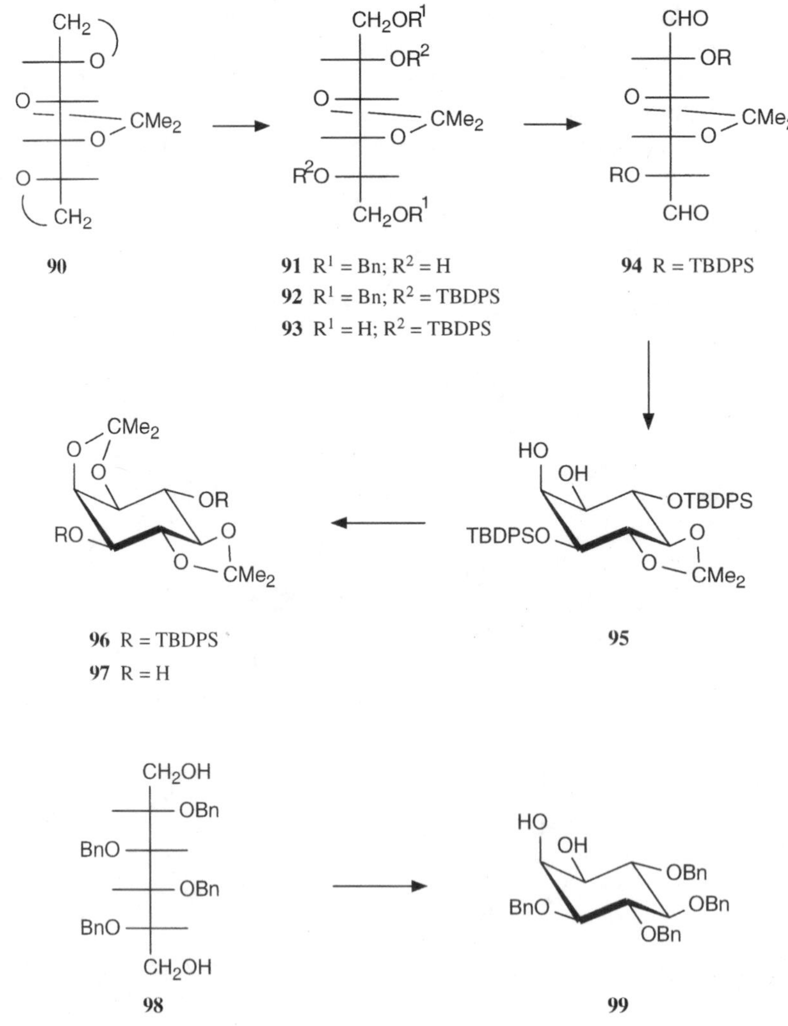

90

91 R^1 = Bn; R^2 = H
92 R^1 = Bn; R^2 = TBDPS
93 R^1 = H; R^2 = TBDPS

94 R = TBDPS

96 R = TBDPS
97 R = H

95

98

99

Scheme 11

([α]$_D$ −24°) as the major product (80%, together with 7% of the *scyllo*-isomer). This was converted into the di-*O*-isopropylidene derivative **96**, which was desilylated with tetrabutylammonium fluoride to give the diol **97** (m.p. 177°; [α]$_D$ −22°—which figures compare favorably with those reported [31] for the enantiomer). A similar approach was used by the Strasbourg-Gif group [32] to prepare 1L-1,4,5,6-tetra-*O*-benzyl-*myo*-inositol (**99**; 56%; m.p. 142°; [α]$_D$ −22°) from the dialdehyde derived from 2,3,4,5-tetra-*O*-benzyl-L-iditol (**98**) in the presence of samarium(II) iodide and *t*-butanol.

H. Syntheses by the Dundee Group

The Dundee group [14] also used the chiral inositol derivative **51** (Scheme 6), and this was converted (Scheme 12) into the allyl ether **100** (m.p. 96°; [α]$_D$ −45°), which on partial acid hydrolysis gave the diol **101** (m.p. 137°; [α]$_D$ +20°), and this gave the tri-*O*-benzyl ether **102**. Acidic hydrolysis of the latter gave the diol **103** (m.p. 120°; [α]$_D$ −35°), and this on tin-mediated *p*-methoxybenzylation gave **104** (m.p. 106°; [α]$_D$ −10°), which was benzylated to give **105**. Deallylation of **105** gave the required alcohol **106** (m.p. 78°; [α]$_D$ −9°). To confirm the absolute configuration of **106**, it was treated with cerium(IV) ammonium nitrate to give the diol **107** (m.p. 152°; [α]$_D$ +14°), which was compared with material prepared previously [33]. The enantiomer of **106** was also prepared by the same route.

The Dundee group also used [34] another approach to get a chiral *myo*-inositol derivative. The camphanate **51** (Scheme 6) was converted via the alcohol **108** (Scheme 13) into the *p*-methoxybenzyl ether **109** (m.p. 155°; [α]$_D$ −57°), and this

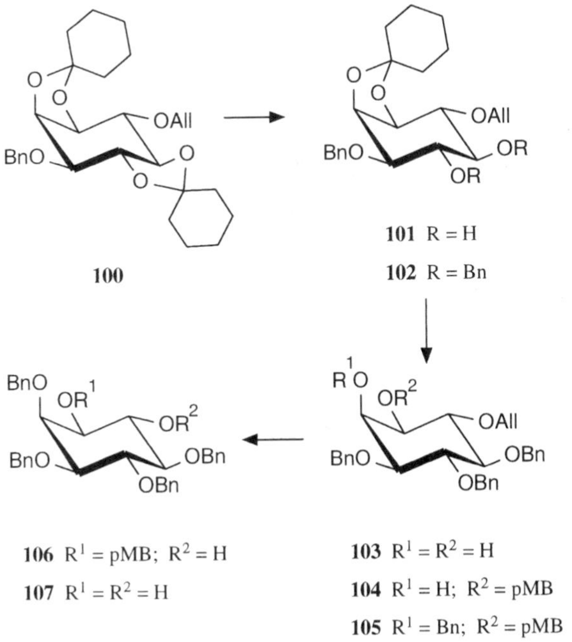
O
O
OAll
BnO
O
O
100
O
O
OAll
BnO
OR
OR
101 R = H
102 R = Bn
BnO
OR1
OR2
BnO
OBn
OBn
106 R1 = pMB; R2 = H
107 R1 = R2 = H
R1O
OR2
OAll
BnO
OBn
OBn
103 R1 = R2 = H
104 R1 = H; R2 = pMB
105 R1 = Bn; R2 = pMB

Scheme 12

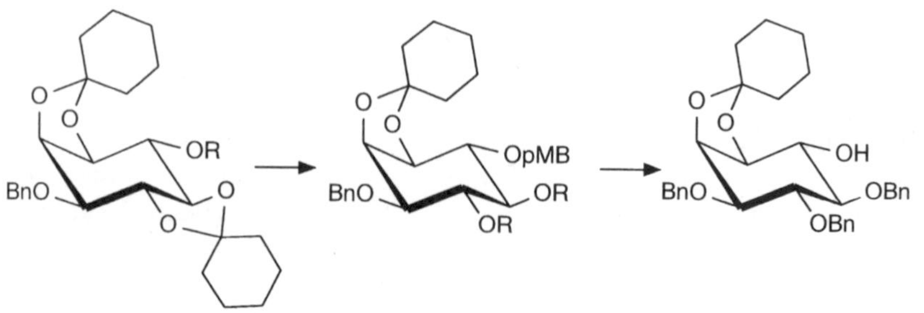

| 108 R = H | 110 R = H | 112 |
| 109 R = pMB | 111 R = Bn | |

Scheme 13

on partial hydrolysis gave the diol **110** (m.p. 147°; $[\alpha]_D$ +0.6°), which was benzylated to give **111**. The p-methoxybenzyl group was removed from **111** by the action of boron trifluoride in dry dichloromethane to give the alcohol **112** (=**50**, Scheme 5; m.p. 98°; $[\alpha]_D$ −18°) in high yield without affecting the cyclohexylidene ketal group. Compound **112** was used for glycosidation, as shown in Section III.H.

I. Syntheses by the Konstanz Group

The Konstanz group [35] started with the racemic dicyclohexylidene-myo-inositol **10** (Scheme 3), and this was treated with bis(tributyltin)oxide in toluene and subsequently with (−)-menthoxycarbonyl chloride to give, regioselectively, the diastereoisomeric mixture of the esters **113** (Scheme 14), which on acid hydrolysis gave the mixture of diastereoisomeric pentaols **114**. These were crystallized to give the known ([21]—see Scheme 9) chiral **78**. This was treated with methoxymethyl chloride in the presence of N-ethyldiisopropylamine (Hünigs base) to give **115**, and this on saponification with potassium carbonate in methanol gave the alcohol **116**.

In subsequent work [36] the Konstanz group described the resolution of the diastereoisomeric mixture **113** by crystallization from light petroleum to give the chiral **117** required for glycosidation (see Section III.I.).

J. Syntheses by the London-Cambridge Group

In a radically different approach to a chiral myo-inositol derivative, the London-Cambridge group [37] started from benzene (Scheme 15). Microbial oxidation of benzene with *Pseudomonas putida* gave cis-1,2-dihydroxycyclohexa-3,5-diene (**118**, [38])—a compound now commercially available from this procedure—and this was converted into the carbonate **119** (m.p. 90° dec.). Oxidation of **119** with 3-chloroperbenzoic acid gave predominantly the epoxide **123** (m.p. 88°) together with some of the stereoisomer. The epoxide ring of **123** was opened regioselectively with R-(+)-sec-phenethyl alcohol in the presence of fluoroboric acid to give the diastereoisomeric mixture of alcohols **122** (67%), which were separated by HPLC to give the chiral alcohol **120** ($[\alpha]_D$ +90°). This was benzylated with benzyl bromide and silver(I) oxide in N,N-dimethylformamide to give **121** (m.p. 98°; $[\alpha]_D$ +103°), and this

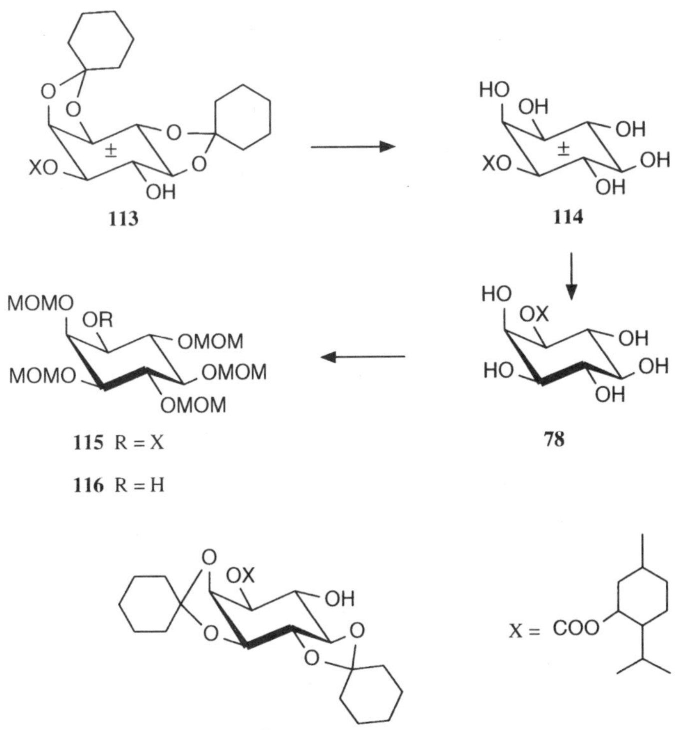

Scheme 14

on saponification with triethylamine in aqueous methanol gave the diol **124** (m.p. 47°; $[\alpha]_D$ +82°). Treatment of the latter with 3-chloroperbenzoic acid gave predominantly the epoxide **125** ($[\alpha]_D$ +99°), which was converted into the isopropylidene derivative **126** (m.p. 115°; $[\alpha]_D$ +134°). This was reduced [39] with lithium aluminum hydride to give predominantly the alcohol **129**, which was oxidized to the ketone **128**, and this on treatment with t-butyldimethylsilyl chloride gave the enol ether **127**. Hydroboration of **127** and subsequent oxidation with hydrogen peroxide gave predominantly the *myo*-inositol derivative **130**. Alkylation of this with *p*-methoxybenzyl chloride and sodium hydride in toluene at 80° gave the ether **131** formed as a result of an alkali-catalyzed silyl migration prior to *p*-methoxybenzylation. Desilylation with tetrabutylammonium fluoride gave the alcohol **132**, ready for glycosidation, as described in Section III.G.

As an alternative approach for investigations on the preparation of a chiral *myo*-inositol derivative suitable for the synthesis of the glycophosphatidylinositol anchor of *Trypanosoma brucei*, the London-Cambridge group [40] treated the symmetrical 2,5-di-*O*-benzoyl-*myo*-inositol (**133**) [41] with (2S,2S′)-2,2′-dimethyl-3,3′,4,4′-tetrahydro-6,6′-bi-2H-pyran (**134**, Scheme 16) under acid-catalyzed conditions to give **135**. Saponification gave the tetraol **136**, which was benzylated to give **137**. This on treatment with 95% trifluoroacetic acid gave the chiral diol **138** (63%; $[\alpha]_D$ −16°).

The required enantiomeric *myo*-inositol derivative **139** was prepared in a similar way from **133** using the isomeric reagent **140** or the analog **141**.

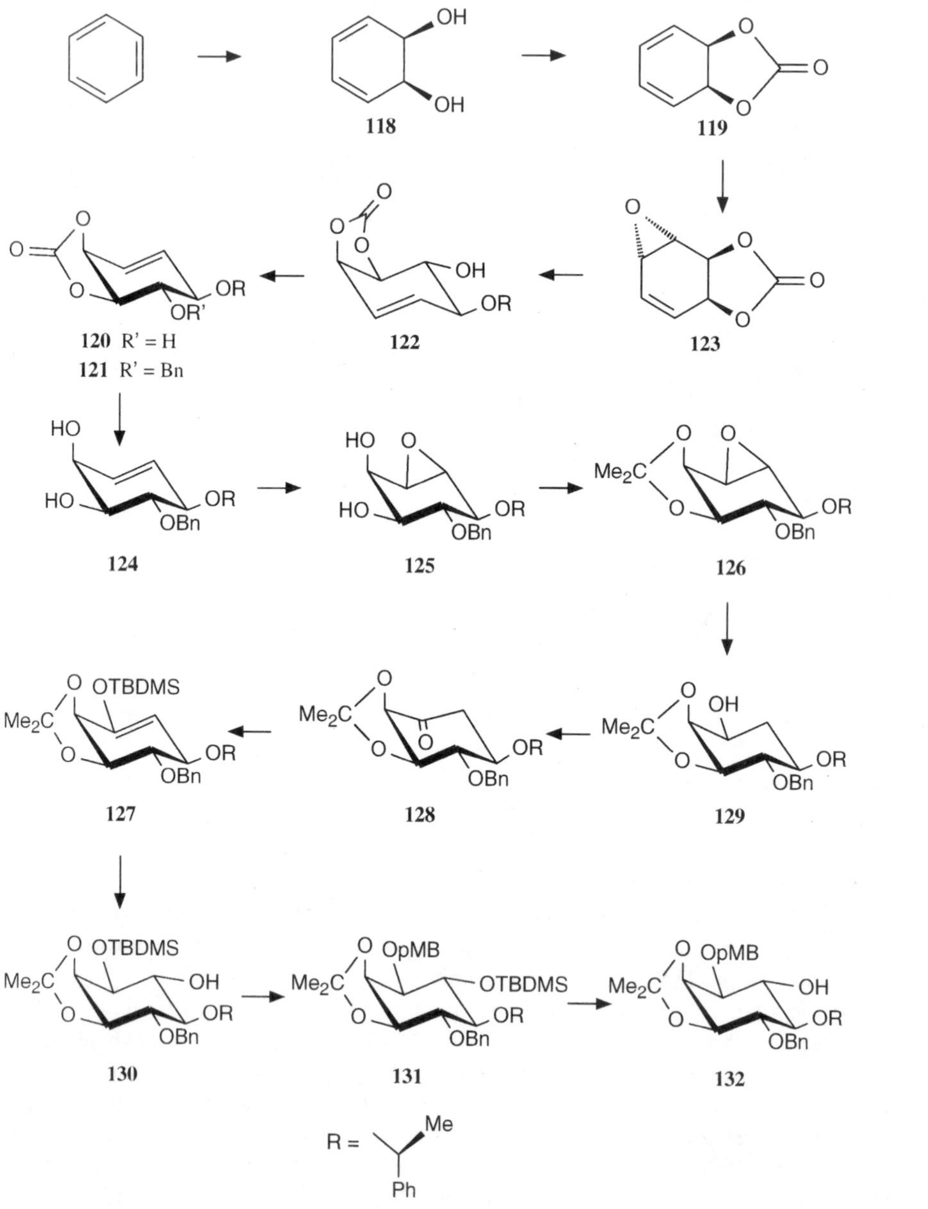

Scheme 15

III. ADDITION OF THE GLUCOSAMINE RESIDUE

A. Syntheses by the Durham Group

The Durham group [4] condensed the pentenyl glucosamine derivative **142** (m.p. 98°; [α]$_D$ +19°) [42,8] with the chiral inositol derivative **9** (Scheme 2) in the presence

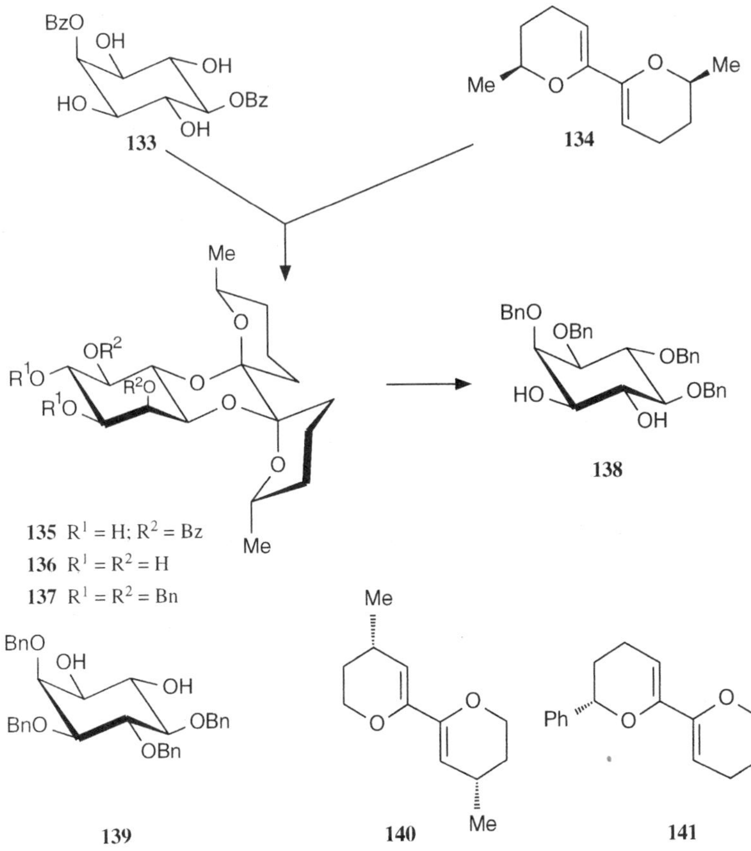

Scheme 16

of the collidine complex of iodine perchlorate to give the disaccharide **143** (Scheme 17). The amino-protecting group was removed with tosyl hydrazine to give **144**, which was converted into the carbobenzyloxy derivative **145** ($[\alpha]_D$ +58°). The benzylidene group was opened by reduction with sodium cyanoborohydride in tetrahydrofuran under acidic conditions to give the disaccharide **146** ($[\alpha]_D$ +75°) required for further glycosidation.

In an alternative procedure to give a more versatile intermediate (in which the azido group was used as a protected form of the amino group in the glucosamine residue), the Durham group [8,43] started from the 2-azido-2-deoxy-1,6-anhydroglucose derivative **147** (Scheme 18) [44]. This was treated with trifluoroacetic acid in acetic anhydride to give the glycosyl acetate **148**, which on treatment with titanium(IV) bromide in dichloromethane-ethyl acetate gave the glycosyl bromide **149** previously described by Paulsen and his co-workers [45]. Glycosidation of the *myo*-inositol derivative **27** (Scheme 3) with the bromide **149** in the presence of silver perchlorate in ether (see [46]) gave the α-linked disaccharide **150** in 63% yield (together with 22% of the β-linked disaccharide), the separation of the two being achieved after saponification to the diol **151**. A rather lengthy procedure was then required to convert **151** into the alcohol 156 required for further glycosylation. The

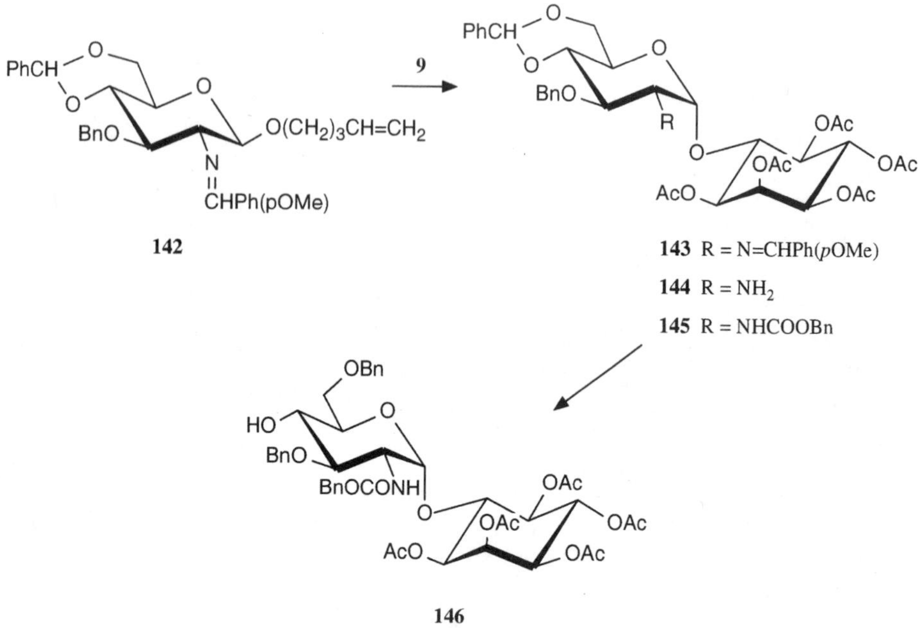

Scheme 17

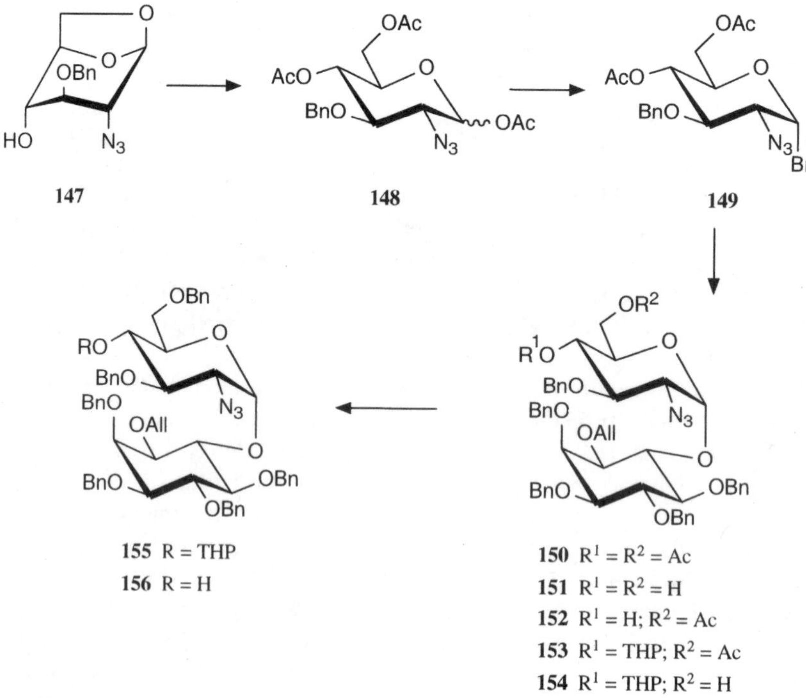

Scheme 18

primary hydroxyl group of **151** was acetylated (to give **152**), and this was tetrahy-dropyranylated to give **153**. Saponification of **153** gave the alcohol **154**, and this was benzylated to give **155**; subsequent acidic hydrolysis gave the required alcohol **156** ($[\alpha]_D$ +41°).

B. Syntheses by the Leiden Group

The Leiden group [12] converted the azido sugar **157** [47] into **158** (Scheme 19) by acetolysis; subsequent regioselective deacetylation [48] gave the free sugar **159**, which was converted [49] into the trichloracetimidate **160**. Condensation of **160** with the chiral inositol derivative **50** (Scheme 5) in the presence of trimethylsilyl triflate gave the disaccharide **161**, which was deacetylated. Benzylation of the product gave **162**, and this on acidic hydrolysis gave the diol **163**.

Compound **160** was also condensed with the racemate **47** of the chiral inositol derivative **50** (Scheme 5), and the diastereoisomeric mixture **161** was converted into the mixture **163**. Crystallization of this mixture gave pure **163**, identical with the material prepared from the chiral inositol derivative **50**.

p-Methoxybenzylation of the diol **163** gave **164**, which was benzylated to give **165**. Mild acidic hydrolysis of **165** to remove the *p*-methoxybenzyl group gave the alcohol **166**, suitable for further glycosidation or phosphorylation.

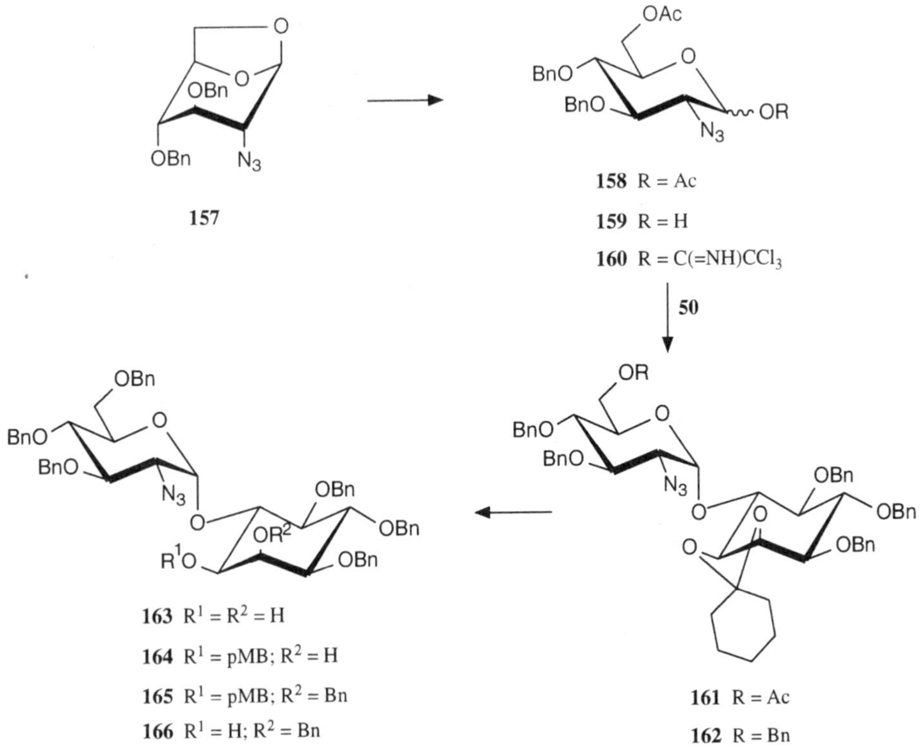

157

158 R = Ac
159 R = H
160 R = C(=NH)CCl₃

| 50

163 R¹ = R² = H
164 R¹ = pMB; R² = H
165 R¹ = pMB; R² = Bn
166 R¹ = H; R² = Bn

161 R = Ac
162 R = Bn

Scheme 19

C. Syntheses by the Charlottesville and Madrid Groups

Although the glucosamine-containing disaccharide **170** (Scheme 20) prepared by the
Charlottesville group [13] is not the exact isomer required for lipid anchor synthesis
(because the glucosamine is attached to the 4-position rather than the 6-position of
D-*myo*-inositol 1-phosphate), it illustrates a different approach for linking the glucosamine to the inositol. The azido-nitro derivative **167** was treated with tetraethylammonium chloride to give **168**, which was condensed with the chiral inositol
derivative **58** (Scheme 6) in the presence of silver triflate to give the α-linked disaccharide **169** (47%; $[\alpha]_D$ +44°). This was desilylated with tetrabutylammonium
fluoride and deacetylated with ammonia in methanol to give **170** ($[\alpha]_D$ +34°). Hydrogenolysis of **170** gave 1D-4-*O*-(2-amino-2-deoxy-α-D-glucopyranosyl)-*myo*-
inositol 1-phosphate **176** {m.p. 178° (dec.), $[\alpha]_D$ +1°, H_2O} because the acidity of the
liberated phosphate was sufficient to remove the cyclohexylidene-protecting groups.

The same disaccharide (with a different optical rotation) was prepared by the
Madrid group [50]. The disaccharide **171** (Scheme 20) was saponified with a catalytic

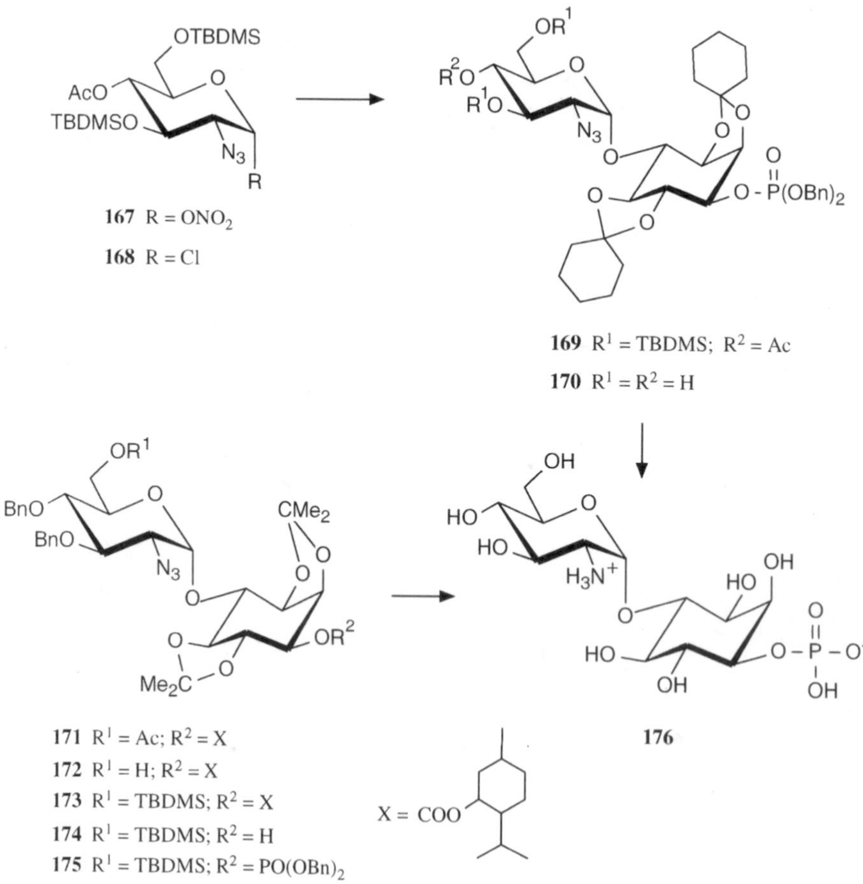

167 R = ONO_2

168 R = Cl

169 R^1 = TBDMS; R^2 = Ac

170 $R^1 = R^2 = H$

171 R^1 = Ac; R^2 = X

172 R^1 = H; R^2 = X

173 R^1 = TBDMS; R^2 = X

174 R^1 = TBDMS; R^2 = H

175 R^1 = TBDMS; R^2 = PO(OBn)$_2$

X = COO

176

Scheme 20

Scheme 20

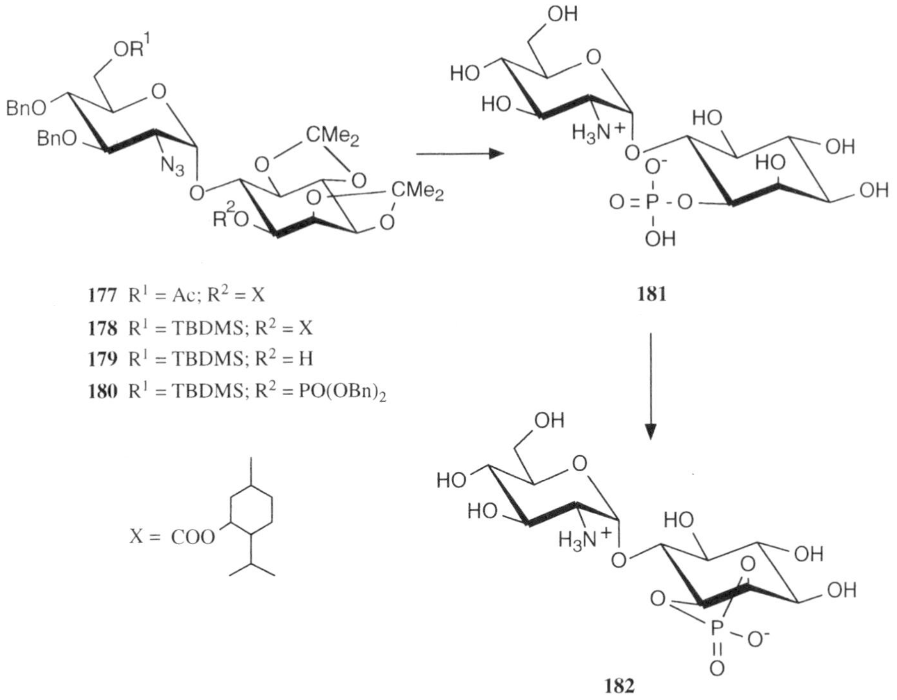

Scheme 21

quantity of sodium methoxide in methanol-dichloromethane, which removed only the 6-*O*-acetyl group on the glucosamine residue to give **172**, and this was silylated to give **173** ($[\alpha]_D$ +40°). Saponification of the menthoxycarbonyl group gave the alcohol **174**, which was phosphorylated using phosphoramidite reagents to give **175** ($[\alpha]_D$ +61°). Deprotection of **175** by desilylation with tetrabutylammonium fluoride and subsequent hydrogenolysis and acid hydrolysis gave **176** ($[\alpha]_D$ +74°, H$_2$O).

In a similar way the Madrid group [50] converted the disaccharide **177** (Scheme 21) via **178** ($[\alpha]_D$ +20°) and **179** into **180** ($[\alpha]_D$ +34°), and this was deprotected as for **175** to give 1D-6-*O*-(2-amino-2-deoxy-α-D-glucopyranosyl)-*myo*-inositol 1-phosphate (**181**; $[\alpha]_D$ +80°, H$_2$O). This was treated with a water-soluble carbodiimide to give the 1,2-cyclic phosphate **182** (=**200**; see Scheme 23). The possible insulin-like biological activity of **182** was discussed [50].

D. Syntheses by the RIKEN Group

The RIKEN group [16] converted the azido derivative **183** [51] (Scheme 22) into the 4,6-*O*-benzylidene derivative **184** ($[\alpha]_D$ −40°) and this was benzylated to give **185** (m.p. 101°; $[\alpha]_D$ −85°). The benzylidene group was opened with borane-trimethylamine-aluminum chloride [52] to give the 6-*O*-benzyl derivative **186** (83%; $[\alpha]_D$ −32°). This was condensed with the thiomannoside **187** in the presence of silver triflate, tetrabutylammonium bromide, and cupric bromide to give the disaccharide **188** (90%; $[\alpha]_D$ +5°). Desilylation of **188** with tetrabutylammonium fluoride gave

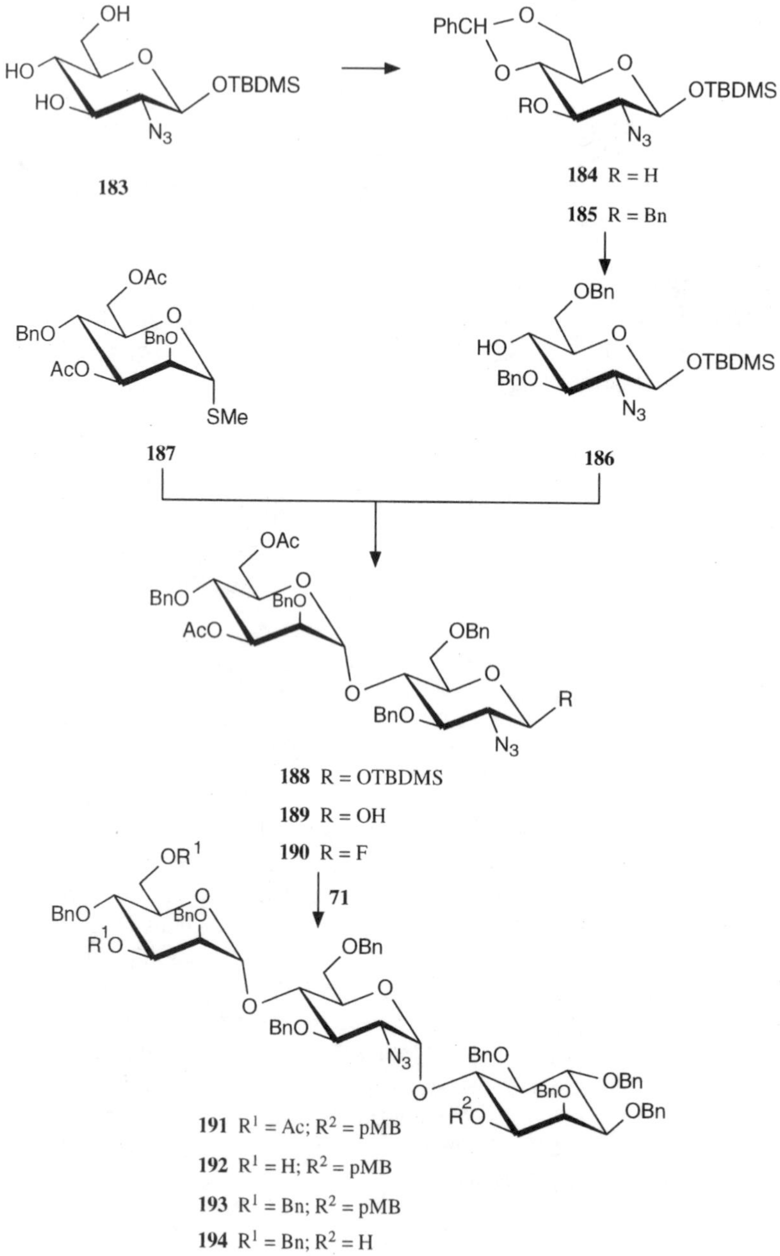

183

184 R = H
185 R = Bn

187

186

188 R = OTBDMS
189 R = OH
190 R = F

71

191 R^1 = Ac; R^2 = pMB
192 R^1 = H; R^2 = pMB
193 R^1 = Bn; R^2 = pMB
194 R^1 = Bn; R^2 = H

Scheme 22

the free disaccharide **189**, which was treated with diethylaminosulfur trifluoride to give the glycosyl fluoride **190** (in 89% yield from **188**). Condensation of the fluoride **190** with the chiral inositol derivative **71** (Scheme 7) in the presence of zirconocene dichloride and silver perchlorate gave the trisaccharide **191** (73%; [α]$_D$ +49°). Saponification of **191** gave the diol **192** ([α]$_D$ +54°), which was benzylated to give **193** ([α]$_D$ +40°), and the *p*-methoxybenzyl group was removed with cerium(IV) ammonium nitrate to give the alcohol **194** ([α]$_D$ +37°), suitable for phosphorylation (see Section V.A.).

E. Syntheses by the Tufts Group

The Tufts group [20] converted the azido sugar **195** [51] (Scheme 23) into the α-glycosyl bromide **196**, using (bromomethylene)dimethylammonium bromide (Vils-

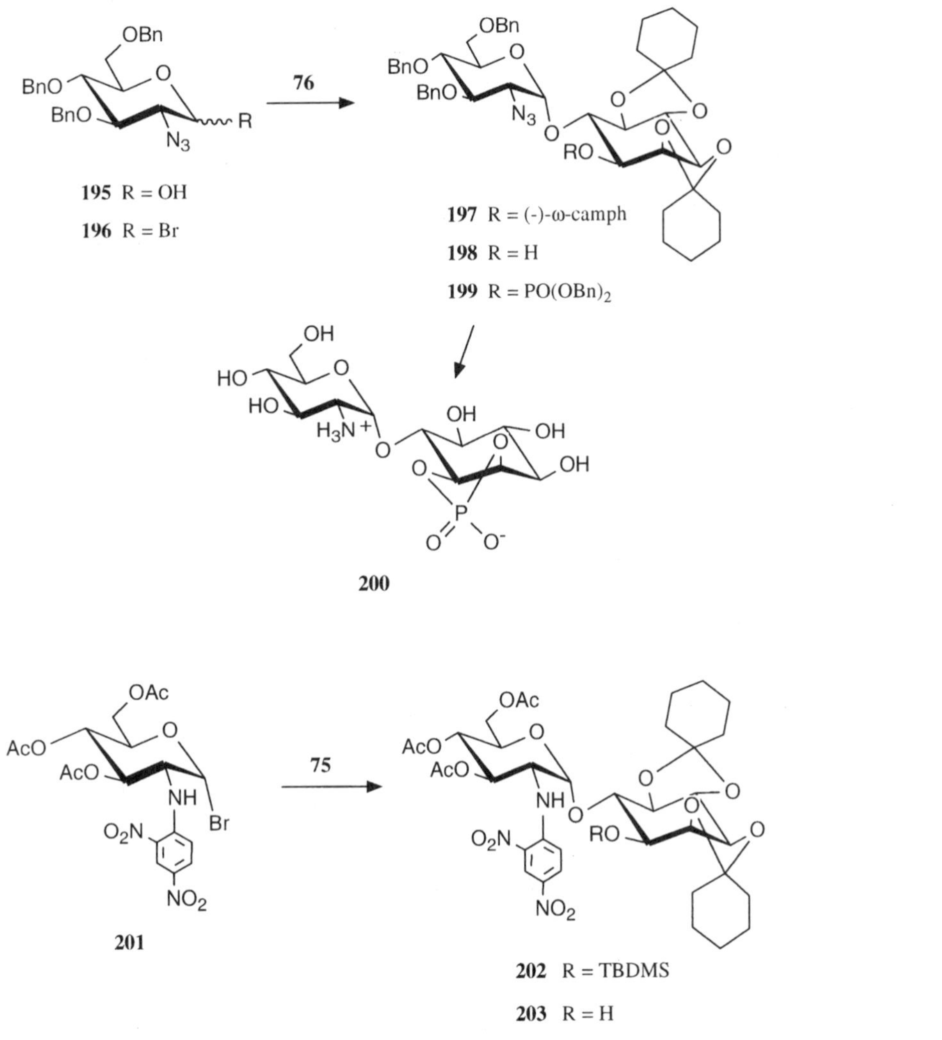

195 R = OH

196 R = Br

197 R = (-)-ω-camph

198 R = H

199 R = PO(OBn)$_2$

200

201

202 R = TBDMS

203 R = H

Scheme 23

meier reagent), and this was condensed with the chiral inositol derivative **76** (Scheme 8) in the presence of silver carbonate, silver perchlorate, and 2,4,6-trimethylpyridine to give the α-linked disaccharide **197** in 19% yield. Saponification of this gave the alcohol **198**, which was phosphorylated by treatment with dibenzyloxy-*N,N*-diiso-propylaminophosphine and subsequent oxidation to give the phosphate **199**. Hydrogenolysis of **199** and subsequent acidic hydrolysis of the cyclohexylidene groups gave the amino-phosphate, which was treated with a water-soluble carbodiimide reagent to give the cyclic phosphate **200**.

In an earlier publication [19] this group described the condensation of the glucosamine derivative **201** [53] with the chiral inositol derivative **75** (Scheme 8) in the presence of silver carbonate, silver perchlorate, and 2,4,6-trimethylpyridine to give a mixture of α- and β-isomers of the disaccharide **202**. Desilylation with tetrabutylammonium fluoride gave the mixture of α- and β-isomers of **203**, and chromatographic separation gave 33% of the α-anomer **203**, which was phosphorylated and deprotected to give a disaccharide phosphate.

In a further approach [19] to **203**, the glucosamine derivative **201** was condensed with the racemic inositol derivative **59** (Scheme 7), as described earlier, to give a mixture of eight isomeric disaccharides, which, after chromatography, gave **203** in 30% yield.

These disaccharides have been used for enzymatic studies concerning the inhibition of phosphatidylinositol phospholipase C [54].

F.   Further Syntheses by the Madrid Group

The Madrid group [22] prepared the azido sugar **205** (Scheme 24) from the mannopyranose **204**, using a published procedure [55], and converted this into the trichloracetimidate **206**. This was condensed with the racemic inositol derivative **29** (Scheme 4), using trimethylsilyl triflate, to give a mixture of diastereoisomers (71%), which was separated by repeated thin layer chromatography to give **207**.

In a further approach [22], **204** was glycosylated with 2,3,4,6-tetra-*O*-acetyl-α-D-galactopyranosyl bromide in the presence of mercury salts to give **208** (Scheme 25) in 51% yield. The benzylidene group was oxidatively cleaved with dichlorodicyanoquinone to give **209** (71%; m.p. 191°; [α]$_D$ −51°). This was converted into the triflate, which was treated with sodium azide in *N,N*-dimethylformamide to give **210** (92%; m.p. 132°; [α]$_D$ −1°). Acetolysis of **210** gave **211**, and this was converted

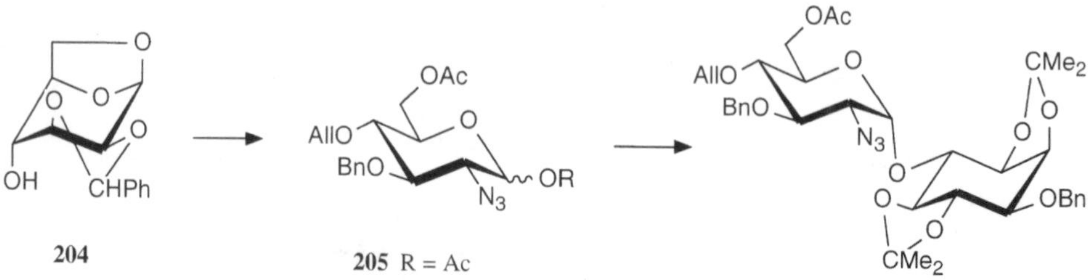

204

205 R = Ac

206 R = C(=NH)CCl$_3$

207

Scheme 24

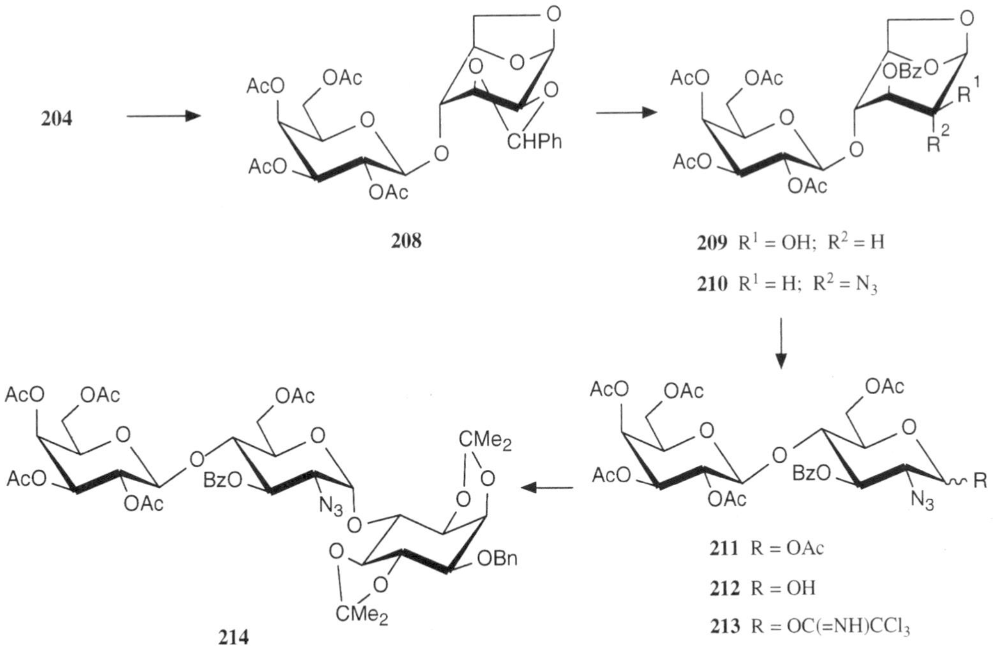

204 →

208

209 $R^1 = OH$; $R^2 = H$

210 $R^1 = H$; $R^2 = N_3$

214

211 R = OAc

212 R = OH

213 R = OC(=NH)CCl$_3$

Scheme 25

into the free sugar **212** by the action of hydrazine hydrate. Compound **212** was treated with trichloroacetonitrile and potassium carbonate to give the trichloracetimidate **213**. Condensation of **213** with the racemic inositol derivative **29** (Scheme 4), in the presence of trimethylsilyl triflate, gave the diastereoisomeric mixture containing **214**. Column chromatography gave **214** (m.p. 105°; [α]$_D$ +102°), which is an isomer of the core structure of the lipid anchor.

In order to prepare the disaccharide **218** (Scheme 26), which is a portion of the lipid anchor and may have some function in the action of insulin, the Madrid group [25] condensed the glucosamine derivative **160** (Scheme 19) with the inositol derivative **89** (Scheme 10) in the presence of trimethylsilyl triflate to give the disaccharide **215** (65%; [α]$_D$ +47°). Saponification of **215** with ammonia in methanol gave **216**. Removal of the allyl group by isomerization to a *trans*-prop-1-enyl group with an iridium catalyst [56], followed by hydrolysis of the prop-1-enyl group with aqueous iodine [57], gave the diol **217**. Hydrogenolysis of the latter gave the disaccharide **218** ([α]$_D$ +73°, H$_2$O).

Compounds related to **218** but containing *chiro*-inositol rather than *myo*-inositol have also been prepared [25,58].

G. Syntheses by the London-Cambridge Group

The London-Cambridge group also prepared [39] **218** from the protected inositol derivative **132** (Scheme 15). Glycosidation of **132** with the trichloracetimidate **219** [59] (Scheme 27) gave the α-linked disaccharide **220** (59% together with 24% of the β-linked disaccharide). The p-methoxybenzyl group was removed with dichlo-

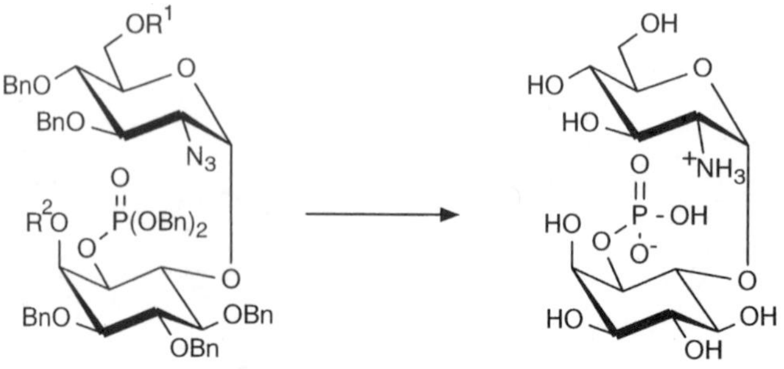

215 R^1= Ac; R^2 = All
216 R^1= H; R^2 = All
217 R^1= R^2 = H

218

Scheme 26

rodicyanoquinone to give the alcohol **221**, and this was phosphorylated using phosphoramidite methodology to give **222**. Catalytic hydrogenolysis of **222** and subsequent acidic hydrolysis of the isopropylidene group gave **218** (Scheme 26) ([α]$_D$
+62°, H$_2$O).

Using a similar procedure, the London-Cambridge group [60] condensed the
myo-inositol derivative **132** (Scheme 15) with the trichloracetimidate **223** [61] in

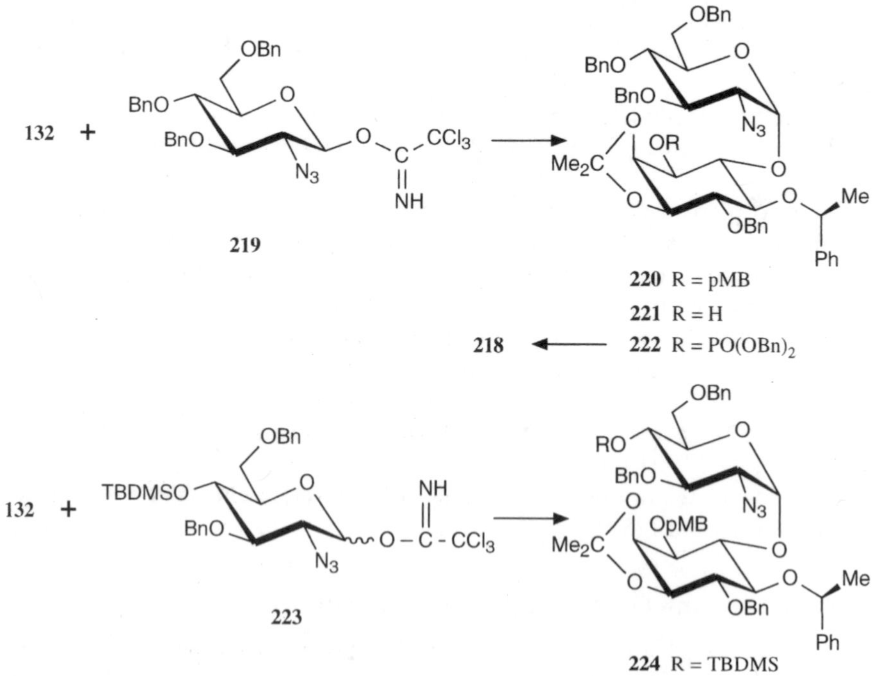

219

220 R = pMB
221 R = H
218 ←——— 222 R = PO(OBn)$_2$

223

224 R = TBDMS
225 R = H

Scheme 27

toluene in the presence of *t*-butyldimethylsilyl triflate [61] to give the α-linked disaccharide **224** (48% together with 24% of the β-linked disaccharide). Desilylation of **224** with tetrabutylammonium fluoride in tetrahydrofuran gave the alcohol **225** suitable for further glycosidation (see Section IV.C.).

H. Syntheses by the Dundee Group

The Dundee group [14] converted the azido sugar **195** (Scheme 23) into the glycosyl fluoride **226** (Scheme 28) by the action of diethylaminosulfur trifluoride, and this was coupled with the chiral inositol derivative **106** (Scheme 12) in the presence of zirconocene dichloride and silver perchlorate to give the disaccharide **227** together with the β-isomer. The mixture was treated with cerium(IV) ammonium nitrate to remove the *p*-methoxybenzyl group, and subsequent chromatography of the product gave the pure α-linked disaccharide **228** (34%; $[\alpha]_D$ +44°).

In a new approach by the Dundee group [34], the inositol derivative **112** (Scheme 13) was glycosylated with **229** (m.p. 104°; $[\alpha]_D$ −3°) (Scheme 28) in the presence of trimethylsilyl triflate, and subsequent hydrolysis of the cyclohexylidene group gave mainly the α-linked glycoside **230** (46%; m.p. 124°; $[\alpha]_D$ +36°; previously prepared by the Leiden group, see **163**, Scheme 19), which was separated by chromatography from the corresponding β-linked glycoside (25%; m.p. 118°; $[\alpha]_D$ −21°). Tin-mediated *p*-methoxybenzylation of **230** gave **231** ($[\alpha]_D$ +26°), and this was benzylated to give **227**, identical with the material prepared previously [14].

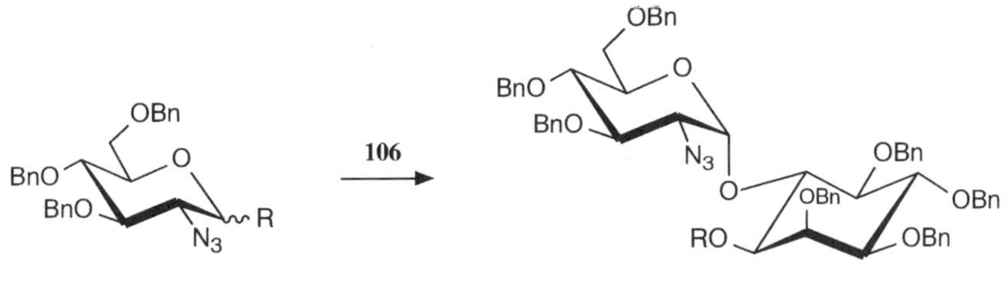

195 R = OH
226 R = F

227 R = pMB
228 R = H

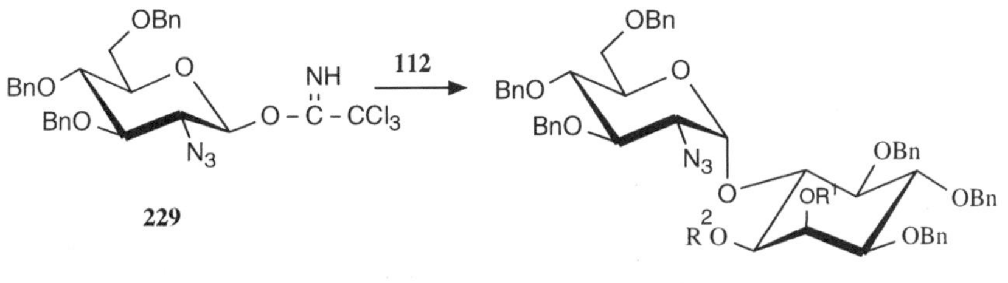

229

230 R¹ = R² = H
231 R¹ = H; R² = pMB

Scheme 28

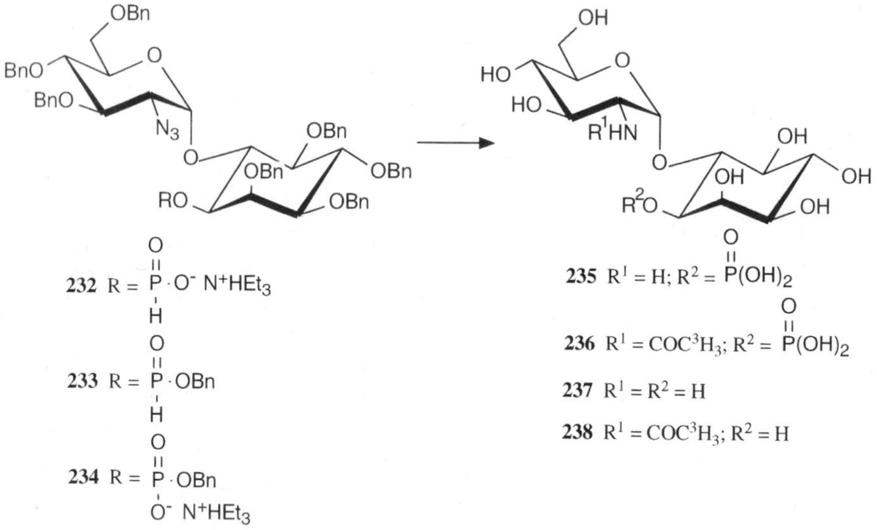

232 R = $\overset{\overset{\displaystyle O}{\|}}{\underset{\underset{\displaystyle H}{|}}{P}} \cdot O^- \ N^+HEt_3$

233 R = $\overset{\overset{\displaystyle O}{\|}}{\underset{\underset{\displaystyle H}{|}}{P}} \cdot OBn$

234 R = $\overset{\overset{\displaystyle O}{\|}}{\underset{\underset{\displaystyle O^- \ N^+Et_3}{|}}{P}} \cdot OBn$

235 $R^1 = H; R^2 = \overset{\overset{\displaystyle O}{\|}}{P}(OH)_2$

236 $R^1 = COC^3H_3; R^2 = \overset{\overset{\displaystyle O}{\|}}{P}(OH)_2$

237 $R^1 = R^2 = H$

238 $R^1 = COC^3H_3; R^2 = H$

Scheme 29

Treatment of **227** with trifluoroacetic acid in dichloromethane removed the *p*-methoxybenzyl group to give **228**.

Treatment [62] of the alcohol **228** (Scheme 28) with phosphorus acid in pyridine gave a phosphonate, which was isolated as the triethylammonium salt **232** ([α]$_D$ +63°; Scheme 29), and this was treated with benzyl alcohol and pivaloyl chloride in pyridine to give the phosphonic diester **233**. Oxidation of the latter with iodine in aqueous pyridine gave a phosphate diester, which was isolated as the triethylammonium salt **234** ([α]$_D$ +45°). Hydrogenolysis of **234** gave the completely deprotected disaccharide **235** ([α]$_D$ +75°, H$_2$O; cf **218**, Scheme 26), which was converted [63] into the $^3H,N$-acetyl derivative **236**, and this was used in enzymatic studies (see also Ref. 64 for further biological studies with these synthetic compounds).

Similar catalytic hydrogenation of **228** (Scheme 28) gave **237** ([α]$_D$ +82°; MeOH), and this was converted into the tritiated acetyl derivative **238** [63] for use in enzymatic studies.

I. Syntheses by the Konstanz Group

The Konstanz group [36] condensed the trichloracetimidate **239** [65] (Scheme 30) with the chiral inositol derivative **117** (Scheme 14) in ether in the presence of trimethylsilyl triflate to give only the α-linked disaccharide **240**. Selective cleavage of the *O*-acetyl groups with sodium methoxide-methanol gave the triol **241**, and benzylation of this with sodium hydride-benzyl bromide-tetrabutylammonium bromide gave **242**. Cleavage of the carbonate group with potassium carbonate in methanol at 40° gave the alcohol **243**.

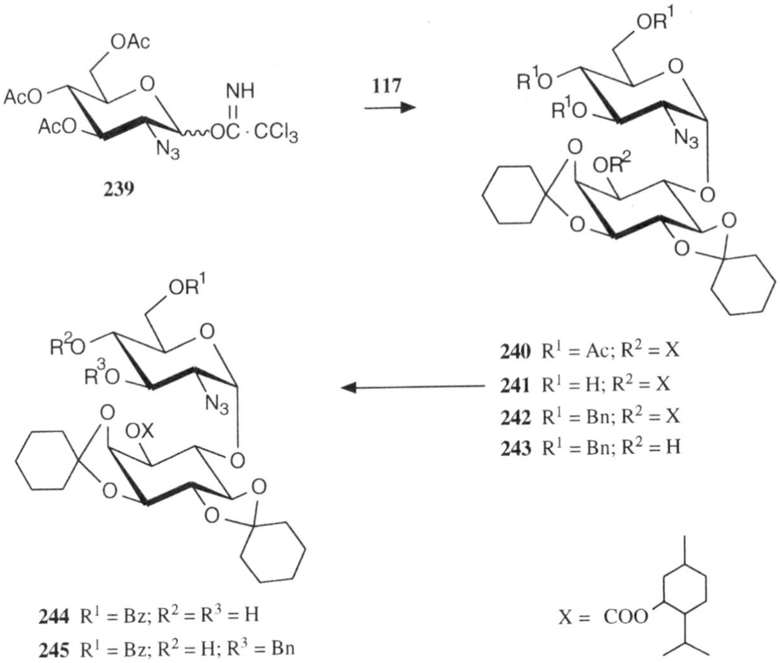

OAc

AcO

AcO

NH
||
OC·CCl₃

N₃

239

117

OR¹

R¹O

R¹O

N₃

OR²

240 R¹ = Ac; R² = X
241 R¹ = H; R² = X
242 R¹ = Bn; R² = X
243 R¹ = Bn; R² = H

OR¹

R²O

R³O

N₃

OX

244 R¹ = Bz; R² = R³ = H
245 R¹ = Bz; R² = H; R³ = Bn

X = COO

Scheme 30

Regioselective 6-O-benzoylation of **241** with benzoyl cyanide at $-70°$ gave **244**. This was regioselectively benzylated at the 3-position of the azido-sugar residue using benzyl bromide-silver oxide to give **245**, suitable for further elaboration into the lipid anchor of *Saccharomyces cerevisiae* (see Section V.D.).

IV. ADDITION OF FURTHER SUGARS OF THE CORE OLIGOSACCHARIDE

A. Syntheses by the Durham Group

The Durham group [4,66] prepared the pentasaccharide α-D-Man$p$-(1 → 2)α-D-Man$p$-(1 → 6)α-D-Man$p$-(1 → 4)α-D-GlcNp-(1 → 6)-D-*myo*-inositol, as outlined in Schemes 31 and 32. 2,3,4,6-Tetra-O-benzoyl-α-D-mannopyranosyl bromide was converted into the orthoester **246**, which on saponification with sodium methoxide and subsequent benzylation gave **247**. Acidic hydrolysis of this and subsequent saponification gave the pentenyl mannoside **251** ([α]$_D$ +48°), which was condensed with 2,3,4,6-tetra-O-benzoyl-α-D-mannopyranosyl bromide in the presence of silver triflate to give the dimannosyl disaccharide **254** ([α]$_D$ −29°).

The pentenyl mannoside **249** was converted [4,8] into **250**, and this was condensed with the disaccharide **146** (Scheme 17) in the presence of N-iodosuccinimide and triflic acid to give the trisaccharide **255**, from which the silyl group was removed to give the alcohol **256** ([α]$_D$ +61°).

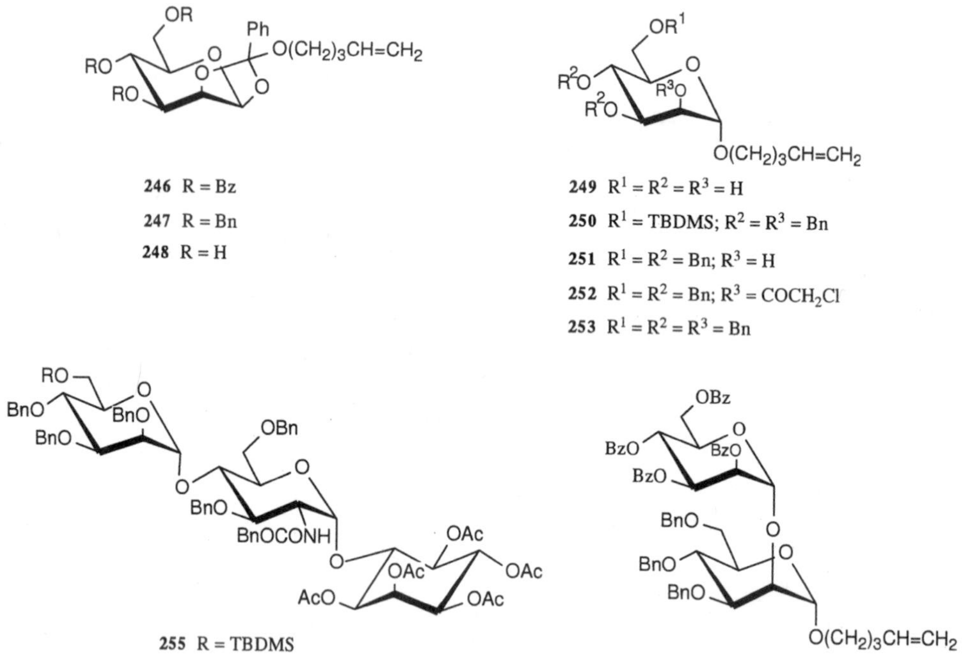

246 R = Bz

247 R = Bn

248 R = H

249 R¹ = R² = R³ = H

250 R¹ = TBDMS; R² = R³ = Bn

251 R¹ = R² = Bn; R³ = H

252 R¹ = R² = Bn; R³ = COCH₂Cl

253 R¹ = R² = R³ = Bn

255 R = TBDMS

256 R = H

254

Scheme 31

Condensation of the pentenyl mannoside **252** with the trisaccharide **256** in the presence of *N*-iodosuccinimide and triflic acid gave the tetrasaccharide **257**, and removal of the chloroacetyl group with thiourea gave the alcohol **258**.

Condensation of the pentenyl disaccharide **254** with the trisaccharide **256** in the presence of *N*-iodosuccinimide and triflic acid gave in high yield the pentasaccharide mixture comprising **259** (Scheme 32) and its β-isomer (at the new glycosidic linkage) in a ratio of 2 to 3 due to the high reactivity of the primary hydroxyl group of **256**. However, condensation of the orthoester **246** with the tetrasaccharide **258** under the same conditions gave the pure α-linked pentasaccharide **259**. Compound **259** was completely deprotected by saponification of the ester groups and hydrogenolysis of the benzyl ethers to give the free pentasaccharide α-D-Man*p*-(1 → 2)α-D-Man*p*-(1 → 6)α-D-Man*p*-(1 → 4)α-D-GlcN*p*-(1 → 6)D-*myo*-inositol.

In an approach to the trimannosyl trisaccharide **270** (Scheme 33), required as the mannan component of the glycophosphatidylinositol lipid anchor, the Durham group [43,66] prepared the triol **248** by saponification of **246** (Scheme 31), and this was selectively *t*-butyldiphenylsilylated at the primary hydroxyl group. Subsequent benzylation gave **260** (Scheme 33), and this on treatment with bromine in dichloromethane gave the mannosyl bromide **262**. This was coupled with **251** (Scheme 31) in the presence of silver triflate to give the disaccharide **263** ([α]$_D$ +8°), and this on saponification gave **264**. Treatment of the orthoester **247** with bromine in dichloromethane gave the mannosyl bromide **265**, and this was condensed with the disaccharide **264** in the presence of silver triflate to give the trisaccharide **266** ([α]$_D$ +3°). This was saponified to give **267**, which was benzylated to give **268**, and

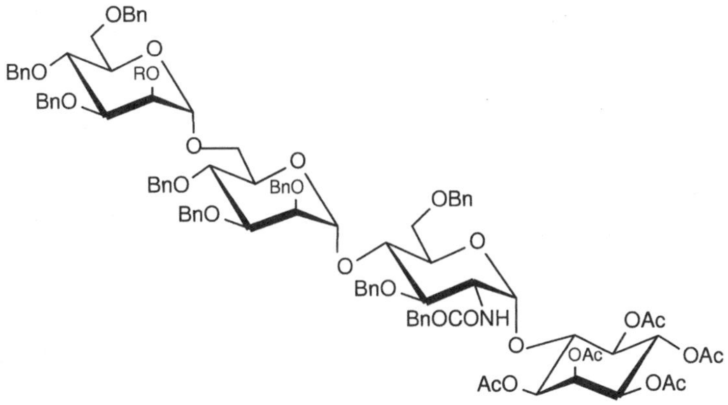

257 R = COCH$_2$Cl

258 R = H

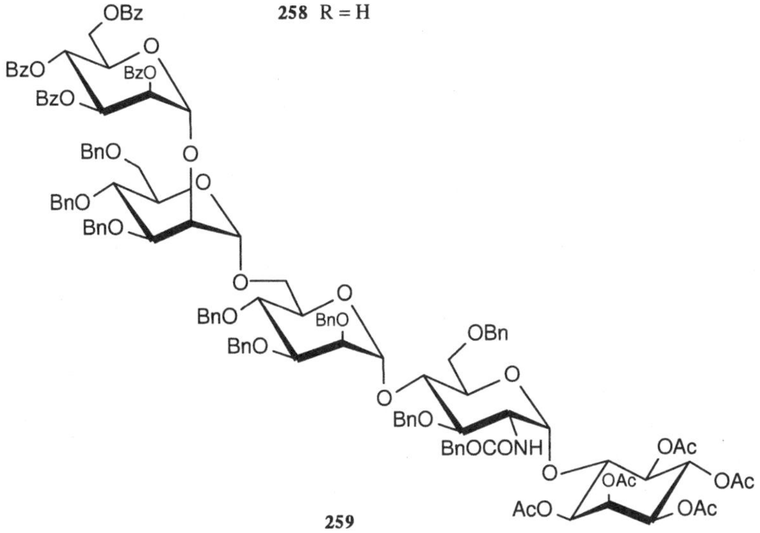

259

Scheme 32

treatment of this with tetrabutylammonium fluoride to remove the silyl group gave the alcohol **269**. This alcohol was treated with chloroacetic anhydride to give the trisaccharide **270** ([α]$_D$ +27°) required for further coupling (see below).

As this procedure for the synthesis of **270** was considered a little unwieldy, the Durham group [66] developed an alternative route taking advantage of the potential of the pentenyl groups. Three new mannosyl units were required for this route and were prepared as follows:

1. The *t*-butyldiphenylsilyl group of **260** (Scheme 33) was replaced with acetyl by desilylation with tetrabutylammonium fluoride and subsequent acetylation to give **261**.
2. The alcohol **251** (Scheme 31) was benzylated to give **253**.
3. The pentenyl group of the alcohol **251** (Scheme 31) was brominated to give **271** (Scheme 34), which was now ready to act as a glycosyl acceptor.

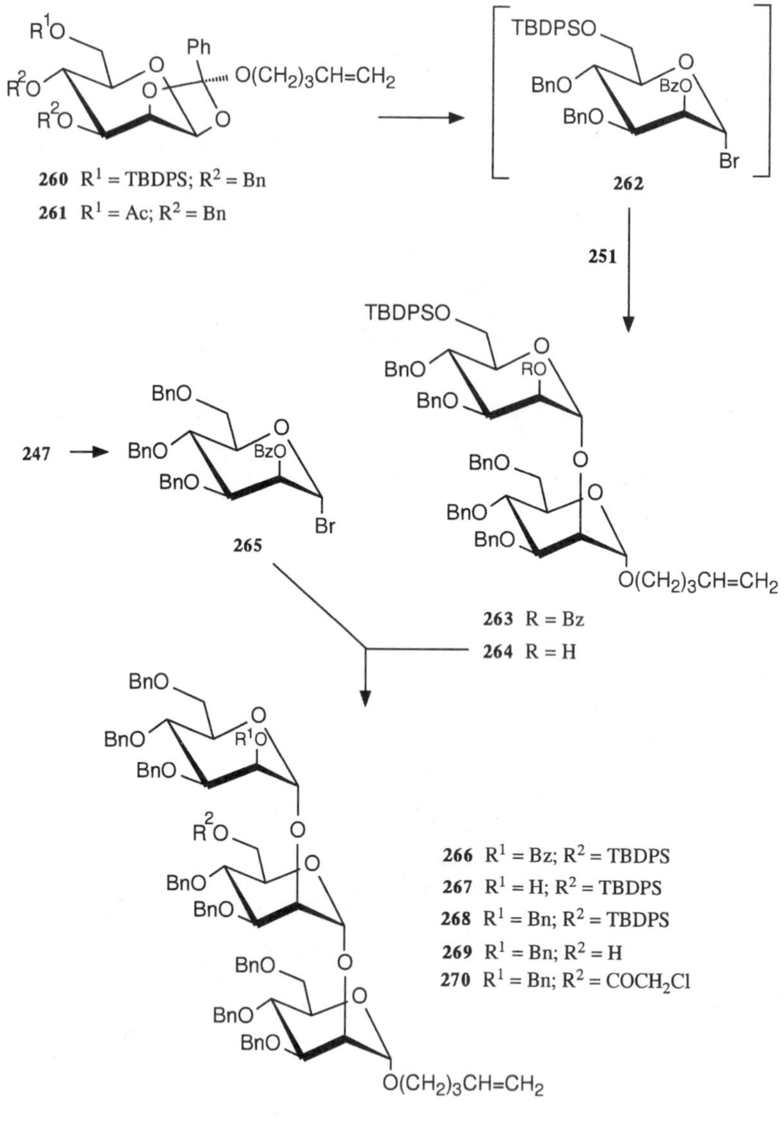

260 R^1 = TBDPS; R^2 = Bn
261 R^1 = Ac; R^2 = Bn

262

247 →

265

263 R = Bz
264 R = H

266 R^1 = Bz; R^2 = TBDPS
267 R^1 = H; R^2 = TBDPS
268 R^1 = Bn; R^2 = TBDPS
269 R^1 = Bn; R^2 = H
270 R^1 = Bn; R^2 = COCH$_2$Cl

Scheme 33

Thus condensation (Scheme 34) of the orthoester **261** with **271** in the presence of *N*-iodosuccinimide and a catalytic amount of triethylsilyl triflate in dichloromethane gave the disaccharide **272** (78%; $[\alpha]_D$ +11°), which was saponified to give the diol **273**. Preferential acylation of the primary hydroxyl group of **273** with chloroacetic anhydride-triethylamine gave the required mono-chloroacetyl derivative **274** (94%; $[\alpha]_D$ +24°), and this was condensed with **253** (Scheme 31) in the presence of *N*-iodosuccinimide and a catalytic amount of triethylsilyl triflate in dichloromethane to give the trisaccharide **275** (75%; $[\alpha]_D$ +8°). Reductive debromination of **275** with zinc and tetrabutylammonium iodide was accompanied by dechloroacetylation to

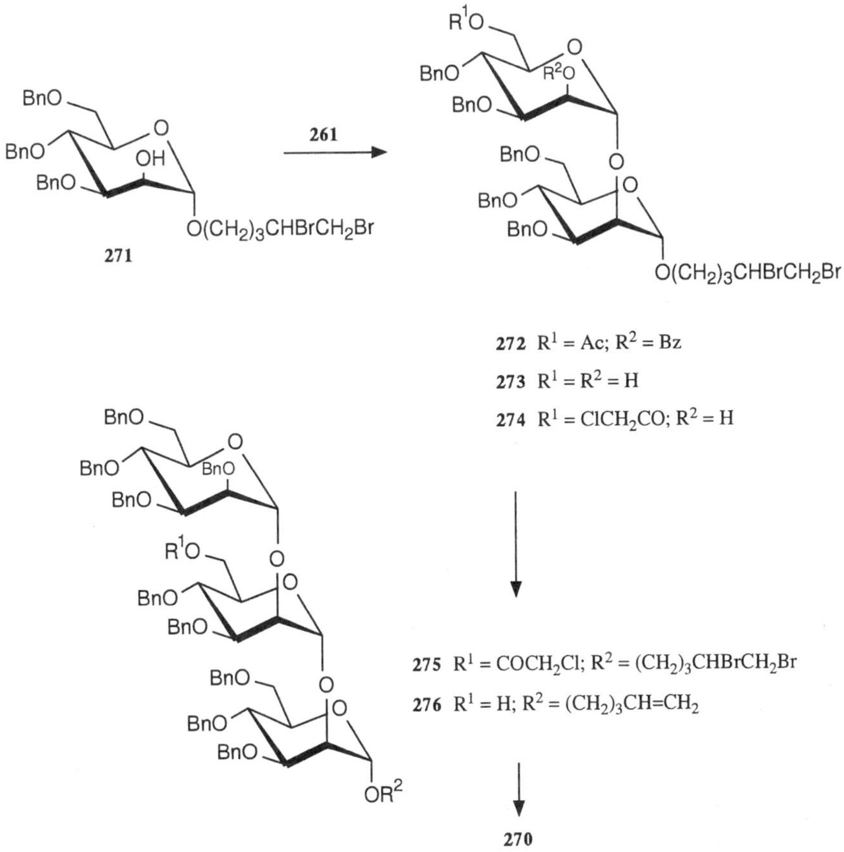

Scheme 34

give **276**, and this on rechloroacetylation gave **270**, identical with the material pre-pared as described earlier (Scheme 33).

To prepare the heptasaccharide derivative **295** (Scheme 36) ready for phos-phorylation to give the Thy-1 anchor, the Durham group [8] required some suitably substituted intermediates **282** and **285** (Scheme 35). These were prepared as follows.

1. The pentenyl α-mannoside **277** [67] was converted into the 4,6-O-benzyli-dene derivative **278** ($[\alpha]_D$ +54°), and this on tin-mediated benzylation gave **279**, which was acetylated to give **280** ($[\alpha]_D$ +17°). Acidic hydrolysis of the benzylidene group and preferential acetylation of the primary hydroxyl group of the diol **281** produced gave **282** ($[\alpha]_D$ +1°), required as a glycosyl acceptor.

2. 2-Amino-2-deoxy-D-galactose was converted [68] into the phthalimido de-rivative **283**, which, on reaction with p-methoxyphenol and trimethylsilyl triflate, gave **284** ($[\alpha]_D$ +30°). This was saponified with sodium methoxide in methanol and the product benzylated to give **285** ($[\alpha]_D$ +63°). Alternatively the 2-amino-2-deoxy-D-glucose derivative **287** [69] was converted into the triflate, and this, on reaction with cesium acetate in methyl sulfoxide, gave the 2-amino-2-deoxy-D-galactose de-

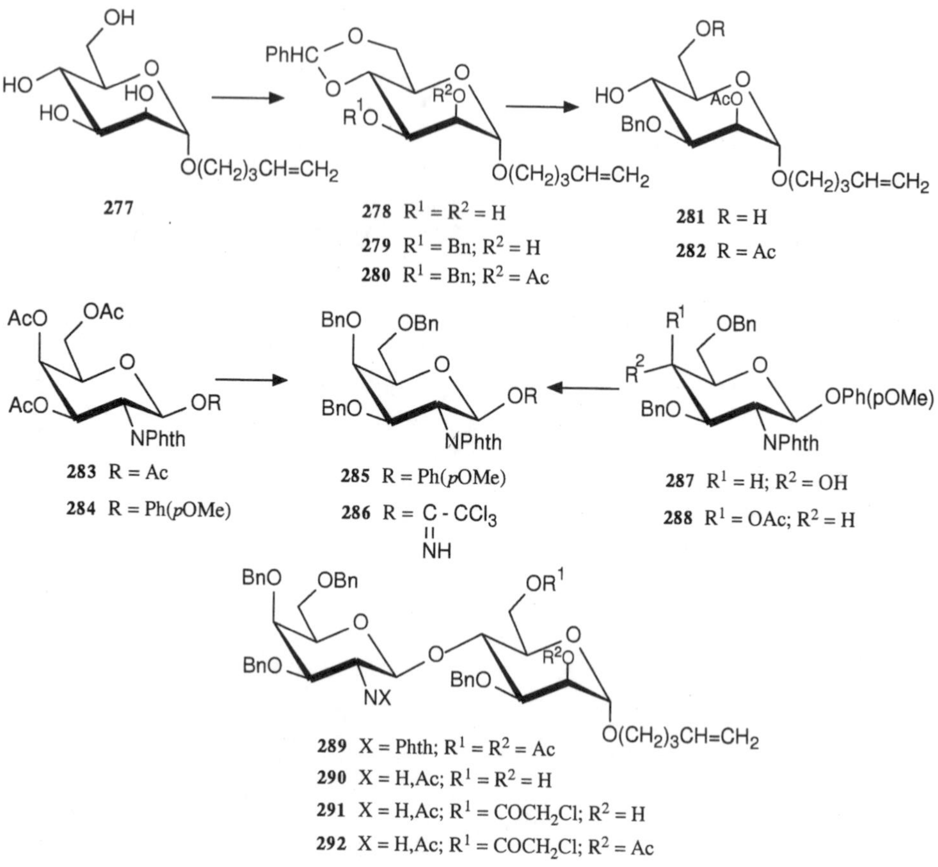

Scheme 35

rivative **288** (78%; $[\alpha]_D$ +76°). This was saponified with sodium methoxide in methanol, and subsequent benzylation gave **285**, required as a glycosyl donor.

For the synthesis of the disaccharide **292**, compound **285** was treated with cerium(IV) ammonium nitrate (to remove the *p*-methoxyphenyl group), and the free sugar was treated with trichloroacetonitrile and diazabicycloundecene to give the crystalline trichloroacetimidate **286** ($[\alpha]_D$ +59°). This was condensed with the alcohol **282** in the presence of trimethylsilyl triflate to give the disaccharide **289** (79%; $[\alpha]_D$ +78°). The phthaloyl and acetyl groups were removed with methylamine in aqueous methanol, and the free amino group was acetylated with acetic anhydride to give **290**. The diimide produced (from reaction of trace amounts of oxygen with hydrazine) caused some saturation of the double bond of the pentenyl group if hydrazine was used to remove the phthaloyl group. The primary hydroxyl group of **290** reacted preferentially with chloroacetic anhydride and triethylamine to give **291**, and the secondary hydroxyl group was acetylated to give **292** (68%; $[\alpha]_D$ +38°), the required disaccharide donor.

Condensation of the donor galactosamine-mannose disaccharide derivative **292** (Scheme 35) with the acceptor glucosamine-inositol disaccharide derivative **156**

(Scheme 18) in the presence of N-iodosuccimimide and triethylsilyl triflate (under precise conditions because of competing silylation of **156**) allowed the Durham group [8,43] to prepare the tetrasaccharide derivative **293** (Scheme 36) (66% from **292**; $[\alpha]_D$ +56°). This was treated with thiourea in chloroform-methanol to remove the chloroacetyl group and give the alcohol **294** (88%; $[\alpha]_D$ +57°).

This was coupled with the trimannan donor **270** (Scheme 33) in the presence of N-iodosuccinimide and triethylsilyl triflate to give the heptasaccharide derivative **295** (39% based on recovered **294**; $[\alpha]_D$ +19°). Compound **295** has three orthogonal, temporary protecting groups (chloroacetyl, acetyl, and allyl), and these were removed independently for the phosphorylation studies described in Section V.C.

B. Syntheses by the RIKEN Group

The RIKEN group [17] prepared the glycoside **296** (Scheme 37) by the reaction of penta-O-acetyl-β-D-galactopyranose with p-methoxyphenol in the presence of trimethylsilyl triflate, and subsequent saponification gave the tetraol **297**. This was converted into the 4,4′-dimethoxytrityl derivative **298**, which was benzylated to give **299**, and this on acidic hydrolysis gave the alcohol **300**. Glycosylation of **300** with the thiogalactoside **301** [70] in the presence of copper(II) bromide and tetrabutylammonium bromide gave the α-linked isomer **302** in 67% yield after chromatographic separation of some β-linked isomer. Treatment of **302** with cerium(IV) ammonium nitrate removed the p-methoxyphenyl group [71], and the product **303** was treated with diethylaminosulfur trifluoride to give the glycosyl fluoride **304** as a mixture of anomers.

The 3,6-diol **192** (Scheme 22) was converted into the 6-O-acetyl derivative by the action of acetyl chloride in pyridine, and this was condensed with the digalactosyl fluoride **304** in the presence of zirconocene dichloride and silver perchlorate to give the α-linked pentasaccharide **305** (Scheme 38) (67% contaminated with some of the β-linked isomer). Deacetylation of the crude mixture gave the crude alcohol **306**, from which the pure compound was isolated by chromatography. The pentasaccharide **306** was glycosylated with the mannosyl chloride **308** [72] in the presence of mercury(II) chloride and mercury(II) cyanide to give the hexasaccharide **307** (R′ = Ac) in 89% yield. This was saponified to give the alcohol **307** (R′ = H), which was condensed with the mannosyl fluoride **309** to give the heptasaccharide **356** (Scheme 45), in which there are two positions that can be deprotected preferentially for phosphorylation, as described in Section V.A.

C. Syntheses by the London-Cambridge Group

For their approach to the lipid anchor of *Trypanosoma brucei*, the London-Cambridge group [60] prepared the galactose derivative **312** (Scheme 39) from the diol **311** [73], using standard procedures. Iodonium dicollidine perchlorate-mediated, chemoselective glycosidation of **312** with the galactose derivative **310** gave predominantly the α-linked digalactosyl derivative **313** (60%, together with 24% of β-linked isomer), and further chemoselective glycosidation of the mannose derivative **314** with **313** in the presence of N-iodosuccinimide-triflic acid gave the α-linked trisaccharide **315** (63%).

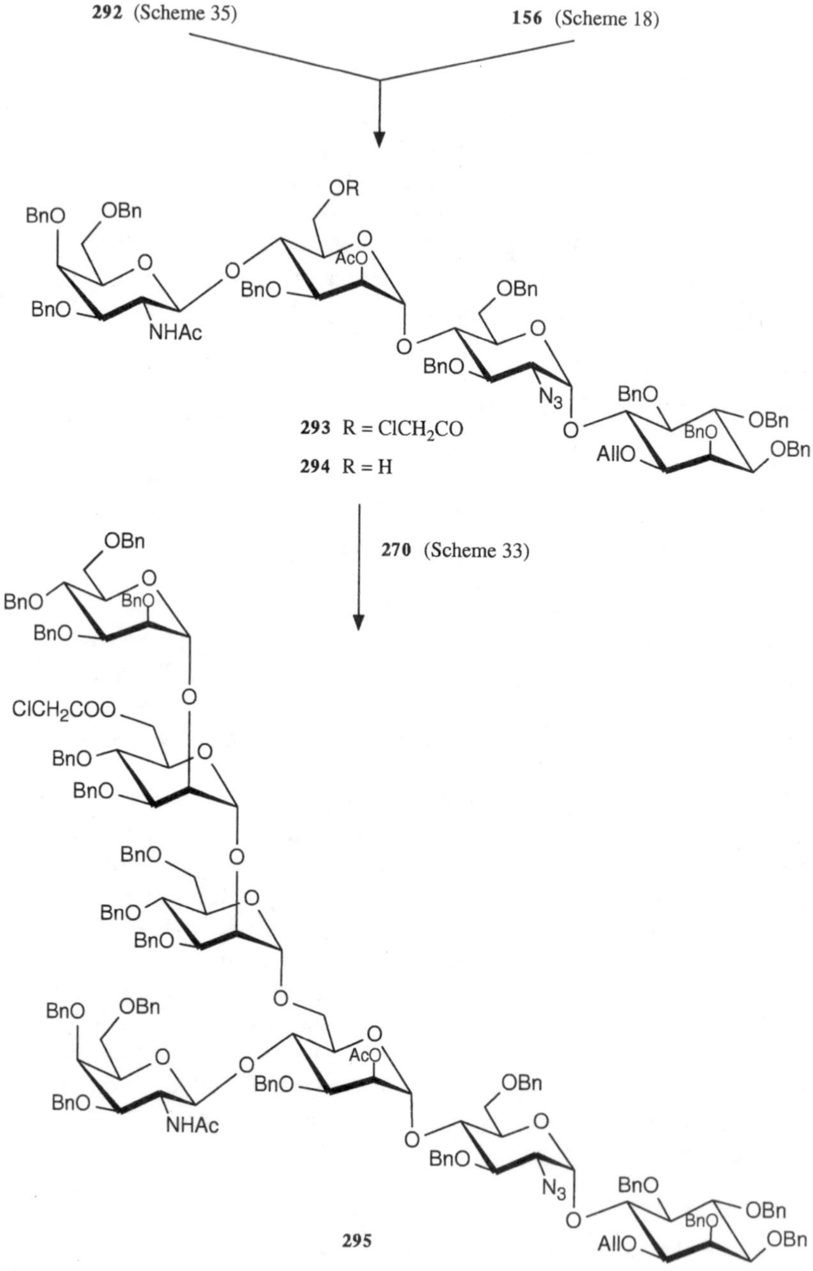

292 (Scheme 35)   **156** (Scheme 18)

293 R = ClCH$_2$CO
294 R = H

270 (Scheme 33)

295

Scheme 36

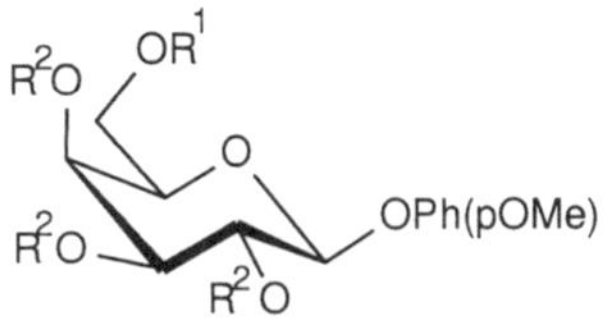

296 $R^1 = R^2 = Ac$

297 $R^1 = R^2 = H$

298 $R^1 = DMT;\ R^2 = H$

299 $R^1 = DMT;\ R^2 = Bn$

300 $R^1 = H;\ R^2 = Bn$

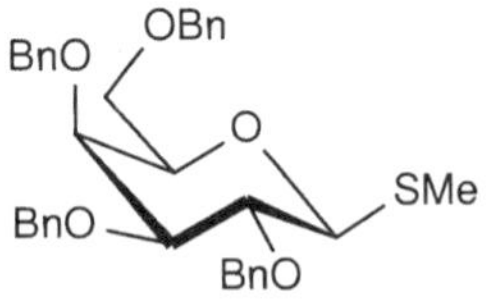

301

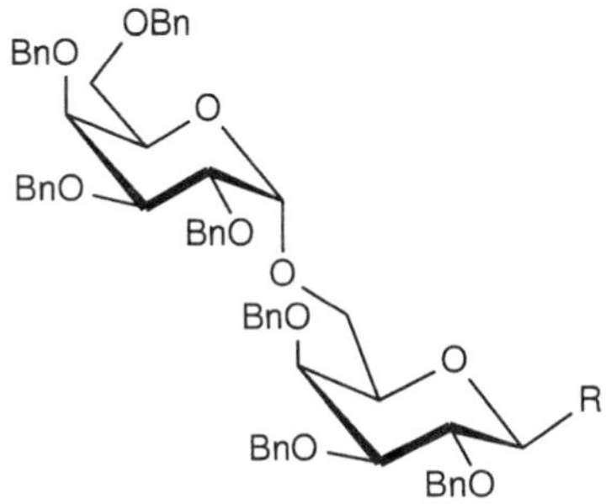

302 $R = OPh(pOMe)$

303 $R = OH$

304 $R = F$

Scheme 37

Glycosidation [60] of the disaccharide **225** (Scheme 27) with **315** (Scheme 39) in the presence of *N*-iodosuccinimide and triflic acid gave the pentasaccharide **316** (41%, Scheme 40), and this has the appropriate protection pattern for further processing to give the GPI anchor of *Trypanosoma brucei*.

D. Syntheses by the Madrid Group

For their preparation of a tetragalactosyl unit **326** (Scheme 41) required in the synthesis of the GPI anchor of *Trypanosoma brucei*, the Madrid group [74] used the sulfoxide glycosylation reaction of Kahne [75]. The phenylthio-β-galactoside **317** [76] was saponified and acetonated to give **318** ($[\alpha]_D$ +7°), which was selectively acetylated with acetyl chloride-*sym*-collidine to give **319** (85%; $[\alpha]_D$ +6°), and this was converted into the *t*-butyldiphenylsilyl derivative **320** (m.p. 98°; $[\alpha]_D$ +26°). Saponification of **320** gave the acceptor molecule **321** (m.p. 145°; $[\alpha]_D$ +29°), whereas oxidation of **320** with 3-chloroperbenzoic acid gave the donor sulfoxide **322** (S, R mixture).

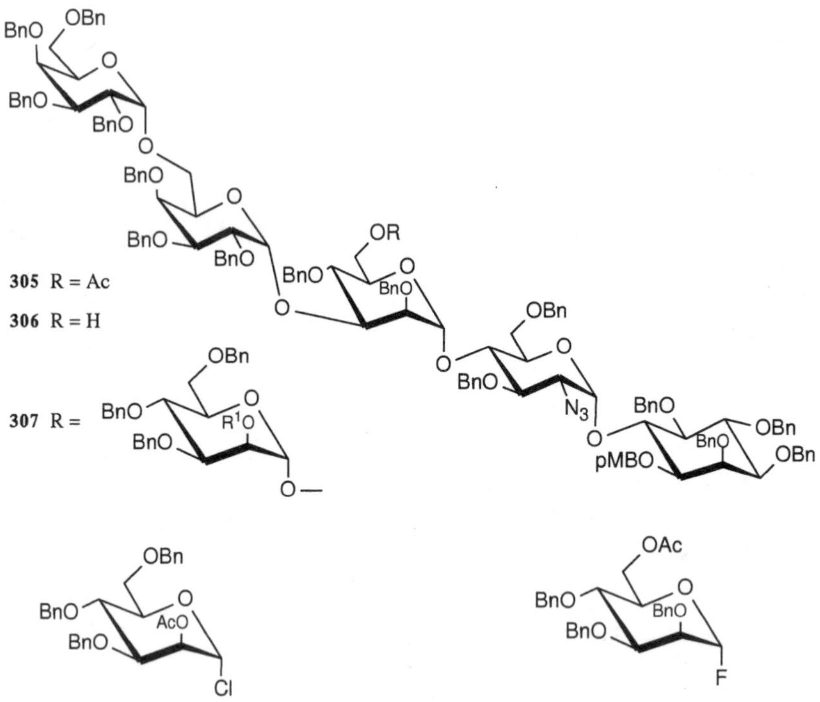

Scheme 38

Condensation of **322** with **321** in the presence of triflic anhydride and 2,6-di-*t*-butyl-4-methylpyridine gave the α-linked disaccharide **323** (63%; $[\alpha]_D$ +51°), which was separated from a little of the β-linked isomer by chromatography. Treatment of **323** with tetrabutylammonium fluoride gave the diol **324** ($[\alpha]_D$ +52°).

Glycosylation of **324** with **322** using the same promoters as described earlier gave the protected tetrasaccharide **325** (40%; $[\alpha]_D$ +62°), which was oxidized to give the sulfoxide **326** required for coupling to a mannose residue for the synthesis of the GPI anchor of *Trypanosoma brucei*.

E. Syntheses by the Konstanz Group

For the synthesis of the trimannoside **344** (Scheme 42) required for the synthesis of the lipid anchor of *Saccharomyces cerevisiae*, the Konstanz group [36] first benzylated the orthoester **327** to give **328**. This was hydrolyzed with acetic acid and the product acetylated to give **331**. The 1-*O*-acetyl group was removed with ammonium carbonate in *N,N*-dimethylformamide to give **332**, which was converted into the trichloracetimidate **333**.

Regioselective 6-*O*-silylation of **327** with *t*-butyldiphenylsilyl chloride in the presence of imidazole gave **329**, and this was benzylated to give **330**. Hydrolysis of **330** with acetic acid and acetylation of the product gave **334**, which was selectively de-*O*-acetylated at the 1-position (as for the preparation of **332**) to give **335**, and this was converted into the trichloracetimidate **336**.

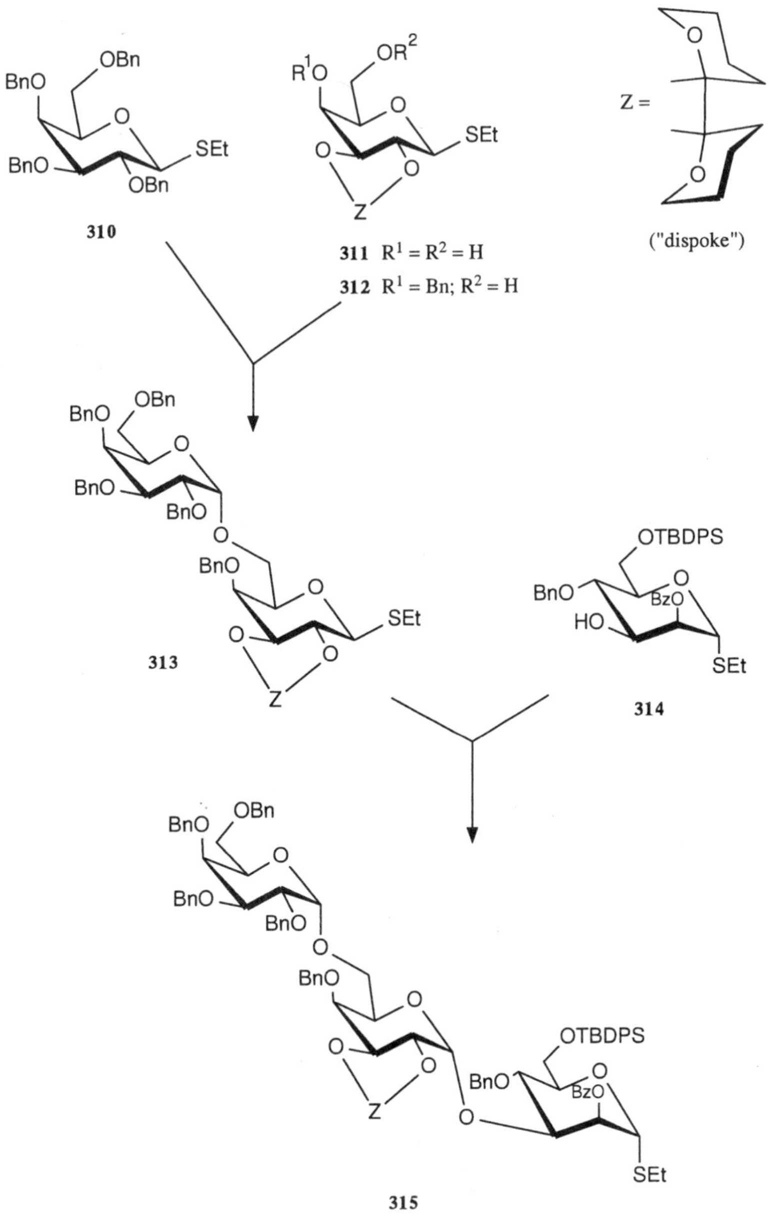

Scheme 39

Reaction of **336** with allyl alcohol in the presence of trimethylsilyl triflate gave **337**, and this on saponification gave **338**, which was allylated to give **339**. The 6-*O*-silyl group was removed with tetrabutylammonium fluoride to give the alcohol **340**. Condensation of the trichloracetimidate **333** with the alcohol **340** gave the α-linked disaccharide **341**, and this was saponified to give the alcohol **342**. Conden-

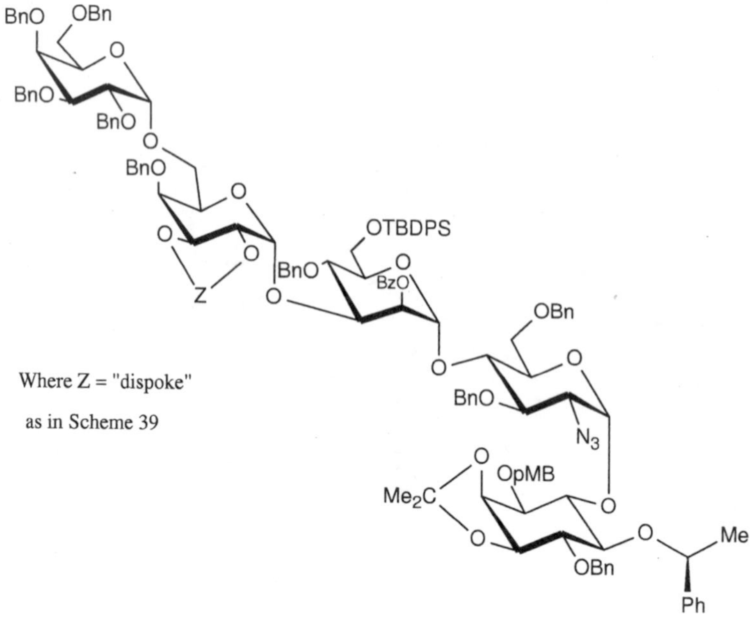

Where Z = "dispoke"
as in Scheme 39

316

Scheme 40

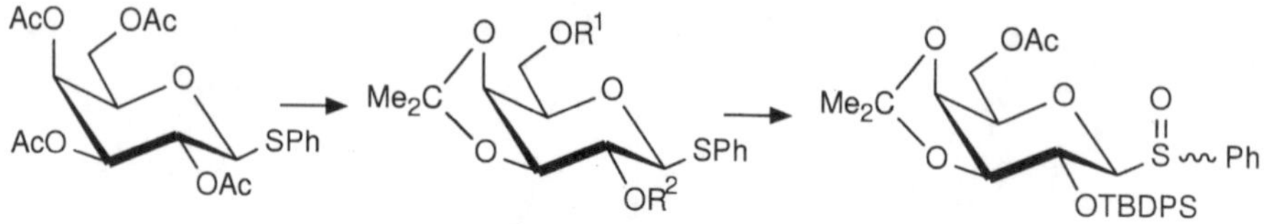

317

318 $R^1 = R^2 = H$
319 $R^1 = Ac; R^2 = H$
320 $R^1 = Ac; R^2 = TBDPS$
321 $R^1 = H; R^2 = TBDPS$

322

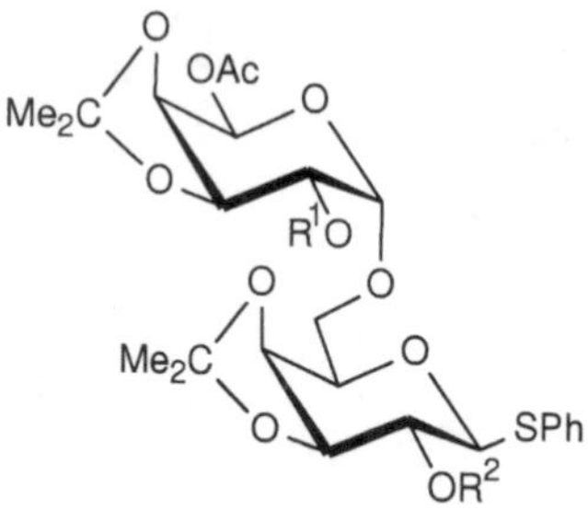

323 $R^1 = R^2 = TBDPS$
324 $R^1 = R^2 = H$

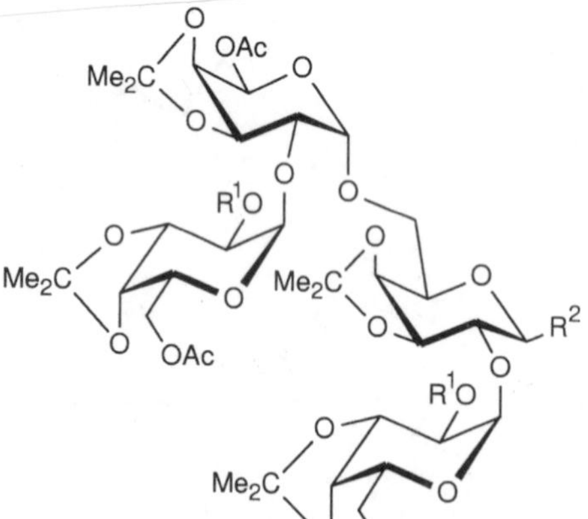

325 $R^1 = TBDPS; R^2 = SPh$
326 $R^1 = TBDPS; R^2 = S \text{~~} Ph$ with $\underset{O}{\overset{\parallel}{}}$

Scheme 41

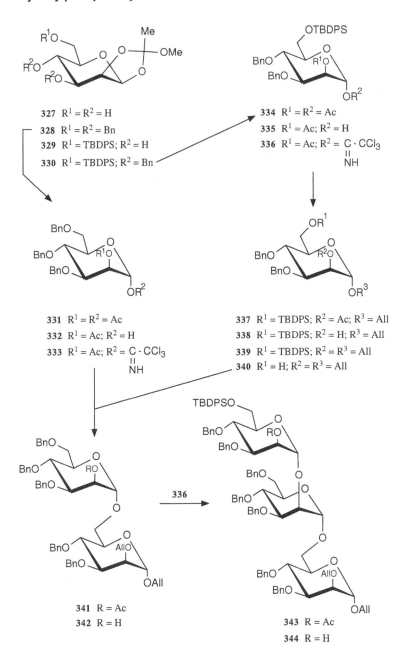

Scheme 42

sation of the trichloracetimidate **336** with **342** gave the α-linked trisaccharide **343**, which was saponified to give the alcohol **344**.

Condensation of the trichloracetimidate **333** with the alcohol **344** in the presence of trimethylsilyl triflate gave the tetramannoside **345** (Scheme 43) [36], and this

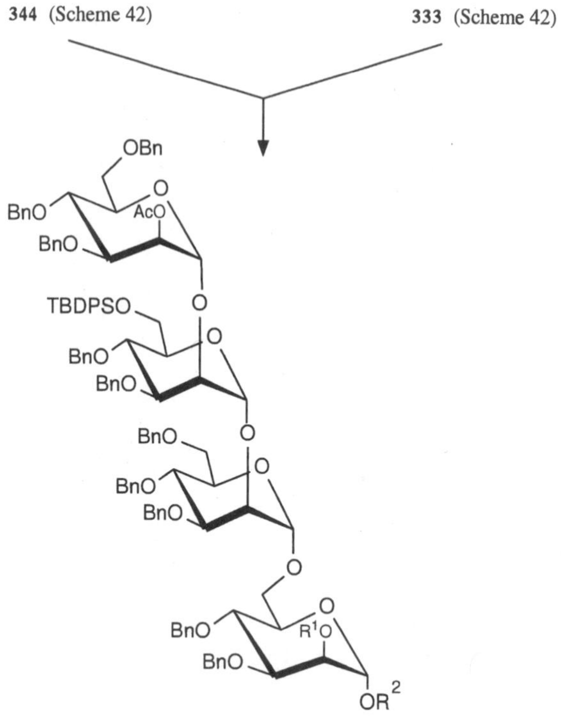

344 (Scheme 42)   333 (Scheme 42)

345 $R^1 = R^2 = All$
346 $R^1 = R^2 = H$
347 $R^1 = R^2 = Ac$
348 $R^1 = Ac; R^2 = H$
349 $R^1 = Ac; R^2 = C - CCl_3$
                     $\|$
                     NH

Scheme 43

was deallylated using tristriphenylphosphinerhodium(I) chloride and subsequent acid hydrolysis to give **346**, which was acetylated to give **347**. Preferential de-O-acetylation of the 1-position gave the free sugar **348**, which was converted into the trichloracetimidate **349**.

Condensation [36] of the tetramannoside trichloracetimidate **349** (Scheme 43) with the disaccharide **245** (Scheme 30) in the presence of trimethylsilyl triflate gave the hexasaccharide **350** (Scheme 44). The acetyl and benzoyl groups were removed by the action of sodium cyanide in methanol, and the triol **351** produced was benzylated to give **352**. The menthoxycarbonyl group was cleaved with potassium carbonate in methanol to give **353**, which was acetylated to give **354**. The silyl group was then removed by the action of tetrabutylammonium fluoride to give the alcohol **355** ready for phosphorylation (see Section V.D.).

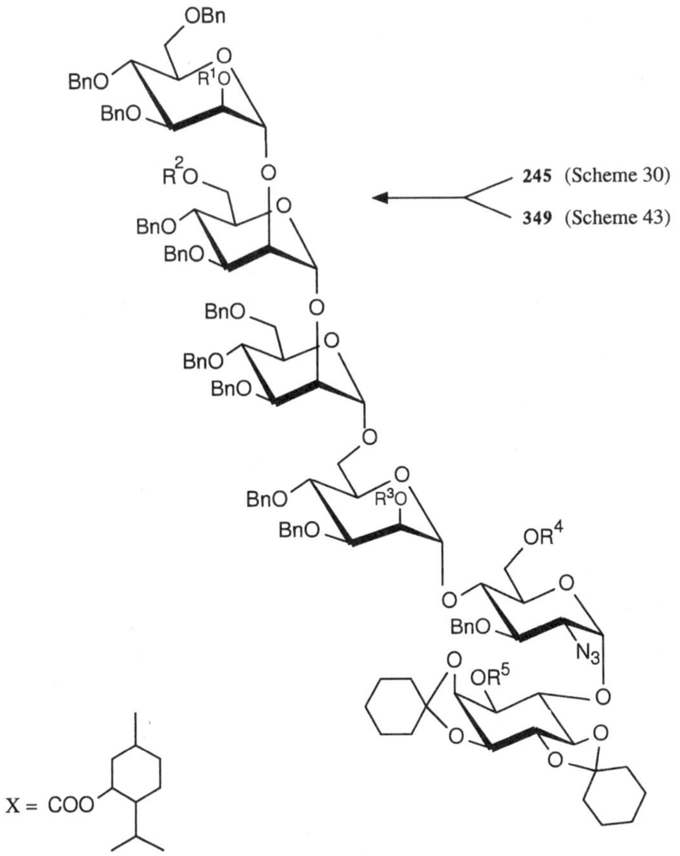

350 $R^1 = R^3 = Ac; R^2 = TBDPS; R^4 = Bz; R^5 = X$
351 $R^1 = R^3 = R^4 = H; R^2 = TBDPS; R^5 = X$
352 $R^1 = R^3 = R^4 = Bn; R^2 = TBDPS; R^5 = X$
353 $R^1 = R^3 = R^4 = Bn; R^2 = TBDPS; R^5 = H$
354 $R^1 = R^3 = R^4 = Bn; R^2 = TBDPS; R^5 = Ac$
355 $R^1 = R^3 = R^4 = Bn; R^2 = H; R^5 = Ac$

Scheme 44

V. ADDITION OF THE PHOSPHATE MOIETIES

A. Syntheses by the RIKEN Group

The RIKEN group [17] continued the synthesis described in Section IV.B. by saponification of the acetyl group in **356** (Scheme 45) to give the alcohol **357**, which was converted into the chloroacetyl derivative **358** by the action of chloroacetic anhydride in pyridine. The p-methoxybenzyl group was removed from **358** by the action of trimethylsilyl triflate in dichloroethane to give the alcohol **359**.

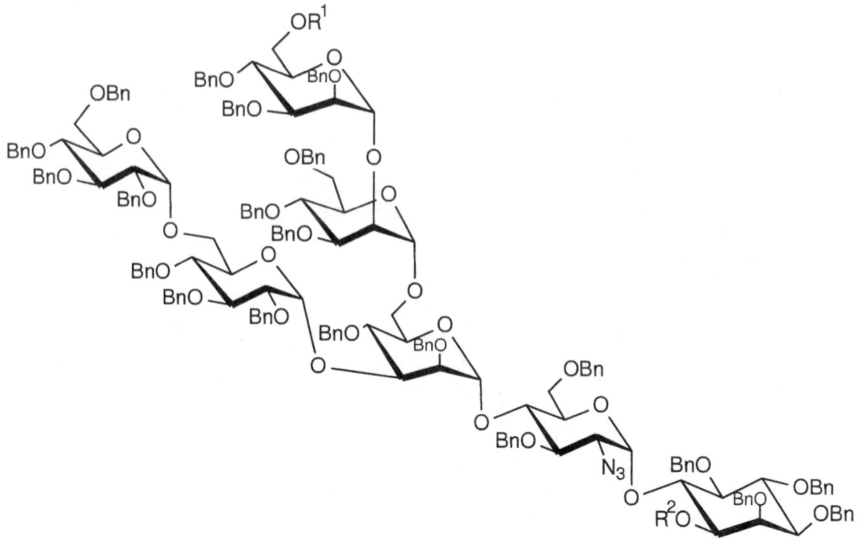

356 $R^1 = Ac$; $R^2 = pMB$

357 $R^1 = H$; $R^2 = pMB$

358 $R^1 = COCH_2Cl$; $R^2 = pMB$

359 $R^1 = COCH_2Cl$; $R^2 = H$

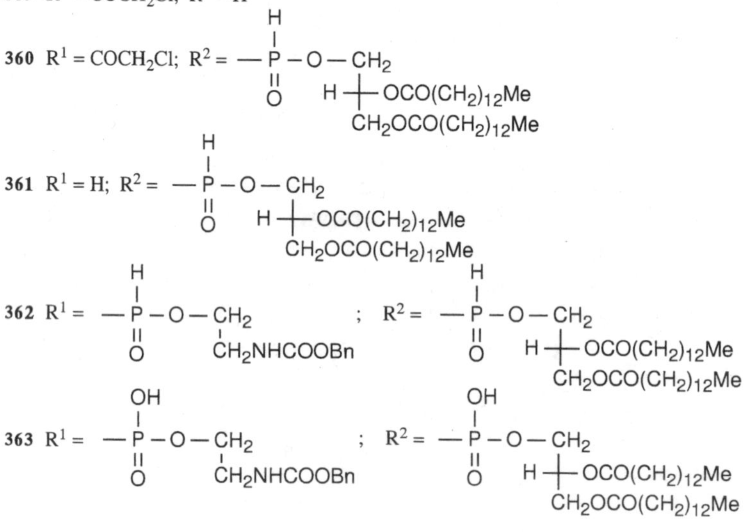

360 $R^1 = COCH_2Cl$; $R^2 = $
$$-\overset{H}{\underset{\overset{\|}{O}}{P}}-O-CH_2$$
$$H\overset{}{+}OCO(CH_2)_{12}Me$$
$$CH_2OCO(CH_2)_{12}Me$$

361 $R^1 = H$; $R^2 = $
$$-\overset{H}{\underset{\overset{\|}{O}}{P}}-O-CH_2$$
$$H\overset{}{+}OCO(CH_2)_{12}Me$$
$$CH_2OCO(CH_2)_{12}Me$$

362 $R^1 = $
$$-\overset{H}{\underset{\overset{\|}{O}}{P}}-O-CH_2$$
$$CH_2NHCOOBn$$
; $R^2 = $
$$-\overset{H}{\underset{\overset{\|}{O}}{P}}-O-CH_2$$
$$H\overset{}{+}OCO(CH_2)_{12}Me$$
$$CH_2OCO(CH_2)_{12}Me$$

363 $R^1 = $
$$-\overset{OH}{\underset{\overset{\|}{O}}{P}}-O-CH_2$$
$$CH_2NHCOOBn$$
; $R^2 = $
$$-\overset{OH}{\underset{\overset{\|}{O}}{P}}-O-CH_2$$
$$H\overset{}{+}OCO(CH_2)_{12}Me$$
$$CH_2OCO(CH_2)_{12}Me$$

Scheme 45

The diacylglycerol phosphate residue was then introduced into **359** using the *H*-phosphonate method. The *H*-phosphonate **364** (Scheme 46) and compound **359** were condensed in the presence of pivaloyl chloride to give **360** in 64% yield. The chloroacetyl group was removed by treatment with thiourea to give **361** in 75% yield. This was coupled with the unstable *H*-phosphonate **366** (prepared from *N*-benzyloxycarbonyl ethanolamine **365** as shown in Scheme 46) in the presence of

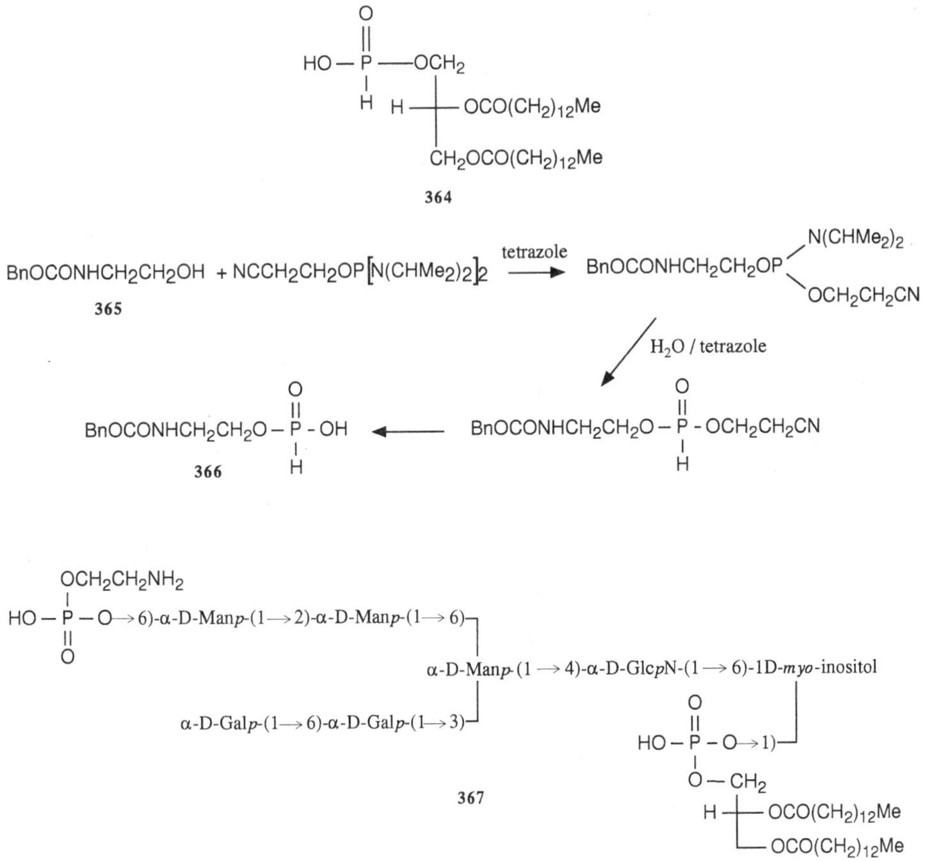

$$
\begin{array}{c}
\overset{O}{\underset{\|}{}} \\
HO - P - OCH_2 \\
\underset{H}{|} \quad H \dashv OCO(CH_2)_{12}Me \\
CH_2OCO(CH_2)_{12}Me
\end{array}
$$

364

BnOCONHCH$_2$CH$_2$OH + NCCH$_2$CH$_2$OP[N(CHMe$_2$)$_2$]$_2$ $\xrightarrow{\text{tetrazole}}$ BnOCONHCH$_2$CH$_2$OP$\underset{OCH_2CH_2CN}{\overset{N(CHMe_2)_2}{<}}$

365

$\swarrow$ H$_2$O / tetrazole

$$
\begin{array}{ccc}
\overset{O}{\underset{\|}{}} & & \overset{O}{\underset{\|}{}} \\
BnOCONHCH_2CH_2O - P - OH & \longleftarrow & BnOCONHCH_2CH_2O - P - OCH_2CH_2CN \\
\underset{H}{|} & & \underset{H}{|}
\end{array}
$$

366

$$
\begin{array}{c}
OCH_2CH_2NH_2 \\
| \\
HO - P - O \rightarrow 6)\text{-}\alpha\text{-D-Man}p\text{-}(1 \rightarrow 2)\text{-}\alpha\text{-D-Man}p\text{-}(1 \rightarrow 6) \\
\| \\
O
\end{array}
$$

α-D-Man$p$-(1 → 4)-α-D-Glc$p$N-(1 → 6)-1D-myo-inositol

α-D-Gal$p$-(1 → 6)-α-D-Gal$p$-(1 → 3)

$$
\begin{array}{c}
O \\
\| \\
HO - P - O \rightarrow 1) \\
| \\
O - CH_2 \\
H \dashv OCO(CH_2)_{12}Me \\
OCO(CH_2)_{12}Me
\end{array}
$$

367

Scheme 46

pivaloyl chloride to give **362** in 40% yield. The bisphosphonate **362** was oxidized with iodine to give the bisphosphate **363**, which represents the fully protected derivative of the required compound. Complete deprotection of **363** was achieved by hydrogenolysis over Pd/C to give the glycosyl phosphatidylinositol anchor **367** of *Trypanosoma brucei*.

In an earlier paper the RIKEN group [16] used a similar *H*-phosphonate approach to condense the *H*-phosphonate **364** (Scheme 46) with the trisaccharide **194** (Scheme 22) in the presence of pivaloyl chloride in pyridine to give the phosphonate **368** (Scheme 47), which was oxidized with iodine to give the phosphate **369**. Deprotection of **369** by hydrogenolysis over Pd/C gave the glycobiosyl phosphatidylinositol **370**.

The work of the RIKEN group was reviewed by Ogawa in his Haworth Memorial Lecture to the Royal Society of Chemistry Carbohydrate Group [77], and the structures of the synthetic molecules **363** and **369** have been revised (see p. 401 of Ref 77). Because of the previous misassignment (see Sections II.D. and II.E.) of the absolute configurations of the *myo*-inositol derivatives used in this work, the synthetic products are stereoisomers (in the *myo*-inositol portion) of the natural products.

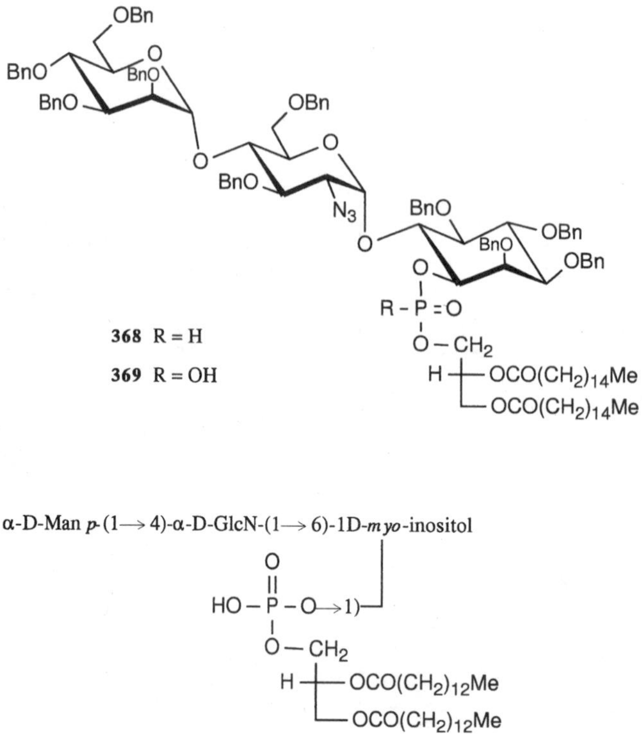

368 R = H

369 R = OH

α-D-Man *p*-(1→4)-α-D-GlcN-(1→6)-1D-*myo*-inositol

370

Scheme 47

B.   Syntheses by the Dundee Group

The Dundee group [14] also used the *H*-phosphonate method for the introduction of the diacylglycerol phosphate residue into the disaccharide **228** (Scheme 28). Thus condensation of **228** with the phosphonate **371** [78] (Scheme 48) in the presence of pivaloyl chloride in pyridine gave the phosphonic diester **372**, which was oxidized with iodine in pyridine to give the phosphoric diester **373** ([α]$_D$ +19°). Hydrogenolysis of **373** over Pd/C gave the glycosyl phosphatidylinositol **375** ([α]$_D$ +21°, Me$_2$SO). Acetylation of **375** [63] with ^3H-labeled acetic anhydride gave the *N*-acetyl derivative, which was used as a substrate in the partial purification of the α-D-Glc*p*NAc-PI de-*N*-acetylase of the bloodstream form of *Trypanosoma brucei*.

Deacylation [62] of **373** (Scheme 48) gave the glycerol phosphate **374**, which on hydrogenolysis gave **376** ([α]$_D$ +54°, H$_2$O), and the free amino group of this was acetylated [63] with ^3H-acetic anhydride. This product was used in enzymatic studies [63] to assess the roles of lipid, phospholipid, and fatty acid moieties in substrate recognition by the *T. brucei* de-*N*-acetylase.

Analogs of **375** (Scheme 48) in which the 2-amino-2-deoxy-D-glucose unit was replaced by 2-deoxy-D-glucose or D-glucose were prepared by the Dundee group [62] and used by them [63] in enzymatic studies. For the 2-deoxy-D-glucose analog **383**, 3,4,6-tri-*O*-benzyl-D-glucal **377** (Scheme 49) was condensed with the *myo*-in-

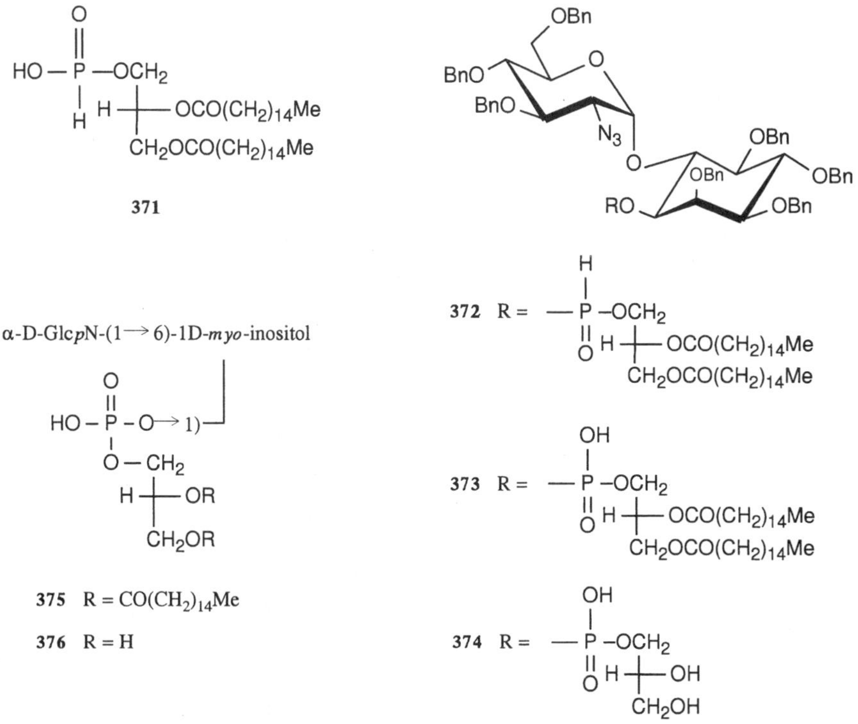

371

α-D-Glc*p*N-(1→6)-1D-*myo*-inositol

375 R = CO(CH$_2$)$_{14}$Me

376 R = H

372 R =

373 R =

374 R =

Scheme 48

ositol derivative **106** (Scheme 12) in the presence of triphenylphosphine hydrogen bromide [79] to give the α-linked disaccharide **378** (56%; [α]$_D$ +38°), and the *p*-methoxybenzyl group was removed with cerium(IV) ammonium nitrate to give **379** (m.p. 110°; [α]$_D$ +71°). The phospholipid component was introduced as for the preparation of compound **373** (Scheme 48), using the *H*-phosphonate method, to give **385**, and this was hydrogenolyzed to give the 2-deoxy-D-glucose analog **383** of **375**, which was used in enzymatic studies [63].

For the synthesis of the D-glucose analogue **384**, the glucosyl trichloracetimi-date **382** [80] was coupled with **106** in the presence of trimethylsilyl triflate to give the α-linked disaccharide **380** (76%; [α]$_D$ +38°), and removal of the *p*-methoxyben-zyl group with trifluoroacetic acid in dichloromethane gave the alcohol **381** ([α]$_D$ +45°). The phospholipid component was introduced as before to give **386**, which was hydrogenolyzed to give the D-glucose analog **384** of **375**, which was also used [63] in enzymatic studies.

C. Syntheses by the Durham Group

For the synthesis of the fully phosphorylated glycan and lipid anchor of rat brain Thy-1, the Durham group [81] manipulated the protected heptasaccharide **295** (Schemes 36 and 50). Dechloroacetylation of **295** with thiourea gave the alcohol **387**, which was phosphitylated with the amidite **394** in the presence of tetrazole,

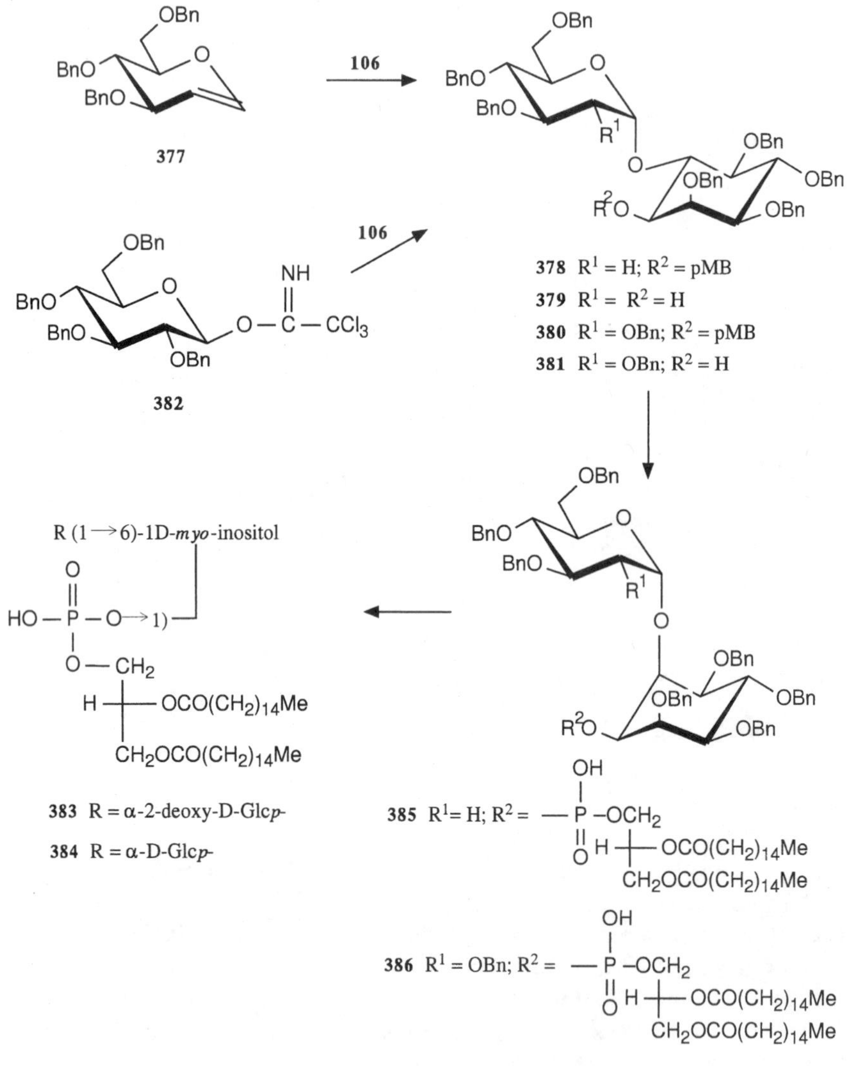

OBn
BnO
BnO
O
377
106
106
NH
O—C—CCl3
OBn
382
R¹
R²O
378 R¹ = H; R² = pMB
379 R¹ = R² = H
380 R¹ = OBn; R² = pMB
381 R¹ = OBn; R² = H
R (1→6)-1D-myo-inositol
HO—P—O→1)
O—CH2
H—OCO(CH2)14Me
CH2OCO(CH2)14Me
383 R = α-2-deoxy-D-Glcp-
384 R = α-D-Glcp-
385 R¹= H; R² = —P–OCH2
386 R¹ = OBn; R² = —P–OCH2
OH

Scheme 49

and subsequent oxidation with 3-chloroperbenzoic acid gave the phosphate diastereoisomers **388**. Deacetylation of **388** with sodium methoxide in methanol gave the alcohol **389**, and repeat phosphorylation with **394** gave a mixture (due to the chirality around the phosphorous atoms) of the isomers **390** (79%). Deallylation of **390** with palladium(II) chloride in acetate buffer gave **391** (66%, mixed isomers).

Phosphorylation of **391** using the amidite **395** gave **392**, which on hydrogenolysis gave the nonlipidic glycan of Thy-1. Phosphorylation of **391** with the amidite **396** gave **393**, which on hydrogenolysis to remove all of the protecting groups gave the GPI anchor corresponding to Thy-1.

In a separate publication [82], the Durham group described the preparation of the phosphoramidite reagents **394–396** (Scheme 50), as well as different protecting

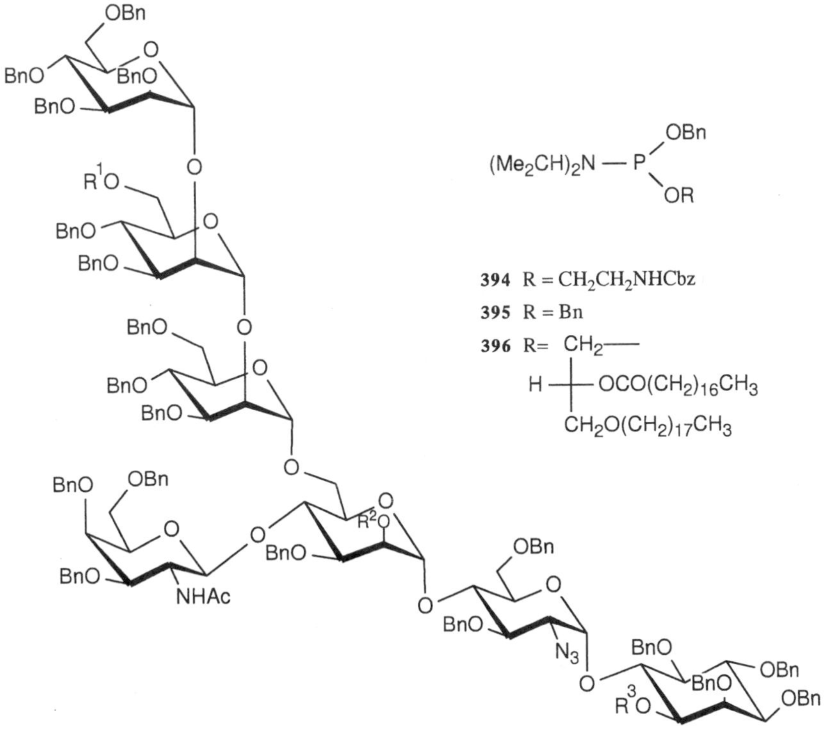

$(Me_2CH)_2N \!-\! P \overset{\displaystyle OBn}{\underset{\displaystyle OR}{\big\langle}}$

394 R = CH_2CH_2NHCbz

395 R = Bn

396 R= $CH_2\!-\!$

$H \!-\!\!\!\!\!-\!\!\!\!\!- OCO(CH_2)_{16}CH_3$

$CH_2O(CH_2)_{17}CH_3$

295 $R^1 = COCH_2Cl$; $R^2 = Ac$; $R^3 = All$ (Scheme 36)

387 $R^1 = H$; $R^2 = Ac$; $R^3 = All$

388 $R^1 = Y$; $R^2 = Ac$; $R^3 = All$ $Y= \ \overset{O}{\overset{\|}{-P}}(OBn)OCH_2CH_2NHCbz$

389 $R^1 = Y$; $R^2 = H$; $R^3 = All$

390 $R^1 = R^2 = Y$; $R^3 = All$

391 $R^1 = R^2 = Y$; $R^3 = H$

392 $R^1 = R^2 = Y$; $R^3 = \ \overset{O}{\overset{\|}{-P}}(OBn)_2$

393 $R^1 = R^2 = Y$; $R^3 = \ \overset{O}{\overset{\|}{-P}}\!-\!OCH_2$

$\qquad\qquad\qquad\qquad\quad BnO \ \ H \!-\!\!\!\!\!-\!\!\!\!\!- OCO(CH_2)_{16}CH_3$

$\qquad\qquad\qquad\qquad\qquad\quad CH_2O(CH_2)_{17}CH_3$

Scheme 50

groups for the amino group of 2-aminoethanol. In this communication the projected attachment of a cysteine residue (the site of attachment of the protein in the GPI lipid anchors) to one of the 2-aminoethanol residues is also discussed.

D. Syntheses by the Konstanz and Moscow Groups

The Konstanz group [35] converted the sphingosine derivative **397** (Scheme 51) into the *t*-butyldimethylsilyl derivative **398** [83,84], and the azido group was reduced to

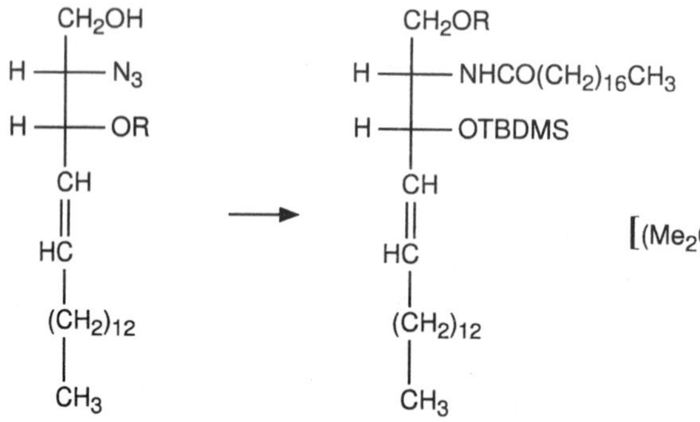

397 R = H
398 R = TBDMS

399 R = H
400 R = P – OCH$_2$CH$_2$CN
                    |
                 N(CHMe$_2$)$_2$

$[(Me_2CH)_2N]_2POCH_2CH_2CN$

401

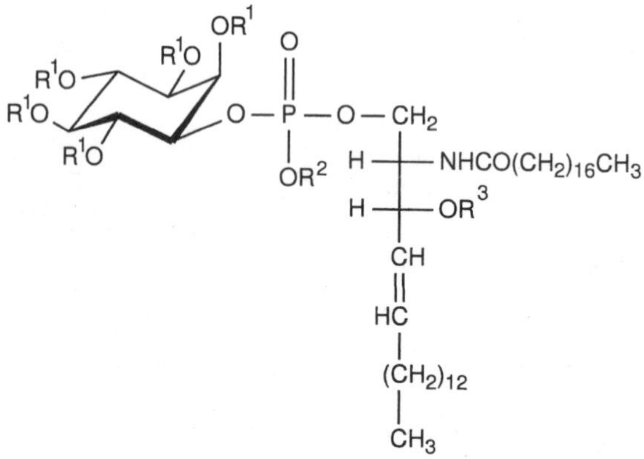

402 R^1 = MOM; R^2 = CH$_2$CH$_2$CN; R^3 = TBDMS
403 R^1 = MOM; R^2 = H; R^3 = TBDMS
404 R^1 = R^2 = R^3 = H

Scheme 51

an amino group by the action of triphenylphosphine in aqueous pyridine. Acylation of the amino group with stearic acid anhydride and triethylamine gave **399**, which was phosphitylated with bis(diisopropylamino)cyanoethoxyphosphine (**401**) in the presence of diisopropylammonium tetrazolide to give the phosphoramidite **400**. This was condensed with the inositol derivative **116** (Scheme 14) in the presence of tetrazole, and subsequent oxidation of the phosphite produced (with *t*-butylhydroperoxide) gave **402**. The cyanoethyl group was removed by the action of dimethylamine in methanol to give **403**, and this on treatment with toluene *p*-sulfonic acid in methanol, which removed the methoxymethyl and silyl groups, gave **404**. This compound (ceramide-1-phosphoinositol) is a component of yeasts and fungi and is related to the GPI anchor of *Saccharomyces cerevisiae*.

A similar approach to the synthesis of a mixture of diastereoisomers of **404** using racemic 2,3,4,5,6-penta-*O*-acetyl-*myo*-inositol and racemic 3-*O*-benzoylcer-amide was described by the Moscow group [85], with the final deprotection being alkaline de-*O*-acylation. The diastereoisomeric mixture of the phosphorothioate an-alogue of **404** was also prepared by the Moscow group [86], using a similar approach.

For the synthesis of the GPI lipid anchor of *Saccharomyces cerevisiae*, the Konstanz group [36] condensed the alcohol **355** (Scheme 44) with the phosphor-amidite **405** (Scheme 52; see also Scheme 46) [17] in the presence of tetrazole, and oxidation of the phosphite produced gave **406**, which was saponified with sodium methoxide in methanol to give the alcohol **407** (Scheme 52), ready for coupling to the phospholipid.

This phytosphingosine-based phospholipid was prepared by the Konstanz group [36] from the azido-phytosphingosine derivative **409** [87], which was converted into the isopropylidene derivative **410** (Scheme 53). The azido group was reduced with

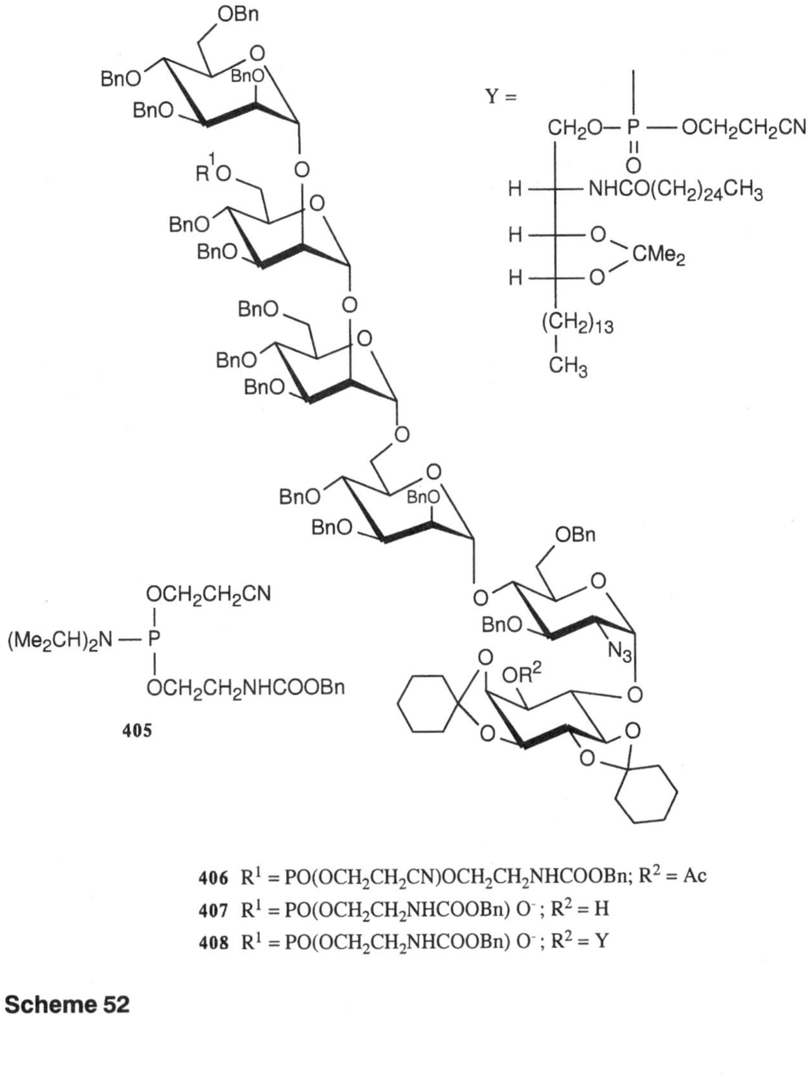

406 R^1 = PO(OCH$_2$CH$_2$CN)OCH$_2$CH$_2$NHCOOBn; R^2 = Ac
407 R^1 = PO(OCH$_2$CH$_2$NHCOOBn) O$^-$; R^2 = H
408 R^1 = PO(OCH$_2$CH$_2$NHCOOBn) O$^-$; R^2 = Y

Scheme 52

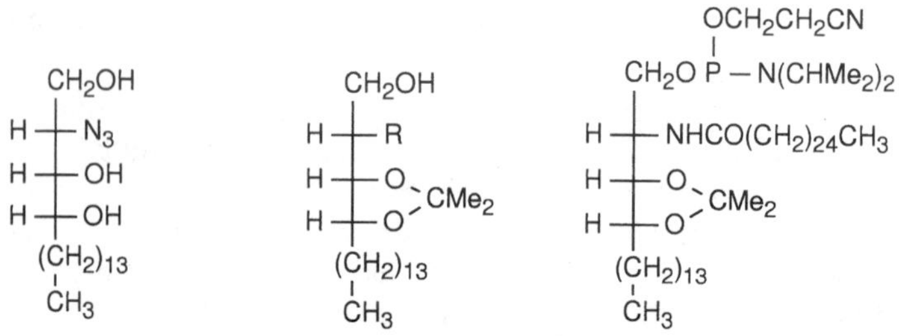

OCH₂CH₂CN type structures...

$$OCH_2CH_2CN$$

CH₂OH
H — N₃
H — OH
H — OH
(CH₂)₁₃
CH₃

409

CH₂OH
H — R
H — O⟩CMe₂
H — O⟩
(CH₂)₁₃
CH₃

410 R = N₃
411 R = NH₂
412 R = NHCO(CH₂)₂₄CH₃

$$CH_2O\ \overset{OCH_2CH_2CN}{\underset{}{P}} - N(CHMe_2)_2$$

H — NHCO(CH₂)₂₄CH₃
H — O⟩CMe₂
H — O⟩
(CH₂)₁₃
CH₃

413

NCCH₂CH₂O
CH₂O——P——
            ‖
            O
H — NHCO(CH₂)₂₄CH₃
H — O⟩CMe₂
H — O⟩
(CH₂)₁₃
CH₃

OBn / BnO / BnO / N₃ (glucosamine and inositol ring system with cyclohexylidene groups)

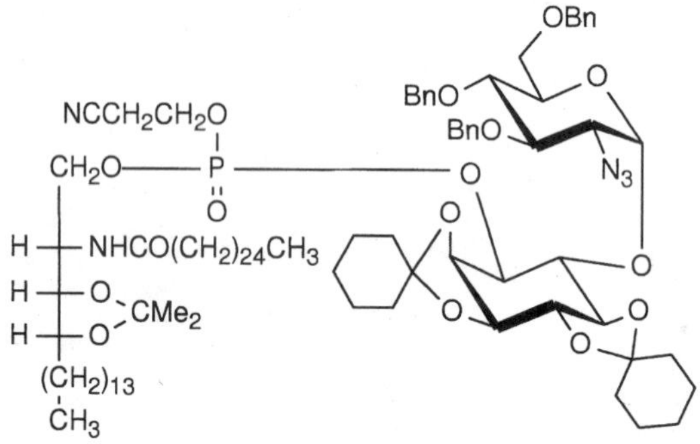

414

Scheme 53

triphenylphosphine to give the amine **411**, which was acylated with hexacosanoic acid in the presence of 2-ethoxy-1-ethoxycarbonyl-1,2-dihydroquinoline to give the ceramide derivative **412**. This was treated with the bisphophoramidite **401** (Scheme 51) in the presence of tetrazole to give the phosphitylating agent **413**.

Condensation of the phosphoramidite **413** (Scheme 53) with the alcohol **407** (Scheme 52) in the presence of tetrazole gave **408**. The cyanoethyl group was cleaved with dimethylamine, and subsequent acid hydrolysis removed the isopropylidene and cyclohexylidene groups. Hydrogenolysis removed all of the benzyl groups and reduced the azido group to give the GPI anchor of *Saccharomyces cerevisiae*.

P-ethanolamine
↓
6

α-D-Man*p*-(1 → 2)α-D-Man*p*-(1 → 2)α-D-Man*p*-(1 → 6)α-D-Man*p*-(1 → 4)α-D-GlcN*p*-(1 → 6)-D-*myo*-inositol-1-*P*-1-ceramide

The Konstanz group [36] also prepared the smaller protected portion **414** (Scheme 53) of the GPI anchor by condensation of the phosphoramidite **413** with the alcohol **243** (Scheme 30) in the presence of tetrazole and subsequent oxidation

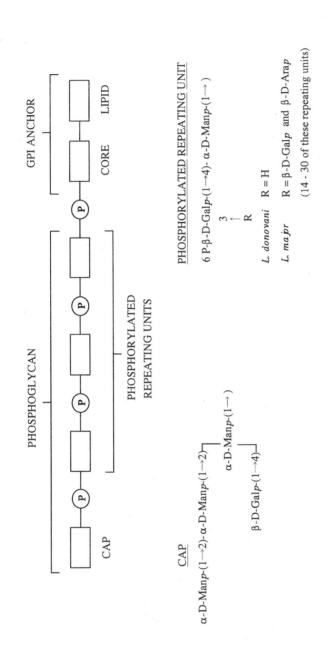

PHOSPHOGLYCAN

CAP

P

PHOSPHORYLATED
REPEATING UNITS

P

P

P

GPI ANCHOR

CORE   LIPID

<u>PHOSPHORYLATED REPEATING UNIT</u>

6 P-β-D-Galp-(1→4)- α-D-Manp-(1→)

3
↑
R

L. donovani   R = H

L. major   R = β-D-Galp and β-D-Arap

(14 - 30 of these repeating units)

<u>CAP</u>

α -D-Manp-(1→2)-α-D-Manp-(1→2)

α-D-Manp-(1→)

β-D-Galp-(1→4)-

<u>GPI ANCHOR</u>

6 P-β-D-Galp-(1→6)- α-D-Galp-(1→3)- β-D-Galf-(1→3)

α-D-Manp-(1→3)- α-D-Manp-(1→4)- α-D-GlcNp-(1→6)-*myo*-Inositol-(1→P-)Lipid

α-D-Glcp-(1→P→6)

Scheme 54

of the phosphite produced. Compound **414** was deprotected as described earlier for **408** to give the phospholipid disaccharide α-D-GlcNp-(1 → 6)D-*myo*-inositol-1-P-1-ceramide.

VI. LIPOPHOSPHOGLYCAN OF *LEISHMANIA*

Because the lipophosphoglycan of *Leishmania* differs from the lipid anchors of *Trypanosoma* and Thy-1, particularly in the phosphorylated repeating units (see Scheme 54), the published synthetic work is being reviewed in a separate section.

A.   Syntheses by the Durham Group

For their work on the synthesis of the lipophosphoglycan of *Leishmania*, the Durham group [88] prepared the protected derivative **424** (Scheme 56) of the tetrasaccharide that forms the cap region of the molecule (see general structure [89] in Scheme 54) by two different routes.

The pentenyl mannoside **415** [67] (Scheme 55) was converted into **416** by tin-mediated benzylation, and subsequent chloroacetylation gave **417**. Reductive opening of the benzylidene acetal using triethylsilane-trifluoroacetic acid [90] gave the 6-O-benzyl ether **418** (84%), and this on bromination gave **419**. Condensation of pent-4-enyl 2,3,4,6-tetra-O-acetyl-β-D-galactopyranoside with **419**, using N-iodosuccinimide-triethylsilyl triflate as condensing agent, gave the disaccharide **420** (87%), and the chloroacetyl group was removed with thiourea to give the alcohol **421**.

The dimannoside **422** (Scheme 56) prepared [91] from **425** was debrominated with zinc dust (or sodium iodide) to give the pent-4-enyl disaccharide **423**, and this was condensed with the disaccharide **421** (Scheme 55) to give the tetrasaccharide **424** (69%).

In an alternative approach to the tetrasaccharide **424**, the donor compound **427** was prepared from **425** by deacetylation (to give **426**) and subsequent chloroacetylation. Compound **427** was condensed with the disaccharide **421**, as described earlier, to give the trisaccharide **428** (63%). The chloroacetyl group was removed with thiourea to give the alcohol **429**, which was condensed with the donor **425** to give the tetrasaccharide **424** (35%).

Because the *Leishmania* phosphoglycan contains a galactofuranose residue (see Scheme 54), the Durham group [92] investigated the synthesis of n-pentenyl galactofuranosides and studied their behavior in glycosidation reactions. D-Galactose was converted into the α,β-mixture of pentenyl galactofuranosides **430** (43%; Scheme 57) by glycosidation with pentenyl alcohol under acidic conditions. The mixture was converted into the peracetyl **431** and perbenzyl **432** derivatives, and the anomers of the latter were separated by chromatography.

Glycosidations with the α,β-mixture **431** under normal conditions gave exclusively β-galactofuranosides **433**, whereas glycosidations with **432** gave a mixture of α,β-glycosides containing predominantly the β-anomer **434**.

B.   Syntheses by the Dundee Group

The Dundee group [93] investigated the synthesis of the phosphorylated repeating units (see Scheme 54) of the *Leishmania* lipophosphoglycan, and for this purpose

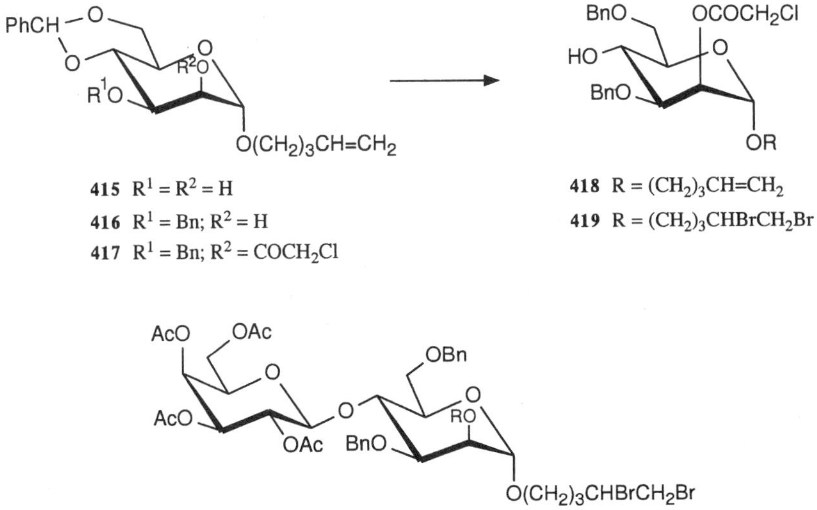

415 $R^1 = R^2 = H$

416 $R^1 = Bn; R^2 = H$

417 $R^1 = Bn; R^2 = COCH_2Cl$

418 $R = (CH_2)_3CH=CH_2$

419 $R = (CH_2)_3CHBrCH_2Br$

420 $R = COCH_2Cl$

421 $R = H$

Scheme 55

they first synthesized the disaccharide units **441** and **452** (Scheme 58) containing *H*-phosphonate groups and also the disaccharide **448** containing a decenyl aglycone suitable for coupling the products to a macromolecule for the preparation of artificial antigens.

For the synthesis of **441** (Scheme 58), the tetra-*O*-benzoyl-mannopyranoside **435** ([α]$_D$ −43°; prepared from D-mannose by treatment with benzoyl chloride in pyridine at −40°) was condensed with tetra-*O*-acetyl-α-D-galactopyranosyl bromide in the presence of silver triflate and 2,4,6-trimethylpyridine to give the β-linked disaccharide **436** (74%; m.p. 109°; [α]$_D$ +26°). De-*O*-acetylation of **436** with hydrogen chloride in methanol [94] gave **437**, which was treated with dimethoxytrityl chloride in pyridine to give **438**, and this on benzoylation gave **439** (72% from **436**; [α]$_D$ +63°). 1-*O*-Debenzoylation of **439** with dimethylamine in acetonitrile [95] gave **440** (77%; [α]$_D$ +32°), which was phosphitylated with tri-imidazoylphosphine [95], and subsequent treatment with triethylammonium bicarbonate gave the *H*-phosphonate **441** (92%; [α]$_D$ +24°).

For the synthesis of **448** (Scheme 58), acetobromomannose was condensed with dec-9-ene-1-ol to give **442**, and this was saponified to give **443** ([α]$_D$ +56°, MeOH). Selective benzoylation [96] of **443** gave **444** (m.p. 87°; [α]$_D$ +13°), and this was condensed with acetobromogalactose to give the β-linked disaccharide **445** (67% together with some α-linked isomer; [α]$_D$ −10°). De-*O*-acetylation of **445** gave **446**, which was converted into **447** by dimethoxytritylation and subsequent benzoylation. Acidic hydrolysis of **447** gave the alcohol **448** (71% from **445**; [α]$_D$ +91°).

For the preparation of **452** (Scheme 58), 1,3,4,6-tetra-*O*-acetyl-β-D-mannopyranose (**449**) was glycosylated with benzobromomannose to give the α-linked disaccharide **450** ([α]$_D$ −44°). This was converted via **451** ([α]$_D$ −55°) into the required

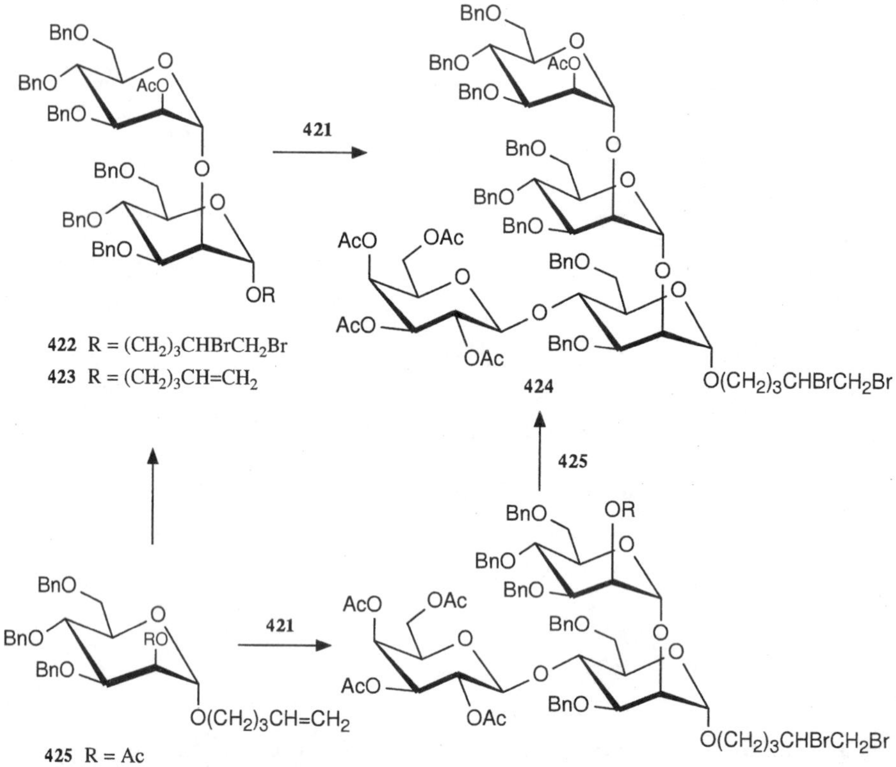

422 R = (CH$_2$)$_3$CHBrCH$_2$Br
423 R = (CH$_2$)$_3$CH=CH$_2$

424

425

425 R = Ac
426 R = H
427 R = COCH$_2$Cl

428 R = COCH$_2$Cl
429 R = H

Scheme 56

H-phosphonate **452** ([α]$_D$ −22°), as described earlier for the conversion of **439** into **441**.

With the building blocks **441**, **448**, and **452** (Scheme 58) assembled, the Dundee group [93] proceeded to link them to produce the required units. Thus condensation of the *H*-phosphonate **441** with the disaccharide **448** in pyridine in the pres-

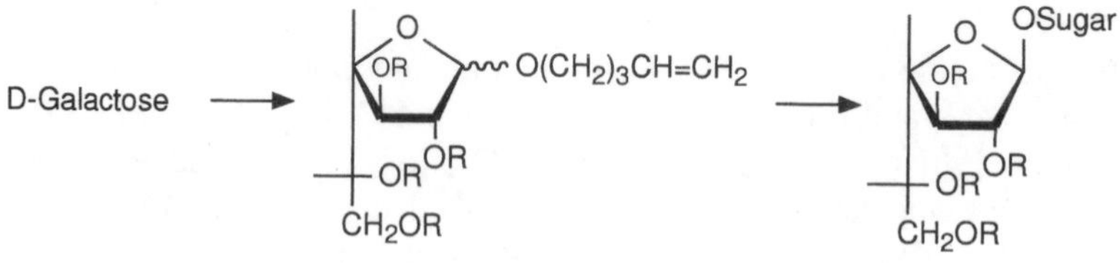

D-Galactose ⟶

430 R = H
431 R = Ac
432 R = Bn

433 R = Ac
434 R = Bn

Scheme 57

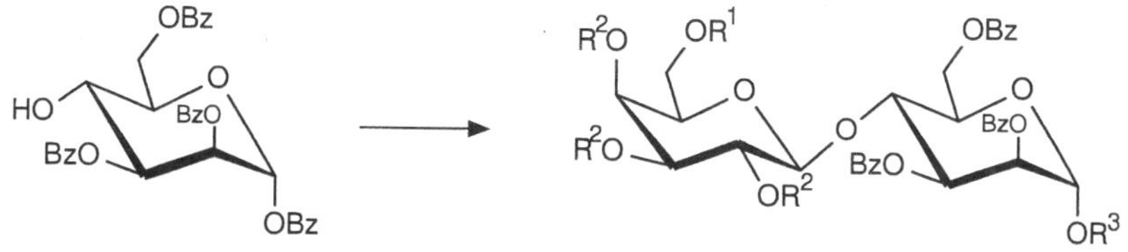

435

436 $R^1 = R^2 = Ac; R^3 = Bz$
437 $R^1 = R^2 = H; R^3 = Bz$
438 $R^1 = DMT; R^2 = H; R^3 = Bz$
439 $R^1 = DMT; R^2 = R^3 = Bz$
440 $R^1 = DMT; R^2 = Bz; R^3 = H$
441 $R^1 = DMT; R^2 = Bz; R^3 = PH(O)O^- \ N^+HEt_3$

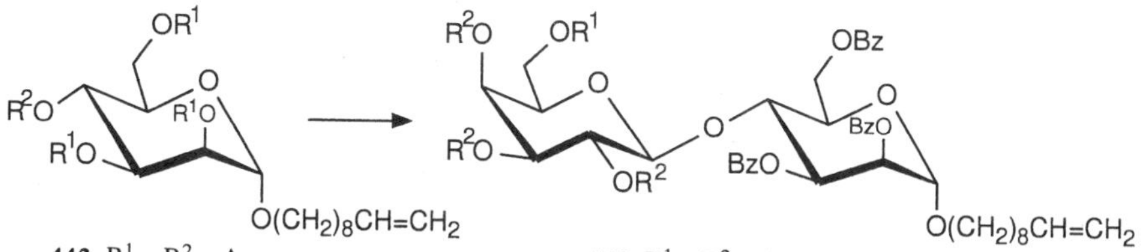

442 $R^1 = R^2 = Ac$ **445** $R^1 = R^2 = Ac$
443 $R^1 = R^2 = H$ **446** $R^1 = R^2 = H$
444 $R^1 = Bz; R^2 = H$ **447** $R^1 = DMT; R^2 = Bz$
                                  448 $R^1 = H; R^2 = Bz$

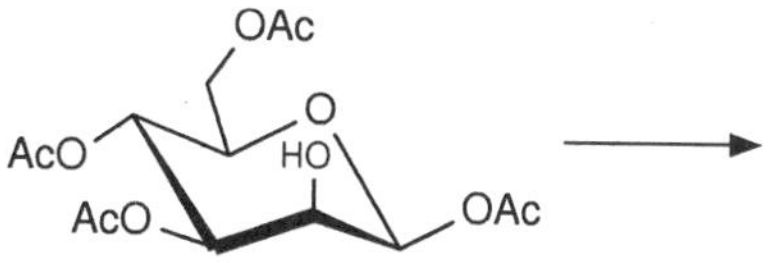

449

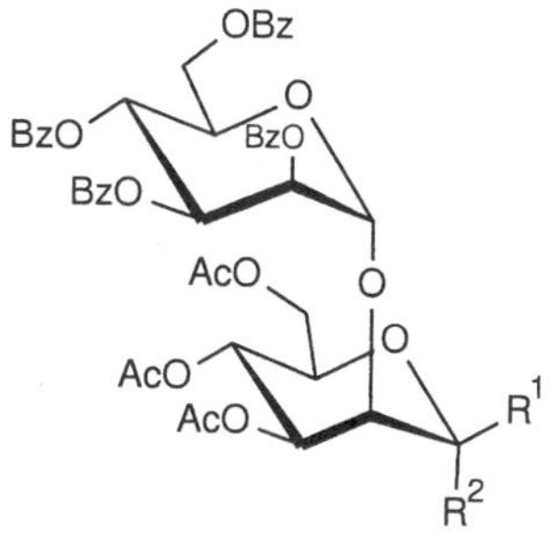

450 $R^1 = OAc; R^2 = H$
451 $R^1 = H; R^2 = OH$
452 $R^1 = H; R^2 = OPH(O)O^- \ NH^+Et_3$

Scheme 58

ence of adamantane-1-carbonyl chloride and subsequent oxidation of the product with iodine in aqueous pyridine gave the tetrasaccharide phosphoric diester **453** (Scheme 59). Dedimethoxytritylation of this with 1% trifluoroacetic acid in dichloromethane gave the alcohol **454** (81%; $[\alpha]_D$ +70°).

The hexasaccharide diphosphate diester **456** ($[\alpha]_D$ +65°) was obtained in a similar way, via **455**, by condensation of the H-phosphonate **441** with **454**. In a similar way the octasaccharide triphosphate diester **457** ($[\alpha]_D$ +31°) was obtained by coupling the disaccharide H-phosphonate **452** with **456**.

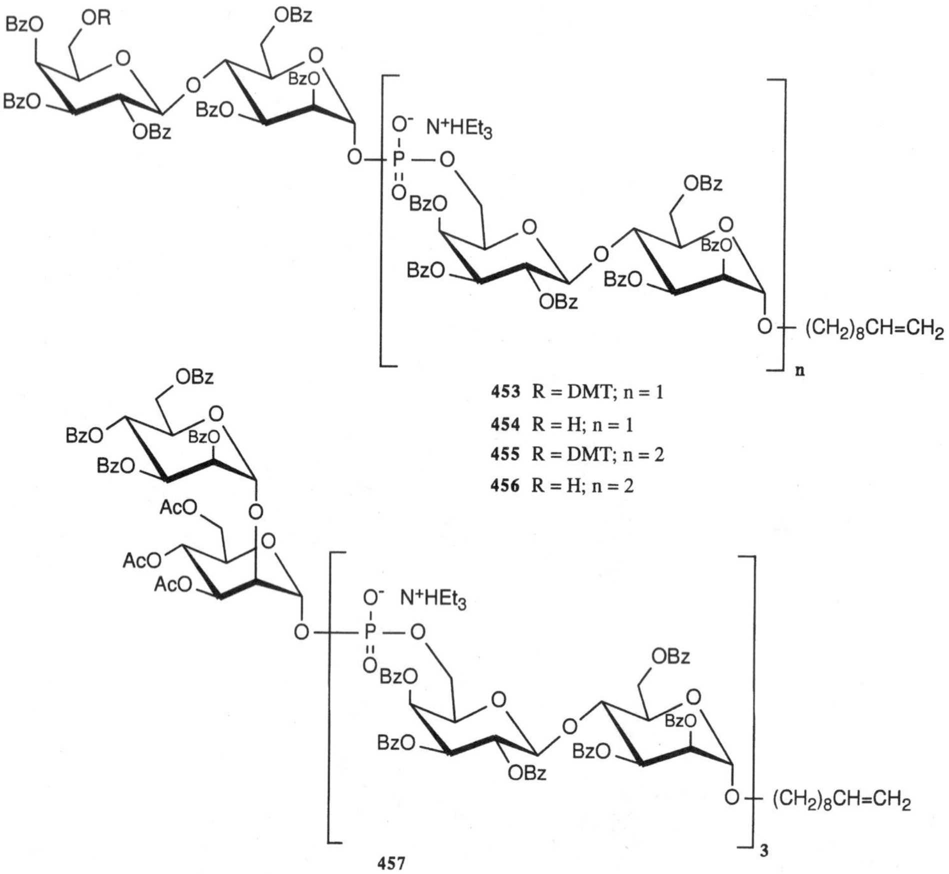

453 R = DMT; n = 1
454 R = H; n = 1
455 R = DMT; n = 2
456 R = H; n = 2

457

Scheme 59

Deacylation of **448**, **454**, **456**, and **457** with sodium methoxide-methanol gave the deprotected molecules suitable for biological studies and for coupling to macromolecular supports.

The Dundee group [97] continued their studies on the synthesis of the lipophosphoglycan of *Leishmania* by carrying out polycondensations of *H*-phosphonate disaccharides to provide molecules related to those of the phosphorylated repeating units (see Scheme 54). Thus **441** (Scheme 58) was dedimethoxytritylated to give the required monomer **458** ($[\alpha]_D$ +67°; Scheme 60). Polycondensation of **458** in pyridine-triethylamine in the presence of pivaloyl chloride (using a high concentration of **458** to avoid the formation of cyclic products) gave the mixture of polymers **459**, which was oxidized with iodine in aqueous pyridine to give **460** ($[\alpha]_D$ +53°). Deacylation of **460** with sodium methoxide in methanol-dioxane-chloroform followed by ion-exchange chromatography, with aqueous ammonium hydrogen carbonate as eluant, gave **461**, with an average degree of polymerization ($\tilde{n}$) of 10.

Copolymerization of **458** (Scheme 60) with **448** (Scheme 58), followed by oxidation, debenzoylation of the product, and ion-exchange chromatography, gave **462** (60%; $\tilde{n}$ = 6).

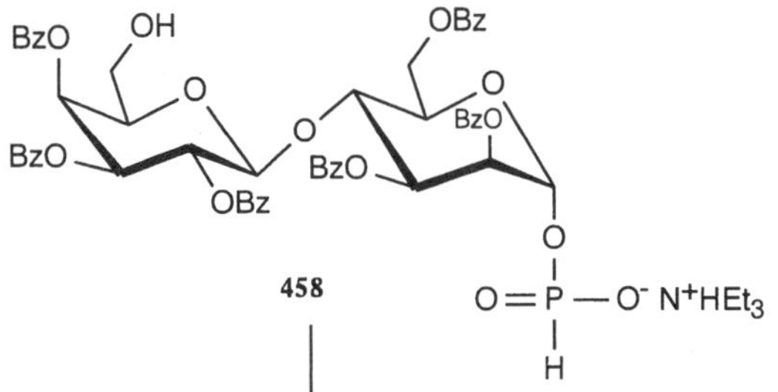

458

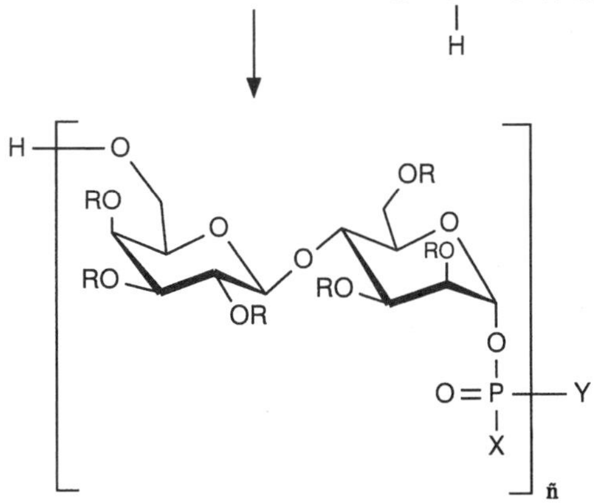

459 R = Bz; X = H; Y = O⁻ N⁺HEt₃

R = Bz; X = H; Y = O^- N^+HEt_3

459 R = Bz; X = H; Y = O^- N^+HEt_3
460 R = Bz; X = Y = O^- N^+HEt_3
461 R = H; X = Y = O^- N^+H_4

462 R = H; X = O^- N^+H_4;

Y =

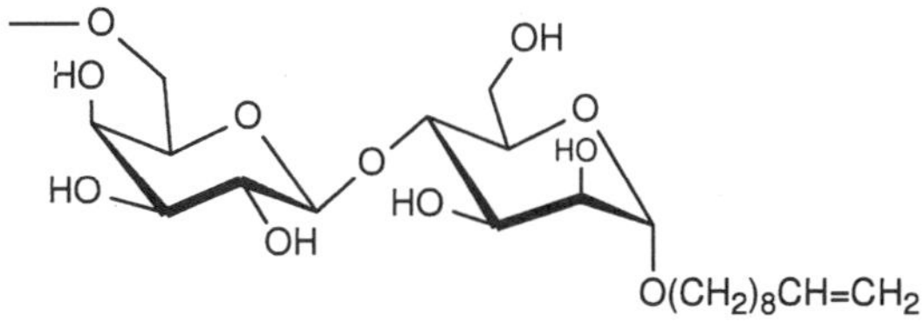

$O(CH_2)_8CH=CH_2$

Scheme 60

LIST OF SYMBOLS FOR SCHEMES

Ac   = $COCH_3$

All   = $CH_2-CH=CH_2$

Bn   = CH_2Ph

Bz   = COPh

DMT   = $CPh[Ph(pOMe)]_2$

pMB   = $CH_2Ph(pOMe)$

MOM   = CH_2OMe

TBDMS   = $SiMe_2Bu^t$

TBDPS   = $SiPh_2Bu^t$

(-)-ω-camph =

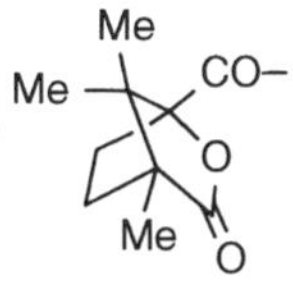

Phth =

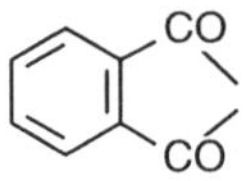

THP =

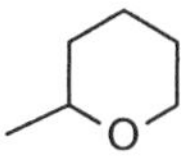

REFERENCES

1. R. Parthasarathy and F. Eisenberg, The inositol phospholipids: A stereochemical view of biological activity, *Biochem. J. 235*: 313 (1986).
2. B. V. L. Potter and D. Lampe, Chemistry of inositol lipid mediated cellular signalling. *Angew. Chem. Int. Ed. Engl. 34*: 1933 (1995).
3. S. L. Beaucage and R. P. Iyer, The synthesis of specific ribonucleotides and unrelated phosphorylated biomolecules by the phosphoramidite method, *Tetrahedron 49*: 10,441 (1993).
4. D. R. Mootoo, P. Konradsson, and B. Fraser-Reid, *n*-Pentenyl glycosides facilitate a stereoselective synthesis of the pentasaccharide core of the protein membrane anchor found in *Trypanosoma brucei. J. Am. Chem. Soc. 111*: 8540 (1989).
5. P. J. Garegg, T. Iversen, R. Johansson, and B. Lindberg, Synthesis of some mono-*O*-benzyl- and penta-*O*-methyl-*myo*-inositols, *Carbohydr. Res. 130*: 322 (1984).
6. D. C. Billington, R. Baker, J. J. Kulagowski, and I. M. Mawer, Synthesis of *myo*-inositol 1-phosphate and 4-phosphate, and of their individual enantiomers, *J. Chem. Soc. Chem. Commun.*: 314 (1987); J. P. Vacca, S. J. deSolms, and J. R. Huff, Total synthesis of D- and L-*myo*-inositol 1,4,5-triphosphate, *J. Am. Chem. Soc. 109*: 3478 (1987).
7. J. P. Vacca, S. J. deSolms, J. R. Huff, D. C. Billington, R. Baker, J. J. Kulagowski, and I. M. Mawer, The total synthesis of *myo*-inositol polyphosphates, (a) *Tetrahedron 45*: 5679 (1989); (b) correction *Tetrahedron 47*: 907 (1991).
8. R. Madsen, U. E. Udodong, C. Roberts, D. R. Mootoo, P. Konradsson, and B. Fraser-Reid, Studies related to synthesis of glycophosphatidylinositol membrane-bound protein anchors. 6. Convergent assembly of subunits, *J. Am. Chem. Soc. 117*: 1554 (1995).
9. C. J. J. Elie, R. Verduyn, C. E. Dreef, D. M. Brounts, G. A. van der Marel, and J. H. van Boom, Synthesis of 6-0-(α-D-mannopyranosyl)-D-*myo*-inositol: A fragment from *Mycobacteria* phospholipids, *Tetrahedron 46*: 8243 (1990).
10. T. Desai, A. Fernandez-Mayoralas, J. Gigg, R. Gigg, and S. Payne, The allyl group for protection in carbohydrate chemistry. Part 22. The synthesis and resolution of (±)-1,5,6-tri-*O*-benzyl-*myo*-inositol, *Carbohydr. Res. 205*: 105 (1990).
11. J. Gigg, R. Gigg, S. Payne, and R. Conant, The allyl group for protection in carbohydrate chemistry. Part 18. Allyl and benzyl ethers of *myo*-inositol. Intermediates for the synthesis of *myo*-inositol trisphosphates, *J. Chem. Soc. Perkin Trans. I*: 423 (1987).
12. R. Verduyn, C. J. J. Elie, C. E. Dreef, G. A. van der Marel, and J. H. van Boom, Stereospecific synthesis of partially protected 2-azido-2-deoxy-D-glucosyl-D-*myo*-inositol: Precursor of a potential insulin mimetic and membrane protein anchoring site, *Recl. Trav. Chim. Pays-Bas 109*: 591 (1990).
13. W. K. Berlin, W.-S. Zhang, and T. Y. Shen, Glycosyl-inositol derivatives III. Synthesis of hexosamine-inositol-phosphates related to putative insulin mediators, *Tetrahedron 47*: 1 (1991).
14. S. Cottaz, J. S. Brimacombe, and M. A. J. Ferguson, Parasite glycoconjugates Part 1. The synthesis of some early and related intermediates in the biosynthetic pathway of glycosyl-phosphatidylinositol membrane anchors, *J. Chem. Soc. Perkin Trans. I*: 2945 (1993).
15. C. Murakata and T. Ogawa, (a) Synthetic study on glycophosphatidyl inositol (GPI) anchor of *Trypanosoma brucei*: Glycoheptaosyl core, *Tetrahedron Lett. 31*: 2439 (1990); (b) Synthetic studies on glycophosphatidylinositol anchor: A highly efficient synthesis of glycobiosyl phosphatidylinositol through *H*-phosphonate approach, *Tetrahedron Lett. 32*: 101 (1991); (c) A total synthesis of GPI anchor of *Trypanosoma brucei, Tetrahedron Lett. 32*: 671 (1991).

16. C. Murakata and T. Ogawa, Stereoselective synthesis of glycobiosyl phosphatidylino-
    sitol, a part structure of the glycosylphosphatidylinositol (GPI) anchor of *Trypanosoma
    brucei, Carbohydr. Res. 234*: 75 (1992).

17. C. Murakata and T. Ogawa, Stereoselective total synthesis of the glycosyl phosphati-
    dylinositol (GPI) anchor of *Trypanosoma brucei, Carbohydr. Res. 235*: 95 (1992).

18. C. E. Dreef, M. Douwes, C. J. J. Elie, G. A. van der Marel, and J. H. van Boom,
    Application of the bifunctional phosphonylating agent bis[6-trifluoromethyl)benzo-
    triazol-1-yl] methylphosphonate towards the preparation of isosteric D-*myo*-inositol
    phospholipid and phosphate analogues, *Synthesis*: 443 (1991).

19. R. Plourde and M. d'Alarcao, Synthesis of a potentially insulin-mimetic phosphodisac-
    charide, *Tetrahedron Lett. 31*: 2693 (1990).

20. R. Plourde, M. d'Alarcao, and A. R. Saltiel, Synthesis and characterization of an insulin-
    mimetic disaccharide, *J. Org. Chem. 57*: 2606 (1992).

21. A. Aguilo, M. Martín-Lomas, and S. Penades, The regioselective synthesis of enantio-
    merically pure *myo*-inositol derivatives. Efficient synthesis of *myo*-inositol 1,4,5-tris-
    phosphate, *Tetrahedron Lett. 33*: 401 (1992).

22. A. Zapata and M. Martín-Lomas, Building blocks for the synthesis of glycosyl-*myo*-
    inositols involved in the insulin intracellular signalling process, *Carbohydr. Res. 234*:
    93 (1992).

23. C. Jaramillo, R. Fernandez de la Pradilla, and M. Martín-Lomas, Synthesis of 1D-1,2-
    anhydro-*myo*-inositol, *Carbohydr. Res. 209*: 296 (1991).

24. C. Jaramillo and M. Martín-Lomas, Approaches to the synthesis of glycosyl phospha-
    tidylinositols. Enantioselective synthesis of optically active *chiro-* and *myo*-inositols,
    Tetrahedron Lett. 32: 2501 (1991).

25. C. Jaramillo, J. L. Chiara, and M. Martín-Lomas, An effective strategy for the synthesis
    of 6-*O*-(2-amino-2-deoxy-α-D-glucopyranosyl)-D-*chiro*- and -D-*myo*-inositol 1-phos-
    phate related to putative insulin mimetics, *J. Org. Chem. 59*: 3135 (1994).

26. D. Semeria, M. Philippe, J.-M. Delaumeny, A.-M. Sepulchre, and S. D. Gero, A general
    synthesis of cyclitols and aminocyclitols from carbohydrates, *Synthesis*: 710 (1983).

27. J. L. Chiara and M. Martín-Lomas, A stereoselective route to enantiomerically pure
    myo-inositol derivatives starting from D-mannitol, *Tetrahedron Lett. 35*: 2969 (1994).

28. L. F. Wiggins, The anhydrides of polyhydric alcohols. Part 2. Derivatives of 1:2-5:6-
    dianhydro mannitol, *J. Chem. Soc.*: 384 (1946); Y. Le Merrer, A. Duréault, C. Greck,
    D. Micas-Languin, C. Gravier, and J.-C. Depezay, Synthesis of diepoxides and diaziri-
    dines, precursors of enantiomerically pure α-hydroxy and α-amino aldehydes or acids,
    from D-mannitol, *Heterocycles 25*: 541 (1987).

29. A. Duréault, M. Portal, and J.-C. Depezay, Enantiospecific syntheses of 2,5-dideoxy-
    2,5-imino-D-mannitol and -L-iditol from D-mannitol, *Synlett*: 225 (1991).

30. J. L. Chiara, W. Cabri, and S. Hanessian, The stereocontrolled formation of cyclic vicinal
    cis-diols via a samarium diiodide pinacol coupling of dialdehydes, *Tetrahedron Lett. 32*:
    1125 (1991).

31. M. Jones, K. K. Rana, J. G. Ward, and R. C. Young, Improved syntheses of inositol
    phospholipid analogues, *Tetrahedron Lett. 30*: 5353 (1989); T. Desai, J. Gigg, R. Gigg,
    E. Martín-Zamora, and N. Schnetz, The synthesis and resolution of (±)-1,4-di-*O*-benzyl-
    2,3-*O*-isopropylidene-*myo*-inositol, *Carbohydr. Res. 258*: 135 (1994).

32. J. P. Guidot, T. Le Gall, and C. Mioskowski, Samarium diiodide-mediated synthesis of
    D-3,4,5,6-tetra-*O*-benzyl-*myo*-inositol, *Tetrahedron Lett. 35*: 6671 (1994).

33. T. Desai, J. Gigg, R. Gigg, and S. Payne, The allyl group for protection in carbohydrate
    chemistry. Part 24. The preparation and phosphorylation of 2,5-and 1D-2,6-di-*O*-benzyl-
    myo-inositol, *Carbohydr. Res. 228*: 65 (1992).

34. S. Cottaz, J. S. Brimacombe, and M. J. Ferguson, Synthesis of a partially protected 1D-6-O-(2-azido-2-deoxy-α-D-glucopyranosyl)-myo-inositol: A useful precursor of glycosylphosphatidylinositols and related compounds, *Carbohydr. Res. 270*: 85 (1995).

35. B. Kratzer, T. G. Mayer, and R. R. Schmidt, Synthesis of D-erythro-sphingomyelin and of D-erythro-ceramide-1-phosphoinositol, *Tetrahedron Lett. 34*: 6881 (1993).

36. T. G. Mayer, B. Kratzer, and R. R. Schmidt, Synthesis of a GPI anchor of yeast (*Saccharomyces cerevisiae*), *Angew. Chem. Int. Ed. Engl. 33*: 2177 (1994).

37. S. V. Ley, F. Sternfeld, and S. Taylor, Microbial oxidation in synthesis: A six-step preparation of (±)-pinitol from benzene, *Tetrahedron Lett. 28*: 225 (1987); S. V. Ley and F. Sternfeld, Microbial oxidation in synthesis: Preparation from benzene of the cellular secondary messenger myo-inositol-1,4,5-trisphosphate (IP₃) and related derivatives, *Tetrahedron Lett. 29*: 5305 (1988); S. V. Ley and F. Sternfeld, Microbial oxidation in synthesis: Preparation of (+)- and (−)- pinitol from benzene, *Tetrahedron 45*: 3463 (1989); S. V. Ley, M. Parra, A. J. Redgrave, F. Sternfeld, and A. Vidal, Microbial oxidation in synthesis: Preparation of 6-deoxy cyclitol analogues of myo-inositol 1,4,5-trisphosphate from benzene, *Tetrahedron Lett. 30*: 3557 (1989); S. V. Ley and A. J. Redgrave, Microbial oxidation in synthesis: Concise preparation of (+)- conduritol F from benzene, *Synlett*: 393 (1990); S. V. Ley, M. Parra, A. J. Redgrave, and F. Sternfeld, Microbial oxidation in synthesis: Preparation of myo-inositol phosphates and related cyclitol derivatives from benzene, *Tetrahedron 46*: 4995 (1990).

38. D. T. Gibson, J. R. Koch, and R. E. Kallio, Oxidative degradation of aromatic hydrocarbons. 1. Enzymatic formation of catechol from benzene, *Biochemistry 7*: 2653 (1968).

39. S. V. Ley and L. L. Yeung, Microbial oxidation in synthesis: Preparation of a potential insulin mimic from benzene, *Synlett*: 997 (1992).

40. P. J. Edwards, D. A. Entwistle, C. Genicot, K. S. Kim, and S. V. Ley, Dispiroketals in synthesis (Part 11): Concomittant enantioselective and regioselective protection of 2,5-dibenzoyl-myo-inositol, *Tetrahedron Lett. 35*: 7443 (1994); R. Downham, P. J. Edwards, D. A. Entwistle, A. B. Hughes, K. S. Kim, and S. V. Ley, Dispiroketals in synthesis (Part 19): Dispiroketals as enantioselective and regioselective protective agents for symmetric cyclic and acyclic polyols, *Tetrahedron: Asymm. 6*: 2403 (1995).

41. Y. Watanabe, M. Mitani, T. Morita, and S. Ozaki, Highly efficient protection by the tetraisopropyldisiloxane-1,3-diyl group in the synthesis of myo-inositol phosphates as inositol 1,3,4,6-tetrakisphosphate, *J. Chem. Soc. Chem. Commun.*: 482 (1989).

42. D. R. Mootoo and B. Fraser-Reid, n-Pentenyl 2-amino-2-deoxy glycosides undergo stereoselective coupling under mild chemospecific conditions, *Tetrahedron Lett. 30*: 2363 (1989).

43. U. E. Udodong, R. Madsen, C. Roberts, and B. Fraser-Reid, A ready convergent synthesis of the heptasaccharide GPI membrane anchor of rat brain Thy-1 glycoprotein, *J. Am. Chem. Soc. 115*: 7886 (1993).

44. H. Hori, Y. Nishida, H. Ohrui, and H. Meguro, Regioselective de-O-benzylation with Lewis acids, *J. Org. Chem. 54*: 1346 (1989).

45. H. Paulsen, A. Richter, V. Sinnwell, and W. Stenzel, Bausteine von Oligosacchariden. Mitteilung 7. Darstellung selektiv Blockierter 2-azido-2-desoxy-D-gluco- und D-galactopyranosylhalogenide: Reaktivität und ¹³C-NMR-spektren, *Carbohydr. Res. 64*: 339 (1978).

46. P. Kováč and K. J. Edgar, Synthesis of ligands related to the O-specific antigen of Type 1 *Shigella dysenteriae*. 3. Glycosylation of 4,6-O-substituted derivatives of methyl 2-acetamido-2-deoxy-α-D-glucopyranoside with glycosyl donors derived from mono- and oligosaccharides, *J. Org. Chem. 57*: 2455 (1992).

47. P. A. M. van der Klein, G. J. P. H. Boons, G. H. Veeneman, G. A. van der Marel, and J. H. van Boom, An efficient route to 3-deoxy-D-manno-2-octulosonic acid (KDO) derivatives via a 1,4-cyclic sulfate approach, *Tetrahedron Lett. 30*: 5477 (1989).

48. G. Excoffier, D. Gagnaire, and J-P. Utille, Coupure sélective par l'hydrazine des groupements acétyles anomères de résidus glycosyles acétylés, *Carbohydr. Res. 39*: 368 (1975).

49. R. R. Schmidt, J. Michel, and M. Roos, Glycosyl imidates 12. Direct synthesis of *O*-α- and *O*-β-glycosyl imidates, *Liebigs Ann. Chem.*: 1343 (1984).

50. A. Zapata, Y. León, J. M. Mato, I. Varela-Nieto, S. Penadés, and M. Martín-Lomas, Synthesis and investigation of the possible insulin-like activity of 1D-4-*O*- and 1D-6-*O*- (2-amino-2-deoxy-α-D-glucopyranosyl)-*myo*-inositol 1-phosphate and 1D-6-*O*-(2-amino-2-deoxy-α-D-glucopyranosyl)-*myo*-inositol 1,2-(cyclic phosphate), *Carbohydr. Res. 264*: 21 (1994).

51. W. Kinzy and R. R. Schmidt, Glycosyl imidates 16. Synthesis of the trisaccharide of the repeating unit of the capsular polysaccharide of *Neisseria meningitidis* (serogroup L), *Liebigs Ann. Chem.*: 1537 (1985).

52. M. Ek, P. J. Garegg, H. Hultberg, and S. Oscarsson, Reductive ring openings of carbohydrate benzylidene acetals using borane-trimethylamine and aluminum chloride. Regioselectivity and solvent dependence. *J. Carbohydr. Chem. 2*: 305 (1983).

53. S. Ogawa and Y. Shibata, Synthesis of biologically active pseudo-trehalosamine:[(1*S*)- (1,2,4/3,5)-2,3,4-trihydroxy-5-hydroxymethyl-1-cyclohexyl] 2-amino-2-deoxy-α-D-glucopyranoside, *Carbohydr. Res. 176*: 309 (1988); P. F. Lloyd and M. Stacey, Reactions of 2-deoxy-2-(2,4-dinitrophenylamino)-D-glucose ("DNP-D-glucosamine") and derivatives, *Tetrahedron 9*: 116 (1960).

54. J. C. Morris, L. Ping-Sheng, T.-Y. Shen, and K. Mensa-Wilmot, Glycan requirements of glycosylphosphatidylinositol phospholipase C from *Trypanosoma brucei*. Glucosaminylinositol derivatives inhibit phosphatidylinositol phospholipase C, *J. Biol. Chem. 270*: 2517 (1995).

55. M. Kloosterman, M. P. de Nijs, and J. H. van Boom, Synthesis of 1,6-anhydro-2-*O*-trifluoromethanesulphonyl-β-D-mannopyranose derivatives and their conversion into the corresponding 1,6-anhydro-2-azido-2-deoxy-β-D-glucopyranoses: A convenient and efficient approach, *J. Carbohydr. Chem. 5*: 215 (1986).

56. J. J. Oltvoort, C. A. A. van Boeckel, J. H. de Koning, and J. H. van Boom, Use of the cationic iridium complex 1,5-cyclo-octadiene-bis[methyldiphenylphosphine]-iridium hexafluorophosphate in carbohydrate chemistry. Smooth isomerization of allyl ethers to 1-propenyl ethers, *Synthesis*: 305 (1981).

57. M. A. Nashed and L. Anderson, Iodine as a reagent for the ready hydrolysis of pro-1-enyl glycosides, or their conversion into oxazolines, *J. Chem. Soc. Chem. Commun.*: 1274 (1982).

58. K. K. Reddy, J. R. Falck, and J. Capdevila, Insulin second messengers: Synthesis of 6-*O*-(2-amino-2-deoxy-α-D-glucopyranosyl)-D-*chiro*-inositol-1-phosphate, *Tetrahedron Lett. 34*: 7869 (1993).

59. W. Kinzy and R. R. Schmidt, Glycosylimidate, 16. Synthese des Trisacchrids aus der "Repeating Unit" des Kapselpolysaccharids von *Neisseria meningitidis* (Serogruppe L), *Liebigs Ann. Chem.*: 1537 (1985).

60. G.-J. Boons, P. Grice, R. Leslie, S. V. Ley, and L. L. Yeung, Dispiroketals in synthesis (Part 5): A new opportunity for oligosaccharide synthesis using differentially activated glycosyl donors and acceptors, *Tetrahedron Lett. 34*: 8523 (1993).

61. R. R. Schmidt, New synthetic methods (56). New methods for the synthesis of glycosides and oligosaccharides—Are there alternatives to the Koenigs–Knorr method?, *Angew. Chem. Int. Ed. Engl. 25*: 212 (1986).

62. S. Cottaz, J. S. Brimacombe, and M. A. J. Ferguson, Parasite glycoconjugates. Part 3. Synthesis of substrate analogues of early intermediates in the biosynthetic pathway of glycosylphosphatidylinositol membrane anchors, *J. Chem. Soc. Perkin Trans. I*: 1673 (1995).

63. K. G. Milne, R. A. Field, W. J. Masterson, S. Cottaz, J. S. Brimacombe, and M. A. J. Ferguson, Partial purification and characterization of the *N*-acetylglucosaminyl-phosphatidylinositol de-*N*-acetylase of glycosylphosphatidylinositol anchor biosynthesis in African Trypanosomes, *J. Biol. Chem. 269*: 16403 (1994).

64. T. K. Smith, S. Cottaz, J. S. Brimacombe, and M. A. J. Ferguson, Substrate specificity of the dolichol phosphate mannose: Glucosaminyl phosphatidylinositol α1-4 mannosyl-transferase of the glycosylphosphatidylinositol biosynthetic pathway of African Trypan-osomes, *J. Biol. Chem. 271*: 6476 (1996).

65. G. Grundler and R. R. Schmidt, Glycosylimidate, 13. Anwendung des Trichloracetimi-dat-Verfahrens auf 2-Azidoglucose- und 2-Azidogalactose-Derivate, *Liebigs Ann. Chem.*: 1826 (1984).

66. C. Roberts, R. Madsen, and B. Fraser-Reid, Studies related to synthesis of glycophos-phatidylinositol membrane-bound protein anchors. 5. *n*-Pentenyl ortho esters for mannan components, *J. Am. Chem. Soc. 117*: 1546 (1995); C. Roberts, C. L. May, and B. Fraser-Reid, Streamlining the *n*-pentenylglycoside approach to the trimannoside component of the Thy-1 membrane anchor, *Carbohydr. Lett. 1*: 89 (1994).

67. B. Fraser-Reid, U. E. Udodong, Z. Wu, H. Ottosson, J. R. Merritt, C. S. Rao, C. Roberts, and R. Madsen, *n*-Pentenyl glycosides in organic chemistry: A contemporary example of serendipity, *Synlett*: 927 (1992).

68. H. Paulsen and A. Bünch, Bausteine von Oligosacchariden. Mitteilung 34. Synthese der Pentasaccharid-Kette des Forssman-Antigens, *Carbohydr. Res. 100*: 143 (1982).

69. T. Nakano, Y. Ito, and T. Ogawa, Synthetic studies on cell-surface glycans: Part 92. Synthesis of sulfated glucuronyl glycosphingolipids; carbohydrate epitopes of neural cell-adhesion molecules, *Carbohydr. Res. 243*: 43 (1993).

70. K. Koike, M. Sugimoto, S. Sato, Y. Ito, Y. Nakahara, and T. Ogawa, Total synthesis of globotriaosyl-*E*- and *Z*-ceramides and isoglobotriaosyl-*E*-ceramide, *Carbohydr. Res. 163*: 189 (1987).

71. R. Johansson and B. Samuelsson, Regioselective reductive ring opening of 4-methox-ybenzylidene acetals of hexopyranosides. Access to a novel protective group strategy, *J. Chem. Soc. Chem. Commun.*: 201 (1984); T. Fukayama, A. A. Laird, and L. M. Hotchkiss, *p*-Anisyl group. A versatile protecting group for primary alcohols, *Tetrahe-dron Lett. 26*: 6291 (1985).

72. T. Ogawa, K. Katano, and M. Matsui, Regio- and stereo-controlled synthesis of core oligosaccharides of glycopeptides, *Carbohydr. Res. 64*: C3 (1978); P. J. Garegg and L. Maron, Improved synthesis of 3,4,6-tri-*O*-benzyl-α-D-mannopyranosides, *Acta Chem. Scand., Ser. B. 33*: 39 (1979).

73. S. V. Ley, R. Leslie, P. D. Tiffen, and M. Woods, Dispiroketals in synthesis (Part 2); A new group for the selective protection of diequatorial vicinal diols in carbohydrates, *Tetrahedron Lett. 33*: 4767 (1992).

74. N. Khiar and M. Martín-Lomas, A highly convergent synthesis of the tetragalactose moiety of the glycosyl phosphatidyl inositol anchor of the variant surface glycoprotein of *Trypanosoma brucei*, *J. Org. Chem. 60*: 7017 (1995).

75. D. Kahne, S. Walker, Y. Cheng, and D. Van Engen, Glycosylation of unreactive sub-strates, *J. Am. Chem. Soc. 111*: 6881 (1989); L. Yan, C. M. Taylor, R. Goodnow, and D. Kahne, Glycosylation on the Merrifield resin using anomeric sulfoxides, *J. Am. Chem. Soc. 116*: 6953 (1994).

76. R. J. Ferrier and R. H. Furneaux, 1,2-*trans*-1-Thioglycosides. *Methods Carbohydr. Chem. 8*: 251 (1980).

77. T. Ogawa, Haworth Memorial Lecture. Experiments directed towards glycoconjugate synthesis, *Chem. Soc. Rev. 23*: 397 (1994).

78. I. Lindh and J. Stawinski, A general method for the synthesis of glycerophospholipids and their analogues via *H*-phosphonate intermediates, *J. Org. Chem. 54*: 1338 (1989).

79. V. Bollitt, C. Mioskowski, S.-G. Lee, and J. R. Falck, Direct preparation of 2-deoxy-D-glucopyranosides from glucals without Ferrier rearrangement, *J. Org. Chem. 55*: 5812 (1990); N. Kaila, M. Blumenstein, H. Bielawska, and R. W. Franck, Face selectivity of the protonation of glycals, *J. Org. Chem. 57*: 4576 (1992).

80. R. R. Schmidt, J. Michel, and M. Roos, Glycosylimidate, 12. Direkte Synthese von *O*-α- und *O*-β-Glycosyl-imidaten, *Liebigs Ann. Chem.*: 1343 (1984).

81. A. S. Campbell and B. Fraser-Reid, First synthesis of a fully phosphorylated GPI membrane anchor: Rat brain Thy-1, *J. Am. Chem. Soc. 117*: 10387 (1995).

82. A. S. Campbell and B. Fraser-Reid, Support studies for installing the phosphodiester residues of the Thy-1 glycoprotein membrane anchor, *Bioorg. Med. Chem. 2*: 1209 (1994).

83. P. Zimmerman and R. R. Schmidt, Synthese von Sphingosinen, 4. Synthese von *erythro*-Sphingosinen über die Azidoderivate, *Liebigs Ann. Chem.*: 663 (1988).

84. B. Kratzer and R. R. Schmidt, An efficient synthesis of sphingosine-1-phosphate, *Tetrahedron Lett. 34*: 1761 (1993).

85. A. Y. Frantova, A. E. Stepanov, A. S. Bushnev, E. N. Zvonkova, and V. I. Shvets, The synthesis of ceramide phosphoinositol, *Tetrahedron Lett. 33*: 3539 (1992).

86. A. Y. Zamyatina, A. E. Stepanov, A. S. Bushnev, E. N. Zvonkova, and V. I. Shvets, Synthesis of ceramide phosphoinositol and its thioanalogue, *Bioorg. Khim. 19*: 347 (1993).

87. R. R. Schmidt and T. Maier, Glycosylimidates. Part 32. Synthesis of D-*ribo*- and L-*lyxo*-phytosphingosine: Transformation into the corresponding lactosyl-ceramides, *Carbohydr. Res. 174*: 169 (1988).

88. A. Arasappan and B. Fraser-Reid, *n*-Pentenylglycoside methodology in the stereoselective construction of the tetrasaccharyl cap portion of *Leishmania* lipophosphoglycan, *J. Org. Chem. 61*: 2401 (1996).

89. M. J. McConville and M. A. J. Ferguson, The structure, biosynthesis and function of glycosylated phosphatidylinositols in the parasitic protozoa and higher eukaryotes, *Biochem. J. 294*: 305 (1993).

90. M. P. DeNinno, J. B. Etienne, and K. C. Duplantier, A method for the selective reduction of carbohydrate 4,6-*O*-benzylidene acetals, *Tetrahedron Lett. 36*: 669 (1995).

91. J. R. Merritt, E. Naisang, and B. Fraser-Reid, *n*-Pentenyl mannoside precursors for synthesis of the nonamannan component of high mannose glycoproteins, *J. Org. Chem. 59*: 4443 (1994).

92. A. Arasappan and B. Fraser-Reid, *n*-Pentenyl furanosides: Synthesis and glycosidation reactions of some *galacto* derivatives, *Tetrahedron Lett. 36*: 7967 (1995).

93. A. V. Nikolaev, T. J. Rutherford, M. A. J. Ferguson, and J. S. Brimacombe, The chemical synthesis of *Leishmania donovani* phosphoglycan fragments, *Bioorg. Med. Chem. Lett. 4*: 785 (1994); A. V. Nikolaev, T. J. Rutherford, M. A. J. Ferguson, and J. S. Brimacombe, Parasite glycoconjugates. Part 4. Chemical synthesis of disaccharide and phosphorylated oligosaccharide fragments of *Leishmania donovani* antigenic lipophosphoglycan, *J. Chem. Soc. Perkin Trans. I*: 1977 (correction p. 2809) (1995).

94. N. E. Byramova, M. V. Ovchinnikov, L. V. Bachinowsky, and N. K. Kochetkov, Selective removal of *O*-acetyl groups in the presence of *O*-benzoyl groups by acid-catalyzed methanolysis, *Carbohydr. Res. 124*: C8 (1983).

95. A. V. Nikolaev, I. A. Ivanova, and V. N. Shibaev, The stepwise synthesis of oligo(glycosyl phosphates) via glycosyl hydrogenphosphonates. The chemical synthesis of oligomeric fragments from *Hansenula capsulata* Y-1842 exophosphomannan and from *Escherichia coli* K51 capsular antigen, *Carbohydr. Res. 242*: 91 (1993).

96. J. M. Williams and A. C. Richardson, Selective acylation of pyranosides 1. Benzoylation of methyl α-D-glycopyranosides of mannose, glucose and galactose, *Tetrahedron 23*: 1369 (1967).

97. A. V. Nikolaev, J. A. Chudek, and M. A. J. Ferguson, The chemical synthesis of *Leishmania donovani* phosphoglycan via polycondensation of a glycobiosyl hydrogenphosphonate monomer, *Carbohydr. Res. 272*: 179 (1995).

8

Analysis of GPI Protein Anchors and Related Glycolipids

Malcolm J. McConville and Julie E. Ralton
University of Melbourne, Parkville, Victoria, Australia

I. INTRODUCTION

Glycosylated phosphoinositides that contain the core structure Manα1-4GlcNα1-6 *myo* inositol-1-PO$_4$-lipid are synthesized by all eukaryotic cells. They were first discovered covalently linked to the carboxyl-terminus of several cell surface glycoproteins and are now known to act as membrane anchors for a wide variety of type-1 proteins. As such, they may be viewed as an alternative to a hydrophobic peptide domain for these proteins. In addition, GPI anchors may have other roles that either enhance or are essential for protein function. In higher eukaryotes, GPI anchors are thought to be involved in (1) targeting proteins to particular membrane domains, such as the apical surfaces of polarized epithelial cells, (2) controlling the rate of endocytosis and turnover of the attached proteins, (3) the selective release of proteins from the cell surface and/or flipping of membrane proteins between cells, and (4) transmembrane signaling in some cell types (i.e., lymphocytes) through their association with membrane microdomains on the cell surface (reviewed in Cross, 1990; McConville and Ferguson, 1993). Interestingly, GPI-anchored proteins are most abundant on the surfaces of single-celled eukaryotes (i.e., yeast and protozoa), which are exposed to relatively harsh external environments. Some of these proteins are expressed in very high copy numbers, and it is thought that the use of GPI anchors may facilitate their surface packing and, in some cases, form a protective glycocalyx. Several protozoa also synthesize GPI structures [commonly referred to as glycoinositol phospholipids (GIPLs)] that are not involved in protein anchoring. These structures are frequently major cell surface components and in many cases are crucial for parasite infectivity and survival (McConville and Ferguson, 1993; Turco and Descoteaux, 1992).

A wide variety of techniques have been used to characterize GPI structures, and several monographs devoted to specific methodologies for GPI analysis have appeared recently (Hooper, 1992; Ferguson, 1992a; Menon, 1994). This chapter will provide a detailed overview of the various methods that have been used to identify and characterize both the protein-linked and the free GPI species.

II. THE STRUCTURE OF GPI GLYCOLIPIDS

Detailed structural studies have now been carried out on more than 20 protein anchors (see Figure 1). These studies indicate the presence of a common backbone sequence EtN-PO$_4$-6Manα1-2Manα1-6Manα1-4GlcNα1-6 *myo* inositol-1-PO$_4$-lipid, in which the terminal ethanolamine residue is amide linked to the C-terminal α-carboxyl group of the mature protein. In most cases, the glycan core is elaborated with one or more side chains, which may comprise Man, Gal, GalNAc, sialic acid, or additional ethanolamine-phosphate residues (Figure 1). As with other types of glycosylation, the GPI anchors of any one protein may comprise a variety of glycoforms, which vary, depending on the cell type and stage of differentiation. Heterogeneity may also occur in the lipid moieties of the protein anchors, which may contain diacylglycerol, monoacylglycerol, alkylacylglycerol, or ceramide, depending on the protein and cell type. These lipids are invariably distinct from those found in the total PI pool, suggesting that only a subpopulation of the cellular PI pool enters the GPI biosynthetic pathway and/or that extensive remodeling of the GPI lipids can occur before or after transfer to the protein. Many GPI anchors contain an additional fatty acid (usually, but not always, palmitic acid) on the inositol ring (Figure 1). This type of acylation has important implications for structural studies as it renders the GPI anchors resistant to cleavage by phosphatidylinositol-specific phospholipase C (PI-PLC) and GPI-specific phospholipase C (GPI-PLC), which are often used to demonstrate the presence of a GPI anchor (Roberts et al., 1988a).

Free GPIs, which are structurally related to the protein anchors, have been found in yeast and mammalian cells and are generally thought to be precursors for the protein anchors (Mayor et al., 1990a,b; Puoti and Conzelmann, 1992; Kamitani et al., 1992). Interestingly, some of these GPIs may be elaborated with substituents that are not found on the corresponding protein anchors (i.e., additional ethanolamine-phosphate groups), suggesting that they may be metabolic end products or that the addition of these substituents is reversible.

Representative structures of abundant protein-free GPIs, or GIPLs, of protozoa belonging to the Order Kinetoplastida are shown in Figure 2. While some of the protozoan GIPLs contain the same trimannose backbone as the protein anchors, many of these structures diverge from the protein anchors beyond the first mannose residue and contain novel glycan extensions (Figure 2). These extensions may comprise linear or branched chains that are made of residues such as Man, Gal, Ara, GalNAc, Xyl, ethanolamine-phosphate, and 2-aminoethylphosphonate. Amongst the most complex of these GPI-related structures is the lipophosphoglycan (LPG) from *Leishmania* parasites. LPG consists of a GPI anchor and a linear phosphoglycan chain made up of [Galβ1-4Manα1-PO$_4$] repeat units. The length of the phosphoglycan chain and the degree to which the disaccharide backbone is substituted with mono- or oligosaccharide branches varies in a species-, strain-, and stage-specific manner (McConville et al., 1990b; 1992; 1995; Ilg et al., 1992; Thomas et al., 1992). Interestingly, many of the abundant Kinetoplastida GIPLs contain common sequence motifs, suggesting that they may have evolved in a common protozoal ancestor (Figure 2). The lipid moieties of these GIPLs are also diverse (including diacylglycerol, alkylacylglycerol, and ceramide) and tend to be different from the lipid moieties of the protein anchors of the same cell. In all cases, these structures are expressed in the cell as a spectrum of biosynthetically related structures that may vary during parasite development.

III. INITIAL IDENTIFICATION OF GPI ANCHORS

A number of approaches have been used to determine whether or not a protein is GPI anchored. One of the simplest methods is to determine if the protein is released from the cell surface after treatment with bacterial PI-specific phospholipase C. Loss of a particular protein can be monitored by fluorescence-activated cell sorter (FACS) analysis of the treated cells or by analysis of metabolically or surface-labeled proteins that are released into the supernatant (Low and Kincade, 1985). Another approach to identify GPI-anchored proteins involves detergent partitioning with Triton X-114. This technique is based on the ability of the nonionic detergent Triton X-114 to partition into two distinct phases above 20°C: a detergent-rich phase and a detergent-depleted or aqueous phase (Bordier, 1981). Amphiphilic proteins, particularly those containing attached lipid (such as a GPI-anchor or peptide-linked myristate or pal-

```
Protein
  |
 C=O
  |
PO4-CH2-CH2-NH
  |
6Manα1-2Manα1-6Manα1-4GlcNα1-6myo-inositol-1-PO4-lipid
 |       |       |        |                |
 R1      R2      R3       R4               R5
```

Protein	Side chain substituents	¹Lipid	Refs
Trypanosoma brucei Variant surface glycoprotein	R3: +/-Galα1-2Galα1-6 ⟍ Galα1-3 / +/-Galα1-2 ⟋ Galα1-3	DAG (14:0 - 14:0)	Ferguson et al.,1985a;b; 1988
T. brucei Procyclic acidic repeat protein	R?: [NANA₅, GlcNAc₉,Gal₉]	MAG (18:0) R5 = 16:O	Ferguson et al., 1993 Ferguson 1992
Leishmania major promastigote surface protease		AAG (24:0, 26:0 - 16:0,18:0)	Schneider et al., 1990
T. cruzi 1G7 antigen (metacyclic stage)	R1: Manα1-2	AAG (16:0 - 16:0,18:0) Cer as minor component	Güther et al., 1992, Heise et al., 1995
mucin glycoprotein	R1: Manα1-2 +/- AEP-6 (in place of EtN-P-6) R4: AEP-6 or EtN-P-6	AAG (16:0 - 16:O) (in epimastigotes) Cer (16:O, 24:0 - 18:0 base) (in metacyclics)	Acosta Serrano et al., 1995 Previato et al., 1995
Tc-85 (trypomastigotes)		*lyso* alkylglycerol (16:0)	Couto et al., 1993
Plasmodium falciparum MSP-1 & -2	R1: Manα1-2	DAG (18:0 - 18:0) R5: 14:0	Gerold et al., 1995
Saccharomyces cerevisiae ggp125 other proteins	R1: +/- Manα1-2Manα1-2 "	DAG (26:0) Cer (26:0 / 18:0 base)	Conzelmann et al., 1990 Frankhauser et al., 1993
Dictyostelium discoideum PsA	R1: +/- Manα1-2 R?: EtN-P-6	Cer (18:1 / 18:0 base)	Haynes et al., 1993 Stadler et al., 1989

Acetylcholinesterase *Torpedo californica* electric organ	R1: Glcα1-2 R3: GalNAcβ1-4 EtN-P-6	DAG (16:0,18:0,18:1)	Bütikofer et al., 1990 Mehlert et al., 1992
Bovine erythrocytes	R2: EtN-P-6	AAG (18:0 - 18:0,18:1)	Roberts et al, 1988a
Human erythrocytes	R2: EtN-P-6	AAG (18:0,18:1 - 22:4, 22:5, 22:6) R5: +/- 18:0	Roberts et al., 1988b
Alkaline phosphatase (human placental)	R?: EtN-P-6	AAG (18:0,18:1 - 16:0,18:0)	Redman et al., 1994b
Folate-binding protein (Human KB cells)	?	AAG (16:0,18:1, 20:1 - 22:0) R5: 16:0,18:0	Luhrs and Slomiany 1989
Thy-1 (rat brain)	R1: Manα1-2 R3: GalNAcβ1-4	AAG	Homans et al., 1988
CD52 (human spleen)	R1: +/- Manα1-2 R3: EtN-P-6	DAG (18:0 - 18:0, 20:4) R5: +/- 18:0	Treumann et al., 1995
CD59 (human urine)	R1: +/- Manα1-2 R3: GalNAcβ1-4 EtN-P-6	?	Nakano et al., 1994
Dipeptidase[2] (human and porcine)	R3: +/-X-GalNAcβ1-4	?	Brewis et al., 1992b
Scrapie prion protein, PrP (hamster brain)	R? NANA-Gal-GalNAc-	?	Stahl et al., 1987; 1992

DAG = diacylglycerol; AAG = alkylacylglycerol; MAG = monoacylglycerol; EtN-P = ethanolamine-phosphate; X = unknown monosaccharide; Cer = ceramide; ? = not determined.

[1] Only the lipid composition of major molecular species of each anchor are given. The composition of alkyl/acyl chains on *sn*-1 position of glycero lipids are listed first, followed (after hyphen) by composition of acyl chains on *sn*-2 position.

[2] Glycans were analyzed after removal of potential sialic acid and ethanolamine-phosphate residues.

Figure 1 Structures of GPI protein anchors. All protein anchors so far characterized contain the conserved backbone sequence shown at the top of the figure. The nature of carbohydrate and noncarbohydrate side chain substituents and variable lipid moieties are indicated for individual protein anchors.

Parasite GPI	Structure	Refs
L. donovani M3	Manα1-2Manα1-6↘ 　　　　　　　　Manα1-4GlcNα1-6 *myo* inositol-1-PO_4-AAG	McConville and Blackwell, 1991
L. mexicana iM4	EtN-P | 6 Manα1-2Manα1-6↘ 　　　　　　　　Manα1-4GlcNα1-6 *myo* inositol-1-PO_4-AAG Manα1-3↗	McConville et al., 1993
L panamensis GPI 6	Manα1-2Manα1-6↘ 　　　　　　　　Manα1-4GlcNα1-6 *myo* inositol-1-PO_4-DAG Galα1-2Galβ1-3↗	McConville, unpublished
Trypanosoma cruzi LPPG	Gal$_f$β1　Gal$_f$β1　　　　　AEP 　|　　　　|　　　　　　| 　3　　　　3　　　　　　6 Manα1-2Manα1-6Manα1-4GlcNα1-6 *myo* inositol-1-PO_4-Cer / AAG	Previato et al., 1990 Lederkremer et al., 1991
Herpetomonas samuelpessoai	AEP | 6 Manα1-6↘ 　　　　　Manα1-4GlcNα1-6 *myo* inositol-1-PO_4-Cer Manα1-3↗	Routier et al., 1995
L. major GIPL-3	Galα1-6Galα1-3Gal$_f$ β1-3Manα1-3Manα1-4GlcNα1-6 *myo* inositol-1-PO_4-AAG	McConville et al., 1990a
Leishmania LPG anchor	[Galβ1-4Man-P-]$_n$　　　　　Glcα1-P 　　|　　　　　　　　　| 　　6　　　　　　　　　6 Galα1-6Galα1-3Gal$_f$ β1-3Manα1-3Manα1-4GlcNα1-6 *myo* inositol-1-PO_4-*lyso*-alkylglycerol	McConville et al., 1990b; 1995, Thomas et al., 1992, Ilg et al., 1992
Endotypanum schaudinni	EtN-P　　　　　EtN-P 　|　　　　　　　| 　6　　　　　　　6 Arap β1-2Galα1-6Galα1-3Gal$_f$ β1-3Manα1-3Manα1-4GlcNα1-6 *myo* inositol-1-PO_4-Cer	Wait et al., 1994
Crithidia fasciculata Araβ1-2	Araβ1-2 | [Galβ1-3]$_n$ Galα1-3Gal$_f$ β1-3Manα1-3Manα1-4GlcNα1-6 *myo* inositol-1-PO_4-Cer	Schneider et al., 1995
Leptomonas samueli	[GalcA$_{0-1}$, Glc$_{0-1}$, Xyl$_{0-5}$]　AEP　　　AEP 　　|　　　　　　　　　|　　　　| 　　4　　　　　　　　　6　　　　6 Manβ1-3Manα1-3Gal$_f$ β1-3Manα1-3Manα1-4GlcNα1-6 *myo* inositol-1-PO_4-Cer	Previato et al., 1992

EtN-P = ethanolamine-phosphate; AEP = aminoethylphosphonate; DAG = diacylglycerol; AAG = alkylacylglycerol; Cer = ceramide
Sequences with homology with the protein anchors are underlined

Figure 2 Structures of selected protein-free GPI glycolipids from Kinetoplastid parasites.

mitate) partition predominantly into the detergent-rich phase. Partitioning of proteins into the detergent phase can be monitored by SDS-polyacrylamide gel electrophoresis (SDS-PAGE), by Western blot analysis if antibodies are available, or by enzymic assays if the protein of interest is an enzyme. To confirm the presence of a GPI anchor it is usually necessary to treat these proteins with either a bacterial PI-specific PLC, the *Trypanosoma brucei* GPI-specific PLC, or the PI-specific PLD from mammalian serum, to remove the lipid moiety and convert the protein to a hydrophilic form that preferentially partitions into the aqueous phase. It is important to note that many GPI anchors are resistant to the PI-PLC/GPI-PLC group of enzymes due to the presence of an extra fatty acid on the 2-position of the inositol headgroup. This is consistent with the putative reaction mechanism of these enzymes, which are thought to use the hydroxyl on the 2-position of the inositol for nucleophilic attack on the phosphodiester linkage, generating inositol-1,2 cyclic phosphate with concomitant release of the lipid (Low, 1992). Prior treatment of GPI-anchored proteins with mild base or hydroxylamine can be used to remove selectively the inositol-linked fatty acid and thus render the protein susceptible to PI-PLC/GPI-PLC digestion (Toutant et al., 1989). Alternatively, GPI-anchored proteins containing an acylated inositol can be digested with serum PI-PLD, which is not affected by this modification. However, the protein product of PI-PLD digestion still retains the fatty acid on the inositol and may thus still partition into the Triton X-114 detergent phase. Other methods that can be used to show conversion of GPI-anchored proteins from an amphiphilic form to a hydrophilic form after lipase treatment include electrophoresis under nondenaturing conditions (Toutant et al., 1989) and the reconstitution of proteins into lipid vesicles (Hooper, 1992).

Evidence for a GPI anchor can also be obtained by the immunochemical detection of the cross-reacting determinant (CRD) that is exposed after PI-PLC digestion. The CRD epitope is recognized by a number of different rabbit polyclonal antibodies and is absolutely dependent on the presence of the inositol 1,2-cyclic phosphate in conjunction with other features of the GPI anchor glycan, such as the amino group on the core GlcN (Zamze et al., 1988; Jager et al., 1990). The CRD epitope is generally detected by Western blot analysis or enzyme-linked immunoadsorbant assays (ELISA). Again, treatment of the protein with hydroxylamine may be necessary to remove any inositol-acyl prior to PI-PLC digestion. This approach has recently been extended with the finding that both the PI-PLC digestion and base treatment (50 mM in 20% 1-propanol, 60 min, room temperature) can be performed on proteins *after* they have been transferred to nitrocellulose membranes, allowing the simultaneous detection of PI-PLC-resistant and PI-PLC-sensitive anchors (Güther et al., 1994). The *T. brucei* GPI-PLC is more reliable in generating the CRD epitope as some of the bacterial PI-PLCs have an additional cyclic phosphodiesterase activity that slowly hydrolyzes the inositol-1,2-cyclic phosphate epitope with concomitant loss of anti-CRD antibody reactivity.

Metabolic labeling can be used to identify the presence of a GPI anchor and is often useful for further structural studies. GPI anchors can be metabolically labeled with radiolabeled fatty acids (i.e., myristate, palmitate), specific monosaccharides (Man, GlcN), inositol, or ethanolamine. While the use of radiolabeled fatty acids may greatly facilitate subsequent structural studies, it is important to bear in mind that there may be considerable heterogeneity in the lipid moieties of GPI structures between different cells and even within the same cell and that this may lead to selective labeling of a subpopulation of structures. The presence of both diacyl-PI and inositolphosphoceramide in the yeast GPI anchors is a striking example of this phenomenon. The incorporated fatty acids may also be positionally restricted. For example, ^{3}H-palmitic acid is only incorporated into the sn-2-position of the alkylacyl-PI moiety of the T. cruzi 1G7 anchor, suggesting that there is little conversion of fatty acids into ether-linked alkyl chains (Heise et al., 1995). Conversely, the fatty alcohol, ^{14}C-hexadecanol, has been used to label selectively the alkyl chains of GPI anchors containing alkylacyl-PI (J. Ralton and M. McConville, unpublished data). Thus for uniform labeling of the GPI lipid moieties it may be necessary to use a variety of lipid precursors, either individually or as a mixture. An advantage of fatty acid/alcohol labeling is that the label can be specifically released (with lipases, nitrous acid deamination, etc.), confirming the specificity of incorporation. Metabolic labeling of GPI-anchored proteins with [^{3}H]GlcN or [^{3}H]Man is complicated by the possible incorporation of label into O- and N-linked sugars, although incorporation of label into the latter can be greatly reduced or eliminated if the labeling is done in the presence of tunicamycin. In contrast, incorporation of [^{3}H]ethanolamine into protein is a relatively specific indicator of the presence of a GPI anchor. Inhibition of [^{3}H]ethanolamine incorporation into protein with mannosamine has been used as an additional criterion for labeling specificity (Lisanti et al., 1991).

Several procedures have been used to label chemically either the lipid or the glycan portion of GPI anchors and nonprotein-linked GPIs. The lipid domain can be labeled selectively with the photoactivated reagent 3-trifluoromethyl-3-(m-[^{125}I]-iodophenyl)-diazirine ([^{125}I]TID) (Roberts and Rosenberry, 1986). On UV irradiation, the [^{125}I]TID generates a reactive carbene that covalently labels all the aliphatic chains in the GPI anchor. This procedure results in the incorporation of a high specific activity label, which is not selective for chain type (i.e., alkyl/acyl). Moreover, the TLC mobility of the TID-labeled lipids is only slightly different from the corresponding unmodified lipid, facilitating analysis of the products of chemical and enzymic treatments of the lipid moiety. For radiolabeling of the glycan moiety, the free amino groups present in all GPIs (i.e., on GlcN and any unsubstituted ethanolamine-phosphate) can be reductively methylated with formaldehyde and CNB^{3}H$_4$ (Roberts and Rosenberry, 1986) or N-[^{3}H]-acetylated with [^{3}H]acetic anhydride (Milne et al., 1994). The former procedure was used to identify the nonprotein-linked GPI glycolipids that occur in low abundance in rat liver plasma membranes (Deeg et al., 1992b). It should be mentioned that both of these labeling techniques prevent subsequent release of the glycan moiety by nitrous acid deamination (which requires a free amino group) for further structural studies (see Section VI.A.). If galactose residues are present (either as terminal residues or in internal positions not substituted on the 6-hydroxyl), they may be labeled by sequential treatment of the GPI with galactose oxidase followed by NaB^{3}H$_4$ reduction (McConville and Bacic, 1989). Some GPI glycolipids contain sialic acid and galactofuranose residues, both of which contain exocyclic glycols that can be radiolabeled by very mild periodate oxidation (10 mM NaIO$_4$, 10 min, 4°C) and NaB^{3}H$_4$ reduction (Lederkremer et al., 1980; McConville and Bacic, 1990).

IV. PURIFICATION OF GPI GLYCOLIPIDS

A. Release of the GPI Moiety from Proteins

For determination of the GPI anchor structure it is often necessary to release the GPI anchor from the bulk of the protein, particularly if *N*- or *O*-linked glycans are present that can interfere with subsequent analyses. Release of a GPI-peptide is commonly achieved by proteolysis of the carboxymethylated/reduced peptide or cyanogen bromide cleavage. Exhaustive pronase digestion normally results in a GPI anchor with 1–4 amino acids still attached, although in some cases, longer peptides (up to 24 amino acids long) may remain (Redman et al., 1994a). It is important to note that the efficiency of proteolysis may be affected by the presence of an intact GPI anchor, which may lead to the formation of micelles and hinder access of the protease to the peptide (Redman et al., 1994a). Following proteolysis and removal of detergent by organic solvent extraction, the GPI-peptide can be purified by exploiting its hydrophobicity using reverse-phase HPLC (Ferguson et al., 1988; Moran et al., 1991) or octyl-Sepharose chromatography (McConville and Bacic, 1989; Schneider et al., 1990). The latter procedure gives higher yields and is used more frequently. GPI-peptides are loaded onto the octyl-Sepharose column in a volatile salt buffer, such as ammonium acetate, and eluted with a linear gradient of 1-propanol. It should be noted that octyl-Sepharose chromatography gives very good separation of glycolipids that differ in the number of acyl or alkyl chains, with inositol-acylated GPI species being eluted later in the propanol gradient than noninositol-acylated species (Sevlever et al., 1995). Some intact GPI-anchored glycoproteins have also been purified by this procedure (Ferguson et al., 1993; Acosta Serrano et al., 1995). After pooling and lyophilization, the GPI peptides are recovered in a detergent- and salt-free form that is suitable for chemical analysis. Alternatively, GPI peptides can be purified by a two-step gel filtration procedure that exploits their tendency to associate with detergent micelles (Roberts et al., 1988a).

   Bütikofer et al. (1995) have used hydrofluoric acid (HF) to release the dephosphorylated GPI glycolipid moiety from the intact glycoprotein. This procedure exploits the finding that the phosphodiester bridge between the ethanolamine and the glycan core is more susceptible to HF than the bond between the inositol and lipid. When the *T. brucei* variant surface glycoprotein was subjected to partial HF hydrolysis (50% HF, 4°C, 18 h), approximately 10% of the anchor was recovered as an intact GPI species. This procedure may be useful for generating material either for structural studies (although information on the location of ethanolamine-phosphate residues is lost) or for assaying the activities of specific glycosyltransferases involved in GPI biosynthesis (Bütikofer et al., 1995).

B. Isolation of Free GPIs

Polar solvent systems that have been used to extract phospholipids and polar glycosphingolipids are also suitable for extracting free GPI glycolipids. The protein anchor precursors are frequently extracted from whole cells in either chloroform-methanol-water (10:10:3 v/v) (Masterson et al., 1989) or by sequential extraction of cells in chloroform-methanol (2:1 v/v) followed by chloroform-methanol-water (10: 10:3 v/v) (Menon et al., 1990). In the latter procedure, most of the mature GPI anchor precursors are extracted in the second step, resulting in significant enrichment over the total lipid pool. A more polar chloroform-methanol-water mixture (1:2:0.8 v/v), initially devised to extract polar glycosphingolipids, has been used to extract the GIPLs of *Leishmania* parasites (McConville and Bacic, 1989; McConville et al., 1990a). After phase partitioning of this extract, most of the GIPLs partition into the aqueous phase, resulting again in significant enrichment. Extremely polar GPI species, such as *Leishmania* lipophosphoglycan, have been extracted from delipidated cell pellets in solvent mixtures such as 9% 1-butanol in water (McConville et al., 1990b) or ethanol-water-diethylether-pyridine-13 M ammonia (15:15:5:1:0.017 v/v) (Orlandi and Turco, 1987). Following their extraction, these free GPI glycolipids have been purified using a combination of octyl-Sepharose chromatography and HPTLC (McConville and Bacic, 1989) or by silica gel chromatography using Iatro-beads (Masterson et al. 1989). If the GPI species are inositol acylated, contaminating phospholipids can be removed largely by digesting the extract with bee venom phospholipase A_2 (which cleaves the phospholipids but not the inositol-acylated GPIs) and chromatography on octyl-Sepharose (Sevlever et al., 1995).

V.   ANALYSIS OF INTACT GPI PEPTIDES AND GPI GLYCOLIPIDS

A.   Compositional Analyses

Compositional analysis of the purified glycoprotein and/or isolated GPI peptides can be used to confirm the presence of a GPI anchor and provide considerable structural information, respectively. Although a variety of analytical techniques are suitable for determining the composition of GPI glycolipids, the quantitative determination of monosaccharides and lipids by GC-MS and amino acids and amino sugars by HPLC is most commonly employed (Ferguson, 1992a). For these analyses, 1–5 nmol of purified glycoprotein/glycopeptide (containing suitable internal standards such as *scyllo*inositol or deuterated *myo*inositol and *nor*leucine) is divided into three aliquots. Approximately 100 pmol of sample is hydrolyzed in 6 M HCl to release *myo*inositol, which is then quantitated as its trimethylsilyl derivative by GC-MS using selected ion monitoring. A second aliquot (~1 nmol sample) is subjected to methanolysis (0.5 M methanolic HCl, 80°C, 3–18 h), to release neutral and phosphorylated monosaccharides. It should be noted that cleavage of the GlcN-*myo*inositol bond does not occur with standard methanolysis conditions, unless the GlcN is chemically *N*-acetylated prior to solvolysis. Prior *N*-acetylation is thus essential if the GlcN is to be quantitated in these analyses. Methanolysis also releases ester-linked fatty acids as their methyl esters and ether-linked alkyl chains as 1-*O*-alkylglycerols. The amide-linked fatty acids and long-chain bases of ceramide lipids are released in low yield. Both the monosaccharide and lipid constituents are derivatized with trimethylsilyl reagents and can be analyzed simultaneously by GC-MS using a fused silica capillary column coated with apolar stationary phases. The recoveries of phosphorylated sugars (e.g., Man-6-PO_4) in these analyses tend to be rather poor, although this can be improved if the trimethylsilyl derivatives are permethylated with diazomethane (to methylate the phosphate groups) then rederivatized with TMS prior to GC-MS analysis (Ferguson et al., 1988). The remaining aliquot (~1 nmol sample) is subjected to 6 M HCl hydrolysis (110°C, 18 h), to release any amino acids, ethanolamine, and glucosamine, which are then detected using standard HPLC protocols with pre- or postcolumn derivatization. These analyses can also be used to identify 2-aminoethylphosphonate (2-AEP), which is a common constituent in the free GPI glycolipids and some of the protein-linked GPIs of protozoan parasites (Figures 1 and 2; Acosta Serrano et al., 1995).

While this scheme can be used to determine the ratio of most constituents in a GPI anchor, some modifications are required for the accurate quantitation of particular lipid constituents. For example, the fatty acids and long-chain bases of inositolphosphoceramides are more effectively released by strong base hydrolysis (1 M NaOH, 100°C, 16 h). Long-chain bases are recovered by direct diethyl ether extraction and quantitated by GC-MS after N-acetylation and conversion to their TMS derivatives. The amide-linked fatty acids are recovered by diethyl ether extraction of the acidified mixture and analyzed by GC-MS as their methyl esters. The quantitation of ester-linked fatty acids is also improved if they are first released by milder base conditions (0.1 M NaOH, 37°C, 4 h), then methylated with diazomethane. Some GPI anchors contain sialic acid residues that are partially degraded during methanolysis. Milder acid hydrolysis conditions or specific enzymes may be used if the type of sialic acid is to be identified (Manzi and Varki, 1993).

The monosaccharide composition can also be determined after acid hydrolysis and high-pH anion-exchange chromatography (HPAEC) (Hardy and Townsend, 1994). This approach is very sensitive, does not require prior derivatization, and can be used to detect phosphorylated monosaccharides directly. Hydrolysis of GPI anchors in 4 M trifluoroacetic acid (100°C, 4 h) releases all neutral and phosphorylated monosaccharides as well as a characteristic disaccharide fragment that contains GlcN and myoinositol-phosphate (Deeg et al., 1992a). This fragment may be readily detected if the samples have been radiomethylated prior to hydrolysis and is a useful diagnostic fragment for the presence of a GPI anchor or free GPI glycolipids (Deeg et al., 1992).

B.   Analysis of Intact GPI Peptides by Mass Spectrometry

Fast-atom-bombardment mass spectrometry (FAB-MS) and, more recently, electrospray-ionization mass spectrometry (ESI-MS) are extremely useful techniques for the rapid, accurate, and sensitive determination of molecular weight of underivatized GPI species. The main advantages of these techniques are that (1) they can be used to analyze GPI mixtures (different glycoforms/molecular species), (2) prior chemical degradation and/or derivatization is not required, and (3) when combined with tandem mass spectrometry, information on the glycan sequence (branch points, location of noncarbohydrate substituents) and the nature of the lipid moiety can also be obtained.

FAB-MS uses a beam of high-energy atoms (usually zenon or argon) to bombard a solution of analyte dissolved in a relatively involatile liquid matrix in the source of the mass spectrometer. The most common matrices used for GPI samples are combinations of thioglycerol, glycerol, and triethanolamine (Roberts et al.,

1988b; McConville et al., 1990a). Dissipation of the kinetic energy of the bombarding atom beam induces volatilization and ionization of the analyte and matrix molecules. Both positive and negative ions are produced that can be detected by appropriate choice of instrument parameters. These include molecular-weight-related ions ($[M - H]^-$, $[M + H]^+$, and $[M + \text{cation adduct}]^+$) and, in some cases, fragment ions that are generated in the ionization process. Since FAB is a relatively soft ionization technique, it is usually necessary to resort to tandem mass spectrometry to obtain extensive sequence information. FAB-MS provides accurate mass measurements to within 0.5 dalton, with an upper working mass range for glycoconjugates of 6 kDa (Dell et al., 1994).

FAB-MS has been used to analyze the GPI peptide derived from human acetylcholinesterase (Roberts et al., 1988b) and the free GPI species of *Leishmania* parasites (McConville et al., 1990a,b; Thomas et al., 1992; Ilg et al., 1992). The acetylcholinesterase GPI peptide was analyzed after reductive methylation of free amino groups and mild base treatment to remove ester-linked fatty acids. Molecular-weight-related ions and useful fragment ions were obtained in both the positive and the negative ion mode. Fragment ions containing both the nonreducing terminus (resulting from A-type cleavage across glycosidic linkages) and the phospholipid tail (resulting from internal ring cleavages through C1-O rupture) predominated (Roberts et al., 1988b). These ions were used to define the $\text{Hex}_3 \cdot \text{HexN} \cdot \text{inositol-PO}_4 \cdot \text{lipid}$ backbone and the location of the peptide-EtN-PO_4-bridge and an extra unsubstituted ethanolamine-phosphate residue. Reductive methylation of residues containing free amino groups (i.e., GlcN and unsubstituted EtN residues) with either $NaCNBH_3$ or $NaCNBD_3$ was used to corroborate some of these assignments. When reduction was carried out with $NaCNBD_3$, fragment ions containing one, two, or three free amino groups were 2, 4, or 6 mass units larger, respectively, than the corresponding ions generated from the $CNBH_3$-reduced material (Roberts et al., 1988).

For free GPI glycolipids, particularly those containing multiple phosphate groups, FAB-MS in the negative ion mode is the most sensitive (McConville et al., 1990a,b; Thomas et al., 1992). While the relative abundance of fragment ions in these mass spectra can be quite variable, structurally informative fragment ions that retained the phospholipid moiety were generated by β-cleavage along each of the glycosidic linkages in the GPI anchors of *Leishmania* lipophosphoglycan (McConville et al., 1990b, Thomas et al., 1992). Permethylation of the *Leishmania* GIPLs greatly increased the sensitivity of the FAB-MS analyses, although information on the fatty acid composition was lost due to cleavage of ester-linked residues during methylation (McConville et al., 1990a). These derivatives were analyzed in the positive ion mode and gave a different range of fragment ions, some of which contained the nonreducing terminus. When permethylation is carried out without prior *N*-acetylation of the core GlcN, GPI species containing two or three methyl groups on the GlcN are produced, which are detected as a characteristic pair of abundant pseudomolecular species differing in mass by 14 mass units. In contrast, only one molecular-weight-related species is seen if samples are *N*-acetylated before permethylation (McConville et al., 1990a).

ESI-MS differs from FAB-MS in that the sample is dissolved in a suitable solvent and injected directly into the electrospray source, where the sample molecules are stripped of solvent, leaving them as singly or multiply charged species whose charges reflect the number of functional groups that can be protonated (positive ion mode) or deprotonated (negative ion mode). Typically, ESI is interfaced with quadrupole mass analyzers that are able to tolerate high pressures and have good resolving power. Although quadrupole mass detectors have a relatively low m/z range, multiple charging means that even very large biomolecules can be detected.

A limited number of studies suggest that ESI-MS may be more suitable for analyzing intact GPI peptide species than FAB-MS (Redman et al., 1994a; Treumann et al., 1995). The presence of multiple positive charges on the peptide and free ethanolamine components of GPI peptides may increase the sensitivity of these analyses. Samples are introduced into the electrospray source in a suitable solvent (acetonitrile-0.2% formic acid, 1:1 v/v, or chloroform-methanol 2:3 v/v) and analyzed in either the positive or the negative ion mode. The spectra of the GPI peptides generally contain ion clusters corresponding to the double- and triple-charged pseudomolecular ions. Higher charged states (i.e., $[M + 4H]^{4+}$) may predominate if the attached peptide is relatively long (Redman et al., 1994a). Multiple charging means that more accurate molecular weight determination can be made from the distribution of multiply charged peaks. For GPI peptides up to 5 kDa, the accuracy of mass measurements is easily within ± 1 mass unit (Redman et al., 1994a). These analyses can be used to define the composition of the glycan, peptide, and lipid moieties and identify the spectrum of anchor glycoforms in a mixture (Redman et al., 1994a). When ESI is coupled with tandem mass spectrometry, fragment ions can be obtained that provide further information on the GPI structure. Fragmentation is usually achieved by collision-induced dissociation (CID) and/or collision-activated dissociation (CAD). When a triple quadrupole mass analyzer is used, the ion of interest (frequently a pseudomolecular ion) is selected in the first quadrupole and introduced into the second quadrupole, also known as the collision cell, together with a collision gas (usually argon) to induce fragmentation, and the resultant daughter ions analyzed in the third quadrupole. ESI-MS-MS on the GPI peptide derived from a *T. brucei* VSG resulted in fragmentation of the GPI moiety, with no detectable fragmentation of the attached peptide, greatly simplifying interpretation of the mass spectra. The predominant fragment ions reflected cleavage along the glycan backbone (readily defining the sequence peptide-EtN-PO_4-Hex), loss of side-chain sugars and the release of dimyristoylglycerol from the diacyl-PI lipid moiety (Redman et al., 1994a).

Under optimal conditions, these analyses can be carried out on less than 1 nmol of material. However, for both FAB-MS and ESI-MS, sensitivity is crucially dependent on the nature of the sample. The presence of residual detergent, other lipids, or salts can greatly reduce secondary ion yields with concomitant loss of sensitivity. If GPI samples are being extracted or chromatographed in organic solvents, it is prudent to carry out all procedures in glassware rather than plastic tubes, to avoid the accumulation of plasticizers.

VI. SEQUENCING OF THE GPI GLYCAN MOIETY

A. Release of C-Terminal Glycopeptide and Neutral Glycans

A variety of strategies have been used to release the glycan moiety of a GPI anchor for further structural characterization. One approach that can be applied to both GPI peptides (Ferguson et al., 1988; Homans et al., 1988; Schneider et al., 1990; Haynes et al., 1993) and intact proteins (Ferguson et al., 1993; Truemann et al., 1995; Acosta Serrano et al., 1995) involves the following steps: (1) nitrous acid deamination and recovery of the released inositolphospholipid by solvent extraction, (2) reduction of the newly formed 2,5-anhydromannose at the reducing terminus of the GPI glycan to 2,5-anhydromannitol (AHM) with either NaB^3H_4 or $NaBH_4$, and (3), treatment of the deaminated glycopeptide/glycoprotein with hydrofluoric acid to remove substituents that are attached to the glycan via phosphodiester linkages, including the ethanolamine-phosphate residues and any attached peptide (Figure 3). These steps release the intact inositolphospholipid moiety, a deaminated/reduced C-terminal glycopeptide and a deaminated/reduced "neutral" glycan, respectively, which are required for elucidation of GPI structure. The use of NaB^3H_4 allows for the incorporation of a high-specific-activity ^3H-label into the reducing terminus of the GPI glycan to facilitate subsequent sequencing analyses.

There are a number of points to note about this procedure:

1. It is important that the released glycans be repurified after NaB^3H_4 reduction to remove contaminating radiochemical impurities. This is normally performed after the reduction step, although it can also be done after HF dephosphorylation (particularly if the starting material is intact protein). At the end of the reduction step, ^3H-labeled samples are desalted on small gel filtration columns (e.g., BioGel-P10 for glycoprotein samples) or by passage down a cation exchange column [i.e., AG50-X12 (H^+) for glycopeptide samples] and coevaporation with methanol. Most of the radiochemical impurities can then be removed by descending paper chromatography in 1-butanol-ethanol-water (4:1:0.6 v/v), where the sample remains at the origin. High-voltage electrophoresis can be used as an additional cleanup step to remove charged radiochemical contaminants (Ferguson et al., 1988).
2. Fine pH control is crucial for optimal deamination of glycoproteins/glycopeptides, particularly where small volumes are being used. It is imperative that the pH of the deamination mixture be checked (using $\sim$1-μL sample on pH paper) to ensure that any salts remaining in the sample do not affect the final pH.
3. Free GPIs and small GPI peptides may be difficult to deaminate because of micelle formation (Menon et al., 1988; McConville and Bacic, 1989). This can be overcome with multiple additions of $NaNO_2$ at 1-h intervals and by increasing the temperature of the reaction (40–55°C). Alternatively, the sample can be delipidated with PI-specific lipases (PI-PLC or PI-PLD depending on the nature of the lipid moiety) or mild base treatment (13 M ammonia: 50% 1-propanol, 1:1 v/v; 18 h, 37°C) to remove acyl chains from diacyl- or alkylacylglycerol-containing GPIs.
4. Standard HF incubation times (i.e., 60 h) result in the partial cleavage of the βGalNAc side chains that occur in some of the mammalian anchors (Homans et al., 1988). If the presence of these structures is suspected, shorter HF hydrolysis times ($\sim$36 h) can be used.

Alternatively, the glycan moiety can be released from intact glycoproteins or GPI peptides with HF, usually after base or lipase treatment to remove the lipid (Güther et al., 1992; Ferguson, 1992). The HF-released glycan fragment is ideal for meth-

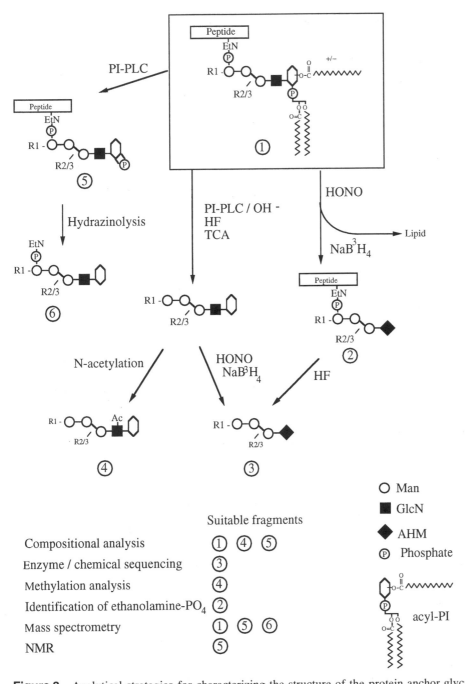

Figure 3 Analytical strategies for characterizing the structure of the protein anchor glycan moiety. Procedures used include: PI-PLC digestion to remove the neutral lipid moiety (note that PI-PLC will not remove the lipid from inositol-acylated anchors); nitrous acid deamination (HONO) followed by NaB^3H_4 reduction, to remove the PI lipid and convert the core GlcN residue to radiolabeled 2,5-anhydromannitol; hydrolysis with hydrofluoric acid to cleave all phosphate and phosphonate bonds (HF); hydrazinolysis to remove attached peptide; *N*-acetylation to convert positively charged GlcN to the neutral GlcNAc species; mild base hydrolysis to remove ester-linked fatty acids.

ylation and FAB-MS analysis, or it can be subjected to nitrous acid deamination/ NaB^3H_4 reduction to generate a labeled neutral glycan for sequencing (Schneider et al., 1990; Güther et al., 1992; Baldwin et al., 1990; Redman et al., 1994b). The protein or peptide components that are present in the HF mixture can be separated from the GPI glycan by TCA precipitation (Güther et al., 1992) or reverse-phase HPLC (Baldwin et al., 1990). It is important to note that the location of peptide bridge and of any unsubstituted ethanolamine-phosphate residues cannot be identified from these fragments.

For NMR and some mass spectrometric analyses, preparations containing the C-terminal glycopeptide have been used. The C-terminal glycopeptides are commonly prepared by PI-PLC treatment (which may also facilitate the purification of the protein), followed by extensive proteolysis, or vice versa (Ferguson et al., 1988). The C-terminal glycopeptide can then be separated from other peptides by reverse-phase HPLC (where the former often elutes in the unbound fraction), gel filtration on BioGel P-30 (Homans et al., 1988), or Concanavalin A chromatography. In some cases, reverse-phase chromatography has been used to partially resolve C-terminal glycopeptides that differ in glycan structure (Stahl et al., 1992). Brewis et al. (1992b) have employed a novel affinity chromatography approach, using anhydrotrypsin-Sepharose, to purify the C-terminal glycopeptide generated by lipase and trypsin digestion. Anhydrotrypsin binds the C-terminus of tryptic peptides, so that only the C-terminal glycopeptide derived from a GPI-anchored protein will elute in the un-retained fraction.

In some cases, proteolysis may result in the release of GPI peptides with peptide chains of variable length. This heterogeneity can greatly reduce the sensitivity of FAB-MS analyses. Hydrazinolysis has been used specifically to remove the peptide moiety from C-terminal glycopeptides while leaving the ethanolamine-phosphate residues intact (Frankhauser et al., 1993). The base conditions of hydra-zinolysis also remove the 1,2-cyclic phosphate residue on the inositol that remains after PI-PLC digestion (Frankhauser et al., 1993).

Strong base hydrolysis (1 M KOH, 100°C, 6 h, or 37°C, 48–72 h) has been used to cleave the phospho-ceramide linkage in the inositol-phosphoceramide lipid moiety of several nonprotein-linked parasite GPIs (Previato et al., 1990, 1992; Routier et al., 1995). The released glycans contain a terminal inositol-phosphate and can either be fractionated by gel filtration (i.e., on BioGel P-4) or by high-pH anion-exchange chromatography (HPAEC) on a Dionex CarboPac column (Routier et al., 1994). This treatment does not remove 2-aminoethylphosphonate or ethanolamine-phosphate substituents, and the derived glycans are used for both FAB-MS and NMR studies.

B. Sequencing of the Glycan Backbone

Sequencing of the glycan backbone is normally performed on the deaminated/reduced glycan that is released after HF hydrolysis. These glycans may then be characterized by their retention times relative to defined standards on a variety of different chromatographic systems, including BioGel P-4 gel filtration, HPAEC and TLC. This approach is particularly powerful if carried out in conjunction with exoglycosidase digestions and partial chemical degradations to monitor the sequential removal of sugar residues. Loss of sugar residues can also be monitored with FAB- or ESI-MS (Stahl et al., 1992).

Chromatographic Analysis

BioGel P-4 chromatography is the most informative for sequencing, because it gives a direct measurement of size (hydrodynamic volume). It can be used preparatively, without internal standards, or analytically by coinjecting a series of glucose oligomer standards with the radiolabeled glycan. The use of internal standards allows the accurate determination of glycan size in terms of glucose units (GU) and hence the number of monosaccharide residues that have been released by individual enzymatic and chemical treatments. The relative elution times of glycans on BioGel P-4 are highly reproducible, facilitating the identification of unknown glycans through co-migration with defined standards (Table 1). In some cases, BioGel P-4 has been used to resolve GPI glycans that contain one or more ethanolamine-phosphate residues (Kamitari et al., 1992). As negatively charged molecules elute in the void volume if the BioGel P-4 is eluted in water (due to a residual net negative charge on the resin), chromatography is performed with a volatile buffer (e.g., ammonium acetate, pyridine-acetate).

Table 1   Chromatographic Properties of the Neutral Glycans from Protein Anchors

Structure	BioGel P-4	HPAEC
AHM	1.7	1.0
Manα 1-4AHM	2.3	1.1
Manα1-6Manα1-4AHM	3.2	2.2
Manα1-2Manα1-6Manα1-4AHM	4.2	2.5
Manα1-2Manα1-2Manα1-6Manα1-4AHM	5.2	3.0
Manα1-2Manα1-6Manα1-4AHM GalNAcB1-4⌋	5.7	3.0
Manα1-2Manα1-2Manα1-6Manα1-4AHM GalNAcβ1-4⌋	6.5	3.5
Glcα1-2Manα1-2Manα1-6Manα1-4AHM	5.1	3.6
Glcα1-2Manα1-2Manα1-6Manα1-4AHM GalNAcβ1-4⌋	6.4	4.2
Manα1-2Manα1-6Manα1-4AHM Galα1-3⌋	5.2	3.6
Manα1-2Manα1-6Manα1-4AHM Galα1-6Galα1-3⌋	6.1	3.8
Manα1-2Manα1-6Manα1-4AHM Galα1-6[Galα1-2]Galα1-3⌋	6.8	4.4
Manα1-2Manα1-6Manα1-4AHM Galα1-2Galα1-6Galα1-3⌋	6.8	4.0
Manα1-2Manα1-6Manα1-4AHM Galα1-2Galα1-6[Galα1-2]Galα1-3⌋	7.6	4.7

Retention times are given relative to dextran oligosacharides coinjected with the deaminated/NaB^{3}H$_4$ reduced GPI glycan.
Adapted from Ferguson, 1992.

HPAEC on a CarboPac (Dionex) column provides excellent resolution of a wide variety of neutral and charged glycan structures (Hardy and Townsend, 1994). Chromatography is performed at high pH (~14), to ionize sugar hydroxyl groups, allowing the glycans to be separated by anion-exchange chromatography using an acetate gradient. HPAEC has been used to analyze the deaminated neutral glycans (Ferguson, 1992), the HF-released glycans that retain the inositol moiety (Puoti and Conzelmann, 1992), and even C-terminal glycopeptides (Deeg et al., 1992a). Although the elution position of glycans on HPAEC is less predictable than on BioGel P-4, closely related structures are usually resolved from each other (Table 1). As a result, HPAEC has been used to purify individual glycans from heterogeneous samples and to make tentative structural assignments based on the coelution of unknown glycans with defined standards. The use of internal standards (e.g., dextran hydrolysate) is imperative for the latter type of analysis to counter the variability in absolute retention times that occur between chromatographic runs. Identification by comigration is particularly powerful if more than one type of chromatography is used. In this respect, no two GPI glycan structures have been found to comigrate on both HPAEC and BioGel P-4 chromatography (Table 2). HPAEC is less commonly used for sequencing because a shift in elution times is sometimes difficult to interpret unless a full range of standards is available.

Table 2 TLC Solvent Systems Useful for Analyzing Intact GPI Glycolipids or Components Derived by Chemical or Enzymic Degradation

Compounds for analysis	Plate	Solvent system
GPI anchor precursors (including GlcN-PI and GlcNAc-PI)	Si60	1. $CHCl_3$/CH_3OH/ 1M NH_4OH (10:10:3) 2. $CHCl_3$/CH_3OH/ 13 M NH_4OH/1 M NH_4OAc/H_2O (180:140:9:9:23)
Non-ethanolamine-containing GPIs and phospholipids	Si60	$CHCl_3$/CH_3OH/HOAc/H_2O (25:15:4:2)
Polar GIPLs	Si60	$CHCl_3$/CH_3OH/0.2%KCl (10:10:3) followed by 1-butanol/pyridine/water (9:4:3)
PI species	Si60	$CHCl_3$/CH_3OH/HOAc/H_2O (25:15:4:2)
Phosphatidic acid and other phospholipids	Si60	$CHCl_3$/CH_3OH/90% formic acid (50:30:7)
Ceramides	Si60	$CHCl_3$/CH_3OH (9:1)
Benzoylated diacyl/alkyl acyl/ alkenylacyl-glycerols	Si60	benzene/2-propanol/ether (50:45:4 or 70:30:2)
Diacyl/alkylacyl glycerol acetates	Si60	hexane/2-propanol (96:4)
	RP-18	$CHCl_3$/CH_3OH/H_2O (15:45:3)
Fatty acid methyl esters (FAMEs)	RP-18	$CHCl_3$/CH_3OH/H_2O (25:75:5) CH_3CN/HOAc (1:1)
Mixture of FAMEs, diacyl-, and monoalkyl-glycerol	RP-60	hexane/diethylether/HOAc (60:30:1)
	Si60	petroleum ether/diethyl ether/HOAc (80:20:1 or 70:30:2)
Deaminated/reduced GPI glycans	Si60	1. 1-propanol/acetone/H_2O (9:6:5 or 5:4:1) 2. 1-propanol/ethanol/H_2O (7:1:2) 3. $CHCl_3$:CH_3OH/H_2O (3 × 10:10:3)
O-methylated mannose derivatives	Si60	1. Benzene/acetone/H_2O/30% NH_3 (100:400:6:3)

Thin-layer chromatography (TLC or high-performance TLC) is also useful for analyzing the products of enzymic or chemical treatments. A wide range of deaminated neutral GPI glycans can be resolved on silica 60 HPTLC sheets developed in 1-propanol-acetone-water (9:6:5 and 5:4:1 v/v) (Schneider et al., 1993). Migration of the glycans in these solvents is mainly dependent on size, although in some cases different isomers (i.e., Manα1-3Manα1-4AHM and Manα1-6Manα1-4AHM) have been resolved. Chloroform-methanol-water (10:10:3 v/v) has also been used to separate a series of neutral GPI glycans and glycans containing 1–3 ethanolamine-phosphate residues on silica TLC (Kamitani et al., 1992). The sensitivity of TLC/HPTLC is high (less than 1000 cpm of labeled glycan can be detected by fluorography or a linear analyzer), and a large number of samples can be sequenced simultaneously.

Lectin-affinity chromatography can also be used to fractionate and analyze the deaminated/reduced GPI glycans. Concanavalin A-Sepharose was used to fractionate and characterize four deaminated/reduced glycans derived from human CD59 anchor (Nakano et al., 1994). Glycans containing no side-chain substituents or an extra α-mannose residue were highly retained on Conanavalin A-Sepharose, whereas glycans with an extra GalNAc side chain were weakly retained. The presence of GalNAc side chains was confirmed by chromatography on *Wistaria floribunda* agglutinin (GalNAc-binding)-agarose (Nakano et al., 1994).

Enzymatic and Chemical Sequencing

A number of exoglycosidases have been used for the sequencing of GPI glycans.

α-Mannosidases.   Jack bean α-mannosidase acts on terminal unsubstituted α-D-mannose (pyranose) residues. Following overnight incubation virtually all α-mannose residues can be released from linear or branched glycans and native GPI structures (in the presence of detergent). This enzyme is particularly useful for locating the position of branch residues (nonmannose-containing glycan side chains, ethanolamine-phosphate residues) along the trimannose backbone of GPI anchors. A number of linkage-specific α-mannosidases are also useful for the sequencing of GPI glycans. These include the *Aspergillus saitoi* (*phoenicis*) α-mannosidase, which specifically cleaves the terminal α-D-Man(1–2)D-Man glycosidic linkage(s) of GPI anchors, and two α-mannosidases from *Xanthomonas manihotis*, which cleave terminal α1–6-linked and α1–2/3-linked Man residues, respectively. It should be noted that the *Xanthomonas* α-mannosidases are only active on linear structures.

Galactosidases.   Some protein anchors (i.e., *T. brucei* VSG) and several parasite GIPLs contain αGal and βGal residues as side chains and as internal residues. These can be removed using the broad-specificity α-galactosidase (from coffee beans) and the β-galactosidases (from jack bean and bovine testis). A number of details about the specificity of the β-galactosidases have emerged from studies on some of the novel parasite glycans (GPI- and *O*-linked). Neither β-galactosidase will digest oligosaccharides terminating in Galβ1–3(Galβ1–2)βGal, or Galβ1–2βGal*f*, while bovine testicular β-galactosidase cannot digest Galβ1–6GlcNAc (Acosta Serrano et al., 1995).

The short β1–4-linked GalNAc branch that occurs on some mammalian anchors (Homans et al., 1988) can be removed with jack bean β-hexosaminidase, which has a broad specificity for any terminal, nonsubstituted β-D-*N*-acetylhexosamine

(GalNAc, GlcNAc). This enzyme has also been used to digest the poly N-acetyllac-
tosamine side chains of the *T. brucei* PARP anchor following digestion with the
Bacteriodes fragilis endo-galactosidase (Ferguson et al., 1993).

Two chemical treatments are also commonly used for sequencing GPI glycans.
The first is partial acetolysis, which selectively cleaves the Manα1–6Man glycosidic
linkage in the conserved glycan backbone. GPI glycans are first O-acetylated in
pyridine:acetic anhydride, then subjected to acetolysis in acetic anhydride–acetic
acid–sulphuric acid (10:10:3 v/v; 37°C, 6–8 h). After recovery of the products by
chloroform-water partitioning, the glycans are de-O-acetylated in ammonia/methanol
(1:1 v/v) (Ferguson, 1992). Although the Manα1–6Man glycosidic linkage is the
most sensitive to acetolysis (>70% cleavage), the Manα1–4AHM bond is also
cleaved to some extent ($\sim$20%) (Mayor et al., 1990a; McConville et al., 1990a). As
a result, shorter acetolysis times are used to reduce the second cleavage and allow
detection of all acetolysis intermediates. It should be noted that some other glycosidic
linkages (i.e., Galβ1–3Hex), which occur in the parasite GPIs are also cleaved by
these acetolysis conditions (M. McConville, unpublished data).

Partial acid hydrolysis has also been used in glycan sequencing. For example,
the galactofuranosidic linkages that occur in several parasite GIPLs are selectively
cleaved with 40 mM trifluoroacetic acid (100°C, 1 h) (McConville et al., 1990a). If
these hydrolysis conditions are extended (100°C, 2–5 h), hexopyranosidic linkages
are partially cleaved, resulting in a nested series of glycan bands when analyzed by
HPTLC. If the glycans are linear, the number of bands generated by this procedure
equals the number of residues in the glycan (additional bands are generated by double
cleavages if the glycans are branched). Partial acid hydrolysis can also provide a
way of exposing internal sequences if the glycan contains residues that are resistant
to available exo- or endoglycosidases.

Methylation Analysis

Methylation analysis is required to identify the linkages that occur within the GPI
glycan structure. These analyses are usually performed on the HF-released neutral
glycans that have either been deaminated and reduced or N-acetylated to convert the
GlcN to 2,5-anhydromannitol or GlcNAc, respectively (Figure 4). The procedure of
Ciucanu and Kerek (1984) with modifications for analysis of 1–5 nmol of material
is commonly used (Ferguson, 1992). The glycans are permethylated in NaOH-DMSO
with three equivalents of methyliodide. After recovery by chloroform-water phase
partitioning, the permethylated glycans are subjected to acid hydrolysis by 0.25 M
H_2SO_4 in 93% acetic acid, 7% water (4 h, 80°C). These acid conditions are used to
hydrolyze the glycosidic linkage between the GlcNAc and inositol; milder conditions
can be used (2 M trifluoroacetic acid, 2 h, 100°C) if the glycan has been deaminated
and lacks other hexosamine residues. The released, partially O-methylated monosaccharides are converted into their alditol acetate derivatives by $NaBD_4$ reduction and
peracetylation with acetic anhydride. The partially methylated alditol acetates may
be identified by their retention times after gas chromatography as well as by electron
ionization (EI)-MS on the basis of their characteristic fragmentation patterns. If analyses are being carried out on small amounts of starting material (<1–2 nmol), many
of the background contaminants that are seen in methylation GC-MS analyses can
be removed by repurifying the permethylated glycan by reversed-phase HPLC

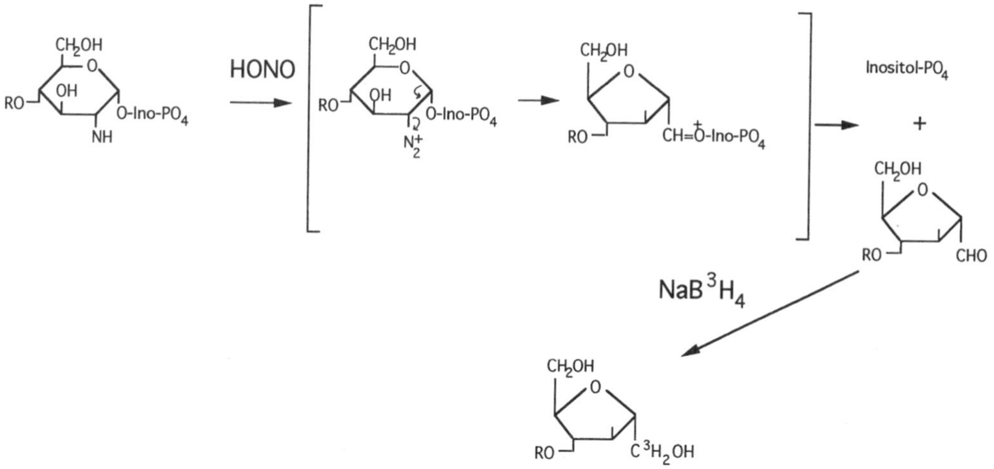

Figure 4 Ring contraction and cleavage of the GlcN-myoinositol linkage during nitrous acid deamination. Loss of nitrogen occurs after formation of an intermediate diazonium ion, with concomitant ring contraction to form 2,5 anhydromannose. Reduction with NaB^3H_4 results in the incorporation of a tritium label into the terminal 2,5 anhydromannitol (AHM) residue of the GPI glycan.

(Güther et al., 1992). This procedure requires that the glycan be radiolabeled for detection and can be used further to fractionate different glycoforms that differ in size. Methylation analysis has also been used to characterize GPI species that have been metabolically labeled with [^{3}H]mannose. The partially O-methylated monosaccharides released by trifluoroacetic acid can be analyzed directly by silica TLC using benzene-acetone-water-30% ammonia (50:200:3:1.5, v/v) as developing solvent (Mayor et al., 1990a).

C. Site of Attachment of Ethanolamine-phosphate and 2-Aminoethyl-phosphonate Residues

GPI anchors are characterized by having one or more ethanolamine-phosphate (or 2-AEP) residues, one of which acts as the bridging residue between the protein and the glycan. The number of these residues can be determined from the compositional analysis, by FAB or ESI mass spectrometry or by conventional anion-exchange chromatography of the deaminated glycans (in which the positively charged amino groups are removed) (Kamitani et al., 1992). Additional modification of these residues by radiomethylation (Roberts and Rosenberry, 1986; Deeg et al., 1992a) or dansylation (Menon et al., 1988) can be used to show whether or not they are substituted with peptide.

The location of the bridging ethanolamine-phosphate is commonly determined by assessing the susceptibility of the backbone mannose residues in the deaminated/ NaB^3H_4-reduced glycopeptide residues to jack bean α-mannosidase before and after the removal of the ethanolamine-phosphate with HF (Ferguson et al., 1988; Homans et al., 1988, Mayor et al., 1990a). This approach has recently been modified to allow the position of additional ethanolamine-phosphate substituents along the glycan backbone to be determined (Treuman et al., 1995). The deaminated/NaB^{3}H$_4$-reduced glycopeptide (or glycoprotein) is partially hydrolyzed (in 0.1 M trifluoroacetic acid, 100°C, 4 h) to generate a series of oligosaccharide fragments that retain any ethanolamine-phosphate substituents. This mixture is digested with α-mannosidase before and after HF dephosphorylation and the products analyzed by TLC (Figure 5). The location of ethanolamine phosphate residues can then be determined by identifying the TLC bands that are resistant to α-mannosidase digestion when the digestion is performed prior to the HF treatment (see Figure 5).

FAB or ESI tandem mass spectrometry of GPI-peptides (see above, Redman et al., 1994a; Roberts et al., 1988b) or, more commonly, the C-terminal glycopeptides (Deeg et al., 1992a; Frankhauser et al., 1993; Stahl et al., 1992) can be extremely useful for locating the position of these residues. Negative ion ESI-MS-MS of the reductively methylated C-terminal glycopeptide of human erythrocyte acetylcholinesterase was used to identify the position of the bridging ethanolamine phosphate and up to two additional ethanolamine-phosphate residues (Figure 1) (Deeg et al., 1992a). For FAB-MS analyses, permethylation of the C-terminal glycopeptide was found to increase the sensitivity of the analyses and generate diagnostic fragment ions. Prior treatment of the glycopeptide with hydrazinolysis and α-mannosidase to remove the peptide component and heterogeneity in the glycan moiety, respectively, also increased sensitivity. As with the intact GPI species, it is important to note that the nature of the fragment ions varies depending on whether the core GlcN residue is N-acetylated prior to methylation (Frankhauser et al., 1993).

D. NMR of GPI Glycans

Nuclear magnetic resonance (NMR) spectroscopy has been an invaluable tool in solving the structures of a variety of GPI anchors (Ferguson et al., 1988; Homans et al., 1988; Schneider et al., 1990; Frankhauser et al., 1993; Mehlert et al., 1992) and nonprotein-linked GPIs (McConville et al., 1990a,b; Thomas et al., 1992; Previato et al., 1990, Lederkremer et al., 1991; McConville and Homans, 1992; Jones et al., 1994; Routier et al., 1994). The nondestructive nature of NMR spectroscopy makes it the method of first choice, particularly for the ab initio structural determination of the complex and novel parasite GPIs. However, such analyses require relatively large amounts of material (~100 nmol) that frequently limit the use of NMR. In nearly all cases, NMR is performed on the delipidated C-terminal glycopeptide or glycan moiety to avoid the problems associated with micelle formation. One- and two-dimensional ^{1}H and ^{31}P NMR techniques have been used to define the anomeric configuration of constituent monosaccharides, glycan sequences, and the location of phosphoryl branch substituents. While the two-dimensional approaches are required for sequence information, one-dimensional NMR is more sensitive and provides a useful fingerprint that can be used to identify the presence of the conserved core and additional side-chain residues. Two-dimensional ^{1}H NMR has also been used, together with molecular orbital calculations and restrained molecular dynamic simulations, to define the conformation of the GPI glycans (Homans et al., 1989; Weller et al., 1994).

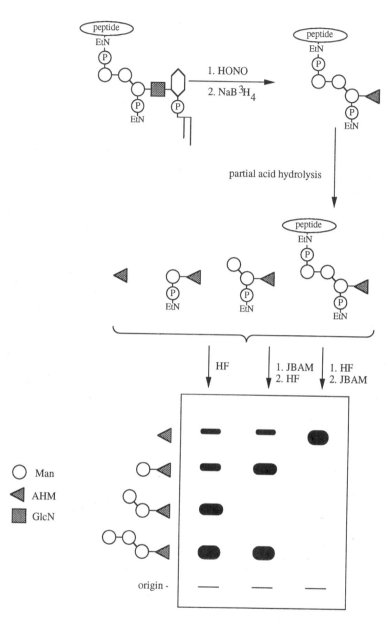

Fluorograph of HPTLC

Figure 5 A microsequencing approach to determine the location of ethanolamine-phosphate residues. Partial acid hydrolysis of the deaminated/NaB³H₄-reduced glycopeptides is used to generate a series of fragments, which form a ladder of bands when analyzed by HPTLC after HF dephosphorylation to remove ethanolamine-phosphate substituents. Treatment of the mixture with jack bean α-mannosidase (JBAM) prior to or after HF hydrolysis can be used to identify the fragments that contain a terminal phosphoryl substituent (and are therefore resistant to prior JBAM digestion) and those that lack a terminal modification (and are therefore digested) (Treuman et al., 1995).

VII.   CHARACTERIZATION OF THE GPI LIPID MOIETY

A.   Release of Lipid Moieties

The GPI lipid moieties can be released using a variety of enzymic or chemical treatments (Figure 6). For detection purposes, it is frequently desirable to incorporate a radiolabel metabolically or chemically into the released moiety. PI-PLC liberates diacyl, alkylacyl, and ceramide from GPI-anchored glycoproteins and GPI peptides that lack an acyl modification on the C(O)2 position of the inositol. It is important to note that some protein anchors are PI-PLC sensitive even though the inositol headgroup is acylated (e.g., human FBP), indicating that more than one type of inositol acylation exists (Luhrs and Slomiany, 1989). Conversely, some protein anchors are resistant to PI-PLC but are not inositol acylated (Haynes et al., 1993; Heise et al., 1995). Protein aggregation, micelle formation, or the presence of other post-translational modifications may contribute to this resistance in the absence of any inositol modification. In contrast, serum PI-PLD is unaffected by inositol acylation and should release phosphorylated 1,2-diacylglycerol or alkylacylglycerol (phosphatidic acid) from protein anchors containing a glycerolipid. Other lipases, such as

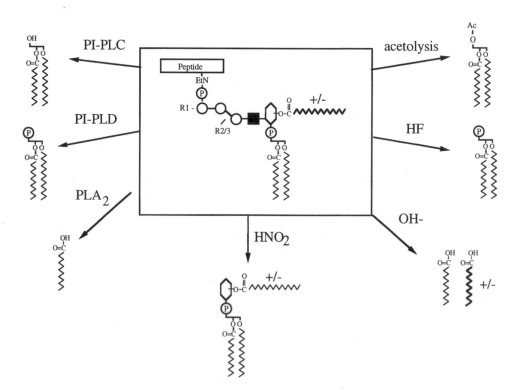

Figure 6   Strategies for releasing the lipid moieties of GPI glycolipids. Only the lipid products of each treatment are shown. Note that PI-PLC does not hydrolyze inositol-acylated GPIs. Also, only the phospholipase A_2 from *Croatalus atrox* is effective against inositol-acylated GPIs (bee venom phospholipase A_2 does not hydrolyze these species).

phospholipase A_2, are useful for determining the position of specific (radiolabeled) fatty acids (McConville and Bacic, 1989; Mayor et al., 1990b).

The most commonly used chemical procedures are nitrous acid deamination and HF treatment. Nitrous acid deamination releases the intact inositolphospholipid, including any fatty acids on the inositol headgroup, while HF releases the neutral glycerolipid or ceramide. Acetolysis also releases the glycerolipid and, like HF hydrolysis, can be used to differentiate between glycero- and inositol-linked fatty acids (Mayor et al., 1990b). All ester-linked fatty acids are released with mild base treatment (using 13 M ammonia in either methanol or 50% 1-propanol, 1:1 v/v or 0.1 M methanolic NaOH), while selective release of the inositol-linked fatty acids can be achieved with hydroxylamine treatment (0.8 M hydroxylamine hydrochloride, 0.1 M triethylamine, pH 10–12 for 2–24 h at 4°C) (Toutant et al., 1989). Stronger base conditions (1 M KOH, 100°C, 6 h) are used to release the ceramide from GPIs with inositol-phosphoceramide lipid moieties. The main advantage of chemical cleavage is that lipids are recovered free of detergent that interfere with subsequent mass spectrometric analyses. The lipids released by any of these enzymic or chemical treatments are readily recovered by solvent partitioning (e.g., in 1-butanol-water; toluene-water; acidified chloroform-water).

B. Mass Spectrometry of Released Lipids

Following nitrous acid deamination, the released inositolphospholipid moiety can be directly analyzed by either FAB-MS or ESI-MS (Roberts et al., 1988b; Redman et al., 1994b; Heise et al., 1995; Treumann et al., 1995). This is the most powerful approach for analyzing the GPI lipid moiety as it provides information on the composition of individual molecular species. This information, which is usually lost when the constituent lipid components are released by acid or base cleavage, is important for understanding the selectivity of biosynthetic enzymes and/or the origin of GPI precursor pools. Negative ion ESI tandem mass spectrometry of the released PI species gives spectra with several groups of diagnostic ions: (a) the $[M\!-\!H]^-$ pseudomolecular ion, (b) the fatty acid carboxylate ions $[RCO_2]^-$ that are usually the base peak, (c) a monoalkylglycerol-phosphate fragment if the PI contains alkyacylglycerol, and (d) low-mass ions corresponding to $[PO_3]^-$ and inositol-1,2-cyclic phosphate (Treumann et al., 1995; Acosta Serrano et al., 1995). Interestingly, the carboxylate ions derived from inositol-linked fatty acids were much weaker than those derived from the glycero lipids, which may be useful in defining the position of specific fatty acids. The ESI-MS spectrum of inositolphospho-ceramides are somewhat less informative, since the stable amide bond within the ceramide prevents the formation of the carboxylate ions, and the major fragment ions are low-mass phosphate and inositol-phosphate ions (Acosta Serrano et al., 1995). FAB tandem mass spectrometry has been used extensively to analyze other classes of phospholipids and may provide more informative fragment ions (Evershed, 1992; Domon and Costello, 1988). Diagnostic daughter ions derived from acylated-PI lipid of human acetylcholinesterase corresponded to loss of palmitoyl-inositol and phosphatidic acid, which localized the palmitoyl group to the inositol ring (Roberts et al., 1988b). Finally, degradative methodologies (such as base or lipase treatment) can be used to corroborate the FAB- or ESI-MS assignments and provide additional information unobtainable by analysis of the intact PI species.

To identify which hydroxyl(s) in the inositol ring are substituted with fatty acid in acylated-PI species, the inositol ring can be selectively degraded by periodate oxidation and the products generated after release of the inositol fatty acid and glycero lipid, by base and HF treatments, respectively, analyzed by GC-MS (Ferguson, 1992b). This procedure was used to show that the palmitoyl group was attached to either the 2- or 3-positions of inositol in the anchor of *T. brucei* PARP.

GC-MS has also been used to analyze the mixtures of diacylglycerols and 1-*O*-alkylglycerols that are released by PI-PLC digestion from some GPIs. These lipids can be analyzed as their acetate or, more commonly, trimethylsilyl- (TMS) or *tert*-butyldimethylsilyl (TBDMS) derivatives (McConville and Bacic, 1989; Evershed, 1992). The TMS and TBDMS derivatives of 1,2-diacylglycerols and 1-*O*-alkylglycerols have excellent GC properties and produce useful electron-impact mass spectra. Individual molecular species can be resolved on either short (for diacylglycerols) or normal (for 1-*O*-alkylglycerols) fused silica capillary columns coated with apolar stationary phases. In contrast, the 1-*O*-alkyl-2-acylglycerol molecular species tend to be analyzed as their acetate derivatives (Roberts et al., 1988a), since recoveries of the TMS derivatives can be very low.

C. TLC and HPLC Analysis of Released Lipids

TLC and high-performance TLC has been used for both the purification and the identification of released lipids. Some solvent systems that have been used to resolve different lipid moieties and derivatives are listed in Table 2. TLC is commonly used to analyze the PI species released after nitrous acid deamination (i.e., to determine whether they are inositol acylated) and the neutral glycerolipids released by phospholipases or HF treatment. For separation of glycerolipids into subclasses (i.e., diacyl, alkylacyl, etc.), the released lipids are analyzed as their benzoate derivatives (Bütikofer et al., 1990; Heise et al., 1995).

Reversed-phase HPLC has been used to characterize the diacyl- and alkylacylglycerol molecular species of several GPI protein anchors (Bütikofer et al., 1990, 1992). Prior to HPLC analysis, the diradyglycerols released by PI-PLC are converted to their benzoate derivatives and separated into their respective subclasses (diacyl, alkylacyl, and alk-1-enyl types) by TLC. The molecular species in each subclass were then analyzed by reversed-phase HPLC using a mobile phase of acetonitrile-2-propanol (8:2 v/v). The elution position on reversed-phase HPLC is dependent mainly on the combined chain length, with retention being modified by presence of double bonds. The lipid composition of individual molecular species can be identified by the retention times relative to defined standards or more effectively by interfacing the HPLC with a mass spectrometer that enables comigrating species to be identified and quantified (Bütikofer et al., 1990, 1992). Fatty acid migration during release or derivatization may complicate these analyses, for the presence of both 1,2- and 1,3-diradylglycerols results in the appearance of doublet peaks in the HPLC profile (Bütikofer et al., 1992).

VIII. CONCLUDING REMARKS

More than 40 GPI structures (including protein anchors and free GPIs) have been partially or completely characterized, and the analytical strategies for dealing with this class of glycolipids are now well developed. While studies on the GPI anchors of *T. brucei* VSG and human Thy-1 required mmolar amounts of starting material and the use of a wide variety of techniques, including NMR, for complete characterization (Ferguson et al., 1988; Homans et al., 1988), it is now possible to characterize a protein anchor with 1–20 nmol of starting material. Two factors that have contributed to the increased sensitivity of these analyses are the refinement of the enzyme and chemical microsequencing strategies and the application of mass spectrometric techniques such as FAB-MS and more recently ESI-MS. Microsequencing can be used to define some of the major structural features of the GPI glycan moiety and can be carried out in most laboratories without access to expensive instrumentation. On the other hand, mass spectrometry can be used to analyze intact GPI or GPI-peptide species without prior degradation and provides information on the composition and sequence of the peptide, glycan, and lipid moieties. Importantly, mass spectrometry can be used to characterize individual molecular species and thus the extent of glycan and lipid heterogeneity. However, it should be emphasized that a combination of techniques is still required for the complete elucidation of GPI structure, because new structural features are constantly encountered in this diverse class of biologically important glycoconjugates.

ABBREVIATIONS

GPI glycosylphosphatidylinositol, GIPL, glycoinositol phospholipid; GPI-peptide, the GPI anchor moiety with a short C-terminal peptide attached; C-terminal glycopeptide, the delipidated GPI-peptide; GC-MS, gas chromatography-mass spectrometry; FAB-MS, fast-atom-bombardment mass spectrometry; ESI-MS, electrospray-ionization mass spectrometry; CID, collision-induced dissociation; PI, phosphatidylinositol; HPAEC, high-pH anion-exchange chromatography; AHM, anhydromannitol; PLC, phospholipase C; PLD, phospholipase D; PLA$_2$, phospholipase A$_2$; PARP, procyclic acidic repeat protein.

ACKNOWLEDGMENTS

This work was supported by an Australian Wellcome Trust Senior Research Fellowship and the Australian National Health and Medical Research Council.

REFERENCES

Acosta Serrano, A., Schenkmann, S., Yoshida, N., Mehlert, A., Richardson, J. M., and Ferguson, M. A. J. (1995). The lipid structure of the GPI-anchored mucin-like sialic acid acceptors of *Trypanosoma cruzi* changes during parasite differentiation from epimastigotes to infective metacyclic trypomastigote forms. *J. Biol. Chem.* 270, 27244–27253.

Baldwin, M. A., Stahl, N., Reinder, L. G., Gibson, B. W., Prusiner, S. B., and Burlingame, A. L. (1990). Permethylation and tandem mass spectrometry of oligosaccharides having free hexosamine: Analysis of the glycoinositol phospholipid anchor glycan from the Scrapie prion protein. *Anal. Biochem.* 191, 174–182.

Bordier, C. (1981). Phase separation of integral membrane proteins in Triton X-114 solution. *J. Biol. Chem. 256*, 1604–1607.

Brewis, I. A., Ferguson, M. A. J., Turner, A. J., and Hooper, N. M. (1992a). Structural determination of the glycolipid anchors of human and porcine membrane dipeptidases. *Biochem. Soc. Trans. 21*, 46S.

Brewis, I. A., Hooper, N. M., and Turner, A. J. (1992b). Identification of the site of attachment of the glycolipid anchor in porcine membrane dipeptidase. *Biochem. Soc. Trans. 21*, 44S.

Bütikofer, P., Kuypers, F. A., Shackleton, C., Brodbeck, U., and Stieger, S. (1990). Molecular species analysis of the glycosylphosphatidylinositol anchor of *Torpedo marmorata* acetylcholinesterase. *J. Biol. Chem. 265*, 18,983–18,987.

Bütikofer, P., Zollinger, M., and Brodbeck, U. (1992). Alkylacyl glycerophosphatidylinositol in human and bovine erythrocytes: molecular species composition and comparison with glycosyl-inositolphospholipid anchors of erythrocyte acetylcholinesterases. *Eur. J. Biochem. 208*, 677–683.

Bütokofer, P., Boschung, M., and Menon, A. K. (1995). Production of a nested set of glycosylphosphatidylinositol structures from a glycosylphosphatidylinositol-anchored protein. *Anal. Biochem. 229*, 125–132.

Ciucanu, I., and Kerek, F. (1984). A simple and rapid method for the permethylation of carbohydrates. *Carbohydr. Res. 131*, 209–217.

Conzelmann, A., Riezman, H., Desponds, C., and Bron, C. (1988). A major 125-kd membrane glycoprotein of *Saccharomyces cerevisiae* is attached to the lipid bilayer through an inositol-containing phospholipid. *EMBO J. 7*, 2233–2240.

Conzelmann, A., Puoti, A., Lester, R. L., and Desponds, C. (1992). Two different types of lipid moieties are present in glycophosphoinositol-anchored membrane proteins of *Saccharomyces cerevisae. EMBO J. 11*, 457–466.

Couto, A. S., Lederkremer, de, R. M., Colli, W., and Alves, M. J. M. (1993). The glycosylphosphatidylinositol anchor of the trypomastigote-specific Tc-85 glycoprotein from *Trypanosoma cruzi*. Metabolic-labeling and structural studies. *Eur. J. Biochem. 217*, 597–602.

Cross, G. A. M. (1990). Glycolipid anchoring of plasma membrane proteins. *Annu. Rev. Cell Biol. 6*, 1–39.

Deeg, M. A., Humphrey, D. R., Tang, S. H., Ferguson, T. R., Reinhold, V. N., and Rosenberry, T. L. (1992a). Glycan components in the glycoinositol phospholipid anchor of human erythrocyte acetylcholinesterase: Novel fragments produced by trifluoroacetic acid. *J. Biol. Chem. 267*, 18,573–18,580.

Deeg, M. A., Murray, N. R., and Rosenberry, T. L. (1992b). Identification of glycoinositol phospholipids in rat liver by reductive radiomethylation of amines but not in H4IIE hepatoma cells or isolated hepatocytes by biosynthetic labeling with glucosamine. *J. Biol. Chem. 267*, 18,581–18,588.

Dell, A., Reason, A. J., Khoo, R.-H., Panico, M., McDowell, R. A., and Morris, H. R. (1994). Mass spectrometry of carbohydrate-containing biopolymers. *Meth. Enzymol. 230*, 108–131.

Domon, B., and Costello, C. E. (1988). Structural elucidation of glycosphingolipids and gangliosides using high performance tandem mass spectrometry. *Biochemistry 27*, 1534–1543.

Evershed, R. P. (1992). Mass spectrometry of lipids. In *Lipid Analysis: A Practical Approach* (Hamilton, J. R., and Hamilton, S., eds.), pp. 263–308, IRL Press, New York.

Ferguson, M. A. J. (1992a). Chemical and enzymic analysis of glycosyl-phosphatidylinositol anchors. In *Lipid Modification of Proteins: A Practical Approach*, (Hooper, N. M., and Turner, A. J., eds.), pp. 191–230, IRL Press, New York.

Ferguson, M. A. J. (1992b). Site of palmitoylation of a phospholipase C-resistant glycosyl-phosphatidylinositol membrane anchor. *Biochem. J. 284*, 297–300.

Ferguson, M. A. J., Halder, K., and Cross, G. A. M. (1985a). *Trypanosoma brucei* variant surface glycoprotein has a *sn*-1,2-dimyristyl glycerol membrane anchor at its COOH terminus. *J. Biol. Chem. 260*, 4963–4968.

Ferguson, M. A. J., Low, M. G., and Cross, G. A. M. (1985b). Glycosyl-*sn*-1,2-dimyristoyl-phosphatidylinositol is covalently linked to *Trypanosoma brucei* variant surface glycoprotein. *J. Biol. Chem. 260*, 14,547–14,555.

Ferguson, M. A. J., Homans, S. W., Dwek, R. A., and Rademacher, T. W. (1988). Glycosyl-phosphatidylinositol moiety that anchors *Trypanosoma brucei* variant surface glycoprotein to the membrane. *Science 239*, 753–759.

Ferguson, M. A. J., Murray, P., Rutherford, H., and McConville, M. J. (1993). A simple purification of procyclic acidic repetitive protein and demonstration of a sialylated glycosyl-phosphatidylinositol membrane anchor. *Biochem. J. 291*, 51–55.

Frankhauser, C., Homans, S. W., Thomas-Oates, J. E., McConville, M. J., Desponds, C., Conzelmann, A., and Ferguson, M. A. J. (1993). Structures of glycosylphosphatidylinositol membrane anchors from *Saccharomyces cerevisiae*. *J. Biol. Chem. 268*, 26,365–26,374.

Gerold, P., Schofield, L., Blackman, M. J., Holder, A. A., and Schwarz, R. T. (1995). Structural analysis of the glycosyl-phosphatidylinositol membrane anchor of the merozoite surface proteins-1 and -2 of *Plasmodium falciparum*. *Mol. Biochem. Parasitol. 75*, 131–143.

Güther, M. L. S., Cardoso de Almeida, M. L., Yoshida, N., and Ferguson, M. A. J. (1992). Structural studies on the glycosylphosphatidylinositol membrane anchor of the *Trypanosoma brucei* 1G7 antigen: Structure of the glycan core. *J. Biol. Chem. 267*, 6820–6828.

Güther, M. L. S., Cardoso de Almeida, M. L., Rosenberry, T. L., and Ferguson, M. A. J. (1994). The detection of phospholipase-resistant and -sensitive glycosyl-phosphatidylinositol membrane anchors by Western blotting. *Anal. Biochem. 219*, 249–255.

Haas, R., Brandt, P. T., Knight, J., and Rosenberry, T. L. (1986). Identification of amine components in a glycolipid membrane-binding domain at the C-terminus of human erythrocyte acetylcholinesterase. *Biochemistry 25*, 3098–3105.

Hardy, M. R. and Townsend, R. R. (1994). High-pH anion exchange chromatography of glycoprotein-derived carbohydrates. *Meth. Enzymol. 230*, 208–225.

Haynes, P. A., Gooley, A. A., Ferguson, M. A. J., Redmond, J. W., and Williams, K. L. (1993). Post-translational modifications of the *Dictyostelium discoideum* glycoprotein PsA: Glycosylphosphatidylinositol membrane anchor and composition of *O*-linked oligosaccharides. *Eur. J. Biochem. 216*, 729–737.

Heise, N., Cardoso de Almeida, M. L., and Ferguson, M. A. J. (1995). Characterization of the lipid moiety of the glycosylphosphatidylinositol anchor of *Trypanosoma cruzi* 1G7 antigen. *Mol. Biochem. Parasitol. 70*, 71–84.

Homans, S. W., Ferguson, M. A. J., Dwek, R. A., Rademacher, T. W., Anand, R., and Williams, A. F. (1988). Complete structure of the glycosyl phosphatidylinositol membrane anchor of rat brain Thy-1 glycoprotein. *Nature 333*, 269–272.

Homans, S. W., Edge, C. J., Ferguson, M. A. J., Dwek, R. A., and Rademacher, T. W. (1989). Solution structure of the glycosylphosphatidylinositol membrane anchor glycan of *Trypanosoma brucei* variant surface glycoprotein. *Biochemistry 28*, 2881–2887.

Hooper, N. M. (1992). Identification of a glycosyl-phosphatidylinositol anchor on membrane anchors. In *Lipid Modification of Protein: A Practical Approach* (Hooper, N. M., and Turner, A. J., eds.), pp. 89–115, IRL Press, New York.

Ilg, T., Etges, R., Overath, P., McConville, M. J., Thomas-Oates, J. E., Homans, S. W., and

Ferguson, M. A. J. (1992). Structure of *Leishmania mexicana* lipophosphoglycan. *J. Biol. Chem. 267*, 6834–6840.

Jager, K., Meyer, P., Stieger, S., and Brodbeck, U. (1990). Production and characterization of antibodies against the cross-reacting determinant of glycosyl-phosphatidylinositol-anchored acetylcholinesterase. *Biochim. Biophys. Acta 1039*, 367–373.

Jones, C., Previato, J. O., Mendonça-Previato, L., and Wait, R. (1994). The use of NMR spectroscopy in the structure determination of a *Leptomonas samueli* glycosyl phosphosphingolipid-derived oligosaccharide. *Brazilian J. Med. Biol. Res. 27*, 219–226.

Kamitani, T., Menon, A. K., Hallaq, Y., Warren, C. D., and Yeh, E. T. H. (1992). Complexity of ethanolamine phosphate addition in the biosynthesis of glycosylphosphatidylinositol anchors in mammalian cells. *J. Biol. Chem. 267*, 24,611–24,619.

Lederkremer, R. M., Casal, O. L., Alves, M. J. M., and Colli, W. (1980). Evidence for the presence of D-galactofuranose in the lipopeptidophosphoglycan from *Trypanosoma cruzi. FEBS Lett. 116*, 25–29.

Lederkremer, de R. M., Lima, C., Tamirez, M. L., Ferguson, M. A. J., Homans, S. W., and Thomas-Oates, J. (1991). Complete structure of the glycan of lipopeptidophosphoglycan from *Trypanosoma cruzi* epimastigotes. *J. Biol. Chem. 266*, 23,670–23,675.

Lederkremer, de R. M., Lima, C. E., Ramirez, M. L., Gonçalvez, M. F., and Colli, W. (1993). Hexadecylpalmitoylglycerol or ceramide is linked to similar glycophosphoinositol anchor-like structures in *Trypanosoma cruzi. Eur. J. Biochem. 218*, 929–936.

Lisanti, M. P., Field, M. C., Caras, I. W., Menon, A. K., and Rodriguez-Boulan, E. (1991). Mannosamine, a novel inhibitor of glycosyl-phosphatidylinositol incorporation into proteins. *EMBO J. 10*, 1969–1977.

Low, M. G. (1992). Phospholipases that degrade glycosylphosphatidylinositol anchors of membrane proteins. In *Lipid Modification of Protein: A Practical Approach* (Hooper, N. M., and Turner, A. J., eds.), pp. 117–154, IRL Press, New York.

Low, M. G., and Kincade, P. W. (1985). Phosphatidylinositol is the membrane-anchoring domain of the Thy-1 glycoprotein. *Nature 318*, 62–64.

Luhrs, C. A., and Slomiany, B. L. (1989). A human membrane-associated folate binding protein is anchored by a glycosyl-phosphatidylinositol tail. *J. Biol. Chem. 264*, 21,446–21,449.

Manzi, A. E., and Varki, A. (1993). Compositional analysis of glycoproteins. In *Glycobiology, A Practical Approach* (Fukuda, M., and Kobata, A., eds.), pp. 27–78, Oxford University Press, New York.

Masterson, W. J., Doering, T. L., Hart, G. W., and Englund, P. T. (1989). A novel pathway for glycan assembly: Biosynthesis of the glycosyl phosphatidylinositol anchor of the trypanosome variant surface glycoprotein. *Cell 56*, 793–800.

Mayor, S., Menon, A. K., Cross, G. A. M., Ferguson, M. A. J., Dwek, R. A., and Rademacher, T. W. (1990a). Glycolipid precursors for the membrane anchor of *Trypanosoma brucei* variant surface glycoproteins. I. Glycan structure of the phosphatidylinositol-specific phospholipase C sensitive and resistant glycolipids. *J. Biol. Chem. 265*, 6164 –6173.

Mayor, S., Menon, A. K., and Cross, G. A. M. (1990b). Glycolipid precursors for the membrane anchor of *Trypanosoma brucei* variant surface glycoproteins. II. Lipid structure of phosphatidylinositol-specific phospholipase C sensitive and resistant glycolipids. *J. Biol. Chem. 265*, 6174–6181.

McConville, M. J., and Blackwell, J. M. (1991). Developmental changes in the glycosylated phosphatidylinositols of *Leishmania donovani*. Characterization of the promastigote and amastigote glycolipids. *J. Biol. Chem. 266*, 15,170–15,179.

McConville, M. J., and Ferguson, M. A. J. (1993). The structure, biosynthesis and function of glycosylated phosphatidylinositols in the parasitic protozoa and higher eukaryotes. *Biochem. J. 294*, 305–324.

McConville, M. J., and Homans, S. W. (1992). Identification of the defect in lipophosphoglycan biosynthesis in a non-pathogenic strain of *Leishmania major. J. Biol. Chem. 267,* 5855–5861.

McConville, M. J., Homans, S. W., Thomas-Oates, J. O., Dell, A., and Bacic, A. (1990a). Structures of the glycosylinositolphospholipids from *Leishmania major.* A family of novel galactofuranose-containing glycolipids. *J. Biol. Chem. 265,* 7385–7394.

McConville, M. J., Thomas-Oates, J. E., Ferguson, M. A. J., and Homans, S. W. (1990b). Structure of the lipophosphoglycan from *Leishmania major. J. Biol. Chem. 265,* 19,611–19,623.

McConville, M. J., Turco, S. J., Ferguson, M. A. J., and Sacks, D. L. (1992). Developmental modification of lipophosphoglycan during differentiation of *Leishmania major* promastigotes to an infectious stage. *EMBO J. 11,* 3593–3600.

McConville, M. J., Collidge, T. A. C., Ferguson, M. A. J., and Schneider, P. (1993). The glycoinositol phospholipids of *Leishmania mexicana* promastigotes. Evidence for the presence of three distinct pathways of glycolipid biosynthesis. *J. Biol. Chem. 268,* 15,595–15,604.

McConville, M. J., Schnur, L. F., Jaffe, C., and Schneider, P. (1995). Structure of *Leishmania* lipophosphoglycan: Inter- and intra-specific polymorphism in Old World species. *Biochem. J. 310,* 807–818.

Mehlert, A., Silman, I., Homans, S. W., and Ferguson, M. A. J. (1992). The structure of the glycosylphosphatidyl inositol anchor from *Torpedo californica* acetylcholinesterase. *Biochem. Soc. Trans. 21,* 43S.

Menon, A. K. (1994). Structural analysis of glycosylphosphatidylinositol anchors. *Meth Enzymol. 230,* 418–442.

Menon, A. K., Mayor, S., Ferguson, M. A. J., Duszenko, M., and Cross, G. A. M. (1989). Candidate glycophospholipid precursor for the glycosylphosphatidylinositol membrane anchor of *Trypanosoma brucei* variant surface glycoprotein. *J. Biol. Chem. 263,* 1970–1977.

Menon, A. K., Mayor, S, Ferguson, M. A. J., Duszenko, M., and Cross, G. A. M. (1988). Candidate glycophospholipid precursor for the glycosylphosphatidylinositol membrane anchor of *Trypanosoma brucei* variant surface glycoprotein. *J. Biol. Chem. 263,* 1970–1977.

Menon, A. K., Schwarz, R. T., Mayor, S., and Cross, G. A. M. (1990). Cell-free synthesis of glycosyl-phosphatidylinositol precursors for the glycolipid membrane anchor of *Trypansoma brucei* variant surface glycoproteins. Structural characterization of putative biosynthetic intermediates. *J. Biol. Chem. 265,* 9033–9042.

Milne, K. G., Field, R. A., Masterson, W. J., Cottaz, S., Brimacombe, J. S., and Ferguson, M. A. J. (1994). Partial purification and characterization of the *N*-acetylglucosaminyl-phosphatidylinositol de-*N*-acetylase of glycosylphosphatidylinositol anchor biosynthesis in African trypanosomes. *J. Biol. Chem. 269,* 16,403–16,408.

Moran, P., Roab, H., Kohr, W. J., and Caras, I. (1991). Glycophospholipid membrane anchor attachment. Molecular analysis of the cleavage/attachment site. *J. Biol. Chem. 266,* 1250–1257.

Nakano, Y., Noda, K., Endo, T., Kobata, A., and Tomita, M. (1994). Structural study on the glycosyl-phosphatidylinositol anchor and the asparagine-linked sugar chain of a soluble form of CD59 in human urine. *Arch. Biochem. Biophys. 311,* 117–126.

Olandi, P. A., Jr., and Turco, S. J. (1987). Structure of the lipid moiety of the *Leishmania donovani* lipophosphoglycan. *J. Biol. Chem. 262,* 10,384–10,391.

Previato, J. O., Gorin, P. A. J., Mazurek, M., Xavier, M. T., Fournet, B., Wieruszesk, J. M., and Mendonça-Previato, L. (1990). Primary structure of the oligosaccharide chain of lipopeptidophosphoglycan of epimastigote forms of *Trypanosoma cruzi. J. Biol. Chem. 265,* 2518–2526.

Previato, J. O., Mendonça-Previato, L., Jones, C., Wait, R., and Fournet, B. (1992). Structural characterization of a novel class of glycophosphosphingolipids from the protozoan *Leptomonas samueli*. *J. Biol. Chem. 267*, 24,279–24,286.

Previato, J. O., Jones, C., Xavier, M. T., Wait, R., Travassos, L. R., Parodi, A. J., and Mendonça-Previato, L. (1995). Structural characterization of the major glycosylphosphatidylinositol membrane anchored-glycoprotein from epimastigote forms of *Trypanosoma cruzi* T-strain. *J. Biol. Chem. 270*, 7241–7250.

Puoti, A., and Conzelmann, A. (1992). Structural characterization of free glycolipids which are potential precursors for glycophosphatidylinositol anchors in mouse thymoma cell lines. *J. Biol. Chem. 267*, 22,673–22,680.

Puoti, A., Desponds, C., Frankhauser, C., and Conzelmann, A. (1991). Characterization of a glycosylphospholipid intermediate in the biosynthesis of glycosylphosphatidylinositol anchors accumulating in the Thy-1-negative lymphoma line SIA-b. *J. Biol. Chem. 266*, 21,051–21,059.

Redman, C. A., Green, B. N., Thomas-Oates, J. E., Reinhold, V. N., and Ferguson, M. A. J. (1994a). Analysis of glycosylphosphatidylinositol membrane anchors by electrospray ionization-mass spectrometry and collision induced dissociation. *Glycoconjugate J. 11*, 187–193.

Redman, C. A., Thomas-Oates, J. E., Ogata, S., Ikehara, Y., and Ferguson, M. A. J. (1994b). Structure of the glycosylphosphatidylinositol membrane anchor of human placental alkaline phosphatase. *Biochem. J. 302*, 861–865.

Roberts, W. L., and Rosenberry, T. L. (1986). Selective radiolabeling and isolation of the hydrophobic membrane-binding domain of human erythrocyte acetylcholinesterase. *Biochemistry 25*, 3091–3098.

Roberts, W. L., Myher, J. J., Kuksis, A., Low, M. G., and Rosenberry, T. L. (1988a). Lipid analysis of the glycoinositol phospholipid membrane anchor of human erythrocyte acetylcholinesterase. Palmitoylation of inositol results in resistance to phosphatidylinositol-specific phospholipase C. *J. Biol. Chem. 263*, 18,766–18,775.

Roberts, W. L., Santikarn, S., Reinhold, V. N., and Rosenberry, T. L. (1988b). Structural characterization of the glycoinositol phospholipid membrane anchor of human erythrocyte acetylcholinesterase by fast atom bombardment mass spectrometry. *J. Biol. Chem. 263*, 18,776–18,784.

Routier, F., Mendonça-Previato, J. O., Jones, C., and Wait, R. (1994). Structures of four oligosaccharides derived from the glycoinositolphospholipid of *Leishmania adleri*. *Brazilian J. Med. Res. 27*, 211–217.

Routier, F. H., da Silveira, E. X., Wait, R., Jones, C., Previato, J. O., and Mendonça-Previato, L. (1995). Chemical characterization of glycosylinositolphospholipids of *Herpetomonas samuelpessoai*. *Mol. Biochem. Parasitol. 69*, 81–92.

Schneider, P., Ferguson, M. A. J., McConville, M. J., Mehlert, A., Homans, S. W., and Bordier, C. (1990). Structure of the glycosyl-phosphatidylinositol membrane anchor of the *Leishmania major* promastigote surface protease. *J. Biol. Chem. 265*, 16,955–16,964.

Schneider, P., Ralton, J. E., McConville, M. J., and Ferguson, M. A. J. (1993). Analysis of the neutral glycan fragments of glycosyl-phosphatidylinositols by thin-layer chromatography. *Anal. Biochem. 210*, 106–112.

Schneider, P., Treumann, A., Milne, K. G., McConville, M. J., Zitmann, N., and Ferguson, M. A. J. (1995). Structural studies on a lipoarabinogalactan of *Crithidia fasciculata*. *Biochem. J. 313*, 963–971.

Sevlever, D., Humphrey, D. R., and Rosenberry, T. L. (1995). Compositional analysis of glucosaminyl(acyl)phosphatidylinositol accumulated in HeLa S3 cells. *Eur. J. Biochem. 233*, 384–394.

Stadler, J., Keenan, T. W., Bauer, G., and Gerisch (1989). The contact site. A glycoprotein of *Dictyostelium discoideum* carries a phospholipid anchor of a novel type. *EMBO J. 8*, 371–377.

Stahl, N., Borchelt, D. R., Hsiao, K., and Prusiner, S. B. (1987). Scrapie prion protein contains a phosphatidylinositol glycolipid. *Cell 51*, 229–240.

Stahl, N., Baldwin, M. A., Hecker, R., Pan, K.-M., Burlingame, A. L., and Prusiner, S. B. (1992). Glycosylinositol phospholipid anchors of the Scrapie and cellular prion proteins contain sialic acid. *Biochemistry 31*, 5043–5053.

Taguchi, R., Hamakawa, N., Harada-Nishida, M., Fukui, T., Nojima, K., and Ikezama, H. (1994). Microheterogeneity in glycosylphosphatidylinositol anchor structures of bovine liver 5'-nucleotidase. *Biochemistry 33*, 1017–1022.

Thomas, J. R., McConville, M. J., Thomas-Oates, J. E., Homans, S. W., Ferguson, M. A. J., Gorin, P. A. J., Greis, K. D., and Turco, S. J. (1992). Refined structure of the lipophosphoglycan of *Leishmania donovani. J. Biol. Chem. 267*, 6829–6833.

Toutant, J.-P., Roberts, W. L., Murray, N. R., and Rosenberry, T. L. (1989). Conversion of human erythrocyte acetylcholinesterase from an amphiphilic to a hydrophilic form by phosphatidylinositol-specific phospholipase C and serum phospholipase D. *Eur. J. Biochem. 180*, 503.

Treumann, A., Lifely, M. R., Schneider, P., and Ferguson, M. A. J. (1995). Primary structure of CD52. *J. Biol. Chem. 270*, 6088–6099.

Turco, S. J., and Descoteaux, A. (1992). The lipophosphoglycan of *Leishmania* parasites. *Annu. Rev. Microbiol. 46*, 65–94.

Wait, R., Jones, C., Reviato, J. O., and Mendonça-Previato, L. (1994). Characterization of phosphoinositol oligosaccharides from parasitic protozoa by fast atom bombardment and collisional activation mass spectrometry. *Brazilian J. Med. Biol. Res. 27*, 203–210.

Weller, C. T., McConville, M., and Homans, S. W. (1994). Solution structure and dynamics of a glycoinositol phospholipid (GIPL-6) from *Leishmania major. Biopolymers 34*, 1155–1163.

Zamze, S. E., Ferguson, M. A. J., Collins, R., Dwek, R. A., and Rademacher, T. W. (1988). Characterization of the cross-reacting determinant (CRD) of the glycosyl-phosphatidylinositol membrane anchor of *Trypanosoma brucei* variant surface glycoprotein. *Eur. J. Biochem. 176*, 527–534.

9

Synthesis of Intermediates in the Dolichol Pathway of Protein Glycosylation

Vladimir N. Shibaev and Leonid L. Danilov
N. D. Zelinsky Institute of Organic Chemistry,
Russian Academy of Sciences, Moscow, Russia

I. INTRODUCTION

The dolichol pathway of glycoprotein glycosylation represents a series of enzymic reactions that occurs during biosynthesis of asparagine-linked carbohydrate chains in glycoproteins. As a result of the process, which takes place in the endoplasmic reticulum membranes of eukaryotic cells, the tetradecasaccharide Glc_3Man_9-$(GlcNAc)_2$, linked to dolichyl diphosphate, is assembled. After transfer of the oligosaccharide block onto a protein molecule, processing of the oligosaccharide chains and addition of terminal monosaccharide residues occur, leading to the large variety of structures present in glycoproteins of this class.

Participation of dolichol-linked sugars as intermediates in the process was first demonstrated in 1970 by Behrens and Leloir [1]. The details of the complicated reaction sequence were clarified toward the beginning of the '80s and discussed in a number of excellent reviews [2–7]. Initially identified in mammalian tissues and cell cultures, the dolichol pathway was shown to be common to all eukaryotic organisms [8,9]. Particular attention was devoted to the investigation of this pathway in yeast (for reviews see Refs. 10–13).

In this case, as well as in other fungi, a derivative of dolichol, dolichyl mannosyl phosphate (Dol-P-Man) was found to serve as a glycosyl donor in the transfer of the first mannosyl residue during the formation of O-linked oligomannose chains, characteristic of glycoproteins of these organisms. By contrast, in higher eukaryotes glycoprotein O-linked carbohydrate chains are formed without involvement of dolichyl-linked sugars. More recently, participation of Dol-P-Man in the biosynthesis of the mannosyl core of glycosyl phosphatidylinositol anchors was demonstrated (for reviews see Refs. 12,14); the process seems to be widely distributed in different organisms. Intermediate formation of dolichyl-linked sugars was suggested for some plant polysaccharide biosynthesis [9] and was shown to occur in the biosynthesis of archaebacterial cell-wall polymers and glycoproteins present in bacterial S-layers [15–17].

For biosynthesis of carbohydrate chains in prokaryotic cell-wall and membrane polymers (including peptidoglycan, teichoic acids, O-specific chains of lipopolysaccharides), as well as of many capsular and exopolysaccharides, the participation of lipid-linked sugar intermediates is also characteristic, although in these cases bacterial undecaprenol, instead of dolichol, serves as a lipid carrier. These processes, first discovered as early as in 1967 [18,19], were intensively studied and are similar in many respects to the dolichol pathway of protein glycosylation (for reviews see Refs. 20–22).

One of the common features of all biochemical pathways mentioned here is a rather low steady-state concentration of dolichyl (polyprenyl) mono- and diphosphate sugar intermediates under conditions normally employed for biochemical experiments. As a rule, these derivatives were identified by chromatographic methods with the use of a radioactive label. Studies of chemical synthesis in this field are greatly stimulated by the requirement of synthetic standards for definite identification of the biosynthetic products and for unequivocal determination of their structure. The availability of chemically defined substances in this series is of great help in unraveling the enzymic reaction sequences within the biosynthetic pathways, and is a prerequisite for assay and purification of the enzymes involved. For studies of substrate specificity, a set of closely related analogs of the natural substrate is necessary; in most cases these may be only obtained by chemical synthesis.

The aim of the present chapter is to summarize the present development of chemical synthetic methods for preparation of dolichyl glycosyl phosphates and diphosphates. The synthesis of analogous derivatives of polyprenols and, in some cases, of short-chain prenols, is also reviewed. In reality, the synthesis of these phosphate or diphosphate diesters is always a multistage project starting from dolichol (polyprenol) and a protected sugar derivative. In this chapter we attempt to discuss all steps of the necessary synthetic work. Accordingly, sections are also included devoted to the availability of natural dolichols and polyprenols and to progress in their chemical synthesis. Prior to the discussion of the diester synthesis, the preparation of the corresponding phosphate monoesters, namely, dolichyl (polyprenyl) mono- or diphosphates and glycosyl phosphates, are considered. In the last section, a short review is given on applications of the synthetic dolichyl phosphates and their glycosylated derivatives in biochemical studies. Previous reviews [23,24] dealing mainly with the chemical synthesis of bacterial polysaccharide biosynthetic intermediates may be also consulted for some points omitted in the present chapter.

II. DOLICHOLS AND POLYPRENOLS IN NATURE

It so happens that dolichol was identified as a component of living tissues a long time before the role of its derivatives in biosynthetic processes was established. The isolation of dolichol from human kidney [25] and pig liver [26,27] was performed in 1960; dolichol was one of the first examples of long-chain, acyclic alcohols composed of isoprene units linked "head-to-tail." It soon became clear that a large group of structurally related alcohols exist in nature, and the term *polyprenol* was coined for these derivatives.

The characteristic feature of the natural polyprenols is the presence of a large number of internal (*Z*)-isoprene units in the hydrocarbon chain, and the existence of these alcohols in natural materials as a mixture of oligomer homologs.

According to the present IUPAC-IUB recommendations on nomenclature [28], the following abbreviations have to be used for different types of isoprene units characteristic of polyprenols and dolichols:

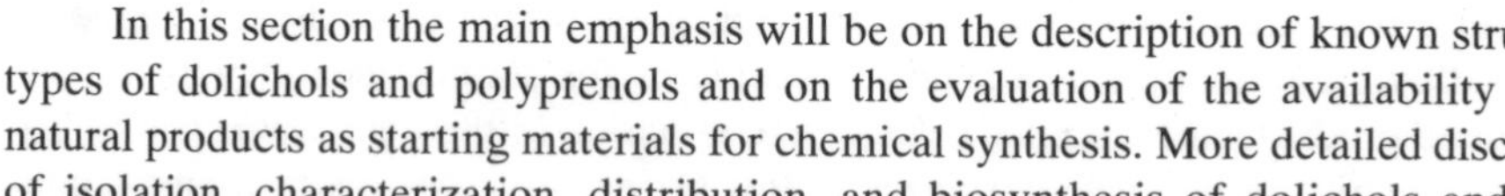

W-   T-   C-   S-

In this section the main emphasis will be on the description of known structural types of dolichols and polyprenols and on the evaluation of the availability of the natural products as starting materials for chemical synthesis. More detailed discussion of isolation, characterization, distribution, and biosynthesis of dolichols and polyprenols may be found in other reviews [29–34].

A. Structure of Dolichols and Their Phosphorylated Derivatives

Contrary to the exact meaning of the term *dolichol* (derived from Greek *dolikos*, meaning "long"), it is used not to characterize the chain length of polyprenols but to define a group of natural α-dihydropolyprenols of the general structure **1**, which differ from analogous allylic polyprenols by the presence of a single chiral center in the molecule. Studies of the degradation products of pig liver dolichol performed in 1980 [35] allowed investigators to establish its (*S*)-configuration.

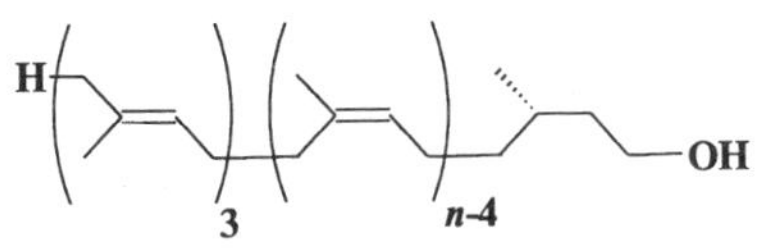

1

In terms of the abbreviations presented earlier, the general structure of dolichols may be represented as $WT_2C_{n-4}S$-OH, dolichols from various sources differing mainly by chain length (the value of n). Although with the present chromatographic techniques it is possible to isolate individual dolichols, mixtures of oligomer homologs characteristic of the selected source are generally used for the different studies.

Pig liver is the most usual source for isolation of dolichols. From 5 kg of the liver, approximately 200 mg of a mixture of oligomer-homologs may be prepared. According to a typical analysis [36], it contains mainly dolichols with $n = 18$ (20%), 19 (44%), and 20 (29%), as well as a detectable quantity of the homologs with $n = 17$ (2%) and 21 (5%). Traces (approximately 1%) of short-chain dolichol-11 with three internal (E)-isoprene units were also detected [37]. A similar picture is characteristic of other mammalian tissues, although the proportion of different oligomers may be slightly altered. Thus, for rat tissues the predominance of only two dolichol homologs with $n = 18$ and 19 was noted. Human tissues usually contain many more dolichols than tissues of laboratory animals. A typical analysis of human liver dolichol shows that approximately 75% of the preparation represents a mixture of almost equal quantities of the oligomers with $n = 19$ and 20, in addition to dolichols with $n = 17$ (1%), 18 (10%), 21 (12%), 22 (3%), and 23 (<1%).

As another readily available source of dolichols, yeast cells were used. Structures of yeast dolichols correspond to structure **1**, with shorter chain-lengths than those of mammalian dolichols. In a preparation from yeast, dolichol-15 (43%) and dolichol-16 (35%) were the main components, and they were accompanied by homologs with $n = 13$ (3%), 14 (14%), and 17 (5%) [36,38]. A similar distribution of oligomers was reported for *Phytopthora cactorum*; in other fungi, such as *Fusarium*, the chain length of dolichols ($n = 18-21$) is similar to that for mammalian dolichols. The same is true for dolichols from algae and higher plants [9]; in this case, dolichols represent only a small fraction of isolated polyprenols. For some fungi the presence of unusual dolichols was reported. Thus, dolichols from *Aspergillus fumigatis* contain two additional dihydroisoprene units at the ω-end of the chain and may be represented as H-$S_2T_2C_{n-5}S$-OH ($n = 19-23$) [39]. The additional modification of the ω-terminal dihydroisoprene unit in the same general structure was reported for dolichols from *A. niger* [40] (see structure **2**).

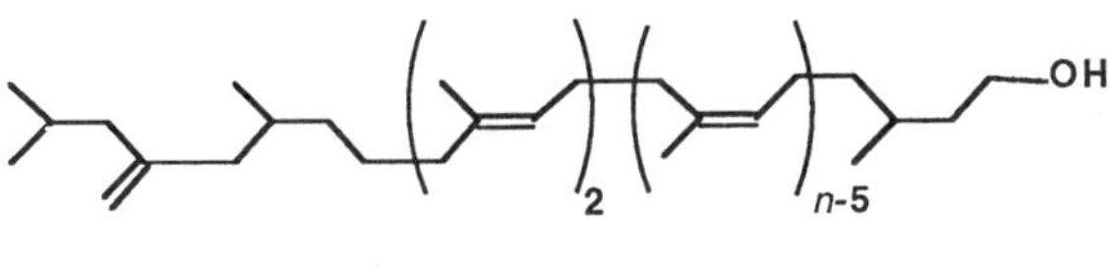

2 ($n = 19 - 23$)

Dolichols of the general structure **1** with shorter chain length ($n = 11, 12$) are characteristic of protozoa [41] and archaebacteria [15], but these sources could hardly be considered convenient for preparative isolation.

In living tissues, dolichols are present mainly as esters with fatty acids (45%–75% of the total dolichol) or as free alcohols. The content of phosphorylated derivatives, which represent a metabolically active form of dolichol, is usually very low, such as 4% of total dolichol reported for rat liver [42]. In contrast, they become the main fraction of total dolichol during the logarithmic phase of growth in yeast and mammalian cell cultures.

Four different types of phosphorylated dolichol derivatives are known, their interconversions are shown in general form in Scheme 1. Dolichyl phosphate is in a central position in this series of enzymic reactions. It may be formed from dolichol through the action of a specific kinase and cytidine 5′-triphosphate (reaction *a*) and dephosphorylated back to dolichol (reaction *b*) by a specific phosphatase. On the other hand, Dol-P serves as a substrate for the enzymic synthesis of dolichyl glycosyl phosphates (reaction *c*) or dolichyl glycosyl diphosphates (reaction *d*) by transfer of glycosyl or glycosyl-phosphate residues from the corresponding nucleoside diphosphate sugars. The enzymes that catalyze these reactions are specific for the structures of the latter derivatives. Dolichyl glycosyl phosphates function mainly as glycosyl donors in glycosylation reactions occurring inside the lumen of endoplasmic reticulum, and as a result of these processes (reaction *e*), dolichyl phosphate is regenerated. In contrast, dolichyl glycosyl diphosphates, derived from monosaccharides, serve as acceptors of monosaccharides in glycosylation reactions leading to dolichyl-diphosphate-linked oligosaccharides (reaction *f*). The latter derivatives participate as oligosaccharyl donors in transfer of a carbohydrate chain onto a final acceptor during the formation of glycoproteins. This glycosylation reaction (*g*) produces dolichyl diphosphate, and its dephosphorylation to dolichyl phosphate (reaction *h*) is necessary to complete the biosynthetic cycle.

In the classical dolichol pathway of protein glycosylation, only guanosine diphosphate mannose (GDP-Man) and uridine diphosphate glucose (UDP-Glc) are able to participate in the reaction (*c*), giving Dol-P-Man and Dol-P-Glc. Reaction *d* is characteristic of uridine diphosphate *N*-acetylglucosamine (UDP-GlcNAc). The resulting Dol-PP-GlcNAc is further converted into oligosaccharide derivatives through a series of enzymic reactions, with UDP-GlcNAc, GDP-Man, Dol-P-Man, and Dol-P-Glc as glycosyl donors.

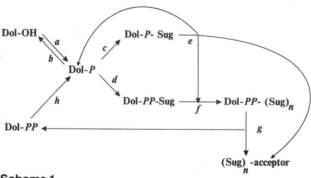

Scheme 1

B. Bacterial Polyprenols and Intermediates of Bacterial Polysaccharide Biosynthesis

The biosynthesis of bacterial polysaccharides and cell-wall polymers in many cases includes intermediate formation of phosphorylated polyprenol derivatives analogous to those shown in Scheme 1 but derived from an allylic undecaprenol with two internal (*E*)-isoprene units **3**.

WT$_2$C$_8$-OH   WTC$_8$-OH   H-S$_4$C$_3$-OH

3   **4**   **5**

In a first study of a bacterial polyprenol [43], the name "bactoprenol" was suggested for undecaprenol formed from mevalonic acid in lactobacilli. On the basis of the available evidence the presence of a dihydroisoprene unit in its structure was suggested. The following investigations [18,19,44,45] demonstrated absence of these units in undecaprenols isolated from a number of bacteria, including *Lactobacillus plantarum*, and this conclusion was confirmed in subsequent studies.

Contrary to the situation observed with dolichols and plant polyprenols (see Section II.C.), for bacterial polyprenols the strong predominance of undecaprenol over other oligomer homologs was observed in most bacteria, although in some cases small amounts of nona-, deca-, and dodecaprenols were found. The notable exception is mycobacteria, where decaprenol was identified as a main component of the polyprenol fraction [46]. It was initially thought to be related to undecaprenol **3**, but in a more recent study [47] the presence of only one internal (*E*)-isoprene unit was demonstrated, and structure **4** for this polyprenol seems the most plausible. In addition, mycobacteria contain another, structurally different polyprenol, octahydroheptaprenol **5**, with dihydroisoprene units situated near the ω-end of the chain [48].

The content of polyprenols in bacterial cells is similar to that of dolichols in pig liver. Thus for *L. plantarum*, isolation of 40 μg of undecaprenol from 1 g of dry bacterial cells was reported [45]. As with other cell cultures, during logarithmic phase of bacterial growth about 75% of total undecaprenol is present as different phosphorylated derivatives, whereas, in stationary phase, the content of free undecaprenol was found to be between 75% and 90%. Although the application of bacterial undecaprenol as a starting material for chemical synthesis is possible in principle, in most studies use was made of structurally related plant polyprenols, which are much more readily available.

C. Polyprenols from Higher Plants

Participation of dolichyl-linked sugars in biosynthesis of *N*-linked carbohydrate chains of glycoproteins in higher plants, and isolation of dolichols from these sources, was already mentioned. Nevertheless the main components of polyprenol mixtures in green leaves and wood are allylic polyprenols present as free alcohols, acetates, and esters of higher fatty acids. Among their phosphorylated derivatives, only diphosphates were identified as end products of biosynthetic reactions in vitro, and no evidence was presented for the specific participation of allylic polyprenyl phosphosugars in the biosynthesis of carbohydrate chains of plant polysaccharides or glycoconjugates. Three general structure types of allylic polyprenols from higher plants are known, as represented by formulas **6–8**. For additional reviews on plant polyprenols, see Refs. 49–52. Due to the longtime efforts of T. Chojnacki and his co-workers, a collection of plant polyprenols was created, from which individual polyprenols of all structural types, together with some of their derivatives, are available (for description see Ref. 53).

$$\text{WT}_2\text{C}_{n\text{-}3}\text{-OH} \qquad \text{WT}_3\text{C}_{n\text{-}4}\text{-OH} \qquad \text{WT}_{n\text{-}1}\text{-OH}$$

$$\textbf{6} \qquad\qquad\qquad \textbf{7} \qquad\qquad \textbf{8}$$

Polyprenols **6** with two internal (E)-isoprene units (as in dolichols and bacterial undecaprenol) and relatively short chain length ($n = 6$–8) were first isolated from birch wood [54,55] and are known under the trivial name *betulaprenols*. The content of betulaprenols, present exclusively as fatty acid esters, in wood is rather high (0.5%–1%), and they were found also in some other nonphotosynthetic tissues of different plants.

The presence of other structural types of polyprenols is characteristic of green leaves. There are a few cases when *all-trans*-polyprenols **8** were isolated, such as solanesol ($n = 9$) from tobacco and potato [56–58] or spadicol ($n = 10$) from *Arum maculatum* [59]. The presence of poly-(Z)-prenols with similar or larger chain length is much more usual. Their content in leaves is between 0.2% and 1%, increasing with the age of the leaves. These polyprenols are undoubtedly the most readily available representatives of the isoprenoids of this group.

Initially, a series of polyprenols **7** ($n = 9$–13), present mainly as free alcohols with a small proportion of fatty acid esters, was isolated from leaves of numerous angiospermous plants. At the beginning it was usual to give different trivial names to these very similar mixtures of oligomer homologs, such as castaprenol [60], ficaprenol [61], moraprenol [62], and cleomeprenol [63], but now this practice seems to have been abandoned. Nevertheless, the terms *betulaprenols* and *ficaprenols* are used sometimes as collective names for polyprenols of the general structures **6** and **7**. For reviews on distribution of these polyprenols see Refs. 52, 64, and 65. In most cases undecaprenol is the main component of the mixtures. It may be considered as a close analog of bacterial undecaprenol, and, in fact, derivatives of ficaprenol and moraprenol were frequently employed in studies on biosynthetic intermediates of bacterial polysaccharides. From some plants, polyprenols of the same structural type were isolated with longer or shorter chain length (for example, $n = 14$–17 in leaves of *Capparis* species [66] but 6–8 in fruits of *Serenoa repens* [67]).

The predominance of betulaprenol-type polyprenols **6** with longer chain length (as free alcohols or acetates) is characteristic of needles of coniferous trees and leaves of other gymnospermous plants [68–71]. For example, polyprenols from the pine tree *Pinus silvestris*, where the main components with $n = 15$ (23%), 16 (42%), and 17 (25%) were found, are similar in chain length to yeast dolichols, whereas ginkgoprenols from *Ginkgo biloba* with $n = 17$ (25%), 18 (36%), and 19 (18%) are more similar to mammalian dolichols. These derivatives were used for partial chemical synthesis of dolichols (Section III.A.). In many gymnosperms the presence of two families of polyprenols was demonstrated [51,52,72–74]: In addition to dolichol-like polyprenols, even longer polyprenols, with $n = 21$–24, were found as the main components.

High-molecular-weight polyprenols were also discovered in leaves of angiosperms belonging to the *Rosaceae* family [75–77]. As a rule, two groups of polyprenols are present. In plants of the *Potentilla* genus, polyprenols **6** with $n = 24$–28 predominate in a higher-molecular-weight group, whereas for the *Rosa* genus mainly polyprenols **7** with $n = 31$–34 were found. In addition to the main components, smaller quantities of higher oligomer homologs are present and may be isolated. The longest individual polyprenol of this series seems to be C_{225}-prenol (**6**, $n = 45$) from *Potentilla aurea*.

Leaves of some plants contain, in addition to normal polyprenols, the analogous derivatives with three dihydroisoprene units at the ω-end of the chain. Such polyprenols H-S_3TC$_{n-4}$-OH (**9**) were isolated, for example, from potato ($n = 5$) [58] and soybeans ($n = 9$–11) [78].

9

D. Short-Chain Prenols

According to present nomenclature, the term *polyprenol* is applied only to compounds that contain more than four isoprene units in their hydrocarbon chain. Structurally related lower isoprenoid alcohols are thus not included in this class. Nevertheless, these compounds, which are widely distributed in nature and readily available through both isolation and synthesis, usually serve as starting materials for the chemical synthesis of polyprenols and as model compounds for the development of methods in the preparation of phosphorylated polyprenol derivatives. Thus a short review of their structure seems necessary.

This series of natural products includes C_{10}-alcohols geraniol **10**, nerol **11**, and citronellol **12**. The last was isolated from different sources, both as (*R*)- and (*S*)-isomers. Among the alcohols composed of three isoprene units, (2*E*,6*E*)-farnesol **13** is the most usual, but the (2*Z*,6*E*)- and (2*Z*,6*Z*)-isomers were also isolated in several cases.

WT-OH	WC-OH	WS-OH
10	**11**	**12**
WTT-OH		WTTT-OH
13		**14**

In the C_{20}-series, only the (2*E*,6*E*,10*E*)-isomer of geranylgeraniol **14** was found in nature. To this series phytol belongs also, readily obtained by saponification of chlorophyll, which represents probably the most widely distributed terpenoid on earth. In addition to the usual isomer of phytol **15**, the isolation of small quantity of its $(2Z)$-isomer was recorded.

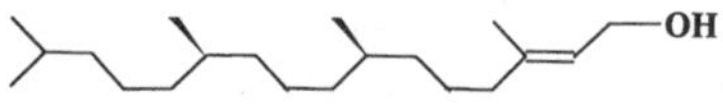

15

The diphosphates of some short-chain prenols serve as biosynthetic intermediates at the first stages of terpenoid biosynthesis, and much effort was directed, starting from the early '60s, into the development of chemical syntheses of these derivatives. The key biosynthetic reaction of isoprenoid chain elongation catalyzed by prenyl transferases is shown in Scheme 2. It consists of interaction of Δ^3-isopentenyl diphosphate **17** with a primer **16**. Stereochemistry of the new isoprene unit in the chain is determined by specificity of the enzyme involved.

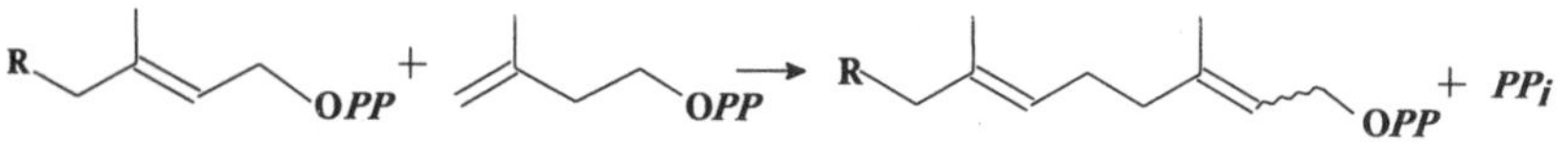

16a–d   **17**
a R = H
b R = W-
c R = WT-
d R = WT$_2$-
Scheme 2

In the first reaction of the sequence, 3,3-dimethylallyl diphosphate **16a** participates, and the reaction leads to incorporation of the (E)-isoprene unit. The analogous stereochemistry was observed for the other biosynthetic reactions common to all terpenoids, where the diphosphates of geraniol **16b**, (E,E)-farnesol **16c** or (E,E,E)-geranylgeraniol **16d** serve as primers.

Further elongation of the chain with (E)-isoprene units was shown to lead to polyprenols of type **8**. At the same time, the existence of the enzymes that catalyze the addition of (Z)-isoprene unit to **16c** or **16d**, and the further elongation of the chain with the same units, is well established. As a result of these reactions, the plant polyprenols **7** are formed from **16d**, whereas **16c** serves as a precursor for plant polyprenols **6**, bacterial undecaprenol **3**, and dolichols **1**. In the last case, incorporation of the dihydroisoprene unit occurs at the later biosynthetic stages. This fact is confirmed, among other observations, by the isolation of small quantities of polyprenols **6**. These are similar in chain length to dolichols from some mammalian tissues, and these polyprenols accumulate in several mutants of cultivated Chinese hamster ovary cells. Isolation of plant polyprenols of type **9** may be explained by the ability of phytyl diphosphate to serve as a primer for the prenyltransferases discussed earlier.

It is necessary to note that although distribution of internal (*E*)- and (*Z*)-isoprene units of polyprenols may be unambiguously determined by thorough analysis of their ^{13}C-NMR spectra, this analysis was actually performed only for readily available polyprenols. In other cases the conclusion regarding an "accumulation" of (*E*)-isoprene units at the ω-end of the hydrocarbon chain was based on the biosynthetic evidence, and sometimes was simply assumed by analogy.

III. SYNTHESIS OF POLYPRENOLS AND DOLICHOLS

During the last 15 years, significant progress was achieved in chemical synthesis of dolichols and polyprenols (for a recent review see Ref. 79). As a result of these studies it became possible to prepare not only the natural products of this series, but also a number of their analogs that may be useful for structural-biological activity studies.

A. Conversion of Natural Polyprenols into Dolichols

Partial chemical synthesis of dolichols from readily available plant polyprenols seems very attractive from the preparative point of view, although it is much more limited in terms of possibilities for the construction of different dolichol analogs.

Selective Hydrogenation of the α-Terminal Isoprene Unit

Selective hydrogenation of the α-terminal isoprene unit in plant undecaprenol WT$_3$C$_7$-OH was first described by Mankowski et al. in 1976 [37]. The reaction was performed in the presence of a platinum catalyst and NaBH$_4$. After chromatographic purification, including separation from other dihydro derivatives, racemic (*30E*)-dolichol-11 was isolated in a yield of about 60%. A series of dolichols with different chain lengths was successfully prepared by this method, and it was used also for incorporation of tritium into the α-terminal dihydroisoprene unit of dolichols (for a review see Ref. 80; a modified procedure is described in Ref. 53).

More recently it became possible to achieve preparation of optically active dolichols through catalytic hydrogenation with a homogeneous catalyst including a chiral ligand, namely, with ruthenium complexes containing 2,2'-bis(diphenylphosphino)-1,1'-binaphthyl (BINAP). For example, a complex **18** was found to be extremely useful in this respect.

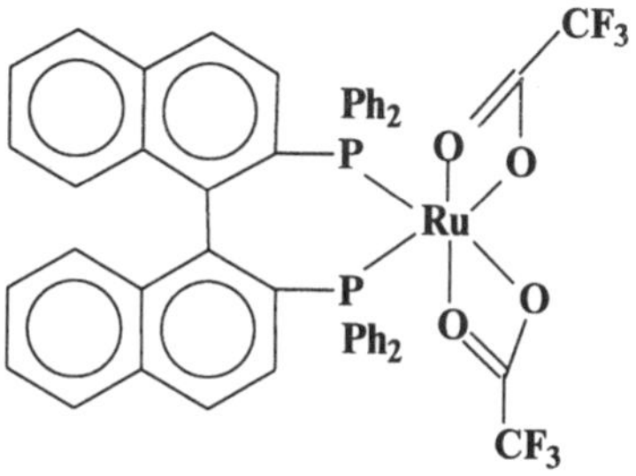

18

During studies of geraniol and nerol hydrogenation [81], a strict correlation was found to exist between the stereochemistry of the α-terminal isoprene unit in the substrate, the configuration of the chiral ligand in the catalyst, and the stereochemical result of the reaction. Thus, (*S*)-citronellol (**19**, R = W-) may be obtained from geraniol (**20**, R = W-) with the (*R*)-BINAP-derived catalyst, or from nerol (**21**, R = W-) with the catalyst containing (*S*)-BINAP as a ligand (Scheme 3).

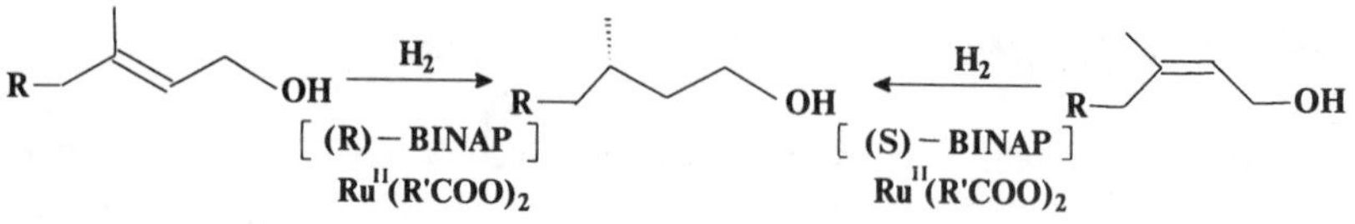

Scheme 3

On the basis of this procedure, an efficient method was developed [82] for the conversion of the readily available ginkgoprenols (**21**, R = WT_2C_{11-17}-) into (*S*)-dolichols of the same chain length, using **18** with (*S*)-BINAP.

Elongation of Polyprenols with a Dihydroisoprene Unit

Another approach to the partial synthesis of dolichols consists of elongation of the polyprenol chain at the α-end with a dihydroisoprene unit through formation of C—C bond. Of two possible retrosynthetic disconnections of the dolichol molecule, only the approach using an anionic C_5-component, shown in Scheme 4, has been applied so far in syntheses.

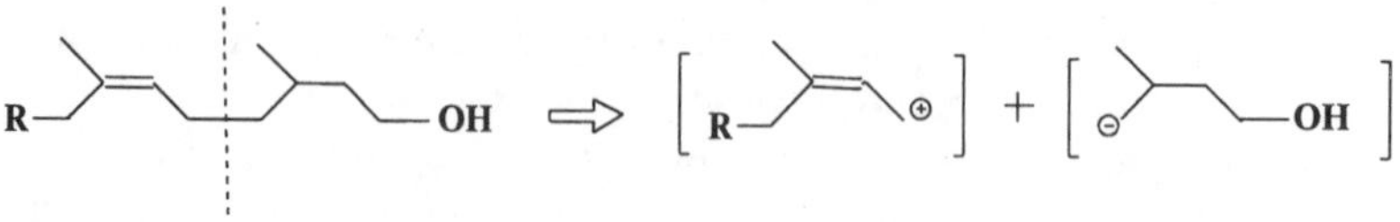

Scheme 4

Thus, interaction of the acetate **22** with (*S*)- or (*R*)-isomers of the magnesium-organic reagent **23** was shown to occur smoothly in the presence of Li_2CuCl_4. After initial studies with neryl acetate (**22**, R = W-) [83], the synthesis of (*S*)- and (*R*)-dolichols was achieved [84,85] through chain elongation of the ginkgoprenol derivative **22** (R = WT_2C_{11-17}-). With the C_4-derivative **24** it became possible to prepare a truncated dolichol derivative that was further converted into [1-^{14}C]dolichol in four steps with the use of labeled potassium cyanide [85].

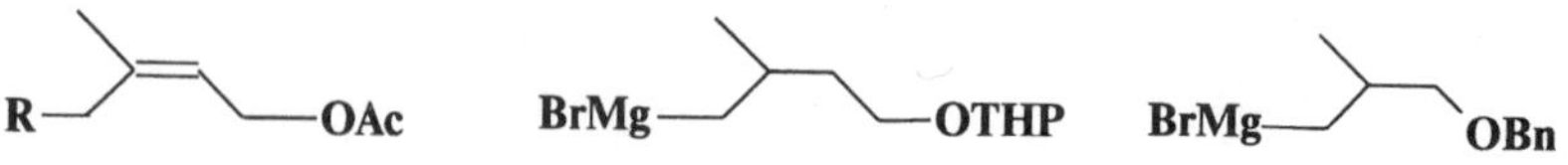

Another procedure for chain elongation is based on the interaction of the bromide **25** and the hydroxysulfone derivative **26**. The initially formed product **27** was desulfonylated by treatment with sodium in liquid ammonia or related reagents. It was used successfully for the synthesis of dolichol-6 (WT$_2$C$_2$S-OH) and (14E)-dolichol-7 (WT$_3$C$_2$S-OH), first as racemates [86] and later, when the chiral sulfone was employed, as (S)-isomers [87]. The starting bromides **25** (R = WT$_2$C-, WT$_3$C-) were prepared from derivatives of (E,E)-farnesol or (E,E,E)-geranylgeraniol through (Z)-C$_5$-homologization (see Section 3.B.).

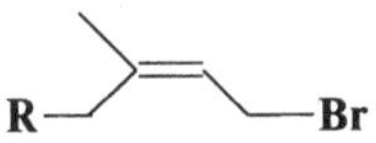

| **25** | **26** | **27** |

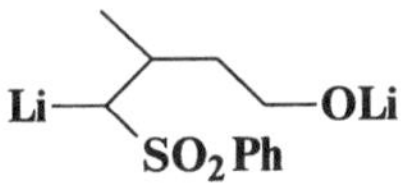

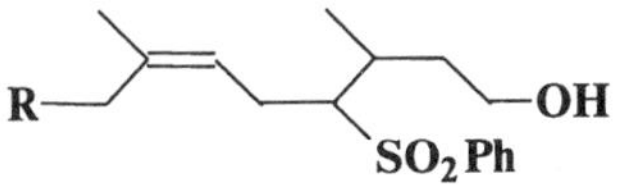

B. Assembly of Polyprenol Chains

A much more flexible approach to the synthesis of polyprenols and dolichols of different structure consists of complete assembly of the polyprenol chain. The easily available derivatives of (E,E)-farnesol or (E,E,E)-geranylgeraniol are usually employed as starting materials, and the main synthetic problem is the construction of oligo-(Z)-isoprene blocks characteristic of these natural products.

With Formation of C—C Bond

The most widely used approach for the preparation of different hexa- and hepta-prenols and their modified derivatives is based on the (Z)-C$_5$-homologization procedure developed by A. M. Moiseenkov and co-workers in the early '80s [88–91]. In a way this approach is analogous to the prenyltransferase reaction that occurs in the biosynthesis of polyprenols (see Scheme 2): The ω-terminal fragment of the chain participates in the reaction as a synthetic equivalent of a carbocation, whereas for chain elongation a synthetic equivalent of a C$_5$-carbanion was used. The structure of the latter determines the stereochemistry of the new isoprene unit in the chain.

For the initial stages of syntheses through this approach, use was made of the bromides **28ab**, readily available from the corresponding alcohols by the action of PBr$_3$ in pyridine. For incorporation of (Z)-isoprene units into the chain, the dilithium derivatives of hydroxysulfonamide **29a** or hydroxysulfone **29b** were employed. Anal-

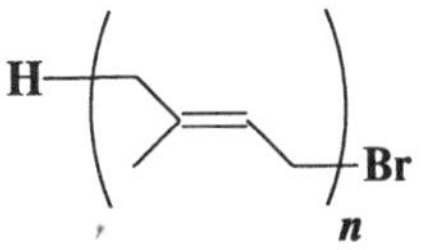

28ab   **29ab**   **30**

a $n = 3$ a X = —N⟨ ⟩O

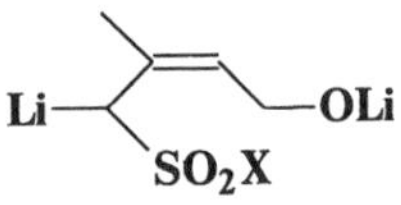

b $n = 4$ b X = Ph

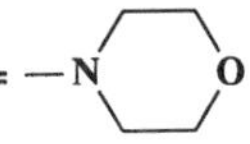

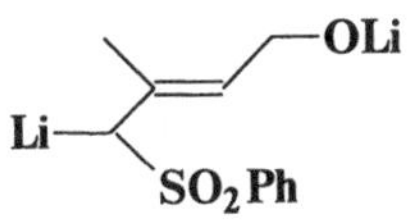

ogous reaction with **30** allowed the addition of an (E)-isoprene unit at the α-end of the chain; **26** (see earlier) may be used for incorporation of the dihydroisoprene unit.

Interaction of the above-mentioned reactants at lowered temperature ($-50°-70°C$) in tetrahydrofuran produced the sulfone (for example, **31**), which may be converted by treatment with Na/NH$_3$ into the alcohol **32**, which, in turn, may be readily transformed into the bromide serving as a starting material for further chain elongation. The limiting factor in these conversions is partial migration of a double bond under the conditions of the desulfonylation leading to the homoallylic alcohol **33**. In many cases this side reaction may be inhibited in the presence of crown ethers such as dibenzo-18-crown-6.

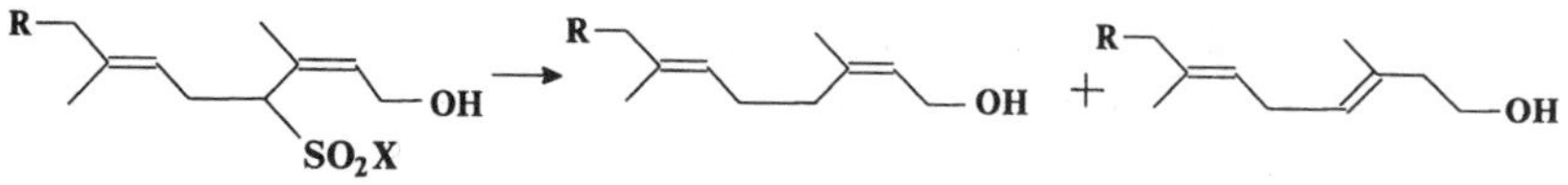

31   **32**   **33**

With this approach, three consecutive cycles of chain elongation with **28a** were performed in a first synthesis of betulaprenol-6 (WT$_2$C$_3$-OH) [88,89]. The corresponding prenols-4 (WT$_2$C-OH) and -5 (WT$_2$C$_2$-OH) were obtained as intermediates, and the latter was employed in the above-mentioned synthesis of dolichol-6. Starting from **28b**, the analogous syntheses of penta-, hexa-, and heptaprenols WT$_3$C$_n$-OH ($n = 1-3$) were achieved [90]. The approach was found especially useful for the preparation of the derivatives with alternating isoprene units of different configuration, such as WT$_3$CT-OH and WT$_3$CTC-OH [91]. With the use of the corresponding C$_5$-synthons, elongation of the normal alkyl chain in n-C$_{16}$H$_{33}$-OH with one and two (Z)-isoprene units were performed [92], as well as (Z)-C$_5$-homologization of solanesol [91], (E)-C$_5$-homologization of moraprenol, and consecutive addition to the latter of dihydroisoprene and (Z)-isoprene units [92].

Elongation of the polyprenol chain through interaction of sulfone derivatives and alkyl halogenides may be used for addition to the chain of several isoprene units in one step. In this approach, first suggested by Sato et al. [93], the tolylsulfones **34ab** were employed as synthetic equivalents of a carbanion for the ω-end of the

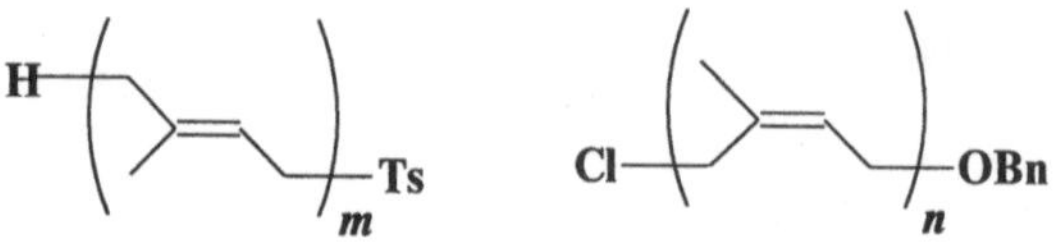

34ab   **35ab**
a $m = 3$ **a** $n = 2$
b $m = 2$ **b** $n = 3$

chain, whereas for chain elongation ω-chlorides derived from nerol **35a** or (Z,Z)-farnesol **35b** served as synthetic equivalents of a cationic fragment.

Interaction of these reactants (treatment with BuLi in THF-HMPA at −78°C, or, in some cases, phase-transfer catalysis reaction with Bu$_4$NBr and 50% aqueous NaOH) leads smoothly to the derivatives **36**. When the latter were treated with Li/EtNH$_2$, reductive desulfonylation occurred, producing the polyprenol **37**, and in a side process the isomer **38** was formed.

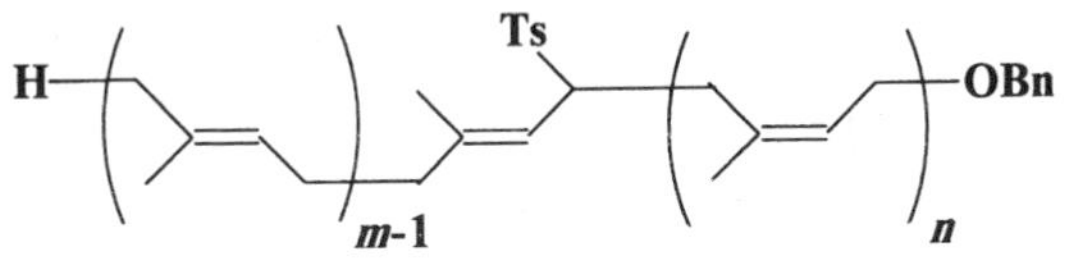

36

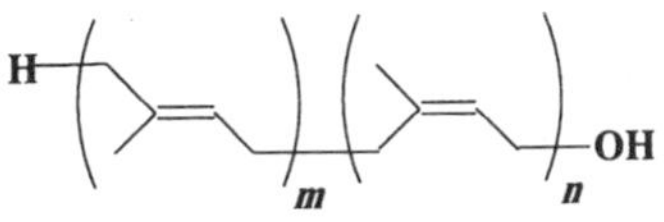

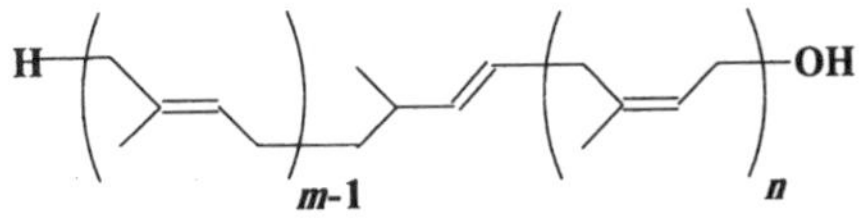

37

38

A series of the reactions shown in Scheme 5 may be used for conversion of **37** into the sulfone that allows further elongation of the polyprenol chain.

Preparation of the chlorides **42ab** from the corresponding benzyl prenols **39ab** requires multistep synthesis, as shown in Scheme 6. It includes selective epoxidation of the ω-terminal isoprene unit, elimination of three-carbon fragment from the ω-end through periodate oxidation to give the aldehydes **40ab**, and incorporation of the functionalized fragment through Wittig reaction with tetra-hydropyranyloxyacetone. The resulting **41ab** were converted further to the chlorides.

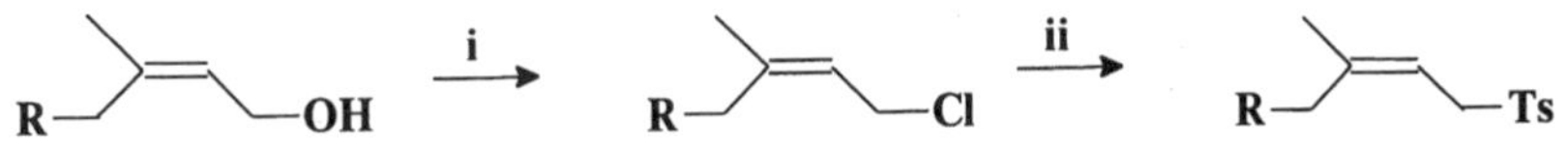

$$R = WT_mC_{n-1}-$$

i MsCl, LiCl, s-collidine/DMF; ii TsNa/DMF

Scheme 5

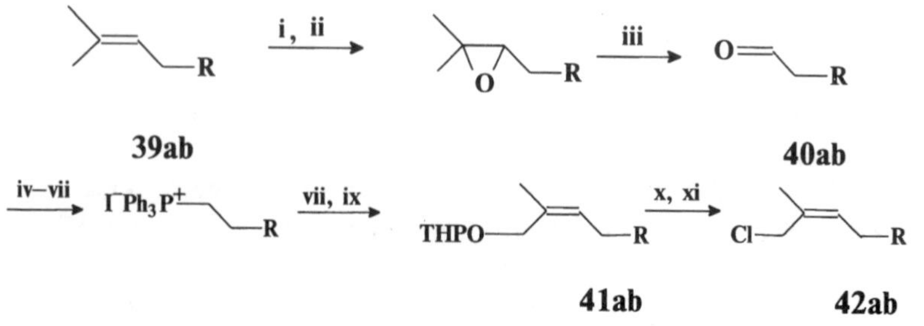

39ab   **40ab**

41ab   **42ab**

$$R = \text{-C-OBn (a)}, \text{-C}_2\text{-OBn (b)}$$

i *N*-Bromosuccinomide;   ii K_2CO_3/NaOH;   iii H_5IO_6;   iv $NaBH_4$;   v TsCl;
vi NaI;   vii Ph_3P;   viii BuLi;   ix $THPOCH_2COCH_3$;   x H^+;
xi MsCl, LiCl

Scheme 6

Starting from the sulfone **34a** and the C_{10}-chloride **35a**, betulaprenol-5 WT_2C_2-OH and betulaprenol-7 WT_2C_4-OH were obtained [93], and the use of the C_{15}-chloride **35b** allowed the preparation of analogous polyprenols with 30, 40, and 45 carbon atoms [94]. In an analogous synthesis of bacterial undecaprenol WT_2C_8-OH [95], the C_{20}-chloride **44** was employed for the chain elongation. It was prepared (Scheme 7) from the chloride **35a** and the sulfone **43**, readily available from **41a**.

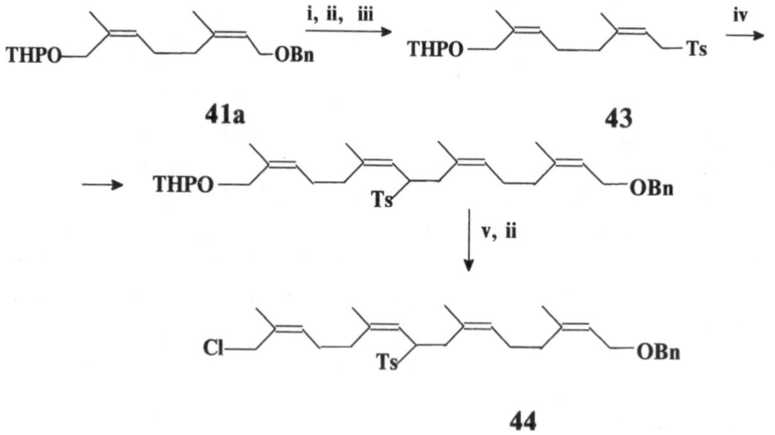

41a   **43**

44

i Na/NH$_3$;   ii MsCl, LiCl;   iii TsNa;   iv **35a**, BuLi;   v H$^+$

Scheme 7

Further elongation of the polyprenol chain with the use of **44** allowed the synthesis of betulaprenol-15 WT_2C_{12}-OH, which served as an intermediate in preparation of dolichol-20 [96].

For synthesis of dolichols through this approach, it is necessary to use in the last stage a building block that contains a dihydroisoprene unit, for example **45**, which may be prepared from citronellol through a sequence of reactions analogous to that shown in Scheme 6 [96,97].

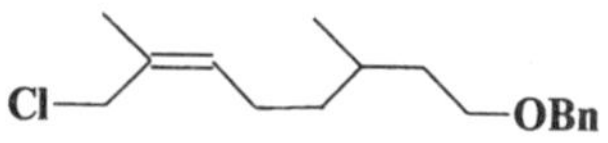

45

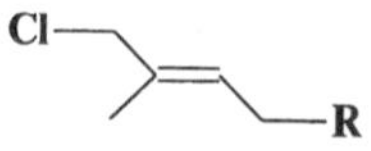

46a-c
a R = -S-OBn
b R = -C-OBn
c R = -T-OBn

Interaction of **45** and **34a** allowed the synthesis of dolichol-5. And after elongation of polyprenols prepared from **34a** and **35a** with the C_{10}-block **45**, dolichols-7, -9, and -11 [97,98] were prepared. In some cases, particularly for the synthesis of dolichols with an even number of isoprene units, the geraniol derivative **34b** was used as a starting sulfone, and in the next stage of the chain elongation an additional (E)-isoprene unit was incorporated from the chlorides **46a–c** prepared from citronellol, nerol, or geraniol derivatives, through selective oxidation of the ω-terminal methyl group with SeO_2. In this way, the preparation of WT_2S-OH, WT_2C-OH, WT_2C_2S-OH, and WT_3C_2S-OH was performed [98].

Finally, the total synthesis of (S)-dolichol-20 was achieved [96] through chain elongation with the use of a C_{25}-block **47**, prepared from **41b** and the (S)-isomer of **45**.

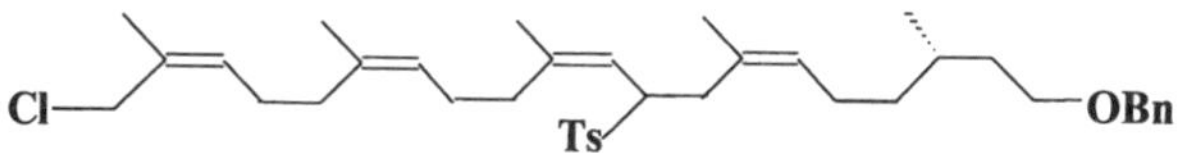

47

In conclusion, the described approach was found to be very useful for total syntheses of natural polyprenols and dolichols, as well as for preparation of a series of their analogs differing in the chain length and stereochemistry of some of the isoprene units.

With Formation of C=C Bond

An alternative approach to polyprenol synthesis was found in some cases to be very beneficial, especially for the preparation of polyprenols and dolichols with 6–9 isoprene units in the chain and for their modified derivatives. It is based on an aldol-type condensation of aldimines **48** and aldehydes **49** [90,99,100], as shown in Scheme 8.

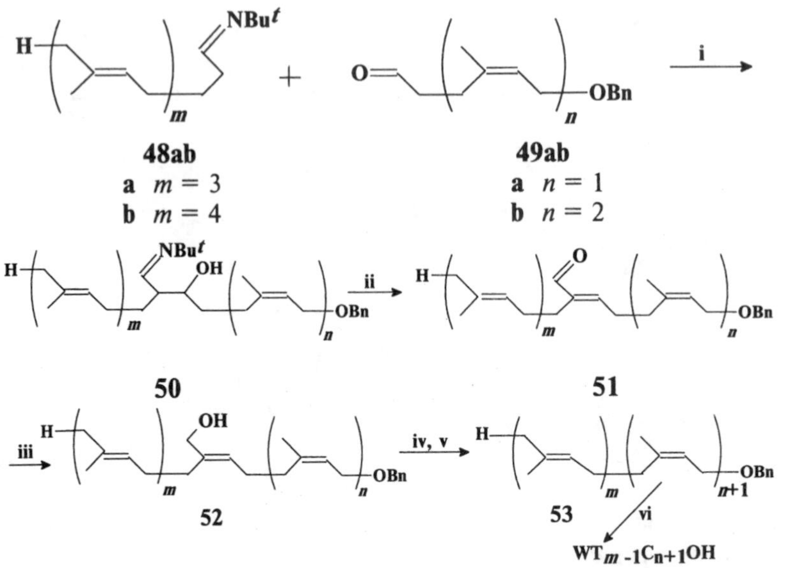

Scheme 8

i Pr_2^i; ii H^+ (pH 4.0–4.5); iii $NaBH_4$; iv Py•SO_3;
v $LiAlH_4$; vi Li/NH_3

The initial product of the condensation, derivative **50**, is stereoselectively trans-formed, after treatment with dilute acid, into the derivative **51** with a thermodynam-ically more stable internal (E)-acrolein unit. It was then converted further, through the allylic alcohol **52**, into the derivative **53** with an internal (Z)-isoprene unit. As a result, in this approach an additional (Z)-isoprene unit arises in the middle of the chain, with three carbon atoms originating from the methylene component **48** and two carbon atoms derived from the aldehyde **49**.

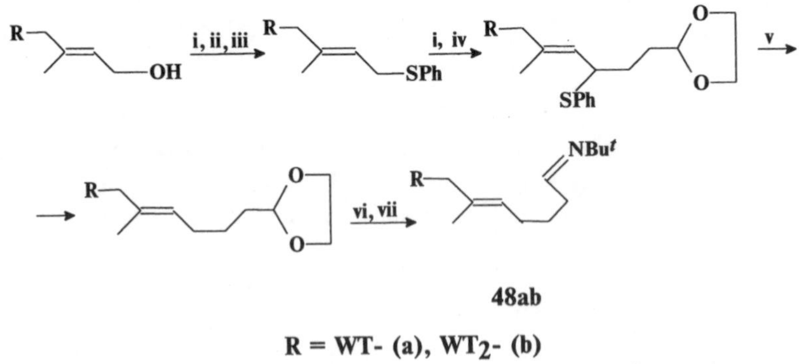

R = WT- (a), WT$_2$- (b)

i BuLi; ii TsCl; iii PhSCl; iv $Br(CH_2)_2CH(OCH_2)_2$;
v Li/NH_3; vi H^+; vii Bu^iNH_2

Scheme 9

To prepare the necessary aldimines **48ab** from (E,E)-farnesol [100] or (E,E,E)-geranylgeraniol [90,99], the alcohols were converted into thioethers, followed by C_3-elongation of the chain through reaction with a 3-bromopropionic aldehyde derivative (Scheme 9).

The analogous reaction sequence may be applied to other prenols. Also, a procedure for (Z)-C_5-homologization of the aldimines was developed [101,102]. It includes the mentioned aldol condensation with the use of **54** as an aldehyde component. The best results were achieved with α-trimethylsilylaldimines, for example, **55**, as a second component of the reaction. The transformation of the initial condensation product **56** into the aldimine **57** was achieved through reactions analogous to those included in Schemes 8 and 9.

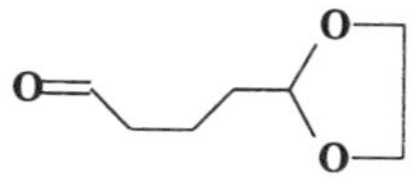

54 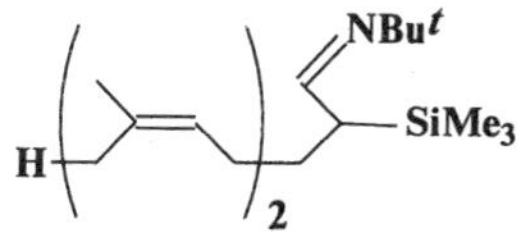   **55**

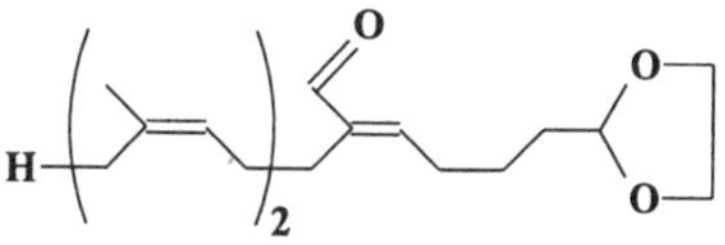

56 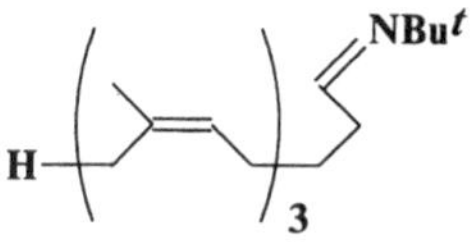   **57**

Several procedures were suggested for the preparation of the aldehyde components of the condensation shown in Scheme 8. These aldehydes may be obtained through chain-shortening of the readily available lower prenols (see **40ab** in Scheme 6) or from low-molecular-weight precursors, through interaction of the glutaric aldehyde derivative **58** with benzyloxyacetaldehyde (for synthesis of **49a**) or with **49a** (for synthesis of **49b**).

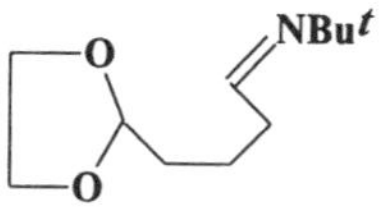

58

In some polyprenol syntheses through this approach, as aldehyde components the aldehydoesters **59**, available as both $(2Z)$- and $(2E)$-isomers, were employed. The higher member of the series may be obtained from the derivatives **60**, isolated from ozonolysis products of rubber [103]. The reaction sequence shown in Scheme 8 is quite similar when **59** is used instead of **49**, except that treatment with LiAlH$_4$ results simultaneously in reduction of the ester and sulfoester groups.

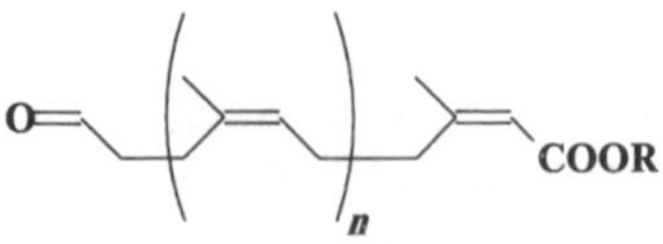

59

$n = 1, 2, 3;$ R = Et, Bu^t

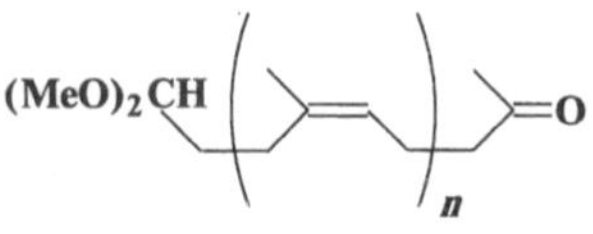

60

$n = 2,3$

The blocks **59** may be converted into blocks **61ab**, suitable for dolichol synthesis, through acetalation and treatment with Li/NH$_3$, resulting in simultaneous reduction of the ester group and the α-terminal double bond, followed by benzylation and mild acid hydrolysis. The chiral aldehyde **62**, for the synthesis of optically active dolichol, was prepared through degradation of the ω-terminal isoprene unit in (S)-citronellol.

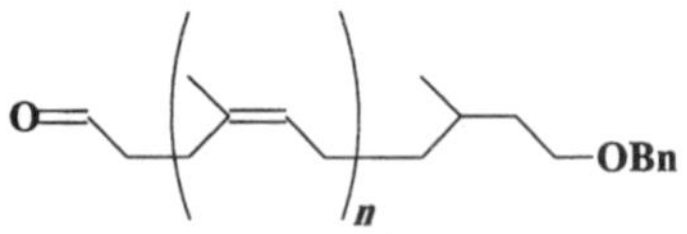

61ab

a $n = 2$
b $n = 3$

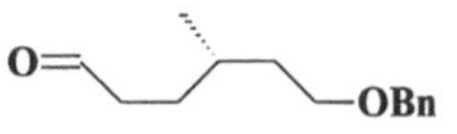

62

The approach was successfully applied to the synthesis of a series of natural and modified polyprenols, including hexaprenol WT$_3$C$_2$-OH [104], heptaprenol WT$_3$C$_3$-OH [90,99,100], and octaprenol WT$_3$C$_4$-OH [105]. Derivatives modified in stereochemistry of the α-terminal isoprene unit WT$_3$C$_n$T-OH ($n = 2, 3$) were prepared [99,100,105] as well as *all-cis*-hexaprenol WC$_5$-OH [106]. The development of methods for the preparation of an aldehyde condensation fragment from low-molecular-weight precursors opens the way for the synthesis of some unusual hexaprenols, such as those with altered positions of hydroxyl groups **63** and **64** [107], or of a derivative devoid of the methyl group in the α-terminal unit **65** [108].

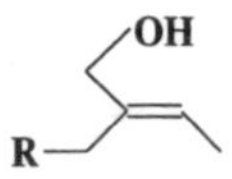

63

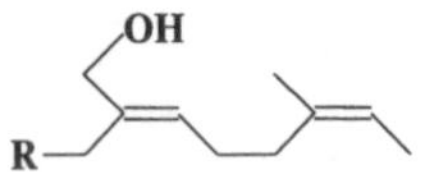

64

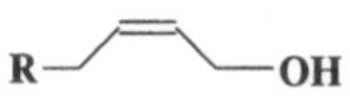

65

R = WT$_3$C-

R = WT$_3$-

R = WT$_3$C-

In the dolichol series this approach allowed syntheses of racemic alcohols WT$_3$C$_3$S-OH and WT$_3$C$_4$S-OH [103], and the (S)-isomer of the latter was also successfully prepared [104].

These examples clearly demonstrate the flexibility of the approach based on aldol condensation in preparation of different analogs of medium-chain-length polyprenols.

IV. SYNTHESIS OF DOLICHYL MONO- AND DIPHOSPHATES AND RELATED DERIVATIVES

The key significance of dolichyl (polyprenyl) mono- and diphosphates in metabolic reactions makes the task of chemical synthesis of these derivatives extremely important. In this series, specific difficulties exist for performance of the phosphorylation reactions. First of all, high hydrophobicity of these alcohols and their low solubility in polar solvents hamper the finding of suitable experimental conditions for efficient conversion into phosphates. Secondly, numerous side reactions are possible in the course of deprotection of the initial phosphorylation products, due to high unsaturation of the substrates. Therefore some established phosphorylation procedures cannot be used. The situation is even more complicated for allylic polyprenols readily subjected to monomolecular nucleophilic substitution at C-1; the facile nature of these reactions results in the instability of both intermediates and end products.

Historically, the first synthetic studies in this field, started in the early '60s, were connected with the preparation of short-chain prenol derivatives. Then followed the development of phosphorylation procedures for readily available plant polyprenols, which were treated as analogs of bacterial undecaprenol. The first syntheses of dolichol derivatives were reported in the middle of the '70s.

A. Dolichyl and Polyprenyl Monophosphates

Generally speaking, two approaches may be used for the preparation of phosphoric acid monoesters. As shown in formula **66**, as a nucleophilic component, a phosphoric acid derivative (retrosynthetic disconnection a) or an alcoholate anion equivalent (the disconnection b) may be applied.

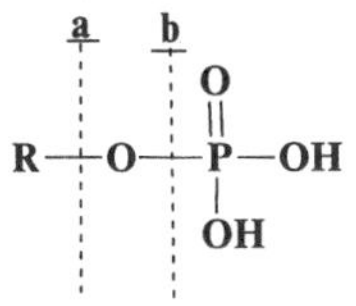

$$\text{R} \!-\!\! \underset{\underset{b}{|}}{\overset{\overset{a}{|}}{\text{O}}} \!-\! \underset{\text{OH}}{\overset{\overset{\text{O}}{\|}}{\text{P}}} \!-\! \text{OH}$$

66

a $[R^+] + [H_2PO_4^-]$ b $[RO^-] + [H_2PO_3^+]$

Only the latter strategy was employed for the syntheses of dolichyl phosphate. In these cases dolichol served as a nucleophilic component. As electrophilic components, different phosphorochloridates or derivatives formed in situ from phosphoric acid salts in the presence of activating reagents were used.

Phosphorochloridates as Reagents

The usual procedures for the synthesis of complicated phosphate monoesters include the reaction of an alcohol with diester phosphorochloridates followed by deprotection of the initially formed phosphate triesters. The first efficient synthesis of dolichyl phosphate [109,110] was achieved with the use of o-phenylene phosphorochloridate **67**, which is one of the most reactive phosphorylating agents of this group. It was applied earlier for the preparation of phosphates of a series of simple alcohols, including 3,3-dimethylallyl phosphate [111]. The reaction was performed in anhydrous dioxane as a solvent in the presence of 2,6-lutidine. The initially formed phosphotriester was transformed rapidly into the diester **68** after the addition of water. Deprotection of the latter to yield dolichyl phosphate required the action of an oxidizing reagent (such as lead tetraacetate) followed by treatment with alkali. In this synthesis a total yield of dolichyl phosphate was 46%.

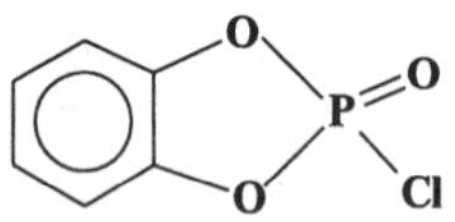

67 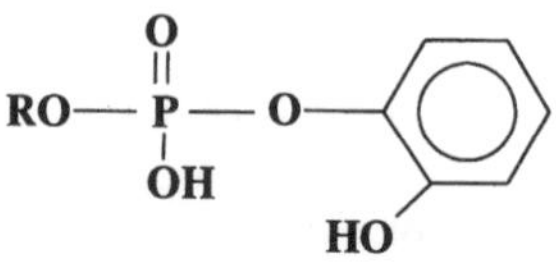   **68**

Prior to application of **67** for the synthesis of dolichyl phosphate, the reagent was successfully employed for the preparation of ficaprenyl phosphate [112,113] and, in more recent time, a slightly modified procedure was used to phosphorylate a series of plant polyprenols with different chain lengths [62,114].

Another related phosphorylating agent successfully applied to the synthesis of dolichyl phosphate [115] is 2-chloromethyl-4-nitrophenyl phosphorodichloridate **69** [116]. Under the reaction conditions, only one chlorine atom is substituted, and the resulting phosphorochloridate may be smoothly converted by treatment with aqueous pyridine, through intermediate formation of **70** and **71**, into dolichyl phosphate, isolated in a yield of 76%.

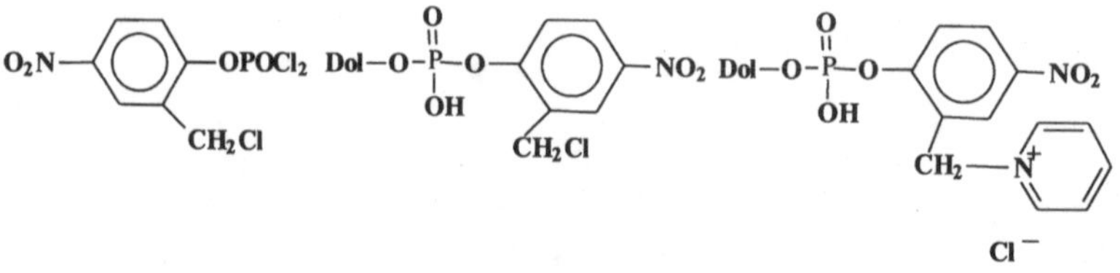

69   **70**   **71**

In recent years, the use of protected phosphorochloridates for the synthesis of dolichyl phosphate was practically discontinued after investigators found experimental conditions providing high yields of these phosphate monoesters with $POCl_3$ as the reagent [117] (Scheme 10).

For this purpose it was found essential to perform the reaction in hexane as a solvent, with the slow addition of dolichol solution to an excess (5 equiv) of $POCl_3$ and triethylamine. The initially formed dichloridate **72** was transformed into the phosphate monoester by treatment with a mixture of triethylamine, acetone, and water.

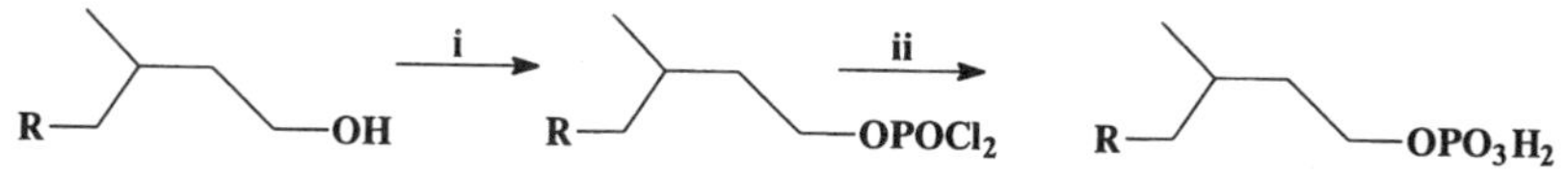

R = WT2C_{II}, n = 6, 14, 15, 16, 17
i POCl3/Et3N; ii H2O/Et3N

Scheme 10

The high efficiency of this simple method was demonstrated on a series of dolichols with different chain lengths. The reaction gave good results even when performed on a very small scale in the preparation of labeled derivatives, and was widely applied recently (see, for example, Refs. 33, 53, 97, and 118–121).

The procedure may not be used for allylic polyprenols, probably due to the easy nucleophilic substitution at C-1 of the intermediate phosphodichloridate with chloride ion formed in the reaction. Nevertheless, the trimethylsilyl ethers of lower prenols, (*E*)-geraniol and (*E,E*)-farnesol, may be successfully phosphorylated with $POCl_3$ [122]. The reaction requires more drastic conditions (reflux of dichloromethane solution for 48 hours), and the possibility of its application to higher prenols remains unexplored.

Use of Phosphoric Acid Salts and Condensing Reagents

A number of activating reagents were suggested for conversion in situ of phosphoric acid derivatives into their mixed anhydrides or imidoyl phosphates, which may serve as synthetic equivalents of the phosphoryl cation in retrosynthetic disconnection b (see formula **66**). Among them, only trichloroacetonitrile has found an application for the synthesis of dolichyl or polyprenyl phosphates.

This procedure for the phosphorylation of allylic alcohols, including lower prenols, was first described by Cramer et al. [123,124] in the early '60s. Bis-(triethylammonium) hydrogen phosphate was used in the reaction. It was suggested that after interaction of the salt with trichloroacetonitrile, the intermediate **73** is formed, which reacts further with an alcohol, giving the monoester. This, in turn, may participate in reaction with **73** or be further activated with the condensing reagent. In addition, **73** may interact with a second molecule of the phosphate salt, giving inorganic pyrophosphate, which may be further converted into inorganic polyphosphates. As a result, a complex mixture of products, including mono- and diphosphates of prenols as well as condensed inorganic phosphates, is formed in the reaction. The content of this mixture may be regulated by a change of the experimental conditions.

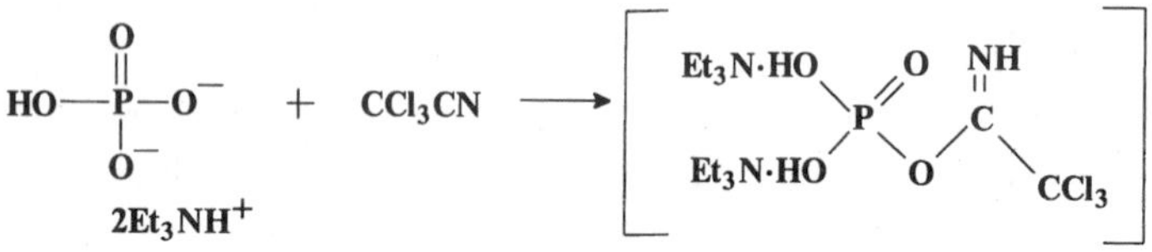

73

Thus, to achieve the best yield of geranyl phosphate (24%) it was recommended [124] to add slowly a solution of the phosphate salt in acetonitrile to a mixture of the prenol and trichloroacetonitrile, the final ratio of alcohol:phosphate:CCl_3CN being 1.3:1:2.5. When an excess of the phosphate salt was used, the predominant formation of the prenyl diphosphates and inorganic polyphosphates was observed (see Section IV.B.). Conditions were found for efficient conversion of farnesol into farnesyl phosphate [112]: The reaction was performed with the ratio of farnesol: phosphate:CCl_3CN equal to 1:3:2.5 for a short time. In the initial reaction, the degree of substrate conversion was low. To achieve a better degree of substrate conversion, the unreacted alcohol was separated and subjected twice more to the same procedure.

Elongation of the hydrocarbon chain of the substrate, resulting in an increase of the hydrophobicity of the molecule, hinders the finding of suitable conditions for the efficient performance of the Cramer reaction. In numerous publications, application of the Cramer reaction for the preparation of dolichyl or polyprenyl phosphates was mentioned, but experimental conditions were generally not described and the yields of the products were not reported (see, for example, Ref. 125). A detailed description of polyprenol phosphorylation was made by Jankowski and Chojnacki [126]. They were able to prepare ficaprenyl phosphate in a yield of 20%, using acetonitrile as solvent and a much larger content of the phosphorylating reagents, in comparison with the lower prenols (the final ratio of alcohol:phosphate:CCl_3CN was 1:2.5:6). For dolichyl phosphate, a rather detailed description of the procedure may be found in Ref. 127: It was found necessary to employ an acetonitrile-toluene mixture (4:6) as solvent, and a very large excess of the phosphate salt (150-fold) and of trichloroacetonitrile (300-fold). Under these conditions, both dolichyl mono- and diphosphates were obtained in 13% yields.

A significant improvement in the synthesis of polyprenyl and dolichyl phosphates through CCl_3CN-mediated condensations was achieved after finding that tetra-n-butylammonium dihydrogen phosphate may be used as a reagent [128,129]. The application of this salt allows the reaction to be performed in such solvents as chloroform or dichloromethane, and a substantial increase of the reaction rate was observed. The optimal conditions for the preparation of the monophosphates were 3–5-fold excess of the phosphate salt in relation to an alcohol and a ratio phosphate: CCl_3CN equal to 1:1.3–1.5. Under these conditions farnesyl, moraprenyl, solanesyl, and dolichyl phosphates were obtained in yields of 80%–87%; less than 7% of the diphosphates were formed (for a modified procedure see a review [23]).

Phosphorylation with tetra-n-butylammonium dihydrogen phosphate and trichloroacetonitrile was successfully applied to a series of modified polyprenols [129], retinol [130], and some hydrophobic alcohols [131]. In all cases the corresponding monophosphates were obtained in high yields. At present this procedure seems to be the simplest and most efficient for the preparation of polyprenyl phosphates and related derivatives. Results obtained during ³¹P-NMR studies of the reaction [132] allowed suggestion of a different mechanism for phosphorylation with trialkylammonium and tetraalkylammonium phosphates as reagents.

Some Special Cases

In some enzymic reactions the dolichyl derivatives may be efficiently substituted with derivatives of short-chain, saturated, branched alcohols, particularly with those of phytanol, which is readily available. In connection with these studies several syntheses of phytanyl phosphate were described. Initially it was prepared through the Cramer reaction [133] or through interaction with diphenyl phosphorochloridate followed by hydrogenolysis of the intermediate phosphotriester [134]. In more recent papers [135–137] on the preparation of phytanyl phosphate and related saturated derivatives, use was made of an efficient phosphorylation method [138] based on the application of a trivalent phosphorus derivative, aminodialkoxyphosphine **74**. Its interaction with dolichol analogs in tetrahydrofuran in the presence of 1H-tetrazole, followed by oxidation of the resulting phosphite-triester with m-chloroperbenzoic acid, smoothly gave the phosphate triester **75**, and the phytanyl derivative was obtained in 90% yield. Conversion of **75** into the monophosphate was readily achieved by treatment with trifluoroacetic acid in dichloromethane.

$$Et_2NP(OBu^t)_2$$

$$RO-\overset{\overset{\textstyle O}{\|}}{\underset{\underset{\textstyle OBu^t}{|}}{P}}-OBu^t$$

74   **75**

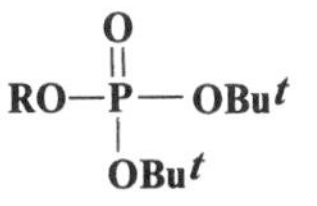

At present it is not clear whether a similar approach may be adapted for dolichol or other polyunsaturated derivatives, although with the use of milder oxidizing reagents it may become feasible.

As another special case phosphorylation of allylic polyprenols through the use of synthetic equivalents of prenyl cation (see disconnection a in formula **66**) may be considered. This approach, widely used for preparation of short-chain prenyl diphosphates (see Section IV.B.), found only limited application in the monophosphate series.

For short-chain prenols, interaction of tetra-n-butylammonium dihydrogen phosphate with bromide **76a** or sulphonium salts **77ab** was investigated [139]. The latter derivatives seem more preferable; 3,3-dimethylallyl phosphate was prepared from **77a** in a yield of 60%, but with the (E)-geranyl derivative **77b** only 26% of the phosphate was obtained. As described in a preliminary note [122], farnesyl phosphate may be successfully prepared from the bromide **76b** with the use of bis-(trimethylsilyl) hydrogen phosphate **78** as a nucleophile.

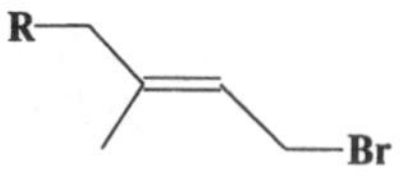

76ab	77ab	78
a R = H	a R = H, R' = $C_{12}H_{25}$-,	
b R = W-	X = ClO_4	
	b R = W-, R' = C_6H_{13},	
	X = MsO	

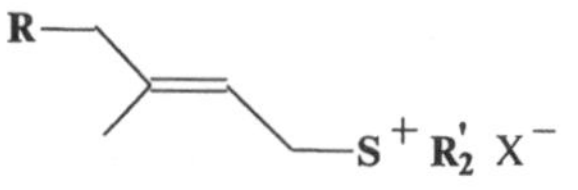

Until now, the only application of this approach in the polyprenol series was based on the use of polyprenyl O-trichloroacetimidates **79**, readily prepared from polyprenols [129,140–142]. Their interaction with an excess of phosphoric acid in tetrahydrofuran led to the formation of polyprenyl phosphates and minute quantities of tertiary-alcohol phosphates **80** (Scheme 11). The products may be separated with ion-exchange chromatography, and the procedure was employed for the synthesis of phosphates of farnesol, moraprenol, solanesol, and some modified polyprenols [129,140,141]. It may not be used for phosphorylation of α-dihydropolyprenols.

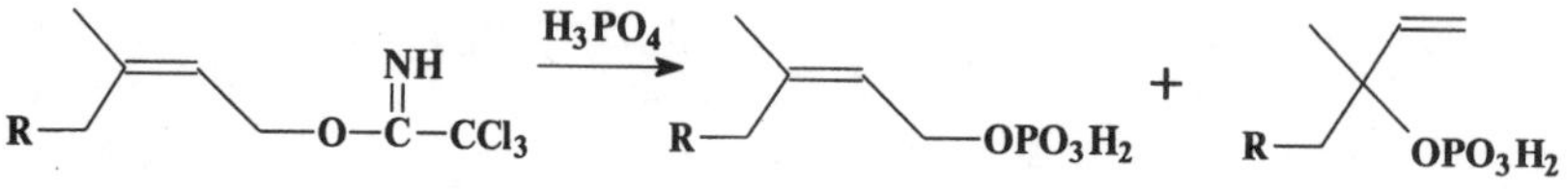

Scheme 11

B. Dolichyl and Polyprenyl Diphosphates

Three different approaches were applied to the synthesis of dolichyl diphosphates and related derivatives. They are based on the activation of phosphoric acid, on the interaction of activated derivatives of dolichyl (or polyprenyl) phosphates with inorganic phosphate, and, finally, on the reaction of activated prenol derivatives with inorganic pyrophosphate. Until now, attention was mainly devoted to the preparation of the lower prenol derivatives, and syntheses of dolichyl diphosphates were described in only a few papers.

Trichloroacetonitrile-Mediated Condensation for the Synthesis of the Diphosphates

As mentioned earlier (Section IV.A.), trichloroacetonitrile-mediated condensation of prenols and phosphoric acid salts lead to a mixture of prenyl mono- and diphosphates, and experimental conditions may be found for the formation of the diphosphates as the main reaction products.

For the preparation of short-chain prenyl diphosphates it was found essential to use a large excess of bis-(triethylammonium) hydrogen phosphate; the reaction may be performed in acetonitrile. The procedure described in 1962 by G. Popjak et al. for the synthesis of farnesyl diphosphate [143,143a] is widely applied. With the final ratio of reactants farnesol:phosphate:CCl_3CN equal to 1:3:5, it was possible to isolate 50% of farnesyl diphosphate and 13.5% of the monophosphate. Several modifications of the procedure were described (for example, Refs. 144 and 145), ensuring the preparation of diphosphates from different C_5-C_{20}-prenols, including citronellol, in similar yields.

The application of similar conditions to the synthesis of dolichyl or polyprenyl diphosphates was less successful (see, for example, Refs. 127 and 146–148). The best result in the small-scale synthesis of dolichyl and undecaprenyl diphosphates with the use of bis-(triethylammonium) hydrogen phosphate (yields about 50%) was achieved when a very large excess of the salt (200-fold) and CCl_3CN (1000-fold) in acetonitrile-toluene mixture, was applied [149].

As mentioned earlier, with tetra-*n*-butylammonium dihydrogen phosphate as a reagent in CCl_3CN-mediated condensation, the formation of polyphosphates including polyprenyl diphosphates is strongly inhibited. Nevertheless, with a reactant ratio moraprenol:phosphate:CCl_3CN equal to 1:4:36, quantitative conversion of the starting alcohol into a mixture of moraprenyl mono- and diphosphates was observed, and both products were isolated in yields of 45% [23,128,129]. Similar conditions were found suitable for the preparation of dolichyl diphosphate [53].

Synthesis Through Activated Derivatives of Dolichyl Monophosphates

This approach toward the synthesis of dolichyl and polyprenyl diphosphates is presented in Scheme 12. In this approach, phosphate salts serve usually as the anionic component. As synthetic equivalents of the prenyl phosphoryl cation the derivatives **81–86** were suggested. These activated derivatives were mostly applied only in the synthesis of dolichyl glycosyl diphosphates (Section VII), but their use for preparation of the diphosphates also seems quite possible.

Dolichyl phosphorodichloridate **81** may be readily prepared by treatment of dolichyl phosphate with oxalyl chloride [119]. The derivatives **82** and **83** are known both in the dolichol and the polyprenol series. The mixed anhydride **82** was easily formed when the monophosphates were treated with diphenyl phosphorochloridate [110,112,150]. For the preparation of the phosphoroimidazolidate **83**, the reaction of the corresponding phosphates with *N,N'*-sulfinyldiimidazole [23,151,152] or *N,N'*-carbonyldiimidazole [153,154] was employed. The possibility of the activation of polyprenyl phosphate as the phosphorobenzimidazolidate was also demonstrated [154].

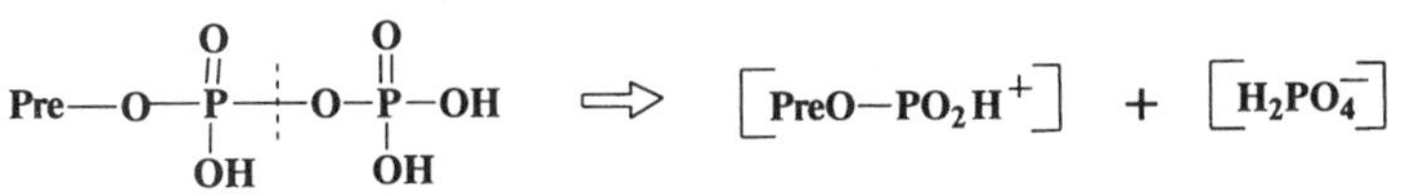

$$\text{Pre}-\text{O}-\overset{\overset{\text{O}}{\|}}{\underset{\underset{\text{OH}}{|}}{\text{P}}}-\text{O}-\overset{\overset{\text{O}}{\|}}{\underset{\underset{\text{OH}}{|}}{\text{P}}}-\text{OH} \implies \left[\text{PreO}-\text{PO}_2\text{H}^+\right] + \left[\text{H}_2\text{PO}_4^-\right]$$

Scheme 12

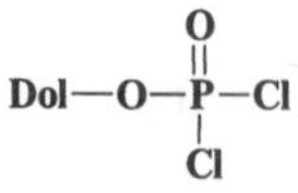

81 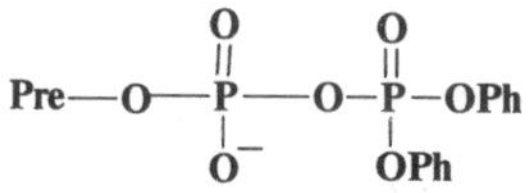　　　　　　　　　　　　　　**82**

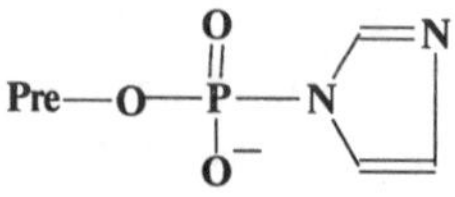

83 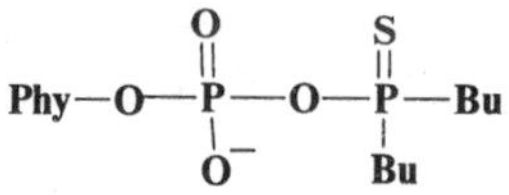　　　　　　　　　　　　　　**84**

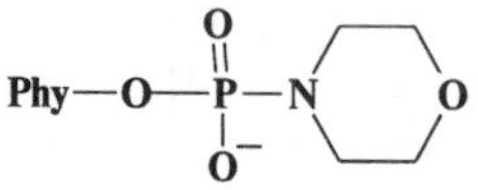

85 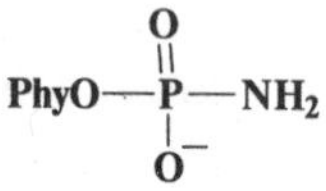　　　　　　　　　　　　　　**86**

Dol - dolichyl, **Pre** - polyprenyl or dolichyl, **Phy** - phytanyl

Derivatives **84–86** were obtained only for phytanol and related saturated alcohols. Preparation of **84** was achieved by interaction of the phosphate with di-(*n*-butyl)thiophosphonyl bromide [135,136]. For synthesis of **85**, the *N,N'*-dicyclohexylcarbodiimide-mediated reaction was used [134]. In the same paper, preparation of phytanyl phosphoramidate **86** through a multistep synthesis was described.

Interaction of dolichyl phosphoroimidazolidate **83** with bis-(tri-*n*-octylammonium) hydrogen [^{33}P]phosphate in tetrahydrofuran-dimethylsulfoxide mixture was employed for the first synthesis of dolichyl [β-^{33}P]diphosphate [152], the desired product being isolated in a yield of 18%. In an application of moraprenyl-derived **83**, moraprenyl diphosphate was obtained in a yield of 24% after reaction with bis-(triethylammonium) cyanoethyl phosphate, followed by treatment of the initially formed diphosphate diester with ammonia in aqueous *n*-butanol [155].

Derivative **82** seems preferable for the synthesis of polyprenyl and dolichyl diphospates [155]. Its interaction with an excess of tetra-*n*-butylammonium dihydrogen phosphate in a dichloromethane-pyridine mixture was found to produce 60%–65% of moraprenyl, solanesyl, and dolichyl diphosphates.

In the phytanol series, derivatives **85** and **86** were successfully employed for the synthesis of the diphosphate, using dioxane diphosphate as reagent [134].

Synthesis Through Activated Prenol Derivatives

An alternative approach to the synthesis of prenyl diphosphates is based on their retrosynthetic disconnection into prenyl cation and diphosphate anion (Scheme 13):

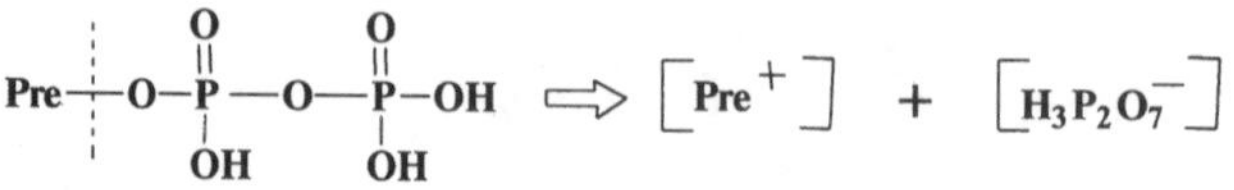

Scheme 13

It was found to be highly successful for the derivatives of lower prenols. Specifically, 3,3-dimethylallyl, geranyl, farnesyl, and geranylgeranyl diphosphates were obtained in 70%–75% yields through interaction of the corresponding chlorides with tris-(tetra-n-butylammonium) hydrogen diphosphate in acetonitrile [156–159]. In some cases the bromides may be used as synthetic equivalents of the prenyl cation; for α-dihydroprenols O-tosylates gave better results.

Until now, attempts to use the analogous approach for polyprenol or dolichol derivatives have been unsuccessful (L. L. Danilov, S. D. Maltsev, and V. N. Shibaev, unpublished results).

C. Analogs with Modified Phosphate or Diphosphate Fragment

Modified derivatives of dolichyl (polyprenyl) mono- and diphosphates with alterations in the phosphate fragment of the molecule may be useful for physicochemical studies and as potential inhibitors of some enzymic reactions. In the dolichol series only two derivatives of this group have so far been synthesized [160]: dolichyl thiophosphate **87** and dolichyl H-phosphonate **88**.

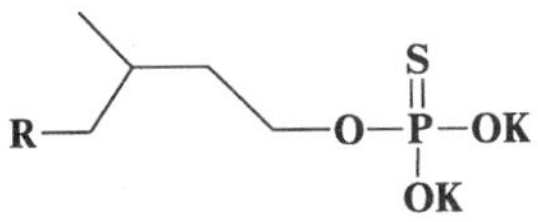

87 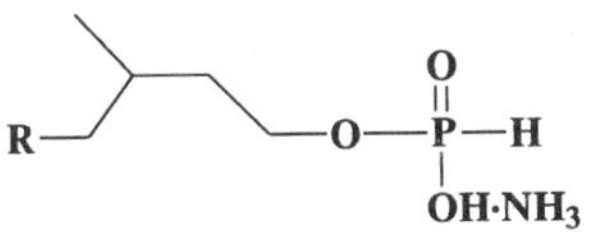 **88**

$$R = W\text{-}, \ WT_2C_{12\text{-}16}\text{-}$$

The former derivative was obtained through interaction of dolichol with $PSCl_3$ in pyridine, followed by treatment with alkali under controlled conditions. For preparation of the latter, reaction of dolichol with triimidazolyl phosphine was employed. The analogous citronellyl phosphate analogs were also prepared.

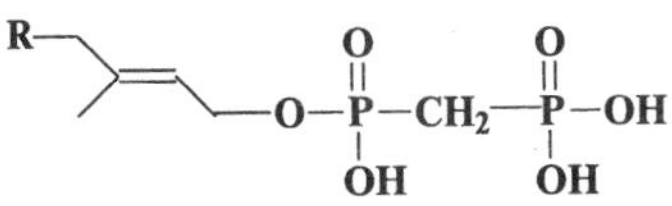

89 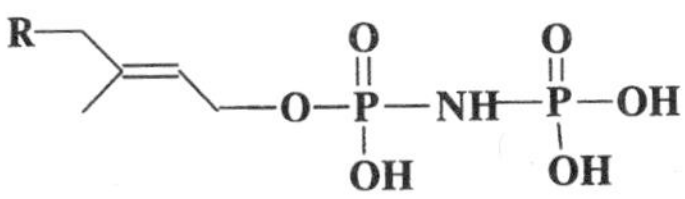 **90**

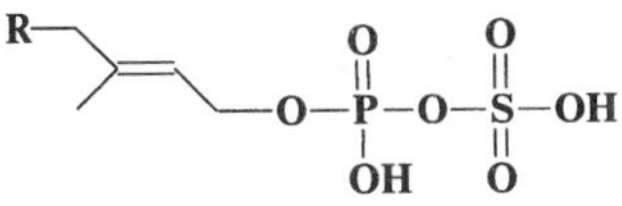

91 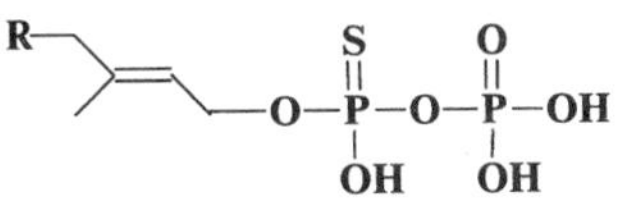 **92**

$$R = H \ (\text{for } 90), \ W\text{-}, \ WT\text{-}$$

A series of modified derivatives of dimethylallyl, geranyl, and farnesyl diphosphates with alterations of the diphosphate fragment were prepared in connection with studies on the mechanism of the enzymic prenyltransferase reaction (see Refs. 158 and 161–163), including analogs **89–92**.

It remains unclear whether similar analogs may be prepared in the polyprenol and dolichol series.

V. SYNTHESIS OF GLYCOSYL PHOSPHATES

For chemical synthesis of glycosylated dolichyl phosphates or diphosphates that serve as biosynthetic intermediates in carbohydrate-chain assembly, it is necessary in most cases to employ glycosyl phosphates or their protected derivatives as starting materials. The preparation of these phosphates includes some novel features in comparison with the synthesis of more usual phosphate monoesters. The most significant difference is a necessity to secure a definite stereochemistry at C-1 of the reaction product. The stereochemical result of the phosphorylation is strongly dependent on the structure of the sugar and, particularly, on the nature of the protecting groups used. It is significantly influenced also by the phosphorylating reagent and the experimental conditions of the reaction.

For the synthesis of the intermediates of the dolichol pathway of protein glycosylation, only three glycosyl phosphates derived from monosaccharides are required, namely, β-D-glucopyranosyl phosphate **93**, β-D-mannopyranosyl phosphate **94**, and 2-acetamido-2-deoxy-α-D-glucopyranosyl phosphate **95**. Also, oligosaccharide derivatives with a residue of **95** at the ''reducing'' end of the carbohydrate chain are necessary for the preparation of dolichyl diphosphate oligosaccharides.

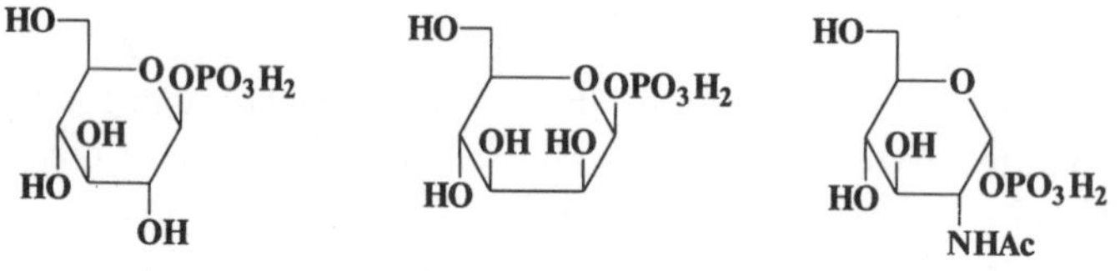

93   **94**   **95**

The polyprenol-linked intermediates of bacterial polysaccharide biosynthesis in some cases contain residues of α-D-glucopyranosyl phosphate, α-D-galactopyranosyl phosphate, or phosphates of oligosaccharides derived therefrom. Moreover, for structure-activity studies of the natural products in this series, the preparation of some analogs with a modified glycosyl phosphate residue may be useful, and data concerning the synthesis of the corresponding glycosyl phosphates are included in this section.

As mentioned earlier (see formula **66** in Section IV.A.), two strategic approaches to the synthesis of phosphoric acid monoesters are possible. For glycosyl phosphates the methods based on the use of synthetic equivalents of a glycosyl cation and phosphate anion are much better developed.

A. Through Synthetic Equivalents of a Glycosyl Cation

A large number of synthetic equivalents for the glycosyl cation are known as a result of intensive studies on the synthesis of glycosides and oligosaccharides (for recent reviews, see Refs. 164 and 165), but a few of them were practically useful for glycosyl phosphate synthesis. As synthetic equivalents of the phosphate anion, phosphoric acid or its salts were initially employed, mainly in acid-promoted phosphorylation reactions. Later, phosphate diesters or their salts were used, with subsequent deprotection of the initially formed phosphate triesters. Dibenzyl and diphenyl phosphates were frequently employed, often giving rise to products with different anomeric configuration.

The structure of any substituent at C-2 of the monosaccharide residue and the stereochemistry at this center are both of utmost importance for the stereochemical result of glycosyl phosphate synthesis.

Oxygen-Containing Substituents at C-2 of the Substrate

Both anomers of glycosyl phosphate derivatives may be prepared from monosaccharide substrates **96ab** with oxygen-containing substituents in an equatorial position (i.e., from hexopyranoses of gluco- or galacto-configuration), the processes occurring as shown in Scheme 14.

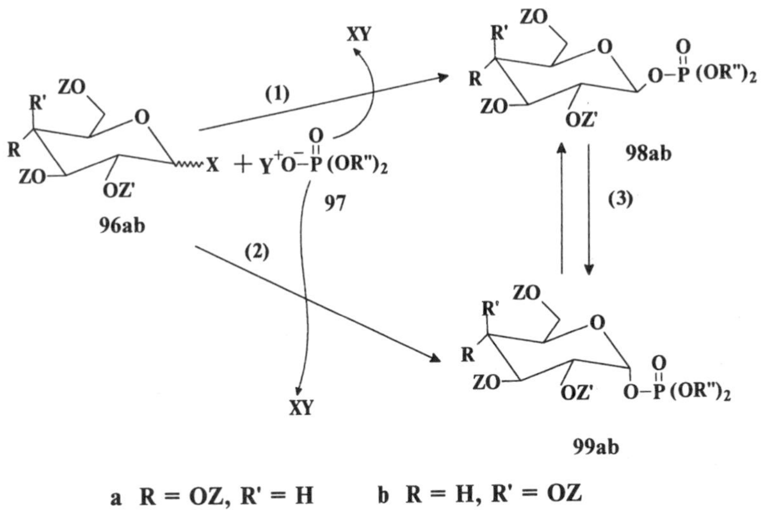

$$\text{XY} \quad (1) \quad \textbf{98ab} \quad (3) \quad \textbf{99ab}$$

$$\textbf{96ab} \quad \sim\!\!X + Y^+O\!-\!\overset{\overset{O}{\|}}{P}(OR'')_2 \quad \textbf{97} \quad (2) \quad \text{XY}$$

$$\textbf{a} \ R = OZ, \ R' = H \qquad \textbf{b} \ R = H, \ R' = OZ$$

Scheme 14

In most cases, substrates with acetyl-protecting groups (Z = Z' = OAc) were used. Due to neighboring-group participation of the acyl group at C-2, nucleophilic substitution at C-1 with anion **97** leads to the 1,2-*trans*(β)-glycosyl phosphates **98ab** (reaction 1) as kinetically controlled products. In a parallel reaction (2), the 1,2-*cis*(α)-anomers **99ab** may be produced. But the main pathway for the formation of these thermodynamically more stable glycosyl phosphates is probably reaction (3), anomerization of **98ab**. The efficiency of this reaction is strongly dependent on the properties of the **97** anion as a leaving group.

The simplest and most widely used procedure for glycosyl phosphate synthesis is the MacDonald method (for reviews, see Refs. 166–169), interaction of β-glycosyl peracetates, for example, **96a** (Z = Z' = Ac, X = β-OAc) with anhydrous phosphoric acid **97** (Y = R'' = H). After melting of the reactants in vacuo for 15 min at 50°, a high yield of β-D-glucopyranosyl phosphate **93** was obtained, whereas after reaction for 2 h only the α-phosphate was isolated. The reaction may be performed also with α-peracetates as substrates, although with a lower efficiency. In the usual procedure, the initially formed products were subjected directly to deacetylation, but their isolation may also be easily achieved.

In pioneering syntheses of **93** or its α-anomer from tetra-O-acetyl-α-D-glucopyranosyl bromide **96a** (Z = Z' = Ac, X = α-Br), phosphoric acid salts were used as reagents (for a review of classical methods of glycosyl phosphate synthesis, see Ref. 170). With trisilver phosphate, formation of the α-phosphate was noted, whereas the β-anomer was the product of the reaction with silver dihydrogen phosphate. Application of phosphate diester salts for this purpose was found much more efficient, and these classical methods are still in use.

In the reaction of **96a** (Z = Z' = Ac, X = α-Br) with silver dibenzyl phosphate (**97**, Y = Ag, R'' = OBn) after reflux for 90 min in benzene solution, the β-phosphotriester **98a** (Z = Z' = Ac, R'' = Bn) was isolated in a yield of 73%. This was further converted into **93** by hydrogenolysis over a palladium catalyst and deacetylation with aqueous alkali. A similar procedure was found useful for the preparation of a series of other 1,2-*trans*-glycosyl phosphates; in some cases, triethylammonium or tetra-*n*-butylammonium dibenzyl phosphates were successfully employed. Recently a procedure was described [171] for the synthesis of **98a** and **98b** (Z = Z' = Ac, R'' = Bn) in high yields from the corresponding bromides under conditions of phase-transfer-catalysis reaction (tetra-*n*-butylammonium hydrogen sulfate in aqueous $NaHCO_3/CH_2Cl_2$ system).

The stereochemistry of the reaction product was found to be different when a substituent at C-2 of the substrate was unable to participate in substitution at C-1. Thus, reaction of 2-O-trichloroacetyl-3,4,6-tri-O-acetyl-β-D-glucopyranosyl chloride (**96a**, Z = Ac, Z' = COCCl$_3$, X = β-Cl) with silver dibenzyl phosphate was shown [172] to give smoothly (93%) the α-triester **99a** (Z = Ac, Z' = COCCl$_3$, R'' = Bn), which is probably formed through reaction (2) (Scheme 14).

Interaction of acetylated glycosyl bromides with silver diphenyl phosphate resulted in predominant formation of the thermodynamically stable products. Thus, when **96a** (Z = Z' = Ac, X = α-Br) was treated with this reagent under conditions quite similar to those used for reaction with silver dibenzyl phosphate (60-min reflux of benzene solution), the α-anomer **99a** (Z = Z' = Ac, R'' = Ph) was produced. This was subjected without purification to hydrogenolysis over PtO_2 and deacetylation to give α-D-glucopyranosyl phosphate, in a total isolated yield of 37%.

Glycosyl O-trichloroacetimidates were found [173–175] to serve as efficient glycosyl cation synthetic equivalents in glycosyl phosphate synthesis. After interaction of dibenzyl phosphate (**97**, Y = H, R'' = Bn) with the acetylated derivative **96a** (Z = Z' = Ac, X = α-O—C(=NH)CCl$_3$) (1 h at room temperature in CH_2Cl_2), the β-phosphotriester **98a** (Z = Z' = Ac, R'' = Bn) was formed. The analogous reaction with the benzylated trichloroacetimidates **96a** (Z = Z' = Bn, X = α- or β-

O—C(=NH)CCl$_3$) gave the phosphates with inverted configuration at C-1 (**98a** or **99a**, Z = Z′ = R″ = Bn, respectively). The β-anomer **98a** was readily transformed into the more stable α-phosphotriester **99a** in the presence of acidic catalysts (for example, HCl/CH$_2$Cl$_2$, p-TsOH, BF$_3$·Et$_2$O), such anomerization occurring readily under the influence of a strong acid impurity present in a commercial preparation of dibenzyl phosphate. The thermodynamically stable α-phosphotriester **99a** (Z = Z′ = Bn, R″ = Ph) was found to be the only product (yield of 89%) in the reaction of benzylated α-trichloroacetimidate with diphenyl hydrogenphosphate.

The use of other potential glycosyl cation synthetic equivalents of the general structure **96** in the preparation of glycosyl phosphates is more limited. Thus, the interaction of ethyl 2,3,4,6-tetra-O-benzoyl-1-thio-β-D-galactopyranoside (**96b**, Z = Z′ = Bz, X = β-SEt) with dibenzyl phosphate promoted with N-iodosuccinimide was shown [176] to give an 80% yield of the β-phosphate **98b** (Z = Z′ = Bz, R″ = Bn). The reaction of an anomeric mixture of benzylated 4-pentenyl glycosides **96a** (Z = Z′ = Bn, X = —O(CH$_2$)$_3$CH=CH$_2$) and dibenzyl phosphate in the presence of N-bromosuccinimide or I(s-collidine)$_2$ClO$_4$ produced a mixture of the α- and β-phosphotriesters **98a** and **99a** (Z = Z′ = R″ = Bn) in ratios from 1:4 to 4:1, depending on the experimental conditions [177].

For the preparation of 1,2-trans-glycosyl phosphates, including **93**, high efficiency of some bicyclic carbohydrate derivatives as synthetic equivalents of glycosyl cations was observed. Thus, the β-phosphotriester **100a** was obtained in a 90% yield through the interaction of the 1,2-anhydro-derivative **101** with dibenzyl phosphate [178]. It is interesting that **100a** is smoothly formed also by treatment of the chloride **100b** with silver dibenzyl phosphate, probably via intermediate formation of **101**.

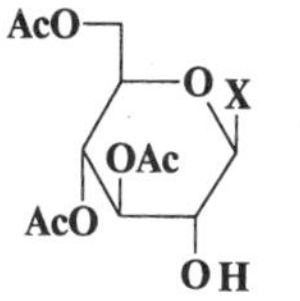

100ab	**101**	**102a-c**
a X = OP(O)(OBn)$_2$		a R = Bu^t
b X = Cl		b R = Ph
		c R = Et

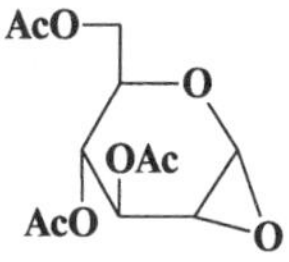

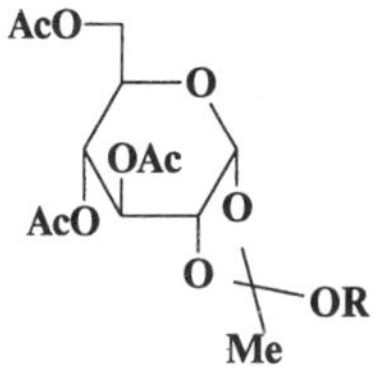

The 1,2-orthoacetate derivatives **102a–c** were successfully employed for the preparation of **93** [179,180]. Thus, after reaction of **102a** with dibenzyl hydrogen phosphate (benzene solution, 30 min, room temperature), the β-phosphotriester **98a** (Z = Z′ = Ac, R″ = Bn) was obtained in a high yield (85%), and the analogous result was achieved with the phenyl orthoester **102b** as a substrate. In the case of the primary or secondary alcohol derivatives, such as **102c**, in a side reaction the 2-O-deacetylated product **100a** was formed. In contrast to other cases mentioned, the interaction of the orthoacetate **102a** with diphenyl hydrogen phosphate under similar conditions also gave the β-phosphate **98a** (Z = Z′ = Ac, R″ = Ph). Similar results were obtained for the D-galactopyranose derivatives [181], although in this case the reaction seems to have been slightly slower.

In the synthesis of glycosyl phosphates from the 1,2-orthoacetates, employment of phosphoric acid as a reagent proved useful. With a small excess of H_3PO_4 (1.2 mol/mol of the orthoester), efficient syntheses of **93** and β-D-galactopyranosyl phosphate were performed [182,183]. Other investigators reported synthesis of **93** [184], β-L-galactopyranosyl phosphate [185], and β-L-fucopyranosyl phosphate [186] under more drastic experimental conditions (5 mol H_3PO_4/mol of the orthoester, addition of P_2O_5 into the reaction mixture). In this case some side processes were observed, particularly formation of a small proportion of the α-anomer, with increasing reaction times [185].

The interaction of **102a** and its D-galacto-analog with H_3PO_3 was recently used [187] to prepare the corresponding β-D-glycosyl H-phosphonates, which are of interest as activated derivatives for the synthesis of dolichyl glycosyl phosphates (see Section VI.A.).

The dependence of phosphorylation stereochemistry on the structure of reactants described above for the glucose and galactose derivatives remains generally correct for other monosaccharides with the same configuration at C-2, although all structural changes leading to an increase in the rate of substitution at C-1 favor more facile formation of the thermodynamically controlled products.

Thus, for deoxysugars preparation of the α-phosphates may be readily achieved. Specifically, analogs of α-D-glucopyranosyl phosphate with a deoxy-unit at C-6, C-4, or C-3 were prepared with the use of the MacDonald reaction [188,189], and through the interaction of silver diphenyl phosphate with the corresponding glycosyl bromides [190–192] or chlorides [193]. The same methods were applied successfully in other cases, for example, in syntheses of α-anomers of D-glucosyl and D-galactosyl phosphates containing a fluorine atom [189,194–196], iodine atom

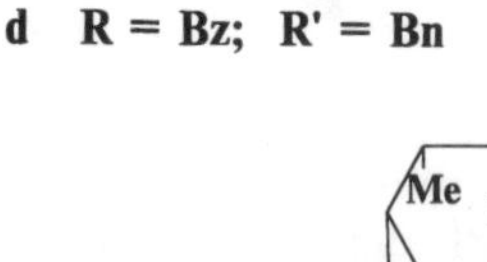

103a-d

a   R = R' = H
b   R = Ac, R' = Bn
c   R = R' = Bn
d   R = Bz; R' = Bn

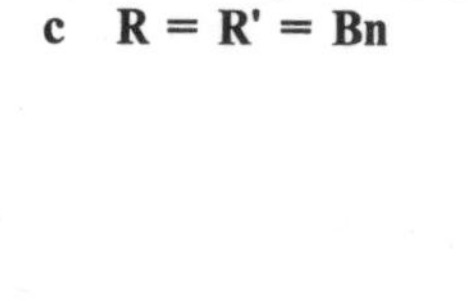

104a-c

a   R = R' = H
b   R = Ac, R' = Bn
c   R = R' = Bn

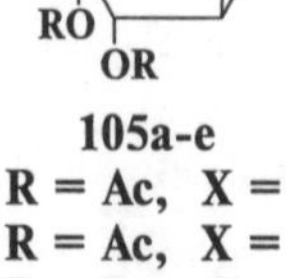

105a-e

a   R = Ac, X = Cl
b   R = Ac, X = -OC(=NH)CCl₃
c   R = Bn, X = -OC(=NH)CCl₃
d   R = Ac, X = Br
e   R = Bz, X = Br

[193] or azidogroup [193]. The preparation of 5-thio-α-D-glucopyranosyl phosphate [197] was achieved only through the use of silver diphenyl phosphate, and not through the MacDonald procedure.

In contrast, the synthesis of β-anomers of deoxyglycosyl phosphates may be difficult. The problem was studied most extensively in the case of β-L-fucopyranosyl phosphate **103a**. The first attempts at its preparation were not successful. Only the α-anomer **104a** was isolated after interaction of 2,3,4-tri-O-acetyl-α-L-fucopyranosyl chloride **105a** with silver dibenzyl phosphate [198] or under standard conditions of the MacDonald procedure [199]. Even when the latter reaction was performed for 1 min, the ratio of **104a** and **103a** in the products was found to be 9:1. The first satisfactory preparation of **103a** through nucleophilic substitution at C-1 was achieved with the use of the 1,2-orthoacetate [186]. Also, the reaction of α-O-trichloroacetimidates **105b** and **105c** with dibenzyl hydrogen phosphate was found [200] to give the β-triesters **103b** and **103c**, although the extreme ease with which they underwent anomerization into the derivatives of α-phosphates **104b** or **104c** under the influence of traces of a strong acid was noted.

In further studies on the interaction between 2,3,4-tri-O-acetyl-α-L-fucopyranosyl bromide **105d** and different salts of dibenzyl phosphate, the strong influence of experimental conditions on the stereochemical result was observed. With tetra-n-butylammonium salt as the reagent (5 min in DMF solution), almost pure β-phosphate **103a** was obtained (50% after deblocking) [201]. With triethylammonium dibenzyl phosphate solution and in the presence of Ag$_2$CO$_3$, only the α-triester **104b** was formed [202].

The use of the bromide **105e** with benzoyl-protecting groups gave excellent yields of the β-phosphate derivative **103d** after treatment with dibenzyl phosphate in the presence of Ag$_2$CO$_3$ [202,203]. Reaction of the same bromide with Bu$_4$N·H$_2$PO$_4$ in the presence of 2,6-lutidine, followed by deblocking, gave a mixture of equal amounts of **103a** and **104a** in a yield of 40% [203]. Also, a good yield of the β-phosphate **103a** was obtained when ethyl 2,3,4-tri-O-benzoyl-1-thio-β-L-fucopyranoside was employed as a substrate [176].

The preparation of **94** and related β-phosphates is much more difficult than the synthesis of **93**. In the case of the manno-configuration, an acyl substituent at C-2 is in the axial position, and its efficient participation in substitution at C-1 leads to the formation of a 1,2-trans(α)-diaxial product that is more stable than the corresponding 1,2-cis(β)-anomer. Accordingly, α-D-mannopyranosyl phosphate **106** may be readily obtained by the MacDonald procedure [168] and through interaction of 2,3,4,6-tetra-O-acetyl-α-D-mannopyranosyl chloride with silver diphenyl or dibenzyl phosphates [204]. The formation of **106** as a single product was observed [174] in the reaction of 2,3,4,6-tetra-O-benzyl-α-D-mannopyranosyl O-trichloroacetimidate with dibenzyl phosphate as well as with 1,2-orthoacetates as substrates [180,182].

106   **107**

Similar methods were successfully employed for preparation of α-L-rhamnopyranosyl phosphate **107**, analogs of **106** with a deoxy-unit at C-6, C-4, or C-3 [205,206], and the analogous derivative of 4-fluoro-4-deoxy-D-mannose [207].

As was shown more recently [208], the synthesis of the β-glycosyl phosphates in this series may be successfully performed through the use of glycosyl halides possessing a 2,3-carbonate-protecting group, such as the L-rhamnopyranose derivative **108**. Its interaction with triethylammonium dibenzyl phosphate followed by deprotection gave smoothly (71%) β-L-rhamnopyranosyl phosphate **109**. The analogous derivative was employed for preparation of β-D-mannopyranosyl phosphate **91** [142].

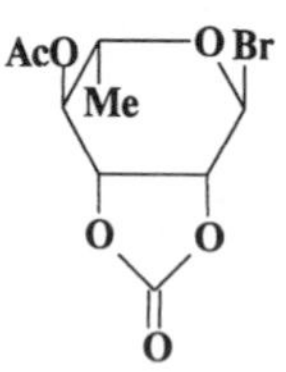

108 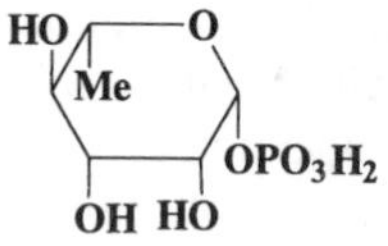   **109**

Nitrogen-Containing Substituents at C-2

Several procedures based on nucleophilic substitution at C-1 were suggested for the synthesis of **95** and related glycosyl phosphates. The synthetic equivalents of the glycosyl cation are often different from those described in the previous subsection.

The only synthetic method that may be treated as completely analogous to those used in the hexose series is the MacDonald procedure, i.e., the interaction of 2-acetamido-2-deoxyhexose peracetates with phosphoric acid. Under standard reaction conditions (3 h, 50°C) or at slightly elevated temperature (45 min, 83°–90°C), the *N*-acetyl-D-glucosamine derivative was shown [209–211] to give anomeric mixtures of the corresponding glycosyl phosphates **95** and **110** in ratios from 1:1 to 2:3. A procedure for isolation of their tri-*O*-acetates was developed [113]. The stereochemical result seems to depend on the structure of the sugar phosphorylated. Thus, after reaction with the *N*-acetylglucosamine derivative (3 h, 65°C), the predominance of the α-glycosyl phosphate was noted [113], whereas from the *N,N'*-diacetyl-chitobiose derivative mainly β-phosphate was formed [212].

Preparation of the α-phosphate **95** as a single reaction product was achieved [213] when dioxane-diphosphate was employed as a reagent and the process was performed at 95°–100°C. The same conditions were successfully applied to the synthesis of 2-acetamido-2-deoxy-α-D-galactopyranosyl phosphate **111**. The analogous 2-acetamido-2-deoxy-α-D-mannopyranose derivative **112** was obtained under the standard condition of the MacDonald reaction in a yield of 14% [214].

The use of glycosyl halides as synthetic equivalents of the glycosyl cation in this series is very limited. The 2-phthalimido-2-deoxy-derivatives **113ab**, widely used

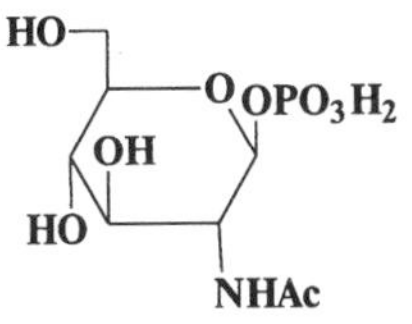

110

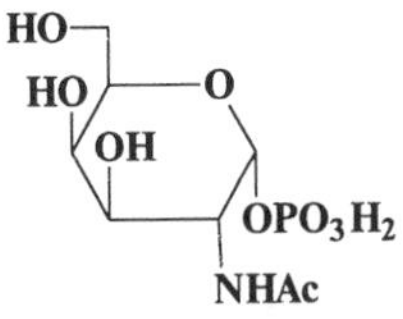

111

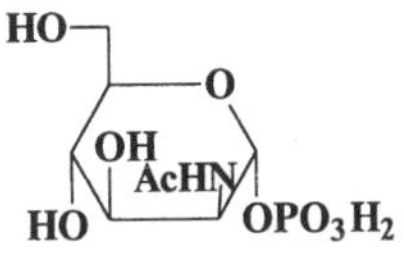

112

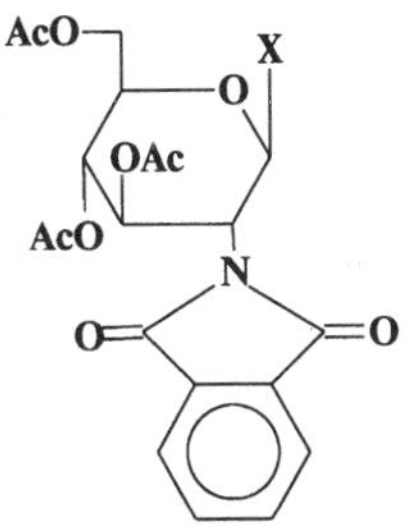

113ab
a X = Br
b X = Cl

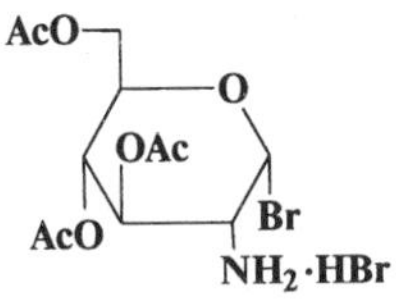

114

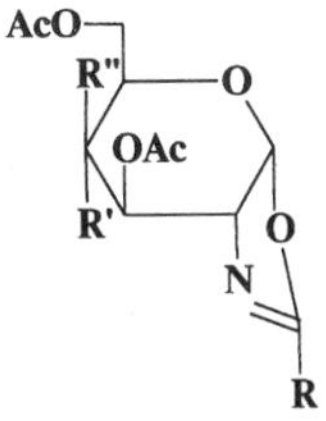

115a-c

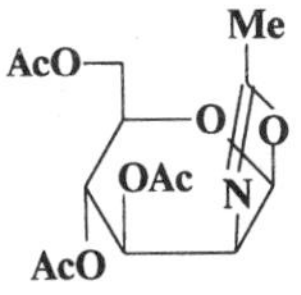

116

a R = Me, R' = OAc, R" = H
b R = Me, R' = H, R" = OAc
c R = Me(CH$_2$)$_{14}$-, R' = OAc, R" = H

in oligosaccharide synthesis, were found [215] to react smoothly with silver dibenzyl phosphate, producing the corresponding β-phosphotriesters (65% and 81%, respectively), which after deblocking were converted into **110**. The first synthesis of **95** was achieved [216,217] through reaction of the bromide **114** with triethylammonium diphenyl phosphate followed by *N*-acetylation and de-*O*-acetylation. When triethyl-ammonium dibenzyl phosphate was used as the reagent in a similar phosphorylation [218], a mixture of the α- and β-triesters (isolated as *N*-benzyloxycarbonyl derivatives) was formed.

The most useful substrates for synthesis of 2-acetamido-2-deoxy-hexopyranosyl phosphates were bicyclic derivatives containing the oxazoline ring, such as **115ab** or **116**. They are readily available from the corresponding glycosyl halides or acetates, and are applied widely in oligosaccharide synthesis for 1,2-*trans*-glycosylation.

In the D-*manno*-series, interaction of **116** with dibenzyl phosphate, followed by hydrogenolysis and deacetylation, was shown [214] to give the expected α-phosphate **112** in a yield of 46%. In a subsequent paper [219] a 61% yield was reported.

In contrast, in the analogous reaction of the D-*gluco*-derivative **115a**, formation of the α-phosphotriester **118** (R = Me, R′ = R″ = Bn) was observed [220]; and in a later, more detailed study, an intermediate formation of the β-phosphate **117** (R = Me, R′ = R″ = Bn) was demonstrated [221] (see Scheme 15).

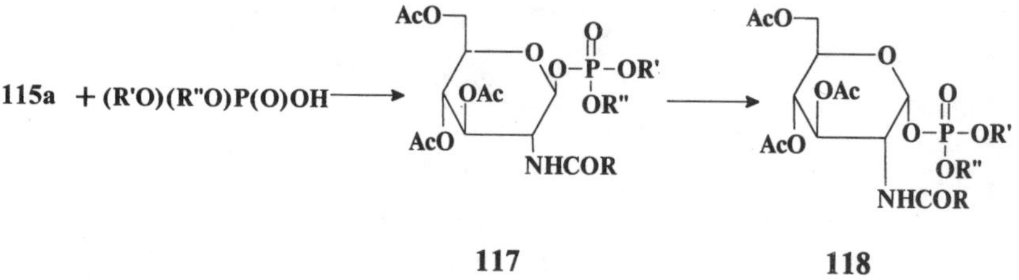

$$115a \ + (R'O)(R''O)P(O)OH \longrightarrow$$

117   118

Scheme 15

The stereochemical result of the reaction was shown to be the same for the benzylated analog of **115a** [222] and for the D-*galacto*-derivative **115b** [219]. On the basis of this reaction, efficient procedures were developed for preparation of **95** and an analogous *N,N′*-diacetylchitobiose derivative [212,223]. The reaction was further applied to the synthesis of fluorine-containing analogs of **95** and **111** [224]. The preparation of some lipid A analogs with higher fatty acid residues linked to the amino-group at C-2 was achieved from the corresponding oxazoline derivatives (such as **115c**) [221,225].

Other synthetic equivalents of the phosphate anion may be successfully used in reaction with oxazoline derivatives of aminosugars. Thus, the α-phosphotriester **118** (R = Me, R′ = R″ = Ph) was smoothly formed after interaction of **115a** with diphenyl phosphate [212]; the reagent was applied to the synthesis of α-phosphates derived from some disaccharides [220,226]. The interaction of **115c** with the phosphomonoester $CBr_3CH_2OPO_3H_2$ was shown [227] to give either a mixture of the diesters **117** and **118** (R = $Me(CH_2)_{14}$—, R′ = —CH_2CBr_3, R″ = H), with the preponderance of the β-anomer (after reaction for 6 h at room temperature), or the pure α-phosphate (66% after 3 days at 35°C).

Other Substituents at C-2

In this series, several glycosyl phosphates derived from 2-deoxy-2-fluorohexoses and 2-deoxyhexoses were prepared, mainly with use of the MacDonald procedure.

High electronegativity of a fluorine substituent at C-2 leads to significant diminishing of the rate of substitution at C-1. As a result, more drastic reaction conditions are usually applied to prepare the thermodynamically stable α-glycosyl phosphates. Thus, for synthesis of **119** (35% yield), the corresponding sugar per-acetate was treated with phosphoric acid for 48 h at 55°C [196] and **120** was prepared in a yield of 36% after reaction for 30 min at 80°C [228]. Slightly more mild conditions (9 h, 55°C) were found acceptable [189] to obtain the 2,6-dideoxy-2,6-difluoro-α-D-glucopyranosyl phosphate in a yield of 61%.

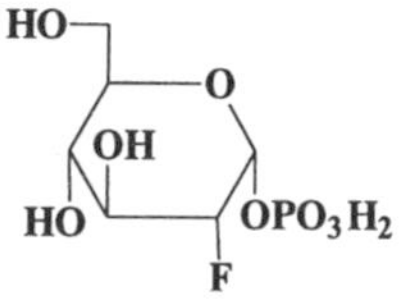

119 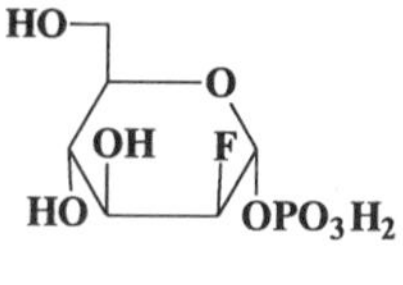   **120** 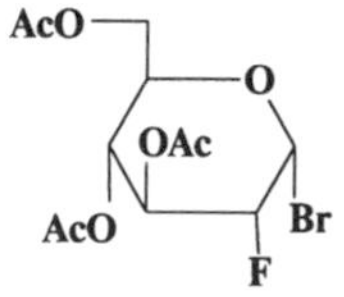   **121**

The bromide **121** was used for the synthesis of the β-anomer of **119** [196] through an unusual reaction sequence, including treatment of the bromide with silver tosylate (−20°C in acetonitrile) followed by reaction with phosphoric acid in tetrahydrofuran (reflux for 90 min). After deacetylation, both anomeric glycosyl phosphates (in approximately equal amounts) were isolated in a total yield of 63%.

In contrast, nucleophilic substitution at C-1 in 2-deoxyhexosyl derivatives proceeded extremely fast, but the resulting product was highly unstable and prone to further conversions.

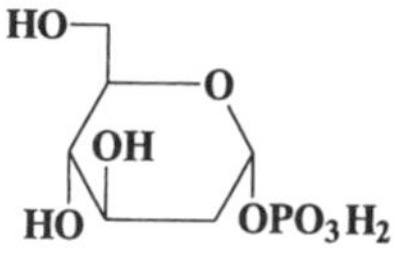

122 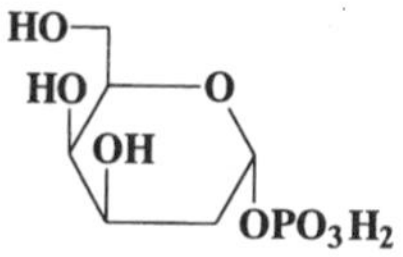   **123** 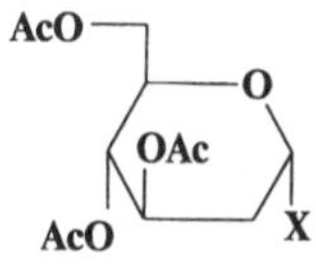   **124a-c**

a X = OP(O)(OBn)$_2$
b X = SP(S)(OEt)$_2$
c X = Cl

The synthesis of 2-deoxy-α-D-*arabino*-hexopyranosyl phosphate **122** was achieved after a short treatment (30 min, 50°C) of the corresponding peracetate with anhydrous phosphoric acid [229–231]. The analogous conditions were successfully applied for preparation of **123** [230] and 2,6-dideoxy-α-L-*arabino*-hexopyranosyl phosphate [232]. At the same time, there were also reports of an inability to obtain **122** through the MacDonald reaction [196,233].

More recently, other approaches to the synthesis of glycosyl phosphates of this series, with the use of dibenzyl phosphate as a reagent, were described [234]. The α-triester **124a** was prepared in high yield (75%), in the presence of I(*s*-collidine)$_2$ClO$_4$ as a promoter, from the dithiophosphate **124b**, readily available by addition of (EtO)$_2$P(S)SH to 3,4,6-tri-*O*-acetyl-D-glucal **125**. The analogous reaction was successfully applied to 2-deoxy-analogs of D-galactose and lactose. Also, **124a** was prepared in lower yields through interaction of the α-chloride **124c** and silver dibenzyl phosphate or by reaction of **125** with dibenzyl phosphate in the presence of Ph$_3$P·HBr. When the latter reaction was performed in the presence of *N*-iodosuccinimide, the derivative **126** was identified as the main product, but attempts to convert it into **122** failed.

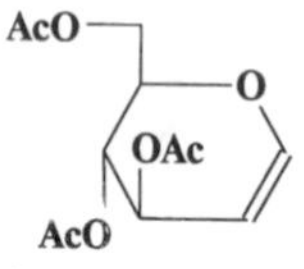

125 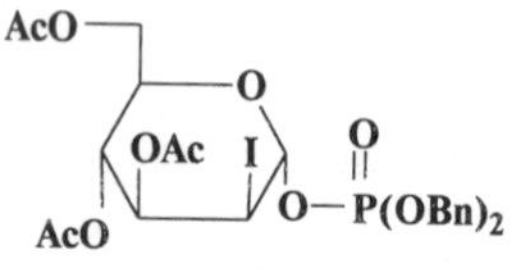   126

B. Through Synthetic Equivalents of Glycosyloxy Anion

The approach to glycosyl phosphate synthesis based on the use of glycosyloxy anion synthetic equivalents, namely, carbohydrate derivatives with free hydroxyl at C-1 or their metal salts, attracted much attention in recent years. In this approach, as synthetic equivalents of the phosphoryl cation, chlorides of phosphoric acid diesters were usually employed. More recently, some derivatives of phosphorous acid were successfully applied for this purpose.

With Derivatives of Phosphoric Acid

In general, the stereochemical result of phosphorylation of the free hemiacetal hydroxyl group in monosaccharide derivatives is dependent on the interrelationship between the rates of the reactions shown in Scheme 16, namely, mutarotation of the starting material (reaction 1), interaction of each of the anomers with a phosphorylating reagent (reactions 2a and 2b), and isomerization of the phosphotriesters (reaction 3; cf. Scheme 14). The last reaction may readily occur when the phosphate diester anion serves as a good leaving group, for example, in the case of diphenyl phosphate derivatives. Accordingly, variation of experimental conditions would allow predominant formation of either of the glycosyl phosphate anomers.

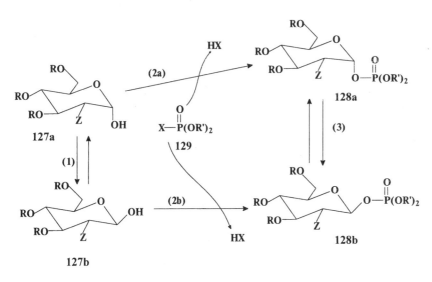

Scheme 16

The influence of monosaccharide structure and reaction conditions on the ste-reochemistry of phosphorylation was studied most thoroughly [235] for diphenyl phosphorochloridate (**129**, X = Cl, R' = Ph) as a reagent, with methylene dichloride as a solvent and 4-dimethylaminopyridine as a base. When an *O*-acetyl substituent is present at C-2 of a monosaccharide in the equatorial position, i.e., for D-*gluco*- (as shown in formulas **127ab**) or D-*galacto*-configurations, reaction (2b) is much faster than reaction (2a). As a result it was possible to perform selective conversion of **127b** (Z = OAc, R = Ac) into **128b** (Z = OAc, R = Ac, R' = Ph) at lowered temperature (such as −25°C). For example, the β-D-galactopyranosyl phosphate de-rivative was obtained in a yield of 50%. In contrast, in the presence of an excess of the free base, which catalyzed reaction (1), and at higher temperature (0°C), when a significant increase in velocity of the reaction (3) occurred, the thermodynamically stable anomer **128a** (Z = OAc, R = Ac, R' = Ph) was formed predominantly. Thus, for example, α-D-galactopyranosyl phosphate derivative was prepared in a yield of 78%. These conditions were used successfully for the preparation of derivatives of α-D-glucopyranose and α-L-fucopyranose. In the latter case, attempts to perform selective synthesis of the β-phosphate at lowered temperature failed, probably due to the higher rate of anomerization for the phosphotriester derivatives of 6-deoxyhexoses.

The nature of the protecting groups in the monosaccharide residue of **127ab** was found to influence the stereochemical result of the phosphorylation. When *O*-benzyl derivatives were used as substrates under identical conditions, a higher pro-portion of the α-phosphotriester was present in the reaction products, whereas with *O*-benzoyl derivatives formation of the β-anomers was favored in comparison with the *O*-acetates.

For the 2-acetamido-2-deoxy-D-glucopyranose derivatives (**127ab**, Z = NHAc, R = Ac), the β-phosphotriester was highly unstable under reaction conditions and rapidly converted into the oxazoline derivative **115a**. The best results for preparation of **128a** (Z = NHAc, R = Ac, R' = Ph) were achieved with the use of a large excess of the phosphorylating reagent and of the base, under lowered temperature ($-25°C$, 62% yield).

When the starting monosaccharide was of the *manno*-configuration, the stereo-chemical result of the phosphorylation reaction was different. In fact, when a mixture of **130a** and **131a** was treated with the reagent under lowered temperature ($-30°C$), the α-phosphate **131b** was formed exclusively (94%), whereas at room temperature in the presence of an excess of 4-dimethylaminopyridine a mixture of **130b** and **131b** (4:1) was formed, and **130b** was isolated in a yield of 77%. These results may be interpreted as a consequence of having a strong preponderance of **131a** in the starting material, rather slow reaction (3) even at room temperature when the C-2 acetoxy-group is present in an axial position, and an increase in this case of the difference in anomer reactivity, i.e., the ratio of the rates for reactions (2b) and (2a), with increasing temperature. The analogous conditions were found suitable for preparation of the α- and β-L-rhamnopyranosyl phosphate derivatives.

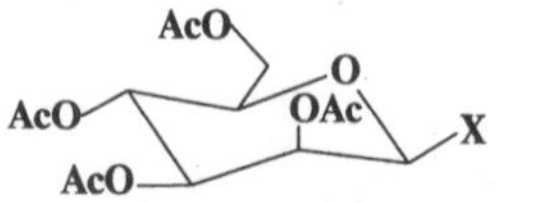

130ab
a X = OH

b X = OP(O)(OPh)$_2$

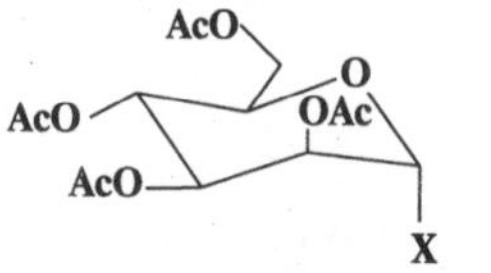

131ab
a X = OH

b X = OP(O)(OPh)$_2$

The reaction of diphenyl phosphorochloridate with metal salts of **127ab** was also used in some cases for the preparation of glycosyl phosphates. Thus after its reaction with the thallium(I) salt of **127ab** (Z = OBn, R = Bn) in acetonitrile, the α-anomer **128a** (Z = OBn, R = Bn, R' = Ph) was formed, whereas the analogous reaction in benzene gave a mixture of **128a** and **128b** (1:2) [236]. The lithium salt of the same monosaccharide substrate in tetrahydrofuran was found [237] to produce only the α-phosphotriester, while the analogous tetra-O-benzoyl derivative gave a mixture of **128a** and **128b** (Z = OBz, R = Bz, R' = Ph) in approximately equal amounts.

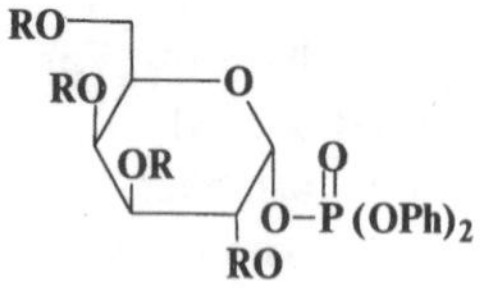

132a–c
a R = Bn
b R = Ac
c R = Bz

Under similar conditions the α-phosphates **132a** [237] and **132b** [237a] were prepared from the lithium salts of 2,3,4,6-tetra-*O*-substituted D-galactose with the *O*-benzyl ethers and *O*-acetates, but in the case of *O*-benzoates a mixture of **132c** and the β-anomer (3:1) was obtained [237].

When other phosphodiester derivatives were used as synthetic equivalents of the phosphoryl cation, the isomerization of the phosphotriesters (reaction 3 in Scheme 16) was of low significance. Among these reagents highly reactive *o*-phenylene phosphorochloridate (**129**, R' + R' = *o*-phenylene; see **64**) was frequently applied. Under the experimental conditions used for reaction with monosaccharide derivatives having a free hemiacetal hydroxyl group (tetrahydrofuran or dioxane as solvent, 2,6-lutidine or 2,4,6-collidine as a base, 0°–20°), the phosphorylation reactions (2a and 2b in Scheme B) are much faster than mutarotation of the starting monosaccharides (reaction 1), and the ratio of the anomers in the products corresponds to that in the starting derivative.

The initially formed phosphotriester was smoothly converted into the phosphodiester **68** by mild hydrolysis in the presence of base. Further transformation of the latter into a glycosyl phosphate was performed by action of an oxidizing agent (bromine water or lead tetraacetate) followed by alkaline treatment.

This reagent was employed for the first syntheses of some β-glycosyl phosphates, which are difficult to prepare through nucleophilic substitution at C-1. Thus from the acetate **130a**, β-D-mannopyranosyl phosphate **94** was first prepared [238]. Isolation of its tetraacetate for further use in the synthesis of a dolichyl glycosyl phosphate was described in Refs. 110 and 239. The method was successfully applied in the *manno*-series for the synthesis of β-L-rhamnopyranosyl phosphate **109** [240,241] and its α-anomer **107** [240], as well as for the preparation of the glycosyl phosphates derived from β-L-mannopyranose [242] and 4,6-dideoxy-L-*lyxo*-hexopyranose [243]. It was found useful for the synthesis of β-D-glucopyranosyl phosphate **93** [240], β-L-fucopyranosyl phosphate **103a**, and its α-anomer **104a** [240,244]. Also, phosphates of α-D-fucopyranose [245], 6-deoxy-α-D-glucopyranose [246], and 2-deoxy-2-palmitoylamino-α-D-glucopyranose [227] were prepared with this method.

The use of phosphorylating reagents derived from dibenzyl phosphate in reactions with monosaccharide derivatives possessing a free hemiacetal hydroxyl group is rather limited. Thus a mixture of **128b** and **128a** (Z = OBn, R = R' = Bn) in a ratio of 3:2 was formed after treatment of **127ab** (Z = OBn, R = Bn) with dibenzyl phosphorofluoridate (**129**, X = F, R' = Bn) in acetonitrile in the presence of cesium fluoride [247].

In contrast, reagents such as dibenzyl phosphorochloridate and tetrabenzyl pyrophosphate were found extremely useful for synthesis of glycosyl phosphates through reactions with lithium salts of alkoxides of **127ab** (R = Ac), particularly in the 2-acylamino-2-deoxy-D-glucose series.

Initially, the method was developed for the synthesis of lipid A fragments, containing residues of higher fatty acids [248]. Interaction of 1-*O*-lithium salts **133a** or **133b** with dibenzyl phosphorochloridate (tetrahydrofuran, 10 min, −60°C) was shown to give the α-phosphotriesters **128a** (Z = NH-acyl, R = Ac, R' = Bn), smoothly converted after deblocking into the α-glycosyl phosphates, in a total yield of 83%–85%. The analogous reaction with the thallium salt **133c** (benzeneacetonitrile, 10 min, −40°C) leads, however, to the β-triester **128b**, the corresponding β-glycosyl phosphate being isolated after deblocking in a yield of 55%. This fast and efficient procedure for the synthesis of 2-acylamino-2-deoxy-α-D-glucopyranosyl phosphates through 1-*O*-lithium salts has been widely used for preparation of lipid A derivatives and analogs (for a review see Ref. 249).

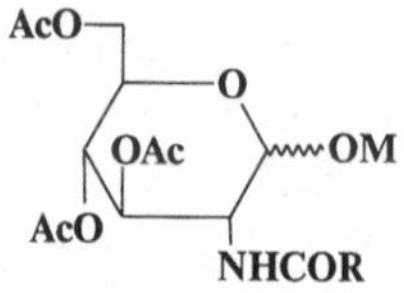

133a-d

a $R = Me(CH_2)_{12}-, M = Li$
b $R = Me(CH_2)_{10}CHOAcCH_2-, M = Li$
c $R = Me(CH_2)_{10}CHOAcCH_2-, M = Tl$
d $R = Me, M = Li$

A similar procedure was successfully applied to obtain *O*-acetylated derivatives of *N*-acetylglucosamine and *N,N'*-diacetylchitobiose α-phosphates [120,136,250], which were used further in the synthesis of the corresponding dolichyl glycosyl diphosphates. For example, phosphate **128a** (Z = NHAc, R = Ac, R' = H) was prepared after treatment of **133d** with tetrabenzyl pyrophosphate (1 h, 0°C) followed by hydrogenolysis, in a yield of 72%. Some deoxy-analogs of 2-acetamido-2-deoxy-α-D-glucopyranosyl phosphate were also synthesized by this method [251].

When the same reaction was applied to 1-*O*-lithium salts of monosaccharides with an oxygen-containing substituent at C-2, the stereochemical result was not so definite. Thus phosphorylation of D-galactose derivative **134a** was found [237a] to give a mixture of the β- and α-phosphates in a ratio of 2.3:1 (total yield 68%), whereas only α-phosphates were isolated after reaction with the derivatives of 6-deoxyhexose **134b** [252] or 3,6-dideoxyhexose **134c** [253]. In the cases of **134d** [252] and **134e** [253], a mixture of anomers, with strong preponderance of the α-phosphate, was formed. The interaction of the 2-deoxy-derivative **134f** with dibenzyl phosphorochloridate was successfully used for the synthesis of 2-deoxy-α-D-*lyxo*-hexopyranosyl phosphate **123** [254].

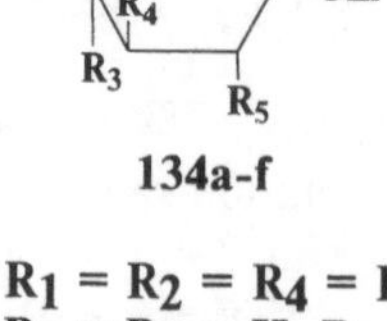

134a-f

a $R_1 = R_2 = R_4 = R_5 = OAc, R_3 = H$
b $R_1 = R_3 = H, R_2 = R_4 = R_5 = OAc$
c $R_1 = R_3 = R_4 = H, R_2 = R_5 = OBz$
d $R_1 = F, R_2 = R_4 = R_5 = OAc, R_3 = H$
e $R_1 = R_2 = R_4 = H, R_3 = R_5 = OBz$
f $R_1 = R_2 = R_4 = OAc, R_3 = R_5 = H$

With Derivatives of Phosphorous Acid

The trivalent phosphorus derivatives are generally much better electrophiles than the corresponding phosphate derivatives, and their use as synthetic equivalents of the phosphoryl cation for the synthesis of glycosyl phosphates was initiated recently. The finding of rather stable reagents and mild oxidation conditions for conversion of the intermediate phosphite esters into phosphates are the main limiting factors in the development of synthetic methods of this group. Until now, the phosphines **135** and **136** were found useful for this purpose.

$$\text{Cl}-\text{P}\overset{\diagup \text{NPr}_2^i}{\diagdown \text{OCH}_2\text{CH}_2\text{CN}} \qquad\qquad \text{Pr}_2^i\text{N}-\text{P}\overset{\diagup \text{OBn}}{\diagdown \text{OBn}}$$

$$\textbf{135} \qquad\qquad\qquad\qquad \textbf{136}$$

The first example of this approach was the synthesis of benzylated α-L-fucopyranosyl phosphate from 2,3,4-tri-*O*-benzyl-α-L-fucopyranose and **135** [255], as shown in Scheme 17. Phosphitylation of the monosaccharide derivative into **137** was found to proceed in 97% yield and the product was further converted into the phosphite triester, which was oxidized with *tert*-butyl hydroperoxide into the phosphate triester. The latter was deblocked to give the glycosyl phosphate in a total yield of 89% from **137**.

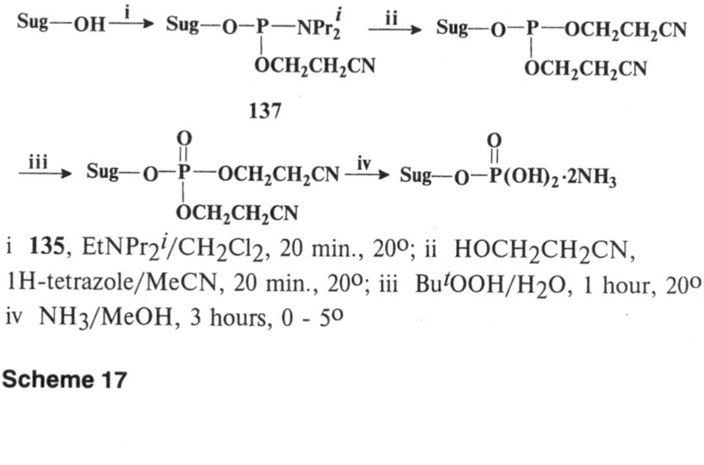

$$\text{Sug}-\text{OH} \xrightarrow{\text{i}} \underset{\underset{\text{OCH}_2\text{CH}_2\text{CN}}{|}}{\text{Sug}-\text{O}-\text{P}-\text{NPr}_2^i} \xrightarrow{\text{ii}} \underset{\underset{\text{OCH}_2\text{CH}_2\text{CN}}{|}}{\text{Sug}-\text{O}-\text{P}-\text{OCH}_2\text{CH}_2\text{CN}}$$

$$\textbf{137}$$

$$\xrightarrow{\text{iii}} \underset{\underset{\text{OCH}_2\text{CH}_2\text{CN}}{|}}{\overset{\overset{\text{O}}{\|}}{\text{Sug}-\text{O}-\text{P}-\text{OCH}_2\text{CH}_2\text{CN}}} \xrightarrow{\text{iv}} \overset{\overset{\text{O}}{\|}}{\text{Sug}-\text{O}-\text{P(OH)}_2 \cdot 2\text{NH}_3}$$

i **135**, EtNPr$_2^i$/CH$_2$Cl$_2$, 20 min., 20°; ii HOCH$_2$CH$_2$CN, 1H-tetrazole/MeCN, 20 min., 20°; iii Bu^{t}OOH/H$_2$O, 1 hour, 20°; iv NH$_3$/MeOH, 3 hours, 0 - 5°

Scheme 17

The same reaction sequence was successfully applied for the synthesis of β-L-rhamnopyranosyl phosphate **106** [256] and 3-acetamido-3-deoxy-α-D-mannopyranosyl phosphate [257] from the corresponding monosaccharide acetates. The derivative **137** was prepared also from 2-acetamido-3,4-di-*O*-benzoyl-2-deoxy-

6-O-(p-anisyldiphenylmethyl)-α-D-glucopyranose [258]. It was not converted into the glycosyl phosphate but was used for the synthesis of its diester.

The application of **136** for the synthesis of glycosyl phosphates was studied more thoroughly. The reaction sequence shown in Scheme 18 was applied to monosaccharide acetates, with a free hydroxyl group at C-1 derived from common hexoses, 6-deoxyhexoses, and 2-acetamido-2-deoxyhexoses [259], and in all cases glycosyl phosphates were obtained in high yields.

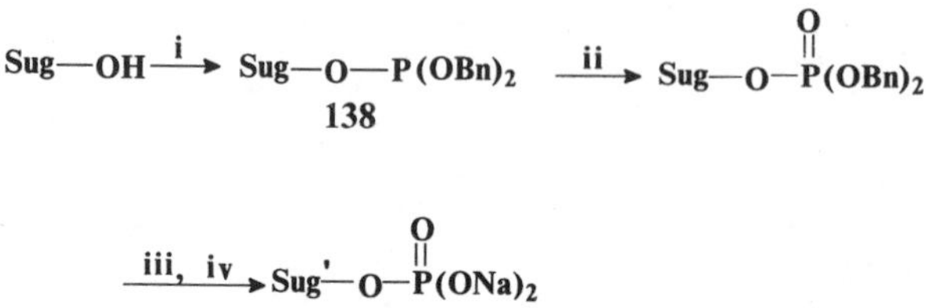

$$Sug-OH \xrightarrow{\ i\ } \underset{\textbf{138}}{Sug-O-P(OBn)_2} \xrightarrow{\ ii\ } Sug-O-\overset{\overset{O}{\|}}{P}(OBn)_2$$

$$\xrightarrow{\ iii,\ iv\ } Sug'-O-\overset{\overset{O}{\|}}{P}(ONa)_2$$

Sug - O-acetylated monosaccharide residue

Sug' - deblocked monosaccharide residue

i **136**, 1,2,4-triazole/CH_2Cl_2 or THF, 1 - 2 hours, 20°; ii 30% H_2O_2/THF; iii H_2/Pd-C in EtOH - 10% $NaHCO_3$; iv 1 N NaOH or 10% Et_3N/MeOH - H_2O, then Dowex-50 (Na)

Scheme 18

From 2-acetamido-2-deoxyhexose derivatives pure α-glycosyl phosphates were formed. In other cases, mixtures of anomeric phosphates were present. The content of the β-anomer in the intermediates **138** was found to be higher than in the starting monosaccharide acetates, this being probably due to higher reactivity of the equatorial hydroxyl group under the reaction conditions. During conversion of **138** into the glycosyl phosphate, the ratio of anomers was not changed.

For monosaccharides of the *manno*-series, the α-anomer predominates in the products (thus, for a D-mannose derivative, the ratio of α- and β-anomers was 3:1), whereas the opposite is true in the case of *gluco*- and *galacto*-configurations: The previously mentioned ratios were 1:2 for D-glucose and 1:4 for D-galactose derivatives. A significant increase in content of the β-anomer observed for the L-fucose derivatives (the ratio of α- and β-anomers was found to be 1:10) made it possible to employ this approach successfully in the synthesis of β-L-fucopyranosyl phosphate [203].

The presented data allow us to conclude that at present, synthetic methods for the preparation of glycosyl phosphates with a variety of different structures are rather well developed. Specifically, among the derivatives necessary for the synthesis of intermediates in the dolichol pathway of protein glycosylation, phosphate **93** may be readily prepared through various methods based on nucleophilic substitution at C-1. The choice of synthetic methods for **94** is more limited, but substitution at C-1 in a glycosyl halide with a 2,3-O-carbonate group and phosphorylation of the 1-OH derivative with *o*-phenylene phosphochloridate are quite suitable for this purpose. Preparation of **95** may be readily achieved through the oxazoline method or by phosphorylation of lithium salts of the monosaccharide acetate.

Some of the methods mentioned in the present discussion may also be readily applied to the synthesis of glycosyl phosphates derived from oligosaccharides. A series of these phosphates was prepared by use of the MacDonald procedure. For derivatives with a residue of GlcNAc at the "reducing" end of the carbohydrate chain, the oxazoline method and phosphorylation of lithium salts were both shown to be quite useful. The known examples of oligosaccharide phosphates are mentioned in Section VII.

VI. SYNTHESIS OF DOLICHYL GLYCOSYL PHOSPHATES AND RELATED DERIVATIVES

As noted already, two dolichyl monophosphate sugars serve as glycosyl donors in enzymic reactions occurring in the lumen of endoplasmic reticulum as part of the dolichol pathway of *N*-glycoprotein biosynthesis, namely, the derivatives of β-D-glucopyranose (**139a**) and β-D-mannopyranose (**140a**).

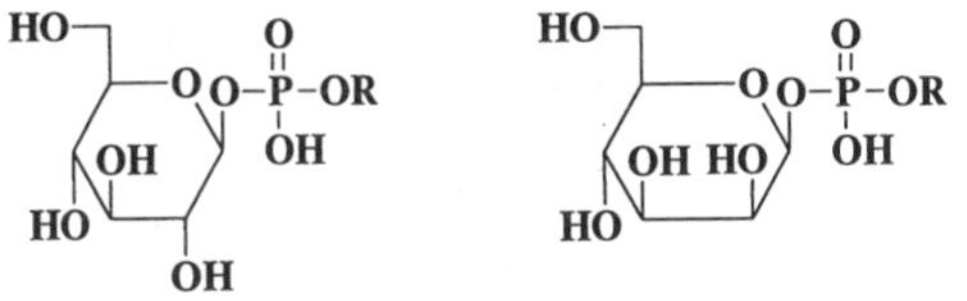

139a-d   **140a-d**

a $R = WT_2C_{12-16}S-$ b $R = WT_2C_8-$ c $R = WS-$

d $R = WT_3C_{6-8}-$

Formation of dolichyl monophosphate D-xylose in hen oviduct membranes was also reported [260]. In bacterial systems, the analogous glycosyl phosphate diesters **139b** and **140b**, derived from bacterial undecaprenol, were identified as biosynthetic intermediates, and in some bacterial strains the presence of other related compounds was demonstrated, including β-D-galactopyranosyl and 2-acetamido-2-deoxy-β-D-glucopyranosyl derivatives, as well as phosphodiesters containing α-anomers of D-glucose and of *N*-acetyl-D-glucosamine residues (see Ref. 20 for review and more recent papers [261–263]). As a new development in this field, the identification of β-D-arabinofuranosyl and D-ribosyl derivatives of polyprenyl phosphate in mycobacteria [47] has to be noted.

Four different approaches to the synthesis of dolichyl (or polyprenyl) glycosyl phosphates are possible, using the retrosynthetic disconnections shown in formula **141**. Approaches c and d require the use of anomerically pure glycosyl phosphates as a starting material. In the case of approaches a and b, it is necessary to find experimental conditions that ensure stereospecific formation of the glycosyl phosphate linkage. As can be seen from the previous section, this may not always be easy.

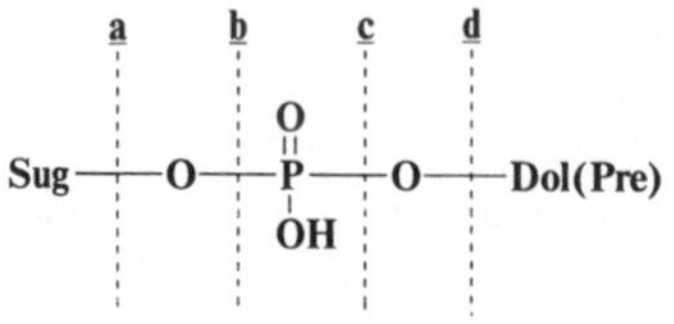

$$\begin{array}{ccc}
\underline{a} & \underline{b} & \underline{c} & \underline{d} \\
\end{array}$$

Sug—O—P—O—Dol(Pre)

with O double bonded to P and OH below P.

141

a $[Sug^+] + [Dol(Pre)OPO_3H^-]$

b $[SugO^-] + [Dol(Pre)OPO_2H^+$

c $[SugOPO_2H^+] + [Dol(Pre)O^-]$

d $[SugOPO_3H^-] + [Dol(Pre)^+]$

Biosynthesis of **139ab** and **140ab** occurs through enzymic glycosylation reactions of dolichyl (or polyprenyl) phosphate with the corresponding nucleoside diphosphate sugars; i.e., it corresponds to the retrosynthetic disconnection $\underline{a}$. In chemical syntheses, approach $\underline{c}$ is more usual; it requires the use of an activated derivative of the glycosyl phosphate as synthetic equivalent of the glycosyl phosphoryl cation.

A. Through Synthetic Equivalents of the Glycosyl Phosphoryl Cation

Stable activated derivatives of glycosyl phosphates, suitable for preparation of the phosphodiesters, are not known, and the generation of reactive species in situ under the action of suitable activating reagents is used in the present methods for synthesis of dolichyl or polyprenyl monophosphate sugars. Until now, only protected derivatives of glycosyl phosphates or glycosyl H-phosphonates (the analogous derivatives of phosphorous acid) have been employed as substrates for the reaction.

From Acetylated Glycosyl Phosphates

The first syntheses of dolichyl and polyprenyl glycosyl phosphates were described in 1973 by Warren and Jeanloz [264,265]. In their work, pyridinium tetra-O-acetyl-α-D-mannopyranosyl phosphate **142** served as a substrate. As condensing reagents, N,N'-dicyclohexylcarbodiimide (DCC) or 2,4,6-triisopropylbenzenesulfonyl chloride (TPS) were employed. These reagents had been applied widely in oligonucleotide synthesis through the so-called "phosphodiester method" (for reviews see Refs. 266 and 267). The phosphodiesters **143a–d**, derived from dolichol, ficaprenol, and their short-chain analogs, citronellol and farnesol, were obtained after interaction of the activated glycosyl phosphate derivatives with the corresponding isoprenoid alcohols in anhydrous pyridine, followed by deacetylation with sodium methoxide in chloroform-methanol mixture.

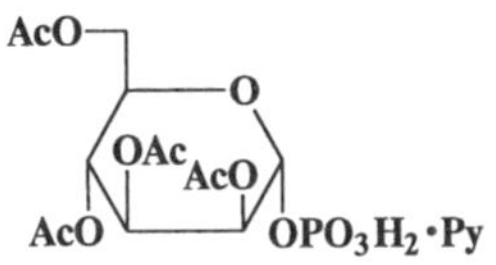

AcO—⌐——⌐—O
      |OAc AcO|
AcO—⌐————⌐—OPO₃H₂·Py

HO—⌐——⌐—O
      |OH HO| O
HO—⌐————⌐—O—P—OR
                   ‖
                   OH

142 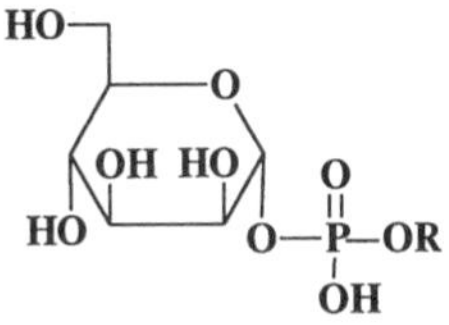   143a-d

a R = WT₂C₁₂₋₁₆S- , b R = WT₂C₈⁻

c R = WS- , d R = WT₂-, (+WTS-)

The analogous procedure with the use of TPS as a condensing reagent was applied to the synthesis of the derivatives of β-D-mannose **140c** [239] and **140a** [110,239], and a similar preparation was described for β-D-glucose phosphodiesters **139a** [110] and **139d** [268], as well as for the α-anomer of this last [268]. Yields of dolichyl glycosyl phosphates were between 40% and 50%, whereas for the derivatives of polyprenols with unsaturated α-terminal isoprene units they decreased to 20%–30%.

In a more recent paper [269], the synthesis of **143a** and **143c** with the use of DCC was described, and the same reagent was employed for the preparation of derivatives that included a [¹⁴C]mannose residue [269,270].

Rather cumbersome purification of the desired reaction products is necessary due to numerous side reactions observed under the conditions of phosphodiester synthesis. Some of them are connected with splitting of the glycosyl phosphate linkage. In the case of ficaprenyl and farnesyl derivatives, destruction of the prenyl-phosphate linkage may also occur. The mechanism of phosphodiester synthesis probably involves the intermediate formation of glycosyl and polyprenyl esters of pyrophosphoric, triphosphoric and tetrapolyphosphoric acids after initial activation of glycosyl phosphates (cf. Ref. 271 for review of investigations of the related reactions in the nucleotide series). In these derivatives, which contain good leaving groups at C-1 of both glycosyl and prenyl residues, efficient nucleophilic substitution at these centers may be expected to occur as a competing reaction to nucleophilic substitution at a phosphorus atom, the latter being necessary for the generation of the phosphodiesters.

Glycosyl H-Phosphonate Synthesis

Similar side reactions may be anticipated to proceed with the use of other synthetic equivalents of glycosyl phosphoryl cations. Nevertheless, when derivatives of trivalent phosphorus were employed for this purpose, substitution at the phosphorus atom was found to be extremely fast, and the degree of side processes was diminished.

In this respect, an application of glycosyl-*H*-phosphonates **144** for the synthesis of glycosyl phosphate diesters seems of interest. Specifically, a series of glycosyl phosphosugars, fragments of numerous biopolymers of different origin, was prepared successfully with the use of these derivatives (cf. Ref. 272 and references therein). The general reaction sequence is shown in Scheme 19. When **144** was treated with trimethylacetyl chloride in the presence of an alcohol in anhydrous pyridine, the initially formed mixed anhydride **145** was rapidly converted into rather unstable *H*-phosphonate diester **146**, which after mild oxidation with iodine in aqueous pyridine gave the desired phosphate diester **147**.

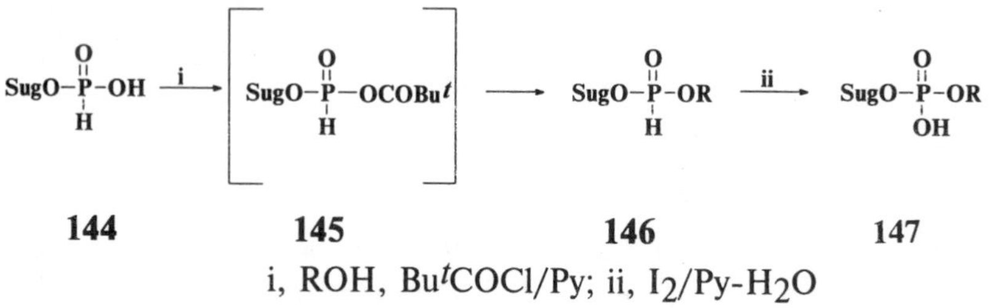

144   **145**   **146**   **147**

i, ROH, Bu^{t}COCl/Py; ii, I$_2$/Py-H$_2$O

Scheme 19

Although some side reactions, including those connected with splitting of the glycosyl phosphate linkage, were detected during a detailed investigation of the process with ^{31}P-NMR spectroscopy [273], their velocity was shown to be very low in comparison with the main reaction of phosphodiester formation, and the products **147** (R = monosaccharide residue) were isolated in high yields.

The glycosyl *H*-phosphonate approach was found to be successful for the synthesis of dolichyl glycosyl phosphates [187]. Initially, tetra-*O*-acetyl-α-D-glucopyranosyl *H*-phosphonate and dolichol were used as reaction substrates. The smooth formation of the phosphate diester was observed and, importantly, no significant alterations in the polyunsaturated chain were noted under conditions of oxidation of **146** to **147**. The procedure was then employed for the synthesis of **139a** and **139c** (their yields were 62% and 48%, respectively); also, their analogs with β-D-galactopyranosyl residues were prepared.

Further use of this prospective method for the preparation of dolichyl glycosyl phosphates is limited by the availability of the starting *H*-phosphonates. While the β-D-glucopyranosyl and β-D-galactopyranosyl derivatives were readily obtained from the 1,2-orthoacetates of the corresponding monosaccharides (see Section V.A.), the synthesis of the β-D-mannopyranosyl derivative has not yet been achieved.

B. Other Approaches

The successful synthesis of citronellyl β-D-glucopyranosyl phosphate **139c** was performed with an approach based on retrosynthetic disconnection a [274]. In this case, *tert*-butyl 1,2-orthoacetyl-3,4,6-tri-*O*-acetyl-α-D-glucopyranose **102a** was used as a synthetic equivalent of the glycosyl cation. As a synthetic equivalent of the prenyl-phosphoryloxy anion, citronellyl benzyl phosphate **148a** was prepared through the interaction of citronellyl bromide with silver dibenzyl phosphate, followed by anionic debenzylation of the resulting phosphotriester with sodium iodide in acetone. The interaction of **102a** and **148a** smoothly gave the corresponding phosphotriester, which was converted, after anionic debenzylation and deacetylation, into **139c** in a total yield of 80%.

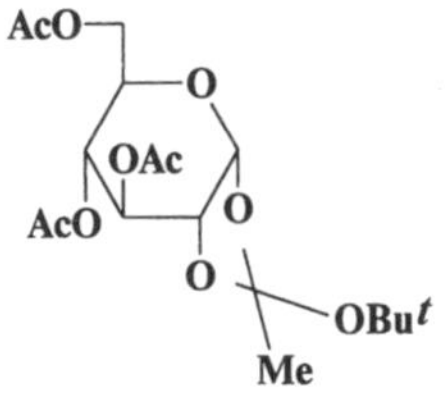

| **102a** | **148ab** |
| | a R = Bn   b R = H |

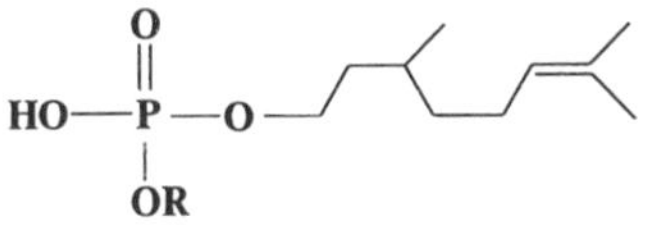

Later it was found [275] that glycosylation of citronellyl phosphate **148b** was also quite feasible under similar conditions, and **139c** was obtained in a yield of 55%. Moreover, glycosylation with **102a** was successful for phosphates of prenols containing an α-terminal isoprene unit, i.e., derivatives of geraniol [275] and moraprenol [276]. Nevertheless, the necessity to use phosphomonoesters, which are unstable as free acids, limits the preparative value of this method.

The most convenient method for preparation of the glycosyl phosphate derivatives of polyprenols of this type is based on retrosynthetic disconnection d̲, which requires the use of a synthetic equivalent of the polyprenyl cation. As already noted, until now only polyprenyl trichloroacetimidates **79** were found useful for this purpose. Interaction of the moraprenyl derivative with mono-(tri-n-octylammonium) salts of protected glycosyl phosphates was found [142] to lead smoothly to protected moraprenyl glycosyl phosphates. In Scheme 20 the reaction with the β-D-mannopyranose derivative is shown, and in other cases peracetates of glycosyl phosphates were used as starting materials.

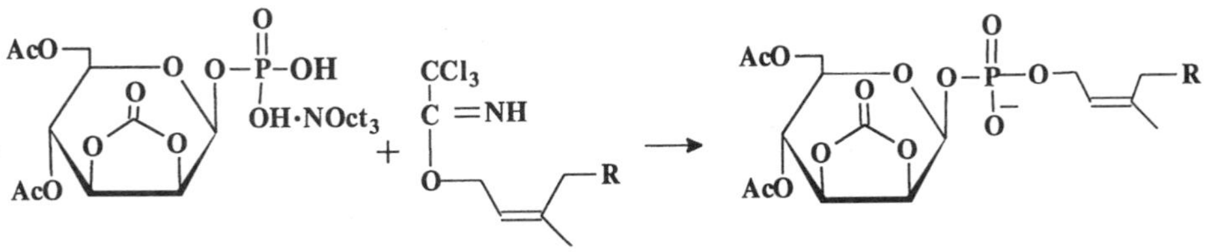

| **149** | **79** | **150** |

$$R = WT_3C_{5\text{-}7}-$$

Scheme 20

Under standard reaction conditions (benzene as a solvent, 8 h at 77°C), the conversion of **149** into **150** could easily be monitored with ^{31}P-NMR spectra, which showed the absence of rearrangement products. The phosphodiester **150** was isolated in a yield of 71% and converted into **140d** after treatment with sodium methoxide in chloroform-methanol mixture.

Similar reaction conditions were found to be suitable for preparation of **139d** and the analogous β-D-galactopyranose derivative, with comparable yields. In the case of the 2-acetamido-2-deoxy-β-D-glucopyranose derivative, the yield increased to 96%. Polyprenyl α-D-glycopyranosyl phosphates were also successfully obtained, although the rate of the reaction was slightly slower, probably due to the lower acidity of the α-glycosyl phosphates in comparison with the β-anomers. Under standard reaction conditions, 50%–60% yields were achieved for the derivatives of α-D-glucopyranosyl, α-D-galactopyranosyl, and 2-acetamido-2-deoxy-α-D-glucopyranosyl phosphates. An anomalously low yield of 38% was observed in the case of the α-D-mannopyranosyl derivative, but it probably may be improved by prolongation of the reaction.

This method is undoubtedly the most convenient one for preparation of the polyprenyl glycosyl phosphates that serve as intermediates in bacterial polysaccharide biosynthesis.

VII. SYNTHESIS OF DOLICHYL GLYCOSYL DIPHOSPHATES AND RELATED DERIVATIVES

In biosynthetic processes, dolichyl and polyprenyl glycosyl diphosphates serve usually as membrane-bound acceptors for enzymic glycosyl transfer. As a result of these glycosylation reactions, the initially formed monosaccharide derivatives are converted into oligosaccharide derivatives, which in turn function as oligosaccharide donors in the transfer of the oligosaccharide chain to suitable acceptors. The first reaction of these biosynthetic cycles consists of the transfer of a glycosyl phosphate residue from the corresponding nucleoside diphosphate sugars to dolichyl or polyprenyl phosphate.

In terms of retrosynthetic analysis, this reaction corresponds to the disconnection a as shown in formula **151**; in other words, nucleoside diphosphate sugars serve as activated forms of glycosyl phosphates.

$$\text{Sug}\text{—O—}\overset{\overset{O}{\|}}{\underset{OH}{P}}\text{—O—}\overset{\overset{O}{\|}}{\underset{OH}{P}}\text{—O——Dol(Pre)}$$

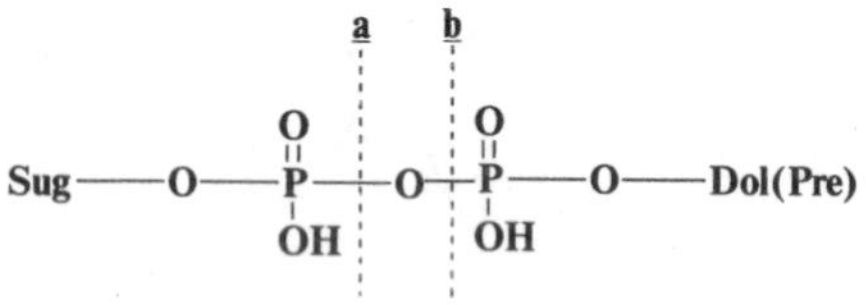

151

a [SugOPO$_2$H$^+$] + [Dol(Pre)OPO$_3$H$^-$]

b [SugOPO$_3$H$^-$] + [Dol(Pre)OPO$_2$H$^+$]

In contrast, in chemical synthesis of these derivatives an approach based on the disconnection b is more usual. In this case, the use was made of glycosyl phosphates or their peracetates as a nucleophilic component, whereas as discussed in Section IV.B, activated dolichyl or polyprenyl phosphate derivatives served as the electrophile.

A. Intermediates in the Biosynthesis of *N*-linked Chains of Glycoproteins

The first intermediates in the dolichol pathway of glycoprotein biosynthesis are dolichyl diphosphate derivatives of *N*-acetylglucosamine **152** and of *N,N'*-diacetylchitobiose **153**. Most of the publications in this field are concerned just with preparation of these compounds.

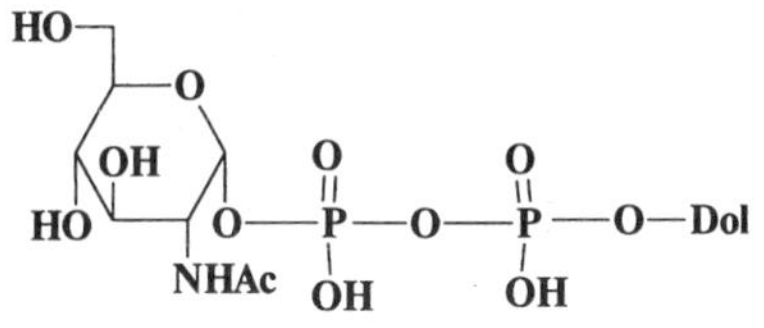

152

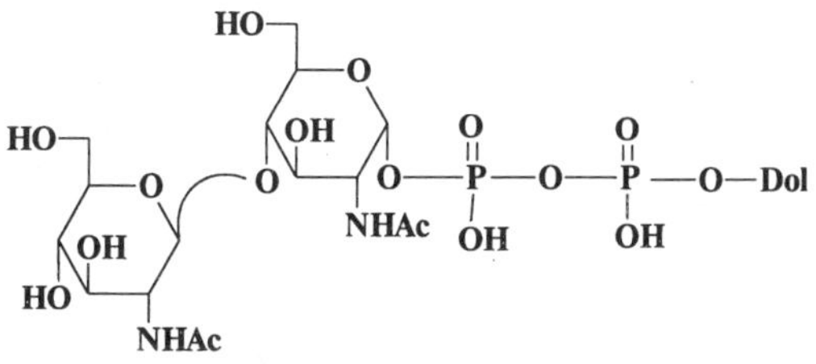

153

Dol = $WT_2C_{12-16}S-$

In all cases, peracetates **154ab** were employed as synthetic equivalents of glycosyl phosphates, whereas different activated derivatives of dolichyl phosphate were used, so comparison of different synthetic procedures is possible.

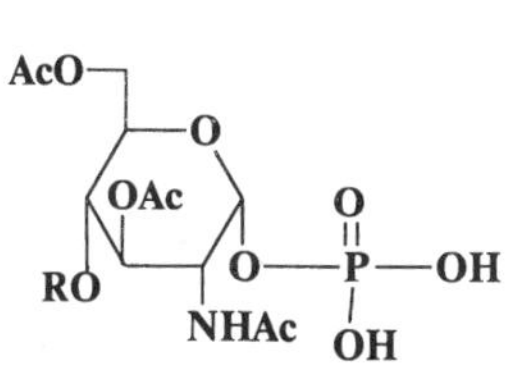

154ab 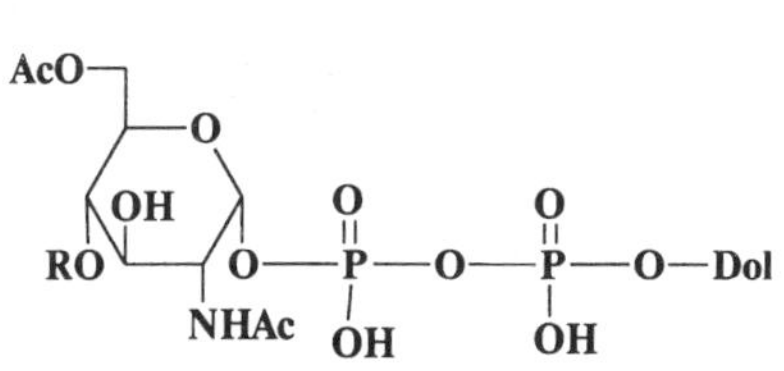   **155ab**

a R = OAc; b R = GlcNAc(OAc)$_3$(β1–)

In the pioneering studies of Warren and Jeanloz [110,150,212] the diphenyl pyrophosphate derivative **82** (Pre = $WT_2C_{12-16}S$—) was used. Its interaction with a 4–5-fold molar excess of acetylated bis-(tri-*n*-butylammonium) glycosyl phosphates

was performed in dichloroethane in the presence of pyridine (5 mol/mol of **82**) for
48 hours at room temperature. The diesters **155ab**, after purification by TLC, were
converted into the desired dolichyl diphosphate sugars by brief treatment with sodium methoxide in chloroform-methanol mixture. The reported [110] yields were
40% of **152** and 73% for **153**. The method was applied successfully to the preparation
of other intermediates of the dolichol pathway, particularly the derivatives of the
trisaccharide **156** [277] and the tetrasaccharides **157** [277] and **158** [118].

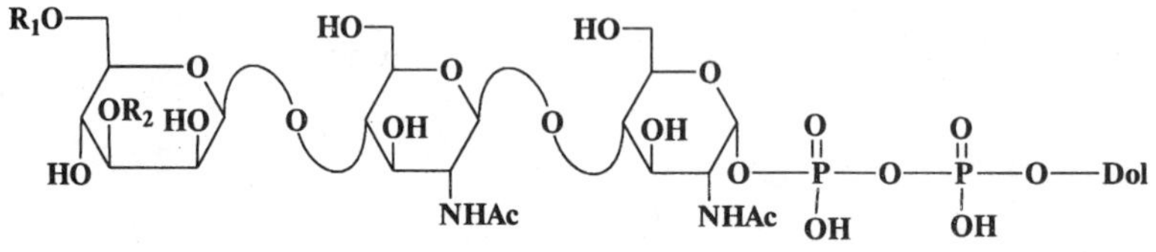

156	$R_1 = R_2 = H$
157	$R_1 = H;\ \ R_2 = Man(\alpha 1\text{-})$
158	$R_1 = Man(\alpha 1\text{-});\ \ R_2 = H$

The synthesis of **152** from dolichyl phosphodichloridate **81** was described by
Imperiali and Zimmerman [119]. The activated derivative was treated with 1.7 equivalents of pyridinium **154a** in anhydrous pyridine (1 hour at room temperature), followed by addition of aqueous triethylamine, giving 60% of **155a**, which was
smoothly deacetylated to **152**.

Conversion of **154ab** into **155ab** by using the phosphoroimidazolidate method
was also described [120]. In this case an approach was applied that corresponds to
the retrosynthetic disconnection a (see formula **151**). The protected glycosyl
phosphates were converted into the imidazolidates by reaction with N,N'-
carbonyldiimidazole (CDI); the products were treated without isolation with an equimolar amount of bis-(tri-n-butylammonium) dolichyl phosphate in DMF-
dichloromethane mixture (48 hours at room temperature). The crude yields of the
pyrophosphates were reported to be 66%, but after extensive purification only 22%
of **155b** was isolated. The latter was deblocked with sodium methoxide to give **153**
quantitatively.

Finally, an application of the mixed anhydride **84** for pyrophosphate synthesis
was studied with analogs of **155b** that contained phytanyl or lauryl residues
[135,136]. In a reaction catalyzed by silver acetate (pyridine as a solvent, 4 days at
room temperature) the desired diesters were obtained in yields of 21% and 40%,
respectively. For the first of these derivatives other synthetic procedures were also
investigated. When **81** was used as an activated derivative, the reported yield was
45% [136], and even better results (60%) were achieved with the phosphoroimidazolidate method [250].

B. Other Dolichyl Glycosyl Diphosphates

Only a few syntheses of these derivatives were reported. The diphenyl pyrophosphate procedure with use of the acetylated glycosyl phosphate was found to be quite useful for the preparation of dolichyl α-D-mannopyranosyl diphosphate **159** and the analogous citronellyl derivative [146], the yield of **159** being 40%. With the same method it was possible to obtain the dolichyl diphosphate derivative of a disaccharide fragment of hyaluronic acid, β-D-GlcA-(1-3)-α-D-GlcNAc [226].

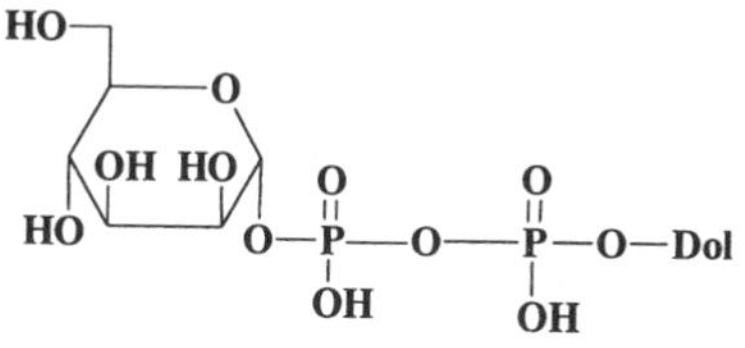

159

From unprotected α-D-[^{14}C]mannopyranosyl phosphate and dolichyl phosphoroimidazolidate **83** (Pre = WT$_2$C$_{12-16}$S—), **159** with a labelled monosaccharide residue was prepared in a yield of 22% after extensive purification [152]. The application of unprotected glycosyl phosphates in this procedure was found to be superior in many other cases, as may be seen from the following section.

C. Polyprenyl Glycosyl Diphosphates

Polyprenyl glycosyl diphosphates, derived from bacterial undecaprenol, serve as important intermediates in the biosynthesis of bacterial polysaccharides, and numerous synthetic studies were performed in this field. In general, the use was made of readily available derivatives of plant polyprenols, which are quite similar in structure to bacterial undecaprenol.

For these derivatives, additional methodologic difficulties exist connected with the high instability of allylic pyrophosphates. What is more significant in this case is that the synthetic targets include not only the derivatives of 2-acetamido-2-deoxyhexoses, but also those of α-D-galactopyranose, as identified in numerous bacterial strains (for review, see Ref. 20). The corresponding polyprenyl diphosphate sugars are highly unstable in alkaline medium due to the easy formation of 1,2-cyclic phosphates. Therefore, deblocking of the corresponding acetylated derivatives is quite difficult.

The diphenyl pyrophosphate method was used in the first syntheses of the diphosphates of this group. The ficaprenyl diphosphate derivatives of α-D-galactopyranose **160** [112] and 2-acetamido-2-deoxy-α-D-glucopyranose **161** [113] were prepared from the peracetates of the corresponding glycosyl phosphates. For the suppression of side reactions, it was found essential to decrease the content of pyridine in the reaction mixture (to 0.1 mol per 1 mol of **82**) and to diminish the reaction time to 20 hours. After deacetylation, **161** was isolated in a yield of 12%, and for **160** it decreased to only 6%.

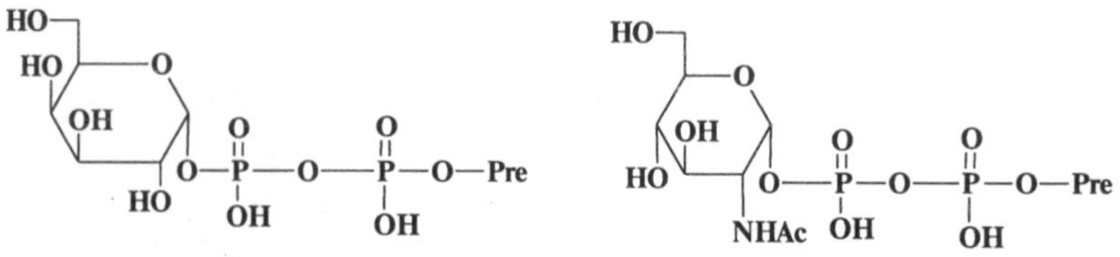

160   161

$$Pre = WT_3C_{6-8}-$$

Much better results were obtained in the synthesis of polyprenyl glycosyl diphosphates through interaction of polyprenyl phosphoroimidazolidate **83** (Pre = $WT_3C_{6-8}-$) with unprotected glycosyl phosphates. Under standard conditions, a 2-fold molar excess of bis-(tri-*n*-octylammonium) or bis-(triethylammonium) glycosyl phosphate was employed, and the reaction was performed in tetrahydrofuran-dimethyl sulfoxide mixture for 16–18 hours at room temperature, or at 37°C [151] (see also a description of the procedure in a review [23]).

With the use of this protocol, the yield of **160** was 47% [151], whereas **161** was obtained in a yield of 63% [213]. In the latter case the resulting polyprenyl glycosyl diphosphate is rather stable under conditions of mild deacetylation, and application of **154a** under similar conditions with subsequent deacetylation gave 81% of **161** [213].

Excellent results were obtained also in the synthesis of other polyprenyl diphosphate sugars, including the derivatives of α-D-glucopyranose [151], α-D-mannopyranose [278], 2-acetamido-2-deoxy-α-D-galactopyranose [213], and α-D-glucopyranuronic acid [153]. For biochemical studies some analogs of polyprenyl α-D-galactopyranosyl diphosphate [183,278] and of polyprenyl 2-acetamido-2-deoxy-α-D-glucopyranosyl diphosphate [279] were prepared. Moreover, the method was found to be applicable to the preparation of the polyprenyl diphosphate derivatives of some oligosaccharides [151,280–286] that are of interest as putative biosynthetic intermediates for O-antigenic polysaccharides of *Salmonella* or *Escherichia coli* or for analogs of these intermediates. In some cases, slight modification of the reaction conditions was necessary. Also, the yields of the products in the oligosaccharide series are lower than those for the monosaccharide derivatives.

It seems that at present the phosphoroimidazolidate method of synthesis is the method of choice for both dolichyl and polyprenyl glycosyl diphosphates. The application of protected or unprotected derivatives in the synthesis is dependent on structure of the monosaccharide residue in the desired product.

VIII. USE OF SYNTHETIC DERIVATIVES IN STUDIES OF THE DOLICHOL PATHWAY OF PROTEIN GLYCOSYLATION

The above-described progress in the development of synthetic methods for dolichyl phosphate and dolichyl phosphate-linked sugars opens the way for the use of these derivatives in a variety of biochemical studies. In the present section a short review of such applications of the different synthetic dolichol derivatives is given; for reviews of analogous studies with bacterial polyprenol derivatives see Ref. 20.

A. Identification of Biosynthetic Intermediates and Studies of the Enzymes Participating in the Process

As noted earlier, dolichyl phosphate serves as a substrate for several important enzymes. Specifically, the dolichol pathway of protein glycosylation includes three enzymic reactions with different nucleoside diphosphate sugars, namely, formation of Dol-P-Man (**140a**), Dol-P-Glc (**139a**), and Dol-PP-GlcNAc (**152**), catalyzed by mannosyltransferase, glucosyltransferase, and *N*-acetylglucosamine-phosphate-transferase, respectively. All these derivatives were prepared by chemical synthesis and used as standards for comparison with the products of the enzymic reactions in chromatographic behavior and hydrolytic stability. These studies represented an important step in unequivocal identification of the biosynthetic products.

Much attention was devoted to the elucidation of the structure of dolichyl phosphate mannose [270,287–289] formed after incubation of GDP-[^{14}C]Man with membrane preparations from different sources. In particular, the anomeric configuration of the mannosyl residue was clarified after detailed, comparative studies of the biosynthetic products with phosphodiesters **140a** and **143a**. The analogous approach was used for the identification of a product, formed from UDP-[^{14}C]Glc, as **139a** [290].

Dolichyl diphosphate *N*-acetylglucosamine formed from UDP-[^{14}C]GlcNAc in calf pancreas microsomes was shown to be identical to a sample of synthetic **152** [291], and further enzymic conversion of the synthetic **152** into **153** was also demonstrated [119,292]. Synthetic dolichyl glycosyl diphosphates, derivatives of oligosaccharides, were shown to serve as glycosyl acceptors in three consecutive glycosylation reactions catalyzed by the enzymes from calf pancreas: (β1–4)-mannosylation of **153** into **156** [293], (α1–3)-mannosylation of **156** into **157** [294], and (α1–6)-mannosylation of **157** into **162a** [295]. In the last case, forma-

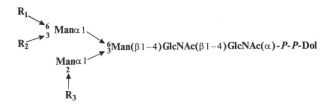

162a-e

a $R_1 = R_2 = R_3 = H$

b $R_1 = R_2 = H$; $R_3 = Man(\alpha 1-$

c $R_1 = R_2 = H$; $R_3 = Man(\alpha 1-2)Man(\alpha 1-$

d $R_1 = R_2 = R_3 = Man(\alpha 1-2)Man(\alpha 1-$

e $R_1 = R_2 = Man(\alpha 1-2)Man(\alpha 1-$;
    $R_3 = Glc(\alpha 1-2)Glc(\alpha 1-3)Glc(\alpha 1-3)Man(\alpha 1-2)Man(\alpha 1$

tion of small quantities of the derivatives of hexasaccharide **162b** and of heptasaccharide **162c** was also observed.

For further elongation of the oligosaccharide chain in dolichyl-linked biosynthetic intermediates, i.e., for the formation of **162d** and **162e**, it is necessary to use dolichyl glycosyl phosphates **140a** and **139a** as glycosyl donors. Chemical synthesis of these derivatives with a labeled hexose residue has not yet been described, and for their preparation enzymic methods are usually employed, starting from nucleoside diphosphate sugars and synthetic polyprenyl phosphates of definite structure (see, for example, Refs. 269 and 296).

At the same time, synthetic dolichyl α-D-[^{14}C]mannosyl phosphate **143a** was studied as an analog of the natural substrate. It was found to be inactive in mannosylation reactions leading to **162d** in pig brain microsomes [269], as an activator of enzymic synthesis of **152** [297], and as a glycosyl donor in biosynthesis of the glycosyl phosphatidylinositol protein anchor [298]. Thus in all these cases strict specificity of the enzymes toward the anomeric configuration of the mannosyl residue was observed.

The undecasaccharide derivative **162e** serves as a natural substrate for oligosaccharyltransferase that catalyzes the formation of the GlcNAc-Asn-linkage in glycoproteins. Synthetic disaccharide derivative **153** was also found to be able to participate in the enzymic reaction [120,299]. Interaction of **153** with synthetic peptides was widely used in recent studies on the mechanism of the oligosaccharyltransferase reaction and on the relation between conformation and substrate properties of the peptide acceptors [299–302].

A search for new biosynthetic pathways connected with the participation of dolichyl-linked sugars may be greatly facilitated by the use of the corresponding synthetic derivatives. Thus, Shabalin et al. were able to demonstrate the transfer of a [^{14}C]mannose residue from synthetic **159** into yeast mannan [303,304]. It was suggested that the reaction is significant for biosynthesis of the outer chain of the yeast cell-wall polymer.

B. Specificity of the Biosynthetic Enzymes Toward the Structure of Dolichol Residue, and Chemical-Enzymic Synthesis of Oligosaccharides

For biosynthetic experiments performed in vitro it has become usual practice to incorporate dolichyl phosphate or a related derivative into incubation mixtures as an exogenous acceptor of glycosyl residues. The availability of a wide range of synthetic polyprenyl phosphates made possible numerous studies on the specificity of different enzymes of the dolichol pathway toward the structure of the polyprenol residue in biosynthetic intermediates (for a review see Ref. 80).

Initially, clear preference for α-dihydropolyprenyl derivatives was shown for the enzymes catalyzing formation of **139a**, **140a**, and **152** [80,305–309], using phosphates of plant polyprenols and their α-dihydroanalogs prepared by selective hydrogenation of the α-terminal isoprene unit. The results of studies on substrate properties of (S)- and (R)-dolichyl phosphates [310–312] clearly demonstrated the sensitivity of the enzymes toward stereochemistry at C-3 of the α-dihydropolyprenyl residue, the natural (S)-stereoisomer being a much better substrate.

Numerous papers were devoted to substrate properties of short-chain analogs of dolichyl phosphate for enzymic transfer of mannose, glucose, and N-acetylglucosamine 1-phosphate residues. In initial studies [305,306,308], α-dihydroundecaprenyl phosphate was shown to serve as an efficient acceptor of glycosyl residues. More detailed investigations [313–316] showed a gradual decrease of the rate of the enzymatic reactions with shortening of the hydrocarbon chain. This resulted in almost complete loss of substrate properties for the derivatives of α-dihydrohexaprenol [313] or α-dihydropentaprenol [315,316].

The substrate specificity of the mannosyl transferase deserves special comment. On the one hand, the enzyme from rat liver microsomes was shown [317] to accept as substrates derivatives of α-dihydropolyprenols with longer chain length than in natural dolichols (up to $n = 32$). On the other hand, these enzymes from different sources were able to use as substrates a large number of different low-molecular-weight hydrophobic phosphates, including the derivatives of retinol [308,318–320], phenol [321], phytanol [133,136,137], and (S)-3-methyloctadecanol [137]. However, phosphates of dihydrocitronellol and tetradecanol were found to be inactive [137].

The significance of the stereochemistry of the internal isoprene units for the substrate properties of dolichyl phosphate is not completely clear. It was reported [315] that phosphates WT_2C_3S—P and WT_3C_2S—P showed similar acceptor properties for glucosyl transfer with an algal enzyme, whereas the efficiency of WT_5S—P was much lower. At the same time, high acceptor efficiency of α-dihydrosolanesyl phosphate was noted in an earlier work [305] on glucosyl transfer with the enzyme from rat liver microsomes.

Analogs of **140a** prepared from modified dolichyl phosphates by enzymic synthesis were employed for studies of the specificity of several enzymes that use Dol-P-Man (**140a**) as a substrate [269,296–298,307,312,313], and to a lesser degree, similar studies were performed for the analogs of **139a** [296]. In general, the observed regularities are similar to those found for modified dolichyl phosphates: The enzymes prefer to use as substrates the derivatives of α-dihydropolyprenols rather than the diesters of polyprenols, and derivatives of (S)-dolichols rather than their (R)-isomers. Increase of the hydrophobicity of the polyprenyl residue was found to lead to better substrate properties of the derivatives. Nevertheless, some of the enzymes studied were shown to be able to use the citronellyl derivative **140c** as a substrate, although its efficiency was rather low [269,297,298].

A study of the substrate properties of synthetic **152**, **153**, and **157**, containing residues of (S)- and (R)-dolichol, was performed for glycosyltransferases from calf liver [322]. The first enzyme, catalyzing conversion of **152** into **153**, was found to show a strong preference for the derivative of the (S)-isomer; but with increase of the oligosaccharide chain length, the influence of stereochemistry of the dolichol residue on the rate of enzymic reactions became less significant. An analogous observation was that when the derivatives of perhydrodolichol and (S)-dolichol were compared, the relative reactivity of the perhydrodolichol was increased with elongation of the oligosaccharide chain. Also, the significance of the hydrophobicity of the dolichol residue seems to be diminished with elongation of oligosaccharide chains. This was shown in studies of dolichol derivatives with different chain length ($n = 6–11$) as mannosyl and glucosyl acceptors during formation of **162d** and **162e** [316]. In the same work the efficiency of the oligosaccharyltransferase reaction was found to be independent of dolichol chain length within the limits mentioned. The latter enzyme showed strict specificity for the stereochemistry of the dolichol residue, accepting only the derivative of the (S)-isomer as a substrate [80].

The relaxed specificity of some glycosyltransferases toward the structure of the dolichol residue characteristic of the oligosaccharide-diphosphate derivatives made possible chemical-enzymic synthesis of some fragments of N-linked carbohydrate chains of glycoproteins using readily available phytanyl N,N'-diacetylchitobiosyl diphosphate as a glycosyl acceptor [136,250,323,324]. It was found to give readily the phytanyl analog of **156**, and the possibility of further elongation of the oligosaccharide chain through enzymic reactions with GDP-Man was also noted.

C. Incorporation of Modified Sugars into Dolichol-Linked Intermediates

One of the obvious advantages of the chemical-enzymic synthesis of complicated oligosaccharides consists of the possibility of using this approach for the preparation of a series of structurally related derivatives. This is possible because of a lack of absolute specificity of glycosyltransferases toward the structures of monosaccharide residues of both donor and acceptor. A number of modified polyprenyl diphosphate oligosaccharides related to intermediates of bacterial polysaccharide biosynthesis were prepared through chemical-enzymic synthesis (for reviews, see Ref. 20 and references in Section VII). For the intermediates of the dolichol pathway of biosynthesis, only limited information on incorporation of modified sugars is available at present.

The formation of several derivatives of this type was demonstrated in studies of the influence of modified monosaccharides on the biosynthesis of the *N*-linked carbohydrate chains in glycoproteins (see a review [5] and more recent publications [207,228,325]). With the use of synthetic analogs of GDP-Man, the formation of several analogs of **156** was observed. The compounds contained 2-deoxy-D-*arabino*-hexose [326], 4-deoxy-D-*lyxo*-hexose [207], or 4-deoxy-4-fluoro-D-mannose [207] residues instead of a D-mannose residue. Also, the identification was described of an analog of **157** where 2-deoxy-D-*arabino*-hexose substituted for the terminal mannose residue [326], and of an analog of **162c** that contains five 4-deoxy-D-*lyxo*-hexosyl residues [207].

In the dolichyl monophosphate sugar series, enzymic synthesis was achieved for analogs of **140a** with 2-deoxy-D-*arabino*-hexose [327,328], 2-deoxy-2-fluoro-D-mannose [228], 4-deoxy-D-*lyxo*-hexose [207], or 4-deoxy-4-fluoro-D-mannose [207] residues. Results of inhibition studies of the enzymic synthesis of **140a** [325] suggest the possibility of forming similar derivatives of 6-deoxy-D-mannose and 3-deoxy-D-*arabino*-hexose. In the enzymic synthesis of **139a**, UDP-2-deoxy-D-*arabino*-hexose was unable to serve as an analog of UDP-Glc [327–329]. However, UDP-2-amino-2-deoxy-D-glucose could substitute for the natural glucosyl donor [330]. The enzymic formation of dolichyl xylosyl phosphate [260], as was shown in a more detailed study [331], is also the result of UDP-D-xylose functioning as an analog of UDP-Glc.

D. Synthetic Dolichyl Derivatives Used in Biophysical Studies

Numerous biophysical studies have been performed in recent years for the elucidation of the influence of dolichol and related derivatives on the structure of biological membranes (for review, see Ref. 34). For such studies (see review [34]), several derivatives of dolichols and polyprenols were prepared that contained spin [332,333], deuterium [334], or fluorescent [335] labels.

ACKNOWLEDGMENTS

The authors greatly appreciate the contributions made to the investigations performed in this group by our co-workers, whose names are shown in the reference list. Our work in recent years has been partly supported by the International Science Foundation (Grants MMG000 and MMG300) and the Russian Foundation for Basic Research (Grants 93-03-18196 and 96-03-32473). We are very grateful to Dr. N. A. Kalinchuk for her help in preparation of the manuscript.

REFERENCES

1. N. H. Behrens and L. F. Leloir, Dolichol monophosphate glucose: An intermediate in glucose transfer in liver, *Proc. Natl. Acad. Sci. USA 66*: 153 (1970).
2. S. C. Hubbard and R. J. Ivatt, Synthesis and processing of asparagine-linked oligosaccharides, *Annu. Rev. Biochem. 50*: 555 (1981).
3. R. Kornfeld and S. Kornfeld, Assembly of asparagine-linked oligosaccharides, *Annu. Rev. Biochem. 54*: 631 (1985).
4. A. J. Parodi and L. F. Leloir, The role of lipid intermediates in the glycosylation of proteins in the eukaryotic cells, *Biochim. Biophys. Acta 559*: 1 (1979).
5. R. T. Schwarz and R. Datema, The lipid pathway of protein glycosylation and its inhibitors: The biological significance of protein-bound carbohydrates, *Adv. Carbohydr. Chem. Biochem. 40*: 287 (1982).
6. A. Kaiden and S. S. Krag, Regulation of glycosylation of asparagine-linked glycoproteins, *Trends Glycosci. Glycotechnol. 3*: 275 (1991).
7. D. J. Cummings, Synthesis of asparagine-linked oligosaccharides: Pathways, genetics, and metabolic regulation, *Glycoconjugates* (H. J. Allen and E. C. Kisailus, eds.), p. 333, Dekker, New York, 1992.
8. R. C. Hughes and T. D. Butters, Glycosylation patterns in cells: An evolutionary marker? *Trends Biochem. Sci. 6*: 228 (1981).
9. R. Pont Lezica, G. R. Daleo, and P. M. Dey, Lipid-linked sugars as intermediates in the biosynthesis of complex carbohydrates in plants, *Adv. Carbohydr. Chem. Biochem. 44*: 341 (1986).
10. W. Tanner and L. Lehle, Protein glycosylation in yeast, *Biochim. Biophys. Acta 906*: 81 (1987).
11. M. A. Kukuruzinska, M. L. E. Bergh, and B. J. Jackson, Protein glycosylation in yeast, *Annu. Rev. Biochem. 50*: 555 (1987).
12. A. Herscovics and P. Orlean, Glycoprotein biosynthesis in yeast, *FASEB J. 7*: 540 (1993).
13. R. B. Trimble and M. F. Verostek, Glycoprotein oligosaccharide synthesis and processing in yeast, *Trends Glycosci. Glycotechnol. 7*: 1 (1995).
14. P. T. Englund, The structure and biosynthesis of glycosyl phosphatidyl inositol protein anchors, *Annu. Rev. Biochem. 62*: 121 (1993).
15. J. Lechner and F. Wieland, Structure and biosynthesis of prokaryotic glycoproteins, *Annu. Rev. Biochem. 58*: 173 (1989).
16. E. Hartmann and H. Koenig, Uridine and dolichol activated oligosaccharides are intermediates in the biosynthesis of the S-layer glycoprotein of *Methanothermus fervidus*, *Arch. Microbiol. 151*: 274 (1989).
17. P. Messner and U. B. Steytr, Crystalline bacterial cell surface layers, *Adv. Microbial Physiol. 33*: 212 (1992).
18. A. Wright, M. Dankert, P. Fennesey, and P. W. Robbins, Characterization of a polyisoprenoid compound functional in O-antigen biosynthesis, *Proc. Natl. Acad. Sci. USA 57*: 1798 (1967).
19. Y. Higashi, J. L. Strominger, and C. C. Sweeley, Structure of a lipid intermediate in cell wall peptidoglycan biosynthesis: A derivative of a C_{55}-isoprenoid alcohol, *Proc. Natl. Acad. Sci. USA 57*: 1878 (1967).

20. V. N. Shibaev, Biosynthesis of bacterial polysaccharide chains composed of repeating units, *Adv. Carbohydr. Chem. Biochem. 44*: 277 (1986).

21. C. Whitfield and M. A. Valvano, Biosynthesis and expression of cell-surface polysaccharides in Gram-negative bacteria, *Adv. Microbial Physiol. 35*: 135 (1993).

22. T. D. H. Bugg and P. E. Brandish, From peptidoglycan to glycoproteins: Common features of lipid-linked oligosaccharide biosynthesis, *FEMS Microbiol. Lett. 119*: 255 (1994).

23. L. L. Danilov and V. N. Shibaev, Phosphopolyprenols and their glycosyl esters: Chemical synthesis and biochemical application, *Studies in Natural Products Chemistry, v. 8* (Atta-ur-Rahman, ed.), p. 63, Elsevier, Amsterdam, 1991.

24. V. N. Shibaev and L. L. Danilov, New developments in the synthesis of phosphopolyprenols and their glycosyl esters, *Biochem. Cell Biol. 70*: 429 (1992).

25. F. W. Hemming, R. A. Morton, and J. F. Pennock, An unsaturated alcohol from human kidney, *Biochem. J. 74*: 38 (1960).

26. J. F. Pennock, F. W. Hemming, and R. A. Morton, Dolichol: A naturally occurring isoprenoid alcohol, *Nature 186*: 470 (1960).

27. J. Burgos, F. W. Hemming, J. F. Pennock, and R. A. Morton, Dolichol: A naturally occurring C_{100} isoprenoid alcohol, *Biochem. J. 88*: 470 (1963).

28. IUPAC-IUB Joint Commission on Biochemical Nomenclature, Prenol nomenclature. Recommendations 1986, *Eur. J. Biochem. 167*: 181 (1987).

29. F. W. Hemming, Lipids in glycan biosynthesis, *Biochemistry of Lipids, Ser. 1, v. 4* (T. W. Goodwin, ed.), p. 39, Butterworths, London, 1974.

30. F. W. Hemming, Biosynthesis of dolichols and related compounds, *Biosynthesis of Isoprenoid Compounds, v. 2* (J. W. Porter and S. L. Spurgeon, eds.), p. 305, Wiley, New York, 1983.

31. F. W. Hemming, Glycosyl phosphopolyprenols, *Glycolipids* (H. Wiegandt, A. Neuberger, and L. L. M. van Deenen, eds.), p. 261, Elsevier, Amsterdam, 1985.

32. J. W. Rip, C. A. Rupar, K. Ravi, and K. K. Carroll, Distribution, metabolism and functions of dolichol and polyprenols, *Progr. Lipid Res. 24*: 269 (1985).

33. W. L. Adair and R. K. Keller, Isolation and assay of dolichol and dolichyl phosphate, *Methods Enzymol. 111*: 201 (1985).

34. T. Chojnacki and G. Dallner, The biological role of dolichol, Biochem. J. 251: 1 (1988).

35. W. L. Adair and S. Robertson, Absolute configuration of dolichol, *Biochem. J. 189*: 441 (1980).

36. I. A. Tavares, N. J. Johnson, and F. W. Hemming, A sensitive quantitative assay method for dolichols, cholesterol and ubiquinone using high-pressure liquid chromatography, *Biochem. Soc. Trans. 5*: 1771 (1977).

37. T. Mankowski, W. Jankowski, T. Chojnacki, and P. Franke, C_{55}-Dolichol: Occurrence in pig liver and preparation by hydrogenation of plant undecaprenol, *Biochemistry 15*: 2125 (1976).

38. W. L. Adair and N. Cafmeyer, Characterization of the *Saccharomyces cerevisiae cis*-prenyl transferase, *Arch. Biochem. Biophys. 259*: 589 (1987).

39. K. J. Stone, P. H. W. Butterworth, and F. W. Hemming, Characterization of the hexahydropolyprenols of *Aspergillus fumigatis* Fresenius, *Biochem. J. 102*: 443 (1967).

40. R. M. Barr and F. W. Hemming, Polyprenols of *Aspergillus niger*. Their characterization, biosynthesis and subcellular distribution, *Biochem. J. 126*: 1193 (1972).

41. P. Loew, G. Dallner, S. Mayer, S. Coven, B. T. Chait, and A. K. Menon, The mevalonate pathway of the bloodstream form of *Trypanosoma brucei*. Identification of dolichols containing 11 and 12 isoprene residues, *J. Biol. Chem. 266*: 19250 (1991).

42. O. Tollbom, C. Valtersson, T. Chojnacki, and G. Dallner, Esterification of dolichol in rat liver. *J. Biol. Chem. 263*: 1347 (1988).

43. K. J. I. Thorne and E. Kodicek, The structure of bactoprenol, a lipid formed by lactobacilli from mevalonic acid, *Biochem. J. 99*: 123 (1966).

44. M. Scher, W. J. Lennarz, and C. C. Sweeley, The biosynthesis of mannosyl-1-phosphorylpolyisoprenol in *Micrococcus lysodeikticus* and its role in mannan synthesis, *Proc. Natl. Acad. Sci. USA 59*: 1313 (1968).

45. D. P. Gough, A. L. Kirby, J. B. Richards, and F. W. Hemming, The characterization of undecaprenol of *Lactobacillus plantarum, Biochem. J. 118*: 167 (1970).

46. K. Takayama and D. S. Goldman, Enzymic synthesis of mannosyl-1-phosphoryl-decaprenol by a cell-free system of *Mycobacterium tuberculosis, J. Biol. Chem. 245*: 6251 (1970).

47. B. A. Wolucka, M. R. McNeil, E. de Hoffmann, T. Chojnacki, and P. J. Brennan, Recognition of the lipid intermediate for arabinogalactan/arabinomannan biosynthesis and its relation to the mode of action of ethambutol on mycobacteria, *J. Biol. Chem. 269*: 23328 (1994).

48. K. Takayama, H. K. Schnoes, and E. J. Semmler, Characterization of the alkali-stable mannophospholipids of *Mycobacterium smegmatis, Biochim. Biophys. Acta 316*: 212 (1973).

49. F. W. Hemming, Polyprenol alcohols (prenols), *Terpenoids in Plants* (J. B. Pridham, ed.), Academic Press, p. 223, New York, 1967.

50. D. R. Threlfall, Polyprenols and terpenoid quinones and chromanols, *Encyclopedia of Plant Physiology. New Series, v. 8* (A. Pirson and M. H. Zimmerman, eds.), p. 288, Springer, New York, 1980.

51. T. Chojnacki, E. Swiezewska, and T. Vogtman, Polyprenols from plants—Structural analogues of mammalian dolichols, *Chem. Scripta 27*: 209 (1987).

52. E. Swiezewska, W. Sasak, T. Mankowski, W. Jankowski, T. Vogtman, I. Krajewska, J. Hertel, E. Skoczylas, and T. Chojnacki, The search for plant polyprenols, *Acta Biochim. Polon. 41*: 221 (1994).

53. *Dolichols, Polyprenols and Derivatives. Collection of Polyprenols*, Techgen, Warsaw, 1995.

54. B. O. Lindgren, Homologous aliphatic $C_{30}-C_{45}$ terpenols in birch wood, *Acta Chem. Scand. 19*: 1317 (1965).

55. A. R. Wellburn and F. W. Hemming, Polyprenols of wood and leaf tissue of the silver birch *Betula verrucosa, Nature 212*: 1364 (1966).

56. R. L. Rowland, P. H. Latimer, and J. A. Giles, Flue-cured tobacco. I. Isolation of solanesol, an unsaturated alcohol, *J. Amer. Chem. Soc. 78*: 4680 (1956).

57. M. Koffer, A. Langemann, R. Ruegg, U. Gloor, U. Schwieter, J. Wuersch, O. Wiss, and O. Isler, Struktur und Partialsynthese des pflanzliches Chinons mit isoprenoider Seitenkette, *Helv. Chim. Acta 42*: 2252 (1959).

58. M. Toyoda, M. Asahina, H. Fukawa, and T. Shimizu, Isolation of new acyclic C_{25}-isoprenyl alcohol from potato leaves, *Tetrahedron Lett.*: 4879 (1969).

59. F. W. Hemming, R. A. Morton, and J. F. Pennock, Constituents of the unsaponifiable lipid fraction from the spadix of *Arum maculatum, Proc. Roy. Soc. London, Ser. B 158*: 291 (1963).

60. A. R. Wellburn, F. W. Hemming, J. Stevenson, and R. A. Morton, The characterization of properties of castaprenol-11, -12 and -13 from the leaves of *Aesculum hippocastanum* (horse chesnut), *Biochem. J. 102*: 313 (1967).

61. K. J. Stone, A. R. Wellburn, F. W. Hemming, and J. F. Pennock, The characterization of ficaprenol-10, -11 and -12 from leaves of *Ficus elastica* (decorative rubber plant), *Biochem. J. 102*: 325 (1967).

62. G. I. Vergunova, I. S. Glukhoded, L. L. Danilov, G. I. Eliseyeva, N. K. Kochetkov, M. F. Troitsky, A. I. Usov, A. S. Shashkov, and V. N. Shibaev, Structure of moraprenol and synthesis of moraprenyl phosphate, *Biorg. Khim. 3*: 1484 (1977).

63. T. Suga and T. Shihibori, Structure and biosynthesis of cleomeprenols from the leaves of *Cleome spinosa, J. Chem. Soc. Perkin Trans. 1*: 2098 (1980).

64. P. J. Dunphy, J. D. Kerr, J. F. Pennock, K. J. Whittle, and J. Feeney, The plurality of long chain isoprenoid alcohols (polyprenols) from natural sources, *Biochim. Biophys. Acta 136*: 136 (1967).

65. A. R. Wellburn and F. W. Hemming, The occurrence and seasonal distribution of higher isoprenoid alcohols in the plant kingdom, *Phytochemistry 5*: 969 (1966).

66. W. Jankowski and T. Chojnacki, Long chain isoprenoid alcohols in leaves of *Capparis* species, *Acta Biochim. Polon. 38*: 265 (1991).

67. G. Jommi, L. Verotta, P. Gariboldi, and P. Grabetta, Constituents of the lipophilic extract of the fruits of *Serenoa repens* (Bart) small, *Gazz. Chim. Ital. 118*: 823 (1988).

68. D. F. Zinkel and B. B. Evans, Terpenoids of *Pinus strobus* cortex tissues, *Phytochemistry 11*: 3387 (1972).

69. K. Hannus and G. Pensar, Polyisoprenols in *Pinus silvestris* needles, *Phytochemistry 13*: 2563 (1974).

70. K. Ibata, M. Mizuno, Y. Tanaka, and A. Kageyu, Long-chain polyprenols in the family *Pinaceae, Phytochemistry 23*: 783 (1984).

71. K. Ibata, M. Mizuno, T. Takigawa, and Y. Tanaka, Long-chain betulaprenol-type polyprenols from leaves of *Gingko biloba, Biochem. J. 213*: 305 (1983).

72. W. Sasak, T. Mankowski, T. Chojnacki, and W. M. Daniewsky, Polyprenols in *Juniperus communis* needles, *FEBS Lett. 64*: 55 (1976).

73. K. Ibata, A. Kageyu, T. Takigawa, M. Okada, T. Nishida, M. Mizuno, and Y. Tanaka, Polyprenols from conifers: Multiplicity of chain length distribution, *Phytochemistry 23*: 2517 (1984).

74. E. Swiezewska and T. Chojnacki, Long-chain polyprenols in gymnosperm plants, *Acta Biochim. Polon. 35*: 131 (1988).

75. E. Swiezewska and T. Chojnacki. The occurrence of unique, long-chain polyprenols in the leaves of *Potentilla* species, *Acta Biochim. Polon. 36*: 143 (1989).

76. E. Swiezewska and T. Chojnacki, Long-chain polyprenols from *Potentilla aurea, Phytochemistry 30*: 267 (1991).

77. E. Swiezewska, T. Chojnacki, W. Jankowski, A. K. Singh, and J. Olsson, The occurrence of long chain polyprenols in leaves of plants of *Rosaceae* family and their isolation by time-extended liquid chromatography, *Biochem. Cell Biol. 70*: 448 (1992).

78. T. Suga, S. Ohta, A. Nakai, and K. Munesada, Glycinoprenols: Novel polyprenols possessing a phytyl residue from the leaves of soybean, *J. Org. Chem. 54*: 3390 (1989).

79. N. Ya. Grigorieva and O. A. Pinkser, New approaches to the synthesis of regular functionalized Z-isoprenoids and their 2,3-dihydro derivatives, *Usp. Khim. 63*: 177 (1994).

80. W. Jankowski, G. Palamarczyk, I. Krajewska, and T. Vogtman, Specificity of cellular processes and enzymes towards polyisoprenoids of different structure, *Chem. Phys. Lipids 51*: 249 (1989).

81. H. Takaya, T. Ohta, N. Sayo, H. Kumobayashi, S. Akutagawa, S. Inoue, I. Kasahara, and R. Noyori, Enantioselective hydrogenation of allylic and homoallylic alcohols, *J. Amer. Chem. Soc. 109*: 1596 (1987).

82. B. Imperiali and J. W. Zimmerman, Synthesis of dolichols via asymmetric hydrogenation of plant polyprenols, *Tetrahedron Lett. 29*: 5343 (1988).

83. S. Suzuki, M. Shiono, and Y. Fujita, Grignard coupling reaction of (Z)-trisubstituted allylic acetates with retention of the double bond stereo- and regiochemistry, *Synthesis*: 804 (1983).

84. S. Suzuki, F. Mori, T. Takigawa, K. Ibata, J. Ninagawa, T. Mishida, M. Mizuno, and Y. Tanaka, Synthesis of mammalian dolichols from plant polyprenols, *Tetrahedron Lett. 24*: 5103 (1983).

85. T. Takigawa, K. Ibata, and M. Mizuno, Synthesis of mammalian dolichols from plant polyprenols, *Chem. Phys. Lipids 51*: 171 (1989).

86. V. A. Koptenkova, V. V. Veselovsky, M. A. Novikova, and A. M. Moiseenkov, Synthesis of dolichol-like prenols, WT_2C_2S-OH and WT_3C_2S-OH, *Izv. Akad. Nauk SSSR. Ser. Khim.*: 817 (1987).

87. V. V. Veselovsky, V. A. Koptenkova, M. A. Novikova, and A. M. Moiseenkov, Synthesis of dolichol-like (S)-hexa- and (S)-heptaprenol, *Izv. Akad. Nauk SSSR. Ser. Khim.*: 2052 (1989).

88. A. M. Moiseenkov, E. V. Polunin, and A. V. Semenovsky, Synthesis of (2Z,6Z,10Z,14E,18E)-farnesylfarnesol, *Tetrahedron Lett. 22*: 3309 (1981).

89. A. M. Moiseenkov, M. A. Novikova, E. V. Polunin, and S. I. Torgova, Synthesis of 2Z,6Z,10Z,14E,18E-farnesylfarnesol, *Izv. Akad. Nauk SSSR. Ser. Khim.*: 1557 (1983).

90. A. V. Semenovsky, N. Ya. Grigorieva, I. M. Avrutov, V. V. Veselovsky, M. A. Novikova, and A. M. Moiseenkov, Two approaches to stereospecific construction of polyprenols with heptaprenol WTTTCCC-OH as an example, *Izv. Akad. Nauk SSSR. Ser. Khim.*: 152 (1984).

91. M. A. Novikova, A. V. Lozanova, V. V. Veselovsky, V. A. Dragan, and A. M. Moiseenkov, Stepwise synthesis of three modified prenols, *Izv. Akad. Nauk SSSR. Ser. Khim.*: 356 (1987).

92. V. V. Veselovsky, V. A. Koptenkova, and A. M. Moiseenkov, Synthesis of tridecaprenol WT_3C_7S-OH and tetrakis-nor-octahydroanalogs of penta- and hexaprenol, *Izv. Akad. Nauk SSSR. Ser. Khim.*: 2296 (1989).

93. K. Sato, O. Miyamoto, S. Inoue, F. Furusawa, and Y. Matsuhashi, Stereospecific synthesis of a cisoid C_{10} isoprenoid building block and some *all-cis*-polyprenols, *Chem. Lett.*: 725 (1983).

94. K. Sato, O. Miyamoto, S. Inoue, F. Furusawa, and Y. Matsuhashi, General method of stereospecific synthesis of natural polyprenols. Synthesis of betulaprenols-6, -7, -8, and -9, *Chem. Lett.*: 1105 (1984).

95. K. Sato, O. Miyamoto, S. Inoue, Y. Matsuhashi, S. Koyama, and T. Kaneko, Stereospecific synthesis of (Z,Z,Z,Z,Z,Z,Z,Z,E,E)-undecaprenol (bacterial prenol) using an *all-cis* diterpene building block, *J. Chem. Soc. Chem. Commun.*: 1761 (1986).

96. S. Inoue, T. Kaneko, Y. Takahashi, O. Miyamoto, and K. Sato, Stereoselective total synthesis of (S)-(−)-dolichol-20, *J. Chem. Soc. Chem. Commun.*: 1036 (1987).

97. L. Jaenicke and H.-U. Siegmund, Total synthesis of chain-length uniform dolichyl phosphates and their fitness to accept hexoses in the enzymatic formation of lipoglycans, *Biol. Chem. Hoppe-Seyler 367*: 787 (1986).

98. L. Jaenicke and H.-U. Siegmund, Synthesis and characterization of dolichols and polyprenols of designed geometry and chain length, *Chem. Phys. Lipids 51*: 159 (1989).

99. N. Ya. Grigorieva, I. M. Avrutov, and A. V. Semenovsky, Novel approach to the stereoselective synthesis of polyprenols via directed aldol condensation, *Tetrahedron Lett. 24*: 5531 (1983).

100. N. Ya. Grigorieva, I. M. Avrutov, O. A. Pinsker, O. N. Yudina, A. I. Lutsenko, and A. M. Moiseenkov, Directed aldol condensation as a stereoselective synthetic way to Z-trisubstituted olefins, *Izv. Akad. Nauk SSSR. Ser. Khim.*: 1824 (1985).

101. N. Ya. Grigorieva, O. N. Yudina, O. A. Pinsker, E. D. Daeva, and A. M. Moiseenkov, Synthesis of Z,Z-trishomofarnesal *tert*-butylaldimine, *Izv. Akad. Nauk SSSR. Ser. Khim.*: 97 (1990).

102. N. Ya. Grigorieva, O. N. Yudina, and A. M. Moiseenkov, Glutaraldehyde derivatives as building blocks for stereoselective (Z)-C_5-elongation of a regular isoprenoid chain, *Synthesis*: 591 (1989).

103. N. Ya. Grigorieva, O. A. Pinsker, V. N. Odinokov, G. A. Tolstikov, and A. M. Moiseenkov, Synthesis of racemic octaprenol WT_3C_5S-OH and nonaprenol WT_3C_4S-OH, *Izv. Akad. Nauk SSSR. Ser. Khim.*: 1546 (1987).

104. N. Ya. Grigorieva, O. A. Pinsker, E. D. Daeva, and A. M. Moiseenkov, Stereoselective synthesis of dolichol-like octaprenol (S)-WT$_3$C$_3$S-OH, *Izv. Akad. Nauk SSSR. Ser. Khim.*: 2325 (1991).

105. N. Ya. Grigorieva, O. A. Pinsker, and A. M. Moiseenkov, Stereospecific synthesis of octaprenols WT$_3$C$_4$-OH and WT$_3$C$_3$T-OH, *Izv. Akad. Nauk SSSR. Ser. Khim.*: 2333 (1991).

106. N. Ya. Grigorieva, O. N. Yudina, E. D. Daeva, and A. M. Moiseenkov, Synthesis of modified hexaprenol WC$_5$-OH from glutaric aldehyde derivatives, *Izv. Akad. Nauk SSSR. Ser. Khim.*: 89 (1990).

107. N. Ya. Grigorieva, O. N. Yudina, and A. M. Moissenkov, Synthesis of two hexaprenols, modified in position of a hydroxyl group, *Izv. Akad. Nauk SSSR. Ser. Khim.*: 2036 (1986).

108. N. Ya. Grigorieva, O. N. Yudina, E. G. Cherepanova, and A. M. Moiseenkov, Synthesis of 3-desmethylhexaprenol WT$_3$C$_2$-OH, *Izv. Akad. Nauk SSSR. Ser. Khim.*: 803 (1990).

109. J. F. Wedgwood, J. L. Strominger, and C. D. Warren, Transfer of sugars from nucleoside diphosphate sugar compounds to endogenous and synthetic dolichyl phosphate in human lymphocytes, *J. Biol. Chem. 249*: 6316 (1974).

110. C. D. Warren and R. W. Jeanloz, Chemical synthesis of dolichyl phosphate and dolichyl glycosyl phosphates and pyrophosphates or "dolichol intermediates," *Methods Enzymol. 50*: 122 (1978).

111. T. A. Khwaja and C. B. Reese, A convenient general procedure for the conversion of alcohols into their monophosphate esters, *J. Chem. Soc. C*: 2092 (1970).

112. C. D. Warren and R. W. Jeanloz, Chemical synthesis of pyrophosphodiesters of carbohydrates and isoprenoid alcohols. Lipid intermediates of bacterial cell wall and antigenic polysaccharide biosynthesis, *Biochemistry 11*: 2565 (1972).

113. C. D. Warren, Y. Konami, and R. W. Jeanloz, The synthesis of P^1-(2-acetamido-2-deoxy-α-D-glucopyranosyl) P^2-ficaprenyl pyrophosphate, *Carbohydr. Res. 30*: 257 (1973).

114. N. A. Kalinchuk, L. L. Danilov, T. N. Druzhinina, V. N. Shibaev, and N. K. Kochetkov, Specificity of enzymes of O-antigen biosynthesis in *Salmonella anatum* towards polyprenyl derivatives of different chain-lenth and saturation, *Bioorg. Khim. 11*: 219 (1985).

115. C. A. Rupar and K. K. Carroll, Preparation of tritiated dolichol and dolichyl phosphate, *Chem. Phys. Lipids 17*: 193 (1976).

116. T. Hata, Y. Mushika, and T. Mukaiyama, A new phosphorylating reagent. I. The preparation of alkyl dihydrogen phosphates by means of 2-chloromethyl-4-nitrophenyl phosphorodichloridate, *J. Amer. Chem. Soc. 91*: 4532 (1969).

117. L. L. Danilov and T. Chojnacki, A simple procedure for preparing dolichyl monophosphates by the use of POCl$_3$, *FEBS Lett. 131*: 310 (1981).

118. C. D. Warren, S. Nakabayashi, and R. W. Jeanloz, Synthesis of tetrasaccharide lipid intermediate P^1-dolichyl P^2-[O-α-D-mannopyranosyl-(1-6)-O-β-D-mannopyranosyl-(1-4)-O-(2-acetamido-2-deoxy-β-D-glucopyranosyl)-(1-4)-O-(2-acetamido-2-deoxy-α-D-glucopyranosyl)] diphosphate, *Carbohydr. Res. 169*: 221 (1987).

119. B. Imperiali and J. W. Zimmerman, Synthesis of dolichyl-linked oligosaccharides, *Tetrahedron Lett. 31*: 6485 (1990).

120. J. Lee and J. K. Coward, Enzyme-catalyzed glycosylation of peptides using a synthetic lipid disaccharide intermediate, *J. Org. Chem. 57*: 4126 (1992).

121. R. W. Keenan, R. G. Martinez, and R. F. Williams, Synthesis of [^{32}P]dolichyl phosphate, utilizing a general procedure for [^{32}P]phosphorus oxychloride preparation, *J. Biol. Chem. 257*: 14817 (1982).

122. J. L. Montero, C. Toiron, J.-L. Clavel, and J.-L. Imbach, Phosphorylation of hydroxy groups via silylated intermediates, *Recl. Trav. Chim. Pays-Bas 107*: 570 (1988).

123. F. Cramer and W. Boehm, Synthese von Geranyl- und Farnesylpyrophosphate, *Angew. Chem. 76*: 775 (1959).

124. F. Cramer, W. D. Rittersdorf, and W. Boehm, Synthese von Phosphorsaureestern und Pyrophosphorsaureestern der Terpenalkohole, *Lieb. Ann. 654*: 180 (1962).

125. R. M. Barr and F. W. Hemming, Polyprenol phosphate as an acceptor of mannose from guanosine diphosphate mannose in *Aspergillus niger, Biochem. J. 126*: 1203 (1972).

126. W. Jankowski and T. Chojnacki, Enzymic formation of polyisoprenol phosphate sugars, *Acta Biochim. Polon. 19*: 51 (1972).

127. J. F. Wedgwood and J. L. Strominger, Enzymatic activities of cultured human lymphocytes that dephosphorylate dolichyl pyrophosphate and dolichyl phosphate, *J. Biol. Chem. 255*: 1120 (1980).

128. L. L. Danilov, S. D. Maltsev, and V. N. Shibaev, Phosphorylation of polyprenols with tetra-*n*-butylammonium phosphate and trichloroacetonitrile, *Bioorg. Khim. 14*: 1287 (1988).

129. L. L. Danilov, T. N. Druzhinina, N. A. Kalinchuk, S. D. Maltsev, and V. N. Shibaev, Polyprenyl phosphates: Synthesis and structure-activity relationship for a biosynthetic system of *Salmonella anatum O*-specific polysaccharide, *Chem. Phys. Lipids 51*: 191 (1989).

130. L. L. Danilov, S. D. Maltsev, and V. N. Shibaev, Efficient synthesis of retinyl phosphate, *Bioorg. Khim. 14*: 1293 (1988).

131. L. L. Danilov, S. D. Maltsev, and V. N. Shibaev, Phosphorylation of nonacosanol and cholesterol with tetra-*n*-butylammonium dihydrogen phosphate and trichloroacetonitrile, *Bioorg. Khim. 16*: 1002 (1990).

132. L. L. Danilov, S. D. Maltsev, and V. N. Shibaev, The mechanism of phosphorylation of alcohols with tetra-*n*-butylammonium dihydrogen phosphate and trichloroacetonitrile, *Bioorg. Khim. 16*: 1423 (1990).

133. A. F. Clark and C. L. Villemez, Artificial mannosyl acceptor for GDP-D-mannose. Lipid phosphate transmannosylase from *Phaseolus aureus, FEBS Lett. 32*: 84 (1973).

134. C. N. Joo, C. E. Park, J. K. G. Kramer, and M. Kates, Synthesis and hydrolysis of monophosphate and pyrophosphate esters of phytanol and phytol, *Can. J. Biochem. 51*: 1527 (1973).

135. S. L. Flitsch, J. P. Taylor, and N. J. Turner, Synthesis of a novel acceptor substrate for a mannosyl transferase, *J. Chem. Soc. Chem. Commun.*: 380 (1991).

136. S. L. Flitsch, H. L. Pinches, J. P. Taylor, and N. J. Turner, Chemo-enzymatic synthesis of a lipid-linked core trisaccharide of *N*-linked glycoproteins, *J. Chem. Soc. Perkin Trans. I*: 2087 (1992).

137. I. B. H. Wilson, J. P. Taylor, M. C. Webberley, N. J. Turner, and S. L. Flitsch, A novel mono-branched lipid phosphate acts as a substrate for dolichyl phosphate mannose synthetase, *Biochem. J. 295*: 195 (1993).

138. J. W. Perich and R. B. Johns, Di-*tert*-butyl *N,N*-diethylphosphoramidite. A new phosphitylating agent for efficient phosphorylation of alcohols, *Synthesis*: 142 (1988).

139. H. Julia, H. Mestdagh, and C. Rolando, Une méthode simple de synthèse des phosphates terpeniques allyliques primaires et tertiares, *Tetrahedron 42*: 3841 (1986).

140. L. L. Danilov, V. N. Shibaev, and K. N. Kochetkov, Phosphorylation of polyprenols via their trichloroacetimidates, *Bioorg. Khim. 9*: 844 (1983).

141. L. L. Danilov, V. N. Shibaev, and N. K. Kochetkov, Phosphorylation of polyprenols via their trichloroacetimidates, *Synthesis*: 404 (1984).

142. S. D. Maltsev, L. L. Danilov, and V. N. Shibaev, Effective synthesis of moraprenyl monophosphate sugars, *Bioorg. Khim. 14*: 69 (1988).

143. G. Popjak, J. W. Cornforth, R. H. Cornforth, R. Ryhage, and D. S. Goodman, Chemical synthesis of $1\text{-}H_2^3\text{-}2\text{-}C^{14}$ and $1\text{-}D_2\text{-}2\text{-}C^{14}\text{-}trans\text{-}trans$-farnesyl pyrophosphate and their utilization in squalene biosynthesis, *J. Biol. Chem. 237*: 56 (1962).

143a. R. H. Cornforth and G. Popjak, Chemical synthesis of substrates of sterol biosynthesis, *Methods Enzymol. 15*: 359 (1969).

144. D. V. Banthorpe, P. V. Christon, C. R. Pink, and D. G. Watson, Metabolism of linaloyl, neryl, and geranyl pyrophosphates in *Artemisia annua*, *Phytochemistry 22*: 2465 (1983).

145. K. R. Keller and R. Thompson, Rapid synthesis of isoprenoid diphosphates and their isolation in one step, using either TLC or flash-chromatography, *J. Chromatogr. 645*: 161 (1993).

146. C. D. Warren and R. W. Jeanloz, Synthesis of P^1-dolichyl P^2-α-D-mannopyranosyl pyrophosphate. The acid and alkaline hydrolysis of polyisoprenyl α-D-mannopyranosyl mono- and pyrophosphate diesters, *Biochemistry 14*: 412 (1975).

147. B. Pelc, Preparation of bactoprenyl pyrophosphate, *Biochim. Biophys. Acta. 208*: 155 (1970).

148. O. Samuel, Z. El Hachini, and H. Azerad, Preparation of $[1\text{-}^3\text{H}]$polyprenyl pyrophosphates, *Biochimie 56*: 1279 (1974).

149. M. G. Scher and C. J. Waechter, Brain dolichyl pyrophosphate phosphatase. Solubilization, characterization and differentiation from dolichyl monophosphate phosphatase activity. *J. Biol. Chem. 259*: 14580 (1984).

150. C. D. Warren and R. W. Jeanloz, Chemical synthesis of P^1-2-acetamido-2-deoxy-α-D-glucopyranosyl P^2-dolichyl pyrophosphate, *Carbohydr. Res. 37*: 252 (1974).

151. L. L. Danilov, S. D. Maltsev, V. N. Shibaev, and N. K. Kochetkov, Synthesis of polyprenyl pyrophosphate sugars from unprotected mono- and oligosaccharide phosphates, *Carbohydr. Res. 88*: 203 (1981).

152. Y. A. Shabalin, A. V. Naumov, V. M. Vagabov, I. S. Kulaev, L. L. Danilov, and V. N. Shibaev, Synthesis of dolichyl $[\beta\text{-}^{33}P]$pyrophosphate and dolichyl pyrophosphate $[^{14}\text{C}]$mannose, *Bioorg. Khim. 11*: 651 (1985).

153. L. L. Danilov and V. N. Shibaev, Synthesis of moraprenyl pyrophosphate α-D-glucuronic acid, *Bioorg. Khim. 14*: 558 (1988).

154. Z. P. Belousova, P. P. Purygin, L. L. Danilov, and V. N. Shibaev, Use of N,N'-carbonyl-bis-azoles for synthesis of polyprenyl pyrophosphate sugars, *Bioorg. Khim. 14*: 379 (1988).

155. S. D. Maltsev, A. V. Trifonov, L. L. Danilov, and V. N. Shibaev, Efficient synthesis of polyprenyl diphosphates, *Bioorg. Khim. 21*: 61 (1995).

156. V. J. Davisson, A. B. Woolside, and C. D. Poulter, Synthesis of allylic and homoallylic isoprenoid pyrophosphates, *Methods Enzymol. 110*: 130 (1985).

157. V. M. Dixit, F. M. Lascovics, W. I. Noall, and C. D. Poulter, Tris-(tetra-n-butylammonium) hydrogen pyrophosphate. A new reagent for the preparation of allylic pyrophosphate esters, *J. Org. Chem. 46*: 1967 (1981).

158. V. J. Davisson, A. B. Woodside, T. R. Neal, K. E. Stremler, M. Muehlbacher, and C. D. Poulter, Phosphorylation of isoprenoid alcohols. *J. Org. Chem. 51*: 4768 (1986).

159. V. J. Davisson, T. M. Zabriskie, and C. D. Poulter, Radiolabelled allylic isoprenoid pyrophosphates: Synthesis, purification and determination of specific activity, *Bioorg. Chem. 14*: 46 (1986).

160. L. L. Danilov, S. D. Maltsev, and V. N. Shibaev, Synthesis of 2,3-dihydropolyprenyl H-phosphonates and thiophosphates, *Bioorg. Khim. 17*: 1292 (1991).

161. T. Gotoh, T. Koyama, and K. Ogura, Farnesyl diphosphate synthase and solanesyl diphosphate synthase reactions of diphosphate-modified allylic analogs: The significance of the diphosphate linkage involved in the allylic substrates for prenyltransferase, *J. Biochem. 112*: 20 (1992).

162. D. S. Mautz, V. J. Davisson, and C. D. Poulter, Synthesis of *O*-geranyl (1-thio)di-phosphate, *Tetrahedron Lett. 30*: 7333 (1989).

163. S. A. Biller and C. Forster, The synthesis of isoprenoid (phosphinylmethyl)-phosphonates, *Tetrahedron 46*: 6645 (1990).

164. K. Toshima and K. Tatsuta, Recent progress in *O*-glycosylation methods and its ap-plication to natural product synthesis, *Chem. Rev. 93*: 1503 (1993).

165. J. Banoub, P. Boullanger, and D. Lafont, Synthesis of oligosaccharides of 2-amino-2-deoxysugars, *Chem. Rev. 92*: 1167 (1992).

166. D. L. MacDonald, Chemical synthesis of aldose 1-phosphates, *Methods Enzymol. 8*: 121 (1966).

167. D. L. MacDonald, Glycosyl phosphates, *Methods Carbohydr. Chem. 6*: 389 (1972).

168. P. M. Barna, Carbohydrate phosphorylation with phosphoric acid, polyphosphoric acid, and anhydrous phosphoric acid, *Synth. Commun. 1*: 207 (1971).

169. Yu. Yu. Kusov and V. N. Shibaev, Synthesis and properties of glycosyl phosphates, *Usp. Biol. Khim. 12*: 182 (1971).

170. E. W. Putman, Aldose 1-phosphates, *Methods Carbohydr. Chem. 2*: 261 (1963).

171. R. Roy, F. D. Tropper, and C. Grand-Maitre, Syntheses of glycosyl phosphates by phase transfer catalysis. *Can. J. Chem. 69*: 1462 (1991).

172. C. L. Stevens and R. E. Harmon, Proof of structure and synthesis of α-D-glucopyranosyl (dihydrogen phosphate), *Carbohydr. Res. 11*: 93 (1969).

173. R. R. Schmidt, M. Stumpp, and J. Michel, α- and β-D-Glucopyranosyl phosphates from *O*-α-D-glucopyranosyl trichloroacetimidates, *Tetrahedron Lett. 23*: 405 (1982).

174. R. R. Schmidt and M. Stumpp, Glycosylphosphate aus Glycosyl(trichloracetimidaten), *Lieb. Ann.*: 680 (1984).

175. M. Hoch, E. Heinz, and R. R. Schmidt, Synthesis of 6-deoxy-6-sulfo-α-D-glucopyranosyl phosphate, *Carbohydr. Res. 191*: 21 (1989).

176. G. H. Veeneman, H. J. G. Broxterman, G. A. van der Marel, and J. H. van Boom, An approach towards synthesis of 1,2-*trans* glycosyl phosphates via iodonium ion assisted activation of thioglycosides, *Tetrahedron Lett. 32*: 6175 (1991).

177. P. Pale and G. M. Whitesides, Synthesis of glycosyl phosphates using the Fraser-Reid activation, *J. Org. Chem. 56*: 4547 (1991).

178. C. L. Stevens and R. E. Harmon, Synthesis of β-D-glucopyranosyl (dihydrogen phos-phate), *Carbohydr. Res. 11*: 99 (1969).

179. L. V. Volkova, L. L. Danilov, and R. P. Evstigneeva, A novel, stereospecific synthesis of β-D-glucopyranosyl phosphate, *Carbohydr. Res. 32*: 165 (1974).

180. L. L. Danilov, L. V. Volkova, and R. P. Evstigneeva, Stereospecific synthesis of 1,2-*trans*-glycosyl phosphates, *Zh. Obsch. Khim. 45*: 2307 (1975).

181. L. L. Danilov, L. V. Volkova, V. A. Bondarenko, and R. P. Evstigneeva, Stereospecific synthesis of 1,2-*trans*-galactosyl and mannosyl phosphates, *Bioorg. Khim. 1*: 905 (1975).

182. Yu. L. Sebyakin, L. V. Volkova, E. E. Rusanova, and R. P. Evstigneeva, Stereospecific synthesis of 1,2-*trans*-galactosyl and -mannosyl phosphates, *Zh. Organ. Khim. 15*: 2228 (1979).

183. S. D. Maltsev, N. N. Yurchenko, L. L. Danilov, and V. N. Shibaev, Synthesis of moraprenyl pyrophosphates of β-D-galactose, β-D-glucose and 4-deoxy-α-D-*xylo*-hexose, *Bioorg. Khim. 9*: 1097 (1983).

184. M. A. Salam and E. J. Behrman, Synthesis of glycosyl phosphates from sugar or-thoesters: Formation of bis(2,3,4,6-tetra-*O*-acetyl-β-D-glucopyranosyl) phosphate and the effect of solvents on the synthesis of β-D-glucopyranosyl phosphate, *Carbohydr. Res. 90*: 83 (1981).

185. P. A. Hebda, E. J. Behrman, and G. A. Barber, The guanosine 5'-diphosphate D-mannose:guanosine 5'-diphosphate L-galactose epimerase of *Chlorella pyrenoidosa*.

Chemical synthesis of guanosine 5'-diphosphate L-galactose and further studies of the enzyme and the reaction it catalyzes, *Arch. Biochem. Biophys. 194*: 496 (1979).

186. J.-H. Tsai and E. J. Behrman, Synthesis of β-L-fucopyranosyl phosphate from fucose orthoacetates, *Carbohydr. Res. 64*: 297 (1978).

187. N. S. Utkina, S. D. Maltsev, L. L. Danilov, and V. N. Shibaev, Synthesis of citronellyl and dolichyl glycosyl phosphates, derivatives of β-D-glucose and β-D-galactose, by the *H*-phosphonate method, *Bioorg. Khim. 21*: 376 (1995).

188. V. N. Shibaev, Yu. Yu. Kusov, S. Kucar, and N. K. Kochetkov, Synthesis of deoxy-glycosyl phosphates, derivatives of 6-, 4- and 3-deoxy-D-glucose, *Izv. Akad. Nauk SSSR. Ser. Khim.*: 430 (1973).

189. S. G. Withers, M. D. Percival, and I. P. Street, The synthesis and hydrolysis of a series of deoxy- and deoxyfluoro-α-D-glucopyranosyl phosphates, *Carbohydr. Res. 187*: 43 (1989).

190. N. K. Kochetkov, E. I. Budowsky, V. N. Shibaev, and Yu. Yu. Kusov, Synthesis of uridine diphosphate 6-deoxyglucose, *Izv. Akad. Nauk SSSR. Ser. Khim.*: 1136 (1969).

191. N. K. Kochetkov, E. I. Budowsky, V. N. Shibaev, and Yu. Yu. Kusov, Synthesis of uridine diphosphate 4-deoxyglucose, *Izv. Akad. Nauk SSSR. Ser. Khim.*: 404 (1970).

192. V. N. Shibaev, Yu. Yu. Kusov, I. V. Komlev, E. I. Budowsky, and N. K. Kochetkov, Synthesis of uridine diphosphate 3-deoxyglucose, *Izv. Akad. Nauk SSSR. Ser. Khim.*: 2522 (1969).

193. B. Leon, S. Liemann, and W. Klaffke, Synthesis of some specifically deoxygenated D-hexopyranosyl phosphates, *J. Carbohydr. Chem. 12*: 597 (1993).

194. P. W. Kent and J. A. Wright, Chemical and enzymic reduction and phosphorylation of 6-deoxy-6-fluoro-α-D-galactose, *Carbohydr. Res. 22*: 193 (1972).

195. J. A. Wright and N. F. Taylor, Synthesis of 3-deoxy-3-fluoro-D-glucose 1- and 6-phosphates, *Carbohydr. Res. 32*: 366 (1974).

196. S. G. Withers, D. J. MacLennan, and I. P. Street, The synthesis and hydrolysis of a series of deoxyfluoro-D-glucopyranosyl phosphates, *Carbohydr. Res. 154*: 127 (1986).

197. S. G. Whistler and J. H. Stark, 5-Thio-D-glucopyranose 1-phosphate and 6-phosphate, *Carbohydr. Res. 13*: 15 (1970).

198. D. H. Leaback, E. C. Heath, and S. Roseman, Preparation and properties of crystalline acetochlorofucose and its conversion into the β-glycosides and α-1-phosphate of L-fucose, *Biochemistry 8*: 1351 (1969).

199. H. S. Prihar, J.-H. Tsai, S. R. Wanamaker, S. J. Duber, and E. J. Behrman, Synthesis of β-L-fucopyranosyl phosphate and L-fucofuranosyl phosphates by the MacDonald procedure, *Carbohydr. Res. 56*: 316 (1977).

200. R. R. Schmidt, B. Wegmann, and K.-H. Jung, Stereospecific synthesis of α- and β-L-fucopyranosyl phosphates and GDP-fucose via trichloroacetimidate, *Lieb. Ann.*: 121 (1991).

201. U. B. Gokhale, O. Hindsgaul, and M. M. Palcic, Chemical synthesis of GDP-fucose analogs and their utilization by the Lewis α(1–4) fucosyltransferase, *Can. J. Chem. 68*: 1063 (1990).

202. K. Adelhorst and G. M. Whitesides, Large-scale synthesis of β-L-fucopyranosyl phosphate and the preparation of GDP-β-L-fucose, *Carbohydr. Res. 242*: 69 (1993).

203. Y. Ichikawa, M. M. Sim, and C.-H. Wong, Efficient chemical synthesis of GDP-fucose, *J. Org. Chem. 57*: 2943 (1992).

204. T. Posternak and J. P. Posternak, Synthese d'ester phosphorique d'interet biologique. III. Synthese des acides α-D-mannose-1-phosphorique, D-mannose-6-phosphorique et α-D-mannose-1,6-diphosphorique. Action de la phosphoglucomutase, *Helv. Chim. Acta 36*: 1614 (1953).

205. S. Kucar, J. Zamocky, J. Zemek, and S. Bauer, Preparation of 3-, 4- and 6-deoxyderivatives of guanosine diphosphate D-mannose, *Chem. Zvesti 32*: 414 (1978).

206. J. Niggeman and J. Thiem, Synthesis of some deoxymannosyl phosphates, *Lieb. Ann.*: 535 (1992).

207. W. McDowell, T. J. Grier, J. R. Rasmussen, and R. T. Schwarz, The role of C-4 substituted mannose in protein glycosylation. Effect of guanosine diphosphate esters of 4-deoxy-4-fluoro-D-mannose and 4-deoxy-D-mannose on lipid-linked oligosaccharide assembly, *Biochem. J. 248*: 523 (1987).

208. V. N. Shibaev, N. S. Utkina, L. L. Danilov, and G. I. Eliseeva, Synthesis of β-L-rhamnopyranosyl phosphate and thymidine 5'-(β-L-rhamnopyranosyl) pyrophosphate, *Bioorg. Khim. 6*: 1778 (1980).

209. P. J. O'Brien, The synthesis of N-acetyl-α- and N-acetyl-β-D-glucosamine 1-phosphates (2-acetamido-2-deoxy-α- and β-D-glucose 1-phosphates), *Biochim. Biophys. Acta 86*: 628 (1964).

210. H. Heymann, R. Turdiu, B. K. Lee, and R. S. Burkulis, A synthesis of uridinediphospho-N-acetylmuramic acid and its use as an acceptor of L-[^{14}C]alanine, *Biochemistry 7*: 1393 (1968).

211. A. H. Olavesen and E. A. Davidson, The chemical synthesis of the 1-phosphate and the uridine diphosphate derivative of N-acetylchondrosine (O-β-D-glucopyranosyluronic acid (1–3)-2-acetamido-2-deoxy-D-galactose), *J. Biol. Chem. 240*: 992 (1965).

212. C. D. Warren, A. Herscovics, and R. W. Jeanloz, The synthesis of P^1-2-acetamido-4-O-(2-acetamido-2-deoxy-β-D-glucopyranosyl)-2-deoxy-α-D-glucopyranosyl P^2-dolichyl pyrophosphate (P^1-di-N-acetyl-α-D-chitobiosyl P^2-dolichyl pyrophosphate), *Carbohydr. Res. 61*: 181 (1978).

213. L. L. Danilov, S. D. Maltsev, and V. N. Shibaev, Synthesis of moraprenyl pyrophosphates of 2-acetamido-2-deoxy-α-D-glucose, 2-acetamido-2-deoxy-β-D-glucose and 2-acetamido-2-deoxy-α-D-galactose through phosphoimidazolidates, *Bioorg. Khim. 12*: 934 (1986).

214. W. L. Salo and H. G. Fletcher, Synthesis of 2-acetamido-2-deoxy-α-D-mannopyranosyl phosphate and uridine 5'-(2-acetamido-2-deoxy-α-D-mannopyranosyl dipotassium pyrophosphate), *Biochemistry 9*: 878 (1970).

215. G. Baluja, B. H. Chase, G. W. Kenner, and A. R. Todd, Nucleosides, XLV. Derivatives of β-2-amino-2-deoxy-D-glucose (β-D-glucosamine) 1-phosphate, *J. Chem. Soc.*: 4678 (1960).

216. F. Maley, G. F. Maley, and H. A. Lardy, The synthesis of α-D-glucosamine-1-phosphate and N-acetyl-α-D-glucosamine-1-phosphate. Enzymatic formation of uridine diphosphoglucosamine, *J. Amer. Chem. Soc. 78*: 5303 (1956).

217. K. Tadano, T. Tsuchiya, T. Suami, and K. L. Rinehart, Synthesis of cytidine, uridine, and adenosine 5'-(2,6-diamino-2,6-dideoxy-α-D-glucopyranosyl diphosphates) (neosamine C-CDP, -UDP, and -ADP), *Bull. Chem. Soc. Jpn. 55*: 3840 (1982).

218. F. Trigalo, D. Charon, and L. Szabo, Chemistry of bacterial endotoxins. IV. Synthesis of anomeric 2-deoxy-[(3R)-3-hydroxytetradecanamido]-D-glucopyranosyl phosphate and pyrophosphate derivatives related to lipid A, *J. Chem. Soc. Perkin Trans. I.*: 2243 (1988).

219. T. Yamazaki, C. D. Warren, A. Herscovics, and R. W. Jeanloz, The synthesis of uridine diphosphate N-acetylhexosamines and uridine 5'-(2-acetamido-2-deoxy-α-D-mannopyranosyluronic acid diphosphate), *Can. J. Chem. 59*: 2247 (1981).

220. A. Ya. Khorlin, S. E. Zurabyan, and T. S. Antonenko, Synthesis of 2-acetamido-2-deoxy-α-D-glycosyl phosphates via 2-methyl-glyco[2',1':4,5]-2-oxazolines, *Tetrahedron Lett.*: 4803 (1970).

221. M. Inage, H. Chaki, S. Kusumoto, and T. Shiba, Chemical synthesis of phosphorylated fundamental structure of lipid A, *Tetrahedron Lett. 22*: 2281 (1981).

222. C. D. Warren, M. A. E. Shaban, and R. W. Jeanloz, The synthesis and properties of benzylated oxazolines derived from 2-acetamido-2-deoxy-D-glucose, *Carbohydr. Res. 59*: 427 (1977).

223. J. E. Heidas, W. J. Lees, P. Pale, and G. M. Whitesides, Gram-scale synthesis of uridine 5′-diphospho-*N*-acetylglucosamine: Comparison of enzymic and chemical routes, *J. Org. Chem. 57*: 146 (1992).

224. R. L. Thomas, S. A. Abbas, and K. L. Matta, Synthesis of uridine 5′-(2-acetamido-2,4-dideoxy-4-fluoro-α-D-galactopyranosyl) diphosphate and uridine 5′-(2-acetamido-2,6-dideoxy-6-fluoro-α-D-glucopyranosyl) diphosphate, *Carbohydr. Res. 184*: 77 (1988).

225. M. Kiso, K. Nishikori, and A. Hasegawa, Synthesis of 2-(acylamino)-2-deoxy-α-D-glucopyranosyl phosphates: Monosaccharide analogs of lipid A, *Agr. Biol. Chem. 45*: 545 (1981).

226. E. Walker-Nasir and R. W. Jeanloz, The synthesis of P^1-[2-acetamido-2-deoxy-3-*O*-(β-D-glucopyranosyluronic acid)-α-D-glucopyranosyl] P^2-dolichyl diphosphate (*N*-acetylhyalobiosyluronic dolichyl diphosphate), *Carbohydr. Res. 68*: 343 (1979).

227. C. A. A. van Boeckel, J. P. G. Hermans, P. Westerduin, J. J. Oltvoort, G. A. van der Marel, and J. H. van Boom, Synthesis of two phosphorylated lipid A derivatives containing α- and β-anomeric phosphates, *Recl. Trav. Chim. Pays Bas 102*: 438 (1983).

228. W. McDowell, R. Datema, P. A. Romero, and R. T. Schwarz, Mechanism of activation of protein glycosylation by the antiviral sugar analogue 2-deoxy-2-fluoro-D-mannose: Inhibition of synthesis of Man(GlcNAc)$_2$-*PP*-Dol by the guanosine diphosphate ester, *Biochemistry 24*: 8145 (1985).

229. V. N. Shibaev, Yu. Yu. Kusov, S. Kucar, and N. K. Kochetkov, Synthesis of 2-deoxy-α-D-*arabino*-hexopyranosyl phosphate, *Izv. Akad. Nauk SSSR. Ser. Khim.*: 922 (1973).

230. S. Kucar, J. Zamocky, and S. Bauer, Synthesis of 2-deoxy-α-D-glucopyranosyl and 2-deoxy-α-D-galactopyranosyl phosphates, *Chem. Zvesti 28*: 115 (1974).

231. R. T. Schwarz and M. F. G. Schmidt, Formation of uridine diphosphate 2-deoxy-D-glucose and guanosine diphosphate 2-deoxy-D-glucose in vitro using animal enzymes, *Eur. J. Biochem. 62*: 181 (1976).

232. V. N. Shibaev, Yu. Yu. Kusov, and V. A. Petrenko, Phosphorylation of 2,6-dideoxy-L-*arabino*-hexose, *Izv. Akad. Nauk SSSR. Ser. Khim.*: 1843 (1975).

233. M. D. Percival and S. G. Withers, Application of enzymes in the synthesis and hydrolytic study of 2-deoxy-α-D-glucopyranosyl phosphate, *Can. J. Chem. 66*: 1970 (1988).

234. J. Niggeman, T. K. Lindhorst, M. Walfort, L. Lanpicher, H. Sajus, and J. Thiem, Synthetic approaches to 2-deoxyglycosyl phosphates, *Carbohydr. Res. 246*: 173 (1993).

235. S. Saseban and S. Neira, Synthesis of glycosyl phosphates and azides, *Carbohydr. Res. 223*: 169 (1992).

236. A. Granata and A. S. Perlin, Use of *O*-thallium(I) salts in the synthesis of phosphate, sulfite and related ester derivatives, *Carbohydr. Res. 94*: 165 (1981).

237. S. Hashimoto, T. Honda, and S. Ikegami, A rapid and efficient synthesis of 1,2-*trans*-β-linked glycosides via benzyl or benzoyl-protected glycopyranosyl phosphates, *J. Chem. Soc. Chem. Commun.*: 685 (1989).

237a. N. S. Utkina, L. L. Danilov, and V. N. Shibaev, Synthesis of glycosyl phosphates via lithium salts of acetylated mono- and oligosaccharides, *Bioorg. Khim. 12*: 1372 (1986).

238. H. S. Prihar and E. J. Behrman, Phosphorylation of the hemiacetal hydroxyl group: synthesis of β-D-mannopyranosyl phosphate, *Carbohydr. Res. 23*: 456 (1972).

239. C. D. Warren, Y. I. Liu, A. Herscovics, and R. W. Jeanloz, The synthesis and chemical properties of polyisoprenyl β-D-mannopyranosyl phosphates, *J. Biol. Chem. 250*: 8069 (1975).

240. H. S. Prihar and E. J. Behrman, Chemical synthesis of β-L-fucopyranosyl phosphate and β-L-rhamnopyranosyl phosphate, *Biochemistry 12*: 997 (1973).

241. V. N. Shibaev, Yu. Yu. Kusov, V. A. Petrenko, and N. K. Kochetkov, Improved synthesis of β-L-rhamnopyranosyl phosphate, *Izv. Akad. Nauk SSSR. Ser. Khim.*: 1852 (1974).

242. V. N. Shibaev, Yu. Yu. Kusov, V. A. Petrenko, and N. K. Kochetkov, Synthesis of thymidine 5'-(β-L-mannopyranosyl) and (4,6-dideoxy-L-*lyxo*-hexopyranosyl) pyrophosphates, *Izv. Akad. Nauk SSSR. Ser. Khim.*: 2588 (1976).

243. V. N. Shibaev, Yu. Yu. Kusov, V. A. Petrenko, and N. K. Kochetkov, Synthesis of 4,6-dideoxy-α- and β-L-*lyxo*-hexopyranosyl phosphates, *Izv. Akad. Nauk SSSR. Ser. Khim.*: 887 (1976).

244. H. A. Nunez, J. V. O'Connor, R. P. Rosevear, and R. Barker, The synthesis and characterization of α- and β-L-fucopyranosyl phosphates and GDP fucose, *Can. J. Chem. 59*: 2086 (1981).

245. T. Faust, C. Theurer, K. Eger, and W. Kreis, Synthesis of uridine 5'-(α-D-fucopyranosyl diphosphate) and (digitoxigenin-3β-yl)-β-D-fucopyranoside and enzymic β-D-fucosylation of cardenolide aglycons in *Digitalis lanata*, *Bioorg. Chem. 22*: 140 (1994).

246. L. D. Liu and H. D. Liu, Synthesis of cytidine diphosphate D-quinovose, *Tetrahedron Lett. 30*: 35 (1989).

247. Y. Watanabe, N. Hyodo, and S. Ozaki, Dibenzyl phosphofluoridate, a new phosphorylating agent, *Tetrahedron Lett. 29*: 5763 (1988).

248. M. Inage, H. Chaki, S. Kusumoto, and T. Shiba, A convenient preparative method of carbohydrate phosphates with butyl lithium and phosphorochloridate, *Chem. Lett.*: 1281 (1982).

249. S. Kusumoto, Chemical synthesis of lipid A, *Bacterial Endotoxic Lipopolysaccharides, v. 1* (D. C. Morrison and J. L. Ryan, eds.), CRC Press, p. 81, Boca Raton, 1992.

250. S. L. Flitsch, D. M. Goodridge, B. Guilbert, M. C. Webberley, and I. B. H. Wilson, The chemoenzymatic synthesis of neoglycolipids and lipid-linked oligosaccharides using glycosyltransferases, *Bioorg. Med. Chem. 2*: 1243 (1994).

251. G. Srivastava and G. Alton, Combined chemical-enzymic synthesis of deoxygenated oligosaccharide analogs: Transfer of deoxygenated D-Glc*p*NAc residues from their UDP-Glc*p*NAc derivatives using *N*-acetylglucosaminyltransferase I, *Carbohydr. Res. 207*: 259 (1990).

252. H. Kodama, Y. Kojihira, T. Endo, and H. Hashimoto, Synthesis of UDP-6-deoxy- and 6-fluoro-D-galactoses and their enzymic glycosyl transfer to mono- and biantennary carbohydrate chains, *Tetrahedron Lett. 34*: 6419 (1993).

253. N. S. Utkina, L. L. Danilov, T. N. Druzhinina, and V. N. Shibaev, Synthesis of cytidine diphosphate 3,6-dideoxyhexoses, glycosyl donors in biosynthesis of O-specific polysaccharides of *Salmonella* serogroups A, B, and C, *Bioorg. Khim. 15*: 1375 (1989).

254. G. Srivastava, O. Hindsgaul, and M. M. Palcic, Chemical synthesis and kinetic characterization of UDP-2-deoxy-D-*lyxo*-hexose ("UDP-2-deoxy-D-galactose"), a donor-substrate for β-(1–4)-D-galactosyltransferase, *Carbohydr. Res. 245*: 137 (1993).

255. P. Westerduin, G. H. Veenenman, J. E. Maregg, G. A. van der Marel, and J. H. van Boom, An approach to the synthesis of α-L-fucopyranosyl phosphoric acid mono- and diesters via phosphite intermediates, *Tetrahedron Lett. 27*: 1211 (1986).

256. G. A. Barber and E. J. Behrman, The synthesis and characterization of uridine 5'-(β-L-rhamnopyranosyl diphosphate) and its role in the enzymic synthesis of rutin, *Arch. Biochem. Biophys. 288*: 239 (1991).

257. W. Klaffke, Synthesis of GDP-3-acetamido-3-deoxy-α-D-mannose and GDP-3-azido-3-deoxy-α-D-mannose, *Carbohydr. Res. 266*: 285 (1995).

258. P. Westerduin, G. H. Veeneman, G. A. van der Marel, and J. H. van Boom, Synthesis of the fragment GlcNAc-α-(1-P-6)-GlcNAc of the cell wall polymer of *Staphylococcus lactis* having repeating *N*-acetylglucosamine phosphate units, *Tetrahedron Lett. 27*: 6271 (1986).

259. M. M. Sim, H. Kondo, and C.-H. Wong, Synthesis and use of glycosyl phosphites—An effective route to glycosyl phosphates, sugar nucleotides and glycosides, *J. Amer. Chem. Soc. 115*: 2260 (1993).

260. C. J. Waechter, J. J. Lucas, and W. J. Lennarz, Evidence for xylosyl lipids as intermediates in xylosyl transfers in hen oviduct membranes, *Biochem. Biophys. Res. Commun. 56*: 343 (1974).

261. K. Yokoyama, Y. Araki, and E. Ito, The function of galactosyl phosphorylpolyprenol in biosynthesis of lipoteichoic acid in *Bacillus coagulans, Eur. J. Biochem. 173*: 453 (1988).

262. A. Shimada, M. Ohta, H. Iwasaki, and E. Ito, The function of β-N-acetyl-D-glucosaminyl monophosphate undecaprenol in biosynthesis of lipoteichoic acids in a group of *Bacillus* strains, *Eur. J. Biochem. 176*: 559 (1988).

263. A. Shimada, J. Tamatukuri, and E. Ito, Function of α-D-glucosyl monophosphorylpolyprenols in biosynthesis of cell wall teichoic acids in *Bacillus coagulans, J. Bacteriol. 171*: 2835 (1989).

264. C. D. Warren and R. W. Jeanloz, Chemical synthesis of ficaprenyl α-D-mannopyranosyl phosphate, *Biochemistry 12*: 5031 (1973).

265. C. D. Warren and R. W. Jeanloz, Chemical synthesis of dolichyl α-D-mannopyranosyl phosphate and citronellyl α-D-mannopyranosyl phosphate. *Biochemistry 12*: 5538 (1973).

266. V. Amarnath and A. D. Broom, Chemical synthesis of oligonucleotides, *Chem. Rev. 77*: 183 (1977).

267. R. I. Zhdanov and S. M. Zhenodarova, Chemical methods of oligonucleotide synthesis, *Synthesis*: 222 (1975).

268. T. Yamazaki, D. W. Laske, A. Herscovics, C. D. Warren, and R. W. Jeanloz, Biosynthesis of α-D-glucosyl polyisoprenyl diphosphate in particulate preparations of *Micrococcus lysodeikticus, Carbohydr. Res. 120*: 159 (1983).

269. J. S. Rush, J. G. Shelling, N. S. Zingg, P. H. Ray, and C. J. Waechter, Mannosylphosphoryldolichol-mediated reactions in oligosaccharide-*PP*-dolichol biosynthesis. Recognition of the saturated α-isoprene unit of the mannosyl donor by pig brain mannosyltransferases, *J. Biol. Chem. 268*: 13110 (1993).

270. A. Herscovics, C. D. Warren, and R. W. Jeanloz, Anomeric configuration of the dolichyl D-mannosyl phosphate formed in calf pancreas microsomes, *J. Biol. Chem. 250*: 8079 (1975).

271. V. F. Zarytova and D. G. Knorre, Intermediate compounds and reactions in oligonucleotide synthesis, *Usp. Khim. 54*: 313 (1985).

272. N. S. Utkina, G. I. Eliseyeva, A. V. Nikolaev, and V. N. Shibaev, The synthesis of glycosyl phosphosugars containing 2-acetamido-2-deoxy-α-D-mannopyranosyl phosphate residues, including a fragment of the capsular antigen from *Neisseria meningitidis* A, *Bioorg. Khim. 19*: 228 (1993).

273. A. V. Nikolaev, I. A. Ivanova, V. N. Shibaev, and A. V. Ignatenko, ^{31}P NMR studies on the reaction of glycosyl hydrogenphosphonates with trimethylacetyl chloride and alcohols, *Bioorg. Khim. 17*: 1550 (1991).

274. L. L. Danilov, L. V. Volkova, and R. P. Evstigneeva, The synthesis of β-D-glucopyranosyl citronellyl phosphate. *Zh. Obsch. Khim. 47*: 2137 (1977).

275. Y. L. Sebyakin, E. E. Rusanova, L. V. Volkova, and R. P. Evstigneeva, Preparation of short-chain analogs of natural polyisoprenyl monophosphate sugars, *Khim. Prir. Soed.* 246 (1982).

276. L. L. Danilov, T. N. Druzhinina, V. N. Shibaev, and N. K. Kochetkov, Chemical synthesis of the intermediates of *Salmonella senftenberg O*-antigen biosynthesis. *Bioorg. Khim. 6*: 468 (1980).

277. C. D. Warren, M.-L. Milat, C. Auge, and R. W. Jeanloz, The synthesis of a trisaccharide and a tetrasaccharide lipid intermediate. P^1-Dolichyl P^2-[O-β-D-mannopyranosyl-(1–4)-O -(2-acetamido-2-deoxy-β-D-glucopyranosyl)-(1-4)-O-(2-acetamido-2-deoxy-α-D-glucopyranosyl)] diphosphate and P^1-dolichyl P^2-[O-α-D-mannopyranosyl-(1–3)-O -β-D-mannopyranosyl-(1–4)-O-(2-acetamido-2-de-oxyβ-D-glucopuranosyl)-(1–4)-O-(2-acetamido-2-deoxy-α-D-glucopyranosyl)] diphosphate, *Carbohydr. Res. 126*: 61 (1984).

278. L. L. Danilov, S. D. Maltsev, V. N. Shibaev, and N. K. Kochetkov, Synthesis of moraprenyl pyrophosphates of α-D-mannose, α-D-talose, α-D-fucose and labeled hexoses, *Bioorg. Khim. 8*: 109 (1982).

279. S. D. Maltsev, L. L. Danilov, and V. N. Shibaev, Synthesis of polyprenyl diphosphate sugars and uridine diphosphate sugars, derivatives of *N*-acetyl-D-quinovosamine and *N*-acetyl-D-fucosamine, *Bioorg. Khim. 17*: 540 (1991).

280. L. L. Danilov, T. N. Druzhinina, V. N. Shibaev, and N. K. Kochetkov, Synthesis of moraprenyl pyrophosphate oligosaccharides, putative substrates for biosynthesis of *Salmonella* O-specific polysaccharides, *Bioorg. Khim. 7*: 1718 (1981).

281. V. I. Torgov, C. A. Panosyan, A. T. Smelyansky, and V. N. Shibaev, Synthesis of moraprenyl pyrophosphate disaccharides, the intermediates in biosynthesis of *Salmonella* O-specific polysaccharides serotypes C_2 and C_3, *Bioorg. Khim. 11*: 83 (1985).

282. V. I. Torgov, C. A. Panosyan, and V. N. Shibaev, Synthesis of moraprenyl pyrophosphate trisaccharide, a precursor in the biosynthesis of main chain of *Salmonella* serogroup C_2 and C_3 O-specific polysaccharides, and its isomer, *Bioorg. Khim. 12*: 559 (1986).

283. V. I. Torgov, C. A. Panosyan, and V. N. Shibaev, Synthesis of di-, tri- and tetrasaccharides, the fragments of branched O-antigenic polysaccharide from *Salmonella kentucky* and their moraprenyl pyrophosphate derivatives, *Bioorg. Khim. 12*: 652 (1986).

284. V. I. Torgov, T. N. Druzhinina, O. A. Nechaev, and V. N. Shibaev, Synthesis of moraprenyl pyrophosphate derivatives of 3-O-α-L-rhamnopyranosyl-(4-O-α-D-glucopyranosyl)-α-D-galactopyranose and 3-O-α-L-rhamnopyranosyl-(4-O-β-D-glucopyranosyl)-α-D-galactopyranose and investigation of their substrate properties in the reaction of biosynthesis of O-antigenic polysaccharides from *Salmonella typhimurium* and *S. anatum*, *Bioorg. Khim. 13*: 947 (1987).

285. O. A. Nechaev, O. V. Sizova, T. N. Druzhinina, V. I. Torgov, and V. N. Shibaev, Transformation of di- and trisaccharide methylglycosides into moraprenyl pyrophosphate derivatives and their use for investigation of substrate specificity of mannosyltransferases from *Salmonella kentucky*, *Bioorg. Khim. 14*: 1290 (1988).

286. V. I. Torgov, V. N. Shibaev, and N. K. Kochetkov, Synthesis of moraprenyl pyrophosphate oligosaccharides, possible biosynthetic precursors of *E. coli* O8 and O9 O-antigens, *Bioorg. Khim. 10*: 946 (1984).

287. J. S. Tkacz, A. Herscovics, C. D. Warren, and R. W. Jeanloz, Mannosyltransferase activity in calf pancreas microsomes. Formation from guanosine diphosphate D-[^{14}C]mannose of a ^{14}C-labeled mannolipid with properties of dolichyl mannopyranosyl phosphate, *J. Biol. Chem. 249*: 6372 (1974).

288. A. Herscovics, C. D. Warren, R. W. Jeanloz, J. F. Wedgwood, I. Y. Liu, and J. L. Strominger, Occurrence of β-D-mannopyranosyl phosphate residue in the polyprenyl mannosyl phosphate formed in calf pancreas and in human lymphocytes, *FEBS Lett.* *45*: 312 (1974).

289. J. S. Tkacz and A. Herscovics, Ozonolytic cleavage of authentic and pancreatic dolichyl mannopyranosyl phosphate: Determination of sugar configuration in the fragments with α- and β-mannosidases, *Biochem. Biophys. Res. Commun. 64*: 1009 (1975).

290. A. Herscovics, B. Bugge, and R. W. Jeanloz, Glucosyltransferase activity in calf pancreas microsomes. Formation of dolichyl D-[^{14}C]glycosyl phosphate and ^{14}C-labeled lipid-linked oligosaccharides from UDP-D-[^{14}C] glucose, *J. Biol. Chem. 252*: 2271 (1977).

291. M. A. Ghalambor, C. D. Warren, and R. W. Jeanloz, Biosynthesis of a P^1-2-acetamido-2-deoxy-D-glucosyl P^2-polyisoprenyl pyrophosphate by calf pancreas microsomes, *Biochem. Biophys. Res. Commun. 56*: 407 (1974).

292. A. Herscovics, C. D. Warren, B. Bugge, and R. W. Jeanloz, Biosynthesis of P^1-di-*N*-acetyl-α-chitobiosyl P^2-dolichyl pyrophosphate in calf pancreas microsomes, *J. Biol. Chem. 253*: 160 (1978).

293. A. Herscovics, C. D. Warren, B. Bugge, and R. W. Jeanloz, Biosynthesis of dolichyl pyrophosphate trisaccharide from synthetic dolichyl pyrophosphate di-*N*-acetylchitobiose and GDP-D-[^{14}C]mannose in calf liver microsomes, *FEBS Lett. 120*: 271 (1980).

294. A. Herscovics, C. D. Warren, and R. W. Jeanloz, Solubilization of α-(1−3)-D-mannosyltransferase from pancreas which utilizes synthetic dolichyl pyrophosphate trisaccharide β-Man-(1−4)-β-GlcNAc-(1−4)-GlcNAc as substrate, *FEBS Lett. 156*: 298 (1983).

295. W. Sasak, C. Levrat, C. D. Warren, and R. W. Jeanloz, Biosynthesis of dolichyl pentasaccharide diphosphate in calf pancreas microsomes, *J. Biol. Chem. 259*: 332 (1984).

296. C. D'Souza-Schorey, K. R. McLachlan, S. S. Krag, and A. D. Elbein, Mammalian glycosyltransferases prefer glycosyl phosphoryl dolichols rather than glycosyl phosphoryl polyprenols as substrates for oligosaccharyl synthesis, *Arch. Biochem. Biophys. 308*: 497 (1994).

297. E. L. Kean, J. S. Rush, and C. J. Waechter, Activation of GlcNAc-*PP*-Dol synthesis by mannosylphosphoryldolichol is stereospecific and requires a saturated α-isoprene unit, *Biochemistry 33*: 10508 (1994).

298. A. W. DeLuca, J. S. Rush, A. Lehrman, and C. J. Waechter, Mannolipid donor specificity of glycosylphosphatidylinositol mannosyltransferase-I (GPIMT-I) determined with an assay system utilizing mutant CHO-K1 cells, *Glycobiology 4*: 909 (1994).

299. B. Imperiali and K. L. Shannon, Differences between Asn-Xaa-Thr-containing peptides: A comparison of solution conformation and substrate behavior with oligosaccharyltransferase, *Biochemistry 30*: 4374 (1991).

300. B. Imperiali, K. L. Shannon, and K. W. Rickert, Role of peptide conformation in asparagine-linked glycosylation, *J. Amer. Chem. Soc. 114*: 7942 (1992).

301. B. Imperiali, K. L. Shannon, M. Unno, and K. W. Rickert, A mechanistic proposal for asparagine-linked glycosylation, *J. Amer. Chem. Soc. 114*: 7944 (1992).

302. J. Lee and J. K. Coward, Oligosaccharyltransferase: synthesis and use of deuterium-labeled peptide substrates as mechanistic probes, *Biochemistry 32*: 6794 (1993).

303. Y. A. Shabalin, V. M. Vagabov, and I. S. Kulaev, Dolichyl diphosphate mannose as a possible intermediate of glycoprotein biosynthesis in yeast, *Dokl. Akad. Nauk SSSR 283*: 720 (1985).

304. Y. A. Shabalin and I. S. Kulaev, Some properties of dolichyl pyrophosphate mannose: polymannose glycosyltransferase from yeast membrane fraction, *Biokhimiya 52*: 368 (1987).

305. T. Mankowski, W. Sasak, and T. Chojnacki, Hydrogenated polyprenol phosphates—Exogenous lipid acceptors of glucose from UDP-glucose in rat liver, *Biochem. Biophys. Res. Commun. 65*: 1292 (1975).

306. T. Mankowski, W. Sasak, E. Janczura, and T. Chojnacki, Specificity of polyprenyl phosphates in the in vitro formation of lipid-linked sugars, *Arch. Biochem. Biophys. 181*: 393 (1977).

307. D. D. Pless and G. Palamarczyk, Comparison of polyprenyl derivatives in yeast glycosyl transfer reactions, *Biochim. Biophys. Acta 529*: 21 (1978).

308. A. Bergman, T. Mankowski, T. Chojnacki, L. M. de Luca, E. Peterson, and G. Dallner, Glycosyl transfer from nucleotide sugars to C_{85}- and C_{55}-polyprenyl and retinyl phosphates by microsomal subfractions and Golgi membranes of rat liver, *Biochem. J. 172*: 123 (1978).

309. K. R. McLachlan and S. S. Krag, Substrate specificity of *N*-acetylglucosamine 1-phosphate transferase activity in Chinese hamster ovary cells, *Glycobiology 2*: 313 (1992).

310. P. Low, E. Peterson, M. Mizuno, T. Takigawa, T. Chojnacki, and G. Dallner, Reaction of optically active *S*- and *R*-forms of dolichyl phosphate with activated sugars, *Biochem. Biophys. Res. Commun. 130*: 460 (1985).

311. T. Chojnacki, G. Palamarczyk, W. Jankowski, I. Krajewska-Rychlik, A. Szkopinska, and T. Vogtman, The enzymic formation of dolichyl phosphate mannose from C-3 enantiomeric dolichyl phosphates, *Biochim. Biophys. Acta 793*: 187 (1984).

312. G. Palamarczyk, T. Vogtman, T. Chojnacki, E. Bause, M. Mizuno, and T. Takigawa, Enantiomeric forms of phospho-2,3-dihydroprenols in mannosyl transfer, *Chem. Scripta 27*: 135 (1987).

313. G. Palamarczyk, L. Lehle, T. Mankowski, T. Chojnacki, and W. Tanner, Specificity of solubilized yeast glycosyl transferases for polyprenyl derivatives, *Eur. J. Biochem. 105*: 517 (1980).

314. P. Low, E. Peterson, M. Mizuno, T. Takigawa, T. Chojnacki, and G. Dallner, Effectivity of dolichyl phosphates with different chain length as acceptors of nucleotide activated sugars, *Biosci. Rep. 6*: 677 (1986).

315. L. Jaenicke, K. van Leyen, and H.-U. Siegmund, Dolichyl phosphate-dependent glycosyltransferases utilize truncated cofactors, *Biol. Chem. Hoppe-Seyler 372*: 1021 (1991).

316. R. Berendes and L. Jaenicke, Short-chain dolichols of defined chain length as cofactors in reactions of the microsomal dolichyl-phosphate cycle and transglycosylations, *Biol. Chem. Hoppe-Seyler 373*: 35 (1992).

317. A. Szkopinska, E. Swiezewska, and T. Chojnacki, On the specificity of dolichol kinase and Dol-P-Man synthase towards isoprenoid alcohols of different chain length in rat liver microsomal membranes, *Int. J. Biochem. 24*: 1151 (1992).

318. L. Lehle, A. Haselbeck, and W. Tanner, Synthesis of retinylphosphate mannose in yeast and its possible involvement in lipid-linked oligosaccharide formation, *Biochim. Biophys. Acta 757*: 77 (1983).

319. J. Stoll, L. Resenberg, D. D. Carson, W. J. Lennarz, and S. S. Krag, A single enzyme catalyzes the synthesis of mannosyl phosphoryl derivative of dolichol and retinol in rat liver and Chinese hamster ovary cells, *J. Biol. Chem. 260*: 232 (1985).

320. L. M. de Luca, C. S. Silverman-Jones, D. Rimoldi, K. E. Greek, and C. D. Warren, Retinoids and glycosylation, *Chem. Scripta 27*: 193 (1987).

321. S. Kato, M. Tsuji, Y. Nakanishi, and S. Suzuki, Phenyl phosphate as a water-soluble analog of dolichyl phosphate in microsomal mannosyltransferase systems, *J. Biochem.* 87: 929 (1980).

322. C. D. Warren, B. Bugge, R. DeGasperi, T. Chojnacki, R. Meuwly, M. Mizuno, and R. W. Jeanloz, The significance of polyprenol structure for protein glycosylation via the dolichol pathway, *Glycoconjugate J.* 5: 315 (1988).

323. S. L. Flitsch, J. P. Taylor, and N. J. Turner, Chemo-enzymatic synthesis of β-mannosyl containing trisaccharide, *J. Chem. Soc. Chem. Commun.*: 382 (1991).

324. L. Revers, I. B. H. Wilson, M. C. Webberley, and S. L. Flitsch, The potential dolichol recognition sequence of β-1,4-mannosyltransferase is not required for enzymic activity using phytanyl-pyrophosphoryl-α-*N,N'*-diacetylchitobiose as acceptor, *Biochem. J.* 299: 23 (1994).

325. W. McDowell and R. T. Schwarz, Specificity of GDP-Man:dolichyl-phosphate mannosyltransferase for the guanosine diphosphate esters of mannose analogues containing deoxy and deoxyfluoro substituents, *FEBS Lett.* 243: 413 (1989).

326. R. Datema and R. T. Schwarz, Formation of 2-deoxyglucose-containing lipid-linked oligosaccharides. Interference with glycosylation of glycoproteins, *Eur. J. Biochem.* 90: 505 (1978).

327. L. Lehle and R. T. Schwarz, Formation of dolichol monophosphate 2-deoxy-D-glucose and its interference with glycosylation of mannoproteins in yeast, *Eur. J. Biochem.* 67: 239 (1976).

328. R. T. Schwarz, M. F. G. Schmidt, and L. Lehle, Glycosylation in vitro of Semliki-forest-virus and influenza-virus glycoproteins and its suppression by nucleotide-2-deoxy-hexose, *Eur. J. Biochem.* 85: 163 (1978).

329. R. Datema, R. T. Schwarz, L. A. Rivas, and R. Pont Lezica, Inhibition of β-1,4-glucan biosynthesis by deoxyglucose. The effect on the glucosylation of lipid intermediates, *Plant Physiol.* 71: 76 (1983).

330. W. McDowell, G. Weckbecker, D. O. R. Keppler, and R. T. Schwarz, UDP-glucosamine as a substrate for dolichyl monophosphate glucosamine synthesis. *Biochem. J.* 233: 749 (1986).

331. C. R. Faltynek, J. E. Silbert, and L. Hof, Xylosylphosphoryldolichol synthesized by chick embryo epiphyses. Not an intermediate in proteoglycan synthesis, *J. Biol. Chem.* 257: 5490 (1982).

332. M. A. McCloskey and F. A. Troy, Paramagnetic isoprenoid carrier lipid. 1. Chemical synthesis and incorporation into model membranes, *Biochemistry 19*: 2056 (1980).

333. M. A. McCloskey and F. A. Troy, Paramagnetic isoprenoid carrier lipid. 2. Dispersion and dynamics in lipid membranes, *Biochemistry 19*: 2061 (1980).

334. M. J. Knudsen and F. A. Troy, Nuclear magnetic resonance studies of polyisoprenols in model membranes, *Chem. Phys. Lipids 51*: 205 (1989).

335. L. W. Jiang, B. A. Mitchell, J. G. Teodoro, and J. W. Rip, Uptake and transport of fluorescent derivatives of dolichol in human fibroblasts, *Biochim. Biophys. Acta 1147*: 205 (1993).

10

Synthesis of Inhibitors of the Glycosidases and Glycosyltransferases Involved in the Biosynthesis and Degradation of Glycoproteins

J. Michael Williams
University of Wales Swansea, Swansea, Wales

I. INTRODUCTION

Increasingly in recent years it has been recognized that the oligosaccharide constituents of glycoproteins can influence their properties, such as solubility, rate of proteolysis, and antigenicity, as well as their intracellular fate, including targeting to lysosomes and insertion into the cellular membrane. These observations have stimulated much interest in the synthesis of potential inhibitors of enzymes involved in glycoprotein biosynthesis and degradation, since, in principle, such inhibitors can be used to modify the structures of the oligosaccharide moieties of glycoproteins and thus assist the study of structure–function relationships. Other potential applications of such inhibitors include interference with the viability of viruses and cancer cells.

The enzymes involved in the formation, modification, and degradation of the oligosaccharide moieties of glycoproteins are the glycosyltransferases and glycosidases (glycoside hydrolases). The biosynthesis of the serine or threonine-linked (O-linked) oligosaccharides of glycoproteins involves the sequential addition of monosaccharides by the appropriate glycosyltransferases. The biosynthesis of the asparagine-bound (N-linked) oligosaccharides is more complicated, and the most studied enzymes in this context are the glycosidases that are involved in the processing of N-linked oligosaccharides. In general, studies of glycosyltransferases are more difficult because they are membrane bound and of low abundance. But progress has been made in the synthesis of inhibitors of these enzymes also.

The biosynthesis of the N-linked oligosaccharide chains of glycoproteins involves the initial formation of a dolichol pyrophosphate-linked oligosaccharide, which is then transferred to the Asn-X-Ser/Thr sequons of newly synthesized polypeptide in the endoplasmic reticulum (ER). Subsequent processing of these oligosaccharides (**1**, Figure 1) involves trimming by glucosidases and mannosidases and extension by various glycosyltransferases [1]. Removal of glucose units involves first α-glucosidase I, which removes the terminal (α1–2)-linked glucose unit to form (**2**), and then α-glucosidase II removes the two (α1–3)-linked glucose units to form (**3**). α-Mannosidases in the ER and Golgi complex next remove the four (α1–2)-linked mannose units to form $Man_5GlcNAc_2$ (**4**). Removal of these four mannose units was originally attributed to α-mannosidase I in the Golgi, but other α-mannosidases, some associated with the ER, have since been discovered (reviewed by Moremen and associates, [2]). A further complication is the possible recycling of proteins between the ER and the Golgi [3]. Glucose-free oligosaccharides may be transiently glucosylated in the ER by the action of a glucosyltransferase [4], and it has been suggested recently that some of the processing involving α-mannosidases takes place in an intermediate compartment between the ER and the Golgi [5].

The α-mannosidases have recently been divided into two classes on the basis of amino acid sequence homologies [2] and inhibitor studies [5]. Class 1 α-mannosidases are specific for (α1–2)-mannose linkages, whereas class 2 enzymes can hydrolyze (1–2)-, (1–3)-, and (1–6)-linkages.

An alternative processing pathway involves an endomannosidase that cleaves Glc_3Man from $Glc_3Man_9GlcNAc_2$ (1) and Glc_2Man from $Glc_2Man_9GlcNAc_2$ (**2**) [6–10]. This enzyme also removes GlcMan from monoglucosylated oligosaccharides [7].

The next processing step after the formation of $Man_5GlcNAc_2$ (**4**) is the addition of one GlcNAc unit by the action of GlcNAc transferase I to form (**6**) followed by removal of one (α1–3)- and one (α1–6)-linked mannose unit to form $GlcNAcMan_3GlcNAc_2$ (**7**). This oligosaccharide is then extended with the addition of GlcNAc, Gal, Fuc, and sialic acid units by the appropriate glycosyltransferases to form bi-, tri- and tetraantennary oligosaccharides. Oligosaccharides prior to (**6**) may be extended to form high-mannose oligosaccharides. Some of the mannosidases mentioned may have a role in glycoprotein catabolism, as do endo-glycosidases such as endo-N-acetyl-β-D-glucosaminidase [5].

Since the glycosidases involved in the processing of N-linked oligosaccharide chains have a much stricter substrate specificity than lysosomal or other glycosidases, they are attractive targets for the development of specific inhibitors. In this review only representative examples of inhibitors in the above context will be discussed. All enzyme inhibitions referred to are competitive unless otherwise stated.

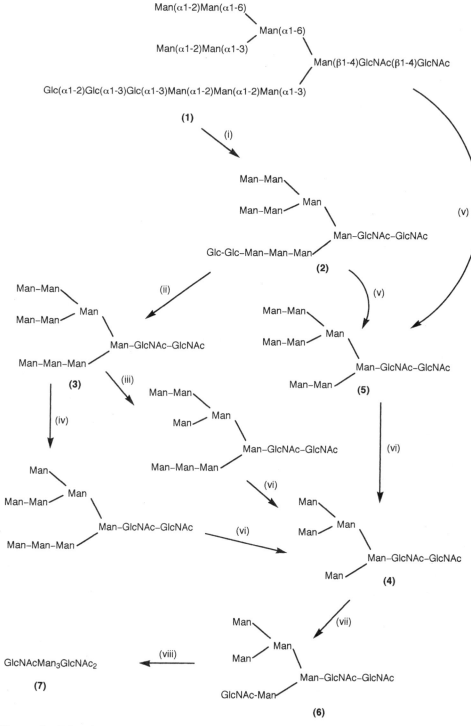

Figure 1 Trimming steps in the processing of *N*-linked oligosaccharides. (i) α-Glucosidase I. (ii) α-Glucosidase II. (iii) ER mannosidase I. (iv) ER mannosidase II. (v) Endomannosidase. (vi) Golgi mannosidase I. (vii) GlcNAc transferase I. (viii) Golgi mannosidase II.

II. INHIBITORS OF GLYCOSIDASES

A. Monocyclic Compounds

Lactones, Lactams, and Related Compounds

Glycosidases may act with inversion or retention of configuration at the anomeric center, but most act with retention. There are two possible mechanisms, originally suggested by Koshland [11], one involving a double displacement, the other involving a glycosyl oxocarbenium ion stabilized by a suitably located negatively charged group. It is considered likely that for most configuration-retaining glycosidases the active site contains two carboxyl groups, one acting as a proton donor, the other stabilizing the glycosyl cation that remains after departure of the aglycone. For a more detailed discussion of the mechanism of glycosidase action, the reader is referred to reviews by Sinnott [12] and Legler [13].

The inhibition of glycosidases by aldonolactones has been known for over 50 years, and the suggestion that such inhibition was due to the resemblance of the structure of the lactone, e.g., (**8**), to that of the glycosyl oxocarbenium ion intermediate (**9**) or to a related transition state was made in 1968 [14]. Studies of the inhibition by lactones are complicated by the interconversion of 1,5- and 1,4-lactones in solution, but the former are generally regarded as better inhibitors than the latter. Aldono-1,5-lactams, which cannot isomerize in the same way as lactones, inhibit glycosidases to about the same extent as the aldonolactones, β-glycosidases being inhibited more strongly. Examples are provided by GlcNAc-lactam (2-acetamido-5-amino-2,5-dideoxy-D-gluconolactam, K_i 1.8 μM for N-acetyl-β-D-glucosaminidase from bovine kidney) [15] and L-fuconolactam (5-amino-5,6-dideoxy-L-galactonolactam, K_i 0.3 mM for bovine kidney α-L-fucosidase) [16], which was synthesized as shown in Scheme 1. The known starting material is readily accessible from D-glucose in four steps, and this synthesis is typical of many syntheses of nitrogen-containing sugar-related inhibitors. An alternative route to these lactams was recently reported that used readily formed aldonamides as intermediates. O-Benzylated aldoses such as 2,3,4,6-tetra-O-benzyl-D-glucose, can be converted into the corresponding amide by oxidation followed by reaction with ammonia. Oxidation of the alcohol group then forms the 5-ulose, which is cyclized by the action of ammonia to form the epimeric hydroxylactams (Scheme 2). Reduction of this epimeric mixture with sodium cyanoborohydride in the presence of formic acid gave the D-gluconolactam. The D-galactonolactam was similarly prepared, but in the mannose series the reduction of the intermediate acyliminium ion was not stereoselective, D-*manno* and L-*gulo* lactams being formed in a 2:3 ratio. Five-membered-ring lactam formation was also possible [17].

(9)   (8)

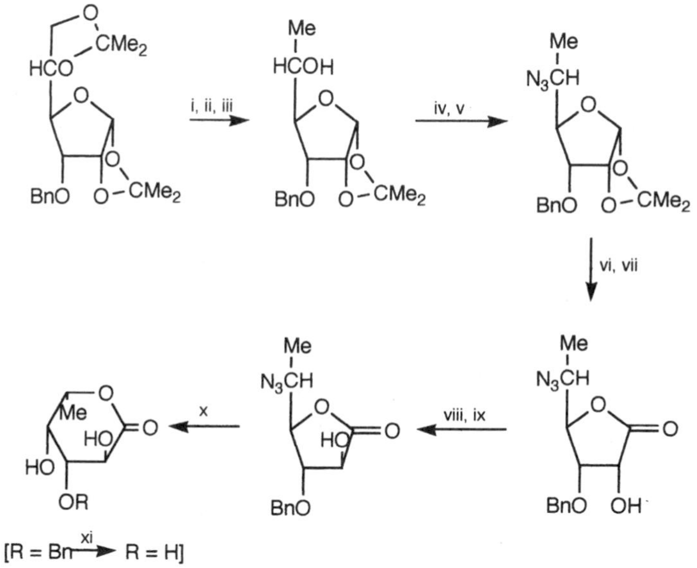

Scheme 1

i) Aq. HOAc
ii) TsCl, pyridine
iii) LiBHEt₃, THF
iv) MsCl, pyridine
v) NaN₃, DMF
vi) Aq. TFA

vii) Br₂, aq. dioxane, BaCO₃
viii) Tfl₂O, pyridine
ix) CF₃COONa, DMF, then aq. MeOH
x) H₂, Pd-C, EtOAc
xi) H₂, Pd-C, EtOH, HCl

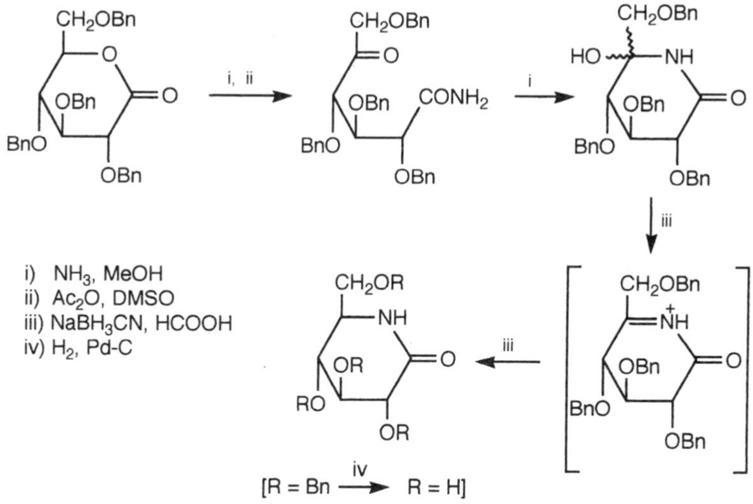

i) NH₃, MeOH
ii) Ac₂O, DMSO
iii) NaBH₃CN, HCOOH
iv) H₂, Pd-C

Scheme 2

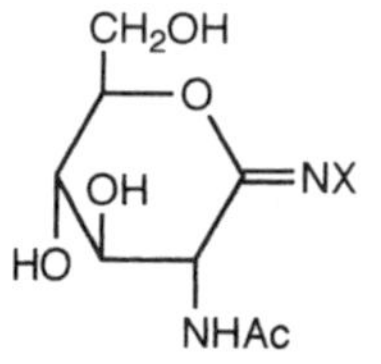

(10) X = OH
(11) X = NHCONHPh
(12) X = OCONHPh

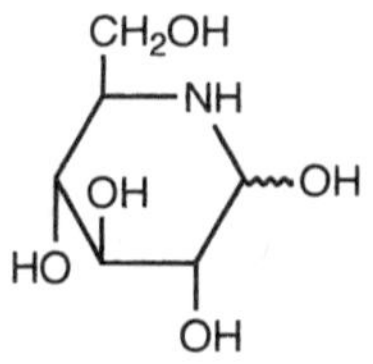

(13)

509

Other compounds related in structure to 2-acetamido-2-deoxy-D-glucono-1,5-lactone and found to be potent inhibitors include the oximinolactone (10) (K_i 0.45 μM for bovine kidney *N*-acetyl-β-D-glucosaminidase) [15], the lactone phenylsemicarbazone (11) (K_i 0.13−6.0 μM for enzymes from different sources) [18], and the phenylcarbamoyl oxime (12) (K_i 40 nM for a fungal enzyme) [19].

Aza-Sugar Analogs: Piperidine Derivatives

The discovery that sugar-related alkaloids and other basic derivatives of saccharides can inhibit glycosidases has stimulated many syntheses of such compounds. Analogs and related compounds have been synthesized in a search for compounds with enhanced inhibitory properties.

The inhibitory properties of glycosylamines have been known for many years. The glycosylamines derived from D-glucose, D-galactose, D-mannose [20], and 2-acetamido-2-deoxy-D-glucose [21] were found to inhibit the glycosidases that act on glycosides, corresponding to the inhibitor glycone. Detailed studies of this inhibition, however, were limited by the rapid anomerization of glycosylamines and their ease of hydrolysis.

The first aza-sugar analog to be reported as an inhibitor was 5-amino-5-deoxy-D-glucopyranose, nojirimycin (13), which was isolated from *Streptomyces* culture filtrates and found to be a powerful inhibitor of β-glucosidases. It also inhibits α-glucosidases (K_i 1.0 μM at pH 4.0 for the enzyme from human liver) [22], but no data on the inhibition of α-glucosidases I and II have appeared. The later discovery that the 1-deoxy derivative, 1,5-dideoxy-1,5-imino-D-glucitol (14) (deoxynojirimycin, DNJ), also a microbial product, was a good inhibitor of α-glucosidases (including α-glucosidases I and II) led to the synthesis of the corresponding aza analogs of other monosaccharides.

Several syntheses of DNJ have been reported. A 12-step synthesis from 1,2:5,6-di-*O*-isopropylidene-α-D-glucofuranose inserted the required nitrogen atom by an azide opening of the known epoxide (15) to form a 6-azido-6-deoxy-L-idofuranose derivative, which was converted into DNJ as shown in Scheme 3 [23]. The reactivity of a trifluoromethanesulfonate was necessary to achieve the nucleophilic displacement at C-2. Some syntheses involve formation of the *N*-containing ring by cyclizations (of an aminoalkene or amino epoxide intermediate) that form mixtures of regioisomers [24,25]. Such regiochemical problems were overcome in a recent approach in which the key step was cyclization of an aldosulose, e.g., D-*xylo*-hexos-5-ulose, by reductive amination (Scheme 4). This did have the potential disadvantage that two epimers could be formed, but the stereoselectivity was very high and the major glucitol product was readily obtained pure by recrystallization [26]. This route is useful for the formation of *N*-alkylated piperidines such as (16) that are otherwise difficult to prepare. Stereoselectivity was not high when the same reaction sequence was applied to the hexos-5-ulose derived from D-mannose, a mixture of D-*manno* and L-*gulo* products being isolated in a ratio of 2:1 [26]. Excellent stereoselectivity was observed in the corresponding cyclization of 2-acetamido-2-deoxy-D-*xylo*-hexos-5-ulose, prepared from the known oxazoline derivative (17), to form the aza-GlcNAc product (18, R = CH$_2$Ph). The debenzylated compound (18, R = H) inhibited *N*-acetyl-β-D-glucosaminidase from bovine kidney (K_i 1.7 μM) [15]; earlier studies had

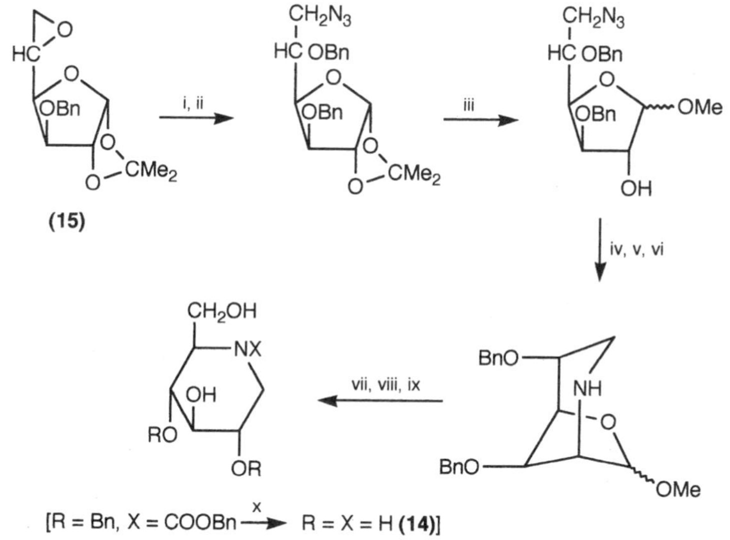

Scheme 3

i)   NaN₃, DMF
ii) BnBr, NaH, Bu₄NI, THF
iii) MeOH, HCl
iv) Tfl₂O
v)   SnCl₂, MeOH
vi) NaOAc, EtOH
vii) BnOCOCl
viii) TFA, aq. dioxane
ix) NaBH₄, EtOH
x)   H₂, Pd-C, HOAc

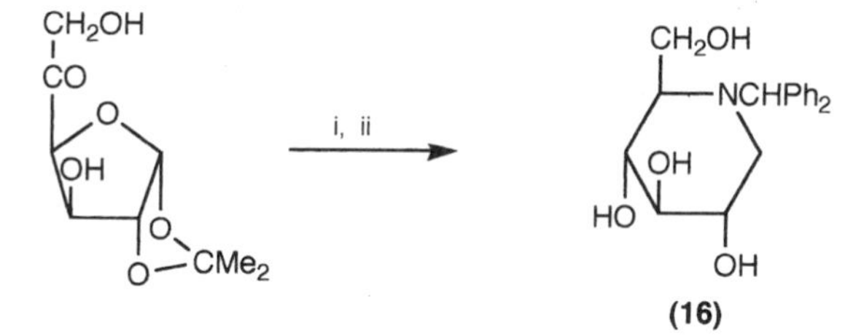

i) Dowex 50W-X8
ii) NaBH₃CN, Ph₂CHNH₂, MeOH

Scheme 4

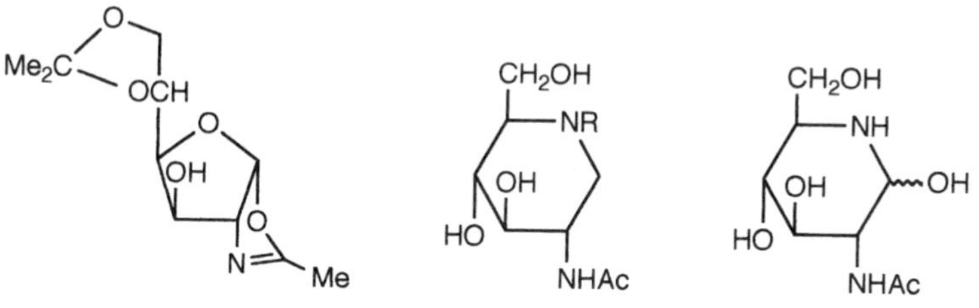

(17)   (18)   (19)

shown that the corresponding aminal, 2-acetamido-5-amino-2,5-dideoxy-D-glucopyr-anose (**19**), was a much more potent inhibitor of this enzyme [21].

N-Alkylation has been found to improve the potency and specificity of DNJ, a value of $K_i = 0.07$ μM being reported for the inhibition of calf liver glucosidase I by the N-methyl derivative [27]. N-Alkylation also increased the inhibition of glu-cosidase I more than that of glucosidase II [28]. N-Alkylation of the parent alkaloid is relatively straightforward, the main methods being nucleophilic displacement of alkyl bromide or triflate by the secondary amine, or reductive amination with the appropriate aldehyde [29].

1,5-Dideoxy-1,5-imino-D-mannitol, deoxymannojirimycin (DMJ), which inhib-its mannosidase I (K_i 5 μM for pig liver enzyme) [30], has been synthesized from sucrose by an ingenious route that utilized both monosaccharide units (Scheme 5) [31]. The diazido sucrose starting material is readily prepared by reaction of sucrose with tetrachloromethane and triphenyl phosphine in pyridine to form the 6,6'-di-chloride followed by azide displacement. The glucose and fructose azides were sep-arated, and the efficiency of the synthesis could be increased by partially converting the azidoglucose into the azidofructose by the action of polymer-supported glucose isomerase. An overall yield of 35% was thus achieved for this four-step synthesis. Another synthesis used as starting material the chiral diol (**20**), readily available from the microbial oxidation of chlorobenzene, (Scheme 6) [32]. A chemoenzymatic ap-proach was used by Wong and co-workers in their synthesis of DMJ [33]. Fructose 1,6-diphosphate aldolase from *E. Coli* was used to catalyze the reaction between dihydroxyacetone phosphate and (R)-3-azido-2-hydroxypropanal, which was formed from the racemate by an enzymatic resolution involving lipase-catalyzed hydrolysis of 3-azido-2-acetoxypropanal diethyl acetal. Dephosphorylation of the product gave 6-azido-6-deoxy-D-*arabino*-hexulose (**21**), which was converted by reductive ami-nation into DMJ in 80% yield. The last step in this sequence, reduction of the cyclic

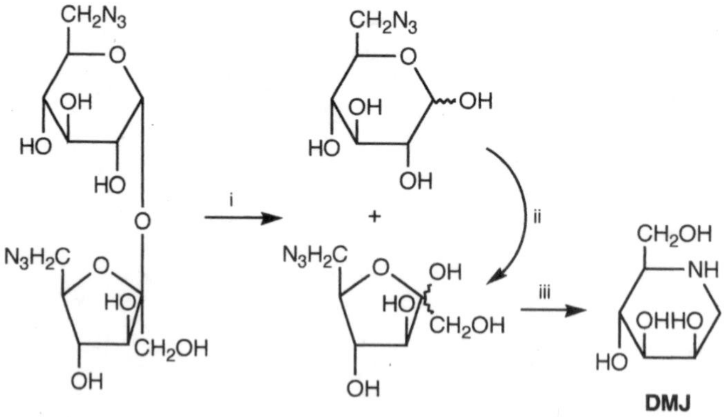

i) H_3O^+

ii) Glucose isomerase

iii) H_2, Pd-C, MeOH, H_2O

Scheme 5

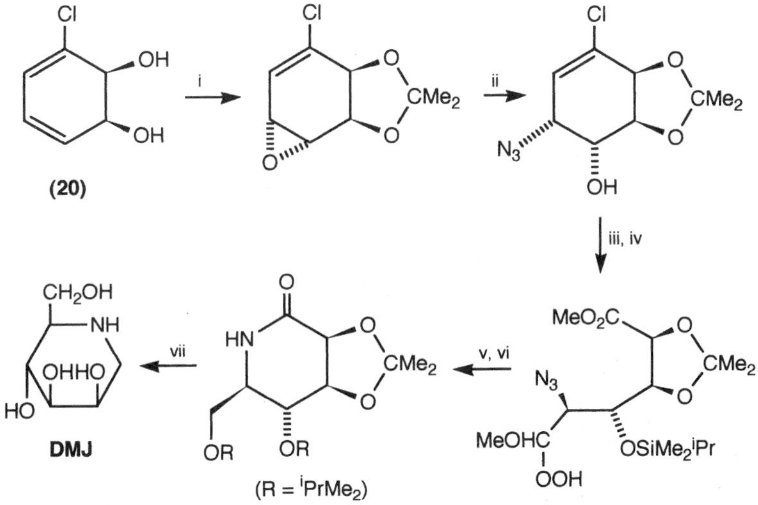

Scheme 6

i) MeC(OMe)₂Me, H⁺;
m-CPBA, phosphate buffer

ii) LiCl, EtO₂CCH₂COMe; NaN₃

iii) ⁱPrMe₂SiCl, imidazole

iv) O₃, MeOH-H₂O, NaHCO₃, -78°C

v) NaBH₄, CeCl₃, -20°C; H₂, Pd-C

vi) ⁱPrMe₂SiCl, DBU

vii) BH₃.Me₂S, THF; aq.TFA

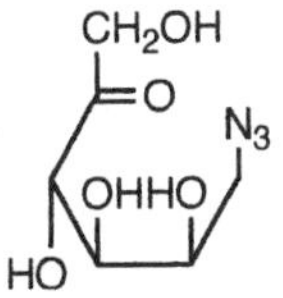

imine intermediate (**22**), was clearly highly stereoselective. In a study of structure–activity relationships, van den Broek and co-workers found that the 3-OH group in DNJ and DMJ was essential for inhibitory activity, the 3-O-methyl derivatives being devoid of activity against the appropriate glycosidases [29].

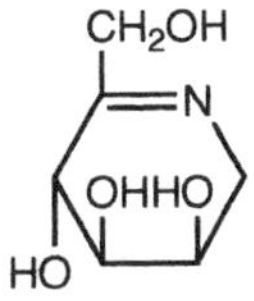

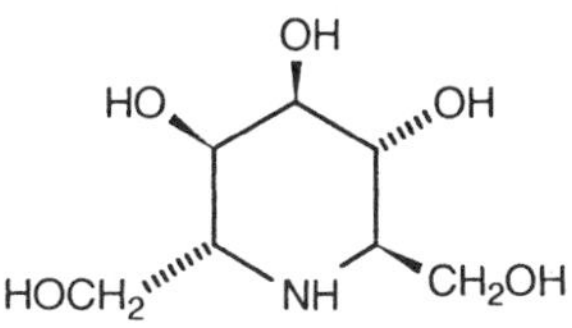

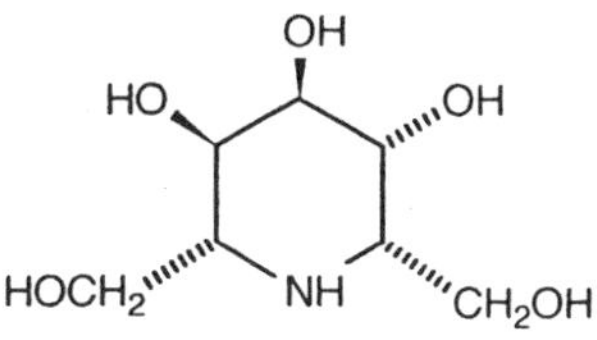

Attempts to improve the specificity of the inhibition by DMJ were made by extending the carbon skeleton. α-Homo-DMJ (**23**) was a weaker inhibitor of lysosomal α-mannosidase, whereas 6-epi-α-homo-DMJ (**24**) was a better inhibitor of α-fucosidase (K_i 4.5 mM) and did not inhibit a-mannosidase [30]. The most powerful inhibitor of a (canine) α-fucosidase is 1,5-dideoxy-1,5-imino-L-fucitol (deoxyfuco-nojirimycin) (K_i 0.04 nM), a short synthesis of which was reported by Fleet and co-workers (Scheme 7) [34]. The starting material, D-lyxono-1,4-lactone, was accessible from D-galactose by oxidative degradation, and the 3,5- and 2,3-O-isopropylidene derivatives (**25, 26**), which were formed in yields of 20% and 60%, respectively, were readily separated by chromatography. Addition of methyl lithium to the azido lactone gave the hexulose derivative (**27**) as a single stereoisomer in 97% yield, and the overall yield in this synthesis was 41%. The inhibition constants for deoxyfu-conojirimycin and β-L-homofuconojirimycin were the same (K_i 10 nM) for human liver α-fucosidase [35]; α-L-homofuconojirimycin (**28**) was a highly selective inhib-itor for α-fucosidases (K_i 11.3 and 5.8 nM for enzymes from human neutrophils and bovine epididymus, respectively), with no activity against α-mannosidase, β-galac-tosidase, or neuraminidase [36]. Its synthesis utilized as a key intermediate the known heptulose derivative (**29**), which was prepared from the corresponding lactone by addition of methyl lithium [35].

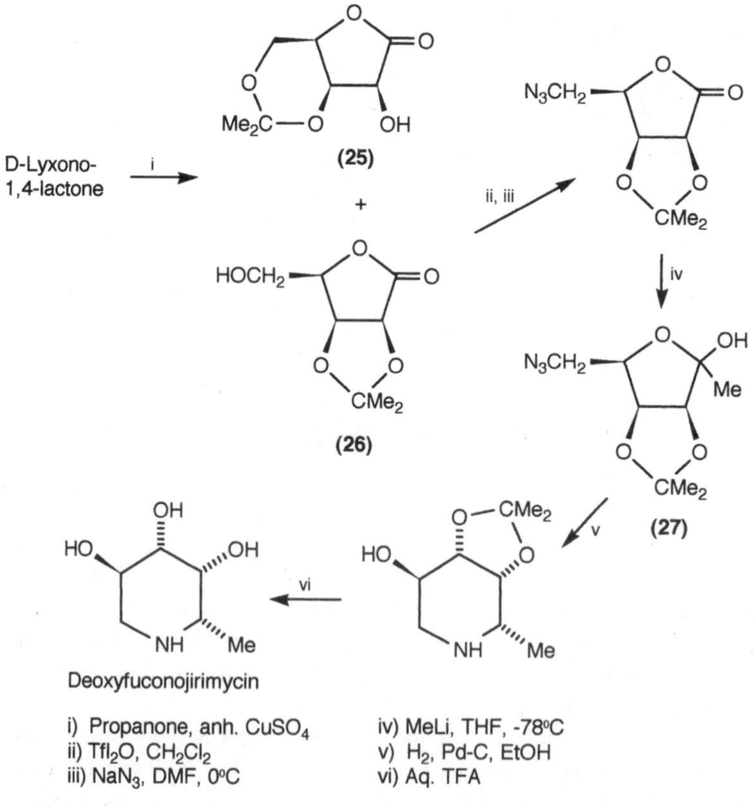

i) Propanone, anh. CuSO$_4$
ii) Tfl$_2$O, CH$_2$Cl$_2$
iii) NaN$_3$, DMF, 0°C
iv) MeLi, THF, -78°C
v) H$_2$, Pd-C, EtOH
vi) Aq. TFA

Scheme 7

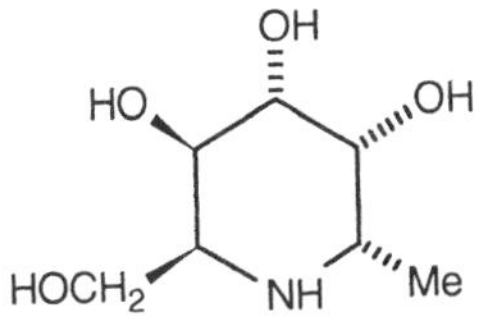

(28)

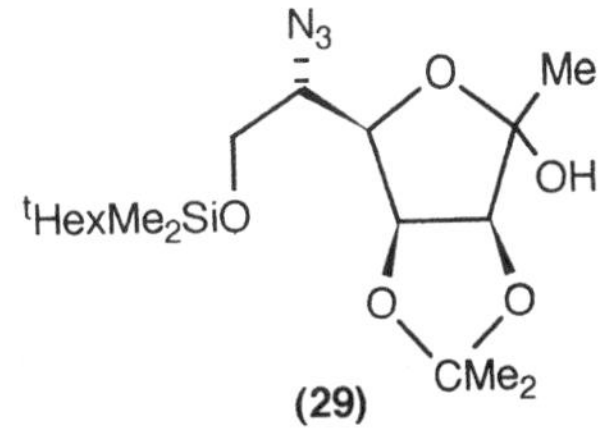

(29)

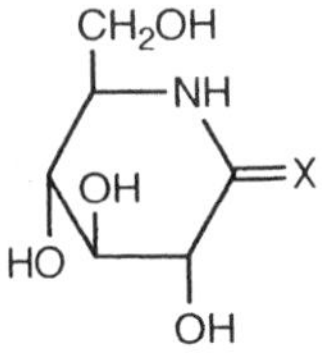

(30) X = O
(31) X = S

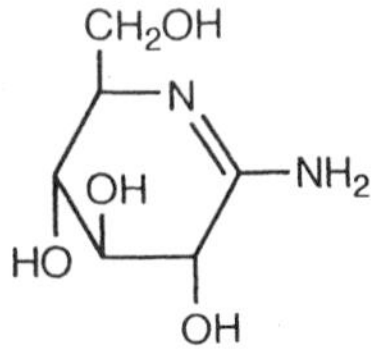

(32)

Amidines and Related Compounds

Cyclic amidine and lactim ether derivatives of monosaccharides were first suggested as potential glycosidase inhibitors in 1971 [37], and the first syntheses of the cyclic amidines corresponding to D-glucose and D-galactose were reported recently [38–40]. The lactam, e.g., (30), was protected as the tetra-O-trimethylsilyl ether, which was converted into the thionolactam (31) by the action of Lawesson's reagent followed by desilylation in acidic methanol. Methanolic ammonia then formed the amidine (32). Attempts to form the corresponding D-mannose amidine gave only the glucose amidine as a result of epimerization. However, with more reactive nucleophiles such as hydrazine and hydroxylamine the mannoamidrazone and mannoamidoxime were successfully prepared [40].

In an analysis of the inhibitory properties of several inhibitors of α-mannosidase it was shown, using semiempirical molecular orbital calculations, that the potent inhibitors were good topographical analogs of the mannopyranosyl cation rather than of α-D-mannopyranose. The lower activity of DMJ compared with swainsonine, for example, is attributed to the mismatch between the chair-shaped DMJ and the half-chair conformation predicted for the mannopyranosyl cation [41]. In a later modeling study, in which molecular mechanics and semiempirical MO calculations were used, Wong and co-workers [42] concluded that a good glycosidase inhibitor should have either a half-chair conformation with a positive charge character around the anomeric carbon and ring heteroatom or a chair conformation with a positive charge character and the same disposition of hydroxyl groups as the corresponding monosaccharide. Ganem and Papandreou [39] argued that the amidines and related compounds discussed above would have a half-chair shape because of the endocyclic double bond. However, this has been challenged by Vasella's group, who cite X-ray and NMR data in support of an exocyclic double bond for the lactam oxime and report theoretical calculations as further evidence [43]. The amidines are indeed potent glycosidase inhibitors that showed some cross-reactivity. D-Mannono-1,5-lactam amidrazone was the first potent inhibitor of the soluble (ER) α-mannosidase (IC$_{50}$ 1 μM)

[44], and in addition to inhibiting the Golgi α-mannosidases I and II (IC_{50} 4 and 0.09 μM, respectively), it also inhibited almond β-glucosidase (K_i 0.2 mM) and bovine liver β-galactosidase (K_i 57 μM) [40].

Endomannosidase Inhibitors

The discovery in 1987 of the Golgi-located endo-mannosidase that cleaves N-linked oligosaccharide chains, removing the terminal Glc_{1-3}Man fragments, accounted for some earlier failures of the inhibition of glucosidases I and II to block the processing of the oligosaccharide chains to form complex units. In an attempt to prepare compounds that inhibit this mannosidase, Spiro and co-workers [45–47] modified the structure of the disaccharide Glc(α1–3)Man. Several inhibitors were thus obtained, the most potent being Glc(α1–3)DMJ [1,5-dideoxy-3-O-(α-D-glucopyranosyl)-1,5-imino-D-mannitol] (IC_{50} 1.7 μM) and 1,5-anhydro-2-deoxy-3-O-(α-D-glucopyrano-syl)-D-*arabino*-hexitol (IC_{50} 3.8 μM). These compounds had a negligible effect on other ER and Golgi processing glycosidases [48]. Glc(α1–3)DMJ was prepared [46] from the DMJ derivative (33) as outlined in Scheme 8 [46]. A similar synthesis was reported by Fleet and co-workers, who found an IC_{50} of 5.6 μM for the inhibition of the tetradecasaccharide substrate, $Glc_3Man_9GlcNAc_2$ [49]. The corresponding Man(α1–3)DMJ had IC_{50} 25 μM.

Aza-Sugar Analogs: Pyrrolidine Derivatives

Polyhydroxylated pyrrolidines have also been found to be inhibitors of glycosidases. Thus 1,4-dideoxy-1,4-imino-D-mannitol was synthesized from the azidomannoside derivative (34), prepared from a protected mannopyranoside via a 4-uloside inter-mediate, as outlined in Scheme 9 [50]. It inhibited lysosomal, neutral and Golgi II

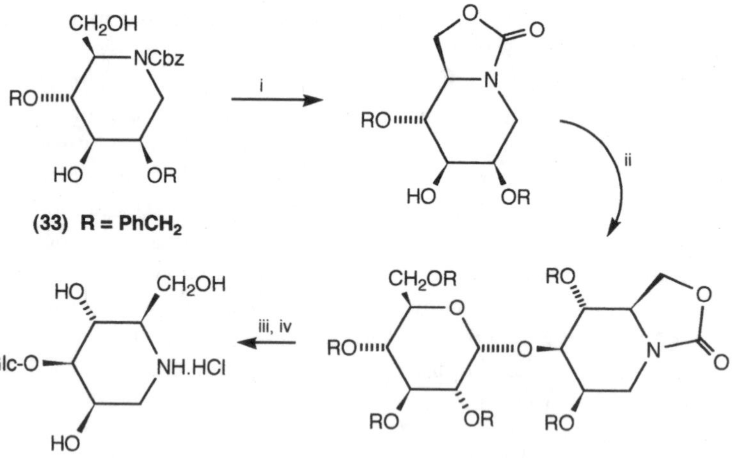

i) NaOMe
ii) 2,3,4,6-tetra-O-benzyl-α-D-glucosyl bromide, NEt_4 Br

iii) KOH, EtOH
iv) H_2, Pd-C, HCl

Scheme 8

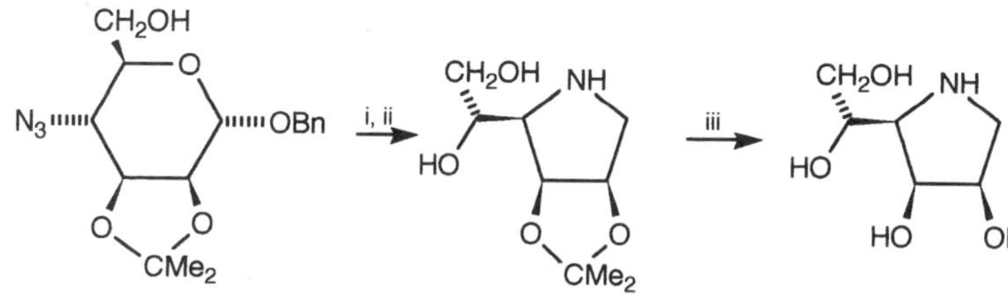

(34)

i) H_2, Pd-C, MeOH
ii) H_2, Pd-C, HOAc
iii) Aq. TFA

Scheme 9

mannosidases [51,52], and the 6-deoxy and 6-deoxy-6-fluoro derivatives were more active against the lysosomal and neutral enzymes [52]. 1,4-Dideoxy-1,4-imino-L-allitol (**34a**) was found to inhibit α-mannosidase of human liver and, to a lesser extent, α-L-fucosidase and N-acetyl-β-D-glucosaminidase. N-Methylation of these pyrrolidine derivatives markedly decreased the inhibition of the glycosidases except N-acetyl-β-D-glucosaminidase, whereas N-benzylation abolished all activity except toward the fucosidase, which was more strongly inhibited (K_i 50 μM).

Other diastereoisomeric 1,4-dideoxy-1,4-iminohexitols have been synthesized from brominated lactones. For example 1,4-dideoxy-1,4-imino-D-talitol (**35**) was prepared from 2,6-dibromo-2,6-dideoxy-D-glucono-1,4-lactone by the action of aqueous ammonia, which gave two products, (**36**) and (**37**), in a 2:1 ratio, followed by ion-exchange chromatography, amide hydrolysis, ethyl ester formation and reduction with sodium borohydride [53]. An efficient synthesis of this compound from D-mannose involved no chromatography and gave an overall yield of 52%; the key cyclization step (Scheme 10) involved a double displacement of the dimethanesul-fonate (**38**) with benzylamine [54]. This pyrrolidine (**35**) specifically inhibited human

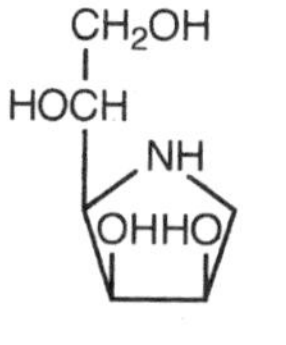

(34a)

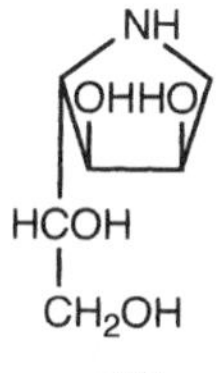

(35)

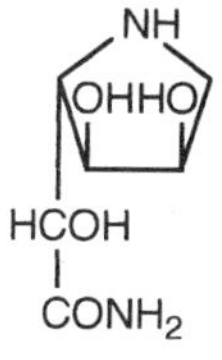

(36)

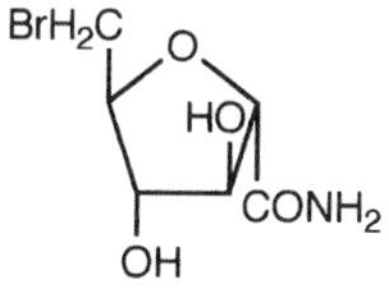

(37)

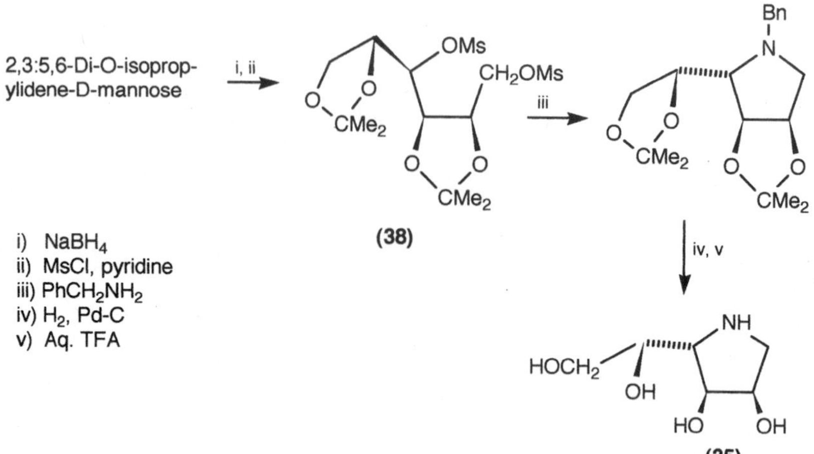

2,3:5,6-Di-O-isopropylidene-D-mannose
i, ii
OMs
CH2OMs
iii
CMe2
(38)
Bn
N
iv, v
NH
HOCH2
OH
HO
i) NaBH4
ii) MsCl, pyridine
iii) PhCH2NH2
iv) H2, Pd-C
v) Aq. TFA

(35)

Scheme 10

liver lysosomal α-mannosidase (K_i 0.12 mM at pH 4). Other pyrrolidine inhibitors include 2,5-dideoxy-2,5-imino-D-mannitol, which inhibits mammalian glucosidase I [55], and pyrrolidine analogs of aldopentoses such as 1,4-dideoxy-1,4-imino-D-arabinitol, which inhibits yeast α-glucosidase (K_i 0.18 μM) [56]. 2,5-Dideoxy-2,5-imino-D-mannitol was synthesized in 85% yield from 5-azido-5-deoxy-D-fructose via an intramolecular reductive amination of the corresponding amine [57]. The azidofructose was prepared in 81% yield from 5-azido-5-deoxy-D-glucose by the action of immobilized glucose isomerase (for a related isomerization see Scheme 5).

A chemoenzymatic approach was also used in the synthesis of 2,5,6-trideoxy-2,5-iminohexitols from 2-butyn-1-ol [58]. Reduction with Lindlar catalyst followed by epoxidation and ring opening by azide ion gave a 6:1 mixture of the azides **(39)** and **(40)**. Resolution of **(39)** was achieved by means of a lipase-catalyzed acetylation with vinyl acetate. The (2*R*,3*R*)-isomer was isolated as the diacetate and the enantiomer as the monoacetate. Hydrolysis and periodate cleavage then formed the enantiomeric 2-azidopropanals **(41,42)**, which were reacted with dihydroxyacetone phosphate in the presence of either fuculose 1-phosphate aldolase or rabbit muscle fructose 1,6-diphosphate aldolase (Scheme 11). Enzymatic phosphate hydrolysis followed by reductive ring closure gave the pyrrolidines **(43–45)** stereoselectively. The diastereoisomer **(46)** was synthesized from 2-azido-3-hydroxypropanal (Scheme 11). All four diastereoisomers **(43–46)** inhibited α-fucosidase from bovine kidney (K_i 1.4, 8, 22, and 4 μM, respectively), illustrating the difficulty of predicting which enzyme will be inhibited by which diastereoisomer [58].

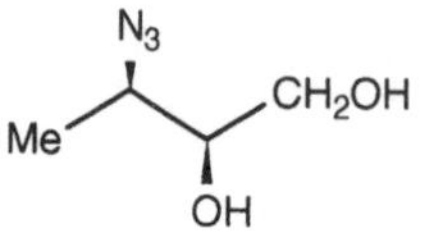

N3
CH2OH
Me
OH

(39)

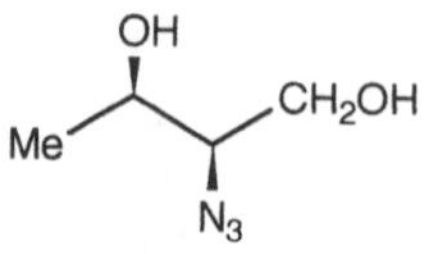

OH
CH2OH
Me
N3

(40)

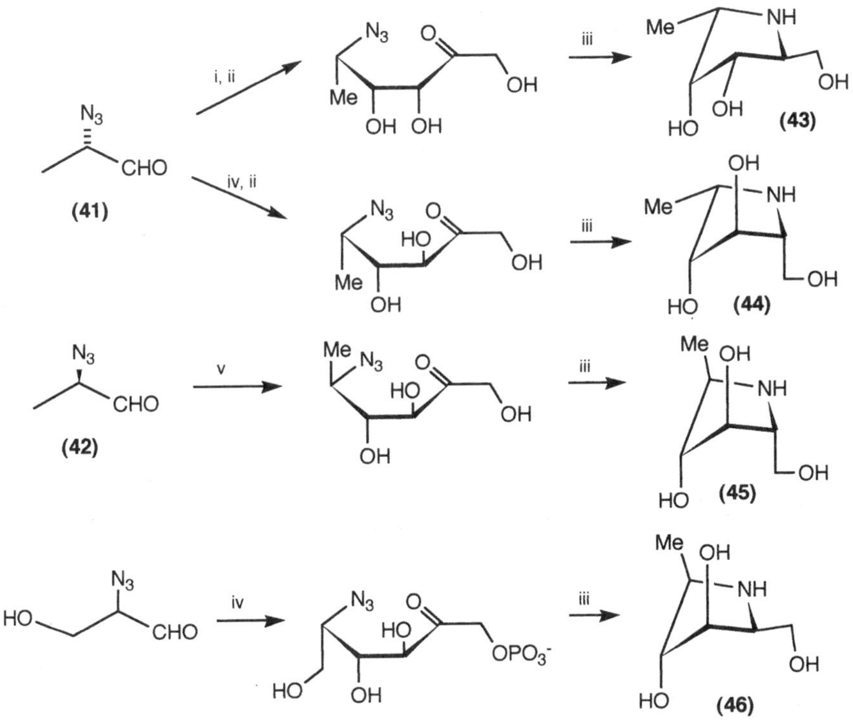

i)   DHAP, fuculose 1-phosphate aldolase
ii) Acid phosphatase
iii) H$_2$, Pd-C, 50 psi
iv) DHAP, fructose 1,6-diphosphate aldolase

Scheme 11

Miscellaneous Inhibitors

That other basic compounds that do not closely resemble a monosaccharide can act as a strong inhibitor of glycosidases is illustrated by manostatin A (**47**), a metabolite of *Streptoverticillium verticillus*. Mannostatin A inhibits α-mannosidase II (IC$_{50}$ 10–15 nM) but not α-mannosidase I [59]. It has been synthesized from D-ribono-lactone via the known carbocyclic intermediate (**48**) in 32% yield (Scheme 12) [60]. An enantioselective synthesis from 5-methylthiocyclopentadiene achieved asymmetric induction using the hydroxamic acid (**49**) [61].

Glycals have long been known to inhibit glycosidases; for example, D-galactal (1,5-anhydro-D-*arabino*-hex-1-enitol) inhibits β-galactosidase from various sources, and the slow formation and release of a 2-deoxy-D-hexosyl-enzyme intermediate has been implicated [for a discussion see ref. 13], except for the sialidase inhibitor (**50**) [62], which is a more effective inhibitor than the aza-analog (**51**) [63]. Modifications of the structure of the glycal (**50**) have been made in attempts to improve the potency and selectivity of the inhibition. Such research, aided by molecular modeling studies of ligand binding to sialidase from influenza virus [64], has led to the development of the most potent sialidase inhibitors known. The modeling studies indicated that a

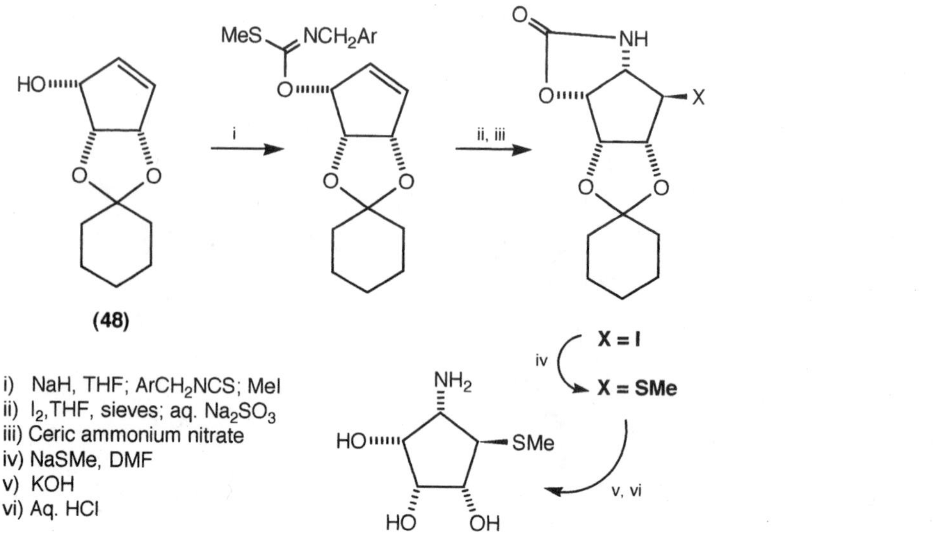

(48)

i)   NaH, THF; ArCH₂NCS; MeI
ii) I₂,THF, sieves; aq. Na₂SO₃
iii) Ceric ammonium nitrate
iv) NaSMe, DMF
v) KOH
vi) Aq. HCl

Scheme 12

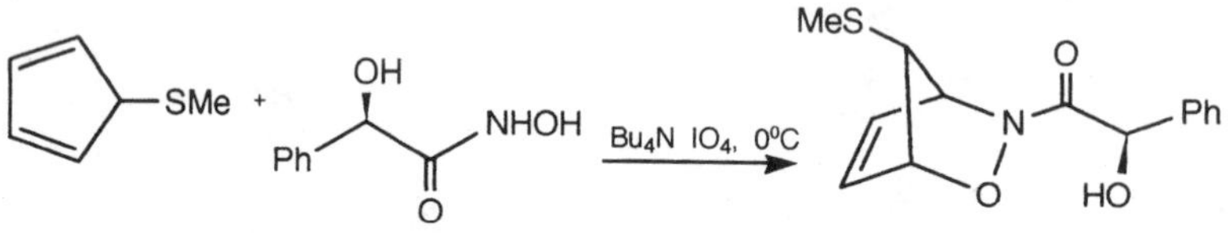

(47)

(49)

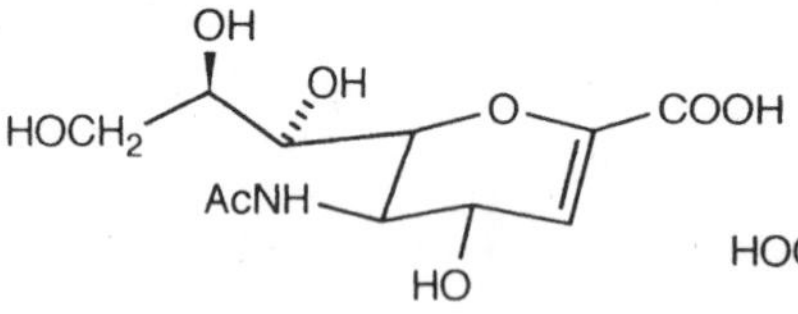

(50)

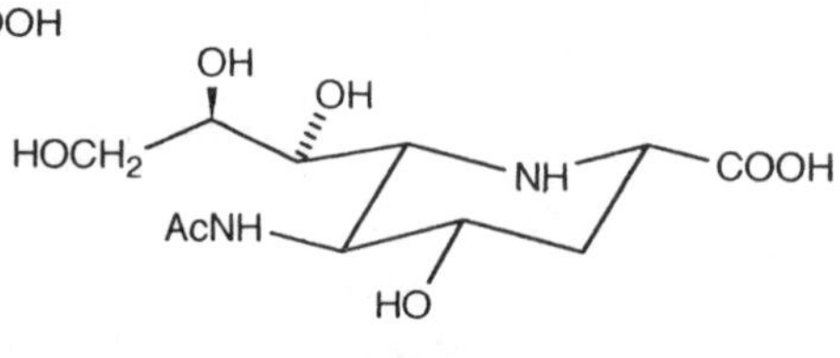

(51)

positively charged group at C-4 of the glycal would enhance binding to the enzyme, and the 4-amino- and 4-guanidino (**52**) derivatives, synthesized according to Scheme 13, were highly potent inhibitors of influenza virus sialidase (K_i 50 and 0.2 nM, respectively) [65,66]. The compounds were much less active than the parent glycal against mammalian sialidases.

The naturally occurring siastatin B (**53**) is an inhibitor of sialidases from *Clostridium perfringens* (IC$_{50}$ 0.23 mM) and *Streptococcus* (IC$_{50}$ 29 μM), and it has been converted into several derivatives in attempts to enhance activity [67]. The four compounds (**54–57**) were better inhibitors of sialidase and had activity comparable to that of the glycal (**50**). For example, (**56**) had IC$_{50}$ 28 μM against the enzyme

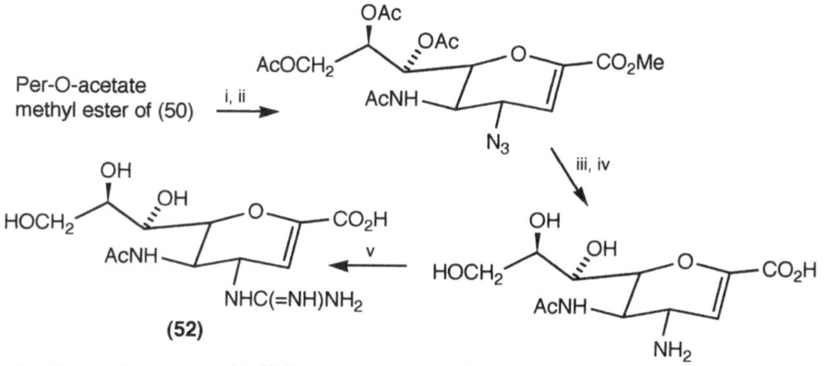

Per-O-acetate
methyl ester of (50) —i, ii→

OAc
OAc
AcOCH$_2$
AcNH
N$_3$
O
CO$_2$Me

iii, iv

OH
OH
HOCH$_2$
AcNH
O
CO$_2$H
NH$_2$

←v—

OH
OH
HOCH$_2$
AcNH
O
CO$_2$H
NHC(=NH)NH$_2$

(52)

i) BF$_3$.Et$_2$O   iv) OH⁻
ii) NaN$_3$ v) HN=C(NH$_2$)SO$_3$H, aq. K$_2$CO$_3$
iii) H$_2$, Pd-C

Scheme 13

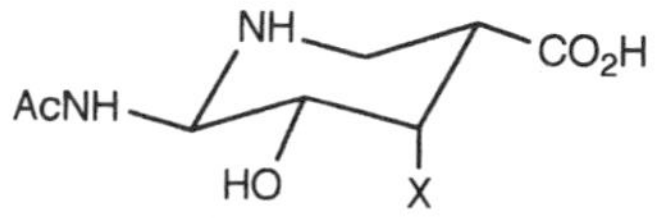

NH
AcNH
HO   X
CO$_2$H

(53) X = OH
(54) X = H

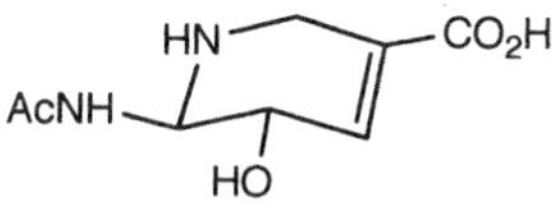

HN
AcNH
HO
CO$_2$H

(55)

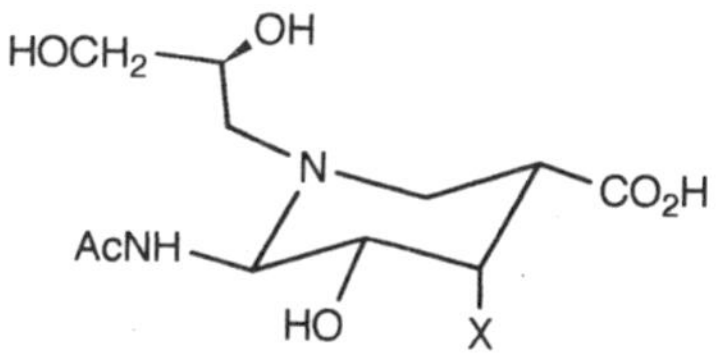

HOCH$_2$
OH
N
AcNH
HO   X
CO$_2$H

(56) X = OH
(57) X = H

CH$_2$OH
OH
OH
HO   NH$_2$
OH

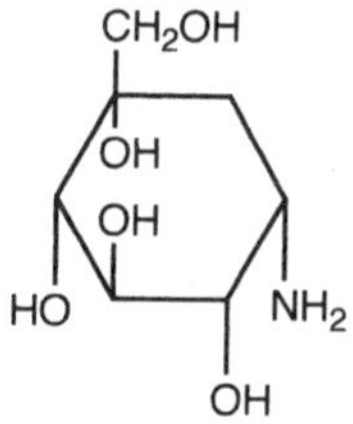

(58)

from *Clostridium perfringens*. Some of the compounds also inhibited yeast α-glucosidase and bovine liver β-glucuronidase.

Carbocyclic analogs ("pseudo sugars" or carba-sugars) were suggested as potential enzyme inhibitors in 1968 [68], and such compounds were subsequently found in nature (reviewed by Suami and Ogawa [69]). Amine derivatives such as valiolamine (58) inhibit α-glucosidases I and II (IC$_{50}$ 50 μM, substrate Glc$_3$Man$_9$GlcNAc$_2$). A synthesis of the penta-*NO*-acetyl derivative is outlined in Scheme 14 [70]. The starting material in this scheme was accessible from D-glucose, but in 14 steps.

B. Bicyclic Compounds

Indolizidines and related compounds

Swainsonine (59) was the first indolizidine alkaloid to show inhibitory properties against α-mannosidase; rat liver α-mannosidase II was strongly inhibited (IC$_{50}$ 0.2

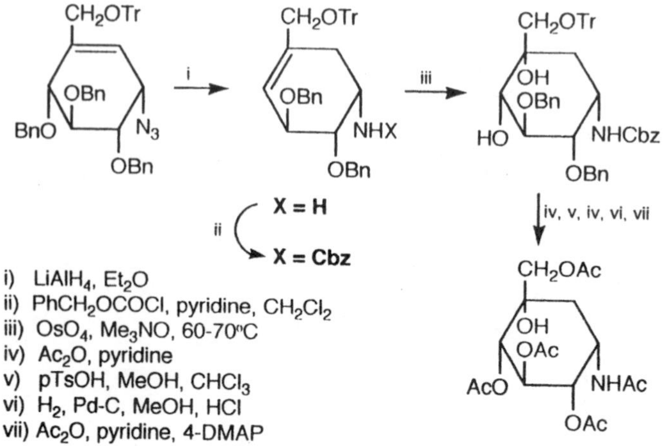
CH2OTr
OBn
BnO
N3
OBn
i
NHX
iii
OH
NHCbz
HO
X = H
ii
X = Cbz
iv, v, iv, vi, vii
CH2OAc
OAc
AcO
NHAc
i) LiAlH4, Et2O
ii) PhCH2OCOCl, pyridine, CH2Cl2
iii) OsO4, Me3NO, 60-70°C
iv) Ac2O, pyridine
v) pTsOH, MeOH, CHCl3
vi) H2, Pd-C, MeOH, HCl
vii) Ac2O, pyridine, 4-DMAP

Scheme 14

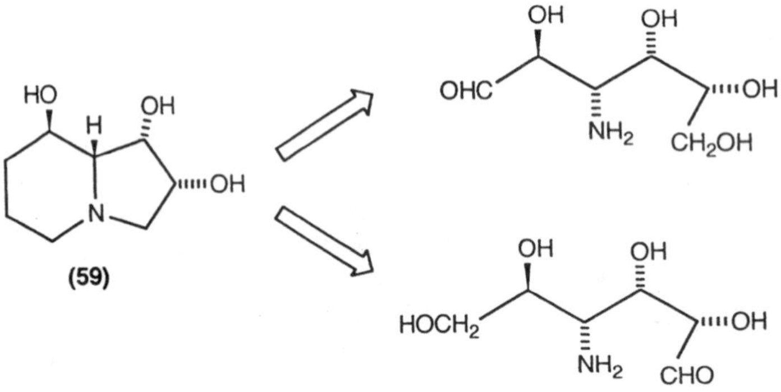
HO
OH
H
OH
N
OHC
NH2
CH2OH
HOCH2
CHO

(59) (60) (61)

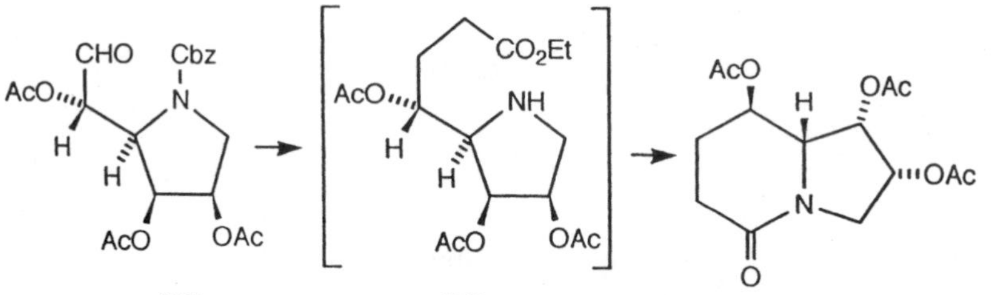
CHO
Cbz
AcO
N
H
AcO
OAc
CO2Et
NH
O

(62) (63)

μM) [71] but not α-mannosidase IA or IB. Human lysosomal α-mannosidase was also strongly inhibited (K_i 70 nM) [72]. Many different approaches have been used in the synthesis of swainsonine. Hexoses have been favored starting materials, and retrosynthetic analysis indicated that the appropriate chiral precursors of swainsonine could be either 3-amino-3-deoxy-D-mannose (**60**) or 4-amino-4-deoxy-D-mannose (**61**). Pyrrolidine ring formation by linking the amino group to the appropriate terminal position of the hexose and introduction of a —CH_2CH_2— unit between the other terminal carbon atom and the nitrogen atom would then give the correct stereochemistry and skeleton.

3-Amino-3-deoxy-D-mannose, readily made from methyl α-D-glucopyranoside by the periodate/nitromethane procedure, was used as starting material by Richardson and co-workers [73]. The pyrrolidine ring was formed by cyclization of a 6-tosyl derivative, and subsequent chain extension at C-1 involved a Wittig reaction of the aldehyde (62) with ethoxycarbonylmethylene triphenylphosphorane. Reduction of the E/Z mixture of alkenes gave the ester (63), which cyclized to form the required lactam together with the product of O → N acetyl migration. In a synthesis of similar length, Fleet and co-workers [74] used as an intermediate a 4-azido-4-deoxy-D-mannopyranoside (64), chain extension at C-6 being achieved by oxidation to the aldehyde and a Wittig reaction with formylmethylene triphenylphosphorane. Reduction and cyclization formed the piperidine ring (Scheme 15), and further reduction removed the benzyl group to release the aldehyde for the second cyclization to form the indolizidine ring system.

A shorter synthesis (11 steps from D-mannose) of swainsonine was reported by Setoi and co-workers [75], who formed the bicyclic ring system by a double cyclization of the epoxide intermediate (65) (Scheme 16). The first enantioselective synthesis of swainsonine from an achiral starting material was reported by Sharpless and co-workers, who used the iterative epoxidation of allylic alcohols starting from N-p-toluenesulfonyl-4-benzylamino-2-buten-1-ol [76]. Although this synthesis was very long (21 steps from N-benzyl-p-toluene sulfonamide; 6.6% overall yield), it had the advantage that the stereochemistry could be controlled at each stage, thus providing, in principle, access to all 16 stereoisomers of swainsonine.

The resemblance in structure between protonated swainsonine (66) and the mannopyranosyl cation (67), a possible intermediate in mannoside hydrolysis, is

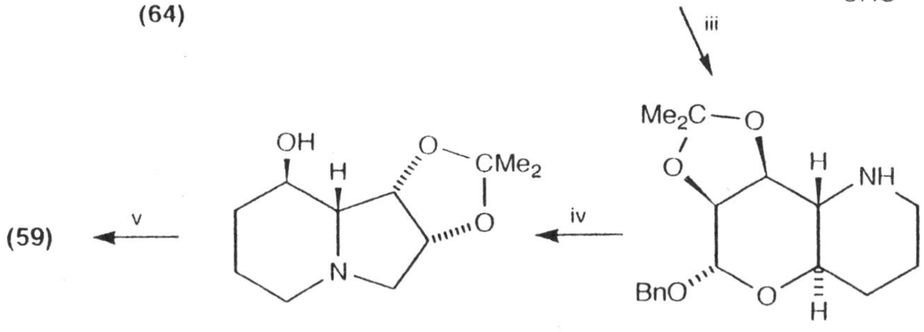

i)   PCC
ii) Ph$_3$P=CHCHO
iii) H$_2$,Pd-C, MeOH
iv) H$_2$, Pd-C, HOAc
v)   Aq. TFA

Scheme 15

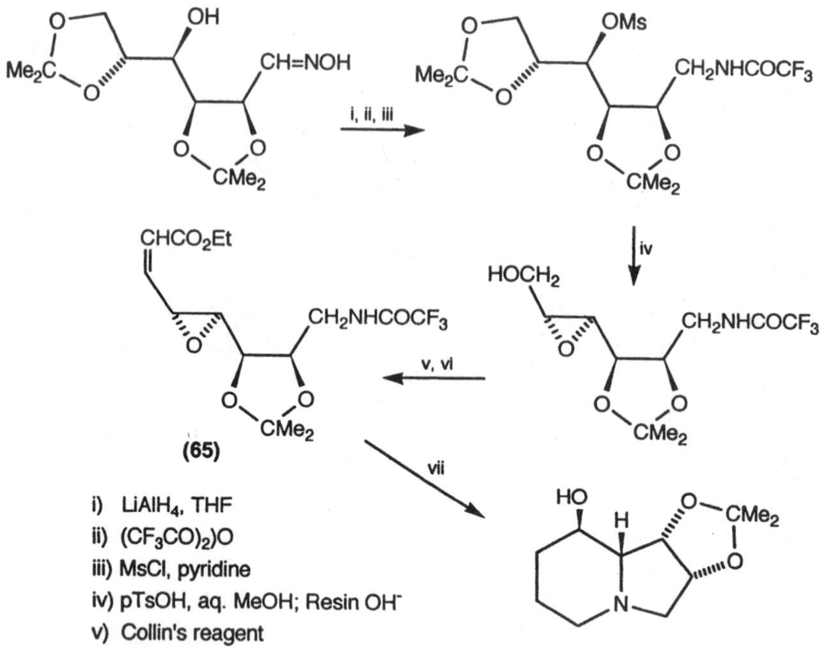

Scheme 16

i)   LiAlH₄, THF
ii) (CF₃CO)₂O
iii) MsCl, pyridine
iv) pTsOH, aq. MeOH; Resin OH⁻
v)   Collin's reagent
vi) Ph₃PCHCO₂Et, THF
vii) NaBH₄, EtOH, CF₃CH₂OH

i) LiAlH$_4$, THF
ii) (CF$_3$CO)$_2$O
iii) MsCl, pyridine
iv) pTsOH, aq. MeOH; Resin OH⁻
v) Collin's reagent
vi) Ph$_3$PCHCO$_2$Et, THF
vii) NaBH$_4$, EtOH, CF$_3$CH$_2$OH

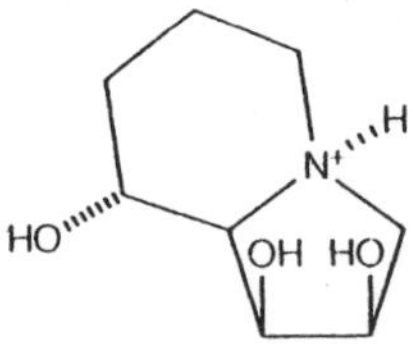

(66) 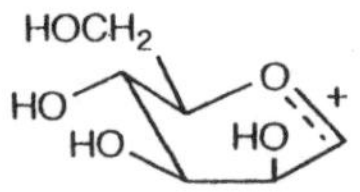   (67)

evident, and it is noteworthy that the corresponding monocyclic analog (a better term than "open-chain analog" [77]), namely 1,4-dideoxy-1,4-imino-D-mannitol, is a weaker (though still potent) inhibitor of lysosomal α-mannosidase (K_i 13 μM) [77].

Epimers of swainsonine have been synthesized to gain some understanding of the specificity of the inhibition. The 8a-epimer (K_i 75 μM) and the 8,8a-diepimer (K_i 2 μM), but not the 8-epimer, the 1,8-diepimer, or the 2,8a-diepimer, were specific inhibitors of lysosomal α-mannosidases in vitro. They also caused the storage of mannose-rich oligosaccharides in human fibroblasts in culture, thus indicating that processing α-mannosidases had also been affected [77].

Castanospermine (**68**) was reported in 1983 to inhibit glucosidase I and to a lower extent glucosidase II in rabbit liver microsomal preparations (substrate: Glc₁₋₃Man₉GlcNAc₂) [78]. An IC₅₀ value of 0.1 μM has been reported for inhibition of pig kidney α-glucosidase [79]. Synthetic approaches to castanospermine and analogs were recently reviewed [80]. Many syntheses use a hexose as starting material, but separation of diastereoisomers formed in reactions of low stereoselectivity can reduce the efficiency and convenience of some routes. Ganem and co-workers used a DNJ derivative **(69)** as an intermediate, the carbon chain of which was extended by a chelation-controlled Sakurai allylation that proceeded with excellent stereoselectivity (Scheme 17) [81].

The greater inhibitory potency of castanospermine than DNJ against α-glucosidase I is noteworthy. Castanospermine may be regarded as a conformationally restricted analog of DNJ [28,82]. It has been pointed out that the comparable inhibitory potency of *N*-alkylated (e.g., *N*-methyl) DNJ may be correlated with a preferred conformation of the 6-OH group, which resembles the 1-OH group of castanospermine in its disposition [28,29].

Castanospermine epimers have different inhibitory properties (reviewed by Winchester [83]). For example, 6-epi-castanospermine does not inhibit lysosomal α-

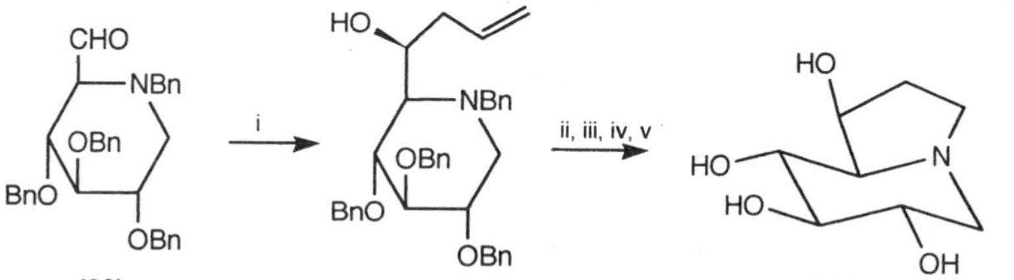

i) AllylSiMe$_3$, TiCl$_4$, -85°C
ii) O$_3$
iii) NaBH$_4$
iv) MsCl (1 equiv.), Et$_3$N
v) H$_2$, Pd-C

Scheme 17

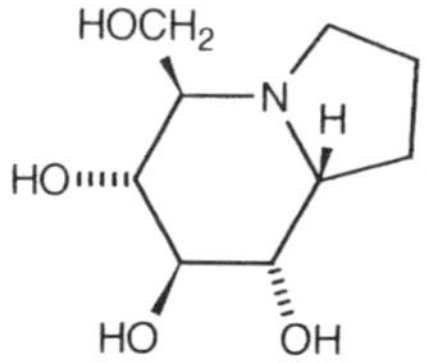

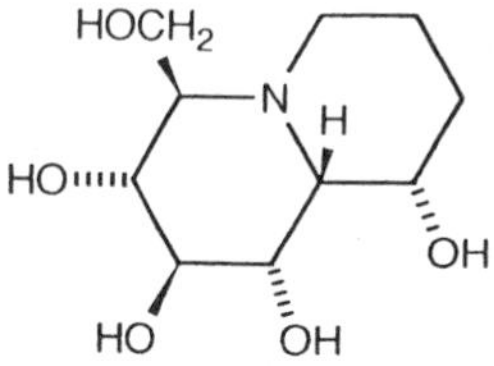

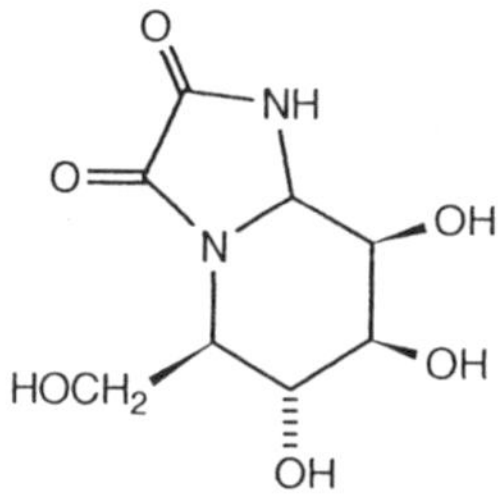

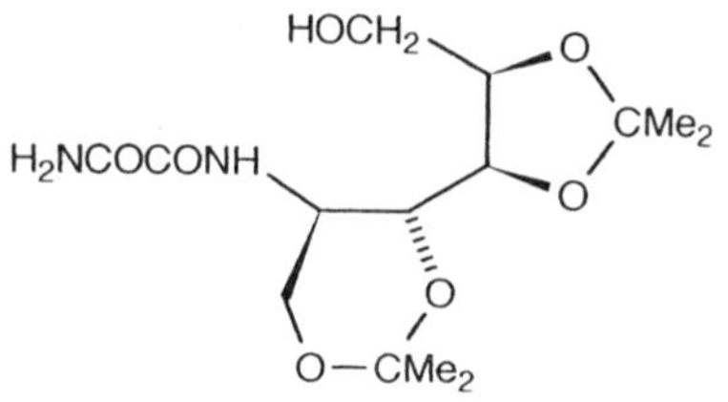

mannosidase but is a good inhibitor of cytosolic or neutral α-mannosidase. 1-Deoxy-6-epi-castanospermine inhibits lysosomal α-mannosidase strongly but not the neutral α-mannosidase. 1-Deoxy-1,8a-diepi-castanospermine, which has four chiral centers identical with α-fucose, is a potent inhibitor of α-fucosidase (K_i 1.3 μM).

The related indolizidine (70) and quinolizidine (71), synthesized from α-homonojirimycin, also inhibited α-glucosidase from pig kidney (IC_{50} 0.3 and 0.15 μM, respectively) [79].

Kifunensine (72), a recently isolated fungal alkaloid, is a potent inhibitor of α-mannosidase I (IC_{50} 20 nM, substrate: Man$_9$GlcNAc) and is useful in that it does not inhibit other (class 2) α-mannosidases [84]. It was synthesized via (73) from 2-amino-2-deoxy-D-mannose [85]. An alternative route (Scheme 18) to the intermediate (73) used the chiral azide (74) prepared from chlorobenzene (see Scheme 6) by a microbial transformation [86].

Tetrazoles such as (75) have been investigated as potential glycosidase inhibitors since they are nonbasic homologs of the corresponding amidines. The D-*gluco* (75) and D-*manno* (76) isomers were synthesized via intramolecular cycloaddition of azidonitrile intermediates [87,88]. The route to the D-*manno* isomer is outlined in Scheme 19. Reduction of the intermediate 5-ulose was highly stereoselective, the ratio of L-*gulo* to D-*manno* products being 93:7 [88]. Both tetrazoles were shown by X-ray crystallography to have half-chair conformations in the solid state. ^{1}H NMR data were interpreted as supporting half-chair ($^4H_3$) and sofa (S_7) conformations for the *gluco* and *manno* isomers, respectively, in solution. The inhibitory properties of these tetrazoles against α- and β-glucosidases and -mannosidases and a β-galacto-

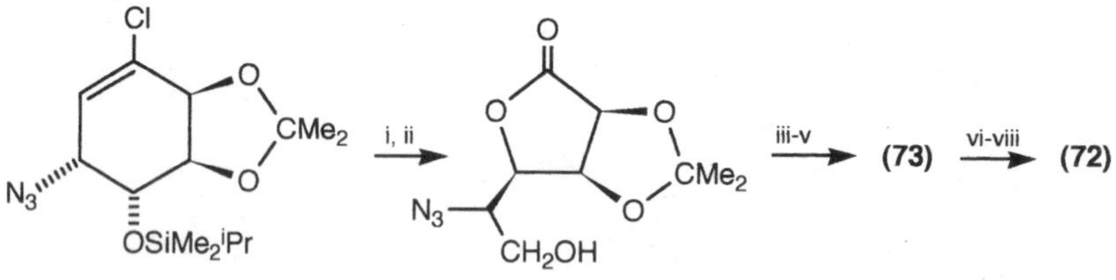

i) O$_3$, aq. MeOH, NaHCO$_3$, -78°C;   v) (MeO$_2$C)$_2$, MeOH; then NH$_3$
   then NaBH$_4$, CeCl$_3$ vi) CrO$_3$.2pyr., CH$_2$Cl$_2$
ii) Bu$_4$NF   vii) NH$_3$, MeOH
iii) DMP, camphorsulfonic acid   viii) 75% Aq. TFA
iv) LiAlH$_4$, Et$_2$O

Scheme 18

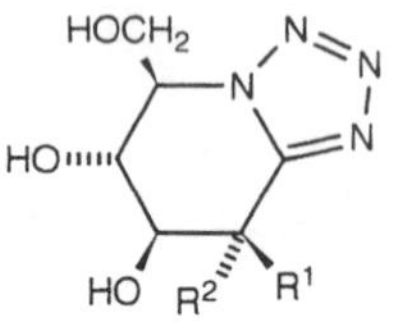

(75) R^1 = H, R^2 = OH   (77)
(76) R^1 = OH, R^2 = H

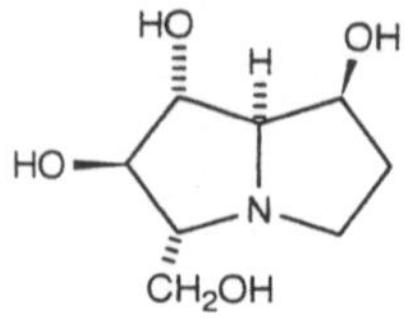

sidase were measured. The importance of the configuration at C-2 was shown by the much poorer inhibition of the β-galactosidase (from bovine liver) by mannonojiritetrazole (**76**) (K_i 14 mM) compared with gluconojiritetrazole (K_i 0.0015 mM). The important feature of these results was the linear correlation between log K_i for each inhibitor–enzyme pair and log (V_m/K_m) for the corresponding substrate–enzyme pair (*p*-nitrophenyl glycosides were used as substrates). This provided the first proof that any glycosidase inhibitor was a transition state analog, and it was suggested by the Vasella group that transition state analog inhibitors for glycosidases should possess a conformation close to that of a flattened chair with an sp^2 hybridized anomeric center and the correct configuration at C-2, C-3, C-4, and C-5.

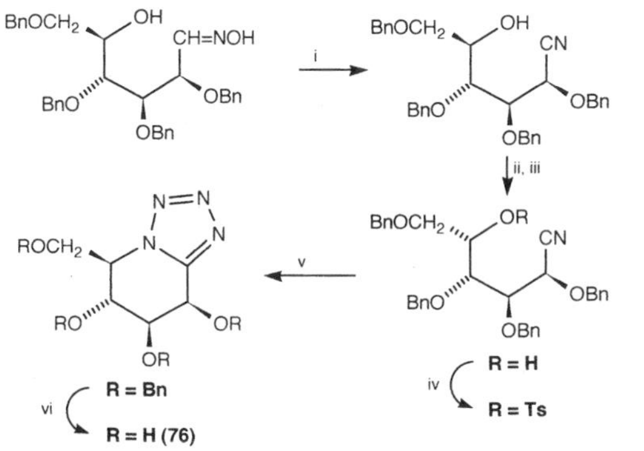

i) CBr_4, PPh_3, MeCN
ii) PCC, 3A mol. sieves, CH_2CCl_2
iii) $NaBH_4$, $CeCl_3$, MeOH, -60 to -40°C
iv) TsCl, pyrididne, 40-50°C
v) NaN_3, DMSO, 110-120°C
vi) H_2, Pd-C, MeOH, HOAc

Scheme 19

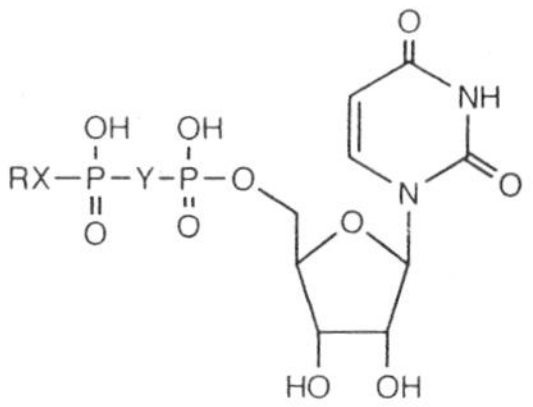

(78) X = -, Y = O
(79) X = O, Y = CH_2
R = α-D-Galactopyranosyl

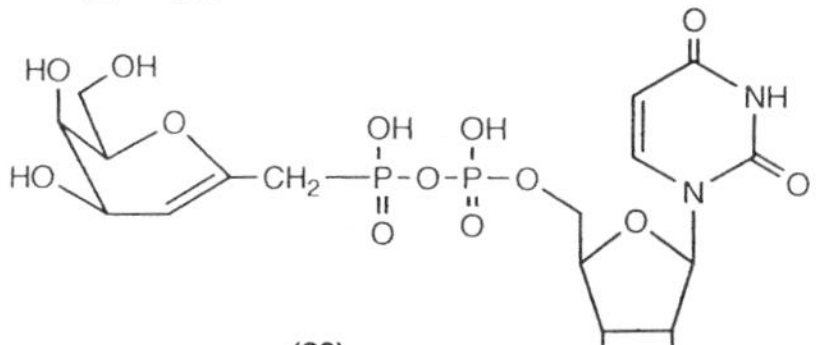

(80)

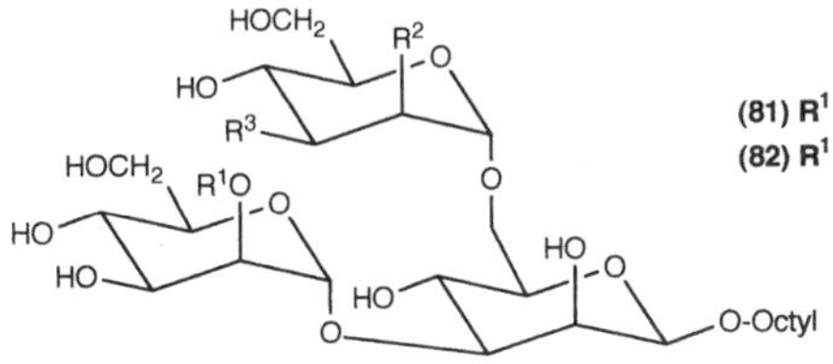

(81) R^1 = H
(82) R^1 = βGlcNAc

Pyrrolizidines

Several hydroxylated pyrrolizidine alkaloids with a carbon branch at C-3 have been isolated and found to have inhibitory properties. Australine (**77**) strongly inhibited α-glucosidase I from mung beans (IC$_{50}$ 20 μM) [89]; inhibition of α-glucosidase II was very low. Australine can be regarded as a conformationally restricted analog of 2,5-dideoxy-2,5-imino-D-mannitol.

III. INHIBITORS OF GLYCOSYLTRANSFERASES

It has been estimated that 100 or more glycosyltransferases are required for the synthesis of the oligosaccharide moieties of glycoproteins and glycolipids. Glycosyltransferases catalyze the transfer of a monosaccharide from its nucleoside diphosphate derivative to an acceptor, which in the glycoprotein context is usually a monosaccharide or oligosaccharide. Early attempts to synthesize inhibitors of glycosyltransferases involved unreactive sugar nucleotide analogs such as the galactosyl phosphonate (**78**), made by coupling α-galactopyranosyl-phosphonate with uridine 5′-phosphoric dibutylphosphinothioic anhydride (IC$_{50}$ 35 μM for the galactosyl transferase from mouse ascites fluid) [90], the galactosyl diphosphonate (**79**) (apparent K_i 97 μM) [91], and the 1,5-anhydro-D-*arabino*-1-hexenitol-1-ylmethylphosphonate (**80**) (K_i 62 μM for the galactosyltransferase from bovine milk) [92]. However, such approaches are unlikely to provide specific glycosyltransferase inhibitors, since any given sugar nucleotide can be used by many enzymes.

A more promising approach appears to involve modification of the natural acceptor. An obvious modification to make is to replace the hydroxyl group undergoing glycosylation by a hydrogen atom, and this approach in Hindsgaul's laboratory has yielded four inhibitors of two fucosyl transferases and two GlcNAc-transferases (Table 1) [93]. However, four other deoxy-acceptors were not inhibitors of the appropriate glycosyltransferases, and it was proposed that for these enzymes [(α1−3/ 4)fucosyltransferase from human milk, (α1−3)fucosyltransferase from human serum, (β1−4)-galactosyltransferase from bovine milk, and (α2−3)sialyltransferase from rat liver] the reactive acceptor hydroxyl group is involved in a critical hydrogen bond interaction with a basic site on the enzyme that removes the proton during the glycosyl transfer reaction [93].

A second and potentially more general strategy, also introduced by Hindsgaul's group, involves the alteration of a specific acceptor "so as to provide a steric impediment to the transfer reaction without removing the reactive hydroxyl group to which transfer occurs" [94]. This was successfully implemented when the 4-OH group of the mannose unit in the trisaccharide acceptor GlcNAc(β1−2)Man(α1− 6)Glcβ-O(CH$_2$)$_7$CH$_3$ was replaced by methoxyl to give an analog that was a good inhibitor of GlcNAc-transferase V (K_i 14 μM) [94]. Similar results have been reported for (β1−2)-GlcNAc-transferase II [95]. In this study use was made of a glycosyl transferase for the synthesis of the required acceptor analogs. Though glycosyltransferases have a high specifity, some variation in acceptor structure is tolerated. For example, some hydroxyl groups not directly involved in the glycosylation can be replaced by other groups without impairment of the acceptor properties [96−99].

Thus, a recombinant (β1−2)GlcNAc-transferase I was used to convert the trisaccharides (**81**), modified at position 2 or 3 of the (α1−6)-linked mannose unit into

Table 1 Data for Deoxygenated Acceptor Analogs as Inhibitors of Glycosyltransferases

Enzyme	Source	Donor	Acceptor deoxy analog[b]	K_i (mM)
(α1–2)Fuc-T[a]	Pig submaxillary	GDP-Fuc	2-DeoxyGal(β1–3)GlcNAcβ-OR[1]	0.80
(α1–4)Fuc-T	Mung bean	GDP-Fuc	Gal(β1–3)(4-deoxy)GlcNAcβ-OR[1]	0.54
(β1–6)GlcNAc-T-V	Hamster kidney	UDP-GlcNAc	GlcNAc(β1–2)(6-deoxy)Man(α1–6)Glcβ-OR[2]	0.063
(β1–6)GlcNAc-T (mucin core-2)	Mouse kidney	UDP-GlcNAc	Gal(β1–3)(6-deoxy)GalNAcα-OR[1]	0.56

[a] T = transferase
[b] R[1] = $(CH_2)_8COOCH_3$
R[2] = $(CH_2)_7CH_3$

the tetrasaccharide (**82**). The problem of product inhibition that can arise in glycosyltransferase-catalyzed reactions was overcome by using a fourfold excess of the
donor, UDP-GlcNAc, and Co^{2+} instead of Mn^{2+}. Four inhibitors of ($\beta1-2$)GlcNAc-
transferase II were thus prepared: the 2'-deoxy analog (R^2 = H, R^3 = OH, K_i 0.13
mM), a 3'-O-pentyl ether (R^2 = OH, R^3 = O-pentyl, K_i 2.4 mM), the 3'-O-(4,4-
azo)pentyl ether (K_i 1.0 mM) and the 3'-O-(5-aminopentyl) ether (K_i 2.4 mM). The
photolabile diazirine analog and the iodoacetamide derivative [R^3 = 3-(5-iodoacetami-
dopentyl), R^2 = OH], which is a substrate but not an inhibitor, offer the prospect of
identifying the active site of the enzyme by a photochemically or thermally induced
reaction, respectively.

Aza-sugar analogs have been found to have some inhibitory properties against
glycosyltransferases. Gal($\beta1-4$)DNJ, synthesized using a ($\beta1-4$)-galactosyltransfer-
ase [100] with DNJ as acceptor, was a moderate inhibitor of a human recombinant
($\alpha1-3$)-fucosyltransferase (IC$_{50}$ 8 mM). The pyrrolidine derivatives (**43**) and (**46**)
also showed some inhibitory activity, but this was much enhanced in the presence
of GDP. It was suggested that this synergistic effect might be due to an interaction
between GDP and the aza-sugar in the active site [58,101].

The proposal that for some glycosyltransferases there is a critical hydrogen
bond between the acceptor hydroxyl group to be glycosylated and a basic residue at
the active site (loc. cit.) of the enzyme stimulated Field and co-workers [102] to
synthesize acceptor analogs that contained amino and other groups close to or in
place of the OH group to be glycosylated. It was argued that the protonated amino
group may interact with the basic group on the enzyme. A series of 4- and 6-
substituted derivatives of benzyl 2-acetamido-2-deoxy-β-D-glucopyranoside were
prepared and assessed as inhibitors of bovine ($\beta1-4$)-galactosyltransferase. The 4-
substituted derivatives were synthesized from benzyl 2-acetamido-2-deoxy-β-D-gal-
actopyranoside as outlined in Scheme 20. Of the azido, amino, and acetamido compounds tested, only the 4-amino compound produced significant inhibition (K_i 0.85
mM).

A similar result was obtained for the glycosyltransferase responsible for the
biosynthesis of the A and B blood group antigens. These enzymes transfer GalNAc
and Gal, respectively, to the 3-OH of the galactose residue in the disaccharide
Fuc($\alpha1-2$)Galβ-OR. A modified disaccharide acceptor (R = octyl) with an amino
group at C-3 of the galactose residue was an inhibitor of both transferases. Inhibition

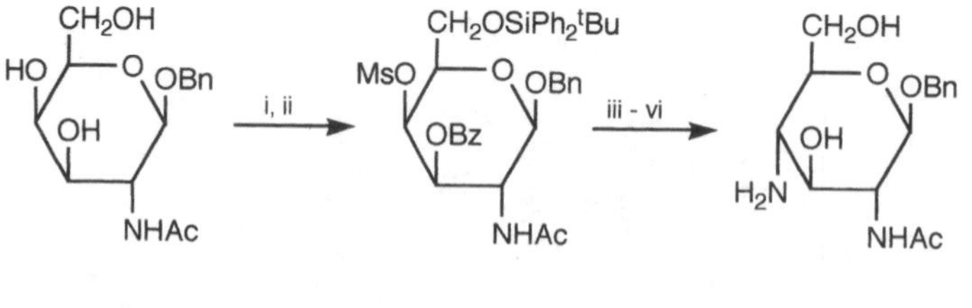

i) TBDPS-Cl, DMF, imidazole
ii) BzCl, pyridine, CH_2Cl_2, -50°C;
     MsCl
iii) NaN$_3$, DMF, 110°C

iv) MeONa
v) Bu$_4$NF, THF
vi) PPh$_3$, MeOH, pyridine, NH$_4$OH

Scheme 20

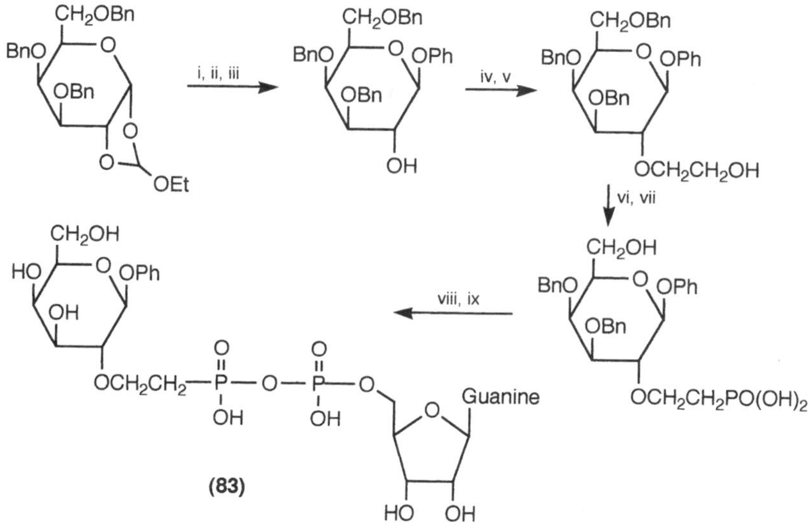

i)   AcBr
ii) NaOPh, aq. acetone
iii) NaOMe, MeOH
iv) NaH, DMF; THPOCH$_2$CH$_2$Br
v)   80% Aq. HOAc

vi)   NIS, DMF, PPh$_3$; P(OMe)$_3$
vii) Me$_3$SiI, CH$_2$Cl$_2$, 0°C
viii) H$_2$, Pd-C, EtOH
ix)   AG-50 WX (pyridinium salt);
      pyridine, GMP-morpholidate

Scheme 21

constants could not be calculated because of a complex mode of inhibition; K_i was estimated at 200 nM for the inhibition of the A transferase. The 3-epimer (gulose in place of galactose) and the 3-deoxy and 3-deoxy-3-fluoro analogs were also found to inhibit both transferases [97,98]. A similar study of the blood group H gene-specific glycosyltransferase using structural analogs of octyl β-D-galactopyranoside, the minimum acceptor substrate for the enzyme, and a cloned (α1–2)-fucosyltrans-ferase identified several C-3 modified analogs as inhibitors with estimated K_i values in the range 0.9–43 mM. Detailed kinetic studies with similar analogs of more potent disaccharide acceptors should help to establish the mechanism of inhibition [99].

Porcine submaxillary gland β-galactoside (α1–2)-fucosyltransferase transfers fucose from GDP-fucose to the 2-OH group of β-D-galactopyranoside. The GDP derivative (**83**) was synthesized (Scheme 21) as a bisubstrate analog of the postulated transition state and was found to inhibit both the membrane-bound and soluble forms of the enzyme (K_i 2.3–16 μM for substrates phenyl β-D-galactopyranoside and Gal(β1–3)GlcNAc-O(CH$_2$)$_8$COOMe) [103].

IV. FUTURE PROSPECTS

A feature of many of the inhibitors that have been identified is that they are not completely specific. For example, castanospermine, in addition to inhibiting lyso-somal α-glucosidase, also inhibits β-glucosidase. 1,4-Dideoxy-1,4-imino-D-lyxitol inhibits both α-mannosidase (jack beans) and α-galactosidase (coffee beans) [56].

Thus a target for future research is the development of highly specific inhibitors, and it is clearly important to assess new potential inhibitors against a wide range of purified enzymes. The ultimate objective must be to crystallize where possible the enzymes of interest, and establish the complete structure by X-ray crystallography. Probing the active site of the enzyme by molecular modeling, as illustrated by the elegant work of von Itzstein and co-workers on sialidase inhibitors (see page 519), can be a very powerful method for inhibitor design. Where crystallization of the enzyme is not possible, other approaches, such as structure–activity relationships, can be used to identify which groups are important for binding to the enzyme. Thus, known inhibitors or the natural substrates can be modified by alkylation of OH groups, replacement of OH by H, or charged groups, as illustrated by examples cited in this review. Irreversible inhibitors can label functional groups in the active site of the enzyme of interest, but the enzyme–inhibitor bond needs to be stable enough to survive peptide degradation so that the labeled amino acids can be identified. Irreversible inhibitors such as bromoconduritol (6-bromo-3,4,5-trihydroxycyclohex-1-ene), reported to inhibit glucosidase II [104], and α-D-mannopyranosylmethyl (p-nitrophenyl)triazene, reported to inhibit α-mannosidase I and lysosomal α-mannosidase [105,106], probably react with a carboxylate group at the active site to form esters. Stable linkages can be formed using more reactive functional groups such as carbene and nitrene precursors (diazo carbonyl compounds, diazirines, azide).

The use of mutated forms of enzymes, in which an amino acid involved in binding the inhibitor (or substrate) is altered, can also provide useful information, as illustrated by the mutant of *Agrobacter* β-glucosidase, studied in connection with the binding of the tetrazole inhibitor (**75**) [88]. For a review of this approach for glycosidases, see ref. 107.

Recent developments in synthesis, such as the use of enzymes and microbes, give the chemist a wide choice of methods for the synthesis of potential inhibitors, and the increasing interest in inhibitors of glycosyltransferases is likely to focus attention on the synthesis of modified oligosaccharides. Here, the combinatorial approach [108], whereby mixtures of oligosaccharides are synthesized in reactions designed to be indiscriminate, i.e., of low regioselectivity, enables many compounds to be screened simultaneously for acceptor or inhibitory activity.

ABBREVIATIONS

Ac$_2$O	acetic anhydride
Bn	benzyl
Bz	benzoyl
Cbz	benzyloxycarbonyl
m-CPBA	meta-chloroperbenzoic acid
DBU	1,8-diazabicyclo[5.4.0]undec-7-ene
DHAP	dihydroxyacetone phosphate
4-DMAP	4-dimethylaminopyridine
DMP	2,2-dimethoxypropane
ER	endoplasmic reticulum
Fuc	L-fucopyranose
Gal	D-galactopyranose

GDP	guanosine 5'-diphosphate
Glc	D-glucopyranose
GlcNAc	2-acetamido-2-deoxy-D-glucopyranose
GMP	guanosine 5'-monophosphate
Man	D-mannopyranose
MsCl	methanesulfonyl chloride
NIS	N-iodosuccinimide
PCC	pyridinium chlorochromate
TFA	trifluoroacetic acid
Tfl$_2$O	trifluoromethanesulfonic anhydride
THP	tetrahydropyranyl
pTsOH	para-toluenesulfonic acid
Tr	trityl (triphenylmethyl)
UDP	uridine 5'-diphosphate

REFERENCES

1. R. Kornfeld and S. Kornfeld, Assembly of asparagine-linked oligosaccharides, *Ann. Rev. Biochem. 54*: 631 (1985).
2. K. W. Moremen, R. B. Trimble, and A. Herscovics, Glycosidases of the asparagine-linked oligosaccharide processing pathway, *Glycobiol. 4*: 113 (1994).
3. H. R. B. Pelham, Recycling of proteins between the endoplasmic reticulum and Golgi complex, *Curr. Opin. Cell Biol. 3*: 585 (1991).
4. M. C. Sousa, M. A. Ferrero-Garcia, and A. J. Parodi, Recognition of the oligosaccharide and protein moieties of glycoproteins by the UDP-glucose:glycoprotein glucosyltransferase, *Biochem. 31*: 97 (1992).
5. P. F. Daniel, B. Winchester, and C. D. Warren, Mammalian α-mannosidases—multiple forms but a common purpose? *Glycobiol. 4*: 551 (1994).
6. W. A. Lubas and R. G. Spiro, Golgi endo-α-D-mannosidase from rat liver, a novel N-linked carbohydrate unit processing enzyme, *J. Biol. Chem. 262*: 3775 (1987).
7. W. A. Lubas and R. G. Spiro, Evaluation of the role of rat liver Golgi endo-α-D-mannosidase in processing N-linked oligosaccharides, *J. Biol. Chem. 263*: 3990 (1988).
8. S. E. H. Moore and R. G. Spiro, Demonstration that Golgi endo-α-D-mannosidase provides a glucosidase-independent pathway for the formation of complex N-linked oligosaccharides of glycoproteins, *J. Biol. Chem. 265*: 13,104 (1990).
9. S. E. H. Moore and R. G. Spiro, Characterisation of the endomannosidase pathway for the processing of N-linked oligosaccharides in glucosidase II-deficient and parent mouse lymphoma cells, *J. Biol. Chem. 267*: 8443 (1992).
10. S. Hiraizumi, U. Spohr, and R. G. Spiro, Ligand affinity chromatographic purification of rat liver Golgi endomannosidase, *J. Biol. Chem. 269*: 4697 (1994).
11. D. E. Koshland, Stereochemistry and the mechanism of enzymic reactions, *Biol. Rev. 28*: 416 (1953).
12. M. Sinnott, Catalytic mechanisms of enzymic glycosyl transfer, *Chem. Rev. 90*: 1171 (1990).
13. G. Legler, Glycoside hydrolases: Mechanistic information from studies with reversible and irreversible inhibitors, *Adv. Carbohydr. Chem. Biochem. 48*: 319 (1990).
14. D. H. Leaback, Inhibition of β-N-acetyl-D-glucosaminidase by 2-acetamido-2-deoxy-D-glucono-1,5-lactone, *Biochem. Biophys. Res. Commun. 32*: 1025 (1968).

15. M. Horsch, L. Hoesch, G. W. J. Fleet, and D. M. Rast, Inhibition of β-*N*-acetylglu-cosaminidase by glycon-related analogues of the substrate, *J. Enzyme Inhibition 7*: 47 (1993).

16. G. W. J. Fleet, N. G. Ramsden, R. A. Dwek, T. W. Rademacher, L. E. Fellows, R. J. Nash, D. St. C. Green, and B. Winchester, δ-Lactams: Synthesis from D-glucose, and preliminary evaluation as a fucosidase inhibitor, of L-fuconic-δ-lactam, *JCS Chem. Commun.* 483 (1988).

17. H. S. Overkleeft, J. van Wiltenburg, and U. K. Pandit, A facile transformation of sugar lactones to azasugars, *Tetrahedron 50*: 4215 (1994).

18. D. R. Wolk, A. Vasella, F. Schweikart, and M. G. Peter, Synthesis and enzyme-inhibition studies of phenylsemicarbazones derived from D-glucono-1,5-lactone and 2-acetamido-2-deoxy-D-glucono-1,5-lactone, *Helv. Chim. Acta 75*: 323 (1992).

19. M. Horsch, L. Hoesch, A. Vasella, and D. M. Rast, *N*-Acetylglucosamino-1,5-lactone oxime and the corresponding (phenylcarbamoyl)oxime. Novel and potent inhibitors of β-*N*-acetyl-glucosaminidase, *Eur. J. Biochem. 197*: 815 (1991).

20. H.-Y. L. Lai and B. Axelrod, 1-Aminoglycosides, a new class of specific inhibitors of glycosidases, *Biochem. Biophys. Res. Commun. 54*: 463 (1978).

21. E. Kappes and G. Legler, Synthesis and inhibitory properties of 2-acetamido-2-deoxy-nojirimycin (2-acetamido-5-amino-2,5-dideoxy-D-glucopyranose) and 2-acetamido-1,2-dideoxynojirimycin (2-acetamido-1,5-imino-1,2,5-trideoxy-D-glucitol), *J. Carbohydr. Chem. 8*: 371 (1989).

22. J. P. Chambers, A. D. Elbein, and J. C. Williams, Nojirimycin—a potent inhibitor of purified lysosomal α-glucosidase from human liver, *Biochem. Biophys. Res. Commun. 107*: 1490 (1982).

23. G. W. J. Fleet, N. M. Carpenter, S. Petursson, and N. G. Ramsden, Synthesis of deoxynojirimycin and of nojirimycin δ-lactam, *Tetrahedron Lett. 31*: 409 (1990).

24. R. C. Bernotas and B. Ganem, Efficient preparation of enantiomerically pure cyclic aminoalditols; total synthesis of 1-deoxynojirimycin and 1-deoxymannojirimycin, *Tetrahedron Lett. 26*: 1123 (1985).

25. R. C. Bernotas and B. Ganem, Total synthesis of (+)-castanospermine and (+)-deoxynojirimycin, *Tetrahedron Lett. 25*: 165 (1984).

26. A. B. Reitz and E. W. Baxter, Pyrrolidine and piperidine aminosugars from dicarbonyl sugars in one step. Concise synthesis of 1-deoxynojirimycin, *Tetrahedron Lett. 31*: 6777 (1990); E. W. Baxter and A. B. Reitz, Expeditious synthesis of azasugars by the double reductive amination of dicarbonyl sugars. *J. Org. Chem. 59*: 3175 (1994).

27. J. Schweden, C. Borgmann, G. Legler, and E. Bause, Characterisation of calf liver glucosidase I and its inhibition by basic sugar analogues, *Arch. Biochem. Biophys. 248*: 335 (1986).

28. A. Tan, L. van den Broek, S. van Boeckel, H. Ploegh and J. Bolscher, Chemical modification of the glucosidase inhibitor 1-deoxynojirimycin; structure-activity relationships. *J. Biol. Chem. 266*: 14,504 (1991).

29. L. A. G. M. van den Broek, D. J. Vermaas, B. M. Heskamp, C. A. A. van Boeckel, M. C. A. A. Tan, J. G. M. Bolscher, H. L. Ploegh, F. J. van Kemenade, R. E. Y. de Goede, and F. Miedema, Chemical modification of azasugars, inhibitors of *N*-glyco-protein-processing glycosidases and of HIV-I infection, *Recl. Trav. Chim. Pays-Bas 112*: 82 (1993).

30. I. Bruce, G. W. J. Fleet, I. C. di Bello and B. Winchester, Iminoheptitols as glycosidase inhibitors: Synthesis of α-homomannojirimycin, 6-*epi*-α-homomannojirimycin and of a highly substituted pipecolic acid, *Tetrahedron 48*: 10,191 (1992).

31. A. de Raadt and A. E. Stutz, A simple convergent synthesis of the mannosidase inhibitor 1-deoxymannojirimycin from sucrose, *Tetrahedron Lett. 33*: 189 (1992).

32. T. Hudlicky, J. Rouden and H. Luna, Rational design of aza sugars via biocatalysis: Mannojirimycin and other glycosidase inhibitors, *J. Org.Chem. 58*: 985 (1993).

33. C. H. von der Osten, A. J. Sinskey, C. F. Barbas III, R. L. Pederson, Y.-F. Wang and, C.-H. Wong, Use of a recombinant bacterial fructose-1,6-diphosphate aldolase in aldol reactions: Preparative syntheses of 1-deoxynojirimycin, 1-deoxymannojirimycin, 1,4-dideoxy-1,4-imino-D-arabinitol and fagomine, *J. Amer. Chem. Soc. 111*: 3924 (1989).

34. G. W. J. Fleet, S. Petursson, A. L. Campbell, R. A. Mueller, J. R. Behling, K. A. Babiak, J. S. Ng, and M. G. Scaros, Short efficient syntheses of the α-L-fucosidase inhibitor, deoxyfuconojirimycin [1,5-dideoxy-1,5-imino-L-fucitol] from D-lyxonolactone, *JCS Perkin Trans I:* 665 (1989).

35. G. W. J. Fleet, S. K. Namgoong, C. Barker, S. Baines, G. S. Jacob, and B. Winchester, Iminoheptitols as glycosidase inhibitors: Synthesis and specific α-L-fucosidase inhibition by β-L-homofuconojirimycin and 1-β-C-substituted deoxymannojirimycins, *Tetrahedron Lett. 30*: 4439 (1989).

36. D. M. Andrews, M. I. Bird, M. M. Cunningham, and P. Ward, Synthesis of and glycosidase inhibition by α-L-homofuconojirimycin. *Bioorg. Med. Chem. Lett. 3*: 2533 (1993).

37. E. T. Reese, F. W. Parrish, and M. Epplinger, Nojirimycin and D-glucono-1,5-lactone as inhibitors of carbohydrases, *Carbohydr. Res. 18*: 381 (1971).

38. M. K. Tong, G. Papandreou, and B. Ganem, Potent, broad-spectrum inhibition of glycosidases by an amidine derivative of D-glucose, *J. Amer. Chem. Soc. 112*: 6137 (1990).

39. B. Ganem and G. Papandreou, Mimicking the glucosidase transition state: Shape/charge considerations, *J. Amer. Chem. Soc. 113*: 8984 (1991).

40. G. Papandreou, M. K. Tong, and B. Ganem, Amidine, amidrazone, and amidoxime derivatives of monosaccharide aldonolactams: Synthesis and evaluation as glycosidase inhibitors, *J. Amer. Chem. Soc. 115*: 11,682 (1993).

41. D. A. Winkler and G. Holan, Design of potential anti-HIV agents. 1. Mannosidase inhibitors, *J. Med. Chem. 32*: 2084 (1989).

42. T. Kajimoto, K. K.-C Liu, R. L. Pederson, Z. Zhong, Y. Ichikawa, J. A. Porco Jr., and C.-H. Wong, Enzyme-catalysed aldol condensation for asymmetric synthesis of aza-sugars: Synthesis, evaluation and modelling of glycosidase inhibitors, *J. Amer. Chem. Soc. 113*: 6187 (1991).

43. R. Hoos, A. B. Naughton, W. Thiel, A. Vasella, W. Weber, K. Rupitz, and S. G. Withers, D-Gluconhydroximo-1,5-lactam and related N-arylcarbamates. Theoretical calculations, structure, synthesis and inhibitory effect on β-glucosidases, *Helv. Chim. Acta 76*: 2666 (1993).

44. Y.-T. Pan, G. P. Kaushal, G. Papandreou, B. Ganem, and A. D. Elbein, D-Mannonolactam amidrazone. A new mannosidase inhibitor that also inhibits the endoplasmic reticulum or cytoplasmic α-mannosidase, *J. Biol. Chem. 267*: 8313 (1992).

45. U. Spohr, M. Bach, and R. G. Spiro, Inhibitors of endo-α-mannosidase. Part 1. Derivatives of 3-O-(α-D-glucopyranosyl)-D-mannopyranose, *Canad. J. Chem. 71*: 1919 (1993).

46. U. Spohr, M. Bach and R. G. Spiro, Inhibitors of endo-α-mannosidase. Part 2. 1-Deoxy-3-O-(α-D-glucopyranosyl) mannojirimycin and conjeners modified in the mannojirimycin unit, *Canad. J. Chem. 71*: 1928 (1993).

47. U. Spohr and M. Bach, Inhibitors of endo-α-mannosidase. Part 3. Conjeners of 1-deoxy-3-O-(α-D-glucopyranosyl)-mannojirimycin modified in the glucose unit, *Canad. J. Chem. 71*: 1943 (1993).

48. S. Hiraizumi, U. Spohr, and R. G. Spiro, Characterisation of endomannosidase inhibitors and evaluation of their effect on N-linked oligosaccharide processing during glycoprotein biosynthesis, *J. Biol. Chem. 268*: 9927 (1993).

49. H. Ardron, T. D. Butters, F. M. Platt, M. R. Wormald, R. A. Dwek, G. W. J. Fleet, and G. S. Jacob, Synthesis of 1,5-dideoxy-3-*O*-(α-D-mannopyranosyl)-1,5-imino-D-mannitol and 1,5-dideoxy-3-*O*-(α-D-glucopyranosyl)-1,5-imino-D-mannitol: powerful inhibitors of endomannosidase, *Tetrahedron Asym. 4*: 2011 (1993).

50. G. W. J. Fleet, P. W. Smith, S. V. Evans, and L. E. Fellows, Design, synthesis and preliminary evaluation of a potent α-mannosidase inhibitor: 1,4-Dideoxy-1,4-imino-D-mannitol, *JCS Chem. Commun.*: 1240 (1984).

51. G. Palamarczyk, M. Mitchell, P. W. Smith, G. W. J. Fleet, and A. D. Elbein, 1,4-Dideoxy-1,4-imino-D-mannitol inhibits glycoprotein processing and mannosidase, *Arch. Biochem. Biophys. 243*: 35 (1985).

52. B. Winchester, S. Al Daher, N. C. Carpenter, I. Cenci di Bello, S. S. Choi, A. J. Fairbanks, and G. W. J. Fleet, The structural basis of the inhibition of human α-mannosidases by azafuranose analogues of mannose, *Biochem J. 290*: 743 (1993).

53. I. Lundt and R. Madsen, Deoxyiminoalditols from aldonolactones; I. Preparation of 1,4-dideoxy-1,4-iminohexitols with D- and L-allo and D- and L-talo configuration: Potential glycosidase inhibitors, *Synthesis*: 714 (1993).

54. G. W. J. Fleet, J. C. Son, D. St. C. Green, I. Cenci di Bello, and B. Winchester, Synthesis from D-mannose of 1,4-dideoxy-1,4-imino-L-ribitol and of the α-mannosidase inhibitor 1,4-dideoxy-1,4-imino-D-talitol, *Tetrahedron 44*: 2649 (1988).

55. A. D. Elbein, M. Mitchell, B. A. Sandford, L. E. Fellows, and S. V. Evans, The pyrrolidine alkaloid, 2,5-dihydroxymethyl-3,4-dihydroxypyrrolidine, inhibits glycoprotein processing, *J. Biol. Chem. 259*: 12,409 (1984).

56. G. W. J. Fleet, S. J. Nicholas, P. W. Smith, S. V. Evans, L. E. Fellows, and R. J. Nash, Potent competitve inhibition of α-galactosidase and α-glucosidase activity by 1,4-dideoxy-1,4-iminopentitols; synthesis of 1,4-dideoxy-1,4-imino-D-lyxitol and of both enantiomers of 1,4-dideoxy-1,4-iminoarabinitol, *Tetrahedron Lett. 26*: 3127 (1985).

57. G. Legler, A. Korth, A. Berger, C. Ekhart, G. Gradnig, and A. E. Stutz, 2,5-Dideoxy-2,5-imino-D-mannitol and -D-glucitol. Two-step bio-organic syntheses from 5-azido-5-deoxy-D-glucofuranose and -L-idofuranose; evaluation as glucosidase inhibitors and application in affinity purification and characterisation of invertase from yeast, *Carbohydr. Res. 250*: 67 (1993).

58. Y.-F. Wang, D. P. Dumas, and C.-H. Wong, Chemo-enzymatic synthesis of five-membered azasugars as inhibitors of fucosidase and fucosyltransferase: An issue regarding the stereochemistry discrimination at transition states, *Tetrahedron Lett. 34*: 403 (1993).

59. J. E. Tropea, G. P. Kaushal, I. Pastuszak, M. Mitchell, T. Aoyogi, R. J. Molyneux, and A. D. Elbein, Mannostatin A, a new glycoprotein-processing inhibitor, *Biochem. 29*: 10,062 (1990).

60. S. Knapp and T. G. M. Dhar, Synthesis of the mannosidase II inhibitor Mannostatin A, *J. Org. Chem. 56*: 4096 (1991).

61. S. B. King and B. Ganem, Enantioselective synthesis of Mannostatin A: A new glycoprotein processing inhibitor, *J. Amer. Chem. Soc. 113*: 5089 (1991).

62. V. Kumar, J. Kessler, M. E. Scott, and B. H. Patwardhan, 2,3-Dehydro-4-epi-*N*-acetylneuraminic acid; a neuraminidase inhibitor, *Carbohydr. Res. 94*: 123 (1981).

63. F. Bamberger, A. Vasella, and R. Schauer, Synthesis of new sialidase inhibitors, 6-amino-6-deoxysialic acids, *Helv. Chim. Acta 71*: 429 (1988).

64. N. R. Taylor and M. von Itzstein, Molecular modelling studies on ligand binding to sialidase from influenza virus and the mechanism of catalysis, *J. Med. Chem. 37*: 616 (1994).

65. M. von Itzstein, W.-Y. Wu, G. B. Kok, M. S. Pegg, J. C. Dyason, B. Jin, T. V. Phan, M. L. Smythe, H. F. White, S. W. Oliver, P. M. Colman, J. N. Varghese, D. M. Ryan, J. M. Woods, R. C. Bethell, V. J. Hotham, J. M. Cameron, and C. R. Penn, Rational

design of potent sialidase-based inhibitors of influenza virus replication, *Nature 363*: 418 (1993).

66. M. von Itzstein, W.-Y. Wu, and B. Jin, The synthesis of 2,3-didehydro-2,4-dideoxy-4-guanidinyl-*N*-acetylneuraminic acid: A potent influenza virus sialidase inhibitor, *Carbohydr. Res. 259*: 301 (1994).

67. T. Kudo, Y. Nishimura, S. Kondo, and T. Takeuchi, Synthesis of the potent inhibitors of neuraminidase, *N*-(1,2-dihydroxypropyl) derivatives of siastatin B and its 4-deoxy analogues, *J. Antiobiotics 46*: 300 (1993).

68. G. E. McCasland, S. Furuta, and L. J. Durham, Alicyclic carbohydrates. XXXII. Synthesis of pseudo-β-DL-gulopyranose from a diacetoxybutadiene. Proton magnetic resonance studies, *J. Org. Chem. 33*: 2835 (1968).

69. T. Suami and S. Ogawa, Chemistry of carba-sugars (pseudo-sugars) and their derivatives, *Adv. Carbohydr. Chem. Biochem. 48*: 21 (1990).

70. M. Hayashida, N. Sakairi, and H. Kuzuhara, Novel synthesis of penta-*N,O*-acetylvaliolamine, *J. Carbohydr. Chem. 7*: 83 (1988).

71. D. R. P. Tulsiani, T. M. Harris, and O. Touster, Swainsonine inhibits the biosynthesis of complex glycoproteins by inhibition of Golgi mannosidase II, *J. Biol. Chem. 257*: 7936 (1982).

72. K.-I. Tadana, M. Morita, Y. Hotta, S. Ogawa, B. Winchester, and I. Cenci di Bello, Synthesis of (−)-2,8a and (−)-8,8a-di-epi-swainsonine and evaluation of their inhibitory activity against several glycosidases, *J. Org. Chem. 53*: 5209 (1988).

73. M. H. Ali, L. Hough, and A. C. Richardson, A chiral synthesis of swainsonine from D-glucose, *JCS Chem. Commun:* 447 (1984); M. H. Ali, L. Hough, and A. C. Richardson, Synthesis of the indolizidine alkaloid swainsonine from D-glucose, *Carbohydr. Res. 136*: 225 (1985).

74. B. P. Bashyal, G. W. J. Fleet, M. J. Gough, and P. W. Smith, Synthesis of swainsonine and 1,4-dideoxy-1,4-imino-D-mannitol from mannose, *Tetrahedron 43*: 3083 (1987).

75. H. Setoi, H. Takeno, and M. Hashimoto, Enantiospecific total synthesis of (−)-swainsonine: new applications of sodium borohydride reduction, *J. Org. Chem. 50*: 3948 (1985).

76. C. E. Adams, F. J. Walker, and K. B. Sharpless, Enantioselective synthesis of swainsonine, a trihydroxylated indolizidine alkaloid, *J. Org. Chem. 50*: 420 (1985).

77. I. Cenci di Bello, G. W. J. Fleet, S. K. Namgoong, K.-I. Tadana, and B. Winchester, Strucure-activity relationship of swainsonine, *Biochem. J. 259*: 855 (1989).

78. Y. T. Pan, H. Hori, R. Saul, B. A. Sandford, J. R. Molyneux, and A. D. Elbein, Castanospermine inhibits processing of the oligosaccharide portion of the influenza viral haemagglutinin, *Biochem. 22*: 3975 (1983).

79. P. S. Lui, R. S. Rogers, M. S. Kang, and P. S. Sunkara, Synthesis of polyhydroxylated indolizidine and quinolizidine compounds—Potent inhibitors of α-glucosidase I, *Tetrahedron Lett. 32*: 5853 (1991).

80. K. Burgess and I. Henderson, Synthetic approaches to stereoisomers and analogues of castanospermine, *Tetrahedron 48*: 4045 (1992).

81. H. Hamana, N. Ikota, and B. Ganem, Chelate selectivity in chelation-controlled allylations. A new synthesis of castanospermine and other bioactive indolizidine alkaloids, *J. Org. Chem. 54*: 5492 (1987).

82. A. Hempel, N. Camerman, D. Mastropaola, and A. Camerman, Glucosidase inhibitors: Structures of 1-deoxynojirimycin and castanospermine, *J. Med. Chem. 36*: 4082 (1993).

83. B. G. Winchester, I. Cenci di Bello, A. C. Richardson, R. J. Nash, L. E. Fellows, N. G. Ramsden, and G. W. J. Fleet, The structural basis of the inhibition of human glycosidases by castanspermine analogues, *Biochem. J. 269*: 227 (1990).

84. A. D. Elbein, J. E. Tropea, M. Mitchell, and G. P. Kaushal, Kifunensine, a potent inhibitor of glycoprotein processing mannosidase I, *J. Biol. Chem. 265*: 15,599 (1990).

85. H. Kayakiri, C. Kasahara, K. Nakamura, T. Oku, and M. Hashimoto, Synthesis of kifunensine, an immunomodulating substance isolated from a microbial source, *Chem. Pharm. Bull. 39*: 1392 (1991).

86. J. Rouden and T. Hudlicky, Total synthesis of (+)-kifunensine, a potent glycosidase inhibitor, *JCS Perkin Trans I:* 1095 (1993).

87. P. Ermert and A. Vasella, Synthesis of a glucose-derived tetrazole as a new β-glucosidase inhibitor. A new synthesis of 1-deoxynojirimycin, *Helv. Chim. Acta 74*: 2043 (1991).

88. P. Ermert, A. Vasella, M. Weber, K. Rupitz, and S. G. Withers, Configurationally selective transition state analogue inhibitors of glycosidases. A study with nojiritetrazoles, a new class of glycosidase inhibitors, *Carbohydr. Res. 250*: 113 (1993).

89. J. E. Tropea, R. J. Molyneux, G. P. Kaushal, Y. T. Pan, M. Mitchell, and A. D. Elbein, Australine, a pyrrolizidine alkaloid that inhibits amyloglucosidase and glycoprotein processing, *Biochem. 28*: 2027 (1989).

90. M. M. Vaghefi, R. J. Bernacki, N. K. Dalley, B. E. Wilson, and R. K. Robins, Synthesis of glycopyranosylphosphonate analogues of certain nucleoside diphosphate sugars as potential inhibitors of glycosyltransferases, *J. Med. Chem. 30*: 1383 (1987).

91. M. M. Vaghefi, R. J. Bernacki, W. J. Hennen, and R. K. Robins, Synthesis of certain nucleoside methylenediphosphonate sugars as potential inhibitors of glycosyltransferases, *J. Med. Chem. 30*: 1391 (1987).

92. R. R. Schmidt and K. Frische, A new galactosyl transferase inhibitor, *Bioorg. Med. Chem. Lett. 3*: 1747 (1993).

93. O. Hindsgaul, K. J. Kaur, G. Srivastava, M. Blaszczyk-Thurin, S. C. Crawley, L. D. Heerze, and M. M. Palcic, Evaluation of deoxygenated oligosaccharide acceptor analogues as specific inhibitors of glycosyltransferases, *J. Biol. Chem. 266*: 17,858 (1991).

94. S. H. Khan, S. C. Crawley, O. Kanie, and O. Hindsgaul, A trisaccharide acceptor analogue for *N*-acetylglucosaminyltransferase V which binds to the enzyme but sterically precludes the transfer reaction, *J. Biol. Chem. 268*: 2468 (1993).

95. F. Reck, M. Springer, H. Paulsen, I. Brockhausen, M. Sarkar, and H. Schachter, Synthesis of tetrasaccharide analogues of the *N*-glycan substrate of β-(1,2)-*N*-acetylglucosaminyltransferase II using trisaccharide precursors and recombinant β-(1,2)-*N*-acetylglucosaminyltransferase I, *Carbohydr. Res. 259*: 93 (1994).

96. E. J. Toone, E. S. Simon, M. D. Bednarski, and G. Whitesides, Enzyme-catalysed synthesis of carbohydrates, *Tetrahedron 45*: 5365 (1989).

97. T. L. Lowary and O. Hindsgaul, Recognition of synthetic deoxy and deoxyfluoro analogues of the acceptor α-L-fucp-(1,2)-β-D-galp-OR by the blood-group A and B gene-specific glycosyltransferases, *Carbohydr. Res. 249*: 163 (1993).

98. T. L. Lowary and O. Hindsgaul, Recognition of synthetic *O*-methyl, epimeric, and amino analogues of the acceptor α-L-fucp-(1,2)-β-D-galp-OR by the blood-group A and B gene-specific glycosyltransferases, *Carbohydr. Res. 251*: 33 (1994).

99. T. L. Lowary, S. J. Swiedler, and O. Hindsgaul, Recognition of synthetic analogues of the acceptor, β-D-galp-OR, by the blood-group H gene-specific glycosyltransferase, *Carbohydr. Res. 256*: 257 (1994).

100. C.-H. Wong, Y. Ichikawa, T. Krach, C. Gautheron-Le Narvor, D. P. Dumas, and G. C. Look, Probing the acceptor specificity of β-1,4-galactosyltransferase for the development of enzymatic synthesis of novel oligosaccharides, *J. Amer. Chem. Soc. 113*: 8137 (1991).

101. C.-H. Wong, D. P. Dumas, Y. Ichikawa, K. Koseki, S. J. Danishefsky, B. W. Weston, and J. B. Lowe, Specificity, inhibition, and synthetic utility of a recombinant human α-1,3-fucosyltransferase, *J. Amer. Chem. Soc. 114*: 7321 (1992).

102. R. A. Field, D. C. A. Neville, R. W. Smith, and M. A. J. Ferguson, Acceptor analogues as potential inhibitors of bovine β-1,4-galactosyltransferase, *Bioorg. Med. Chem. Lett. 4*: 391 (1994).

103. M. M. Palcic, L. D. Heerze, O. P. Srivastava, and O. Hindsgaul, A bisubstrate analogue inhibitor for α(1,2)-fucosyltransferase, *J. Biol. Chem. 264*: 17,174 (1989).

104. R. Datema, P. A. Romero, G. Legler, and R. T. Schwarz, Inhibition of formation of complex oligosaccharides by the glucosidase inhibitor bromoconduritol, *Proc. Natl. Acad. Sci. USA, 79*: 6787 (1982).

105. W. McDowell, A. Tlusty, R. Rott, J. N. BeMiller, J. A. Bohn, R. W. Meyers, and R. T. Schwarz, Inhibition of glycoprotein oligosaccharide processing *in vitro* and in influenza-virus-infected-cells by α-D-mannopyranosylmethyl-*p*-nitrophenyltriazene, *Biochem. J. 255*: 991 (1988).

106. P. A. Docherty and N. N. Aronson, α-D-Mannopyranosylmethyl-*p*-nitrophenyltriazene inhibition of rat liver α-D-mannosidases, *Biochim. Biophys. Acta 914*: 283 (1987).

107. B. Svensson and M. Sogaard, Mutational analysis of glycosylase function, *J. Biotechnol. 29*: 1 (1993).

108. O. Hindsgaul, Chemical Strategies for the Discovery of Tight-Binding Carbohydrate Ligands, lecture presented at the meeting of the Royal Society of Chemistry Carbohydrate Group, September 1994, University of Manchester Institute of Science and Technology, England.

11

Use of Permethylation with GC-MS for Linkage and Sequence Analysis of Oligosaccharides: Historical Perspectives and Recent Developments

Steven B. Levery
University of Georgia, Athens, Georgia

I. INTRODUCTION

Primary structure elucidation of oligosaccharides, polysaccharides, or glycoconjugates requires determination of the following characteristics (see, for example, Refs. 1–4):

1. The glycosyl composition, i.e., identity (including absolute configuration [D- or L-], ring form [furanose or pyranose], and anomeric configuration [α- or β-]) and ratio of the monosaccharide units that are glycosidically linked to each other within the carbohydrate
2. The sequence of the glycosyl residues
3. The positions of all glycosidic linkages between sugars
4. The identity, points of attachment, and stereochemistry (where appropriate) of any noncarbohydrate moieties, including aglycones, as well as any nonglycosidic linkages between sugars (e.g., interresidue esterification or amidation)

The combination of permethylation with GC-MS forms the basis of two important techniques for carbohydrate structure determination, namely (a) methylation linkage analysis, used for determining the *positions of glycosidic attachment* of all monosaccharide units in a carbohydrate polymer, as well as confirming the *identity* and *ring form* of each of the component sugars, and providing, in many cases, a means of determining the points of attachment (if not the identity) of noncarbohydrate moieties or nonglycosidic linkages; and (b) GC-MS of permethylated oligosaccharides, which, like a number of other mass spectrometric techniques in use for this purpose (direct probe EI-, DCI-, FAB-, ESI-, and MALDI-MS or -MS/MS), can provide the *composition* and *sequence* of the isobaric monosaccharide units constituting an intact (nondepolymerized) glycan (but not their precise identities), and, moreover, can provide this information on each of the components of a mixture of oligosaccharides. Neither technique is capable, by itself or in combination, of determining the anomeric or absolute configuration of the component sugars, but must be used in conjunction with other methods, such as NMR or glycosidase digestion, to provide a complete primary structure determination. Nevertheless, methylation linkage analysis in particular has been an essential part of many elucidation strategies, and will probably remain so in the near future, because it is as yet the only completely general, unambiguous method of determining linkage structure. Although mass spectrometry of permethylated oligosaccharides and glycoconjugates is likewise in widespread use, the GC-MS variation has not been one of the more popular alternatives, due to a number of practical limitations that, however, as work described in this chapter may show, have recently been overcome to a great extent.

II. PERMETHYLATION REACTIONS

Purdie and co-workers [5,6] were the first to report formation of alkyl ethers involving the nonanomeric hydroxyl groups of a sugar. They established that the major product from prolonged treatment of α-methyl glucoside with methyl iodide and silver oxide in methanol was a trimethyl α-methyl glucoside (proposed to be the 1,2,3,4-tetra-O-methyl derivative of glucose); refluxing of this intermediate with silver oxide in pure methyl iodide then yielded the fully methylated sugar, which could be converted into 2,3,4,6-tetra-*O*-methyl glucose upon hydrolysis. The reaction was subsequently applied to a number of sugars, including sucrose [7–9]. Later, Haworth [10] was able to prepare many of the same derivatives with a procedure employing dimethyl sulfate and sodium hydroxide. With some sugars (e.g., sucrose), the method of Purdie and Irvine [5] had to be used to complete the alkylation. On the other hand, sugars are susceptible to oxidation by silver oxide, particularly at elevated temperatures. For this reason, ethyl or higher alkyl sugars could not be synthesized except by Haworth's [10] method.

Among the earliest to attempt methylation of polysaccharides were Denham and Woodhouse [11], who treated different types of cellulose repeatedly with methyl sulfate and sodium hydroxide (similar to Haworth). While they came nowhere near complete permethylation, their paper contains the following interesting suggestion:

> If cellulose, for example, were methylated, subsequent cleavage of the molecule should yield one or more of the methylated dextroses in which the methoxyl groups represent hydroxyl groups of the original cellulose complex, so that the nature of the linkings would be determined to an extent dependent on the degree to which the cellulose had been methylated before the cleavage of the molecule.

This early statement contains, *in nuce*, a clear representation of the strategy that would evolve decades later into routine methylation linkage analysis, as well as a prescient glimpse at one of the major hurdles along that evolutionary path—the necessity of achieving virtually complete methylation for the results to be analytically meaningful.

In early practice, methylation of an oligo- or polysaccharide was usually initiated by the method of Haworth [10] and then completed by exhaustive treatment with silver oxide and methyl iodide. Generally, the low solubility of sugars in methyl iodide (or methyl iodide and methanol) precludes the direct application of Purdie and Irvine's [5] reaction. A later improvement by Kuhn [12] used dimethylformamide as solvent for this reaction, which allowed direct application to a wider variety of sugars. Another improvement, avoiding the oxidizing effects of silver oxide, was the use of barium oxide as the base in the Kuhn modification [13]. Still later, the higher solubility of many polysaccharides in dimethylsulfoxide was used to advantage by methylation in this solvent using the combination of methyl iodide with silver oxide and drierite or barium oxide and/or barium hydroxide [14–16]; or dimethyl sulfate with barium oxide and/or barium hydroxide, or sodium hydroxide pellets [14,17,18]. Despite all these improvements, ''complete'' permethylation (adequate for the purposes of linkage analysis) could take days, sometimes even weeks, to achieve.

A major breakthrough allowing routine linkage analysis was the Hakomori permethylation [19], based on application of the Corey–Chaykovsky base [20–22]. The Hakomori method immediately replaced the older methylations. The procedure utilizes methylsulfinyl carbanion, prepared from NaH and DMSO, as a strong base to deprotonate all labile sites (OH, NH, COOH) of a carbohydrate. The presence of sufficient excess of methylsulfinyl carbanion can be verified by reaction of a small aliquot with triphenylmethane [23,24] (the general validity of this test has been questioned [25]). Methylation is then effected quickly and uniformly with methyl iodide. Under these conditions, all hydroxy groups are converted to methyl ethers, carbonyl groups to methyl esters; *N*-acyl groups, including acetamides of sugars and long-chain amides of ceramide aglycones, are *N*-methylated. In addition, *O*-acyl groups are replaced by *O*-methyl groups. Information on naturally *O*-methylated sugars is also lost, unless isotopically labeled methyl iodide is used. Alternatively, the location of natural O-methyl groups can be identified simply by applying the alditol acetate monosaccharide protocol, skipping the permethylation; naturally methylated sugars can then be identified by GC-MS in the same manner as PMAAs from linkage analysis (Section III). Acetal and ketal (such as naturally occurring pyruvate) substituents are retained. Since these are lost during the subsequent acid depolymerization, their points of attachment can be determined by comparison of results of linkage analysis on both mild-acid-pretreated and native samples. A similar strategy can be used to locate terminal sugars labile to weak acid treatment, such as Fuc and NeuAc residues. More generally, monosaccharide residues can be removed sequentially by exoglycosidases, with permethylation of an aliquot following each reaction (see, for example, Ref. 26). Naturally occurring O-acyl groups can be located by a ''negative imaging'' technique, whereby the glycoconjugate is first acetalated with methyl vinyl ether [27] or dihydropyran [28] in dimethylsulfoxide in the presence of an acidic catalyst such as p-toluenesulfonic acid. After permethylation of the protected material and acid depolymerization, which consequently removes the acetal protecting groups, the former sites of O-acyl groups will be marked by O-methyl groups [29,30]. The rather special case of internal esterification (ganglioside lactones) will be discussed in detail in Section IV.

Polysaccharides and glycoconjugates having uronic acid residues can be permethylated after reduction of the carboxyl group by the carbodiimide/sodium borohydride method [31,32]; alternatively, the reduction can be performed with LiAlH$_4$ after permethylation [30], and the carbohydrate remethylated. Reduction with a deuterated reagent is recommended [29,30,33] to facilitate mass spectrometric distinction of residues formerly present as uronates.

Numerous adaptations of the Hakomori procedure have been developed to cope with differing types of samples and personal preferences [29,30,33–37]. Microscale methylations have been developed [4,38–44] for minute quantities of sample (100 ng–20 μg) to keep pace with reduced detection limits for PMAAs by GC-MS. A number of alternative procedures have introduced the use of dimsyl anion (methylsulfinyl carbanion) made with bases other than NaH, such as KH [25,45], and butyllithium [43,44,46]. Permethylations have also been performed with potassium *tert*-butoxide in DMSO [47], although its effectiveness has been disputed [25,48]. The most recent refinements have incorporated the production of dimsyl anion by butyllithium into a comprehensive procedure capable of producing a linkage analysis on as little as 10 ng of a sample of glycoprotein glycan [43] (see Section III.A.). Their success with such small amounts of material is ascribed at least in part to the fact that, unlike sodium and potassium hydride, which are normally stored and shipped in an oil dispersion, butyllithium is available in high purity and therefore contributes much less chemical background to the analysis [43].

Twenty years after the introduction of the Hakomori [19] procedure, Ciukanu and Kerek [48] found that, in many cases, a simple mixture of NaOH and MeI in DMSO is an adequate permethylation reagent for carbohydrates. Furthermore, the permethylation appears to be unimpeded by the presence of even relatively large amounts of water [49]. The method is so simple and the results so convincing, one can't help wondering why it was not reported earlier. It has been suggested (Hakomori, personal communication) that previous, unreported observations of the inadequacy of NaOH/MeI/DMSO permethylations could have been due to the inferior quality of sodium hydroxide available prior to the 1970s. One might also wonder why NaOH appears to be superior to potassium *tert*-butoxide, when the latter is considered a stronger base [50].

Adaptations of this procedure have now superceded, to considerable extent, the Hakomori permethylation [37,49,51−56], although it is unclear whether it will be an adequate replacement in all situations. The production of artifacts in linkage analysis of oligoglycosyl alditols was noted, and this was ascribed to the formation of an oxidizing agent and the consequent conversion of some sugar alkoxides to carbonyls during the NaOH-catalyzed permethylation [57]. But this problem appears to be suppressed by the simple expedient of delaying the introduction of MeI some time after NaOH addition [58]. We have noticed (unpublished) some problems in per- methylating gangliosides, which may be due to saponification of the sialic acid methyl esters by leftover sodium hydroxide during workup; others have not reported encountering this difficulty with gangliosides (see, for example, Refs. 52 and 53), but some particular attention may need to be paid to the workup procedure in these cases. Gunnarsson's [53] procedure includes neutralization with HAc prior to chlo- roform extraction, while the procedure of Larson and co-workers [52], like the orig- inal [48], does not. For a further discussion of advantages and disadvantages of the Ciukanu vs. Hakomori procedures, see Hanisch [59]; for another discussion of per- methylations in general, see Carpita and Shea [36]; for pre-Hakomori reviews of permethylation reactions, see Bouveng and Lindberg [60] and Wallenfels et al. [14].

An interesting procedure for permethylating carbohydrates under mild, non- basic conditions was introduced by Prehm [61], employing methyltrifluorometha- nesulfonate in trimethyl phosphate, with 2,6-di-(*tert*-butyl)-pyridine acting as a pro- ton scavenger. The brief publication described methylation of a number of core oligosaccharides derived from bacterial lipopolysaccharides, as well as complete li- popolysaccharides following their conversion to triethylammonium salts. The pro- cedure was recommended particularly for carbohydrates possessing alkali-labile sub- stituents. In one example of its use, the distribution pattern of acetyl groups in commercial cellulose acetates with DS (degree of substitution) 2.5 was determined [62]; following Prehm methylation, the method of Ciukanu and Kerek was used to substitute higher alkyls for the acetyl groups. When completeness of Prehm meth- ylation was checked by subsequent use of a stronger but related methylating agent, trimethyloxonium tetrafluoroborate [63], no increase in methyl content was observed [62]. Trimethyloxonium tetrafluoroborate, in the presence of 2,6-di-(*tert*-butyl)-pyr- idine, was recommended [62] as a most effective methylating reagent for partially derivatized carbohydrates, provided the sample is soluble in dichloromethane. More recently, a number of examples of the use of Prehm methylation for solving glyco- conjugate structural problems have appeared; it has been used, for example, in the detailed characterization of a succinoglycan from *Rhizobium meliloti* [63a], and of acylated phosphatidylinositol mannosides from *Mycobacterium tuberculosis* [63b].

III.   METHYLATION LINKAGE ANALYSIS BY GC-MS

A.   Derivatization Schemes

In general terms, the standard strategy is to (1) derivatize all free hydroxyl groups with a simple, acid-stable group (methyl); (2) obtain monosaccharide units by acid depolymerization; (3) derivatize all newly freed hydroxyl groups with a second simple group that will simultaneously label those positions formerly engaged in glycosyl linkages and produce compounds with good characteristics for analysis, generally by GC-MS (acetate or trimethylsilyl), although other more classical techniques (e.g., paper and thin-layer chromatographies) are still in use. In our laboratory, depolymerization of permethylated glycosphingolipids is accomplished by an ''acetolysis'' procedure similar to that described previously for alditol acetates (0.5 N H_2SO_4 in 95% acetic acid; sealed tube, 80°C, 7–8 h). Numerous other reagents have been used for hydrolysis, but for work with 2-*N*-acetamido sugars this type of reagent is prescribed for avoiding de-*N*-acetylation and consequent inhibition of depolymerization [64]. Other hydrolysis conditions that have been used for various types of polysaccharides, oligosaccharides, and glycoconjugates include 90% formic acid at 100°C for 2 hours followed by 0.25 M sulfuric acid at 100°C for 12 hours [30], 2M trifluoroacetic acid at 121°C for 1 hour [1,25,39,65], and mixtures of formic and trifluoroacetic acids at 100°C for varying lengths of time [49]. These reagents are quite effective where no 2-*N*-acetamido sugars are expected, but usually produce lower yields than acetolysis when such sugars are present [49].

The monosaccharides are then reduced (with $NaBH_4$ or $NaBD_4$) and peracetylated to produce partially methylated alditol acetates (PMAAs). This method eliminates the stereochemical multiplicity associated with the reducing forms of sugars, yielding only one derivative for each sugar and linkage type in the original carbohydrate (see Scheme 1). It has to be kept in mind that sugars adopting the pyranoside form in the original sample will always have *O*-Ac groups at C-5 and C-1 in their PMAA products, while furanosides will end up with *O*-Ac groups at C-4 and C-1. It is thus impossible to distinguish by this method alone a pyranoside with a substituent linked at C-4 from a furanoside with a C-5 substituent. Additional information is required to discriminate these cases. Mammalian glycoconjugates so far appear to incorporate pyranosides exclusively. But in dealing with carbohydrates from other organisms, such as bacteria, this can't be assumed. Note that linkage analysis normally does not allow deduction of the anomeric configuration of sugars in a polysaccharide. Whether or not the derivatization scheme includes a reduction step, this information is lost upon acid-catalyzed depolymerization, due to the possibility for epimerization at C-1. Some exceptions to this rule occur when employing the method of reductive cleavage (see Section F), where, in some cases, indirect evidence can be obtained from the differential susceptibility of α- and β-anomeric forms under certain depolymerization conditions. Finally, it is worth emphasizing that no information concerning sugar sequence can be derived from methylation linkage analysis procedures having only monosaccharide derivatives as the end product, except for identification of nonreducing terminal residues and, in the case of a reduced oligosaccharide, the reducing end residue.

Derivatives of KDO and sialic acids are not obtained using the PMAA scheme described above. However, specialized conditions permitting linkage analysis of oligosaccharides containing 3-deoxy-D-*manno*-2-octulosonic acid (KDO) residues as partially methylated 3-deoxyoctitol acetates have been established. Analysis of these derivatives will also be discussed separately (Section D). Sialic acids, on the other hand, are normally converted by acidic methanolysis to partially methylated methyl ketoside methyl esters, which can be analyzed by GC-MS as their *O*-trimethylsilyl or *O*-acetyl derivatives (see Section E).

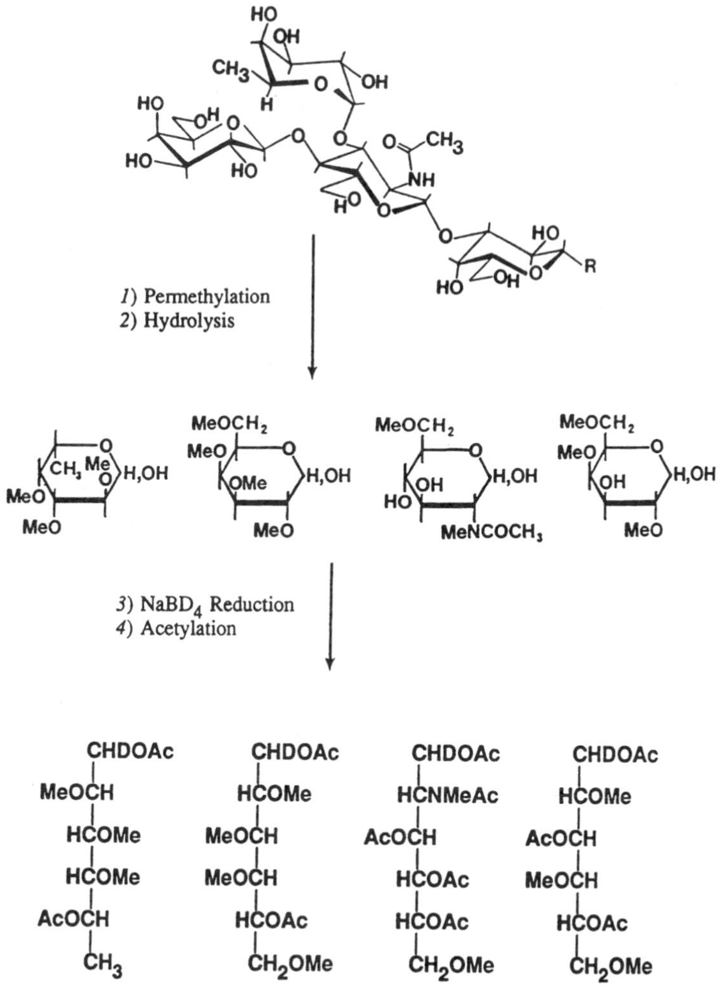

1) Permethylation
2) Hydrolysis

3) NaBD$_4$ Reduction
4) Acetylation

Scheme 1 Stepwise conversion of Galβ1→4(Fucα1→3)GlcNAcβ1→3Galβ1→R to partially methylated alditol acetates.

Over the years, a large number of protocols for producing and analyzing PMAAs have been published. These have been prompted by the variety of oligosaccharide and glycoconjugate types as well as by the desire to apply new technological developments as they become available [24,25,29,30,33–37,66–70; for microscale adaptations, see 4,38–44,49]. Aside from the issue of depolymerization, improving the acetylation step has also received some attention. The acetylation has generally been performed by heating with acetic anhydride, with catalysis by either pyridine [4,30,33,35,42,43,66,67] or sodium acetate (present following destruction of borohydride by acetic acid) [38,39,41,65]. With these reagents, one must be careful to remove completely all residual borate following the reduction step, because complexation of sugar hydroxyl groups by borate interferes with the acetylation.

This is usually accomplished by repeated addition and evaporation of methanol acidified with a bit of HAc or HCl, a procedure that removes borate as its volatile methyl ester. The possibility of using a borate-specific resin [71] for this step has not yet seen practical application in methylation analysis. Removal of borate may be unnecessary if 1-methylimidazole is used as catalyst [36,72–75]; however, as with pyridine catalysis, the appearance of artifact peaks in the GC-MS analysis may be a problem when using 1-methylimidazole [75]. It has also been found that not all partially methylated alditols are completely acetylated with this catalyst [25]. It was suggested that perchloric acid catalysis provided more complete acetylation [25]. Another recent improved procedure employed acetylation at room temperature using 4-dimethylaminopyridine as catalyst, with a considerable reduction in artifact formation [49].

In early attempts at analysis of complex carbohydrates, derivatization and GC-MS detection of 2-acetamido-2-deoxyhexoses were problematic. Difficulties with these materials were eventually overcome by a number of laboratories [35,70,76–78]. The special case of 2-acetamido-2-deoxy-hexitols obtained in linkage analysis of glycoprotein glycans that have been reduced prior to methylation (such as mucin structures isolated by NaOH-NaBH$_4$ treatment) was studied by Hase and Rietschel [79] and by Finne and Rauvala [80]. The formation of by-products containing putative N-acetylacetamido groups was noted when an acetolysis procedure was used for depolymerization, attributable to de-N-methylation prior to the acetylation step [79]. The extent of formation appeared to be dependent on the substitution pattern. For this reason, a methanolysis procedure was recommended [80]. Later, it was shown by Caroff and Szabo [81] that the by-products are in fact the result of de-O-methylation, primarily at C-1 (a more reasonable finding). This was confirmed by Nilsson [82], who also found small, but significant amounts of de-O-acetylation at C-3, as expected but not detected by Caroff and Szabo [81]. Therefore, the recommendation for methanolysis in this case appears to stand [82]. Unfortunately, if PMAA analysis is desired, this means a separate treatment for the reducing sugar and for the remainder of the saccharide. However, it should be noted that in all of these studies, quite harsh conditions for acetolysis were employed, involving either elevated temperatures (100°C) or extended reaction times (16 hours or more) or both. It might be wondered if the question ought not to be reexamined using milder temperatures and shorter times for acetolysis. In particular, it should be noted that only the product of de-O-methylation at C-3 causes real confusion as to linkage structure. In the microanalysis procedure of Geyer and Geyer [43], acetolysis was employed, with de-O-methylation products appearing in only trace amounts for a reduced N-glycan. In this case, detection of the partially methylated reduced amino-sugar seemed to be influenced to a far greater extent by the occurrence of selective degradation of amino-sugar derivatives on the GC column [43].

A novel cleanup procedure in which PMAAs of hexosamines were separated from those of hexoses was reported by Lowe and Nilsson [83]. This protocol was designed particularly to enable linkage analysis to be obtained directly from intact glycoprotein samples, without the necessity of first isolating their glycopeptides or free oligosaccharides. The cleanup procedure, involving stepwise elution from a small silica gel column, eliminated major interferences from peptide-derived contaminants in the GC analysis of the PMAAs.

So far, the most sensitive overall results appear to have been achieved by Geyer and Geyer [43], using a protocol incorporating many sample-handling refinements and technical advances developed since the early 1980s: micromethylation with lithium methylsulfinyl carbanion (see Section II), gas chromatography on bonded-phase fused silica capillary columns (Section B) with sample introduction using an all-glass moving needle injector, and detection of PMAAs with ammonia chemical ionization mass spectrometry and selected ion monitoring (Section C). Detection of PMAAs from as little as 10 ng of a reduced glycoprotein glycan was possible, with qualitatively useful results acquired with 50 ng of sample.

An interesting variant of the PMAA procedure has been employed for sequence analysis of polysaccharides [1,84]. In this procedure, a polysaccharide is first permethylated, then *partially* depolymerized by formolysis or trifluoroacetolysis (uronic acid–containing polymers are reduced by $LiAlD_4$ after permethylation). The resulting mixture of oligosaccharide fragments is then reduced with $NaBD_4$ (to alditols) and *perethylated*. The mixture of partially methylated, partially ethylated oligoalditol fragments is then fractionated by HPLC, with subsequent analysis of fractions by on-line (CI)-MS, GC-MS, direct probe EI-MS, or proton NMR. The analysis is completed by further depolymerization, reduction, and acetylation to partially methylated, partially ethylated, acetylated monosaccharide alditols (PMPEAAs), which are subjected to GC-MS for linkage analysis. Note that this procedure not only provides sequence information but also enables one to distinguish between furanosyl and pyranosyl residues, since the second alkylation clearly establishes at the reducing and nonreducing ends of each oligoalditol fragment which carbons are or were in glycosyl linkages and which were or are linked within a ring. The ring-form information can be obtained from GC-MS of the final mixture of PMPEAAs even if the intermediate HPLC fractionation and analysis of the partially methylated, partially ethylated oligoalditol fragments is omitted [1,85]. Although the method is in principle generally applicable, it may not be suitable in practice for samples available in very limited amounts, since it requires not only additional hydrolysis, reduction, and alkylation steps, but that optimal conditions for the partial hydrolysis be established by prior study. These conditions can't be assumed to be invariant from sample to sample [84].

A number of viable alternatives to the PMAA scheme have been developed, including methods in which no borohydride or borodeuteride reduction step is employed, yielding methylated monosaccharide derivatives in their ring forms, either as free reducing sugars following hydrolysis [86] or as methyl glycosides following methanolysis. Although many early separation experiments were conducted on partially methylated methyl glycosides without further derivatization (see, for example, Wallenfels et al. [14]; early papers of Fournet and Montreuil also utilize this technique [87,88]), partially methylated monosaccharide derivatives are generally either per-*O*-trimethylsilylated or *O*-acetylated prior to analysis by GC-MS [86–95]. The production of multiple peaks by these methods can be considered either an advantage (more than one peak for identification) or disadvantage (complex chromatograms). A potential advantage of this approach, as demonstrated by Fournet et al. [95] for a sialoglycopeptide, is the concurrent detection of the methylated derivatives of sialic acid, which are normally destroyed in the PMAA protocol (see Section E). Another method utilized methanolysis, followed by hydrolysis and reduction, with per-*O*-trimethylsilylation of the resulting alditols [96,97] (these references describe GC analysis only). A further alternative to the PMAA scheme is conversion of the partially methylated monosaccharides to aldonitrile derivatives by reaction with hydroxylamine followed by dehydration. The 1-carbon is now included in a nitrile function. Following acetylation, the partially methylated aldonitriles can be analyzed by GC-MS [98–100]. These alternative methods will not be discussed in detail here (for summaries of GC retention characteristics and EI fragmentation patterns of partially methylated, partially acetylated, methyl glycoside derivatives, see, for example, Refs. 40 and 95).

In a rather different approach pioneered by Gray [101], depolymerization is accomplished by a reductive cleavage step, using triethylsilane as the reducing agent in the presence of an acid catalyst such as boron trifluoride etherate, trimethylsilyl trifluoromethanesulfonate, or trimethylsilyl methanesulfonate and boron trifluoride etherate [102–104]. This procedure produces stable monosaccharide derivatives in which the stereochemistry at C-1 has been washed out by formation of 1,5- and 1,4-anhydroalditols (for pyranose and furanose forms, respectively). The ring form is thus preserved relative to the structure in the original polysaccharide. Developments in this pioneering technique will be discussed separately (Section F).

B. Gas Chromatographic Analysis of Partially Methylated Alditol Acetates

The first practical GC separation of partially methylated pentitol and hexitol acetates was reported by Björndal et al. [66]. In the years following, a large body of retention-time data was systematically compiled for pentitol, hexitol, and deoxyhexitol PMAAs [29,30,33,105,106] on various stationary phases (ECNSS-M, OV-17, OV-225, OS-138) in wide-bore packed glass (support-coated open tubular; SCOT) columns (for a more complete review of PMAA separations prior to 1974, see Dutton [69]). Jansson et al. [33] reported greatly improved resolution with a relatively nonpolar SP-1000 glass capillary (wall-coated open tubular; WCOT) column. Effective separations of partially O-methylated 2-deoxy-2-N-methylacetamidohexitol acetates were reported [35,70,76,77,78,80] on wide-bore SCOT columns (ECNSS-M, OV-17, OV-210, OV-225). Further continuous improvements in GC separation of PMAAs using glass capillary SCOT (e.g., Silar 10C) columns [74], glass and fused silica WCOT (OV-101, OV-225, OV-275, SE-30, SE-54, SP-1000, SP-2100, CP-Sil88, Silar 9CP, Silar 10C, Dexsil 410) columns [38,83,84,107–113], and then bonded-phase fused silica capillary (DB-1 [BP-1], DB-5, DB-210, DB-225, BP-75) columns, were reported throughout the 1980s [38,39,41,109,111,114–122] (see in addition the numerous examples compiled in Ref. 43; note that in Ref. 116, the methods section mistakenly refers to a column of OV-101, but the GC-MS figure clearly identifies the column as DB-5; in Ref. 109, the reference to a DB-5 column "coated with OV-101" is clearly in error—the column was simply DB-5). For further discussion of GC columns and an extensive compilation of retention data, see Carpita and Shea [36].

Despite the excellent separations of PMAAs achieved with modern gas chromatographic instrumentation, unambiguous identifications are best made by combination of GC with mass spectrometric analysis, as recommended [29,30,33,67]. Mass spectrometry can be particularly helpful in identifying PMAAs not previously encountered [67].

C. Identification of Partially Methylated Alditol Acetates by Mass Spectrometry

Under electron impact ionization, the fragmentation of partially methylated alditol acetates is orderly, predictable, and easy to interpret [67,123–125; reviewed, 4,29,30,33,36,42,126,127]. Molecular ions are generally not observed. The base peak is usually found at m/z 43 ($[CH_3-C\equiv O]^+$). Primary fragments are formed by cleavage between carbon atoms of the alditol chain. Preferences for cleavage between pairs of carbon atoms are in the order: both methoxylated > one methoxylated + one acetoxylated > both acetoxylated. The charge is retained preferentially on the fragment(s) possessing the methoxyl group. The presence of a deoxy group will inhibit cleavage from adjacent carbon atoms, especially from one that is acetoxylated. Secondary fragments are formed by one or more neutral losses of methanol (—32 a.m.u.), acetic acid (—60), ketene (—42), and, in some cases, formaldehyde (—30). Mass spectra of partially methylated alditol acetates derived from 2-acetamido-2-deoxyhexopyranose residues (O-methyl-2-N-methylacetamido-2-deoxyhexitol acetates) are distinguished by a characteristic primary fragment at m/z 158 made up of C-1 and C-2 bearing the N-methylacetamido group, and a highly abundant secondary ion at m/z 116 from elimination of ketene [35,70,76–78] (preliminary observations were reported in reviews by Björndal et al. [29] and Lindberg [30]). These fragmentation rules are summarized with examples in Schemes 2–4. In the special case of O-methyl-2-N-methylacetamido-2-deoxyhexitol acetates derived from prereduced glycoprotein oligosaccharides, which have a methoxy group at C-1, the characteristic primary C-1—C-2 fragment is found at m/z 130, with the abundant secondary fragment occurring at m/z 88 [79,80]. The masses of fragments including this end of the molecule are adjusted accordingly.

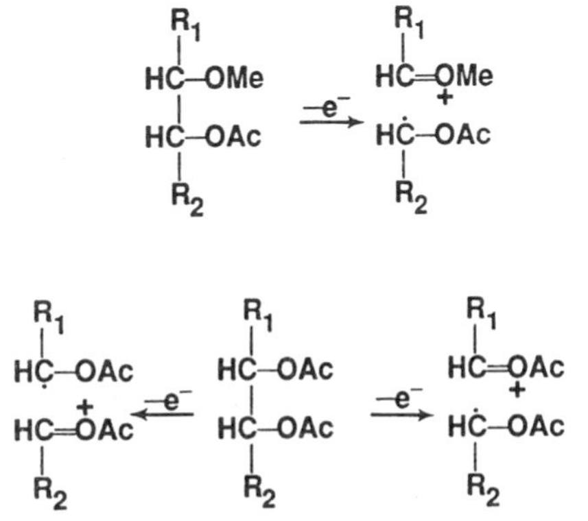

Scheme 2　Origin of primary fragments of partially methylated alditol acetates under electron impact ionization. (Adapted from Ref. 33.)

CH₂OAc   CH₂
$$\text{CH}_2\text{OAc} \quad\quad \text{CH}_2$$
$$\text{HC}-\text{OMe} \quad\xrightarrow{-\text{AcOH}}\quad \text{HC}-\text{OMe}$$
$$\overset{+}{\text{HC}}=\text{OMe} \quad\quad \overset{+}{\text{HC}}=\text{OMe}$$
$$m/z\ 161 \quad\quad\quad m/z\ 101$$

$$\text{H}\overset{+}{\text{C}}=\text{OMe} \quad\quad \text{H}\overset{+}{\text{C}}=\text{OMe}$$
$$\text{HC}-\text{OAc} \quad\xrightarrow{-\text{AcOH}}\quad \text{CH}$$
$$\text{CH}_2\text{OMe} \quad\quad \text{HCOMe}$$
$$m/z\ 161 \quad\quad\quad m/z\ 101$$

$$\xrightarrow{-\text{MeOH}}$$

$$\text{H}\overset{+}{\text{C}}=\text{OMe} \quad\quad \text{H}\overset{+}{\text{C}}=\text{OMe}$$
$$\text{C}-\text{OAc} \quad\xrightarrow{-\text{CH}_2\text{CO}}\quad \text{C}=\text{O}$$
$$\text{CH}_2 \quad\quad\quad \text{CH}_3$$
$$m/z\ 129 \quad\quad\quad m/z\ 87$$

Scheme 3 Origin of secondary fragments of partially methylated alditol acetates under electron impact ionization. (Adapted from Ref. 33.)

Due to the magnitude of energy deposition associated with electron impact ionization, it is not possible to distinguish stereoisomers by differences in their mass spectra. Mass spectra are characteristic for substitution patterns only. Fortunately, they can in most instances be distinguished by their GC retention times [29,30,33,35,70,78]. It was realized from the beginning [29,30,66] that certain pairs

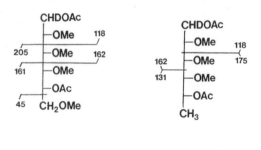

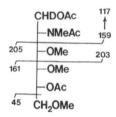

Scheme 4 Characteristic fragmentation of 2,3,4,6-tetra-O-methyl-hexitol acetate, 2,3,4-tri-O-methyl-6-deoxyhexitol acetate, and 3,4,6-tri-O-methyl-N-methyl-2-acetamido-2-deoxyhexitol acetate under electron impact ionization.

of methylated sugars could yield alditols with the same substitution pattern (for example, a 3- and a 4-O-methyl hexose), and in some cases these could be identical, or enantiomorphic, and therefore inseparable by GC. Such ambiguities are resolved mass spectrometrically if the reduction step is carried out with borodeuteride, which unambiguously labels C-1 of the alditol [29,30,67,123].

Applications of CI-MS to detection of PMAAs were examined starting in the mid- to late 1970s [128–133] and continued throughout the 1980s and early '90s [38,41,43,108,109,113,115–120,134–137]. Under methane or isobutane CI conditions, hexose and deoxyhexose PMAA spectra are dominated by MH^+, $[MH - 32]^+$, and $[MH - 60]^+$, with the latter two considerably more abundant with methane CI and the former more abundant with isobutane. For 2-deoxy-2-N-methylacetamido-derived PMAAs, the spectra in methane or isobutane CI are dominated by MH^+ and $[MH - 60]^+$ ions. Adducts with $C_2H_5^+$ and $C_3H_5^+$ ($[M + 29]^+$ and $[M + 41]^+$) when using methane, and with $C_3H_3^+$ ($[MH + 39]^+$) when using isobutane, are noticeable for all types of PMAAs under these conditions. The relative intensities of the major ions have been shown to be sensitive to reagent gas pressure, source temperature, and, to a limited but useful extent, substitution pattern and configuration [129,135] (see, for example, Table 1; Figure 1). Helium CI produces pseudo-EI (charge-transfer) spectra [132]. The introduction of small amounts of helium, either in the form of GC eluent or as part of a reagent gas mixture, produces spectra with mixed CI and EI character [132,134] (see Figure 2). The production of fragments similar to those in EI spectra can be useful in the identification process, but must be traded for a concomitant loss in intensity of molecular weight related ions, which are the most useful for selected ion monitoring (see below).

Ammonia CI detection of PMAAs was introduced by Geyer et al. [38] and has been used extensively in the characterization of glycoprotein glycans, glycopeptides, and glycolipids (see numerous references compiled in Ref. 43). Ammonia CI spectra of PMAAs are characterized by abundant $[M + 18]^+$ (ammonia adduct), accompanied by a small (usually less than 10%) $[MH - 60]^+$, for hexose derivatives; and abundant MH^+ and $[M + 18]^+$ pairs for hexosamine derivatives [38,43].

In practical operation, the different PMAAs can be divided conveniently into groups by molecular weight and retention time, with monodeoxy hexose and pentose derivatives eluting first, followed by hexose derivatives and finally hexosamine derivatives. Within these groups, increasingly substituted sugars, having increasing numbers of acetyl versus methyl groups attached, are separated by step increases in

Table 1 Relative Intensities of the M + 1, M + 1 − 32, and M + 1 − 60 Ions Obtained from 2,3,4,6-Tetra-O-methyl Hexitol Acetates (MW = 323 D) Under Isobutane CI Conditions

	M + 1 m/z 324	M + 1 − 32 m/z 292	M + 1 − 60 m/z 264
Gal	100	17 (5)	87 (18)
Glc	9.0 (1.8)	1.3 (0.4)	100
Man	2.6 (0.9)	0.9 (0.2)	100

Excerpted from Ref. 129. Numbers in parentheses are standard deviations over 5–7 spectra.

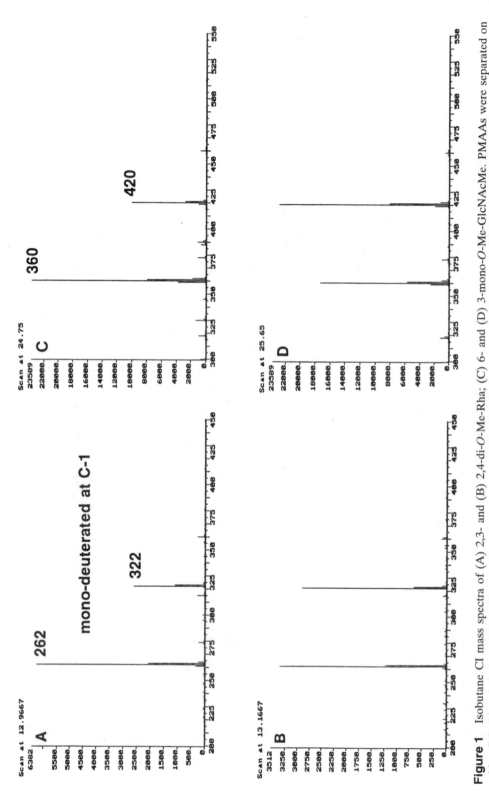

Figure 1 Isobutane CI mass spectra of (A) 2,3- and (B) 2,4-di-*O*-Me-Rha; (C) 6- and (D) 3-mono-*O*-Me-GlcNAcMe. PMAAs were separated on a 30 m DB-5 bonded-phase fused silica capillary column, temperature programed from 140°–250°C at 4°/min. Analyses were performed on a Hewlett-Packard 5890 Series II gas chromatograph interfaced to an Extrel ELQ 400 mass spectrometer, mass range *m/z* 50–500 scanned 60/min. (From Ref. 135.)

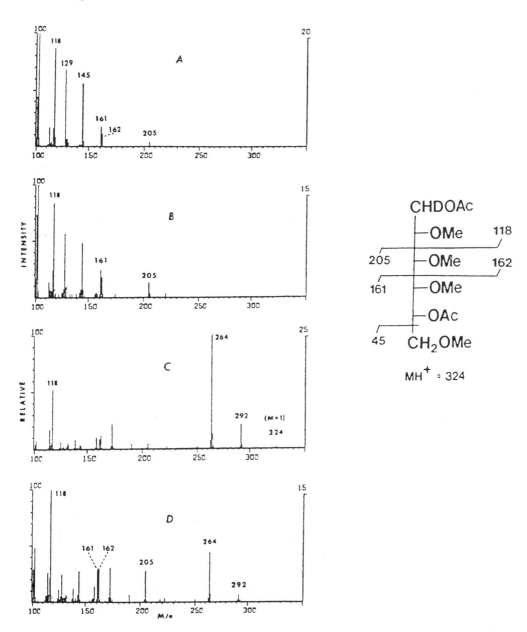

Figure 2 Mass spectrum of 2,3,4,6-tetra-*O*-methyl hexitol acetate (MW = 323 D), representing a nonreducing terminal residue, under different mass spectrometric conditions: (A) EI; (B) CI (He, 0.2 torr); (C) CI (methane, 1 torr); (D) CI (methane, 0.8 torr + He, 0.2 torr). (From Ref. 132.)

molecular weight of 28 a.m.u., and are characterized by generally increasing retention times. In methane and isobutane CI, there is a convenient overlap in mass between $[MH - 32]^+$ of one group and $[MH - 60]^+$ of the next-heaviest group. PMAA analysis by CI is ideally suited to mass chromatography (enhancement of total ion current chromatograms by reconstruction with selected ions) [130–133] (see also Refs. 113 and 115–118) and especially to selected ion monitoring (SIM; also called mass fragmentography), with considerable advantage in selectivity and sensitivity, as suggested by Laine [130,132,133] and successfully demonstrated [38,41,43 (and references cited therein), 119,120,136,137]. Superiority in sensitivity over EI detection has been clearly demonstrated [43]; the claim of an advantage of ammonia over methane or isobutane CI is probably also correct, due to the relative lack of fragmentation in the former [43].

The loss of information obtained from characteristic fragment ions should be compensated for by the use of carefully prepared groups of characterized PMAA standards, to be injected either before or after unknown samples or, in certain cases, coinjected. Analysis on one column is often sufficient for simple glycoconjugates; analysis on two columns (e.g., DB-5 and DB-225 [41]; DB-1 and DB-210 [43]) is generally recommended to remove any ambiguities in identification due to retention-time overlap [110,111]. The ratios of the major ions in isobutane CI spectra can also be of use in distinguishing PMAA isomers, as demonstrated for hexose derivatives by McNeil and Albersheim [129]. This approach may also work for hexosamine derivatives [135].

More recently, the use of an ion-trap type of detector has been recommended [4,42] for PMAA analysis with excellent sensitivity. Interestingly, some molecular weight related ions are detected in the EI spectra, due to CI processes taking place in the trap, while abundant EI fragments are retained in the methane CI spectra [42], similar to observations by Laine [132,134] using a He/methane CI gas mixture.

D. Linkage Analysis of Oligosaccharides Containing KDO Residues

Linkage analysis of KDO residues is complicated by the variety of functional groups found on adjacent carbons (1-carboxyl, 2-keto, 3-deoxy, 4-8 hydroxy). Thus, modifications to the procedures outlined in the previous sections (A–C) have to be made for carbohydrates containing KDO residues, such as bacterial lipopoly- and lipooligosaccharides (LPS and LOS), certain acidic bacterial capsular polysaccharides, and some plant cell wall constituents.

Linkages to a reducing KDO can be determined by a sequence consisting of reduction (of the reducing-end C-2 keto group) with sodium borohydride or borodeuteride, permethylation, reduction (of the newly formed carboxylic methyl ester at C-1), hydrolysis, borohydride reduction, and peracetylation [138–141]. The resulting partially methylated, partially acetylated 3-deoxyoctitol(s) can be analyzed by GC-MS, with interpretation of the EI fragmentation pattern according to the rules already outlined. In this case, primary fragments are observed from cleavage between C-3 deoxy/C-2 methoxy pairs but not between C-3 deoxy/C-4 methoxy pairs [138–141]. Note that the nonstereospecific reduction of the keto group yields two diastereomeric products, having the D-*glycero*-D-*talo*- and D-*glycero*-D-*galacto*- configurations. Using this approach, Prehm et al. [138] established the existence of a Hep1→5KDO linkage in a fragment of LPS from *E. coli* B by detection of a 1,5-

di-*O*-acetylated product. Although their GC method (ECNSS-M on a packed SCOT column) did not resolve the diastereomeric pair of products, this was not essential to the conclusion. York et al. [139] used a similar approach to elucidate the structure of a Rha1→5KDO disaccharide fragment released from plant cell wall rhamnogalacturonans. In this work, the authors separated the 1,5-di-*O*-acetyl-2,4,6,7,8-penta-*O*-methyl-3-deoxyoctitol diastereomers into two peaks (DB-1 bonded-phase fused silica capillary); in addition, a potential source of ambiguity in their EI fragmentation was resolved by synthesis and GC-MS analysis of the 1,4-di-*O*-acetylated pair of isomers, having different retention times and EI fragmentation; finally, the authors obtained isobutane CI spectra for both pairs of isomers, observing characteristic $[MH]^+$, $[MH - 32]^+$, $[MH - 60]^+$, and $[M + 39]^+$ ions as expected [129].

Prehm et al. [138] used a modification of this sequence to analyze linkages to internal KDO residues (e.g., in an LPS core oligosaccharide from mutant *E. coli* BB12, having two KDO residues only linked to lipid A): After permethylation, the product was carboxyl-reduced with calcium borohydride in THF and hydrolyzed, and the products of hydrolysis were carbonyl-reduced with sodium borohydride, acetylated, and analyzed by GC-MS. Two 3-deoxyoctitol derivatives were detected, corresponding to terminal and 4-linked KDO residues, establishing the presence of a KDO2→4KDO→lipid A sequence in the LPS. Later, however, it was claimed [140] that the procedure could not be satisfactorily reproduced. In order to fill a clear need for a systematic approach to KDO linkage analysis, Tacken et al. [140] developed new protocols, incorporating a number of refinements at critical points in the reaction sequence, as well as synthesizing an extensive set of partially methylated, partially acetylated 3-deoxyoctitol standards with defined structures, obtaining for each compound electron impact and ammonia chemical ionization spectrometric data, along with retention times (SE-54 bonded-phase fused silica capillary column) for all of the D-*glycero*-D-*talo*- and D-*glycero*-D-*galacto*- diastereomers. Derivatives having an *O*-methyl group at C-1 were synthesized by interpolating a second methylation step after carboxyl reduction, while derivatives having an *O*-acetyl group at C-1 resulted from omitting this step. These protocols and standards were then used in the structural analysis of the inner core region of *O*-deacylated LPS (LPS-OH) from a number of *Salmonella minnesota* rough mutants [141]. The authors also used a modified two-step hydrolysis procedure: Carboxyl-reduced and permethylated LPS-OH yielded an intermediate disaccharide fragment after a mild first hydrolysis, carbonyl reduction, and acetylation, which could be analyzed by GC-EI- and -CI-MS; after trideuteriomethylation, a second hydrolysis, carbonyl reduction, and acetylation produced monosaccharide derivatives that were then analyzed further by GC-MS and compared with synthetic standards. In a third procedure, LPS was partially hydrolyzed, and the resulting mono-, di-, and trisaccharides carbonyl reduced, methylated, and analyzed by GC-MS (see Section V). The data, taken together, were used to confirm a tetrasaccharide core structure, produced by one mutant, consisting of a KDO2→4KDO2→4KDO2→ homotrimer substituted by an L-glycero-D-manno-heptopyranosyl unit on O-5 of the reducing KDO [141].

E.   Analysis of Partially Methylated Neuraminic Acids by GC-MS

As mentioned previously, derivatives of sialic acids are not obtained using the PMAA derivatization scheme. Their partially methylated methyl ketosides obtained by methanolysis can be analyzed by GC-MS as O-trimethylsilyl or O-acetyl derivatives. The gas chromatographic retention times and EI mass spectra of several partially methylated derivatives of the methyl ester methyl ketoside of N-acetyl neuraminic acid have been reported [142–146]. A number of these publications have compared the properties of both O-acetyl and O-trimethylsilyl derivatives [142,143,145]. An extensive GC-EIMS study of the O-trimethylsilyl derivatives was reported by Bruvier et al. [146]. The O-acetyl derivatives are conveniently prepared (by acetylation with acetic anhydride-pyridine) and analyzed, and the method is easily extended to analysis of N-glycolyl neuraminic acid at the nonreducing terminal position [144]. It has been found that while fully methylated N-acyl neuraminic acids are stable to methanolysis, derivatives bearing a free hydroxy group in the 8-position of the glyceryl side chain are quickly de-N-acylated, in common with underivatized neuraminic acids [147–149]. This would mean that such derivatives will always end up N-acetylated, which creates no confusion if they were originally N-acetylated, but is misleading if they were N-glycolylated. This is a problem because sialic acids are frequently linked to each other as α-(2→8) ketosides. The only answer at this time appears to be the use of extremely mild conditions for methanolysis (0.05 N HCl, 80°C, 1 h [150]) when N-glycolyl neuraminic acids are suspected to be present. Unfortunately, the cleavage products are not obtained quantitatively under this set of conditions. The subject has been studied extensively by Inoue and Matsumura, who have chosen ammonia CI-GC-MS for characterizing an array of derivatives from synthetic and natural sources [149], particularly polysialoglycoproteins isolated from fish eggs [151–153]. Ammonia CI spectra of these derivatives are characterized by MH^+ (base peak), accompanied by losses of one and two molecules of MeOH ([MH − 32]$^+$ and [MH − 64]$^+$), with [MH − 15]$^+$ observed at modest abundance (4–32%), and occasional detection of the ammonia adduct [M + 18]$^+$ [149]. Methane CI spectra for the terminal and 8-O-substituted NeuAc derivatives have also been published [117]. Again, characteristic fragments appear to be MH^+, [MH − 32]$^+$ and [MH − 64]$^+$ (base peak), with smaller contributions of [M + 29]$^+$ and [M + 41]$^+$ adduct ions (see Figure 3). These conditions are fairly well suited to SIM, with detection in the pmol range easily attainable [41]. It is helpful that the β-ketoside is formed almost exclusively during methanolysis (≈95%). An advantage of the O-acetylated over the O-trimethylsilylated derivatives is their exceptional stability; samples can be stored for years at −20°C with no noticeable degradation.

F. Linkage Analysis by Reductive Cleavage

Reductive cleavage was introduced in 1982 by Rolf and Gray [101] to deal with some of the weaknesses in the standard linkage analysis strategy, notably the lack of a facile method for distinguishing ring forms. As mentioned previously (Section A), a key feature of reductive cleavage is the preservation of ring configurations of monosaccharide residues in the form of partially methylated 1,5- and 1,4-anhydroalditols. A second potentially desirable feature is that by choosing different reagents for depolymerization, either total or selective partial cleavage may be achieved, yielding small, sequenceable oligomeric products along with inferences about the anomeric configurations and positions of both the cleaved and uncleaved glycosidic linkages. However, under certain circumstances it is conceivable that the lack of a completely general catalyst capable of cleaving every linkage could be considered a

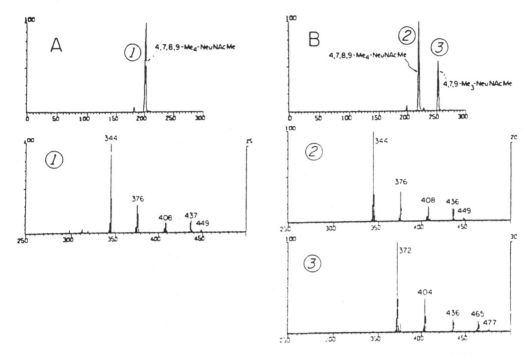

Figure 3 Limited mass chromatograms and methane CI (300 μ) mass spectra of fully or partially methylated sialic acid methyl ester methyl glycosides, following methanolysis and acetylation. Derivatives from (A) bovine brain GD$_{1a}$ (NeuAcα2→3Galβ1→3GalNAcβ1→ 4[NeuAcα2→3]Galβ1→4Glc1→1Cer); and (B) GD$_3$ (NeuAcα2→8NeuAcα2→3Galβ1→ 4Glc1→1Cer) from human melanoma. Derivatives were separated on a DB-5 column (split-less injection; temperature program, 50°–230°C in 5′40″, followed by 230°C isothermal). The mass range m/z 40–500 was scanned at the rate of 30/min on a Finnigan 3300/6110 GC-MS. The figure is a summation of ion chromatograms: *abscissa*, scan number; *ordinate*, sum of ion intensities at m/z 408, 376, 344, and 436, 404, 372, which represent MH$^+$, [MH − MeOH]$^+$, and [MH − 2MeOH]$^+$ for the 4,7,8,9-tetra-O-Me and 4,7,9-tri-O-Me derivatives, respectively. M/z pairs 436 and 448, 464 and 476, represent [M + C$_2$H$_5$]$^+$ and [M + C$_3$H$_5$]$^+$ adduct ions for the 4,7,8,9-tetra-O-Me and 4,7,9-tri-O-Me derivatives, respectively. (From Ref. 117.)

disadvantage. Furthermore, in some cases reagents have failed to yield the expected products (see below), indicating that considerably more work needs to be done before the chemistry has been defined adequately enough for confident use on unknown, complex glycoconjugates.

Developments in the reductive cleavage method prior to 1992, including the application to a sizable number of polysaccharide model systems, have been reviewed [102–104,126]. Mass spectra of anhydroalditols, under both CI and EI conditions, have been discussed by Gray [103] and Wait [126]. Fragmentation patterns of partially methylated, partially acetylated anhydroalditol derivatives under EI conditions appear to be similar to those of partially methylated, partially acetylated methyl glycosides [126,154], and will not be discussed further in this section. The emphasis here will be on newer developments not already reviewed elsewhere, with selected examples from previous work provided for background, particularly where these relate to problems addressed in the more recent publications.

In the original version of the method [101], depolymerization was achieved with a mixture of boron trifluoride etherate and trifluoroacetic acid in triethylsilane. However, the use of trifluoroacetic acid was considered optional, and in later studies was abandoned by Gray, although retained by others [154,155]. In addition to the use of boron trifluoride etherate as catalyst [102], cleavage is commonly promoted by a 5:1 mixture of trimethylsilyl methanesulfonate and boron trifluoride etherate [103,156] or by trimethylsilyl trifluoromethanesulfonate [102,157], in each case with triethylsilane as reducing agent. In the latter case, the subsequent acetylation step can be performed in situ prior to workup for GC-MS. Of these reagents, boron trifluoride etherate is the most selective. For example, it was very ineffective in catalyzing the reductive cleavage of 1→6 linkages to D-mannopyranose residues in yeast mannans [102,158] and to D-glucopyranose residues in pullulan [102,157]; it effectively catalyzed the reductive cleavage of permethylated amylose but not of cellulose [102,157,159]. In experiments on the *Klebsiella* K2 capsular polysaccharide (→4[GlcAα1→3]Manβ1→4Glcα1→3Glcβ1→)$_n$, it cleaved both of the β1→4 linkages but neither of the α1→3 linkages [104,160]; this was expected in the case of the α-GlcA residue but not for the α-Glc [104].

A significant barrier to the adoption of reductive cleavage methodology for structural problems of mammalian complex-type glycoconjugates has probably been the relatively slow development of a general, effective way to deal with linkages involving aminosugars [104,154,155]. Van Langenhove and Reinhold [154] isolated methyl 4-O-acetyl-3,6-di-O-methyl-2-deoxy-2-(N-methylacetamido)-D-glucopyranoside from Man1→3Man1→4GlcNAc following permethylation, reductive cleavage, and acetylation, and Edge et al. [155] isolated the expected product from the reduced GalNAc of O-glycosidically linked oligosaccharides released from rat erythrocyte membrane sialoglycoproteins, but failed to detect any anhydroalditol or other products from internally linked β-GlcNAc residues. A step forward in this area was taken by Bennek et al. [161], who showed that the product of reductive cleavage of a permethylated 2-acetamido-2-deoxy-β-D-glucopyranosyl residue was the reducing sugar, due to formation of an intermediate oxazolinium ion. This intermediate was hydrolyzed during the subsequent workup and the product recovered in the aqueous layer, separate from the dichloromethane-soluble anhydroalditol products from neutral sugars. The α-anomer, on the other hand, was found to be stable to reductive cleavage [161], indicating the probable origin of the methyl glycoside product obtained from the reducing end GlcNAc of a trisaccharide [154] (any α-methyl glycoside product yielded upon permethylation of the free oligosaccharide would have survived intact the subsequent reductive cleavage into monosaccharide derivatives). In addition, it was found that reduction of the 2-N-methylacetamido group with lithium aluminum hydride rendered the β-anomer resistant to cleavage as well, leading to production of a disaccharide anhydroalditol from a complex-type glycan [161] (this stability is also observed under acetolysis conditions). While this work provided some measure of success, the necessity for separate analysis of neutral and acetamido sugar derivatives was admittedly awkward [104].

More recently, taking advantage of the known glycosyl-donor properties of oxazolinium ion derivatives of 2-acetamido sugars, a modified workup procedure was examined [162] that allowed for recovery of partially methylated acetamido sugar derivatives in the same organic phase as the anhydroalditol derivatives after acetylation. This was accomplished by quenching the reductive cleavage reaction with alcohol, producing, from the intermediate oxazolinium ions, alkyl glycosides instead of the free sugars. Model studies with methyl 3,4,6-tri-*O*-methyl-2-deoxy-2-(*N*-methylacetamido)-α,β-D-glucopyranoside, in which the intermediate oxazolinium ion was quenched with methanol-d_4, confirmed that the α-glycoside was completely resistant to reductive depolymerization under three different conditions, while the β-glycoside was quantitatively cleaved, as shown by both GC-CI-MS and NMR analysis of the products. Furthermore, quenching of the intermediate (from cleavage of the β-glycoside) with racemic 2-butanol produced two diastereomeric *sec*-butyl glycosides separable by GC, while an optically pure 2-butanol yielded a single diastereomeric product in 96% excess; the procedure thus facilitates determination of absolute configuration of an *N*-methylated, partially *O*-methylated, 2-deoxy-2-acetamido sugar residue directly after reductive cleavage.

These procedures were then applied to a polysaccharide having a trisaccharide repeating unit previously established as [→2)-β-D-Glc*p*-(1→3)-α-L-Fuc*p*NAc-(1→3)-β-D-Fuc*p*NAc-(1→]ₙ, providing a suitable test for the methodology [163]. While the analysis was ultimately successful, confirming both the structure of the polysaccharide and to great extent, the utility of the method, an unexpected lack of selectivity of cleavage with respect to the α-L- and β-D-Fuc*p*NAc residues in the fully methylated polysaccharide led to the initial conclusion that the original proposal for the structure was incorrect. Confirmation of the presence of the two forms of Fuc*p*NAc was instead provided by NMR study of the intact polysaccharide [163]. In addition, (*S*)-2-butanol showed significant kinetic selectivity in the quenching reaction, yielding the expected products, but in nonstoichiometric ratios. These results, while pointing to the need for further study of reaction conditions before reductive cleavage can be relied on exclusively as a means for elucidating *complete* structures, nevertheless provided more information than conventional linkage analysis. The reliability of the method can be expected to increase with greater experience.

The application of linkage analysis to sialosylated oligosaccharides is another area of particular concern for the structure elucidation of glycoprotein and glycolipid glycans. Following permethylation and reductive cleavage of reduced *O*-glycosidically linked oligosaccharides with the triethylsilane/boron trifluoride etherate/trifluoroacetic acid reagent, Edge et al. [155] were able to detect by GC-MS a pair of anhydroalditol products (α- and β-) of terminal NeuAc residues. The products of the residues to which they were linked were also observed. The NeuAc derivatives had mass spectra identical with those obtained from reductive cleavage of permethylated *N*-acetylneuraminic acid and neuraminyllactose. A more systematic study by Srivastava and Gray [164] showed that per-*N*,*O*-methylated-*N*-acetylneuraminic acid α- and β-methyl glycosides were resistant to reductive cleavage by the selective reagent combination of boron trifluoride etherate and borane dimethylsulfide (Me₂S·BH₃) but were cleaved almost quantitatively following borohydride reduction of the methyl ester group. The expected anhydroalditol derivatives (α- and β-) from permethylated, reduced NeuAc were detected by GC-MS following acetylation. The resistance of the unreduced glycosides to reductive cleavage was attributed to the electron-withdrawing effect of the C-1 methoxycarbonyl group. Uncharacterized artifacts related to NeuAc that were noticed when reductive cleavage was performed with the tri-ethylsilane/boron trifluoride etherate/trifluoroacetic acid reagent [155] were not observed under the conditions described by Srivastava and Gray [164]. The authors then applied the new reagent combination to a simple model oligosaccharide, 3'-sialyllactose (NeuAcα2→3Galβ1→4Glc). Both permethylated 3'-sialyllactose and its carboxyl-reduced derivative were treated with boron trifluoride etherate and borane dimethylsulfide (Me$_2$S·BH$_3$), and the products acetylated and analyzed by GC-MS. From the unreduced compound, the fully methylated terminal disaccharide anhydroalditol arising from cleavage of the Galβ1→4Glc linkage was observed, along with the anhydroalditol expected from a 4-linked glucopyranosyl residue. From the carboxyl-reduced compound, four products were obtained, one each from the 3-linked galactopyranose and the 4-linked glucopyranose residues, and two from the nonreducing terminal NeuAc residue (as observed by Edge et al. [155], both α- and β-anhydroalditol products were obtained). These studies demonstrated that simple sialosylated glycans can be analyzed by reductive cleavage; there seems to be little reason to believe that the method can't eventually be successfully applied to glycans of more complex type (containing both NeuAc and HexNAc residues and possibly polysialosyl units).

Limited compilations of retention times and mass spectra of synthetically prepared anhydroalditol standards were published up to 1990 [103,154,155,165,166]. More recently, Gray and co-workers have begun more systematic construction of a database of anhydroalditol chromatographic and spectral properties, with a view toward synthesis and characterization of all possible combinations of ring form and linkage position for the commonly encountered monosaccharide residues. So far, publications have dealt with derivatives of fucopyranosyl, galactopyranosyl, mannopyranosyl, glucopyranosyl, xylofuranosyl, fucofuranosyl, and ribofuranosyl residues [167–173]. In addition to electron impact mass spectra and GC retention times of all partially methylated anhydroalditols as their acetates, LC capacity factors and proton NMR data of all of these compounds as benzoates have been included. The GC-MS standard data compiled in this work have the advantage that each compound was fully characterized by NMR before debenzoylation and acetylation, so there can be no mistake about its identity.

With solutions to some of the problems with reductive cleavage more clearly in sight, it can be anticipated that the method will be more widely applied in the near future.

IV. CHARACTERIZATION OF GLYCOCONJUGATES CONTAINING SIALIC ACID INNER ESTERS

A. The Lactone Problem

Although the tendency of sialic acid residues to lactonize spontaneously and reversibly has not for some time been a matter of dispute, thirty years of conjecture and argument concerning the possible physiological significance of this reaction has given rise to a small field all its own, enjoying an almost cultlike status among interested glycobiologists. McGuire and Binkley first noted [174] evidence for intramolecular esterification of some of the NeuAc residues in colominic acid (a

[NeuAcα2→8]$_n$ homopolymer produced by *E. coli*) and proposed that these linkages occur between the C-1 carboxyl of one residue and C-9 of the residue to which it is glycosidically linked. In the wake of these observations, it became immediately apparent to such pioneering workers in the ganglioside field as Kuhn [175] and Wiegandt [176] that the same structural modification might be found in sialic acid–containing glycosphingolipids. This possibility was subsequently confirmed by McCluer and Evans [177,178], who demonstrated the formation of a different type of intramolecular ester, occurring between the sialic acid and the aglyconic residue in the terminal NeuAcα2→3Galβ1→ disaccharide of a simple ganglioside, G$_{M3}$ (some observations of NeuAcα2→8NeuAcα2→ lactone behavior in the disialoganglioside G$_{D3}$ are briefly alluded to in a paper by Wiegandt [179]). McCluer and Evans also showed that lactonization of gangliosides could be induced by treatment with glacial acetic acid at room temperature [178]. Using this method, lactones (inner esters) have been prepared from a number of gangliosides [180–184]. In addition, ganglioside lactones have been isolated and characterized following extraction from freshly prepared rat, mouse, and human brain tissue [185,186]. G$_{M3}$ lactone has been radiolabelled in situ on B16 melanoma cells [187], and monoclonal antibodies reacting with lactones of NeuAcα2→8NeuAcα2→3Galβ1→-containing gangliosides have been reported [188–190].

A recent study demonstrated a rapid pH-dependent equilibrium for lactone formation in aqueous media [191]. The apparent low-energy barrier has made it difficult to establish that glycoconjugates isolated with lactonized sialic acids actually represent compounds occurring in vivo, not artifacts introduced during tissue preparation, extraction, or purification. The same low barrier obviously applies to the reverse reaction, resulting in an extreme lability in alkaline media, with consequential difficulties in isolating lactonized glycoconjugates intact, as well as in the subsequent elucidation of their precise structures. Conventional linkage analysis employing base-catalyzed permethylation cannot be used to determine the site of esterification, nor can protection schemes that utilize acid catalysis, such as reaction with methyl vinyl ether [27] or dihydropyran [28], since in the former case any lactones present will be destroyed, and in the latter case they may form spontaneously or perhaps isomerize. These circumstances have hindered the debate over any putative role that such structures might play in cellular processes. As discussed in the next two sections, there has been a clear need to develop analytical methods to characterize unambiguously the structures of inner ester-containing glycoconjugates, and to establish that these are not artifacts promoted by the conditions of isolation.

Although all work so far published has dealt explicitly with the formation of inner esters either from gangliosides or from NeuAc-containing homopolymers, it should be apparent that any effective analytical methods should be applicable as well to glycoprotein glycans terminating in sialosyl and polysialosyl groups. For example, those of N-CAM are known [192,193] to be extended by [NeuAcα2→8]$_n$ NeuAcα2→3Galβ1→ structures of considerable length, and, therefore, with considerable potential for inner esterification. The polysialosyl glycoproteins found in the unfertilized eggs of Salmonidae fishes [151–153,194–196] constitute another well-characterized system in which lactonization could potentially play a biological role. (For a discussion of the occurrence, structure, and biological significance of polysaccharides and glycoconjugates containing polysialic acids, see the recent review by Troy [197]).

Until recently, the most convincing evidence for the location of esterification sites has come from 1- and 2-D ^{1}H-NMR spectra of synthetic ganglioside lactones, for which the best indications are that NeuAc residues attached to O-3 of Gal are lactonized to the C-2 hydroxyl of that sugar [180,181,183,184], while NeuAc residues attached to O-8 of another NeuAc are lactonized to the C-9 hydroxyl of that sugar [182,184]. In parallel with this, ^{13}C-NMR spectral data on internally esterified [NeuAcα2→8]$_n$ and [NeuAcα2→9]$_n$ homopolymers have been interpreted to indicate that lactonization is to the C-9 hydroxyl in the former and to the C-8 hydroxyl in the latter [198,199]. However, in all of these cases, acceptance of the interpretations must be based on the judgment that the chemical shifts of all relevant nuclei have been reliably assigned and that the observation of a characteristic downfield "lactonization shift" is a completely accurate indicator of primary structure. Given the known possibility of anomalous shift effects that can be caused by changes in saccharide secondary structure, these interpretations may not be completely unambiguous.

In addition to NMR spectroscopy, fast atom bombardment (FAB) mass spectrometry, in both positive- and negative-ion modes, has been used to characterize ganglioside lactones [181,182,186,200,201]. These methods give reliable information on saccharide sequences and on the presence of lactones in that sequence, but provide no indication of linkage structure. Following reduction of sialic acid esters using NaBH$_4$, or ammonolysis to produce sialic acid amides, the sialic acid derivatives released by methanolysis have been analyzed by high-performance liquid chromatography (HPLC) [185], by GC [202], and by GC-MS [181,182,183,186,203]. None of these methods provide information on linkage structure or sequence. As described in the next two sections, new GC-MS methods have been recently introduced [204,205] to assist in characterizing more precisely glycoconjugates containing internal esters of sialic acids.

B. Identification of Lactonized NeuAc Residues

Positive-ion fast atom bombardment mass spectra of permethylated gangliosides exhibit abundant fragments at intervals indicative of the presence of one or more sialic acid residues in the glycan sequence [206,207]. There is no difference for ganglioside lactones since base-catalyzed permethylation opens the internal ester ring to produce the same sialic acid methyl ester derivative. However, if the ganglioside lactone is ammonolyzed prior to permethylation, the sialic acid–containing fragments will exhibit a 13 amu mass shift (the difference between CONMe$_2$ and COOMe) for each residue that was previously esterified [190,201]. In analogy to existing methods of sialic acid linkage analysis by GC-MS (see Section III.E.), subsequent methanolysis and acetylation could be expected to yield volatile derivatives characteristic for the previously lactonized sialic acids, identify them as originating from internal or terminal residues, and, in the former case, allow one to determine the point of glycosylation. This has been shown to be the case for artificially prepared lactones of the disialogangliosides GD$_3$ and GD$_{1b}$ [205].

When disialogangliosides such as GD$_3$ and GD$_{1b}$, containing the sequence NeuAcα2→8NeuAcα2→3Galβ1→, are treated with glacial acetic acid and lyophilized [178], both singly charged and neutral products can be isolated, corresponding to the mono- and dilactone of each, respectively [182,184]. Following ammonolysis, permethylation, methanolysis, and acetylation (Scheme 5), GC-MS analysis showed

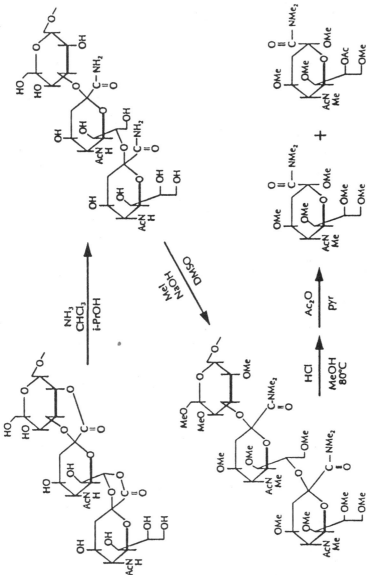

Scheme 5 Reaction sequence used for the production of NeuAc dimethylamide derivatives from GD$_3$-L2. (From Ref. 205.)

two major products from the dilactones (3β, 4β; Figure 4, Panel C) in place of those normally found in methylation analysis of the corresponding unlactonized gangliosides (1β, 2β; Figure 4, Panel A).

Mass spectral analysis of the peaks under both CI (Figure 5) and EI (Figure 6) conditions indicated that these new products corresponded to terminal and internal NeuAc residues in which the usual methoxycarbonyl group had been replaced by an N,N-dimethylamide group. The CI spectra clearly displayed pseudomolecular ions for the two new derivatives (MH$^+$ at m/z 421 and 449) that were 13 amu higher in mass than those for the corresponding methyl ester methyl glycosides (Figure 5, Panels C and D versus Panels A and B). Further indication of the substitution of OMe with NMe$_2$ came from abundant fragments corresponding to neutral losses, which in the former case are observed at [MH − 32]$^+$ and [MH − 64]$^+$ (neutral loss of one and two MeOH), and in the latter case at [MH − 32]$^+$ and [MH − 73]$^+$ (neutral loss of MeOH or of Me$_2$NCHO). The same pattern of neutral losses were observed in positive-ion FAB mass spectra of the intact permethylated compounds [190,201]. EI spectra of the new products (Figure 6, Panels C and D) were dominated by [M − 72]$^+$ ions found at m/z 348 and 376, respectively, corresponding to loss of the C-1 fragment (Scheme 6, Fragment B). These ions are also found in EI spectra of the methyl ester analogs [142,144,145], although in lower abundance (Figure 6, Panels A and B), attesting to the stabilizing effect of the nitrogen on the fragment lost from the N,N-dimethylamide derivatives.

GC-MS of the products from monolactonized disialogangliosides GD$_3$ and GD$_{1b}$ (Figure 4, Panel B) confirmed previous reports [182,184,186,201] that for gangliosides containing the NeuAcα2→8NeuAcα2→3Galβ1→ sequence the lactone ring is formed much more rapidly between the two NeuAc residues than between NeuAc and Gal. This was indicated by the observation of considerable amounts of derivative 3β, with a concomitant decrease in the peak for 1β, with virtually none of the derivative corresponding to the internal NeuAc residue being detected as 4β (see Scheme 6 for structures and fragmentation of all derivatives).

The structure of the singly charged intermediate product is of considerable interest to studies of possible biological function of sialic acid lactonization. For example, monoclonal antibodies have been reported that react with lactones and their parent disialogangliosides [188,189], while others are specific for the lactones and do not react with the parent ganglioside [190]. In the latter case, although the antibody reacted with the dilactone as well as the monolactone, it also cross-reacted with lactones formed from polysialic acids, as might be expected if the minimum epitope is formed from the disialosyl unit of a disialosyl ganglioside.

The GC-MS method described could provide information about lactonization of sialic acids occurring in either gangliosides, glycoproteins, or oligosaccharides on a micro scale without resorting to more expensive methods requiring FAB-MS or NMR equipment. Used in conjunction with those methods, it would provide complementary or confirmatory information. A protocol including ammonolysis, permethylation, FAB-MS, followed by methanolysis, acetylation, and GC-MS, could potentially allow detection and characterization of sialic acid lactones formed in living cells in situ in much smaller quantities than possible by any other methods. Complete characterization of sialic acid inner esters would require one further step, unambiguous confirmation of the site of esterification on the aglyconic residue. This is addressed by another GC-MS method.

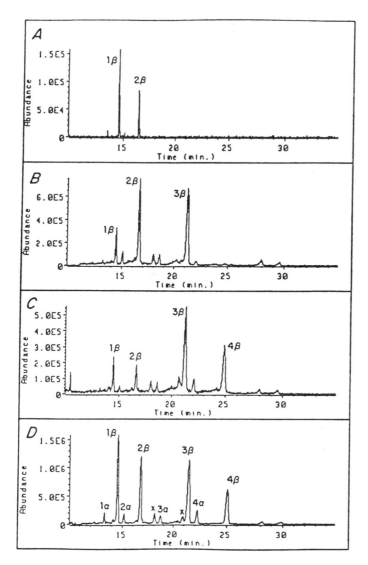

Figure 4 Gas chromatography–electron impact mass spectrometry of partially *O*-acetylated, *O*-methylated *N*-acetyl-*N*-methyl-β-methyl glycoside-methyl esters and -N^1,N^1-dimethylamides produced from methanolysis and acetylation of: (*A*) permethylated GD$_3$; (*B*) ammonolyzed/permethylated GD$_{1b}$-L1; (*C*) ammonolyzed/permethylated GD$_{1b}$-L2. Panel (*D*) shows results from coinjection of derivatives in (*A*) and (*C*), to illustrate identity of products 1 and 2. Traces are the summation of *m/z* 348 and 376 ion intensities. Peaks identified were the α- and β-methyl glycoside derivatives 1–4 as drawn in Scheme 6. Virtually identical results, since (*B*)–(*D*) were obtained from GD$_3$-L1 and -L2. (From Ref. 205.)

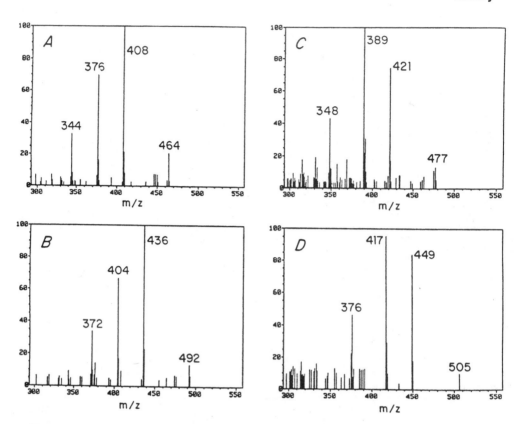

Figure 5 Iso-butane chemical ionization mass spectra of derivatives 1β–4β [panels (A)–(D)]. (From Ref. 205.)

C. Location of Internally Esterified Hydroxyl Groups

The key to determining unambiguously which aglyconic sugar hydroxyl is internally esterified is to employ a protecting group that can be introduced under conditions that will not cleave ester linkages. Furthermore, in order to be suitable for use on a putative natural lactone, the conditions should be such as not to induce artifactual formation of a lactone ring. Finally, the protecting group should be stable in strongly alkaline media, able to survive permethylation procedures that will introduce a methyl ether at any site previously acylated, but easily removed by the acid treatments used for subsequent depolymerization of carbohydrates. Any such method would be suitable for locating acylation sites in general, but the requirement for essentially neutral conditions, to avoid induction of lactonization, eliminates many traditional protection schemes, such as reaction with dihydropyran [27] or methylvinyl ether [28], which require acidic catalysis. The method of Prehm [61] might be appropriate for a direct methylation approach. But as far as this author is aware, no one has yet reported its application to this problem, although it was used in the synthesis of a partially methylated, partially acetylated 3-deoxy-octitol derivative via the intermediate 1,4-lactone of carbonyl-reduced KDO [139] (see Section D). In any

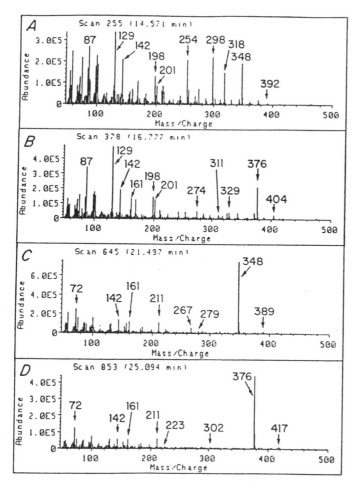

Figure 6 Electron impact mass spectra of derivatives 1β–4β [panels (A)–(D)]. (From Ref. 205.)

case, a suitable alternative strategy has recently been reported [204,205] employing methoxyethoxymethyl (MEM) chloride, a reagent first described by Corey et al. [208]. The formation of MEM ethers from MEMCl can be catalyzed by diisopropylethylamine (iPr$_2$EtN), a nonnucleophilic base, enabling protection of hydroxyl groups in the presence of lactones [209].

Thus, the inner ester of GM$_3$ ganglioside, formed by the method of McCluer and Evans [178], was completely converted to its per-O-MEM derivative ("per-MEMOrized") by MEMCl and iPr$_2$EtN in THF [204]. Following permethylation, depolymerization, reduction, and acetylation according to the normal scheme for production of PMAAs, analysis by GC-MS allowed identification of 2-mono-O-Me-1,3,4,5,6-penta-O-Ac-galactitol in the products, along with glucitol hexaacetate and a small amount of galactitol hexaacetate from residual unlactonized material (Figure 7, Panel B). By contrast, a sample of unlactonized GM$_3$ produced only glucitol and

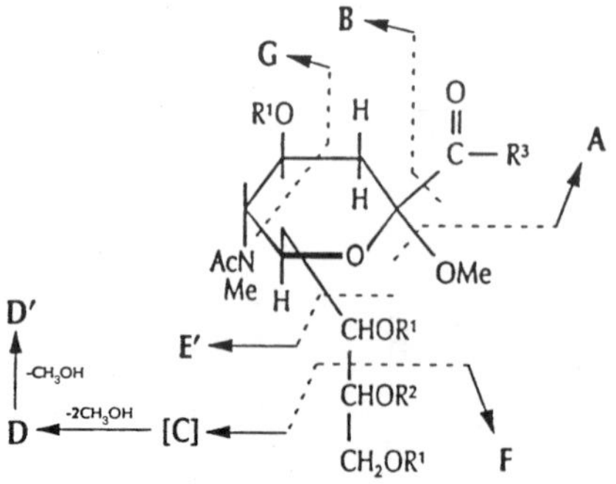

$\underline{1}$ R¹ = R² = Me ; R³ = OMe ; M = 407

$\underline{2}$ R¹ = Me ; R² = Ac ; R³ = OMe ; M = 435

$\underline{3}$ R¹ = R² = Me ; R³ = NMe$_2$; M = 420

$\underline{4}$ R¹ = Me ; R² = Ac ; R³ = NMe$_2$; M = 448

Scheme 6   Characteristic fragment ions produced by NeuAc derivatives 1–4. (From Ref. 205.)

galactitol hexaacetates (Figure 7, Panel A). As summarized in Scheme 7, these results provided the first unequivocal confirmation of previous assumptions [178,180] that in GM$_3$ lactone, it is the 2-hydroxyl group of the Gal residue that is esterified by the NeuAcα2→3 carboxylate.

A similar scheme (see Scheme 8) was used to confirm that in the dilactone of GD$_3$ the 2-hydroxyl group of the Gal residue is similarly esterified by the inner NeuAc carboxylate, while treatment of the permethylated intermediate according to protocols for sialic acid linkage analysis (methanolysis followed by acetylation or trimethylsilylation) demonstrated that the inner NeuAc is in turn esterified at C-9 (rather than C-7) by the α2→8-linked outer NeuAc residue [205], as had been previously proposed [174,182,184]. For example, GC-MS of the per-*O*-trimethylsilylated, partially-*O*-methylated NeuAc methylester β-methyl glycosides produced from GD$_3$-L2 allowed identification of the 9-*O*-Me-4,7,8-tri-*O*-TMS along with the 4,7,8,9-tetra-*O*-TMS derivative (Figure 8, Panel B). Application of the method to unlactonized GD$_3$ produced only the latter derivative (Figure 8, Panel A).

Although some GC-MS strategies had already been used for detection of internally esterified NeuAc residues [181–183,186,203], the methods described in this section were the first to address specifically the problems of linkage structure. Particularly, the methods producing derivatives of NeuAc could be expected to work well in laboratories already performing routine sialic acid linkage analysis by GC-MS.

V.   GC-MS OF PERMETHYLATED OLIGOSACCHARIDES

Early experiments with GC-MS of permethylated oligosaccharides were fairly successful but were limited as to the size of molecule that could be analyzed. These

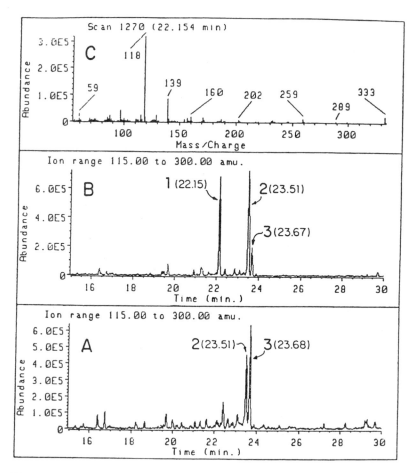

Figure 7 GC-MS of alditol acetate derivatives obtained following MEM derivatization, permethylation, hydrolysis, reduction, and acetylation of GM₃ (*A*) and GM₃ lactone (*B*). Derivatives identified are (1) 2-mono-*O*-Me-Gal; (2) glucitol hexaacetate; and (3) galactitol hexaacetate. Panel (*C*) shows the EI mass spectrum of 2-mono-*O*-Me-Gal obtained from the sample of GM₃ lactone. Temperature program: 140°–250°C at 4°C/min, with splitless injection. EI mass spectra were acquired every 0.95 s, mass range 50–500 amu. (From Ref. 204.)

early studies established that, similar to EI spectra obtained by direct insertion probe, glycosyl sequence information could be obtained from fragments derived from cleavages on both sides of the glycosidic oxygen, with charge retention on either the nonreducing or the reducing end (for a summary of early investigation, see Lönngren and Svensson [127]). Furthermore, in the case of prereduced oligosaccharides, the positions of certain linkages to the reducing sugar could often be inferred from ions originating from fragmentation of the alditol moiety. Kärkkäinen [210,211] analyzed a series of permethylated, mostly hexose-containing di- and trisaccharide alditols by GC-MS and concluded that 1→6-linked alditols could be readily distinguished from 1→3- and 1→4-linked alditols; the latter two types could also be distinguished if borodeuteride was used in the reduction step. Separation of most derivatives was

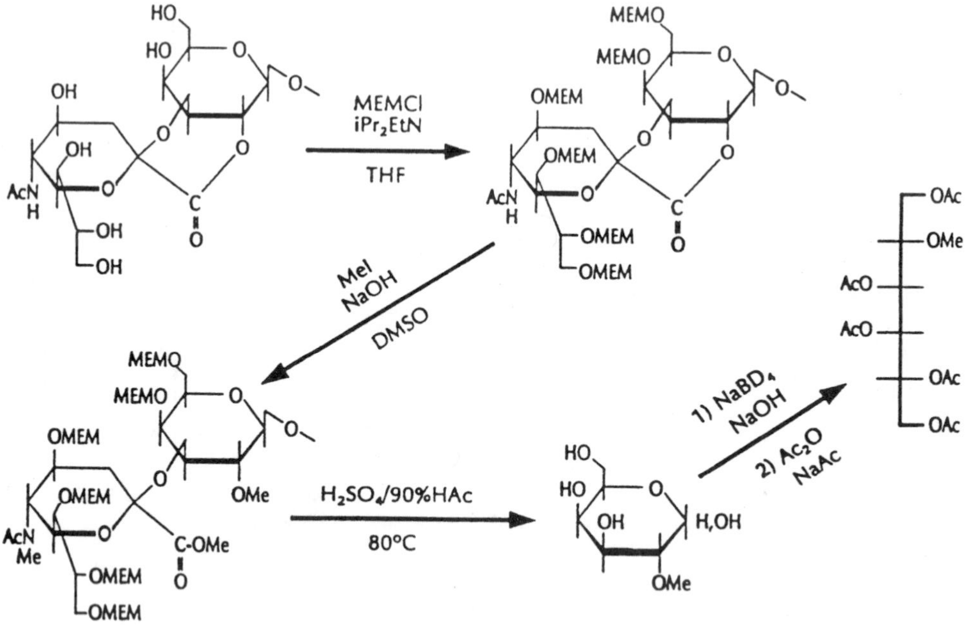

Scheme 7 Reaction sequence used for production of sugar derivatives carrying an *O*-Me label indicating position of inner esterification. Application to GM₃ lactone. (From Ref. 204.)

demonstrated with a variety of common stationary phases [210,211]. Later, Mononen et al. [212] analyzed a series of permethylated hexopyranosyl-2-acetamido-2-deoxy-hexitol isomers by GC-MS. They systematically recorded retention times on three different stationary phases, along with electron impact mass spectra, for six different compounds, and were able to distinguish 1→3-, 1→4-, and 1→6-linked isomers using the appearance or absence, or differences in intensity, of characteristic ions. In this case, of course, the acetamido group establishes a natural label for the 2-carbon. It was not possible to identify sugars beyond isobaric type, or to specify anomeric configuration, from the EI mass spectra, but clear differences in retention times for all six derivatives could be discerned when using a more polar stationary phase [212]. Such a difference in retention time is illustrated for an anomeric pair, Galα or β1→3GalNAc-itol, in Figure 9, Panel A. In addition, some configuration-related differences in the relative abundance of certain ions (*m/z* 378, 422) may also be observed (Figure 9, Panels B and C; Levery and Bendiak, unpublished data).

A crucial problem for further development was the limited temperature range of the GC columns employed. Thus, it was possible to analyze permethylated small urinary oligosaccharides [213–217] or fragments of larger molecules [68,218], but generally nothing larger than a tetrasaccharide could be expected to pass through the GC columns employed. Using a packed GC column, Fournet et al. [219] performed GC-MS analysis on a series of permethylated di-, tri-, and tetrasaccharide alditols obtained by partial acetolysis of glycoprotein glycans. Using permethylation and GC-MS on packed columns, Funakoshi and Yamashina [220] analyzed the structures of *O*-glycosidically linked glycans released from the plasma membranes of an ascites

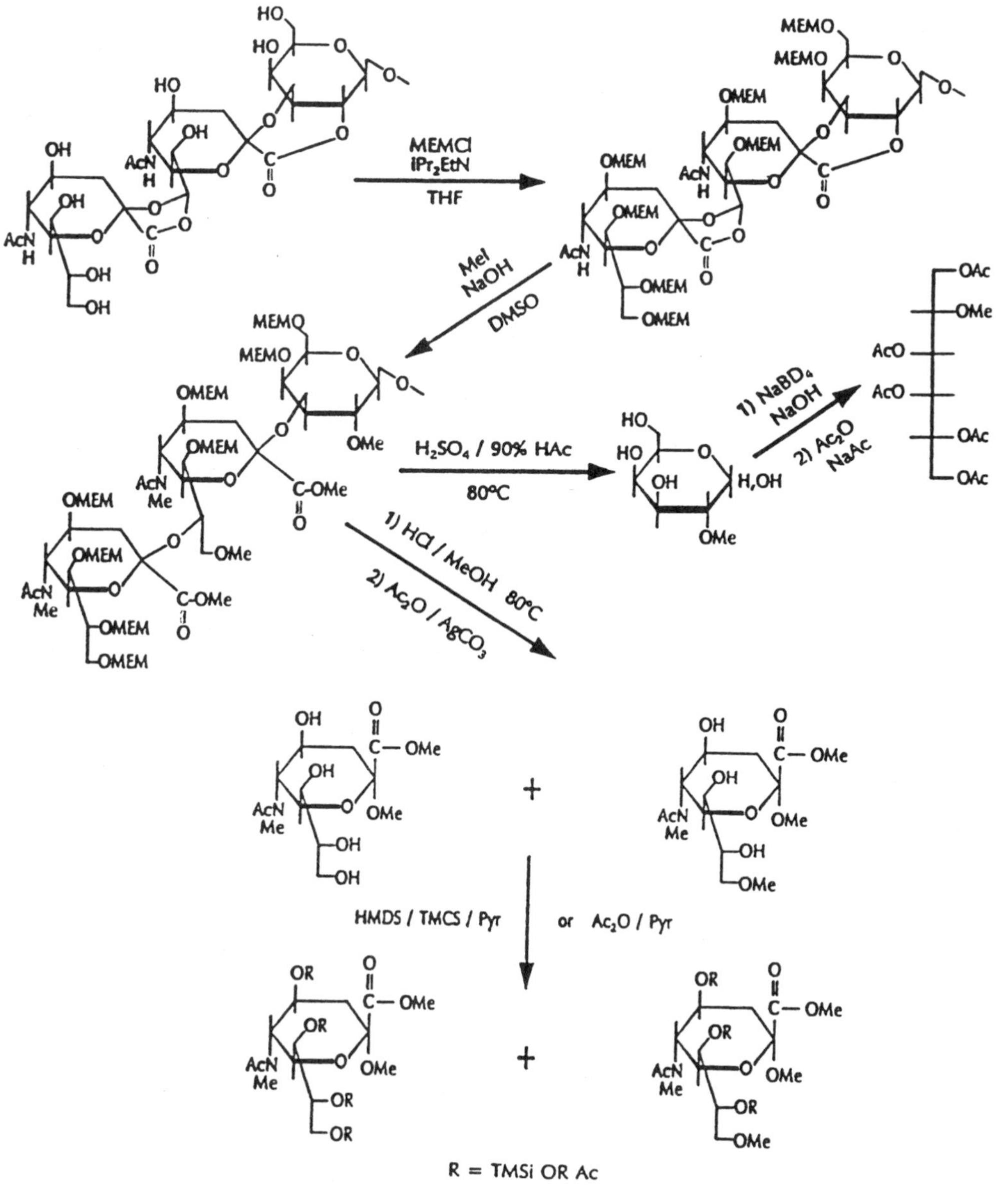

Scheme 8 Reaction sequence used for production of sugar derivatives carrying an *O*-Me label indicating position of inner esterification. Application to GD$_3$ lactone. (From Ref. 205.)

hepatoma by reductive elimination; Pierce-Cretel et al. [221] analyzed *O*-linked structures from the hinge region of human milk secretory immunoglobulins, and van Halbeek et al. [222] analyzed oligosaccharides released from hog submaxillary gland mucin glycoproteins. In each case, the largest compounds analyzed by GC-MS were permethylated tetrasaccharide alditols; in the latter paper, analysis of permethylated

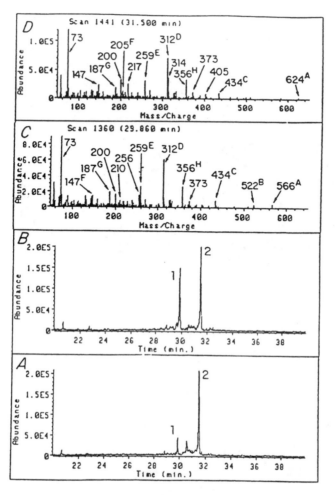

Figure 8 Gas chromatography−mass spectrometry of per-*O*-trimethylsilylated, partially -*O*-methylated NeuAc methyl ester β-methyl glycosides produced from (*A*) GD₃ and (*B*) GD₃-L2 following per-*O*-methoxyethoxymethylation, permethylation, methanolysis, and re-*N*-acetylation. Electron impact mass spectra of derivatives of peaks marked *1* and *2* in (*B*) are reproduced in Panels (*C*) and (*D*) and are identified as the 9-*O*-Me-4,7,8-tri-*O*-trimethylsilyl and 4,7,8,9-tetra-*O*-trimethylsilyl derivatives, respectively. Capital letter superscripts refer to fragments named according to schemes in van Halbeek et al. [145] and Bruvier et al. [146]. From Levery et al. [205].

sialic acid−containing di- and trisaccharide alditols was reported. Kurosaka et al. [223] performed GC-MS analysis on permethylated mucin-type disaccharides from human rectal adenocarcinoma, and Tsuji and Osawa [224] analyzed several oligosaccharides from bovine submaxillary mucin, the largest being a tetrasaccharide. Schmidt and Jann [225] used GC-MS analysis on carbonyl-reduced permethylated rhamnosyl-KDO-containing di- and trisaccharides from the capsular polysaccharide antigen of a urinary-tract-invasive *E. coli* strain. As mentioned previously in con-

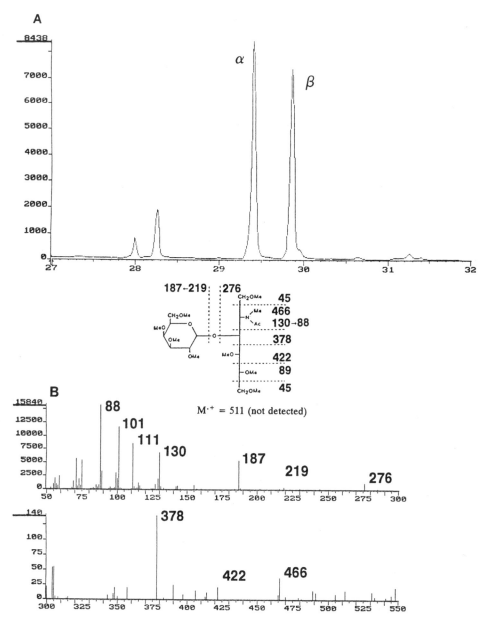

Figure 9 (A) Separation of permethylated Gal1→3GalNAc-itol isomers on a 30 m DB-5 bonded-phase fused silica capillary. GC-MS was performed on a Hewlett-Packard 5890 series II GC (splitless injection; temperature program 140°–320°C @ 4°/min) interfaced to an Extrel ELQ 400 mass spectrometer operating in EI mode with selected ion monitoring. Thirteen relevant masses were acquired with a cycle time of 0.5 s. α, Galα1→3GalNAc-itol; β, Galβ1→3GalNAc-itol. About 1 nmol of each compound was injected. (B) Electron impact mass spectrum of permethylated Galα1→3GalNAc-itol with fragmentation scheme. Mass spectrum was obtained during GC-MS using a Hewlett-Packard 5890 series II GC (splitless injection; temperature program 140°–320°C @ 8°/min) interfaced to an Extrel ELQ 400 mass spectrometer (the mass range m/z 50–600 was acquired at 1.2 s/scan).

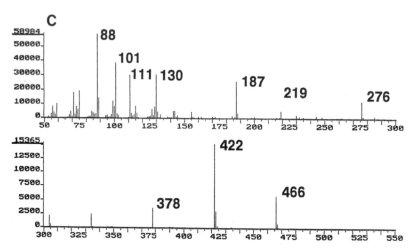

Figure 9 *Continued* (C) Electron impact mass spectrum of permethylated Galβ1→ 3GalNAc-itol. Mass spectrum was obtained during GC-MS using a Hewlett-Packard 5890 series II GC (splitless injection; temperature program 140°–320°C @ 8°/min) interfaced to an Extrel ELQ 400 mass spectrometer (the mass range *m/z* 50–600 was acquired at 1.2 s/scan).

nection with linkage analysis of reducing KDO-containing oligosaccharides (Section III.D.), a pair of diastereomeric products is obtained in each case from the nonstereospecific reduction of the keto group. The permethylated products, having either the D-*glycero*-D-*talo*- or the D-*glycero*-D-*galacto*- configuration at the reducing end, are usually separable by GC while exhibiting virtually identical EI mass spectra. As with the permethylated alditols mentioned earlier, fragments can be observed in these spectra indicative of both overall sequence and linkage to the reduced KDO residue.

A distinct advantage was realized with the introduction of bonded-phase fused silica capillary columns. Thus, Sweeley and co-workers [226] were able to separate a large number of oligosaccharides up to pentasaccharides on a DB-1 column, but did not analyze these by mass spectrometry. GC-MS analysis was also performed on permethylated disaccharides previously derivatized with ABEE [227]. Brade, Rietschel, and co-workers [141,228–230] have used a chemically bonded SE-54 fused silica capillary for GC-MS of carbonyl-reduced permethylated KDO-containing di-, tri-, and tetrasaccharide fragments isolated from lipopolysaccharides of various enterobacterial strains. Mass spectra of these permethylated oligosaccharides were studied under EI conditions [141,228–230], as well as in CI mode using either ammonia [141,229,230] or methane [230]. A rhamnosyl-KDO disaccharide isolated from a plant cell wall rhamnogalacturonan was shown [139], in part by its EI spectrum, after carbonyl reduction, permethylation, and GC-MS, to be identical to one isolated from the capsular polysaccharide antigen of a urinary-tract-invasive *E. coli* [225].

The development of high-temperature capillary GC, using thermostable thin-film bonded-phase columns, has considerably expanded the scope of GC-MS analysis to include permethylated oligosaccharides with up to 11 sugar residues (molecular masses to 2300 Da [231–237]; for a recent review of this group's work, see Karlsson et al. [238]). The success of the technique also depends upon the introduction of higher temperature GC-MS interfaces with direct introduction of the capillary column into the ion source, and has benefited as well from the availability of higher mass range spectrometers, developed, to a great extent, in response to the revolution in soft ionization of biomolecules.

GC-MS of permethylated derivatives has been applied to free oligosaccharides [232,234], oligosaccharides released chemically from mucin glycoproteins [231,232,234–237], and oligosaccharides released enzymatically from glycosphingolipids [233,235]. In the case of mucin alditols containing sialic acid residues, poor yields were obtained on methylation. But by using a protocol in which the compounds were first converted to lactones (concurrent with peracetylation in acetic anhydride/pyridine), treated with ammonia in dry methanol to convert sialic acids to their corresponding amides (with concurrent ammonolysis of all O-acetyl groups), and then permethylated, dimethylamide derivatives (analogous to those produced from ganglioside lactones [201], although the two protocols were developed independently) were obtained conveniently in high yield [236]. These oligosaccharide alditol derivatives have good chromatographic properties and provide intense EI mass spectra [236]. As observed in FAB-MS of the analogous permethylated ganglioside derivatives [190,201], the terminal sialic acid dimethylamide provides an abundant pair of characteristic ions at m/z 389 and 357. In addition, facile loss of the C-1 fragment from the sialic acid residue yields an abundant [M − 72]$^+$ ion, as noted with GC-EI-MS of the monosaccharide derivatives [205] (see Section IV), in this case providing a useful molecular mass related marker [236]. Interestingly, compounds having a sialic acid linked 2→6 to the GalNAc-itol residue were also able to be derivatized this way. NeuAc linked 2→6 to a monosaccharide in its ring form does not readily form a lactone, but the flexibility of the alditol, and the consequent availability of HO-5, allows internal esterification to take place [236].

The methodology described was recently used in the characterization of glycopeptides released from two different glycosylated domains of insoluble mucin complex from rat small intestine [239]. More recently, this group has developed an improved, more general protocol for analyzing O-linked oligosaccharides released from mucins, in which alditols were fractionated into neutral, sialic acid–containing, and sulfated species prior to analysis [240]. The protocol incorporated a different method for producing dimethylamide derivatives that is not dependent on having lactonizable sialic acids. The mixture of oligosaccharide alditols was applied to an anion exchange column, from which the neutral species were eluted unretarded and permethylated directly. The sialic acid–containing species were methylesterified on-column, after which they could also be eluted as a neutral fraction, subsequently treated with methylamine to make N-methylamide derivatives, and finally permethylated as before to produce N,N-dimethylamides. These two fractions were analyzed by GC-MS, while the sulfated fraction, following separate elution from the column, was analyzed by high-performance anion exchange chromatography. The method successfully identified 21 neutral, and 28 sialylated (see Figure 10), oligosaccharide species previously found to be present in mucin glycopeptides of porcine small intestine [234,236].

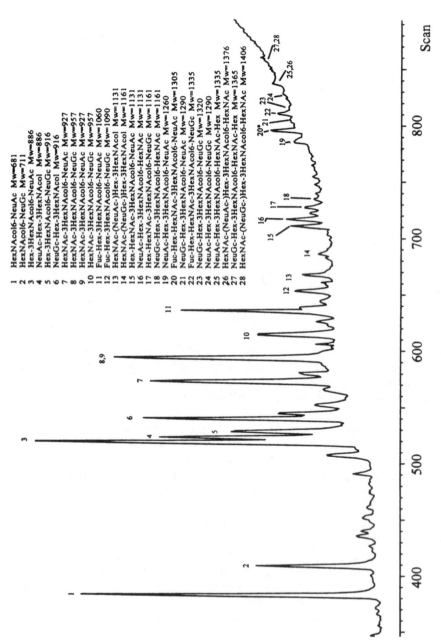

1 HexNAcol6-NeuAc Mw=681
2 HexNAcol6-NeuGc Mw=711
3 Hex-3HexNAcol6-NeuAc Mw=886
4 NeuAc-Hex-3HexNAcol Mw=886
5 Hex-3HexNAcol6-NeuGc Mw=916
6 NeuGc-Hex-3HexNAcol Mw=916
7 HexNAc-3HexNAcol6-NeuAc Mw=927
8 HexNAc-3HexNAcol6-NeuGc Mw=957
9 HexNAc-3HexNAcol6-NeuAc Mw=927
10 HexNAc-3HexNAcol6-NeuGc Mw=957
11 Fuc-Hex-3HexNAcol6-NeuAc Mw=1060
12 Fuc-Hex-3HexNAcol6-NeuGc Mw=1090
13 HexNAc-(NeuAc-)Hex-3HexNAcol Mw=1131
14 HexNAc-(NeuGc-)Hex-3HexNAcol Mw=1161
15 Hex-HexNAc-3HexNAcol6-NeuAc Mw=1131
16 NeuAc-Hex-3HexNAcol6-HexNAc Mw=1131
17 Hex-HexNAc-3HexNAcol6-NeuGc Mw=1161
18 NeuGc-Hex-3HexNAcol6-HexNAc Mw=1161
19 NeuAc-Hex-3HexNAcol6-NeuAc Mw=1260
20 Fuc-Hex-HexNAc-3HexNAcol6-NeuAc Mw=1305
21 NeuGc-Hex-3HexNAcol6-NeuAc Mw=1290
22 Fuc-Hex-HexNAc-3HexNAcol6-NeuGc Mw=1335
23 NeuGc-Hex-3HexNAcol6-NeuGc Mw=1320
24 NeuAc-Hex-3HexNAcol6-NeuGc Mw=1290
25 NeuAc-Hex-3HexNAcol6-HexNAc-Hex Mw=1335
26 HexNAc-(NeuAc-)Hex-3HexNAcol6-HexNAc Mw=1376
27 NeuGc-Hex-3HexNAcol6-HexNAc-Hex Mw=1365
28 HexNAc-(NeuGc-)Hex-3HexNAcol6-HexNAc Mw=1406

Figure 10 Total ion chromatogram and interpreted structures of permethylated sialic acid–containing oligosaccharides as their dimethyl amides obtained from mucin glycopeptides in porcine small intestine. (From Ref. 240.)

ABBREVIATIONS

ABEE	4-aminobenzoic acid ethyl ester
DCI	desorption chemical ionization
EI	electron impact ionization
ESI	electrospray ionization
FAB	fast atom bombardment
GC	gas chromatography
HPLC	high-performance liquid chromatography
KDO	3-deoxy-D-*manno*-2-octulosonic acid
LPS	lipopolysaccharide
LOS	lipooligosaccharide
MALDI	matrix-assisted laser desorption ionization
MEM	methoxyethoxymethyl
MS	mass spectrometry
N-CAM	neural cell-adhesion molecule
NMR	nuclear magnetic resonance
PMAA	partially methylated alditol acetate
PMPEAA	partially methylated, partially ethylated alditol acetate
SCOT	support-coated open tubular
WCOT	wall-coated open tubular

REFERENCES

1. M. McNeil, A. G. Darvill, P. Åman, L. -E. Franzén, and P. Albersheim, Structural analysis of complex carbohydrates using high-performance liquid chromatography, gas chromatography, and mass spectrometry, *Methods Enzymol. 83*: 3 (1982).
2. C. C. Sweeley and H. A. Nunez, Structural analysis of glycoconjugates by mass spectrometry and nuclear magnetic resonance spectroscopy, *Ann. Rev. Biochem. 54*: 765 (1985).
3. R. Laine, Glycoconjugates: Overview and strategy, *Methods Enzymol. 193*: 539 (1990).
4. C. G. Hellerqvist and B. Sweetman, Mass spectrometry of carbohydrates, *Methods of Biochemical Analysis, Vol 34: Biomedical Applications of Mass Spectrometry* (C. H. Suelter and D. T. Watson, eds.), Wiley, New York, 1990, p. 91.
5. T. Purdie and J. C. Irvine, The alkylation of sugars, *J. Chem. Soc. (London) 83*: 1021 (1903).
6. T. Purdie and R. C. Bridgett, Trimethyl α-methylglucoside and trimethyl glucose, *J. Chem. Soc. (London) 83*: 1037 (1903).
7. J. C. Irvine and A. Cameron, The alkylation of galactose, *J. Chem. Soc. (London) 85*: 1071 (1904).
8. T. Purdie and J. C. Irvine, Synthesis from glucose of an octamethylated disaccharide. Methylation of sucrose and maltose, *J. Chem. Soc. (London) 87*: 1022 (1905).
9. J. C. Irvine and A. M. Moodie, The alkylation of mannose, *J. Chem. Soc. (London) 87*: 1465 (1905).
10. W. N. Haworth, A new method of preparing alkylated sugars, *J. Chem. Soc. (London) 107*: 8 (1915).
11. W. S. Denham and H. Woodhouse, The methylation of cellulose, *J. Chem. Soc. (London) 103*: 1735 (1913).

12. R. Kuhn, H. Trischmann, and I. Löw, Zur permethylierung von zuckern und glykosiden, *Angew. Chem. 67*: 32 (1955).

13. R. Kuhn, H. H. Baer, and A. Seeliger, Zur permethylierung von *N*-acetylglucosaminderivaten, *Liebigs Ann. Chemie. 611*: 236 (1958).

14. K. Wallenfels, G. Bechtler, R. Kuhn, H. Trischmann, and H. Egge, Permethylation of oligomeric and polymeric carbohydrates and quantitative analysis of the cleavage products, *Angew. Chem. Internatl. Ed. Engl. 2*: 515 (1963).

15. H. C. Srivastava, S. N. Harshe, and P. P. Singh, Methylation of carbohydrates: Part I. Use of dimethylsulphoxide in the methylation reaction, *Indian J. Chem. 1*: 304 (1963).

16. H. C. Srivastava, S. N. Harshe and P. P. Singh, Methylation of carbohydrates: Part II. A convenient method for the methylation of polysaccharides, *Tetrahedron Lett.* 1869 (1963).

17. R. Kuhn and H. Trischmann, Permethylierung von inosit und anderen kohlenhydraten mit dimethylsulfat, *Chem. Ber. 96*: 284 (1963).

18. H. C. Srivastava, P. P. Singh, S. N. Harshe, and K. Virk, Methylation of carbohydrates: Part III. Methylation of polysaccharides with dimethyl sulfate, *Tetrahedron Lett.* 493 (1964).

19. S. Hakomori, A rapid permethylation of glycolipid and polysaccharide catalyzed by methylsulfinylcarbanion in dimethylsulfoxide, *J. Biochem. (Tokyo) 55*: 205 (1964).

20. E. J. Corey and M. Chaykovsky, Methylsulfinylcarbanion, *J. Am. Chem. Soc. 84*: 866 (1962).

21. M. Chaykovsky and E. J. Corey, Some new reactions of methylsulfinyl and methylsulfonyl carbanion, *J. Org. Chem. 28*: 254 (1963).

22. E. J. Corey and M. Chaykovsky, Methylsulfinyl carbanion (CH_3-SO-CH_2^-). Formation and applications to organic synthesis, *J. Am. Chem. Soc. 87*: 1345 (1965).

23. H. Rauvala, Use of triphenylmethane as an indicator of complete methylation of glycolipids and glycopeptides, *Carbohydr. Res. 72*: 257 (1979).

24. H. Rauvala, J. Finne, T. Krusius, J. Kärkkäinen, and J. Järnefelt, Methylation techniques in the structural analysis of glycoproteins and glycolipids, *Adv. Carbohydr. Chem. Biochem. 38:* 389 (1981).

25. P. J. Harris, R. J. Henry, A. B. Blakeney, and B. A. Stone, An improved procedure for the methylation analysis of oligosaccharides and polysaccharides, *Carbohydr. Res. 127*: 59 (1984).

26. G. K. Ostrander, S. B. Levery, H. L. Eaton, M. E. K. Salyan, S. Hakomori, and E. H. Holmes, Isolation and characterization of four major neutral glycosphingolipids from the liver of the English sole (*Parophrys vetulus*). Presence of a novel branched lacto-ganglio-*iso*-globo hybrid structure, *J. Biol. Chem. 263*: 18,716 (1988).

27. A. N. DeBelder and B. Norrman, The distribution of substituents in partially acetylated dextran, *Carbohydr. Res. 8*: 1 (1968).

28. J. C. Promé, C. Lacave, A. Ahibo-Coffy, and A. Savagnac, Séparation et étude structurale des espèces moléculaire de monomycolates et de dimycolates de α-D-tréhalose présents chez *Mycobacterium phlei*, *Eur. J. Biochem. 63*: 543 (1976).

29. H. Björndal, C. G. Hellerqvist, B. Lindberg, and S. Svensson, Gas-flüssigkeits-chromatographie und massenspektrometrie bei der methylierungsanalyse von polysacchariden, *Angew. Chem. 82*: 643 (1970); for English translation, see, Gas-liquid chromatography and mass spectrometry in methylation analysis of polysaccharides, *Angew. Chem. Intl. Ed. Engl. 9*: 610 (1970).

30. B. Lindberg, Methylation analysis of polysaccharides, *Methods Enzymol. 28B*: 178 (1972).

31. R. L. Taylor, J. E. Shively, H. E. Conrad, and J. A. Cifonelli, Uronic acid composition of heparins and heparan sulfates, *Biochemistry 12*: 3633 (1973).

32. R. L. Taylor, J. E. Shively, and H. E. Conrad, Stoichiometric reduction of uronic acid carboxyl groups in polysaccharides, *Methods Carbohydr. Chem. 7*: 149 (1976).

33. P.-E. Jansson, L. Kenne, H. Liedgren, B. Lindberg, and J. Lönngren, A practical guide to methylation analysis, *Univ. Stockholm Chem. Commun. 8:* 1 (1976).

34. P. A. Sandford and H. E. Conrad, The structure of the *Aerobacter aerogenes* A3 (s1) polysaccharide. I. A reexamination using improved procedures for methylation analysis, *Biochemistry 5:* 1508 (1966).

35. W. Stoffel and P. Hanfland, Analysis of amino sugar containing glycosphingolipids by combined gas-liquid chromatography and mass spectrometry, *Hoppe-Seyler's Z. Physiol. Chem. 354:* 21 (1973).

36. N. C. Carpita and E. M. Shea, Linkage structure of carbohydrates by gas chromatography–mass spectrometry (GC-MS) of partially methylated alditol acetates, *Analysis of Carbohydrates by GLC and MS* (C. J. Biermann and G. D. McGinnis, eds.), CRC Press, Boca Raton, FL, 1989, p. 157.

37. A. Dell, Preparation and desorption mass spectrometry of permethyl and peracetyl derivatives of oligosaccharides, *Methods Enzymol. 193:* 647 (1990).

38. R. Geyer, H. Geyer, S. Kühnhardt, W. Mink, and S. Stirm, Methylation analysis of complex carbohydrates in small amounts: Capillary gas chromatography–mass fragmentography of methylalditol acetates obtained from *N*-glycosidically linked glycoprotein oligosaccharides, *Anal. Biochem. 133:* 197 (1983).

39. T. J. Waeghe, A. G. Darvill, M. McNeil, and P. Albersheim, Determination, by methylation analysis, of the glycosyl-linkage compositions of microgram quantities of complex carbohydrates, *Carbohydr. Res. 123:* 281 (1983).

40. J. Montreuil, S. Bouquelet, H. Debray, B. Fournet, G. Spik, and G. Strecker, Glycoproteins, *Carbohydrate Analysis: A Practical Approach* (M. F. Chaplin and J. F. Kennedy, eds.), IRL Press, Oxford, 1986, p. 143.

41. S. B. Levery and S. Hakomori, Micro-sale methylation analysis of glycolipids using capillary gas chromatography–chemical ionization mass fragmentography with selected ion monitoring, *Methods Enzymol. 138E:* 13 (1987).

42. C. G. Hellerqvist, Linkage analysis using Lindberg method, *Methods Enzymol. 193:* 554 (1990).

43. R. Geyer and H. Geyer, Saccharide linkage analysis using methylation and other techniques, *Methods Enzymol. 230:* 86 (1994).

44. J. Paz Parente, P. Cardon, J. Montreuil, B. Fournet, and G. Ricart, A convenient method for methylation of glycoprotein glycans in small amounts by using lithium methylsulfinyl carbanion, *Carbohydr. Res. 141:* 41 (1985).

45. L. R. Phillips and B. A. Fraser, Methylation of carbohydrates with dimsyl potassium in dimethyl sulfoxide, *Carbohydr. Res. 90:* 149 (1981).

46. A. B. Blakeney and B. A. Stone, Methylation of carbohydrates with lithium methylsulfinyl carbanion, *Carbohydr. Res. 140:* 319 (1985).

47. J. Finne, T. Krusius, and H. Rauvala, Use of potassium *tert*-butoxide in the methylation of carbohydrates, *Carbohydr. Res. 80:* 336 (1980).

48. I. Ciukanu and F. Kerek, A simple and rapid method for the permethylation of carbohydrates, *Carbohydr. Res. 131:* 209 (1984).

49. K. R. Anumula and P. B. Taylor, A comprehensive procedure for preparation of partially methylated alditol acetates from glycoprotein carbohydrates, *Anal. Biochem. 203:* 101 (1992).

50. J. March, *Advanced Organic Chemistry: Reactions, Mechanisms, and Structure,* McGraw-Hill, New York, 1968, pp. 219–221.

51. A. Isogai, A. Ishizu, J. Nakano, S. Eda, and K. Kato, A new facile methylation method for cell-wall polysaccharides, *Carbohydr. Res. 138:* 99 (1985).

52. G. Larson, H. Karlsson, G. C. Hansson, and W. Pimlott, Application of a simple methylation procedure for the analyses of glycosphingolipids, *Carbohydr. Res. 161:* 281 (1987).

53. A. Gunnarsson, *N*- and *O*-alkylation of glycoconjugates and polysaccharides by solid base in dimethylsulphoxide/alkyl iodide, *Glycoconj. J. 4*: 239 (1987).

54. E. H. Holmes and S. B. Levery, Preparative *in vitro* generation of lacto-series type 1 chain glycolipids catalyzed by β1→3-galactosyltransferase from human colonic adenocarcinoma colo 205 cells, *Arch. Biochem. Biophys. 274*: 14 (1989).

55. E. H. Holmes and S. B. Levery, Biosynthesis of fucose containing lacto-series glycolipids in human colonic adenocarcinoma colo 205 cells, *Arch. Biochem. Biophys. 274*: 633 (1989).

56. C. E. Costello and J. E. Vath, Tandem mass spectrometry of glycolipids, *Methods Enzymol. 193*: 738 (1990).

57. W. S. York, L. L. Kiefer, P. Albersheim, and A. G. Darvill, Oxidation of oligoglycosyl alditols during methylation catalyzed by sodium hydroxide and iodomethane in methyl sulfoxide, *Carbohydr. Res. 208*: 175 (1990).

58. P. W. Needs and R. R. Selvendran, Avoiding oxidative degradation during sodium hydroxide/methyl iodide-mediated carbohydrate methylation in dimethyl sulfoxide, *Carbohydr. Res. 245*: 1 (1993).

59. F.-G. Hanisch, Methylation analysis of complex carbohydrates: Overview and critical comments, *Biol. Mass Spectrom. 23*: 309 (1994).

60. H. O. Bouveng and B. Lindberg, Methods in structural polysaccharide chemistry, *Adv. Carbohydr. Chem. 15*: 53 (1960).

61. P. Prehm, Methylation of carbohydrates by methyl trifluoromethanesulfonate in trimethyl phosphate, *Carbohydr. Res. 78*: 372 (1980).

62. P. Mischnick, Determination of the substitution pattern of cellulose acetates, *J. Carbohydr. Chem. 10*: 711 (1991).

63. P. J. Stang, M. Hanack, and L. R. Subramanian, Perfluoroalkanesulfonic esters: Methods of preparation and applications in organic chemistry, *Synthesis* 85 (1982).

63a. B. B. Reinhold, S. Y. Chan, T. L. Reuber, A. Marra, G. C. Walker, and V. N. Reinhold, Detailed structural characterization of succinoglycan, the major exopolysaccharide of *Rhizobium meliloti* Rm 1021, *J. Bacteriol. 176*: 1997 (1994).

63b. K.-H. Khoo, A. Dell, H. R. Morris, P. J. Brennan, and D. Chatterjee, Structural definition of acylated phosphatidylinositol mannosides from *Mycobacterium tuberculosis*: Definition of a common anchor for lipomannan and lipoarabinomannan, *Glycobiology 5*: 117 (1995).

64. C. G. Hellerqvist and B. Lindberg, Structural studies of the common-core polysaccharide of the cell-wall lipopolysaccharide from *Salmonella, Carbohydr. Res. 16*: 39 (1971).

65. P. Albersheim, D. J. Nevins, P. D. English, and A. Karr, A method for the analysis of sugars in plant cell-wall polysaccharides by gas–liquid chromatography, *Carbohydr. Res. 5*: 340 (1967).

66. H. Björndal, B. Lindberg, and S. Svensson, Gas–liquid chromatography of partially methylated alditols as their acetates, *Acta Chem. Scand. 21*: 1801 (1967).

67. H. Björndal, B. Lindberg, and S. Svensson, Mass spectrometry of partially methylated alditol acetates, *Carbohydr. Res. 5*: 433 (1967).

68. B. Lindberg and J. Lönngren, Methylation analysis of complex carbohydrates: General procedure and application for sequence analysis, *Methods Enzymol. 50C*: 3 (1978).

69. G. G. S. Dutton, Applications of gas–liquid chromatography to carbohydrates: Part II. *Adv. Carbohydr. Chem. Biochem. 30*: 9 (1974).

70. K. Stellner, H. Saito, and S. Hakomori, Determination of aminosugar linkages in glycolipids by methylation: Aminosugar linkages of ceramide pentasaccharides of rabbit erythrocytes and of Forssman antigen, *Arch. Biochem. Biophys. 155*: 464 (1973).

71. K. B. Hicks, G. L. Simpson, and A. G. W. Bradbury, Removal of boric acid and related compounds from solutions of carbohydrates with a boron-selective resin (IRA-743), *Carbohydr. Res. 147*: 39 (1986).

72. C. C. Chen and G. D. McGinnis, The use of 1-methylimidazole as a solvent and catalyst for the preparation of aldonitrile acetates of aldoses, *Carbohydr. Res. 90*: 127 (1981).

73. G. D. McGinnis, Preparation of aldonitrile acetates using *N*-methylimidazole as catalyst and solvent, *Carbohydr. Res. 108*: 284 (1982).

74. A. B. Blakeney, P. J. Harris, R. J. Henry, and B. A. Stone, A simple and rapid preparation of alditol acetates for monosaccharide analysis, *Carbohydr. Res. 113*: 291 (1983).

75. A. Fox, S. L. Morgan, and J. Gilbart, Preparation of alditol acetates and their analysis by gas chromatography (GC) and mass spectrometry (MS), *Analysis of Carbohydrates by GLC and MS* (C. J. Biermann and G. D. McGinnis, eds.), CRC Press, Boca Raton, FL, 1989, p. 87.

76. S.-S. J. Sung, W. J. Esselman, and C. C. Sweeley, Structure of a pentahexosylceramide (Forssman hapten) from canine intestine and kidney, *J. Biol. Chem. 248*: 6528 (1973).

77. G. O. H. Schwarzmann and R. W. Jeanloz, Separation by gas–liquid chromatography, and identification by mass spectrometry, of the methyl esters of 2-deoxy-2-(*N*-methylacetamido)-D-glucose, *Carbohydr. Res. 34*: 161 (1974).

78. T. Tai, K. Yamashita, and A. Kobata, Synthesis and mass fragmentographic analysis of partially *O*-methylated 2-*N*-methylglucosamines, *J. Biochem. (Tokyo) 78*: 679 (1975).

79. S. Hase and E. T. Rietschel, Methylation analysis of glucosaminitol and glucosaminylglucosaminitol disaccharides. Formation of 2-deoxy-2-(*N*-acetylacetamido)-glucitol derivatives, *Eur. J. Biochem. 63*: 93 (1976).

80. J. Finne and H. Rauvala, Determination (by methylation analysis) of the substitution pattern of 2-amino-2-deoxyhexitols obtained from *O*-glycosylic carbohydrate units of glycoproteins, *Carbohydr. Res. 58*: 57 (1977).

81. M. Caroff and L. Szabo, *O*-demethylation of per-*O*-methyl derivatives of 2-amino-2-deoxyhexitols during acid hydrolysis and acetolysis, *Carbohydr. Res. 84*: 43 (1980).

82. B. Nilsson, *O*-demethylation of methylated 2-acetamido-2-deoxy-D-hexitols during acid hydrolysis, *Glycoconj. J. 2*: 335 (1985).

83. M. E. Lowe and B. Nilsson, A method of purification of partially methylated alditol acetates in the methylation analysis of glycoproteins and glycopeptides, *Anal. Biochem. 136*: 187 (1984).

84. B. S. Valent, A. G. Darvill, M. McNeil, and P. Albersheim, A general and sensitive chemical method for sequencing the glycosyl residues of complex carbohydrates, *Carbohydr. Res. 79*: 165 (1980).

85. A. G. Darvill, M. McNeill, and P. Albersheim, General and facile method for distinguishing 4-linked aldopyranosyl residues from 5-linked aldofuranosyl residues, *Carbohydr. Res. 86*: 309 (1980).

86. A. De Bettignies-Dutz, G. Reznicek, B. Kopp, and J. Jurenitsch, Gas chromatographic-mass spectrometric separation and characterization of methyl trimethylsilyl monosaccharides obtained from naturally occurring glycosides and carbohydrates, *J. Chromatogr. 547*: 299 (1991).

87. B. Fournet and J. Montreuil, Procédé d'identification et de préparation par chromatographie en phase gazeuse des éthers di-, tri, et tétraméthyliques de l'α-méthyl-D-mannoside, *J. Chromatogr. 75*: 29 (1973).

88. B. Fournet, J. M. Dhalluin, Y. Leroy, J. Montreuil, and H. Mayer, Analytical and preparative gas–liquid chromatography of the fifteen methyl ethers of methyl α-D-galactopyranoside, *J. Chromatogr. 153*: 91 (1978).

89. P. A. J. Gorin and A. J. Finlayson, Synthesis and chromatographic properties of partially *O*-methylated 2-deoxy-2-methylamino-D-glucoses; standards for methylation studies on polysaccharides, *Carbohydr. Res. 18*: 269 (1971).

90. P. A. J. Gorin, Synthesis and chromatographic properties of 2-deoxy-2-methylamino-D-galactose and its methyl ethers, *Carbohydr. Res. 18*: 281 (1971).

91. A. Stoffyn, P. Stoffyn, and J. W. Orr, Methylation of 2-acetamido-2-deoxy-D-hexoses, *Carbohydr. Res. 23*: 251 (1972).

92. S. K. Kundu, R. W. Ledeen, and P. A. J. Gorin, Determination of position of substitution on 2-acetamido-2-deoxy-D-galactosyl residues in glycolipids, *Carbohydr. Res. 39*: 179 (1975).

93. S. K. Kundu, R. W. Ledeen, and P. A. J. Gorin, Determination of position of substitution on 2-acetamido-2-deoxy-D-glucosyl residues in glycolipids, *Carbohydr. Res. 39*: 329 (1975).

94. B. Fournet, Y. Leroy, J. Montreuil, and H. Mayer, Analytical and preparative gas–liquid chromatography of methyl α-D-mannoside monomethyl ethers, *J. Chromatogr. 92*: 185 (1974).

95. B. Fournet, G. Strecker, Y. Leroy, and J. Montreuil, Gas–liquid chromatography and mass spectrometry of methylated and acetylated methyl glycosides. Application to the structural analysis of glycoprotein glycans, *Anal. Biochem. 116*: 489 (1981).

96. A. A. Akhrem, G. V. Avvakumov, and O. A. Strel'chyonok, Methylation analysis in glycoprotein chemistry: Low bleeding columns for the gas chromatographic analysis of methylated sugar derivatives, *J. Chromatogr. 176*: 207 (1979).

97. A. A. Akhrem, G. V. Avvakumov, I. V. Sidorova, and O. A. Strel'chyonok, Methylation analysis in glycoprotein chemistry: General procedure for quantification of the products of solvolysis of permethylated glycopeptides and glycoproteins, *J. Chromatogr. 180*: 69 (1979).

98. B. A. Dmitriev, L. V. Backinowsky, O. S. Chizhov, B. M. Zolotarev, and N. K. Kochetkov, Gas–liquid chromatography and mass spectrometry of aldonitrile acetates and partially methylated aldonitrile acetates, *Carbohydr. Res. 19*: 432 (1971).

99. F. R. Seymour, R. D. Plattner, and M. E. Slodki, Gas–liquid chromatography–mass spectrometry of methylated and dueteriomethylated per-*O*-acetyl-aldonitriles from D-mannose, *Carbohydr. Res. 44*: 181 (1975).

100. F. R. Seymour, M. E. Slodki, R. D. Plattner, and A. Jeanes, Six unusual dextrans: Methylation structural analysis by combined G.L.C.-M.S. of per-*O*-acetyl-aldonitriles, *Carbohydr. Res. 53*: 153 (1977).

101. D. Rolf and G. R. Gray, Reductive cleavage of glycosides, *J. Am. Chem. Soc. 104*: 3539 (1982).

102. G. R. Gray, Reductive cleavage of permethylated polysaccharides, *Methods Enzymol. 138*: 26 (1987).

103. G. R. Gray, Linkage analysis using reductive cleavage method, *Methods Enzymol. 193*: 573 (1990).

104. G. R. Gray, Characterization of complex carbohydrates by the reductive cleavage method, *Frontiers in Carbohydrate Research-2* (R. Chandrasekaran, ed.), Elsevier, New York, 1992, p. 154.

105. J. Lönngren and Å. Pilotti, Gas–liquid chromatography of partially methylated alditols as their acetates II, *Acta Chem. Scand. 25*: 1144 (1971).

106. G. M. Bebault, G. G. S. Dutton, and R. H. Walker, Separation by gas–liquid chromatography of tetra-*O*-methylaldohexoses and other sugars as acetates, *Carbohydr. Res. 23*: 430 (1972).

107. N. Shibuya, Gas–liquid chromatographic analysis of partially methylated alditol acetates on a glass capillary column, *J. Chromatogr. 208*: 96 (1981).

108. S. Hakomori, E. Nudelman, S. Levery, D. Solter, and B. B. Knowles, The hapten structure of a developmentally regulated glycolipid antigen (SSEA-1) isolated from human erythrocytes and adenocarcinoma: A preliminary note, *Biochem. Biophys. Res. Commun. 100*: 1578 (1981).

109. S. Hakomori, E. Nudelman, S. B. Levery, and R. Kannagi, Novel fucolipids accumulating in human adenocarcinoma. I. Glycolipids with di- or trifucosylated type 2 chain, *J. Biol. Chem. 259*: 4672 (1984).

110. R. Geyer, H. Geyer, S. Kühnhardt, W. Mink, and S. Stirm, Capillary gas chromatography of methylhexitol acetates obtained upon methylation of *N*-glycosidically linked glycoprotein oligosaccharides, *Anal. Biochem. 121*: 263 (1982).

111. J. A. Lomax, A. H. Gordon, and A. Chesson, A multiple-column approach to the methylation analysis of plant cell-walls, *Carbohydr. Res. 138*: 177 (1985).

112. P. J. Harris, A. Bacic, and A. E. Clarke, Capillary gas chromatography of partially methylated alditol acetates on a SP-2100 wall-coated open-tubular column, *J. Chromatogr. 350*: 304 (1985).

113. J. Klok, H. C. Cox, J. W. De Leeuw, and P. A. Schenck, Analysis of synthetic mixtures of partially methylated alditol acetates by capillary gas chromatography, gas chromatography–electron impact mass spectrometry and gas-chemical ionization mass spectrometry, *J. Chromatogr. 253*: 55 (1982).

114. A. B. Blakeney, P. J. Harris, R. J. Henry, B. A. Stone, and T. Norris, Gas chromatography of alditol acetates on a high-polarity bonded-phase vitreous silica column, *J. Chromatogr. 249*: 180 (1982).

115. S. Hakomori, E. Nudelman, R. Kannagi, and S. B. Levery, The common structure in fucosyllactosaminolipids accumulating in human adenocarcinomas, and its possible absence in normal tissue, *Biochem. Biophys. Res. Commun. 109*: 36 (1982).

116. S. Hakomori, E. Nudelman, S. B. Levery, and C. M. Patterson, Human cancer-associated gangliosides defined by a monoclonal antibody (IB9) directed to sialosylα2-6 galactosyl residue: A preliminary note, *Biochem. Biophys. Res. Commun. 113*: 791 (1983).

117. E. D. Nudelman, S. Hakomori, R. Kannagi, S. Levery, M.-Y. Yeh, K. E. Hellström, and I. Hellström, Characterization of a human melanoma-associated ganglioside antigen defined by a monoclonal antibody, 4.2, *J. Biol. Chem. 257*: 12,752 (1982).

118. E. G. Bremer, S. B. Levery, S. Sonnino, R. Ghidoni, S. Canevari, R. Kannagi, and S. Hakomori, Characterization of a glycosphingolipid antigen defined by the monoclonal antibody MBr1 expressed in normal and neoplastic epithelial cells of human mammary gland, *J. Biol. Chem. 259*: 14,773 (1984).

119. H. Clausen, S. B. Levery, E. Nudelman, S. Tsuchiya, and S. Hakomori, Repetitive A epitope (type 3 chain A) defined by blood group A_1-specific monoclonal antibody TH-1: Chemical basis of qualitative A_1 and A_2 distinction, *Proc. Natl. Acad. Sci. USA 82*: 1199 (1985).

120. H. Clausen, S. B. Levery, J. M. McKibbin, and S. Hakomori, Blood group A determinants with mono- and difucosyl type 1 chain in human erythrocyte membranes, *Biochemistry 24*: 3578 (1985).

121. E. T. Oakley, D. F. Magin, G. H. Bokelman, and W. S. Ryan, Jr., Preparation, separation and identification of partially methylated alditol acetates for use as standards in methylation analysis, *J. Carbohydr. Chem. 4*: 53 (1985).

122. A. Bacic, P. J. Harris, E. W. Hak, and A. E. Clarke, Capillary gas chromatography of partially methylated alditol acetates on a high-polarity bonded-phase vitreous-silica column, *J. Chromatogr. 315*: 373 (1984).

123. H. Björndal, B. Lindberg, Å. Pilotti, and S. Svensson, Mass spectra of partially methylated alditol acetates. Part II. Deuterium labelling experiments, *Carbohydr. Res. 15*: 339 (1970).

124. H. B. Borén, P. J. Garegg, B. Lindberg, and S. Svensson, Mass spectra of partially methylated alditol acetates. Part III. Labelling experiments in the mass spectrometry of partially methylated deoxyalditol acetates, *Acta Chem. Scand. 25*: 3299 (1971).

125. K. Axberg, H. Björndal, Å. Pilotti, and S. Svensson, Mass spectra of partially methylated alditol acetates. Part IV. Deuterium labelling experiments on some higher fragments, *Acta Chem. Scand. 26*: 1319 (1972).

126. R. Wait, Structural analysis of carbohydrates by mass spectrometry, *Adv. Carbohydr. Anal. 1*: 335 (1991).

127. J. Lönngren and S. Svensson, Mass spectrometry in structural analysis of natural carbohydrates, *Adv. Carbohydr. Chem. Biochem. 29*: 41 (1974).

128. R. A. Hancock, K. Marshall, and H. Weigel, Structure of the levan elaborated by *Streptococcus salivarus* Strain 51: An application of chemical-ionization mass-spectrometry, *Carbohydr. Res. 49*: 351 (1976).

129. M. McNeil and P. Albersheim, Chemical-ionization mass spectrometry of methylated hexitol acetates, *Carbohydr. Res. 56*: 239 (1977).

130. R. A. Laine, L. Hodges, and A. M. Cary, Carbohydrate composition at high sensitivity using CH_4-chemical ionization GLC-mass spectrometry of the alditol acetates, *J. Supramol. Struct. Suppl., 15*: 31 (1977).

131. T. C-Y. Hsieh, K. Kaul, R. A. Laine, and R. L. Lester, Structure of a major glycophosphoceramide from tobacco leaves, PSL-I: 2-Deoxy-2-acetamido-D-glucopyranosyl($\alpha1\rightarrow4$)-D-glucuronopyranosyl($\alpha1\rightarrow2$)myoinositol-1-*O*-phosphoceramide, *Biochemistry 17*: 3575 (1978).

132. R. A. Laine, Chemical ionization GC-mass spectrometry in the structural analysis of saccharide chains, *27th International Congress of Pure and Applied Chemistry, Helsinki, 1979* (A. Varmavuori, ed.), Pergamon Press, New York, 1980, p. 193.

133. R. A. Laine, Enhancement of detection for partially methylated alditol acetates by chemical ionization mass spectrometry, *Anal. Biochem. 116*: 383 (1981).

134. K. Barr, R. A. Laine, and R. L. Lester, Carbohydrate structures of three novel phosphoinositol-containing sphingolipids from the yeast *Histoplasma capsulatum*, *Biochemistry 23*: 5589 (1984).

135. C. E. Roberts and S. B. Levery, Analysis of partially methylated alditol acetates of 6-deoxyhexoses and 2-deoxy-2-acetamidohexoses by gas chromatography-isobutane chemical ionization mass spectrometry, *38th ASMS Conference on Mass Spectrometry and Allied Topics, 1990, Tucson, AZ, Collected Abstracts*, p. 1373.

136. R. Kannagi, S. B. Levery, and S. Hakomori, Hybrid type glycolipids (lacto-ganglio series) with a novel branched structure: Their presence in undifferentiated murine leukemia cells and their dependence on differentiation, *J. Biol. Chem. 259*: 8444 (1984).

137. H. Clausen, S. B. Levery, E. Nudelman, M. Baldwin, and S. Hakomori, Further characterization of type 2 and type 3 chain blood group A glycosphingolipids from human erythrocyte membranes, *Biochemistry 25*: 7075 (1986).

138. P. Prehm, S. Stirm, B. Jann, and K. Jann, Cell wall polysaccharide from *Escherichia coli* B, *Eur. J. Biochem. 56*: 41 (1975).

139. W. S. York, 3-deoxy-D-*manno*-2-octulosonic acid (KDO) is a component of rhamnogalacuronan II, a pectic polysaccharide in the primary cell walls of plants, *Carbohydr. Res. 138*: 109 (1985).

140. A. Tacken, H. Brade, F. M. Unger, and D. Charon, G.L.C.-M.S. of partially methylated and acetylated derivatives of 3-deoxyoctitols, *Carbohydr. Res. 149*: 263 (1986).

141. A. Tacken, E. T. Rietschel, and H. Brade, Methylation analysis of the heptose/3-deoxy-D-*manno*-2-octulosonic acid region (inner core) of the lipopolysaccharide from *Salmonella minnesota* rough mutants, *Carbohydr. Res. 149*: 279 (1986).

142. A. K. Bhattacharjee and H. J. Jennings, Determination of the linkages in some methylated, sialic acid-containing, Meningococcal polysaccharides by mass spectrometry, *Carbohydr. Res. 51*: 253 (1976).

143. J. Haverkamp, J. P. Kamerling, J. F. G. Vliegenthart, R. W. Veh, and R. Schauer, Methylation analysis determination of acylneuraminic acid residue type 2→8 glyco-

sidic linkage. Application to GTlb ganglioside and colominic acid, *FEBS Lett. 73*: 215 (1977).

144. H. Rauvala, and J. Kärkkäinen, Methylation analysis of neuraminic acids by gas chromatography–mass spectrometry, *Carbohydr. Res. 56*: 1 (1977).

145. H. van Halbeek, J. Haverkamp, J. P. Kamerling, J. F. G. Vliegenthart, C. Versluis, and R. Schauer, Sialic acids in permethylation analysis: Preparation and identification of partially *O*-methylated derivatives of methyl *N*-acetyl-*N*-methyl-β-D-neuraminate methyl glycoside, *Carbohydr. Res. 60*: 51 (1978).

146. C. Bruvier, Y. Leroy, J. Montreuil, B. Fournet, and J. P. Kamerling, Gas–liquid chromatography and mass spectrometry of methyl ethers of methyl *N*-acetyl-*N*-methyl-β-D-neuraminate methyl glycoside, *J. Chromatogr. 210*: 487 (1981).

147. S. Inoue and G. Matsumura, Identification of the *N*-glycolylneuraminyl-(2→8)-*N*-glycolylneuraminyl group in a trout-egg glycoprotein by methylation analysis and gas–liquid chromatography-mass spectrometry, *Carbohydr. Res. 74*: 361 (1979).

148. S. Inoue and G. Matsumura, Stability of *N*-acyl groups of neuraminic acid residues in 2→8-linked polymers toward methanolysis used in methylation analysis, *FEBS Lett. 121*: 33 (1980).

149. S. Inoue, G. Matsumura, and Y. Inoue, Stability of *N*-acyl groups in methylated α2→8-linked oligosialosyl chains toward methanolysis. Analysis by chemical ionization–mass spectrometry, *Anal. Biochem. 125*: 118 (1982).

150. R. K. Yu and R. W. Ledeen, Gas–liquid chromatographic assay of lipid-bound sialic acids: Measurement of gangliosides in brain of several species, *J. Lipid Res. 11*: 506 (1970).

151. S. Inoue and M. Iwasaki, Characterization of a new type of glycoprotein saccharide containing polysialosyl sequence, *Biochem. Biophys. Res. Commun. 93*: 162 (1980).

152. M. Iwasaki, S. Inoue, K. Kitajima, H. Nomoto, and Y. Inoue, Novel oligosaccharide chains on polysialoglycoproteins isolated from rainbow trout eggs. A unique carbohydrate sequence with a sialidase-resistant sialyl group, DGalNAcβ1→4 (NeuGc2→3)DGalNAc, *Biochemistry 23*: 305 (1984).

153. M. Shimamura, T. Endo, Y. Inoue, S. Inoue, and H. Kambara, Fish egg sialoglycoproteins: Structures of new sialooligosaccharide chains isolated from the eggs of *Onchorhynchus keta* (Walbaum). Fucose-containing units with oligosialyl groups, *Biochemistry 23*: 317 (1984).

154. A. van Langenhove and V. N. Reinhold, Determination of polysaccharide linkage and branching by reductive depolymerization. Gas–liquid chromatography and gas–liquid chromatography–mass spectrometry reference data, *Carbohydr. Res. 143*: 1 (1985).

155. A. S. B. Edge, A. van Langenhove, V. N. Reinhold, and P. Weber, Characterization of *O*-glycosidically linked oligosaccharides of rat erythrocyte membrane sialoglycoproteins, *Biochemistry 25*: 8017 (1986).

156. J.-G. Jun and G. R. Gray, A new catalyst for reductive cleavage of methylated glycans, *Carbohydr. Res. 163*: 247 (1987).

157. D. Rolf, J. A. Bennek, and G. R. Gray, Analysis of linkage positions in D-glucopyranosyl residues by the reductive-cleavage method, *Carbohydr. Res. 137*: 183 (1985).

158. J. U. Bowie, P. V. Trescony, and G. R. Gray, Analysis of linkage positions in *Saccharomyces cerevisiae* D-mannans by the reductive-cleavage method, *Carbohydr. Res. 125*: 301 (1984).

159. D. Rolf, J. A. Bennek, and G. R. Gray, Reductive cleavage of glycosides. Stereochemistry of trapping of cyclic oxonium ions, *J. Carbohydr. Chem. 2*: 373 (1983).

160. C. K. Lee and G. R. Gray, A general strategy for the chemical sequencing of polysaccharides, *J. Am. Chem. Soc. 110*: 1292 (1988).

161. J. A. Bennek, M. J. Rice, and G. R. Gray, Analysis of linkage positions in 2-acetam-ido-2-deoxy-D-glucopyranosyl residues by the reductive-cleavage method, *Carbohydr. Res. 157*: 125 (1986).

162. A. J. D'Ambra and G. R. Gray, An improved procedure for the analysis of linkage positions in 2-acetamido-2-deoxy-D-glucopyranosyl residues by the reductive-cleav-age method, *Carbohydr. Res. 251*: 115 (1994).

163. A. J. D'Ambra and G. R. Gray, Analysis by the reductive-cleavage method of a polysaccharide containing 2-acetamido-2,6-dideoxy-D- and L-galactopyranosyl resi-dues, *Carbohydr. Res. 251*: 127 (1994).

164. V. Srivastava and G. R. Gray, Structural analysis of sialic acid–containing carbohy-drates by the reductive-cleavage method, *Carbohydr. Res. 248*: 167 (1993).

165. J. U. Bowie and G. R. Gray, Synthesis and mass spectra of the partially methylated and partially ethylated anhydro-D-mannitol acetates derived by reductive cleavage of permethylated and perethylated *Saccharomyces cerevisiae* D-mannans, *Carbohydr. Res. 129*: 87 (1984).

166. S. G. Zeller, A. J. D'Ambra, M. J. Rice, and G. R. Gray, Synthesis and mass spectra of 4-*O*-acetyl-1,5-anhydro-2,3,6-tri-*O*-ethyl-D-glucitol and the positional isomers of 4-*O*-acetyl-1,5-anhydro-di-*O*-ethyl-*O*-methyl-D-glucitol and 4-*O*-acetyl-1,5-anhydro-*O*-ethyl-di-*O*-methyl-D-glucitol, *Carbohydr. Res. 182*: 53 (1988).

167. L. E. Elvebak II, T. Schmitt, and G. R. Gray, Authentic standards for the reductive-cleavage method. The positional isomers of partially methylated and acetylated or benzoylated 1,5-anhydro-D-fucitol, *Carbohydr. Res. 246*: 1 (1993).

168. L. E. Elvebak II, C. Abbott, S. Wall, and G. R. Gray, Authentic standards for the reductive-cleavage method. The positional isomers of partially methylated and acet-ylated or benzoylated 1,5-anhydro-D-galactitol, *Carbohydr. Res. 269*: 1 (1995).

169. N. Wang and G. R. Gray, Authentic standards for the reductive-cleavage method. The positional isomers of partially methylated and acetylated or benzoylated 1,4-anhydro-D-xylitol, *Carbohydr. Res. 274*: 45 (1995).

170. N. Wang, L. E. Elvebak II, and G. R. Gray, Synthesis and characterization of authentic standards for the reductive-cleavage method. The positional isomers of partially meth-ylated and acetylated or benzoylated 1,4-anhydro-L-fucitol, *Carbohydr. Res. 274*: 59 (1995).

171. L. E. Elvebak II, H. J. Cha, P. McNally, and G. R. Gray, Authentic standards for the reductive-cleavage method. The positional isomers of partially methylated and acet-ylated or benzoylated 1,5-anhydro-D-mannitol, *Carbohydr. Res. 274*: 71 (1995).

172. L. E. Elvebak II and G. R. Gray, Authentic standards for the reductive-cleavage method. The positional isomers of partially methylated and acetylated or benzoylated 1,5-anhydro-D-glucitol, *Carbohydr. Res. 274*: 85 (1995).

173. C. R. Rozanas, N. Wang, K. Vidlock, and G. R. Gray, Synthesis and characterization of authentic standards for the analysis of ribofuranose-containing carbohydrates by the reductive-cleavage method, *Carbohydr. Res. 274*: 99 (1995).

174. E. J. McGuire and S. B. Binkley, The structure and chemistry of colominic acid, *Biochemistry 3*: 247 (1964).

175. R. Kuhn and H. Müldner, Über glyko-lipo-sialo-proteide des gehirns, *Naturwissen-schaften 51*: 635 (1964).

176. H. Wiegandt, Ganglioside, *Ergeb. Physiol. Biochem. Exptl. Pharm. 57*: 190 (1966).

177. J. E. Evans and R. H. McCluer, Synthesis and characterization of sialosyllactosylcer-amide inner ester, *Fed. Proc. 30*: 1133 (1971).

178. R. H. McCluer and J. E. Evans, Ganglioside inner esters, *Adv. Exp. Med. Biol. 19*: 95 (1972).

179. H. Wiegandt, Gangliosides of extraneural organs, *Hoppe-Seyler's Z. Physiol. Chem. 354*: 1049 (1973).

180. R. K. Yu, T. A. W. Koerner, S. Ando, H. C. Yohe, and J. H. Prestegard, High-resolution proton NMR studies of lactones. III. Elucidation of the structure of ganglioside GM3 lactone, *J. Biochem. (Tokyo) 98*: 1367 (1985).

181. S. Sonnino, G. Kirschner, G. Fronza, H. Egge, R. Ghidoni, D. Acquotti, and G. Tettamanti, Synthesis of GM1-ganglioside inner ester, *Glycoconj. J. 2*: 343 (1985).

182. D. Acquotti, G. Fronza, L. Riboni, S. Sonnino, and G. Tettamanti, Gangioside lactones: 1H-NMR determination of the inner ester position of GD1b-ganglioside lactone naturally occurring in human brain or produced by chemical synthesis, *Glycoconj. J. 4*: 119–127 (1987).

183. G. Fronza, G. Kirschner, D. Acquotti, R. Bassi, L. Tagliavacca, and S. Sonnino, Synthesis and structural characterization of the dilactone derivative of GD1a ganglioside, *Carbohydr. Res. 182*: 31 (1988).

184. S. Ando, R. K. Yu, J. N. Scarsdale, S. Kusunoki, and J. H. Prestegard, High-resolution proton NMR studies of gangliosides: Structure of two types of GD3 lactones and their reactivity with monoclonal antibody R24, *J. Biol. Chem. 264*: 3478 (1989).

185. S. K. Gross, M. A. Williams, and R. H. McCluer, Alkali-labile, sodium borohydride-reducible ganglioside sialic acid residues in brain, *J. Neurochem. 34*: 1351 (1980).

186. L. Riboni, S. Sonnino, D. Acquotti, A. Malesci, R. Ghidoni, H. Egge, S. Mingrino, and G. Tettamanti, Natural occurrence of ganglioside lactones: Isolation and characterization of GD1b inner ester from adult human brain, *J. Biol. Chem. 261*: 8514 (1986).

187. G. A. Nores, T. Dohi, M. Taniguchi, and S. Hakomori, Density-dependent recognition of cell surface G_{M3} by a certain anti-melanoma antibody, and G_{M3} lactone as a possible immunogen: Requirements for tumor-associated antigen and immunogen, *J. Immunol. 139*: 3171 (1987).

188. T. Tai, I. Kawashima, N. Tada, and S. Ikegami, Different reactivities of monoclonal antibodies to ganglioside lactones, *Biochem. Biophys. Acta 958*: 134 (1988).

189. T. Tai, I. Kawashima, N. Tada, and T. Fujimori, Mouse monoclonal antibodies to ganglioside GD2: Characterization of the fine specificities, *Gangliosides in Cancer* (H. F. Oettgen, ed.), Verlagsgesellschaft, Weinheim, 1989, p. 149.

190. B. Bouchon, S. B. Levery, H. Clausen, and S. Hakomori, Production and characterization of a monoclonal antibody (BBH5) directed to ganglioside lactone, *Glycoconj. J. 9*: 27 (1992).

191. R. Bassi, L. Riboni, S. Sonnino, and G. Tettamanti, Lactonization of GD1b ganglioside under acidic conditions, *Carbohydr. Res. 193*: 141 (1989).

192. S. Hoffman, B. C. Sorkin, P. C. White, R. Brackenbury, R. Mailhammer, U. Rutishauser, B. A. Cunningham, and G. M. Edelman, Chemical characterization of a neural cell adhesion molecule purified from embryonic brain membranes, *J. Biol. Chem. 257*: 7720 (1982).

193. J. Finne, Occurrence of unique polysialosyl carbohydrate units in glycoproteins of developing brain, *J. Biol Chem. 257*: 11966 (1982).

194. K. Kitajima, H. Nomoto, Y. Inoue, M. Iwasaki, and S. Inoue, Fish egg polysialoglycoproteins: Circular dichroism and proton nuclear magnetic resonance studies of novel oligosaccharide units containing one sialidase-resistant *N*-glycolylneuraminic acid residue in each molecule, *Biochemistry 23*: 310 (1984).

195. Evidence for unique homologous peptide sequences around the glycosylated seryl and threonyl residues in polysialoglycoproteins isolated from the unfertilized eggs of the Pacific salmon *Onchorynchus keta*, *Biochemistry 24*: 5470 (1985).

196. Polysialoglycoproteins of *Salmonidae* fish eggs. Complete structure of 200-kDa polysialoglycoprotein from the unfertilized eggs of rainbow trout (*Salmo gairdneri*), *J. Biol. Chem. 261*: 5262 (1986).

197. F. A. Troy, II, Polysialylation: From bacteria to brains, *Glycobiology 2*: 5 (1992).

198. M. R. Lifely, A. S. Gilbert, and C. Moreno, Sialic acid polysaccharide antigens of *Neisseria meningitidis* and *Escherichia coli*: Esterification between adjacent residues, *Carbohydr. Res. 94*: 193 (1981).

199. M. R. Lifely, A. S. Gilbert, and C. Moreno, Rate, mechanism, and immunochemical studies of lactonisation in serogroup B and C polysaccharides of *Neisseria meningitidis*, *Carbohydr. Res. 134*: 229 (1984).

200. H. Egge, J. Peter-Katalinić, G. Reuter, R. Schauer, R. Ghidoni, S. Sonnino, and G. Tettamanti, Analysis of gangliosides using fast atom bombardment mass spectrometry, *Chem. Phys. Lipids 37*: 127 (1985).

201. S. B. Levery, M. E. K. Salyan, C. E. Roberts, B. Bouchon, and S. Hakomori, Strategies for characterization of ganglioside inner esters. I. Fast atom bombardment mass spectrometry, *Biomed. Env. Mass. Spectrom. 19*: 303 (1990).

202. M. R. Lifely and F. H. Cottee, Formation and identification of two novel anhydro compounds obtained by methanolysis of *N*-acetylneuraminic acid and carboxyl-reduced, meningococcal B polysaccharide, *Carbohydr. Res. 107*: 187 (1982).

203. S. K. Gross, J. E. Evans, V. N. Reinhold, and R. H. McCluer, Identification of the internal acetal 5-acetamido-2,7-anhydro-3,5-dideoxy-D-*glycero*-D-*galacto*-nonulopyranose, *Carbohydr. Res. 41*: 344 (1975).

204. S. B. Levery, C. E. Roberts, M. E. K. Salyan, and S. Hakomori, A novel strategy for unambiguous determination of inner esterification sites of ganglioside lactones, *Biochem. Biophys. Res. Commun. 162*: 838 (1989).

205. S. B. Levery, C. E. Roberts, M. E. K. Salyan, B. Bouchon, and S. Hakomori, Strategies for characterization of ganglioside inner esters. II. Gas chromatography/mass spectrometry, *Biomed. Env. Mass Spectrom. 19*: 311 (1990).

206. H. Egge and J. Peter-Katalinić, Fast atom bombardment mass spectrometry for structural elucidation of glycoconjugates, *Mass Spectrom. Rev. 6*: 331 (1987).

207. A. Dell, F.A.B.-mass spectrometry of carbohydrates, *Adv. Carbohydr. Chem. Biochem. 45*: 19 (1987).

208. E. J. Corey, J.-L. Gras, and P. Ulrich, A new general method for protection of the hydroxyl function, *Tetrahedron Lett.* 809 (1976).

209. E. J. Corey and R. H. Wollenburg, Useful stereoselective and position-selective transformations of Brefeldin A and derivatives at carbons 4, 7, and 15, *Tetrahedron Lett.* 4701 (1976).

210. J. Kärkkäinen, Analysis of disaccharides as permethylated disaccharide alditols by gas-liquid chromatography-mass spectrometry, *Carbohydr. Res. 14*: 27 (1970).

211. J. Kärkkäinen, Structural analysis of trisaccharides as permethylated trisaccharide alditols by gas-liquid chromatography-mass spectrometry, *Carbohydr. Res. 17*: 11 (1971).

212. I. Mononen, J. Finne, and J. Kärkkäinen, Analysis of permethylated hexopyranosyl-2-acetamido-2-deoxyhexitols by g.l.c.-m.s., *Carbohydr. Res. 60*: 371 (1978).

213. A. Lundblad, P. K. Masson, N. E. Nordén, S. Svensson, P.-A. Öckerman, Determination of a mannose-containing trisaccharide in the urine of patients and heterozygotes of mannosidosis, *Biomed. Mass Spectrom. 2*: 285 (1975).

214. G. Lennartson, A. Lundblad, S. Sjöblad, S. Svensson, and P.-A. Öckerman, Quantitation of a urinary tetrasaccharide by gas chromatography and mass spectrometry, *Biomed. Mass Spectrom. 3*: 51 (1976).

215. P. Hallgren and A. Lundblad, Structural analysis of nine oligosaccharides isolated from the urine of a blood group O, nonsecretor, woman during pregnancy and lactation, *J. Biol. Chem. 252*: 1014 (1977).

216. P. Hallgren and A. Lundblad, Structural analysis of oligosaccharides isolated from the urine of a blood group A, secretor, woman during pregnancy and lactation, *J. Biol. Chem. 252*: 1023 (1977).

217. P. Hallgren, B. S. Lindberg, and A. Lundblad, Quantitation of some urinary oligo-saccharides during pregnancy and lactation, *J. Biol. Chem. 252*: 1034 (1977).

218. B. Nilsson, N. E. Nordén, and S. Svensson, Structural studies on the carbohydrate portion of fetuin, *J. Biol. Chem. 254*: 4545 (1979).

219. B. Fournet, J.-M. Dhalluin, G. Strecker, J. Montreuil, C. Bosso, and J. Defaye, Gas−liquid chromatography and mass spectrometry of oligosaccharides obtained by partial acetolysis of glycans of glycoproteins, *Anal. Biochem. 108*: 35 (1980).

220. I. Funakoshi and I. Yamashina, Structure of *O*-glycosidically linked sugar units from plasma membranes of an ascites hepatoma, AH 66, *J. Biol. Chem. 257*: 3782 (1982).

221. A. Pierce-Cretel, M. Pamblanco, G. Strecker, J. Montreuil, and G. Spik, Heterogeneity of the glycans *O*-glycosidically linked to the hinge region of secretory immunoglob-ulins from human milk, *Eur. J. Biochem. 114*: 169 (1981).

222. H. van Halbeek, L. Dorland, J. Haverkamp, G. A. Veldink, J. F. G. Vliegenthart, B. Fournet, G. Ricart, J. Montreuil, W. D. Gathmann, and D. Aminoff, Structure deter-mination of oligosaccharides isolated from A$^+$, H$^+$, A$^-$H$^-$ hog-submaxillary-gland mucin glycoproteins, by 360-MHz ^{1}H-NMR spectroscopy, permethylation analysis and mass spectrometry, *Eur. J. Biochem. 118*: 487 (1981).

223. A. Kurosaka, H. Nakajima, I. Funakoshi, M. Matsuyama, T. Nagayo, and I. Yamash-ina, Structures of the major oligosaccharides from a human rectal adenocarcinoma glycoprotein, *J. Biol. Chem. 258*: 11594 (1983).

224. T. Tsuji and T. Osawa, Carbohydrate structures of bovine submaxillary mucin, *Car-bohydr. Res. 151*: 391 (1986).

225. M. A. Schmidt and K. Jann, Structure of the 2-keto-3-deoxy-D-*manno*-octonic-acid-containing capsular polysaccharide (K12 antigen) of the urinary-tract-infective *Escherichia coli* O4:K12:H$^-$, *Eur. J. Biochem. 131*: 509 (1983).

226. W.-T. Wang, F. Matsuura, and C. C. Sweeley, Chromatography of permethylated oli-gosaccharide alditols, *Anal. Biochem. 134*: 398 (1983).

227. W.-T. Wang, N. C. LeDonne, Jr., B. Ackerman, and C. C. Sweeley, Structural char-acterization of oligosaccharides by high-performance liquid chromatography, fast atom bombardment−mass spectrometry, and exoglycosidase digestion, *Anal. Biochem. 141*: 366 (1984).

228. H. Brade, U. Zahringer, E. T. Rietschel, R. Christian, G. Schulz, and F. M. Unger, Spectroscopic analysis of a 3-deoxy-D-*manno*-2-octulosonic acid (KDO)-disaccharide from the lipopolysaccharide of a *Salmonella godesberg* Re mutant, *Carbohydrate Res. 134*: 157 (1984).

229. H. Brade and E. T. Rietschel, α-2→4-Interlinked 3-deoxy-D-*manno*-octulosonic acid disaccharide. A common constituent of enterobacterial lipopolysaccharides, *Eur. J. Biochem. 145*: 231 (1984).

230. H. Brade, H. Moll, and E. T. Rietschel, Structural investigations on the inner core region of lipopolysaccharides from *Salmonella minnesota* rough mutants, *Biomed. Mass Spectrom. 12*: 602 (1985).

231. H. Karlsson, I. Carlstedt, and G. C. Hansson, Rapid characterization of mucin oli-gosaccharides from rat small intestine with gas chromatography-mass spectrometry, *FEBS Lett. 226*: 23 (1987).

232. H. Karlsson and G. C. Hansson, Gas chromatography and gas chromatography/mass spectrometry for the characterization of complex mixtures of large oligosaccharides, *J. High Res. Chromatogr. & Chromatogr. Commun 11*: 820 (1988).

233. G. C. Hansson, Y.-T. Li, and H. Karlsson, Characterization of glycosphingolipid mixtures with up to ten sugars by gas chromatography and gas chromatography−mass spectrometry as permethylated oligosaccharides and ceramides released by ceramide glycanase, *Biochemistry 28*: 6672 (1989).

234. H. Karlsson, I. Carlstedt, and G. C. Hansson, The use of gas chromatography and gas chromatography-mass spectrometry for the characterization of permethylated oligosaccharides with molecular mass up to 2300, *Anal. Biochem. 182*: 438 (1989).

235. G. C. Hansson and H. Karlsson, High-mass gas chromatography–mass spectrometry of permethylated oligosaccharides, *Methods Enzymol. 193*: 733 (1990).

236. G. C. Hansson, J.-F. Bouhours, H. Karlsson, and I. Carlstedt, Analysis of sialic acid-containing mucin oligosaccharides from porcine small intestine by high-temperature gas chromatography-mass spectrometry of their dimethylamides, *Carbohydr. Res. 221*: 179 (1991).

237. G. C. Hansson and H. Karlsson, Gas chromatography and gas chromatography-mass spectrometry of glycoprotein oligosaccharides, *Methods in Molecular Biology, Vol. 14* (E. F. Hounsell, ed.), Humana Press, Totawa, 1993, p. 47.

238. H. Karlsson, N. Karlsson, and G. C. Hansson, High-temperature gas chromatography-mass spectrometry of glycoprotein and glycosphingolipid oligosaccharides, *Molecular Biotechnology 1*: 165 (1994).

239. I. Carlstedt, A. Herrmann, H. Karlsson, J. Sheehan, L.-Å. Fransson, and G. C. Hansson, Characterization of two different glycosylated domains from the insoluble mucin complex of rat small intestine, *J. Biol. Chem. 268*: 18771 (1993).

240. N. G. Karlsson, H. Karlsson, and G. C. Hansson, Strategy for the investigation of O-linked oligosaccharides from mucins based on the separation into neutral, sialic acid- and sulfate-containing species, *Glycoconj. J. 12*: 69 (1995).

12

Structural Determination of Protein-Bound Oligosaccharides by Mass Spectrometry

David J. Harvey
Glycobiology Institute, University of Oxford, Oxford, England

I. INTRODUCTION

Mass spectrometry has, until recently, only been useful for the examination of compounds having a relatively low molecular weight. In recent years, however, methods have become available that allow large and fragile molecules to be ionized, and mass spectrometry is increasingly being accepted as a biochemical tool, particularly in the field of protein research. Predictions have even been made that most structural work in this area will be performed by mass spectrometry within a few years. Glycoproteins and their constituent carbohydrates are no exception to this trend. This chapter aims to review work in the application of mass spectrometry to the structural analysis of these compounds.

II. GENERAL STRUCTURE OF GLYCOPROTEINS AND THEIR CONSTITUENT CARBOHYDRATES

Proteins may be modified by the addition of carbohydrates in three ways [1]: attachment to asparagine by an amide bond, attachment to serine or threonine through an ether bond, or attachment to the carboxy terminus in the form of a glycosyl-phosphatidylinositol lipid anchor. The first two types of carbohydrate are generally known as N-linked and O-linked, respectively. This chapter is concerned with the application of mass spectrometry to the structural elucidation of these two structural types, with emphasis on the N-linked structures.

   N-linked oligosaccharides are synthesized predominantly in the endoplasmic reticulum and Golgi apparatus [2]. First, the oligosaccharide moiety of the glycolipid dolichol-pyrophosphate-$(Glc)_3(Man)_9(GlcNAc)_2$, where Glc = glucose, Man = mannose, and GlcNAc = N-acetylglucosamine, is cotranslationally attached to asparagine in an asparagine-X-serine (or threonine) motif, where X is not proline. A series of glycan-processing enzymes then removes the three glucose residues and four of the mannoses to give $(Man)_5(GlcNAc)_2$ (commonly known as ''Man-5''). Next, a GlcNAc residue is transferred to the 2-position of the mannose on the 3-linked antenna, after which two more mannoses can be trimmed off the 6-linked antenna to give $GlcNAc(Man)_3(GlcNAc)_2$. At this point GlcNAc is also transferred to the 6-linked antenna, after which (for a biantennary glycan) both antennae are extended by the addition of galactose, then (usually) sialic acid. The carbohydrate chains of the resulting compounds are known as ''complex'' oligosaccharides. If, for some reason, only the 3-linked antenna is extended, the chains are called ''hybrid.'' Further antennae can be added to the mannose residues at the 3- or 6-positions, giving tri- and tetra-antennary structures. Although a GlcNAc residue can also be added to the 4-position of the mannoses, further extension at this point is uncommon. This residue is usually referred to as a ''bisecting GlcNAc.'' Fucose (Fuc) can be added to the reducing-terminal GlcNAc residue or to the 3- or 2-positions of the outer-arm GlcNAc or galactose residues, respectively. Many other, less common modifications occur, particularly in organisms other than mammals; these can include addition of xylose and glucuronic acid, the addition of phosphate or sulfate and variations in the sialic acid structure.

   Although these N-linked structures are complex in the general sense of the word, they all contain the common core structure of $(Man)_3(GlcNAc)_2$. This property considerably facilitates their structural determination. *O*-Linked oligosaccharides, on the other hand, tend to be smaller, but structural determination is more difficult on account of their more varied compositions.

Unlike proteins or nucleic acids, whose primary structure is simply a linear chain of monomer units, carbohydrates contain a number of additional structural features that must be determined. First, many of the constituent monosaccharides are isobaric and, as such, are not readily identified directly by mass spectrometry. Second, chains can branch, leading to additional isomeric structures. And third, any of the hydroxy groups of a monosaccharide unit can be involved in bonding, or linkage as it is more commonly called, to the 1-hydroxy group of the next monosaccharide in the chain. Finally, the monosaccharide ring structure is important. Because a reducing monosaccharide can exist in both open- and closed-chain forms, there is the possibility that ring closure can give rise to different-size rings, depending on which hydroxy group is involved in ring formation. The rings may exist in different conformations, such as chair or boat; and when ring closure occurs, a new anomeric site is generated.

Further complications arise when the structure of the intact glycoprotein is addressed. Many glycoproteins have several *N*- and *O*-sites, each of which may or may not be occupied by a sugar. Furthermore, each site is usually occupied by a number of different oligosaccharides, in varying proportions.

No one mass spectrometric technique is able to address all of these structural problems. Indeed, until recently only relatively small oligosaccharides could be examined by mass spectrometry at all. Thus, several different techniques are necessary for complete structural definition. Even then, mass spectrometry is usually combined with other methods for a satisfactory analysis.

III. MASS SPECTROMETERS AND IONIZATION METHODS

Early forms of mass spectrometry were capable of examining only small sugars, such as monosaccharides. As the field has progressed, larger and larger molecules have become accessible, so today carbohydrates and glycoproteins with masses of over 200 kDa can be examined.

A. General Features of Mass Spectrometers

Mass spectrometry involves the production of ions from the sample molecules and the measurement of their motion in magnetic or electric fields in order to determine their mass. As many of the ions have short lifetimes, a number of structurally related fragment ions are produced; a plot of their relative abundance against mass [or, more accurately, mass:charge ratio (m/z), since the charge is not always unity] constitutes a mass spectrum.

Although many types of mass spectrometer exist [3], they all posses certain common features. Because they rely for their operation on the production and movement of ions in the gas phase, they contain a vacuum system and some method of sample introduction. Samples can either be evaporated outside the vacuum and introduced as a vapor or be introduced as a solid and evaporated in situ. Vapor can be introduced either in the form of a continuous stream or as the effluent from a chromatograph, either gas (gas chromatography/mass spectrometry, or GC/MS) or liquid (LC/MS), with a suitable interface. Methods for ion formation, dealt with later, are very sample-dependent, and it is progress in this area that has led to the main advances in applications to carbohydrate and glycoprotein mass spectrometry.

Ionization Methods

Electron-Impact (EI) Ionization.   Electron-impact (EI) ionization involves the bombardment of vaporized sample molecules with a beam of electrons, which has causes ionization by removal of an electron. The resulting "odd-electron" ions tend to be unstable and readily fragment to a large number of products. Derivatization of carbohydrates, in the form of methylation, acetylation, or trimethylsilylation [4] is necessary in order to vaporize them, but molecules as complex as phosphated monosaccharides can be examined [5,6]. GC/MS usually employs this ionization method and is used extensively to obtain composition and linkage information from carbohydrates, as described later. But its application to polysaccharides is limited to those with less than about seven constituent monosaccharides. Karlsson et al. [7] have reported a method using especially prepared columns for permethylated oligosaccharides, but the temperatures of over 400°C necessary for elution of compounds with more than six residues were reported to cause decomposition of the sugars.

Chemical Ionization (CI).   In response to the fact that many molecular ions generated by EI are weak or nonexistent, the technique of chemical ionization (CI) became popular in the 1970s. Molecules are ionized by collision in a relatively high-pressure ion source, with "reagent" ions generated from gases such as methane, isobutane, and ammonia. The resulting molecular ions from the sample are usually in the form of adducts such as MH^+ or MNa^+ or, in the case of negative ions, $[M - H]^-$. Because the energy transferred to the sample in this so-called "soft" ionization technique is less than in EI and because the ions are "even electron," fragmentation is minimal.

Fast Atom Bombardment (FAB) Mass Spectrometry.   Fast atom bombardment (FAB) mass spectrometry was introduced in 1981 [8] as a technique for ionizing molecules of intermediate mass (up to about 10 kDa). The sample, dissolved in a solvent of low volatility, is placed on a metal probe, introduced into the vacuum system of the mass spectrometer, and bombarded either with a beam of neutral argon or xenon atoms or with a beam of cesium ions. More accurately, the ion bombardment technique, known as liquid secondary-ion mass spectrometry (LSIMS), is somewhat more sensitive than FAB for larger molecules. Glycerol was the first matrix used for FAB studies; but for oligosaccharides, thioglycerol or a 1:1 mixture of glycerol and thioglycerol usually gives superior results.

Electrospray Ionization.   Electrospray mass spectrometry has made a significant contribution to our ability to examine intact glycoproteins. In this technique, a solution of the sample molecules is ejected from a capillary needle that carries a high electrical potential, on the order of 3–5 kV. This produces a spray of fine, charged droplets from which the solvent is evaporated to leave charged sample ions of the type MX_n^{n+}, where X = H or a metal, or $[M - H_n]^{n-}$. Larger molecules attract several charges, depending on factors such as the proton affinity and the number of chargeable sites, to give an envelope of ions differing by one charge state each. Deconvolution of the spectrum using a computer leads to a very accurate measurement of the molecular weight. Furthermore, because the ions are multiply charged, they appear in the mass spectrum at relatively low m/z values, allowing quite modest instruments such as quadrupoles to be used for their measurement even though masses of up to 100 kDa may be involved.

Matrix-Assisted Laser Desorption/Ionization (MALDI). The use of lasers without matrices to ionize large biological molecules, such as glycolipid lipid-A, predates MALDI by several years [9–12], but the incorporation of a matrix greatly extended the scope and ease of use of the technique. To obtain a spectrum, the sample, typically in the low picomole range, is mixed, in solution, with a large excess (about 5000-fold) of a suitable matrix dissolved in 1–2 μL of solvent, allowed to dry, and introduced into the mass spectrometer. Ionization is by irradiation with a laser, usually at 337 nm or 266 nm, and detection is usually with a time-of-flight (TOF) analyzer, although a magnetic sector instrument fitted with an array detector can be used [13].

The function of the matrix is to dilute the sample, minimize cluster formation, absorb the laser energy, and ionize the sample. Ionization mechanisms are not well understood, although Vertes et al. [14] have proposed a hydrodynamic model of the MALDI process in which a plume of matrix and sample molecules is generated from the target surface after heating by the laser beam. Subsequent rapid expansive cooling of the plume prevents thermal degradation of the sample. Most matrices are crystalline compounds, although a few liquids, such as m-nitrobenzyl alcohol, are effective. Matrices for positive-ion work are usually low molecular weight organic acids with low volatility and the ability to absorb energy at the laser frequency. Intimate contact between the matrix and the sample is essential, and it is believed that the sample must be entrained within the crystal lattice for successful ionization to occur. Proteins and glycoproteins are usually most successfully ionized with sinapinic acid (3,5-dimethoxy-4-hydroxycinnamic acid) [15], whereas lighter compounds give better responses with α-cyano-4-hydroxycinnamic acid [16] or 2,5-dihydroxybenzoic acid (2,5-DHB) [17].

Plasma Desorption. This technique is also performed with a TOF instrument but generally has lower resolution than MALDI, and spectra take considerably longer to acquire. Nevertheless, it has given results with intractable compounds such as heparin fragments. The sample, on a thin metal target, is placed into the spectrometer and bombarded with fission fragments from ^{252}californium. Fission of this isotope ejects two fragments in diametrically opposed directions; one fragment ionizes the sample, and the other starts the timing clock for the TOF process.

Mass Separation

Ion separation was originally performed with large magnets whose fields cause ions to traverse a path with a radius of curvature that is a function of the mass-to-charge (m/z) ratio of the ions. These instruments are large, have mass ranges of 1000 to about 10,000, and can achieve high resolutions. However, they are now generally being replaced by other instrumentation that tends to be smaller and more easy to control by data systems. Quadrupole instruments are of lower performance as far as mass range and resolution are concerned but are more easily automated and coupled to chromatographs. Mass separation is achieved electrically by means of alternating electric fields. A related instrument, the ion trap, also relies on electric fields for ion separation but has the capability of storing ions, thus allowing their fragmentation to be studied. Another storage device, the ion cyclotron resonance mass spectrometer, can achieve very high resolutions and sensitivity and is probably the instrument of the future. The only other instrument commonly encountered in carbohydrate work is the TOF analyzer. As its name implies, ion separation is a function of flight time in an evacuated tube. Ideally it is coupled to ion sources that emit pulses of ions. It has the advantage over other instruments of an almost unlimited mass range and is, thus, used extensively for high-mass work. In its linear form, resolution is low except with the recently developed delayed-extraction ion source, which can give resolutions approaching that of a large magnetic sector instrument [18]. An electric mirror, or reflectron, incorporated into a TOF instrument can also be used to increase resolution or to study fragmentations occurring during the ion's flight to the detector by a process known as "post source decay" (PSD).

Combination Instruments

These instruments, consisting of two mass spectrometers in series, can be used with most types of ion source. The first mass spectrometer selects a parent ion, usually a molecular ion, from the total ion cloud and injects it into a collision cell filled with an inert gas such as helium, argon, or xenon. Collision between the ions and the gas molecules induces fragmentation, the energy of which is a function of the initial ion velocity and the mass of the target gas. Fragment, or product, ions can then be analyzed by the second mass spectrometer. Such instruments can consist of many combinations of analyzer for the first or second stages, with magnetic sector instruments generally being employed as the first stage if high-energy collisions are required. Triple quadrupole instruments are probably the most common MS/MS instruments (the center quadrupole acts as the collision cell) but only provide low-energy collisions. The recent combination of a TOF analyzer to receive the product ions from a magnetic sector instrument appears to be a particularly attractive combination and is suitable for obtaining MS/MS spectra from MALDI ion sources [19]. An obvious advantage of MS/MS-type instruments is that they are able to select and fragment single components of mixtures, thus minimizing the amount of pre-mass spectrometric sample preparation.

B. Ionic and Molecular Masses Used in Mass Spectrometry

Three types of ionic masses are used in mass spectrometry, depending on the mass of the compound being examined and the resolution of the instrument. Nominal masses are commonly used in low-mass work with low-resolution magnetic sector and quadrupole instruments. They are obtained by addition of the masses of the most abundant isotope of each element rounded to the nearest whole number and based on the mass of carbon as 12.0000. Monoisotopic masses, which are necessary at higher resolution, are obtained from the sum of the accurate atomic weights of the most abundant isotope, whereas average or chemical masses, which can be several mass units higher than monoisotopic masses, are obtained when resolution is not sufficient to resolve the isotopes. Low-resolution instruments such as linear TOF mass spectrometers usually return the average mass, whereas higher resolution instruments can return either monoisotopic or average masses, depending on the resolution and mass range being investigated. Some older forms of data system used to record mass spectra return nominal masses under all conditions.

C. Fragmentation of Carbohydrates

Polysaccharides show two general types of fragmentation, cleavages between the monosaccharide rings and cleavages across the rings (Figure 1). The former fragmentation, known as glycosidic cleavage, is more prominent at low energies, whereas the cross-ring cleavages, which yield more information on linkage, become more abundant as the energy rises.

The most common glycosidic fragmentations occur between the 1-carbon atom and its linking oxygen, to give Y-type ions if the charge is retained by the reducing terminus and B ions if the charge resides at the other end of the molecule. Z and C ions result from cleavages on the other side of the linking oxygen (Figure 1). The nomenclature used to describe these fragmentations is that proposed by Domon and Costello in 1988 [20]. B-type ions are thought to involve oxonium species and to be formed following protonation of the linking oxygen. Y-type ions, on the other hand, involve a hydrogen transfer from the eliminated fragment. The hydrogen from the 2-position is thought to be involved in the case of derivatized oligosaccharides [21–23] but has been shown to be a hydroxylic hydrogen from free sugars [24]. So-called "internal" fragments, formed by elimination of residues from both ends of the molecules, are abundant in many spectra. Although many of these are undoubtedly formed by consecutive reactions, there is evidence to suggest that at least some arise from concerted mechanisms [25].

Cross-ring cleavages are generally the result of retro-aldol or retro-ene rearrangements and are particularly abundant at high energy [26,27] and in spectra produced by infrared laser desorption [28]. Orlando et al. [29] have observed that they appear to be more prominent in the spectra of MNa^+ species than in the spectra of carbohydrates ionized as MH^+, although, in general, MNa^+ species appear to be more reluctant to ionize than MH^+ ions. These cross-ring fragment ions are classified as X ions if the charge resides at the reducing end and as A ions for nonreducing-end ions. Cleavages across the branching mannose residue are particularly valuable for determination of antennae structure. Lemoine et al. [26] have identified another ion, termed a W ion, in high-energy spectra that originated from cleavage of the 6-arm from this mannose residue and that is also useful in defining these structural features. Linkage position has frequently been shown to affect the relative abundance of many of these ions in small oligosaccharides [24,30–32], but few such studies have been reported for larger sugars.

IV. EXAMINATION OF INTACT GLYCOPROTEINS BY MASS SPECTROMETRY

Large molecules such as glycoproteins were inaccessible by mass spectrometry until recently, when the techniques of electrospray, MALDI, and, to a lesser extent, PD, were developed.

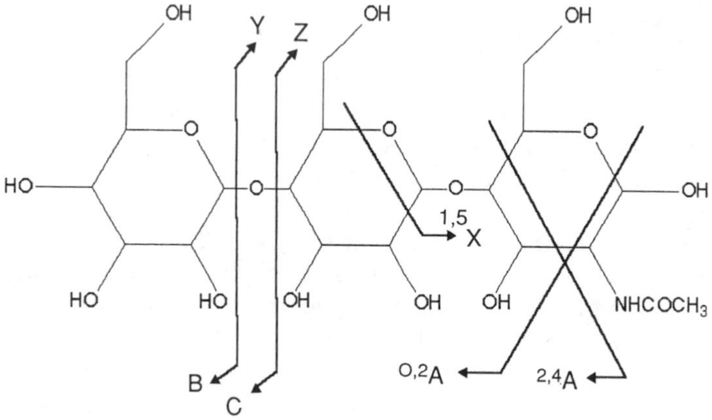

Figure 1 Formation of some major fragment ions from oligosaccharides. (Nomenclature from Ref. 20.)

Table 1 Residue Masses of Monosaccharides Commonly Found in Oligosaccharides

Monosaccharide	Formula	Monoisotopic mass	Chemical mass
Pentose	$C_5H_8O_4$	132.0423	132.12
Deoxyhexose	$C_6H_{10}O_4$	146.0579	146.14
Hexose	$C_6H_{10}O_5$	162.0528	162.14
Hexuronic acid	$C_6H_8O_6$	176.0321	176.13
Na salt	$C_6H_7O_6Na$	198.0140	198.11
N-Acetylaminohexose	$C_8H_{13}NO_5$	203.0794	203.19
N-Acetylneuraminic acid	$C_{11}H_{17}NO_8$	291.0954	291.26
Na salt	$C_{11}H_{16}NO_8Na$	313.0773	313.24

The oligosaccharide masses are obtained by addition of the residue masses together with the mass of one molecule of water (18.0106 monoisotopic, 18.02 chemical). [Alternatively, the full masses of the monosaccharides less $n - 1$ water molecules (where n = number of monosaccharides) may be used]. For the mass of the $[M + Na]^+$ ion, an additional 22.9899 (monoisotopic) or 23.00 (chemical) mass units should be added.

A. Examination of Intact Glycoproteins by Electrospray

Proteins and glycoproteins give exceptionally good electrospray spectra, although the complexity of some glycoprotein structures sometimes reduces the mass range that can be attained on account of instrument resolution. Nevertheless, the method can be used to give a measurement of the total glycoform population and, if the mass of the protein is known, to yield the mass of the sugars. Because *N*- and *O*-linked sugars are composed of relatively few isobaric structures, it is then a relatively simple matter to calculate the composition of the sugar in these terms (see Table 1) and even to propose a probable structure. For example, a neutral mammalian *N*-linked sugar of nominal mass 1641 reduces to a composition of (hexose)$_5$(*N*-acetylamino-hexose)$_4$, for which a biantennary structure is the most likely. It must be emphasized, however, that this proposed structure must be confirmed by additional information, such as a composition and linkage analysis by GC/MS and by techniques such as MS/MS and NMR.

Duffin et al. [33] have examined the glycoform profile of hen ovalbumin by electrospray and found reasonable correlation between the spectrum and the reported oligosaccharide structures. The complexity of the glycoform population, however, resulted in a complex spectrum that almost reached the resolution limit of the mass spectrometer even though the mass of the protein was only 44 kDa. Electrospray has also been used in the study of carbohydrate-deficient glycoprotein syndrome, both to characterize the unglycosylated and both mono- and di-glycosylated serum trans-ferrins and to examine the mixture of the three glycoforms. Resolution was high enough for complete separation of the glycoforms of this 75-kDa glycoprotein [34].

B. Examination of Intact Glycoproteins by MALDI

Proteins and glycoproteins generally give abundant $[M + H]^+$ ions accompanied by lower amounts of dimeric and doubly charged ions when ionized by MALDI. Tri- and tetrameric ions are occasionally seen. Multiply charged ions are particularly abundant with α-cyano-4-hydroxycinnamic acid.

Poor resolution generally restricts the use of MALDI on linear TOF instruments to studies of glycoproteins with masses below about 20 kDa. Above this mass, even structures differing by the mass of a monosaccharide residue may not be resolved. Early results from ribonuclease B (around 15 kDa), a glycoprotein containing five glycoforms each differing by one mannose residue, illustrate the problem [35]. Above about 20 kDa, only broad peaks are produced, with no resolution of glycoforms. However, MALDI can be used at much higher masses to determine the presence or absence of *N*-linked glycosylation, as illustrated by the study of the carbohydrate-deficient glycoprotein syndrome recently reported by Wada et al. [36]. Human serum transferrin from normal controls was shown to produce an ion at 79.6 kDa, with sinapinic acid as the matrix, whereas patients with the syndrome produced transfer-rins with additional peaks, appearing as shoulders on the main peak, at 77.4 and 75.2 kDa, corresponding to species lacking one and two sialylated biantennary oli-gosaccharides, respectively. Resolution, however, was inferior to that obtained in the electrospray spectra in the above study [34].

Although TOF instruments fitted with a reflectron are capable of improving resolution, problems occur with glycoproteins containing sialic acid, where fragmentation within the reflectron has been reported [37]. Such fragmentation is strongly dependent on the matrix and is particularly abundant with α-cyano-4-hydroxycinnamic acid. Less fragmentation is obtained with 2,5-DHB, but the best results have been obtained from 3-hydroxypicolinic acid [37].

V. DETERMINATION OF GLYCOSYLATION SITE OCCUPANCY BY MASS SPECTROMETRY

Identification of the sites in glycoproteins that are occupied by oligosaccharides generally depends on chemical or enzymatic cleavage of the protein chain, separation of the resulting peptides and glycopeptides by reverse-phase high-performance chromatography (HPLC), and detection by mass spectrometry. A variety of mass spectrometric techniques can be used, either off-line as in the case of MALDI, or on-line as with electrospray [38] or dynamic-FAB mass spectrometry. These methods have recently been reviewed by Carr et al. [39]. Conditions can usually be found that induce cleavage between the glycosylation sites to leave a mixture of peptides and glycopeptides containing a single glycosylation site. Assuming that the peptide sequence of the glycopeptide is known, the masses of the oligosaccharides can be obtained by difference.

Medzihradszky et al. [40] have used this approach to examine tryptic fragments of bovine fetuin by microbore liquid chromatography–electrospray mass spectrometry and identified the oligosaccharides specific to Asn-91, -138, and -158. Several O-linked sites were also identified. Strong positive-ion spectra were recorded at the 20-pmol level, and glycopeptides bearing triantennary oligosaccharides with up to four sialic acids were detected.

Identification of the glycopeptides in tryptic mixtures, however, is sometimes difficult, and a number of techniques have been developed to overcome the problem. For example, a two-dimensional plot of m/z values against time from a series of repetitively scanned spectra of tryptic digest obtained from mutant and wild-type tissue plasminogen activator has been used by Guzzetta et al. [41]. The presence of glycopeptides was revealed by negatively sloping families of peaks in the resulting contour plot.

Enzymatic removal of the sugars with endoglycosidases such as protein-N-glycosidase-F (PNGase-F) or endoglycosidase-H (endo-H) and revaluation of the mixture by LC/MS reveals the glycopeptides by their shift in the chromatogram [42]. Both enzymes leave a modified protein; thus, partial site occupancy can be determined. PNGase-F cleaves the sugar at the amide bond to leave an aspartic acid residue rather than asparagine, and this can be detected by a mass increment of 1 Da [43]. The extent of acid formation gives an indication of site occupancy. An interesting extension of this technique was recently published by Gonzalez et al. [44], who performed their incubations in a water containing 40% $H_2^{18}O_2$. The labeled oxygen was incorporated into the asparagine, giving doublet peaks separated by two mass units for the freed aspartic acid in the resulting spectra. Endo-H cleaves the sugar between the two GlcNAc residues of the core and, thus, leaves the reducing-terminal GlcNAc residue, together with any substituents, attached to the peptide.

Carr et al. [39] have described a method whereby the glycoprotein is first alkylated (pyridylethylation or carboxymethylation) and then cleaved with enzymes such as trypsin, endoproteinase Lys-C, endoproteinase Asp-N, or *Staphylococcus aureus* V8 protease and then examined by FAB mass spectrometry. As described below, glycoproteins give weak or nonexistent signals when examined by this technique but can be located following incubation with PNGase-F.

A better detection procedure, however, makes use of the fact that molecules can be fragmented, both in multianalyzer instruments and, for use in single-stage instruments, in the spray from the electrospray needle, by suitable adjustment of the ion source voltages [39,45,46]. Under conditions producing fragmentation, glycosidic cleavages occur to give ions at low mass from each constituent monosaccharide, usually its residue mass plus hydrogen. These fragments are most abundant in the spray when the voltage on the orifice is high. The mass spectrometer is, thus, scanned repetitively with a high orifice voltage at the low-mass end of the scan, reducing to normal for the rest of the mass range so as not to distort the spectrum. Single-ion chromatograms of the diagnostic monosaccharide ions, typically m/z 204 from HexNAc and 366 from Hex-HexNAc, are then plotted, and the chromatograms are compared with the total-ion chromatogram. Glycopeptides are recognized by coincidence of peaks in both chromatograms. The method was originally used to study the 25 potential glycosylation sites of soluble-complement receptor type 1, a 240-kDa glycopeptide.

Hunter and Games [47] have investigated these methods for locating glycopeptides in peptide mixtures obtained from bovine alpha-1-acid glycoprotein following digestion by cyanogen bromide and endoproteinase Glu-C. Collisional activation with argon in a triple-stage quadrupole was compared with the production of ions in the supersonic regions of the electrospray. The latter method was found to be more versatile and to give the highest sensitivity. The ion at m/z 204 derived from HexNAc proved to be generally applicable to this type of analysis. But to avoid possible confusion with ions of the same mass derived from peptides, the use of m/z 366 from HexHexNAc was recommended.

The orifice-voltage stepping method has now become a relatively routine method for locating glycopeptides in proteolysis digests by electrospray. Schindler et al. [48] have used the technique to identify eight different groups of glycopeptides in HPLC traces from lecithin:cholesterol acetyltransferase. Sialylated bi- and triantennary sugars were found at the four N-linked sites, and two O-linked sugars were found in a fifth glycoprotein. Kragten et al. [49] have examined N-linked oligosaccharides from murine polymeric immunoglobulin A. Asn-49 from the J-chain was found to carry bi- and triantennary sugars with and without fucose and with either N-acetyl or N-glycolyl neuraminic acid. One of the two glycosylation sites of the heavy chain was found to contain two high-mannose structures, whereas the other site carried a mixture of oligosaccharides. Identification of the sugars was made by comparing their masses, after subtraction of the peptide mass, with compositions from the CarbBank database [50]. This process, of course, assumes that no sugars of unreported structure are present.

Treuheit et al. [51] have used the technique of enzymatic digestion and mass profiling by MALDI for site analysis and to propose structures for N-linked oligosaccharides from the β-subunit of (Na, K)-ATPase from lamb and dog kidney. Tryptic digestion and HPLC fractionation were first used to obtain the three glycopeptides containing the N-linked glycosylation sites. Masses of the molecular ions obtained by negative-ion MALDI-TOF analysis, with 2,5-DHB as the matrix, combined with the results from lectin-binding studies, indicated that the major oligosaccharides were tetraantennary, with additional lactosamine extensions to some of the antennae.

Examination of site occupancy in a hybrid plasminogen activator (mass = 42,926 Da) has been made by MALDI using 2,5-DHB as the matrix by Muller et al. [52]. The glycoprotein was split with plasmin into A and B chains, each containing one N-linked glycosylation site, and the extent and type of glycosylation were revealed by the mass difference between the peptides following deglycosylation with PNGase-F. Further investigations of the fucosylated, sialylated bi-, tri-, and tetraantennary oligosaccharides were made with electrospray mass spectrometry using the glycopeptides. A comparison between MALDI and PD for the detection of glycopeptides for site analysis has strongly favored MALDI. Of the two methods, only MALDI gave results from Chinese hamster ovary (CHO) recombinant IL-4 glycopeptides [53].

A. Detection of O-Glycosylation

O-linked sugars do not appear to have received as much attention as the N-linked carbohydrates, mainly because of the lack of suitable glycosidases. These glycans are usually released by β-elimination, which, from threonine, leaves an unsaturated propyl substituent. Rademaker et al. [54] have reduced the double bond in this alkyl chain with sodium borodeuteride to give 2-aminobutyric acid in order to label the site of carbohydrate attachment with deuterium. Of the four glycopeptides tested, only three underwent reduction, but the fourth was still identified by the presence of the double bond.

VI. DETERMINATION OF OLIGOSACCHARIDE STRUCTURE

Glycoproteins are generally too large and too complex for mass spectrometry to be used directly to examine the detailed structure of the attached oligosaccharides. Thus, they need to be broken into smaller pieces. This can be achieved either by cleaving the peptide chain, as above, to give glycopeptides, or by releasing the oligosaccharide.

A. Examination of Glycopeptides

Glycopeptides obtained by protease cleavage can be separated by HPLC and then examined by mass spectrometry to reveal the glycoform population as well as to define the site occupancy. Kroon et al. [55], for example, have recently used this method to derive oligosaccharides from murine imminoglobulin-2 glycopeptides for subsequent examination by MALDI mass spectrometry. Carbohydrate masses and compositions were obtained by subtracting the peptide mass and were found to agree closely with those of the expected glycans.

Exoglycosidase Digestion of Glycopeptides

Much information on the structure of the carbohydrate portion of the glycoproteins can be obtained by a modification of the classical procedure in which the oligosaccharide is sequentially cleaved by specific exoglycosidases, with the occurrence of a reaction indicating the nature of the monosaccharide at the nonreducing terminus, together, in many cases, with its linkage. MALDI has proved to be particularly valuable in this respect because of its speed. Any problems associated with low resolution are generally of little significance for determining mass losses of mono-saccharides, since these amount to hundreds of mass units.

The use of MALDI to examine the products of such an analysis was first demonstrated by Sutton et al. [56,57] for the glycoprotein recombinant human tissue inhibitor of metalloproteinases (TIMP). The well-characterized glycoproteins fetuin, alpha1-acid glycoprotein, and tissue plasminogen activator (TPA) were included in the study as controls. The glycoprotein was first hydrolyzed with trypsin, and the two glycopeptides were identified by mass difference following incubation with PNGase-F. Two hundred pmol of each of the two glycopeptides was then incubated with a sialidase from *Arthrobacter ureafaciens*, and the resulting glycopep-tides containing neutral oligosaccharides were examined by MALDI with α-cyano-4-hydroxycinnamic acid as the matrix. The measured masses indicated mainly core-fucosylated bi-, tri-, and tetraantennary structures. Successive incubations with galactosidases, hexosaminidases, mannosidases, and fucosidases enabled the structures of these compounds to be established. Similar methods have recently been used to examine the oligosaccharides of baculovirus-expressed mouse interleukin-3 peptides [58].

Huberty et al. [59] have examined recombinant human macrophage colony-stimulating factor (rhM-CSF), a 223 amino acid glycoprotein with two *N*- and several *O*-linked carbohydrate attachment sites, using similar exoglycosidase digestion. The glycoprotein was first digested with the lysine-specific protease Achro K, and the resulting 15 peptides and glycopeptides were separated by reverse-phase HPLC using a Vydac C$_{18}$ column. Glycopeptides were examined by both FAB and MALDI mass spectrometry using the 2,5-DHB matrix. Structural information was obtained using both exo- and endoglycosidase digestions. Thus, incubation with the endoglycosidase PNGase-F released the intact biantennary *N*-linked oligosaccharides, whose masses were determined by the difference in peptide mass before and after digestion. Alternative incubations of the intact glycopeptides with exoglycosidases such as neur-aminidase were used to define the structures further. Information on the *O*-linked oligosaccharides was obtained with enzymes such as *O*-glyconase. Postsource decay was used, as described below, to obtain information on the detailed structure of the *N*-linked oligosaccharides.

In another application of the technique Ogonah et al. [60] identified the high-mannose sugars (Man)$_5$ and 7(GlcNAc)$_2$ from human gamma-interferon produced in baculovirus-infected insect cells. The glycoprotein was cleaved with trypsin, and the glycans were found attached to Asn-25 and -97. In addition, the glycans at Asn-97 were shown to have fucose at the reducing terminus.

Oligosaccharide profiles from glycoproteins have also been obtained by complete digestion of the protein with pronase. The approach leaves at the reducing terminus of the oligosaccharide an amino acid whose basic amino group can be protonated in electrospray, FAB, or MALDI spectra. However, this approach removes data on site specificity. Wang et al. [61] used pronase digestion to study *N*-linked sugars from ovalbumin and proposed several new structures on the basis of positive- and negative-ion FAB spectra. Pronase digestion was also used by Bock et al. [62]

in their study of the oligosaccharides from the crystalline surface-layer glycoprotein from *Thermoanaerobacter thermohydrosulfuricus* L111-69. Much of the structural analysis was performed by NMR and GC/MS, but MALDI was used to obtain the oligosaccharide profile. Twelve linear oligosaccharides averaging 60 monosaccharide residues were found (mean mass around 9.5 kDa), with the linkage being between galactose and tyrosine. This was the first report of such a linkage in glycoproteins. The matrix was 2.5-DHB, and ions of the type $[M - H + 2Na]^+$ were formed as the result of salt formation at the tyrosine carboxylic acid group. Although pronase apparently digested this protein without much difficulty, such complete digestion does not appear to be universally true, with some peptides being particularly resistant. In such systems, the incubation mixtures can become contaminated by peptide fragments produced by autodigestion of the pronase.

B. Examination of Released Oligosaccharides

Most work on carbohydrate structure has been performed on the sugars themselves after cleavage from the protein by either chemical or enzymatic methods.

Chemical Methods of Carbohydrate Release

Chemical release methods usually take the form of β-elimination, used extensively for *O*-linked oligosaccharides, hydrazinolysis [63,64], the preferred method for *N*-linked sugars, or reduction with lithium aluminum hydride to release both types of sugar. β-Elimination employing a mild base is performed in the presence of a reducing agent to prevent base-catalyzed degradation of the sugar, but gives a product that cannot then be derivatized at the reducing terminus for improving HPLC sensitivity or mass spectrometric detection.

   Hydrazinolysis may be used to release both *N*- and *O*-linked oligosaccharides, with selective release of *O*-linked sugars generally being favored at lower temperatures. Although the method generally achieves a more quantitative release of oligosaccharides than does the use of enzymes, it is particularly prone to more artefact production. Perhaps the most serious of these problems results from the fact that this reagent cleaves amide bonds reasonably unspecifically. While this is useful for release of the oligosaccharide, problems arise because the acyl group is cleaved from acylamino sugars. Normally, therefore, following hydrazinolysis, amino sugars are reacetylated under mild conditions, with the assumption that all acylamino sugars originally contained acetyl groups. However, this is not always the case. For example, the GPI anchors frequently contain unacylated sugars, and the sialic acids in the glycoproteins, such as human alpha-1-acid glycoprotein, contain about 50% glycolylamino rather than acetylamino groups [47]. Hydrazinolysis can also produce additional compounds as the result of alterations to the reducing terminus. Thus, incomplete removal of the intermediate hydrazide gives a compound that, after acetylation, produces a compound appearing at 56 mass units above that of the parent sugar. Also present is a compound containing an acetylamino group at the reducing terminus, giving an ion with a 41-mass-unit increment. The incorporation of one such acetylamino group effectively produces a carbohydrate that appears to contain one more acetylamino monosaccharide and one less unsubstituted monosaccharide than is actually the case. Further problems can occur through overreaction with hydrazine, which can cleave an acetylamino sugar from the reducing terminus. Loss of the acetylamino group from the reducing GlcNAc residue has also been found.

Enzymatic Methods of Carbohydrate Release

Although many of the preceding problems can be overcome by using PNGase-F release, the enzyme is not totally effective, and some structures, particularly those containing a reducing terminal 1–3 linked fucose, may remain intact. However, such sugars may be released with PNGase-A, as demonstrated by Hasham et al. [65] in a study of core fucosylation of honey bee venom phospholipase-A. Because of the bulky nature of many glycoproteins, they must first be denatured and, in some cases, cleaved into smaller fragments before PNGase is effective. Denaturing, however, may cause degradation of sensitive structures such as sialic acids. In addition, PNGase-F releases oligosaccharides at different speeds and, thus, might not yield a representative product.

Endoglycosidase-H cleaves the sugar between the core GlcNAc residues and results in loss of information on the reducing-terminal GlcNAc and its possible substituents. The enzyme is also rather selective and only releases oligomannose and hybrid-type sugars. Nevertheless, it has been used, for example, by Ziegler et al. [66], combined with MALDI mass spectrometry, to record the molecular ions of $(Hex)_{10}GlcNAc$ and $(Hex)_{13}GlcNAc$ from the yeast *Schizosaccharomyces pombe* in a study of general glycosidation mechanisms in this species. In this study, only the sugars at the nonreducing termini were being investigated, and loss of information at the reducing end of the molecule was immaterial.

C. Structure Determination of Released Carbohydrates

Detection of the products of oligosaccharide release is often carried out using gel filtration with Bio-Gel P4 columns to measure the hydrodynamic volume of the oligosaccharide. However, this process is time consuming (typically several hours), has relatively poor resolution, and requires that the oligosaccharide be radio- or fluorescently labeled at the reducing terminus. In addition, gel filtration columns are only effective with neutral sugars; thus, the sialic acids, sulfate, and phosphate groups often present on *N*-linked structures must first be removed. HPLC methods are now becoming increasingly common for this type of analysis and do not require prior removal of the acidic functions. Indeed, sugars can now be fractionated on the basis of charge number. Mass spectrometry, particularly MALDI, is also applicable to both neutral and acidic oligosaccharides and can give results more rapidly and with higher precision than the earlier methods. Techniques such as gel filtration are, however, still used extensively for fractionation of oligosaccharides prior to such structural analysis.

Possibly the best approach to structural determination of oligosaccharides is to isolate enough material for NMR investigation. However, this is rarely possible, and the information must be obtained by more sensitive techniques. Isolation for these studies is not always necessary, because techniques such as MS/MS allow mixtures to be examined directly. Combination methods such as LC/MS and LC/MS/MS are less widely used for free oligosaccharide mixtures than for glycopeptides, but HPLC itself is becoming increasingly important for oligosaccharide separation prior to off-line mass spectrometric examination. As with chromatography of other compounds, the behavior of oligosaccharides on different chromatographic columns can yield valuable information on structure and can provide isomeric information not available by mass spectrometry.

Compositional Analysis by Methanolysis and GC/MS

It is difficult to determine the type of constituent monosaccharide in polysaccharides by direct mass spectrometric examination; consequently, this information must be obtained by other means. Most commonly this involves cleavage of the polysaccharide into its constituent monosaccharides and examination by GC/MS. Hydrolysis with aqueous acids, although efficient, is problematical on account of the acid lability of some monosaccharides [67,68]. Unwanted side reactions such as deacylation can also occur. Consequently, methanolysis, a milder procedure, is often used in its place. The technique involves heating the oligosaccharide with methanol containing hydrogen chloride to produce methyl glycosides, which are then converted into their trimethylsilyl ethers for GC/MS evaluation [69]. The method provides both a retention time and a mass spectrum to aid identification. All commonly occurring monosaccharides are separable on GLC columns of low polarity, and, although some sugars give multiple GLC peaks, identification is usually unambiguous.

Determination of Sequence, Branching, and Linkage

Once the constituent monosaccharide composition is known, a variety of other mass spectrometric techniques can be used to obtain information about how these sugars are linked in the oligosaccharide. These methods include FAB, electrospray, MALDI, and further uses of GC/MS.

Examination of Released Oligosaccharides by FAB. *N*-linked oligosaccharides and Asn-peptides can be examined directly by FAB mass spectrometry [70]. But in general they do not perform well under these conditions. For the technique to be successful, a layer of sample molecules should form on the matrix surface, where it can be desorbed by the atom or ion beam. Oligosaccharides and glycopeptides are generally too hydrophilic for this process to occur and are generally not seen when mixtures of peptides and glycopeptides are examined. In order to overcome this problem, the oligosaccharides are invariably derivatized to reduce the hydrophilicity. Permethylation and peracetylation tend to be the most widely used methods [71–73], but derivatization of the reducing terminus by reductive amination with compounds such as long-chain amines has also been used. Permethylation needs somewhat vigorous conditions and is rarely quantitative, thus leading to rather complicated spectra. The usual reagent is methyl iodide, catalyzed either by the methylsulfenyl carbanion [Hakomori method, 74] or, more recently, by solid sodium hydroxide in DMSO [75].

Reducing-terminal derivatization exploits the unique properties of the aldehyde group of the reducing sugar in order to introduce a single group, most commonly for imparting UV-absorbing or fluorescent properties for chromatographic detection. However, the reaction can also be used to modify the carbohydrate in ways that improve mass spectrometric detection or fragmentation. Reductive amination is the most common technique. The procedure first involves reaction of the sugar with a primary amine in the presence of an acid catalyst in order to form the Schiff base.

Stabilization is achieved by concomitant reduction, usually with sodium borohydride or sodium cyanoborohydride. 2-Aminopyridine is possibly the most well known of these fluorophores [76], although it has also been used for mass spectrometric studies [77].

A popular reducing-terminal derivative for FAB mass spectrometry is that formed from *p*-aminobenzoic acid ethyl ester (ABEE). This derivative was synthesized by Webb et al. in 1988 [78] specifically to combine a fluorophore with a hydrophobic group for enhanced FAB performance [79]. A further systematic study of homologous esters was carried out in 1991. It was generally observed that FAB sensitivity increased with the length of the alkyl ester chain, with gains up to 40-fold being obtained [22,80]. Although the decyl and tetradecyl esters were most effective, yields were lower. The best compromise was achieved with hexyl and octyl esters. The derivatives gave particularly clear negative-ion FAB spectra containing prominent Y-type cleavage ions that provided sequence and branching information.

Zhang et al. [81] have added a chromophore and a hydrophobic group in a two-stage reaction. The reducing sugar was first reacted with *n*-octylamine by reductive amination, and the amine was then reacted further with 2,4-dinitrofluorobenzene to give the *N,N*-(2,4-dinitrophenyl)octylamine derivative. Abundant negative ions were formed under FAB conditions, and a detection limit of 1 pmol was obtained.

High sensitivity is of paramount importance for much carbohydrate identification work. In addition to the use of derivatization, improved detection systems can be fitted to the mass spectrometer. Array, or focal-plane, detectors that capture a range of ions simultaneously have been used for carbohydrate analysis and have been reported to give sensitivity increases up to 100-fold [81]. However, such increases are generally only realized with MS/MS studies or with ionization techniques that give a low background. FAB spectra are generally too noisy for the efficient application of array detection.

In addition to producing abundant molecular ions, usually MH^+, which gives compositional information on the oligosaccharide, FAB also catalyzes a considerable amount of fragmentation [73]. Fragment ions are usually produced by glycosidic cleavage, with Y and B ions being particularly abundant. They provide a considerable amount of information on constituent monosaccharide sequence and branching. Fragmentation is particularly abundant adjacent to HexNAc residues, resulting, in the case of permethylated complex *N*-linked oligosaccharides, in very abundant ions of the type $[\text{Gal-GlcNAc-H}]^+$ (*m/z* 464). These ions tend to eliminate methanol if there is a free methoxy group in the 3-position of the GlcNAc residue. But other than this fragmentation, there is little in the spectrum that yields much more information on linkage. High-energy MS/MS spectra of MNa^+ ions generated by FAB, however, contain a wealth of such information [26].

Linkage information has been obtained directly from FAB spectra following additional chemical modification, such as periodate oxidation [82,83]. This reaction causes oxidative cleavage of the carbon–carbon bond between carbons bearing *cis*-hydroxy groups with the formation of a dialdehyde. Reduction of this aldehyde with sodium borodeuteride followed by methylation gives a product whose mass differs from that of the original permethylated oligosaccharide. For hexoses, the presence of 1- and 4-linkages causes cleavage between the 2- and 3-carbon atoms, resulting in a shift of the residue weight of the hexose moiety from 204 mass units to 208 units. Although the presence of linkages at the 1- and 2-positions also causes a similar mass shift due to cleavage of the 3–4 bond, the linkage can be recognized by the presence of a prominent ion due to loss of methanol in the spectrum of the unoxidized oligosaccharide. Hexoses containing 1- and 3-linkages produce no periodate cleavage product because of the absence of a *cis*-diol group, whereas linkage at the 1- and 6-positions causes cleavages between both C-3 and C-4 and between C-4 and C-5, resulting in the loss of the carbon atom at C-3. A terminal hexose gives rise to a prominent ion at m/z 179. Other mass shifts are listed in Ref. 83.

Periodate oxidation, coupled with FAB mass spectrometry, has been used on numerous occasions for structural determinations of native N- and O-linked oligosaccharides. Thus, Taguchi et al. [84] have used it to identify large, bisected pentaantennary N-linked oligosaccharides from the fish *Fundulus heteroclitus*, and Angel et al. [85] have characterized polylactosamine extensions to tri- and tetraantennary N-linked glycans from murine glycophorin. Isomeric high-mannose isomers have also been characterized as their oxidized alditols [86] and as products following trifluoracetolysis to remove the reducing-terminal GlcNAc residues [87]. Manzi et al. [88] have used periodate oxidation, coupled with the formation of acetates and ABEE derivatives and detection by FAB, to differentiate positional isomers of sialic acids. Both the acid and lactone forms of the acid were produced. Lactone formation was minimal from the 4-substituted acids, suggesting this position as the site of lactone formation. Linsley et al. [89] employed the technique to identify a large number of O-linked and bi-, tri-, and tetraantennary N-linked glycans from erythropoietin and noted that examination of the compounds as released glycans was not as informative as examination of the glycopeptides.

Fast atom bombardment mass spectrometry in general has had a considerable impact on the analysis of large sugars, with the identification and confirmation of several novel structural types. Novel tyvelose (3,6-di-deoxy-D-arabinohexose) containing tri- and tetraantennary oligosaccharides have been identified from the intracellular parasite *Trichinella spiralis* [90]. The antennae contained mainly Tyv1–3GalNAc1–4GlcNAc chains and were identified by a combination of FAB mass spectrometry, GC/MS linkage analysis, and exoglycosidase digestions. The presence of GalNAc itself in the antenna of this type of oligosaccharide was confirmed in 1991 from glycans present at Asn-184 and Asn-448 in tissue plasminogen activator extracted from Bowes melanoma [91]. Other applications include the characterization of N-linked glycans from the human transferrin receptor [92], bi- and triantennary N-linked oligosaccharides from the light chain of human glycoprotein IIb [93], hybrid sugars from frog rhodopsin [94], and xylose-containing glycans from the plant *Nicotiana alata* [95].

Anionic sugars such as sialic acids generally present no particular problems when analyzed by FAB mass spectrometry, with the presence of the acid groups promoting formation of negative ions. Sulfates, on the other hand, are more problematical. Dell et al. [96] have reported that, although sulfated oligosaccharides give good negative-ion FAB spectra, from which the position of the sulfate can be deduced, the sulfate group is not seen in spectra recorded in the positive-ion mode. Li et al. [97], on the other hand, have reported that, although sulfate loss is observed in both positive- and negative-ion modes, molecular ions are, nevertheless, seen in both ion modes from trisaccharide structures. The presence of the sulfate caused several ions to be observed, depending on whether or not the sulfate formed a metal salt. Fast atom bombardment mass spectrometry has, however, contributed to the analysis of these anionic compounds; Siciliano et al. [98], for example have identified sulfated GalNAc residues at the nonreducing end of *N*-linked sugars from bovine pro-opiomelanocortin.

Examination of Released Oligosaccharides by Electrospray. Electrospray has had much less impact on the examination of neutral oligosaccharides than the desorption techniques, such as FAB or MALDI, mainly because the neutral oligosaccharides lack a site with any appreciable proton affinity. Thus, ion formation usually requires the presence of a metal such as sodium or potassium, which is often added to the electrospray in the form of sodium acetate [33]. Addition of acetic or formic acid can produce protonation but not with the exclusion of the metal-adducted species. Doubly charged ions are usually more abundant than their singly charged analogs, with spectra being dominated by species such as MH_2^{2+}, $MHNa^{2+}$, and MNa_2^{2+}. Ionization appears to be critically dependent in the solvent composition. Thus, the singly charged MH^+ ion was reported by Suzukisawada et al. [77] to be formed when 20 nM triethylamine/acetic acid at pH 4 was used as the buffer, whereas a 100 nM ammonium/acetic acid buffer at the same pH favored formation of the doubly charged MH_2^{2+} ion. Sodium phosphate 5 nM at pH 3.8, on the other hand, produced MNa^{2+} ions as the major species. Sialic acids give good negative-ion spectra on account of the ready loss of protons from the carboxylic acid functions.

The ready coupling of electrospray with liquid chromatography and HPLC, together with the high solubility of oligosaccharides in the normal electrospray solvents, provides great impetus for development in this area. Dionex high-pH ion-exchange chromatography is frequently used for oligosaccharide analysis, and effluents from these columns have also been examined [99]. Anionic micromembrane suppression was used to remove the high base concentrations, and the resulting solution was examined by ionspray (pneumatically assisted electrospray). Either ammonium or lithium acetate was added in order to produce the necessary cations. The resulting ammonium adducts of oligosaccharides released with endo-H from ribonuclease B gave singly and doubly charged ions and substantial fragmentation, whereas the lithium adducts yielded mainly molecular ions.

One approach that has been used to improve the detection of oligosaccharides in electrospray experiments parallels that seen for FAB mass spectrometry, namely, derivative formation at the reducing terminus, either with reagents that provide a proton accepting site or that supply a constitutive charge. Reductive amination again appears to be the method of choice and has been used by Yoshino et al. [100] in the form of 4-aminobenzoic acid 2-(diethylamino)ethyl ester to introduce an amino group with a reported 5000-fold increase in sensitivity. Maltohexose could be measured with this derivative at the 10-fmol level. An even more promising technique is to derivatize in such a manner as to produce a charged molecule prior to the electrospray step. Thus, Hogeland et al. [101] have derivatized maltohexose and *N*-linked oligosaccharides from baculovirus-expressed mouse interleukin-3 with *n*-hexylamine, and the charge was introduced by exhaustive permethylation that quaternized the amino group at the reducing terminus. Okamoto et al. [102] used trimethyl(*p*-aminophenyl)ammonium chloride to introduce a charged trimethylammonium group in a single stage by the method originally used by Dell et al. [72] for preparation of similar derivatives for FAB mass spectrometry. Sugars gave mixtures of M^+ and MNa^{2+} ions, but increases in sensitivity of 5000-fold over that of the free sugars were reported; detection limits were in the region of 20 fmol. Prevention of negative-ion formation by these derivatives contributes to the high sensitivity. Other derivatives of the type used for introducing fluorophores for HPLC detection were also examined in this publication, but these derivatives did not produce such a dramatic increase in sensitivity.

Several other fluorophore-derivatized oligosaccharides have, however, been investigated as electrospray-enhancing derivatives by other investigators, with good results. Thus 2-aminopyridine derivatives have been reported to give a detection limit of 1 pmol for biantennary oligosaccharides isolated from bovine IgG [103]. In-source fragmentation in the spray region was observed to produce glycosidic cleavage ions from which the sequence of the constituent monosaccharides could be deduced. 2-Aminobenzamide derivatization [104] is becoming popular for fluorophore introduction. But under electrospray conditions this derivative again gives mainly doubly charged ions of the types MNa_2^{2+}, $MHNa^{2+}$, and MH_2^{2+}, with the relative abundance of the last ion increasing if acetic or formic acid is added to the electrospray solvent. Fragmentation of the doubly charged ions from these derivatives by collisional activation gives mainly singly charged ions formed by glycosidic cleavages.

An improved technique for the preparation of charged derivatives that avoids the removal of the excess of reagents required by the reductive amination procedure was recently introduced by Naven and Harvey [105]. N-Linked oligosaccharides were reacted with Girard's reagent T (carboxymethyltrimethylammonium chloride) to form the corresponding substituted hydrazone. Only a small excess of reagent was required, which did not require removal before analysis. Again, mainly doubly charged ions were formed, and detection limits were in the low femtomole range by electrospray and in the mid-femtomole range by MALDI.

Examination of Released Oligosaccharides by MALDI.   The first MALDI matrix to be used for free oligosaccharides, 3-amino-4-hydroxybenzoic acid, was reported by Mock et al. in 1991 [106]. However, this matrix has now been superseded by 2,5-DHB [107]. In order to obtain the best signals, it is necessary to recrystallize the matrix from water before use. However, the reagent-grade matrix, as usually supplied, works well with most samples. When the matrix:sample solutions evaporate, 2,5-DHB tends to crystallize from the periphery of the target spot in the form of long needles that point toward the center of the target. In many cases the center is devoid of crystals. To overcome this problem, it has been reported that recrystallization of the dried target from ethanol leads to a much more even distribution of crystals and a stronger signal [108,109]. In addition to aiding crystallization, it is thought that ethanol also enables more of the sample molecules to become trapped into the crystal lattice. This is because, when the original aqueous:organic solvent is evaporating, the residual liquid becomes enriched in the aqueous phase, thus keeping the sugars in solution until late into the evaporation process. With only ethanol present, the lower solubility of the sugars probably aids their uptake by the crystal. Recent work on milk oligosaccharides [110] has supported this theory. It was noticed that, in mixtures containing both oligosaccharides and proteins, strong oligosaccharide signals could be obtained from the microcrystalline area at the center of the target, whereas ionization of the larger crystals that had formed earlier gave mainly signals for the proteins. Other investigators have added their sample to predried crystalline surfaces in order to aid the even distribution of crystals.

A modified form of 2,5-DHB, namely, 1,4-dihydroxy-2-naphthoic acid, has been found to have matrix properties similar to that of 2,5-DHB and to crystallize in a much more even layer from aqueous solvents [108]. However, this matrix does not produce such a strong signal. Karas et al. [111] have achieved some 2- to 3-fold improvements in the signal obtained from dextrans, with molecular weights between 500 and 2000 Da, by incorporating 2-hydroxy-5-methoxybenzoic acid into the 2,5-DHB matrix. The rationale behind this experiment was that the methoxy analog causes disruption of the crystal lattice, thus facilitating the release of the sample ions.

Several other compounds have been investigated as suitable matrices for oligosaccharides. Most isomeric dihydroxybenzoic acids produce signals, but of considerably reduced intensity to that produced by the 2,5-dihydroxy isomer [108,111]. The reason for this is not clear, although it may be speculated that, because the 2,5-isomer is the only isomer capable of photochemical decarboxylation to give the stable *p*-benzoquinone, then this photochemical reaction is responsible for the ionization. Other matrices, such as α-cyano-4-hydroxycinnamic acid, esculetin (6,7-dihydroxycoumarin), and 2-(4-hydroxyphenylazo)benzoic acid (HABA) also ionize sugars, but generally produce weaker signals. Sinapinic acid appears to be ineffective. 3-Aminoquinoline has recently been reported to give results superior to those of 2,5-DHB for the ionization of plant inulins [112]. In particular, the spectral peaks produced from this matrix appeared sharper than those produced by 2,5-DHB, and the base line was lower. However, to date, the general applicability of this matrix for oligosaccharide analysis has not been determined. Another matrix that has proved to be very satisfactory is 2,5-DHB containing 25% 1-hydroxyisoquinoline [113]. This matrix provided both high resolution and sensitivity, reduced the relative abundance of matrix ions, and was remarkably tolerant to the presence of buffers.

Reducing-terminal derivatization has not been exploited so extensively for MALDI spectra as for FAB and electrospray. Naven and Harvey [105], however, have synthesized several amine-containing derivatives by reductive amination and found increases in sensitivity of about 10-fold. Best results were obtained by reacting the sugars with Girard's reagent T. Ashton et al. [114] have reacted sugars with 1-phenyl-3-methylpyrazoline-5-one to give a product containing two pyrazoline molecules and have used the derivatives for detection of carbohydrates in humanized IgG antibodies expressed in CHO cells. Increases up to 100-fold by use of a charged derivative in MALDI were recently reported by Whittal et al. [115] for examination of oligosaccharides in serum. Derivatization was achieved with tetramethylrhodamine, with linkage through a methoxycarbonyloctyl chain [116].

Although MALDI is reasonably tolerant to the presence of small amounts of buffer salts and other contaminants [117], larger amounts cause problems and can inhibit crystal formation. Several recent micropurification techniques have, thus, been developed for cleanup of MALDI samples. Salts may be removed by drop dialysis on a 500-Da cutoff membrane. A small section of membrane is placed or floated onto the surface of water, and 1–2 µL of the oligosaccharide solution is placed on top. This solution is removed after about 15 min and analyzed as normal. An alternative technique employs a Nafion membrane in its hydrogen form [118]. The membrane is again placed on a water surface, with the oligosaccharide solution on top. The membrane not only removes sodium but also abstracts peptides by adsorption onto the surface. It has also been reported that the membrane, in its salt form, can be used to prepare oligosaccharide solutions containing only the metal of choice. Thus potassium can be substituted for sodium, a procedure that yields greater subsequent MALDI sensitivity.

In contrast to proteins, neutral oligosaccharides give only molecular-ion peaks, produced by the addition of alkali metal ions when examined with TOF mass spectrometers. Although the MNa^+ ion is the only abundant ion, MK^+ ions are present if the matrix contains a significant amount of potassium. Generally, sufficient sodium is present in the matrix solution to achieve ionization, and no addition of sodium salts is required. However, with recrystallized 2,5-DHB it is sometimes advantageous to add small amounts of sodium acetate in order to effect efficient ion formation. However, an excess of salt is detrimental. No multimeric ions have been reported in MALDI spectra of neutral oligosaccharides, but weak fragment ions are occasionally seen in strong spectra recorded with linear TOF instruments. It has recently been noticed that these ions become much more abundant with instruments employing delayed extraction, presumably because of the longer lifetime of the ion before acceleration.

Quantitative Aspects of MALDI. Sample consumption with MALDI is minimal, but amounts in the femtomole to low picomole range are currently required for successful sample handling. Only a small fraction of this amount is actually ionized, and it has been reported that MALDI is 10–100 times as sensitive as FAB mass spectrometry for detection of glycopeptides [56]. The original carbohydrate matrix, 3-amino-4-hydroxybenzoic acid, does not appear to give a quantitative response with oligosaccharides, because a saturation effect has been reported [119]. However, if a chemically related internal standard is included in the sample mixture and peak ratios are measured, oligosaccharides may be quantified over a large concentration range. Saturation has not been found with 2,5-DHB, and the signal produced from oligosaccharides reflects the sample amount over several decades of concentration. With sugars of similar structure, there is a good correlation between sample concentration and signal intensity [120], although with high-mannose oligosaccharides a slight drop in signal intensity has been found with the larger sugars [119]. β(1,2)-Cyclic glycans have also been reported to give a quantitative response and to give profiles from both 2,5-DHB and α-cyano-4-hydroxycinnamic acid that closely matched profiles obtained by techniques such as HPLC [121].

There appears to be little difference in the relative signal intensity produced by *N*- and *O*-linked sugars of different structure, although it has been noticed that, on a linear TOF instrument not employing matrix suppression facilities, there is a pronounced drop in signal intensity with decreasing mass for sugars with mass below about 1000 Da [120]. This effect is thought to be due to detector saturation by the ions from the matrix (m/z 100–200 region). With mixtures, the maximum signal that can be recorded from an individual sugar is lower than that from the sugar in isolation, because the ion current produced with each laser shot is distributed over each of the components of the mixture.

Structural Determination of Neutral Oligosaccharides by MALDI. The absence of fragment, adduct, or multiply charged ions and the good quantitative relationship between peak height and sample concentration makes MALDI an ideal technique for profiling oligosaccharide mixtures. For example, Figure 2 shows the profile of underivatized oligosaccharides obtained from human IgG recorded from

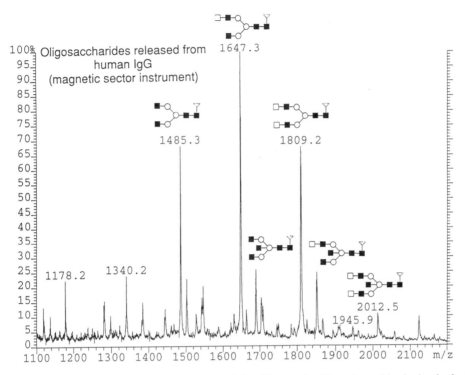

Figure 2 Positive-ion MALDI spectrum of the oligosaccharides released by hydrazinolysis from human IgG and recorded on a magnetic sector instrument, with 2,5-DHB as the matrix. Symbols used to identify the monosaccharides are: D-Galactose (□), D-mannose (○), L-fucose (6-deoxy-β-L-galactopyranose) (▽), 2-acetamido-2-deoxy-D-glucose (■). (Reproduced from Ref. 108 with permission from J. Wiley and Sons, Ltd.)

2,5-DHB with a magnetic sector instrument. All reported structural types are present, and structures can be confirmed by exoglycosidase digestion, as shown in Figure 3.

MALDI can also be used to confirm structures of oligosaccharides separated by chromatography. Thus, high-mannose oligosaccharides from ribonuclease have been separated by gel filtration from PNGase-F-released mixtures and confirmed with a mass accuracy of ±0.5 Da of the calculated value [122].

Fragmentation of MALDI-generated molecular ions is becoming increasingly important for structural characterization. Although little or no fragmentation of oligosaccharides is seen in spectra recorded with linear TOF spectrometers, there is a considerable amount of postsource decay observable with a reflectron. For example, Huberty et al. [59] studied N-linked oligosaccharides from the glycoprotein rhM-CSF and recorded predominantly glycosidic cleavages in the constituent biantennary oligosaccharide of the glycoheptapeptide containing the N-linked site. Positive-ion spectra recorded in the normal reflectron mode gave a base peak corresponding to the asialo-biantennary sugar, demonstrating very rapid fragmentation by loss of sialic acid. In the negative-ion spectra, recorded under similar conditions, the base peak contained one sialic acid, but this was ascribed to the negative charge being carried by the sialic acid group. Further sialic acid loss would have removed the charge site.

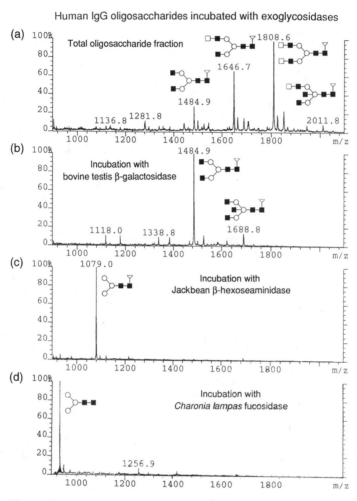

Figure 3 Positive-ion MALDI spectra (2,5-DHB matrix) of oligosaccharides from human IgG successively digested with exoglycosidases. (a) Undigested oligosaccharides, (b) digestion with bovine testis β-galactosidase, (c) digestion with jackbean β-hexosaminidase, (d) digestion with *Charonia lampas* fucosidase. (Symbols explained in legend for Figure 2.) (Reproduced from Ref. 108 with permission from J. Wiley and Sons, Ltd.)

Postsource decay of bi- and triantennary oligosaccharides recovered from human urine and thought to have been released from glycoproteins with endoglycosidase-H (endo-H), and hence lacking the asparagine-linked GlcNAc residue, has been reported by Spengler et al. [123]. Again, the major fragmentations were the result of glycosidic cleavage, mainly B, C, and Y ions, but there were also some cross-ring cleavages, particularly of the terminal GlcNAc moiety. This pattern of PSD fragmentation was also found by Harvey et al. [124] in a study of several intact N-linked oligosaccharides with varying structure and was reported to be similar to that obtained by in-source decay on a magnetic sector instrument [108,124].

Production of high-energy CID spectra of $[M + Na]^+$ ions from oligosaccharides has recently been achieved on a tandem magnetic sector instrument fitted with an orthogonal TOF analyzer [124]. Fragmentation was extensive (Figure 4), with more cross-ring cleavages revealing the linkage than were seen in either postsource or in-source decay spectra.

Structural Determination of Acidic Oligosaccharides by MALDI. Negatively charged oligosaccharides, such as those containing sialic acid, give weak MALDI signals in the positive ion mode but frequently produce more abundant ions ($[M -$ H]$^-$) in the negative-ion mode. In the positive-ion mode, sialic acids usually appear as the sodium salt, although peaks corresponding to the free acid can sometimes be seen [108]. The ability of sialic acids to form sodium salts under these conditions is

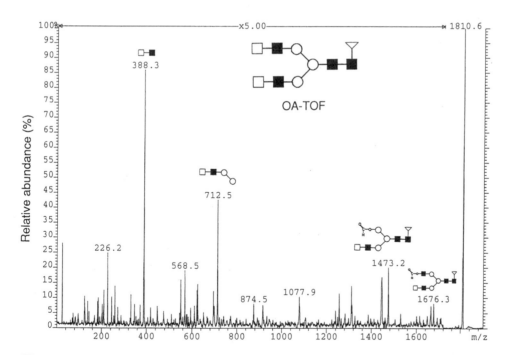

Figure 4 Fragment ions formed by high (800-keV) voltage from the biantennary oligosaccharide $(Gal)_2(GlcNAc)_4(Man)_3Fuc$ and recorded with an orthogonal TOF analyzer. Only one of the two possible structures is shown for the two $^{1,5}X$ ions at m/z 1676.3 and 1473.2. The ion at m/z 388.3 is a Y-type ion, and that at m/z 712.5 is an internal cleavage fragment. (Symbols explained in legend for Figure 2.)

fortuitous, since the residue mass difference between sialic acid (291 Da) and two deoxyhexose residues (292 Da) is only 1 mass unit, an interval that is difficult to determine with a TOF mass spectrometer. Stahl et al. [111] have recently reported that 3-aminoquinoline is a superior matrix to 2,5-DHB for examination of sialylated oligosaccharides. It gives sharper peaks, with a better signal:noise ratio, than either 2,5-DHB or α-cyano-4-hydroxycinnamic acid, but loss of sialic acid by fragmentation can be a problem when spectra are examined on a reflectron TOF instrument.

Talbo and Mann [125] have recently studied postsource decay of a sialylated biantennary oligosaccharide and, again, reported prominent sialic acid loss in both positive- and negative-ion modes. The negative-ion spectra also contained a prominent loss of "about 100" mass units, which probably represents the O,2A cleavage of the reducing terminus discussed earlier. Because fragmentation of neutral sugars is more prominent in positive-ion mode, decomposition of the desialylated ion at m/z 1663 was then studied in this mode and was shown to fragment mainly by glycosidic cleavage.

When examined with a magnetic sector instrument, sialylated oligosaccharides are again observed to undergo extensive fragmentation, mainly by loss of sialic acid but also by decarboxylation [108]. Glycolipids containing sialic acid behave in a similar manner [126]. Thus, in the positive-ion mode, fetuin oligosaccharides, of which a triantennary oligosaccharide is the major species, eliminate most of the sialic acid, to give major ions corresponding to the unsialylated sugars. Much stronger spectra are obtained in the negative-ion mode from these compounds. Recently, a method has been developed for stabilizing sialic acids to MALDI conditions and enabling good positive-ion spectra to be obtained. The sodium salt of the acidic sugar was reacted with methyl iodide in DMSO, without the use of a catalyst, to form the methyl ester. The reaction was complete in about one hour, and the product could be examined directly by MALDI because all reagents were volatile [127].

Linkage Analysis by GC/MS.   None of the preceding methods for examination of intact oligosaccharides give full information on linkage. Mass spectrometry, however, still provides the best way to determine this property, using a method first developed in 1968 [128] and commonly referred to as *methylation analysis* [129]. Many variants exist but all rely essentially on derivatization of the oligosaccharide with a reagent such as methyl iodide, cleavage to the constituent monosaccharides, and derivatization of the newly introduced hydroxyl groups with another reagent. The products, which are specifically labeled at the points of linkage, are then examined by GC/MS. The most popular of these derivatization techniques involves initial permethylation, either by the method developed by Hakomori [74] or by the more recent and milder method of Ciucanu and Kerek [75]. The initial derivatization is followed by acid hydrolysis to release the monosaccharides, which are then reduced with sodium borohydride to their alditols in order to avoid multiple GLC peak formation as the result of equilibration between the anomeric forms. Finally, the alditols are acetylated to produce partially methylated alditol acetates, or PMAAs. Most PMAAs give unique GC/MS properties, but the method is unable to distinguish between certain sugars, e.g., distinguishing 4-linked aldohexopyranose from 5-linked aldohexofuranose. The reducing terminal sugar can be specifically labeled with deuterium if the oligosaccharide is reduced with sodium borodeuteride before methylation. The review by Hellerqvist [129] lists ions diagnostic to specific sugars.

One of the disadvantages of this method is the rather laborious multiderivatization steps. In an attempt to simplify the reaction and to overcome the inability to distinguish between the furanose and pyranose forms just mentioned, the so-called "reductive cleavage" method has been developed. In this method, the hydrolysis and reduction steps are essentially combined by reaction of the permethylated oligosaccharide with triethylsilane [130] catalyzed with either trimethyl trifluoromethane sulfonate or a mixture of trimethylsilylmethane sulfonate and borontrifluoride etherate. Acetic anhydride can be added to the reaction mixture to achieve the acetylation stage in essentially a single step. The main difference between this procedure and the conventional methylation analysis technique is that the sugars are examined as partially methylated anhydroalditols, thus preserving the sugar ring [131].

In an attempt to improve the low sensitivity of the conventional linkage analysis method, Patoprsty et al. [132] have recently used chemical, rather than electron-impact, ionization to ionize PMAAs, with pyridine as the reagent gas. The sugars formed essentially only the pyridine adduct, which provided easy identification of the molecular weight and gave increased sensitivity. However, no diagnostic fragment ions were present in the spectra.

Determination of Anomericity by Mass Spectrometry. This is a topic not easily addressed by mass spectrometry, and the problem is generally more easily answered by other techniques, such as exoglycosidase digestion. However, Khoo and Dell [133] have utilized the ability of chromium trioxide in acetic acid to oxidize β-pyranoses to keto esters. α-Pyranoses remain essentially intact under these conditions. Sugars were deuteroacetylated and heated with the reagent for 2 hours at 50°C before being examined by FAB mass spectrometry, with thioglycerol as the matrix. Oxidation produced a mass increment of 14 mass units, and the difference in mass recorded before and after oxidation defined the number of β-anomers. Their position can be located by changes in the fragmentation pattern. Deuteroacetylation was used in preference to acetylation in order to avoid ambiguities due to the coincidence in mass between three stages of oxidation (42-mass-unit increment) and an acetyl group. The method was fairly specific, but it was noted that 1–6 linked sugars underwent oxidation at the 6-position.

VII. CONCLUSIONS

Mass spectrometry is playing an ever-more-prominent role in the structural elucidation of glycoproteins and oligosaccharides. With the advent of techniques such as electrospray and MALDI, complex structures are directly accessible, and molecular-weight information, sometimes leading directly to a composition, can be rapidly obtained. More emphasis is now being placed on the use of fragmentation to provide structural information, particularly from compounds present in mixtures, and both on-line LC/MS and MS/MS techniques are expected to undergo considerable development over the next few years. One of the most important areas of development will be the problem of sensitivity. Although the modern techniques, and, indeed, mass spectrometry in general, are sensitive, they do not approach the sensitivity that can be attained with, for example, fluorescence detection. It is probably in this area that most effort should be directed in the next few years so that, ultimately, structural information can be obtained directly from functionally significant constituents of single cells.

ABBREVIATIONS

ABEE	aminobenzoic acid ethyl ester
Asn	asparagine
ATP	adenosine triphosphate
CHO	Chinese hamster ovary
CID	collision-induced decomposition
CI	chemical ionization
Da	daltons
DHB	dihydroxybenzoic acid
DMSO	dimethylsulfoxide
El	electron impact
endo-H	endoglycosidase-H
FAB	fast atom bombardment
Gal	galactose
GC/MS	gas chromatography/mass spectrometry
GLC	gas–liquid chromatography
Glc	glucose
GlcNAc	N-acetylglucosamine
GPI	glycosylphosphatidylinositol
HABA	2-(4-hydroxyphenylazo)benzoic acid
Hex	hexose
HexNAc	N-acetylhexosamine
HPLC	high-performance liquid chromatography
IgG	immunoglobulin-G
LC/MS	liquid chromatography/mass spectrometry
LSIMS	liquid secondary-ion mass spectrometry
Lys	lysine
MALDI	matrix-assisted laser desorption/ionization
Man	mannose
MS/MS	mass spectrometry/mass spectrometry
m/z	mass-to-charge ratio
NMR	nuclear magnetic resonance
PD	plasma desorption
PMAAs	partially methylated alditol acetates
PNGase	protein-N-glycosidase
PSD	postsource decay
rhM-CSF	recombinant human macrophage colony-stimulating factor
TIMP	tissue inhibitor of metalloproteinase
TOF	time of flight
TPA	tissue plasminogen activator
UV	ultraviolet

ACKNOWLEDGEMENT

The author wishes to thank R. A. Dwek, Director of the Glycobiology Institute, for his encouragement and support.

REFERENCES

1. M. Fukuda, Cell surface carbohydrates: Cell-type specific expression, *Molecular Glycobiology* (M. Fukuda and O. Hindsgaul, eds.), IRL Press (Oxford University Press), Oxford, 1994, p. 1.
2. R. J. Sturgeon, The glycoproteins and glycogen, *Carbohydrate Chemistry* (J. F. Kennedy, ed.), Oxford University Press, Oxford, 1988, p. 263.
3. J. R. Chapman, *Practical Organic Mass Spectrometry*, Wiley, Chichester, 1982.
4. C. C. Sweeley, R. Bentley, M. Makita, and W. W. Wells, Gas–liquid chromatography of trimethylsilyl derivatives of sugars and related substances, *J. Am. Chem. Soc. 85*: 2497 (1963).
5. D. J. Harvey, M. G. Horning, and P. Vouros, The mass spectra of the trimethylsilyl derivatives of glycerophosphoric acids. Inter- and intramolecular rearrangements of siliconium ions, *J. Chem. Soc, Perkin Trans I*, 1074 (1972).
6. D. J. Harvey and M. G. Horning, Characterization of the trimethylsilyl derivatives of sugar phosphates and related compounds by gas chromatography and gas chromatography–mass spectrometry, *J. Chromatogr. 75*: 51 (1973).
7. H. Karlsson, I. Carlstedt, and G. Hansson, The use of gas chromatography and gas chromatography–mass spectrometry for the characterization of permethylated oligosaccharides with molecular mass up to 2300, *Anal. Biochem. 182*: 438 (1989).
8. M. Barber, R. S. Bordoli, R. D. Sedgwick, and A. N. Tyler, Fast atom bombardment of solids (FAB): A new ion source for mass spectrometry, *Chem. Commun.*: 325 (1981).
9. U. Seydel, B. Lindnar, H.-W. Wollenweber, and E. T. Rietschel, Structural studies on the lipid component of enterobacterial lipopolysaccharides by laser desorption mass spectrometry. Location of acyl groups on the lipid A backbone, *Eur. J. Biochem. 145*: 505 (1984).
10. Z. Lam, G. G. S. Dutton, M. B. Comisarow, D. A. Weil, and A. Bjarnason, Structural information obtained by negative-ion laser desorption ionization–Fourier transform–ion cyclotron resonance (LDI/FT/ICR) mass spectrometry of bacterial capsular polysaccharides, *Carbohydrate Res. 180*: C1 (1988).
11. Z. Lam, M. B. Comisarow, G. G. S. Dutton, H. Parilis, L. A. S. Parolis, A. Bjarnason, and D. A. Weil, Laser desorption-ionization Fourier-transform ion cyclotron resonance mass spectrometry of carbohydrates, Part 1. Bacterial oligosaccharides containing neuraminic acid or pyruvic acid acetal groups, *Anal. Chim. Acta. 241*: 187 (1990).
12. Z. Lam, L. Beynon, M. B. Comisarow, G. S. Dutton, and A. Bjarnason, Laser desorption-ionization Fourier-transform ion cyclotron resonance (LDI/FT/ICR) MS of bacterial oligosaccharides, *Biochem. Soc. Trans. 19*: 922 (1991).
13. R. S. Bordoli, K. Howes, R. G. Vickers, R. H. Bateman, and D. J. Harvey, Matrix-assisted laser desorption mass spectrometry on a magnetic sector instrument fitted with an array detector, *Rapid Commun. Mass Spectrom. 8*: 585 (1994).
14. A. Vertes, G. Irinyi, and R. Gijbels, Hydrodynamic model of matrix-assisted laser desorption mass spectrometry, *Anal. Chem. 65*: 2389 (1993).
15. R. C. Beavis and B. T. Chait, Cinnamic acid derivatives as matrices for ultraviolet laser desorption mass spectrometry of proteins, *Rapid Commun. Mass Spectrom. 3*: 432 (1989).

16. R. C. Beavis, T. Chaudhary, and B. T. Chait, Alpha-cyano-4-hydroxycinnamic acid as a matrix for matrix-assisted laser desorption mass spectrometry, *Org. Mass Spectrom.* *27*: 156 (1992).

17. K. Strupat, M. Karas, and F. Hillenkamp, 2,5-Dihydroxybenzoic acid: A new matrix for laser desorption-ionization mass spectrometry, *Int. J. Mass Spectrom. Ion Proc.* *111*: 89 (1991).

18. M. L. Vestal, P. Juhasz, and S. A. Martin, Delayed extraction matrix-assisted laser desorption time-of-flight mass spectrometry, *Rapid Commun. Mass Spectrom. 9*: 1044 (1995).

19. R. H. Bateman, M. R. Green, G. Scott, and E. Clayton, A combined magnetic sector–time-of-flight mass spectrometer for structural determination studies by tandem mass spectrometry, *Rapid Commun. Mass Spectrom. 9*: 1227 (1995).

20. B. Domon and C. E. Costello, A systematic nomenclature for carbohydrate fragmentations in FABMS/MS spectra of glycoconjugates, *Glycoconjugate J. 5*: 397 (1988).

21. B. L. Gillece-Castro and A. L. Burlingame, Oligosaccharide characterization with high-energy collision-induced dissociation mass spectrometry, *Methods Enzymol. 193*: 689 (1990).

22. L. Poulter and A. L. Burlingame, Desorption mass spectrometry of oligosaccharides coupled with hydrophobic chromophores, *Methods Enzymol. 193*: 661 (1990).

23. A. Dell and J. E. Thomas-Oates, Fast atom bombardment mass spectrometry (FABMS): Sample preparation and analytical strategies, *Analysis of Carbohydrates by GLC and MS* (C. J. Biermann and G. D. McGinnis, eds.), CRC Press, Boca Raton, FL, 1989, p. 217.

24. G. E. Hofmeister, Z. Zhou, and J. A. Leary, Linkage position determination in oligosaccharides: MS/MS study of lithium-cationized carbohydrates, *J. Am. Chem. Soc. 113*: 5964 (1991).

25. V. Kovacik, J. Hirsch, P. Kovac, W. Heerma, J. Thomas-Oates, and J. Haverkamp, Oligosaccharide characterization using collision-induced dissociation fast atom bombardment mass spectrometry. Evidence for internal monosaccharide residue loss, *J. Mass Spectrom. 30*: 949 (1995).

26. J. Lemoine, B. Fournet, D. Despeyroux, K. Jennings, R. Rosenberg, and E. de Hoffmann, Collision-induced dissociation of alkali-metal cationized and permethylated oligosaccharides: Influence of the collision energy and of the collision gas for the assignment of linkage, *J. Am. Soc. Mass Spectrom. 4*: 197 (1993).

27. V. N. Reinhold, B. B. Reinhold, and C. E. Costello, Carbohydrate molecular weight profiling, sequence, linkage, and branching data: ES-MS and CID, *Anal. Chem. 67*: 1772 (1995).

28. B. Spengler, J. W. Joice, and R. J. Cotter, Infrared laser desorption mass spectrometry of oligosaccharides: Fragmentation mechanisms and isomer analysis, *Anal. Chem. 62*: 1731 (1990).

29. R. Orlando, C. A. Bush, and C. Fenselau, Structural analysis of oligosaccharides by tandem mass spectrometry—collisional activation of sodium adduct ions, *Biomed. Environ. Mass Spectrom 19*: 747 (1990).

30. B. Domon, D. R. Müller, and W. J. Richter, Tandem mass spectrometry in structural characterization of oligosaccharide residues, *Int. J. Mass Spectrom. Ion. Proc. 100*: 301 (1990).

31. E. Yoon and R. A. Laine, Linkage-position determination in a novel set of permethylated neutral trisaccharides by collision-induced dissociation and tandem mass spectrometry, *Biol. Mass Spectrom. 21*: 479 (1992).

32. A. R. Dongre and V. H. Wysocki, Linkage position determination of lithium-cationized disaccharides by surface-induced dissociation tandem mass spectrometry, *Org. Mass Spectrom. 29*: 700 (1994).

33. K. L. Duffin, J. K. Welply, E. Huang, and J. D. Henion, Characterization of *N*-linked oligosaccharides by electrospray and tandem mass spectrometry, *Anal. Chem. 64*: 1440 (1992).

34. K. Yamashita, T. Ohkura, H. Ideo, K. Ohno, and M. Kanai, Electrospray ionization mass spectrometric analysis of serum transferrin isoforms in patients with carbohydrate-deficient glycoprotein syndrome, *J. Biochem (Tokyo) 114*: 766 (1993).

35. D. J. Harvey, The role of mass spectrometry in glycobiology, *Glycoconjugate J. 9*: 1 (1992).

36. Y. Wada, J. Gu, N. Okamoto, and K. Inui, Diagnosis of carbohydrate-deficient glycoprotein syndrome by matrix-assisted laser desorption time-of-flight mass spectrometry, *Biol. Mass Spectrom. 23*: 108 (1984).

37. M. Karas, U. Bahr, K. Strupat, F. Hillenkamp, A. Tsarbopoulos, and B. N. Pramanik, Matrix dependence of metastable fragmentation of glycoproteins in MALDI TOF mass spectrometry, *Anal. Chem. 67*: 675 (1995).

38. M. J. Huddleston, M. F. Bean, and S. A. Carr, Collisional fragmentation of glycopeptides by electrospray ionization LC/MS and LC/MS/MS—Methods for selective detection of glycopeptides in protein digests, *Anal. Chem. 65*: 877 (1993).

39. S. A. Carr, J. R. Barr, G. D. Roberts, K. R. Anumula, and P. B. Taylor, Identification of attachment sites and structural classes of asparagine-linked carbohydrates in glycoproteins, *Methods Enzymol. 193*: 501 (1990).

40. K. F. Medzihradszky, D. A. Maltby, S. C. Hall, C. A. Settineri, and A. L. Burlingame, Characterization of protein *N*-glycosylation by reversed-phase microbore liquid chromatography-electrospray mass spectrometry, complementary mobile phases and sequential exoglycosidase digestion, *J. Am. Soc. Mass Spectrom. 5*: 350 (1994).

41. A. W. Guzzetta, L. J. Basa, W. S. Hancock, B. A. Keyt, and W. F. Bennett, Identification of carbohydrate structures in glycoprotein peptide maps by the use of LC/MS with selected ion extraction with special reference to tissue plasminogen activator and a glycosylation variant produced by site directed mutagenesis, *Anal. Chem. 65*: 2963 (1993).

42. J. P. Liu, K. L. Volk, E. H. Kerns, S. E. Klohr, M. S. Lee, and I. E. Rosenberg, Structural characterization of glycoprotein digests by microcolumn liquid chromatography-ionspray tandem mass spectrometry, *J. Chromatogr. 632*: 45 (1993).

43. P. Ferranti, P. Pucci, G. Marino, I. Fiume, B. Terrana, C. Ceccarini, and A. Malorni, Human alpha-fetoprotein produced from Hep G2 cell line: Structure and heterogeneity of the oligosaccharide moiety, *J. Mass Spectrom. 30*: 632 (1995).

44. J. Gonzalez, T. Takao, H. Hori, V. Besada, R. Rodriguez, G. Padron, and Y. Shimonishi, A method for determination of *N*-glycosylation sites in glycoproteins by collision-induced dissociation analysis in fast atom bombardment mass spectrometry—identification of the positions of carbohydrate-linked asparagine in recombinant alpha-amylase by treatment with peptide-*N*-glycosidase-F in *O*-18-labelled water, *Anal. Biochem. 205*: 151 (1992).

45. M. J. Huddleston, M. F. Bean, and S. A. Carr, Collisional fragmentation of glycopeptides by electrospray ionization LC/MS and LC/MS/MS. Methods for selective detection of glycopeptides in protein digests, *Anal. Chem. 65*: 877 (1993).

46. S. A. Carr, M. J. Huddleston, and M. F. Bean, Selective identification of *N*- and *O*-linked oligosaccharides in glycoproteins by liquid chromatography–mass spectrometry, *Protein Sci. 2*: 183 (1993).

47. A. P. Hunter and D. E. Games, Evaluation of glycosylation site heterogeneity and selective differentiation of glycopeptides in proteolytic digests of bovine alpha-1-acid glycoprotein by mass spectrometry, *Rapid Commun. Mass Spectrom. 9*: 42 (1995).

48. P. A. Schindler, C. A. Settiner, X. Collet, C. J. Fielding, and A. L. Burlingame, Site specific detection and structural characterization of the glycosylation of human plasma

proteins lecithin:cholesterol acetyltransferase and apolipoprotein D using HPLC/electrospray mass spectrometry and sequential glycosidase digestion, *Protein Sci. 4*: 791 (1995).

49. E. A. Kragten, A. A. Bergwerff, J. van Oostrum, D. R. Müller, and W. J. Richter, Site-specific analysis of the *N*-glycans on murine polymeric immunoglobulin A using liquid chromatography/electrospray mass spectrometry, *J. Mass Spectrom. 30*: 1679 (1995).

50. S. Doubet and P. Albersheim, CarbBank, *Glycobiology 2*: 505 (1992).

51. M. J. Treuheit, C. E. Costello, and T. L. Kirley, Structure of the complex glycans found on the beta-subunit of (Na.K)-ATPase, *J. Biol. Chem. 268*: 13914 (1993).

52. D. Müller, B. Domon, M. Karas, J. van Oostrum, and W. J. Richter, Characterization and direct glycoform profiling of a hybrid plasminogen activator by matrix-assisted laser desorption and electrospray mass spectrometry: Correlation with high-performance liquid chromatographic and nuclear magnetic resonance analysis of the released glycans, *Biol. Mass Spectrom. 23*: 330 (1994).

53. Tsarbopoulos, M. Karas, K. Strupat, B. N. Pramanik, T. L. Nagabhushan, and F. Hillenkamp, Comparative mapping of recombinant proteins and glycoproteins by plasma desorption and matrix-assisted laser desorption/ionization mass spectrometry, *Anal. Chem. 66*: 2062 (1994).

54. G. J. Rademaker, J. Haverkamp, and J. Thomas-Oates, Determination of glycosylation sites in *O*-linked glycopeptides: A sensitive mass spectrometric protocol, *Org. Mass Spectrom. 28*: 1536 (1993).

55. D. J. Kroon, J. Freedy, D. J. Burinsky, and B. Sharma, Rapid profiling of carbohydrate glycoforms in monoclonal antibodies using MALDI/TOF mass spectrometry, *J. Pharmaceut. Biomed. Anal. 13*: 1049 (1995).

56. C. W. Sutton and J. S. Cottrell, Characterisation of carbohydrates and glycoconjugates by matrix assisted laser desorption mass spectrometry. In *Newer Methods in Glycoprotein and Glycolipid Characterization* (D. Cumming and V. N. Reinhold, eds.), Academic Press, London, 1993.

57. C. W. Sutton, J. A. O'Neil, and J. S. Cottrell, Site-specific characterization of glycoprotein carbohydrates by exoglycosidase digestion and laser desorption mass spectrometry, *Anal. Biochem. 218*: 34 (1994).

58. Y. K. E. Hogeland, Jr. and M. L Deinzer, Mass spectrometric studies on the *N*-linked oligosaccharides of baculovirus-expressed mouse interleukin-3, *Biol. Mass Spectrom. 23*: 218 (1994).

59. M. C. Huberty, J. E. Vath, W. Yu, and S. A. Martin, Site-specific carbohydrate identification in recombinant proteins using MALDI-TOFMS, *Anal. Chem. 65*: 2791 (1993).

60. O. W. Ogonah, R. B. Freedman, N. Jenkins, and R. C. Rooney, Analysis of human interferon-gamma glycoforms produced in baculovirus infected insect cells by matrix-assisted laser desorption spectrometry, *Biochem. Soc., Trans. 23*: S100 (1995).

61. T. H. Wang, T. F. Chen, and D. F. Barofsky, Mass spectrometry of L-β-aspartamido carbohydrates isolated from ovalbumin, *Biomed. Environ. Mass Spectrom. 16*: 335 (1987).

62. K. Bock, J. Schuster-Kolbe, E. Altman, G. Altmaier, B. Stahl, R. Christian, U. B. Sleytr, and P. Messner, Primary structure of the *O*-glycosidically linked glycan chain of the crystalline surface layer glycoprotein of *Thermoanaerobacter thermohydrosulfuricus* L111-69−Galactosyl tyrosine as a novel linkage unit, *J. Biol. Chem. 269*: 7137 (1994).

63. T. Patel, J. Bruce, A. Merry, C. Bigge, M. Wormald, A. Jaques, and R. Parekh, Use of hydrazine to release in intact and unreduced form both *N*-linked and *O*-linked oligosaccharides from glycoproteins, *Biochemistry 32*: 679 (1993).

64. T. P. Patel and R. B. Parekh, Release of oligosaccharides from glycoproteins by hydrazinolysis, *Methods Enzymol. (Guide to Techniques in Glycobiology) 230*: 57 (1994).

65. S. M. Hasham, A. J. Reason, H. R. Morris, and A. Dell, Core fucosylation of honeybee venom phospholipase A2, *Glycobiology 4*: 105 (1994).

66. F. D. Ziegler, T. R. Gemmill, and R. B. Trimble, Glycopeptide synthesis in yeast. Early events in *N*-linked oligosaccharide processing in *Schizosaccharomyces pombe, J. Biol. Chem. 269*: 12527 (1994).

67. C. J. Biermann, Hydrolysis and other cleavage of glycosidic linkage, *Analysis of Carbohydrates by GLC and MS* (C. J. Biermann and G. D. McGinnis, eds.), CRC Press, Boca Raton, FL, 1989, p. 27.

68. R. K. Merkle and I. Poppe, Carbohydrate composition analysis of glycoconjugates by gas–liquid chromatography/mass spectrometry, *Methods Enzymol. (Guide to Techniques in Glycobiology) 230*: 1 (1994).

69. M. A. J. Fergusson, GPI Membrane anchors: Isolation and analysis, *Glycobiology, A Practical Approach* (M. Fukuda and A. Kobata, eds.), IRL Press (Oxford University Press), Oxford, 1993, p. 349.

70. J. P. Kamerling, W. Heerma, J. J. G. Vliegenthart, B. N. Green, I. A. S. Lewis, G. Strecker, and G. Spik, Fast atom bombardment mass spectrometry of carbohydrate chains derived from glycoproteins, *Biomed. Mass Spectrom. 10*: 420 (1983).

71. A. Dell, Preparation and desorption mass spectrometry of permethyl and peracetyl derivatives of oligosaccharides, *Methods Enzymol. 193*: 647 (1990).

72. A. Dell, N. H. Carman, P. R. Tiller, and J. E. Thomas-Oates, Fast atom bombardment mass spectrometric strategies for characterizing carbohydrate-containing biopolymers, *Biomed. Environ. Mass Spectrom. 16*: 19 (1987).

73. A. Dell, K.-H. Khoo, M. Panico, R. A. McDowell, A. T. Etienne, A. J. Reason, and H. R. Morris, FAB-MS and ES-MS of glycoproteins, *Glycobiology: A Practical Approach* (D. Rickwood and B. D. Hames, eds.), IRL Press (Oxford University Press), Oxford, 1993, p. 197.

74. S.-I. Hakomori, A rapid permethylation of glycolipid and polysaccharide catalysed by methylsulfenyl carbanion in dimethylsulfoxide, *J. Biochem (Tokyo) 55*: 205 (1964).

75. I. Ciucanu and F. Kerek, A simple and rapid method for the permethylation of carbohydrates, *Carbohydr. Res. 131*: 209 (1984).

76. H. Tahemoto, S. Hase, and T. Kenaka, Microquantitative analysis of neutral and amino sugars as fluorescent pyridylamino derivatives by high-performance liquid chromatography, *Anal. Biochem. 145*: 245 (1985).

77. J. Suzukisawada, Y. Umeda, A. Kondo, and I. Kato, Analysis of oligosaccharides by on-line high-performance liquid chromatography and ion-spray mass spectrometry, *Anal. Biochem. 207*: 203 (1992).

78. J. W. Webb, K. Jiang, B. L. Gillece-Castro, A. L. Tarentino, T. H. Plummer, J. C. Byrd, S. J. Fisher, and A. L. Burlingame, Structural characterization of intact, branched oligosaccharides by high performance liquid chromatography and liquid secondary ion mass spectrometry, *Anal. Biochem. 169*: 337 (1988).

79. L. Poulter, J. P. Earnest, R. M. Stroud, and A. L. Burlingame, Cesium ion liquid secondary ion mass spectrometry of membrane-bound glycoproteins: Structural and topological considerations of acetylcholine receptors from Torpedo californica, *Biomed. Environ. Mass Spectrom. 16*: 25 (1988).

80. L. Poulter, R. Karrer, and A. L. Burlingame, Normal-alkyl para-aminobenzoates as derivatizing agents in the isolation, separation and characterization of submicrogram quantities of oligosaccharides by liquid secondary ion mass spectrometry, *Anal. Biochem. 195*: 1 (1991).

81. Y. Zhang, R. A. Cedergren, T. J. Nieuwenhuis, and R. I. Hollingsworth, N,N-(2,4-dinitrophenyl)octylamine derivatives for the isolation, purification and mass spectrometric characterization of oligosaccharides, *Anal. Biochem.* *208*: 363 (1993).

82. A. S. Angel and B. Nilsson, Analysis of glycoprotein oligosaccharides by fast atom bombardment mass spectrometry, *Biomed. Environ. Mass Spectrom.* *19*: 721 (1990).

83. A. S. Angel and B. Nilsson, Linkage positions on glycoconjugates by periodate oxidation and fast atom bombardment mass spectrometry, *Methods Enzymol.* *193*: 587 (1990).

84. T. Taguchi, K. Kitijimi, Y. Moto, S. Inoue, K.-H. Khoo, H. B. Morris, A. Dell, R. A. Wallace, K. Selmar, and Y. Inoue, A precise structural analysis of a fertilization-associated carbohydrate-rich glycopeptide isolated from the fertilized eggs of euryhaline kill fish (*Fundulus heteroclitus*). Novel pentaantennary N-glycan chains with a bisecting N-acetylglucosaminyl residue, *Glycobiology 5*: 611 (1995).

85. A. S. Angel, G. Gronberg, H. Krotkiewski, E. Lisowska, and B. Nilsson, Structural analysis of the N-linked oligosaccharides from murine glycophorin, *Arch. Biochem. Biophys.* *291*: 76 (1991).

86. A. S. Angel, P. Lipniunas, K. Erlansson, and B. Nilsson, A procedure for the analysis by mass spectrometry of the structure of oligosaccharides from high-mannose glycoproteins, *Carbohydrate Res.* *221*: 17 (1991).

87. P. Lipniunas, A. S. Angel, K. Erlansson, F. Linch, and B. Nilsson, Mass spectrometry of high-mannose oligosaccharides after trifluoroacetolysis and periodate oxidation, *Anal. Biochem.* *200*: 58 (1992).

88. A. E. Manzi, A. Dell, and A. Varki, Studies of naturally occurring modifications of sialic acids by fast-atom bombardment mass spectrometry—Analysis of positional isomers by periodate cleavage, *J. Biol. Chem.* *265*: 8094 (1990).

89. K. B. Linsley, S. Y. Chan, S. Chan, B. B. Reinhold, P. J. Lisi, and V. N. Reinhold, Applications of electrospray mass spectrometry to erythropoietin N- and O-linked glycans, *Anal. Biochem.* *219*: 207 (1994).

90. A. J. Reason, L. A. Ellis, J. A. Appleton, N. Wisnewski, R. B. Grieve, D. McNeil, D. L. Wasson, H. R. Morris, and A. Dell, Novel tyvelose-containing tri- and tetraantennary N-glycans in the immunodominant antigens of the intracellular parasite *Trichinella spiralis*, *Glycobiology 4*: 593 (1994).

91. A. L. Chan, H. R. Morris, M. Panico, A. T. Etienne, M. E. Rogers, P. Gaffney, L. Creighton-Kempsford, and A. Dell, A novel sialylated N-acetylgalactosamine-containing oligosaccharide is the major complex-type structure present in Bowes melanoma tissue plasminogen activator, *Glycobiology 1*: 173 (1991).

92. G. Orberger, R. Geyer, S. Stirm, and R. Tauber, Structure of the N-linked oligosaccharides of the human transferrin receptor, *Eur. J. Biochem.* *205*: 257 (1992).

93. A. J. Reason, A. Dell, H. R. Morris, M. E. Rogers, J. J. Calvete, and J. Gonzalez-Rodriguez, Characterization of the N-linked oligosaccharides of the light chain of human glycoprotein IIb by FABMS, *Carbohydrate Res.* *221*: 169 (1991).

94. K. L. Duffin, G. W. Lange, J. K. Welply, R. Florman, P. J. O. Brien, A. Dell, A. J. Reason, H. R. Morris, and S. J. Fliester, Identification and oligostructure analysis of rhodopsin glycoforms containing galactose and sialic acid, *Glycobiology 3*: 365 (1995).

95. J. R. Woodward, D. Craik, A. Dell, K.-H. Khoo, S. L. A. Munro, A. E. Clarke, and A. Back, Structural analysis of the N-linked glycan chains from a stylar glycoprotein associated with expression of self-incompatibility in *Nicotiana alata*, *Glycobiology 2*: 241 (1992).

96. A. Dell, H. R. Morris, F. Greer, J. M. Redfern, M. E. Rogers, G. Weisshaar, J. Hiyama, and A. G. C. Renwick, Fast-atom-bombardment mass spectrometry of sulfated oligosaccharides from ovine lutropin, *Carbohydr. Res.* *209*: 33 (1991).

97. T. Li, Y. Ohashi, S. Nunomura, T. Ogawa, and Y. Nagai, Fast atom bombardment and electrospray ionization mass spectrometry of sulfated Lewis(x) trisaccharides, *J. Biochem (Tokyo) 30*: 1277 (1995).

98. R. A. Siciliano, R. R. Morris, R. A. McDowell, P. Azadi, M. E. Rogers, H. P. J. Bennett, and A. Dell, The Lewis x epitope is a major non-reducing structure in the sulphated *N*-glycans attached to Asn-65 of bovine pro-opiomelanocortin, *Glycobiology 3*: 225 (1993).

99. J. J. Conboy and J. Henion, High-performance anion-exchange chromatography coupled with mass spectrometry for the determination of carbohydrates, *Biol. Mass Spectrom. 21*: 387 (1992).

100. K. Yoshino, T. Takao, H. Murata, and Y. Shimonishi, Use of the derivatizing agent 4-aminobenzoic acid 2-(diethylamino)ethyl ester for high-sensitivity detection of oligosaccharides by electrospray ionization mass spectrometry, *Anal. Chem. 67*: 4028 (1995).

101. K. E. Hogeland, B. Arbogast, and M. L. Deinzer, Liquid secondary ion-mass spectrometric analysis of permethylated *n*-hexylamine-derivatised oligosaccharides: Application to baculovirus-expressed mouse interleukin-3, *J. Am. Soc. Mass Spectrom. 3*: 345 (1992).

102. M. Okamoto, K.-I. Takashashi, and T. Doi, Sensitive detection and structural characterisation of trimethyl(*p*-aminophenyl)ammonium-derivatized oligosaccharides by electrospray ionization mass spectrometry and tandem mass spectrometry, *Rapid Commun. Mass Spectrom. 9*: 841 (1995).

103. J. Gu, T. Hiraga, and Y. Wada, Electrospray ionization mass spectrometry of pyridylaminated oligosaccharide derivatives, sensitivity and in-source fragmentation, *Biol. Mass Spectrom. 23*: 212 (1994).

104. J. C. Bigge, T. P. Patel, J. A. Bruce, P. N. Goulding, S. M. Charles, and R. B. Parekh, Nonselective and efficient fluorescent labelling of glycans using 2-aminobenzamide and anthranilic acid, *Anal. Biochem. 230*: 229 (1995).

105. T. J. P. Naven and D. J. Harvey, Cationic derivatization of oligosaccharides with Girard's T reagent for improved performance in matrix-assisted laser desorption/ionization and electrospray mass spectrometry, *Rapid. Commun Mass Spectrom. 10*: 829 (1996).

106. K. K. Mock, M. Davey, and J. S. Cottrell, The analysis of underivatized oligosaccharides by matrix-assisted laser desorption mass spectrometry, *Biochem. Biophys. Res. Commun. 177*: 644 (1991).

107. B. Stahl, M. Steup, M. Karas, and F. Hillenkamp, Analysis of neutral oligosaccharides by matrix-assisted laser desorption-ionisation mass spectrometry, *Anal. Chem. 63*: 1463 (1991).

108. D. J. Harvey, P. M. Rudd, R. H. Bateman, R. S. Bordoli, K. Howes, J. B. Hoyes, and R. G. Vickers, Examination of complex oligosaccharides by matrix-assisted laser desorption mass spectrometry on time-of-flight and magnetic sector instruments, *Org. Mass Spectrom. 29*: 753 (1994).

109. D. J. Harvey, Matrix-assisted laser desorption/ionisation mass spectrometry of oligosaccharides and glycoconjugates, *J. Chromatogr. 720*: 429 (1996).

110. B. Stahl, S. Thurl, J. Zeng, M. Karas, F. Hillenkamp, M. Steup, and G. Sawatzki, Oligosaccharides from human milk as revealed by matrix-assisted laser desorption/ionization mass spectrometry, *Anal. Biochem. 223*: 218 (1994).

111. M. Karas, H. Ehring, E. Nordhoff, B. Stahl, K. Strupat, F. Hillenkamp, M. Grehl, and B. Krebs, Matrix-assisted laser desorption/ionisation mass spectrometry with additives to 2,5-dihydroxybenzoic acid, *Org. Mass Spectrom. 28*: 1476 (1993).

112. J. O. Metzger, R. Wolsh, W. Tuszynski, and R. Angermann, New type of matrix for matrix-assisted laser desorption mass spectrometry of polysaccharides and proteins, *Fresenius' J. Anal. Chem. 349*: 473 (1994).

113. M. D. Mohr, K. D. Bornsen, and H. M. Widmar, Matrix-assisted laser desorption-ionization mass spectrometry: Improved matrix for oligosaccharides, *Rapid Commun. Mass Spectrom. 9*: 809 (1995).

114. D. S. Ashton, C. R. Beddell, D. J. Cooper, and A. C. Lines, Determination of carbo-heterogeneity in the humanised antibody CAMPATH 1H by liquid chromatography and matrix-assisted laser desorption-ionization mass spectrometry, *Anal. Chim. Acta 306*: 43 (1995).

115. R. M. Whittal, M. M. Palcic, O. Hindsgaul, and L. Li, Direct analysis of enzymatic reactions of oligosaccharide in human serum using matrix-assisted laser desorption ionization mass spectrometry, *Anal. Chem. 67*: 3509 (1995).

116. J. Y. Zhao, N. J. Dovichi, O. Hindsgaul, S. Gosselin, and M. M. Palcic, Detection of 100 molecules of product in a fucosyltransferase reaction, *Glycobiology 4*: 239 (1994).

117. K. K. Mock, C. W. Sutton, and J. S. Cottrell, Sample immobilization protocols for matrix-assisted laser-desorption mass spectrometry, *Rapid Commun. Mass Spectrom. 6*: 233 (1992).

118. K. O. Bornsen, M. D. Mohn, and H. M. Widmer, Ion exchange and purification of carbohydrates on a Nafion (RTM) membrane as a new sample pretreatment for matrix-assisted laser desorption-ionization mass spectrometry, *Rapid Commun. Mass Spectrom. 9*: 1031 (1995).

119. D. J. Harvey, Quantitative aspects of the matrix-assisted laser desorption mass spectrometry of complex oligosaccharides, *Rapid Commun. Mass Spectrom. 7*: 614 (1993).

120. T. J. P. Naven and D. J. Harvey, Effect of structure on the signal strength of oligosaccharides in matrix-assisted laser desorption/ionization mass spectrometry on time-of-flight and magnetic sector instruments, *Rapid Commun. Mass Spectrom. 10*: 1361 (1996).

121. D. Garozzo, E. Spina, L. Sturiale, G. Montaudo, and R. Rizzo, Quantitative determination of beta (1-2) cyclic glycans by matrix-assisted laser desorption mass spectrometry, *Rapid Commun. Mass Spectrom. 8*: 358 (1994).

122. D. Fu, L. Chen, and R. A. O'Neil, A detailed structural characterization of ribonuclease B oligosaccharides by ^{1}H NMR spectroscopy and mass spectrometry, *Carbohydrate Res. 261*: 173 (1994).

123. B. Spengler, D. Kirsch, R. Kaufmann, and J. Lemoine, Structure analysis of branched oligosaccharides using post-source decay in matrix-assisted laser desorption ionization mass spectrometry, *J. Mass Spectrom 30*: 782 (1994).

124. D. J. Harvey, T. J. P. Naven, B. Küster, R. H. Bateman, M. R. Green, and G. Critchley, Comparison of fragmentation modes for the structural determination of complex oligosaccharides ionized by matrix assisted laser desorption/ionization mass spectrometry, *Rapid Commun. Mass Spectrom. 9*: 1556 (1995).

125. G. Talbo and M. Mann, Aspects of the sequencing of carbohydrates and oligonucleotides by matrix-assisted laser desorption/ionization post source decay, *Rapid Commun. Mass Spectrom. 10*: 100 (1996).

126. D. J. Harvey, Matrix-assisted laser desorption/ionisation mass spectrometry of sphingo- and glycosphingo-lipids, *J. Mass Spectrom. 30*: 1311 (1995).

127. A. Powell and D. J. Harvey, Stabilization of sialic acids in *N*-linked oligosaccharides and gangliosides for analysis by positive-ion matrix-assisted laser desorption-ionization mass spectrometry, *Rapid Commun. Mass Spectrom. 10*: 1027 (1996).

128. C. G. Hellerqvist, B. Lindberg, S. Svensson, T. Holme, and A. A. Lindberg, Structural studies on the *O*-specific side-chains of the cell wall lipopolysaccharide from *Salmonella typhimurium* 395 MS, *Carbohydrate Res. 8*: 43 (1968).

129. C. G. Hellerqvist, Linkage analysis using Lindberg method, *Methods Enzymol. 193*: 554 (1990).
130. D. Rolf and G. R. Gray, Reductive cleavage of glycosides, *J. Am. Chem. Soc. 104*: 3539 (1982).
131. G. R. Gray, Linkage analysis using reductive cleavage method, *Methods Enzymol. 193*: 573 (1990).
132. V. Patoprsty, V. Kovacik, and S. Karacsonyi, Enhancement of methylation analysis of complex carbohydrates using pyridine as reagent in chemical ionization mass spectrometry, *Rapid Commun. Mass Spectrom. 9*: 840 (1995).
133. K.-H. Khoo and A. Dell, Assignment of anomeric configurations of pyranose sugars on oligosaccharides using a sensitive FAB-MS strategy, *Glycobiology 1*: 83 (1990).

13

Approaches to the Structural Determination of Oligosaccharides and Glycopeptides by NMR

Elizabeth F. Hounsell and David Bailey
University College London, London, England

I. INTRODUCTION

NMR spectroscopy holds a unique position in the analysis of oligosaccharides, glycoproteins, and lower molecular weight glycoconjugates in that it alone can give both unambiguous structural characterization and conformational information. NMR spectroscopists exploit the inherent physicochemical properties of atoms and molecules: The atoms involved in carbohydrates have or can be replaced by nuclei with magnetic spin, most importantly the protons ^{1}H, ^{13}C, ^{15}N, ^{31}P, and ^{33}S. To simplify spectra, NMR "active" atoms can be exchanged with NMR "inactive" atoms, e.g., ^{2}H for ^{1}H, as with ^{2}H$_2$O (D$_2$O) exchange of hydroxyl groups carried out for the spectra discussed in this chapter. On the other hand, NMR active atoms can be incorporated into oligosaccharide compounds by chemical synthesis using isotopically enriched starting materials, e.g., ^{13}C for ^{12}C. Bacteria can also be fed ^{15}N compounds with which they can provide biosynthetically enriched polysaccharides and proteins. The glycosidic ring structure offers additional advantages when we exploit the pulse sequences primarily designed for protein analysis, as follows.

2D ^{1}H–^{1}H *correlated spectroscopy* (COSY) yields signals from protons that are *directly* coupled. This makes it possible to "walk around" the glycosidic ring from the C-1 proton, assigning each individual proton via $^3J_{H,H}$ coupling.

The *double quantum-filtered* (DQF) COSY experiment provides data similar to those available from COSY spectra, but the pulse sequence incorporates a "quantum filter" that reduces the signal intensity of uncoupled nuclei (singlets) and also gives better resolution.

Triple quantum-filtered (TQF) experiments use a "spin filter" conceptually similar to that employed in DQF-COSY experiments; however, this results in spectra whose chief signals are those involving three or more mutually coupled spins. With oligosaccharides this reveals the hydroxymethylene systems in hexopyranosides (H-6, H-6′, H-5), pentoses (H-5, H-5′, H-4), and sialic acids (H-9, H-9′, H-8 and H-3$_{eq}$, H-3$_{ax}$, H-4).

Heteronuclear multiple-bond correlation (HMBC) is useful for long-range ^{1}H–hetereonuclear connectivities. With the carbohydrate part of a glycopeptide, sequencing is possible by using the ^{13}C–^{1}H couplings between glycosidic bonds; when used with ^{15}N-labeled peptide, this experiment correlates the amide N with the Cα proton.

Proton assignment using overcrowded 2D ^{1}H–^{1}H correlated experiments can be supplemented by using the *heteronuclear multiple-quantum coherence* (HMQC) experiment or the *heteronuclear single-quantum coherence* (HSQC) experiment. Correlations between carbon and directly attached protons are obtained. Since ^{13}C chemical shifts have better dispersion, this allows easier spectral assignment of both nuclei.

Total correlation spectroscopy (TOCSY) allows us to see correlations between every spin in a coupled system and not just between those giving rise to $^3J_{H,H}$ couplings, as in COSY/DQF-COSY experiments. With carbohydrates (using D$_2$O as solvent), all protons showing $^3J_{H,H}$ coupling *and* inside the area bounded by glycosidic linkages should be correlated. With peptides, the barrier to spin coupling is the carbonyl part of the peptide linkage.

In *nuclear overhauser effect spectroscopy* (NOESY), protons spatially close but not physically linked are correlated. Quantitative intensities for the cross-peaks obtained can then be processed to generate proton–proton distance constraints. These constraints can be added into distance geometry packages to give "structures" consistent with the NOE data that can be explored by molecular dynamics.

Like the NOESY experiment, *rotating-frame nuclear overhauser effect spectroscopy* (ROESY) is also used to measure "through space" correlations. It has found use when applied to molecules smaller than proteins, e.g., small peptides and carbohydrates. Molecules like these can give reduced or no correlation when examined using a NOESY experiment. ROESY can be used to obtain qualitative proton–proton distance information.

The present chapter will concentrate on the methods required to provide chemical shift assignments, which are necessary before conformational information can be obtained. They provide diagnostic structural profiles, including the positions of linkage and acyl or ester substitutions of hydroxyl groups. Profiling of *O*- and *N*-protein glycosylation can be achieved by comparing the chemical shift fingerprint to a preexisting database. Further de novo characterization of novel oligosaccharide sequences can also be achieved by the clues gained from chemical shifts, coupling constants, and NOE data. In addition measurements of $^1H-^1H$ coupling constants are indicative of monosaccharide type (e.g., Gal, Fuc, Man) and their ring conformation (i.e., the different chair $^1C_4$, $^4C_1$, boat or skew forms). The number of different oligosaccharides identified in plants, yeast, algae, fungi, microorganisms, fish, amphibians, birds, and mammals is constantly expanding as the methods for analysis are improved (reviewed in Refs. 1–3). In addition to the characterization of the oligosaccharides, data are now accumulating on the molecules to which they are attached in nature. The conjugates include proteoglycans, *O*- and *N*-linked glycoproteins, glycolipids, glycophosphatidylinositol anchors, mycobacterial glyco(peptido)lipids, and bacterial lipopolysaccharides. These are discussed here in the context of their importance as, variously, antigens and markers of differentiation and oncogenesis which are targets for immunotherapy and otherwise of potential interest in the pharmaceutical industry.

II. *O*-LINKED GLYCOSYLATION

There are several types of glycosylation linked through the hydroxyl group of serine (Ser) or threonine (Thr) amino acids. Limited NMR data exist for glycoproteins having the single GlcNAcβ found in nuclear and cytoplasmic proteins [4], the Glc, Xyl, or Fuc residues found linked to EGF-(epidermal growth factor-)like domains of certain proteins [5] and larger oligosaccharides in the latter series [6]. However, Xyl is also the linkage monosaccharide for the glycosaminoglycans (GAGs) of proteoglycans for which a large database is now accumulating, due to the importance of the interactions of GAG chains in anticoagulation, extracellular matrix function, and cell signaling [7–11]. *O*-Linked chains released from mucin-type glycoproteins as their alditols (terminating in reduced GalNAc, GalNAcol) have provided a useful source of material for NMR characterization of these important tumor- and disease-associated antigens [reviewed in Refs. 1–3]. Data are also accumulating on *O*-linked glycopeptides and glycosides, either isolated from natural sources or chemically synthesized, from which conformational details are being gathered leading to an understanding of structure/function relationships during tumorigenesis and specific recognition by bacterial and plant lectins in pathogenesis [3].

A. Proteoglycans

By definition, proteoglycans have long polysulfated oligosaccharide chains (GAGs) attached to a protein core. Although the GAGs are made up of a disaccharide repeat, –amino sugars–uronic acids–, the diversity of structures made possible by variable epimerization, acylation, and sulfation are just beginning to be appreciated. Specific functions are now being assigned to different sequences; hence, speculation also abounds as to the qualitative and quantitative controls in their biosynthesis. The majority of detailed structural studies by ^{1}H-NMR have been carried out on oligosaccharides released by sequence-specific enzymes, e.g., for heparin, heparinase [12–17], and heparatinase I and II [18]. The oligosaccharides have been used to probe the different specificities of these enzymes, which serve to show that variation in patterns of O- and N-sulfation of glucosamine and ratios of glucuronic to iduronic acids and their sulfation are recognized in biological systems. This is reflected in the characterization of specific heparin oligosaccharides with high affinity for e.g., antithrombin III, coagulation factor Xa, heparin cofactor II, lipoprotein lipase, and basic fibroblast growth factor [13,15,19–26], although there is still some controversy about the exact recognition sequences.

The variability in heparin structure is shown in Tables 1–3 with averaged ^{1}H-NMR data from the above sources. Figure 1 illustrates the structures from one study [16] in Haworth projection. Figures 2–4 show typical 1D ^{1}H–NMR spectra and ^{1}H–^{1}H TOCSY experiments for three of these oligosaccharides. The most easily assigned signal in the ^{1}H-NMR spectrum is that for H-4 of the uronic acid, which has a C-4—C-5 double bond (Δ4,5) formed by the eliminase enzymes used in the preparation (Figure 1). This is the most downfield signal found, at around 6 ppm (omitted in Figures 2 and 3 to reduce the spectral width), from which in a TOCSY experiment one can trace connectivity to the H-1 of Δ4,5 in the region downfield of the HOD peak. The approximate order of the chemical shifts of the H-1 protons in this region is δ GlcNS$_{ig}$ > $\Delta_{4,5}$ > GlcNS$_r$ > GlcNS$_{ii}$ > GlcNS$_\Delta$ > GlcNAc (where ig is the internal linked to GlcA, $\Delta_{4,5}$ is the nonreducing terminal hexuronic acid, r is the reducing end, ii is the internal linked to IdoA, and Δ is adjacent to the unsaturated nonreducing terminus). The HOD peak is present because of residual water in the sample, which cannot be completely removed on repeated evaporation with D$_2$O (also used to exchange the hydroxyl groups) but can be reduced by presaturation in the NMR experiment. The H-1 and H-5 of iduronic acids are nearest to the HOD signal. Just upfield of the HOD signal are the H-1 of glucuronic acid and the H-2 of the unsaturated hexuronic acid. From these, the majority of signals can be traced from their connectivity around the ring, which in a COSY experiment will give C-1—C-2, C-2—C-3, etc., and in a TOCSY experiment optimally will give C-1—C-2—C-3—C-4—C-5—C-6, if this sequence is not interrupted by poor magnetization

Table 1 Abbreviated Nomenclature for Glycosaminoglycan Chains[a]

△	4,5 unsaturated hexuronic acid			◇	Ido (2S)	◆	GlcA
○	GlcNS (6S,3S)	●	GlcNAc	X	Xylose	◈	IdoA
■	GalNAc	□	Gal	S	Sulfate	◐	GlcNS

[a]This relates to the nomenclature suggested [3] for oligosaccharide chains [see Table 4 and Figures 8 and 9]: Circular signs relate to glucose-type glycosidic rings, and square signs relate to galactose-type. Here, however, △ is not fucose but Greek capital delta for the unsaturated hexuronic acid formed by eliminase enzymes. Because this is usually 2-O-sulfated, a logical extension to the nomenclature is that ◇ is the noneliminated hexuronic acid IdoA(2S). ◆ is used for glucuronic acid (rather than the sialic acid N-glycolylneuraminic acid of mucins and N-linked chains, which has so far not been found in glycosaminoglycan chains). Glc is also not found in glycosaminoglycans; therefore, ○ can now be used to designate GlcNS. Another subtlety to help compare structures is to distinguish nonsulfated IdoA from IdoA(2S) and nonsulfated GlcNS from 6-O-sulfated GlcNS by half filling in the diamond and circle, representing less O-sulfation.

Table 2 References to the Chemical Shifts for Heparin Hexasaccharide Sequences

Ratio[a]		References
8/1/–	6S · 6S · 6S △—○—◇—○—◑—● hexa 2 2S · 2S	Lankjaer et al. 1995 [21]
8/–/–	6S · 6S △—○—◇—○—◇—◑ hexa 3 2S · 2S · 2S	Lankjaer et al., 1995 [21]
8/–/1	6S · 6S · 6S △—○—◇—○—◆—○ hexa 4 2S · 2S	Lankjaer et al., 1995 [21]; Chai et al., 1995 [16] F3-8; Horne and Gettins, 1992 [14]; Hexamer 1
9/–/–	6S · 6S · 6S △—○—◇—○—◇—○ hexa 5 2S · 2S · 2S	Lankjaer et al., 1995 [21]; Chai et al., 1995 [16] F3-9; Horne and Gettins, 1992 [14]; Hexamer 2
7/1/1	6S · 6S · 6S △—○—◑—●—◆—○ F3-15 2S · 3S	Chai et al., 1995 [16]; Linhardt et al., 1986 [19]; Petitou et al., 1988 [13]; Tsuda et al., 1996 [17]
6/1/1	6S · 6S · 6S △—○—◑—●—◆—○ 2S	Lindhardt et al., 1992 [20]
7/1/1	6S · 6S · 6S △—○—◇—●—◆—○ b-24 2S · 2S	Tsuda et al., 1996 [17]
7/1/1	6S · 6S △—○—◇—●—◆—○ F4.5 2S · 2S	

[a]Ratio of the number of sulfate groups/N-acetamido groups/GlcA residues.

transfer between adjacent protons where the coupling constants $^3J_{H,H}$ are smaller than the spectral linewidths.

Figures 2 and 3 are illustrative of the subtle changes in the spectrum from analysis of two tetrasaccharides varying only in the addition of a sulfate group on the reducing-end monosaccharide. Comparison with Figure 4 shows the additional signals from an oligosaccharide having N-acetylglucosamine and glucuronic acid in addition to N-sulfated glucosamine and iduronic acid and also a trisulfated reducing-end glucosamine residue. These experiments were carried out at 22°C (295 K) in D_2O as solvent, with the addition of acetone as an internal standard that, with this temperature and solvent, resonates at 2.225 ppm with respect to DSS (4,4-di-methyl-4-silapentane-1-sulfonate). Increasing the temperature has the effect of revealing close upfield signals; decreasing the temperature reveals signals downfield of the water. Changing the temperature can also help in assigning other protons with similar chemical shift and in sharpening the signals in larger oligosaccharides. Native, high molecular weight heparins are usually analyzed at 30°–60°C by both ^{1}H- and ^{13}C-

Table 3 Averaged ^{1}H-NMR Chemical Shifts of Monosaccharides[a] in the Oligosaccharide Sequences of Heparin (ppm at 22°–30°C with Respect to Acetone at 2.225 ppm in D$_2$O)

	2S 6S △(—○—)	2S 6S △—)○(—◆—)	2S 6S 2S △—)○(—◇—)	6S 6S (—○—)◆(—○—)	6S 6S (—○—)▲(—●—)
H-1	5.50	5.35	5.40	5.16	5.02
H-2	4.62	3.28	3.29	4.34	3.74
H-3	4.31	3.67	3.65	4.28	4.14
H-4	6.00	3.83	3.84	4.28	4.08
H-5	—	3.96	4.05	5.12	4.80
H-6, H-6'	—	4.27, 4.37	4.23, 4.34	—	—

	6S 2S 6S (—○—)◇(—○—)	6S 2S 6S (—○—)◇(—●—)	2S 6S 2S (—◇—)○(—◇—)	2S 6S (—◇—)○(—◇—)
H-1	5.19	5.13	5.39	5.58
H-2	4.33	4.33	3.29	3.28
H-3	4.22	4.08	3.65	3.65
H-4	4.10	4.20	3.78	3.78
H-5	5.17	5.03	4.02	3.97
H-6, H-6'	—	—	4.21, 4.32	4.26, 4.37

	6S (—◆—)●(—●—)	2S 6S (—◇—)●(—●—)	6S 2S 6S (—○—)◇(—○α/β)
H-1	5.38	5.38	5.23
H-2	3.91	3.93	4.32

H-3	3.76	3.78	4.24
H-4	3.83	3.73	4.10
H-5	4.07	4.18	4.85
H-6	4.2–4.35	4.24, 4.36	—

	6S (—○—)◆(—○α/β	6S 3S, 6S (—●—)◆(—○α/β	6S 6S (—●—)◆(—○α/β
H-1	4.62	4.63	4.59
H-2	3.39	3.39	3.35
H-3	3.85	3.70	3.71
H-4	3.81	3.82	3.75
H-5	3.83	3.90	3.83

	6S (—◆—)●○α/β	2S 6S (—◇—)○α/β	2S 6S (—◇—)○α/β	6S (—◆—)○α/β	3S, 6S (—◆—)○α/β
H-1	5.40	5.45	5.46	5.46	5.46
H-2	3.94	3.24	3.26	3.26	3.42
H-3	3.78	3.71	3.72	3.81	4.46
H-4	3.74	3.75	3.74	3.86	4.00
H-5	4.03	4.15	3.95	4.05	4.21
H-6	4.28, 4.35	4.27–4.39	3.90, 3.95	4.30–4.36	4.46–4.31

[a]For nomenclature, see Table 1, plus S, sulfate.

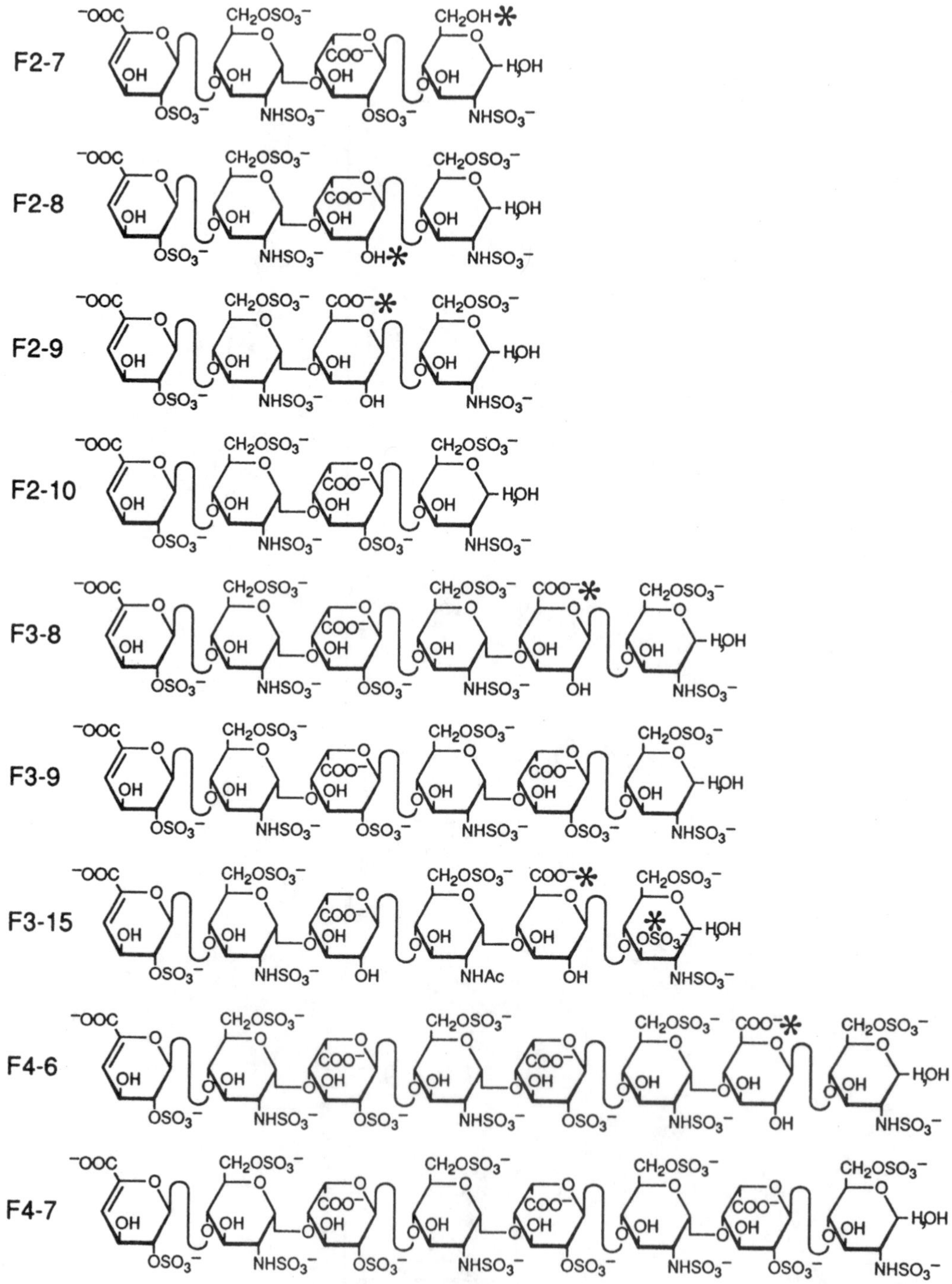

Figure 1 The structure of the most abundant oligosaccharides isolated from porcine intestinal heparin. The common monosaccharides are 6-sulfated, 2-N-sulfated glucosamine, and 2-sulfated iduronic acid. Differences are shown by *. (From Ref. 16.)

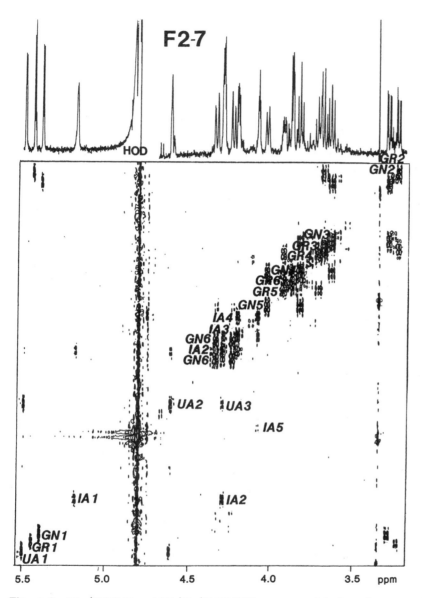

Figure 2 The ^{1}H-NMR and 2D ^{1}H–^{1}H TOCSY spectrum of the heparin tetrasaccharide F2-7. Abbreviations: UA, uronic acid ($\triangle$4,5); GR, reducing and GlcNS; GN, internal GlcNS; IA, iduronic acid. (From Ref. 16, Figure 1.)

NMR. For oligosaccharides that are usually not available in large amounts, ^{1}H-detected ^{13}C-hetereonuclear experiments, such as HMQC, HMBC, and HSQC, are applicable. The HMBC experiment can be used to supply $^3J_{C,H}$ coupling constants to obtain conformational information across the glycosidic bond. From the coupling constants $^3J_{H,H}$ information can be obtained about the ring geometry that is important in proteoglycans due to the relative flexibility of the iduronic acid ring compared to other monosaccharides (Figure 5). Several studies have used ^{1}H- and ^{13}C-NMR and

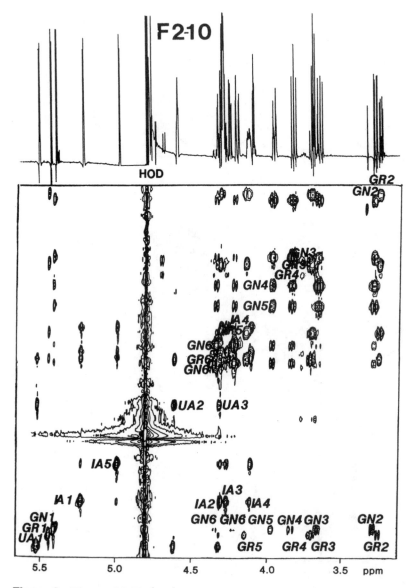

Figure 3 The partial 2D ^{1}H–^{1}H TOCSY spectrum of the heparin tetrasaccharide F2-10. Abbreviations as in Figure 2. (From Ref. 16, Figure 1.)

relaxation to provide conformational information on functionally active, chemically synthesized heparin oligosaccharides or heparin fragments and their chemically modified derivatives [24–30]. Chemical synthesis of different heparin oligosaccharides has been a paradigm in rational drug design [31], which has resulted in new, potent anticoagulant agents.

In other areas where heparin binding to growth factors can mediate cell signaling, such as cell regulation, inhibition of proteoglycan synthesis or action may be an appropriate therapeutic strategy. In vitro, chlorate has been used to inhibit sulfa-

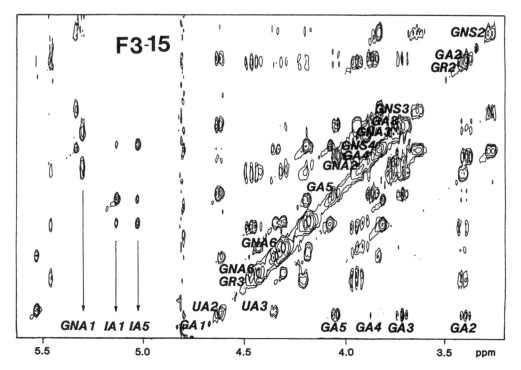

Figure 4 The partial ¹H-NMR and 2D ¹H−¹H TOCSY spectrum of the heparin hexasaccharide F3-15. Abbreviations as in Figure 2 and GA, glucuronic acid; GNA, GlcNAc (as compared to GlcNS). (From Ref. 16, Figure 1.)

tion and β-xylosides to inhibit biosynthesis. Xylose is the linkage sugar to serine (Ser) in the protein core of heparin (Hep), heparan sulfate (HS), chondroitin sulfate (CS), and dermatan sulfate (DS). Several core-region sequences of Hep, CS, and DS obtained by chemical synthesis or isolation have now been characterized [32−35]. Additional NMR data on CS and DS chains are available [36,37], together with

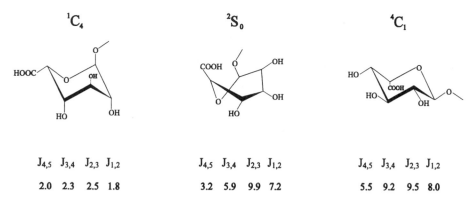

Figure 5 The ⁴C₁, ¹C₄, and ²S₀ conformations for iduronic acid, and their typical coupling constants.

Table 4   Core-Region Sequences of Heparin, Chondroitin Sulfate, Dermatan Sulfate, and Keratan Sulfate[a]

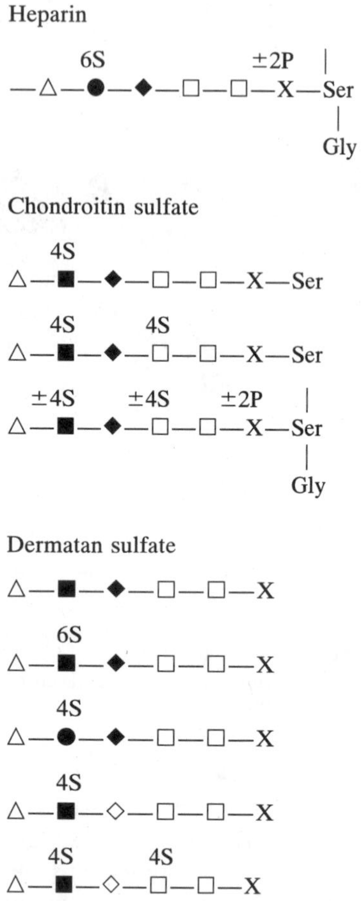

Heparin

<pre>
   6S   ±2P |
—△—●—◆—□—□—X—Ser
                              |
                             Gly
</pre>

Chondroitin sulfate

<pre>
   4S
△—■—◆—□—□—X—Ser

   4S   4S
△—■—◆—□—□—X—Ser

  ±4S   ±4S   ±2P |
△—■—◆—□—□—X—Ser
                          |
                         Gly
</pre>

Dermatan sulfate

<pre>
△—■—◆—□—□—X

   6S
△—■—◆—□—□—X

   4S
△—●—◆—□—□—X

   4S
△—■—◇—□—□—X

   4S   4S
△—■—◇—□—□—X
</pre>

conformational information [38]. Core-region sequences of Hep, CS, DS, and a fourth proteoglycan, keratan sulfate (KS), are shown in Table 4. Keratan sulfate has GAG chains linked to the classical N-linked cores of glycoproteins (Type I KS) and to mucin-type O-linked cores (Type II KS), which are considered below. Keratan sulfate is also unique in lacking uronic acid and is in fact sulfated poly-N-acetyllactosamine[-3Galβ1–4GlcNAcβ1-]$_n$. Since the first characterization by ^{1}H-NMR spectroscopy of this sequence as having sulfate on the majority of Gal residues and on all the GlcNAc residues [39], other chain terminating and core sequences (Table 4) have been extensively characterized by ^{1}H-NMR [40–42]. As discussed previously [1,39], the addition of the sulfate group at C-6 has a large deshielding effect on the signals for the two H-6 protons, moving them downfield by 0.46 and 0.44 ppm, with progressively lesser downfield shift differences for H-5, H-4, H-3, and H-2 (e.g., on the order of 0.2, 0.04, 0.015, and 0.03 ppm, respectively, for GlcNAc). Short sulfated sequences are also found as components of nonproteoglycan-type glycoproteins e.g.,

Table 4 Continued

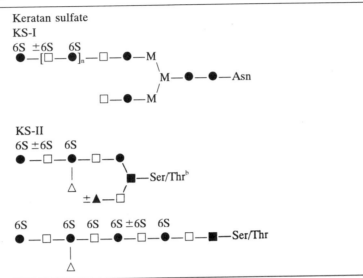

Keratan sulfate
KS-I

KS-II

[a]Nomenclature is as in Table 1, plus: X, xylose; M, mannose; P, phosphate; S, sulfate. NMR data are given in Refs. 32–35 and 39–42.
[b]NMR data given for the oligosaccharide alditol after release from cartilage by KS by β-elimination and concomitant reduction.

sulfate esters at C-6 of GlcNAc in *N*- and *O*-linked chains, at C-6 and C-3 of Gal in *O*-linked chains, and at C-4 of GalNAc in the nonreducing-end sequence GalNAcβ1–4GlcNAc of, e.g., glycopeptide hormones. There is considerable evidence that specifically sulfated sequences of *N*- and *O*-linked chains function in cell signaling (e.g., in the immune system), and sulfation of *O*-linked sequences close to the core has also been implicated in the tissue and mucin changes that occur in gastrointestinal cancers and inflammatory bowel diseases. Other structural variations in the protein–carbohydrate interface of gastrointestinal and respiratory tract mucins are described next.

B. Oligosaccharide Alditols Released from GalNAc-Ser/Thr Cores

Knowledge about the extent of the diversity of the oligosaccharides linked through GalNAc to Ser/Thr hydroxyl groups has been achieved largely by release of the oligosaccharide chains from mucin-type glycoproteins using a β-elimination reaction catalyzed by mild alkaline conditions (Figure 6). For GalNAc residues substituted at C-3 (the majority—see Ref. 3), the β-elimination can continue back through the attached oligosaccharide, leading to degradation ("peeling"). This is stopped by concomitant reduction on alkaline release, leading to the GalNAcol at the end of the chain, prevalent in studies of mucin-type oligosaccharides. The most common core-region sequences (on which numerous NMR studies have been carried out [reviewed in Refs. 1–3]) have Galβ linked at C-3 in the absence (core 1) or presence (core 2) of GlcNAcβ linked at C-6, or GlcNAcβ linked at C-3 in the absence (core 3) or presence (core 4) of GlcNAcβ at C-6. Core 5 has GalNAc linked to C-3 [43,44] and

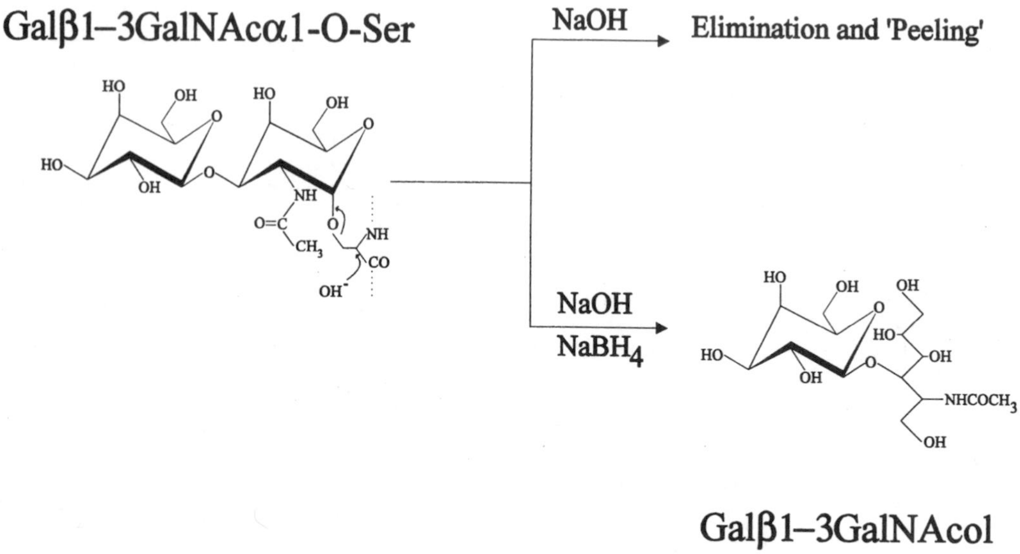

Galβ1–3GalNAcα1-O-Ser

NaOH → Elimination and 'Peeling'

NaOH
NaBH₄ →

Galβ1–3GalNAcol

Figure 6 β-Elimination of the Galβ1–3GalNAcα1-Ser O-glycopeptide linkage.

core 6 GlcNAcβ linked at C6 [43]. Complete ¹H-NMR and coupling-constant data for this last, novel, linear 6-linked sequence has been described [45]. The equivalent structure having GlcNAc linked to C-6 of Gal (rather than GalNAc) in the absence of a substituent at C-3 has also been characterized by ¹H-NMR [46], and extended backbone sequences, [Galβ1–4GlcNAcβ1–6]$_n$Galβ1–, have also been identified [47]. More generally found in both mucins and free oligosaccharides (see Section III) are the linear sequences Galβ1–3/4GlcNAcβ1–3Galβ1– and the branch Galβ1–3/4GlcNAcβ1–3[Galβ1–4GlcNAcβ1–6]Galβ1–.

The Gal/GalNAc residues of the mucin core sequences are commonly substituted by sialic acidα2–3, the GlcNAc/GalNAc by sialic acidα2–6, and the Gal/GlcNAc/GalNAc by Fucα (giving blood group type 1 to type 3 A, B, and H antigens and the Leᵃ, Leᵇ, Leˣ, and Leʸ structures [1–3,48–52]). The sialic acid in human tissues is almost exclusively N-acetylneuraminic acid (Neu5NAc or NeuAc), but in other species N-glycolylneuraminic (Neu5NGc or NeuG) and KDN (Neu5OH) occur in sequences that are species specific [3,53]. In humans, the blood group and related antigens are carried on proteins expressed by the MUC genes, which are important targets for cancer detection and immunotherapy. These highly glycosylated proteins are either secreted as mucins (e.g., MUC2–5) or retained in the epithelial cell membrane (e.g., MUC1) [54]. Oligosaccharides linked via GalNAc-Ser/Thr also occur on membrane-bound nonmucin-type glycoproteins (i.e., those having relatively low molecular weight and percentage of glycosylation), where they have been implicated as influencing the overall conformation of the protein (reviewed in Ref. 3). The oligosaccharides here are primarily of the core 1 type, either disialylated on Gal and GalNAc or monosialylated at Gal or GalNAc. However, O-linked oligosaccharides with the different core regions and extensions described above are also found on nonmucin-type serum and membrane glycoproteins [1,3]. The consensus amino acid sequence predisposing toward O-linked glycosylation is not as clearcut as for

N-linked, but there are some indicators that can be used to predict occupied sites [55].

The NMR chemical shift data can be extrapolated from one type of oligosaccharide chain to another using the computerized searching of available databases [56–60]. That originally set up for ^{1}H-NMR [56] deals extensively with the mucin oligosaccharides and their motifs as also found on *N*-linked chains. ^{13}C-NMR databases are used largely for bacterial polysaccharides [57,58] because of the restraints on amount of material. The original data for *N*-linked chains [59] can now be accessed through the World Wide Web and the search programs are being expanded to include the use of neural networks [60].

C. Studies of *O*-Linked Glycopeptides

In order to understand the effects of *O*-linked glycosylation on glycoprotein conformation and function, studies have addressed the NMR characterization and conformations of oligosaccharides having GalNAc at the reducing end (rather than GalNAcol), as well as glycosides, glycopeptides, and glycoproteins [reviewed in Ref. 3]. Of necessity the glycoside or protein glycosylation studied so far is of the cores (without the further extensions discussed above), which would be expected to have the most effect on protein conformation. In contrast, the more distant oligosaccharides have their own conformational motifs, which are recognized by anticarbohydrate antibodies and carbohydrate-binding proteins in the absence of protein. *O*-glycopeptides having intact GalNAc-Ser/Thr with one to several amino acids and with or without galactose and sialic acid have been studied from two sources, (i) those occurring naturally, presumably arising from glycoprotein catabolism (isolated mostly from human urine) and (ii) those prepared by chemical synthesis either as part of solution- or solid-phase synthesis methods [61–70]. Table 5 compares the ^{1}H-NMR data for a representative glycopeptide from each type of source. The first are our data for a sialylated trisaccharide dipeptide (TSDP) obtained from human urine (Figure 7), and the second, an octapeptide [68] having three disaccharides on adjacent Thr residues. The data can be used to compare the affect of sialylation, length of peptide, and contiguous oligosaccharides on chemical shift variations in the monosaccharides and amino acids.

III. GLYCOPROTEIN-CHAIN TERMINATING SEQUENCES ALSO FOUND ON LACTOSE-BASED CORES

The disaccharide lactose (Galβ1–4Glcα/β) is a major component of the milk of all mammals. This is the substrate for a large number of glycosyltransferase enzymes that give multiple oligosaccharides that are particularly abundant in the milk of humans. The glycosyltransferases either are the same or have the same specificity as those that glycosylate *O*- and *N*-linked glycoprotein chain cores. Thus human milk oligosaccharides are a rich source of material to provide NMR chemical shifts and conformational information for these sequences occurring at the end of oligosaccharide chains. Work on a computerized database of chemical shifts [56] has shown that oligosaccharide chains can be viewed as being made up of domains that can act

Table 5 ^{1}H-NMR Chemical Shifts for the $\alpha2-3$-Sialylated T Glycopeptide (TSDP)[a] shown in Figure 7 and for T on Adjacent Thr Residues of an Octapeptide[b]

Monosaccharide/Amino Acid	TSDP (ppm)	—NH—Thr—Thr—Thr—CO— (ppm)		
NeuAc H-3$_{eq}$	2.755			
H-3$_{ax}$	1.784			
H-4	3.66			
H-5	3.88			
H-6–H-9	ND			
NAc	2.028			
Gal H-1	4.531	4.42	4.43	4.45
H-2	3.529	3.55	3.54	3.52
H-3	4.064	3.61	3.61	3.61
H-4	3.923	3.90	3.90	3.90
H-5	ND	3.61	3.61	3.61
H-6, H-6^1	ND	3.73–3.79		
GalNAc H-1	4.963	5.03	4.89	4.82
H-2	4.362	4.27	4.23	4.24
H-3	4.058	4.00	3.98	4.01
H-4	4.238	4.23	4.21	4.22
H-5	ND	4.06	3.99	4.05
H-6, H-6^1	ND	3.78	3.75	3.77
NAc		2.04	2.03	2.07
Ser αCH (4.3)[c]	4.311	—	—	—
βCH$_2$ (3.9)	4.212	—	—	—
βCH$_2$ (3.9)	3.931	—	—	—
Leu αCH (4.2)	4.268	—	—	—
βCH$_2$ (1.6)	1.63	—	—	—
γCH (1.6)	1.63	—	—	—
δCH$_3$ (0.9)	0.92/0.91	—	—	—
Thr αCH (4.3)	—	4.68	4.81	4.75
βCH (4.3)	—	4.33	4.38	4.40
δCH$_3$ (1.3)	—	1.30	1.27	1.34

[a]Chemical shift with respect to acetone at 2.225 ppm in D$_2$O at 295 K.
[b]Data for NH$_2$-Thr-Ser-Ile-Pro-Thr*-Thr*-Thr*-Pro-OH, where * is Galβ1–3GalNAcα1– taken from Ref. 66 in solution 30% CD$_3$CO$_2$D in H$_2$O at 300 K with respect to acetone at 2.03 ppm.
[c]Textbook data given for nonglycosylated amino acids.

as relatively conformationally distinct entities. Thus, for example, we and others have fully characterized sequences within the oligosaccharides

GalNAcα1–3[Fucα1–2]Galβ1–3[Fucα1–4]GlcNAcβ1–3Galβ1–4Glcα/β [71,72]

NeuAcα2–3Galβ1–3[NeuAcα2–6]GlcNAcβ1–3Galβ1–4Glcα/β [73,74]

which have the nonreducing terminal penta- and tetrasaccharides, respectively, found, for example, on the complex N-linked chains of the receptor for epidermal growth factor on A431 cells [75] and of rat serotransferrin [76] (the lactose sequence being replaced by the mannose- and glucosamine-containing core of N-linked chains). The ^{1}H-NMR spectra of these oligosaccharides, called ALe^b heptasaccharide and disi-

NeuAcα2–3Galβ1–3GalNAcα1–O–Ser-Leu

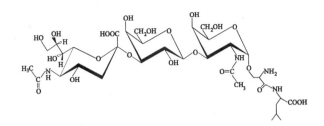

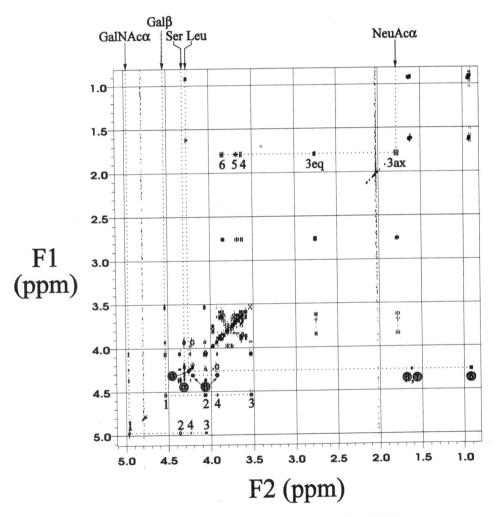

Figure 7   The 2D ¹H–¹H TOCSY of the sialylated T glycopeptide (TSDP).

alyllacto-N-tetraose (DSLNT), have been published previously, with full chemical shift assignments, including ROESY data [71,73]. They are just two of the many examples showing the diversity of oligosaccharides based on the lactose core and their relevance to cell-surface and secreted N-linked glycoproteins.

We have recently characterized a partial monosialylated derivative of the disialylated motif shown above [74]. The 1D ^{1}H-NMR spectrum and 2D TOCSY are shown in Figure 8. The oligosaccharide is based on the GlcNAcβ1–3Galβ1–4Glc linear sequence lacking the nonreducing terminal Galβ1–3 or Galβ1–4 of the classical lacto-N-tetraose (LNT) and lact-N-neo-tetraose (LNNT) oligosaccharides. Larger oligosaccharides are generally built up by branching at the Gal residue, i.e., GlcNAcβ1–6[GlcNAcβ1–3]Galβ. However, as discussed in Section II.B., O-linked oligosaccharides having linear GlcNAcβ1–6GalNAc and GlcNAcβ1–6Gal are also known to occur. We have also extensively characterized by ^{1}H-NMR [74] oligosaccharides having the galactose residue of reducing-end lactose linked only at the C-6 position by GlcNAc (e.g., Figure 9). This oligosaccharide is more likely to have been formed by breakdown of a larger branched oligosaccharide than by a specific synthesis (unlike that in the O-linked chains discussed above). The large majority of oligosaccharides now isolated from milk have branched galactose. Early work on the characterization of tri- to hexasaccharides [77] has now been supplemented with the identification of many higher molecular weight oligosaccharides, which have been characterized in detail by ^{1}H- and ^{13}C-NMR studies (e.g., Refs. 78–84). Similar oligosaccharides have also been characterized in glycosphingolipids of the lactosamine type [1,2,73], and additional sequences have been identified on globosides of the Forssman blood group P family [85].

IV. FROM MAMMALS TO MICROORGANISMS—GLYCO(PEPTIDO)LIPID CONJUGATES, POLYSACCHARIDES, AND GLYCOPROTEINS IN IMMUNE ACTIVATION

In addition to the globoside and lactosamine series of glycosphingolipids discussed above, other lipid conjugates with distinct structures are found in many organisms, e.g., mammalian and parasite glycophosphatidylinositol (GPI) anchors [86], mycobacterial phenolic glycolipids [87–90], lipooligosaccharides [91,92], lipoarabinomannans [93–97], and glycopeptidolipids [98,99] and bacterial lipid A [100,101]. Of these the structures and NMR data of the nonpeptide-containing conjugates have been reviewed previously [2,102]. Data are now accumulating for the lipoarabinomannans that are associated with the mycobacterial cell walls through mannose inositol phosphate (MIP) linkages and peptidoglycan [93–96]. Peptidoglycan is a polymer of glucosamine, lactosylated glucosamine (muramic acid), and amino acids also present in bacterial cell walls. Similar amino acids are also found in the lower molecular weight glycopeptidolipids (mycolic acids) that are unique to mycobacteria and are the main immune determining factor. Muramyl peptides have been known for some time to be potent immune stimulators. Recently, peptide conjugates and phosphorylated monosaccharides have been implicated, by ourselves and others [103], as the γδ superantigens that activate subsets of T cells in a response that is independent of the major histocompatibility complex (MHC). Posttranslationally modified peptides have recently been identified as eliciting MHC-restricted immune

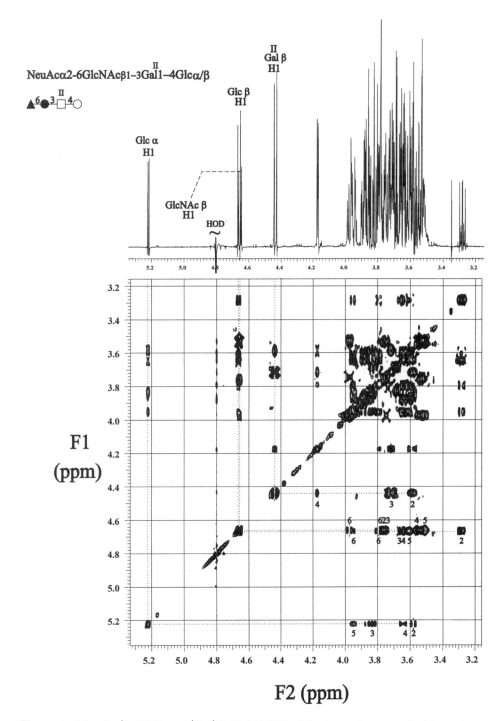

Figure 8 The 1D ¹H-NMR and ¹H–¹H 2D TOCSY of the internal tetrasaccharide portion of DSLNT.

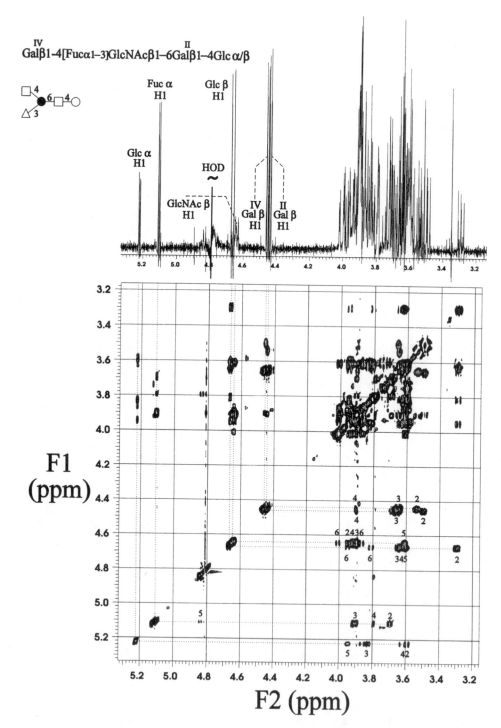

Figure 9 The 1D ¹H-NMR and ¹H–H 2D TOCSY of the pentaoligosaccharide having the Le^x trisaccharide linked via a β1–6 glycoside link to lactose.

responses, and the possibility of recognition of glycopeptides is beginning to be explored [104]. Thus, microorganisms display distinct biosynthetic pathways for modification of proteins and lipids. Oligosaccharide conjugates are targets for novel immunotherapeutic strategies against such scourges as tuberculosis and leprosy and also against common bacterial infections, particularly those of childhood.

The lipopolysaccharide (LPS) capsule antigen sequences linked through lipid A in bacteria are major antigens that elicit an immune response in mammals. Information on the structure and conformation of these is being accumulated in order to design more active antigens that will elicit both a B-cell (antibody) and T-cell (MHC-restricted) response in vaccination. NMR has been essential for these studies not least because the bacteria incorporate many different monosaccharides, for example, sugars of the L-series rather than the more common D and novel acylation. As stated in the Introduction, the bacteria can be more readily cultured in isotopically enriched media than in mammalian cells to obtain ^{15}N and ^{13}C for triple-probe experiments to gain much required conformational information [105]. The structures of all the LPS oligosaccharide sequences characterized so far are too numerous to describe here. NMR data can be obtained by indexing the journal *Carbohydrate Research*, in which the majority have been published, and by accessing X-ray data deposited with Cambridge Crystallographic. X-ray data include those of oligosaccharides complexed with antibodies and lectins [106].

A second aspect of our desire to know the structure and conformation of these structures is to understand the cross-reactivities between bacterial polysaccharides and oligosaccharide sequences of mammalian glycoproteins. A classic study for which extensive NMR data are available is that of the polysialic acid sequence of meningococcal LPS, which mimics the sialylated N-linked chains of the neural-cell-adhesion molecule NCAM [107]. In the human juvenile brain, the same polysialylated sequences as LPS are present, and these diminish in the adult to give shorter oligosaccharides.

V. GLYCOPROTEINS

Unlike the regular repeating oligosaccharides that make up bacterial LPS, the oligosaccharides of glycoproteins are heterogeneous, with any one molecule having several glycoforms. This, together with their large hydrodynamic volume, obviates against high-resolution X-ray crystallographic studies. Strategies include expressing recombinant proteins either in bacteria, where N-linked glycosylation is not added (although bacteria do produce glycoproteins), or in yeast or insect cell lines, which modify proteins with only high-mannose chains. The latter chains can be clipped by the enzyme endo-H between the two GlcNAc residues of the chitobiosyl core, leaving a single GlcNAc to indicate the previous position of glycosylation. For studies of the structure, functions, and dynamics of the oligosaccharides of glycoproteins and GPI-anchored (glyco)proteins it is therefore necessary to resort to NMR spectroscopy.

In the past, structural and conformational studies using NMR on *intact* glycoproteins have concentrated separately on the protein [108,109] or on the carbohydrate part of the molecule [110,111]. Homonuclear 2D techniques (COSY, DQF-COSY, TOCSY, NOESY) can be used to assign the protein part of the glycoprotein.

But assigning the oligosaccharide resonances by using data from these experiments is more difficult, because the Cα backbone proton signals tend to be superimposed on the anomeric signals of the constituent glycan sugars, and oligosaccharides exhibit few NOEs. The greater chemical shift dispersion of ^{13}C can be used to separate the majority of carbohydrate carbon resonances (which with intact glycoproteins is δ_H 3.4 → 5.2 and δ_C 61 → 105) from those of the aliphatic and aromatic carbons of the protein, as has been demonstrated for both N-linked [112,113] and O-linked [114,115] glycoproteins.

The protein structure of human CD59, a 77 amino acid cell-surface glycoprotein that has a fold topology similar to that of snake venom neurotoxins, has recently been structurally characterized by a combination of homonuclear 2D experiments and $^{13}C-^{1}H$ HMQC [116]. Partial assignments for the complex N-glycan have been made, and a $^{31}P-^{1}H$ HSQC spectrum has also been used in an attempt to characterize the phosphorylated sequences in the GPI anchor. Another glycoprotein characterized using standard homonuclear 2D spectroscopy and $^{13}C-^{1}H$ HMQC at natural abundance is the 13.6-kDa adhesion domain of human CD2 [117]. $^{13}C-^{1}H$ DEPT-HMQC spectra were also used in the analysis, to give probably the most comprehensive set of assignments yet for an intact glycoprotein.

Another innovation that should bring better characterization of intact glycoproteins is the recent use of pulsed field gradients [118]. These magnetic field gradients can be used in a variety of standard pulse sequences useful in protein and oligosaccharide structure determination to replace the ''phase cycling'' schemes [119–121]. The carbohydrate chains of the α-subunit of the intact glycoprotein human chorionic gonadotropin [113] have been characterized using gradient-enhanced $^{1}H-^{13}C$ HSQC and HSQC-TOCSY experiments, even with a moderately concentrated sample (5 mM) and a 60-hour acquisition time. Using the above techniques should make better assignments of intact glycoproteins easier and faster.

ACKNOWLEDGMENTS

The authors wish to thank David Renouf and Michael Davies for helpful discussions and for providing the backup necessary to make the work possible, and Gail Evans for her help in the preparation of this chapter. We are also grateful to the MRC and EC for funding and Dr. J. Feeney and Dr. P. Driscoll for use of the NMR facilities at, respectively, the MRC Biomedical NMR Centre and University College London, Department of Biochemistry and Molecular Biology.

REFERENCES

1. E. F. Hounsell, Physicochemical analyses of oligosaccharide determinants of glycoproteins, *Adv. in Carbohydr. Chem. and Biochem. 50*: 311–350 (1994).
2. E. F. Hounsell, ^{1}H-NMR in the structural and conformational analysis of oligosaccharides and glycoconjugates, *Progress in Nuclear Magnetic Spectroscopy 27*: 1–36 (1995).
3. E. F. Hounsell, M. J. Davies, and D. V. Renouf, Protein O-glycosylation structure and function, *Glycoconj. J. 13*: 19–26 (1996).

4. G. W. Hart, W. G. Kelly, M. A. Blomberg, E. P. Roquemore, L-Y. D. Dong, L. Kreppel, T-Y. Chou, D. Snow, and K. D. Greis, Glycosylation of nuclear and cytoplasmic proteins is as abundant and as dynamic as phosphorylation, In *44th Mosbach Colloquium: Glyco and Cell Biology* (Weiland F. Reutter and W. Heidelberg, eds.), Springer-Verlag, New York, (1994), pp. 91–103.

5. R. J. Harris and M. W. Spellman, *O*-Linked fucose and other post-translational modifications unique to EGF modules, *Glycobiology 3*: 219–224 (1993).

6. N. L. Stults and R. D. Cummings, *O*-Linked fucose in glycoproteins from Chinese hamster ovary cells, *Glycobiology 3*: 589–596 (1993).

7. J. T. Gallagher, The extended family of proteoglycans: Social residents of the pericellular zone, *Current Opinion in Cell Biol. 1*: 1201–1218 (1989).

8. D. Carrie, C. Caranobe, S. Saivin, G. Houin, M. Petitiou, J. C. Lormeau, C. Van Boeckel, D. Meuleman, and B. Boneu, Pharmacokinetic and antithrombotic properties of two pentasaccharides with high affinity to antithrombin III in the rabbit: Comparison with CY216, *Blood 84*: 2571–2577 (1994).

9. M. Lyon, J. A. Deakin, K. Mizuno, T. Nakamura, and J. T. Gallagher, Interaction of hepatocyte growth factor with heparan sulfate, *J. Biol. Chem. 269*: 11,216–11,223 (1994).

10. H. van Tilbeurgh, A. Roussel, J-M. Lalouel, and C. Cambillau, Lipoprotein lipase, molecular model based on the pancreatic lipase X-ray structure: Consequences for heparin binding and catalysis, *J. Biol Chem. 269*: 4626–4633 (1994).

11. T. Spivak-Kroizman, M. A. Lemmon, I. Dikic, J. E. Ladbury, D. Pinchasi, J. Huang, M. Jaye, G. Crumley, J. Schlessinger, and I. Lax, Heparin-induced oligomerization of FGF molecules is responsible for FGF receptor dimerization, activation and cell proliferation, *Cell 79*: 1015–1024.

12. Z. M. Merchant, Y. S. Kim, K. G. Rice, and R. J. Linhardt, Structure of heparin-derived tetrasaccharides, *Biochem. J. 229*: 369–377 (1985).

13. M. Petitou, J-C. Lormeau, B. Perly, P. Berhault, V. Bossennec, P. Sie, and J. Choay, Is there a unique sequence in heparin for interaction with heparin Cofactor II? *J. Biol. Chem. 268*: 8685–8690 (1988).

14. A. Horne and P. Gettins, ¹H-NMR spectral assignments for two series of heparin-derived oligosaccharides, *Carbohydr. Res. 225*: 43–57 (1992).

15. S. Yamada, K. Yoshida, M. Sugiura, and K. Sugahara, Sulfated disaccharide prepared from chondroitin sulfate and heparan sulfate/heparin by bacterial eliminase digestion, *J. Biochem. 112*: 440–447 (1992).

16. W. Chai, E. F. Hounsell, C. J. Bauer, and A. M. Lawson, Characterization by LSI-MS and ¹H-NMR spectroscopy of tetra-, hexa-, and octa-saccharides of porcine intestinal heparin, *Carbohydr. Res. 269*: 139–156 (1995).

17. H. Tsuda, S. Yamada, Y. Yamane, K. Yoshida, J. J. Hopwood, and K. Sugahara. Structures of five sulfated hexasaccharides prepared from porcine intestinal heparin using bacterial heparinase, *J. Biol. Chem. 271*: 10,495–10,502 (1996).

18. H. B. Nader, M. A. Porcionatto, I. L. S. Tersariol, M. A. S. Pinhal, F. W. Oliveira, C. T. Moraes, and C. P. Dietrich, Purification and substrate specificity of heparitinase I and heparitinase II from *Flavobacterium heparinum*, *J. Biol. Chem. 265*: 16,807–16,813 (1990).

19. R. J. Linhardt, J. G. Rice, Z. M. Merchant, Y. S. Kim, and D. L. Lohse, Structure and activity of a unique heparin-derived hexasaccharide, *J. Biol. Chem. 261*: 14,448–14,454 (1986).

20. R. J. Linhardt, H-M. Wang, D. Loganathan, and J-H. Bae, Search for the heparin antithrombin III-binding site precursor, *J. Biol. Chem. 267*: 2380–2387 (1992).

21. A. Larnkjaer, A. Nykjaer, G. Olivecrona, H. Thøgersen, and P. B. Østergaard, Structure of heparin fragments with high affinity for lipoprotein lipase and inhibition of lipopro-

tein lipase binding to α_2-macroglobulin-receptor/low-density-lipoprotein-receptor-related protein by heparin fragments. *Biochem J. 307*: 205–214 (1995).

22. A. Walker, J. E. Turnbull, and J. T. Gallagher, Specific heparan sulfate saccharides mediate the activity of basic fibroblast growth factor. *J. Biol. Chem. 269*: 931–935 (1994).

23. M. Maccarana, B. Casu, and U. Lindahl, Minimal sequence in heparin/heparan sulfate required for binding of basic fibroblast growth factor. *J. Biol. Chem. 269*: 3908 (1994).

24. G. Torri, B. Casu, G. Gatti, M. Petitou, J. Choay, J. C. Jacquinet, and P. Sinay, Mono- and bidimensional 500-MHz ^{1}H-NMR spectra of a synthetic pentasaccharide corresponding to the binding sequence of heparin to antithrombin-III: Evidence for conformational pecularity of the sulfated iduronate residue. *Biochem. Biophys. Res. Commun. 128*: 134–140 (1985).

25. M. Petitou, P. Duchaussoy, I. Lederman, J. Choay, J-C. Jacquinet, P. Sinay, and G. Torri, Synthesis of heparin fragments: A methyl α-pentaoside with high affinity for antithrombin III, *Carbohydr. Res. 167*: 67–75 (1987).

26. M. Ragazzi, D. R. Ferro, B. Perly, P. Sinay, M. Petitou, and J. Choay, Conformation of the pentasaccharide corresponding to the binding site of heparin for antithrombin III, *Carbohydr. Res. 195*: 169–185 (1990).

27. M. J. Forster and B. Mulloy, Molecular dynamics study of iduronate ring conformation, *Biopolymers 33*: 575–588 (1993).

28. B. Mulloy, M. J. Forster, C. Jones, A. F. Drake, E. A. Johnson, and D. B. Davies, The effect of variation of substitution on the solution conformation of heparin: A spectroscopic and molecular modelling study, *Carbohydr. Res. 255*: 1–26 (1994).

29. M. Hricovíni and G. Torri, Dynamics in aqueous solutions of the pentasaccharide corresponding to the binding site of heparin for antithrombin III studied by NMR relaxation measurement, *Carbohydr. Res. 268*: 159–175 (1995).

30. M. Hricovini, M. Guerrini, G. Torri, S. Piani, and F. Ungarelli, Conformational analysis of heparin epoxide in aqueous solution an NMR relaxation study. *Carbohydr. Res. 217*: 1–13 (1995).

31. P. D. J. Grootenhuis and C. A. A. van Boeckel, Constructing a molecular model of the interaction between antithrombin III and a potent heparin analog, *J. Am. Chem. Soc. 113*: 2743–2747 (1991).

32. K. Sugahara, I. Yamashima, P. De Waard, H. Van Halbeek, and J. F. G. Vliegenthart, Structural studies on sulfated glycopeptides from the carbohydrate–protein linkage region of chondroitin 4-sulfate proteoglycans of swarm rat chondrosarcoma, *J. Biol. Chem. 263*: 10,168–10,174 (1988).

33. K. Sugahara, S. Yamada, K. Yoshida, P. De Waard, H. Van Halbeek, and J. F. G. Vliegenthart, A novel sulfated structure in the carbohydrate–protein linkage region isolated from porcine intestinal heparin, *J. Biol. Chem. 267*: 1528–1533 (1992).

34. S. Rio, J-M. Beau, and J-C. Jacquinet, Synthesis of sulfated and phosphorylated glycopeptides from the carbohydrate–protein linkage region of proteoglycans, *Carbohydr. Res. 255*: 103–124 (1994).

35. K. Sugahara, Y. Ohkita, Y. Shibata, K. Yoshida, and A. Ikegami, Structural studies on the hexasaccharide alditols isolated from the carbohydrate–protein linkage region of dermatan sulfate proteoglycans of bovine aorta: Demonstration of iduronic acid–containing components, *J. Biol. Chem. 270*: 7204–7212 (1995).

36. K. Sugahara, Y. Takemura, M. Sugiura, Y. Kohno, K. Yoshida, K. Takeda, K. H. Khoo, H. R. Morris, and A. Dell, Chondroitinase ABC–resistant sulfated trisaccharides isolated from digests of chondroitin/dermatan sulfate chains, *Carbohydr. Res. 255*: 165–182 (1994).

37. K. Sugahara, K. Shigeno, M. Masuda, N. Fujii, A. Kurosaka, and K. Takeda, Structural studies on the chondroitinase ABC–resistant sulfated tetrasaccharides isolated from various chondroitin sulfate isomers, *Carbohydr. Res. 255*: 145–163 (1994).

38. S. M. T. D'Arcy, S. L. Carney, and T. J. Howe, Preliminary investigation into the purification, NMR analysis, and molecular modelling of chondroitin sulphate epitopes, *Carbohydr. Res. 255*: 41–59 (1994).

39. E. F. Hounsell, J. Feeney, P. Scudder, P. W. Tang, and T. Feizi, ¹H-NMR studies at 500 MHz of a neutral disaccharide and sulfated di-, tetra-, hexa- and larger oligosaccharides obtained by endo-β-galactosidase treatment of keratan sulphate. *Eur. J. Biochem. 157*: 375–384 (1986).

40. G-H. Tai, T. N. Huckerby, and I. A. Niedusynski, N.M.R. spectroscopic studies of fucose-containing oligosaccharides derived from keratanase digestion of articular cartilage keratan sulfates. Influence of fucose residues on keratanase cleavage, *Biochem. J. 291*: 889–894 (1993).

41. G-H. Tai, T. N. Huckerby, and I. A. Niedusynski, 600-MHz ¹H-NMR study of a fucose-containing heptasaccharide derived from a keratanase digestion of bovine articular cartilage keratan sulphate, *Carbohydr. Res. 255*: 303–309 (1994).

42. G. M. Brown, T. N. Huckerby, and I. A. Nieduszynski, Oligosaccharides derived by keratanase II digestion of bovine articular cartilage keratan sulfates, *Eur. J. Biochem. 224*: 281–308 (1994).

43. E. F. Hounsell, A. M. Lawson, J. Feeney, H. C. Gooi, N. J. Pickering, M. S. Stoll, S. C. Lui, and T. Feizi, Structural analysis of the *O*-glycosidically linked core-region oligosaccharides of human meconium glycoproteins which express oncofoetal antigens, *Eur. J. Biochem. 148*: 367–377 (1985).

44. A. V. Savage, C. M. Donoghue, S. M. D'Arcy, C. A. M. Koeleman, and D. H. van den Eijnden, Structure determination of five sialylated trisaccharides with core types 1, 3 or 5 isolated from bovine submaxillary mucin, *Eur. J. Biochem. 192*: 427–432 (1990).

45. J. Feeney, T. A. Frenkiel, and E. F. Hounsell, Complete ¹H-NMR assignments for two core-region oligosaccharides of human meconium glycoproteins, using ID and 2D methods at 500 MHz, *Carbohydr. Res. 152*: 64–73 (1986).

46. E. F. Hounsell, A. M. Lawson, J. Feeney, G. C. Cashmore, D. P. Kane, M. Stoll, and T. Feizi, Identification of a novel oligosaccharide backbone structure with a galactose residue monosubstituted at C-6 in human foetal gastrointestinal mucins, *Biochem. J. 256*: 397–401 (1988).

47. F-G. Hanisch, G. Uhlenbruck, J. Per-Katalinic, H. Egge, J. Dabrowski, and U. Dabrowski, Structures of neutral *O*-linked polylactosaminoglycans on human skim milk mucins—A novel type of linearly extended poly *N*-acetyllactosamine backbones with Galβ(1–4)GlcNAcβ(1–6) repeating units. *J. Biol. Chem. 264*: 872–883 (1989).

48. N-U. Din, R. W. Jeanloz, G. Lamblin, P. Roussel, H. van Halbeek, J. H. G. M. Mutsaers, and J. F. G. Vliegenthart, Structure of sialyloligosaccharides isolated from bonnet monkey (*Macaca radiata*) cervical mucus glycoproteins exhibiting multiple blood group activities, *J. Biol. Chem. 261*: 1992–1997 (1986).

49. V. K. Dua, B. N. N. Rao, S-S. Wu, V. E. Dube, and C. A. Bush, Characterization of the oligosaccharide alditols from ovarian cyst mucin glycoproteins of blood group a using high-pressure liquid chromatography (HPLC) and high-field ¹H-NMR spectroscopy, *J. Biol. Chem. 261*: 1599–1608 (1986).

50. J. Breg, H. van Halbeek, J. F. G. Vliegenthart, G. Lamblin, M-C. Houvenaghel, and P. Roussel, Structure of sialyl-oligosaccharides isolated from bronchial mucus glycoproteins of patient (blood group O) suffering from cystic fibrosis, *Eur. J. Biochem. 168*: 57–68 (1987).

51. E. F. Hounsell, N. J. Jones, H. C. Gooi, T. Feizi, A. S. R. Donald, and J. Feeney, MHz ¹H-NMR and conformational studies of fucosyl-oligosaccharides recognized by monoclonal antibodies with specifities related to Le[a], Le[b], and SSEA-1, *Carbohydr. Res. 178*: 67–78 (1988).

52. W. Chai, E. F. Hounsell, G. C. Cashmore, J. R. Rosankiewicz, C. J. Bauer, J. Feeney, T. Feizi, and A. M. Lawson, Neutral oligosaccharides of bovine submaxillary mucin —a combined mass spectrometry and ^{1}H-NMR study, *Eur. J. Biochem. 203*: 257–268 (1992).

53. G. Strecker, J. M. Wiersceski, Y. Plancke, and B. Boilly, Primary structure of 12 neutral oligosaccharides-alditols released from the jelly coats of the anuran *Xenopus laevis* by reductive β-elimination, *Glycobiol. 5*: 137–146 (1995).

54. D. Baeckstrom, N. Karlsson, and G. C. Hansson, Purification and characterization of sialyl-Le^a-carrying mucins of human bile; evidence for the presence of MUC1 and MUC3 apoproteins, *J. Biol Chem. 269*: 14,430–14,437 (1994).

55. J. E. Hansen, O. Lund, J. Engelbrecht, H. Bohr, J. O. Nielsen, J. E. S. Hansen, and S. Brunak, Prediction of O-glycosylation of mammalian proteins: Specificity patterns of UDP-GalNAc:polypeptide N-acetylgalactosaminyltransferase, *Biochem. J. 308*: 801–813 (1995).

56. E. F. Hounsell and D. J. Wright, Computer-assisted interpretation of ^{1}H-NMR spectra in the structural analysis of oligosaccharides, *Carbohydr. Res. 205*: 19–29 (1990).

57. K. Hermansson, P.-E. Jansson, L. Kenne, G. Widmalm, and F. Lindh, A ^{1}H- and ^{13}C-NMR study of oligosaccharides from human milk. Application of the computer program CASPER, *Carbohydr. Res. 235*: 69–81 (1992).

58. G. M. Lipkind, A. S. Shashkov, N. E. Nifant'ev, and N. K. Kochetkov, Computer-assisted analysis of the structure of regular branched polysaccharides containing 2,3-disubstituted rhamnopyranose and mannopyranose residues on the basis of ^{13}C-NMR data, *Carbohydr. Res. 237*: 11–22 (1992).

59. J. A. van Kuik, K. Hard, and J. F. G. Vleigenthart, A ^{1}H-NMR database computer program for the analysis of the primary structure of complex carbohydrates, *Carbohydr. Res. 235*: 53–68 (1992).

60. P. Albersheim and J. F. G. Vliegenthart, Workshop to establish databases of carbohydrate spectra, *Glycoconj. J. 13*: i–iii (1996).

61. H. U. Linden, R. A. Klien, H. Egge, J. Peter-Katalinic, J. Dabrowski, and D. Schindler, Urine of patients with hereditary deficiency in α-N-acetylgalactosaminidase activity, *Biol. Chem. Hoppe-Seyler 370*: 661–672 (1989).

62. Y. Hirabayashi, Y. Matsumoto, M. Matsumoto, T. Toida, N. Iida, T. Matsubara, T. Kanzaki, M. Yokota, and I. Ishizuka, Isolation and characterization of major urinary amino acid O-glycosides and a dipeptide O-glycoside from a new lysosomal storage disorder (Kanzaki disease), *J. Biol. Chem. 265*: 1693–1701 (1989).

63. H. Paulsen, Syntheses, conformation and X-ray structure analyses of the saccharide chains from the core region of glycoproteins, *Angew. Chem. Int. Ed. Engl. 29*: 823–839 (1990).

64. H. Paulsen, A. Pollex-Krüger and V. Sinnwell, Konformationsalytische untersuchungen von N-terminalen O-glycopeptidsequenzen des interleukin-2, *Carbohydr. Res. 214*: 199–226 (1991).

65. G. Pèpe, D. Siri, Y. Oddopn, A. A. Pavia, and J-P. Reboul, Conformational analysis of the amino termini (5 residues) of human glycophorin A_M and A_N: differentiation of the structural features of the T_N and T antigenic determinants in relation to their specificity, *Carbohydr. Res. 209*: 67–81 (1991).

66. Y. Mimura, Y. Yamamoto, Y. Inoue, and R. Chûjô, N.M.R. study of interaction between sugar and peptide moieties in mucin-type model glycopeptides, *Int. J. Biol. Macromol. 14*: 242–249 (1992).

67. A. Pollex-Krüger, B. Meyer, R. Stuike-Prill, V. Sinnwell, K. L. Matta, and I. Brockhausen, Preferred conformation and dynamics of five core structures of mucin type O-glycan determined by NMR spectroscopy and force field calculations, *Glycoconj. J. 10*: 365–380 (1993).

68. H. Paulsen, S. Peters, T. Bielfeldt, M. Meldal, and K. Bock, Synthesis of the glycosyl amino acids N^{α}-Fmoc-Ser [Ac$_4$-β-D-Gal p-(1 → 3)-Ac$_2$-α-D-GalN$_3p$]-OPfp and N^{α}-Fmoc-Thr[Ac$_4$-β-D-Gal$p$-(1 → 3)-Ac$_2$-α-D-GalN$_3p$]-OPfp and the application in the solid-phase peptide synthesis of multiply glycosylated mucin peptides with T$''$ and T antigenic structures, *Carbohydr. Res. 268*: 17–34 (1995).

69. J. Rademann and R.R. Schmidt, Solid-phase synthesis of a glycosylated hexapeptide of human sialophorin, using the trichloroacetimidate method, *Carbohydr. Res. 269*: 217–225 (1995).

70. W. M. Macindoe, H. Ijima, Y. Nakahara, and T. Ogawa, Stereoselective synthesis of a blood group A type glycopeptide present in human blood mucin, *Carbohydr. Res. 269*: 227–257 (1995).

71. G. Strecker, J-M. Wieruszeski, J-C. Michalski, and J. Montreuil, Complete analysis of the ^{1}H- and ^{13}C-NMR spectra of four blood-group A active oligosaccharides, *Glycoconj. J. 6*: 271–284 (1989).

72. E. F. Hounsell, Structural and conformational studies of glycoprotein and oligosaccharide recognition determinants, *NATO ASI Series 87*: 246–262 (1994).

73. S. Sabesan, K. Bock, and J. C. Paulson, Conformational analysis of sialyoligosaccharide, *Carbohydr. Res. 218*: 27–54 (1991).

74. D. Bailey, M. J. Davies, F. H. Routier, C. J. Bauer, J. Feeney, and E. F. Hounsell, ^{1}H NMR analysis of novel sialylated and fucosylated lactose-based oligosaccharides having linear GlcNAcβ1–6Gal and NeuAcα2–6GlcNAc sequences, *Carbohydr. Res.* in press.

75. H. C. Gooi, E. F. Hounsell, I. Lax, R. M. Kris, T. A. Libermann, J. Schlessinger, J. D. Sato, T. Kawamoto, J. Mendelsohn, and T. Feizi, The carbohydrate specificities of the monoclonal antibodies 29.1, 455 and 3C1B12 to the epidermal growth factor receptor of A431 cells. *Biosci. Rep. 5*: 83–94 (1985).

76. G. Spik, B. Coddeville, G. Strecker, J. Montreuil, E. Regoeczi, P. A. Chindemi, and J. R. Rudolph, Carbohydrate microheterogeneity of rat serotransferrin—Determination of glycan primary structures and characterization of a new type of trisialylated diantennary glycan, *Eur. J. Biochem. 195*: 397–405 (1991).

77. Y. Tachibana, K. Yamashita, and A. Kobata, Oligosaccharides of human milk: Structural studies of di- and tri-fucosyl derivatives of lacto-*N*-octaose and lacto-*N*-neooctaose, *Arch. Biochem. Biophys. 188*: 83–89 (1978).

78. V. K. Dua, K. Goso, V. E. Dube, and C. A. Bush, Characterization of lacto-*N*-hexaose and two fucosylated derivatives from human milk by high-performance liquid chromatography and proton NMR spectroscopy, *J. Chrom. 328*: 259–269 (1985).

79. J-M. Wieruszeski, J-C. Michalski, J. Montreuil, and G. Strecker, Sequential ^{1}H and ^{13}C resonance assignments for an octa- and decasaccharide of the *N*-acetyllactosamine type by multiple-step relayed correlation and hetero-nuclear correlation nuclear magnetic resonance, *Glycoconj. J. 6*: 183–194 (1989).

80. H. Kitagawa, H. Nakada, A. Kurosaka, N. Hiraiwa, Y. Numata, S. Fukui, I. Funakoshi, T. Kawasaki, and I. Yamashina, Three novel oligosaccharides with the sialyl-Le^a structure in human milk: Isolation by immunoaffinity chromatography, *Biochem. 28*: 8891–8897 (1989).

81. G. Strecker, J-M. Wieruszeski, J-C. Michalski, and J. Montreuil, Primary structure of human milk nona- and decasaccharides determined by a combination of fast atom bombardment mass spectrometry and ^{1}H/^{13}C-nuclear magnetic resonance spectroscopy. Evidence for a new core structure, *iso*-lacto-*N*-octaose, *Glycoconj. J. 6*: 169–182 (1989).

82. N. Platzer, D. Davoust, M. Lhermitte, C. Bauvy, D. M. Meyer, and C. Derappe, Structural analysis of five lactose-containing oligosaccharides by improved, high-resolution, two-dimensional ^{1}H-N.M.R. spectroscopy, *Carbohydr. Res. 191*: 191–207 (1989).

83. G. Grönberg, P. Lipniunas, T. Lundgren, F. Lindh, and B. Nilsson, Isolation and structural analysis of three new disialylated oligosaccharides from human milk, *Arch. Biochem. Biophys. 278*: 297–311 (1990).

84. S. Haeuw-Fievre, J-M. Wieruszeski, Y. Plancke, J-C. Michalski, J. Montreuil, and G. Strecker, Primary structure of human milk octa-, dodeca- and tridecasaccharides determined by a combination of ¹H-NMR and FAB-MS: Evidence for a new core structure, the *para*-lacto-*N*-octaose, *Eur. J. Biochem. 215*: 361–371 (1993).

85. G. Grönberg, U. Nilsson, K. Bock, and G. Magnusson, Nuclear magnetic resonance and conformational investigations of the pentasaccharide of the Forssman antigen and overlapping di-, tri-, and tetra-saccharide sequences, *Carbohydr. Res. 257*: 35–54 (1994).

86. J. R. Thomas, R. A. Dwek, and T. W. Rademacher, Structure, biosynthesis and function of glycosylphosphatidylinositols, *Biochem. 29*: 5413–5422 (1990).

87. S. W. Hunter and P. J. Brennan, A novel phenolic glycolipid from *Mycobacterium leprae* possibly involved in immunogenicity and pathogenicity, *J. Bacteriol. 147*: 728–735 (1981).

88. M. Daffé and P. Servin, Scalar, dipolar-correlated and *J*-resolved 2D-NMR spectroscopy of the specific phenolic mycoside of *Mycobacterium tubercolosis*, *Eur. J. Biochem. 185*: 157–162 (1989).

89. A. Vercellone and G. Puzo, New-found phenolic glycolipids in *Mycobacterium bovis* BCG, *J. Biol. Chem. 264*: 7447–7454 (1989).

90. K. Bock, T. Hvidt, J. Marino-Albernas, and V. Verez-Bencomo, An NMR and conformational analysis of the terminal trisaccharide from the serologically active glycolipid of *Mycobacerium leprae* in different solvents, *Carbohydr. Res. 200*: 33–45 (1990).

91. S. W. Hunter, R. C. Murphy, K. Clay, M. B. Goren, and P. J. Brennan, Trehalose-containing lipooligosaccharides, *J. Biol. Chem. 258*: 10,481–10,487 (1983).

92. M. Daffe, M. McNeil, and P. J. Brennan, Novel type-specific lipooligosaccharides from *Mycobacterium tuberculosis, Biochem. 30*: 378–388 (1991).

93. M. McNeil, M. Daffe, and P. J. Brennan, Evidence for the nature of the link between the arabinogalactan and peptidoglycan of mycobacterial cell walls, *J. Biol. Chem. 265*: 18,200–18,206 (1990).

94. D. Chatterjee, C. M. Bozie, M. McNeil, and P. J. Brennan, Structural features of the arabinan component of the lipoarabinomannan of *Mycobacterium tubercolosis J. Biol. Chem. 266*: 9652–9660 (1991).

95. D. Chatterjee, S. W. Hunter, M. McNeil, and P. J. Brennan, Lipoarabinomannan multiglycosylated form of the mycobacterial mannosylphosphatidylinositols, *J. Biol. Chem. 267*: 5228–6233 (1992).

96. A. Venisse, J-M. Berjeaud, P. Charand, M. Gilleron, and G. Puzo, Structural features of lipoarabinomannan from *mycobacterium bovis* BCG, *J. Biol. Chem. 268*: 12,401–12,411 (1993).

97. A. Lemassu and M. Daffé, Structural features of the exocellular polysaccharides *Mycobacterium tuberculosis, Biochem. J. 297*: 351–357 (1994).

98. L. M. L. Marin, M-A. Lonéelle, D. Promé, M. Daffe, G. Lanéelle, and J-C. Promé, Glycopeptidolipids from *mycobacterium fortuitum*: A variant in the structure of C-mycoside, *Biochem. 30*: 10,536–10,542 (1991).

99. M. Rivière and G. Puzo, Use of ¹H-NMR ROESY for structural determination of *O*-glycosylated amino acids from a serine-containing glycopeptidolipid antigen, *Biochem. 31*: 3575–3580 (1992).

100. H. Paulsen and M. Brenken, Synthese von oligosacchariden der inneren core und lipoid-A-region von lipopolysacchariden, *Liebigs Ann. Chem.*: 1113–1126 (1991).

101. S. Müllrt-Leonnies, O. Holst, and H. Brade, Chemical structure of the core region of *Escherichia coli* J-5 lipopolysaccharide, *Eur. J. Biochem. 224*: 751–760 (1994).

102. P. J. Brennan, Structure of mycobacteria: Recent developments in defining cell wall carbohydrates and proteins, *Rev. of Inf. Dis. 11*: S420–S430 (1989).

103. B. Schoel, S. Sprenger, and S. H. E. Kaufmann, Phosphate is essential for stimulation of Vγ9Vδ2 T lymphocytes by micobacterial low molecular weight ligand, *Eur. J. Biochem. 24*: 1886–1892 (1994).

104. J. S. Haurum, G. Arsequell, A. C. Lellouch, S. Y. C. Wong, R. A. Dwek, A. J. McMichael, and T. Elliott, Recognition of carbohydrate by major histocompatibility complex class I–restricted, glycopeptide-specific cytotoxic T lymphocytes, *J. Exp. Med. 180*: 739–744 (1994).

105. G. P. Reddy, U. Hayat, C. A. Bush, and J. G. Morris, Jr., Capsular polysaccharide structure of a clinical isolate of *Vibrio vulnificus* strain B02316 determined by heteronuclear NMR spectroscopy and high-performance anion-exchange chromatography, *Anal-Biochem. 214*: 106–115 (1993).

106. D. R. Bundle, H. Baumann, J. R. Brisson, S. M. Gagne, A. Zdanov, and M. Cygler, Solution structure of a trisaccharide-antibody complex: Comparison of NMR measurements with a crystal structure, *Biochem. 33*: 5183–5192 (1994).

107. R. Yamasaki and B. Bacon, Three-Dimensional structural analysis of the group B polysaccharide of *Neisseria meningitidis* 6275 by two-dimensional NMR: The polysaccharide is suggested to exist in helical comformations in solution, *Biochem. 30*: 851–857 (1991).

108. N. C. Veitch, R. J. P. Williams, R. C. Bray, J. F. Burke, S. A. Sanders, R. N. F. Thorneley, and A. T. Smith, Structural studies by proton-NMR spectroscopy of plant horseradish peroxidase C, the wild-type recombinant protein from *Escherichia coli* and two protein variants, Phe41 → Val and Arg38 → Lys. *Eur. J. Biochem. 207*: 521–531 (1992).

109. B. Kieffer, P. C. Driscoll, I. D. Campbell, A. C. Willis, P. A. van der Merwe, and S. J. Davis, Three-dimensional solution structure of the extracellular region of the complement regulatory protein CD59, a new cell-surface protein domain related to snake venom neurotoxins, *Biochem. 33*: 4471–4482 (1994).

110. R. L. Brockbank and H. J. Vogel, Structure of the oligosaccharide of hen phosvitin as determined by two-dimensional 1H-NMR of the intact glycoprotein, *Biochem. 29*: 5574–5583 (1990).

111. J. P. M. Lommerse, L. M. J. Kroon-Batenburg, J. P. Kamerling, and J. F. G. Vliegenthart, Conformational analysis of the xylose-containing *N*-glycan of pineapple stem bromelain as part of the intact glycoprotein, *Biochem 34*: 8196–8206 (1995).

112. K. Dill, E. Berman, and A. A. Pavia, Natural-abundance, ^{13}C-nuclear magnetic resonance-spectral studies of carbohydrates linked to amino acids and proteins, *Adv. Carb. Chem. Biochem. 43*: 1–49.

113. T. de Beer, C. W. E. M. van Zuylen, K. Hård, R. Boelens, R. Kaptein, J. P. Kamerling, and J. F. G. Vliegenthart, Rapid and simple approach for the NMR resonance assignment of the carbohydrate chains of an intact glycoprotein; application of gradient-enhanced natural abundance ^{1}H–^{13}C HSQC and HSQC-TOCSY to the α-subunit of human chorionic gonadotropin, *FEBS Letts. 348*: 1–6 (1994).

114. T. A. Gerken and N. Jentoft, Structure and dynamics of porcine submaxillary mucin as determined by natural abundance carbon-13 NMR spectroscopy, *Biochem. 26*: 4689–4699 (1987).

115. T. A. Gerken, K. J. Butenhof, and R. Shogren, Effects of glycosylation on the conformation and dynamics of *O*-linked glycoproteins: Carbon-13 NMR studies of ovine submaxillary mucin, *Biochem. 28*: 5536–5543 (1989).

116. C. M. Fletcher, R. A. Harrison, P. J. Lachmann, and D. Neuhaus, Sequence-specific ^{1}H-NMR assignments and folding topology of human CD59, *Protein Science 2*: 2015–2027, (1993).

117. D. F. Wyss, J. S. Choi, and G. Wagner, Composition and sequence-specific resonance assignments of the heterogeneous *N*-linked glycan in the 13.6-kDa adhesion domain of human CD2 as determined by NMR on the intact glycoprotein, *Biochem. 34*: 1622–1634 (1995).

118. T. H. Norwood, Magnetic field gradients in NMR: Friend or foe? *Chem. Soc. Rev. 23*: 59–66 (1994).

119. R. E. Hurd and B. K. John, Gradient-enhanced proton-detected heteronuclear multiple-quantum coherence spectroscopy, *J. Magnetic Res. 91*: 648–653 (1991).

120. J. Ruiz-Cabello, G. W. Vuister, C. T. W. Moonen, P. van Gelderen, J. S. Cohen, and P. C. M. van Zul, Gradient-enhanced heteronuclear correlation spectroscopy. Theory and experimental aspects, *J. Magnetic Res. 100*: 282–302 (1992).

121. S. Medvedeva, J-P. Simorre, B. Brutscher, F. Guerlesquin, and D. Marion, Extensive ^{1}H-NMR resonance assignment of proteins using natural abundance gradient-enhanced ^{13}C–^{1}H correlation spectroscopy. *FEBS Lett. 333*: 251–256 (1993).

14

Conformational Studies of Glycoprotein Glycans Using NMR and Molecular Dynamics

Trevor J. Rutherford
University of St. Andrews, Fife, Scotland

I. INTRODUCTION

Carbohydrate recognition is understood to have an essential role in regulating biological systems [1]. Determining the three-dimensional structure of oligosaccharides is an important step toward understanding the mechanism of recognition between carbohydrate ligands and their receptors. Modeling studies have recently had an impact on the rational design of carbohydrate analogs as therapeutic inhibitors of influenza virus sialidase [2] and selectin ligand binding [3]. Inhibition of other glycosidases, glycosyl transferases, and carbohydrate receptors are current objectives for pharmaceutical development.

The balance between enthalpy and entropy changes is the key to ligand binding specificity, and to model the recognition process it is necessary to characterize the flexibility of the binding surfaces in free solution [4]. However, even when conformational models are available, the design of potent inhibitors is far from straightforward, largely because of the difficulty of modeling the effects of solvation [5,6]. Carbohydrates are highly solvated in aqueous solution, and upon binding to proteins the solvation shells are displaced and may be exchanged with polar groups on the binding surfaces. Unless models are available for the restructuring of solvent molecules, conformational models cannot be used to predict binding affinities [6]. However, dynamic models can assist in the characterization and design of conformationally restricted ligands, which currently form part of the strategy toward enhancement of oligosaccharide binding affinity [4,7].

High-resolution models of carbohydrate–protein complexes and unligated receptors have been obtained by X-ray diffraction. (For reviews of carbohydrate–protein complexes, see Refs. 8 and 9). These crystal structures provide detailed models of the binding interactions, and often give precise locations for the bound water molecules that usually mediate the interactions. Diffraction techniques are, however, limited by the difficulty of crystallizing branched oligosaccharides and glycopeptides. For almost all crystallographic studies of glycoproteins there are no high-resolution coordinates for the oligosaccharide moiety, due to conformational disorder of the carbohydrate [10]. The coordinates of a biantennary complex-type N-glycan of a legume glycoprotein were first reported in 1991 [11], and the glycosidic torsion angles were found to correspond closely with those determined by nuclear magnetic resonance (NMR) spectroscopy for the free glycan in aqueous solution [12].

NMR spectroscopy, circular dichroism (CD) [13], and fluorescence energy transfer [14,15] are useful and sensitive conformational probes for oligosaccharides in solution. CD is particularly useful for high-resolution structures of saccharides that have one or two different linkage types (i.e., disaccharides [16,17] or homopolymers [18]), but is difficult to apply to the π-electron-containing residues that are common to glycoprotein glycans (e.g., GlcNAc, GalNAc, and Neu5Ac). Fluorescence energy transfer has resolved conformations that differ in their end-to-end distances in fluorophore-containing oligosaccharide derivatives [14,15]. NMR is the principal source of experimental evidence of the solution conformation and dynamics for all linkage types of oligosaccharide and their derivatives in solution.

A general feature of carbohydrate NMR, compared with that of other biomolecules, is the relatively small amount of available evidence for conformation. Glycosidic linkages tend to adopt extended conformations, with a dearth of observable interglycosidic through-space NMR contacts from which to estimate interatomic distances. Hence, the conformational equilibria are underdefined by the experimental data, and it is impractical to use distance-geometry techniques (as commonly applied to peptides) to model the atom coordinates. Carbohydrate modeling protocols therefore rely heavily upon the identification of low energy structures from theoretical molecular mechanics calculations [19].

This chapter is intended as an introduction to the methodology for investigating the conformations of glycoprotein glycans in solution by NMR and molecular mechanics, and discusses recent developments for improving the quantity of NMR contacts and the reliability of interpretation. Branched, complex-type N-linked (Asn-linked) glycans released from glycoproteins and the trisaccharide sialyl lactose are used to illustrate the procedures, although the techniques are equally applicable to intact glycopeptides or polysaccharides.

II. CONVENTIONS FOR NOMENCLATURE

Monosaccharide residues adopt a relatively rigid ring geometry, and the conformation of oligosaccharides is defined by the relative orientation of adjacent residues. In IUPAC nomenclature, the glycosidic torsion angle (Figure 1) ϕ is defined as O5–C1–Ox–Cx, and ψ as C1–Ox–Cx–C(x − 1) [20], where x is the number of the linkage carbon atom on the aglycon. When the glycosylated carbon of the aglycon is exocyclic (i.e., 1→6 linkages in hexopyranoses), the additional degree of freedom about the exocyclic C—C bond is characterized by the angle ω = Ox–Cx–C(x − 1)–C(x − 2). The angles are zero when the first and last bonds are *syn-periplanar*, when looking along the central bond, the angle becomes positive when the atom at the rear is rotated clockwise [21].

For NMR studies it is common to use the ϕ^H and ψ^H IUPAC nomenclature [22], where the glycosidic torsion angles are defined by H1–C1–Ox–Cx and C1–Ox–Cx–Hx, respectively. For 1→6 linkages, ω^H = O6–C6–C5–H5 and ψ^H = C1–O6–C6–C5. As a note of caution, the "H" superscript is almost invariably omitted in the literature, and it is worth checking which convention is in use. Throughout

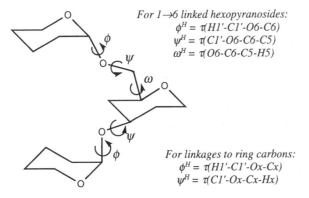

For 1→6 linked hexopyranosides:
$\phi^H = \tau(H1'\text{-}C1'\text{-}O6\text{-}C6)$
$\psi^H = \tau(C1'\text{-}O6\text{-}C6\text{-}C5)$
$\omega^H = \tau(O6\text{-}C6\text{-}C5\text{-}H5)$

For linkages to ring carbons:
$\phi^H = \tau(H1'\text{-}C1'\text{-}Ox\text{-}Cx)$
$\psi^H = \tau(C1'\text{-}Ox\text{-}Cx\text{-}Hx)$

Figure 1 Definition of the ϕ, ψ, and ω glycosidic dihedral angles that determine oligosaccharide conformations.

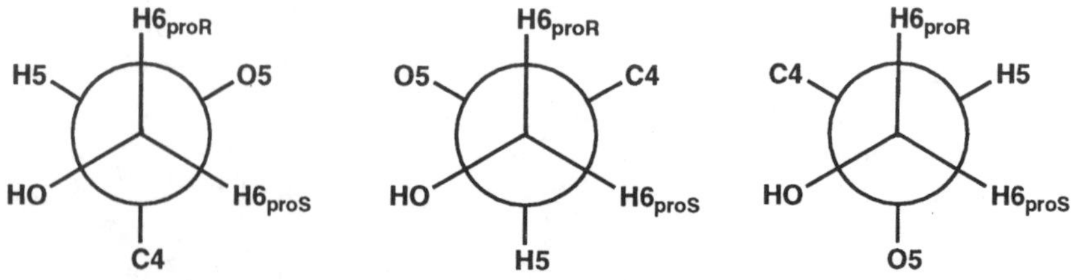

$\omega^H = +60°$ *trans-gauche* (tg) $\omega^H = -60°$ *gauche-trans* (gt) $\omega^H = 180°$ *gauche-gauche* (gg)

Figure 2 Definition of dihedral angles for hydroxymethyl rotamers in hexopyranoses.

this chapter the ϕ^H, ψ^H, and ω^H designations are used to define the linkage torsion angles.

The C5–C6 bond rotamer is commonly defined by the *trans* or *gauche* staggered orientation of O6 with respect to two ring atoms, first with respect to O5, then to C4. Hence, orientations with ω^H = 60°, −60°, or 180° are equivalent to *trans-gauche* (*tg*), *gauche-trans* (*gt*), or *gauche-gauche* (*gg*), respectively (Figure 2).

III. NMR SPECTROSCOPY

Purified oligosaccharides are typically available in small quantities, and most NMR techniques for carbohydrates use exclusively the high-sensitivity ^{1}H nucleus. ^{1}H NMR spectra of oligosaccharides exhibit extreme resonance overlap within an unresolved peak envelope between 3.5 and 4.0 ppm. Relieving the peak overlap is a significant challenge, and requires a combination of experiments that spread the peaks into two or more frequency dimensions. Although there are a huge number of NMR pulse schemes, only relatively few are routinely required for carbohydrate analysis. These, with some interesting recent developments, are outlined below.

A. Sample Preparation

For high-resolution conformational analysis, the sample must be purified to homogeneity, although polydispersity of high molecular weight (MW), regular repeating polymers can be tolerated. With a high-field spectrometer (>400 MHz ^{1}H) two-dimensional (2D) ^{1}H-NMR spectra can be recorded for ~500 μM carbohydrates. Millimolar concentrations are usually required for accurate quantification of ^{1}H–^{1}H distances, whereas 3D experiments and studies of ^{13}C at natural isotopic abundance require concentrations of 10 mM or more. Sample volumes for optimum resolution and lineshape vary according to the design of the NMR probe (typically 0.5–0.7 mL in a 5-mm-i.d. tube and 2.0–3.5 mL in a 10-mm tube).

For most biological samples, aqueous solvents provide the most suitable model of the native environment and are preferred for conformational studies. The observation of exchangeable protons in H_2O or D_2O/H_2O mixtures can be used to significant advantage [23,24]. In D_2O, the solvent peak (HOD) obscures peaks in the structural reporter group region (4–6 ppm), but its intensity is minimized by three cycles of dissolution and evaporation/lyophilization from D_2O. The ^1H chemical shift of the HOD resonance is temperature-dependent, and it is worthwhile recording spectra at different temperatures to reveal peaks that might otherwise be hidden by the solvent. For D_2O at neutral pH and low salt concentration, $\delta_{HOD} = 4.65$ ppm at 37°C, falling at 0.010 ppm/C, relative to internal trimethylsilylpropionate (TSP), with increasing temperature. The pH of isolated neutral glycan solutions need not normally be buffered, but phosphate buffer (or deuterated tris if Ca^{2+} is present in the solution) may be required for studies involving peptides. All other proton-containing material that is commonly found in biological extracts (e.g., glycerol or acetate) must be removed from the solution.

Before measuring NMR relaxation rates, paramagnetic metal ions, such as iron or manganese, must be removed (e.g., by passage over cation exchange resin). Dissolved oxygen is paramagnetic and should also be removed, by up to five cycles of freeze–pump–thaw in an inert atmosphere [25]. In high-MW samples, where cross-relaxation rates are high, the "leakage" of magnetization due to dissolved oxygen is less significant than for small molecules, and it may not be necessary to degas the sample [26].

B. Resonance Assignments in Oligosaccharides

A complete assignment of proton resonance frequencies is a prerequisite for conformational analysis, since the chemical shift is the characteristic label for distinguishing each nucleus when measuring internuclear distances. Relatively few peaks can be unambiguously assigned from an unedited one-dimensional spectrum, but the overcrowding can be relieved by correlating the frequencies of adjacent spins in two or three frequency dimensions.

Two-Dimensional NMR Experiments

The application of NMR spectroscopy in two dimensions to carbohydrate analysis has been reviewed [27–33]. In essence, 2D NMR pulse schemes (illustrated schematically in Figure 3a) incorporate an incremental time delay (t_1) between pulses, to map the chemical shifts and/or J-couplings. Magnetization is then transferred (or a state of coherence is generated) between spins that are connected either by J-coupling (through covalent bonds) or by dipolar-coupling (through space). Subsequently, the chemical shifts of the connected spins are mapped during a second time period (t_2). The amplitude of each component of the free induction decay (FID) signal, which is acquired during t_2, is encoded with chemical shift/coupling information in both time dimensions (hence, 2D NMR). Resonance frequencies during t_1 are mapped by acquiring a series of FIDs (typically 128–1024), with equal increments to t_1, and are converted to the familiar frequency spectrum by two-dimensional Fourier transformation. The value of the t_1-increment determines the spectral width in the transformed t_1 dimension. Most 2D experiments are no more difficult to set up than a 1D experiment, but require an accurately calibrated pulse width for a 90° rotation and values for the number of FIDs and the spectral widths in both dimensions.

Each 2D NMR technique employs different selection procedures to effect the coherence transfer step, and a combination of experiments provides complementary

(a)

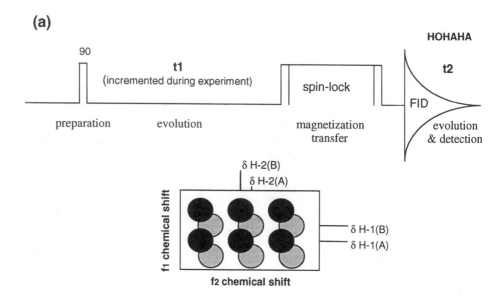

(b)

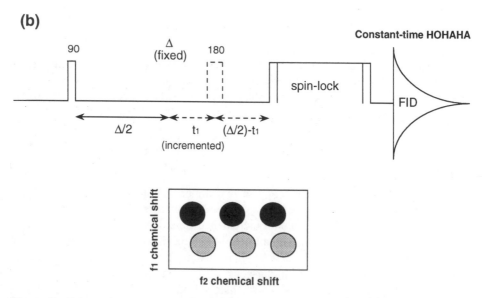

Figure 3 Schematic representation of radiofrequency pulses for 2D HOHAHA experiments. (a) Magnetization is initially prepared by rotation through 90° into the transverse plane. After mapping chemical shifts and J-couplings during an incremental delay (t_1), magnetization is transferred between spins, and the chemical shifts and couplings of the connected spins are mapped during t_2. Two overlapping H-1/H-2 cross-peaks for separate β-GlcNAc residues are shown schematically, and are characterized by chemical shifts and spin couplings in f_1 and f_2. (b) In constant-time HOHAHA, the incremental t_1 period is fitted into a constant time delay, and the resulting loss of J-splitting in f_1 improves the cross-peak resolution.

information for resonance assignment. Virtually all oligosaccharide resonances can be assigned using the combined information from the experiments described below.

COSY. In a COSY (correlation spectroscopy) spectrum (Figure 4), the two frequency axes, labeled f_1 and f_2, mark the chemical shifts of the signal evolving during the t_1 and t_2 time periods, respectively. Because every nucleus has a component of magnetization that evolves at the same frequency in both dimensions, the diagonal has an equivalent appearance to the one-dimensional spectrum. The off-diagonal cross-peaks connect the frequencies of two spins that have a direct J-coupling. COSY is a homonuclear technique that is very commonly applied to $^1H-^1H$ correlation, but it is equally applicable to $^{13}C-^{13}C$ correlation of uniformly ^{13}C-labeled saccharides [34].

There is negligible $^1H-^1H$ J-coupling between adjacent sugar rings, so each residue behaves as an isolated ^1H spin system. Because there is also negligible J-

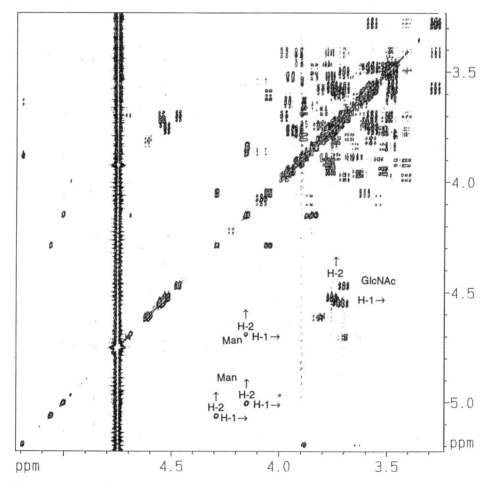

Figure 4 COSY spectrum of a complex-type "bisected" biantennary N-glycan, in D_2O, released from hen ovomucoid by hydrazinolysis. All H-1/H-2 cross-peaks and Man H-2/H-3 are resolvable, but severe overlap in the peak envelope region prevents assignment of most resonance frequencies.

coupling between H-1 and H-5 in hexopyranoses, the H-1 is easily recognized as
the downfield resonance with only one cross-peak. (The only common exception to
this is H-5 of uronic acids). H-1 resonances are a convenient starting point for a
unidirectional sequential assignment around each residue's spin system.

In Figure 4, H-1 through H-3 are readily assignable for the mannose residues
of a pentaantennary N-glycan. The GlcNAc H-1/H-2 connectivities are evident at
~4.6/3.7 ppm, but most cross peaks between resonances in the main peak envelope
are unassignable due to the many coincident ^{1}H chemical shifts. Even for small
oligosaccharides, and some disaccharides, full assignments cannot be obtained from
a COSY spectrum alone. Spectra of N-glycans can be assigned only with recourse
to complementary experiments that relieve the severe overlap problem.

The multiplicity and approximate values of J-coupling constants of each cross-
peak indicate the configuration of the corresponding carbon centers. An axial–axial
^{1}H–^{1}H coupling through three bonds ($^3J_{HH}$) is approximately 8–10 Hz; axial–equatorial or equatorial–equatorial couplings are in the range ~1–4 Hz (Table 1). Thus,
the residue type can readily be identified from the cross-peak fine structure: Distinc-
tively narrow cross-peaks are observed for H-2 of mannose and H-4 of galactose.
Multiplets with a line splitting greater than 10 Hz can be assigned to hydroxymethyl
protons (H-6 and H-6′ of hexopyranoses). Stereospecific assignment of hydroxymethyl resonances is not trivial, but it has been achieved by synthesis of chirally
deuterated monosaccharides [35].

For any J-coupling constants that are smaller than the line width, there can be
no COSY-type coherence transfer. Therefore, there is frequently no observable cross-
peak connecting galactose H-4 and H-5 ($^3J_{H4-H5}$ = 0.8 Hz, Table 1). It is very rare to
resolve $^4J_{HH}$ interglycosidic coupling.

With uniformly ^{13}C-labeled oligosaccharides, HCCH-COSY [36] is an alternative correlation experiment that has no direct magnetization transfer between protons, although the spectrum can be plotted with an identical appearance to that of
the conventional COSY. The pulse scheme was designed to transfer magnetization
sequentially from H-1, for example, to C-1, from C-1 to C-2, and from C-2 to H-2.
Uniform ^{13}C-labeling is essential because of the negligible occurrence at natural

Table 1 $^3J_{H,H}$ Spin-Coupling Constants (Hz) for Common Monosaccharide Residues

Residue	H1–H2	H2–H3	H3–H4	H4–H5	H5–H6	H5–H6′	H6–H6′
α-D-Glc	3.8	9.8	8.9	9.8	2.5	5.4	−11.9
β-D-Glc	7.9	9.3	9.0	9.7	2.2	5.7	−12.2
α-D-GlcNAc	3.5	10.6	8.9	10.1	—	—	—
β-D-GlcNAc	8.4	10.2	8.8	—	2.0	5.6	−12.2
α-D-Man	1.9	3.4	9.7	9.8	2.2	5.8	−11.8
β-D-Man	1.1	3.3	9.6	9.7	2.3	6.2	−12.2
α-D-Gal	3.7	10.3	3.2	1.0	6.2	6.2	—
β-D-Gal	7.8	9.9	3.5	1.0	4.4	7.8	−11.6
α-L-Fuc	3.8	10.3	3.4	~1.0	6.6		
β-L-Fuc	7.9	9.9	3.5	~1.0	6.5		

Data for free monosaccharide in D$_2$O at 30°C, measured at 500 MHz with 0.1 Hz/pt resolution. Values
for Glc, Gal, and Man are mostly consistent with the early report by De Bruyn et al. [166].

abundance of the ^{13}C–^{13}C J-coupling that is required for the intermediate transfer step. The transfer requires resolvable $^1J_{CH}$ (~140–170 Hz) and $^1J_{CC}$ (~50 Hz) couplings and is therefore not hampered by unresolved ^1H–^1H coupling.

COSY-45 and Double-Quantum Filtered COSY. The broad diagonal peaks in COSY spectra are 90° out of phase with the cross-peaks; thus, if the cross-peaks are phased to give pure absorptive lineshapes, the diagonal peak lineshape is dispersive (Figure 5). In these ''phase-modulated'' COSY spectra, all peaks cannot be satisfactorily phased simultaneously and are thus displayed in absolute value, without discrimination of the sign of the peaks. Diagonal peaks in phase-modulated spectra are broad and may obscure cross-peaks for pairs of spins that are J-coupled and have almost coincident chemical shifts. Such spin pairs are said to be ''strongly coupled,'' and their COSY cross-peaks occur close to the diagonal. Two variants of the COSY technique provide identical information to the phase-modulated experiment but improve the resolution of peaks near the diagonal.

The first method, COSY-45 [37], is also a phase-modulated experiment, which employs a 45° rotation pulse to effect coherence transfer. A full treatment of the effect of the 45° pulse is beyond the scope of this chapter, but can be found in textbooks of NMR theory [38]. The observable effect in the transformed spectrum is to narrow the diagonal peak, but at the expense of sensitivity. COSY-45 is preferable to the standard COSY experiment (''COSY-90'') for saccharides concentration of more than ~10 mM.

The second alternative is to record an experiment with a double-quantum filter (DQF-COSY [39]). The DQF-COSY selects for magnetization that can evolve as double-quantum coherence, evolving simultaneously under the influence of two coupled spins. The advantages of DQF-COSY are that all peaks can be phased with absorptive lineshape, and the intensity of methyl signals, HOD, and diagonal peaks are lower than in COSY (relative to the cross-peak intensity). However, the sensitivity is half that of unfiltered COSY spectra.

It is worth noting that cross-peaks in all COSY spectra have antiphase peak splitting for the active coupling (i.e., the coupling that gives rise to the coherence transfer for the cross-peak) and therefore have zero integrated intensity. *Approximate* values and signs of J-coupling constants can be measured (with caution) from phase-sensitive spectra: Where two cross-peaks overlap, the spectrum may be distorted by the cancellation of lines with opposite phase. Where two lines in a COSY multiplet are poorly resolved, the apparent separation of the lines does not equal the true J-

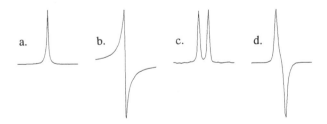

a.   b.   c.   d.

Figure 5 (a) Absorptive and (b) dispersive lineshapes of NMR peaks. (c) and (d) show pure absoptive lineshapes for doublets with in-phase and antiphase peak splitting, respectively.

coupling constant; thus, J-values measured from COSY spectra have large error limits.

Triple-Quantum Filtered COSY. From the spectrum in Figure 4 it is clear that even with only a few saccharide rings it is rarely possible to assign all ^{1}H resonances from a COSY spectrum alone. Some of the overlap can be resolved by editing out cross-peaks. In triple-quantum-filtered COSY (TQF-COSY) experiments [40], cross-peaks are generated only where there are three mutually coupled spins. Hence, in hexopyranoses, TQF-COSY selects only H-5, H-6, and H-6′ connectivities; all other COSY cross-peaks are eliminated by the filter. The characteristic geminal coupling between hydroxymethyl protons (~12 Hz) distinguishes them from H-5. TQF-COSY is not necessary for assigning most saccharides, but can be effective if H-5 and H-6 are poorly resolved from other signals.

In the sequence $90°_{\phi 1}-t_1-90°_{\phi 1}-\Delta-90°_{\phi 2}-t_2$ (acquire), the phase of the first two pulses, $\phi 1$, is cycled in steps of 60°, and the resulting free-induction decays are alternately added and subtracted by inverting the phase of the receiver. Triple-quantum terms are effectively cycled at $3 \times 60°$; hence, only states of pure triple-quantum coherence (generated only in three-spin systems) are added in successive scans. Multiples of 48 scans are required per t_1 increment to complete the phase-cycling scheme of the triple-quantum filter. TQF-COSY provides a prime example of the advantage of pulsed field gradients (vide infra) for the selection of coherence transfer pathways, since the total experiment time can be reduced by a factor of 48 compared with the phase-cycled version.

Relayed COSY. An alternative to removing peaks by filtration (as in TQF-COSY) is to correlate the well-resolved H-1 spins in the structural reporter group region with spins further along the coupling network. In relayed COSY [41] (also known as "RELAY" or "RECSY"), the H-1 shows cross-peaks both to H-2 and to H-3 (as illustrated for fucosyl lactose, Galβ(1−4)[Fucαa(1−3)]Glc, in Figure 6c). Similarly, H-2 correlates with H-1, H-3, and H-4, and so on. Comparison with the COSY (Figure 6b) distinguishes which of the cross-peaks arise by direct J-coupling and which arise by relayed coherence transfer.

The relay pulses generate phase distortions, and relayed COSY peaks can be adequately displayed only in absolute value. By inserting additional relay elements into the pulse sequence, the H-1 resonance can be correlated with resonances further along the coupling network (e.g.. double-relayed COSY [41]; Figure 6d). Sensitivity is lost at each relay step, and three or more relays are less efficient than equivalent HOHAHA-type transfers (below). The complementary information from COSY, TQF-COSY, and relayed COSY for the complete assignment of N-glycans has been well illustrated [42]. More recently, relayed COSY has largely been replaced by HOHAHA experiments [43].

HOHAHA (TOCSY). Homonuclear Hartmann−Hahn spectroscopy [44] (HOHAHA), also known as total correlation spectroscopy (TOCSY), can correlate each resonance with all other resonances in the same spin system. For oligosaccharides, each sugar ring forms a separate ^{1}H spin system (with virtually no resolvable interglycosidic ^{1}H−^{1}H J-coupling), and slices taken through the 2D HOHAHA spectrum at the chemical shift of each of the well-resolved H-1 resonances are equivalent to a 1D spectrum selective for a single residue (Figure 6e). Clearly, this is of enormous value in relieving severe resonance overlap, and HOHAHA experiments are routinely used for resonance assignment.

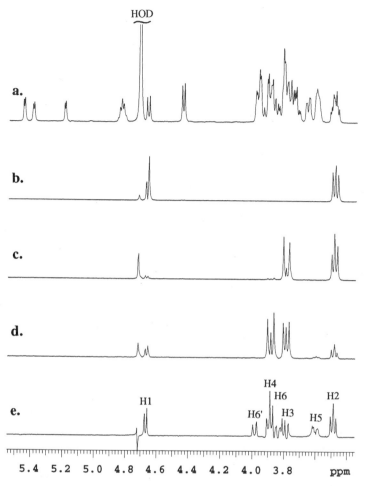

Figure 6 (a) Reference 1D ^{1}H-NMR spectrum of fucosyl lactose, Galβ(1–4)[Fucα(1–3)]Glc, in D$_2$O at 30°C. (b–e) Single rows (parallel to f_2) from 2D NMR spectra, taken at the chemical shift of βGlc H-1. (b) COSY, showing the H-1 diagonal peak and the correlation to H-2. (c) Relayed COSY, showing an additional correlation to H-3. (d) Double-relayed COSY, with an additional correlation to H-4. (e) HOHAHA with long mixing time (120-ms MLEV-17 spin-lock), showing the complete β-Glc subspectrum.

Hartmann–Hahn magnetization transfer occurs between coupled spins that experience the same effective field strength. The typical spread of chemical shifts for ^{1}H nuclei is ±5 ppm. Thus, in a 500-MHz NMR spectrometer, ^{1}H nuclei precess (spin) in the B_0 field at 500 MHz ± 2500 Hz. In a weaker magnetic field, the spread of frequencies is considerably lower; in the field from the transmitter (B_1), spins precess typically at ~10 kHz ± 0.05 Hz (i.e., ±5 ppm). Clearly, in the B_1 field all spins have virtually identical precession frequency, and thus there is minimal difference in the effective field strength at different nuclei. Though the magnetization is "spin-locked" in the B_1 field, all coupled spins are strongly coupled, the Hartmann–Hahn condition is satisfied, and magnetization is mixed throughout the spin system. For short spin-lock times ($\sim$10 ms) there is transfer between only immediately adjacent ^{1}H spins, equivalent to the COSY experiment. For longer spin-lock times ($>$100 ms), transfer can occur to H-1 from all spins up to H-6 and H-6'. There is no transfer between spins that have unresolved couplings.

There are practical difficulties associated with applying long, continuous transmitter pulses for spin-locking. The power level must be attenuated to avoid overheating the transmitter coils, but a continuous ''soft'' pulse irradiates unevenly across the full spectral width. Resonance offset effects are largely removed (and sensitivity is enhanced) by using composite spin-lock pulse trains (e.g., MLEV [45] or DIPSI [46]), which repeat a cycle of short, soft pulses that vary in duration and/or phase.

Acquiring a series of HOHAHA spectra with different spin-lock mixing times can greatly simplify the ^{1}H resonance assignment. The intensity of each cross-peak is an oscillatory function of the spin-lock time, and at long mixing times spectra are distorted. Unlike COSY, all lines can be phased to pure absorption, and all J-couplings produce in-phase line splitting. Thus, overlapping multiplets do not exhibit the cancellation that is observed in COSY, and a HOHAHA experiment with short spin-lock (10 ms) has a significant advantage over COSY if the resolution is low (e.g., in 3D experiments).

Galactose and fucose residues lack H-4/H-5 connectivities in HOHAHA spectra (due to unresolved J-coupling), and their full assignment often requires through-space correlations in NOESY or ROESY spectra (vide infra), e.g., between H-3 and H-5.

Constant-Time Experiments. All of the experiments described above have an incremental time delay for t_1, and the resulting multiplets in the f_1 dimension are characterized by the chemical shifts and J-couplings. Figure 3a shows a schematic representation of a conventional 2D pulse scheme and the resulting H-1/H-2 cross-peaks of two residues (with the β-glucose configuration) with virtually degenerate spin systems. A 1D ^{1}H spectrum selective for each residue could be obtained if the H-1 resonances were resolved, but in a HOHAHA experiment there is insufficient resolution for separating the two spin systems.

In the constant-time [37] variant of the HOHAHA experiment [47], the incremental t_1 delay is replaced with a fixed evolution period. Chemical-shift evolution is mapped by incrementing the position of a 180° inversion pulse inserted into the fixed delay, Δ (Figure 3b). However, the inversion pulse has no effect on the J-coupling terms, and since Δ is constant there is no modulation of signal due to J-coupling. The coupling term attenuates the intensity of each multiplet by a constant, and optimum sensitivity is obtained for $\Delta = n/J$, where n is an integer and J is the coupling constant. The initial position of the inversion pulse is at $\Delta/2$, and t_1 is incremented in equal steps during the experiment to map the chemical shifts. The cross-peaks are characterized only by the chemical shift in the f_1 frequency dimension (Figure 3b) and are effectively f_1-decoupled.

The ^{1}H-NMR spectrum of the pentaantennary N-glycan in Figure 7 has virtually degenerate spin systems for seven of its β-GlcNAc residues. Six of the GlcNAc H-1 resonate within a 55-Hz envelope, and in conventional HOHAHA or COSY spectra the resolution of H-1 peaks is insufficient for extracting data selective for each residue. Unless the H-1 resonances can be resolved, the spectrum is unassignable by homonuclear methods. The H-1 peaks are split only by coupling with H-2, and in this instance all GlcNAc $^3J_{\text{H1-H2}}$ values are identical; therefore, the sensitivity of a constant-time experiment can be optimized for all GlcNAc residues. Slices from the

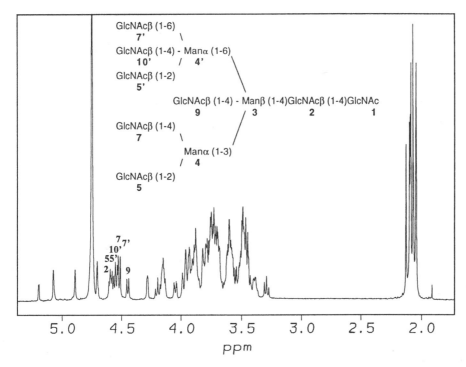

Figure 7 The 1D ^{1}H-NMR spectrum of the ''bisected'' pentaantennary N-glycan from hen ovomucoid in D_2O exhibits severe overlap for the β-GlcNAc H-1 resonances (~4.6 ppm). (Reprinted with permission from Ref. 47. Copyright 1995 Academic Press.)

constant-time HOHAHA of the pentaantennary glycan (Figure 8) show selective 1D subspectra for each of the GlcNAc residues that were separated on the basis of resolved H-1 frequencies. Despite the narrow H-1 chemical shift dispersion, the slices show minimal ''crosstalk'' from resonances in adjacent spin systems, and all of the signals in the pentaantennary glycan were unambiguously assigned [47].

The constant-time technique can also be applied for ''f_1-decoupling'' of COSY [48] or other 2D experiments and can be implemented in 3D NMR to resolve any further ambiguities [47].

Heteronuclear Shift Correlation. ^{13}C-NMR has not yet been developed as extensively as ^{1}H-NMR for the conformational analysis of oligosaccharides, but ^{13}C–^{1}H J-coupling constants [49] and the sensitive conformation dependence of ^{13}C chemical shifts [50] are complementary to ^{1}H-NMR data. Several 2D pulse sequences are available for assigning ^{13}C resonances by correlation with spin-coupled proton resonances. Most early heteronuclear correlation techniques involved detection of signal from the ^{13}C nuclei [51], although the sensitivity is enhanced by starting with ^{1}H magnetization, transferring to ^{13}C nuclei during t_1 and returning to ^{1}H during t_2 [52,53]. These ''inverse-detected'' (i.e., detection of the ^{1}H signal that is associated with the decoupler circuitry) experiments require switching the receiver to the same reference phase as the decoupler channel, and an inverse-detection probe with receiver coils that are optimized for detecting high-resolution ^{1}H signals. The benefit over normal-mode experiments is threefold [53]. First, starting from the high γ nu-

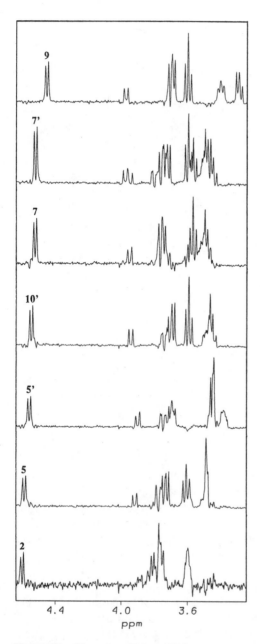

Figure 8 Slices (parallel to f_2) from a constant-time HOHAHA spectrum of the bisected pentaantennary N-glycan, illustrating complete ^{1}H subspectra selective for each of the β-GlcNAc residues. The subspectra were separated on the basis of their H-1 chemical shifts, and were resolvable due to the absence of H-1/H-2 J-splitting in f_1. (Reprinted with permission from Ref. 47. Copyright 1995 Academic Press.)

cleus (^{1}H) generates greater polarization of the energy level populations (hence greater signal intensity) than starting from lower-γ nuclei. Second, ^{1}H-detected experiments exploit the greater sensitivity of the receiver coils to high-frequency signals. Third, there are normally more data points in f_2 than in f_1; hence, the most crowded resonances are mapped in the highest resolution dimension.

HMQC (heteronuclear multiple quantum coherence [52]) and HSQC (heteronuclear single-quantum coherence [54]) experiments produce ^{1}H-detected shift correlation spectra with an identical format (Figure 9a). In HMQC, peaks are split by ^{1}H–^{1}H J-coupling in both f_1 and f_2. The splitting is not resolved in f_1, due to low digital resolution, but the f_1 (^{13}C) linewidths are broadened and peak intensities decreased. HSQC evolves solely under ^{13}C shifts during t_1, and direct comparison with HMQC [55,56] has shown that HSQC often has better resolution in f_1 and better sensitivity than HMQC. The J_{CH} peak splittings in f_2 can be removed by decoupling ^{13}C during acquisition; but when used without decoupling, high-resolution HSQC-based experiments are particularly suited to the measurement of long-range ^{1}H–^{13}C J-coupling constants (vide infra).

Immediately prior to t_2 in HMQC, the magnetization is evolving under ^{1}H chemical shifts. This magnetization can be further transferred between connected ^{1}H spins prior to detection, by means of relay or spin-lock sequences. In 2D HMQC-HOHAHA [55,57] (or the HSQC analog), each ^{13}C nucleus correlates with several protons in the spin system of the directly bonded proton (Figure 9b). A slice taken at the chemical shift of a resolved ^{13}C can reveal the entire ^{1}H spectrum for a single residue, as illustrated for the β-Glc spin system of fucosyl lactose in Figure 9b (which is equivalent to the slice shown in Figure 6e).

HMBC (heteronuclear multiple bond correlation, a derivative of HMQC designed to correlate via long-range C-H coupling) or selective-INEPT (insensitive nuclei enhanced by polarization transfer) experiments are useful aids for resonance assignment by correlating coupled C–H pairs that are separated by two or more bonds [57–60] (e.g., correlating across glycosidic linkages or between acetyl carbons and ring protons).

NMR in Three or More Dimensions

In the previous sections resonance overlap was relieved by editing the peaks into a second frequency dimension. Each slice through the 2D spectrum contains an edited subset of the peaks in the full 1D spectrum (Figure 10). Similarly, overlap in a 2D spectrum can be relieved by expansion to a third frequency dimension, such that a plane through the 3D data set contains a subset of the cross-peaks observable in a 2D spectrum (Figure 10c). For example, in the 3D version of the HMQC-HOHAHA experiment [61], 2D HOHAHA planes are edited by the chemical shift of an attached heteronucleus. Hence, a slice taken at the chemical shift of a resolved C-1 resonance, for example, is a 2D ^{1}H–^{1}H HOHAHA subspectrum selective for one saccharide residue. Three-dimensional methods are invaluable for relieving resonance overlap in oligosaccharides.

In a 3D experiment, the evolution of magnetization is mapped during three time periods: a signal acquisition period and two independently incremented delays. The pulse sequence is a concatenation of two 2D pulse sequences, with (at least) two separate coherence transfer steps. The total acquisition time is several times

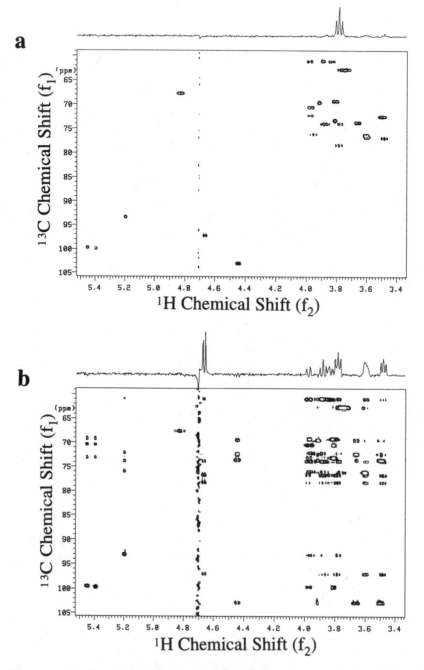

Figure 9 $^1H-^{13}C$ shift correlation spectra of fucosyl lactose in D_2O, with rows (parallel to f_2) taken at the chemical shift of β-Glc C-3. (a) HSQC; (b) 2D HSQC-HOHAHA (100-ms MLEV-17 spin-lock). The 1D slice in (b) shows the complete β-Glc ^1H subspectrum (c.f., Figure 6e).

a.

chemical shift

b.

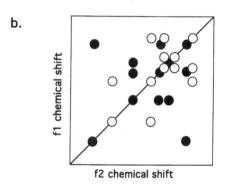

f1 chemical shift

f2 chemical shift

c.

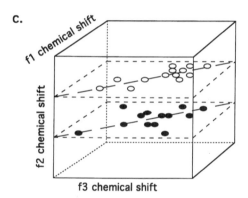

f1 chemical shift

f2 chemical shift

f3 chemical shift

Figure 10 Schematic representation of NMR spectra in (a) one, (b) two, and (c) three frequency dimensions. Cross-peaks are illustrated for two separate spin systems, which are differentiated by open or filled circles. The 2D experiment can be viewed as several 1D subspectra separated by chemical shift. Similarly, the planes in the 3D cube contain a subset of the cross-peaks in the full 2D spectrum.

longer than for a 2D experiment, and would typically require 24–72 hours. Processed data are most conveniently displayed as individual 2D planes extracted from the 3D cube.

In the HOHAHA-COSY experiment [28], magnetization is transferred around the spin system by the initial HOHAHA step (with ~100-ms spin-lock time), and each resonance is labeled with the frequencies of all spins within the same residue. The subsequent step generates COSY-type connectivities between the resonances. All resonances produce a peak on the diagonal of the 3D cube, and an f_1/f_3 plane taken at the f_2 chemical shift of an H-1 comprises a COSY spectrum selective for one residue. The third dimension contains relatively few data points (typically 64), and overlapped antiphase COSY multiplets can partially cancel due to the low digital resolution. Replacing the COSY step with a short (12-ms) spin-lock HOHAHA [47] obtains identical information to HOHAHA-COSY. But because the HOHAHA steps produce in-phase multiplets, unresolved cross-peaks do not cancel.

The remaining assignment ambiguities are almost always resolvable with analogous ROESY-COSY (or ROESY-HOHAHA) experiments [24,62], which use both J-coupling and dipolar-coupling (ROESY—vide infra) for magnetization transfer.

It is straightforward to extend the dimensionality further, to produce 4D experiments, with three incremental delays and an acquisition period. Again data are displayed as 2D planes, but the cross-peaks form a smaller subset of data in the constituent 3D experiment.

Reducing the Dimensionality of NMR Experiments

A significant limitation for 4D and some 3D NMR experiments is the requirement for a large, continuous block of magnet time. A full 4D experiment is usually prohibitively long. But the dimensionality can be reduced by replacing one incremental delay with a semiselective pulse or chemical-shift-selective filter [63], to select a single frequency point in one of the dimensions [64]. Shaped Gaussian pulses [65] or BURP (band-selective radiofrequency pulses [66]) excite with uniform phase and amplitude across a selected narrow bandwidth of the spectrum, and can be tailored to extract a single frequency point in any dimension. A "pseudo"-4D spectrum has peaks defined in terms of four frequencies, but is acquired with only two independently incremented delays and an acquisition period (c.f., 3D). The pseudo-4D HOHAHA-HOHAHA-COSY experiment uses BURP excitation for the first frequency discrimination, and was shown to resolve three almost coincident chemical shifts within the same GlcNAc residue of a complex biantennary N-glycan [64]. Each plane exhibits a single COSY correlation, and the selective pulse ensures that the magnetization originates from a single H-1 spin and evolves within only one residue.

Several pseudo-3D and pseudo-2D experiments have been proposed for saccharides [24,67–71], which resolve overlap problems with the minimum investment of machine time. Selective 1D HOHAHA spectra [57], irradiating any resolved multiplet, can obtain a 1D subspectrum selective for one spin system, without needing to acquire a full 2D HOHAHA. A spin-lock time course can be obtained within a few minutes, to observe resonance frequencies successively along the chain.

Pulsed Field Gradients

Pulsed-field-gradient techniques were recently introduced for time-efficient suppression of artifacts and selection of coherence transfer pathways in NMR experiments (reviewed by Keeler et al. [72]). Mixtures of magnetization components are generated during NMR experiments, and selection of the desired components traditionally uses phase cycling, i.e., a systematic variation of the phase of the transmitter pulses and the phase of the receiver, so that in successive scans the desired signal components are added and all unwanted components are cancelled. A full phase cycle for the 2D COSY experiment requires 16 scans per t_1 increment, but pulsed field gradients can achieve the same selection in a single scan. If sample concentrations are sufficiently high (>10 mM), a high-resolution gradient-enhanced 2D NMR experiment can be acquired in less than 10 minutes [73]; a typical 3D data set (e.g., $512 \times 256 \times 64$ data points) can be acquired in under a day.

Linear field gradients (e.g., 10–50 gauss-cm^{-1}) are applied for a few milliseconds, during which the transverse magnetization acquires a spatially dependent phase. The field gradients exploit the differences in rates of evolution of magnetization with different coherence order: ^{1}H–^{1}H double-quantum coherence (DQC) evolves under the sum of two chemical shift terms and dephases at twice the rate of single-quantum coherence (SQC). If the coherence order is changed from DQC to SQC in between two gradient pulses, the magnetization can be refocused only if the second gradient pulse is in the opposite sense for exactly twice the duration of the first. Thus it becomes possible to refocus only the terms that evolve as chosen coherence orders during two time intervals (e.g., t_1 and t_2), and the unwanted terms remain dephased and have no net contribution to the measured signal.

Pulsed field gradients are easily implemented within NMR experiments and can be used for "single-pulse" (per t_1 data point) homo- and heteronuclear experiments [72], for excellent tailored water suppression [74] and for purging artifacts such as the zero-quantum coherence that distorts cross-peaks in 2D NOESY spectra [75].

C. Conformationally Sensitive NMR Parameters

Chemical shift assignments are used to interpret the conformationally sensitive NMR parameters, all of which originate from one of three effects: Dipolar coupling (from which NOE, T_1, and T_2 are derived) gives information about internuclear distances and mobilities; J-coupling (also called "spin–spin coupling" or "scalar coupling") gives information about bond torsion angles, and the chemical shift provides qualitative information on steric interactions. The origins and applications of these effects are outlined below.

Nuclear Overhauser Effects

The nuclear Overhauser effect, or NOE, is widely applied as a probe of internuclear distances up to 5Å in biological molecules [26,76]. Interpretation of NOE is far from straightforward. In order to appreciate the many caveats, it is worthwhile looking at its origin.

The Origin of the NOE.   Each nucleus behaves as a small magnetic dipole, and the through-space dipole–dipole coupling of nuclear spin energy levels gives rise to nuclear relaxation and to the associated parameters NOE, T_1, and T_2. (Other mechanisms can also contribute to T_1 and T_2 relaxation). The rate of cross-relaxation between two nuclei (σ) depends upon both the internuclear distance and the mobility of the vector connecting the nuclei.

NOEs arise when perturbation in the equilibrium- (Zeeman-) energy-level populations of one spin induces a change in the populations for a nearby nucleus. NOEs appear as intensity changes in 1D spectra and are expressed quantitatively as the fractional change in the intensity of the remote spin compared with the intensity in a control experiment. In 2D (NOESY) and higher dimensional spectra, NOEs are quantified from cross-peak volumes. The peak integral is proportional to the inverse sixth power of the internuclear distance and is detectable for distances up to ~5 Å.

Coupling between nuclear dipoles depends upon the orientation of the inter-nuclear vector with respect to the external magnetic field. For molecules tumbling rapidly in solution, averaging the dipolar-coupling over all possible orientations, there is no net line splitting (unlike J-coupling), but the fluctuating dipole–dipole terms create oscillating magnetic fields that permit energy transfer (cross-relaxation) between nuclei.

Figure 11 shows an energy-level diagram for two dipolar-coupled spins, I and S. The symbol W denotes the probability for a given energy-level transition. Transitions labelled W_1I and W_1S link the low- and high-energy states of a single spin, I or S, respectively. The frequency for a single spin transition is the Larmor frequency (typically 400–600 MHz $\pm$ 5 ppm for ^{1}H in a high-field spectrometer). Following perturbation of spin S, relaxation occurring via the double-quantum and zero-quantum transitions, with probabilities W_2 and W_0, respectively, gives rise to NOEs. Both transitions correspond to a simultaneous flip of two spins; for W_2, spins flip with the same sense, and for W_0 with the opposite sense. The balance between W_0 and W_2 probabilities are linked to the rate of fluctuation of the dipolar fields. Whereas W_2 is most efficient within magnetic fields fluctuating at approximately twice the Larmor spin frequency, W_0 is most efficient with low- (almost zero-) frequency fluctuations.

In slowly tumbling molecules, W_0 processes are the dominant cross-relaxation mechanism, causing a reduction in the intensity of the remote I spin (negative NOE). In the limit, spin I saturates and its peak disappears from the spectrum (NOE = -100%). For rapidly tumbling molecules, W_2 outweighs W_0, resulting in an increased population for the I spin. The observed positive NOE for small molecules and negative NOE for large molecules results from the competing influences of W_0 and W_2 processes. At an intermediate molecular weight, W_2 and W_0 have almost equal and opposite effects and the maximum observable ^{1}H–^{1}H NOE is close to zero, regardless of the internuclear distance.

NOE Evolution in the Rotating Frame (ROESY). ROESY [77] (2D rotating-frame Overhauser spectroscopy) and the 1D variant, CAMELSPIN [78], increase

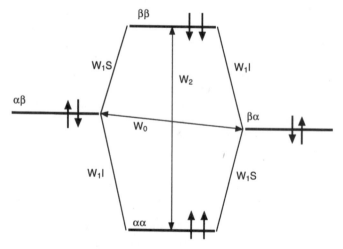

Figure 11 Energy-level diagram for two dipolar-coupled spins.

the intensity of NOEs for molecules within the "intermediate" molecular weight range ($\sim$1000–3000 Da). They are often the methods of choice for NOE measurement in oligosaccharides. In this modified NOE experiment, NOEs evolve in the B_1 (transmitter) field rather than the B_0 field. Since the "ROE" evolves in a weak field (spin-locked in the "rotating frame"—c.f., HOHAHA), the tumbling of all molecules is fast in comparison with the resonance frequencies, and W_2 processes always dominate over W_0. At the end of the spin-lock ROE evolution period, the spins revert to rotating in the B_0 field during detection, and benefit from the high resolution and sensitivity available at high field strengths.

    Conformational Analysis. When enhanced by NOE, the remote spin (I) also relaxes toward thermal equilibrium by transferring the energy it has gained to its dipolar-coupled spins (spin-diffusion), and the distance specificity of the NOE is compromised. For quantifying distances it is more convenient to consider rates of cross-relaxation, σ, rather than the absolute magnitude of the measured NOE. In the earliest stages of NOE buildup from the perturbed spin S, the NOE on spin I is affected by only one distance-dependent term, namely, σ_{IS}. NOEs arising via direct dipolar-coupling initially build up linearly (with rate σ_{IS}) on a plot of NOE intensity vs. evolution time. Quantitative distance information is available from a single experiment, with a mixing time in the linear NOE buildup region. When the "initial rate approximation" (or "isolated spin-pair approximation," ISPA) is valid, all enhancements behave as though they are in a two-spin system and the NOE intensity is directly proportional to r^{-6}. Hence,

$$r_{IS} = r_{\mathrm{ref}} \left(\frac{\sigma_{\mathrm{ref}}}{\sigma_{IS}} \right)^{1/6}$$

(r_{ref}, the distance between S and a reference nucleus with measurable σ_{ref}, is estimated from intraresidue H–H distances in models). However, there are several caveats when using this approximation:

1.   The isolated spin-pair approximation is valid only for evolution times within the linear initial buildup region of the NOE time course.
2.   The internuclear vector connecting the reference NOE spins and the I-S vector must have the same effective correlation time (τ_e). τ_e characterizes the rate of mobility of the vector connecting two nuclei and is inversely proportional to the rate of reorientation. τ_e equates roughly with the time taken to rotate through 1 radian, and is not the same for all H–H vectors in molecules that tumble anisotropically or have internal flexibility.
3.   NOEs are heavily weighted by the conformation with the closest contact in flexible molecules (by virtue of the r^{-6} dependence); it is the NOE that is averaged rather than the distance. A relatively strong NOE does not necessarily imply that two spins are close for the majority of the time.
4.   The accuracy of the calculated distance relies upon that of the reference distance and the reliability of peak integration routines. Due to the r^{-6} dependence of σ, small inaccuracies in measured NOE have negligible effects on calculated internuclear distances.
5.   Inaccuracies are introduced by noninstantaneous saturation; hence, excitation pulses of long duration (such as semiselective shaped pulses) may result in intensity distortion.

6. Strong coupling mixes spin energy levels, and the resonance lines do not belong to a single spin. The NOE is dispersed over all resonance lines in the strongly coupled system, and interpretation is significantly more complicated [26]. Strong coupling is a particularly relevant problem for oligosaccharides.

For mixing times where the isolated spin-pair approximation is not valid, cross-relaxation rates between all spins that contribute to the indirect relaxation pathways must be considered (as reviewed by Borgias et al. [79]). Complete relaxation matrix calculations require a model for the length and correlation time of each H-H vector. Several authors have accounted for multispin effects in carbohydrates, using a matrix of σ values between all spin pairs in a given geometry to simulate the NOE time course [80–85]. In principle there is no need for relaxation matrix calculations if the ISPA is valid. With long mixing times, cross-peak volumes are larger and long-range effects easier to quantify [12,85], but full relaxation matrix calculations are essential for accurate analysis.

It is important to note that indirect NOEs in small molecules have opposite sign to direct NOEs. In a three-spin system, opposing effects through direct and indirect pathways can reduce, or even cancel, the observed NOE (the "three-spin effect"). Therefore, a lack of an observed NOE between two protons is not, on its own, sufficient evidence that they are far apart [25], and the use of such information as modeling constraints should be regarded with great caution. In large molecules, both direct and indirect NOE have negative sign, and cancellation would not normally occur.

Measurement of NOE.   Resonance lines in the structural reporter group region can often be irradiated selectively prior to acquiring a normal 1D spectrum. NOEs evolve during the presaturation period (typically 100–500 ms, depending upon the molecular weight of the analyte) and are observed by subtracting an identical spectrum obtained with off-resonance preirradiation. To minimize subtraction errors, the 1D NOE difference spectrum can be acquired directly by interleaving on- and off-resonance transients, with inversion of the receiver phase to achieve subtraction. Difference experiments are particularly sensitive to small changes in peak heights and are a precise means for quantifying NOE. The presaturation can be replaced with an inversion pulse to perturb one spin selectively, followed by an NOE evolution period. A gradient-enhanced version of this 1D NOESY experiment is effective for eliminating subtraction artifacts, enabling accurate quantification of small NOEs [86].

Methods for relieving overlap in NOE and ROE spectra are analogous to those for resonance assignment, e.g., 2D NOESY, relayed NOESY [87] (employing a through-space followed by through-bond transfer), NOESY-HOHAHA [88], or NOESY-HMQC in two or more dimensions.

Stochastic variation of the NOESY mixing time (e.g., $\pm 5\%$) converts unwanted zero-quantum terms that contribute to cross-peak intensities into a small random variation in signal amplitude, which appears as t_1 noise. More complicated purging pulses, using inhomogeneous fields, are remarkably effective for filtering zero-quantum artifacts without producing t_1 noise [75].

In ROESY experiments, HOHAHA transfer occurs between resonances that are equidistant either side of the transmitter offset frequency. HOHAHA artifacts are minimized by using a weak spin-lock field (e.g., 2 kHz) and by placing the transmitter offset at one edge of the spectrum. The offset dependence of ROE cross-peak intensity can easily be eliminated by modifying the continuous spin-lock pulse [89].

Transferred NOE: Conformations of Rapidly Exchanging Ligands. Bound-state conformations of carbohydrate ligands that associate with high-MW receptors, with a dissociation rate, k_{off}, that is fast compared with spin-relaxation ($k_{off} > 10^2$ s^{-1}) can be detected indirectly by transferred NOE (TRNOE; reviewed by Ni [90]). Many carbohydrate–protein complexes associate with an appropriate exchange rate (equivalent to ~millimolar K_d), and a "pool" of unbound ligand can be generated, with large negative NOEs that evolved while associated with the slowly tumbling protein. As the complex dissociates, the negative NOEs from the bound state are "transferred" to the free ligand. σ values in the unbound state are generally smaller than in the bound state, and produce positive NOEs. The contribution of the positive NOE to peak intensity is evaluated by recording spectra at different protein: ligand ratios. The dissociation rate is crucial, because with a high k_{off} insufficient NOE develops during the lifetime of the complex, whereas with lower k_{off} spin-diffusion leads to loss of distance specificity. The influence of indirect relaxation pathways on measured NOEs is substantial, and relaxation matrix calculations are essential for quantifying H–H distances [81]. Interpreting TRNOE data is generally more difficult than for NOE of the ligand in free solution, particularly if the ligand associates with multiple sites on the receptor.

Receptor concentrations can be below the limit of detection by NMR, since the signals from the bound state are not observed directly. Thus TRNOE can be measured even if the receptors are sparingly soluble or unavailable in large quantities. Since many carbohydrate–protein associations form relatively weak complexes, TRNOE has become an important tool for probing the mechanisms of carbohydrate recognition [81,91–93].

Heteronuclear NOE. Heteronuclear NOEs develop either by perturbing ^{1}H and observing the enhancement of ^{13}C signals, {^{1}H}–^{13}C NOE or in "reverse" [94] (^{1}H–{^{13}C} NOE). ^{13}C nuclei relax predominantly via a dipolar interaction with their directly bonded ^{1}H and thus present relatively little distance information for conformational analysis. {^{1}H}–^{13}C NOE are large (up to ~198%), and their most common application is to enhance the sensitivity of ^{13}C-detected signals [26]. Long-range effects (up to ~4 Å) can be observed by 1D or 2D HOESY (heteronuclear Overhauser effect spectroscopy) techniques [95,96], but there are few reports of their application to carbohydrate analysis.

^{1}H–{^{13}C} NOE are small and are seldom measured. With the dearth of experimental evidence for carbohydrate conformations there has been recent interest in developing the use of heteronuclear NOE for labeled sugars [94]. ^{1}H–{^{13}C} three-spin effects contain both distance and angular information [94] that may receive more attention for conformational analysis.

T_1 and T_2 Relaxation-Time Constants

Several attempts have been made to account for the amplitude, and the time scale, of internal molecular motions. Lipari and Szabo [97,98] modified the spectral density function for each internuclear vector, $\mathbf{J}(\omega)$ (which characterizes the mobility dependence of cross-relaxation rates), with an additional term containing a generalized order parameter, S^2. For a rigid body, $S^2 = 1.0$, for uniformly distributed internal motion, $S^2 = 0.0$. Thus, S describes the spatial restriction of the *movement* of the internuclear vector, without the need for a detailed description of the motions involved (i.e., it is "model-free"). It is inapplicable if the length of the vector is varying, and is particularly suited to analysis of ^{13}C relaxation data, since the ^{13}C nucleus relaxes via its directly bonded protons only, and the C–H bond length is virtually invariant.

Two exponential time constants describe the time course of the relaxation following an excitation pulse: T_1 (the spin-lattice relaxation-time constant) for the recovery of magnetization along the B_0 axis, and T_2 (the spin–spin relaxation-time constant) for the loss of coherent signal from the transverse plane. The spectral density expression for T_2 relaxation has an additional term for low-frequency motions, $J(0)$, and T_2 can be considered to be more sensitive to overall molecular tumbling rates of large molecules than T_1, which is more sensitive to internal motions [99]. ^{13}C T_1 and T_2 are routinely used to probe internal dynamics and overall tumbling rates of oligosaccharides that are available in sufficient quantity.

The field strength dependence of T_1, T_2, and NOE values varies according to the correlation time (τ_c) and S^2 value, and can therefore be a useful probe for distinguishing fixed from flexible H–H distances [100]. The time scale of internal motions cannot be determined unequivocally from studies of field strength dependence.

It is not easy to interpret T_1 measurements for small saccharides in terms of a motional model because the internal and overall tumbling motions are coupled (i.e., it may be impossible to distinguish which atomic movements correspond to an internal motion and which correspond to overall tumbling). By tethering the glycan at one end (e.g., by covalent attachment to a large mass, such as a protein), the internal motions become uncoupled from overall tumbling motions, and variations in T_1 values along the carbohydrate chain reflect the extent of internal flexibility. Restricted internal motions of a high-mannose-type N-glycan attached to ribonuclease B were recently characterized using ^{13}C T_1 measurements [101]. 2D HMQC spectra modified with relaxation delays [102] were used to quantify the relaxation-time course. Data were fitted to a single exponential curve of peak intensity vs. relaxation delay. Traditionally, T_1 and T_2 are calculated from 1D inversion recovery (for T_1) or CPMG (Carr–Purcell–Meiboom–Gill; for T_2) techniques [103].

J-Coupling Constants

J-coupling (also called scalar-coupling or spin–spin-coupling) is a through-bond effect that determines the fine structure of an NMR peak. J-coupling constants through three bonds (^3J) correlate *approximately* with the dihedral angle, θ, according to the generalised Karplus relationship [104],

$$^3J = A \cos^2\theta + B \cos \theta + C$$

where A, B, and C are empirically derived constants (for $^3J_{HH}$, *approximate* values of A, B, and C are 4.2, -0.5, and 4.5, respectively, but these vary according to the composition and orbital hybridization of the spin system) [104]. $^3J_{HH}$ for the monosaccharides most commonly found in glycoproteins are shown in Table 1. The coupling constants indicate that all common monosaccharides with the D– absolute configuration adopt predominantly the $^4C_1$ chair conformer (i.e., looking down on the ring with the atom numbers increasing in a clockwise direction, C-4 is above the

$^3J_{H3\text{-}H4} = 3.3\ Hz$ $^3J_{H4\text{-}H5} = 0.8\ Hz$

Figure 12 Newmann diagrams illustrating the orientation of electronegative substituents around the C-3—C-4 and C-4—C-5 bonds in D-galactopyranose. $^3J_{H4-H5}$ is lower than $^3J_{H3-H4}$ since both H-4 and H-5 are *antiperiplanar* to oxygen atoms.

plane of the ring and C-1 is below). The Karplus equation provides a rough correlation of J with θ, but does not account for the observed difference in $^3J_{H3-H4}$ (~3.3 Hz) and $^3J_{H4-H5}$ (~0.9 Hz) in galactose, for example. The torsion angles H3–C3–C4–H4 and H4–C4–C5–H5 are $+55°$ and $-55°$, respectively, and the differences in ^3J arise not from the dihedral angles but from differences in electron density around H3 through H5. The orientation of electronegative substituents on C-3, C-4, and C-5 are illustrated in Figure 12. Haasnoot and co-workers have parametrized a Karplus curve for monosaccharides, with a correction for substituent electronegativities [105,106].

$$^3J_{HH} = 13.24\ \cos^2\theta - 0.91\ \cos\theta + \Sigma\{0.53 - 2.41\ \cos^2(\xi_i \cdot \phi + 15.5|\Delta\chi_i|)\}\Delta\chi_i$$

$\xi = 1$ or -1 according to the orientation with respect to the geminal coupled proton, and $\Delta\chi_i$ is the difference in electronegativity of the substituents (as calculated by Huggins [107] and recently revised by Altona et al. [108]) relative to hydrogen.

C-5–C-6 bond rotamer populations of hexapyranose rings are readily determined from the magnitude of the H-5–H-6$_{proR}$ and H-5–H-6$_{proS}$ J-coupling constants. The prochiral C-6 protons of glucose and galactose have been assigned stereospecifically by selective deuteration, and the measured J-coupling constants indicate that *gg* and *gt* rotamers are favored for glucose; all three staggered rotamers of galactose are significantly populated (Table 2) [35]. The observed $^3J_{H5-H6}$ are a simple, weighted average of the values for the contributing rotamers, and they provide valuable information on the orientation of 1–6 linkages. Homans et al. have used $^3J_{H5-H6}$, for example, to indicate the almost exclusively *gg* orientation of the

Table 2 Hydroxymethyl Rotamer Populations (%) for α-D-Glucopyranose and α-D-Galactopyranose in D$_2$O, Determined from $^3J_{H5-H6}$ Spin-Coupling Constants

	gg	*gt*	*tg*
α-Glc	56	44	0
α-Gal	21	54	25

Similar values were calculated for the β-anomers and the corresponding methyl glycosides. (From Ref. 35.)

Manα(1−6)Man linkage in complex *N*-glycans with a "bisecting-GlcNAc" on the 4-position of Man(3) [109].

Interglycosidic heteronuclear ${}^{3}J_{CH}$ ($A = 5.7$, $B = -0.6$, $C = 0.5$ [49]) are a powerful supplement to NOE data in oligosaccharide analysis [49,110,111]. The anomeric ${}^{1}J_{CH}$ value is also conformationally sensitive [112–115], and part of the mechanism for the variation has the same origin as the *exo*-anomeric effect (an electron delocalization from lone pairs on the glycosidic linkage oxygen, which depends upon the orientation about the ϕ glycosidic torsion angle [116]).

$${}^{1}J_{CH} = A \cos 2\theta + B \cos \theta + C \sin 2\theta + D \sin \theta + E + a\varepsilon$$

Values for the coefficients A–E depend upon the anomeric configuration and whether the carbon is on the anomeric or aglyconic side of the linkage. The final term is a correction for the influence of solvent dielectric, ε. The ${}^{1}J_{CH}$ vs. ϕ curve has only one maximum and is therefore complementary to the ${}^{3}J_{CH}$ curve, which has maxima at 0° and 180° (Figure 13). ${}^{1}J_{CH}$ can easily be measured from high-resolution 1D ${}^{1}H$-coupled ${}^{13}C$ spectra. There are, however, wide error limits on the calibration of the ${}^{1}J_{CH}$ curve, which has not yet gained widespread use.

It might reasonably be expected that ${}^{13}C-{}^{13}C$ J-coupling constants will provide additional information for isotopically enriched oligosaccharides. Although empirical rules have been proposed for correlating conformation with carbon–carbon couplings through one or more bonds [117,118], it has not yet been possible to parametrize a ${}^{13}C-{}^{13}C$ equivalent to the Karplus equation.

${}^{1}J_{CH}$ (~150 Hz) and ${}^{1}J_{CC}$ (~50 Hz) are easily measurable from 1D spectra, but ${}^{3}J_{CH}$ are usually measured by heteronuclear correlation. Several recent improvements have been proposed for measuring interglycosidic couplings in saccharides, most of which are based upon measuring active ${}^{1}H-{}^{13}C$ couplings from antiphase pure absorptive line-splittings in HSQC-type spectra. The most severe problem arises when one ${}^{1}H-{}^{1}H$ coupling has similar magnitude to the ${}^{13}C-{}^{1}H$ coupling, causing the partial cancellation of overlapping lines from in-phase and antiphase splitting. 1D experiments with selective excitation can produce particularly high digital resolution [119–121].

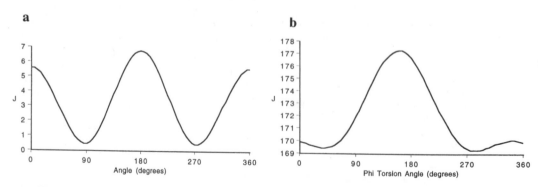

Figure 13 Karplus-type curves giving approximate correlations between the dihedral angle and J-coupling constants. (a) ${}^{3}J_{CH}$ for COCH fragment, (b) ${}^{1}J_{C1-H1}$ in aqueous solvent ($\varepsilon = 80$).

Table 3 ¹H-NMR Chemical Shifts (ppm) for Common Monosaccharide Residues

Residue	H-1	H-2	H-3	H-4	H-5	H-6	H-6'
α-D-Glc	5.23	3.54	3.72	3.42	3.84	3.76	3.84
β-D-Glc	4.64	3.25	3.50	3.42	3.46	3.72	3.90
α-D-GlcNAc	5.21	3.88	3.75	3.49	3.86	3.77	3.85
β-D-GlcNAc	4.72	3.65	3.56	3.46	3.46	3.75	3.91
α-D-Man	5.18	3.94	3.86	3.68	3.82	3.74	3.86
β-D-Man	4.89	3.95	3.66	3.60	3.38	3.75	3.91
α-D-Gal	5.22	3.78	3.81	3.95	4.03	3.69	3.69
β-D-Gal	4.53	3.45	3.59	3.89	3.65	3.64	3.82
α-D-GalNac	5.28	4.19	3.95	4.05	4.13	3.79	3.79
β-D-GalNac	4.68	3.90	3.77	3.98	3.72	3.82	3.84
α-L-Fuc	5.20	3.77	3.86	3.81	3.20	1.21	
β-L-Fuc	4.55	3.46	3.63	3.74	3.79	1.26	

Measured in D_2O solution at 70°C, from internal TSP = 0.00 ppm. Methyl signals for the N-acetylated residues were observed at 2.06 ppm. (Reprinted from Ref. 123 with permission.)

Conformation-Dependent Chemical Shifts

Chemical shifts originate from electron shielding of the nucleus from the external magnetic field and are therefore sensitive to small changes in local electron density. A force component along a bond axis that affects the bond length causes chemical-shift perturbation [122]. Chemical-shift changes relative to the corresponding resonances in the free monosaccharides (''glycosylation shifts'') are in part due to inductive effects (through bonds) and in part conformation-dependent effects, i.e., indicative of steric proximity with adjacent sugar rings. Reference data for ¹H and ¹³C chemical shifts of the most common monosaccharides (from Jansson et al. [123]) are given in Tables 3 and 4, respectively.

Table 4 ¹³C-NMR Chemical Shifts (ppm) for Common Monosaccharide Residues

Residue	C-1	C-2	C-3	C-4	C-5	C-6
α-D-Glc	92.99	72.47	73.78	70.71	72.37	61.84
β-D-Glc	96.84	75.20	76.76	70.71	76.76	61.84
α-D-GlcNAc	91.77	55.00	71.74	71.26	72.51	61.78
β-D-GlcNAc	95.85	57.86	74.81	71.06	76.82	61.85
α-D-Man	94.94	71.69	71.25	67.94	73.34	61.99
β-D-Man	94.55	72.13	74.03	67.69	77.00	61.99
α-D-Gal	93.18	69.35	70.13	70.28	71.30	62.04
β-D-Gal	97.37	72.96	73.78	69.69	75.93	61.84
α-D-GalNac	91.95	51.16	68.40	69.56	71.36	62.11
β-D-GalNac	96.29	54.80	72.01	68.85	75.98	61.89
α-L-Fuc	93.12	69.09	70.30	72.80	67.10	16.33
β-L-Fuc	97.15	72.73	73.93	72.35	71.64	16.33

Measured in D_2O solution at 70°C, from internal dioxane = 67.40 ppm. Methyl signals for the N-acetylated residues were observed at 22.8–23.1 ppm; carbonyl carbons were assigned at 175 ppm. (Reprinted from Ref. 123 with permission.)

The inductive effect of glycosylating a monosaccharide changes the chemical shift of the linkage carbon by ~4 ppm (''α-effect'') and of the β-carbon by -1.5 ppm [50]. A β-^{13}C glycosylation shift that is significantly different from -1.5 ppm indicates that groups attached to the β-carbon atom exhibit steric proximity with neighboring residues [124–127]. The magnitude and direction of the glycosylation shift are a complicated function of the steric interactions and orientations of protons, hydroxyls, and other pendant groups. For such highly flexible molecules as carbohydrates it is not yet possible to interpret glycosylation shifts quantitatively for distance estimates, but their use as a qualitative guide can be an invaluable complement to NOE and J-coupling data.

Glycosylation shifts of linkage ^{13}C resonances are related to steric interactions of the attached proton, and they correlate approximately with the ϕ glycosidic torsion angle [128–130].

Each individual conformer contributes to the observed signal. Lines with separate chemical shifts coalesce to a single peak, with average chemical shift, when the exchange rate exceeds ~$2\nu_o$ (where ν_o is the Larmor frequency). Oligosaccharide spectra invariably exhibit just one multiplet for each resonance, indicating that there is usually no conformational exchange on a time scale slower than ~$2\nu_o$.

IV. MOLECULAR MODELING CALCULATIONS

A. Molecular Mechanics

The reliability of oligosaccharide models is heavily dependent upon the accuracy of energy calculations; thus, considerable research effort has been devoted to ab initio calculations of molecular orbitals and the parametrization of molecular mechanics (MM) force fields. Ab initio studies to calculate the energy of the system from first principles provide the greatest accuracy of all theoretical modeling procedures [131]. The calculations are useful for characterizing forces within monosaccharide derivatives in the gas phase [132,133], but are too computationally intensive to model larger numbers of atoms adequately. Molecular mechanics calculations use the approximations of classical physics to provide a tractable solution for the study of large molecules. Molecular mechanics force fields treat atoms as solid spheres held together by springlike bonds, and the calculated potential energy for the strain on the model is analogous to that for a taut spring. Force fields are parametrized with a set of force constants and optimum bond lengths and bond angles, which are derived empirically from crystal structures or from ab initio calculations. The terms contributing to the total energy (V_T) vary according to the force field, but the general form is

$$V_T = V_{\text{van der Waals}} + V_{\text{bond stretch}} + V_{\text{bond angle}} + V_{\text{torsion angle}} + V_{\text{electrostatic}} + V_{\text{H-bond}}$$

The literature on carbohydrate MM calculations is complicated by a lack of standardization in the force fields and methodology. In the examples given below, all energy calculations use the AMBER force field [134], with a parameter set that is optimized for oligosaccharides [135]. Many force fields use extended atoms (e.g., approximating a hydroxyl group as a single ''hydroxyl atom'') to increase computational efficiency. AMBER, however, is an all-atom force field, and hydrogen bonding is acccounted for by assigning each atom with a partial charge that is designed for condensed phase (solution) calculations [135,136].

It must be emphasized that each force field is a caricature of thermodynamic properties and only approximates the behavior of the physical system. Before embarking on conformational modeling it is essential to decide what problems need to be answered and hence what levels of approximation are acceptable. A crude model of the natural substrate, for example, can generate ideas from which to plan a strategy for synthesizing enzyme inhibitors. Alternatively, more rigorous procedures are required to characterize the extent of flexibility within the ligand. In interpreting the results of the study it is important not to lose sight of the levels of approximation that have been used. Though it is easy to get results from any of the many proprietry MM software packages, it is not so easy to obtain conclusions that are meaningful. The conclusions can only be meaningful if care is taken to avoid overinterpreting the results. Whenever possible, the validity of a model should be evaluated by systematic comparison with experimental observations.

Given the huge number of degrees of freedom available to even a small molecule, it is impossible to calculate energies for all possible structures. The main objective of MM calculations is to derive a low energy structure (e.g., by minimization or simulated annealing) or a distribution of structures (e.g., by Monte Carlo sampling or molecular dynamics simulation) from the search of a wide range of conformational space.

B. Simulated Annealing

Low energy states are easily identified by energy minimization (iterative structure refinement down a potential energy gradient), although the existence of multiple energy wells presents a recurrent problem. It is virtually impossible to establish that the model is in the global energy minimum. Simulated annealing [137] (SA) provides one solution, by giving the system sufficient energy to traverse energy barriers during short periods of molecular dynamics and relaxing into the lowest energy state by gradually withdrawing thermal energy. Empirically, a cooling rate of ~10 K/ps is found to be a suitable compromise between the requirements for slow cooling rates and short computation times [138]. Simulated annealing samples an enormous number of conformations, although it is not certain that the result from any single simulation reaches the global minimum-energy conformation. However, when several SA runs are performed with different "random" atom coordinates for the starting structures (Figure 14a) and all converge to a consensus conformation (Figure 14b), it increases the confidence that the result depicts the true minimum energy structure. Pseudorandom starting structures are easily generated by running molecular dynamics at an arbitrary high temperature (e.g., 700 K).

C. Simulating Internal Flexibility

Relaxed (or adiabatic) potential energy maps are calculated by a grid search of ϕ and ψ torsion angles while optimizing all other internal geometries. They are a highly favored form for displaying the accessible conformational freedom around glycosidic linkages [139]. Boltzmann-weighted populations can be estimated from the relative

a.

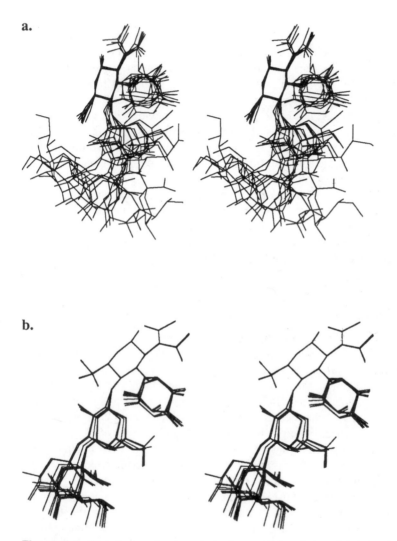

b.

Figure 14 Superimposed stereoviews for models of the sialyl Lewis^x tetrasaccharide, Neu5Acα2−3Galβ1−4[Fucα1−3]GlcNAc. (a) Ten pseudorandom starting geometries for annealing simulations, generated by free MD at 700 K. (b) Nine annealed models from (a), depicting the consensus, global minimum-energy conformation [152].

energies at each point on the potential surface, but the relative populations are force-field dependent. Maps contain no entropic information and may be difficult to interpret for highly branched structures where the conformational freedom of each linkage is not independent. Alternative strategies for sampling relative populations involve exploring the degrees of freedom by rapidly generating new conformers.

Monte Carlo Sampling

In Monte Carlo calculations, random structural changes are made to a low-energy-model structure, and the new potential energy is calculated. If the new energy is lower than its previous value, the trial structure is retained; if it is higher, the trial structure is retained if it passes the Metropolis selection criterion [140]:

$$\xi_3 < \exp\left(\frac{-\Delta E}{kT}\right)$$

where ξ_3 is a random number between 0 and 1, ΔE is the energy difference between the trial structure and the previous structure, k is the Boltzmann constant, and T is the absolute temperature. If retained, the new structure becomes the starting point for a subsequent round of modification. After several thousand steps, ensemble average H–H distances ($<r^{-6}>$) converge on the running average value [84], and the simulation depicts the population probabilities across the ϕ, ψ surface. The extent of the flexibility is controlled by manipulating the temperature scaling factor.

The Metropolis Monte Carlo (MMC) procedure is efficient at crossing energy barriers and can sample a wide range of conformational space. The relative populations of conformers are consistent with experimental observations [84,141–142]. Metropolis Monte Carlo is relatively inefficient at exploring conformational space for molecules with a large number of covalent bonds [143], and some protocols treat individual monosaccharide residues as rigid bodies to overcome this problem [84]. To a first approximation, most monosaccharides with the D– absolute configuration adopt the $^4C_1$ chair conformation almost exclusively (as indicated by ^1H J-coupling constants), and it may be argued that the rigid residue approximation is valid since distortions of the ring geometry are minor. However, it is the *energy* of conformers that determines the relative population and, with an r^{-12} factor in the van der Waals repulsion term, it is important to allow for even minor changes in ring conformations by "relaxing" the internal coordinates during the calculation. Thermal fluctuations frequently draw the structure from the lowest energy conformation, to occupy the periphery of energy wells. Since rigid residue force fields produce unrealistically steep-sided potential energy wells, the rigid residue approximation is likely to be a limitation in dynamics simulations.

Molecular Dynamics Simulations

Molecular dynamics simulations are now routinely used for modeling biological macromolecules and can simulate both the range and the relative rates of exchange between different conformational states [144,145]. An initial set of atom velocities (based upon, for example, a Maxwell–Boltzmann distribution at a chosen temperature) is assigned to a low-energy starting structure, the force on each atom is evaluated from potential energy gradients, and the Cartesian coordinates of each atom are periodically updated by solving Newton's equations of motion. Potential energy in a nonoptimized starting structure is converted to kinetic energy, and such excess thermal energy must be removed by coupling to a thermal bath. Since the time step between successive updates, Δt, must be short in comparison with the fastest motions ($\sim 10^{-15}$ s), simulations are limited to 10^{-12}–10^{-8} s of real time. Simulations of a few hundred picoseconds are sufficient for sampling most motions in oligosaccharides [83,135,146–148].

Simulations can be performed in vacuo or with the explicit inclusion of solvent molecules [149], although in the latter the accessible time scale of the simulation is reduced by the additional computation required to refine the solvent atom coordinates. Explicit solvent molecules may also decrease the rate at which conformational space is sampled [101,135,147]. (Brady and Schmidt, however, have reported that fluctuation between competing intramolecular and solvent–solute hydrogen bonds can increase the rate of internal motions [150]). The area of conformational space sampled in vacuo may be similar to that sampled with explicit solvent, although with solvent molecules a single conformation can persist for several hundred picoseconds [135,138]. For examining the extent of flexibility, in vacuo simulations may be an acceptable compromise, since the effective time scale of the simulation is greater than with explicit solvent.

The major advantage of MD simulations is the sampling of all degrees of freedom simultaneously, although this may be at the expense of computational efficiency compared with rigid residue MMC [143]. Molecular dynamics is inefficient at crossing energy barriers, for modeling slow conformational transitions such as hydroxymethyl group rotations or the flexibility of sialic acid linkages [85,151,152].

D. NOE-Derived Distance Restraints in Molecular Dynamics

The two limitations of MD simulations are the systematic errors introduced by approximations in the force fields and the statistical errors resulting from the short simulation times [145]. No MM force field is perfectly parametrized. Most carbohydrate force fields are adequately parametrized around the global energy minimum, but it is difficult to parametrize for conformations in regions away from the global minimum. Thus it is not yet possible to predict carbohydrate flexibility accurately and to reproduce NOE or J-coupling data from MD simulations that rely solely upon the force field parameters. Even by extending the duration of the simulation it is found that areas of conformational space are oversampled during the MD simulation and imply NOEs that are not observed experimentally. These errors are inherent in the MM approximation, arising inter alia from the approximations of the force field, solvation effects on electrostatic terms, and intra- or intermolecular H-bonding.

NOE-derived distance restraints are a pragmatic approach to modeling flexibility in carbohydrates in the absence of perfect force-field parameters [137,153]. NOE restraints have acheived wider acceptance for modeling peptides or nucleic acids than for modeling carbohydrates, presumably because carbohydrates are believed to be flexible. NOE is a time-averaged bulk property and gives no direct indication of the individual conformers that contribute to the equilibrium [154]. Clearly, if the restraint energies are large, the model can depict a "virtual conformation," a time average of several different conformers that is "devoid of physical meaning" [154]. With care, restraints can allow sufficient flexibility to explore the individual contributing conformers and produce models with excellent agreement with experiment. The following example illustrates the application of restrained MD to the analysis of a glycoprotein glycan.

E. Example—Restrained MD Simulations for a Biantennary *N*-Glycan

For the study of the biantennary *N*-glycan, NOE restraints were applied as a biharmonic pseudopotential [155]. Restraints were classified as strong (no penalty for an H–H distance of 1.8–2.7 Å), medium (1.8–3.3 Å), or weak (1.8–5.0 Å) according to the relative strength of the measured NOE. The restraint potential wells have a wide flat area and the energy penalty at the sides are sufficiently weak (10.0 kcal/mol/Å^2 force constant) that other force-field terms do all of the work when the models are close to the global minimum. As the conformation deviates, the restraints pull the conformation back into the energy well.

The restrained MD simulations for the biantennary *N*-linked glycan used 14 NOE restraints, with a maximum of 10.0 kcal/mol per restraint. Each ϕ vs. ψ trajectory (Figure 15a) occupies the lowest energy region of conformational space for the linkage. Because the time average of the total forcing potential from the 14 NOE restraints is 0.83 kcal/mol, the restraints are not forcing the molecule to adopt a high-energy virtual conformation. Given the wide error limits for the experimental data, NMR provides insufficient evidence to confirm that the trajectories represent the true extent of the flexibility. However, since the models occupy the lowest energy regions it would be difficult to envisage substantially different distributions that are equally consistent with experimental data. Moreover, the extent of internal mobility is consistent with the T_1 and T_2 measurements for *N*-glycans tethered to a protein [101].

Flexibility in restrained MD is at worst underestimated and therefore represents the minimum extent of flexibility that is consistent with experimental data. The maximum extent could be estimated from unrestrained ("free") MD simulations (Figure 15b). Additional energy minima in Figure 15b (compared with Figure 15a) imply several ROE that are not observed experimentally. It has been shown that these conformers could be populated for up to 5%–10% of the time before a measurable NOE would arise, but in the free MD these conformers are oversampled [155].

It might be argued that the NOE back-calculated from restrained models will inevitably agree with experimental NOE (or ROE), since the models are biased by experimental values during the calculation. However, it is worth noting that even in the presence of the tightest constraints (1.8–2.7 Å without energy penalty), the measured relative ROE can vary between ~0.6 and ~7.2, and yet back-calculated ROE values are in the range that is observed experimentally (Table 5). There are also three pieces of experimental data that are not used to bias the MD calculations, which can be used as independent evidence to verify that the models approximate the solution dynamics:

1. NOE—Valid models show no close time-averaged effective H–H distances that would imply the existence of an NOE that is not observed experimentally.
2. Interglycosidic J-coupling constants—Back-calculated $^3J_{CH}$ show close agreement with experimental values.

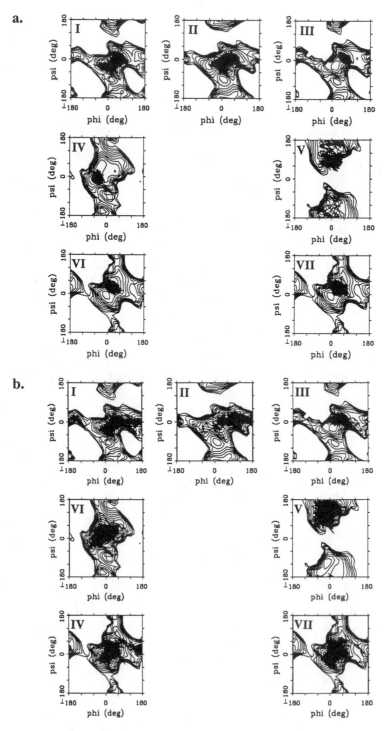

Figure 15 φ, ψ Trajectories from 500 ps of (a) restrained and (b) free molecular dynamics simulations for the bisected biantennary *N*-glycan, superimposed on a relaxed-residue, AMBER φ, ψ iso-energy contour surface (10 kcal/mol). The trajectories depict restricted flexibility at (I) GlcNAcβ1–4GlcNAc, (II) Manβ1–4GlcNAc, (III) GlcNAcβ1–4Man, (IV) Manα1–3Man, (V) Manα1–6Man, (VI) GlcNAcβ1–2Man (3 arm), and (VII) GlcNAcβ1–2Man (6 arm). (Reprinted with permission from Ref. 155. Copyright 1994 American Chemical Society.)

Table 5 Time-Averaged Effective H–H Distances (Å) Back-Calculated from Restrained MD Trajectories of the Bisected Biantennary N-Glycan

GlcNacβ1–2Manα1–6
  5'   4'   \
           GlcNAcβ1–4Manβ1–4GlcNacβ1–4GlcNAc
          9   /   3   2   1
GlcNacβ1–2Manα1–3
  5   4

			H–H Distance			Relative ROE[b]	
		Restraint[a]	$\langle r^{-6} \rangle^{-1/6}$	$(\langle r^{-3} \rangle^2)^{-1/6}$	r_{Tropp}	Calc.[c]	Expt.
H-1(2)	H-4(1)	S	2.27	2.30	2.58	1.63	1.46
H-1(3)	H-4(2)	S	2.21	2.24	2.35	2.86	1.66
H-1(4)	H-3(3)	S	2.28	2.30	2.37	2.72	2.63
H-5(4)	H-2(3)	S	2.54	2.58	2.69	1.27	1.32
H-1(5)	H-4(4)	S	2.36	2.40	2.50	1.97	2.02
H-1(5)	H-2(4)	S	2.31	2.34	2.44	2.28	2.16
H-1(5')	H-1(4')	S	2.42	2.46	2.74	1.14	2.13
H-1(5')	H-2(5')	S	2.33	2.35	2.57	1.67	2.04
H-1(9)	H-4(3)	S	2.59	2.60	2.66	1.36	1.44
H-2(4)	H-2(9)	M	3.02	3.10	3.19	0.46	0.52
H-1(4')	H-6(3)	S	2.23	2.26	2.43	2.34	1.28
H-1(4')	H-6'(3)	S	3.10	3.24	3.64	0.21	1.28
H-1(9)	H-6(3)	S	4.01	4.02	4.11	0.10	1.33
H-1(9)	H-6'(3)	S	2.49	2.53	2.62	1.49	1.33

[a]W, weak; M, medium; S, strong.
[b]Relative to the intraresidue ROE H-1(4)—H-2(4) = 1.00.
[c]Calculated from the ratio of Tropp-averaged distances.
(Reprinted with permission from Ref. 155. Copyright 1994 American Chemical Society.)

3. Conformation-dependent chemical shifts—"Glycosylation shifts" can be rationalized qualitatively in terms of steric proximity of groups on adjacent residues.

Time-dependent NOE restraints [156] and J restraints [157], using time-averaged rather than instantaneous values to drive the MD trajectory toward the target value, were recently introduced to model flexibility in polypeptides and nucleic acids. Time-dependent restraints may produce a more realistic picture of flexibility and reproduce conformational fine detail better than instantaneous restraints [158], but to date there are no reports of time-dependent restraints applied to oligosaccharide analysis.

F. Back-Calculating NOE from Molecular Dynamics Trajectories

In addition to radial averaging ($\langle r^{-6} \rangle$, where $\langle \quad \rangle$ denotes a time average) of the NOE intensity, it is also necessary to consider the often-neglected angular dependence of the dipole–dipole interaction with respect to the direction of the B_0 field [159]. Angular averaging significantly reduces the observed cross-relaxation rate [160] and thus increases the effective measured distances for slowly tumbling flexible molecules [138,160].

If significant overall molecular reorientation occurs during the lifetime of the conformational state (i.e., if $\tau_i \gg \tau_c$, where τ_c and τ_i are the correlation times for overall and internal motions, respectively), the angular terms average away and need not be calculated explicitly. Therefore, a rapidly tumbling molecule approximates to a rigid body, and the appropriate form of radial averaging for back-calculating effective distances from MD data is $\langle r^{-6} \rangle^{-1/6}$ [159]. If there is negligible molecular reorientation during the lifetime of the conformational state, it is necessary to account explicitly for angular anisotropy of the dipolar coupling. Thus, for oligosaccharides with internal motions that are fast compared with overall tumbling motions ($\tau_c \gg \tau_i$), the effective distance is calculated by both radial averaging ($[\langle r_i^{-3} \rangle^2]^{-1/6}$) and angular averaging (given by $\langle Y_{2m}(\Omega_i, \phi) \rangle$, the normalized second-order spherical harmonics) [159,160].

Assuming that angular and radial terms of internal dynamic averaging are uncoupled, the angular term is approximated by the S^2-order parameter determined from MD simulations [160]. Because S^2 is an equilibrium orientational distribution, the duration of the MD must be sufficiently long for the correlation function to reach an equilibrium plateau value.

The Tropp formalism was originally developed for the multisite conformational exchange of methyl groups in protein sidechains [159] and is also appropriate for carbohydrate mobility [138]. Tropp-averaged distances back-calculated from MD data have shown close agreement with experimentally measured time-averaged distances in oligo- [138,155] and polysaccharides [127]. Table 5 shows a comparison of relative ROE calculated from $\langle r^{-6} \rangle$, $\langle r^{-3} \rangle^2$, and Tropp average distances for the biantennary N-glycan. Clearly, the averaging algorithm has a significant effect on the calculated effective distance, since $\langle r^{-6} \rangle$ and r_{Tropp} can vary by more than 0.5 Å. Given the r^{-6}-dependence of the ROE, small inaccuracies in the modeled distances translate to a much larger discrepancy in ROE. Agreement between experimental and back-calculated ROE to within a factor of 2 is considered acceptable agreement (equivalent to $\pm \sim 15\%$ error limits on the distance).

V. FUTURE PROSPECTS

Refining the methodology for conformational analysis is still an active area of research. There are frequent reports of novel techniques for the acquisition and processing of NMR data and refined force-field parameters. Hence, there are some significant recent developments that are not yet in routine use but that will almost certainly be commonplace in the near future.

The number of studies incorporating stable isotope-labeled oligosaccharides is rapidly increasing. Milligram quantities of labeled oligomers are becoming available by efficient enzymatic synthesis, using commercially available glycosyl transferases and regiospecifically or universally labeled monosaccharides [161]. Heteronuclear correlation experiments for ^{13}C-labeled samples offer prospects for resolving the overlap in ^{1}H-NOE spectra and acquiring additional experimental constraints for characterizing conformations. The applications of $^{13}C-^{13}C$ J-coupling constants [114] and $\{^{13}C\}-^1H$ NOE [94], for example, are currently being investigated in carbohydrates. Isotope-edited NMR experiments have enabled the selective detection of signals from labeled ligands in their native environment or complexed with proteins. For example, ^{13}C-labeled capsular polysaccharides have been observed while attached to bacteria [34], dynamic parameters have been measured for lectin-bound glycolipids (^{13}C-labeled) inserted into lipid bilayers [162], and the interactions of tritiated carbohydrate ligands have been inferred from ^3H-NMR (reviewed by Wemmer and Williams [163]). Selective deuteration removes peaks from the ^1H spectrum to aid ^1H signal assignment [35] or to simplify ^1H relaxation pathways [100].

NOEs to hydroxyl proton resonances are extremely useful as long-range sensors of carbohydrate pendant group interactions [164], and increase the number of constraints for conformational analysis. Until recently, hydroxyl and amide proton resonances of carbohydrates were observed only in nonaqueous solvents, which may not be a good representation of physiological conditions. Poppe and van Halbeek have recently observed NOEs to hydroxyl proton resonances of carbohydrates in H_2O solution to supplement the dearth of interglycosidic NOEs between ring protons [23]. By cooling, hydroxyl proton exchange rates can be slowed to a few exchanges per second for NOE measurement [165]. Defining the distances and orientations (from $^3J_{HOCH}$) of hydroxyl group interactions is an important development and will inevitably receive more attention in the future.

VI. SUMMARY

NMR techniques for probing the conformations of carbohydrates and their analogs are now well established, and traditional resonance assignment strategies have been simplified by the recent introduction of high-resolution and time-efficient homonuclear and heteronuclear through-bond correlation techniques. There is active interest in the development of novel approaches to supplement the relatively small number of observable NOEs and interglycosidic $^{13}C-^1H$ J-coupling constants. Modeling procedures rely heavily upon the accuracy of molecular mechanics energy calculations, although no consensus has been reached on the optimum energy force field or modeling protocol. Molecular dynamics simulations have an advantage of sampling conformer distributions over all degrees of freedom simultaneously, and *with care* they can give models that are consistent with all available experimental observations. In the absence of perfect force-field parameters, NOE-derived distance restraints can compensate for the effects that are most difficult to parametrize. Since there are wide error limits on the interpretation of NOE for flexible molecules, it is not possible to confirm that any model depicts the true conformational distribution in solution. However, if the model occupies the lowest energy conformational space and fits all experimental data to within reasonable error limits, there is some confidence that it is a good *model* from which to generate ideas for understanding recognition processes.

REFERENCES

1. A. Varki, Biological roles of oligosaccharides: All of the theories are correct, *Glycobiology 3*: 97 (1993).
2. M. von Itzstein, W.-Y. Wu, G. B. Kok, M. S. Pegg, J. C. Dyason, B. Jin, T. V. Phan, M. L. Smythe, H. F. White, S. W. Oliver, P. M. Colman, J. N. Varghese, D. M. Ryan, J. M. Woods, R. C. Bethell, V. J. Hotham, J. M. Cameron, and C. R. Penn, Rational design of potent sialidase-based inhibitors of influenza virus replication, *Nature 363*: 418 (1993).
3. B. N. N. Rao, M. B. Anderson, J. H. Musser, J. H. Gilbert, M. E. Schaefer, C. Foxall, and B. K. Brandley, Sialyl Lewis X mimics derived from a pharmacophore search are selectin inhibitors with anti-inflammatory activity, *J. Biol. Chem. 269*: 19,663 (1994).
4. J. P. Carver, Oligosaccharides—How can flexible molecules act as signals?, *Pure Appl. Chem. 65*: 763 (1993).
5. H. Beierbeck, L. T. J. Delbaere, M. Vandonselaar, and R. U. Lemieux, Molecular recognition .14. Monte-Carlo simulation of the hydration of the combining site of a lectin, *Can. J. Chem. 72*: 463 (1994).
6. M. C. Chervenak and E. J. Toone, A direct measure of the contribution of solvent reorganization to the enthalpy of ligand binding, *J. Am. Chem. Soc. 116*: 10533 (1994).
7. J. Goddat, A. A. Grey, M. Hricovini, J. Gruschow, J. P. Carver, and R. N. Shah, Synthesis of di- and tri-saccharides with intramolecular NH-glycosidic linkages: Molecules with flexible and rigid glycosidic bonds for conformational studies, *Carbohydr. Res. 252*: 159 (1994).
8. F. A. Quiocho, Carbohydrate-binding proteins: Tertiary structures and protein–sugar interactions, *Annu. Rev. Biochem. 55*: 287 (1986).
9. D. R. Bundle and N. M. Young, Carbohydrate–protein interactions in antibodies and lectins, *Curr. Opinion Struct. Biol. 2*: 666 (1992).
10. J. Deisenhofer, Crystallographic refinement and atomic models of a human F_c fragment and its complex with fragment B of protein A from *Staphylococcus aureus* at 2.9 Å and 2.8 Å resolution, *Biochemistry 20*: 2361 (1981).
11. B. Shaanan, H. Lis, and N. Sharon, Structure of a legume lectin with an ordered N-linked carbohydrate in complex with lactose, *Science 254*: 862 (1991).
12. S. W. Homans, R. A. Dwek, and T. W. Rademacher, Tertiary structure in N-linked oligosaccharides, *Biochemistry 26*: 6553 (1987).
13. W. C. Johnson, The circular dichroism of carbohydrates, *Adv. Carbohydr. Chem. Biochem. 45*: 73 (1987).
14. K. G. Rice, P. G. Wu, L. Brand, and Y. C. Lee, Interterminal distance and flexibility of a triantennary glycopeptide as measured by resonance energy transfer, *Biochemistry 30*: 6646 (1991).
15. P. G. Wu, K. G. Rice, L. Brand, and Y. C. Lee, Differential flexibilities in 3 branches of an N-linked triantennary glycopeptide, *Proc. Nat. Acad. Sci. (USA) 88*: 9355 (1991).
16. E. R. Arndt and E. S. Stevens, Vacuum ultraviolet circular dichroism studies of simple saccharides, *J. Am. Chem. Soc. 115*: 7849 (1993).
17. E. S. Stevens, A potential energy surface of methyl 3-O-(α-D-mannopyranosyl)-α-D-mannopyranoside in aqueous solution: Conclusions derived from optical rotation, *Biopolymers 34*: 1395 (1994).
18. E. R. Arndt and E. S. Stevens, A conformational study of agarose by vacuum UV CD, *Biopolymers 34*: 1527 (1994).
19. A. D. French and J. W. Brady, *Computer Modelling of Carbohydrate Molecules*, American Chemical Society, Washington, 1989.

20. IUPAC-IUB Joint Commission on Biochemical Nomenclature, Symbols for specifying the conformation of polysaccharide chains, *Eur. J. Biochem. 131*: 5 (1983).

21. W. Klyne and V. Prelog, Description of steric relationships across single bonds, *Experientia 16*: 521 (1960).

22. IUPAC-IUB Joint Commission on Biochemical Nomenclature, Abbreviations and symbols for the description of the conformation of polypeptide chains, *Eur. J. Biochem. 17*: 193 (1970).

23. L. Poppe and H. van Halbeek, Nuclear magnetic resonance of hydroxyl and amido protons of oligosaccharides in aqueous solution: Evidence for a strong intra-molecular hydrogen bond in sialic acid residues, *J. Am. Chem. Soc. 113*: 363 (1991).

24. L. Poppe and H. van Halbeek, NOE measurements of carbohydrates in aqueous solution by double-selective pseudo-3D TOCSY-ROESY and TOCSY-NOESY. Application to gentiobiose, *J. Magn. Reson. 96*: 185 (1992).

25. A. E. Derome, *Modern NMR Techniques for Chemistry Research*, Pergamon Press, Oxford, 1987.

26. D. Neuhaus and M. P. Williamson, *The Nuclear Overhauser Effect in Structural and Conformational Analysis*, VCH, New York, 1989.

27. C. A. Bush, High-resolution NMR in the determination of structure in complex carbohydrates, *Bull. Magn. Reson. 10*: 73 (1988).

28. S. W. Homans, Oligosaccharide conformations: Application of NMR and energy calculations, *Prog. NMR Spectrosc. 22*: 55 (1990).

29. C. Jones, Nuclear magnetic resonance spectroscopy methods for the structural analysis of polysaccharides and glycoprotein carbohydrate chains, *Adv. Carbohydr. Analysis 1*: 145 (1991).

30. A. S. Serianni, Nuclear magnetic resonance approaches to oligosaccharide structure, *Glycoconjugates: Composition, Structure, and Function* (H. J. Allen and E. C. Kisailus, eds.), Dekker, New York, 1992.

31. H. C. Siebert, R. Kaptein, and J. F. G. Vliegenthart, Study of oligosaccharide–lectin interaction by various nuclear magnetic resonance (NMR) techniques and computational methods, *Lectins and glycobiology* (H.-J. Gabius and S. Gabius, eds.), Springer-Verlag, Berlin, 1993.

32. H. van Halbeek, ^{1}H nuclear magnetic resonance spectroscopy of carbohydrate chains of glycoproteins, *Methods Enzymol. 230*: 132 (1994).

33. H. van Halbeek, NMR developments in structural studies of carbohydrates and their complexes, *Current Opin. Struct. Biol. 4*: 697 (1994).

34. D. N. M. Jones and J. K. M. Sanders, Assignment of the ^{13}C NMR spectrum of the *Klebsiella* K3 serotype polysaccharide by COSY spectroscopy, *J. Chem. Soc., Chem. Commun.* 167 (1989).

35. Y. Nishida, H. Hori, H. Ohrui, and H. Meguro, ^{1}H NMR analyses of rotameric distribution of C5–C6 bonds of D-glucopyranoses in solution, *J. Carbohydr. Chem. 7*: 239 (1988).

36. A. Bax, G. M. Clore, P. C. Driscoll, A. M. Gronenborn, M. Ikura, and L. E. Kay, Practical aspects of proton–carbon–carbon–proton three-dimensional correlation spectroscopy of ^{13}C-labeled proteins, *J. Magn. Reson. 87*: 620 (1990).

37. A. Bax and R. Freeman, Investigation of complex networks of spin–spin coupling by 2-dimensional NMR, *J. Magn. Reson. 44*: 542 (1981).

38. S. W. Homans, *A Dictionary of Concepts in NMR*, Clarendon Press, Oxford, 1992.

39. A. Wokaun and R. R. Ernst, Selective detection of multiple quantum transitions in NMR by two-dimensional spectroscopy, *Chem. Phys. Lett. 52*: 407 (1977).

40. U. Piantini, O. W. Sorenson, and R. R. Ernst, Multiple-quantum filters for elucidating NMR coupling networks, *J. Am. Chem. Soc. 104*: 6800 (1982).

41. G. Wagner, Two-dimensional relayed coherence transfer spectroscopy of a protein, *J. Magn. Reson. 55*: 151 (1983).

42. S. W. Homans, R. A. Dwek, D. L. Fernandes, and T. W. Rademacher, Multiple-step relayed correlation spectroscopy: Sequential resonance assignments in oligosaccharides, *Proc. Natl. Acad. Sci. USA 81*: 6286 (1984).

43. S. W. Homans, R. A. Dwek, J. Boyd, N. Soffe, and T. W. Rademacher, A method for the rapid assignment of ^{1}H-NMR spectra of oligosaccharides using ^{1}H homonuclear Hartmann–Hahn spectroscopy, *Proc. Natl. Acad. Sci. USA 84*: 1202 (1987).

44. A. Bax, D. G. Davis, and S. K. Sarkar, An improved method for two-dimensional heteronuclear relayed-coherence-transfer NMR-spectroscopy, *J. Magn. Reson. 63*: 230 (1985).

45. A. Bax and D. G. Davis, MLEV-17-based two-dimensional homonuclear magnetization transfer spectroscopy, *J. Magn. Reson. 65*: 355 (1985).

46. A. J. Shaka, C. J. Lee, and A. Pines, Iterative schemes for bilinear operators: Application to spin decoupling, *J. Magn. Reson. 77*: 274 (1988).

47. T. J. Rutherford and S. W. Homans, Proton resonance assignments in oligosaccharides containing multiple monosaccharide residues of the same type, *J. Magn. Reson., Ser. B 106*: 10 (1995).

48. M. E. Girvin, Increased sensitivity of COSY spectra by use of constant-time t_1 periods (CT COSY), *J. Magn. Reson. 108*: 99 (1994).

49. I. Tvaroska, M. Hricovini, and E. Petrakova, An attempt to derive a new Karplus-type equation of vicinal proton–carbon coupling constants for COCH segments of bonded atoms, *Carbohydr. Res. 189*: 359 (1989).

50. A. S. Shashkov, G. M. Lipkind, Y. A. Knirel, and N. K. Kochetkov, Stereochemical factors determining the effects of glycosylation on the ^{13}C chemical shifts in carbohydrates, *Magn. Reson. Chem. 26*: 735 (1988).

51. H. Kessler, C. Greisinger, and G. Zimmermann, Comparison of four pulse sequences for the routine application of proton–carbon shift correlation, *Magn. Reson. Chem. 25*: 579 (1987).

52. L. Mueller, Sensitivity-enhanced detection of weak nuclei using heteronuclear multiple-quantum coherence, *J. Am. Chem. Soc. 101*: 4481 (1979).

53. J. Cavanagh, C. A. Hunter, D. N. M. Jones, J. Keeler, and J. K. M. Sanders, Practicalities and applications of reverse heteronuclear shift corelation: Porphyrin and polysaccharide examples, *Magn. Reson. Chem. 26*: 867 (1988).

54. G. Bodenhausen and D. J. Reuben, Natural abundance nitrogen-15 NMR by enhanced heteronuclear spectroscopy, *Chem. Phys. Lett. 69*: 185 (1980).

55. T. J. Norwood, J. Boyd, J. E. Heritage, N. Soffe, and I. D. Campbell, Comparison of techniques for ^{1}H-detected heteronuclear ^{1}H-^{15}N spectroscopy, *J. Magn. Reson. 87*: 488 (1990).

56. E. R. P. Zuiderweg, A proton-detected heteronuclear chemical-shift correlation experiment with improved resolution and sensitivity, *J. Magn. Reson. 87*: 346 (1990).

57. L. Lerner and A. Bax, Applications of new, high-sensitivity, ^{1}H-^{13}C-NMR-spectral techniques to the study of oligosaccharides, *Carbohydr. Res. 166*: 35 (1987).

58. R. A. Byrd, W. Egan, M. F. Summers, and A. Bax, New NMR-spectroscopic approaches for structural studies of polysaccharides: Application to the *Haemophilus influenzae* type A capsular polysaccharide, *Carbohydr. Res. 166*: 47 (1987).

59. E. Altman, J.-R. Brisson, D. R. Bundle, and M. B. Perry, Structural analysis of the O-chain of the phenol-phase soluble lipopolysaccharide from *Haemophilus pleuropneumoniae* serotype 2, *Can. J. Biochem. Cell Biol. 65*: 876 (1987).

60. T. J. Rutherford, C. Jones, D. B. Davies, and A. C. Elliott, NMR assignment and conformational analysis of the antigenic capsular polysaccharide from *Streptococcus pneumoniae* type 9N in aqueous solution, *Carbohydr. Res. 265*: 79 (1994).

61. P. C. Driscoll, G. M. Clore, D. Marion, P. T. Wingfield, and A. M. Gronenborn, Complete assignment for the polypeptide backbone of interleukin 1b using three-dimensional heteronuclear NMR spectroscopy, *Biochemistry 29*: 3542 (1990).

62. S. W. Homans, Homonuclear three-dimensional NMR methods for the complete assignment of proton NMR spectra of oligosaccharides—Application to Galβ1–4(Fucα1–3)GlcNAcβ1–3Galβ1–4Glc, *Glycobiology* 2: 153 (1992).

63. L. D. Hall and T. J. Norwood, A chemical shift selective filter, *J. Magn. Reson. 76*: 548 (1988).

64. T. J. Rutherford and S. W. Homans, Reducing the overlap problem in the proton NMR spectra of oligosaccharides by application of pseudo-four-dimensional homonuclear HOHAHA-HOHAHA-COSY, *Glycobiology 2*: 293 (1992).

65. C. Bauer, R. Freeman, T. A. Frenkiel, J. Keeler, and A. J. Shaka, Gaussian pulses, *J. Magn. Reson. 58*: 442 (1984).

66. H. Geen and R. Freeman, Band-selective radiofrequency pulses, *J. Magn. Reson. 93*: 93 (1993).

67. L. D. Hall and T. J. Norwood, An alternative design strategy for one-dimensional correlation experiments, *J. Magn. Reson. 87*: 331 (1990).

68. S. W. Homans, Simplification of COSY spectra of oligosaccharides by application of a 2D analogue of 3D HOHAHA-COSY, *J. Magn. Reson. 90*: 557 (1990).

69. D. Uhrin, J.-R. Brisson, and D. R. Bundle, Pseudo-3D NMR spectroscopy, application to oligo- and polysaccharides, *J. Biomol. NMR 3*: 367 (1993).

70. S. Holmbeck, P. J. Hajduk, and L. E. Lerner, Streamlined pseudo-3D TOCSY-NOESY and TOCSY-NOESY experiments, *J. Magn. Reson. 102*: 107 (1993).

71. D. Uhrin, J.-R. Brisson, G. Kogan, and H. J. Jennings, 1D analogs of 3D NOESY-TOCSY and 4D TOCSY-NOESY-TOCSY: Applications to polysaccharides, *J. Magn. Reson., Series B 104*: 289 (1994).

72. J. Keeler, R. T. Clowes, A. L. Davis, and E. D. Laue, Pulsed-field gradients: Theory and practice, *Methods Enzymol. 239*: 145 (1994).

73. T. A. Carpenter, L. D. Colebrook, L. D. Hall, and G. K. Pierens, Application of gradient-selective COSY and double-quantum filtered gradient-selective COSY experiments to carbohydrates: 2-Deoxy-D-*arabino*-hexose ("2-deoxy-D-glucose"), *Carbohydr. Res. 241*: 267 (1993).

74. T. L. Hwang and A. J. Shaka, Water suppression that works: Excitation sculpting using arbitrary waveforms and pulsed-field gradients, *J. Magn. Reson., Series A, 112*: 275 (1995).

75. A. L. Davis, G. Estcourt, J. Keeler, E. D. Lave, and J. J. Titman, Improvement of z filters and purging pulses by the use of zero-quantum dephasing in inhomogeneous B_1 or B_0 fields, *J. Magn. Reson., Series A, 105*: 167 (1993).

76. K. Wuthrich, *NMR of Proteins and Nucleic Acids*, Wiley, New York, 1986.

77. A. Bax and D. G. Davis, Practical aspects of two-dimensional transverse NOE spectroscopy, *J. Magn. Reson. 63*: 207 (1985).

78. A. Bothner-By, R. L. Stephens, J. T. Lee, C. D. Warren, and R. W. Jeanloz, Structure determination of a tetrasaccharide: Transient nuclear Overhauser effects in the rotating frame, *J. Am. Chem. Soc. 106*: 811 (1984).

79. B. A. Borgias, M. Gochin, D. J. Kerwood, and T. L. James, Relaxation matrix analysis of 2D NMR data, *Prog. NMR Spectrosc. 22*: 83 (1990).

80. M. J. Forster, C. Jones, and B. Mulloy, NOEMOL: Integrated molecular graphics and the simulation of nuclear Overhauser effects in NMR spectroscopy, *J. Mol. Graphics 7*: 196 (1989).

81. V. L. Bevilacqua, Y. M. Kim, and J. H. Prestegard, Conformation of β-methylmelibiose bound to the ricin-B chain as determined from transferred nuclear Overhauser effects, *Biochemistry 31*: 9339 (1992).

82. B. R. Leeflang and L. M. J. Kroon-Batenburg, CROSREL—Full relaxation matrix analysis for NOESY and ROESY NMR spectroscopy, *J. Biomol. NMR 2*: 495 (1992).

83. G. Widmalm, R. A. Byrd, and W. Egan, A conformational study of α-L-Rha*p*-(1-2)-α-L-Rha*p*-(1-OMe) by NMR nuclear Overhauser effect spectroscopy (NOESY) and molecular dynamics calculations, *Carbohydr. Res. 229*: 195 (1992).

84. T. Peters, B. Meyer, R. Stuike-Prill, R. Somorjai, and J.-R. Brisson, A Monte Carlo method for conformational analysis of saccharides, *Carbohydr. Res. 238*: 49 (1993).

85. C. Mukhopadhyay, K. E. Miller, and C. A. Bush, Conformation of the oligosaccharide receptor for E-selectin, *Biopolymers 34*: 21 (1994).

86. J. Stonehouse, P. Adell, J. Keeler and A. J. Shaka, Ultrahigh-quality NOE spectra, *J. Am. Chem. Soc. 116*: 6037 (1994).

87. G. Wagner, Two-dimensional relayed coherence transfer-NOE spectroscopy, *J. Magn. Reson. 57*: 497 (1984).

88. G. W. Vuister, P. de Waard, R. Boelens, J. F. G. Vliegenthart, and R. Kaptein, The use of 3D NMR in structural studies of oligosaccharides, *J. Am. Chem. Soc. 111*: 772 (1989).

89. C. Griesinger and R. R. Ernst, Frequency offset effects and their elimination in NMR rotating-frame cross-relaxation spectroscopy, *J. Magn. Reson. 75*: 261 (1987).

90. F. Ni, Recent development in transferred NOE methods, *Prog. NMR Spectrosc. 26*: 517 (1994).

91. V. L. Bevilacqua, D. S. Thomson, and J. H. Prestegard, Conformation of methyl β-lactoside bound to the ricin-B chain—Interpretation of transferred nuclear Overhauser effects facilitated by spin-simulation and selective deuteration, *Biochemistry 29*: 5529 (1990).

92. R. M. Cooke, R. S. Hale, S. G. Lister, G. Shah, and M. P. Weir, The conformation of the sialyl-Lewis-X ligand changes upon binding to E-selectin, *Biochemistry 33*: 10591 (1994).

93. D. R. Bundle, H. Baumann, J.-R. Brisson, S. M. Gagne, A. Zdanov, and M. Cygler, Solution structure of a trisaccharide–antibody complex—Comparison of NMR measurements with a crystal structure, *Biochemistry 33*: 5183 (1994).

94. G. Batta, K. E. Kover, and J. Gervay, Three-spin effects in ^{1}H-{^{13}C} (reverse) heteronuclear NOE and INEPT-(R) HOESY spectra of [1-^{13}C]Me α,β-D-glucopyranoside, *J. Magn. Reson., Series B 103*: 185 (1994).

95. C. Yu and G. C. Levy, Two-dimensional heteronuclear NOE (HOESY) experiments: Investigation of dipolar interactions between heteronuclei and nearby protons, *J. Am. Chem. Soc. 106*: 6533 (1984).

96. G. Batta, K. E. Kover, and Z. Madi, Quantitative ^{13}C-{^{1}H} NOE spectroscopy at natural abundance. The role of indirect effects, *J. Magn. Reson. 73*: 477 (1987).

97. G. Lipari and A. Szabo, Model-free approach to the interpretation of nuclear magnetic resonance relaxation in macromolecules. 1. Theory and range of validity, *J. Am. Chem. Soc. 104*: 4546 (1982).

98. G. Lipari and A. Szabo, Model-free approach to the interpretation of nuclear magnetic resonance relaxation in macromolecules. 2. Analysis of experimental results, *J. Am. Chem. Soc. 104*: 4559 (1982).

99. L. H. Zang, M. R. Laughlin, D. L. Rothman, and R. G. Shulman, ^{13}C-NMR relaxation times of hepatic glycogen in vitro and in vivo, *Biochemistry 29*: 6815 (1990).

100. M. Hricovini, R. N. Shah, and J. P. Carver, Detection of internal motions in oligosaccharides by ^{1}H relaxation measurements at different magnetic fields, *Biochemistry 31*: 10018 (1992).

101. T. J. Rutherford, J. Partridge, C. T. Weller, and S. W. Homans, Characterization of the extent of internal motions in oligosaccharides, *Biochemistry 32*: 12715 (1993).

102. L. E. Kay, D. A. Torchia, and A. Bax, Backbone dynamics of proteins as studied by ^{15}N inverse detected heteronuclear NMR spectroscopy: Application to *Staphylococcal* nuclease, *Biochemistry 28*: 8972 (1989).

103. J. K. M. Sanders and B. K. Hunter, *Modern NMR Spectroscopy: Guide for Chemists*, Oxford University Press, Oxford, 1987.

104. M. Karplus, Vicinal proton coupling in nuclear magnetic resonance, *J. Am. Chem. Soc. 85*: 2870 (1963).

105. C. A. G. Haasnoot, F. A. A. M. de Leeuw, and C. Altona, The relationship between proton–proton NMR coupling constants and substituent electronegativities I—An empirical generalization of the Karplus equation, *Tetrahedron 36*: 2783 (1980).

106. C. A. G. Haasnoot, F. A. A. M. de Leeuw, and C. Altona, Prediction of anti and gauche vicinal proton–proton coupling constants for hexapyranose rings using a generalized Karplus equation, *Bull. Soc. Chim. Belg. 89*: 125 (1980).

107. M. L. Huggins, Bond energies and polarities, *J. Am. Chem. Soc. 75*: 4123 (1953).

108. C. Altona, R. Francke, R. de Haan, J. H. Ippel, G. J. Daalmans, A. J. A. Hoekzema, and J. van Wijk, Empirical group electronegativities for vicinal NMR proton–proton couplings along a C–C bond: solvent effects and reparameterization of the Haasnoot equation, *Magn. Reson. Chem. 32*: 670 (1994).

109. S. W. Homans, R. A. Dwek, J. Boyd, M. Mahmoudian, W. G. Richards, and T. W. Rademacher, Conformational transitions in *N*-linked oligosaccharides, *Biochemistry 25*: 6342 (1986).

110. G. K. Hamer, F. Balza, N. Cyr, and A. S. Perlin, A conformational study of methyl β-cellobioside-d_8 by ^{13}C-NMR spectroscopy: Dihedral angle dependence of ^{3}J(CH) in ^{13}C–O–C–H arrays, *Can J. Chem. 56*: 3109 (1978).

111. B. Mulloy, T. A. Frenkiel, and D. B. Davies, Long-range carbon–proton coupling constants: Application to conformational studies of oligosaccharides, *Carbohydr. Res. 184*: 39 (1989).

112. D. B. Davies, M. MacCross, and S. S. Danyluk, Variation of ^{1}J(C1′,H1′) with glycosidic bond conformation of pyrimidine nucleosides, *J. Chem. Soc., Chem. Commun.* 536 (1984).

113. I. Tvaroska and F. R. Taravel, One-bond carbon–proton coupling constants: Angular dependence in α-linked oligosaccharides, *Carbohydr. Res. 221*: 83 (1991).

114. J. M. Duker and A. S. Serianni, ^{13}C-substituted sucrose—^{13}C–^{1}H and ^{13}C–^{13}C Spin-coupling constants to assess furanose ring and glycosidic bond conformations in aqueous solution, *Carbohydr. Res. 249*: 281 (1993).

115. A. S. Serianni, P. B. Bondo, C. A. Podlasek, W. A. Stripe, and T. J. Church, ^{13}C–^{1}H spin-coupling constants in oligonucleotides—Potential of ^{1}J$_{CH}$, ^{2}J$_{CH}$ and ^{3}J$_{CH}$ as conformational probes, *Biophys. J. 66*: A 236 (1994).

116. I. Tvaroska and T. Bleha, Anomeric and *exo*-anomeric effects in carbohydrate chemistry, *Adv. Carbohydr. Chem. Biochem. 47*: 45 (1989).

117. M. J. King-Morris and A. S. Serianni, ^{13}C-NMR studies of [1-^{13}C]aldoses: Empirical rules correlating pyranose ring configuration and conformation with ^{13}C chemical shifts and ^{13}C–^{13}C spin couplings, *J. Am. Chem. Soc. 109*: 3501 (1987).

118. I. Carmichael, D. M. Chipman, C. A. Podlasek, and A. S. Serianni, Torsional effects of the one-bond ^{13}C–^{13}C spin-coupling constant in ethylene glycol: Insights into the behavior of ^{1}J(C–C) in carbohydrates, *J. Am. Chem. Soc. 115*: 10863 (1993).

119. L. Poppe and H. van Halbeek, Selective, inverse-detected measurements of long-range ^{13}C,^{1}H coupling constants. Application to a disaccharide, *J. Magn. Reson. 93*: 214 (1991).

120. D. Uhrin, A. Mele, J. Boyd, M. R. Wormald, and R. A. Dwek, New methods for measurement of long-range proton–carbon coupling constants in oligosaccharides, *J. Magn. Reson. 97*: 411 (1992).

121. D. Uhrin, A. Mele, K. E. Kover, J. Boyd, and R. A. Dwek, One-dimensional inverse-detected methods for measurement of long-range proton–carbon coupling constants. Application to saccharides, *J. Magn. Reson., Ser. A 108*: 160 (1994).

122. D. M. Grant and B. V. Cheney, Carbon-13 magnetic resonance. VII: Steric perturbation of the carbon-13 chemical shift, *J. Am. Chem. Soc. 89*: 5315 (1967).

123. P.-E. Jansson, L. Kenne, and G. Widmalm, Computer-assisted structural analysis of polysaccharides with an extended version of CASPER using ^{1}H-NMR and ^{13}C-NMR data, *Carbohydr. Res. 188*: 169 (1989).

124. R. U. Lemieux, K. Bock, L. T. J. Delbaere, S. Koto, and V. S. Rao, The conformations of oligosaccharides related to the ABH and Lewis human blood group determinants, *Can. J. Chem. 58*: (1980).

125. N. K. Kochetkov, O. S. Chizhov, and A. S. Shashkov, Dependence of ^{13}C chemical shift on the spatial interactions of protons, and its application in structural and conformational studies of oligo- and polysaccharides, *Carbohydr. Res. 133*: 173 (1984).

126. H. Baumann, B. Erbing, P.-E. Jansson, and L. Kenne, NMR and conformational studies of some 3-*O*-, 4-*O*- and 3,4-di-*O*-glycopyranosyl-substituted methyl α-D-galactopyranosides, *J. Chem. Soc., Perkin Trans. I*: 2153 (1989).

127. T. J. Rutherford, C. Jones, D. B. Davies, and A. C. Elliott, Molecular recognition of antigenic polysaccharides: A conformational comparison of capsules from *Streptococcus pneumoniae* serogroup 9, *Carbohydr. Res. 265*: 97 (1994).

128. K. Bock, A. Brignole, and B. W. Sigurskjold, Conformational dependence of ^{13}C nuclear magnetic resonance chemical shifts in oligosaccharides, *J. Chem. Soc., Perkin Trans. II*: 1711 (1986).

129. R. P. Veregin, C. A. Fyfe, R. H. Marchessault, and M. G. Taylor, Correlation of ^{13}C chemical shifts with torsional angles from high-resolution ^{13}C-CP-MAS NMR studies of crystalline cyclomalto-oligosaccharide complexes, and their relation to the structures of starch polymorphs, *Carbohydr. Res. 160*: 41 (1987).

130. M. J. Gidley and S. M. Bociek, ^{13}C CP/MAS NMR studies of amylose inclusion complexes, cyclodextrins and the amorphous phase of starch granules: Relationships between glycosidic linkage conformation and solid-state ^{13}C chemical shifts, *J. Am. Chem. Soc. 110*: 3820 (1988).

131. W. J. Hehre, L. Radom, P. v. R. Schleyer, and J. A. Pople, *Ab Initio Molecular Orbital Theory*, Wiley, New York, 1986.

132. I. Tvaroska and J. P. Carver, *Ab initio* molecular orbital calculation on carbohydrate model compounds .1. The anomeric effect in fluoro and chloro derivatives of tetrahydropyran, *J. Phys. Chem. 98*: 6452 (1994).

133. I. Tvaroska and J. P. Carver, *Ab initio* molecular orbital calculation of carbohydrate model compounds. 2. Conformational analysis of axial and equatorial 2-methoxytetrahydropyrans, *J. Phys. Chem. 98*: 9477 (1994).

134. U. C. Singh, P. Weiner, J. Caldwell, and P. A. Kollman, *AMBER 3.0*, University of California, San Francisco, CA, 1986.

135. S. W. Homans, A molecular mechanical force field for the conformational analysis of oligosaccharides: Comparison of theoretical and crystal structures of Manα(1–3)Manβ(1–4)GlcNAc, *Biochemistry 29*: 9110 (1990).

136. S. N. Ha, A. Giammona, M. Field, and J. W. Brady, A revised potential-energy surface for molecular mechanics studies of carbohydrates, *Carbohydr. Res. 180*: 207 (1988).

137. G. M. Clore, A. T. Brunger, M. Karplus, and A. M. Gronenborn, Application of molecular dynamics with interproton distance restraints to three-dimensional protein structure determination. A model study of Crambin, *J. Mol. Biol. 191*: 523 (1986).

138. S. W. Homans and M. J. Forster, Application of restrained minimization, simulated annealing and molecular dynamics simulations for the conformational analysis of oligosaccharides, *Glycobiology 2*: 143 (1992).

139. A. D. French, Comparisons of rigid and relaxed conformational maps for cellobiose and maltose, *Carbohydr. Res. 188*: 206 (1989).

140. N. Metropolis, A. W. Rosenbluth, M. N. Rosenbluth, A. H. Teller, and E. Teller, Equation of state calculations by fast computing machines, *J. Chem. Phys. 21*: 1087 (1953).

141. L. Poppe, R. Stuike-Prill, B. Meyer, and H. van Halbeek, The solution conformations of sialyl-α-(2–6)-lactose studied by modern NMR techniques and Monte Carlo simulations, *J. Biomol. NMR 2*: 109 (1992).

142. T. Weimar, B. Meyer, and T. Peters, Conformational analysis of α-D-Fuc-(1–4)-β-D-GlcNAc-OMe: One-dimensional transient NOE experiments and Metropolis Monte-Carlo simulations, *J. Biomol. NMR 3*: 399 (1993).

143. A. E. Howard and P. A. Kollman, An analysis of current methodologies for conformational searching of complex molecules, *J. Med. Chem. 31*: 1669 (1988).

144. J. A. McCammon and S. Harvey, *Dynamics of Proteins and Nucleic Acids*, Cambridge University Press, Cambridge, 1987.

145. M. Karplus and G. A. Petsko, Molecular dynamics simulations in biology, *Nature 347*: 631 (1990).

146. S. N. Ha, L. J. Madsen, and J. W. Brady, Conformational analysis and molecular dynamics simulations of maltose, *Biopolymers 27*: 1927 (1988).

147. C. J. Edge, U. C. Singh, R. Bazzo, G. L. Taylor, R. A. Dwek, and T. W. Rademacher, 500-picosecond molecular dynamics in water of the Man(1–2)Man glycosidic linkage present in Asn-linked oligomannose-type structures on glycoproteins, *Biochemistry 29*: 1971 (1990).

148. Z.-Y. Yan and C. A. Bush, Molecular dynamics simulations and the conformational mobility of blood group oligosaccharides, *Biopolymers 29*: 799 (1990).

149. L. Madsen, S. N. Ha, V. H. Tran, and J. W. Brady, Molecular dynamics simulations of carbohydrates and their solvation, *Computer Modelling of Carbohydrate Molecules* (A. D. French and J. W. Brady, eds.), American Chemical Society, Washington, D.C., 1990.

150. J. W. Brady and R. K. Schmidt, The role of hydrogen bonding in carbohydrates— Molecular dynamics simulations of maltose in aqueous solution, *J. Phys. Chem. 97*: 958 (1993).

151. H. C. Siebert, G. Reuter, R. Schauer, C. W. von der Lieth, and J. Dabrowski, Solution conformations of GM3 gangliosides containing different sialic acid residues as revealed by NOE-based distance mapping, molecular mechanics and molecular dynamics calculations, *Biochemistry 31*: 6962 (1992).

152. T. J. Rutherford, D. G. Spackman, P. J. Simpson, and S. W. Homans, 5 Nanosecond molecular dynamics and NMR study of conformational transitions in the sialyl-Lewis-X antigen, *Glycobiology 4*: 59 (1994).

153. J. N. Scarsdale, P. Ram, J. H. Prestegard, and R. K. Yu, A molecular mechanics–NMR pseudoenergy approach to the solution conformation of glycolipids, *J. Comput. Chem. 9*: 133 (1988).

154. O. Jardetsky, On the nature of molecular conformations inferred from high-resolution NMR, *Biochim. Biophys. Acta 621*: 227 (1980).

155. T. J. Rutherford and S. W. Homans, Restrained vs. free dynamics simulations of oligosaccharides: Application to solution dynamics of biantennary and bisected biantennary N-linked glycans, *Biochemistry 33*: 9606 (1994).

156. A. E. Torda, R. M. Scheek, and W. F. van Gunsteren, Time-dependent distance restraints in molecular dynamics simulations, *Chem. Phys. Letters 157*: 289 (1989).

157. A. E. Torda, R. M. Brunne, T. Huber, H. Kessler, and W. F. van Gunsteren, Structure refinement using time-averaged J-coupling constant restraints, *J. Biomol. NMR 3*: 55 (1993).

158. D. A. Pearlman and P. A. Kollman, Are time-averaged restraints necessary for nuclear magnetic resonance refinement? A model study for DNA, *J. Mol. Biol. 220*: 457 (1991).

159. J. Tropp, Dipolar relaxation and the nuclear Overhauser effects in nonrigid molecules: The effect of fluctuating internuclear distance, *J. Chem. Phys. 72*: 6035 (1980).

160. C. B. Post, Internal motional averaging and three-dimensional structure determination by nuclear magnetic resonance, *J. Mol. Biol. 224*: 1087 (1992).

161. Y. Ichikawa, Y. C. Lin, D. P. Dumas, G. J. Shen, E. Garcia-Junceda, M. A. Williams, R. Bayer, C. Ketcham, L. E. Walker, J. C. Paulson, and C. H. Wong, Chemical-enzymatic synthesis and conformational analysis of sialyl Lewis X and derivatives, *J. Am. Chem. Soc. 114*: 9283 (1992).

162. B. J. Hare, F. Rise, Y. Aubin, and J. H. Prestegard, ^{13}C-NMR studies of wheat-germ-agglutinin interactions with N-acetylglucosamine at a magnetically oriented bilayer surface, *Biochemistry 33*: 10,137 (1994).

163. D. E. Wemmer and P. G. Williams, Use of nuclear magnetic resonance in probing ligand–macromolecule interactions, *Methods Enymol. 239*: 739 (1994).

164. J. Dabrowski and L. Poppe, Hydroxyl and amido groups as long-range sensors in conformational analysis by nuclear Overhauser enhancement—A source of experimental evidence for conformational flexibility of oligosaccharides, *J. Am. Chem. Soc. 111*: 1510 (1989).

165. L. Poppe and H. van Halbeek, NMR spectroscopy of hydroxyl-protons in supercooled carbohydrates, *Nature Struct. Biol. 1*: 215 (1994).

166. A. De Bruyn, M. Anteunis, and G. Verhegge, Glucose, galactose, mannose and their Me-osides in D_2O. A 300-MHz study, *Acta Ciencia Indica 1*: 83 (1975).

15

Glycopeptides in the Immunotherapy of Cancer

R. Rao Koganty, Mark A. Reddish, and B. Michael Longenecker
Biomira Inc., Edmonton, Alberta, Canada

I. INTRODUCTION

The cell surface has many highly specific and functional macromolecules, such as glycoproteins, glycolipids, and proteoglycans. The structural differences in the expression of these molecules contribute to the functional differences between cancer and normal cells. Most epithelial cells express glycoproteins with extensive *O*-linked glycans called *mucins*. The complex and highly branched carbohydrate structures of these mucins completely obscure the protein core, whose sequence is often dominated by clusters of serines and threonines, which account for up to 60% of all amino acids. The magnitude of glycosylation may have some significance in the survival of the cell itself. In normal mucins, the protein core is inaccessible due to the presence of a dense cover of large carbohydrate structures. In the transformed cell, the mucins are characterized by patterns of glycosylation that are altered both in size and distribution of the carbohydrate structures [1]. The patterns observed in these highly repetitive glycosylations appear to have profound effects on the behavior of malignant cell types. Parts of the protein core become exposed in such mucins, resulting in immune responses [2] such as those seen in breast cancer patients. The epitopes reactive with antibodies HMFG-1, HMFG-2, and SM-3 were detected in patients with breast cancer, who also have antibodies that react with human milk fat globule membrane (HMFG) mucin, which is designated as MUC-1. Springer [3] termed such epitopes created by malignant mucins as "autoantigens." These epitopes are characteristically either carbohydrate or peptide in origin. However, a combination of both of these epitopes could create a hybrid or a multiple epitope that might generate immune responses. Normal or precancer epithelial cells, which express these epitopes, generate immune responses to produce antibodies that can be detected by their respective epitopes [2–4].

The presence of a distinct immunogenic epitope [5] within the core protein suggests that the MUC-1 mucin may serve as a target for immunotherapy of cancer, through its exposed protein core, that results in the loss of tolerance, followed by the development of immune responses, as has been observed in breast cancer patients. Cancer establishes and spreads through metastasizing cells and disables the immune system through poorly understood mechanisms. Recent evidence, however, points to high serum concentrations of circulating cancer-associated mucins that are implicated in immune suppression [6]. We have employed an immunotherapeutic strategy that involves stimulating the immune system through the use of synthetic mucin antigens and to modulate the T-helper response between humoral and cell-mediated immunity. Immune responses against immunosuppressive serum-circulating mucins may aid in their clearance from blood and, thereby, restore normal immune function.

II. MEMBRANE-BOUND GLYCOPROTEINS AS TARGETS FOR THE IMMUNOTHERAPY OF CANCER

We have focused our approach on immunotherapy of cancers using membrane-bound mucin-type glycoproteins as the target. The glycosylation patterns and structures of the carbohydrates of mucins from both normal and cancer cells have been a subject of great interest for over two decades. The results of the numerous investigations relating to the carbohydrates that were chemically cleaved and purified from mucins, structural analyses by high-resolution proton (^{1}H) and carbon (^{13}C) nuclear magnetic resonance (NMR) spectroscopy [7–17] and immunohistochemical analyses [17,18–31] yielded valuable information about the basic structural differences between tumor-associated and normal mucins. Core glycosylation appears to follow a pattern that is dependent on the species- and tissue-specific pools of glycosyltransferases within the cell. Cell transformation adds a new dimension to the existing diversity in posttranslational glycosylation of the core protein. Membrane-bound glycoproteins such as mucins can be excellent targets for immunotherapy of cancer. MUC-1, a cell-surface mucin expressed by epithelial cells, is the best studied of the mucins [32,33] and is widely expressed on most adenocarcinomas.

The core protein of MUC-1 mucin consists of three distinct regions. The mucin is anchored in the cell membrane, with a large glycosylated portion extending into the cell surface. The highly hydrophobic transmembrane domain anchors the mucin in the lipid membrane, with the N-terminal tail extending into cytoplasm [34]. It is the highly conserved extracellular domain that displays significant deficiencies in glycosylation patterns that are associated with cancer. This region is characterized by a tandemly repeating 20-amino-acid sequence designated by the single-letter amino acid code known as GVTSAPDTRPAPGSTAPPAH. The tandem repeat has five possible sites of glycosylation (three threonines and two serines), which represents 25% of all amino acids in the sequence. It is not known if all sites are glycosylated even in the normal mucins. The number of tandem repeats varies significantly among individuals, usually between 30–120 repeats [32–37]. Additionally, differences in size and distribution of carbohydrate structures cause further variations in mucins that are known as *glycoforms* [38,39]. Due to this heterogeneity, MUC-1 was originally called *polymorphic epithelial mucin* (PEM). All the carbohydrate structures in the tandem repeat region are *O*-linked to serines and threonines. Since MUC-1 is expressed by a wide variety of cancers, such as breast, ovarian, pancreatic, colorectal, and lung, it is an ideal target for actively stimulating the immune system to destroy the cancer cells [40].

The MUC-1 mucin is believed to extend a great distance (280 nM) from the surface of most epithelial carcinomas, five times further than any other glycoprotein. There is considerable evidence that this mucin molecule acts as the first and perhaps most important contact with cells of the immune system [40,41]. Cancer-associated-mucin–derived segments in the form of glycopeptides can potentially function as vaccines that are mimics of cancer-associated mucin epitopes and effectively stimulate anticancer immune responses. However, in order to design such a ''glycopeptide'' vaccine, the sites of glycosylation as well as the size of the carbohydrate structures present must be determined. It is a tedious exercise to synthesize all possible combinations of glycopeptides and test them for their protective abilities. In our laboratory, we have employed a monoclonal antibody generated against a cancer-associated mucin as well as another generated against a synthetic unglycosylated peptide and compared their binding to various glycosylated synthetic vaccine candidates. We have then used this information to determine the extent and specificity of glycosylation required to mimic the cancer-associated mucin.

III. SYNTHETIC GLYCOPEPTIDES AS ANTIGENS AGAINST CANCER: A SURVEY

The core protein sequences of the mucins are often dotted with clusters of serines and threonines, most of which are glycosylated with α-linked *N*-acetylgalactosamine (Tn antigen). Consequently, the carbohydrates also exist as clusters. It has been suggested that smaller carbohydrate epitopes, such as Tn antigen (αGalNAc-Ser or αGalNAc-Thr), in the form of linear clusters of two or more, are needed to function as an effective epitope [42,43]. It has been demonstrated that such a cluster is required to bind to MLS 128, an anti-Tn antibody raised against LS180 cells, a human colorectal cancer cell line. A minimal binding segment isolated from a protease digest of ovine submaxillary mucin (OSM) was found to contain a sequence STT, glycosylated with three Tn structures as a linear cluster [42–44]. Glycopeptides have been receiving attention since the early 1980s, though most of the work is limited to the chemical and enzymatic synthesis or to the synthesis of clusters of glycopeptides [45–61]. Immune response to glycopeptides has been a subject of only recent interest [45,46,51,62,63]. This interest is driven by the idea that glycopeptides might be the most suitable antigens to generate anticancer immune responses. There has not been a great deal of immunological data available to this day, probably because small synthetic segments are typically not very immunogenic by themselves. The problem of enhancing the immunogenicity has been solved mostly by going back to the concept of spacers, conjugation to carrier proteins, and their use in conjunction with adjuvants, which are usually bacterial-cell-wall preparations. Bacterial-cell-wall–derived synthetic lipopeptide, tripalmitoyl-S-glycerylcysteinylserine (P3CS) (Figure 1), has been the most successful structurally well-defined immune stimulant thus far known in the literature [45,46,62,63] in terms of generating immune responses to an epitope chemically linked to it. In fact, a glycopeptide synthesized as an integral part of this molecule generated immune responses in mice [45,46].

Though there is no known clinical investigation of a glycopeptide as a cancer vaccine, there are a few reports of their use in animal models and in immunological studies. A dimeric Tn-glycopeptide synthesized as an integral part of P3CS has been used to immunize mice without the help of carrier protein or an adjuvant [45,46]. This is perhaps the first time ever that a structurally well-defined synthetic small molecule has generated immune responses in animal models as a single self-adjuvanted immunogen (Figure 2). High serum titers of both IgM and IgG classes of antibodies have been detected following two immunizations of 100 μg each of this immunogen, seven days apart. The high titers of the IgG class of antibodies is attributed to the involvement of helper T cells.

Antigen processing and presentation by class I and class II proteins of the major histocompatibility complex (MHC) is a major part of the immune system's functions. Cytotoxic T lymphocytes (CTL) recognize short peptide segments of approximately nine amino acids that are intracellularly processed and presented by class I MHC molecules. It has long been thought that glycopeptides may not be among the repertoire of MHC class I restricted T cell epitopes, and, consequently, there were no car-

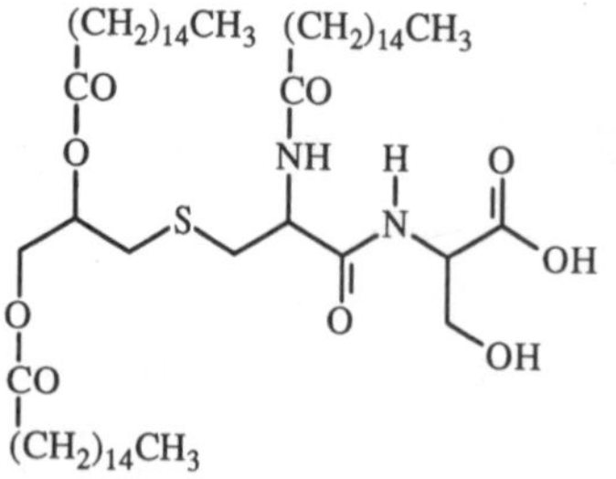

Figure 1 Tripalmitoyl-S-glycerylcysteinylserine (P$_3$CS), an *E. Coli* cell-wall-derived lipopeptide that has been used as a conjugated adjuvant.

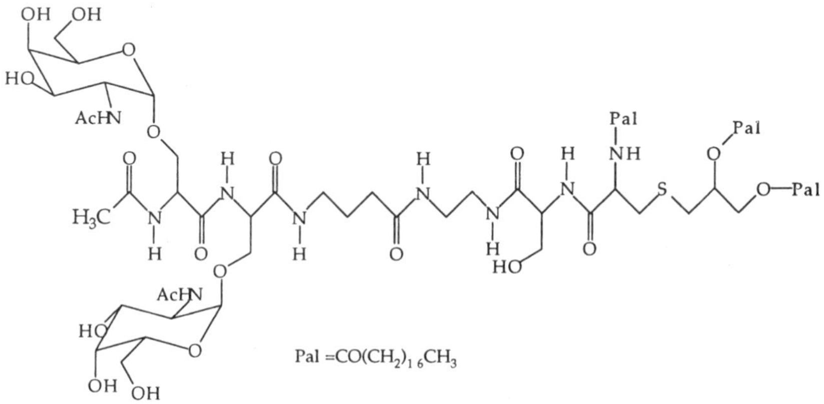

Figure 2 Dimeric Tn glycopeptide with P_3CS. A self-adjuvanted antigen as a well-defined single molecule that has generated immune responses.

bohydrate-specific cytotoxic T lymphocytes. In a recent report [64], a K^b-restricted CTL epitope, FAPGNYPAL, derived from *Sendai* virus nucleoprotein has been mutated with serine at G4. Glycosylated versions of the peptide in which the serine is glycosylated with *N*-acetylglucosamine or *N*-acetylgalactosamine were synthesized. Mice were immunized with either of the glycopeptides mixed with incomplete Freund's adjuvant. The spleen cells taken from the immunized mice were stimulated using RMA-S cells, which were precultured to express empty K^b and D^b molecules and preincubated with the corresponding glycopeptide and washed. The results of this study demonstrate that MHC class I molecules can bind glycopeptides and that the glycoepitope of the presented glycopeptide may determine the CTL's antigen specificity. This is an important finding for the immunotherapy of cancers, where cell-mediated immunity is regarded as an important weapon to fight tumors. Though there has not been any reported identification of class I bound glycopeptides, it has been suggested that this might be due purely to technical difficulties in purification and identification of such molecules, which might be present only in femtogram quantities or less among a pool of many other class I bound peptides.

Many reports of glycopeptide design and synthesis and their possible applications are outside the domain of cancer immunotherapy and, hence, beyond the scope of this chapter. Nevertheless, they exemplify the growing interest in this important area of vaccine design.

IV. STRUCTURAL FEATURES OF THE EPITOPES OF MUC-1 MUCIN

The biosynthesis of mucin-associated oligosaccharides is initiated with the attachment of an αN-acetylgalactosamine to serines and threonines in the peptide core. From there on, the diversity approaches a much higher magnitude with the attachment of carbohydrates like *N*-acetylglucosamine, galactose, fucose, sialic acid, and repeating units such as *N*-acetyllactosamine. The variations in the pool of glycoforms that arise from different tissues and species may originate from differences in the occurrences and functions of the various associated glycosyltransferases. This is true

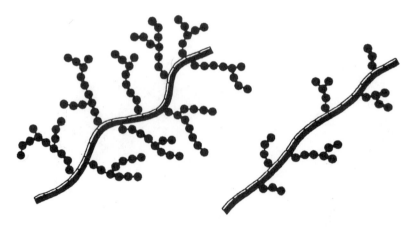

Normal Mucin   Cancer Mucin

Figure 3 Normal and cancer-associated mucins with long and truncated carbohydrate chains, respectively.

in both normal and malignant tissue, and results in significant differences observed in the size and distribution of the carbohydrate structures expressed on the tumor-associated mucins, such as the cell-surface-bound MUC-1. The biosynthesis and structure of O-linked mucin-bound carbohydrates has been the subject of several research and review articles [65–67].

The extracellular tandem-repeat region of MUC-1 mucin displays significant differences between tumor-associated core-glycosylation and that of the normal cell. Figure 3 displays the resultant differences in terms of glycosylation and the consequent exposure of the protein core epitopes.

The truncated versions of the normal carbohydrate structures that are present on cancer-associated MUC-1 mucin are well known as the Thompsen–Friedenreich (TF) and related antigens (Figure 4). Antibodies against these structures have been detected in humans, both normal and patients with cancer.

Linking of N-acetylgalactosamine is a primary step in glycosylations. Further biosynthesis through a battery of glycosyltransferases, produced intracellularly by their respective encoding genes, extends this structure to more complex and branched polylactosamine-based structures which terminate usually with sialylation and/or fucosylation [65,66]. As can be seen from Figure 3, the cancer-associated mucins in general are incompletely and sparsely glycosylated versions of normal mucins, exposing the core protein. In fact, several small carbohydrate structures, such as Tn, TF, STn, have been identified as predominantly cancer-associated and regarded as indicators of prognosis [68,69]. These differences make the cancer mucin a twofold target for immunotherapy of cancer, through its active stimulation of the immune system against both protein core and the truncated carbohydrate structures that constitute a significant proportion of the known MUC-1–associated epitopes on cancer cells.

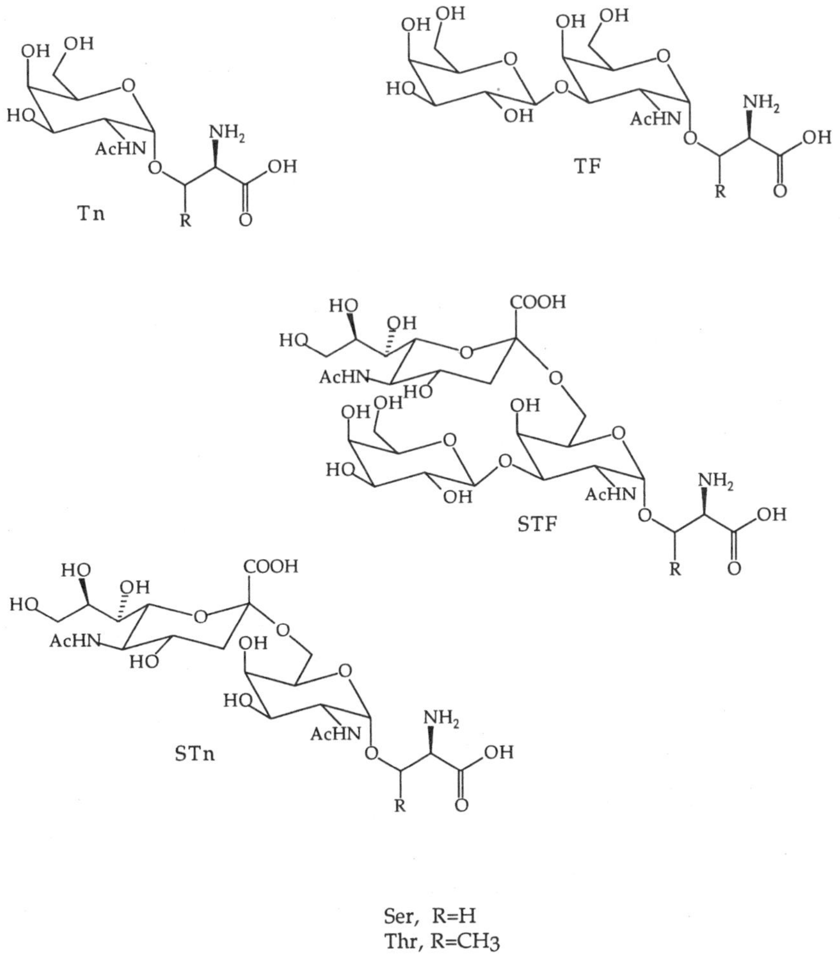

Figure 4 Thomsen–Friedenreich (TF) and related carbohydrate structures that are expressed on cancer-associated mucins.

V. RECENT DEVELOPMENTS IN THE SYNTHESIS AND PRODUCTION OF GLYCOPEPTIDES

So far, glycopeptide synthesis is limited by the accessibility of glycosylated amino acids, particularly the O-glycosylated serines and threonines. In mucins, the predominant O-linked primary carbohydrate is αN-acetylgalactosamine. In addition to their accessibility, the techniques for the routine, automated synthesis of glycopeptides are not well established, in terms of protecting groups for amino acid side chains as well as for carbohydrates. Cleavage conditions for glycopeptides from solid supports, as well as removal of blocking groups, need to be revisited in the light of the sensitivity of glycosidic linkages to the strong acid environment that is used for unglycosylated peptides.

Several bypass techniques are in vogue currently. It is not uncommon to synthesize small glycopeptides in solution, where the conditions are more amenable to the carbohydrate part than they are in the solid-phase mode. Certain protecting groups, such as pentamethylchroman sulfonyl (pmc), which requires a strongly acidic environment for its removal, may be avoided. Enzymatic extension of small molecules to complex carbohydrate structures following conventional synthesis of preliminary glycopeptides with monosaccharides seems to be a popular approach. This combination of synthetic and enzymatic processing overcomes the difficulty of the initial lack of site specificity of transferases in glycosylating a multisite peptide. It is widely regarded that enzymes perform complex carbohydrate synthesis more efficiently than has been possible through chemical synthesis [70,71]. In a single step, for example, an enzyme can bring about stereospecific attachment of a sialic acid to a specific position of a carbohydrate substrate, often in a matter of minutes. Chemical synthesis of the same would need several manipulative steps using blocking groups to protect other sites from reacting, often taking several days to achieve the same linkage. While the enzymatic approach is growing fast, the feasibility is still in its infancy, probably due to the lack of widespread demand for complex oligosaccharides. Column-type immobilized enzymes are now being developed for large-scale "flow-through" production of both carbohydrates and glycopeptides. Complex, high molecular weight glycopeptides can be purified using high-efficiency reverse-phase column chromatography. Structural characterization has become fairly routine with the advent of multidimensional nuclear magnetic resonance (NMR) spectroscopy and analytical mass spectrometry. Critical glycosyl donors for transferases are now being commercially produced. Technological advancements in coupled multienzyme syntheses that perform complex glycopeptide synthesis look promising. In spite of many advancements in enzymatic syntheses, production of glycopeptides and their purification on a commercial scale has not advanced. Enzymatic glycosylation of a multisite peptide is not yet clearly defined in terms of site specificity. Chemical synthesis, though at a small scale, still offers promise for the production of structurally well-defined glycopeptides.

The recent upsurge in enzymatic developments is attributable to a certain lack of interest in the development of chemical production methods. During the last two decades carbohydrate chemistry has not grown beyond small-scale experiments which often fail to translate into larger scale production. Given that, carbohydrate and glycopeptide production is expensive compared to conventional organic pharmaceuticals; the main reason for the lack of industrial-scale methods may be attributable to the lack of need. There has not been any approved synthetic glycopeptide-based vaccine or drug in commercial production. Commercial-scale processes make economic sense if and when there is need for such development. An example is the number of reactions [72] that are used to glycosylate serines and threonines to obtain Tn-antigen (Figure 5). Though this method employs a highly reactive 2-azido precursor for the formation of an α-glycosidic linkage, the synthesis using a circuitous route makes it unsuitable for commercial production. There has been a need for large-scale production of these glycosylated amino acids in order to produce the mucin-type glycopeptides that are being investigated as vaccines against cancer. Consequently, research into newer glycosyl donors led to a simplified approach (Figure 6) that makes it possible not only to use commercially available raw material but also to produce these glycosylated amino acids [73] in a configuration that can be adopted for the synthesis of glycopeptides.

The synthetic scheme in Figure 6 has been developed for large-scale production of glycopeptides. N-Acetylgalactosamine has not been extensively used synthetically

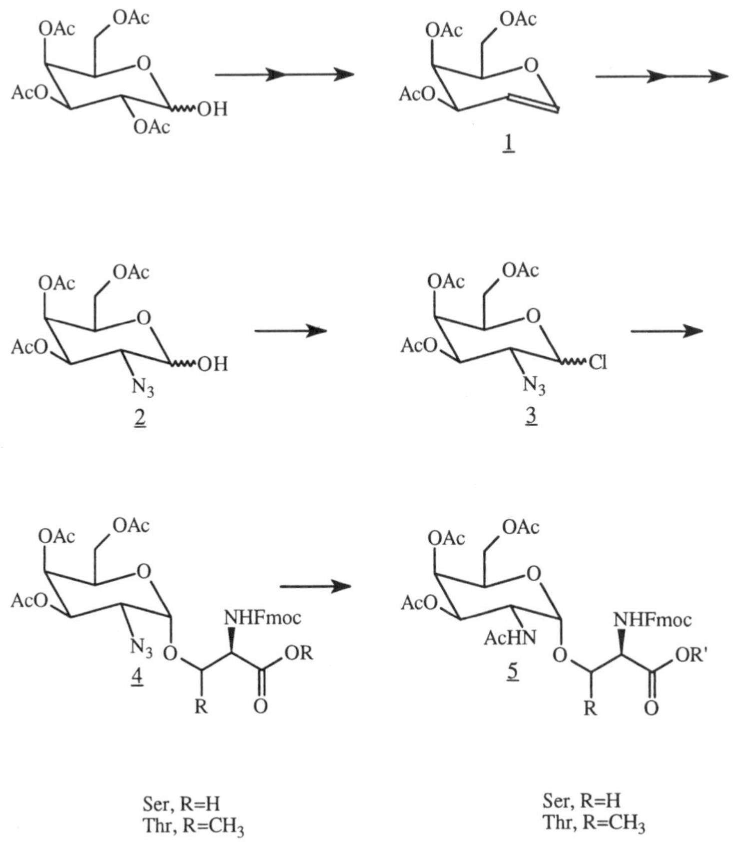

Figure 5 Scheme for the synthesis of Tn serine using 2-azido galactose derivative as a precursor.

since its synthetic manipulation is considered to be very difficult due to the presence of a neighboring acetamido group that participates in any reaction involving the anomeric carbon atom. Formation of oxazoline derivatives as a major by-product has been commonly observed, which makes it almost impossible to create a glycosidic bond with N-acetylgalactosamine. Creation of an exclusive α-glycosidic bond has never been possible. Lemieux [72] devised an elegant scheme (Figure 5) to circumvent this problem by creating a precursor to the N-acetyl group. 3,4,6-Triacetyl galactal is synthesized from galactose by acetylation and treatment with dry, gaseous hydrogen bromide. The resultant peracetylated bromogalactose undergoes elimination under basic conditions to form the unsaturated galactal **1**. Introduction of the azido group at the 2-position is a complex chemical process using ceric ammonium nitrate and sodium azide as the reagents in dry acetonitrile as solvent. 2-Azido galactosyl nitrate derivative is formed among a complex mixture of other by-products. Hydrolysis of the 2-azido galactosyl nitrate gives 2-azido-3,4,6-triacetyl galactose **2** in a modest overall yield. Structure **2** can readily be converted into a donor **3** by treatment with dry hydrogen chloride. The 2-azido donor **3** (Figure 5) forms a precursor α-glycoside **4** in a glycosylation reaction catalyzed usually by silver or mer-

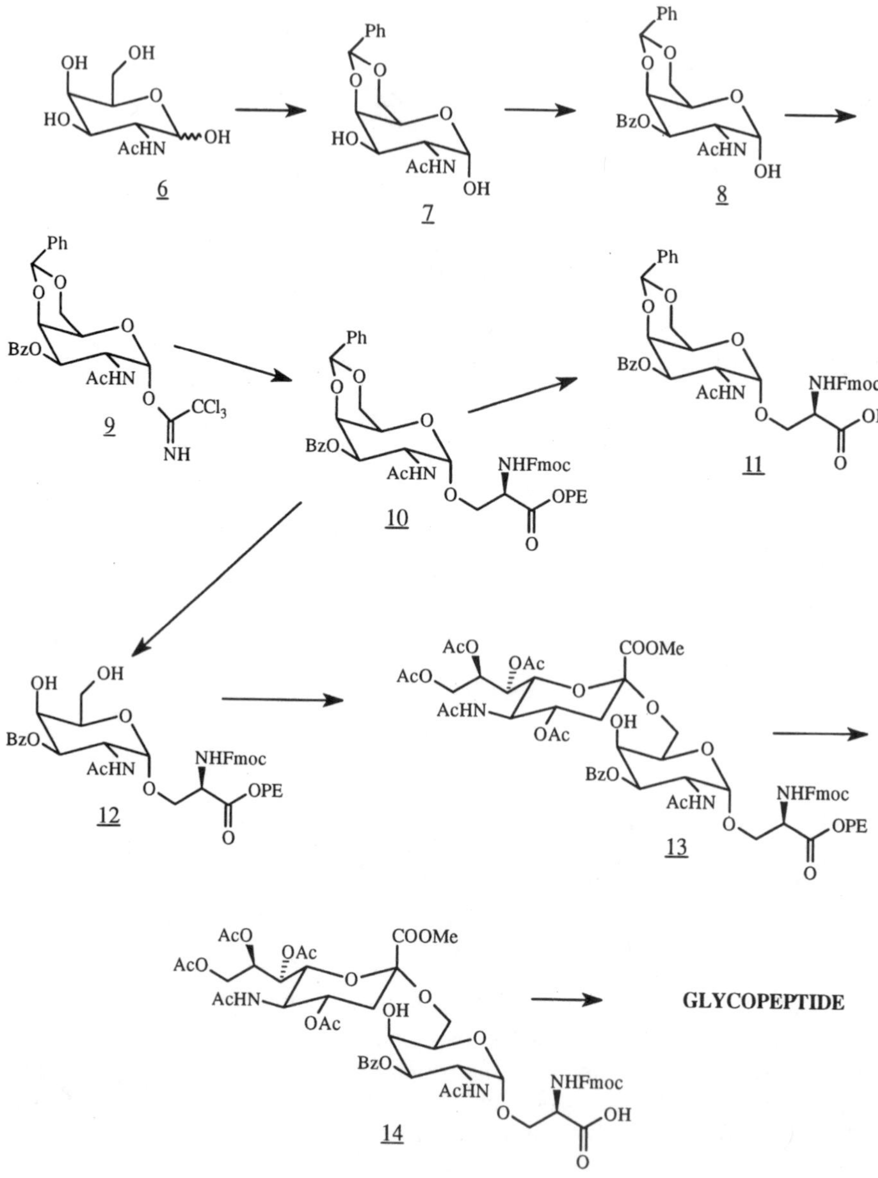

Figure 6 Scheme for the synthesis of Sialyl-Tn-serine using *N*-acetylgalactosamine derivative as a donor (PE = phenacyl ester).

curic salts. The yield at this stage is quite impressive, since there is no participating group involved at the 2-position. The 2-azido group can easily be converted to an *N*-acetamido group by catalytic hydrogenation in acetic acid at low pressure using 5% palladium-on-charcoal to obtain blocked glycosylated Tn-serine **5**. In spite of the elegance and the ease of formation of the α-linkage, further development of this process to extend **5** to larger carbohydrate structures adds more complex steps to the synthesis. For example, synthesis of **13** (Figure 6) from **5** may add more deblocking and blocking steps that are complex in terms of selectivity and that may reduce the yields of the final product further.

The synthesis of 2-azido galactose makes this route lengthy and uneconomical for bulk production. Commercially available *N*-acetylgalactosamine **6** can be converted to its 4,6-benzylidene acetal **7** in bulk scale. The solubility of *N*-acetylgalactosamine is not a factor in this reaction, since the use of benzaldehyde dimethylacetal with a catalytic amount of *p*-toluene sulfonic acid in dimethylformamide (DMF) overcomes the initial solubility problem as the product quickly goes into solution. Purification of the final product is achieved by simple crystallization from ethanol. The 3-position of the benzylidene derivative is selectively protected by acetylation or benzoylation in pyridine to obtain 1-OH compound **8**. The 1-OH is easily converted to a stable trichloroacetimidate donor **9**, by treating it with trichloroacetonitrile in dry methylene chloride with a catalytic amount of diazabicyclo undecene (DBU). The simple and fast access to this donor makes this glycosylated amino acid synthesis adaptable to bulk production. The glycosylation is straightforward and quick, with moderate yields of α-glycosides. In a reaction typically catalyzed with boron trifluoride etherate (BF$_3$ · OEt$_2$) in dry tetrahydrofuran (THF) as solvent, **9** reacts with suitably protected serine or threonine to form predominantly α-glycoside **10**. The major by-product is a recyclable 1-OH compound **8**. The carboxy-protecting phenacylester (PE) group of **10** can be removed selectively by reductive cleavage in zinc/acetic acid to obtain **11**, which can be used in solid-phase glycopeptide synthesis. Synthetic extension of **10** is facilitated by deblocking the benzylidene-protecting group in acetic acid give 4,6-diol, **12**. The 6-position of **12** is selectively sialylated using 2-chloro, peracetylated sialic acid methylester as donor to obtain α2−6 sialyl Tn derivative, **13**, in moderate yield. The phenacylester deblocked version, **14**, can be used in solid phase glycopeptide synthesis [73]. This process is suitable for commercial production of glycopeptides, since the building blocks, such as **11** and **14**, are stable and can be commercially produced and stored. The formation of side products, particularly during the α-glycosylation with **9**, is minimal, which facilitates the handling of bulk-scale purification.

Peptide synthesis has advanced technically with the development of solid-phase methods during the '60s. Large quantities of amino acids or glycosylated amino acids, often at fourfold excess, are used in the solid coupling process. The coupling efficiency, though typically high at the initial stages of the synthesis, gradually falls with increasing peptide length. Consequently, solid-phase synthesis suffers from an increasing number of deletions (2^n, where n = number of couplings) that occur with an increase in the length of the peptide, not to mention the associated purification problems. The synthesis of glycopeptides with longer sequences of several protected amino acids results in very poor yields of the desired glycopeptide on any scale. The processes that are currently in use for peptide synthesis are too severe for glycopeptides, particularly for those involving sialylated glycosyl amino acids. Most of the existing techniques need improvement, and newer techniques have to be developed to make the production of glycopeptides commercially feasible. With the interest in glycopeptides increasing, as possible vaccines and drugs for therapy of disease states such as AIDS, cancer, and bacterial and viral infections, there is a need for the development of commercial-scale production methods.

VI. ANIMAL MODELS FOR THE IMMUNOTHERAPY OF CANCER

Human clinical testing must be preceded by extensive animal model studies, not only to demonstrate the concepts but also to prove that such concepts are translated into efficacious therapy of the disease under study. To this end, we have developed several relevant murine-tumor models for the evaluation of immunotherapeutic vaccines.

One of the earliest carbohydrate antigens that has been identified as tumor associated is known as the Thomsen–Friedenreich (TF) antigen (see Figure 4). In order to demonstrate that the immune system can be directed to fight cancer, it is necessary to test the principle in an animal model, in which, for example, tumor-bearing mice are immunized with an antigen that is expressed by the tumor cells, and then to show the effect of immune responses on the tumor burden of the mouse. The TF antigen is widely expressed both on human carcinomas and on the murine mammary carcinoma. A well-known and widely studied mucin, epiglycanin, is rich in TF antigen and expressed on the murine mammary carcinoma cell line, TA3Ha [6,74]. TF is also expressed on human-breast-cancer–associated MUC-1 mucin, and has been the subject of clinical studies in human ovarian immunotherapy trials [75]. This constitutes an excellent prospect for a relevant animal model to demonstrate the active specific immunotherapy (ASI) of human cancers. TA3Ha is a lethal and highly proliferative cell line that at a dose of even a few hundred cells establishes cancer and grows rapidly in mice and kills all untreated control groups within three weeks.

The main problem lies in increasing the immunogenicity of small carbohydrate or small glycopeptide molecules. Ordinarily such small molecules are not immunogenic. They have to be conjugated chemically to an immunogenic carrier protein such as keyhole limpet hemocyanin, better known as KLH. Keyhole limpet hemocyanin is a highly immunogenic protein that is purified from the hemolymph of the giant keyhole limpet, *Megathura crenulata*. Small molecules chemically attached to this carrier generate immune responses, both humoral and cell mediated, when administered along with an adjuvant to stimulate the immune system, such as DE-TOX™ or Freund's adjuvant. Figure 7 shows an ideal animal model in which the vaccine-treated mice survived the longest and the untreated control group died within three weeks after being given tumor cells. The prolonged survival is attributed to the prevention of spread, and retarded growth, of the tumor rather than cure and elimination of cancer completely.

This disease-relevant animal model was used as the rationale to design and conduct human clinical trials. Indeed, TF-KLH–immunized ovarian cancer patients generated high titers of IgG specific to the TF carbohydrate structure. Though the antigen TF-KLH is a neoglycoconjugate and not a true glycopeptide derived from epiglycanin, and only the carbohydrate portion is a true antigen, this concept raises prospects for the true glycopeptide antigen. Another way to study human tumors in animal models is to use transfected mouse carcinoma cell lines that grow in mice and express human tumor-associated antigens such as MUC-1. Immune responses generated against a MUC-1–derived peptide protected mice against tumor growth when challenged with E3 cells, a murine mammary adenocarcinoma cell line, transfected with MUC-1 gene [5]. The antibodies generated against the MUC-1–derived peptide bound to both E3 cells as well as MCF-7, a human breast carcinoma cell line that expresses MUC-1 mucin. It is often thought that a segment of the mucin

ASI ANIMAL MODEL

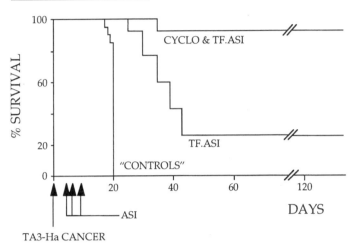

Figure 7 Survival curves of CAF1 mice challenged with the murine adenocarcinoma cell line TA3/Ha followed by treatments with synthetic TF-KLH vaccine in RIBI adjuvant. The effects of combining low-dose cyclophosphamide pretreatment (day 3) with TF-KLH vaccination show significant and syngeristic actions resulting in >90% survival at 120 days following tumor challenge. The effects of cyclophosphamide in this protocol have been shown to be mediated through the abrogation of suppressor T cell activities [6] that relate to the production of the immunosuppressive mucin epiglycanin. Control groups in this experiment include PBS, adjuvant only, and cyclophosphamide only. Each experimental group consists of 14 mice (ASI = active specific immunotherapy).

molecule that carries a cluster of small carbohydrate structures may be needed as an antigen that truly represents the cancer cell surface, both structurally and chemically. This may be true in generating humoral immune responses against small carbohydrate structures, since the latter may not constitute a distinct epitope that is recognized by an antibody [42–44]. Immune responses to a vaccine based on synthetic trimeric sialyl-Tn cluster glycopeptide were found to be much more pronounced than to the vaccine based in monomeric sialyl-Tn [76].

VII. DESIGNING VACCINES AGAINST CANCER

In developing vaccines for immunotherapy of cancer it is desirable that the target be specific; that is, the immune responses should be specific to the epitopes found on tumor cells and not on normal cells, though this exclusivity may not be entirely feasible. However, chemically defined structures are the most promising therapeutic candidates since the immune responses generated against them are reasonably specific compared to that of the whole mucin molecule [77]. The vaccine must be a structural and compositional mimic of the tumor-associated epitope. It is imperative that the segments of mucin that are predominantly cancer associated be identified. Cancer-associated MUC-1 contains tandem repeats of 20-amino-acid sequences that are sparsely glycosylated with small carbohydrate structures that are shown in Figure

2. Systematic investigations were carried out, to locate the tumor-specific glyco-sylation sites, by measuring the total activity of the pool of tumor cell enzymes on synthetic MUC-1 peptides [78,79]. It was discovered that lysates from both human pancreatic cancer cells and breast cancer cells glycosylated the MUC-1 tandem repeat GV**TS**APD**T**RPAPG**S**TAPPAH (glycosylated sites in bold) at identical sites with N-acetylgalactosamine. The consistency in occurrence of this mucin model throughout the tumor cell population becomes significant because the therapeutic value of the synthetic structure depends on its ability to target the whole population of tumor cells. It is probably unlikely that a unique epitope distinguishes all tumor cells from normal cells. Multiepitopic vaccines that cover a wide range of tumor antigens may be a viable option to protect against tumor antigen heterogeneity in the population of tumor cells. A major concern is the conformation of the protein core of the mucin molecule, because this undergoes changes resulting from variations in glycosylation, in terms both size and distribution of the carbohydrate structures. Small carbohydrate structures that are widely distributed on tumor-associated mucins induce only subtle conformational changes of the protein core, as evidenced by a recent study [77]. In mucins characterized by tandem repeats, each epitope within the tandem repeat may be significantly influenced by the changes in local conformation. Unglycosylated versions of the MUC-1 tandem-repeat segment (Figure 8) did not undergo significant changes in conformation when small carbohydrate structures were introduced at spe-cific sites.

Figure 9 displays a stereoscopic projection of the PDTRP portion of the 16-amino-acid sequence GVTSAPDTRPAPGSTA of the 20-amino-acid tandem repeat. T3 and S4 of the sequence are glycosylated with αN-acetylgalactosamine as shown in Table 1.

As indicated in Figure 9, the minor (small carbohydrates) glycosylation-related conformational changes in the epitopic region or elsewhere in the tandem-repeat sequence seem to be insignificant. Antibody-binding studies indicate, however, that these subtle changes do result in significant changes in the relative affinities for specific antibodies. The binding patterns of two well-characterized monoclonal an-tibodies that were raised to either the native glycosylated mucin or to a synthetic

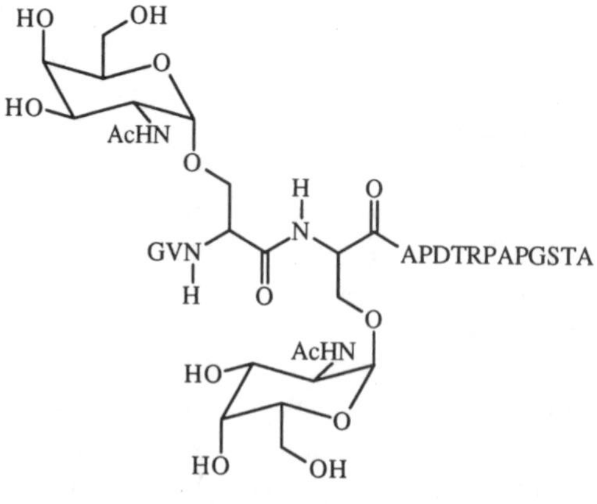

Figure 8   MUC-1–derived synthetic glycopeptide carrying two N-acetylgalactosamine (Tn) structures.

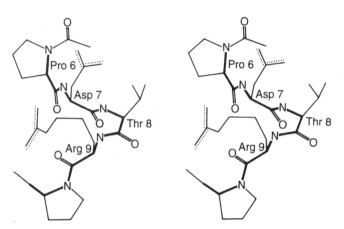

Figure 9 Stereoscopic projection of optimized conformation of PDTRP epitope from synthetic peptides and glycopeptides.

peptide antigen show significant characteristic peptide and glycopeptide reactivities (Table 1). B27.29 is a monoclonal antibody raised against MUC-1 cancer-associated mucin [80], while BCP-8 is a monoclonal antibody raised against synthetic unglycosylated peptide [81]. It is not surprising that B27.29 binds better to the glycosylated peptides, as shown by the inhibition of its binding to the MUC-1 mucin solid phase (Figure 10a) and to the synthetic unglycosylated peptide (BP1–7) as solid phase (Figure 10b). The monoclonal antibody BCP-8 shows greater affinity toward the unglycosylated peptide as indicated by the inhibition of its binding by various peptides (Table 1) to the mucin as solid phase (Figure 11a) and to the peptide as solid phase (Figure 11b). Both the antibodies, BCP-8 and B27.29, bind to the cancer-associated MUC-1 mucin, with the latter displaying stronger affinity. Since B27.29 also shows greater affinity to the diglycosylated peptide, it appears to describe a glycopeptide as its antigen. Though the observed differences as shown in Figure 10 are minor, the consistency reveals the glycopeptide characteristic of the MAb B27.29. It is desirable to analyze the structural features of synthetic peptides and glycopep-

Table 1 Synthetic Peptides and Glycopeptides of MUC-1 Mucin

Peptide	Sequence
BP1-7	GVTSAPDTRPAPGSTA
PB1-47	GVTS(Tn)APDTRPAPGSTA
BP1-48	GVT(Tn)SAPDTRPAPGSTA
BP1-52	GVT(Tn)S(Tn)APDTRPAPGSTA

BP1-7 is a synthetic peptide derived from the 20-amino-acid tandem-repeat domain of MUC-1, a surface-expressed adenocarcinoma-associated mucin. BP1-47, BP1-48, and BP1-52 are the corresponding glycosylated synthetic peptides with α-linked N-acetylgalactosamine (Tn) on T^3 and/or S^4 of the 16-amino-acid sequence.

a.

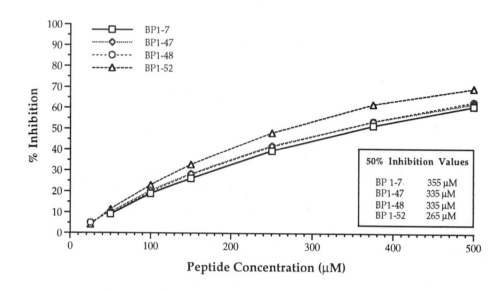

b.

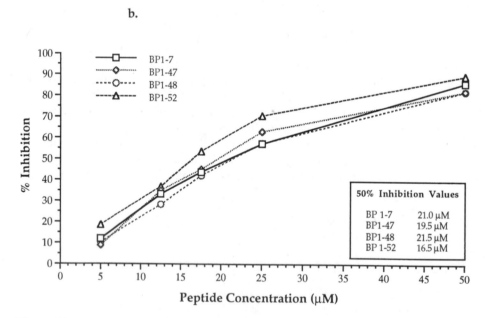

Figure 10   Binding-inhibition studies with MAb B27.29 [79] on either an adenocarcinoma-derived MUC-1 mucin solid phase (a) or a synthetic peptide solid phase (b). Comparisons between synthetic peptide (BP1-7) and various glycopeptide structures show that only the doubly glycosylated structure (BP1-52) has significantly increased affinity for the anti-native mucin antibody. Data presented are expressed as percent binding inhibition relative to the no-peptide controls and represent the mean of three assays, each performed in triplicate.

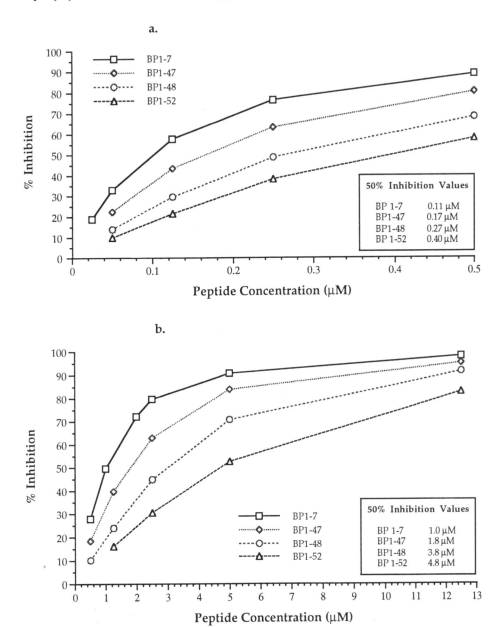

Figure 11 Binding-inhibition studies with MAb BCP8 [80] on either an adenocarcinoma-derived MUC-1 mucin solid phase (a) or a synthetic peptide solid phase (b). Comparisons between synthetic peptide inhibition (BP1-7) and various glycopeptide structures demonstrates the high relative affinity of this synthetic peptide-specific antibody for the nonglycosylated structure. The lowest relative affinity is observed with the doubly glycosylated structure (BP1-52). Data presented are expressed as percent inhibition relative to the no-inhibitor controls and represent the mean of three assays, each performed in triplicate.

tides and their similarities to the native glycoprotein and to use the information to
design a vaccine whose immune responses may be specific to tumor cells.

Glycosylation of serines and threonines with very large carbohydrate structures
is more likely to bring about significant changes in the protein core conformation in
addition to affecting recognition by the immune system. In an example of a synthetic
segment of a mucin that carries three successive large-sized carbohydrate structures,
such as trimeric sialyl-Tn serine tripeptide, the core peptide appears to assume a
near-spiral configuration in order to accommodate three large disaccharide molecules
in a succession in the available space around the tripeptide backbone (Figure 12).
The probable linearity of the serine core is significantly influenced by extensive
glycosylation, particularly on successive sites, an occurrence commonly associated
with the mucins.

In the case of MUC-1, the unglycosylated repeating segments, dominated by
the presence of prolines and ionic side chains such as aspartate and guanidinium of
arginine, will have greater influence on the population of conformers, irrespective of
the number of tandem repeats. That the influence of side chains is uniform in preserving local conformation is evident from our recent investigation [77]. This is
contrary to the earlier belief that a minimum number of repeats is necessary to

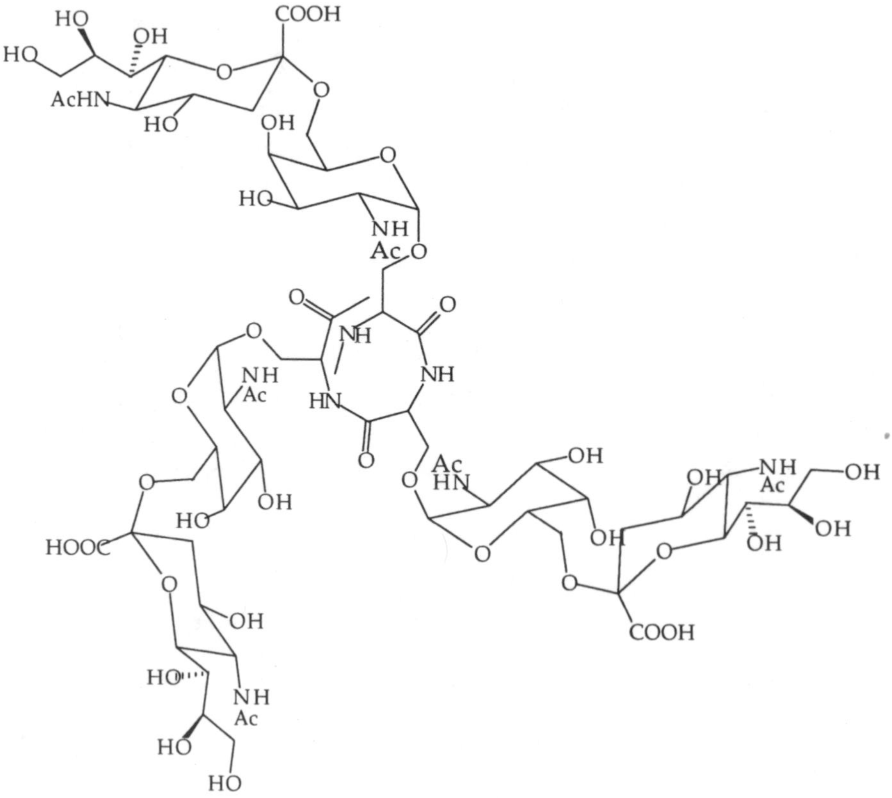

Figure 12 Predicted arrangement of sialyl-Tn carbohydrate structures placed on successive
serines.

preserve such local conformation [82,83]. At the same time, the native mucins are heavily but diversely glycosylated, causing significant diversity in the distribution of glycoforms. These may have a wide distribution of global conformations, in addition to changes in local conformation resulting from the size and distribution of carbohydrate structures.

Though the mucin is a very large molecule that displays a large number of epitopes in a paradigm of tandem repeats and glycoforms, it becomes a practical target if an average representative antigen, in the form of a glycopeptide, can be effectively designed. In order to design a glycopeptide as a vaccine, the uncertainties that are associated with the target mucin in describing the size, number, location of the carbohydrate structures must be overcome. The immune response generated against such a rationally designed glycopeptide may be successful in directing the immune system to destroy the cancer cell targets.

VIII. STRATEGIES FOR RATIONAL DESIGN OF MUC-1 GLYCOPEPTIDE VACCINES

The immune system responds to antigenic challenge via essentially two pathways. These major response pathways are described by the helper T-cell subsets that respond and the regulatory cytokines that they secrete. The T-helper (TH1) pathway is characterized by a predominantly cellular immune response, with CD4+ T cells secreting IL-2 and gamma interferon (IFN) and tumor necrosis factor (TNF). Frequently this pathway results in the generation of true $CD8^+$ CTL activities. (Recently IL-12 has been shown to play a key role early in the TH1 pathway, resulting in a preferential selection of TH1 over TH2 responses.) The alternative T-cell regulatory pathway, termed TH2, is characterized by secretion of IL-4 and IL-5 by $CD4^+$ cells and results in strong antibody production [84,85].

The therapeutic potential of each of these pathways of immune response must be evaluated in the development of a vaccine and of clinical strategies. Various disease settings may benefit from these characteristically different immune responses. This is most evident in the area of infectious disease vaccines. Traditional vaccines that are typically derived from killed organisms and adjuvanted in aluminum hydroxide induce high antibody titers that are effective in blocking the disease transmission. Live, attenuated virus vaccines, such as vaccinia, result in the generation of cell-mediated immune responses. In the case of HIV, it is becoming increasingly evident that antibody responses are not therapeutically effective in treating disease [86,87], but antibodies may offer some protection from viral challenge, particularly if secretory IgA is induced and viral challenge is at mucosal sites. Cancer, however, is a pleiotropic disease that is more a process of incremental changes over long periods of time. Varied clinical states of the disease may require different immunotherapeutic strategies. For instance, in the case of colorectal cancer, it has been thought for some time that at the time of surgery, tumor cells released from the primary site and spilled into the bloodstream and/or the peritoneal cavity may be the source of postsurgical metastasis. In such cases, where the disease spreads through the bloodstream, a strong antibody response may offer clinical benefit through a "sterilization effect" in serum. In an experiment of inducing artificial metastasis in preimmunized and nonimmunized control groups of mice, we have obtained further

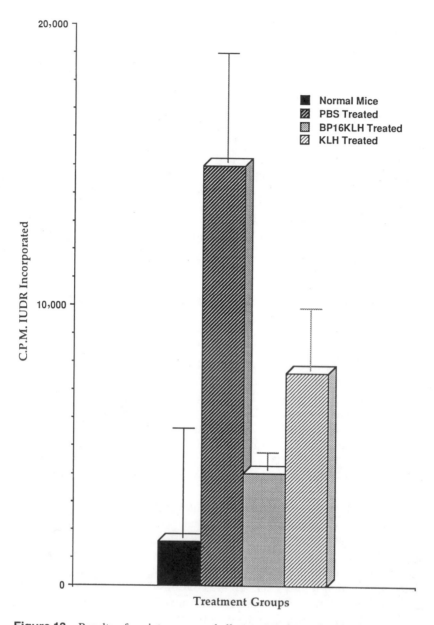

Figure 13 Results of an intravenous-challenge experiment in CB6F1 mice pretreated (day −3 and day −17) with MUC-1 peptide conjugate vaccine (50 μg BP1-7-KLH in DETOX™-B SE adjuvant) or controls, which include KLH DETOX™ or PBS (n = 10 per group). Intravenous tumor challenge was 3 days after the second vaccine priming, and 410.4 murine adenocarcinoma cells (400,000) that had been transfected with the human MUC-1 gene. At day 35 after tumor challenge, mice were injected with 3 μCurie of [125I] IUDR (5-iodo-2-deoxyuridine). After 2 hours, mice were sacrificed and their lungs surgically excised, washed in TCA, and counted in a gamma counter to determine radioactive incorporation into lung tissues. Non-tumor-challenged mice (normal mice) were included to determine normal lung uptake. Data presented are the mean of CPM with standard errors, for the ten mice in each group. Tumor-bearing PBS-treated mice serve to note the uptake into tumors for comparison with vaccine-treated mice.

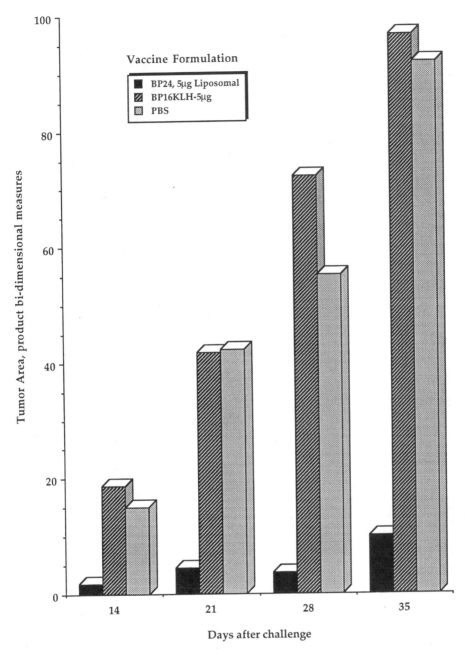

Figure 14 The kinetics of subcutaneous tumor growth in vaccine-primed and non-pretreated control CB6F1 mice. Mice were vaccine pretreated either with a synthetic peptide liposomal formulation that includes monophosphoryl lipid A (MPL, Ribi ImmunoChem Inc., Hamilton, MT) or with a peptide-KLH conjugate construct adjuvanted in DETOX™-B SE. Significant suppression of tumor growth is noted with the liposomal formulation that was shown to induce strong TH1-type immune responses, as characterized by no specific antibody responses and strong gamma interferon production to specific antigen challenge. In contrast, there were no significant differences noted for the peptide conjugate vaccine construct that induced the TH2 response. Animals in this group were noted to have specific antibody titers in excess of 1/5120 and no antigen-specific gamma interferon responses.

evidence of the sterilization effect. The antigen is a 16-amino-acid MUC-1 sequence, GVTSAPDTRPAPGSTA, conjugated to KLH and administered in DETOX™-B SE adjuvant (Ribi ImmunoChem Research, Hamilton, MT). The cell line chosen for this purpose is a murine carcinoma cell line transfected to express human MUC-1. When these cells are given intravenously to induce artificial metastasis to both preimmunized and nonimmunized control groups, the metastatic growth of tumors is retarded considerably in the preimmunized group whereas tumors establish and grow faster in the control group. Results of the experiments on the effects of preimmunization on artificial metastasis are shown in Figure 13. Mice immunized with this construct consistently generated strong IgM and IgG antibody responses, while only weak T-cell proliferative responses were observed.

Using the same MUC-1⁺ tumor cell line in a subcutaneous tumor challenge that resembles more the metastatic soft-tissue spread of breast cancer, these antibody responses are not significantly immunotherapeutic. In contrast, a MUC-1 vaccine construct that generates a TH1-mediated response, including gamma interferon and CTL activity in an absence of antibody, is clearly beneficial (Figure 14 shows 5 μg BP24 vs. 5 μg BP16-KLH PBS, KLH, etc.). Thus, depending on the disease setting, relevant animal models must be established that most closely mimic the clinical situation to be addressed.

IX. BINDING OF MONOCLONAL ANTIBODIES GENERATED AGAINST SYNTHETIC MUCIN MIMICS NATURAL MUCIN EPITOPES

A set of Tn and STn structures constituting single nonpeptide analogs and mono-, di-, and trilinked serine glycopeptides were used as immunogens in the form of KLH conjugates. Standard hybridoma fusions resulted with monomeric-STn-disaccharide–specific MAbs [88] and multimeric-serine-cluster–specific MAbs. The MAb B195.3, which was raised to the STn disaccharide linked via a two-carbon linker arm to KLH, does not appear to be influenced by the presence or absence of the *O*-glycosidic residues, exclusively to serine or threonine, in its binding to natural STn on mucins.

The MAbs B231 (anti-Tn) and B239 (anti-STn) were raised to synthetic Tn and STn trimers, respectively, following immunization with the synthetic mucin like Ser-Ser-Ser–based glycopeptides. In contrast to B195, the reactivity of these cluster-specific MAbs appears to be directed to the linkage-specific primary carbohydrate, *N*-acetylgalactosamine, as well as to its 6-*O*-linked sialic acid, on a succession of serines on the glycopeptide core. The significant cross-reactivity of the MAb B239 with dimeric and trimeric Tn clusters may be of importance in diagnosis and therapy of cancers, since these epitopes are expressed on cancer-associated mucins. Of note for the MAbs of the B239 STn series is the strong cross-reactivity to the nonsialylated Tn analogs. All of the MAbs screened in these fusions are reactive with clusters of two STn or Tn serine units; a third STn(Tn) serine does not significantly affect the binding curves in these hapten studies, suggesting that the epitopes are constituted by two serines and minimally the *N*-acetylgalactosamine core group (Figure 15).

In contrast to the monomeric-disaccharide–specific MAb B195.3, these mucin-core-reactive MAbs show far less sensitivity in binding to neuraminidase-treated

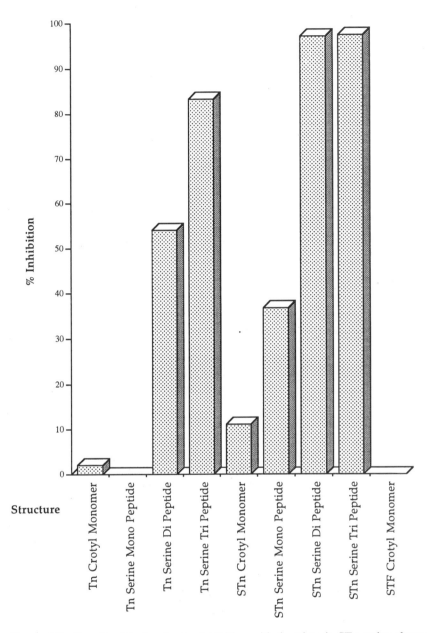

Figure 15 Results of testing hapten inhibition with the trimeric-STn-serine-cluster–specific MAb B239.6. The solid-phase antigen is the native mucin ovine submaxillary mucin (OSM), which is characterized by repetitive STn epitopes. MAb 239.6 (175 ng/mL) was preincubated (1 hour) with the described Tn and STn hapten constructs (100 nM/mL) prior to incubation (1 hour) with the mucin solid phase. Bound antibody was detected after washing by incubating with peroxidase-labeled goat antimouse IgG and substrate. Data are presented as the percent inhibition relative to the no-inhibitor controls and represent the mean of two experiments, each conducted in triplicate.

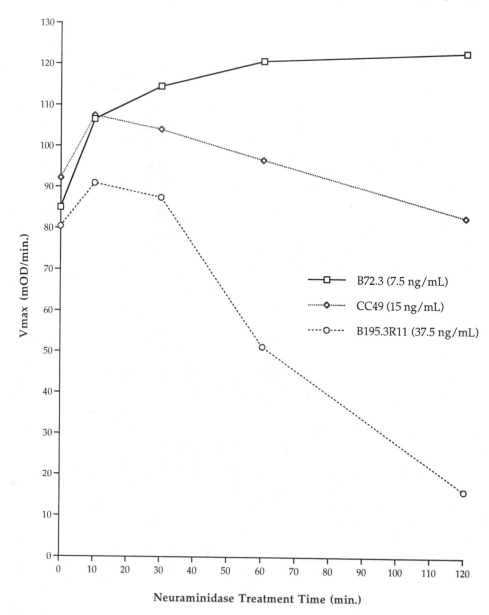

Figure 16 Results of solid-phase binding assays on neuraminidase-treated ovine submaxillary mucin antigen. Monoclonal antibodies B72.3, CC49, and B195.3R11 were tested for binding onto an OSM solid phase that had been digested with neuraminidase for various periods of time (0–120 minutes). Data are expressed as the ELISA-rate kinetic reading of milli-OD generated per minute using the kinetic ELISA reader Vmax from Molecular Devices.

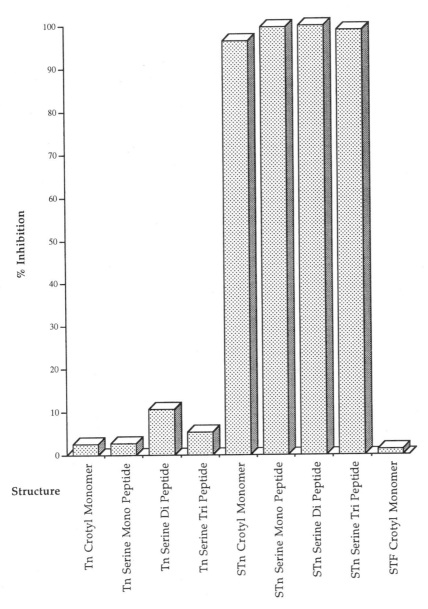

Figure 17 Results of studies of hapten inhibition with the Stn-crotyl-monomer–specific monoclonal antibody B195.3R11 [87]. The solid-phase antigen is ovine submaxillary mucin, a native mucin source of multiple STn epitopes. The MAb (37.5 ng/mL) was preincubated (1 hour) with the described Tn and STn haptens, each at 250 nmol/mL, prior to incubation with the antigen solid phase. Bound antibody was detected after washing by incubating with peroxidase-labeled goat antimouse IgG and subsequent to washing the addition of the appropriate colorimetric substrate. Data are presented as the percent inhibition relative to the no-inhibitor controls and represent the mean of two experiments, each performed in triplicate.

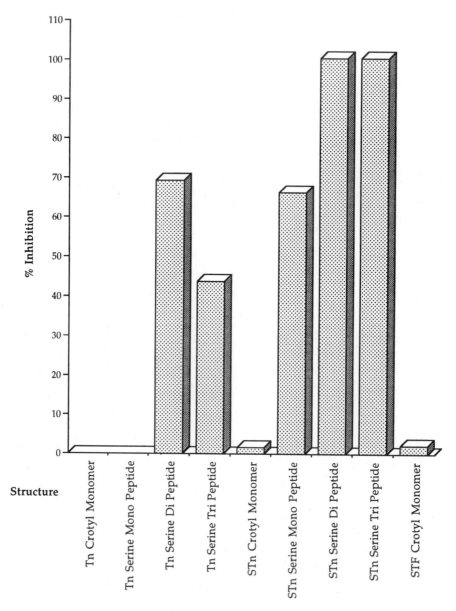

Figure 18 Results of studies of hapten inhibition with the STn-crotyl-monomer–specific monoclonal antibody B72.3 [90]. The solid-phase antigen is ovine submaxillary mucin, a native mucin source of multiple STn epitopes. The MAb (7.5 ng/mL) was preincubated (1 hour) with the described Tn and STn haptens, each at 250 nmol/mL, prior to incubation with the antigen solid phase. Bound antibody was detected after washing by incubating with peroxidase-labeled goat antimouse IgG and subsequent to washing the addition of the appropriate colorimetric substrate. Data are presented as the percent inhibition relative to the no-inhibitor controls and represent the mean of two experiments, each performed in triplicate.

native mucin antigens (Figure 16). Both B72.3 and CC49 seem to exhibit more independent of sialic acid binding compared to MAb B195.3. Kinetic removal of sialic acid through neuraminidase treatment of ovine submaxillary mucin (OSM) did not abrogate the binding of either B72.3 or CC49, whereas B195.3 showed remarkable loss of binding with the removal of sialic acid, which is further confirmed by the lack of inhibition of B195.3 by Tn-Crotyl or Tn-Serine multimers (Figure 17). Though in a previous report by Kieldsen et al. [89] it has been suggested that the specificity of B72.3 is mostly to sialyl-Tn, the authors also demonstrated that the binding of B72.3 to OSM solid phase is inhibited by mM concentrations of N-acetylgalactosamine (the concentration of B72.3 is not defined) as an unlinked re-

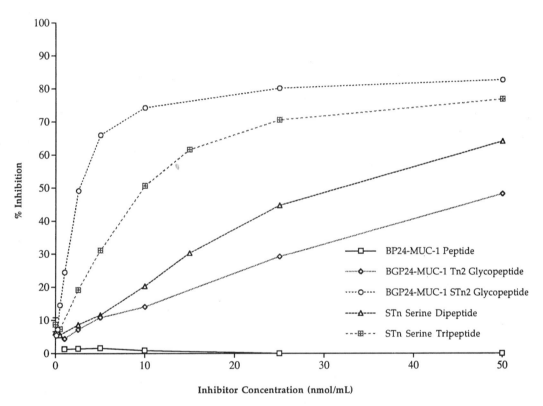

Figure 19 Results of studies of synthetic hapten inhibition with the MAb CC49 that was generated against the human-adenocarcinoma–derived mucin TAG 72 [90]. The MAb (37.5 ng/mL) was preincubated (1 hour) with the described peptide and glycopeptide structures prior to incubation with the OSM mucin solid phase. Bound antibody was detected by peroxidase-conjugated goat antimouse IgG. Data presented are expressed as percent inhibition relative to the no-inhibitor controls and represent the mean of four assays, each conducted in triplicate. Peptide (BP24) and glycopeptides based on the human MUC-1 tandem-repeat sequence are compared to the theoretical mucin segments serine di- or tripeptides, each residue substituted with the STn disaccharide. The MUC-1 glycopeptide substituted at the serine-threonine site with STn disaccharides shows significantly greater relative affinity compared to the serine di- or triglycopeptide structures.

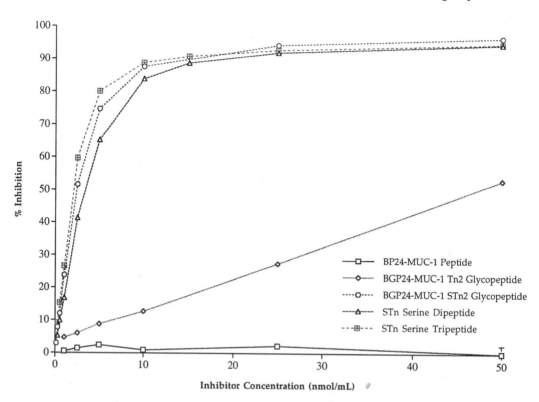

Figure 20 Results of studies of synthetic hapten inhibition with the MAb B239.6 that was generated against the theoretical mucin segment STn-serine tripeptide. Monoclonal antibody was preincubated (1 hour) with the described peptide and glycopeptide structures prior to incubation with OSM mucin solid phase. Bound antibody was detected with peroxidase-conjugated goat antimouse IgG. Data presented are expressed as percent inhibition relative to the no-inhibitor controls and represent the mean of four assays, each conducted in triplicate. This MAb does not discriminate between theoretical mucin segments STn-serine dipeptide (or tripeptide) and the MUC-1 24-amino-acid sequence substituted at the serine-threonine site. This is in contrast to the TAG-72−specific MAb CC49 (see Figure 19) that demonstrates enhanced inhibition with the MUC-1 glycopeptide-based STn cluster.

ducing structure. We believe that *N*-acetylgalactosamine α-linked to an aglycon such as serine or threonine will be an appropriate inhibitor as a Tn structure. We have not observed such inhibition with monomeric Tn-serine (Figure 18) which was done only at micromolar concentration. However, dimeric and trimeric Tn-serine at similar concentrations demonstrate significant inhibition binding of B72.3 OSM solid phase. This is not an abnormal observation, since B72.3 is a TAG-72 mucin antibody and mucins are known to express carbohydrates in clusters. Hence it appears that B72.3 is more a cluster-specific antibody with significant cross-reactivities with both multimeric Tn and STn structures. The data presented in Figure 17 also suggest that the MAb B195.3 shows true STn specificity with significant sialic-acid−dependent binding. A monoclonal antibody such as B195.3 may be useful in specifically detecting all STn epitopes, both clustered and monomeric.

A comparison of the antisynthetic mucin mimic MAbs and the well-character-ized anti-native mucin-specific monoclonals B72.3 and CC49 indicates that these native mucin-specific MAbs are similar to the B231- and B239-series MAbs in that they appear to be core polypeptide-*O*-N-acetylgalactosamine dimer-reactive (Figure 17). The B72.3 MAb does show binding affinity with the monomeric STn structure (STn-*O*-Serine), but not to the crotyl linked disaccharide (Figure 18). Thus, B72.3 is specific to a true glycopeptide epitope. Both the MAb B72.3 and CC49 appear to be strongly cross-reactive with the Tn-Serine dimeric structure. Immunohistology results with these MAbs must be interpreted with caution keeping in mind their cross-reactivity to nonsialated Tn clusters.

The MUC-1 mucin repeat sequence consists of two potential dimeric gly-cosylation sites, each of which is a mixed amino acid dimer of either Thr-Ser or Ser-Thr. A group of MUC-1 backbone-based STn or Tn dimers were synthesized that represent the ST site in the 24-amino-acid core peptide sequence, TAPPAHGVTS-APDTRPAPGSTAPP (single-letter amino acid code). These larger structural glyco-peptides made to mimic the MUC-1 core mucin appear to represent true mimics of the CC49 epitope, as evidenced by the nanomolar inhibitions observed with the synthetic MUC-1 glycopeptide containing sialyl-Tn dimeric structure at the serine threonine site (Figure 19).

In contrast to CC49, the serine-cluster–specific MAb B239.6 shows a greater relative affinity to the synthetic tri-serine STn structure as opposed to the mixed-amino-acid- (Ser-Thr) cluster specificity of CC49 (Figure 20).

X. ANALYSIS OF GLYCOPEPTIDE VACCINES IN MURINE IMMUNE MODELS USING T-CELL IMMUNE RESPONSE ANALYSIS

Because it is frequently a goal in vaccine design to achieve a cellular immune re-sponse, T-cell response analysis must be included in vaccine design. Analyzing drain-ing lymph node T-cell population following subcutaneous immunization, we have compared various peptide and glycopeptide MUC-1 vaccine structures. Shown in Figures 21 and 22 are the T-cell proliferative responses and the gamma interferon responses of draining lymph node T-cells from C57/Bl mice immunized with a 24-amino-acid MUC-1 peptide or the same 24-amino-acid glycopeptide containing a Tn-Ser/Thr dimeric structure.

The synthetic glycopeptide construct generated strong T-cell responses com-parable to those generated against the peptide structure.

These studies on MAb binding and T-cell immunogenicity indicate that it is possible to construct glycopeptide immunogens designed to mimic true tumor-associated epitopes. These analytical and synthetic technologies are extendible to the rational design of defined synthetic immunogens of pharmaceutical quality for any variety of vaccine needs.

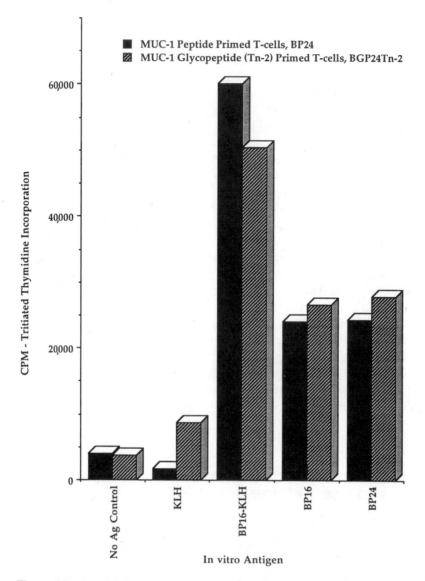

Figure 21 Results of an experiment on draining lymph node T-cell blastogenesis comparing the recall responses of peptide-primed and glycopeptide-primed C57/B1 mice. Mice were primed with 5 mg of the defined structures adjuvanted with monophosphoryl lipid A (MPL, Ribi ImmunoChem Inc., Hamilton, MT) by subcutaneous injection. Draining lymph node T-cell populations were prepared on day 10 and incubated in vitro for 5 days, at which time tritiated thymidine was added for incorporation into DNA of responding % cells. Data are expressed as counts per minute incorporated from triplicate cultures, and test groups are compared to no-peptide or irrelevant-peptide control cultures.

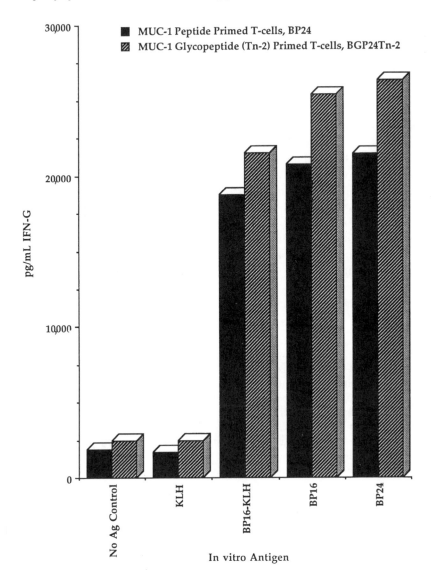

Figure 22 Results of an experiment on draining lymph node T-cell gamma interferon production comparing the recall responses of peptide-primed and glycopeptide-primed C57/B1 mice. Mice were primed with 5 mg of the defined structures adjuvanted in MPL by subcutaneous injection. Draining lymph node T-cells were obtained on day 10 by passage through nylon wool and incubated in irradiated APCs and peptides for 4 days prior to culture supernatant harvest. Triplicate supernatants were tested for the presence of gamma interferon by reference to a standardized two-site ELISA technique, results of which are presented in picograms/mL of supernatant.

XI. CONCLUSIONS

The immunotherapy of cancers must deal with the antigens produced by all cells while exploiting only the features that distinguish them as cancer associated. A glycopeptide that is designed to mimic the cancer-associated macromolecule both in physical and biological properties should display the same antigenic properties as its parent glycoprotein. Such glycopeptides, acting as immunogens, should direct the immune system to the target cancer cell that displays the parent molecule. Vaccine design under these guidelines offers hope for the immunotherapy of cancers as well as other diseases where the respective target antigens are well characterized. Glycopeptides, combining the distinct structural features of both peptide core and carbohydrates, expand the repertoire of epitopes beyond those based on carbohydrate or peptide alone. Though chemically and structurally challenging, synthetic glycopeptides with inherent structural definition and specificity have a lot to offer as vaccines and as adhesion inhibitors of bacteria, viruses, and inflammatory invasion of the body's own immune cells.

REFERENCES

1. S. R. Hull, A. Bright, K. L. Carraway, M. Abe, D. F. Hayes, and D. W. Kufe, Glycosaccharide differences in the DF_3 sialo-mucin antigen from normal human milk and BT-20 human breast carcinoma cell line, *Cancer Commun. 1*: 261 (1989).
2. J. Burchell, J. Taylor-Papadimitriou, M. Boshell, S. Gendler, and T. Duhig, A short sequence within the amino acid tandem repeat of a cancer associated mucin contains immunodominant epitopes, *Int. J. Cancer 44*: 691 (1989).
3. G. F. Springer, T and Tn, General carcinoma autoantigens. *Science 224*: 1198 (1984).
4. J. Burchell, S. Gendler, J. Taylor-Papadimitriou, A. Girling, A. Lewis, R. Millis, and D. Lamport, Development and characterization of breast cancer reactive monoclonal antibodies directed to the core protein of the human milk mucin. *Cancer Res. 47*: 5476 (1987).
5. L. Ding, E.-N. Lalani, M. Reddish, R. Koganty, T. Wong, J. Samuel, M. B. Yacyshyn, A. Meikle, P. Y. S. Fung, J. Taylor-Papadimitriou, and B. M. Longenecker, Immunogenicity of synthetic peptides related to core peptide sequence encoded by the human MUC-1 gene: Effect of immunization on the growth of murine mammary adenocarcinoma cells transfected with the human MUC-1 gene, *Cancer Immunol. Immunother. 36*: 9 (1993).
6. P. Y. S. Fung and B. M. Longenecker, Specific immunosuppressive activity of epiglycanin, a mucin-like glycoprotein secreted by a murine mammary adenocarcinoma, TA3-Ha, *Cancer Res. 51*: 1170 (1990).
7. J. Breg, H. van Halbeek, J. F. G. Vliegenthart, A. Klein, G. Lamblin, and P. Roussel, Primary structure of neutral oligosaccharides derived from respiratory mucus glycoproteins of a patient suffering from bronchiectasis, determined by combination of 500-MHz proton NMR spectroscopy and quantitative sugar analysis. 2. Structure of 19 oligosaccharides having the GlcNAcβ(1–3)GalNAc-ol core (type 3) or the GlcNAcβ(1–3)[GlcNAcβ(1–6)]GalNAc-ol core (type 4), *Eur. J. Biochem. 171*: 643 (1988).
8. A. Klein, G. Lamblin, M. Lhermitte, P. Roussel, J. Breg, H. van Halbeek, and J. F. G. Vliegenthart, Primary structure of neutral oligosaccharides derived from respiratory mucus glycoproteins of a patient suffering from bronchiectasis, determined by combination

of 500-MHz proton NMR spectroscopy and quantitative sugar analysis. 1. Structure of 16 oligosaccharides having the Galβ(1–3)GalNAc-ol core (type 1) or the Galβ(1–3)[GlcNAcβ(1–6)]GalNAc-ol core (type 2), *Eur. J. Biochem. 171*: 631 (1988).

9. G. Lamblin, A. Boersma, M. Lhermitte, P. Roussel, J. H. G. M. Mutsaers, H. van Halbeek, and J. F. G. Vliegenthart, Further characterization, by a combined high-performance liquid chromatography/proton NMR approach, of the heterogeneity displayed by the neutral carbohydrate chains of human bronchial mucins, *Eur. J. Biochem. 143*: 227 (1984).

10. H. van Halbeek, J. Breg, J. F. G. Vliegenthart, A. Klein, G. Lamblin, and P. Roussel, Isolation and structural characterization of low-molecular-mass monosialyl oligosaccharides derived from respiratory-mucus glycoproteins of a patient suffering from bronchiectasis, *Eur. J. Biochem. 177*: 443 (1988).

11. E. C. Yurewicz, F. Matsuura, and K. S. Moghissi, Structural characterization of neutral oligosaccharides of human midcycle cervical mucin, *J. Biol. Chem. 257*: 2314 (1982).

12. E. C. Yurewicz, F. Matsurra, and K. S. Moghissi, Structural studies of sialylated oligosaccharides of human midcycle cervical mucin, *J. Biol. Chem. 262*: 4733 (1987).

13. M. D. G. Oates, A. C. Rosbottom, and J. Schrager, Further investigation into the structure of human gastric mucin: The structural configuration of the oligosaccharide chains, *Carbohydr. Res. 34*: 115 (1974).

14. C. A. Bush, M. M. Panitch, V. K. Dua, and T. E. Rohr, Carbon nuclear magnetic resonance spectra of oligosaccharides isolated from human milk and ovarian cyst mucin, *Anal. Biochem. 145*: 124 (1985).

15. V. K. Dua, B. N. N. Rao, S. Wu, V .E. Dube, and C. A. Bush, Characterization of the oligosaccharide alditols from ovarian cyst mucin glycoproteins of blood group A using high pressure liquid chromatography (HPLC) and high field proton NMR spectroscopy, *J. Biol. Chem. 261*: 1599 (1986).

16. J. H. G. M. Mutsaers, H. van Halbeek, J. F. G. Vliegenthart, A. M. Wu, and E. A. Kabat, Typing of core and backbone domains of mucin-type oligosaccharides from human ovarian cyst glycoproteins by 500 MHz proton NMR spectroscopy, *Eur. J. Biochem. 157*: 139 (1986).

17. M. Fukuda, S. R. Carlsson, J. C. Klock, and A. Dell, Structures of O-linked oligosaccharides isolated from normal granulocytes, chronic myelogenous leukemia cells, and acute myelogenous leukemia cells, *J. Biol. Chem. 261*: 12,796 (1986).

18. F. G. Hanisch, H. Egge, J. Peter-Katalinic, and G. Uhlenbruck, Primary structure of a major sialyl-oligosaccharide alditol from human amniotic mucins expressing the tumor-associated sialyl Lewis-X antigenic determinant, *FEBS Lett. 200*: 42 (1986).

19. D. K. Podolsky, Oligosaccharide structures of human colonic mucin, *J. Biol. Chem. 260*: 8262 (1985).

20. D. K. Podolsky, Oligosaccharide structures of isolated human colonic mucin species, *J. Biol. Chem. 260:* 15,510 (1985).

21. F. G. Hanisch, G. Uhlenbruck, C. Dienst, M. Stottrop, and E. Hippauf, Ca 125 and Ca 19–9: Two cancer-associated sialylsaccharide antigens on a mucus glycoprotein from human milk, *Eur. J. Biochem. 149*: 323 (1985).

22. F.-G. Hanisch, G. Uhlenbruck, J. Peter-Katalinic, H. Egge, J. Dabrowski, and U. Dabrowski, Structures of neutral O-linked polylactosaminoglycans on human skim milk mucins, *J. Biol. Chem. 264*: 872 (1989).

23. A. M. Wu, E. A. Kabat, B. Nilsson, D. A. Zopf, F. G. Gruezo, and J. Liao, Immunochemical studies on blood groups. LXXI. Purification and characterization of radioactive 3H-reduced di- to hexasaccharides produced by alkaline beta-elimination-borohydride 3H reduction of Smith degraded blood group A active glycoproteins, *J. Biol. Chem. 259*: 7178 (1984).

24. A. Kurosaka, H. Nakajima, J. Funakoshi, M. Matsuyama, T. Nagayo, and I. Yamashina, Structures of the major oligosaccharides from a human rectal adenocarcinoma glyco-protein, *J. Biol. Chem. 258*: 11,594 (1983).

25. D. C. Gowda, V. P. Bhavanandan, and E. A. Davidson, Structures of *O*-linked oligo-saccharides present in the proteoglycans secreted by human mammary epithelial cells, *J. Biol. Chem. 261*: 4935 (1986).

26. S. R. Carlsson, H. Sasaki, and M. Fukuda, Structural variations of *O*-linked oligosac-charides present in leukosialin isolated from erythroid, myeloid, and T-lymphoid cell lines, *J. Biol. Chem. 261*: 12,787 (1986).

27. A. Leppanen, A. Korvuo, K. Puro, and O. Renkonen, Glycoproteins of human terato-carcinoma cells (PA1) carry both anomers of *O*-glycosyl-linked D-galactopyranosyl-(1–3)-2-acetamido-2-deoxy-α-D-galactopyranosyl group, *Carbohydr. Res. 153*: 87 (1986).

28. J. Amano, R. Nishimura, M. Mochizuki, and A. Kobata, Comparative study of the mucin-type sugar chains of human chorionic gonadotropin present in the urine of pa-tients with trophoblastic diseases and healthy pregnant women, *J. Biol. Chem. 263*: 1157 (1988).

29. L. A. Cole, The *O*-linked oligosaccharide structures are strikingly different on pregnancy and choriocarcinoma hCG, *J. Clin. Endocrinol. Metab. 65*: 811 (1987).

30. N. Jentoff, Why are proteins *O*-glycosylated? *Trends Biochem. Sci. 15*: 291 (1990).

31. F.-G. Hanisch, G. Uhlenbruck, H. Egge, and J. Peter-Katalinic, A B72.3 second gener-ation monoclonal antibody (CC49) defines the mucin carried carbohydrate epitope Galβ(1–3) [NeuAca(2–6)]GalNAc, *Biol. Chem. 370:* 21 (1989).

32. S. J. Gendler, C. A. Lancaster, J. Taylor-Papadimitriou, T. Duhig, N. Peat, J. Burchell, L. Pemberton, E.-N. Lalani, and D. Wilson, Molecular cloning and expression of the human tumour-associated polymorphic epithelial mucin, *J. Biol. Chem. 265*: 15,286 (1990).

33. M. Ligtenberg, H. Vos, A. Gennissen, and J. Hilkens, Episialin, a carcinoma-associated mucin, is generated by a polymorphic gene encoding splice variants with alternative amino termini, *J. Biol. Chem. 265*: 5573 (1990).

34. D. H. Wreschner, M. Hareuveni, I. Tsarfaty, N. Smorodinsky, J. Horev, J. Zaretsky, P. Kotkes, M. Weiss, R. Lathe, A. Dion, and I. Keydar, Human epithelial tumor antigen cDNA sequences, *Eur. J. Biochem. 189*: 463 (1990).

35. J. Siddiqui, M. Abe, D. Hayes, E. Shani, E. Yunis, and D. Kufe, Isolation and sequencing of a cDNA coding for the human DF3 breast carcinoma-associated antigen, *Proc. Natl. Acad. Sci. USA 85*: 2320 (1988).

36. M. Lan, S. Batra, W.-N. Qi, R. Metzgar, and M. Hollingsworth, Cloning and sequencing of a human pancreatic tumour mucin cDNA, *J. Biol. Chem. 265*: 15,294 (1990).

37. C. Lancaster, N. Peat, T. Duhig, D. Wilson, J. Taylor-Papadimitriou, and S. Gendler, Structure and expression of the human polymorphic epithelial mucin gene: An expressed VNTR unit, *Biochem. Biophys. Res. Commun. 173*: 1019 (1990).

38. F. G. Hanish, G. Uhlenbruck, J. Peter-Katalinic, H. Egge, J. Dabrowski, and U. Da-browski, Structures of neutral *O*-linked polylactosaminoglycans on human skim milk mucins. A novel type of linearly extended poly-*N*-acetyl-lactosamine backbones with Galβ(1–4) GlcNAcβ(1–6) repeating units, *J. Biol. Chem. 265*: 872 (1989).

39. S. R. Hull, A. Bright, K. L. Carraway, M. Abe, D. F. Hayes, and D. W. Kufe, Oligo-saccharide differences in the DF3 sialomucin antigen from normal human milk and the BT-20 human breast carcinoma cell line, *Cancer Commun. 1*: 261 (1989).

40. D. L. Barnd, M. S. Lan, R. S. Metzgar, and O. J. Finn, Specific, major histocompatibility complex-unrestricted recognition of tumor-associated mucins by cytotoxic T cells, *Proc. Natl. Acad. Sci. USA 86*: 7159 (1989).

41. K. R. Jerome, D. L. Barnd, K. M. Bendt, C. M. Boyer, J. Taylor-Papadimitriou, I. F. C. McKenzie, R. C. Bast Jr., and O. J. Finn, Cytotoxic T lymphocytes derived from patients

with breast adenocarcinoma recognize an epitope present on the protein core of a mucin molecule preferentially expressed by malignant cells, *Cancer Res. 51*: 2908 (1991).

42. H. Nakada, M. Inoue, Y. Numata, N. Tanaka, I. Funakoshi, S. Fukui, A. Mellors, and I. Yamashina, Epitopic structure of Tn glycophorin A for an anti-Tn antibody (MLS128), *Proc. Natl. Acad. Sci. USA 90*: 2495 (1993).

43. Y. Numata, H. Nakada, S. Fukui, H. Kitagowa, K. Ozaki, H. Inoue, T. Kowasaki, I. Funakoshi, and I. Yamashina, Elucidation of an essential structure recognized by an anti-GalNAc α-Ser(Thr) monoclonal antibody (MLS 128), *Biochem. Biophys. Res. Commun. 170*: 980 (1990).

44. H. Nakada, Y. Numata, H. Inoue, N. Tanaka, H. Kilagowa, I. Funakoshi, S. Fukui, and I. Yamashina, A monoclonal antibody directed to Tn antigen, *J. Biol. Chem. 266*: 12,402 (1991).

45. T. Toyokuni, S.-I. Hakomori, and A. K. Singhal, Synthetic carbohydrate vaccines: Synthesis and immunogenicity of Tn antigen conjugates, *Bioorganic & Medicinal Chem. 2*: 1119 (1994).

46. T. Toyokuni, B. Dean, S. Cai, D. Boivin, S.-I. Hakomori, and A. K. Singhal, Synthetic vaccines: Synthesis of a dimeric Tn antigen-lipopeptide conjugate that elicits immune responses against Tn-expressing glycoproteins, *J. Am. Chem. Soc. 116*: 395 (1994).

47. Y. Nakahara, H. Iijima, and T. Ogawa, Synthesis of human M blood group antigenic glycopeptide, *Tetrahedron Lett. 35*: 3321 (1994).

48. C. Unverzagt, S. Kelm, and J. C. Paulson, Chemical and enzymatic synthesis of multivalent sialoglycopeptides, *Carbohyrate Res. 251*: 285 (1994).

49. M. Meldal, T. Bielfeldt, S. Peters, K. J. Jensen, H. Paulsen, and K. Bock, Susceptibility of glycans to β-elimination in Fmoc-based *O*-glycopeptide synthesis, *Int. J. Peptide Protein Res. 43*: 529 (1994).

50. H. Paulsen, T. Bielfeldt, S. Peters, M. Meldal, and K. Bock, Application of the azido glycopeptide synthesis strategy for the multiple column solid phase synthesis of mucin *O*-glycopeptides, *Liebigs Ann. Chem.* 381 (1994).

51. F.-Y. Dupradeau, M. R. Stroud, D. Boivin, L. Li, S.-I. Hakomori, A. K. Singhal, and T. Toyokuni, Solid phase synthesis and immunoreactivity of penta-*O*-(*N*-acetyl-α-D-galactosaminyl)-MUC1 eicosapeptide, a glycosylated counterpart of the highly immunogenic tandem repeat sequence of carcinoma-associated mucin, *Bioorganic Medicinal Chem. 4*: 1813 (1994).

52. H. Paulsen, T. Bielfeldt, S. Peters, M. Meldal, and K. A. Bock, A new strategy for the solid-phase synthesis of *O*-glycopeptides via 2-azido-glycopeptides, *Liebigs Ann. Chem.* 369 (1994).

53. H. Kunz, Glycopeptides of biological interest: A challenge for chemical synthesis, *Pure Appl. Chem. 65*: 1223 (1993).

54. S. Peters, T. Bielfeldt, M. Meldal, K. Bock, and H. Paulsen, Solid phase peptide synthesis of mucin glycopeptides, *Tetrahedron Lett. 33*: 6445 (1992).

55. B. Lüning, T. Norberg, G. Rivera-Baeza, and J. Tejbrant, Solid phase synthesis of the fibronectin glycopeptide V(Galβ3GalNAcα)THPGY, its β analogue, and the corresponding unglycosylated peptide, *Glycoconjugate J. 8*: 450 (1991).

56. S. Peters, T. Bielfeldt, M. Meldal, K. Bock, and H. Paulsen, Multiple column solid phase glycopeptide synthesis, *Tetrahedron Lett. 32*: 5067 (1991).

57. M. Meldal and K. J. Jensen, Pentafluorophenyl esters for the temporary protection of the α-carboxy group in solid phase glycopeptide synthesis, *J. Chem. Soc. Chem. Commun.* 483 (1990).

58. B. G. de la Torre, J. L. Torres, E. Bardaji, P. Clapes, N. Xaus, X. Jorba, S. Calvet, F. Albericio, and G. Valencia, Improved method for the synthesis of *O*-glycosylated Fmoc amino acids to be used in solid-phase glycopeptide synthesis (Fmoc = fluoren-9-ylmethoxycarbonyl), *J. Chem. Soc. Chem. Commun.* 965 (1990).

59. B. Lüning, T. Norberg, and J. Tejbrant, Solid phase synthesis of mono- and di-saccha-ride-containing glycopeptides, *J. Chem. Soc. Chem. Commun.* 1267 (1989).

60. B. Lüning, T. Norberg, and J. Tejbrant, Synthesis of mono- and disaccharide amino-acid derivatives for use in solid phase peptide synthesis, *Glycoconjugate J. 6*: 5 (1989).

61. H. Paulsen, G. Merz, and U. Weichert, Solid-phase synthesis of *O*-glycopeptide sequences, *Angew. Chem. Int. Ed. Engl. 27*: 1365 (1988).

62. G. Jung, K. H. Wiesmuller, G. Becker, H. J. Büwring, and W. G. Bessler, Increased production of specific antibodies by presentation of the antigen determinants with co-valently coupled lipopeptide mitogens, *Angew. Chem. Int. Ed. Engl. 24*: 872 (1985).

63. K. Deres, H. Schild, K. H. Weismuller, G. Jung, and H. G. Rammensee, In vivo priming of virus-specific cytotoxic T lymphocytes with synthetic lipopeptide vaccine, *Nature 342*: 561 (1989).

64. J. S. Haurum, G. Arsequell, A. C. Lellouch, S. Y. C. Wong, R. A. Dwek, A. J. Mc-Michael, and T. Elliott, Recognition of carbohydrate by major histocompatibility complex class I-restricted, glycopeptide-specific cytotoxic T lymphocytes, *J. Exp. Med. 180*: 739 (1994).

65. H. Schachter and C. A. Tilley, Biosynthesis of human blood group substances, *Biochemistry of Carbohydrates* (D. J. Manners, ed.), University Park Press, Baltimore, 1978, p. 85.

66. T. A. Bayer, J. E. Sadler, J. I. Rearick, J. C. Pausen, and R. L. Hill, Glycosyltransferasese and their use in assessing oligosaccharide structure and structure-function relationships, *Adv. Enzymol. 52*: 23 (1981).

67. H. Schacter and I. Brockhausen, Biosynthesis of serine (threonine)-*N*-acetylgalac-tosamine-linked carbohydrate moities, *Glycoconjugates: Composition, Structure and Function* (H. J. Allen and E. C. Kisailus, eds.), Dekker, New York, 1992, p. 263.

68. S. H. Itzkowitz, E. J. Bloom, W. A. Kokal, G. Modin, S.-I. Hakomori, and Y. S. Kim, Sialosyl-Tn: A novel mucin antigen associated with prognosis in colorectal cancer patients, *Cancer 66*: 1960 (1990).

69. H. Kobayashi, T. Toshihiko, and Y. Kawashima, Serum sialosyl Tn as an independent predictor of poor prognosis in patients with epithelial ovarian cancer, *Clin. Oncol. 10*: 95 (1992).

70. Y. Ichikawa, G. C. Look, and C. H. Wong, Enzyme-catalyzed oligosaccharide synthesis, *Anal. Biochem. 202*: 215 (1992).

71. C. H. Wong, R. L. Halcomb, Y. Ichikawa, and T. Kajimoto, Enzymes in organic synthesis: Applications to the problems of carbohydrate recognition (Part 1), *Angew. Chem. Int. Ed. Engl. 34*: 412 and *34*: 521 (Part 2) (1995).

72. R. U. Lemieux and R. M. Ratcliff, The azidonitration of tri-*O*-acetyl-D-galactal, *Can. J. Chem. 57*: 1244 (1979).

73. J. E. Yule, T. C. Wong, S. S. Gandhi, D. Qiu, M. A. Riopel, and R. R. Koganty, Steric control of *N*-Acetylgalactosamine in glycosidic bond formation, *Tetrahedron Letters 36*: 6839 (1995).

74. D. Van den eijnden, N. A. Evans, J. F. Codington, V. Reinhold, C. Silber, and R. W. Jeanloz, Chemical structure of epiglycanin—the major glycoprotein of TA3-Ha ascites cell, *J. Biol. Chem. 254*: 12,153 (1979).

75. G. D. McLean, M. B. Bowen-Yacyshyn, J. Samuel, A. Meilke, G. Stuart, J. Nation, S. Peppema, M. Jerry, R. Koganty, T. Wong, and B. M. Longenecker, Active immunization of human ovarian cancer patients against a common carcinoma (Thomsen−Friedenreich) determinant using a synthetic carbohydrate antigen, *J. Immunother. 11*: 292 (1992).

76. S. Zhang, L. A. Walberg, S. Ogata, S. H. Itzkowitz, R. R. Koganty, M. A. Reddish, S. S. Gandhi, B. M. Longenecker, K. O. Lloyd, and P. O. Livingston, Immune sera and

monoclonal antibodies define two figurations for the sialyl Tn tumor antigen, *Cancer Res. 55*: 3364 (1995).

77. X. Liu, J. Sejbal, G. Kotovych, R. R. Koganty, M. A. Reddish, L. Jackson, S. S. Gandhi, A. J. Mendonca, and B. M. Longenecker, Structurally defined synthetic cancer vaccines: Analysis of structure, glycosylation and recognition of cancer associated mucin, MUC-1 derived peptides, *Glycoconjugate J.* (in press) (1995).

78. I. Nishimori, N. R. Johnson, S. D. Sanderson, F. Perini, K. P. Mountjoy, R. Cerny, M. L. Cross, O. Finn, and M. A. Hollingsworth, The influence of acceptor substrate primary amino acid sequence on the activity of human UDP-GalNAc: polypeptide *N*-acetylgalactosaminyltransferase: Studies with the MUC1 tandem repeat, *J. Biol. Chem. 269*: 16,123 (1994).

79. N. Nishimori, F. Perini, K. P. Mountjoy, S. D. Sanderson, N. R. Johnson, R. Cerny, M. L. Gross, D. R. Fontenot, and M. A. Hollingsworth, *N*-Acetylgalactosamine glycosylation of MUC1 tandem repeat peptides by pancreatic tumor cell extracts, *Cancer Res. 54*: 3738 (1994).

80. M. A. Reddish, N. Helbrecht, A. F. Almeida, R. Madiyalakan, M. R. Suresh, and B. M. Longenecker, Epitope mapping of CA27.29 within the protein core of the malignant breast carcinoma associated mucin antigen MUC-1, *J. Tumor Marker Oncol. 7*: 19 (1992).

81. P.-X. Xing, J. Prenzoska, K. Quelch, and I. F. C. McKenzie, Second generation anti-MUC1 peptide monoclonal antibody, *Cancer Res. 52*: 2310 (1992).

82. Y. Kotera, J. D. Fontenot, G. Pecher, R. S. Metzgar, and O. J. Finn, Humoral immunity against a tandem repeat epitope of human mucin MUC-1 in sera from breast, pancreatic and colon cancer patients, *Cancer Res. 54*: 2856 (1994).

83. J. D. Fontenot, N. Tjandra, D. Bu, C. Ho, R. C. Montelaro, and O. J. Finn, Biophysical characterization of one-, two-, and three-tandem repeats of human mucin (MUC-1) protein core, *Cancer Res. 53*: 5386 (1993).

84. T. R. Mosmann, and R. L. Coffman, Two types of mouse helper T-cell clones, *Immunology Today 8*: 223 (1987).

85. D. F. Fiorentino, M. W. Bond, and T. R. Mosmann, Two types of mouse T helper cell Iv. Th2 clones secrete a factor that inhibits cytokine production by Th1 clones, *J. Exp. Med. 170*: 2081 (1989).

86. T. R. Mosmann, Cytokine patterns during the progression to AIDS, *Science 265*: 193 (1994).

87. M. Clerici, and G. M. Shearer, A TH1 to TH2 switch is a critical step in the etiology of HIV infection, *Immunology Today 14*: 107 (1993).

88. B. M. Longenecker, M. Reddish, R. R. Koganty, and G. D. MacLean, Immune responses of mice and human breast cancer patients following immunization with synthetic sialyl-Tn conjugated to KLH plus DETOX™ adjuvant, *Specific Immunotherapy of Cancers with Vaccines* (J. C. Bystryn, S. Ferrone, and P. Livingston, eds.), Annals of New York Academy of Sciences, Vol. 690, 1993, p. 276.

89. T. Kjeldsen, H. Clausen, S. Hirohashi, T. Ogawa, H. Iijima, and S.-I. Hakomori, Preparation and characterization of monoclonal antibodies directed to the tumor-associated *O*-linked sialosyl-2→6 α-*N*-acetylgalactosaminyl (sialosyl-Tn), *Cancer Research 48*: 2214–2220 (1988).

90. D. Colcher, P. Horan Hand, M. Nuti, and J. Schlom, A spectrum of monoclonal antibodies reactive with human mammary tumor cells, *Proc. Natl. Acad. Sci. USA 78*: 3199–3203 (1981).

Index